Recent Advances in Actinide Science

Recent Advances in Actinide Science

Edited by

Iain May, Rebeca Alvares and Nicholas Bryan
Centre for Radiochemistry Research, School of Chemistry,
The University of Manchester, Manchester, UK

RSC Publishing

The proceedings of the eighth Actinide Conference, Actinides 2005, held at the University of Manchester, UK on 4–8 July 2005.

The cover graphic is a depiction of part of the polar, three-dimensional network structure of $NpO_2(IO_3)$ showing the alignment of the stereochemically active lone pair of electrons on the iodate anions along the c-axis. This compound also contains neptunium oxide sheets formed from cation-cation interactions between neptunyl units. This graphic was provided by Prof. Thomas E. A. Albrecht-Schmitt of Auburn University.

Special Publication No. 305

ISBN-10: 0-85404-678-X
ISBN-13: 978-0-85404-678-2

A catalogue record for this book is available from the British Library

Published by The Royal Society of Chemistry,
Thomas Graham House, Science Park, Milton Road,
Cambridge CB4 0WF, UK

Registered Charity Number 207890

For further information see our web site at www.rsc.org

Printed by Henry Lings, Dorchester, Dorset, UK

Preface

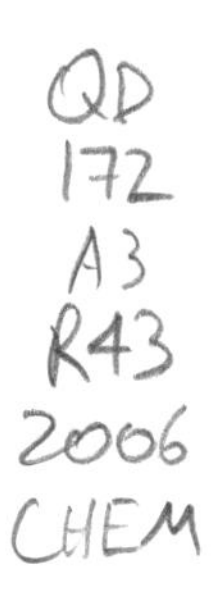

This book consists of 231 papers based on contributions to the international conference, ACTINIDES 2005 (4th – 8th July). Held in Manchester in the UK, this conference was organised by the Centre for Radiochemistry Research (CRR) through The University of Manchester.

The ACTINIDES conference series started in Baden-Baden (1975) and this first meeting was followed by Asilomar (1981), Aix-en-Provence (1985), Tashkent (1989), Santa Fe (1993), Baden-Baden (1997) and Hayama (2001), before coming to the UK for the first time in Manchester (2005). The ACTINIDES conference covers a wide range of science and technology from basic academic research through to the full range of nuclear applications. ACTINIDES 2005 was organised by Francis Livens, with assistance from Christine Martyniuk, Nick Bryan, Steve Faulkner, Iain May and the CRR research group, and with the full support of Professor Gerry Lander (ITU, Germany), Chair of the International Advisory Committee. Financial support was provided by ACTINET, AWE, EPSRC and Nexia Solutions.

Over 300 delegates from 21 countries attended the conference, which was opened by the President and Vice-Chancellor of The University of Manchester, Professor Alan Gilbert. Overview lectures were given by Graham Fairhall (Nexia Solutions, UK), Brian Bowsher (AWE, UK) and Jacques Bouchard (CEA, France) before the technical programme got underway. This programme consisted of 5 plenary lectures, numerous invited and contributed lectures and two extensive poster sessions.

This book is split into 6 chapters which cover the major topic areas of the conference: i) analysis, the environment and biotransformations, ii) coordination and organometallic chemistry, iii) heavy elements, iv) nuclear fuels, materials and waste forms, v) separation and solution chemistry and vi) spectroscopy, magnetism and superconductivity. Contributions to two additional themed sessions, based on the work of the UK actinide chemistry network and of the ACTINET European network for actinide sciences, are incorporated within the appropriate technical chapters.

Due to its wide range of topics and vast number of papers, this book provides a comprehensive overview of the current status of international actinide research. Until the publication of the Proceedings of ACTINIDES 2009, it remains the most complete, up-to-date reference volume for actinide science.

Dr Rebeca Alvarez
Centre for Radiochemistry Research and
School of Earth, Atmospheric and Environmental Science
The University of Manchester

Dr Nicholas Bryan
Centre for Radiochemistry Research
School of Chemistry
The University of Manchester

Dr Iain May
Centre for Radiochemistry Research
School of Chemistry
The University of Manchester

Contents

Analysis, the Environment and Biotransformations

Coordination and Organometallic Chemistry

Heavy Elements

Nuclear Fuels, Materials and Waste Forms

Separation and Solution Chemistry

Spectroscopy and Magnetism

Analysis, the Environment and Biotransformations

BIOTRANSFORMATION OF ACTINIDES

T. Ohnuki

Advanced Science Research Center, Japan Atomic Energy Research Institute,
2-4 Shirakata, Tokai, Ibaraki, 319-1195 Japan

1 INTRODUCTION

The presence of actinides in nuclear reactors and radioactive wastes is a major environmental concern, due to their long radioactive half-lives, their high energy radiation emissions, and their chemical toxicity. In order to estimate the potential impact of actinides on human beings, the mobility of actinides has been examined in terms of its interactions with soils and subsoils composed of abiotic and biotic components, principally minerals and bacteria.[1-7] Among the biotic components, some microorganisms have cells whose surfaces sorb actinides.[6-11] The high capacity of microbial surfaces to bind actinides may affect the migration of actinides in the environment. However, we have only limited knowledge of the role of microorganisms in the migration of actinides in the environment.

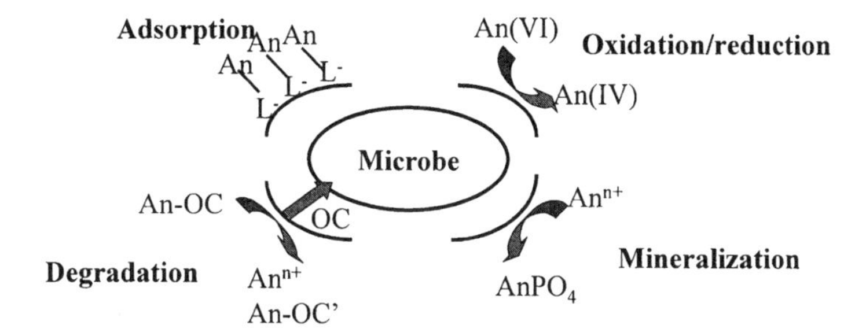

Figure 1 *Schematic interaction of actinides with a microorganism. An are actinides, L ligands on cell surfaces, OC organic carbon.*

The interaction of actinides with microorganisms involves (i) adsorption, (ii) oxidation/reduction, (iii) degradation of actinide-organic component complexes and (iv) mineralization (Fig. 1). The interaction of actinides with microorganisms results in changes in their chemical state (biotransformation). The microbiology research group of JAERI is conducting basic scientific research on microbial interactions with actinides. Fundamental research on microbial transformations of actinides involves elucidating the mechanisms of dissolution and precipitation of various chemical forms, such as ions, oxides, and organic and inorganic complexes of actinides by aerobic or anaerobic microorganisms under relevant microbial process conditions. In the present report, recent findings from the heavy elements microbiology research group of JAERI are summarized.

2 ADSORPTION OF Pu(VI) AND U(VI) BY LICHEN AND YEAST

Lichens are symbiotic organisms consisting of fungal (mycobiont) and photosynthetic (photobiont) components, the latter of which may be a green alga or cyanobacterium. Lichens occur worldwide[12] and on account of their dominance in certain terrestrial ecosystems (especially arctic/Antarctic tundra regions), play a major role in plant ecology and the cycling of some elements, such as C, N and P,[12,13] and radionuclides.[14] The ability of lichens to accumulate metals has led to their use in monitoring Pu fall-out from accidents, e.g. from Chernobyl[15] and nuclear weapons tests.[16] The concentration of Pu in reindeer lichen (*Cladonia* subgen. *Cladina*) has been studied,[16,17] because *Cladonia* species form the first link of the lichen-reindeer-man subarctic food chain. These findings suggest that lichen affects the long-term migration of Pu in the environment. However, the accumulation mechanisms have not yet been identified.

Actinides can exist in different oxidation states of III, IV, V and VI in solution, and their chemical behaviour depends on the oxidation state. Actinides (V) and (VI) are more mobile than actinides (III) and (IV) in groundwater. Ohnuki et al. have studied the uptake of plutonium (VI) and uranium (VI) by lichen biomass of foliose lichen *Parmotrema tinctorum* to elucidate the migration behaviour of Pu and U in the terrestrial environment.[18]

In accumulation experiments, discs 1 cm diam., weighing *ca* 9 mg cut from the outer margin of *Parmotrema tinctorum* (Nyl.) Hale were exposed to a Pu (VI) or U (VI) solution of 4.0×10^{-4} mol L^{-1} for 96 h at pH 3, 4 and 5. The oxidation states of Pu and U in the solutions were determined by UV/VIS absorption spectroscopy. The oxidation states of Pu and U accumulated by *P. tinctorum* were determined by UV/VIS absorption spectroscopy on extracts from the Pu/U-accumulated samples with 50% H_3PO_4 solutions.

Plutonium and uranium uptake by *P. tinctorum* after 96 h incubation averaged 0.040 and 0.055 g g_{dry}^{-1}, respectively. SEM observations showed that the accumulated Pu is evenly distributed on the upper and lower surfaces of *P. tinctorum*. On the contrary, U(VI) was accumulated in medullary layers as well as both cortexes (Fig. 2). Interestingly, U in the region of the algal layer was below the detection limit. UV/VIS absorption spectroscopy demonstrated that a fraction of Pu(VI) in the solution was reduced to Pu(V), and the accumulated Pu on the surface was reduced to Pu(IV). Meanwhile, U(VI) maintained the oxidation state of VI in the solution and on the lichen.

Plutonium (VI) was reduced to Pu(V) in the exudates solution, in which *P. tinctorum* was immersed in a 0.01 M NaCl solution for 96 h and then *P. tinctorum* was removed. These results indicate that the exudates reduce Pu from VI to V in solution. Melanin can act as an electron donor.[19] Approximately 5% Pu(VI) was reduced to Pu(IV) in the solution containing the exudates within 480 h contact. This reduction was caused by the exudates of organic substances released from *P. tinctorum*. Metabolites other than melanin may have a similar potential to melanin to be electron donors.[20] These findings suggest that the organic substances on the cortical cortex directly transfer electrons from their functional groups to the sorbed Pu(VI) and Pu(V) to Pu(IV). This direct electron transfer is one of the possible mechanisms for the reduction of Pu(VI) and Pu(V). Since the solubility of Pu(IV) hydroxides is very low, reduced Pu(VI) does not penetrate to medullary layers, but is probably precipitated as Pu(IV) hydroxides on the cortical lichen surface.

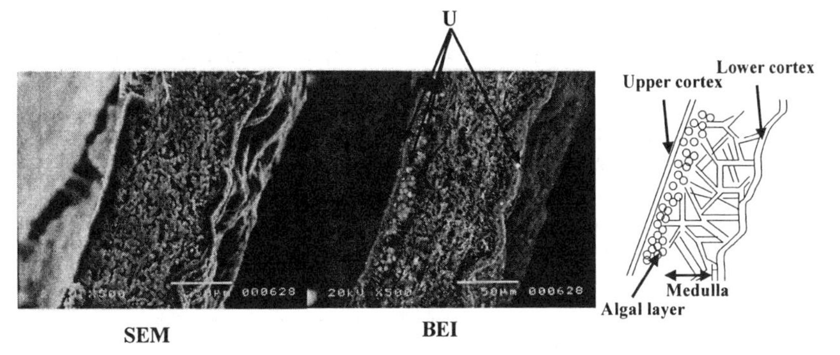

Figure 2 *SEM and backscattered electron images (BEI) of the cross section of the U(VI) accumulated P. tinctorum. Brighter area shows higher U concentration in BEI.*

Similar adsorption experiments have been carried out to examine the interactions of U(VI) and Pu(VI) with *Saccharomyces cerevisiae* and thereby elucidate the accumulation mechanism of actinides(VI) by microorganisms. In the accumulation experiments, pre-cultured *S. cerevisiae* was exposed to $4x10^{-4}$ M U(VI) or Pu(VI) in 0.01 M NaCl solution with an initial pH between 3 and 5. Concentrations and oxidation states of U and Pu in the solutions were measured at predetermined intervals. Oxidation states of the sorbed Pu and U were determined by UV/VIS spectrometry and XANES, respectively.

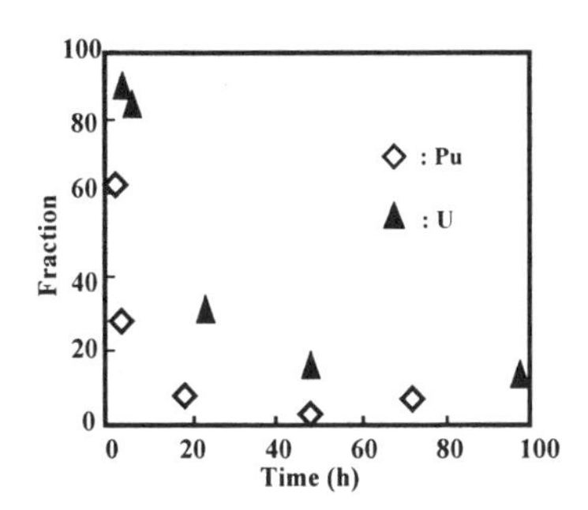

Figure 3 *Time course of sorption of Pu(VI) and U(VI) by yeast Saccharomyces cerevisiae.*

Time courses of sorption of Pu(VI) and U(VI) by the yeast cells (Fig. 3) showed abrupt decreases in solution fraction with time after exposure to the yeast. Interestingly, the accumulation rate of Pu was higher than that of U. More than 90% of Pu and U were accumulated by the yeast under experimental conditions. These results indicated that the cell surfaces of the yeast also have high affinities for Pu and U.

Only one peak around 570 nm (Pu(V)) was distinguished in the UV/VIS spectrum of the Pu solution at 2 h after the exposure (Fig. 4a). No peak around 831 nm (Pu(VI) appeared. No peak for Pu was distinguished in the UV/VIS spectrum of the Pu solution at

24 h. These results indicated that the oxidation state of Pu in the solution changed from VI to V within 2 h. A sharp peak appeared around 831 nm, and small peaks around 642 and 667 nm (Pu(IV)) were distinguished in the UV/VIS spectrum of the extract from the Pu-accumulated *S. cerevisiae* at 2 h after the exposure (Fig. 4b). Peak intensity around 642 and 667 nm increased with the exposure time; by contrast, peak intensity around 831 nm decreased. It is known that Pu(V) in the acid solution changes its oxidation state to IV and VI by disproportionation, i.e.,

$$2Pu(V)O_2^{+} + 4H^{+} = Pu(VI)O_2^{2+} + Pu^{4+} + 2H_2O$$

Thus, the oxidation state of Pu was V at 2 h, and changed to IV with increasing time.

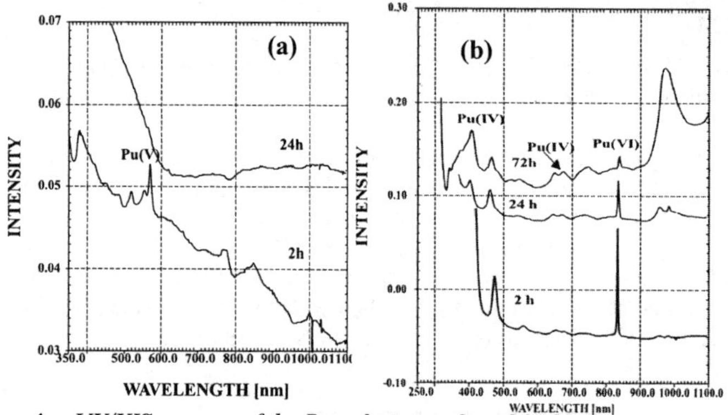

Figure 4 *UV/VIS spectra of the Pu solution at 2 and 24 h after exposure to yeast (a), and the extracts with 50% H_3PO_4 solutions from the Pu-accumulated yeast for 2, 24 and 72 h (b).*

These results showed that Pu was highly accumulated in microorganisms after their exposure to it. Even though the oxidation state of Pu kept at VI in the solution without microorganisms, it changed from VI to V in solution after exposure to microorganisms, probably by released exudates. Direct electron transfer from functional groups of organic substances to the sorbed Pu(VI) and Pu(V) is one of the possible mechanisms for the reduction of Pu(VI) and Pu(V). The Pu(VI) reduction occurs by a two step reaction from VI to IV through V, and differs from that of U(VI), where no reduction occurred.

3 ADSORPTION OF Ln(III) AND An(III) ON CELL SURFACES OF MICROORGANISMS

Various microbial species have different cell-surface characteristics. *Chrorella vulgaris* is an autotrophic unicellular alga having cellulose as cell-wall components. *Bacillus subtilis* is a Gram-positive bacterium whose cell surface is composed of peptidoglycan and teicoic acid. *Pseudomonas fluorescens* is a Gram-negative bacterium. Its cell wall is more complex in both chemical and structural terms than that of a Gram-positive bacterium, and its external outer membrane is composed of protein, phospholipid and lipopolysaccharide. *Halomonas* sp., *Halomonas salinarum* and *Halomonas halobium* are halophic bacteria having highly hydrophobic cell surfaces. The association of Eu(III) and Cm(III) with these microorganisms was studied to elucidate the effects of cell surface structure on adsorption.

Ozaki et al. have studied the kinetics of Eu(III) and Cm(III) sorption by various kinds of bacteria at pH 3-5.[21,22] The amounts of organic carbon exuded from *C. vulgaris*, *Halomonas* sp. and *H. halobium* were determined by measurements of the dissolved organic carbon (DOC). The kinetics of the adsorption of Eu(III) and Cm(III) on the cell surfaces of *C. vulgaris* indicated that adsorption reached a maximum in a very short time, and then decreased with increasing time. This decrease reflects the presence of exuded organic carbon, which desorbed Eu(III) and Cm(III) from the cell surfaces.[22] A similar tendency was observed for *Halomonas* sp. at pH 5 and *H. halobium* at pH 4. These imply that exudates from bacteria enhance the mobility of actinides and lanthanides.

The kinetics of Eu(III) and Cm(III) sorption showed no significant difference between Eu(III) and Cm(III) for *C. vulgaris*, *B. subtilis*, *P. fluorescens*, and *H. balobium*. However, a difference in the sorption kinetics of Eu(III) and Cm(III) by *Halomonas* sp. was observed. Higher amounts of Cm(III) than Eu(III) were accumulated at 20 min after exposure to *Halomonas* sp., suggesting a difference in the affinity of Eu(III) and Cm(III) to the functional groups of its cell surfaces. A slight higher fraction of Cm(III) was accumulated by *H. halobium* than Eu(III). These results suggest that halophilic bacteria have different sorption sites for actinides from non-halophilic bacteria.

4 EFFECTS OF METABOLITES ON THE SORPTION ACTINIDES

As mentioned above, exudates from microorganisms reduce the sorption of Eu(III) by bacteria. Naturally occurring chelating substances also have the potential to reduce the sorption of actinides and lanthanides by forming complexes in the environment. Siderophores, produced by microorganisms, access insoluble cations and form complexes, not only with Fe but also with actinides, causing their solubility to increase.[23] Yoshida et al.[24-26] have studied the effects of desferrioxamine (DFO) B on the sorption of trivalent and tetravalent lanthanides and actinides by soil bacteria of *P. fluorescens* and *B. subtilis*.

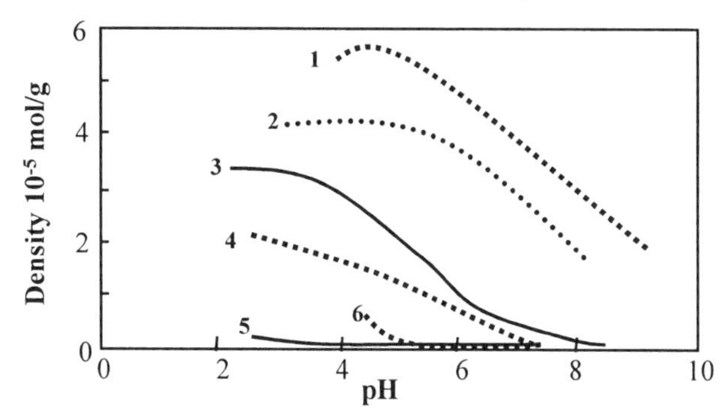

Figure 5 *Sorption density of Th(IV) and Pu(IV) on P. fluorescens and B. subtilis in the presence of DFO. Initial concentrations of Fe(III), Eu(III), Th(IV), Pu(IV), and DFO were 20 μM. 1: Eu on P. fluorescens, 2: Th on P. fluorescens, 3: Th on B. subtilis, 4: Pu on P. fluorescens, 5: Pu on B. subtilis, 6: Fe on P. fluorescens. Data are plotted based on refs.*[24,25]

The sorption density of Pu(IV) and Th(IV) on bacterial cells, and Fe(III) and Eu(III) on *P. fluorescens* in the presence of DFO (Fig. 5) indicated that sorption of Pu(IV) on *P. fluorescens* increased from 3 to 19 μM g^{-1} with a decrease of pH from 7.3 to 3.0, while sorption of Pu(IV) on *B. subtilis* and Fe(III) on *P. fluorescens* was smaller than 3 μM g^{-1} at about pH 3-8. Sorption of Th(IV) on *P. fluorescens* increased from 21 to 42 μM g^{-1} with a decrease of pH from 7.5 to 4.0, and sorption of Th(IV) on *B. subtilis* increased from 3 to 31 μM g^{-1} with a decrease of pH from 7.8 to 3.3. On the contrary, adsorption of DFO on both species was negligible at 3 hours after contact of the 1:1 Th(IV)-DFO complex with *P. fluorescens* or *B. subtilis* cells at pH 5.5. No DFO was sorbed on *P. fluorescens* cells from Eu(III)-DFO complexes. These results indicate that Th(IV)-, Pu(IV)- and Eu(III)-dissociate by contact with cells, after which the metals are sorbed.

Stability constants of the metal-DFO complexes decrease in the order of Pu(IV) (log K = 30.8)[27] > Fe(III) (log K = 30.6)[28] > Th(IV) (log K = 26.6)[29] > Eu(III) (log K = 15).[30] Adsorption density of Eu(III), Th(IV), and Pu(IV) on *P. fluorescens* cells decreased in the order Eu(III) > Th(IV) > Pu(IV), which corresponds to the increasing order of the stability constant of the DFO complexes. Adsorption of hydrated Eu(III) on *P. fluorescens* cells does not change significantly at pH 3 - 8,[8, 31] indicating that the affinity of *P. fluorescens* cell surfaces with metal ions is not changed significantly at these pHs. These facts indicate that pH dependence of adsorption density of metal ions on cells is dominated by the stability of the metal-DFO complexes.

Yoshida et al.[26] investigated the influence of DFO on the sorption behavior of 11 rare-earth elements (REEs), La, Ce, Pr, Nd, Sm Eu, Gd, Tb, Dy, Ho, and Er on *P. fluorescens* cells at neutral pH. *Pseudomonas fluorescens* cell suspensions (1.3 g (dry wt.) L^{-1}) in the suspensions were incubated for 30 minutes in 10 ml of 0.1 M Tris-HCl solutions containing 1.0 mg L^{-1} of each REE (La, 7.20 μM; Ce, 7.14 μM; Pr, 7.10 μM; Nd, 6.93 μM; Sm, 6.65 μM; Eu, 6.58 μM; Gd, 6.36 μM; Tb, 6.29 μM; Dy, 6.15 μM; Ho, 6.06 μM; Er, 5.98 μM) and 0.5 mM DFO in air at room temperature. The oxidation state of Ce was determined by X-ray absorption near edge structure (XANES) spectroscopy in fluorescence mode at the BL27B line at the High Energy Accelerator Research Organization (Tsukuba, Japan).

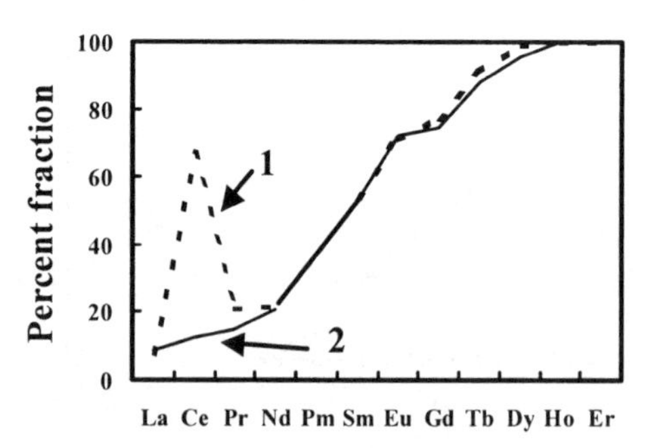

Figure 6 *Percent fraction of REEs complexed with DFO in the solutions after exposure to P. fluorescens cells. 1: without hydroxylammonium and 2: with hydroxylammonium.*

In the presence of DFO, the percent fraction of REEs in the solution after exposure to *P. fluorescens* cells (Fig. 6) showed a tendency to increase with increasing atomic number, except for Ce. The adsorption of Ce was significantly lower than those of the neighbouring REEs, La and Pr. On the contrary, no Ce anomaly in the sorption was distinguished in the solution with hydroxylammonium. XANES analysis of Ce in the Ce-DFO complex showed that Ce was in the tetravalent state. Adding hydroxylammonium reduced the tetravalent Ce in the complex to its trivalent form and erased the Ce anomaly (Fig. 6). These results show that DFO can oxidize Ce(III) to Ce(IV). Cyclic voltammetry revealed that the redox potential of the Ce(IV)/Ce(III) couple in the DFO complex was much lower than the standard redox potential, and that the stability of Ce(IV)-DFO is much higher than that of Ce(III)-DFO. These findings suggest that the observed Ce anomaly is due to the oxidation of the Ce(III)-DFO complex to the more stable Ce(IV)-DFO complex, and that naturally occurring organic ligands can contribute to this Ce anomaly in the natural environment.

5 SORPTION OF U(VI) BY THE MIXTURES OF MICROORGANISM AND CLAY

Soils and subsoils are composed of abiotic and biotic components, principally minerals and bacteria. It is therefore important to elucidate the role of microorganisms on the accumulation of actinides in such mixtures. Ohnuki et al. assessed the accumulation of uranium (VI) by a bacterium, *Bacillus subtilis*, suspended in a slurry of kaolinite clay, to elucidate the role of microbes on the mobility of U(VI).[32] Various mixtures of bacteria and the koalinite were exposed to solutions of 8×10^{-6} M and 4×10^{-4} M-U(VI) in 0.01 M NaCl at pH 4.7. After 48 h, the mixtures were separated from the solutions by centrifugation, and treated with a 1 M CH_3COOK for 24 h to determine the associations of U within the mixture. The mixture exposed to 4×10^{-4} M U was analyzed by a transmission electron microscope (TEM) equipped with EDS.

The accumulation of U by the mixture increased with an increase in the amount of *B. subtilis* cells present at both U concentrations. Treatment of kaolinite with CH_3COOK removed approximately 80% of the associated uranium. However, in the presence of *B. subtilis* the amount of U removed was much less. TEM-EDS analysis confirmed that most of the U removed from solution was associated with *B. subtilis*. XANES analysis of the oxidation state of uranium associated with *B. subtilis*, kaolinite, and with the mixture containing both revealed that it was present as U(VI). These results suggest that the bacteria have a higher affinity for U than the kaolinite clay mineral under the experimental conditions tested, and that they can immobilize significant amounts of uranium.

Acknowledgements.

The author thanks Drs. T. Ozaki, F. Sakamoto, N. Kozai, T. Yoshida, T. Nankawa and Y. Suzuki for their contribution to the research in Heavy Elements Microbiology Research Group, ASRC, JAERI. The author also thanks Dr. A. J. Francis of Brookhaven National Laboratory for fruitful discussions.

References

1 K. V. Ticknor, *Radiochim. Acta.*, 1994, **64**, 229.
2 T. D. Waite, J. A. Davis, T. E. Payne, G. A. Waychunas and N. Xu., *Geochim. Cosmochim. Acta,* 1994, **58,** 5465.
3 E. R. Sylwester, E. A. Hudson, and P. G. Allen., *Geochim. Cosmochim. Acta,* 2000, **64**, 2431.
4 A. J. Dent, J. D. Ramsay and S. W. Swanton., *J. Colloid Interface Sci,* 1992, **150**, 45.
5 A. J. Francis, J. B. Gillow, C. J. Dodge, R. Harris, T. J. Beveridge, and H. W. Papenguth, *Radiochim. Acta,* 2004, **92,** 481.
6 D. A. Fowle, J. B. Fein, and A. M. Martin, *Sci. Technol,* 2000, **34,** 3737.

7 J. H. Haas, T.J. Dichristina, and R. Wade Jr, *Chem. Geol.*, 2001. **180**, 33.
8 Y. Suzuki, T. Nankawa, T. Yoshida, T. Ozaki, T. Ohnuki, A. J. Francis, S. Tsushima, Y. Enokida, and I. Yamamoto, *J. Nucl. Radiochem. Sci.*, in press.
9 S. L. Brantley, L. Liermann and M. Bau, *Geomicrobiology J.,* 2001, **18**, 37.
10 P. J. Panak, and H. Nitsche, *Radiochim. Acta,* 2001, **89**, 499.
11 S. G. John, JC. E. Ruggiero, L. E. Hersman, C. S. Tung and M. P. Neu, *Environ. Sci. Technol.,* 2001, **35**, 2942.
12 T. H. Nash, Nutrients, elemental accumulation and mineral cycling. In: T. H. Nash, (ed.), *Lichen biology*, 1996. 136.
13 J. M. H. Knops, T. H. Nash, V. L. Boucher, and W. H. Schlesinger, *Lichenologist,* 1991, **23**, 309.
14 G. Shaw, and J. N. B. Bell, Plants and Radionuclides. In: M. E. Farago (ed.) *Plants and Chemical Elements: Biochemistry, Uptake, Tolerance and Toxicity.* VCH. Weinheim, 1994. New York. 180.
15 J. Paatero, T. Jaakkola, and S. Kulmala, *J. Environ. Radioactivit*, 1998, **38**, 223.
16 W. G. Hanson, *Health Phys,* 1972, **22,** 39.
17 E. Holm, and R. B. R. Persson, *Radiat. Environ. Biophys.,* 1978, **15**, 261.
18 T. Ohnuki, H. Aoyagi, Y. Kitatsuji, M. Samadfam, Y. Kimura and O. W. Purvis, *J. Environ. Radioact.*, 2004, **77**, 339.
19 R. D. Lillie, in *Pigments in Pathology* (*Histochemistry of melanins*) (ed. Wolman, M.) Academic Press, New York, 327 (1969).
20 C. Saiz-Jiminez and F. Shafizadeh, *Curr. Microbiol.,* 1984, **10**, 281.
21 T. Ozaki, J. B. Gillow, T. Kimura, T. Ohnuki, Z. Yoshida and A. J. Francis, *Radiochim. Acta,* 2004, **92**, 741.
22 T. Ozaki, T. Kimura, T. Ohnuki, Z. Yoshida and A.J. Francis, *Environ. Toxicol. Chem.,* 2003, **22**, 2800.
23 J. R. Brainard, B. A. Strietelmeier, P. H. Smith, P. J. Langston-Unkefer, M. E. Barr and R. R. Ryan, *Radiochim. Acta,* 1992, **58-59**, 357.
24 T. Yoshida, T. Ozaki, T. Ohnuki and A.J. Francis, *J. Nucl. Radiochem. Sci.*, 2005, (in press).
25 T. Yoshida, T. Ozaki, T. Ohnuki and A. J. Francis, *Radiochim. Acta,* 2004, **92**, 749.
26 T. Yoshida, T. Ozaki, T. Ohnuki and A. J. Francis, *Chem. Geol.*, 2004, **212**, 239.
27 N. V. Jarvis and R. D. Hancock, *Inorg. Chim. Acta,* 1991, **182**, 229.
28 G. Schwarzenbach and K. Schwarzenbach, *Helv. Chim. Acta,* 1963, **46**, 1390.
29 D. W. Whisenhunt, M. P. Neu, Z. Hou, J. Xu, D. C. Hoffman, and K. N. Raymond, *Inorg. Chem.*, 1996, **35**, 4128.
30 J. R. Brainard, B. A. Strietelmeier, P. H. Smith, P. J. Langston-Unkefer, M. E. Barr, and R. R. Ryan, *Radiochim. Acta,* 1992, **58/59**, 357.
31 T. Yoshida, T. Ozaki, T. Ohnuki, A. J. Francis, Eu(III) adsorption by *Pseudomonas fluorescens* in the presence of organic ligands, *Proceedings of the international Symposium on Radioecology and Environmental Dosimetry*, ed. J. Inaba, H. Tsukada, and A. Takeda, Institute of Environmental Sciences, Aomori, Japan, 2003, p296.
32 T. Ohnuki, T. Yoshida, T. Ozaki, M. Samadfam, N. Kozai, K. Yubuta, T. Mitsugashira, T. Kasama, and A. J. Francis, Interactions of Uranium with Bacteria and Kaolinite Clay, *Chem. Geol.*, 2005, (in press).

MICROBIAL TRANSFORMATIONS OF ACTINIDES IN TRANSURANIC- AND MIXED-WASTES: IMPLICATIONS FOR RADIOACTIVE-WASTE DISPOSAL

A.J. Francis

Environmental Sciences Department
Brookhaven National Laboratory, Upton, New York 11973, USA

1 INTRODUCTION

The presence of actinides (U, Np, Pu, Am), organic- (cellulose, plastics, rubber, chelating agents) and inorganic- (nitrate and sulfate) compounds in transuranic (TRU) and mixed wastes are a major concern, because of their potential for migration from the waste repositories and contamination of the environment. The primary causes of this disquiet are the toxicity of the actinide elements and the long half-lives of their isotopes. The radionuclides may be present in TRU wastes in various forms, such as elemental, oxide, co-precipitates, ionic, inorganic-, and organic-complexes, and also as naturally occurring minerals, depending on the process and waste stream.

The actinides existing in various oxidation states that are the ones of concern are III (Am, Pu, U), IV (Pu, U), V (Np, Pu), and VI (Pu, U). Significant aerobic- and anaerobic-microbial activity is expected to occur in the waste because of the presence of electron donors and acceptors. The actinides initially may be present as soluble- or insoluble-forms but, after disposal, may be converted from one to the other by microorganisms.[1] The direct enzymatic or indirect non-enzymatic actions of microbes could alter the speciation, solubility, and sorption properties of the actinides, thereby increasing or decreasing their concentrations in solution. Dissolution of radionuclides reflects changes in the Eh and pH of the local environment caused by the microorganisms, by their production of CO_2, or of extra cellular metabolic products, such as organic acids, and sequestering agents such as siderophores. Immobilization or precipitation of radionuclides is due to changes in the Eh of the environment, enzymatic reductive precipitation (reduction from a higher to lower oxidation state), biosorption, bioaccumulation, biotransformation of radionuclide-organic and -inorganic complexes, and bioprecipitation.[1] Free-living bacteria suspended in the groundwater fall within the colloidal size range and may have a strong radionuclide sorbing capacity, giving them the potential to transport radionuclides in the subsurface. Further, gases generated from the biodegradation of TRU wastes, thereby pressurizing the containment areas, with subsequent reduction in the volume of the waste and subsidence in the repository. Microbial corrosion of the waste canisters also is a major concern.

Although the physical-, chemical-, and geochemical-processes affecting dissolution, precipitation, and mobilization of actinides in TRU waste have been investigated, we have limited information on microbial transformations of actinides in wastes and its impact on their environmental mobility and stability.

2 MICROORGANISMS IN RADIOACTIVE WASTES, NATURAL ANALOGUE SITES, BACKFILL MATERIALS, AND WASTE-REPOSITORY SITES.

Microorganisms have been detected in low-level radioactive wastes, transuranic wastes, Pu-contaminated soils, and in waste-repository sites being considered for the disposal of nuclear waste.[2] Pu-238, 239, 240 (gross alpha activity 1.7 x 10^5 pCi/L) was found in leachates collected from the low-level radioactive-waste disposal sites at West Valley, NY, and Maxey Falts, KY. Several aerobic- and anaerobic- bacteria were isolated from the leachate samples, among them, *Bacillus* sp., *Pseudomonas* sp., *Citrobacter* sp., and *Clostridium* sp. The radioactive- and organic-chemicals present in the leachate were not toxic to the bacteria, which metabolized them, producing tritiated- and carbon-14 methane.[1] Metabolically active microbes were identified at the Los Alamos National Laboratory's (LANL) TRU waste burial site containing ^{239}Pu contaminated soil and flammable waste.[3] Even extreme environments, such as the hypersaline groundwaters at the Waste Isolation Pilot Plant (WIPP) site, and the extremely low-nutrient granodiurite pore-waters in Switzerland, harbour microorganisms able to interact with actinides in wastes.[4]

2.1 Analogue sites.

A survey of the distribution of microbial population in water- and subsoil-samples from deep mines designated for disposal of high-level radioactive waste in Europe showed the presence of a variety of organisms representing autotrophic and heterotrophic groups including both native species, and those introduced by mining operations.[5]

2.1.1 Pocos de Caldos, Brazil. Core materials and groundwater samples collected at various depths from boreholes at the Osamu Utsumi open-pit uranium mine and Morro do Ferro Th/REE ore body, Pocos de Caldos, Brazil, contained microbes. The groundwater had higher numbers of bacteria than the solid material, and was dominated by the sulphur-cycle bacteria.[6]

2.1.2 Oklo, Africa. Many different bacteria inhabit the Oklo site in Gabon, the only known natural fission reactor that has plutonium, suggesting that they are likely to thrive in constructed nuclear-waste repositories.[7]

2.1.3 Cigar Lake, Canada. The high-grade uranium deposit at Cigar Lake is a natural analogue site for disposing of nuclear waste. Microorganisms are present in the ore zone and groundwater samples, and seems able to withstand the radiation. Anaerobic bacteria were 10 times more abundant than the aerobes, in agreement with the prevailing reducing chemical environment in the deposit, thereby reducing the potential dissolution and mobilization of redox-sensitive radionuclides. Sulphate-reducing, iron-reducing and denitrifying bacteria are common in both groundwaters and rocks. The latter were the predominant group in all the water samples tested; an adequate supply of metabolizable carbon was present in the samples, although denitrification was limited by the availability of nitrate.[8, 9]

2.1.4 Yucca Mountain Site. The proposed geologic repository at Yucca Mountain (YM), Nevada is located approximately 300 meters below the surface and 300 meters above the groundwater table in an unsaturated region of welded volcanic tuff. Conditions anticipated after waste emplacements include the eventual infiltration of water from surface precipitation. The repository is expected to remain aerobic throughout its evolution. Microorganisms, both native and introduced, potentially could degrade engineered barrier systems (e.g., waste packages, rock bolts, steel sets, or drip shields). Over time, temperatures are predicted to fall, and the water to increase during the repository's operation. Characterization of the microbial communities in the vadose zones focused on correlating the types, numbers, and their interactions with extant geological features, as well as the availability of water and carbon.[10,11] These studies showed that bacteria are present in the vadose zones thus far examined, and, further, that Thiobacilli and sulfate-reducing bacteria, which can corrode metals, exist in the rock and soil of the Nevada Test Site and Yucca Mountain.[12, 13] The presence of an active microbial population in deep subsurface environments clearly suggests that under appropriate conditions they could play a significant role in transforming and transporting radionuclides.

3 MICROBIAL GENERATION OF GAS FROM BIODEGRADATION OF TRU WASTE AT THE WASTE ISOLATION PILOT PLANT (WIPP).

Seventy-percent of the TRU waste disposed of in the WIPP repository is cellulose; its biodegradation under the repository's hypersaline conditions can produce CO_2 and also affect the solubility of the actinides.[14] Gas generation from the microbial degradation of the organic constituents of TRU waste (mixed cellulose and electron-beam irradiated plastics and rubbers) under conditions expected at the WIPP repository showed aerobic-, denitrifying-, fermentative-, sulfate-reducing-, and methanogenic-activity (Gillow et al., 2005, *in preparation*). The rate of total gas production in anaerobic samples was as follows: 0.002 ml g^{-1} cellulose day^{-1} without nutrients; 0.004 ml g^{-1} cellulose day^{-1} in the presence of nutrients; and, 0.01 ml g^{-1} cellulose day^{-1} with excess nitrate. Correspondingly, the rate of CO_2 production was 0.009 μmol g^{-1} cellulose day^{-1}, 0.05 μmol g^{-1} cellulose day^{-1}, and 0.2 μmol g^{-1} cellulose day^{-1} (Figure 1). The brine contained lactic-, acetic-, propionic-, butyric-, and valeric-acid produced by cellulose degradation. Carbon dioxide was produced in samples incubated at 70% relative humidity, and adding bentonite, a potential backfill material that contains a significant amount of iron, enhanced gas production. Gas was not generated from the biodegradation of electron-beam irradiated plastic and rubber. Experimental findings from actual and simulated TRU waste-degradation studies showed that microorganisms produce far more gas than released by physical and chemical means, including corrosion.[15] Microbial activity is expected to be the dominant process responsible for degrading TRU waste and to have the greatest impact on performance assessment and compliance.

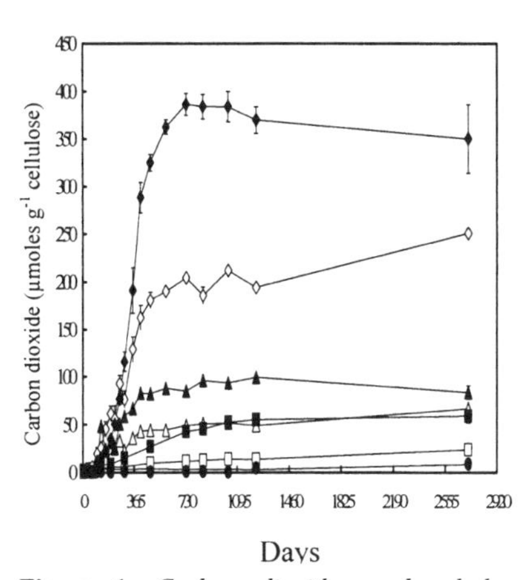

Figure 1. *Carbon dioxide produced by anaerobic samples inundated with brine: unamended (○); unamended and inoculated (□); amended and inoculated (Δ); amended, inoculated, plus excess nitrate (◊). Closed symbols are samples with bentonite (Gillow and Francis 2005, in preparation).*

4 MICROBIAL TRANSFORMATIONS OF URANIUM IN WASTES

We investigated uranium-waste samples collected from the West End Treatment Facility, at the U.S. Department of Energy, Y-12 Plant, Oak Ridge, TN. The sludge was generated from a uranium-process waste stream after biodenitrification of nitric-acid uranium wastewater; the sediment was taken from a contaminated pond (New Hope Pond) that received uranium-process wastewater. Both sediment- and sludge- samples contained varying levels of major elements, Al, Ca, Fe, Mg, K, and Na, along with toxic metals, As, Cd, Cr, Co, Cu, Pb, Mn, Hg, Ni, U, and Zn. The concentrations of uranium in sediment- and sludge-samples were 920 and 3100 ppm, respectively. In addition to those elements listed above, X-ray fluorescence analysis of the sediment at the National Synchrotron Light Source, Brookhaven National Laboratory, revealed the presence of titanium, gallium, bromine, strontium, rubidium, yttrium, and zirconium.

Sludge and sediment samples amended with glucose increased their production of total gas, CO_2, H_2, CH_4, and of organic acids, viz., acetic-, butyric-, propionic-, formic-, pyruvic-, lactic-, isobutyric-, valeric-, and isocaproic-acids. A significant amount of the gas produced was due to glucose fermentation by anaerobic bacteria, as well as from the

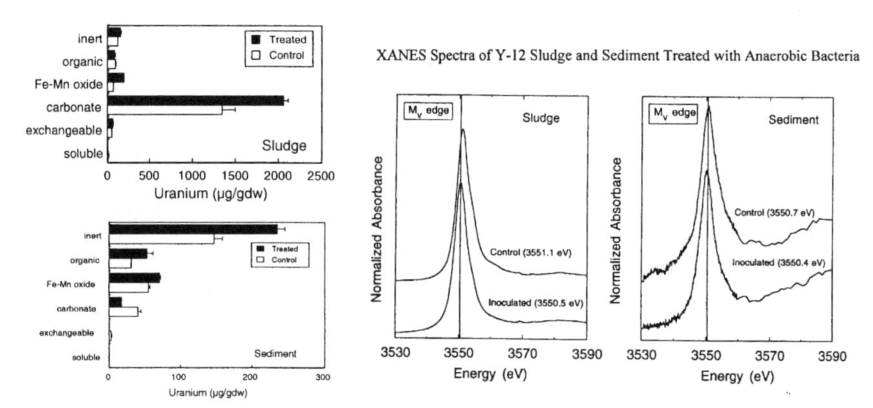

Figure 2. *Mineralogical association and speciation of uranium before and after microbial action.*[16]

dissolution of $CaCO_3$ in the sludge by the organic acids, and lowering of the pH. Only amended samples showed a decrease in sulfate concentration. This treatment removed a large fraction of soluble non-toxic metals, such as Ca, K, Mg, Mn^{2+}, Na, and Fe^{2+}, and enriched and stabilized Cd, Cr, Cu, Ni, Pb, U, and Zn with the remaining solid phase due to the direct and indirect actions of the bacteria.[16] Analysis of the mineralogical association of the metals in the wastes before and after microbiological action demonstrated that many of the metals were redistributed with stable mineral phases, such as organic and silicate fractions. Uranium was predominantly associated with the carbonate fraction, and to a lesser extent with the oxide-, organic-, and inert-fractions; after microbial activity, its concentration increased for all three fractions (Figure 2). The metals associated with the exchangeable-, carbonate-, and iron-oxide-fractions were indirectly solubilized via the production of organic-acid metabolites, whereas the dissolution of iron oxides, and metals co-precipitated with iron oxides, reflected the direct enzymatic reduction of iron. Uranyl ion associated with the exchangeable, carbonate, and iron-oxide fractions was released into solution by the direct and indirect actions of the bacteria, and subsequently, was reduced enzymatically to insoluble U(IV). X-ray absorption near edge spectroscopic (XANES) analysis of uranium in the untreated (control) and treated sludge and sediment samples showed a partial reduction of U(VI) → U(IV) in the treated samples (Figure 2).

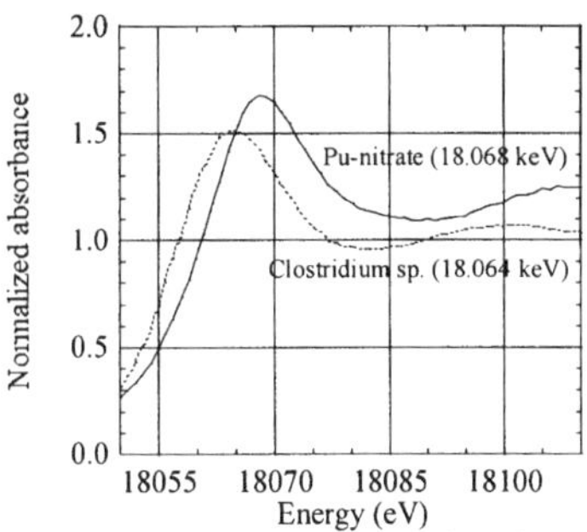

Figure 3. *XANES analysis of Pu before (-) and after (---) anaerobic microbial activity (Francis, Dodge, and Gillow, manuscript in preparation).*

5 MICROBIAL TRANSFORMATIONS OF PLUTONIUM

The solution chemistry of plutonium, which exists in several oxidation states (III, IV, V, VI, VII), is highly complex. It undergoes disproportionation reactions and can exist simultaneously as Pu(IV), Pu(V), and Pu(VI) under oxidizing conditions. Plutonium has a high ionic charge, and tends to undergo hydrolysis, leading to the formation of polymers in systems with pH>2. Soil pH, its organic-matter content, redox conditions, and mineralogy affect the chemical speciation of Pu. Chemical characterization of Pu at contaminated sites shows that its environmental form varies according to site, and depends on the waste stream. For example, at Rocky Flats, CO, the predominant form appears to be $PuO_2(s)$; at the NTS, Pu was found to be associated with mineral colloids.[17] Pu generally is considered to be relatively immobile; however, the transport of Pu, albeit at very low concentrations, was observed at several DOE sites (i.e., Rocky Flats, LANL, and NTS).

A slight increase in microbial activity (respiration) can alter the oxidation state of Pu(VI) to Pu(IV) because of the very small differences in the reduction potential between Pu(VI), Pu(V), and Pu(IV).[2] In general, Pu concentrations > 10^{-5}M seem toxic to most of the microorganisms studied.[2,18,19] This is due to radiation effects rather than metal toxicity, and is modulated by plutonium's chemical form and solubility.

The dissolution of Pu by microorganisms is brought about by their production of organic acids, such as citric acid, extracellular metabolites, and siderophores.[18,20-22] Its immobilization, on the other hand, may be caused indirectly, by changing the Eh, and facilitating abiotic precipitation of Pu, reduction from higher to lower oxidation state, biosorption, and bio-precipitation reactions.[23-25] Keith-Roach *et al.*[26] reported that minimum Pu and Am concentrations in water correspond to maximum biomass due to bioaccumulation. To date, we have only limited information on the interactions of Pu with microbial surfaces.

We investigated the biotransformation of ^{242}Pu-nitrate by the anaerobic bacterium *Clostridium* sp. Adding 1×10^{-7} M ^{242}Pu (IV)-nitrate had no effect upon its growth and metabolism of glucose. Ninety percent of the added Pu in uninoculated growth medium (control) was removed by 0.4μm filtration, and most likely existed as $Pu(OH)_4$ at pH 6.2 due to hydrolysis and polymerization. The growth of *Clostridium* sp. lowered the Eh (-180 mV), and the pH from 6.2 to 2.8, concomitant with the production of the organic acids, acetic and butyric (12 and 17 mM, respectively), and of carbon dioxide (225 μmol). After 14h of growth, 70% of the Pu passed through a 0.4 μm filter and 55% through a 0.03 μm filter. These results suggest that Pu (IV) is solubilized by *Clostridium* sp. due to its reduction to Pu(III). Solvent extraction by thenyltrifluoroacetone (TTA) confirmed a decrease in the polymeric form of Pu and an increase in the soluble fraction, suggesting the presence of Pu^{3+}. XANES analysis of the culture at the Pu L_{III} edge (18.057 keV) confirmed that the oxidation state was Pu^{3+} (Figure 3). The Eh of the medium was low and the CO_2 concentration high, so favouring the reduction of Pu from the tetravalent to the trivalent state. We observed an increase in α- and β-activity from NTS soil and Pu from Rocky Flats soil due to enhanced microbial activity.

6 NEPTUNIUM

The chemical characteristics of pentavalent actinides, for example NpO_2^+, are similar to those of simple monovalent cations: i.e., low ligand-complexing abilities with a high environmental mobility. In general, the pentavalent species of all actinides are unstable, except for Np(V) which can be the commonest form in some natural waters.[27] Neptunium exists in solution primarily as the Np(V) species that forms stable carbonate complexes. Assessments of the importance of neptunium were on the assumption that it will persist in the subsurface as a very mobile, non-sorptive, neptunyl species (NpO_2^+). However, studies demonstrated that NpO_2^+ could be biologically reduced to insoluble Np^{4+} under anaerobic conditions. Although Np^{4+} is easily oxidized in solution, it is stabilized in the presence of complexing ligands.[28] *Shewenella putrefaciens* reduced Np^{5+} to Np^{4+} that then was precipitated from solution as Np^{4+} phosphate.[29]

7 AMERICIUM

The principle oxidation state of Am is +3, with the $Am(OH)_3$ species being extremely surface-reactive. The presence of soluble carbonate complexes, specifically $AmCO_3^+$, $Am(CO_3)_2^-$,and $Am(CO_3)_3^{3-}$ is well established.[30] Microbial activity may convert $Am(OH)^{2+}$ species to Am(III)-carbonate complexes due to carbon dioxide production; likewise, biodegradation of Am(III)-organic complexes may precipitate Am as the hydroxide species. Sorption of Am to biomass has been reported. Bacteria isolated from sediments and grown with ^{241}Am in minimal medium produced exometabolites, which formed soluble complexes with Am.[31] Thus, the potential exists for the dissolution of actinides in wastes by microorganisms, thereby increasing their bioavailability and mobility.

8 BIOTRANSFORMATION OF ACTINIDE-ORGANIC COMPLEXES

Synthetic and natural chelating agents are present in TRU and mixed wastes because they are widely used for decontaminating nuclear reactors and equipment, in cleanup operations, and in separating radionuclides. Natural organic complexing agents, such as humic- and fulvic-acids, and microbially produced complexing agents, such as citrate, and siderophores, as well as synthetic chelating agents, can affect the environmental mobility of actinides. Biotransformation of actinide-organic complexes should result in the breakdown of the organic ligand with the precipitation of the actinide.

8.1 Biotransformation of uranium complexed with citrate.

The type of complex formed between metals and citric acid plays an important role in determining its biodegradability.[32] The bidentate complexes of Fe(III)-, Ni-, and Zn-citrate were readily biodegraded under aerobic conditions, whereas the complexes involving citric acid's hydroxyl group, the tridentate Al-, Cd- and Cu-citrate complexes, and the binuclear U-citrate complex were recalcitrant. The presence of the free hydroxyl group of citric acid is the key determinant in affecting biodegradation of the metal-citrate complex.

The presence of organic ligands affects the reductive precipitation of uranium complexed with organic ligands. Malonate, oxalate, and citrate altered the rate of U reduction by sulfate-reducing *(Desulfovibrio desulfuricans)* and the facultative iron-reducing *(Shewanella alga)* bacteria that decreased in the presence of multi-dentate ligands.[33, 34] The anaerobic bacterium *Clostridium sphenoides*, which utilizes citric acid as the sole carbon source, metabolized equimolar Fe(III)-citrate, whilst degrading citric acid and reducing Fe(III) to Fe(II); the U(VI)-citrate complex was unaffected. However, in the presence of excess citric acid or added glucose, U(VI)-citrate was reduced to U(IV)-citrate complexes. Similarly, *Clostridium* sp., which ferments glucose but not citrate, reduced U(VI)-citrate only when supplied with glucose; the reduced U(IV) was not precipitated out, but present in solution as U(IV)-citrate complexes. EXAFS analysis showed that the 1:1 U(VI):citric acid complex consisted of a binuclear di-μ-OH core with five-fold coordination of oxygen in the equatorial plane. In contrast, U(IV) was present as a bi-ligand mononuclear U(IV)-citrate complex; uranium exhibited eight-fold coordination with oxygen, and tridentate coordination to citric acid (*Francis and Dodge, manuscript in preparation*). These results demonstrate that complexed uranium is readily accessible for microorganisms as an electron acceptor, despite their inability to metabolize the organic ligand complexed to the actinide, and that the presence of this ligand can affect the reductive precipitation of uranium.

8.2 Biotransformation of Pu-citrate.

We investigated the biotransformation by *Pseudomonas fluorescens* of $^{242}Pu^{4+}$ (10^{-8} to 10^{-5} M) in the presence of excess citric acid. Analysis of ^{242}Pu-citrate by electrospray ionization-mass spectrometry (ESI-MS) indicated the presence of a bi-ligand Pu-cit$_2$ complex. Citric acid was metabolized by *P. fluorescens* at a rate of 4.9 μM/h, but fell to 4.0 and 3.8 μM/h when 10^{-8} and 10^{-6} M of Pu, respectively, were present. The pH of the medium containing citric acid alone increased from 6.5 to 7.1, while that of the samples containing Pu increased from 6.8 to 7.9. Initially, Pu added to the growth medium as Pu-citrate remained in solution as Pu^{4+}. Solvent extraction by TTA and microfiltration (0.03 μm) of the medium after biodegradation indicated the presence of polymeric Pu.

9 CONCLUSIONS

There are rather few studies on the microbial effects on actual radioactive wastes. We can only extrapolate from information derived from work using defined single compounds, to binary and ternary systems, to the actual wastes. Microbial transformations of various uranium compounds have been extensively studied, whereas very limited studies have been conducted with other actinides, such as Np, Pu, Am, and Cm.

Under appropriate conditions, the microorganisms present in TRU wastes, Pu-contaminated soils, low-level radioactive wastes, backfill materials, natural analogue sites, and waste-repository sites can affect the chemical speciation of actinides by rendering them mobile or stable. Microbial degradation of organic constituents of the wastes undoubtedly will cause the generation of gas, which might compromise the integrity of the repository.

Gaining a fundamental understanding of the mechanisms of microbial transformations of various chemical forms of actinides under various microbial process conditions, such as aerobic-, anaerobic- (denitrifying, fermentative, and sulphate- reducing), and repository-conditions would be invaluable in assessing the microbial impact on the behavior of radionuclides during on-site storage, shallow land-burial, and disposal in deep geological formations, as well the long-term stewardship of contaminated sites.

Acknowledgments
I thank C.J. Dodge and J.B. Gillow for their contribution to this work. This research was supported by the Environmental Management Science Program (EMSP), Environmental

Remediation Sciences Division, Office of Biological and Environmental Research, Office of Science, U.S. Department of Energy, under Contract No. DE-AC02-98CH10886.

References

1 A.J. Francis, *Experientia,* 1990, **46**, 840.

2 A.J. Francis, *In Plutonium in the Environment* A. Kudo, (Ed) Elsevier Science Ltd., Co., UK. 2001, pp 201-219.

3 B.J. Barnhart, E.W.Campbell, E. Martinez, D.E. Caldwell and R. Hallett, *Potential microbial impact on transuranic wastes under conditions expected in the waste isolation pilot plant (WIPP).* Los Alamos National Laboratory, 1980, LA-8297-PR.

4 J.B. Gillow, M. Dunn, A.J. Francis, and H.W. Papenguth, *Radiochim. Acta*, 2000, **88**, 769.

5 J.M. West, N. Christofi, and I.G. McKinley, *Radioactive Waste Management and Nuclear Fuel Cycle,* 1985, **6**, 79.

6 J.M. West, I.G. Mckinley, and A. Vialata, J. *Geochem. Explor*., 1992, **45**, 439.

7 K.J Pedersen, Arlinger, L. Hallbeck, and C. Pettersson, *Molecular Ecology,* 1996, **5**, 427.

8 A.J. Francis, G. Joshi-Tope, J.B. Gillow, and C.J. Dodge, *Enumeration and characterization of microorganisms associated with the uranium ore deposit at Cigar Lake*, Canada. 1994, BNL Report No. 49737.

9 J.J. Cramer.and J.A.T. Smellie. *Final Report of the AECL/SKB Cigar Lake Analog Study.* 1994, AECL-10851, COG-93-147, SKB TR 94-04.

10 D.L. Haldeman, and P.S. Amy, *Microbial Ecology,* 1993, **25**, 185.

11 T.L Kieft, W.P. Kovacik Jr., D.B. Ringelberg, D.C. White, D.L. Haldeman, P.S. Amy, and L.E. Hersman, *Appl. Environ. Microbiol*., 1997, **63**, 3128.

12 P.R Castro, *Capability of native Yucca Mountain microorganisms to corrode 1020 carbon steel*. 1997. M.S. thesis. University of Nevada, Las Vegas.

13 B.J. Pitonzo, *Characterization of microbes implicated in microbially influenced corrosion from the proposed Yucca Mountain repository*. 1996, Ph.D., Dissertation, University of Nevada, Las Vegas.

14 A.J. Francis, J.B. Gillow, and M.R. Giles, *Microbial Gas Generation Under Expected Waste Isolation Pilot Plant Repository Conditions.* 1997. SAND96-2582. Albuquerque, NM: Sandia National Laboratories.

15 M.A. Molecke. *Gas generation from transuranic waste degradation: An interim assessment.* 1979, SAND 79-0117.

16 A.J. Francis, C.J. Dodge, and J.B. Gillow, US Patent No. 5,047,152. 1991.

17 A.B Kersting. D.W. Efurd, D.L. Finnegan, D.K. Rokop, D.K. Smith, and J.L. Thomson., *Nature*, 1999, **397**, 56.

18 R. E. Wildung, and T.R. Garland, *In Transuranic Elements in the Environment*, W.C. Hanson (Ed.). DOE/TIC-22800. TIC/U.S. DOE, 1980 pp 300.

19 R. E.Wildung, and T.R.Garland, *Appl. Environ. Microbiol*, 1982, **43,** 418.

20 W. F. Beckert, and F.H.F. Au, *In Transuranium Nuclides in the Environment,* IAEA-SM-199/72, 1976 pp. 337-345.

21 J.R. Brainard, B.A. Strietelmeier, P.H. Smith, P.J. Langston-Unkefer, M.E. Barr, and R.R.Ryan, *Radiochim. Acta*, 1992, **58/59**, 357.

22 M.P. Neu, C.E. Ruggiero, and A.J. Francis. Bioinorganic Chemistry of Plutonium and Interactions of Plutonium with microorganisms and Plants. In "*Advances in Plutonium Chemistry 1967-2000*" D. Hoffman (Ed), 2002 pp 169-211. ANS, La Grange Park Illinois and University Research Alliance, Amarillo, Texas.

23 P.A. Rusin, L. Quintana, J.R. Brainard, B.A. Strietelmeier, C.D. Tait, S.A. Ekberg, P.D. Palmer, T.W. Newton, and D.L. Clark, *Environ. Sci. Technol*, 1994, **28,** 1686.
24 J.E. Banaszak, B.E. Rittmann, and D.T.J. Reed, *Radioanalytical Nuc Chem,* 1999, **241**, 385.
25 P.J. Panak, C.H. Booth, D.L. Caulder, J.J. Bucher, D.K. Shuh, and H. Nitsche, *Radiochim. Acta*, 2002, **90**, 315.
26 M.J. Keith-Roach, J.P. Day, L.K. Fifield, N.D. Bryan and F.R. Livens, *Environ. Sci. Technol.*, 2000, **34**, 4273.
27 D.M. Nelson, and M.B. Lovett, *Nature,* 1978, **276**, 599.
28 K.H. Liese, and A. Muhlenweg, *Radiochem.Acta,* 1988, **43**, 27.
29 J.R. Lloyd, P. Yong, and L.E. Macaskie, *Environ. Sci. Technol.*, 2000, **34**, 1297.
30 D.L. Clark, D.E. Hobart, and M.P. Neu, *Chem. Rev.*, 1995, **95**, 25.
31 E.A. Wurtz, T. Sibley, and W.R. Schell, *Health Phys.*, 1986, **50**, 79.
32 A.J. Francis; C.J. Dodge, and J.B. Gillow, 1992, *Nature*, **356**, 140.
33 R. Ganesh, K.G. Robinson, G.R. Reed, and G. S. Saylor, *Appl. Environ. Microbiol.*, 1997, **63**, 4385.
34 R. Ganesh, K.G. Robinson, L. Chu, D. Kucsmas, and G.R. Reed, *Wat. Res.*, 1999, **33**, 3447.

A BRIEF REVIEW OF THE ENZYMATIC REDUCTION OF Tc, U, Np, AND Pu BY BACTERIA

G.A. Icopini, H. Boukhalfa and M.P. Neu

Actinide, Catalysis, and Separations Chemistry (C-SIC), Chemistry Division, Los Alamos National Laboratory, Mail Stop J-514, Los Alamos, New Mexico, 87545, U.S.A.

1 INTRODUCTION

The oxidized species of Tc, U, Np, and Pu are generally more mobile in subsurface environments than are their reduced species. This has led to the development of remedial strategies that involve the reduction of oxidized species and *in situ* stabilization of these radionuclides as precipitated phases. Subsurface reducing conditions are created naturally by anaerobic bacteria, which can cause metal reduction either directly by enzymatic reduction or indirectly by interaction with the reduced species generated by metabolic activity. Even though radionuclides are not required for cell function, their speciation and solid/solution phase distribution can be affected by environmental bacteria. The biological reduction of U and Tc has been well documented; contrastingly, little is known for Np or Pu. Here we review the current literature on the biological reduction of Tc, U, Np, and Pu by anaerobic bacteria, as well as, the stability of the reduction products.

2 MICROBIAL REDUCTION AND STABILIZATION OF RADIONUCLIDES

In subsurface environments dissolved and free oxygen can quickly be depleted. The degree of anoxia is driven by microbial activity, which is controlled by the concentration of organic carbon and other nutrients. Microorganisms derive the greatest energy gain by using O_2 as a terminal electron acceptor. When O_2 is not available, microorganisms utilize other electron acceptors starting with the most energetically favourable and progressing to the least favourable. These processes create reducing environments characterized by the reduction of a dominant terminal electron acceptor (TEA) and may produce zones of progressively more reducing conditions (NO_3^- reduction < Mn^{4+} and/or Fe^{3+} reduction < SO_4^{2-} reduction < fermentation). In contaminated groundwater environments, organic plumes create gradients with distinct zones of anoxia characterized by the dominant TEA.

These reducing environments are dynamic and can vary dramatically over macroscopic and microscopic scales. In studies with microelectrodes, the redox potential adjacent to wetland plant roots varied from as much as +317 at the root surface to -54 mV at a distance of approximately 5 mm.[1] These interfaces between redox zones are very reactive, controlling the redox behavior of even the major elements. Solobolev and Roden (2002) have demonstrated that metal-oxidizing bacteria and metal-reducing bacteria can co-exist at these interfaces and rapidly cycle Fe between the Fe(III) and Fe(II) states. The heterogeneous and transient nature of these reducing environments significantly impact metal and radionuclide speciation.

Dissimilatory metal-reducing bacteria (DMRB), which derive energy by using oxidized metals as terminal electron acceptors, may be particularly important. In natural environments, these organisms typically utilize Fe and Mn solid minerals that have reduction potentials depending on their mineral structure and related thermodynamic stability (Table 1). These potentials are higher than the potential of c-type cytochromes of anaerobic bacteria ranging from –0.233[2] to –0.400 V[3] making their reduction by DMRB thermodynamically feasible. Several radionuclide contaminants also have reduction potentials accessible to DMRB [4] (Table 1). At neutral pH and in the absence of organic chelators, radionuclides predominantly form hydoxo and oxyhydroxo species. Their redox potentials, although shifted towards lower potentials (harder to reduce) as compared to their standard reduction in acid, remain within the accessible range.

Technicium exists primarily in the IV and VII oxidation states. Tc(VII), as TcO_4^- the predominate form under oxic conditions, minimally adsorbs to geologic materials and is therefore very mobile.[5,6] Conversely, Tc(IV) forms relatively insoluble oxide solids in the absence of complexing ligands. Metal-reducing,[7,8] sulfate-reducing,[9,10] and other anaerobic bacteria[11,12] have been shown to couple Tc(VII) reduction with organic carbon or H_2 oxidation for maintenance or growth. Since the reduced species are relatively insoluble, the *in situ* remediation of Tc by biological processes is being considered.

Table 1. *Reduction potentials of predominant species at neutral pH for selected metals, calculated at pH 7. E_0 is the standard reduction potential and E is the effective reduction potential.*

Redox system	Main species at pH 7	Reduction reaction	E_0	E
Iron Fe(III/II)	$Fe(OH)_3$	$Fe(OH)_3 + 3H^+ + e^- = Fe^{2+} + 3H_2O$	0.933	-0.047[a]
	α-FeOOH(s)	$FeOOH(s) + 3H^+ + e^- = Fe^{2+} + 2H_2O$	0.712	-0.173[a]
Manganese Mn(IV)/Mn(II)	MnO_2	$MnO_2 + 4H^+ + 2e^- = Mn^{2+} + 2H_2O$	1.23	0.737[b]
Plutonium Pu(VI)/Pu(IV)	$PuO_2(OH)_2$	$PuO_2(OH)_2 + 2H^+ + 2e^- = PuO_2(s) + 2H_2O$	1.387	0.797[c]
	$(PuO_2)_2(OH)_4$	$(PuO_2)_2(OH)_4 + 4H^+ + 4e^- = 2PuO_2(s) + 4H_2O$	1.328	0.826[c]
	PuO_2CO_3	$PuO_2CO_3 + H^+ + 2e^- = PuO_2(s) + HCO_3^-$	0.762	0.673[d]
	$PuO_2(CO_3)_2^{2-}$	$PuO_2(CO_3)_2^{2-} + 2H^+ + 2e^- = PuO_2(s) + 2HCO_3^-$	0.609	0.609[d]
Uranium U(VI)/U(IV)	$UO_2(OH)_2$	$UO_2(OH)_2 + 2H^+ + 2e^- = UO_2(s) + 2H_2O$	0.794	0.204[c]
	$(UO_2)_2(OH)_2^{2+}$	$(UO_2)_2(OH)_2^{2+} + 2H^+ + 4e^- = 2UO_2(s) + 2H_2O$	0.394	0.239[c]
	$(UO_2)_3(OH)_5^+$	$(UO_2)_3(OH)_5^+ + 5H^+ + 6e^- = 3UO_2(s) + 5H_2O$	0.589	0.185[c]
	$(UO_2)_4(OH)_7^+$	$(UO_2)_4(OH)_7^+ + 7H^+ + 8e^- = 4UO_2(s) + 7H_2O$	0.597	0.279[c]
	UO_2CO_3	$UO_2CO_3 + H^+ + 2e^- = UO_2(s) + HCO_3^-$	0.142	0.053[d]
	$UO_2(CO_3)_2^{2-}$	$UO_2(CO_3)_2^{2-} + 2H^+ + 2e^- = UO_2(s) + 2HCO_3^-$	-0.053	-0.141[d]
	$UO_2(CO_3)_3^{4-}$	$UO_2(CO_3)_3^{4-} + 3H^+ + 2e^- = UO_2(s) + 3HCO_3^-$	-0.208	-0.327[d]
Neptunium Np(V)/Np(IV)	NpO_2^+	$NpO_2^+ + e^- = NpO_2$	0.684	
	$NpO_2(OH)$	$NpO_2(OH) + H^+ + e^- = NpO_2 + H_2O$	1.350	0.583[c]
	$NpO_2(CO_3)^-$	$NpO_2(CO_3)^- + H^+ + e^- = NpO_2 + HCO_3^-$	0.415	0.238[d]
	$NpO_2(CO_3)_2^{3-}$	$NpO_2(CO_3)_2^{3-} + 2H^+ + e^- = NpO_2 + 2HCO_3^-$	0.292	0.115[d]
Technetium Tc (VII)/Tc (IV)	TcO_4^-	$TcO_4^- + 4H^+ + 3e^- = TcO_2(s) + 2H_2O$	0.738	0.068[e]

[a]calculated using existing thermodynamic values[20] and fixing $[Fe^{2+}]$ at 1×10^{-6} M. [b]calculated using existing thermodynamic values[20] and fixing $[Mn^{2+}]$ at 1×10^{-3} M; [c]calculated using existing thermodynamic values[21] and fixing oxidized species at 1×10^{-6} M. [d]calculated by fixing oxidized species at 1×10^{-6} M and 1 mM carbonate using existing thermodynamic values[21]. [e]Calculated using $E = 0.738 - 0.0788\ pH + 0.0197\ \log(TcO_4^-)$[22] and fixing $[TcO_4^-]$ at 1×10^{-6} M.

Uranium typically exists as either U(IV) or U(VI). The reduction of U(VI) species can be accomplished via direct enzymatic electron transfer(s) or indirect reduction. Direct enzymatic reduction of U(VI) to U(IV) and subsequent precipitation of U(IV) was first reported in 1991;[13] growth of both *Geobacter metallireducens* GS15 and *Shewanella putrefaciens* CN32 can be supported by U(VI) as the sole terminal electron acceptor. Other DMRB and sulphate-reducing bacteria have also been shown to reduce U(VI) rapidly in cell suspension (high cell density, non-growth conditions) with the addition of an appropriate electron donor.[14-17] The resultant biogenic U(IV) precipitate has been identified by XRD as being very fine-grained uraninite, UO_2.[18] XAS characterization of the biogenic solid was consistent with uraninite. The study also revealed that the U-U distance in the solid was slightly shorter than it is in bulk uraninite, reportedly corresponding to a surface stress that increases the solubility of biogenic uraninite 10^9 fold over the solubility of well crystallized uraninite.[19]

The biogeochemistry of Np has been little studied, possibly due to its high apparent toxicity.[23, 24] Bacterial reduction of Np(V) to Np(IV) has been reported under non-growth conditions, although the reduction mechanism is not known. Reduction was speculated to have occurred as a result of interactions of Np with ascorbic acid or via interaction with *Shewanella putrefaciens*; however, reduction alone did not remove Np from solution.[25] Only with the addition of a phosphate-producing bacteria was Np precipitated. The biological reduction of Np(V) with the subsequent precipitation of Np(IV) has also been observed in an anaerobic consortium of sulfate-reducing bacteria.[26] In addition to direct reduction of Np by bacteria, indirect reduction of Np(V) by interaction with reduced Fe or S species may also occur. The adsorption of Np(V) onto mackinawite (FeS) has been shown to yield Np(IV).[27]

Plutonium redox chemistry is more complex than that of other radionuclides. The bacterial reduction of Pu(V/VI) may lead to the precipitation of a solid Pu(IV) oxide (highly insoluble), or a relatively more soluble Pu(III) species. The bioreduction of a Pu(IV) hydrous oxide to Pu(III) has been inferred from increased Pu solubility with the addition of metal-reducing *Bacillus* strains.[28] The increase in Pu concentration was attributed to a reductive dissolution of PuO_2 species to produce Pu(III) that was complexed to NTA and reoxidized to Pu(IV).[28] The increased Pu solubility was observed for cell suspension at Pu concentrations too dilute for Pu to have been the sole terminal electron acceptor. Without spectroscopic or other identification of Pu(III), it is not known if the reduction resulted from direct enzymatic reduction or indirect reduction, or if the solubility increase was due to a different mechanism. The adsorption of Pu(VI) on to bacterial cells under aerobic conditions has been shown to result in the production of Pu(V),[29] however it is unlikely that this reduction is an enzymatically-driven process, as apposed to fortuitous interaction with reduced surface sites.

We have observed direct enzymatic reduction of Pu(V) and (VI) in cell suspension cultures of *Shewanella putrefaciens* CN32, *Shewanella oneidensis* MR1, and *Geobacter metallireducens* GS15. In these experiments 0.5 mM Pu(VI) is provided as the sole terminal electron acceptor in monocultures. The Pu(VI) was rapidly reduced, with 99% of the Pu removed from solution within 30 minutes as determined using LSC and UV/vis/NIR spectroscopy. Pu(V), which formed in solutions initially containing Pu(VI) and is the most stable oxic species under these conditions, is present only in controls after 20 hours. Upon the reduction of Pu(VI) a green precipitate that had diffuse reflectance spectrum indicative of a Pu(IV) solid was observed. In control samples no solid formed and Pu(V) remains in solution as PuO_2^+, based on optical spectroscopic data.

Metal-reducing bacteria can also affect Pu solubility via indirect mechanisms, such as the production of reductants. Fe(II), Mn(II), and sulfide are common inorganic reductants that are produced by bacteria and are well known to reduce plutonium.[30,31] Some bacteria produce redox active organic molecules or use humic acids to shuttle electrons during the indirect reduction of iron.[32,33] Electron shuttles serve to increase the reduction rate of iron, but are also indiscriminate reducers of other metals and radionuclides. Humic acids have been shown to reduce Pu(V), with increased reduction occurring in the presence of divalent cations and sunlight.[34] A recent study reported the reductive dissolution of PuO_2 by both Fe(II) and hydroquinone.[31] Thus, the effects of direct and indirect bioreduction on Pu speciation could be significant.

The fate of biogenically reduced Tc, U, Np, and Pu species is of the utmost importance for *in situ* remedial strategies. The biological reduction of Tc(VII) appears to result in the precipitation of very fine-grained TcO_2 (0.2-0.001 μm), TcO_2 adsorption onto the cell surface, or deposition in the periplasmic space of the bacteria.[7,8] The biological reduction of U appears to produce very fine-grained uraninite crystals ($\leq$ 2 nm) or the deposition of U(IV) in the periplasmic space of bacteria.[19,35] Similarly, we have observed the formation of fine-grained Pu(IV) colloids produced from the biological reduction of Pu(VI). Biological reduction of these radionuclides generally appears to produce very fine-grained solid phases near cell surfaces.

It has been proposed that fine-grained uraninite crystals would be mobile in porous sediments and therefore susceptible to reoxidation and/or uptake by other biota.[19] This mechanism could also facilitate Tc and Pu transport.[36] Conversely, in cell cultures nanocrystaline uraninite has been found to agglomerate and/or associate with extracellular polymeric substances (EPS) that may limit their mobility.[19] However, nanocrystalline phases are not necessarily a direct result of biological activity and may instead be the by-product of highly saturated conditions.[37] Fine-grained magnetite, typically attributed to biological activity, can be produced abiologically under appropriate pH and high saturation conditions.[32] Detailed studies of the mobility of biogenically reduced species of Tc, U, Np, and Pu, as well as, the characterization of the *in situ* products of biological reduction are required to understand fully this possible transport mechanism.

The deposition of Tc(IV) and U(IV) within or on the cell and the associated EPS may also be an imperfect mechanism for immobilization of U and Tc, as the bacteria and EPS may become mobile in porous environments. Long-term immobilization of radionuclides adsorbed onto or encapsulated by EPS depends on the stability of the EPS and the mobility of microorganisms. EPS provides a rich source of organic carbon and may be utilized and degraded by other microorganisms. The fate of radionuclides associated with EPS will be influenced greatly by this degradation. Pu(IV) transport via organic rich colloids has been demonstrated[38] and it is possible that these colloids began as EPS material. Bacteria themselves have also been proposed to function as biocolloids[39,40] and thus facilitate radionuclide migration. However, because bacteria are living organisms, their transport in the subsurface is more complex than is the case for abiotic colloids. Not only are bacteria subject to the same physicochemical phenomena as are abiotic colloids, but there are also a number of strictly biological processes that affect their transport (e.g., temporal surface-property changes due to metabolic changes; predation organisms, nutrient availability). Definitive experiments that couple the physicochemical and biological mechanisms underlying bacterial interaction and transport in subsurface environments are lacking.

3 SUMMARY

A diverse array of anaerobic bacteria can influence the environmental speciation of radionuclides by directly or indirectly reducing the oxidized species to less mobile species. Dissimilatory metal-reducing and sulphate-reducing bacteria are the most common species capable of utilizing oxidized radionuclides as terminal electron acceptors. The reduced species produced are often precipitated in the periplasmic space, on the cell surface, on to extracellular material, or as discrete, fine-grained, secondary minerals. For reduction to be an effective method for the *in situ* immobilization of radionuclides the reduction products must be stable and immobile. The stability of the cellular and extracellular associated reduced species will be largely dependant on the degradation of this organic material, which has been little studied. The fine-grained precipitates have the potential to be more mobile and the potential to be more reactive than comparable materials with less surface area. More detailed studies on the distribution, physical form, and fate of reduced radionuclides in subsurface environments is needed to evaluate fully the potential of bioreduction for remedial purposes.

Acknowledgements.

Preparation of this paper and research reported herein was supported by Environmental Remediation and Sciences Division, Office of Biological and Environmental Research of the U.S. Department of Energy. Los Alamos National Laboratory is operated by the University of California for the U. S. Department of Energy.

References

1 A. N. Bezbaruah and T. C. Zhang, *Biotechnology and Bioengineering*, 2004, **88**, 60.
2 A. I. Tsapin, K. H. Nealson, T. Meyers, M. A. Cusanovich, J. VanBeuumen, L. D. Crosby, B. A. Feinberg, and C. Zhang, *Journal of Bacteriology*, 1996, **178**, 6386.
3 P. Bianco and J. Haladjian, *Biochimie (Paris)*, 1994, **76**, 605.
4 J. K. Fredrickson, J. M. Zachara, D. W. Kennedy, M. C. Duff, Y. A. Gorby, S. M. W. Li, and K. M. Krupka, *Geochimica Et Cosmochimica Acta*, 2000, **64**, 3085.
5 K. H. Lieser and C. Bauscher, *Radiochimica Acta*, 1987, **42**, 205.
6 R. E. Wildung, T. R. Garland, K. M. McFadden, and C. E. Cowan, in 'Technetium sorption in surface soils', ed. G. Desmetand and C. Myttenaere, London,England, 1984.
7 R. E. Wildung, Y. A. Gorby, K. M. Krupka, N. J. Hess, S. W. Li, A. E. Plymale, J. P. McKinley, and J. K. Fredrickson, *Applied and Environmental Microbiology*, 2000, **66**, 2451.
8 J. R. Lloyd, V. A. Sole, C. V. G. Van Praagh, and D. R. Lovley, *Applied and Environmental Microbiology*, 2000, **66**, 3743.
9 J. R. Lloyd, H. F. Nolting, V. A. Sole, and K. Bosecker, *Geomicrobiology Journal*, 1998, **15**, 45.
10 A. Abdelouas, M. Fattahi, B. Grambow, L. Vichot, and E. Gautier, *Radiochimica Acta*, 2002, **90**, 773.
11 J. R. Lloyd, J. A. Cole, and L. E. Macaskie, *Journal of Bacteriology*, 1997, **179**, 2014.
12 T. V. Khijniak, N. N. Medvedeva-Lyalikova, and M. Simonoff, *FEMS Microbiology Ecology*, 2003, **44**, 109.
13 D. R. Lovley, E. J. P. Phillips, Y. A. Gorby, and E. R. Landa, *Nature (London)*, 1991, **350**, 413.
14 D. R. Lovley and E. J. P. Phillips, *Environmental Science and Technology*, 1992, **26**, 2228.
15 D. R. Lovley and E. J. P. Phillips, *Applied and Environmental Microbiology*, 1992, **58**, 850.
16 R. K. Sani, B. M. Peyton, J. E. Amonette, and G. G. Geesey, *Geochimica et Cosmochimica Acta*, 2004, **68**, 2639.

17 J. R. Spear, L. A. Figueroa, and B. D. Honeyman, *Applied and Environmental Microbiology*, 2000, **66**, 3711.
18 Y. A. Gorby and D. R. Lovley, *Environmental Science & Technology*, 1992, **26**, 205.
19 Y. Suzuki, S. D. Kelly, K. M. Kemner, and J. F. Banfield, *Nature*, 2002, **419**, 134.
20 U. Schwertmann and R. M. Taylor, in 'iron oxides', ed. J. B. Dixon and S. B. Weed, Madison, Wis., 1977.
21 R. J. Lemire, J. Fuger, H. Nitsche, P. Potter, R. H. M., J. Rydberg, S. K., C. J. Sullivan, J. W. Ullman, P. Vitorge, and H. Wanner, 'Chemical thermodynamics of neptunium and plutonium', ed. O. N. E. Agency, Elsevier, 2003.
22 E. A. Bondietti and C. W. Francis, *Science*, 1979, **203**, 1337.
23 C. E. Ruggiero, H. Boukhalfa, J. H. Forsythe, J. G. Lack, L. E. Hersman, and M. P. Neu, *Environmental Microbiology*, 2005, **7**, 88.
24 J. E. Banaszak, D. T. Reed, and B. E. Rittmann, *Environmental Science and Technology*, 1998, **32**, 1085.
25 J. R. Lloyd, P. Yong, and L. E. Macaskie, *Environmental Science & Technology*, 2000, **34**, 1297.
26 B. E. Rittmann, J. E. Banaszak, and D. T. Reed, *Biodegradation*, 2002, **13**, 329.
27 L. N. Moyes, M. J. Jones, W. A. Reed, F. R. Livens, J. M. Charnock, J. F. W. Mosselmans, C. Hennig, D. J. Vaughan, and R. A. D. Pattrick, *Environmental Science and Technology*, 2002, **36**, 179.
28 P. A. Rusin, L. Quintana, J. R. Brainard, B. A. Strietelmeier, C. D. Tait, S. A. Ekberg, P. D. Palmer, T. W. Newton, and D. L. Clark, *Environmental Science and Technology*, 1994, **28**, 1686.
29 P. J. Panak and H. Nitsche, *Radiochimica Acta*, 2001, **89**, 499.
30 T. W. Newton, in 'Redox Reactions of Plutonium Ions In Aquesous Solutions', ed. D. C. Hoffman, La Grange Park, Illinois, 2002.
31 D. Rai, Y. A. Gorby, J. K. Fredrickson, D. A. Moore, and M. Yui, *Journal of Solution Chemistry*, 2002, **31**, 433.
32 D. K. Newman and R. Kolter, *Nature*, 2000, **405**, 94.
33 R. A. Royer, W. D. Burgos, A. S. Fisher, R. F. Unz, and B. A. Dempsey, *Environmental Science and Technology*, 2002, **36**, 1939.
34 C. Andre and G. R. Choppin, *Radiochimica Acta*, 2000, **88**, 613.
35 J. K. Fredrickson, J. M. Zachara, D. W. Kennedy, C. X. Liu, M. C. Duff, D. B. Hunter, and A. Dohnalkova, *Geochimica et Cosmochimica Acta*, 2002, **66**, 3247.
36 M. Dai, K. O. Buesselerb, and S. M. Pike, *Journal of Contaminant Hydrology*, 2005, **76**, 167.
37 D. Faivre, P. Agrinier, N. Menguy, P. Zuddas, K. Pachana, A. Gloter, J. Y. Laval, and F. Guyot, *Geochimica et Cosmochimica Acta*, 2004, **68**, 4395.
38 K. A. Roberts, P. H. Santschi, G. G. Leppard, and M. M. West, *Colloids and Surfaces A: Physicochemical and Engineering Aspects*, 2004, **244**, 105.
39 A. J. Francis, J. B. Gillow, C. J. Dodge, M. Dunn, K. Mantione, B. A. Strietelmeier, M. E. PansoyHjelvik, and H. W. Papenguth, *Radiochimica Acta*, 1998, **82**, 347.
40 J. B. Gillow, M. Dunn, A. J. Francis, D. A. Lucero, and H. W. Papenguth, *Radiochimica Acta*, 2000, **88**, 769.

ACTINIDES – A CHALLENGE FOR NUCLEAR FORENSICS

M. Wallenius, S. Abousahl, K. Luetzenkirchen, K. Mayer, H. Ottmar, I. Ray,
A. Schubert and T. Wiss

European Commission, Joint Research Centre, Institute for Transuranium Elements,
P.O.Box 2340, 76125 Karlsruhe, Germany

1 INTRODUCTION

Most isotopes of the actinides decay through emission of an alpha particle, which may lead to a high radiotoxicity. The odd-numbered isotopes of uranium and plutonium are known to be fissile; hence, they serve as base material for nuclear explosive devices. This material is consequently called nuclear material. Access to nuclear materials is, however, limited; on the one hand, because the materials are self-protecting (radiotoxicity), and the handling of these materials is usually associated with dedicated facilities and physical protection, and on the other hand, because fissile materials are subject to strict controls, as required in international treaties (Non-Proliferation Treaty, Euratom Treaty). In some instances, the physical protection and the control of this material have not been properly implemented or were circumvented and nuclear material can be diverted. The material might then be offered on the "black market", with intended uses ranging from simply gaining money by selling it, to the use in nuclear explosive devices or for use in so-called "dirty bombs". In any case, the material poses a threat, and efforts are being undertaken to improve the detection possibilities for uncovering the illicit trafficking of nuclear materials. Moreover, the identification of the origin of the material is enabled through nuclear forensic analysis. Knowledge of the origin of the material allows one to improve physical protection and control mechanisms, thus providing a more sustainable way to prevent future thefts or diversions of material. In the following, we will give an overview of the methodologies used, the experience gathered and the on-going development of the detection and categorisation of material, in nuclear forensic investigations and in the attribution of material. Selected examples illustrate the challenges associated with this work.

2 DETECTION AND CATEGORISATION OF NUCLEAR MATERIALS

Gamma spectrometry techniques (occasionally supported by neutron detection) are commonly applied for the detection and first in-field categorisation of illicit nuclear materials. Recent development work is aimed mainly at an enhancement of hand-held gamma monitors used for this purpose, for example by adding capabilities for isotope identification. The detection of uranium - the type of nuclear material so far mostly encountered in illicit trafficking – generally poses the largest problems, because of, (i) its low specific activity, (ii) the relatively low gamma energies from ^{235}U, and (iii) the absence of a notable neutron signal, which makes uranium detection and identification difficult and sometimes even impossible, especially if the uranium is highly enriched and concealed under shielding. In order to quantify the effect of shielding, we have measured uranium gamma spectra with a typical hand-held monitor equipped with a NaI detector using Pb-shielding of variable thickness.[1] The spectra were also calculated by Monte-Carlo simulation. With a 5-mm Pb-shielding, the intensity of the prominent ^{235}U gamma-lines at 186 keV and 205 keV are reduced by a factor of 200 and 850, respectively. However, the bremsstrahlung continuum, which is essentially produced by the electrons from the ^{234m}Pa decay (E_{max}=2.29 MeV), a decay product of ^{238}U, is reduced to a much lesser extent (~ 4 times). Thus, even if the characteristic gamma-rays of ^{235}U cannot be detected due to absorption by shielding material, it is still possible to deduce the presence of uranium from the bremsstrahlung continuum. It has been shown that the evaluation of the spectral shape in the range from 300-1100 keV can provide information on highly enriched uranium, if present in a sufficiently large quantity (≥300 g).

3 NUCLEAR FORENSIC METHODOLOGY

Nuclear forensic investigations start after categorisation of the nuclear material. In most cases, this is done on the spot, but this result generally needs to be confirmed, and a more accurate measurement of the composition (e.g. U enrichment) will be performed using high resolution gamma spectrometry. The main information on the material (e.g. U/Pu content, isotopic composition, impurities) is obtained using several radiometric and mass spectrometric techniques (Table 1). However, in the case of non-typical material, dedicated analysis schemes, as well as the search of information from open sources, are required in order to deduce the origin. To illustrate the methodology, we present the case of uranium pellets and powder seized in Poland.

Table 1 *Information that can be obtained from nuclear material*

Parameter	Information	Analytical technique*
Appearance	Material type (e.g. powder, pellet)	Optical Microscopy
Dimensions (pellet)	Reactor type	- (database)
U, Pu content	Chemical composition	Titration, KED, IDMS
Isotopic composition	Enrichment ⇒ intended use; reactor type	HRGS, TIMS, ICP-MS, SIMS
Impurities	Production process; geolocation	ICP-MS, GDMS
Age	Production date	AS, TIMS, ICP-MS
$^{18}O/^{16}O$ ratio	Geolocation	TIMS, SIMS
Surface roughness	Production plant	Profilometry
Microstructure	Production process	SEM, TEM

*See chapter 3.2. and below for an explanation of the abbreviations.

3.1 Macroscopic inspection

The samples received contained two uranium pellets and a uranium powder (Fig 1). Both of the pellets showed identical geometry, they had a central hole and they were dished. They were weighed and their dimensions were measured (Table 2).

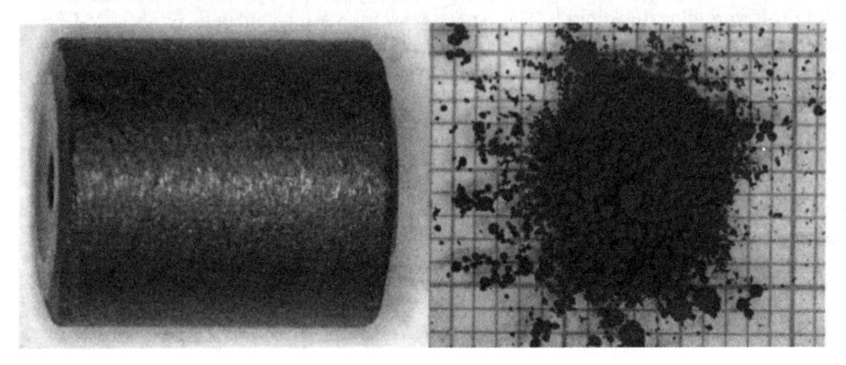

Figure 1 *Uranium pellet and powder*

Table 2 *Macroscopic data for the pellets*

Pellet no.	Weight (g)	Height (mm)	Diameter (mm)	Central hole (mm)
1	14.1779	13.71	11.44	2.1
2	15.3418	14.81	11.44	2.1

3.2 Isotopic analysis

Both pellets were measured individually by high resolution gamma spectrometry (HRGS) using the MGAU code for spectrum deconvolution to obtain a first indication of the isotopic composition.[2] The measured ^{235}U enrichment was 2.1 w-%. As the pellets had identical geometry and isotopic compositions of U, only one of them was dissolved and used for further destructive analysis. The gamma spectrometry analysis of the powder indicated that it was depleted uranium, with ^{235}U enrichment of 0.26 w-%.
In order to quantify the less abundant isotopes (^{234}U and ^{236}U), mass spectrometry techniques are needed. The complete isotopic composition of uranium was determined by thermal ionisation mass spectrometry (TIMS). The results are given in Table 3.

Table 3 *Uranium isotopic composition (w-% ± 1s) as determined by TIMS*

Sample	^{234}U	^{235}U	^{236}U	^{238}U
Pellet	0.0142 ± 0.0005	2.0007 ± 0.0005	0.0060 ± 0.0026	97.9791 ± 0.0010
Powder	0.0014 ± 0.0001	0.2661 ± 0.0001	0.0063 ± 0.0028	99.7262 ± 0.0010

3.3 Chemical analysis

The uranium content after sample dissolution was determined by three different methods, namely by potentiometric titration, by K-Edge Densitometry (KED) and by Isotope Dilution Mass Spectrometry (IDMS).[3,4,5] The analysis showed that the uranium content in

the pellets was 88.0 w-%, which corresponds to the stoichiometry of UO_2, whereas the U-content of the powder was 84.9 w-% corresponding to U_3O_8.
Impurities in the samples were determined by Sector-Field ICP-MS. The main impurities and their concentrations are listed in Table 4.

Table 4 *Main impurities in the pellet and in the powder (μg/g U_xO_y)*

Element	Pellet	Powder	Element	Pellet	Powder
Al	5.5	11.6	Na	20.1	-
Ca	136	-	Ni	5.2	-
Cr	6.0	0.7	P	11.2	74
Fe	33	7.0	Pb	13.8	0.3
K	6.5	-	Zn	7.3	-
Mg	13.0	-			

The age of the material is an important parameter in determining the time of the production of the material and thus in identifying the production campaign or batch. The radioactive decay of the uranium isotopes provides a unique chronometer, which is inherent to the material. This clock is reset to zero each time the decay products (daughter nuclides) are chemically separated from the uranium. Thus, the age of uranium material can be determined from its parent/daughter ratios, $^{234}U/^{230}Th$ and $^{235}U/^{231}Pa$.[6]
The amounts of ^{234}U and ^{230}Th were determined using isotope-dilution alpha-spectrometry, i.e. relative measurements against a known amount of spike isotopes (^{228}Th and ^{233}U). The age of the pellets was 14.2 ± 0.2 years (the U/Th separation date was 03.03.2005), thus they had been produced at the end of 1990. The age of the powder was not determined, because of the low abundance of ^{234}U (and ^{235}U), which makes the age determination for depleted uranium material very difficult.

3.4 Microscopic investigations

The examination of samples by Scanning Electron Microscopy (SEM) and Transmission Electron Microscopy (TEM) plays a vital role in the examination of samples of illicit materials for the following reasons:
1. It is essential to ascertain, particularly in the case of powder samples, whether the material consists of one component, or is a mixture of several distinct components. This is an essential prerequisite to any chemical or mass spectrometric analysis, where a multicomponent system must be separated prior to the examination.
2. SEM enables the material microstructure to be examined and important microstructural characteristics such as particle size and morphology, grain size and porosity to be measured. These are the constituents of the "microstructural" fingerprint of the sample. These characteristics provide information on the possible source of the illicit sample and enable comparisons to be made with other samples or with information stored in the database. In addition, chemical analysis on a fine scale can be made using Energy Dispersive X-ray Analysis (EDX).
3. TEM provides microstructural information on grain size and porosity on a much finer scale (down to 1.0 nm) than SEM, which also constitutes an important element of the microstructural fingerprint. EDX can be used for fine scale chemical analysis, and crystallographic information to identify the phases present.
Figure 2 shows an example of an SEM micrograph, taken in back-scattered electron mode, of the surface of the inner hole in the LEU Pellet No: 2. This clearly images the individual

UO_2 grains, enabling a grain size distribution to be constructed. This micrograph was recorded using a Philips XL40 SEM, operating at 30 kV accelerating potential.

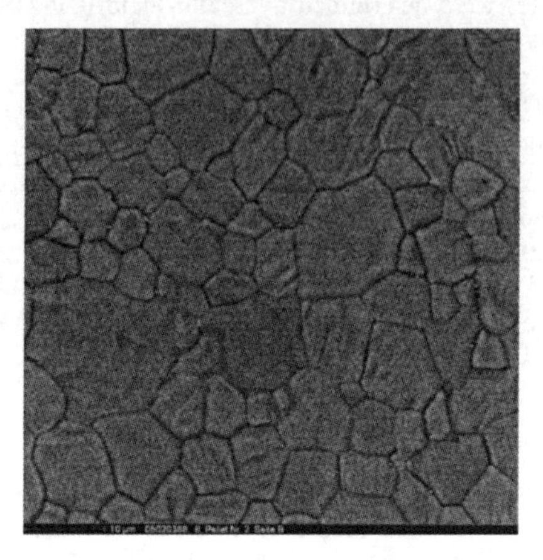

Figure 2 *SEM micrograph showing the grain size distribution in the pellet*

4 SOURCE ATTRIBUTION

A relational database has been established at ITU, which contains data from several commercial nuclear fuel manufacturers (including most of Western Europe and Russia).[7] The data include, e.g., dimensions of pellets, ^{235}U enrichment and typical impurities. Besides the commercial reactor fuels, the database also contains information on several research reactor fuels and information acquired from the open literature. Additionally, the results of old findings are introduced in the database for a comparison with future cases.
In the present case, the database gave an unambiguous answer. Already the pellet dimensions and enrichment were enough to identify the reactor type, which is a RBMK-1500, a Russian type water-cooled, graphite-moderated reactor. There are two models of RBMK reactors, namely 1000 and 1500. The 1000 is the older model, and is more widely distributed. For the 1500 model, there is only one reactor in the world, and this is Ignalina Unit 2 in Lithuania. This reactor started in August 1987, and it is still operational. Furthermore, there is only one manufacturer for this type of fuel, namely MZ Electrostal in Moscow, Russia. The measured impurities were below the maximum values given in the manufacturer specifications, and they also agree with the experimental data from the old findings of the same fuel. The last confirmation parameter was the age, which fitted with the production data of the manufacturer (start of fuel production: December 1989). The information contained in the nuclear materials database thus proved to be essential for the attribution of the material.
For the powder, the situation is different. The database does not contain any information about depleted uranium. However, the uranium must originate from the tailings of the uranium enrichment process. The input material for the enrichment process is fresh uranium, because the ^{236}U abundance in the sample is too low to come from reprocessed uranium. Most likely, this small ^{236}U contribution originates from a contamination (as natural uranium does not contain ^{236}U).
There was a two years difference between the seizures, because the powder was confiscated in 1993 and the pellets in 1995. Thus, even if the origin of the pellets was

found (Ignalina, theft of a fresh fuel assembly in 1992), the powder probably has nothing to do with this incident.

5 CONCLUSIONS

In the last fifteen years we have seen the emergence of a new and potentially hazardous form of smuggling: that of radioactive sources and of nuclear materials, the latter being actinide elements. This has triggered the development of a new discipline in science, helping to support law enforcement authorities in combating illicit trafficking and dealing with criminal environmental issues: nuclear forensics. Existing analytical techniques as used in material science, in nuclear material safeguards and in environmental analysis, were adapted to the specific needs of nuclear forensic investigations. Characteristic parameters (e.g. isotopic composition, chemical impurities, macro- and microstructure) can be combined to give a "nuclear fingerprint", pointing to the origin of the material. Further research is being carried out, aimed at identifying other useful material characteristics in order to reduce the ambiguities often remaining in the interpretation of the data and in source attribution. New methodologies need to be developed, validated and implemented in order to determine parameters with good precision and accuracy. The availability of an up-to-date reference for nuclear material is essential in order to identify the origin and the intended use of the material, or to exclude certain origins.
Significant progress has been achieved in a relatively short time in this new and fascinating discipline. However, the hazards involved with nuclear smuggling and the potential relation with nuclear terrorism are the driving forces for deploying and further improving this methodology.

References

1 S. Abousahl, P. van Belle, P. Ragan, L. Pylkite, J. Bagi and R. Arlt, *Proceedings of the 27th ESARDA Symposium*, London, UK, 10.-12. May 2005.
2 S. Abousahl, A. Michiels, M. Bickel, R. Gunnink and J. Verplancke, *Nucl. Instrum. Methods Phys. Res. A*, 1996, **368**, 443.
3 W. Davies and W. Gray, *Talanta*, 1964, **11**, 1203.
4 H. Ottmar and H. Eberle, *KfK Report 4590,* Forschungszentrum Karlsruhe, 1991.
5 K.G. Heumann, *Int. J. Mass Spectrom. Ion Proc.*, 1992, **118-119**, 575.
6 M. Wallenius, A. Morgenstern, C. Apostolidis and K. Mayer, *Anal. Bioanal. Chem.*, 2002, **374**, 379.
7 J. Dolgov, Y.K. Bibilashvili, N.A. Chorokhov, L. Koch, R. Schenkel and A. Schubert, *Proceedings of the Russian International Conference on Nuclear Material Protection, Control and Accounting*, Obninsk, Russia, 9.-14. March 1997, 116.

NON-EXCHANGEABLE BINDING OF RADIONUCLIDES BY HUMIC SUBSTANCES: A NATURAL CHEMICAL ANALOGUE

N.D. Bryan, D.M. Jones, R.E. Keepax and D.H. Farrelly

Centre for Radiochemistry Research, School of Chemistry, University of Manchester, Oxford Road, Manchester, U.K., M13 9PL.

1 INTRODUCTION

Humic substances control the biological and chemical availability of many pollutants, and particularly metal ions.[1] Because they coordinate metal ions so readily, it has been clear for some time that humic substances can affect transport and bioavailability. However, it has been shown that, as well as binding metals exchangeably, like any simple ligand (e.g. EDTA, citrate), they also have a ***non-exchangeable*** binding mode. Initial uptake is into the exchangeable fraction, but, over time, a significant proportion is transferred into the non-exchangeable. Transfer between exchangeable and non-exchangeable may be described with a spectrum of first order rate constants, ranging from instantaneous to very slow. However, there is a clearly defined fraction, at the slow end of the spectrum of rates that may be described with a single first order desorption rate constant, k_b.[2] This fraction represents a significant proportion of the bound metal ion, typically 10-50% depending upon solution conditions, pH and I etc., concentrations and metal ion chemistry. At present, no component that desorbs more slowly has been found beyond this fraction.

Radionuclide bound in the exchangeable fraction might be transported with the humic, but will desorb instantaneously, if a sink (e.g. mineral surface) of sufficient affinity or concentration is encountered, i.e. it may be stripped from the humic and immobilised. However, once a metal ion is bound non-exchangeably, then it will remain entrapped within the humic structure, regardless of the affinity of any competing sinks, and will effectively transport with the humic.[3] In fact, the only controls on the transport of the non-exchangeable metal will be any sorption of the humic itself and the metal ion desorption rate. Transport calculations have shown that the non-exchangeable fraction is the most significant contributor to environmental impact, and that, in many systems, only this fraction will be transported to any significant extent.[3] In the case of the exchangeable fraction, binding strength is dependent upon metal ion identity. In the case of the non-exchangeable interaction, however, the first order rate constants for the distinct fraction show little dependence upon metal ion identity (Table 1). This behaviour would suggest that the kinetic effect is controlled by some property of the humic itself. Unfortunately, humic substances are very complex, polydisperse materials and their molecular weight/size distributions and structures are not well understood.[4]

Table 1 *Values of desorption rate for some systems from Keepax et al 2002*[3]

Metal Ion (and conditions)	k_b (s^{-1})
Eu(III), pH 4.5	1.2×10^{-6}
Eu(III), pH 6.5	5.0×10^{-7}
Co(II)	1.3×10^{-6}
Am(III)	1.2×10^{-6}

1.1 Ion exchange competition

Metal ion desorption kinetics are determined by an ion exchange technique.[2,5] Samples of humic substance and metal ion are added to a mass of resin (e.g Dowex, Chelex, cellulose phosphate), establishing a competition:

$$HA - M_{non-exch.} \underset{k_b}{\overset{k_f}{\rightleftarrows}} HA - M_{exch.} \overset{HA}{\Leftrightarrow} M_{free} \overset{Resin}{\Leftrightarrow} Resin - M$$

where M_{free}, $HA\text{-}M_{exch}$, $HA\text{-}M_{non\text{-}exch}$, and Resin-M are free, exchangeably bound, non-exchangeably bound and resin bound metal fractions, respectively, HA is the humic, and $\Leftrightarrow$ and $\rightleftarrows$ represent instantaneous and slow reactions, respectively. The resin has a very high affinity for the metal ion, such that any free or exchangeably bound metal ion is instantaneously sorbed onto the resin. Hence, any metal ion remaining in solution is non-exchangeably bound. The solution is analysed at regular intervals to monitor the reduction in $HA\text{-}M_{non\text{-}exch}$ with time. Values of k_b are obtained by fitting. Solution radionuclide concentrations are determined radiometrically.

1.2 The natural anthropogenic sample

Most investigations to date have involved 'synthetic' metal-humate complexes. That is, metal ions and humics are mixed in the laboratory, and allowed to equilibrate, before the desorption kinetics are measured. However, it is also important to study the desorption of metal ions incorporated via natural processes to determine whether parameters determined in synthetic experiments may be applied with validity in the field.

A soil sample was collected from the bank of the Esk Estuary, West Cumbria, UK National Grid Reference SD113964. The site is approximately 12 miles South of the Sellafield nuclear reprocessing plant. Sites adjacent to Sellafield provide a rare opportunity to study the kinetics of a humic acid sample 'naturally' enriched with anthropogenic actinides, for example Am and Pu. Essentially, all the Am and Pu in the soil in the area around the site arise from authorised low level radioactive waste disposals from Sellafield. Peak discharges were in the 1970s, and so the majority of the radionuclide inventory has been equilibrating for 25 – 30 years.

2 EXPERIMENTAL

In these experiments, the term 'natural' corresponds to the Am and Pu already present in the soil from the Sellafield discharges. The radiotracers added to the humic acid in the laboratory are defined as 'artificial'.

The soil sample was sieved through a 2.0 mm mesh, and NaOH (0.5 M) added, in the ratio 5.0 g soil (wet weight): 10.0 ml NaOH. The pH of the mixture was checked, to make sure that it was above pH = 10. The mixture was stirred overnight at room temperature, covered, and allowed to settle for 24 hours. The solution was decanted, centrifuged for 45 minutes (3000 rpm), and the supernatant collected.

The resulting humic acid solution (400.0 ml) was adjusted to pH = 6 ± 0.1 using AnalaR HNO_3. Cellulose phosphate sodium salt ion exchanger (25.0 g) was added to the humic acid solution in a sterile 1000 ml container. The sample was laid on its side and left to shake at 25°C in a thermostated water bath. At regular intervals aliquots of the solution (100.0 ml) were removed and analysed using gamma ray spectrometry to measure the concentration of solution phase ^{241}Am before being returned to the system. Aliquots of the solution (3.0 ml) were also removed and kept for analysis of ^{239,240}Pu. This time, the aliquots were not returned to the experiment. A standard radiochemical separation was used to isolate the Pu, which was determined using alpha-spectrommetry. Note, the technique does not allow resolution of ^{239}Pu and ^{240}Pu, and therefore, the activity is the sum of the two, ^{239,240}Pu.

A spike of radioactive tracer, either ^{241}Am (2.0 kBq) or ^{152}Eu (4 kBq) was added to the same humic solution, which had already been used in the previous natural desorption experiment, and allowed to equilibrate for 7 days (Note, most of the 'natural' ^{241}Am had already been removed during the preceding experiment. Cellulose phosphate sodium salt (25.0 g) was added to the solution and the desorption experiment was repeated.

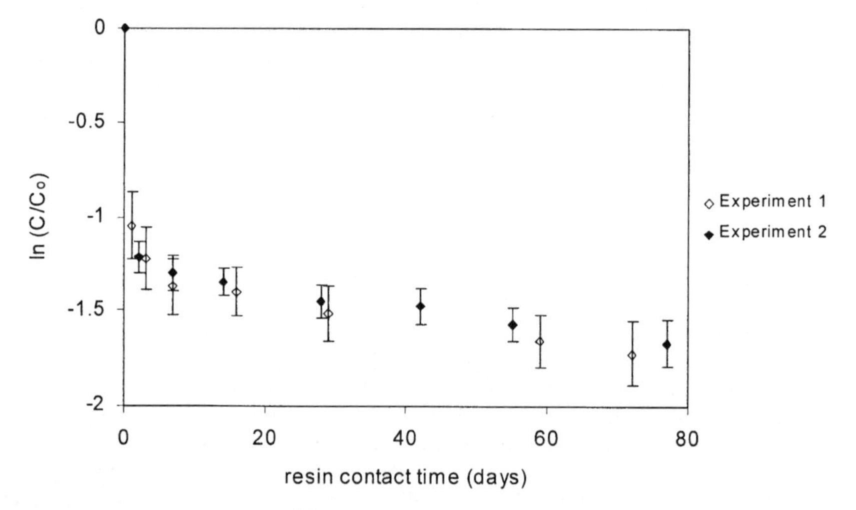

Figure 1 *Desorption of natural ^{241}Am from Esk humic acid at pH = 6 ± 0.1.*

3 RESULTS

The desorption of the 'natural' Am from the sample was repeated twice, starting from scratch with two separate samples of soil from the same site, and proceeding through two separate extraction and desorption series. The results are shown in Figure 1: the data are plotted as the natural log of the concentration of radionuclide remaining in solution at a given contact time (C) divided by the concentration prior to addition of the resin (C_o). The gradient of such a plot gives the first order desorption rate constant with which the radionuclide is desorbing at any given time. The results from the two experiments are clearly concordant. Therefore, even if there is some inherent heterogeneity in the soil from the site, it does not seem to affect the humic desorption kinetics.

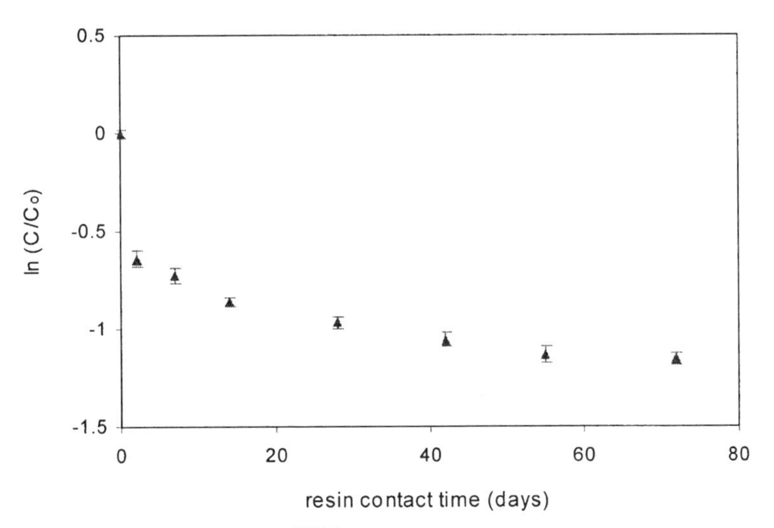

Figure 2 *Desorption of natural* $^{239,240}Pu$ *from Esk humic acid sample at pH = 6 ± 0.1.*

Generally, the form of the plots is very similar to those for completely synthetic batch experiments (metal ion and humic mixed in the lab). There is a sharp drop at the start as any exchangeable isotope is removed, followed by a period when isotope fractions with intermediate rates desorb, and finally the plots show the linear shape that is indicative of the distinct, longest lived, non-exchangeable fraction. Extrapolation of this straight line gives $(C/Co)_{t=0}$, the fraction of radionuclide in the distinct fraction prior to the onset of desorption. Figure 2 shows the data for the 'natural' $^{239,240}Pu$ (one experiment only). Once again, the same general exchangeable, intermediate and slow non-exchangeable behaviour is evident. Figure 3 shows the 'artificial' Am and Eu desorption data compared with the mean result from the two 'natural' Am experiments. The fitted first order desorption rate constants and $(C/C_o)_{t=0}$ values for all the systems are given in Table 2.

4 DISCUSSION

The natural desorption rates are significantly lower than those obtained from standard synthetic batch experiments, which are typically of the order of 10^{-7} s^{-1},[2] compared with 10^{-8} s^{-1} here. The automatic assumption might be that the very long equilibration time is responsible. However, the desorption rates for the isotopes in the 'artificial' systems (only equilibrated for a few days) are the same. The only other significant difference between these experiments and the normal batch type experiments is the humic concentration, 300ppm, compared to 10ppm typically. It is likely that this is responsible for the difference. All of the isotopes, regardless of the origin (natural or artificial) or chemistry (Eu, Am, Pu), show the same desorption rates. This would confirm that it is some property of the humic itself that is responsible for the magnitude of the desorption rate, and more than that, it must be some process that operates on a timescale of days rather than years, at least in this particular case. No evidence has yet been found for any 'irreversible' or pseudo-irreversible behaviour.

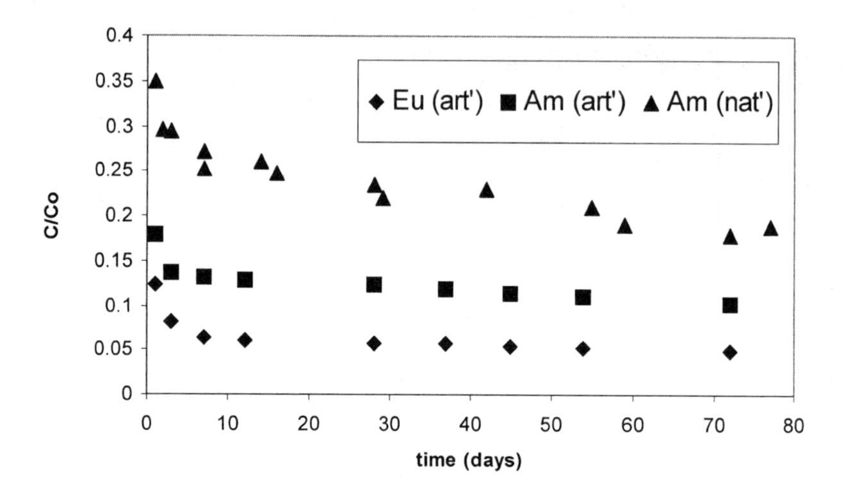

Figure 3 *Desorption of natural (nat') Am and artificial (art') Am and Eu from Esk humic.*

The $(C/C_o)_{t=0}$ values are different for the various systems: the 'natural' systems have significantly higher values. There are several possible factors that could explain the difference. Previous synthetic laboratory batch experiments typically have $(C/C_o)_{t=0}$ for Eu of the order of 5 - 20% at a total concentration of 1×10^{-6} mol dm^{-3},[5] comparable with the 'artificial' $(C/C_o)_{t=0}$ values here, despite 4-5 orders of magnitude difference in total concentration. Therefore, it is unlikely that the difference between the 'artificial' and 'natural' (Table 2) is due to the difference in concentration, which is only 2 orders of magnitude. Hence, the higher $(C/C_o)_{t=0}$ values for the 'natural' systems are almost certainly due to the difference in equilibration times.

Table 2 *Total atom concentrations and kinetic data for Esk humic systems. N.B. there is a range of atom concentrations for the $^{239,240}Pu$ system*, because the two isotopes have different half-lives, but the relative concentrations of the two are unknown. Therefore, the concentration of Pu lies in this range*

Isotope (system)	Total atom concentration (mol dm^{-3})	First order rate k_b (s^{-1}) (2σ error)	$(C/C_o)_{t=0}$ (%) (2σ error)
'natural' ^{241}Am	1.8×10^{-12}	5.9×10^{-8} (18 %)	28 (1 %)
'natural' $^{239,240}Pu$	$5.3 - 19.4 \times 10^{-11}$*	6.1×10^{-8} (40 %)	44 (10 %)
'artificial' ^{241}Am	1.6×10^{-10}	4.9×10^{-8} (11 %)	14 (3 %)
'artificial' ^{152}Eu	1.0×10^{-11}	4.9×10^{-8} (26 %)	7 (2 %)

The fact that these 'artificial' systems give similar $(C/C_o)_{t=0}$ values to the synthetic systems, despite the vast difference in concentration is very significant. At the higher concentrations, there are two possible reasons for only a fraction of the metal entering the non-exchangeable: there could be a large (-ve) free energy difference between the exchangeable and non-exchangeable, but only sufficient sites for a fraction, or there could be a small free energy difference between the two, and the metal distributes itself according to that difference. If there is enough capacity for 10% of the metal to bind non-exchangeably at $1x10^{-6}$ mol dm^{-3}, then there must be a vast excess capacity at the atom concentrations in Table 2, but still only a fraction binds. Therefore, the distribution between exchangeable and non-exchangeable is the result of the free energy difference between the fractions. In fact, given that only a minority binds, the non-exchangeable must have a slightly higher free energy than the exchangeable.

The k_b and $(C/C_o)_{t=0}$ values determined here are the most suitable to be used in coupled chemical transport calculations of Pu and Am migration, rather than those determined in artificial lab studies, at least for the humic at this site. However, such lab studies can provide suitable values of the desorption rate constant, provided that the conditions are similar.

Acknowledgements

The authors wish to thank the European Commission for funding this work as part of the 5th framework HUPA project and the 6th framework FUNMIG Integrated Project.

References

1 A. Piccolo, *Advances in Agronomy*, 2002, **75**, 57.
2 S.J. King, P. Warwick, A. Hall and N.D. Bryan, *Physical Chemistry Chemical Physics,* 2001, **3**, 2080.
3 R.E. Keepax, D.M. Jones, S.E. Pepper and N.D. Bryan, The effects of humic substances on radioactivity in the environment. in Keith-Roach M. and Livens F.R. (eds.) *Microbes and Radionuclides*, 2002, (pp 143- 177), Elsevier, Amsterdam.
4 M.N. Jones and N.D. Bryan, *Advances in Colloid and Interface Science*, 1998, **78**, 1.
5 J.M. Monsallier, W. Schussler, G. Buckau, T. Rabung, J.I. Kim, D. Jones, R. Keepax and N. Bryan, *Analytical Chemistry*, 2003, **75**, 3168.

ON THE MULTI-SPECTROSCOPIC APPROACH FOR RADIONUCLIDE/MINERAL INTERACTION MODELLING IN AQUEOUS SOLUTIONS

R. Drot and E. Simoni

University Paris XI, Nuclear Physics Institute, Radiochemistry Group, 91406 Orsay, France

1 INTRODUCTION

Since sorption phenomena strongly affect the migration of radionuclides,[1] the safety assessment of deep geological disposal needs to estimate the retention capacity of mineral phases over a geological time-scale. To reach that goal, it is necessary to know accurately the interaction mechanisms at the aqueous/mineral interface. Spectroscopic techniques appear to be very powerful tools for characterizing surface species.[2-6] Some of them are typically surface-sensitive, like X-ray photoelectron spectroscopy (XPS) and diffuse reflectance infrared Fourier transform spectroscopy (DRIFT) and give information about the binding between the metal ion and the surface hydroxyl groups. Time-resolved fluorescence spectroscopy (TRLFS), which could be restrictive since the probed ion has to be fluorescent, is nevertheless rather sensitive and appears well adapted for investigating dilute systems, such as metal ions sorbed on a surface as a monolayer. X-ray absorption spectroscopy (EXAFS) can also provide direct evidence of the structural environment of the sorbed species. Since the solid has been previously well characterized (using XRD, elementary analysis, BET method …), it is possible, by coupling spectroscopic techniques, to define all of the components involved in the sorption equilibria and then to perform data fitting using these experimental constraints.[7-9] To illustrate this methodology, several examples will be presented. In the first step, the identification of surface reactive sites, as well as the determination of the sorbed species will be performed by coupling TRLFS and XPS for uranium(VI) sorbed onto Na-Montmorillonite.[10] Then, an example of the use of DRIFT spectroscopy will be presented to investigate the effect of a complexing agent (NO_3^-) on the U(VI)/ZrO_2 system.[11] The application of EXAFS for the determination of the local structure of the sorbed actinide will also be illustrated using the U(VI)/$ZrSiO_4$ system.[11] Finally, some new developments of this methodology will be presented with the U(VI)/TiO_2 system. For all these examples, only the main conclusions will be presented for illustrative purposes. The details of the experimental protocols and setup can be found in the cited literature.

2 EXPERIMENTAL IDENTIFICATION OF SORPTION EQUILIBRIA

2.1 Nature of the reactive sorption sites and sorbed species

The identification of the exact nature of the reactive surface sites is one of the main points for surface complexation modelling. By using the sorbed metal ion as a structural probe, it is possible to identify the reactive surface sites using a spectroscopic investigation. The U(VI)/Na-Montmorillonite system is an illustration of this point.[10] Clay minerals are alumino-silicate substrates, and constitute rather complex systems, since they present several kinds of reactive surface sites.[5,12,13]

2.1.1 Spectroscopic investigation of the U(VI)/Na-Montmorillonite system.[10] This system was investigated for a wide range of total uranium concentrations in solution (from 10^{-6} to 10^{-3} M) and for different ionic strengths (0.1 M and 0.5 M). Examples of sorption edges representative of the two experimental conditions are shown in Figure 1.

Measurements combining TRLFS and XPS were carried out. The spectroscopic characteristics (lifetime from TRLFS and binding energy from XPS) of the sorbed species, which define the local uranyl ion environment on the substrate, are reported in Table 1. The classical uranyl emission spectra are not reported here. For pH values lower than 3, only one lifetime value was obtained (10 μs), correlated to one binding energy (383.7 eV). When the pH was increased, other surface species appeared, which were characterized by other spectroscopic parameters. Therefore, in order to identify clearly the nature of the reactive sorption sites, the spectroscopic results obtained for this system have been compared to reference systems: U(VI) sorbed onto an amorphous silica and onto γ-alumina (Table 2). By comparison with the Montmorillonite data, these systems have allowed the separate study of the sorption sites expected for the clay: "aluminol type" and "silanol type" sites. For all the systems under study, a non-complexing background electrolyte has been used ($NaClO_4$) to give a "uranyl speciation in solution" that was as simple as possible.

Table 1 *Spectroscopic characteristics of U(VI) sorbed species onto Na-Montmorillonite as a function of pH (from reference 10).*

system	*Experimental conditions*	*Decay times (μs)+/-10%*	*U $4f_{7/2}$ binding energies +/-0.3 (eV)*
U(VI) / Montmorillonite	pH < 3	10	383.7
	4 < pH < 5	10 + 55 + 190	383.7 + 382.6
	pH > 5	55 + 190 + 400	382.6 + 381.9

2.1.2 Spectroscopic investigation of the reference systems (Table 2). (i) For the U(VI)/alumina system, it was found that for pH values lower than 5, only one surface species was observed: 45 μs; 381.8 eV. When the pH was increased, another species appeared, characterized by: 120 μs; 380.2 eV. By comparing these results to the U(VI) speciation in solution (not shown here), it has been concluded that the first species (45 μs; 381.8 eV) arises from the sorption of free UO_2^{2+} ion onto "aluminol type" sites. The second one has been interpreted in terms of interaction between "aluminol type" sites and a hydrolyzed U(VI) species.
(ii) For the U(VI)/silica system, when the sorption pH was lower than 5, two fluorescence lifetimes were identified: 65 μs and 180 μs. Above pH 6, 180 μs was still observed, but 65 μs was not, and a new lifetime appeared (400 μs), indicating the presence

of a new surface species. Between pH 5 and 6, the three fluorescence lifetimes were necessary to explain the fluorescence decay. The XPS measurements were very strongly correlated with these results. For samples prepared around pH 4, the XPS U $4f_{7/2}$ spectrum was fitted using two components: 382.2 eV and 383.3 eV. For a sample prepared at pH 6.4, two components were observed: 383.3 eV and 381.8 eV. The presence of these three surface complexes has been explained by solution speciation taking into account the amount of dissolved silicates released by the silica ($[Si]_{solution} \sim 2\times10^{-3}$ M). Since UO_2^{2+} and $UO_2H_3SiO_4^+$ were the main aqueous species at the lower pH values, the surface species (65 μs; 382.2 eV) and (180 μs; 383.3 eV) were attributed to the sorption of $UO_2H_3SiO_4^+$ and UO_2^{2+} onto "silanol type" sites, respectively. The third lifetime (400 μs), only observed for higher pH values, was attributed to the sorption of hydrolyzed uranyl species.

Table 2 *Spectroscopic characteristics of U(VI) sorbed onto amorphous silica and γ-alumina and corresponding attributions (from reference 10).*

E_B (eV) \ τ (μs)	45	65	120	180	400
380.2	--	--	$(UO_2)_n(OH)_m^{(2n-m)+}$ on Al_2O_3	--	--
381.8	UO_2^{2+} on Al_2O_3	--	--	--	$(UO_2)_n(OH)_m^{(2n-m)+}$ on SiO_2
382.2	--	$UO_2H_3SiO_4^+$ on SiO_2	--	--	--
383.3	--	--	--	UO_2^{2+} on SiO_2	--

2.1.3 Identification of the U(VI)/Na-Montmorillonite surface species. By comparing the U(VI)/montmorillonite system with the reference systems, the following interpretation was proposed:

(i) The surface species characterized by (10 μs; 383.7 eV) was observed for low pH values and only for montmorillonite. Therefore, it corresponds to a "clay specific" surface complex. Several studies of the U(VI)/smectite system have shown that for low pH values, an exchange process occurs in the interlayer space *via* the formation of a uranium outer-sphere complex.[5,12,13] For such a complex, the decay time is expected to be quite short, as for the uranyl ion in aqueous solution (between 2 and 10 μs). Hence, this first surface species was attributed to the sorption of a uranyl ion in the interlayer space. This conclusion is supported by the macroscopic data, since for pH values lower than 3 the U(VI) sorption rate depends on the ionic strength (as shown in Figure 1), which is a characteristic of outer-sphere complex formation.

(ii) Since the surface species presenting the longer decay time for U(VI) sorbed onto montmorillonite at high pH values (400 μs; 381.9 eV) is also observed with silica for pH > 6 (400 μs; 381.8 eV), it is proposed that this surface species corresponds to the sorption of a polynuclear U(VI) species onto "silanol type" sites.

(iii) The other surface species obtained for montmorillonite, characterized by 55 μs and 190 μs, were observed over the range of pH values investigated. Since the longer one (190 μs) was very close to 180 μs (U(VI)/silica system), it has been attributed to the sorption of free UO_2^{2+} onto "silanol" edge sites of the clay. Finally, 55 μs was very close (within the uncertainties) to the value obtained for the sorption of UO_2^{2+} onto alumina (45 μs) and to the value for $UO_2H_3SiO_4^+$ sorption onto silica (65 μs). However, since the amount of dissolved silicates arising from the dissolution of montmorillonite was determined to be approximately 1.5×10^{-4} M, the amount of $UO_2H_3SiO_4^+$ aqueous complex was less than 10%, and thus, the U(VI)-silicate complex was not expected to be sorbed. Hence, the signal at 55 μs was assigned to the sorption of UO_2^{2+} onto "aluminol" edge sites. One important fact in this study is that only two binding energies were observed for pH values higher than 6 when U(VI) was sorbed onto montmorillonite (381.9 eV and 382.6 eV). The lower binding energy (correlated to 400 μs) was attributed to an interaction between U(VI) species and "silanol" edge sites. Consequentely, 382.6 eV was correlated with both species characterized by decay times of 55 and 190 μs. This was supported by the fact that for pH values between 4 and 5, the signal at 382.6 eV was obtained, whilst 55 μs and 190 μs were also measured.

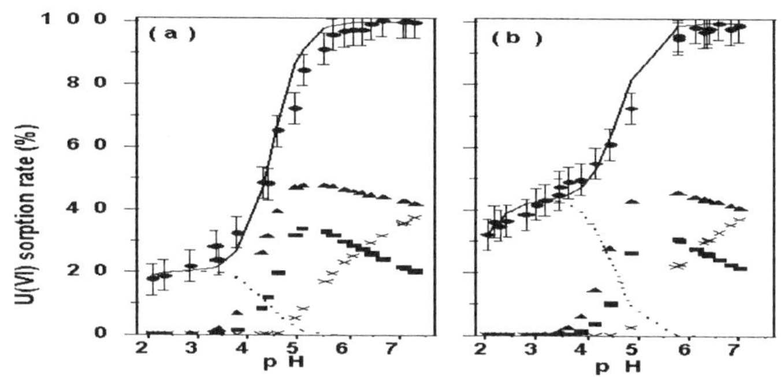

Figure 1 *Sorption modelling data (CCM) of U(VI)/Na-Montmorillonite system for [U(VI)]=10^{-4} M; μ=0.5 M (a) and [U(VI)]=10^{-4} M; μ=0.1 M (b). Experimental points (circles), calculated curve (full line), exchange sites (dots), "aluminol" edge sites (triangles), "silanol" edge sites (rectangles and crosses) (from reference 10).*

2.2 Identification of ternary surface complexes

This point can be illustrated by the investigation of the U(VI)/ZrO_2 system.[11] The experiments were carried out for increasing pH values (from 3 to 6) and considering KNO_3 (0.5 M) as background electrolyte in order to check if a uranyl-nitrate complex was sorbed on the substrate. Strong differences were observed between the different DRIFT spectra recorded at different pH values, which correspond to different amounts of sorbed U(VI) on the surface: the intensity of the nitrate bands (1400 and 1000 cm^{-1}) increased when the sorption pH was raised (increasing concentration of U(VI) sorbed). Moreover, a splitting of the nitrate bands was observed, which is characteristic of a low symmetry arising from a covalent bond between an oxygen of the nitrate group and the U(VI) ion. To check out this hypothesis, different washings of the sample as well as different hydration times were considered, which always led to similar behaviour in the DRIFT spectra. This investigation concluded that a ternary U(VI)/nitrate/ZrO_2 surface species was formed under these experimental conditions. The same kind of experiments were performed on the U(VI)/$ZrSiO_4$ system.[11] For this substrate, no evidence for the formation of a ternary surface complex was found, indicating that the surface oxygen atoms of zirconium oxide and zirconium silicate present strong reactivity differences towards aqueous uranyl-nitrate complexes.

2.3 Local structure of the surface species

The U(VI)/$ZrSiO_4$ (pH=3) system will be used to illustrate the determination of the local structure of sorbed species.[11] First, EXAFS experiments were carried out on both wet and dried sorbed samples in order to check whether the drying step of a sample (necessary to perform XPS measurements for example) does or does not affect the structure of the sorbed species. Whatever the samples, the Fourier transform spectra were similar, which clearly showed that there was no effect from the drying step on the structure of the sorbed species. The best fit to the data was obtained with a structural model consisting of three oxygen coordination shells and one silicon coordination shell. The obtained results are reported in Table 3.

Table 3 *EXAFS best fit parameters obtained for U(VI) sorbed onto zircon at pH=3 (from reference 11).*

Attribution	*Number*	*Distance +/- 0.03 (Å)*
Axial oxygen atoms	2.3	1.81
Equatorial oxygen atoms	3.1	2.36
Equatorial oxygen atoms	1.9	2.54
Silicon atom	1.7	2.76
Zirconium atom	1.2	3.50

Considering the distances between the sorbed uranium and the surface silicon and zirconium atoms and the splitting of the equatorial shell, it was concluded that an inner-sphere surface complex was formed between U and the surface oxygen atoms of the silicate groups. Moreover, the absence of U-U backscattering indicated the formation of a mononuclear complex. Finally, by comparing the distances of both equatorial oxygen shells to the results obtained for uranyl in solution, it was concluded that the two oxygen atoms at 2.36 Å belong to the surface and the three others (at 2.54 Å) belongs to OH groups or water molecules, which completed the coordination shell of the sorbed U(VI).

3 SURFACE COMPLEXATION MODELLING

Since all the chemical components involved in the sorption mechanisms are experimentally determined, the macroscopic sorption data can be fitted on the basis of these experimental constraints to determine accurate sorption constants. The U(VI)/Na-Montmorillonite will be used as example here.[10] One surface complexation model that is appropriate to the experimental conditions considered here (ionic strength higher than 0.1M) is the constant capacitance model. It needs only the inner-capacitance value as an adjustable parameter, the sorption equilibria being defined by the above spectroscopic investigation. The sorption equilibria and corresponding constants were:[10]

$$2\equiv XNa + UO_2^{2+} \Leftrightarrow (\equiv X)_2UO_2 + 2\,Na^+ \qquad \log K^0_{exch.}=3.0$$

$$\equiv Al(OH)_2 + UO_2^{2+} \Leftrightarrow \equiv Al(OH)_2UO_2{}^{2+} \qquad \log K^0_{Al}=14.9$$

$$\equiv Si(OH)_2 + UO_2^{2+} \Leftrightarrow \equiv SiO_2UO_2 + 2\,H^+ \qquad \log K^0_{Si1}=-3.8$$

$$\equiv Si(OH)_2 + 3\,UO_2^{2+} + 5\,H_2O \Leftrightarrow \equiv SiO_2(UO_2)_3(OH)_5^- + 7\,H^+ \qquad \log K^0_{Si2}=-20.0$$

Note that the modelling was performed considering the surface species experimentally determined. Bidentate complexes were only considered in agreement with other work.[5] Figure 1 presents, as an example, the results of the surface complexation modelling of the

U(VI)/Na-Montmorillonite system for two experimental conditions, with C=1 F/m^2 (the acidity constants have been determined elsewhere).

4 CONCLUSION

It has been shown that the use of complementary spectroscopic techniques is a very powerful tool for investigating mineral/actinide interactions in aqueous solutions. It allows the definition of the species involved in the retention processes, which allows the modelling of the macroscopic sorption data on the basis of experimental constraints. Nevertheless, in order to reduce the number of fitted parameters in the surface complexation modelling procedure (such as the surface acidity constants), further developments are now required. In this way, we are now performing investigations on the U(VI)/TiO_2 system, applying the methodology described here. Both powder and single crystal forms of titanium oxide are being studied.[14] The goal of this work is to try to account for the surface properties of the powder from the properties of each of the crystallographic planes that constitute this powder. For instance, using the CD-MUSIC model, it is possible to calculate, without any data fitting, the intrinsic acidity constants of the solid powder with only crystallographic considerations. Finally, quantum chemistry calculations are also performed on the same system in order to try to determine accurately the nature of the bonding between the orbitals of the uranium and the surface oxygen atoms, and to compare the corresponding charge densities with the ones deduced from the CD-MUSIC model.

References

1 G. De Marsily, *Radiochim. Acta*, 1998, **44/45**, 159.

2 D.E. Morris, C.J. Chisholm-Brause, M.E. Barr, S.D. Conradson and P.G. Eller, *Geochim. Cosmochim. Acta*, 1994, **58**, 3613.

3 R. Drot, E. Simoni, M. Alnot and JJ. Ehrhardt, *J. Colloid Interface Sci.*, 1998, **205**, 410.

4 K.H. Chung, R. Klenze, K.K. Park, P. Paviet-Hartmann and J.I. Kim, *Radiochim. Acta*, 1998, **82**, 215.

5 E.R. Sylwester, E.A. Hudson and P.G. Allen, *Geochim. Cosmochim. Acta*, 2000, **58**, 3613.

6 C.C. Fuller, J.R. Bargar, J.A. Davis and M.J. Piana, *Environ. Sci. Technol.*, 2002, **36**, 1062.

7 R. Drot and E. Simoni, *Langmuir*, 1999, **15**, 4820.

8 C. Lomenech, R. Drot and E. Simoni, *Radiochim. Acta*, 2003, **91**, 453.

9 E. Ordoñez-Regil, R. Drot and E. Simoni, *J. Colloid Interface Sci.*, 2003, **263**, 391.

10 A. Kowal-Fouchard, R. Drot, E. Simoni and JJ. Ehrhardt, *Environ. Sci. Technol.*, 2004, **38**, 1399.

11 C. Lomenech, E. Simoni, R. Drot, JJ. Ehrhardt and J. Mielczarski, *J. Colloid Interface Sci.*, 2003, **261**, 221.

12 A.J. Dent, J.D.F. Ramsay and S.W. Swanton, *J. Colloid Interface Sci.*, 1992, **150**, 45.

13 C.J. Chisholm-Brause, J.M. Berg, R.A. Matzner and D.E. Morris, *J. Colloid Interface Sci.*, 2001, **233**, 38.

14 C. Den Auwer, R. Drot, E. Simoni, S.D. Conradson, M. Gailhanou and J. Mustre de Leon, *New J. Chem.*, 2003, **27**, 648.

COMPACT ACCELERATOR MASS SPECTROMETRY: A POWERFUL TOOL TO MEASURE ACTINIDES IN THE ENVIRONMENT

L. Wacker[1], L.K. Fifield[2], S. Olivier[3], M. Suter[1] and H.-A. Synal[4]

[1] Institute of Particle Physics, ETH Hönggerberg, Zürich, Switzerland
[2] Department of Nuclear Physics, Australian National University, Canberra, Australia
[3] Department of Chemistry and Biochemistry, University of Bern, Switzerland
[4] Paul Scherrer Institute c/o ETH Hönggerberg, Zürich, Switzerland

1 INTRODUCTION

In contrast to other mass spectrometry techniques, accelerator mass spectrometry (AMS) has the advantage of avoiding molecular interferences, thus generally having lower detection limits. AMS is mainly used to determine relatively light radioisotopes. Therefore, current AMS facilities are usually not optimised for heavier radioisotopes. However, there are some AMS facilities working at accelerator voltages of 3 – 11 MV, that have already been used to determine low actinide concentrations in environmental samples.[1-5]
A drawback is the complexity of such measuring systems and the resulting costs per measured sample. New compact AMS systems are much simpler and more cost-efficient. The main difference is that a compact AMS system works at accelerator voltages of less than 0.6 MV. The result is an instrumental design that is much more compact. The accelerator and magnets are significantly smaller, which makes a simpler ion optical setup of the machine possible. Only such systems can make the AMS technique more widely used for the measurements of the actinides.
For the last 4 years, we have systematically investigated our compact AMS facility for its potential to measure heavy radioisotopes.[6,7] It is now possible to measure actinides routinely with an automated system. Here, two examples will demonstrate how well our new generation compact AMS facility compares with a much more complex large AMS facility.
We have re-analysed the same ice samples from a glacier in the Altai Mountains that were already prepared and measured for Pu isotopes on the large 15MV tandem accelerator at ANU in Canberra, Australia.[8]

2 PROCEDURE

2.1 Sample preparation

For the AMS measurements, the actinide isotopes have to be separated from the main matrix elements. For isotope ratio measurements of all actinides except uranium, a yield tracer for the chemical separation and the measurement has to be added to the sample. Because there is no bulk material available after the separation, 0.5 mg iron oxide is added before it is mixed with about 10 mg metal powder and then it is pressed for the measurement. A negatively charged oxide beam is extracted from the sample using a Cs sputter source.

2.2 Measurement

The compact AMS system at ETH Zürich[9,10] with its 600 kV tandem accelerator has already been employed for actinide measurements. Several improvements of the instrumental setup were achieved compared to the very first test measurements. Here we will only give a short outline of the instrumental settings.
Negatively charged actinide oxide beams are extracted from a Cs sputter source and a first mass separation is performed with a sector field magnet. Then, the ions are accelerated up to about 330 keV before the oxide molecules are destroyed by stripping in argon gas. With an efficiency of about 15%, 3+ ions are obtained that are accelerated further, before a second mass separation is performed on the high-energy side by a magnet and an electrostatic analyser. Any possible isobaric interference with molecules is avoided by selecting the 3+ charge state. This is in contrast to conventional mass spectrometry that analyses singly charged ions. The final ion detection is performed using a gas detector designed for low energies.[7,11]
The Pu isotopes are measured in a sequence, one after the other. The time for a sequence in which the isotopes 239, 240 and 242 are measured is set to 1½ minutes, and it takes only about 1 s to switch between different isotopes. The measurement of Pu isotopes on the compact AMS facility is fully automated and runs overnight without supervision.

3 RESULTS AND DISCUSSION

The ^{239}Pu concentrations and ^{240}Pu/^{239}Pu ratios measured on the large AMS system (working at 4 MV) and on the new compact AMS system (working at only 0.3 MV) are compared in Table 1. The results are in good agreement according to a χ^2-test: a χ^2 of 1.2 when comparing the ^{239}Pu concentrations and a χ^2 of 0.7 when comparing the ^{240}Pu /^{239}Pu ratios were obtained. While the detection limit of 1 fg for the compact AMS facility is nearly as good as for the large facility (about 0.5 fg), the precision is significantly higher on the compact facility. The reason for the higher precision is twofold.
First, the transmission of the small system is a factor of 5 higher than for the larger machine, due to higher stripping efficiency to the 3^+ charge state and better ion optical transmission. Hence, for a given ion source output, counting statistics are better on the small system.
Secondly, the reproducibility was better on the small AMS system. The reason for this is the short time of only 1½ minutes to measure one sequence. This is more than 5 times shorter than on the large system. Hence, the compact system is less affected by variations in ion source output on time-scales of the order of a minute.

4 CONCLUSIONS

The technical limitations of compact AMS facilities, imposed by the simple and compact design, have been overcome for measurement of the heaviest, long-lived radionuclides. The performance of our compact AMS facility is even better than a large one. Only the detection limit is slightly higher. The new compact system runs more stably with a higher counting rate for a given sample. This results in a significantly higher measurement precision. Furthermore, it runs fully automated.

Table 1 *The measurements of Pu isotopes on the large and the compact AMS facilities are compared. Pu was extracted from ice samples, and each sample was analysed on both systems. The results are given with 1-σ-errors.*

	Large AMS (4 MV)[8]		Compact AMS (0.3 MV)	
Date[8]	^{239}Pu (10^{-8} atoms/kg)	$^{240}Pu/^{239}Pu$	^{239}Pu (10^{-8} atoms/kg)	$^{240}Pu/^{239}Pu$
1983-1986	0.18 ± 0.02	0.29 ± 0.08	0.20 ± 0.03	0.14 ± 0.04
1981-1983	0.28 ± 0.02	0.16 ± 0.04	0.22 ± 0.02	0.13 ± 0.02
1978-1981	2.1 ± 0.16	0.13 ± 0.04	1.93 ± 0.09	0.16 ± 0.01
1975-1978	3.8 ± 0.3	0.20 ± 0.05	3.70 ± 0.20	0.21 ± 0.01
1972-1975	3.9 ± 0.22	0.16 ± 0.03	3.77 ± 0.09	0.17 ± 0.01
1969-1972	12 ± 0.9	0.19 ± 0.05	11.4 ± 0.4	0.17 ± 0.01
1966-1969	13 ± 1.0	0.20 ± 0.05	12.0 ± 0.3	0.18 ± 0.01
1962-1966	79 ± 6.2	0.20 ± 0.05	66.2 ± 1.9	0.20 ± 0.01
1959-1962	39 ± 3.0	0.18 ± 0.05	37.3 ± 1.1	0.17 ± 0.01
1955-1959	43 ± 3.4	0.22 ± 0.06	43.9 ± 1.5	0.21 ± 0.01
1952-1955	4.2 ± 0.33	0.15 ± 0.04	3.87 ± 0.14	0.14 ± 0.01
1949-1952	0.18 ± 0.02	0.08 ± 0.02	0.19 ± 0.02	0.03 ± 0.02
1945-1949	0.11 ± 0.01	0.21 ± 0.06	0.10 ± 0.01	0.15 ± 0.05

References

1 L.K. Fifield, R.G. Cresswell, M.L. di Tada, T.R. Ophel, J.P. Day, A.P. Clacher, S.J. King and N.D. Priest, *Nucl. Instr. and Meth. B*, 1996, **B 117**, 295.

2 X.-L. Zhao, L.R. Kilius, A.E. Litherland and T. Beasley, *Nucl. Instr. and Meth. B*, 1997, **126**, 297.

3 M.A.C. Hotchkis, D. Child, D. Fink, G.E. Jacobsen, P.J. Lee, N. Mino, A.M. Smith and C. Tuniz, *Nucl. Instr. and Meth. B*, 2000, **172**, 659.

4 T.A. Brown, A.A. Marchetti, R.E. Martinelli, C.C. Cox, J.P. Knezovich and T.F. Hamilton, *Nucl. Instr. and Meth. B*, 2004, **223-224**, 788.

5 S. Winkler, I. Ahmad, R. Golser, W. Kutschera, K.A. Orlandini, M. Paul, A. Priller, P. Steier, A. Valenta and C. Vockenhuber, *Nucl. Instr. and Meth. B*, 2004, **223-224**, 817.

6 L.K. Fifield, H.-A. Synal and M. Suter, *Nucl. Instr. and Meth. B*, 2004, **223–224**, 802.

7 L. Wacker, E. Chamizo, L.K. Fifield, M. Stocker, M. Suter and H.A. Synal, *Nucl. Instr. and Meth. B*, 2005, In Press.

8 S. Olivier, S. Bajo, L . K. Fifield, H. W . Gäggeler, T. Papina, P. H. Santschi, U. Schotterer, M. Schwikowski and L. Wacker, *Environ. Sci. Technol.*, 2004, **38**, 6507.

9 H.-A. Synal and S. Jacob, M. Suter, *Nucl. Instr. and Meth. B*, 2000, 172, 1.

10 M. Stocker, R. Bertschinger, M. Döbeli, M. Grajcar, S. Jacob, J. Scheer, M. Suter and H.-A. Synal, *Nucl. Instr. and Meth. B*, 2004, **223-224**, 104.

11 M. Döbeli, C. Kottler, M. Stocker, S. Weinmann, H.-A. Synal, M. Grajcar and M. Suter, *Nucl. Instr. Meth. B*, 2004, **219-220**, 415.

INTERACTION MECHANISMS OF URANIUM WITH BACTERIAL STRAINS ISOLATED FROM EXTREME HABITATS

M. Merroun, M. Nedelkova, A. Rossberg, C. Hennig, A.C. Scheinost and S. Selenska-Pobell

Institute of Radiochemistry, Forschungszentrum Rossendorf, P. O. Box 510119, 01314 Dresden, Germany.

1 INTRODUCTION

Water pollution due to toxic heavy metals and radionuclides remains a serious environmental and public problem. Strict environmental regulations on the discharge of heavy metals make it necessary to develop various efficient technologies for their removal. The main techniques that have been used to reduce the heavy metal content of effluents include chemical precipitation, ion exchange, adsorption onto activated carbon, membrane processes, and electrolytic methods. These methods have been found to be limited, because they often involve high operational costs or may also be insufficient to meet strict regulatory requirements, as for chemical precipitation.[1] The use of microorganisms for the removal of heavy metals and actinides has recently been investigated due to their cost effectiveness at moderate metal concentrations.[2] Understanding the role of bacteria in the migration of U and other heavy metals as well as studies on the underlying mechanisms are of great importance for the development of bioremediation strategies. Optimisation of U(VI) bioremediation strategies requires an extensive understanding of the speciation of uranium associated with bacteria. In this paper we study the interaction mechanisms of bacterial strains isolated from an extreme habitat (Siberian deep-well radioactive disposal sites) as a function of the pH of the U solution using Transmission Electron Microscope (TEM) and XAS spectroscopy.

2 METHODS AND RESULTS

2.1 Isolation and Cultivation of the Bacterial Strains

An enrichment culture technique was used to isolate different bacterial strains from a ground water sample, collected at a depth of 290 to 324 m below the land surface from the S15 monitoring well, located near the nuclear waste repository site Tomsk-7 in Siberia, Russia. Microscopic examination and 16S rDNA sequence analysis were used to characterize these bacterial isolates. One of the isolated bacteria Iso12 was affiliated to the species *Microbacterium oxydans*. The cells of this strain tolerate a concentration of U up to 2 mM and accumulate high amounts (up to 240 mg U/g dry biomass at pH 4.5 and a solution concentration of 0.25 mM).

2.2 EXAFS Analysis of the Uranium Complexes formed by the Bacterial Isolate at Different pH Values

The k^3-weighted χ spectra determined from EXAFS analyses of the uranium species formed at pH 2 and 4.5 are presented in Figure 1, along with the best fits of the data. Quantitative fit results indicate that the adsorbed U(VI) has the common linear trans-dioxo structure: two axial oxygens at about 1.76-1.77 Å, and an equatorial shell of 4 to 5 oxygens at 2.28-2.33 Å. The U-O_{eq1} bond distance is within the range of previously reported values for phosphate bound to uranyl.[3, 4, 5] The FT spectra of the two samples contain an FT peak at about 2.3 Å. After correcting for the scattering phase shift, this distance is typical for carbonate groups coordinated to U(VI) in a bidentate fashion.[6] Indeed, carbon atoms at 2.86 to 2.91 Å provide a good fit to the 2.3 Å FT peak. However, infrared spectroscopy of the same samples indicated that the carboxyl groups are not implicated in the interaction of these bacteria with uranium. Oxygen neighbours (O_{eq2}) provide a good fit to the residual EXAFS spectrum corresponding to this shell. Thus, we interpret this peak in the FT as oxygen neighbours. The fourth FT peak, which appears at 3 Å (radial distance of 3.59-3.62 Å) is a result of the back-scattering from phosphorous atoms. This distance is typical for monodentate coordination of U(VI) by phosphate.[3, 4, 5] At pH 4.5, a sixth shell corresponds to U at a distance of 5.20 Å. This distance is also present in meta-autunite,[5] suggesting that a similar inorganic uranyl phosphate phase was precipitated by the bacterial cells at pH 4.5, probably due to the release of inorganic phosphates by the cells as result of the acid phosphatase activity. At pH 2, however, this U-U peak is absent, but we could fit a contribution of U-C at a distance of 3.74 Å. This C atom most probably originates from the organic phosphate molecules of the cell surface which are implicated in the binding of uranium. In addition, The U-O_{eq1} bond distance of 2.33 ± 0.02 Å and the coordination number of 5.3 ± 0.4 found for the U/bacterial complexes at pH 2 are similar to those found in the uranium complexes formed by fructose 6-phosphate at pH 4.[7]

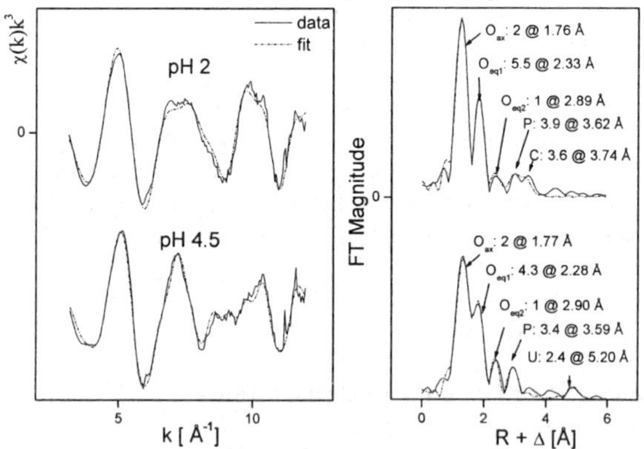

Figure 1 *U L_{III}-edge EXAFS spectra and their corresponding FT of the U complexes formed by M. oxydans Iso12 at pH 2 and 4.5.*

2.3 TEM and EDX Studies

TEM observation of *M. oxydans* Iso12 cells exposed to 0.5 mM U solution at pH 2 and 4.5 (Fig. 2) revealed significant differences on the accumulation profile of the bacterial isolate at both pH values. At pH 4.5, we note the presence of electron dense extracellular accumulations and also on the cell surface. In the case of pH 2, small amounts of uranium are bound only to the bacterial cell surfaces. There was no extracellular precipitation of U. These results are in line with those achieved by EXAFS spectroscopy. The U accumulating at both pH values at the cell surfaces corresponds most likely to the organic phosphate uranyl species. The shape of the extracellular precipitates at pH 4.5, however, is in line with the meta-autunite-like phase observed by EXAFS.

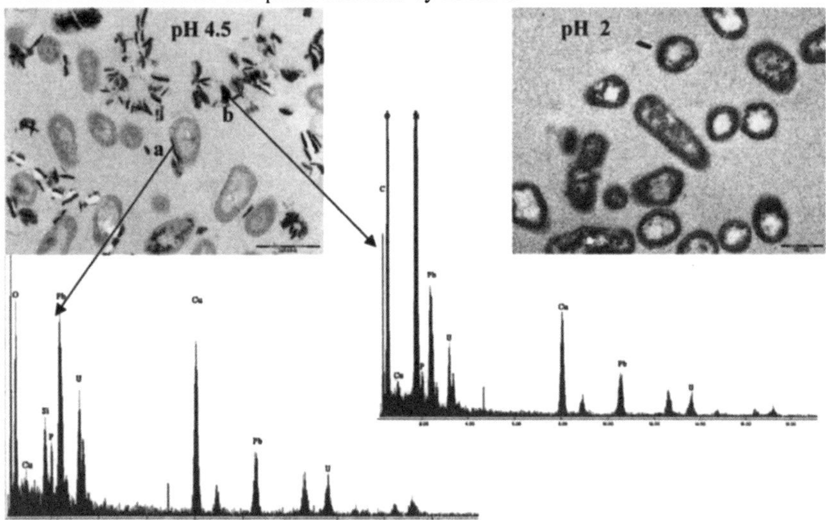

Figure 2 *Transmission electron micrographs of thin section of M. oxydans Iso12 cells treated with uranium at pH 2 and 4.5 (upper) and Energy-disperisve X-ray spectra of the uranium accumulates (bottom).*

3 CONCLUSIONS

The cultured *M. oxydans* isolate Iso12 interacts effectively with U. The EXAFS and TEM studies demonstrate that the interaction mechanisms depend on the pH: At pH 4.5, the cells of this bacterium precipitate U(VI) extracellulary as a meta-autunite-like phase, while at pH 2 U(VI) is bound by organic phosphate complexes at the cell surface.

References

1 Z. Reddad, C. Gerente, Y. Andres, and P. Le Cloirec, *Environ. Sci. Technol.*, 2002, **36**, 2067.
2 N. Renninger, R. Knopp, H. Nitsche, D. S. Clark, and J. D. Keasling, *Appl. Environ. Microbiol.*, 2004, **70**, 7404.
3 M. Merroun, J. Raff, A. Rossberg, C. Hennig, T. Reich, and S. Selenska-Pobell, *Appl. Environ. Microbiol.*, 2005, (in press).
4 M. Merroun, C. Hennig, A. Rossberg, T. Reich, and S. Selenska-Pobell, *Radiochim. Acta*, 2003, **91**, 583.
5 C. Hennig, P. Panak, T. Reich, A. Roßberg, J. Raff, S. Selenska-Pobell, W. Matz, J. J. Bucher, G. Bernhard, and H. Nitsche, *Radiochim. Acta*, 2001, **89**, 625.
6 A. Coda, A. D. Giusta, V. Tazzoli, *Acta Cryst.*, 1981, **B37,** 1496.
7 K. Koban, G. Geipel, A. Rossberg, and G. Bernhard, *Radiochim. Acta*, 2004, **92**, 903.

HIGH SENSITIVITY PLUTONIUM ISOTOPIC ANALYSIS IN ENVIRONMENTAL AND MONITORING SAMPLES

D.P. Child, M.A.C. Hotchkis and M.L. Williams

ANSTO, Lucas Heights, NSW 2234 Australia

1 INTRODUCTION

The nuclear safeguards system administered by the International Atomic Energy Agency (IAEA) has been developing environmental sampling regimes (e.g. wide area environmental sampling - WAES) requiring detection limits for actinides down to environmental levels.[1] High sensitivity isotopic analyses of actinides in environmental samples are now performed routinely by a number of different techniques, such as Thermal Ionisation Mass Spectrometry (TIMS)[2] and Sector Field Inductively Coupled Plasma Mass Spectrometry (SF-ICP-MS).[3,4] AMS offers the advantage of high sensitivity isotopic analysis completely free from molecular interference yielding detection limits down to femtograms of analyte in the case of plutonium.[5,6] At ANSTO, we have developed the capability to isolate chemically and purify plutonium from environmental samples, and to analyse ^{239}Pu:^{240}Pu isotopic ratios on the ANTARES AMS facility in samples containing femtogram levels of plutonium.

2 ANALYTICAL METHODS

2.1 Sample Preparation

For analysis in the AMS ion source, the element of interest must be purified and concentrated into a small pellet of material of no more than a few milligrams. For actinides at low levels, the most suitable form for AMS has been found to be as a co-precipitate with iron oxide.[5,7] For purification, a combination of oxidative high temperature sample combustion, acid leaching, iron hydroxide co-precipitation and ion exchange is used. Full details of the methods used and quantification of our limits of detection are reported elsewhere;[8] here we present a summary and some initial results.

Prior to processing, soil and sediment samples are checked for homogeneity, and if necessary coned and quartered to ensure homogenous representation of the sample. For AMS determination of plutonium concentrations and isotopic analysis (^{239,240}Pu), the technique of isotope dilution is used. Additions of 1pg ^{242}Pu per sample were used in the present work. Approximately 1g of sample is then placed into a tube furnace and combusted at 900 °C for 1 hour under an oxygen flow of 20ml/min.

2.2 Plutonium Purification

Our ion exchange methods are based upon those developed by Horwitz et al.[9] Combustion residues are transferred into Teflon beakers with 50mL of aqua regia, and heated at 100°C for 1 hour to leach all particulate derived plutonium from the samples. The samples are cooled and diluted with Milli-Q water to 100mL. Plutonium is then co-precipitated by the addition of NH_4OH. 50mg of iron (as iron nitrate) is added to samples with little native iron or aluminium.

Precipitates are collected by centrifugation, and washed first with 6M NaOH (to dissolve bulk aluminium) and then Milli-Q water. The remaining precipitate is dissolved in 5mL of 2M HNO_3. 1mL of 0.6M ferrous sulfamate is added to reduce all plutonium to Pu (III). 0.5mL of 1M $NaNO_2$ is then added to oxidise it to Pu (IV).

Samples are added to 0.5g of TEVA resin column,s conditioned with 2M HNO_3. The columns are then washed with 20mL of 2M HNO_3 and the eluent allowed to go to waste. Pu (IV) is retained and uranium is eluted. Columns are next conditioned to the chloride form by the addition of 2mL 9M HCl, and then washed with 20mL of 5M HCl. This wash step will remove americium, thorium and residual uranium. Finally, plutonium is eluted with 20mL of 0.1M HCl. Pu is co-precipitated with 1mg of iron and the precipitate is collected, washed with Milli-Q water and dried ready to be loaded into the AMS sample holders. The typical final sample mass is ~1.5mg of iron oxide in which the plutonium oxide is evenly dispersed.

2.3 AMS Instrumentation and analysis method

The samples are measured by Accelerator Mass Spectrometry (AMS) on the ANTARES accelerator at ANSTO, using the 'actinides beamline' previously described for analysis of $^{236}U/^{238}U$ isotopic ratios.[7] The measurement conditions for Pu isotopes are similar to those used for ^{236}U; the ions are injected into the accelerator as PuO^-, and the 5+ charge state selected for analysis after acceleration using a terminal voltage of 4MV. However, the method for obtaining isotopic ratios of Pu is different, as it is necessary to count atoms of each isotope (mass 239, 240, 242) sequentially in the same ionisation chamber. This requires changing the magnetic field of the main analysers between each isotope measurement; this is relatively slow compared to the technique of Hotchkins *et. al.*[7] and limits the efficiency and precision of the measurements.

3 RESULTS

We have analysed a number of reference materials certified for plutonium content. These were prepared and analysed using the method described above. The results of the analyses are given in Table 1 where they are compared to published values.

In most cases, the measured activities are consistent with previously reported values for the reference materials. In addition, there was good agreement between replicate analyses of IAEA-135, demonstrating good internal reproducibility. The small size of samples we use, 0.5-2 g, means that there is a risk that our results are affected by sample inhomogeneity. This is especially the case for very low activity samples where the activity may reside in only a few hot particles per gram of sample. In the case of IAEA-375, our analysis of the 0.49 g sample represents successful detection of just 23 fg and 7 fg of ^{239}Pu and ^{240}Pu, respectively. With such a low concentration of plutonium, we would expect plutonium particles to be very small and few in this sample, and therefore the risk of inhomogeneity high. However, since good agreement within error is seen between replicate analyses and between our results and published values, this suggests the plutonium in the samples is homogeneously distributed.

Table 1 *Plutonium content of a collection of reference materials comprising radionuclide bearing environmental materials*

Reference material	*Sample mass (g)*	*^{239}Pu activity (Bq/kg)*	*^{240}Pu activity (Bq/kg)*	*$^{239+240}$Pu activity (Bq/kg)*	*240/239 atom ratio*	*Reference*
IAEA-135	0.48-0.64	126±19	86±17	212±25	0.186±0.022	T.W.
Irish Sea				(205-225.8)[a]		certificate
sediment		127±8	96±6	223±5	0.207±0.006	[10]
				245±1.4	0.212±0.004	[6][b]
IAEA-375	2.06	0.113±0.010	0.124±0.020	0.237±0.022	0.296±0.054	T.W.
Russian	0.49	0.108±0.025	0.119±0.049	0.226±0.055	0.299±0.140	T.W.
soil				(0.26-0.34)[a]		certificate
				0.240±0.040		[6][b]
IAEA-300	1.08	2.41±0.09	1.51±0.11	3.92±0.14	0.170±0.014	T.W.
Baltic Sea				(3.09-3.90)[a]		certificate
sediment					0.180	[3]
					0.190±0.020	[4]

**results from 4 samples combined; [a]range; [b]mean value; T.W this work*

For two samples, isotopic information for Pu is reported here for the first time. The high 240:239 ratio for IAEA-375 reflects contamination from the Chernobyl accident. However, the ratio is lower than that of the Chernobyl source material, 0.563,[2] suggesting that the soil contains a mixture of this with global fall-out.

References

1 E. Kuhn, D. Fischer and M. Ryjinski, in Proc. 42nd INMM Annual Meeting, Indian Wells, USA, July 2001.

2 L.W. Cooper, J.M. Kelley, L.A. Bond, K.A. Orlandini and J.M. Grebmeier, *Marine Chem.*, 2000, **69**, 253.

3 S. Sturup, H. Dahlgaard and S.C. Nielsen, *J. Anal. At. Spectrom.*, 1998, **13**, 1321.

4 C.K. Kim, C.S. Kim, B.U. Chang, S.W. Choi, C.S. Chung, G.H. Hong, K. Hirose and Y. Igarashi, *Science of the Total Environment*, 2004, **318**, 197.

5 L.K. Fifield, R.G. Creswell, M.L. di Tada, T.R. Ophel, J.P. Day, A.P. Clacher, S.J. King and N.D. Priest, *Nucl. Inst. & Meth. B*, 1996, **117,** 295.

6 J.E. McAnninch, T.F. Hamilton, T.A. Brown, T.A. Jokela, J.P. Knezovich, T.J. Ognibene, I.D. Proctor, M.L. Roberts, E. Sideras-Haddad, J.R. Southon and J.S. Vogel, *Nucl. Inst. & Meth., B*, 2000, **172**, 711.

7 M.A.C. Hotchkis, D.Child, G.E. Jacobsen, P.J. Lee, N. Mino, A.M. Smith and C. Tuniz, *Nucl. Inst. & Meth. B*, 2000, **172**, 659.

8 D.P. Child et al., to be published

9 E. P. Horwitz, R. Chiarizia, M.L. Dietz and H. Diamond, *Analytica Chim. Acta*, 1993, **281**, 361.

10 S.H. Lee, J. Gastaud , J.J. La Rosa, L. Liong Wee Kwong, P.P. Povinec, E. Wyse, L. K. Fifield, P.A. Hausladen, L.M. Di Tada and G.M. Santos, *J. Radioanal. Nuc. Chem.*, 2001, **248**,757.

SPECIATION OF THE OXIDATION STATES OF NP AND PU IN AQUEOUS SOLUTIONS BY CE-ICP-MS AND CE-RIMS

N.L. Banik, R.A. Buda, S. Bürger, J.V. Kratz, B. Kuczewski and N. Trautmann

Institut für Kernchemie, Johannes Gutenberg-Universität, 55128 Mainz, Germany

1 INTRODUCTION

For the redox speciation of neptunium and plutonium in natural groundwaters, the online coupling of Capillary Electrophoresis (CE) with Inductively Coupled Plasma Mass Spectrometry (ICP-MS) has been developed.[1] Depending on the size and on the effective electrical charge, the IV and V oxidation states of neptunium, and III, IV, V and VI of plutonium are separated by CE, based on the different migration times through the capillary, and are detected by ICP-MS. The detection limit is 20 ppb, i.e.10^9 atoms for one oxidation state with a reproducibility of the retention times of $\leq 1\%$. In this contribution, some experimental details are reported, and an application to the study of the redox kinetics of plutonium oxidation states in the presence of humic substances is presented. In order to improve the sensitivity of the method, the coupling of CE to Resonance Ionization Mass Spectrometry (RIMS) has also been developed. RIMS is a highly sensitive and selective technique for isotopically resolved ultra-trace analysis of long-lived radionuclides. By using a multiple resonant laser excitation and ionization of the element of interest, it provides an excellent element and isotope selectivity and a detection limit of $>10^6$ atoms for plutonium isotopes.[2] The first applications of this new speciation method are presented.

2 CE-ICP-MS

The CE system used in this work is homemade and adapted to meet the special experimental requirements for application in a radiochemical laboratory. Hydrodynamic injection is applied to introduce the sample into the capillary with a pressure continuously adjustable from 20 to 1000 mbar. A fused silica capillary (i.d. 50 μm, o.d. 363 μm) is applied. Before use, the fused-silica capillary is purged for 10 min with 0.1 M HCl, 0.1 M NaOH, and again 0.1 M HCl, and then for 20 min with water and the electrolyte solution. Figure 1 shows the scheme of the CE equipment. The interface from the CE to ICP-MS is made of PEEK as a four way fitting. The “make-up” solution, which is used to fill up the small flow coming out of the CE capillary to the sample uptake of the nebulizer, was conveyed through a PTFE tubing, i.d. 1.5 mm, to the interface. The “make-up” solution reservoir is fixed at the same level as the electrolyte vial at the injection side to prevent a flow by the siphoning effect in the CE capillary. A MicroMist AR 30-I-FM02 nebulizer is

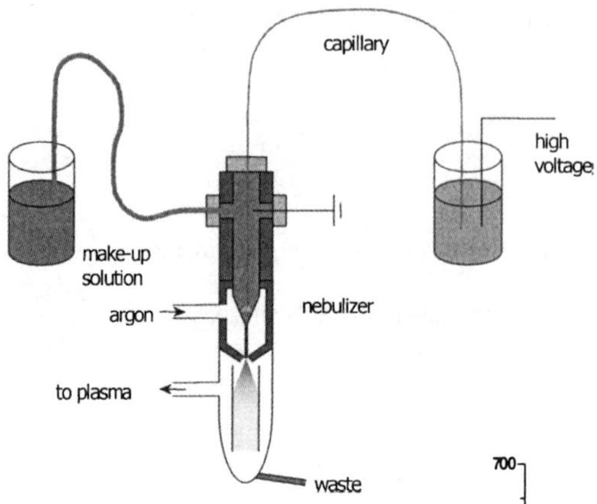

Figure 1. *Schematic of the CE-ICP-MS coupling*

used connected to a Cinnabar small-volume cyclonic spray chamber. The "make-up" flow rate is 560 µL/min, the CE voltage + 30 kV, the CE current 12 – 45 µA, the sample injection takes 10 s with 100 mbar (~44 nL ± 5% sample volume), and the electrolyte buffer is 1 M AcOH at pH 2.47. The distance between the end of the CE capillary and the inlet into the nebulizer capillary influences the chromatographic resolution and best results are obtained at a distance of 1 – 3 mm, see Figure 2 as an example. The order of the migration times depends on the degree of complexation with acetate and the resulting order of mobilities is: Pu(III)>Th(IV)~Np(V)>Pu(VI)> Pu(V)>U(VI)~Np(IV)>Pu(IV). This shows that Th(IV), Np(V), and U(VI) are of limited use as oxidation state analogues for plutonium IV – VI.

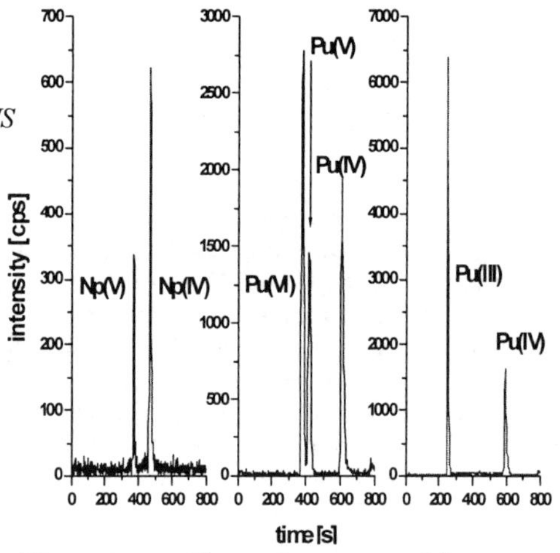

Figure 2. *Electropherograms of the different oxidation states of neptunium and plutonium*

Figure 3 shows the redox kinetic of plutonium resulting from a sequence of measurements with the CE-ICP-MS of the same sample in which Pu(VI) at pH 1.7 was put into contact with natural groundwater from Gorleben. The humic substances contained in this groundwater reduce Pu(VI) rapidly (<10 min) into Pu(V), which is subsequently reduced further into a mixture of Pu(IV) and Pu(III) and an unknown Pu(IV) species which is probably of colloidal constitution. At higher pH values, the reduction occurs even faster. This example shows the applicability of CE-ICP-MS to a direct oxidation state speciation of plutonium in natural groundwater.

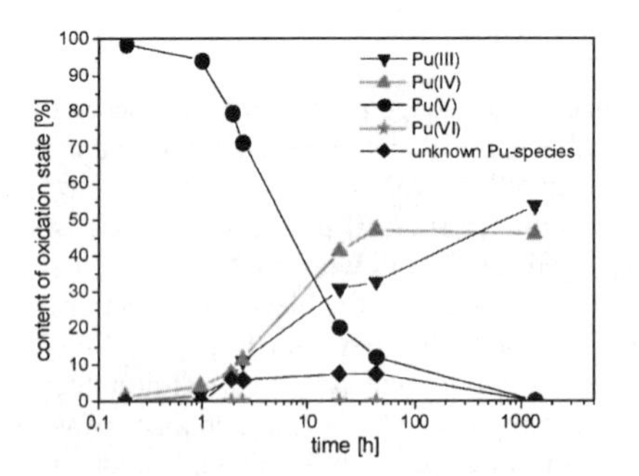

Figure 3. *Reduction of Pu(VI) to lower oxidation states by Gorleben groundwater*

3 CE-RIMS

In order to improve the sensitivity of the method, an attempt has been made to couple CE to RIMS.[2] In contrast to the on-line coupling of CE with ICP-MS, samples for RIMS have to be taken off-line and filaments[2] have to be produced by electrodeposition. From these filaments, the plutonium is atomized by thermal evaporation into vacuum. The atomic beam is simultaneously crossed by laser light of 420.76 nm, 847.28 nm, and 767.53 nm, by which the plutonium atoms are resonantly excited into a Rydberg state, which is subsequently field ionized. The plutonium ions are mass-selectively detected in a reflectron time-of-flight (TOF) mass spectrometer by a channel plate detector.[2] The detection limit is of the order of 10^6 – 10^7 atoms. The experimental scheme for the CE separation is illustrated in Figure 4. In the upper part of the Figure, the electropherogram of the plutonium species III, VI, V, and IV is shown together with the signal of a caesium marker, while in the lower part, it is indicated, where during the CE run, the high voltage was switched off in order to be able to take appropriate samples from the capillary containing the oxidation states III, VI/V, and IV. Note that, due to the very similar migration times for Pu(VI) and Pu(V), a combined fraction of the oxidation states VI/V is collected off-line. Table 1 shows a comparison in which the fractional content of the species VI/V, and IV determined by CE-ICP-MS in a plutonium solution is compared to the results obtained by CE-RIMS in the same plutonium solution after 15 fold dilution. The agreement is satisfactory. In addition, the count rates with RIMS demonstrate that, through the off-line coupling of CE with RIMS, an improvement by about three orders of magnitude in the detection limit of the speciation with CE can be achieved.

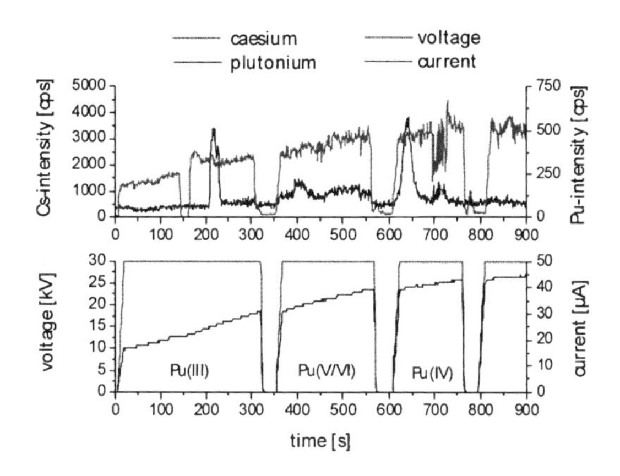

Figure 4. *Calibration of the migration times for the fractionated sample collection for CE-RIMS*

Table 1. *Fractional content of plutonium species in a sample measured with CE-ICP-MS and CE-RIMS*

	CE-ICP-MS (1×10^{11} atoms)	CE-RIMS (6×10^9 atoms)
Pu(VI/VI)	(19.2 ± 2) %	(15.2 ± 5) %
Pu(IV)	(80.8 ± 3) %	(84.8 ± 5) %

References

1 B Kuczewski, C.M. Marquardt, A. Seibert, H. Geckeis, J.V. Kratz and N. Trautmann, *Anal. Chem.*, 2003, **75**, 6769.

2 C. Grüning, G. Huber, P. Klopp, J.V. Kratz, P. Kunz, G. Passler, N. Trautmann, A. Waldek and K. Wendt, *Int. J. Mass Spectrom.*, 2004, **235**, 171.

SOFT X-RAY SPECTROMICROSCOPY OF ACTINIDE PARTICULATES

H.J. Nilsson[1,2], T. Tyliszczak[1], R.E. Wilson[1,3], L. Werme[2,4] and D.K. Shuh[1]

[1]Chemical Sciences Division, The Glenn T. Seaborg Center, Lawrence Berkeley National Laboratory, One Cyclotron Road, Berkeley, CA 94720 USA
[2]SKB, Box 5864, S-102 40, Stockholm, Sweden
[3]Nuclear Sciences Division, Lawrence Berkeley National Laboratory, One Cyclotron Road, Berkeley, CA 94720 USA
[4]Department of Physics, Uppsala University, Box 530, S-751 21, Uppsala, Sweden

1 INTRODUCTION

The Advanced Light Source Molecular Environmental Science (ALS-MES) beamline 11.0.2 scanning transmission x-ray microscope (STXM) has been used to perform soft x-ray spectromicroscopy investigations of solid oxide particulates of uranium, neptunium, and plutonium. The recent progress in the development and application of soft x-ray spectromicroscopy to fundamental scientific studies of actinide materials is described. The motivation for the use of soft x-ray spectromicroscopy is to investigate actinide absorption edges that are not commonly investigated, and to resolve spatially areas within small particles where there are chemically different properties. The K-edges of several important light elements are also located in the soft x-ray region. Of particular relevance for actinide materials is the oxygen K-edge, since actinide oxides are particularly complex and have yet to be fully understood. As soft x-ray STXM spectromicroscopy becomes more fully developed and the actinide spectroscopy understood from a fundamental perspective, this technique will have relevance in several areas of actinide science. A clear area of application will be in the field of actinide environmental chemistry investigating colloids, interface science, biogeochemical interactions, and the transport of actinides, which are related to particulates.

2 EXPERIMENTAL

All soft x-ray STXM spectromicroscopy measurements were performed at the ALS-MES beamline at the Lawrence Berkeley National Laboratory (LBNL). The beamline uses an elliptical polarized undulator to produce x-rays from 80 eV to 2160 eV and the ALS-MES STXM resides on a branchline optimized for spectromicroscopy, both of which have been described previously.[1] This results in excellent spatial resolution of 30 nm and an energy resolution of up to 7500 ($E/\Delta E$). Operation of the STXM under ambient atmosphere facilitates experiments using radioactive materials. STXM samples generally consist of solid phase particles in the size range of 50 nm to 2000 nm sandwiched between two Si_3N_4 windows that are attached to an indexed Al holder. To date, several solid state actinide compounds have been investigated using the ALS-MES STXM. The initial actinide particle spectromicroscopy studies have focused on nominal AnO_2 materials obtained from common sources. The Pu material attached to a particle characterized in this investigation was obtained from an insoluble precipitate isolated following the preparation of a ^{242}Pu stock solution after dissolution and centrifugation.

3 RESULTS

The near edge x-ray absorption fine structure (NEXAFS) spectra of UO_2, NpO_2, and PuO_2 are shown in Figure 1. The $4d_{5/2}$ edge spectra from the three actinide dioxides are quite similar in appearance. Previous measurements of U_3O_8 and UO_3 have yielded NEXAFS spectra shifted to higher energies by 0.9 eV and 1.1 eV, respectively.[2] Thus, actinide oxidation states can be determined from the measurements of charge state shifts at the $4d_{5/2}$ edges, although the magnitudes of the shifts are small. This also indicates that a mixture of actinide oxidation states could be identified by actinide $4d_{5/2}$ peak broadening.

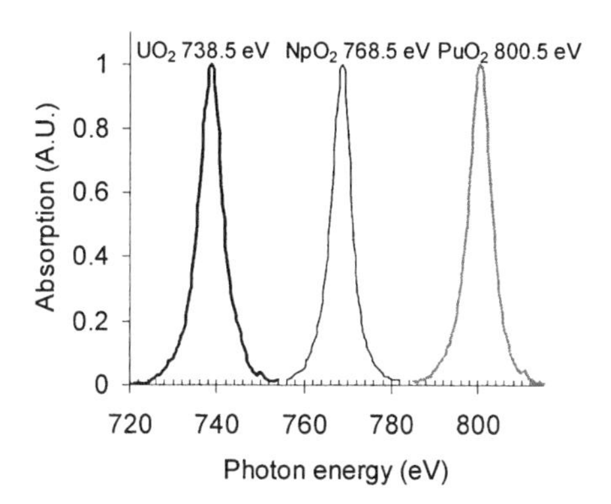

Figure 1 *Spectromicroscopy NEXAFS actinide $4d_{5/2}$ spectra from UO_2, NpO_2, and PuO_2 obtained from particles with dimensions of 100 nm to 500 nm. The spectra have been smoothed, background subtracted, and the $4d_{3/2}$ edges have been removed.*

Figure 2 shows a Pu elemental map (left) and NEXAFS spectra (middle, right) obtained from the precipitated solid described in the experimental section. The Pu elemental map is generated from the on-resonance absorption at 800 eV contrasted to the absorption below the threshold, at 792 eV (left). A comparison of the Pu $4d_{5/2}$ NEXAFS spectra collected from a 40 nm diameter Pu-rich region of the particle located by the elemental map, to that of PuO_2, shows that the Pu material associated with the particles is not PuO_2 (middle). There is a distinct energy shift compared to the PuO_2 spectrum, indicating that there is a clear difference in Pu oxidation state.

The composition of the particles was identified by NEXAFS spectromicroscopy. The right panel of Figure 2 shows the Si K-edge spectra obtained from a bulk-like region of the particle compared to the absorption background obtained from a Si-mirror in the beamline. The Si K-edge spectrum obtained from the particles is characteristic of the signature of a well-known Si material, SiO_2.[3]

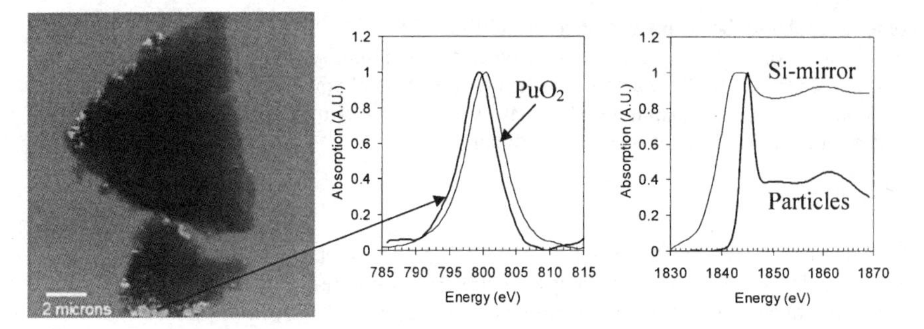

Figure 2 *Pu elemental map (left) showing Pu (light) on the particles (dark) generated from a precipitate obtained during the preparation of a Pu stock solution. The middle and right plots are the NEXAFS spectra collected from the particles at the Pu $4d_{5/2}$ edge and the Si K-edge, respectively. The spectra are compared to a reference PuO_2 spectrum and the Si absorption from beamline mirrors.*

4 CONCLUSIONS

Soft x-ray STXM spectromicroscopy is beginning to develop into a useful tool for investigating actinide particulates. The spectroscopy at the actinide $4d_{5/2}$ edges can provide chemical information about the oxidation states of actinides, although work remains to elucidate more fully the fundamental charge state shifts with reference materials. The absorption at the actinide $4d_{5/2}$ edges provides a suitable response to image the morphologies of sub-micron actinide particles down to the 30 nm spatial resolution level, and can derive actinide elemental maps with the same dimensions. Spectromicroscopy has successfully characterized actinide particles of unknown composition and has shown that the Pu associated with the SiO_2 particles in this investigation is not primarily PuO_2.

Acknowledgements

This work and the ALS are supported by the Director, Office of Science, Office of Basic Energy Sciences, Chemical Sciences, Geosciences, and Biosciences Division of the U.S. Dept. of Energy under Contract Number DE-AC03-76SF00098 at LBNL. This work was funded in part by the Swedish Nuclear Fuel and Waste Management Company, SKB AB.

References

1 T. Tyliszczak, T. Warwick, A.L.D. Kilcoyne, S. Fakra, D.K. Shuh, T.H. Yoon, G.E. Brown Jr., S. Andrews, V. Chrembolu, J. Strachan and Y. Acremann, *Proc. 2003 Synchr. Rad. Instr. Meet.*, San Francisco, AIP Conf. Proc., 2004, **705**, 1356.
2 H. Bluhm, K. Andersson, T. Araki, K. Benzerara, G.E. Brown Jr., J.J. Dynes, S. Ghosal, H.-Ch. Hansen, J.C. Hemminger, A.P. Hitchcock, G. Ketteler, E. Kneedler, J.R. Lawrence, G.G. Leppard, J. Majzlam, B.S. Mun, S.C.B. Myneni, A. Nilsson, H. Ogasawara, D.F. Ogletree, K. Pecher, D.K. Shuh, M. Salmeron, B. Tonner, T. Tyliszczak and T.H. Yoon, *J. Electron Spectros.*, in press, 2005.
3 S. Schuppler, S.L. Friedman, M.A. Marcus, D.A. Adler, Y.-H. Xie, T.D. Harris, W.L. Brown, Y.J. Chabal, L.E. Brus and P.H. Citrin, *Phys. Rev. Lett.*, 1994, **72**, 2651.

URANIUM COLLOID ANALYSIS BY SINGLE PARTICLE INDUCTIVELY COUPLED PLASMA - MASS SPECTROMETRY

C. Degueldre[1], P.-Y. Favarger[2], R. Rossé[1] and S. Wold[3]

[1] Department of Nuclear Energy and Safety, Paul Scherrer Institut, 5232 Villigen-PSI, Switzerland
[2] Institut Forel, University of Geneva, 1290 Versoix, Switzerland
[3] Department of Chemistry, Royal Institute of Technology, Stockholm, Sweden

1 INTRODUCTION

Single particle analysis is currently carried out after filtration by scanning electron microscopy (SEM).[1] Single particle analysis has also been performed over the last decade utilising optical single particle counting (SPC).[2] These techniques have been applied to the analysis of natural water e.g. Degueldre *et. al.*[3] The utilization of inductively coupled plasma - mass spectrometry (ICP-MS) for single particle analysis was first discussed for colloid bearing solutions by the author.[1] Independently, Nomizu *et al.*[4] successfully tested this approach for airborne particles. The feasibility of single particle analysis of TiO_2, Al_2O_3, clay colloids,[5] ZrO_2 colloids[6] and ThO_2 colloids[7] from suspension in water was studied by ICP-MS. These recent studies concerned colloids and the determination of the size resolution limit (< 100 nm). In this work, single particle ICP-MS analysis was tested for yellow cake particles in diluted suspensions, focusing on uranium particles (0.1 – 10.0 μm). This uranium particle specific work may be of use for solutions loaded with yellow cake colloidal particles, e.g. Miekeley *et. al.*[8]

2 THEORETICAL BACKGROUND

In single particle mode, the ICP-MS is adapted for the injection of individual particles in the nebulised water during each analysis time slot of the MS. A colloid suspension of MO_x (molecular weight M_{MOx}) and of concentration N_{col} is injected at a flow rate q_{col} in the stream of water. The diluted suspension is sprayed with a nebulisation yield η_{neb}. Each single particle generates an ion flash, a function of the particle size d_{col} with flash frequency, f, a function of N_{col}. A fraction η_A of the isotope ions $^{A}M^{+}$ produces the ion flash that passes through the cone hole, and is detected by the mass spectrometer with a counting yield η_c. However, the elemental ion $^{A}M^{j+}$ is produced together with other ions (MO^+, *etc.*). The intensity of signal, s_A, is given by the expression:

$$s_A = \xi \cdot d_{col}^3 \qquad (1)$$

(with $\xi = \pi \cdot \eta_A \cdot \eta_C \cdot \rho \cdot N_{Av} / 6 \cdot M_{MOx}$), where ρ is the colloid density and N_{Av} the Avogadro constant. The colloid concentration, N_{col}, in the original suspension is diluted and the fraction η_{neb} is found in the argon. The signal frequency, f, induced by the colloids is then simply given by:

$$f = \eta_{neb} \cdot N_{col} \cdot q_{col} \quad (2)$$

Equations (1) and (2) allow calculation of the metal size distribution (N_{col} vs d_{col}) in the colloid phase on the basis of the signal distribution (f vs s_A) or vice versa.

3 EXPERIMENTAL

The yellow cake (U_3O_8) stock fine was produced by manual milling of a technical powder (Wimmis, Spietz, Switzerland) in an agate mortar. The suspensions were shaken manually for 2 min and diluted 100 times with MilliQ water. Single particle counting (SPC) light scattering investigations were carried out with a commercial single particle monitor (HSLIS-M50) and a single particle spectrometer (HVLIS-C200), both units from Particle Measuring Systems, Inc. (PMS), Boulder, Colorado, were used on-line sequentially as described in Degueldre *et. al.*[3] The ICP-MS unit was a HP (Hewlett Packard) 4500 series 100, as reported earlier.[6]

4 RESULTS AND DISCUSSION

SPC investigations by light scattering were carried out on the initial colloid suspensions (see Fig.1). The distribution follows somewhat the expected Pareto power law:

$$\delta N_{col}/\delta d_{col} = A \cdot d_{col}^{-b} \quad (3)$$

with b ~4. Some particle classes were found around 100, 300 and more specifically 750 nm. This last particle class (i.e., 750±150 nm) is compared with the distribution measured for 802±10 nm latex colloids (Nanosphere starndard).

Yellow cake colloids were tested for $^{238}U^+$, $^{235}U^+$ and $^{254}[UO]^+$. The signals for $^{238}U^+$ were measured as a function of time after injection of the colloids. The signals induced by the U_3O_8 particles were between 10 and 10^5 counts per 10 ms for $^{238}U^+$ (with ξ = 1.0x10^{-5} nm^{-3}). The detection limit for a particle can be estimated in terms of the smallest amount of U, i.e. 10 fg U or approximately 80 nm size. The U_3O_8 colloids were also analysed for the isotope ^{235}U with an isotopic abundance of 0.72%. Isobaric interferences are absent. The background at mass 235 was found to be 0-1 count per 10 ms. The record given in Fig.2 presents typical signals obtained for mass 235 (s_{235}). It is comparable to that registered for $^{254}[^{238}U^{16}O]^+$ (s_{254}). The record at mass 235 avoids the shut down of the detector gate for 800 nm colloids. The signal distribution for the particle class around 750 nm presented in Fig.3 was calculated for ξ = 7.25x10^{-8} nm^{-3} and an average d_{col} of 750±165 nm for a comparable data fit with the SPC result. Yellow cake colloids were finally analysed for mass 254, the ion $^{254}[^{238}U^{16}O]^+$. Potential isobaric interferences were absent as discussed above. The distribution of the ICP-MS signals (Fig.4) shows the signal analysis obtained for mass 254. The signal distribution for the particle class around 750 nm presented was calculated for ξ = 7.03x10^{-8} nm^{-3} and an average d_{col} of 750±165 nm for a comparable data fit with the SPC result.

Colloid size distribution analysis, currently carried out by SEM, however, requires sample preparation, observation and particle counting, which is time consuming e.g. 20 hours. The utilization of SPC is much more rapid (e.g. 20 min run), because it may be carried out on line, however this analysis provides numbers and sizes. The use of ICP-MS in single particle mode makes analysis (isotopic and size distribution) possible in 20 s, demonstrating the potential of this technique.

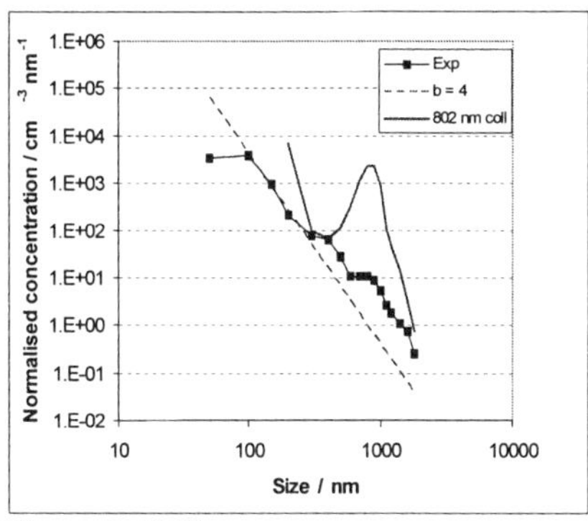

Figure 1: *SPC measurements.*

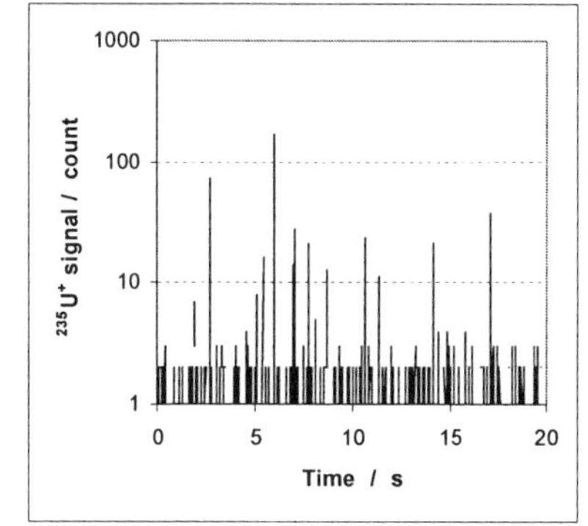

Figure 2: *U_3O_8 colloid ICP-MS ^{235}U signal.*

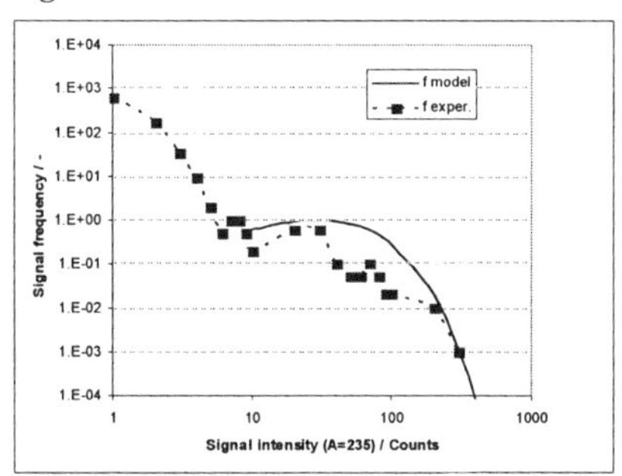

Figure 3: *U_3O_8 colloids, signal analysis: $f(s_{235})$ vs s_{235}. Model conditions: ξ = 7.2 x10^{-8} nm^{-3} and Gaussian d_{col} = 750±165 nm*

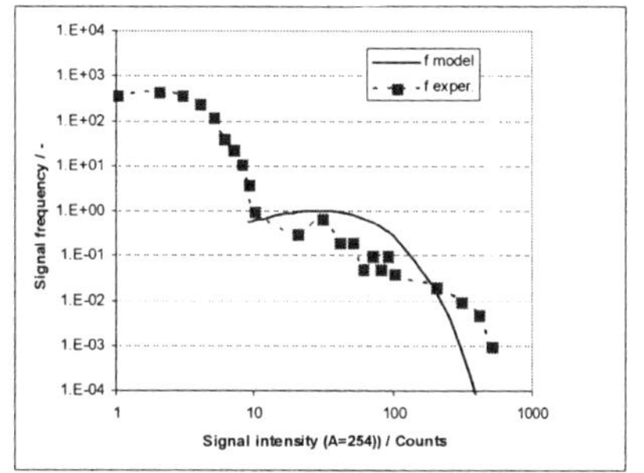

Figure 4: *U_3O_8 colloids, signal analysis: $f(s_{254})$ vs s_{254}. Model conditions: ξ = 7.0 x10^{-8} nm^{-3} and Gaussian d_{col} = 750±165 nm.*

References

1 J. McCarthy and C. Degueldre. *Characterisation of environmental particles*, Lewis Publishers, Chelsea MI, Vol 2 (1993) Chapter 6, 247-315.

2 C. Degueldre, H.-R. Pfeiffer, W. Alexander, B. Wernli and R. Brütsch, *Appl. Geochem.*, 1996, **11**, 677.

3 C. Degueldre, I. Triay, J.I. Kim, M. Laaksoharju, P. Vilks and N. Miekeley, *Appl. Geochem.*, 2000, **15**, 1043.

4 T. Momizu, S. Kaneco, T. Tanaka, T. Yamamoto and H. Kawaguchi, *Anal. Sci.*, 1991, **9**, 843.

5 C. Degueldre and P.-Y. Favarger, *Coll. Surf. A,* 2003, **217**,137.

6 C. Degueldre, P.-Y. Favarger and C. Bitea, *Anal. Chim. Acta,* 2004, **518**, 137.

7 C. Degueldre and P.-Y. Favarger, *Talanta,* 2004, **62**, 1051.

8 N. Miekeley, H. Countinho de Jesus, C.L. Porto da Silveira and C. Degueldre, *J. Geochem. Exploration,* 1992, **45**, 409.

STATUS OF THE EUROPEAN NETWORK FOR ACTINIDE SCIENCES, ACTINET

P.Chaix

CEA, Nuclear Energy Division, Centre d'études de Saclay, 91191 Gif sur Yvette cedex, France

1 INTRODUCTION

The European Union currently produces 35% of its electricity from nuclear fission. Furthermore, as stated in the Green Paper "Toward a European strategy for the security of energy supply" published in 2000 by the European Commission, the European Union should maintain a diversity of its sources of energy supply, and 'must retain its leading position in the field of civil nuclear technology'.

Research plays an important role in this context. In particular, one major issue requiring intensive research and development remains a broadly agreed approach to waste management. Research and development is also needed to explore new concepts for nuclear energy generation that make better use of fissile material and generate less waste. All these issues require expertise and improved knowledge on processes involving actinides.

However, the safety requirements for handling alpha-emitting compounds have gradually made research very costly, so that many radiochemistry laboratories have restricted their activities. This trend has dramatically reduced basic research in actinide sciences in Europe, so that at present, only few research institutions are able to maintain the necessary research capacity, and none of them alone covers the full spectrum required, with the expertise, technical competence and tools at the scale required by the technical challenges.

Because of its strategic importance, research in actinide sciences must be supported and revitalized. This can only be envisaged by the strengthening of links between nuclear research institutes and academic radiochemistry laboratories. This networking will not only facilitate the coordination and utilization of the available facilities for actinide science research, but will also consolidate and optimise research programming and training capacities in Europe.

ACTINET started in March 2004 as a consortium of 27 European research organisations, including both large R&D organisations (CEA, ITU, SCK-CEN, FZK...) and university labs, thus bringing together major experimental facilities, training experience, academic and applied research capacities. ACTINET is supported by the European Commission as a Network of Excellence under the 6th Framework Programme (more information can be found at http//www.actinet-network.org). New organisations can (and will) enter the network.

The network is coordinated by CEA. Its major orientations are decided by a Governing Board where each member organisation has one representative, and it is managed by an Executive Committee. A Scientific Advisory Committee provides scientific evaluation and guidance.

2 POOLED FACILITIES

Currently, European facilities are scattered among several institutes, some of them are redundant, while others are either missing or difficult to gain access to by academic researchers. A global European infrastructure policy is necessary. It should proceed step-by-step to improve the accessibility of the major facilities to the scientific community, to optimise at the European scale the utilisation of existing experimental facilities, and finally to coordinate the future deployment of new facilities and instruments.

The first priority objective within the ACTINET Network is to open parts of the major facilities of some European institutes, for the benefit of a joint programme of research involving all members of the Network.

In the same spirit as the pooling of experimental facilities, another objective is to bring together the tools and expertise available in European actinide modelling and simulation in a long lasting structure (the Theoretical User Laboratory, ThUL) and to make these available to the whole European community of actinide scientists. The ThUL aims to reinforce the community of European theoreticians involved in actinide science, and should strongly couple their interests with the concerns of experimentalists.

The main ACTINET pooled facilities are listed below.

The LN1 laboratory, part of the ATALANTE facility at CEA-Marcoule (France) is dedicated to molecular chemistry of all actinides in solution, solid state and at interfaces. It brings together selected techniques to obtain structural information, speciation and thermodynamic properties. The LN1 Laboratory was commissioned in April 2005. The LECA in CEA-Cadarache provides a set of shielded analytical equipment able to characterize highly irradiated fuels and materials, including electron probe micro-analysis, scanning electron microscope, and secondary ion mass spectroscopy. The Analytical Platform of the Physics and Chemistry Department in CEA-Saclay provides a large set of analytical tools, and the instruments for characterisation of the chemical retention of inactive tracers or radionuclides on natural and industrial solid phases.

ITU is one of a few European centres focused on actinide research, and provides instruments for the study of the solid-state properties of actinide compounds with particular emphasis on metals and alloys, instruments for the study of thermodynamics, thermophysics and radiation damage with particular emphasis on oxides, and instruments for solid-liquid interface chemistry.

INE in FZK provides an analytical platform with radio-analytical methods, trace and isotope analysis, surface and solid-state analysis, and a wide spectrum of speciation methods. INE also provides access to the FZK synchrotron radiation source, ANKA, dedicated for the study of actinides.

SCK-CEN provides access to the LHMA hot cell facility (shielded cells for the post irradiation examination of fuel rods and core components), solid state research tools for nuclear samples, chemical and radiochemical analysis tools.

FZR provides access to the Rossendorf beam-line, at the ESRF in Grenoble (France), and to a pool of laser spectroscopic methods at the Rossendorf Laser Laboratory, complementing the instrumentation provided by CEA and FZK.

Finally PSI will provide the microXAS beam-line at the Swiss Light Source (SLS), which will be operational in 2005.

3 JOINT PROJECTS

Offering access to up-to-date major experimental tools must be accompanied by the definition of shared ambitious research programmes, and by improved mobility between the institutions involved, in particular between academic institutions and national laboratories, in order to reduce the fragmentation of the European community of actinide sciences. The objective of the network is to stimulate and support joint research projects proposed by member organisations. The support essentially consists in access to pooled facilities and funding for the mobility of researchers, information, and radioactive samples.

Three calls for collaborative research projects have already been published within the network. After each call, proposals are selected by the Executive Committee after evaluation by the Scientific Advisory Committee. Access is given to the pooled facilities and support is given for mobility, accommodation, sample transports etc.

The proposals must involve at least two member organisations from two different countries, in order to support exchanges of researchers and of information at the European scale. Sixty-four proposals have been received at the first two calls, twenty-eight have been selected and have received support from the network, ranging from instrumentation to quantum chemistry, from solution chemistry to the physics of irradiated UO_2 (twenty-four of these twenty-eight projects make use of the pooled facilities).

Enhanced mobility and enhanced infrastructure availability for joint research programmes allows the next generation of actinide scientists and engineers to gain hands-on experience as part of their training. The network also supports a stronger participation of national laboratories in training at universities, as well as a stronger use of facilities for teaching and training.

Two calls for education and training projects have been published up to now. Six proposals have been received at the first call, four of them have been approved, funded, and are being organised:

- the “workshop for young researchers in actinide sciences”, part of the Actinides 2005 conference in Manchester;
- a workshop on the characterisation of solid-water interface reaction of metals and actinides on clays and clay minerals in spring 2006 at FZK;
- a short course with tutorial on aqueous - solid solution systems involving actinides (thermodynamics and experimental aspects) in November 2005 at PSI;
- the establishment of an ACTINET sorption board.

Furthermore, ACTINET organises each year an ACTINET Summer School (AnSS). The organisation alternates between ITU and CEA. The first AnSS was organised in June 2004 in Avignon (France) on "Thermodynamics and Kinetics of Liquid-Liquid Extraction". The second AnSS will take place in Karlsruhe on “Actinide Sciences and Applications”, and will include lectures, visits to the laboratories of ITU and the neighbouring German research centre FZK, and laboratory demonstrations during which the participants can learn and experience the challenges of working with actinides (information available at http//www.actinet-network.org, and at http://itu.jrc.eu.int).

Finally, a winter school, ‘theory for experimentalists’ is scheduled for early 2006 within the framework of the Theoretical User Lab.

USING SURFACE IMAGING AND ANALYTICAL TECHNIQUES TO STUDY THE CALCITE SURFACE IN CONTACT WITH AQUEOUS UO_2^{2+} IONS.

L.J.C. Butchins[1,2], F.R. Livens[1,2], D.J. Vaughan[2] and P. Wincott[2]

[1]Centre for Radiochemistry Research, School of Chemistry, The University of Manchester, Manchester M14 9PL, UK
[2]Williamson Research Centre for Molecular Environmental Science, School of Earth Sciences, The University of Manchester, Manchester, M14 9PL, UK

1 INTRODUCTION

The study of solid-solution interactions is important in understanding the control of radionuclide behaviour in the environment and in waste repositories. Radiologically hazardous and toxic elements can be taken out of aqueous solution by coprecipitation, surface complexation or adsorption, or released by dissolution or desorption. Carbonate minerals may dominate behaviour in sediment - water systems due to their moderate solubility, and calcite is the most commonly found carbonate mineral. The UO_2^{2+} ion is the thermodynamically favoured form of uranium in oxic systems, and is relatively mobile in the environment, especially in the presence of CO_3^{2-}.[1] There have been many studies of the uptake of metal ions by calcite, and they tend to show that coprecipitation and the formation of solid solutions are important mechanisms.[2,3] There have also been some investigations of the incorporation of uranium in the calcite structure[4,5] and also some spectroscopic studies of U(VI) sorption on the calcite surface.[6] A wide array of uranium and carbonate containing species may be formed, but the uptake mechanisms are not fully understood. In this study, atomic force microscopy (AFM) and X-ray photoelectron spectroscopy (XPS) have been used, in addition to traditional uptake experiments, to provide a fuller understanding of the mechanisms underlying uranium uptake onto calcite.

2 METHODS

2.1 Uptake experiments

Uranyl nitrate solutions containing ^{232}U tracer (20 Bq), together with ^{238}U to supplement the U concentration, were contacted with crushed, sieved Iceland Spar calcite samples for 24 hours, and uptake onto the solid from solution was measured by scintillation counting of the equilibrated solutions.

2.2 Atomic Force microscopy (AFM)

Rhombohedral blocks of Iceland Spar calcite were freshly cleaved along the 1014 plane. The interaction of the surface in contact with various uranyl nitrate solutions was imaged using AFM. A Digital Instruments Nanoscope III was used for studies in contact (in solution) and tapping modes (in air).

2.3 X-ray Photoelectron Spectroscopy (XPS)

Samples were left in contact with uranium containing solutions for the required periods of time, then rinsed briefly in DI water, and dried under a nitrogen stream. XPS spectra were obtained using a Kratos Axis Ultra automated system. Spectra were taken using monochromatic Al K_{α} radiation at 20 or 80 eV pass energy and 225 W. Data were quantified using Scofield ionisation cross-sections and analysed in CasaXPS.

3 RESULTS

Uranium has a high affinity for calcite over a wide range of initial solution concentrations (Figure 1) with $20 < R_d < 90$.

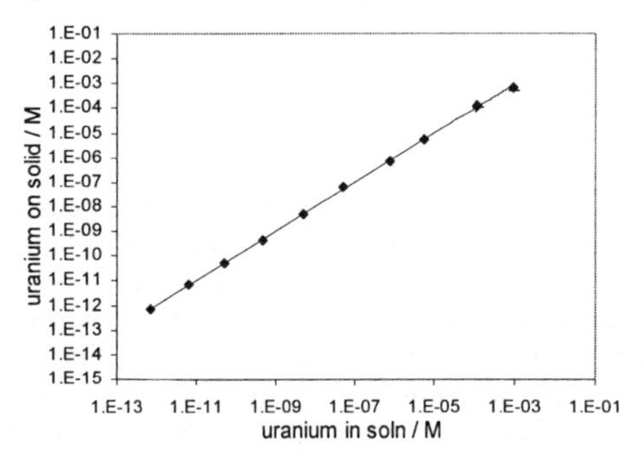

Figure 1 *Sorption isotherm for uranium uptake on calcite. Error bars are ± 1 σ, n = 3 and are too small to show on the plot.*

The XPS spectra showed that there was only one uranium species present on the calcite surface (Figure 2). The side bands at 4eV higher binding energy are satellites. A sample that was contacted with a uranium-containing solution for 7 days showed atomic ratios of Ca 4: U 1: C 4 (originating from the carbonate only): O 12, consistent with the presence of uranium atoms incorporated in the $CaCO_3$ surface structure.

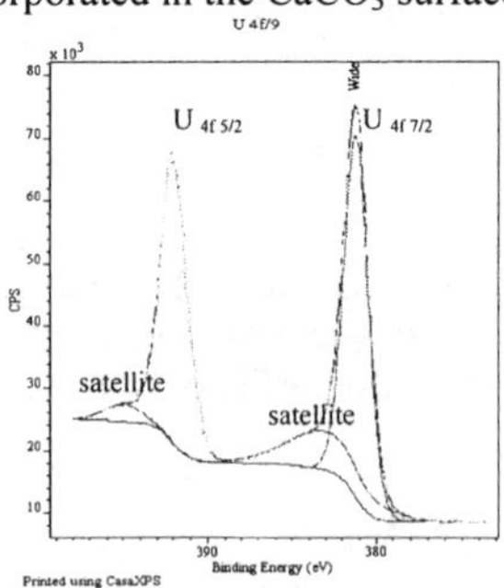

Figure 2 *Region of XPS spectrum showing U 4f peaks.*

AFM images of the $10\bar{1}4$ calcite surface left in contact with uranium-containing solutions showed a range of features surrounding etch pits, as well as varying morphologies of surface growths (Figures 3 and 4).

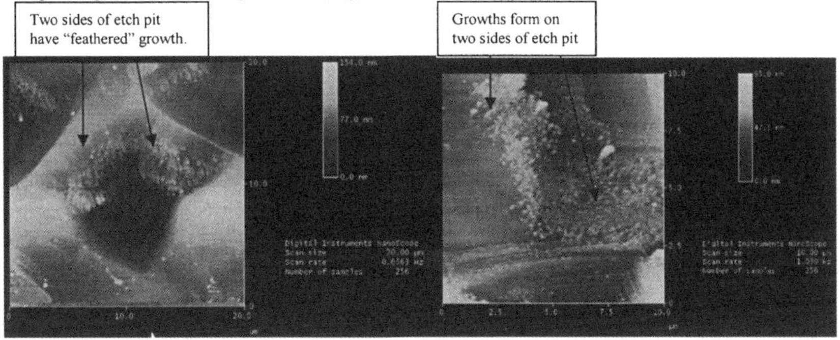

Figure 3 a (left) *AFM image of an etch pit on the calcite surface in contact with 30 ppm (1.26 x 10^{-4} M) UO_2^{2+} solution.* **b** (right) *Corner of an etch pit formed on the calcite surface in contact with 120 ppm (5.04 x 10^{-4} M) uranium solution. The fine crystallites are not observed in the absence of uranium.*

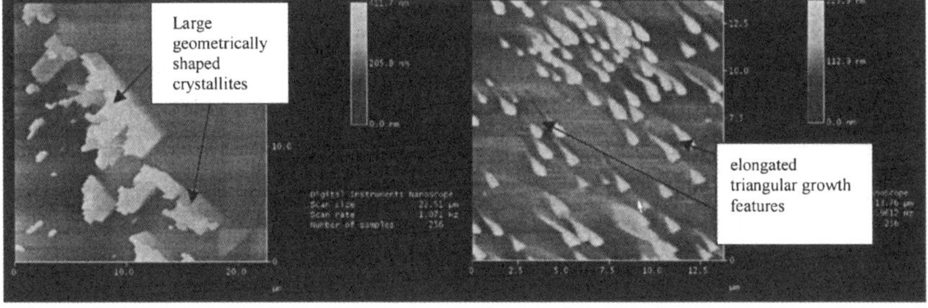

Figure 4 *AFM images of two different growth morphologies formed on the calcite surface when in contact with 1200 ppm (5.04 x 10^{-3} M) UO_2^{2+} solution.*

We can conclude that there seems to be one uranium chemistry on the surface (from XPS) but with a variety of morphologies, with reaction centred around etch pits at low concentration, but more generally over the surface at higher concentration, where two different morphologies are seen.

References

1 D.L. Clark, D.E. Hobart and M.P. Neu, *Chem. Rev.,* 1995, **95**, 25.
2 J.M. Astilleros, C.M. Pina, L. Fernandez-Diaz and A. Putnis, *Chem. Geol.,* 2002, **193**, 93.
3 S.L.S. Stipp, L.Z. Lakshtanov, J.Y. Jenson and J.A. Baker, *J. Contam. Hydrol.,* 2003, **61**, 33.
4 S.D. Kelly, M.G. Newville, L. Cheng, K.M. Kemner, S.R. Sutton, P. Fenter, N.C. Sturchio and C. Spotl, *Environ. Sci. Technol.,* 2003, **37**, 1284.
5 R.J. Reeder, M. Nugent, C.D. Tait, D.E. Morris and S.M. Heald, *Geochim. Cosmochim. Acta,* 2001, **65**, 3491.
6 E.J. Elzinga, C.D. Tait, R.J. Reeder, K.D. Rector, R.J. Donohoe and D.E. Morris, *Geochim. Cosmochim. Acta,* 2004, **68**, 2437.

PRODUCTION OF HIGH-PURITY ^{236}Pu FROM ^{237}Np BY A PHOTONUCLEAR REACTION

T. Kubota[1], A. Kudo[2], T. Kauri[3] and Y. Mahara[1]

[1]Kyoto University Research Reactor Institute, Kumatori, Osaka 590-0494, Japan
[2]Department of Environmental Policy, Kibi International University, Takahashi, Okayama 716-8505, Japan
[3]National Research Council of Canada, M-12, Ottawa, Canada K1A 0R6

1 INTRODUCTION

Neutron deficient ^{236}Pu is a useful yield tracer for determining plutonium concentrations in the environment, because anthropogenic plutonium released to the environment has undergone neutron capture, and consequently its concentration of ^{236}Pu is negligible.

^{236}Pu is produced through bombardment of ^{238}U with deuterons[1] or protons[2], bombardment of ^{235}U with alpha particles[3], and irradiation of ^{237}Np with photons by a LINAC[4]. The methods using uranium require highly enriched isotopes. Additionally, the products contain a considerable amount of ^{238}Pu as an impurity, and hence are not suitable for application as yield tracers without electro-magnetic separation. In contrast, ^{237}Np, the only meta-stable and long half-life (=2.14x10^{6} y) nuclide, can be purified by simple chemical separation. Irradiation of ^{237}Np with photons generated from the bombardment of heavy metals with high-energy electrons from a LINAC always leads to the emission of neutrons, which are captured by ^{237}Np to yield ^{238}Pu.

In this paper, high-purity ^{236}Pu, produced with decreased incidence of neutron capture is used to determine the concentration of ^{238}Pu and $^{239+240}$Pu in surface water at the west pacific coastal region of Shirahama, Japan.

2 METHOD AND RESULTS

2.1 Purification of Neptunium and Plutonium

A stock solution of ^{237}Np, produced from uranium with a LINAC method at Tohoku University, contained ^{236}Pu and ^{238}Pu impurities. The amount of plutonium produced in neptunium is quite low and it is desirable to reduce the activity of ^{237}Np in yield tracing.

To separate the neptunium and plutonium, a cation exchange resin (DOWEX 50Wx8, #200-400 mesh) was used, and different concentrations of acid (HNO_3) were applied to vary adsorption onto the resin for separation. Our first attempt to separate plutonium, as Pu(IV), was conducted by solvent extraction methods (PUREX and TTA-benzene). The very small amount of plutonium in the neptunium matrix was hardly separated in both methods, due to an undesirable change in its chemical form. In moderate acid (0.01 M to 1 M), Pu(III) and Np(V) are stable, which is a favourable property for separation using a cation exchanging resin. The oxidation states of Np(V) and Pu(III) were adjusted with hydroxylammonium chloride. To investigate adsorption, the resin was loaded with neptunium containing plutonium in an atomic ratio of 10^{-8} Pu/Np.

First, the resin was washed with 0.1 M HNO_3, eluting 60% neptunium and no plutonium, and was then washed with 2.5 M HNO_3, eluting 17% neptunium and 47% plutonium. These results infer that the purification of neptunium can be successfully performed using cation resin, and that repeated adsorption and desorption can reduce the amount of neptunium in the irradiated sample.

Next, the irradiated sample was prepared in 0.1 M HNO_3, which was loaded onto the resin and washed with 8 M HNO_3 and then 0.1 M HNO_3. The combined elutes were prepared in 0.1 M HNO_3 and then loaded again onto the same resin. This cycle was repeated 16 times with the last eluant a 0.1 M HNO_3 – 0.2% H_2O_2 mixture, resulting in an overall decontamination factor of 9.4×10^3 and a plutonium recovery of 46% (Fig. 1).

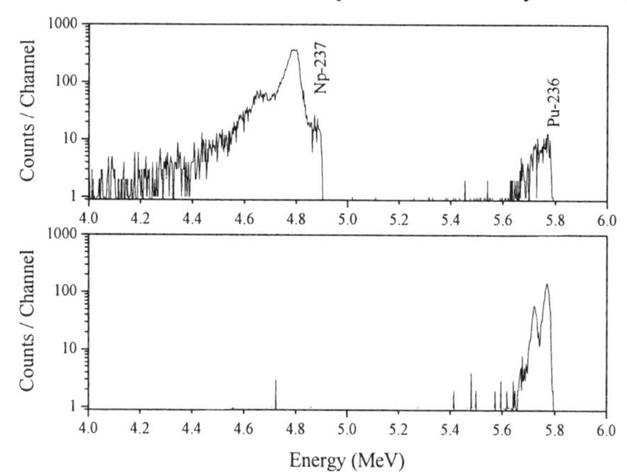

Figure 1 *(upper) Alpha spectrum of initial dried sample. (lower) Results for purified plutonium electrodeposited onto a stainless disk.*

2.2 Production of ^{236}Pu

To prepare an irradiated sample, a small volume of concentrated neptunium solution was dried in a quartz tube. The photon generator consisted of four platinum sheets (0.5 mm thick) in an aluminum casing (40 mm x 40 mm x 40 mm) connected to a cooling line, with a gap between each sheet of 2 mm. When thicker platinum sheets are used, more photons are generated from the accelerated electrons, however, the photons lead to undesirable neutron emission through (γ, n) reaction in the sheets. For the irradiation, the sample was arranged behind the photon generator in the electron beam line.

^{237}Np is transformed through a (γ, n) reaction to the ground and excited states of ^{236}Np. The latter, with a half-life of 22.5 hours, decays to ^{236}Pu (β^-, 48%) and ^{236}U (EC, 52%). Neutron capture by ^{237}Np yields ^{238}Np ($T_{1/2}$ = 2.17 days), which decays into ^{238}Pu. The irradiated neptunium sample was subject to separation after cooling and aging for two days.

The neptunium sample and the platinum sheets were cooled by water to try to avoid destruction of the irradiation system; however, larger volumes of coolant decreased the purity of ^{236}Pu. This effect was investigated by using three sizes of cooling bath for the sample: 1) a large bath (300 mm x 300 mm x 200 mm), 2) a small bath (54 mm x 34 mm x

30 mm), and 3) a cylindrical bath (inner diameter of 28 mm), all made of aluminum. Although the cooling condition of the sample was varied, that of the platinum was unchanged because insufficient cooling could result in melting of the sheets.

Using the large bath, small bath and cylindrical bath gave activity ratios of ^{238}Pu to ^{236}Pu of 0.5%, 0.04% and 0.01%, respectively. The best result is shown in Fig. 2.

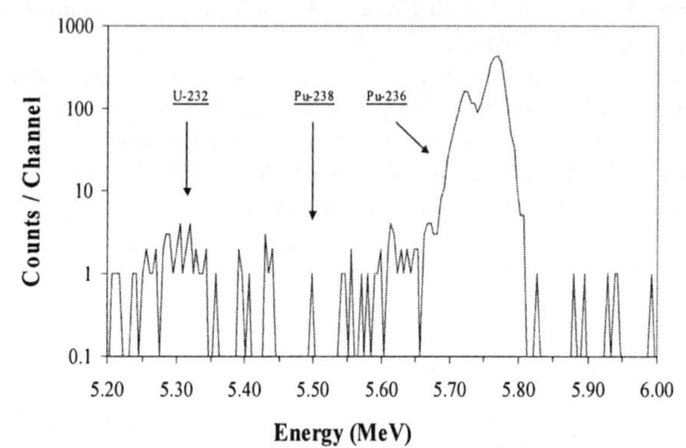

Figure 2 *Alpha spectrum of high purity ^{236}Pu containing ^{238}Pu impurity with an activity ratio of 0.01%. The activity of target ^{237}Np and product ^{236}Pu was 26 kBq and 130 Bq, respectively.*

2.3 Plutonium concentration in North Pacific Surface Water

Water samples were collected at Seto Marine Biological Laboratory, Field Science Education and Research Center, Kyoto University during November 2002. Plutonium isotopes were concentrated by coprecipitation with $Ca(OH)_2$ and $CaCO_3$ from a 700-L water sample. The precipitate, incinerated at 600°C for 3 days, was dissolved in HNO_3, after which the plutonium isotopes were separated from other elements (iron, uranium and thorium) using an anion exchange resin. The activity of plutonium electrodeposited on a stainless disk was determined by alpha spectrometry.

The average concentration of $^{239+240}Pu$ in coastal waters at Shirahama (33°41'N, 135°20'E) was determined as 1.2 μBq/L, which is lower than the concentration (2-8 μBq/L) of surface seawater collected in the western Northern Pacific (21°-35°N) in 1997[5]. The ratio of $^{238}Pu/^{239+240}Pu$ is 0.03, which is consistent with the value of world-wide fallout (=0.04) in the Northern Hemisphere.[6]

References

1 L. F. Bellido, V.J. Robinson and H.E. Sims, *Radiochimica Acta*, 1993, **62**, 123.
2 L. F. Bellido, V.J. Robinson and H.E. Sims, *Radiochimica Acta*, 1994, **64**, 11.
3 S.N. Dmitriev, Y.T. Oganessian, G.Y. Starodub, S.V. Shishkin , G.V. Buklanov, Y.P. Kharitonov, A.F. Novgorodov, Y.V. Yushkevich, D. Newton and R.J. Talbot, *Appl. Radiat. Isot.*, 1995, **46**, 307.
4 H. Yamana, T. Yamamoto, K. Kobayashi, T. Mitsugashira and H. Moriyama, *J. Nucl. Sci. Technol.*, 2001, **38**, 859.
5 K. Hirose, M. Aoyama, T. Miyao and Y. Igarashi *J. Radioanal. Nucl. Chem.*, 2001, **248**, 771.
6 E. Holm, *Appl. Radiat. Isot.*, 1995, **46**, 1225.

DECONTAMINATION OF SOILS IN LIQUID CO_2

A.A. Murzin, V.A. Kamachev, D.N.Shafikov, A.Yu. Shadrin, V.V. Bondin, S.I. Bychkov and I.G. Efremov

Khlopin Radium Institute, St. Petersburg, Russia
Mining and Chemical Combine, Zeleznogorsk, Russia

1 INTRODUCTION

Supercritical fluid extraction is a promising low-waste method for soil decontamination.[1,2] This decontamination method envisages the use of complexone solutions in supercritical carbon dioxide (SC-CO_2) at 200-300 atm. and 60-80 °C. The main drawback of the method is the rather high operating pressures. The operating pressure in the apparatus may be reduced by using complexone solutions not in supercritical carbon dioxide, but in liquid CO_2 (60-70 atm. and 20-25 °C). The objective of the present work is to demonstrate the possibility for the decontamination of real radioactively contaminated soils. The area occupied by the assembled facility, including the operator's position, is equal to 2,5 m^2; the total room capacity is 5,5 m^3. The facility enables the treatment in one cycle of up to 5 kg of soils and sorbents or up to 3 kg (dry weight) of work outfit (overalls, gloves, underwear etc). During the experiments the sample to be decontaminated with a previously compiled chart of radioactive contamination was placed into extractor 5, complexone solution was also introduced, and the extractor was hermetically sealed and filled with liquid CO_2. Hexafluoro-acetylacetone (HFA) was received from Fluorochem Co. (Great Britain).

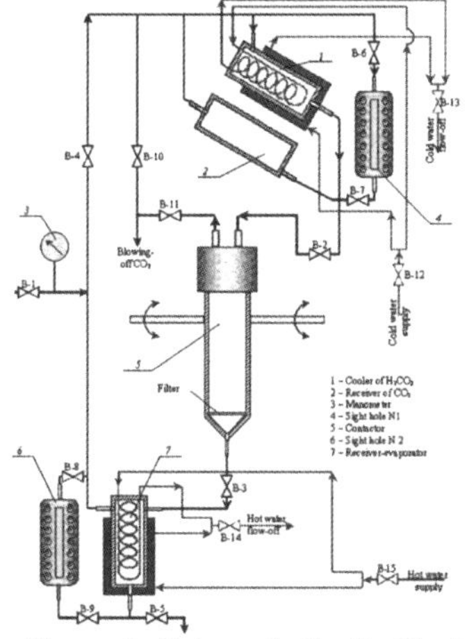

Figure 1. *Enlarged pilot facility for decontamination in liquid CO_2*

2 REAGENTS, EQUIPMENT AND EXPERIMENTAL PROCEDURES

Preliminary investigations were conducted with the use of a laboratory setup for supercritical fluid extraction. The possibility of soil decontamination was demonstrated at an enlarged pilot facility in liquid CO_2. A block diagram of the facility with an extraction chamber volume of 5L is shown in Fig. 1. Tributylphosphate (TBP), di-2-ethylhexyl-phosphoric acid (D2EHPA), octyl alcohol (OctOH), polyethylene glycol (PEG-600) and all inorganic reagents were obtained from Vecton Co. (Russia). Perfluorovaleric acid (PFVA) was obtained from the State Institute of Applied Chemistry. Soil samples were taken at the "Mayak" PA site. The soil samples were dried at 60 °C for 4 hours and screened through a 0,5 mm-sieve to separate coarse fragments.

3 RESULTS AND DISCUSSION

It is known that uranium and transuranium elements may be removed from soils by HFA and TBP solutions in supercritical and liquid CO_2.[2] It is also known that cesium is removed from soil by PVFA and PEG-600 solutions in liquid and supercritical CO_2.[3] The preliminary investigations of the decontamination of real contaminated soils by the compositions listed above have made it possible to refine the experimental conditions for operation at the enlarged pilot facility (Table 1). Depending on soil type, a single treatment operation allows the removal of 50 - 90 % of trivalent rare-earth and transplutonium elements. Two- and three-fold treatment of soil under these conditions enables the removal of up to 99% Am and REE, and affects the plutonium recovery degree only slightly.

Table 1 *Decontamination of real contaminated soil samples (500 μL PFVA, 500 μL OctOH, 100 μL H_2O, 60 mg PEG-600 70 atm., 25 °C)*

Soil sample	Cs recovery, %
Loam, black, sandy (humus layer)	19 ± 12
Sand, light, quartz	53 ± 8

The laboratory experiments have revealed the possibility of the combined removal of An-Ln and Cs–Sr fractions. However, the efficiency of An and REE removal by HFA, TBP, pyridine and H_2O solutions in liquid CO_2 is much higher. Therefore, it is of interest to investigate the possibility of removing these fractions by two successive treatment operations initially with HFA, TBP, pyridine and H_2O solution in liquid CO_2 and subsequently with PFVA, TBP, octanol and PEG solution in liquid CO_2.

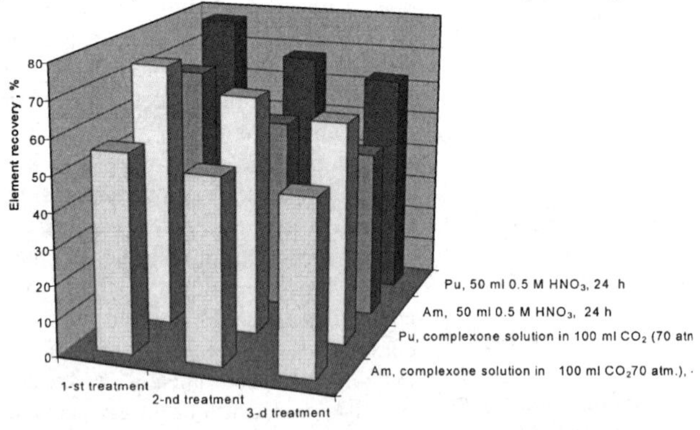

Figure 2. *Leaching of Pu from loam sample by aqueous HNO_3 solution and complexone solutions (0,2 ml TBP, 0,2 ml D2EHPA, 0,2 ml octanol and 0,2 ml H_2O) in liquid CO_2*

For comparison of soil decontamination degrees in CO_2, a soil sample (loam, humus layer) was treated with aqueous solutions by the procedure described in Chirkst et al [4] and Pavlotskaya.[5] Fig. 2 contains data on the leaching of plutonium and americium by aqueous HNO_3 solutions and TBP-D2EHPA-octanol solution in carbon dioxide.

The data obtained have shown that both methods give the same results, but in the case of treatment in the CO_2 medium, it takes 12 hours to remove all labile forms of Am and Pu, instead of 72 hours in the other case. Also, the secondary liquid waste volume is reduced by a factor of more than 60. Similar experiments were conducted to compare the efficiency of cesium removal. It was found that about 45% of Cs may be removed from this soil sample in a single treatment operation with TBP-PFVA-PEG solution in carbon dioxide. Repeat treatment does not lead to any increase of cesium recovery. Treating the soil sample with aqueous HNO_3 solution also enables the removal of about 40% Cs. So, the use of carbon dioxide as a medium for soil decontamination allows the same degree of recovery of cesium in comparison with aqueous solutions; the treatment duration therewith may be decreased no less than three times, and the volume of secondary radioactive liquid waste - by a factor of 60 at least.

To demonstrate the possibility of soil decontamination, some loam samples were treated with HFA and pyridine solutions in CO_2 or with solutions of PFVA, TBP, octanol and PEG-600 in CO_2 one time and then by these solutions in succession. The experiments revealed that radiocontamination of soils may be decreased approximately by a factor of 1.5 (Table 2). In such a manner, the decontamination factor of 1.8 for Am, Cs and γ-activity was achieved. For treating 500 g of soil, 120 ml of organic compounds were applied.

Table 2 *Decontamination of loam sample at enlarged pilot facility (70 atm., 25°C, soil №7, 500 g, contamination 80 μSv/h of total sample)*

Composition	Decontamination factor		Residual contamination, μSv/h
	Am	Cs	
30 ml HFA, 30 ml Py, 30 ml H_2O	1,5	1,3	58
10 ml PFVA, 10 ml TBP, 10 ml OctOH, 10 ml PEG-600, 10 ml H_2O	1,5	1,6	48
Successively by the above compositions	1,8	1,8	42

3 CONCLUSION

With the use of real radioactively contaminated soils, it has been demonstrated that decontamination by complexone solutions in liquid carbon dioxide is more efficient than soil treatment with solutions of salts or acids. The complexone solutions in liquid carbon dioxide allow the joint recovery of U, Pu, Am and Cs, an increase in the proportion of cesium recovered, a reduction in the time of the decontamination process by a factor of 6, and a reduction of the secondary radioactive waste volume by a factor of more than 50. The next stage of this study will concern technology optimization and equipment design. This study was done under the financial support of ISCT (project 2055).

References

1 K. Park, M. Koh, H. Kim and H. Kim, Proceedings of Supergreen 2004, October 2004, Tiyanjing, China, on CD
2 A. Murzin, A. Shadrin , V. Babain et. al., *Radiokhimia*, 1, 1997, p. 44-48 (in Russian)
3 A. Shadrin, A. Murzin, V. Babain et al. Proceedings of International Conference Spectrum'2002, Reno, Nevada, USA, August 04-08, 2002, on CD
4 D. Chirkst, K. Chaliyan, A. Chaliyan, *Radiokhimia*, 5, 1994, p. 459-461 (In Russian).
5 F. Pavlotskaya, Migration of global fallout radioactive products in soil, M., Atomizdat, 1974, p. 215 (In Russian)

A NOVEL HYDROGEN URANYL PHOSPHATE-BASED BIO-INORGANIC ION EXCHANGER FOR THE REMOVAL OF ^{60}Co, ^{85}Sr AND ^{137}Cs FROM AQUEOUS SOLUTIONS

M. Pateson-Beedle[1], L.E. Macaskie[1], C.H. Lee[2], J.A. Hriljac[3], K.Y. Jee[2] and W.H. Kim[2]

[1]School of Biosciences, The University of Birmingham, Edgbaston, Birmingham, UK.
[2]Nuclear Chemistry Research Division, Korea Atomic Energy Research Institute, P.O. Box 105, Yusong, Daejeon 305-600, South Korea
[3]School of Chemistry, The University of Birmingham, Edgbaston, Birmingham, UK

1 INTRODUCTION

The biomineralisation process for the production of hydrogen uranyl phosphate ($H_3OUO_2PO_4.3H_2O$, HUP) relies upon the liberation of inorganic phosphate at a high concentration at the surface of *Serratia* sp. cells, *via* the enzymatic cleavage of a phosphate donor molecule, e.g. glycerol-2-phosphate.[1] Previous studies, using HUP supported on *Serratia* sp. cells, have demonstrated the cation-exchange intercalation of Ni^{2+} into the interlamellar spaces of HUP.[2] This study evaluates the use of HUP, biomanufactured on a *Serratia* sp. biofilm immobilised onto polyurethane reticulated foam in a packed-bed reactor, as an ion exchanger for the removal of Cs, Sr and Co from aqueous solutions. Simulated radioactive waste and water from a pool, sited in a post irradiation examination facility, at the Korea Atomic Energy Research Institute (KAERI) were tested.

2 METHODS AND RESULTS

2.1 Preparation of HUP-Based Bio-Inorganic Ion Exchanger

Serratia sp. (NCIMB 40259) cells, grown in a carbon-limiting continuous culture, in an air-lift fermenter containing polyurethane foam[3], adhere to the foam matrix and grow in a hydrated polymeric matrix of their own synthesis to form a biofilm (Figure 1a). The biofilm, visualised using cryo- and environmental-scanning electron microscopy (cryo-SEM and ESEM), showed clusters of biomass attached to the foam structure as well as monolayers of single cells;[4] the average thickness of the biofilm was estimated as 26 μm.[5] Also, it was shown using magnetic resonance imaging (MRI) that the biofilm coating of the foam is a non line-of-sight method, i.e. all surfaces are covered.[4]

Polyurethane foam coated with biofilm (cubes: 88 *per* reactor, 125 mm^3 or rolled-up rectangular piece: 1 *per* reactor, length = 13.5 cm, width = 6.8 cm and depth = 0.4 cm) were packed in cylindrical glass columns (length = 9.0 cm and internal diameter = 1.5 cm). The reactors were challenged with a solution of $UO_2(NO_3)_2.6H_2O$ (1 mM), disodium glycerol-2-phosphate (5 mM) in sodium citrate buffer (2 mM), pH 6.0 and flow rates (ca. 26 mL/h) were controlled by an external peristaltic pump. The uranium deposited *per* reactor was in the range of 60 to 125 mg. UO_2^{2+} contents in reactor outflows were determined as described previously[3]. On progressive deposition of uranyl phosphate on the biofilm, the cells are transformed to a crystalline appearance, forming interweaving fibrils composed of many cells joined together (Figure 1b). The HUP deposits were removed, washed with deionised water, acetone, dried and examined using a D5000 Siemens Diffractometer. The cell parameters of the phosphate deposits (in Å), defined using the programme 'Chekcell', were $a = 6.9731(1)$ and $c = 17.3603(6)$, which are in reasonable agreement with the literature for $(UO_2)(HPO_4).4(H_2O)$ $a = 6.995$ and $c = 17.491$[6]; the system is tetragonal and the space group *P4/ncc*.

Figure 1 *(a) Polyurethane foam coated with Serratia sp. biofilm visualised using cryo-SEM. Inset: photograph of a cube (1 cm^3) of polyurethane foam coated with biofilm. (b) HUP deposits on Serratia sp. biofilm visualised using ESEM.*

2.2 Removal of Co, Sr and Cs from Solutions Using Packed-Bed Reactors

Reactors containing HUP (~107 mg U), supported on biofilm-cubes, challenged with $Co(NO_3)_2.6H_2O$ (1 mM, pH 6), removed 100% of the metal at an estimated molar ratio of Co:U of 1:1.7 (by assay of the outflows), indicating nearly column saturation at breakthrough and suggesting by this ratio the formation of $Co(UO_2PO_4)_2.nH_2O$. Reactors containing ~102 mg of U and challenged with $Sr(NO_3)_2$ (1 mM), pH 6 retained 100% Sr^{2+} to an estimated molar ratio of Sr:U of 1:1.8, indicating saturation of the HUP host and probable formation of $Sr(UO_2PO_4)_2.nH_2O$. Reactors containing ~61 mg U and challenged with CsCl (1 mM) pH 6 retained 100% of Cs^+ (estimated molar ratio of Cs:U was 2.8:1). The Co^{2+}, Sr^{2+} or Cs^+ contents of reactor outflows were estimated using Nitroso-R salt, Arsenazo III or phosphomolybdic acid, respectively, as described previously[7].

Similar reactors were used for the removal of ^{137}Cs (Isotope Products Laboratory, 0.0904 mCi in 0.1 M HCl, USA), ^{85}Sr and ^{60}Co (Analytics, 0.1 mCi in 0.1 M HCl, USA) from simulated radioactive waste solutions. Radionuclides were analysed using a Gamma spectrometer. The composition of the simulated radioactive waste solution was ^{137}Cs, ^{85}Sr and ^{60}Co at 0.333 mM each, pH 5.4, the reactor temperature was 25 °C and the flow rate was 10 mL/h. Figure 2a shows 100% removal of ^{137}Cs, ^{85}Sr and ^{60}Co together from the simulated radioactive waste solution using the bioinorganic ion exchanger. The breakthrough capacities were Cs:0.11 mmol, Sr:0.055 mmol and Co:0.055 mmol.

MRI of similar bioreactors (using non-radioactive surrogates) showed channelling between the foam cubes and down the side of the reactor (the fluid has higher velocities)[8], i.e. contact between the flow and the HUP-biofilm is reduced. Therefore, the foam cubes were substituted by a rectangular piece of foam containing HUP (124 mg U) supported on biofilm, rolled-up in the reactor, to overcome fluid channelling. This was used for the removal of ^{137}Cs and ^{60}Co from water from a pool, where spent nuclear fuels are stored and cooled, sited in a post irradiation examination facility at KAERI. The flow rate was 4.6 mL/h and the initial radioactivity of the water was ^{137}Cs 1.7810 Bq/mL (0.26 pM) and ^{60}Co 0.1146 Bq/mL (0.71 pM). The reactor was effective in the removal of > 97% of the ^{137}Cs and > 80% of ^{60}Co from the pool water, during 7 days, without saturation (Figure 2b).

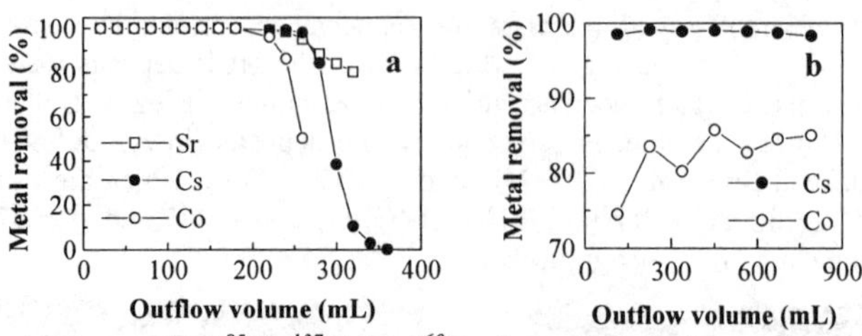

Figure 2 *(a) Removal of ^{85}Sr, ^{137}Cs and ^{60}Co from simulated waste using a reactor containing HUP supported on biofilm-cubes (b) Removal of ^{137}Cs and ^{60}Co from a pool water at KAERI, using a reactor packed with HUP supported on rolled-up biofilm-foam.*

3 CONCLUSIONS

We have shown that using HUP as a bioinorganic ion exchanger it was possible to remove 100% of ^{137}Cs, ^{85}Sr and ^{60}Co from aqueous solutions until the reactor reached saturation. Also, we have shown that the ion exchanger was effective in the removal of > 97% of the ^{137}Cs and > 80% of ^{60}Co from pool water, at KAERI, where spent nuclear fuels are stored for the post irradiation examination.

Acknowledgements

This work was supported by BBSRC (grant no. 6/E1464), Hanil Nuclear Company and KAERI. The authors thank Recticel (Belgium) for the biofilm support (Filtren TM30).

References

1 L.E. Macaskie, R.M. Empson, A.K. Cheetham, C.P. Grey and A.J. Skarnulis, *Science*, 1992, **257**, 782.
2 K.M. Bonthrone, G. Basnakova, F. Lin and L.E. Macaskie, *Nature Biotechnol.*, 1996, **14**, 635.
3 M. Paterson-Beedle and L.E. Macaskie, 'Use of PhoN Phosphatase to Remediate Heavy Metals' in *Methods in Biotechnology: Microbial Products and Biotransformations*, ed., J.L. Barredo, Humana Press, Totowa, 2004, Vol. 18, Chapter 25, pp. 413.
4 L.E. Macaskie, P. Yong, M. Paterson-Beedle, A.C. Thackray, P.M. Marquis, R.L. Sammons, K.P. Nott and L.D. Hall, *J. Biotechnol.*, 2005, **118**, 187.
5 V.J.M. Allan, M.E. Callow, L.E. Macaskie and M. Paterson-Beedle, *Microbiol.,* 2001, **148**, 277.
6 B. Morosin, *Acta Cryst.*, 1978, **B34**, 3732.
7 M. Paterson-Beedle and L.E. Macaskie, 'Removal of Co, Sr and Ce From Aqueous Solutions Using Native Biofilm of *Serratia* sp. and Biofilm Pre-coated with Hydrogen Uranyl Phosphate' in *Biohydrometallurgy: A Sustainable Technology in Evolution*, eds., M. Tsezos, A. Hatzikioseyian and E. Remoundaki, National Technical University of Athens, Athens, 2004, ISBN 960-88415-0-X, pp. 1155.
8 K.P. Nott, F. Heese, L.E. Macaskie, M. Paterson-Beedle and L.D. Hall, *AIChE J.*, 2005, **51**, 3072

EFFECT OF CLOSO-DODECACARBORANE ION EXCHANGERS ON THE POTENTIOMETRIC RESPONSE OF IONOPHORE-BASED ION-SELECTIVE ELECTRODES

S.M. Peper and W.H. Runde

Chemistry Division, Los Alamos National Laboratory, P.O. Box 1663, MS J514, Los Alamos, New Mexico 87545, USA

1 INTRODUCTION

Regulated environmental contaminants that are specifically related to nuclear processing facilities include fission products (e.g., Cs^+ and Sr^{2+}) and actinides (e.g., UO_2^{2+}). In order to ensure regulatory compliance, analytical methods must be available that can provide sensitive, selective, and reliable measurements of these inorganic ions. While possessing the advantages of low cost and in-situ monitoring capabilities, one detection platform that is well suited to meet these criteria is the membrane-based ion-selective electrode (ISE). Over the years, several electrode configurations have been reported for the detection of Cs^{+1} and UO_2^{2+}.[2,3] Today, these sensors are typically composed of a plasticized poly (vinyl chloride) (PVC) matrix containing a lipophilic ionophore and an ion exchanger.

Until recently, the best ion exchanger suitable for ISE use was tetrakis-[3,5-bis(trifluoromethyl)phenyl]borate ($TFPB^-$) (Figure 1).[4]

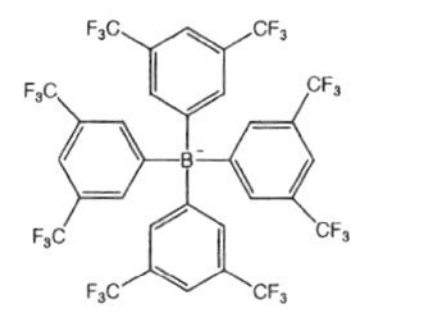

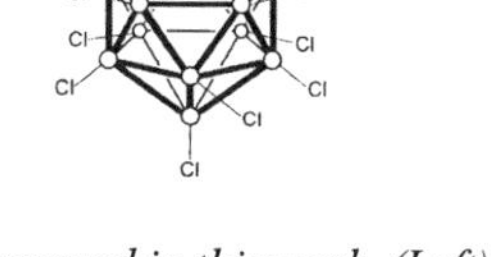

Figure 1 *Chemical structures of the ion exchangers discussed in this work. (Left) $TFPB^-$; (Right) undecachlorinated closo-dodecacarborane, UCC^-. Open circles represent boron atoms; shaded circle represents carbon.*

Although extensively used for sensing applications at near-neutral pH, tetraphenylborates are not suited for acidic samples due to their rapid decomposition.

In order to fabricate durable electrodes for uncomplexed UO_2^{2+}, which exists in solution at pH ≤ 3, a more robust ion exchanger must be employed. Very recently, it has been reported that in the presence of calixarene-type ionophores *closo*-dodecacarborane (carborane) ion exchangers yield ISEs with comparable performance to analogous electrodes containing $TFPB^-$.[5] In addition, carboranes exhibited excellent lipophilicity and improved chemical stability in acidic media, relative to $TFPB^-$.[6]

2 METHOD AND RESULTS

2.1 Response and selectivity of Cs^+-selective ISEs

In this work, Cs^+-selective PVC electrodes plasticized with 2-nitrophenyl octyl ether (NPOE) and containing calix[6]arene-hexaacetic acid hexaethylester (Cs-I) as ionophore and either $TFPB^-$ or UCC^- are compared (Figure 2). For all corresponding electrodes, the interfering ions evaluated responded in a Nernstian manner, suggesting that electrodes containing UCC^- respond in complete analogy to conventional $TFPB^-$-based electrodes.

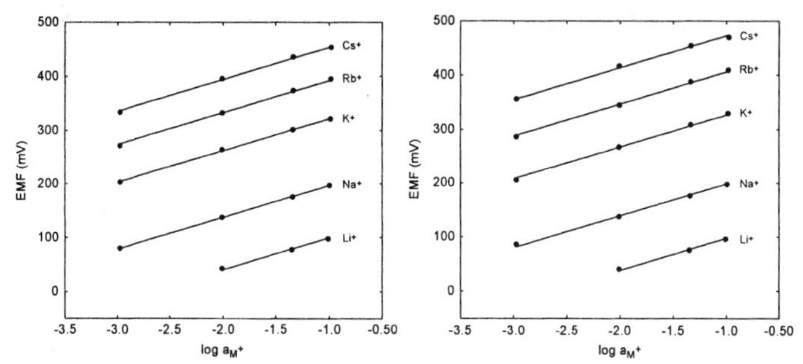

Figure 2 *Mean potentiometric response (N=4) of PVC- NPOE (1:2) ion-selective membranes containing Cs-I (20 mmol/kg) and either $TFPB^-$ (10 mmol/kg) (left) or UCC^- (10 mmol/kg) (right). Solid lines denote a theoretical response slope of 59.16 mV/decade at 25°C.*

Unbiased thermodynamic selectivity coefficients were determined using the modified separate solution method (Figure 3). Interestingly, ISEs containing UCC^- as the ion exchanger exhibit a slight improvement in selectivity over Na^+, Ba^{2+}, Sr^{2+}, Ca^{2+}, and Mg^{2+}. This suggests a weaker interaction between UCC^- and the primary ion-ionophore complex, with respect to $TFPB^-$.

2.2 Response and selectivity of UO_2^{2+}-selective ISEs

In addition to Cs^+, NPOE-plasticized PVC electrodes containing a UO_2^{2+}-selective ionophore, N,N′-Diheptyl-N,N′-6,6-tetramethyl-4,8-dioxaundecanediamide (ETH 295), and $TFPB^-$ were studied. ETH 295 has previously been reported as a useful ionophore for coated-wire[2] and liquid-contact[3] ISEs. Our work evaluates the ligand in the presence of an ion exchanger, $TFPB^-$, which is known to expand the concentration measuring range of ISEs by suppressing coextraction of lipophilic sample anions with primary cations.[7] After calculating the optimized ion exchanger concentration, ion-selective membranes were prepared that contained 200 mol% $TFPB^-$, which corresponds to the optimal ion exchanger concentration for a divalent primary ion and divalent interfering ions that both complex the ionophore with a 1:1 stoichiometry. Surprisingly, at pH 3, instead of exhibiting a response characteristic of the divalent uranyl (VI) cation, a monovalent cation response occurred. Previously, this phenomenon has been reported for ion exchanger-free membranes, UO_2^{2+}-selective membranes composed of a much less polar plasticizer, 1-chloronaphthalene.[2,3] As a result, the electrodes demonstrated an ion exchanger-type selectivity pattern, which is based on the hydration enthalpies of the extracted cations and not on the coordination ability of the ionophore (Figure 3).

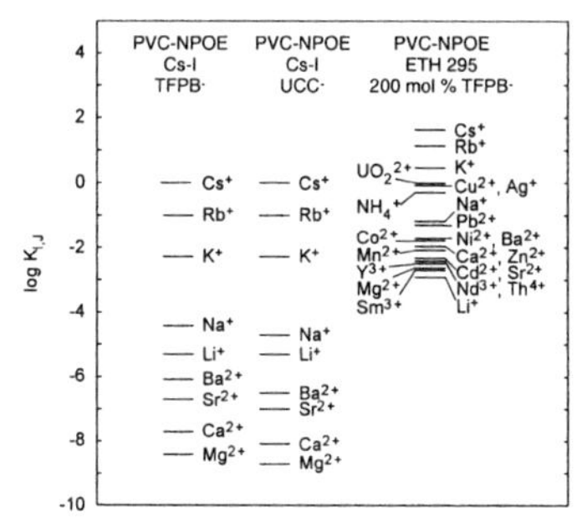

Figure 3 Mean *potentiometric selectivity coefficients (N=4),* $\log K_{I,J}^{pot}$, *for the* Cs^+-*selective ISEs shown in Figure 2 and of PVC-NPOE (1:2) ion-selective membranes containing ETH 295(10 mmol/kg) as ionophore and* $TFPB^-$ *(20 mmol/kg) as ion exchanger.* $\log K_{I,J}^{pot}$ *is defined as the difference in extrapolated electrode potentials at 1 M sample activity for an interfering ion, J, and a primary ion, I, divided by 59.16 mV.*

3 CONCLUSION

Membrane-based ISEs containing a Cs-I and UCC^- as ion exchanger possess similar response characteristics and comparable selectivity to analogous ISEs containing $TFPB^-$ ion exchanger. NPOE-plasticized PVC electrodes based on ETH 295 as a neutral ionophore for free UO_2^{2+} and containing 200 mol% $TFPB^-$ do not function as expected, due to an uncharacteristic monovalent cation response.

Acknowledgements
S.M. Peper is a Los Alamos National Laboratory Director's Postdoctoral Fellow, supported by the Laboratory Directed Research and Development (LDRD) Programme.

References

1 A. Cadogan, D. Diamond, M. R. Smyth, G. Svehla, M. A. McKervey, E. M. Seward and S. J. Harris, *Analyst*, 1990, **115**, 1207.
2 P. A. Bertrand, G. R. Choppin, L. F. Rao and J.-C. G. Bunzil, *Anal. Chem.*, 1983, **55**, 364.
3 J. Senkyr, D. Ammann, P. C. Meier, W. E. Morf, E. Pretsch and W. Simon, *Anal. Chem.*, 1979, **51**, 786.
4 E. Bakker and E. Pretsch, *Anal. Chim. Acta*, 1995, **309**, 7.
5 S. Peper, M. Telting-Diaz, P. Almond, T. Albrecht-Schmitt and E. Bakker, *Anal. Chem.*, 2002, **74**, 1327.
6 S. Peper, Y. Qin, P. Almond, M. McKee, M. Telting-Diaz, T. Albrecht-Schmitt and E. Bakker, *Anal. Chem.*, 2003, **75**, 2131.
7 W. E. Morf, D. Ammann and W. Simon, *Chimia*, 1974, **28**, 65.

GEOCHEMICAL AND MICROBIAL CONTROLS ON DECOMPOSITION AND DISPERSION OF DEPLETED URANIUM (DU) IN THE ENVIRONMENT: AN OVERVIEW OF CURRENT RESEARCH AND THE DEVELOPMENT OF METHODS FOR HANDLING DU MICROPARTICULATES

R. Alvarez[1,5], N.D. Bryan[2], M. Fomina[3], G.M. Gadd[3], S. Handley[4], M.J. Keith-Roach[4], F.R. Livens[2,5], J.R. Lloyd[1,5], P.L. Wincott[1,5] and D.J. Vaughan[1,5]

[1]School of Earth, Atmospheric and Environmental Sciences, University of Manchester.
[2]Centre for Radiochemistry Research, School of Chemistry, University of Manchester.
[3]Division of Environmental and Applied Biology, Biological Sciences Institute, University of Dundee.
[4]Department of Earth, Ocean and Environmental Sciences, University of Plymouth.
[5]Williamson Research Centre for Molecular Environmental Science, University of Manchester.

1 INTRODUCTION

Depleted uranium (DU) is commonly utilised in armour-piercing munitions, armour cladding, aircraft counterweights, in steels as an alloy and for radiation shielding.[1] Conflicts in which DU has been utilised by the military include: 1991 Gulf War I in Iraq/Kuwait; 1995 in Bosnia-Herzegovina; 1999 in Kosovo/Serbia; and recently in 2003 Gulf War II in Iraq.[2] Additionally, the UK has been involved with DU test firing since the early 1960s at Eskmeals in Cumbria, England and also at Kirkudbright in Dumfries and Galloway, Scotland.[3] When a DU projectile impacts on armour it keeps its form better than any projectile made of tungsten or steel, and tends to sharpen itself on impact, presumably due to DU's low melting point and pyrophoric properties.[4] The UK military expended 1.9 tonnes of tank-fired DU during the recent conflict in Iraq.[5] CHARM 3 (120 mm anti-tank penetrator rounds) are one type of DU ammunition used by the UK military.[6] CHARM 3 penetrators consist of DU alloyed with approximately 0.75 % titanium (to increase the fissile strength of the munition). They also contain a coating of aluminium to assist in handling, and small quantities of transuranic elements and other impurities introduced during the manufacturing process.[1] In general, these types of penetrators contain approximately 4.9 kg of DU.[4] It is estimated that 10 – 35 % of the DU penetrator is aerosolised on impact or when the DU catches fire.[7]

2 THE PRESENT STATUS OF RESEARCH ON THE FATE OF DU

The vast majority of DU entering the environment does so as large pieces of DU alloy, and any dusts which are produced from impacting targets are rapidly diluted to levels which are indistinguishable from the natural uranium background.[8] After impact, the DU ultimately deposits on the ground and other surfaces, and is generally in the form of partially oxidised fragments of various sizes.[7] Penetrators which hit soft targets such as sand or soil are expected to penetrate to depths of more than 50 cm and stay intact for long periods of time.[7]

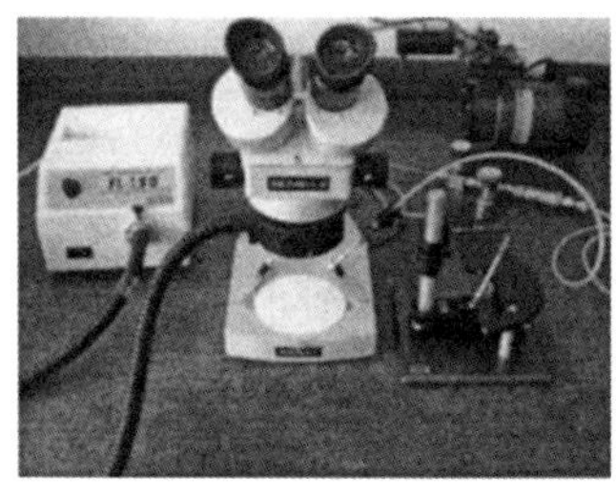

Figure 1 *Micromanipulator, microscope and vacuum-driven needle assembly for the transport of micron-sized DU particles onto substrates for imaging and analysis.*

The environmental fate of DU following corrosion and oxidation of fragments is not well known.[2,9] As yet, no contamination of groundwater by DU has been detected in target regions such as Kosovo.[4] Although DU deposited as oxide is expected to be immobile,[4] simulated predictions, using data obtained from sites in Kosovo, indicate that significant risks may be likely to occur 15 to 30 years after the war.[10] The weathering and possible potential contamination of groundwater from DU penetrators/fragments found in terrestrial environments can vary. Solubilisation rates of DU may be sufficiently high in quartz sand, granite or acidic volcanic rock. The adsorption of DU onto soil particles, particularly clay particles and organic matter is expected to reduce the mobility of DU by minimising re-suspension.[7] In studies where DU penetrators were allowed to weather in a natural environment in arid soils for approximately 22 years, the results showed that the uranium movement occurred primarily through the dissolution and re-precipitation of uranium oxides.[2] With increasing depth in the soil, the uranium oxides were found to be coated with amorphous silica and/or were found in aggregates with clays, which thus limited the mobility of uranium.[2] The uranium oxide phases evident in these studies include schoepite, $(UO_2)_8O_2(OH)_{12}.12H_2O$ and metaschoepite, $(UO_2)_8O_2(OH)_{12}.10H_2O$.[2] Interestingly the titanium, which is associated with the DU penetrators, showed some evidence for mobility in the soil.[2] In short term studies involving the artificial weathering of DU in aqueous media, UO_2 was the initial phase which formed in water, followed by schoepite and then studtite, $UO_4.4H_2O$ after four months of weathering.[11] In calcium phosphate-rich solution, DU metal initially formed uranyl phosphate hydrate, $(UO_2)_3PO_4.4H_2O$, after one week and schoepite after eleven weeks.[11] Silica-rich solutions produced a range of uranyl oxides from UO_2 to UO_3, including becquerelite, $Ca(UO_2)_6O_4(OH)_6.8H_2O$.[9,11] Further studies by the same authors involving the assessment of the alteration of CHARM 3 DU penetrator metal in soil columns revealed that corrosion occurred more rapidly in loamy clay than in sand. Metaschoepite and uranyl phosphate hydrate were evident in loamy clay, and in sand UO_2 and metaschoepite formed.[9] In addition to assessing the geochemical conditions around DU fragments, it is important to study the influences imparted by soil microorganisms, since they provide the potential for the metal to bind directly to the cell surface and hence alter the surrounding soil chemistry, leading to microbially induced corrosion. Very few studies have been published with regards to the microbial influence on the corrosion of DU.[12]

3 PROPOSED STUDY

Our proposed study addresses both the geochemical and microbial controls on the decomposition and dispersion of DU in the environment through a multidisciplinary approach involving both laboratory model systems and field samples to gain a predictive understanding of these processes. The effects of the chemical and microbial processes involved in DU decomposition and transformation will be defined and the effects of DU on bacterial and fungal communities will be explored with both culturing and molecular methods. State-of-the art imaging and analytical techniques will be used to characterise the nano- and macro-particulate DU decomposition products. The dispersion of DU and its decomposition products will also be measured in column experiments and interpreted quantitatively using a mechanistic model.

4 PRELIMINARY PREPARATORY RESULTS

To facilitate the analysis of various sizes of DU particulates and concurrently conform to the service requirements of imaging and analytical instrument manufacturers so as not to jeopardise the safety of servicing personnel, a mechanism was designed by which particulates of less than 100 microns can be selected amongst a scatter of DU decomposition products. With the aid of a stereo microscope (Meiji, EMZ 5) with illuminator (Meiji, FL150), a micromanipulator (Research Instruments, D10P/S10-TSE6) and an in-house vacuum-driven needle assembly (see Figure 1), particulates ranging from approximately 50 to over 100 microns can be selectively picked up and dropped down onto a sample mount secured in place with surface adhesive. At a maximum magnification of 67.5, the resolution for the measurement of particle sizes is 22 ± 11 microns. The Radioactive Substances Act 1993 exempts activities of 0.4 Bq g^{-1} or less from registration and hence the maximum acceptable mass of DU particulates per gram of sample mount to be placed in an instrument is 0.03 mg, which equates to either a single DU particle with 120 micron diameter or a number of smaller particulates, depending on their size and shape.

Acknowledgements

The authors wish to acknowledge Dr Stephen Sestak, PSSRI, Open University, UK for his assistance with the design of the vacuum-driven needle assembly. Funding support from NERC is gratefully acknowledged.

References

1 E. R. Trueman, S. Black, and D. Read, *Sci. Total Environ.*, 2004, **327**, 337.
2 B. J. Buck and A. L. Brock, *Soil Sediment Contam.*, 2004, **13**, 545.
3 T. Carter, '*Comparison of Kirkudbright and Eskmeals Environmental Monitoring Data with Generalised Derived Limits for Uranium*', Unclassified, DRPS report 167/2002, Defence Science and Technology Laboratory UK, 2002.
4 M. Keller, B. Anet, M. Burger, E. Schmid, A. Wicki, and C. Wirz, '*Environmental and Health Effects of Depleted Uranium*', 00437/KMA/LEI, Spiez Laboratory, Swiss Defence Procurement Agency, Spiez, 2001.
5 Ministry of Defence, '*Depleted Uranium*', 2004, http://www.mod.uk/issues/depleted_uranium/middle_east_2003.htm.
6 Ministry of Defence, '*Depleted Uranium*', 2004, http://www.mod.uk/issues/depleted_uranium/.
7 A. Bleise, P. R. Danesi, and W. Burkart, *J. Environ. Radioactiv.*, 2003, **64**, 93.
8 UNEP, '*Depleted uranium in Kosovo: post-conflict environmental assessment*', United Nations, 2001.
9 E. R. Trueman, S. Black, S. Read, and M. E. Hodson, *Geochim. Cosmochim. Acta,* 2004, **68**, A493.
10 M. Durante and M. Pugliese, *J. Environ. Radioactiv.*, 2003, **64**, 237.
11 E. R. Trueman, S. Black, S. Read, and M. E. Hodson, *Geochim. Cosmochim. Acta,* 2003, **67**, A493.
12 D. B. Ringelberg, C. M. Reynolds and L. Karr, *Soil Sediment. Contam.*, 2004, **13**, 563.

FIBROUS "FILLED" SORBENTS FOR THE PRECONCENTRATION OF ACTINIDES AND TECHNETIUM

G.V. Myasoedova, N.P. Molochnikova and I.G. Tananaev

Vernadsky Institute of Geochemistry and Analytical Chemistry RAS, 19 Kosygin str., Moscow, 119991, Russia

1 INTRODUCTION

The need to develop selective methods for isolating and separating radionuclides is connected with the growing interest in their determination in technological solutions for the reprocessing of nuclear materials and in environmental applications. The most dangerous radionuclides are uranium, plutonium, americium, neptunium and technetium, which can be found in different environments and technological solutions. Important requirements for the determination of radionuclides in these contexts are their preliminary isolation and separation from other elements. The most effective are the sorption methods that provide a high degree of extraction and selectivity using complexing sorbents. Different sorbents with functional groups that may form complexes with radionuclides in the presence of salts of alkaline and alkaline earth elements are used for the selective extraction of radionuclides from solutions, even at very low concentrations. Kinetic properties are an important sorbent characteristic for the concentration of trace elements. Better kinetic properties are possessed by fibrous sorbents, compared to traditional granule sorbents. For example, some fibrous sorbents are effective at radionuclide preconcentration from different solutions.

Fibrous "filled" sorbents are promising sorption materials to concentrate radionuclides in aqueous media. These sorbents are composed of two polymers: a thin, porous polyacrylonitrile fiber and a small-dispersed filler. The filler, polymeric and mineral ion exchanger, complexing sorbents and other sorption material, is held strongly inside the fiber. Small filler particles and fiber diameters provide good kinetic properties for these sorbents. Concentration and separation by means of fibrous "filled" sorbents may be carried out in 10-15 minutes. They are stable in acidic and alkaline solutions, and may be used for concentrating in static and dynamic sorption regimes. The sorbents can be used again after the desorption of the radionuclides. These sorbents are produced in the form of non-woven materials, and can be used as filters, disks and can be placed into a column or cartridge. Fibrous "filled" materials with different fillers, intended for sorption concentration and separation are produced in Russia by a special technique (Tver-city). This work demonstrates the sorption properties of some fibrous "filled" sorbents with different fillers, and their potential use for concentrating and separating radionuclides from solutions of varying composition.

2 METHODS AND RESULTS

Table 1 shows the materials used for the preparation of the sorbents. The content of the filler in the fiber was 50%. The density of the fibrous material was 80-100 g/m^3.

Table 1 *Titles of fibrous "filled" sorbents and fillers*

Sorbents*	***Filler***
POLYORGS 33-n, 34-n and 35-n	Complexing sorbents with amidoxime and hydrazidine groups
POLYORGS 4-n and 17-n	Complexing sorbents with 3(5)-methylpyrazole(4-n) and 1,3(5)-dimethylpyrazole groups(17-n)
AV-17-n	Strong basic anion-exchanger

*n denotes fibrous "filled" sorbent

$^{233}U(VI)$, $^{239}Pu(IV)$, $^{241}Am(III)$, $^{243}Am(III)$, $^{237}Np(V)$, $^{239}Np(V)$, and $^{99}Tc(VII)$ were used in the experiments. The sorption of 10 ppm radionuclide solution was carried out under static conditions. The (V : m) ratio = (volume of solution, V, (ml) to mass of sorbent, m, (gram)) ranged between 100 and 2000, and the phase contact time between 10 min and 24 h. The radionuclide concentrations before and after sorption were determined by radiometric measurements. It is known that complexing sorbents with amidoximes and hydrazidine groups exhibit selectivity for transition metals in the presence of high concentrations of alkaline and alkaline earth element salts. The potential of fibrous "filled" sorbents with similar groups for radionuclide concentration was studied from model solutions of the following compositions: **I** - 0.5M NaCl, pH=6; **II** - Ca-0.043; Mg-0.08; Na-0.045; Cl - 0.2 mg/l; pH=6.5; **III** – Ca - 42; Mg-10; Na-54; K-39; Cl-4.6; SO_4-29; HCO_3^- - 159 mg/l.

Table 2 shows the experimental data for radionuclide sorption by these fibrous "filled" sorbents with amidoximes and hydrazidine groups.

Table 2 *Degree of radionuclide extraction from model solutions by fibrous "filled" sorbents V:m=100; contact time 2 h*

Model solution	*Sorbent*	*Sorption degree, %*			
		^{241}Am	^{239}Pu	^{233}U	^{239}Np
I	POLYORGS 33-n	99	93	96	97
II		99	95	96	99
III		99	98	96	99
I	POLYORGS 34-n	96	92	91	-
II		99	99	91	98
III		96	92	91	99

The fibrous "filled" sorbents POLYORGS 33-n and 34-n possess good kinetic properties. The maximum sorption of radionuclides under static conditions at V:m=100 is achieved by 10-15 min; and at V:m=1000 by 30min. This behaviour allows their use for concentrating radionuclides from aqueous salt solutions under dynamic sorption regimes also.[2,3]

It is known that strong basic anion-exchangers can be used for the concentration of plutonium from HNO_3 solutions in the form of anion complexes. We have studied the efficacy of fibrous sorbents, filled with anion-exchanger, AV-17, and complexing sorbents containing groups of heterocyclic amines (Table 1), which in acid solution can show both complexing and anion-exchange properties. The sorption of actinides by the "filled" fibrous sorbents, POLYORGS AV-17-n, POLYORGS 4-n and 17-n, showed that Pu(IV) was sorbed most effectively from 3-7 M HNO_3; the distribution coefficient was found to be 10^3 ml/g. Plutonium concentration by these sorbents from 5 M HNO_3 takes place with 10 min contact time under static conditions. It was established that the concentration of 3-6 M $NaNO_3$ in solution does not affect the sorption degree of plutonium. Under the same conditions, Am(III), U(VI), Np(V) and Tc(VII) were not sorbed; the sorption degree for Np(IV) was found to be ~60%. After concentration, plutonium can be completely desorbed by 0.5-1 M HNO_3, and the sorbents can be used repeatedly. The data show the potential of these sorbents to concentrate Pu(IV) from nitric acid solutions of complex composition and for its separation from other actinides.[4]

Sorption of Tc(VII) was studied with using fibrous "filled" sorbents AV-17-n and POLYORGS 17-n. It was found that these sorbents sorbed Tc from acidic, alkaline and salt-containing solutions. In batch experiments the distribution coefficients are >10^3 ml/g from 0.1-1 M NaOH, 0.1-1 M HNO_3 and 0.1-0.5M $NaNO_3$. The experimental data, obtained for the extraction of Tc(VII) from various model salt solutions are shown in Table 3.

Table 3 *Degree of Tc(VII) extraction (%) from saline solutions by fibrous "filled" sorbents. V:m= 200; time of phase contact 2 h at 20^{0}C; [Tc(VII)] = 100 ppm*

Sorbents	*Sea water**	*I**	*II**	*Karachay lake waters*
POLYORGS -17-n	99	99	99	83
AV-17-n	91	99	98	81

* *simulated sea water, groundwater of areas PA Mayak (**I**) and Yucca Mountain (**II**)*

3 CONCLUSION

This study of the sorption of actinides from solutions of various composition by fibrous materials filled with anion exchangers and complexing sorbents, shows their effectiveness and good kinetic properties. These sorbents are very useful for concentrating and separating actinides from solutions.

References

1. G.V. Myasoedova, I.I. Antokolskaya, O.P. Shvoeva et al., *Solv. Extr. & Ion Exch.*, 1988, **6**, 301.
2. G.V. Myasoedova, V.A. Nikashina, N.P. Molochnikova, and L.V. Lileeva, *Zh. analit. khimiya.* (Russ), 2000, **55**, 611.
3. I.B. Medvedeva, S.I. Rovny, G.V. Myasoedova, and N.P. Molochnikova, *Radiokhimiya* (Russ), 2001, **43**, 359.
4. N.P. Molochnikova, G.V. Myasoedova, and I.G. Tananaev, *Radiokhimiya* (Engl.), 2003, **45**, 546.

Np(V) SORPTION TO GOETHITE IN THE PRESENCE OF NATURAL AND HYDROQUINONE-ENRICHED HUMIC ACIDS

A.B. Khasanova[1], St.N. Kalmykov[1], N.S. Shcherbina[1], A.N. Kovalenko[1], I.V. Perminova[1] and S.B. Clark[2]

[1]Lomonosov Moscow State University, Department of Chemistry, Leninskie Gory 1-3, 119992 Moscow, Russia
[2]Washington State University, P.O. Box 644630, Pullman, WA, USA

1 INTRODUCTION

The sorption of actinides by mineral surfaces is considered as a geochemical barrier in a radioactive waste repository system. However, minerals could be present as colloidal micro- and nano-particles with enhanced mobility in the environment. Therefore, the molecular level description of sorption to mineral surfaces is required for prediction of the migration of the actinides.

The ubiquitous presence of organic matter, in particular of humic substances (HS), is a significant factor for surface processes modeling. Estimated reduction potentials of HS are 0.5-0.7 V.[1] This is consistent with the redox potential of hydroquinone, which probably suggests that these structures are responsible for the redox properties of humic materials. Moreover, direct electrochemical evidence exists on the quinoid nature of the redox-active units of humics.[2] It has been shown that HS can reduce Pu(VI/V) to Pu(IV) and Np(VI) to Np(V).[3,4] However, only one paper has been published on the redox interactions of Np(V) with humic acids.[5] This could result from the high structural heterogeneity of natural HS, which translates into reactive properties that are highly variable between humic fractions and sources of humic materials. Hence, to quantify adequately the impact of HS on the transport of actinides for safety assessment, not only the quantity, but also the quality of the humics present in the system should be taken into consideration. To obtain a comprehensive assessment of this phenomenon, it should be evaluated over a broad range of environmental parameters.

The objective of this study was to evaluate the impact of humic substances that differed in their content of quinonoid moieties on the sorption of Np(V) by a well-characterized, synthesized goethite (α-FeOOH) sample over a broad pH range.

2 METHOD AND RESULTS

2.1 Sorption of Np(V) on goethite at different pH values

The sorption of ^{237}Np(V) onto goethite was studied over a broad pH range. For this purpose, a colloidal sample of goethite α-FeOOH was synthesized according to Atkinson et al.[6] and characterized by powder-XRD, SEM, potentiometric titration, and BET surface analysis. Batch sorption experiments were carried out in 50 mL polypropylene tubes under N_2-atmosphere in a glove-box in the absence of UV-light at 25±1 °C. The necessary pH values were adjusted using NaOH- and $HClO_4$-CO_2 free solutions. The redox speciation of Np(V) was determined using solvent extraction (TTA and HDEHP)[7] and X-ray photoelectron spectroscopy (XPS).[8]

Equilibrium in the Np(V)-goethite system was achieved in a week. The dependence upon pH of Np(V) sorption on the goethite surface is shown in Figure 1. Preliminary experiments have shown a negligibly small effect of ionic strength on Np(V) sorption. This could be indicative of the formation of inner-sphere complexes with the surface hydroxyl groups of goethite. The Np redox-speciation data showed a lack of redox transformations of Np(V), both at the solid phase and in solution.

2.2 Sorption of Np(V) on goethite in the presence of structurally different HS

The next set of sorption experiments was devoted to the study of Np(V) sorption in the presence of humic materials that differed in their content of quinoid moieties. For this purpose, leonardite humic acid and its hydroquinone-enriched derivative were used. Leonardite humic acid was isolated from the commercially available potassium humate Powhumus (Humintech Ltd, Duesseldorf). It was modified using formaldehyde condensation with hydroquinone, as described by Perminova et al.[9] The condensation was conducted at a monomer to humic ratio of 100 mg per 1 g of HS. The content of functional groups in the parent (CHP) and modified humic material (HQ100) was, 3.8 and 3.5 mmol/g of carboxylic groups, respectively, and 1.0 and 3.3 mmol/g of phenolic hydroxyls, respectively. The redox capacity of the CHP and HQ100 accounted for 0.6 and 1.3 mmol/g.[9] The concentration of humic materials in the solution during the sorption process was monitored by UV-absorbance measurements using a CARY 50 spectrophotometer (Varian, USA).

The sequestration of CHP and HQ100 from the aqueous phase onto the goethite surface at different pH values is shown in Figure 1. For both humic materials tested, the pH dependences of sorption onto the goethite surface were very similar: sorption increased drastically with a decreasing pH, and a very steep slope was observed at pH < 6.

Sorption of Np(V) in the presence of humic materials is shown in Figure 1. It should be noted that in the presence of humic materials, equilibrium sorption of Np(V) onto the goethite surface was achieved much more slowly compared to the binary system, taking twenty days compared to seven days. As can be seen from Figure 1, the pH dependences in the range of neutral and alkaline pH in the presence of both humic materials were very close to those obtained in the absence of HS. In the range of pH < 6, however, substantial differences were observed in the sequestration of Np(V), in particular, in the presence of the hydroquinone-enriched humic material. In the case of the modified humic material, the sequestration curve of Np(V) repeated the curve of the humic sample, suggesting the presence of the total pool of Np in the form of humic complexes under these conditions. At the same time, for the parent humic material only a slight increase in the sequestered amount of Np onto the surface of goethite was observed.

Such a substantial difference in the impact of the humic materials on the sorption of Np(V) can be explained by the different redox properties of these samples. The modified sample was enriched with hydroquinone moieties, and had a reducing capacity twofold higher than the parent humic material. Given an increase in the reducing potential of hydroquinone moieties with decreasing pH, the observed correlation between sequestration of humic material and Np at pH < 6 can be explained by reduction of Np(V) to Np(IV). In its turn, the much higher affinity of Np(IV) for binding to humic polyanions provides for its enhanced sequestration in the range of acidic pH.

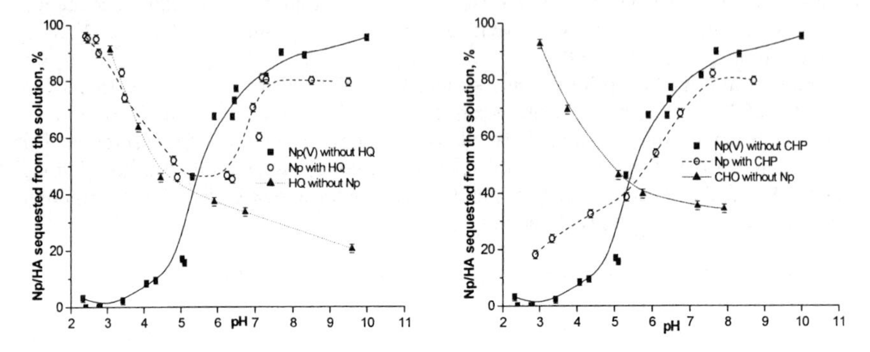

Figure 1 *pH dependence of Np sorption onto goethite in the presence of humic materials: $I(NaClO_4)$ = 0.1 M, $c(Np(V))_T$ = $5.8{\cdot}10^{-7}$ M, c(α-FeOOH) = 0.22 g/L, c(CHP) = 43 mg/L, c(HQ100) = 44 mg/L.*

3 CONCLUSION

The use of humic materials with controlled redox properties has allowed us to demonstrate the important role of the structural features of HS in their interaction with actinides and with Np(V) in particular. The presence of hydroquinone-enriched humic material governed the sorption of Np(V) onto the goethite surface in the range of acidic pH. The results show that interactions with humic substances govern the solid-water distribution of Np in natural systems.

Acknowledgements

The work was supported by US DOE: grants RUC2-20008 and RUC2-20006 administrated through CRDF.

References

1 R.K. Skogerboe and S.A. Wilson, *Anal.Chem.* 1981, **53**, 228.
2 J.T. Nurmi and P.G. Tratnyek, *Environ. Sci. Technol.*, 2002, **36**, 617.
3 C. Andre and G.R. Choppin, *Radiochim. Acta,* 2000, **88**, 613.
4 K.S. Nash, A.M. Fried and J.C. Sullivan, *Environ. Sci. Technol.* 1981, **15**, 834.
5 R. Artinger, C.M. Marquardt, J.I. Kim, A. Seibert, N. Trautmann and J.V. Kratz *Radiochim. Acta,* 2000, **88**, 609.
6 R.J. Atkinson, A.M. Posner and J.P. Quirk, *J.Phys.Chem.*, 1967, **71**, 550.
7 A. Morgenstern and G.R. Choppin, *Radiochim.Acta*, 2002, **90**, 69.
8 Yu. A. Teterin, S. N. Kalmykov, A. P. Novikov, Yu. A. Sapozhnikov, L. Dj. Vukcevic, A. Yu. Teterin, K. I. Maslakov, I. O. Utkin, A. B. Khasanova and N. S. Shcherbina, *Radiochemistry*, 2004, **46**, 545.
9 I.V. Perminova, A.N. Kovalenko, et al. *Environ.Sci.Technol.*, 2005, **39**, 8518.

SORPTION OF Np(IV) AND Np(V) BY DEPLETED URANIUM DIOXIDE

O.N. Batuk[1], St.N. Kalmykov[1], E.V. Zakharova[2], Yu.A. Teterin[3], A.V. Mikhaylina[1], B.F. Myasoedov[4] and V.I. Shapovalov[5]

[1]Radiochemistry Division, Chemistry Department, Lomonosov Moscow State University, Moscow 119992, Russia.
[2]Institute of Physical Chemistry RAS, Moscow, Russia,
[3]Science centre "Kurchatovskiy institute", Moscow, Russia,
[4]Vernadsky Institute of Geochemistry and Analytical Chemistry RAS, Moscow, Russia,
[5]Russian Federal Nuclear Centre – VNIIEF, Sarov, Russia.

1 INTRODUCTION

Depleted uranium (DU) is a waste product of the U enrichment process that typically contains about 99.8 mass % of ^{238}U, 0.2 % of ^{235}U and a very small amount of ^{234}U. The worldwide stocks of DU are estimated at around 1.2 Mt of U, primarily in the form of uranium hexafluoride (UF_6).[1] DU is considered to be low-level waste, but its conversion to stable oxide forms with emplacement in a near-surface disposal facility is not appropriate.[2]

The current civilian application of DU is limited to its use in counterweights or ballasts in aircrafts, radiation shields in medical equipment, as chemical catalysts and heavy concrete casks (DUCRETE™). Large quantities of DU could be used for MOX fuel production for fast breeder reactors.[2,3]

There are a number of beneficial uses of DU in deep repositories. These may include the use of DU in the form of UO_2 for heavy concrete casks as shielding overpack, as a component of floor material (drift "invert"), or as fill and backfill material.[3] In "fill" technology, DU in the form of UO_2 is placed in the voids of spent nuclear fuel waste containers for storage, transport and disposal.[4] The suitability of using DU in deep repositories should be tested, *i.e.* its chemical stability, and its sorption and diffusion properties.

This paper describes the capability of different depleted uranium dioxide samples to sorb Np(IV) and Np(V) from model aqueous solutions.

2 METHODS AND RESULTS

The two samples of depleted uranium dioxide used in this study were prepared from UF_6 at different temperatures 625°C (sample 1) and 800°C (sample 2) in reducing media. Morphology and average particle size were similar for the samples (ranging from 1.5 to 2 μm) as determined by dynamic light scattering and scanning and transmission electron microscopy. The samples were nonporous, and had low values of free surface area (0.5-1.5 m^2/g), as determined from BET analysis.

Partial oxidation of U(IV) to U(VI) takes place as a result of the diffusion of oxygen into the fluorite type crystal structure of UO_2. This is required for charge neutralization, and leads to a decrease of the lattice constant. The bulk composition of the samples was studied by powder X-ray diffraction, while surface composition was studied by X-ray photoelectron spectroscopy. The surface acidic properties were studied by potentiometric titration. The partial oxidation of U(IV) in the surface layer results in a decreasing point of zero charge (pH_{pzc}). The main characteristics of the samples are presented in Table 1.

Table 1 *DU dioxide sample characteristics*

Sample	Lattice constant[a], Å	Oxygen coefficient *x* in UO_X [b]	pHpzc [c]
1	5.4661(2)	2.28±0.23	6.5
2	5.4718(5)	2.17±0.22	7.0

[a] *determined by XRD,* [b] *determined by XPS,* [c] *determined by potentiometric titration.*

Sorption of Np(IV) and Np(V) by depleted uranium dioxide samples was studied in deionized water (DW) under aerobic and anaerobic conditions (N_2 atmosphere) and in J-13, which is a reference water from the unsaturated zone of the proposed nuclear waste repository at Yucca Mountain. J-13 is an oxidized water, low in ionic strength with carbonates and hydroxides as complexing ligands.[5] At neutral pH values, equilibrium was reached in about 2 and 24 hours for Np(IV) and Np(V), respectively for all solutions.

The pH dependence of Np(V) and Np(IV) sorption was studied in DW, as presented in Figure 1. In general, the sorption of Np(IV) increased with increasing of pH. However, for all samples a decrease in sorption at pH between 4.5 and 6.5 was observed. The redox speciation of Np was studied using a solvent extraction technique[6] after desorption by 1M HCl. Partial oxidation of Np(IV) to Np(V) was observed in this pH interval, as presented in Table 2.

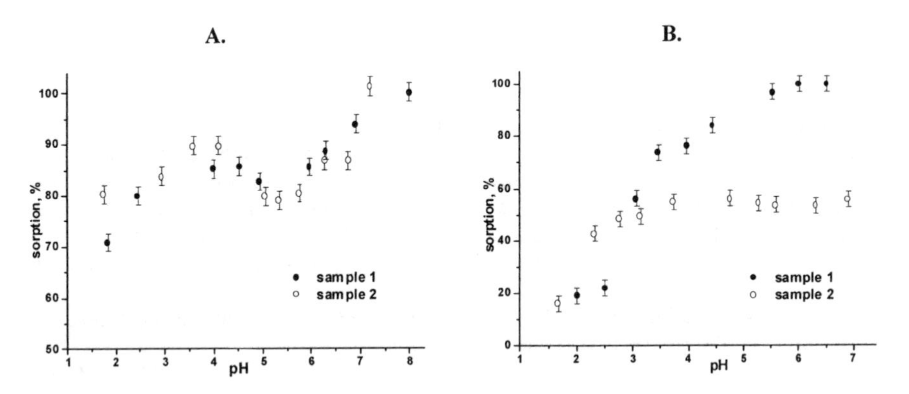

Figure 1 *Dependence of Np sorption by DU dioxide samples upon pH: A - Np(IV) concentration of 1.2×10^{-10} M, B - Np(V) concentration of 4.0×10^{-7} M. The solid phase concentration is 2 g/L.*

The sorption of Np(V) depended on degree of oxidation of the samples and was significantly higher for sample 1 than for sample 2 (Table 3).

Table 2 *Neptunium redox speciation after desorption from sample 2. Sorption experiments were performed in DW with Np concentration of 1.2×10^{-10} M in tetravalent form (mixture of ^{237}Np and ^{239}Np)*

pH value	Np(IV), %	Np(V), %
3.24	93	7
4.40	91	9
5.35	59	41
6.27	87	13
7.33	86	14

Table 3 *Sorption of Np(IV) and Np(V) at fixed pH values in DW and J-13*

Sample	*pH*	*Conditions*	*Sorption, %*	*Sample*	*pH*	*Conditions*	*Sorption, %*
		Np(IV)				Np(V)	
1	6.95	DW (N_2)	95±3	1	6.02	DW (N_2)	100±3
1	6.20	J-13	79±3	1	6.05	DW (air)	100±3
				1	7.25	J-13	65±3
2	7.05	DW (N_2)	76±3	2	6.59	DW (N_2)	56±3
2	7.23	J-13	82±3	2	7.09	J-13	41±3

3 CONCLUSION

Considering only the sorption properties of depleted uranium dioxide samples, it could be postulated that more oxidized samples provide a more effective barrier at a nuclear waste site. However, other properties, *e.g.* solubility, thermal and radiation stability should be studied in separate experiments before a final decision is made.

Acknowledgements

We acknowledge Dr. M.J. Haire and Dr. R.G. Wymer (Oak Ridge National Laboratory) for collaboration in this study as well as ISTC for financial support (project 2694).

References

1 *Management of depleted uranium.* A joint report by the OECD Nuclear Energy Agency and the International Atomic Energy Agency, OECD, 2001.
2 M.J. Haire and A.G. Croff, *Depleted uranium in repositories*, 1998, ORNL/CP-95280.
3 M. Betti, *J. of Environmental Radioactivity*, 2003, **64**, 113.
4 C.W. Frosberg, Depleted uranium oxides and silicates as spent nuclear fuel waste package fill material. *Proc. Materials Research Soc.* 1996 Meeting: Symposium II, Scientific Basis for Nuclear Waste Management XX, Materials Research Society, Pittsburgh, Pennsylvania, Dec, 1996.
5 D. Wes Efurd, W. Runde, J.C. Banar, D. Janecky, J.P. Kaszuba, P.D. Palmer, F.R. Roensch and C.D. Tait, *Environ. Sci. Technol*, 1998, **32**, 3893.
6 G.R. Choppin and A.H. Bond, *J. of Analytical Chemistry*, 1996, **12**, 1240.

THE RECOMMENDED VALUES OF ^{239}Pu and ^{240}Pu CONCENTRATIONS IN IAEA STANDARD REFERENCE MATERIAL ESTIMATED BY THREE MEASURING METHODS: ISOTOPE DILUTION-THERMAL IONIZATON MASS SPECTROMETRY, INDUCTIVELY COUPLED PLASMA MASS SPECTROMETRY AND ALPHA SPECTROMETRY

T. Shinonaga[1], H. Aigner[1], D. Klose[1], M. Hedberg[1], W. Raab[1], D. Donohue,[1] P. Spindler[2], and H. Froeschl[2]

[1]Safeguards Analytical Laboratory, IAEA Agency's Laboratories Seibersdorf, Wagramer Strasse 5, PO Box 100, A-1400 Vienna, Austria
[2]ARC Seibersdorf research GmbH., Chemical Analytics, A-2444 Seibersdorf, Austria

1 INTRODUCTION

The determination of the content and isotopic ratios of Pu nuclides in environmental samples is one of the most important tasks of the Clean Laboratory, Safeguards Analytical Laboratory, International Atomic Energy Agency (IAEA). Recently, more precise and prompt analysis of Pu in various types of sample is increasingly important. The Pu content has normally been reported as $^{239+240}Pu$ measured by alpha spectrometry in soil.[1] The mass spectrometry technique has enabled us to measure ^{239}Pu and ^{240}Pu independently, and these results provide us more specific information on Pu in the environment where the samples were collected.[2] However, independent reference values of ^{239}Pu and ^{240}Pu are still rare. In order to respond to the requirements to validate analytical methods for trace amounts of Pu nuclides and to obtain the recommended value of ^{239}Pu- and ^{240}Pu concentration in the IAEA 315/Marine sediment[3], as an example of a standard material, the quantitative determination of Pu was performed by three measuring methods: isotope dilution-thermal ionization mass spectrometry (ID-TIMS), inductively coupled plasma mass spectrometry (ID-ICP-MS) and alpha spectrometry. The determination of ^{239}Pu and ^{240}Pu ($^{239+240}Pu$ by alpha spectrometry) was carried out with samples from randomly selected bottles using each method. ^{238}Pu was also measured by alpha spectrometry. The obtained data were statistically analysed using the SAS® procedure VARCOMP program.

2 METHOD AND RESULTS

2.1. Sample

The sample used in this study was marine sediment collected near the Tarapur Atomic Power Station, Bombay, India. About 140 kg of sediment was collected, wet sieved to remove large grains, frozen and ground with a mortar and pestle. The obtained grains were sieved with a 0.21 mm diameter sieve, and then mixed in a revolving drum 4 times for 8 hours. The bottles were filled with 100 g of sediment and labelled as IAEA-315.

2.2 Analytical Methods

2.2.1. Sample Preparation and Chemical Separation The sediment sample from a randomly selected bottle was heated at 600 °C for 6 h. About 10 – 14 g (dry weight) of sample was taken from the bottle, weighed and put in a Teflon beaker for digestion. About 6 ng (0.9 Bq) of ^{242}Pu tracer (CBNM IRM-044) was added to the sample. This combined sample was decomposed with 100 ml of 48% HF for several hours and then dried. The residue with a few grams of boric acid was dissolved in 100 ml of aqua regia and boiled for several hours. The solution was dried to wet residue and the residue was converted to the nitric form using 15 M HNO_3. The residue was boiled in 100 ml of 1 M HNO_3 for 1 h. The solution with the residue was transferred into a plastic bottle and centrifuged. The liquid phase was taken from the bottle, and the residue in a plastic bottle was shaken with 40 ml of 8 M HNO_3 and centrifuged. The liquid phase was added to the 1 M HNO_3 solution and the combined solution was finally adjusted to 8 M HNO_3 for the Pu purification.

Qualitative analysis by X-ray diffraction showed that the residue consisted of Ca-, Mg-, Na- and Al-fluorides. However, in order to check whether Pu remains in the residue/solution, the residue was dissolved with 3 M HNO_3 again and the solution was measured with a low background scintillation counter (EG&G, 1220 Quantulus^{TM}). The gross alpha spectrum of the solution showed that there was no significant amount of alpha emitting nuclides in the residue and/or solution compared with the blank measurement.

The chemical separation was performed using ion exchange resin AG 1×8 and AG MP-1 resin (Bio-Rad) for the 1st- and the 2nd purification steps, respectively, for all three techniques. For the ICP-MS measurements, the fraction was diluted with 5 ml of 1 M HNO_3. The Pu fraction as NdF_3 (PuF_4) for the alpha measurements was obtained by the micro co-precipitation method. The typical chemical yield measured by ICP-MS was between 60 – 90 %. For the TIMS measurement, the 3rd–step purification procedure with AG MP-1 resin as described above for the 2nd purification was added. The final fraction was converted to the nitric form and dissolved in 1 μl of 3M HNO_3 and loaded on a rhenium single-filament along with graphite.

2.2.2. TIMS, ICP-MS and alpha measurements The isotopic ratios of Pu were measured by a ThermoElectron Triton instrument using the ion counting detection and peak jump measurements.

The ICP-MS instrument used in this study was a Perkin Elmer SCIEX ELAN 6100 with cross-flow nebulizer.

The alpha spectrometer used in this study was a Canberra, Model 7401 with Ø 25 mm of passivated, implanted, planar silicon detector and GENIE2000 alpha-spectroscopy software. The typical measuring time of the samples was ca. 24 h.

2.3. Data analysis

The results shown in Table 1 are from a preliminary calculation.

The estimates of individual standard uncertainties for the effects "method", "bottle", "portion" (variation among sub-samples within the bottles) and "random" are based on variance component analyses (the procedure VARCOMP of the statistical software SAS® was used), using conventional partitioning of the sum of squares as well as maximum likelihood estimators, where applicable and suggested by the experimental scheme.

The proposed combined standard uncertainties were of the order of 10% (coefficient of variation) applying to the average concentrations within one single analytical portion. The results for the total specific activity from ^{239}Pu plus ^{240}Pu calculated from the individual isotope assays determined by the isotope dilution mass spectrometric methods (TIMS and ICP-MS) were in good agreement with the values directly measured by alpha spectrometry. The results obtained in this study provide a very satisfying mutual verification of the measurement results by two truly independent methods.

Table 1 *Results of preliminary statistical evaluation of measurement results.*

Nuclide	Method	Univariate statistics			Variance component analysis: Standard uncertainties attributable to individual effects as CV[%]				Combined Standard uncertainty
		n	xq [content] (ratio)	CV %	Method	Bottle	portion	random	U(xq) CV %
	TIMS	42	[16.4]	11	-	2	11		
^{239}Pu[ng/kg]	ICPMS	42	[17.6]	10	-	3	10		
	both	84	[17.0]	11	4	3	9	5	11
	TIMS	42	[3.42]	9	-	1	9		
^{240}Pu[ng/kg]	ICPMS	42	[3.51]	14	-	NVE	15		
	both	84	[3.46]	12	NVE	NVE	2	12	12
^{240}Pu:^{239}Pu (mol/mol)	TIMS	42	(0.209)	10	-	NVE	11		
	ICPMS	42	(0.199)	11	-	NVE	12		
	both	84	(0.204)	11	3	NVE	3	10	11
	TIMS	42	[66]	9	-	3	8		
^{240}Pu+^{239}Pu [Bq/kg]	ICPMS	42	[70]	10	-	2	10		
	Alpha	53	[68]	10	-	NVE	10		
	All 3	137	[68]	10	2	1	7	6	10
^{238}Pu[Bq/kg]	Alpha	53	[9.4]	17	-	NVE	18		17

n: Number of analysis., NVE: Negative Variance Estimate; The effect is not estimable because of other factors dominating the total fluctuation.

3 CONCLUSION

The concentrations of ^{239}Pu and ^{240}Pu in the reference standard sample, IAEA 315, were determined by three different measuring techniques. The combined standard uncertainties from "method", "bottle", "portion" and "random" were the order of 10 %. The main contribution of the uncertainty is from the material heterogeneity, which is assumed to be due to the presence of hot particles, which cannot be removed from the samples. In conclusion, based on this study, the concentration of 17.0 ng (39 Bq)/kg, 3.5 ng (29 Bq)/kg and 68 Bq/kg for ^{239}Pu, ^{240}Pu and $^{239+240}$Pu , respectively, can be proposed as recommended values and 0.015 ng (9.4 Bq)/kg for ^{238}Pu as an information value for the IAEA 315 standard reference material.

References

1 NIST, National Bureau of Standard Certificate, Standard Reference Material 4354 (1986).
2 Y. Muramatsu, S. Uchida, K. Tagami, S. Yoshida and T. Fujikawa, *J. Anal. Atom. Spec.*, 1999, **14**, 859.
3 S. Ballestra, J. Gastaud, P. Parsi and D. Vas, IAEA-MEL, B.P. No. 800.

A SEQUENTIAL SEPARATION PROCEDURE FOR PU, AM, U AND NP ISOTOPES IN SOILS AND LIQUID SAMPLES

M.H. Lee, K.Y. Jee, Y.J. Park and W. H. Kim

Nuclear Chemistry Research Division, Korea Atomic Energy Research Institute, P.O. Box 105, Yusung, Taejeon, Korea

1 INTRODUCTION

Some method[1,2] of preconcentrating the actinides and of isolating them from both the large quantities of inactive substances present and any potentially interfering elements usually precedes analysis, due to the low concentrations of the actinides encountered in typical samples. The determination of low levels of radionuclides, such as Pu, Am, U and Np isotopes requires lengthy and tedious chemical processes, which include techniques such as an ion exchange, liquid-liquid extraction, and precipitation for separating and preconcentrating the nuclides. Recently, several studies have been reported using combined procedures for the determination of the radionuclides in environmental samples with extraction chromatographic materials such as TRU Spec, TEVA Spec resins and Diphonix.[3-4] However, these combined methods for the nuclides in soil or sediment samples are limited, because it is difficult to purify the radionuclides, due to the major ion contents, particularly of soils. Also, these combined methods are focused on only certain actinide isotopes, such as Pu and Am, but not Np and U isotopes.

In this study, an extraction chromatography method using anion exchange resin and TRU Spec resin was developed for rapidly and reliably determining low levels of Pu, Np, Am and U isotopes in environmental and radioactive waste samples. The analytical method developed for Pu, Np, Am and U was validated by application to IAEA-Reference samples.

2 METHOD AND RESULTS

2.1 Sample decomposition

A total of 20 g of soil was weighed into a porcelain dish, and ashed in a muffle furnace with a gradual heating program up to 600 °C to eliminate organic matter. To compensate for the chemical recovery, ^{242}Pu, ^{243}Am and ^{232}U, as yield tracers, were added into the soil sample. In the nitric acid dissolution method, the calcined samples were leached with 30 ml 8 M HNO3 with a stirring on a hot plate. In the HNO3 + HF method, the samples were dissolved with 10 ml concentrated HNO3 and 10 ml HF (48 %), and evaporated to dryness. The sample solution was filtered with a 0.45 μm membrane filter. In the 9 M HCl leaching solution, a few drops of concentrated HNO3 were added to give a matrix of 9 M HCl - 0.1 M HNO3.

2.2 Pu and Np separation with anion exchange resin

The sample solution in 8 M HNO_3 medium was passed through a pre-conditioned anion exchange column with 8 M HNO_3 at the rate of 0.5 ml/minute. The column was then washed with 20 ml of 8 M HNO_3. The columns were washed with 20 ml of 9 M HCl to desorb Th. This effluent was evaporated to dryness and reserved for a subsequent separation of Am from U. Finally, Pu and Np were eluted with 20 ml of 0.36 M HCl / 0.01 M HF.[5] The sample solution with 9 M HCl - 0.1 M HNO_3 media was also passed through a pre-conditioned anion exchange column with 9 M HCl - 0.1 M HNO_3 at the rate of 0.5 ml/minute. The column was then washed with 20 ml of 9 M HCl - 0.1 M HNO_3, followed by 5 ml of 8 M HNO_3 and 5 ml of 4 M HNO_3 to elute the U and Am. The column was then washed with 10 ml of 9 M HCl. Pu and Np were eluted with 20 ml of 0.36 M HCl / 0.01 M HF and held for electrodeposition.

2.3 Separation of Am from U with TRU resin

The U / Am residue that was separated from the Pu was dissolved in 30 ml of 2 M HNO_3. TRU Spec columns, conditioned with 30 ml of 2 M HNO_3, were used to separate the Am from the U. The samples were loaded onto the column followed by a washing with 20 ml of 2 M HNO_3. The column was then washed with 4 ml of 9 M HCl and the Am fraction was eluted with 20 ml of 4 M HCl. Uranium was eluted with 20 ml of a 0.1 M ammonium oxalate solution. Americium was separated from rare earth elements with anion exchange resin.[4] Each residue was dissolved in 1 ml of concentrated HCl and evaporated to a dryness. Pu, Am and U isotopes were electroplated on stainless steel platelets and measured by an alpha spectrometer.[6]

3 RESULTS AND DISCUSSION

3.1 Purification of Pu and Np from U and Am with anion exchange resin

With the anion exchange from the 8 M HNO_3 media, a small amount of U was often detected in the final Pu fraction during routine analyses. To overcome the limitations of this conventional Pu separation method in nitric acid, Pu can alternatively be separated from other elements in a hydrochloric acid media.

The concentrations of $^{239,240}Pu$ in the IAEA-375 and IAEA-326 reference materials were consistent with the values reported by the IAEA. Though there was no significant difference in the chemical recoveries for Pu between the 9 M HCl / 0.1 M HNO_3 media and the 8 M HNO_3 media, the Pu separation method in the chloride system has some advantages when compared with a nitrate-based system; the concentration of HCl and the temperature of the operation are not critical, and a complete loading can be obtained whether the Pu is in the (IV) or the (VI) state. The Pu complex with HCl was more strongly retained on the resin than the Pu complex with HNO_3.[7] The more rapid reaction rate of the chlorocomplex with the resin also ensures quantitative adsorption of the plutonium in the presence of competing complexes.

In the loading solution (8 M HNO_3, 9 M HCl), the oxidation states of Pu and Np were adjusted to IV and VI with $NaNO_2$, respectively. It was reported that Np (V) in nitric acid higher than 3 M and in the presence of NO_2^- oxidizes to Np (VI).[8] Under the same conditions, Pu exists as Pu (IV). Because the distribution coefficient of Np (VI) on the anion exchange resin in nitric acid media is similar to that of U (VI), most of the Np was eluted into the loading and washing solution with U, so that only a small amount of Np was detected in the eluting solution. However, in 9 M HCl media, Np (VI) was so strongly adsorbed on the anion resin that most of the Np was detected in the eluting solution. There was no ^{237}Np detected in the IAEA reference soil or in radioactive waste samples. For detecting the level of ^{237}Np in environmental or radioactive samples, mass analytical techniques with TIMS or ICP-MS are preferred rather than a radiation measurement.

3.2 Purification of Am and U with TRU resin

The concentrations of ^{241}Am and ^{238}U in the IAEA-375 and IAEA-326 materials were within the confidence interval, though in the acid leaching method, the concentrations of ^{238}U were a little lower than the other decomposition method, due to an incomplete destruction of the soil matrix. Also, the chemical recoveries for Am and U isotopes with TRU resin in the 8.0 M HNO_3 media were lower than those in the 9.0 M HCl media. This suggests that the Fe, which leached from the soil, interfered in the adsorption of Am and U isotopes onto the TRU resin, even where ascorbic acid is added to the solution to reduce Fe (III). Therefore, in nitric acid, it is necessary to remove Fe with an oxalate coprecipitation before loading the solution containing Am and U isotopes onto the TRU resin. However, in 9 M HCl, because much of the Fe strongly adsorbed onto the anion exchange resin, only a small portion of Fe was eluted in the wash solution, and so it is not necessary to remove Fe before loading the solution onto the TRU resin.

The activity concentrations of ^{241}Am were measured with high chemical recovery in the radioactive waste solution, though the activity concentrations of ^{238}U were not measured in the radioactive waste solution. In addition, it is possible to save on analytical time for Am isotopes in routine liquid samples, because anion-exchange separation of rare earth elements (REEs) in acid-methanol can be omitted, since the concentration of REEs is very low in the liquid samples.

4. CONCLUSION

Sequential separation of Pu, Np, Am and U Isotopes in the soil and radioactive waste samples investigated in this study is rapid and reliable. The activity concentrations of ^{239,240}Pu, ^{241}Am and ^{238}U measured in the IAEA-375 and IAEA-326 were close to the reference values reported by the IAEA. The chemical recoveries of the Pu, Np, Am and U isotopes for the acid leaching methods were similar to those for the fusion method. In the Pu and Am separation with an anion exchange resin and TRU resin, a hydrochloric acid media is preferred rather than a nitric acid media. ^{237}Np was not detected in the IAEA reference soil and radioactive waste samples.

References

1 A. Yamato, *J. Radioanal. Chem.*, 1982, **75**, 265.
2 W. C. Burnett, D. R. Corbett, M. Schultz, E. P. Horwitz, R. Chiarizia, M. Dietz, A. Thakkar and M. Fern, *J. Radioanal. Nucl. Chem.*, 1997, **226**, 121.
3 E. P. Horwttz, R. Chiarizia, M. L. Dietz and H. Diamond, *Anal. Chim. Acta*, 1993, **281**, 361.
4 P. E. Warwick, I. W. Croudace and J. S. Oh, *Anal. Chem.*, 2001, **73**, 3410.
5 J. Korkisch, '*Handbook of Ion Exchange Resins: Their Application to Inorganic Analytical Chemistry*', Vol. II, CRC Press, Inc, 1989.
6 M. H. Lee and C. W. Lee, *Nucl. Instr. and Meth. A*, 2000, **447**, 593.
7 M. Chilton and J. J. Fardy, *J. Inorg. Nucl. Chem.*, 1969, **31**, 1171.
8 T. H. Siddall and E. K. Dukes, *J. Am. Chem. Soc.*, 1950, **81**, 790.

COMBINATION OF ALPHA TRACK ANALYSIS, FISSION TRACK ANALYSIS WITH SEM-EDX AND SIMS TO STUDY SPATIAL DISTRIBUTION OF ACTINIDES

I.E. Vlasova[1], St.N. Kalmykov[1], S.B. Clark[2], S.G. Simakin[3], A.Yu. Anokhin[3] and Yu.A. Sapozhnikov[1]

[1]Radiochemistry Division, Chemistry Department, Lomonosov Moscow State University, Moscow 119992, Russia.
[2]Chemistry Department, Washington State University, P.O. Box 644630, Pullman, WA, USA
[3]Laboratory for Microparticles Analysis, Moscow 117218, Russia

1 INTRODUCTION

The distribution of actinides among colloidal matter, soil and bottom sediment particles is usually non-uniform, due to either preferential sorption by different minerals or the presence of so-called "hot particles" – particles with a high specific radioactivity. Determination of spatial distribution of actinides is necessary to predict their environmental fate, and together with sequential extraction data, is essential for development of remediation strategies.[1] Alpha track analysis (ATA) and fission track analysis (FTA) could be used to detect the presence of "hot particles" with high sensitivity, and to determine their size.[2,3] However, the combination of ATA and FTA with submicron analytical methods, e.g. scanning electron microscopy with energy dispersed X-ray spectrometry (SEM-EDX)[4] and/or secondary ion mass spectrometry (SIMS),[5] is required to study the key structural features and isotopic composition of single particles.

A procedure that combines ATA, FTA, SEM-EDX and SIMS to detect the spatial distribution of fissile actinides (^{235}U and ^{239}Pu) in environmental samples and to determine the structural features of actinide bearing particles, their major element composition and isotope ratios was developed. The procedure was used to study "hot particles" of different radioactivity from soil collected 1.5 km from Chernobyl Nuclear Power Plant (NPP).

2 METHODS AND RESULTS

Soil samples were thinly spread on one side of a piece of double tape (about 1.5 x 1.5 cm^2) that was placed on a plastic support. In order to localize actinide bearing particles, an SEM finder grid with electrodeposited ^{239}Pu was used. The SEM finder grid was placed on the top of the soil sample and then covered by solid-state nuclear track detector (SSNTD). ^{239}Pu from finder grid produced a grid-like track image on the SSNTD used for both ATA and FTA. This technique allowed the finding of single particles of interest for further analysis using SEM-EDX and SIMS. The ATA and FTA were performed according to the procedure described previously.[4]

According to the ATA and FTA analysis of soil collected 1.5 km from the Chernobyl NPP, almost all alpha emitting radionuclides were concentrated in about 30 "hot particles" that produced star-like clusters of tracks. The distribution of fission tracks produced by ^{235}U or ^{239}Pu in general fitted the distribution of alpha tracks, except for a few low-activity particles that produced only ATA images. However, the ratio of fission tracks to alpha tracks for single particles varied significantly. This corresponds to the different isotopic compositions of the individual particles. For 68 hours exposition, the corresponding sensitivity of ATA is about 3×10^{-4} Bq/particle. For FTA, the sensitivity of ^{239}Pu or ^{235}U determination is as low as 1×10^{-5} Bq/particle for irradiation in a fluence of 1×10^{15} neutrons·cm^{-2}.

The morphology, size and major element distributions in selected particles were studied using the SEM-EDX technique (Philips XL-30 TMP). Several "hot particles" that produced intense alpha track and/or fission track images were localized using the SEM finder grid with electrodeposited ^{239}Pu, and examined by SEM-EDX. The backscattered electron (BE) mode was used to trace the distribution of actinides within the particle, since the image contrast is proportional to the atomic number of the scattering atom. Information on the size and morphology of particles was obtained using the secondary electron (SE) image, while major element composition was determined by EDX.

The examined particles were between 200 nm to 15 μm and of irregular shape. Figure 1 shows ATA, FTA and SEM images and EDX spectrum of a typical "hot particle". The examined particle was chosen from an agglomerate of "hot particles" using ATA and FTA. According to the SEM-EDX this was a UO_2 fuel particle. Besides UO_2 fuel particles, several U-Zr oxide particles were found, according to the classification of Yanase et al.[6]

Highly radioactive particles, e.g. fuel particles that consist of actinide oxides could be found directly by BE-SEM. Particles with lower actinide content or natural particles with sorbed or surface precipitated actinides could not be found without preliminary track analysis. Most of the "hot-particles" from about 30 found in the sample were not detected by BE-SEM without track analysis.

According to the $^{235}U/^{238}U$ ratio determined by SIMS, two types of "hot particles" were identified: one with a uranium isotope ratio close to natural and the other with an enriched uranium $^{235}U/^{238}U$ ratio of 0.013-0.017.

2 CONCLUSION

A procedure that combines ATA, FTA, SEM-EDX and SIMS was developed to study the spatial distribution of fissile actinides (^{235}U and ^{239}Pu) in different environmental samples (colloidal matter, soil, bottom sediments, aerosols). ATA and FTA enable highly-sensitive non-destructive determination of the spatial distribution of fissile actinides, while the key structural features of single particles and their major element concentrations are determined by SEM-EDX. The isotope ratios of actinides in single particles are determined by SIMS. The localization of actinide-bearing particles is achieved by using an SEM finder grid with electrodeposited ^{239}Pu that produces a grid-like image on SSNTD and is used through the whole procedure.

Acknowledgements

The research was supported through a joint research programme of US DOE and Russian Academy of Sciences (project RC0-20003-SC14 and RUC2-20008-MO-04).

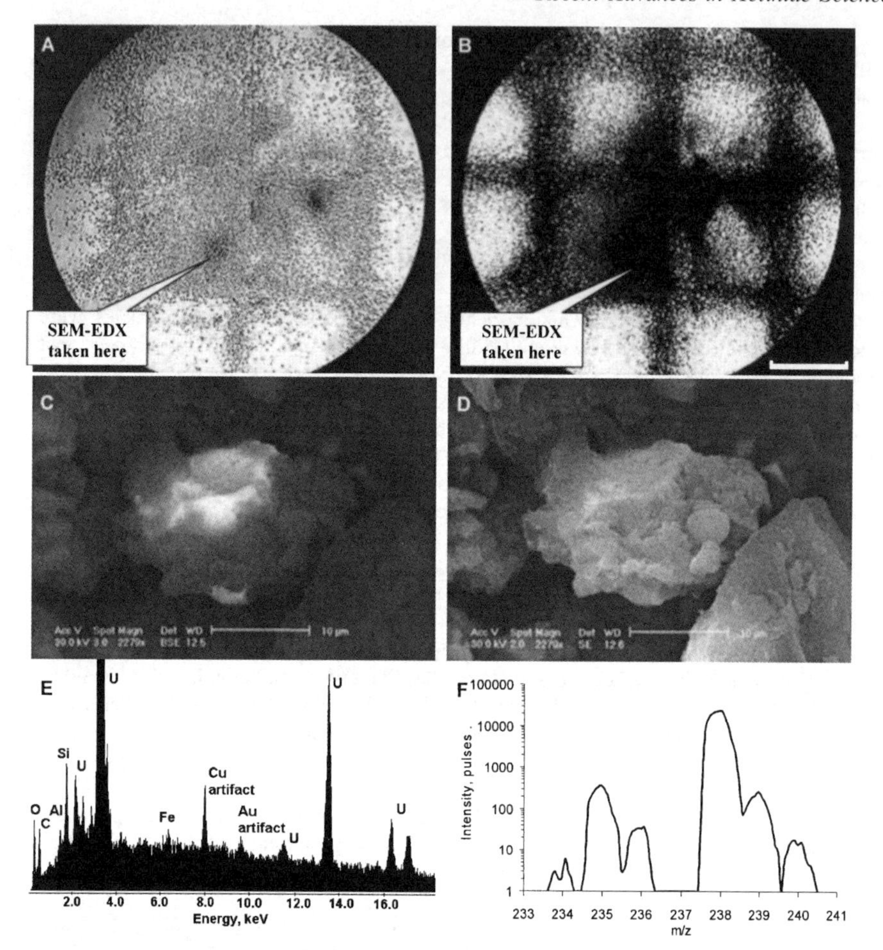

Figure 1. *Alpha track image (A) and fission track image (B) of an agglomerate of "hot particles" (scale bar is 500 μm), SEM images of one of these particles in backscattered mode (C) and secondary electron mode (D), EDX spectrum (E) and SIMS data (F) of this particle.*

References

1 S.M. Loyland Asbury, S.P. Lamont and S.B. Clark, *Environ. Sci. Technol.*, 2001, **35**, 2295.

2 L.L. Kashkarov, G.V. Kalinina and V.P. Perelygin, *Radiat. Meas.*, 2003, **36**, 529.

3 O.G. Povetko and K.A. Higley, *J. Radioanalyt. Nuc. Chem.*, 2001, **248**, 561.

4 I.E. Vlasova, S.N. Kalmykov, L.L. Kashkarov, S.B. Clark and R.A. Aliev, 2004. URL:*http://www.scgis.ru/russian/cp1251/h_dgggms/1-2004/geoecol-11e.pdf*

5 M. Betti, G. Tamborini and L. Koch, *Anal. Chem.*, 1999, **71**, 2616.

6 N. Yanase, H. Isobe, T. Sato, Y. Sanada, T. Matsunaga and H. Amano, *J. Radioanalyt. Nuc. Chem.*, 2002, **252**, 233.

ASPECTS OF UO_3 DISSOLUTION KINETICS

A.J. Heydon[1], N.D. Bryan[1] and J. Small[2]

[1]Centre for Radiochemistry Research, School of Chemistry, The University of Manchester, Manchester, M13 9PL, UK.
[2]Nexia Solutions Ltd, Risley, Warrington, WA3 6AS, UK

1 INTRODUCTION

The slow dissolution of metal oxides often follows zero-order kinetics. Surface properties are thought to influence the dissolution rates, rather than reaction steps involving transport of reactants or products. Whilst UO_2 has been studied; less is reported concerning the dissolution kinetics of the trioxide, UO_3.

An early investigation into UO_3 dissolution kinetics across the pH range 0.5-3 using batch reactors found that dissolution rate increased with decreasing pH, with the dissolution rate proportional to $[H^+]^{0.5}$.[1] The dissolution rate was found to be proportional to lower functions of $[H^+]$ in the pH range 0-0.5.

The aim of this research is to examine the dissolution kinetics of UO_3 at higher pH than has previously been studied.

2 METHODS AND RESULTS

UO_3 of natural isotopic composition was pretreated to remove fines by adding to approximately 10 g of UO_3, 100 ml of 0.1 M $HClO_4$ in a polypropylene container. The resulting suspension was stirred magnetically for 24 hours. The remaining solid UO_3 (ca. 8 g) was filtered under gravity using a Whatman 541 filter paper (20-25 μm cut-off), and left to dry for a further 24 hours.

A series of pH 6.10 (± 0.02) buffered leachate solutions were prepared by dissolving the non-complexing tertiary amine-based buffer 2-(N-morpholino)ethanesulfonic acid (MES) to yield approximately $5x10^{-3}$, $7.5x10^{-3}$, $1x10^{-2}$, $5x10^{-2}$, and 0.1 M solutions in ca. 200 ml of distilled water and adjusting to pH 6.10 with 1 M NaOH. The Henderson-Hasselbalch equation was used to calculate the concentration of the protonated form of MES at pH 6.10, and the concentration of $NaClO_4$ required to yield I=0.1 solutions then calculated. The concentration of MES was varied to examine whether the buffer interacts with the UO_3 surface. 120 ml of each buffer solution was then added to 50 (± 5) mg of UO_3 in polypropylene containers and the containers shaken on a plate shaker. Aliquots were then taken periodically, filtered through a Millipore 0.22 μm PES membrane syringe filter and analysed by LSC. The dissolution profile featured a [U] maximum occurring between 24 and 75 hours, followed by a decrease to a stable concentration of $3\text{-}5\times10^{-6}$ M. No correlation could be found between MES concentration and initial dissolution rates or [U] maximum. This suggests that MES does not interact with UO_3.

The dissolution profile obtained can be interpreted by formation of a secondary phase,[2,3]

$$\frac{d[\mathrm{U}]}{dt} = R_{diss\mathrm{UO}_3} - R_{ppt\mathrm{UO}_3} + R_{diss2^\circ} - R_{ppt2^\circ} \tag{1}$$

where, $R_{dissUO3}$ is the rate of UO_3 dissolution, R_{pptUO3} is the rate of UO_3 precipitation, R_{diss2° is the rate of secondary phase dissolution and R_{ppt2° is the rate of secondary phase precipitation. Assuming that the dissolution rates are zero order and proportional to surface area, and that the precipitation rates are proportional to the product of the surface area and solution U concentration; a FORTRAN program that predicts solution ^{238}U concentrations in Bq l^{-1} was used to model the dissolution and precipitation processes. This used the following equation:

$$\frac{d[\mathrm{U}]}{dt} = k_{diss\mathrm{UO}_3} a_{\mathrm{UO}_3} - k_{ppt\mathrm{UO}_3} a_{\mathrm{UO}_3}[\mathrm{U}] + k_{diss2^\circ} - nk_{ppt2^\circ}[\mathrm{U}]^n \tag{2}$$

where, $k_{dissUO3}$, k_{pptUO3}, k_{diss2°, and, k_{ppt2° are the rate coefficients for dissolution and precipitation of UO_3 and secondary phase, respectively; a_{UO3} and a_{2° are the surface areas of the UO_3 and secondary phase, respectively; [U] is the solution concentration of U in Bq l^{-1} of ^{238}U, and n is an integer that reflects whether nucleation occurs via mononuclear or multinuclear mechanisms. Unfortunately, we have not yet been able to perform any surface area measurements on the pretreated UO_3. For the modelling, a_{UO3} is assumed to be 1 m^2, and hence all other rates used in the model can be considered relative to the surface area of UO_3. As saturation of UO_3 is not expected under these conditions, k_{pptUO3} is assumed to be equal to 0 m^{-2} h^{-1}. This approach is similar to Transition State Theory that relates rates of precipitation to oversaturation indices of secondary phases. However, given that the identity of the secondary phase is unknown, an approach using solution concentrations is preferred.

The surface area of the secondary phase, a_{2°, is calculated based on the difference of the rates of formation of secondary phase, R_{ppt2°, and the dissolution of secondary phase, R_{diss2°:

$$\frac{da_{2^\circ}}{dt} = nk_{ppt2^\circ} a_{2^\circ}[\mathrm{U}]^n - k_{diss2^\circ} a_{2^\circ} \tag{3}$$

Secondary phase formation was assumed to commence after 3 hours with an initial surface area of 1 x 10^{-5} m^2. Iteration yielded the following for n=1,

$$\frac{d[\mathrm{U}]}{dt} = 2.5 \cdot a_{\mathrm{UO}_3} - 0 \cdot a_{\mathrm{UO}_3} \cdot [\mathrm{U}] + 3\times10^{-2} \cdot a_{2^\circ} - 3\times10^{-3} \cdot a_{2^\circ} \cdot [\mathrm{U}] \tag{4}$$

By keeping $k_{dissUO3}$, k_{pptUO3}, and, k_{diss2° constant; n and k_{ppt2° were varied to investigate alternative descriptions of the system. The outcomes of the modelling are shown in Figure 1, where the concentrations are plotted as mol l^{-1}. From Figure 1 it is evident that n=1 yields the best description of the system. Solubilities predicted by the PHREEQC model[4] indicate that the secondary phase may be either schoepite ($(UO_2)_8O_2(OH)_{12}.12H_2O$) or hydrated UO_3 ($UO_3.2H_2O$). In the near future we hope to perform XRD analysis to identify the secondary phase.

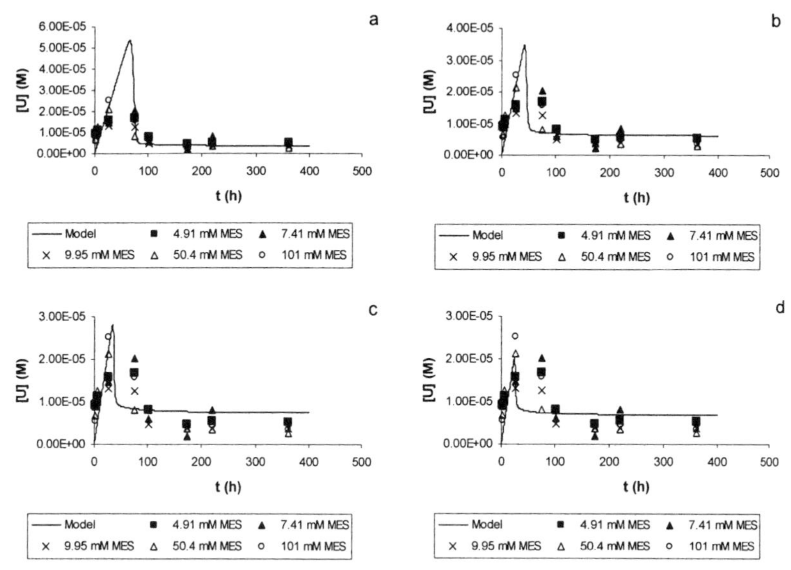

Figure 1 *Modelling of dissolution of* UO_3 *at pH 6.1 with subsequent secondary phase precipitation. Graphs a, b, c, and d are the best possible descriptions obtained for n=1, 2, 3, and 4, respectively.* $k_{ppt2°}$ *values for Figures b-d are:* 5×10^{-5} $l\ Bq^{-1}\ h^{-1}\ m^{-2}$, $1 \times 10^{-6}\ l^2\ Bq^{-2}\ h^{-1}\ m^{-2}$, *and,* $5 \times 10^{-8}\ l^3\ Bq^{-3}\ h^{-1}\ m^{-2}$, *respectively*

3 CONCLUSION

Secondary phase formation from UO_3 dissolution at mildly acidic pH has been reported for the first time. The kinetics of secondary phase formation have been modelled to yield rate laws, although further data are required around the [U] maximum to verify the rate laws proposed. Future work includes XRD analysis of the secondary phase, microscopic investigation into whether the secondary phase exists as a surface precipitate, and modelling of the data using Transition State Theory.

References

1. P.D. Scott, D. Glasser and M.J. Nicol, *Journal of the Chemical Society Dalton Transactions*, 1977, 1939.
2. D.E. Giammar and J.G. Hering, *Geochimica et Cosmochimica Acta*, 2002, **66**, 3235.
3. D.E. Giammar and J.G. Hering, *Environmental Science & Technology*, 2004, **38**, 171.
4. D.L. Parkhurst and C.A.J. Appelo, User's Guide to PHREEQC *A Computer Program for Speciation, Batch-Reaction, One-Dimensional Transport and Inverse Geochemical Calculations*, United States Geological Survey Water-Resources Investigations Report, 1999, 99-4259.

DETERMINING THE RADIATION DAMAGE EFFECT ON GLOVEBOX GLOVE MATERIALS

M.E. Cournoyer, J.J. Balkey, M.M. Maestas and R.M. Andrade

Nuclear Material Technology Division, Los Alamos National Laboratory, Los Alamos, New Mexico, 87545

1 INTRODUCTION

The Nuclear Material Technology (NMT) Division has the largest inventory of glove box gloves at Los Alamos National Laboratory. The minimization of unplanned breaches in the glovebox, e.g., glove failures, is a primary concern in the daily operations in NMT Division facilities, including the Plutonium Facility (PF-4) at TA-55 and Chemical and Metallurgy Research (CMR) Facility.[1] Glovebox gloves in these facilities are exposed to elevated temperatures and exceptionally aggressive radiation environments (particulate ^{239}Pu and ^{238}Pu). Predictive models are needed to estimate glovebox glove service lifetimes, i.e. change-out intervals. Towards this aim, aging studies have been initiated that correlate changes in mechanical (physical) properties with degradation chemistry. This present work derives glovebox glove change intervals based on previously reported mechanical data of thermally aged hypalon glove samples.[2]

Specifications for 30 mil tri-layered hypalon/lead glovebox gloves (TLH) and 15 mil hypalon gloves (HYP) have already been established.[3] The relevant mechanical properties are shown in Table 1.

Table 1. *Mechanical Property Specifications*

Property	*HYP*	*TLH*
Tensile Strength (psi)	1900	1200
Ultimate Elongation (%)	500	300

Tensile strength is defined as the maximum load applied in breaking a tensile test piece divided by the original cross-sectional area of the test piece (Also termed maximum stress and ultimate tensile stress). Ultimate elongation is the elongation at time of rupture (Also termed maximum strain). The specification for the tensile test and ultimate elongation are the minimum acceptable values. In addition, the ultimate elongation must not vary 20% from the original value. In order to establish a service lifetimes for glovebox gloves in a thermal environment, the mechanical properties of glovebox glove materials were studied.

2 RESULTS

Plots of the normalized strain (Ultimate elongation divided by its unaged value, $\varepsilon/\varepsilon_0$) are shown in Figures 1 and 2 for the hypalon and tri-layered hypalon. Note: All points below 80% are out of specification.

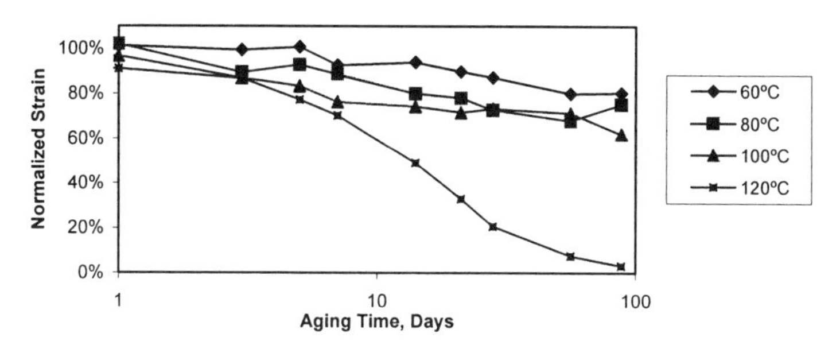

Figure 1 *Normalized Strain for Thermally Aged Hypalon.*

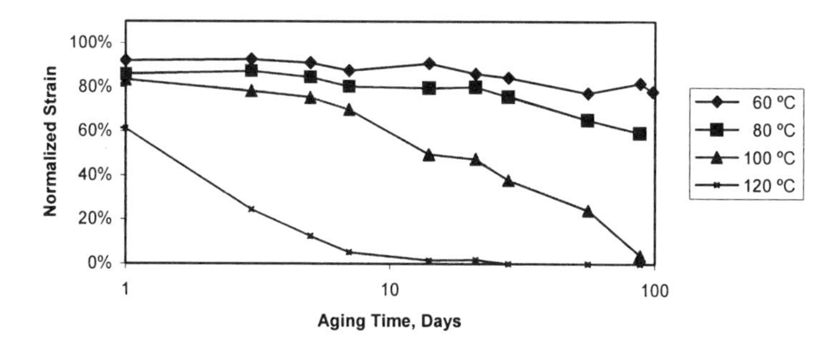

Figure 2 *Normalized Strain for Thermally Aged Tri-Layered Hypalon.*

Based on the specification in Table 1 and the results shown in Figures 1 and 2, the recommended change-out intervals for hypalon and hypalon tri-layer glovebox gloves from North Safety Products are listed in Table 2. Recommended change-out intervals for room temperature ^{238}Pu and ^{239}Pu applications have been discussed in a previous report.[1]

Table 2. *Recommended Change-out Intervals for North Glovebox Gloves in a Thermal Environment*

Temperature	*Glove Material*	
	Hypalon	**Tri-Layered**
20°C	3 years	3 years
60°C	5 day	28 days
80°C	1 day	28 days
100°C	1 day	1 day
120°C	< 1 day	< 1 day

3 DISCUSSION

The lead-loaded glovebox glove made from Hypalon® (hereafter referred to as hypalon) is the workhorse of NMT Division programmatic operations due to its superior properties. The Hypalon material is resistant to interactions with alcohols and strong acids and bases. Glovebox gloves materials stored at ambient temperature and used as control samples meet the acceptance criteria. In general, both materials become out of specification, because the aging causes the strain to drop below the minimum specification. The tensile properties of the hypalon/lead-neoprene/hypalon tri-layered material degrade more quickly and more extensively compared to the hypalon material, although it is the latter that falls out of specification sooner. Changes to glovebox glove maintenance procedures should be made to reflect this. Additionally, there is the issue of ^{238}Pu heat sources that can reach temperatures between 85 and 450°C. If the hot surfaces of the heat sources or storage containers are touched, then the glovebox gloves should be immediately inspected and replaced if visual discoloration in the area where the hot surfaces touched the glove is noticed. It should be noted that the change-out intervals recommendation derived in this study are formulation-specific; they should not be assumed to apply to other formulations of the same generic material classes. Formulation differences, particularly with regard to the identity and amount of stabilizing additives, can exert a strong influence on the thermal resistance of a given material. The next paper in this serial will report on the radiological effects of ^{238}Pu and ^{239}Pu.

4 CONCLUSIONS

We have derived a general methodology for calculating glovebox glove change-out intervals from tensile data. Under various temperatures, stress and strain were measured. The resulting data are compared to product specifications. Once the tensile data are out of specification, the recommended change-out date is reached. The information from this study represents an important baseline in gauging the acceptable standards for polymeric gloves used in a laboratory glovebox environment.

References

1 M.E Cournoyer and J.J. Balkey, *Minimizing Glovebox Glove Breaches*, LA-UR-03-9078, Los Alamos National Laboratory, 2003.
2 K.V. Wilson.; B.L. Smith, J.M. MacDonald, J.R.Schoonover, J.M. Castro, M.E. Smith, M.E. Cournoyer, R. Marx and W.P. Steckle, *Polymer Degradation and Stability*, 2004, **84**, 439.
3 NMT-SPEC-001, *Procurement of Arm-Length Dry Box Gloves Lead-Loaded Neoprene, Hypalon* and NMT-SPEC-006, *Procurement of Hypalon Arm-Length Dry Box Gloves.*

PLUTONIUM-HUMATE COMPLEXATION IN NATURAL GROUNDWATER BY XPS

D. Schild[1], C.M. Marquardt[1], A. Seibert[1] and Th. Fanghänel[1,2]

[1]Forschungszentrum Karlsruhe, Institut für Nukleare Entsorgung, Postfach 3640, D-76021 Karlsruhe, Germany
[2]Physikalisch-Chemisches Institut, Ruprecht-Karls-Universität, Im Neuenheimer Feld 253, D-69120 Heidelberg, Germany

1 INTRODUCTION

The chemical behaviour of plutonium in natural groundwater is controlled by numerous reactions, including solubility, redox potential, hydrolysis, sorption, formation of polymers, and complexation with organic and inorganic ligands. Knowledge of the plutonium redox speciation and the kinetics of the transfer reactions among the various oxidation states is a prerequisite for modelling the migration of plutonium in aquifers. Complexation of plutonium with humic substances (HS) can have a significant influence on its migration behaviour.[1] In this study, plutonium in well defined oxidation states is added to purified Aldrich humic acid solution and to natural groundwater from Gorleben (GoHy-532), a potential site for a high active waste repository in Germany. The Pu-binding in humate complexes was investigated by X-ray photoelectron spectroscopy (XPS). Complexation of Th(IV), considered as an analogue to Pu(IV), with humic acid has been previously analysed.[2]

2 EXPERIMENTAL

Samples were prepared at anoxic conditions in a glove box (99% Ar, 1% CO_2, < 10 ppm O_2). The humate/fulvate was prepared by mixing ^{242}Pu in a defined oxidation state (III, IV or VI) with GoHy-532 groundwater (pH 6.8, (60-70) mg/L HS) or Aldrich humic acid (HA) solution, with final Pu concentrations in the range of $4x10^{-5}$ M to $1.5x10^{-4}$ M. The background electrolyte was 0.01-0.05 M $NaClO_4$. The Aldrich humic acid was purified according to Kim et al.,[3] with a proton exchange capacity (PEC) of $5.4x10^{-3}$ eq/g.

For the XPS analysis, the humic colloids were separated after 1 to 90 days of reaction from solution by ultrafiltration using an Amicon YM10 filter with a nominal molecular weight cut-off of 10 kDa. The colloids on the filter were washed once with Milli-Q water to remove unbound Pu and most of the $NaClO_4$. The humic acid was spread onto Al-foil and dried. The samples were transferred into the XPS, PHI 5600, without air contact by use of a transfer vessel. The energy scale of the spectrometer was calibrated by the Cu $2p_{3/2}$, Ag $3d_{5/2}$, and Au $4f_{7/2}$ lines of pure and Ar^+ sputter cleaned metal foils.[4] Binding energies of elemental lines were charge referenced to the C 1s line, assigned to aliphatic bonding at a binding energy (BE) of 284.8 eV. Excitation by monochromatic Al K_α X-rays (1486.6 eV) minimizes X-ray induced alteration of $NaClO_4$ and Pu oxidation states.

3 RESULTS AND DISCUSSION

Nearly all of the plutonium is removed from the Pu spiked natural groundwater GoHy-532 by ultrafiltration. Besides the elemental lines of C, O and Pu (1 at-%) traces of N, S, Ca, Si and Fe were observed by XPS survey spectra. Because the Pu concentration in solution was well above the solubility limit of Pu(IV),[5] formation of Pu(IV)-polymers due to hydrolysis reaction are expected besides Pu-humate complexation. From the size distribution of the Pu in ultrafiltration experiments[6] it was concluded that the Pu has reacted with HS or formed Pu-polymers that are strongly sorbed on HS. Probably HS prevents Pu(IV)-polymer growth and precipitation of $Pu(OH)_4$(am). These colloids can be considered as a Pu(IV)-polymer-HS conglomerate.

The full-width-at-half-maximum (FWHM) of about 2.0 eV of the Pu $4f_{7/2}$ elemental line, acquired by high resolution scans, allows resolution of the chemical shifts of the different Pu oxidation states, spaced about 1.6 eV from each other. However, besides Pu(IV), low concentrations of Pu(V) or Pu(VI) are hardly resolved, as their 4f lines are superimposed by tailings on the high binding energy side of the asymmetric Pu(IV) 4f lines.

Addition of Pu(VI) to purified Aldrich humic acid solution or natural GoHy-532 groundwater results in fast reduction to Pu(IV) within 1 day at neutral pH. Pu(III) addition leads to Pu(III) and Pu(IV). If Pu(IV) is added, no change of the oxidation state is observed.

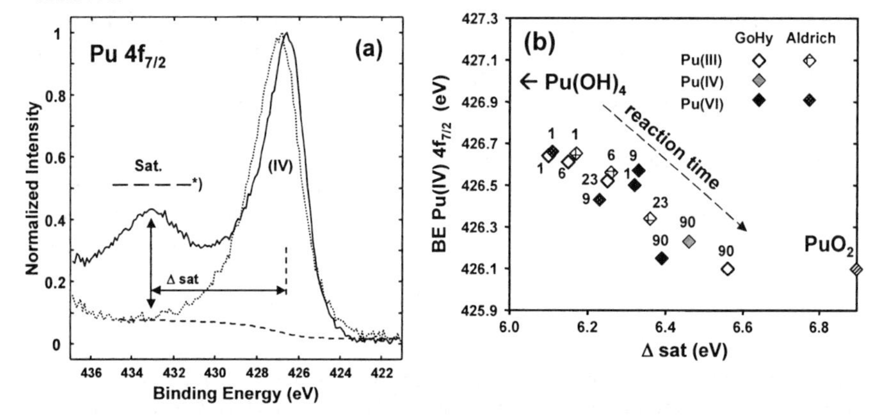

Figure 1 **(a)** *Pu $4f_{7/2}$ spectra of Pu(VI) added to GoHy-532 after 1 day of reaction. Only Pu(IV) can be detected (full line). Also shown Pu(IV)-oxalate (dotted line, BE 426.8 eV) with Shirley background (broken line) and satellite intensity of PuO_2 *) for comparison.*

(b) *Pu $4f_{7/2}$ (IV) binding energy vs. satellite spacing Δ_{sat}. Days of reaction are given as numbers at the symbols. Data for $Pu(OH)_4$ and PuO_2 are given for comparison.*

Pu $4f_{7/2}$ spectra of all samples show a satellite at the high binding energy side but with different intensity and spacing, Δ_{sat}, in relation to the Pu $4f_{7/2}$ main line of Pu(IV). The satellite is assigned to an interatomic shake-up transition of O 2p to Pu 5f with charge transfer energy of Δ_{sat}, typically in the range of 6.5-6.9 eV observed for Pu(IV) compounds.

Calculation of the 4f spectrum of PuO_2 bulk by using the impurity Anderson model leads to analogue satellites.[7] However, Pu(IV)-oxalate, analysed as a reference for the complexation with carboxylic groups like Pu complexation in HS does not show satellites (Fig. 1a) of the Pu 4f elemental lines. Hence, we conclude that the 4f satellites of the colloid samples originate from Pu-O-Pu interaction in the Pu(IV)-polymer fraction. The satellite intensities, in relation to the $4f_{7/2}$ main line intensities of Pu(IV), are lower than in the case of bulk PuO_2 and may be used to estimate the fraction of Pu(IV)-polymers. Thus, 70 at-% of Pu is in the polymeric state and only 30 at-% is complexed by HS after 90 days of reaction (in the GoHy-532 experiments).

Plutonium separated from solution by ultrafiltration after different times of reaction, up to 90 days, show Pu 4f spectra with different satellite intensities, Δ_{sat} and Pu $4f_{7/2}$ binding energies of Pu(IV) (Fig. 1b). These data are well correlated with reaction time and are located between the Δ_{sat} and binding energies of $Pu(OH)_4$ [8] and PuO_2.[9] This time-dependent behaviour indicates ageing/transformation of Pu(IV)-polymers from $Pu(OH)_4$ towards PuO_2, presumably by condensation of OH groups of the polymers.

4 CONCLUSION

The satellite intensity in relation to its Pu(IV) 4f main line may be used to estimate the fraction of Pu(IV)-polymers present. No distinction between Pu-polymers and Pu-polymers bound to HS is provided by XPS, but we suppose from the size distribution that mainly Pu-polymer-HS occurs. Nevertheless, we can conclude from our experiments that at concentrations higher than the solubility of tetravalent plutonium and near neutral pH, the Pu-polymer formation dominates the reaction path of plutonium in solutions rich in humic substances. Although the HS covers the Pu-polymers, the Pu(IV) changes from an amorphous $Pu(OH)_4$ towards a more crystalline PuO_2 with reaction time. The oxidation states Pu(VI) and (V) have no relevance under anaerobic conditions and are reduced to the tetravalent oxidation state. The role of Pu(III) as a relevant species remains unclear, because oxygen intrusion into the samples during the experiments cannot be excluded.

References

1 R. Artinger, B. Kuczewski, C.M. Marquardt, Th. Schäfer, A. Seibert and Th. Fanghänel, *Wissenschaftliche Berichte, FZKA 6969*, ed. G. Buckau, Forschungszentrum Karlsruhe, Germany, 2004, 45.
2 D. Schild and C.M. Marquardt, *Radiochim. Acta*, 2000, **88**, 587.
3 J.I. Kim, G. Buckau, G.H. Li, H. Duschner and N. Psarros, *J. Anal. Chem.*, 1990, **338**, 245.
4 M.P. Seah, I.S. Gilmore and G. Beamson, *Surf. Interface Anal.*, 1998, **26**, 642.
5 V. Neck and J.I. Kim, *Radiochim. Acta,* 2001, **89**, 1.
6 C.M. Marquardt, A. Seibert, R. Artinger, M.A. Denecke, B. Kuczewski, D. Schild and Th. Fanghänel, *Radiochim. Acta* 2004, **92**, 617.
7 A. Kotani and H. Ogasawara, *Physica B,* 1993, **186-188**, 16.
8 J.D. Farr, R.K. Schulze and B.D. Honeyman, *Radiochim. Acta*, 2000, **88**, 675.
9 D. Courteix, J. Chayrousse, L. Heintz and R. Baptist, *Solid State Commun.*, 1981, **39**, 209.

URANIUM SORPTION AND SOLUBILITY UNDER CONDITIONS RELEVANT FOR THE RADIOACTIVE WASTE REPOSITORY IN MORSLEBEN / GERMANY

C. Nebelung and L. Baraniak

Forschungszentrum Rossendorf, Institute of Radiochemistry, P.O. Box 510119, 01314 Dresden, Germany

1 INTRODUCTION

Mobilization and retention of uranium was studied to describe the migration behaviour of uranium in the frame of risk assessment for the Morsleben radioactive waste disposal. Mobilization was considered as dissolution of U_3O_8 by the site-typical brines and retention was caused by the sorption of uranium (VI) onto backfilling materials (near field) and overlaying rock (far field).

2 METHOD AND RESULTS

2.1 Experiments

In all experiments, uranium was in contact with various simulated brines: 5.35 m NaCl and a mixture from 3.8 m $MgCl_2$, 0.8 m KCl and 0.4 m Na_2SO_4. They were subsequently conditioned with site specific solids: (1) backfilling materials, a saline concrete SC (54% NaCl, 23% industrial ashes, 13% water, 10% concrete, all in wt%) and a magnesia binder MB (60% NaCl, 19% crashed slate, 15% calcined dolomite, 6% MgO) and (2) overlaying rock, a salt-rich grey clay "Grauer Salzton" (NaCl, quartz, chlorite, kaolinite, illite). Aliquots of each brine were conditioned either with SC or MB. The resulting four solutions were used for solubility and sorption experiments. Another set of four identically treated brines, and an untreated $MgCl_2$ brine were in a second step conditioned with GS for sorption experiments on GS. NaCl in the solids dissolved almost completely during conditioning. Uranium dissolution experiments were carried out by equilibration of 5 mg $^{238}U_3O_8$ with 10 mL of conditioned brines (with SC or MB or GS). The dissolved uranium was determined by ICP-MS after stepwise filtration through 450-nm and 100-kD membranes. Uranium sorption onto the backfilling materials (in brines conditioned either with SC or MB) and onto the overlaying clay (in brines conditioned with SC and GS, MB and GS, or only GS) was investigated under inert gas with $1.4\times10^{-6}mol\cdot L^{-1}$ $^{234}UO_2^{2+}$ and at two solid-to-liquid ratios of 6 and 10 $mL\cdot g^{-1}$. Aliquots filtered twice by 450-nm and 100-kD membranes were taken after 14, 40, 100, 160, 290, and 410 days. The $^{234}U^{*}$ activity therein was measured by a liquid scintillation counter (LS). For all experiments, three replicates were made (error bars in the figures indicate a confidence level of 95% of the mean value calculated with student's t). The composition of all solids (used fraction

< 1 mm) and all solutions, XRD data, particle size distribution, the pH and redox potential are given in the final report.[1]

2.2 Solubility of uranium

The solubility of uranium oxide (Figure 1) reached its maximum at the time of the first sampling (after 80 days or before). Uranium concentration in the NaCl brines decreased during the following 350 days. A nearly constant solubility was observed in the $MgCl_2$ brines. There the dissolved uranium concentration ranged from 2.6×10^{-6} to 7.8×10^{-6} mol·L^{-1}. In NaCl brines (conditioned with MB) 7.8×10^{-6} mol·L^{-1} was found. But in the case of conditioning with SC, the solubility decreased below the detection limit of ICP-MS (4.2×10^{-9} mol·L^{-1}). The reason may be that the redox potential at -150 mV and the pH, 10.4, promoted the reduction of UO_2^{2+} to the sparingly soluble $UO_2{\cdot}xH_2O$ or $U(OH)_4$.[2, 3] The origin of the negative redox potential is unknown. In the $MgCl_2$ brine conditioned with SC no reducing conditions were observed (pH 5.7 and positive redox potential). Colloidal uranium particles were only found in $MgCl_2$ brines conditioned with GS. The measured solubility was one to two orders of magnitude smaller than noted in other papers.[4]

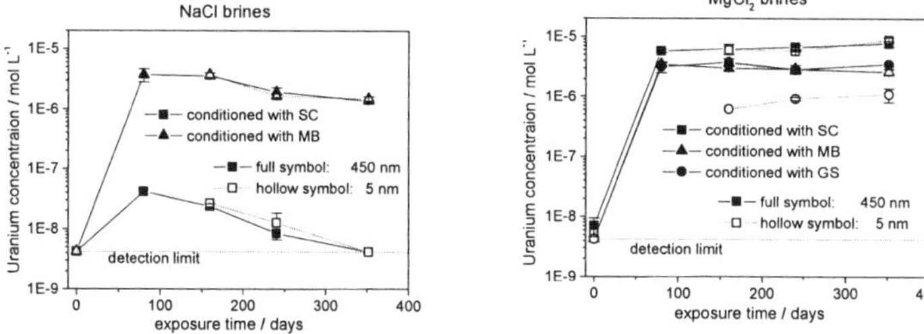

Figure 1 *Solubility of U_3O_8 in conditioned brines of NaCl and $MgCl_2$*

2.3 Adsorption of uranium

U sorption on SC and MB in NaCl brines reached 99 to 100%. In $MgCl_2$ brines values of 96 to 98% were reached (Table 1, Figure 2). In the case of the ^{234}U distribution between SC and NaCl brine (pH 10.3 and Eh -200 mV) the uranium concentration in the liquid phase decreased also (see 2.2) below the detection limit of LS with 6.5×10^{-10} mol·L^{-1}.

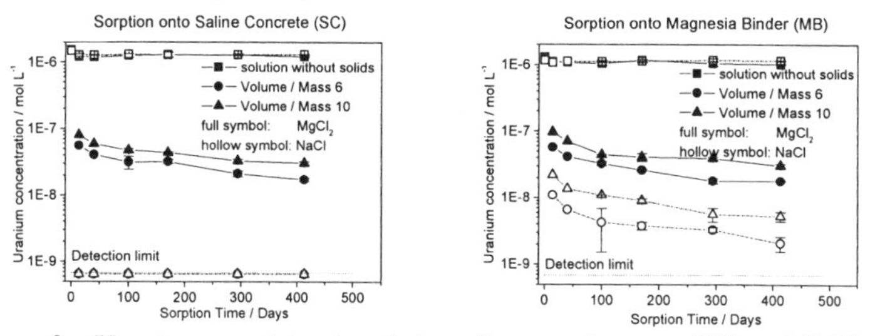

Figure 2 *Uranium remaining in solution after sorption onto (SC) and (MB)*

The sorption coefficients for uranium on alternative backfilling materials in brines were smaller in most cases, only in brown filter coal ashes in $MgCl_2$ was it higher.[4] Uranium sorption on the clay amounted to 93 - 98% (Table 2, Figure 3). In general, steady state was reached after about 300 days. In case of the $MgCl_2$ brine conditioned with SC, the uranium concentration decreased continuously during the experiment. The sorption coefficients ranged between 165 and 392 mL·g^{-1}. The sorption on the GS was somewhat lower than on the backfilling materials.

Table 1 *Sorption coefficient: Sorption of uranium onto SC and MB after 410 days*

Volume to mass / mL·g^{-1}	Sorption coefficient / mL·g^{-1} (sorbed fraction / %)			
	SC / $MgCl_2$	SC / NaCl	MB / $MgCl_2$	MB / NaCl
10	522 (97.4)	>22000[1)] (100)	378 (96.2)	1970 (99.4)
6	524 (98.3)	>13000[1)] (100)	461 (98.0)	2942 (99.8)

[1)] Uranium in solution below the detection level

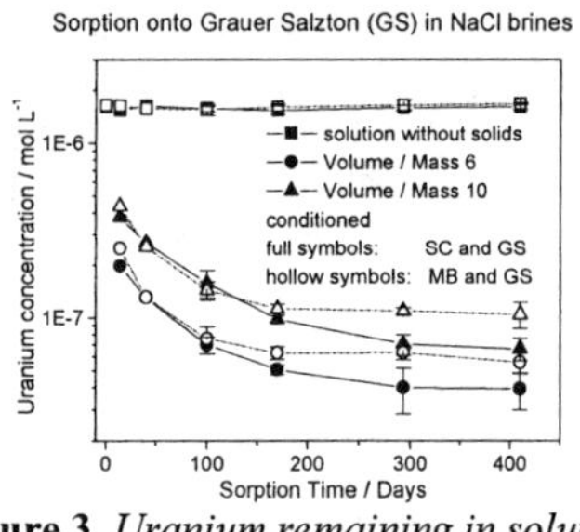

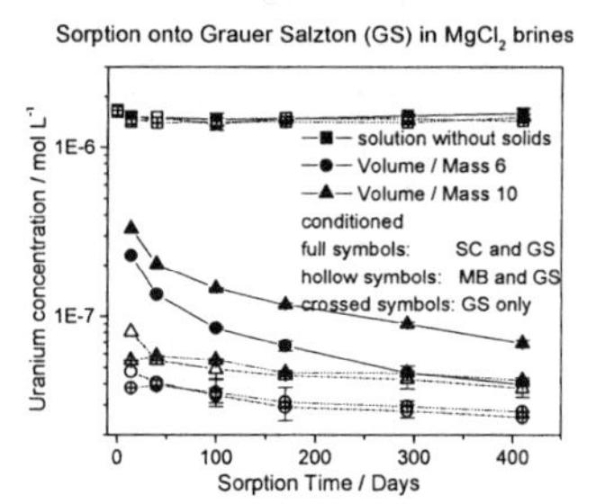

Figure 3 *Uranium remaining in solution after sorption onto clay "Grauer Salzton" GS*

Table 2 *Adsorption coefficient: Sorption of uranium onto GS after 410 days*

Volume to mass / mL·g^{-1}	Sorption coefficient / mL·g^{-1} (sorbed fraction / %)				
	GS / NaCl (SC, GS)[1)]	GS / NaCl (MB, GS)[1)]	GS / $MgCl_2$ (SC, GS)[1)]	GS / $MgCl_2$ (MB, GS)[1)]	GS / $MgCl_2$ (GS)[1)]
10	255 (95.1)	165 (93.3)	251 (94.3)	392 (96.9)	377 (96.8)
6	293 (97.3)	197 (96.3)	298 (96.7)	369 (97.9)	374 (97.9)

[1)] Material for conditioning of the brines

3 SUMMARY

The solubility of U_3O_8 in conditioned NaCl and $MgCl_2$ brines was in the range of 1.5×10^{-6} to 7.8×10^{-6} mol·L^{-1}. Only for NaCl brine conditioned with salt concrete was no dissolved U found, most probably due to a reduction of U(VI) and precipitation of $U(OH)_4$. This process disturbed also the respective sorption experiment. The adsorption coefficient on the backfilling materials in case of the $MgCl_2$ brines was about 500 mL·g^{-1}. In the case of sorption on magnesia binder from NaCl brines it was above 2000 mL·g^{-1}. Smaller coefficients were found for uranium sorption onto the clay (165 to 392 mL·g^{-1}).

Acknowledgement
The project was supported by the BfS under the contract No.: 9M 212230-62

References

1 P. Vejmelka et al. (2004) „Nuklidmigration im Deckgebirge des ERAM (DGL), Abschlußbericht" BfS Auftrag Nr. 9M 212 230-62.
2 V. Neck and J.I. Kim, *Radiochim. Acta*, 2001, **89**, 1.
3 Chemical Thermodynamics of Uranium, Vol. 1, OECD North Holland Amsterdam, London, New York, Tokyo 1992.
4 P. Vejmelka (1999) „Nuklidmigration im Deckgebirge des Endlagers für radioaktive Abfälle Morsleben, Abschlußbericht" BfS Auftrag Nr. 9M 212 230-61.

THE APPLICATION OF COLLOIDAL PHOTOCATALYSIS IN ACTINIDE PHOTOREDOX CHEMISTRY

G. Le Gurun[1], C. Boxall[2] and R.J. Taylor[1]

[1]Nexia Solutions Ltd, Sellafield B170, Seascale, Cumbria CA20 1PG, UK.
[2]Centre for Materials Science, Physics Astronomy & Mathematics Department, University of Central Lancashire, Preston PR1 2HE, UK.

1 INTRODUCTION

Colloidal semiconductor particles may act as efficient photocatalysts for a range of environmentally and industrially useful reactions. The primary step in these reactions is the absorption of ultra-band gap energy photons by the particles, which generates electron-hole (e^-_{CB}, h^+_{VB}) pairs within the semiconductor lattice. The valence band holes/conduction band electrons may recombine or diffuse to the semiconductor surface where they may either reduce/oxidise particle lattice sites or undergo interfacial electron transfer with a surface adsorbed substrate or species in solution.[1] Photocatalytic treatment of metal ion species by such particles has a number of important commercial applications in precious metal recovery[2] and in the removal of heavy elements from effluent streams.[3] The purpose of the work reported in this communication is to investigate the potential of heterogeneous photocatalysis as a mean of controlling the valence state of key actinide and metal species in such a way as to improve the performance of nuclear fuel reprocessing and decontamination technologies. For example, colloidal photocatalysis could be used to convert the valence state of actinides to insoluble species, which may then be selectively removed from solution. In this paper, we report some results obtained for the photoreduction of species such as cerium, a lanthanide whose thermodynamic E_h-pH diagram most closely approximates that of plutonium and uranium, as well as a kinetic analysis of the cerium system.

2 METHODS AND RESULTS

SnO_2 was used as the colloidal semiconductor photocatalyst, because (i) thermodynamic and experimental data[4,5] indicate that it is stable in highly acidic media, i.e. it will not undergo acidic dissolution between pH –2 and 1; (ii) at these pH, the SnO_2 conduction band edge lies comfortably within SnO_2 region of stability on the Sn-H_2O potential-pH predominance diagram, indicating that the material will not undergo photoreductive dissolution; (iii) thermodynamic analysis indicates that at pH 0, photogenerated SnO_2 conduction band electrons can reduce Ce(IV) and U(VI) into Ce(III) and U(IV), respectively. An excess of ethanol was used to scavenge photogenerated valence band holes in order to improve the separation of the (e^-_{CB}, h^+_{VB}) pairs and to simplify the kinetic analysis of the photoreduction. The reaction vessel was surrounded by 12 UV lamps emitting at 312 nm. Photocatalytically induced changes in solution composition were followed in real time by photopotentiometry and UV-Vis spectroscopy for the Ce(IV) and the U(VI) systems respectively. Photopotentiometry involves measuring the potential difference, E, between a reference electrode (saturated calomel electrode) and a working electrode (gold wire) as a function of time, under illumination. Equation 1, derived using the mixed potential theory, allows the transformation of the recorded potential, E, vs time plots into $[Ce^{3+}]$ vs time traces, as shown in Figure 1.

$$[Ce^{4+}] = \frac{k^0_{Ce}.[Ce_T].\exp\left\{\frac{\alpha_{a,Ce}.F}{RT}(E-E^0_{Ce})\right\} + k^0_{Au}.\exp\left\{\frac{\alpha_{a,Au}F}{RT}(E-E^0_{Au})\right\} - k^0_{Au}.[Au^+].\exp\left\{\frac{-\alpha_{c,Au}F}{RT}(E-E^0_{Au})\right\}}{k^0_{Ce}.\exp\left\{\frac{\alpha_{a,Ce}.F}{RT}(E-E^0_{Ce})\right\} + k^0_{Ce}.\exp\left\{\frac{-\alpha_{c,Ce}F}{RT}(E-E^0_{Ce})\right\}} \quad (1)$$

where k^0_{Ce} and k^0_{Au} are the standard electrochemical rate coefficients for the cerium and gold couples, respectively, $\alpha_{a,Ce}$, $\alpha_{a,Au}$, are the anodic charge transfer coefficients for the cerium and gold couples, respectively, $\alpha_{c,Ce}$ and $\alpha_{c,Au}$ are the cathodic charge transfer coefficients for the the cerium and gold couples, respectively, $[Ce^{4+}]$ and $[Ce^{3+}]$ are the solution concentrations of Ce(IV) and Ce(III), $[Au^+]$ is the surface concentration of $Au(H_2O)^+_{ads}$ on the gold working electrode, E^0_{Ce} and E^0_{Au} are the standard potentials of the the cerium and gold couples, respectively, F, R and T are the Faraday constant, the gas constant and the temperature, respectively. Figure 2 illustrates the $[U^{4+}]$ versus time traces obtained using the decreasing values of the maximum absorbance of UO_2^{2+} at $\lambda = 420$ nm.

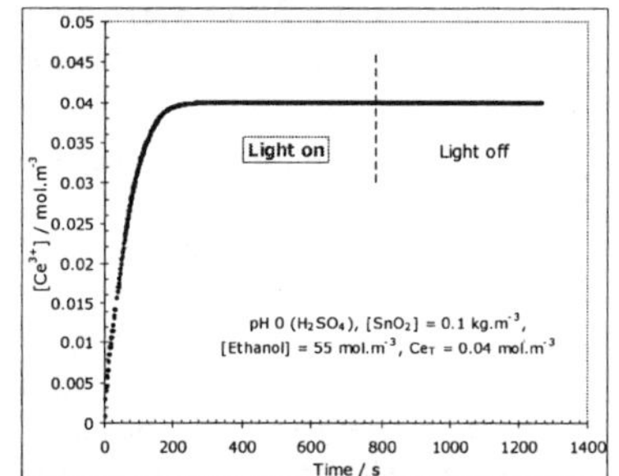

Figure 1 – *Evolution of $[Ce^{3+}]$ as a function of time during and after illumination.*

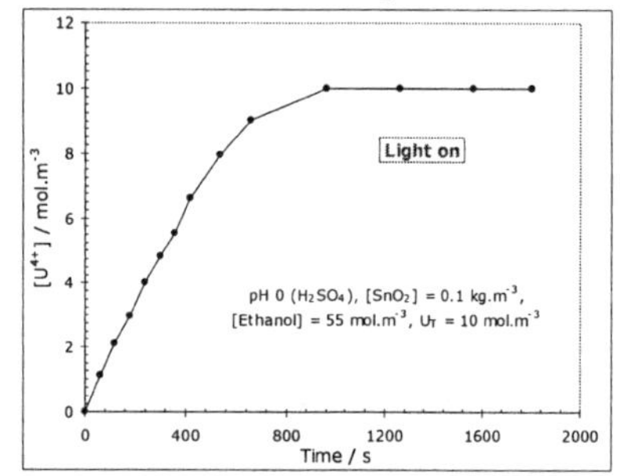

Figure 2 – *Evolution of $[U^{4+}]$ as a function of time during illumination.*

A kinetic analysis of the cerium/SnO_2 system leads to equation 2, which gives a calculated time, t, as a function of $[Ce^{4+}]$ and total cerium, Ce_T, present in the mixture.

$$t = \frac{\frac{k_{VC}}{k_R}}{g\frac{k_{VC}}{k_R}+k_{RVC}}\left(Ce_T-[Ce^{4+}]_t\right) + \left(\frac{g\frac{k_{VC}}{k_R}+k_{RVC}+\frac{k_{VC}}{k_R}Ce_T k_{RVC}}{\left(g\frac{k_{VC}}{k_R}+k_{RVC}\right)^2}\right)\ln\left(\frac{g\frac{k_{VC}}{k_R}Ce_T}{\left(g\frac{k_{VC}}{k_R}+k_{RVC}\right)[Ce^{4+}]_t - k_{RVC}Ce_T}\right) \quad (2)$$

where g is the rate of photogeneration of (e^-_{CB}, h^+_{VB}) pairs, k_R the rate coefficient for recombination of e^-_{CB}, h^+_{VB} pairs, k_{VC} the rate coefficient for reduction of Ce^{4+} by e^-_{CB} and k_{RVC} the rate coefficient for electron reinjection into the particles. Using curve fitting techniques, we extracted values of these parameters, which are shown in Table 1. Figure 3 shows the potential as a function of time recorded under illumination (A) and the theoretical potential vs time trace (B) calculated for $Ce_T = 0.12$ mol.m^{-3}. The very good agreement between recorded and theoretical curves establishes the validity of the mathematical model.

Table 1 – *Kinetic rate parameters determined experimentally, except for k_R[6].*

Parameters	**g**	**k_R**	**k_{VC}**	**k_{RVC}**
Values / units	1.48×10^{-3} mol.m^{-3}.s^{-1} ± 0.2×10^{-3} mol.m^{-3}.s^{-1}	10^{-7} s^{-1} *(from* [6]*)* ± 2×10^{-8} s^{-1}	5.8×10^{8} mol^{-1}.m^{3}.s^{-1} ± 8×10^{7} mol.m^{-3}.s^{-1}	2.7×10^{8} s^{-1} ± 10^{7} .s^{-1}

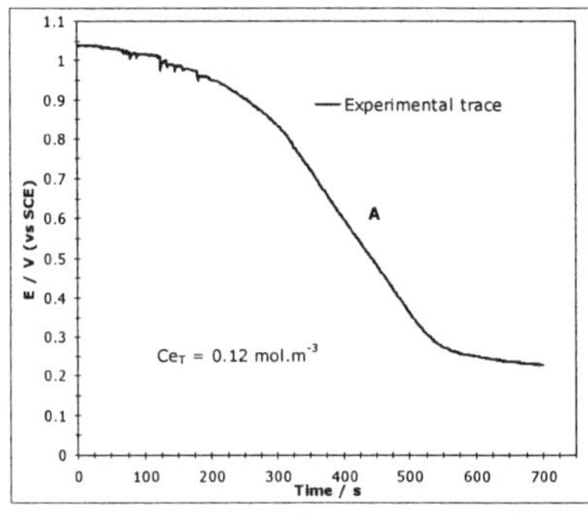

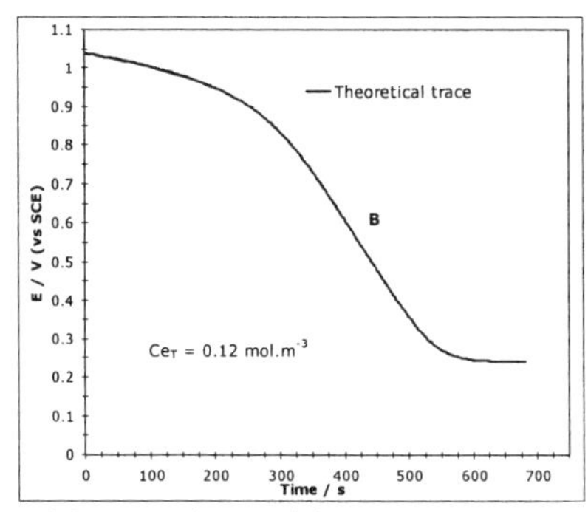

Figure 3 – *Experimental E vs t (A) and theoretical E vs t (**B**) traces during illumination, for the cerium system.*

3 CONCLUSION

We have shown that semiconductor photocatalysis is an efficient means to control the valence state of lanthanide/actinide ions. It is fast, and in the case of the cerium system, is irreversible. The quantum light efficiency, ϕ, for the generation of "useful" charge carrier pairs (i.e. those that can potentially be used in a valence control process) per photon absorbed, was found to be ~ 0.25. This is high compared to values reported for similar processes in the literature, and is a direct consequence of the charge transfer mechanism in operation. As the reduction of U(VI) involves two successive steps, the kinetic modelling of the uranium system is more complex and is still under investigation. Further work includes the application of heterogeneous photocatalysis to actinide mixtures and the use of immobilised colloidal particles in order to eliminate the need for post photocatalytic liquid-solid separation processes.

References

1 S.R. Morrison, *Electrochemistry at Semiconductor and Oxidised Metal Electrodes*, Plenum Press, 1980.
2 R.J. Kriek, W.J. Engelbrecht and J.J. Cruywagen, *J. South Afric. Inst. Min. Metal.*, 1995, 75.
3 N. Serpone, Y.K. Ah-You, T.P. Tran, R. Harris, E. Pelizzetti and H. Hidaka, *Solar Energy*, 1987, **39**, 491.
4 M. Pourbaix, *Atlas of electrochemical equilibria in aqueous solutions*, 2nd Ed., N.A.C.E., 1974.
5 G. Le Gurun, *The application of colloidal photocatalysis in actinide photoredox chemistry*, PhD thesis, 2003. Available in April 2006.
6 D.P. Colombo,Jr., K.A. Roussel, J. Saeh., D.E. Skinner, J.J. Cavaleri and R.M. Bowman, *Chem. Phys. Lett.*, 1995, **232**, 207.

STUDY OF THE CHEMICAL SCHEME OF β-TUPD SOLID SOLUTION ALTERATION

E. du Fou de Kerdaniel[1], N. Clavier[1], N. Dacheux[1], R. Drot[1], R. Podor[2] and E. Simoni[1]

[1] Groupe de Radiochimie, IPN Orsay, Université Paris-Sud-11, 91406 Orsay, France
[2] LCSM, Université H. Poincaré – 54506 Vandoeuvre lès Nancy, France

1 INTRODUCTION

In order to immobilize some radionuclides coming from spent nuclear fuels and/or dismantled nuclear weapons in an underground repository, several specific phosphate matrices like monazites, britholites and β-thorium phosphate - diphophosphate (β-TPD) are studied. β-TPD appears to be a promising material for the immobilization of large amounts of tetravalent actinides, such as uranium, neptunium and plutonium, leading to β-$Th_{4-x}An_x(PO_4)_4P_2O_7$ solid solutions (β-TAnPD). Synthesis and sintering experiments were undertaken on β-TUPD solid solutions showing a high variety of chemical routes and an interesting sintering capability.[1] One of the main properties required for such materials is their high chemical durability during leaching tests. In this field, β-TUPD pellets were leached in several acidic media (HNO_3, HCl, H_2SO_4) at various temperatures in order to collect the experimental data necessary for the determination of the normalized dissolution rates as well as for the identification of the mechanism of dissolution. Consequently, the normalized leaching rates were determined through several leaching experiments, then the neoformed phases were characterized using several techniques, such as Electron Probe MicroAnalysis (EPMA), SEM observations, IR and μ-Raman spectroscopies, as well as X-ray Photoemission Spectroscopy (XPS) and Time Resolved Laser Induced Fluorescence Spectroscopy (TRLIFS).

2 RESULTS AND DISCUSSION

In order to study the successive steps occurring during the dissolution of β-TUPD solid solutions, the behaviours of uranium and thorium were examined during leaching tests. Due to its rapid oxidation to uranyl, uranium was preferentially released in the leachate. On the contrary, thorium was quickly precipitated in neoformed phosphate-based phases which acted afterwards as a protective layer inducing diffusion processes. This observation was correlated with the two different dissolution rates observed for uranium : $2.1\ 10^{-5}$ g $m^{-2}day^{-1}$ during the first days of leaching and then $9.9\ 10^{-6}$g $m^{-2}day^{-1}$ in the second part of the dissolution curves (10^{-3} M HNO_3, $T = 323$ K).

In order to confirm this hypothesis, EPMA and SEM were undertaken on leached samples. According to the SEM observations and EPMA results, the bulk material seemed to be unaffected by the dissolution which occurred preferentially inside the grain boundaries at the surface of the pellets without any modification of the chemical composition of the solid, showing that the dissolution was stoichiometric.

As expected from the thorium behaviour during leaching tests, neoformed phases were found at the surface of the leached pellets (Fig 1). The first appeared as "spider webs" and seemed to be amorphous. The second was composed of ovoid plates of 2 μm to 8 μm in length and formed bigger aggregates of 10 μm to 15 μm. The mole ratios Th/P and U/(U + Th) determined from EPMA experiments on these phases were found to be 0.69 ± 0.01 and 0.02 ± 0.01, respectively, showing a significant depletion of uranium in the neoformed phase (compared to 0.36 of the raw sample), which appeared in good agreement with the formation of pure TPHPH.[2]

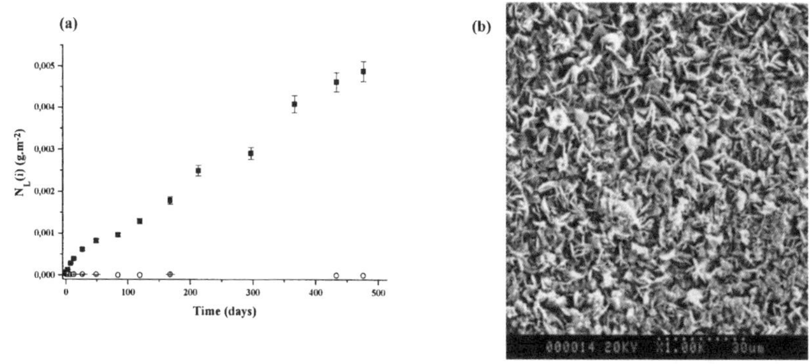

Figure 1 *Evolution of $N_L(U)$ (■) and $N_L(Th)$ (○) during the dissolution of β-TUPD (x = 2) in 10^{-3}M HNO_3 at T = 323 K (a) and SEM observations of leached β-TUPD (x = 1.6) (10^{-1}M HNO_3, T = 363 K, t = 10 months) (b).*

In order to confirm the precipitation of TPHPH onto the surface of the pellets of β-TUPD samples, μ-Raman analyses were performed. The observation of all the bands associated with both P-O bonds and P-O-(H) edge and the absence P-O-P bridge characteristics for diphosphate groups agreed well with the formation of TPHPH. Moreover, the presence of a continuous background in the 800-1200 cm^{-1} region was characteristic of an amorphous phase which was certainly associated with the "spider webs" observed by SEM. The spectra recorded on the uncovered surface revealed no differences with the unleached samples, confirming that the dissolution did not significantly modify the solid.

As was observed previously, uranium (IV) belonging to the solid is released in solution and oxidized into uranyl, which diffuses through the layer. TRLIFS experiments were also undertaken on the samples to identify the species present on the surface. Even though spectrofluorimetry is not a sensitive surface technique, it is of interest for this work, since uranium (VI) was used as a structural probe, on the basis of the dependence of its fluorescence on the uranium environment.[3]

The first results were obtained with spectrofluorimetry. No noticeable differences between the uranyl spectra were observed in intensity or in the peak locations. No fluorescence was observed on the unaltered compounds, showing the absence of uranium (VI) in these samples. On the contrary, the leached samples exhibited a fluorescence signal confirming the presence of uranium (VI) on the surface of the solids.

The decay curves were fitted with two decay times for the β-TUPD leached pellet as well as for uranyl adsorbed on the TPHPH surface. For both samples, the decay times were found to be around 80 μs and 300 μs. These first results seemed to show that TPHPH formed on the surface of the pellet, and that uranium (VI) diffused through this layer.

XPS measurements and further TRLIFS experiments, coupling these two spectroscopic techniques, are now undertaken in order to confirm the EPMA, SEM and μ-Raman results.

On the basis of all the results, the following dissolution mechanism was proposed (Fig 2). The dissolution occurs preferentially inside the grain boundaries and induces the precipitation of a neoformed phase, identified as TPHPH. This layer controls the release of the ions in solution *via* diffusion processes. Uranium (IV) is oxidized to the uranyl form and is released in the leachate, while thorium quickly precipitates as TPHPH in the gelatinous hydrated phase usually formed. This thermodynamically unstable gelatinous phase is progressively transformed into the well crystallized TPHPH with increasing leaching time. The formation of such layers at the beginning of the dissolution of β-TUPD could certainly delay the release of radionuclides to the biosphere.

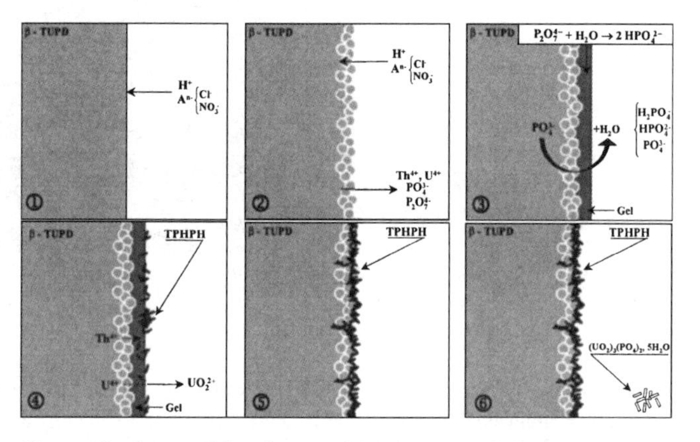

Figure 2 *Chemical scheme of dissolution of β-TUPD sintered samples.*

References

1 N. Clavier, N. Dacheux, P. Martinez, E. du Fou de Kerdaniel, L. Arranda and R. Podor, *Chem. Mater.*, 2004, **16**, 3557.

2 V. Brandel, N. Dacheux, M. Genet and R. Podor, *J. Solid State Chem.*, 2001, **159**, 139.

3 A. Kowal-Fouchard, R. Drot, E. Simoni, J. J. Ehrardt, *Environ. Sci. Technol.*, 2004, **38**, 1399.

SOLUBILITY STUDIES OF URANIUM(IV) BY LASER-INDUCED BREAKDOWN DETECTION (LIBD)

K. Opel, S. Hübener, S. Weiß, H. Zänker and G. Bernhard

Forschungszentrum Rossendorf e.V., Institute of Radiochemistry, PF 510119, D-01314 Dresden, Germany

1 INTRODUCTION

Uranium, among other actinide elements, plays an important role in the long term safety assessment of nuclear waste repositories. For geochemical modeling, the correct determination of the thermodynamic solubility data of uranium is essential. The solubilities of U(IV) species are of particular interest under the redox conditions of deep groundwaters. However, the results of studies on U(IV) solubility in the last decades are scattered by some orders of magnitude. Several reasons for overestimating the solubility have been discussed in the literature:[1-3 and ref. therein]

- U(IV) is readily reoxidized to U(VI) when reducing conditions are not maintained carefully. U(VI) is more soluble than U(IV) by many orders of magnitude, higher solubilities would be produced in this case.
- Tetravalent actinides show a strong tendency to hydrolyze and to form hydroxo complexes even at low pH. Thus, not all U(IV) is present as U^{4+} in the solution but some is in the form $[U(OH)_x]^{(4-x)+}$. The solubility product is overestimated if hydrolysis is neglected or if not all relevant hydrolyzed species are taken into account.
- Small colloidal particles have to be regarded as part of the solid phase. If insufficient methods of phase separation are applied in solubility studies, colloids are misattributed to the liquid phase, resulting in higher solubilities.
- U(IV) initially forms amorphous hydrous oxides, i.e. $UO_2{\cdot}xH_2O(am)$. Aging of these amorphous solids leads to phases with higher crystallinity, and finally to $UO_2(cr)$. Crystalline phases are less soluble than amorphous solids. If crystalline UO_2 fractions are involved, the solubility product of $UO_2{\cdot}xH_2O(am)$ is underestimated.

In the present study, the pH of U(IV) solutions is increased by coulometric titration according to the procedure described by Bundschuh et al. for the generation of thorium colloids[4]. The onset of colloid formation, i.e. the oversaturation of the solution with respect to $UO_2{\cdot}xH_2O(am)$, is detected by laser induced breakdown detection, a method which is capable of detecting colloids of some nanometres in diameter down to the ppt range of mass concentration. This approach avoids the problem of overestimating U(IV) solubility due to traces of U(VI), which was crucial to earlier studies, since the LIBD does not show any reaction to U(VI). For the determination of the solubility product of $UO_2{\cdot}xH_2O(am)$ the speciation of U(IV) at the pH of the onset of colloid formation is calculated in order to take hydrolyzed species into account.

2 EXPERIMENTAL

Solutions of $UO_2(ClO_4)_2$ in 0.2 M $HClO_4/NaClO_4$ are electrochemically reduced. The progression of the reduction is first tracked by UV-Vis spectroscopy showing the increasing U(IV) absorption bands and later by time-resolved laser fluorescence spectroscopy (TRLFS) proving the absence of U(VI).

After the uranium reduction, the pH of the solution is increased by coulometric titration. Samples are transferred into the LIBD setup (Fig. 1) by a peristaltic pump and checked for the presence of colloidal particles. The LIBD uses a pulsed Nd:YAG laser (Soliton DIVA II) as the light source, the breakdown events are detected acoustically by a self-made piezoelectric detector and optically by a CCD-camera (Pulnix TM-1040).
The pH at the onset of colloid formation, pH_{coll}, and the effective uranium concentration $[U(IV)]_{tot}$ determined by ICP-MS are used to calculate the solubility product of $UO_2{\cdot}xH_2O$ (am).

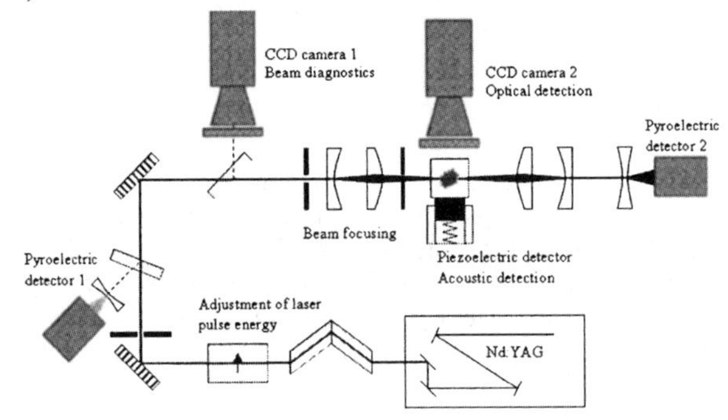

Figure 1 *Setup of the system for laser-induced breakdown detection*

3 RESULTS

Fig. 2 shows the dependence of the breakdown probability (BP) with the pH for two solutions with varying uranium concentrations, measured at laser pulse energies of 1.5 mJ. The background level is marked by a dashed line.

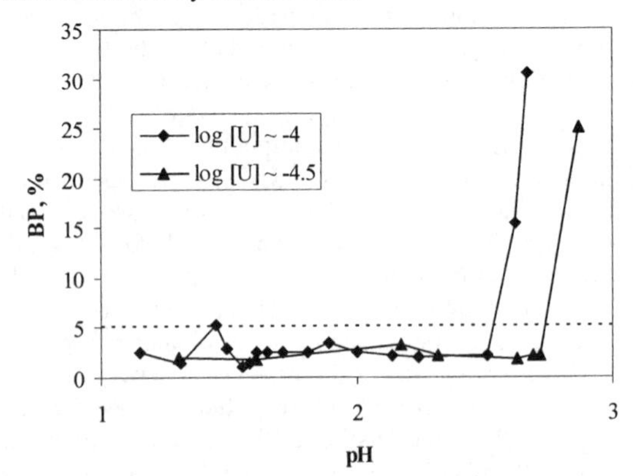

Figure 2 *Breakdown probability as a function of pH for coulometric titrations*

The pH values of the onset of colloid formation, the measured uranium concentrations and the resulting solubility products, $K°_{sp}$, at zero ionic strength calculated by the procedure described below are given in Table 1.

Table 1 *Results of the coulometric titrations*

$[U(IV)]_{tot}/mol\ l^{-1}$	pH_{coll}	$log\ K°_{sp}$
1.06×10^{-4}	2.57	-54.27
3.37×10^{-5}	2.80	-54.34

The solubility product is calculated using equation (1), where γ_i are the activity coefficients of the ionic species. They are determined by applying the specific ion interaction theory explained in detail in the NEA-TDB[5].

$$K_{sp}^{0} = [U^{4+}][OH^{-}]^{4}\gamma_{U^{4+}}\gamma_{OH^{-}}^{4} \quad (1)$$

The concentration of hydroxide ions is calculated from pH_{coll}, the ionic product of water in 0.2 M perchlorate media and the activity coefficients of H^+ and OH^-. For the determination of $[U^{4+}]$, the hydrolysis species are taken into consideration by using their respective formation constants, $\beta°_x$, given in the review by Neck et al.[1]

$$\beta°_x = \frac{[U(OH)_x^{(4-x)+}]}{[U^{4+}][OH^-]^x}\cdot\frac{\gamma_{U(OH)_x^{(4-x)+}}}{\gamma_{U^{4+}}\gamma_{OH^-}^{x}} \quad (2)$$

The concentration of free U^{4+} can be calculated from the total uranium concentration and the uranium speciation under the prevailing conditions according to equation (3).

$$[U^{4+}] = [U(IV)]_{tot} - \sum[U(OH)_x^{(4-x)+}] \quad (3)$$

The mean value of the two calculated solubility products (Table 1) is $log\ K°_{sp} = -54.31$. Thus, the result of this first determination of $UO_2{\cdot}xH_2O$ (am) solubility by LIBD confirms earlier studies based on solubility measurement by phase separation (ultrafiltration) as far as the investigators succeeded in preventing the access of oxygen.[6] They disprove the results of earlier studies in which the investigators obviously failed to exclude oxygen painstakingly.[7] The effect of the particle size on the solubility product, which has been described for analogue investigations of amorphous thorium dioxide,[4] has not yet been taken into consideration. The validation of the presented result by titrations of solutions with other uranium concentrations is in progress.

References

1 V. Neck and J.I. Kim, *Radiochim. Acta*, 2001, **89**, 1.
2 J. Bruno, *Acta Chem. Scand.*, 1989, **43**, 99.
3 K. Fujiwara, H. Yamana, T. Fujii and H. Moriyama, *Radiochim. Acta*, 2003, **91**, 345.
4 T. Bundschuh, R. Knopp, R. Müller, J.I. Kim, V. Neck and T. Fanghänel, *Radiochim. Acta*, 2000, **88**, 625.
5 I. Grenthe, J. Fuger, R.J.M. Konings, R.J. Lemire, A.B. Muller, C. Nguyen-Trung and H. Wanner, *Chemical Thermodynamics, Vol. 1. Chemical Thermodynamics of Uranium*, 1992, Elsevier, Amsterdam.
6 D. Rai, A.R. Felmy, S.M. Sterner, D.A. Moore, M.J. Mason and C.F. Novak, *Radiochim. Acta*, 1997, **79**, 239.
7 J. Bruno, I. Casas, B. Lagerman and M. Munoz, *Mat. Res. Soc. Symp. Proc.*, 1987, **84**, 153.

RESONANCE IONISATION MASS SPECTROMETRY FOR ELEMENT- AND ISOTOPE-SELECTIVE TRACE ANALYSIS OF ACTINIDES IN ENVIRONMENTAL MICRO-PARTICLES

N. Erdmann[1], F.M. Engelberger[1], J.V. Kratz[1], N. Trautmann[1], G. Passler[2], M. Betti[3], P. Lievens[4], R.E. Silverans[4] and E. Vandeweert[4]

[1]Institut für Kernchemie, Universität Mainz, Fritz-Straßmann-Weg 2, D-55128 Mainz, Germany
[2]Institut für Physik, Universität Mainz, Staudinger Weg 7, D-55128 Mainz, Germany
[3]Institute for Transuranium Elements, European Commission JRC, Postfach 2340, D-76125 Karlsruhe, Germany
[4]Laboratorium voor Vaste-Stoffysica en Magnetisme, K.U. Leuven, Celestijnenlaan 200D, B- 3001 Leuven, Belgium

1 INTRODUCTION

Over the last few decades, detection techniques have been developed for ultra-trace analyses of long-lived radioisotopes. Among them is resonance ionisation mass spectrometry (RIMS), which is highly element-selective, due to the specific laser excitation/ionisation process, combined with good mass separation and high detection efficiency. RIMS has been successfully applied to the analysis of long-lived anthropogenic radioisotopes after their chemical separation from bulk environmental samples.[1] Recently, the analysis of micron-sized particles containing high concentrations of actinides has become of great interest for risk assessments of contaminated areas, nuclear forensic analyses, and IAEA and EURATOM safeguards programmes.[2] Information on actinide isotopic compositions is important to deduce the origin, age, and history of the particles. The state-of-the-art standard technique for their analysis, secondary ion mass spectrometry (SIMS),[3] suffers from isobaric interferences (e.g. $^{238}U/^{238}Pu$, $^{241}Am/^{241}Pu$). Post-ionising the sputtered neutrals from actinide-containing micro-particles in an element-selective manner has been suggested as an elegant method to eliminate this problem.

2 METHOD AND RESULTS

To optimize the sample preparation procedures for the highest emission of actinide atoms during ion beam sputtering, evaluation studies were performed at K.U. Leuven. The setup[4] consists of an UHV chamber in which an ion gun (d.c. current ~0.7 μA, spot diameter ca. ~2.5 mm) directs Ar^+ ions with a kinetic energy of 15 keV onto a centrally located target foil at 45° incidence. The ion gun diameter is too large for the analysis of single μm-sized particles; however, basic studies can be performed with this system. The plume of sputtered particles is intersected, parallel to the foil, by two overlapping laser beams from a dye laser system and an optical parametric oscillator, pumped by independent pulsed Nd:YAG lasers operating at 10 Hz. These laser systems generate linearly polarized laser light with laser pulses of ~6 ns tuneable in the wavelength range from 225 to 1600 nm and pulse energies from 3 mJ in the UV up to 50 mJ in the visible range, and with a line width of about 1 cm^{-1}. The atoms sputtered in a polar angle interval of ~10° around the surface normal are photo-ionised and subsequently detected with a time-of-flight mass spectrometer.

The samples consisted of uranium oxide (U_3O_8) powder deposited on foils. The oxide powder was suspended in isopropyl alcohol, and, for sample preparation, several droplets were pipetted onto metal foils, and the alcohol was evaporated. Different foil materials - C, Re, Ta, Ti, and Zr - were used. A small number of samples were studied by scanning electron microscopy (SEM) prior to the RIMS analysis, showing that the U_3O_8 particles resemble small platelets below 1 µm in size. Their distribution on the metal substrates was slightly inhomogeneous, with higher concentrations of particles along the borders of the droplets.

For resonance ionisation of uranium, a two-colour, two-step ionisation scheme (λ_1 = 324.6 nm, λ_2 = 513.5 nm), starting from the atomic ground state and leading to an autoionising state was investigated in a previous study.[5] Using this same excitation scheme, resonant ionisation of uranium atoms sputtered from the uranium oxide particles was studied.

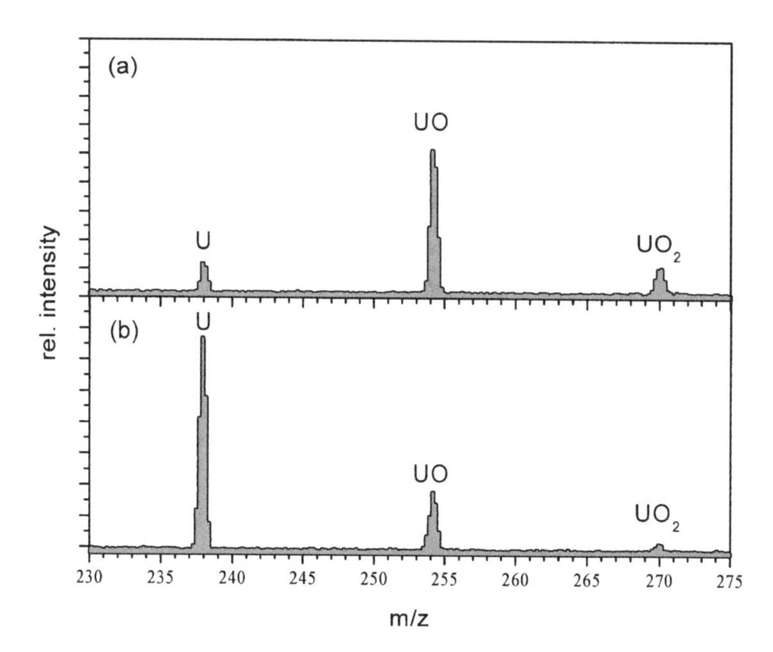

Figure 1. *Mass spectra of uranium and its oxides from U_3O_8 particles, obtained by resonant laser ionisation of the neutrals after sputtering with Ar ions: (a) U_3O_8 deposited on Ti, (b) U_3O_8 deposited on Ti, after sputter-cleaning and bake-out.*

The laser pulse energies were adjusted to saturate the uranium atomic transitions. Under these conditions, peaks of U^+, UO^+ and UO_2^+ were present in the mass spectra, UO and UO_2 being non-resonantly ionised mainly by the laser light at 513.5 nm. This non-resonant process was not saturated and as such very sensitive to small variations in the laser intensity. In principle, only qualitative comparisons are possible. However, it was carefully checked that the laser pulse energies were comparable during the course of all experiments.

When continuously bombarding the sample with the Ar ion beam, the uranium as well as the oxide signals increased with time, reaching saturation after 15 – 20 min. The ratio for atomic uranium to uranium oxide ions was observed to be different for the different substrates, Ti, Ta, and Zr, producing higher yields of atomic uranium than Re or C. A high atomic yield is required to enable the analysis of small quantities of actinides with RIMS. Ti, Ta, and Zr are known to act as reducing materials for the generation of atomic beams in RIMS analysis of bulk samples.[7] To improve the yield of the atomic species, the substrates with particles were baked out for 10 min at 800°C under UHV conditions in a separate chamber. This procedure led to a much higher atomic uranium signal, compared to UO^+ and UO_2^+, and an even stronger increase was obtained for samples that were sputter-cleaned with the Ar ion gun for several minutes prior to the bake-out procedure. Figure 1 shows the mass spectra obtained from U_3O_8 particles on Ti, with and without the sputter-cleaning/bake-out procedure. One can see that about an order of magnitude improvement in the atomic signal can be gained by suitable sample preparation.

3 CONCLUSION AND OUTLOOK

For resonance ionisation of sputtered neutral uranium from uranium oxide particles, it was found that the sample substrate and preparation had a strong influence on the yield of neutral atoms. A new system for single particle analysis is currently being set up, using a commercial TOF-SIMS instrument (TOF-SIMS III, Ion-Tof, Münster, Germany) with a pulsed gallium liquid metal ion gun that allows a sub-µm focus, small enough for micro-analysis. It will be coupled to a high repetition rate Nd:YAG pumped Ti:Sapphire laser system[6] for 3-colour, 3-step resonant ionisation of the sputtered neutrals with high elemental selectivity. The laser repetition rate allows for an optimum coupling to the pulse repetition rate of the ion gun. In this way, traces of actinides in micro-particles could be rapidly analysed without isobaric interferences.

References

1 N. Trautmann, G. Passler, and K.D.A. Wendt, *Anal. Bioanal. Chem.,* 2004, **378**, 348.
2 D.L. Donohue and R. Zieser, *Anal. Chem.,* 1993, **65**, 359A.
3 M. Betti, G. Tamborini, and L. Koch, *Anal. Chem.,* 1999, **71**, 2616.
4 E. Vandeweert, P. Lievens, V. Philipsen, J. Bastiaansen, and R.E. Silverans, *Phys. Rev. B,* 2001, **64**, 195417.
5 N. Erdmann, M. Betti, F. Kollmer, A. Benninghoven, C. Grüning, V. Philipsen, P. Lievens, R.E. Silverans, and E. Vandeweert, *Anal. Chem.,* 2003, **75**, 3175 G.
6 Grüning, G. Huber, P. Klopp, J.V. Kratz, P. Kunz, G. Passler, N. Trautmann, A. Waldek, and K. Wendt, *Int. J. Mass Spectrom.,* 2004, **235**, 171.
7 B. Eichler, S. Hübener, N. Erdmann, K. Eberhardt, H. Funk, G. Herrmann, S. Köhler, N. Trautmann, G. Passler, and F.-J. Urban, *Radiochim. Acta,* 1997, **79**, 221.

THE ROMANIAN URANIUM INDUSTRY- ITS PAST AND FUTURE IN THE CONTEXT OF THE STRENGTHENING AND INTEGRATION OF THE EUROPEAN RESEARCH AREA

I.C.Popescu, P.D. Georgescu, F. Aurelian, St. Petrescu and E. Panturu

Research and Development National Institute for Metals and Radioactive Resources-ICPMRR, 70 Blvd. Carol I, sector 2, 020917, Bucharest, Romania

1 INTRODUCTION

Uranium, a very intriguing and controversial element, was for a long time an important topic of interest for the scientific community, at the beginning for military purposes, and subsequently because of its capacity to offer an environmental friendlier alternative to the electrical energy supplied by the fossil coal burning.

The uranium industry underpins an apparently simple concept, known as the "nuclear fuel cycle", which refers to: uranium mining, processing and refining, fuel production, burned fuel reprocessing and waste treatment.

Economic factors (continuous urban development, the balance between demand and supply, purchasing price, capital costs, etc), political considerations (economic and political strategies developed by the national governmental authorities), societal evolution (social conflicts, "green" attitudes, etc.) and the features of the exploited resources (the ore bodies' nature, their uranium content and so on) influence the development of the uranium industry.

2 BACKGROUND TO THE URANIUM INDUSTRY

Information concerning the activities of European countries is not fully available in all cases, due to the national security policies. Therefore, it is very difficult to capture a realistic image of this matter. Figure 1, which shows the data provided by the World Uranium Mining database, is a good example. In addition, it is important to include the former Soviet Union's contribution, as an economic and political power, to the uranium industry evolution.

The most important European uranium resources and uranium ore processing plants for nuclear fuel output were in France, the former East Germany, Spain, Czech Republic, Hungary and Romania. After 1990, these uranium extraction and processing activities were closed in almost all European countries, and emphasis changed to environmental rehabilitation activities in the areas affected by uranium exploration, exploitation and ore processing.

Romanian activities began[1] in the 1950's, when uranium-bearing ore bodies were found in three main regions, namely Banat, the Apuseni Mountains and the Eastern Carpathians.

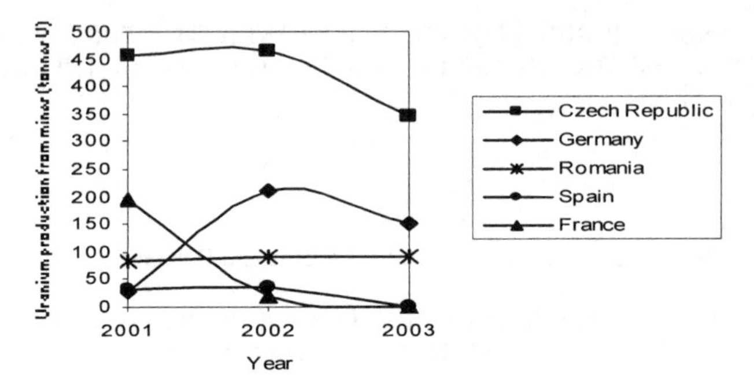

Figure 1 *Uranium production from mines for a few European countries*

Between 1952 and 1962, the uranium ore was exported directly to the former Soviet Union[1] in order to be processed at Sillamäe in Estonia. In 1962[1] the uranium ore exports to the Soviet Union ceased, and the ore was stored outside the mining pits, but from 1978, when the Feldioara plant was commissioned, both the stored ore and the newly extracted material began to be processed. During this time, uranium ore processing[3] was considered a strategic objective, and it was protected by secrecy. The transition from the 'command' economy to a market driven one dramatically affected uranium mining activity, so state intervention was essential. It was concretized in fundamental restructuring measures.

3 ROMANIAN RESEARCH AND DEVELOPMENT ACTIVITY TARGETS AND THEIR HARMONIZATION WITH THE EUROPEAN STANDARDS

Romanian research scientists were assigned to develop[2] extraction technology flow sheets and projects to process Romanian uranium ores, to purify yellowcake and to produce nuclear pure uranium dioxide powder. However, as an accession country, Romania had to conform to European instructions and requirements concerning national economic policy, so a strategy was developed to harmonize the uranium industry's specific problems with the Romanian nuclear energy program and the EU requirements. The uranium demand and costs evolution in the home market was also analyzed, and compared to the situation overseas. Romania was determined to reduce its uranium mining activity significantly due to public concern for radioactive contamination in the environment, so almost all uranium mining activities ceased, and the ecological rehabilitation process commenced. Since then, substantial efforts have been devoted to solving this issue. Unfortunately, the enormous financial resources required were and are very hard to obtain from the national research programmes.

One solution could be the EURATOM programme, but its research is focused on the issues related to the nuclear field grouped around four main research areas, as shown in Figure 3. Another idea is the sustained development of national nuclear electrical power, which is necessary to cover the electrical energy consumption (see Figure 4) generated by continuous urban development. If within the next twenty years all of the nuclear power plant units (NPP) at Cernavoda are completed and commissioned, then the contribution of nuclear energy to the entire electric power output will be about fifty percent (see Figure 4). In this situation, fossil fuel consumption will be reduced by up to 30%, and greenhouse gas emissions will be decreased as well.

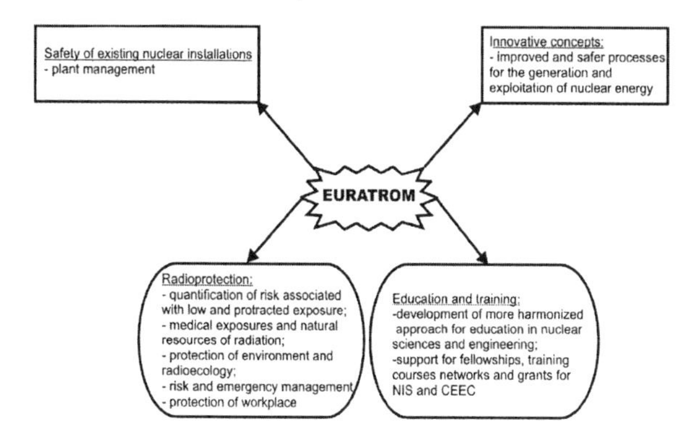

Figure 3 *EURATOM's main research areas*[4]

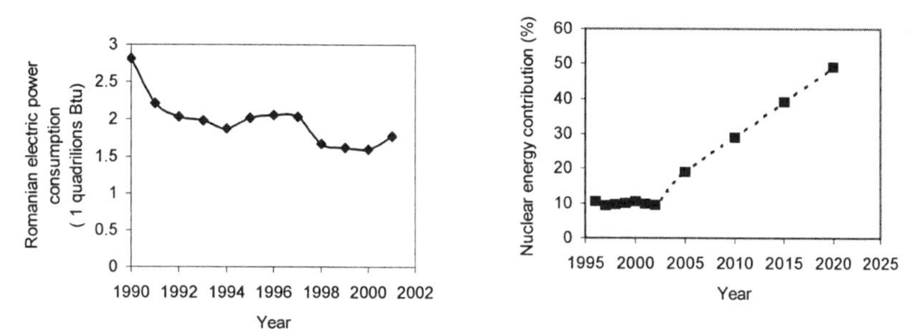

Figure 4 *Romanian*[5] *electric power consumption (left) and nuclear energy*[4] *contribution to the entire power output (dashed line shows its trend after the commissioning of the other three Cernavoda NPP, namely G3, G4 and G5 (right))*

4 CONCLUSION

The conclusion is a very simple one. Only the sustained development of the national nuclear energy output can be the answer to the uranium industry crisis, and the solution to the global warming process caused by the fossil fuel and oil combustion.

References

1 Rapport établi conjointement par l'Agence de l'Organisation de Coopération et de Développement Économiques et l'Agence Internationale de l'Energie Atomique, *Livre Rouge, Uranium 2001 Resources, production et demand*, 2001, IAEA
2 Information available on Research and Development National Institute for Metals and Radioactive Resources-ICPMRR, www.icpmrr.ro
3 Information available on the report entitled: *Part I Uranium, WEC Survey of Energy Resources 2001,*: http://www.worldenergy.org/wec-geis/publications/reports/ser/uranium/uranium.asp
4 Information available on : http://www.cordis.lu/fp6-euratom/activities.htm
5 Lynch, R., *An Energy Overview of Romania,* http://www.fe.doe.gov /international/CentralEastern%20Europe/romnover.html

URANIUM AND NEPTUNIUM SORPTION ONTO MAGNETITE UNDER SUBOXIC CONDITIONS

A.J. Hynes, N.D. Bryan and F.R. Livens

Centre for Radiochemistry Research, School of Chemistry, University of Manchester, Oxford Road, Manchester M13 9PL, UK

1 INTRODUCTION

To make risk assessments for the disposal of radioactive waste it is necessary to understand the aqueous behaviour of those elements most hazardous on release. Uranium and neptunium are two major actinides that, due to their long half lives and risks from their associated daughter nuclides, are of concern.

Much research has taken place concerning the sorption of radionuclides onto iron oxides that result from the corrosion of carbon steel waste containers.[1-8] These iron oxide minerals provide a barrier to groundwater actinide migration, with their effectiveness depending on pH, Eh, and carbonate concentration. Both uranium and neptunium exhibit a range of oxidation states and, under aerobic conditions, it is the trans dioxo uranyl and neptunyl cations, UO_2^{2+} and NpO_2^{+}, that dominate.[9] Their transport and solubility are also affected by hydrolysis and by complexation with the carbonates and bicarbonates present in natural waters[10]. These aqueous species can have strong formation constants, thus competing with the mineral surface for the actinyl centre.

Under reducing conditions, for example resulting from iron corrosion, and at the depths of deep waste repositories, actinide hydrolysis can take place, producing soluble tetravalent actinide hydroxide and oxide complexes, along with precipitates such as $Np(OH)_4$ and $U(OH)_4$. Sorption of nuclides can thus be enhanced under reducing conditions by precipitation of these insoluble species. Magnetite (Fe_3O_4), a mixed iron oxide of spinel structure, is a product of steel corrosion under low oxygen conditions. It provides a barrier to radionuclide migration and under suboxic conditions is an efficient sorption surface for surface complexation/precipitation of uranium and neptunium species.

2 METHODS

Magnetite was synthesised using a conventional method[11] and its structure confirmed using X-ray powder diffraction. Sorption experiments were carried out in 0.01M $NaClO_4$ solution for a magnetite concentration of 10g/L, with either an Ar or Ar/1% CO_2 atmosphere. Sorption of UO_2^{2+} (initial solution concentrations 30 pM and 1 μM) and NpO_2^{+} (initial solution concentration 1 μM) at pH values of 4, 5, 6, 7, 8 and 9 was measured after 2 hours equilibration, which preliminary experiments showed was adequate to reach a steady state. Following the sorption period, the supernatant was filtered through a 0.2 μm membrane, and the uranium/neptunium content quantified. Inductively coupled plasma mass spectrometry was used to analyse 1 μM uranium (^{238}U), and low-level liquid scintillation counting (Quantulus) to quantify 1 μM neptunium (^{237}Np) and 30pM uranium (^{232}U) in alpha/beta discrimination mode.

3 RESULTS AND DISCUSSION

Distribution coefficient values (Rd) were plotted against pH for both U and Np (Figures 1 and 2). The results show a marked effect of CO_2 on Rd values for both Np and U. In general, an increase in pH results in increased sorption, with a more dramatic increase between pH 4 and 5 for U after which sorption plateaus with slight fluctuations between pH 6 and 8. This could possibly be a result of uranyl hydrolysis, and PHREEQE speciation modelling predicts $UO_2(OH)_2$ as a major species. This result is in agreement with that for a magnetite concentration of 2g/L over 4 hours.[2] Inclusion of CO_2 reduces sorption in the case of U, but results in increased Rd values between pH 5 and pH 8 inclusive for Np, although the Rd is the same for both atmospheres at pH 4 and 9. The Rd values for both U and Np start to decrease at higher pH with 1% CO_2 present.

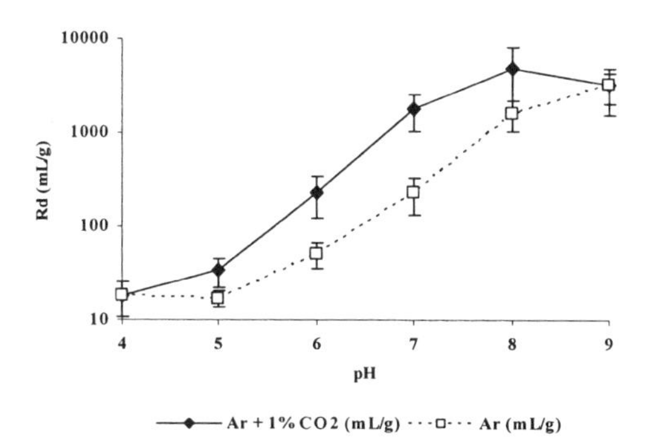

Figure 1 *Distribution coefficients (Rd) for 1 μM neptunium sorption onto synthetic magnetite, I=0.01M $NaClO_4$, suboxic conditions, with and without 1% CO_2.*

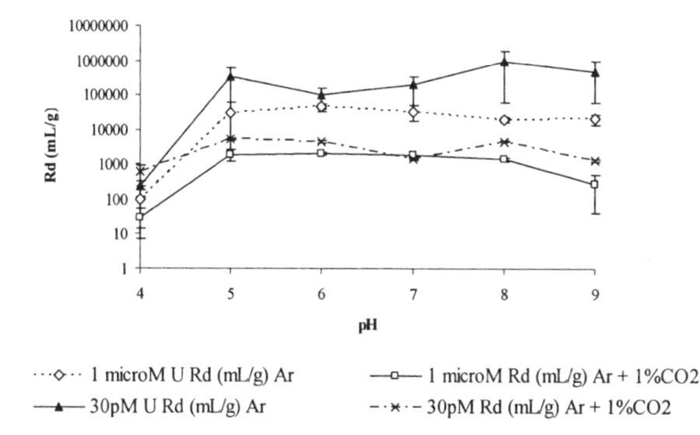

Figure 2 *Distribution coefficients (Rd) for 1 μM and 30 pM uranium sorption onto synthetic magnetite, I=0.01M $NaClO_4$, suboxic conditions, with and without 1% CO_2.*

It is expected that UO_2^{2+} forms more stable carbonate complexes in solution than NpO_2^+, due to its greater acidity, and this is illustrated by higher formation constants. Sorption of carbonate species could thus be more favourable for neptunium. It is also possible that insoluble sodium salts of the form $Na_{2n-1}NpO_2(CO_3)_n$, n=1-3, are precipitating on the magnetite surface, with increasing pH, perhaps leading to formation of ternary surface complexes.[10] PHREEQE speciation modelling, in the absence of magnetite, predicts a rise in neptunium carbonato species from pH 6, reflecting the sorption trend. Above pH~7.25 the major species is the tris-carbonato neptunyl species, $NpO_2(CO_2)_3^{5-}$.
Lowering the U concentration to 30 pM leads to increased sorption, perhaps reflecting the presence of strong surface binding sites at low concentrations. Upon saturation, sorption onto weaker sites follows. The experimental results are presently being analysed using surface complexation modelling.

4 CONCLUSIONS

Magnetite provides an effective sorption surface for U and Np under suboxic conditions, with a marked increase for both U and Np on increasing the pH from 4 to 9. The inclusion of carbonate into the system results in higher Rd values for Np and lower values for U, reflecting stronger carbonate complexation of UO_2^{2+} than NpO_2^+, and thus more favourable binding of NpO_2^+ to the oxide surface. Reducing the U concentration by a factor of $3x10^5$ gives much larger Rd values, particularly in the absence of CO_2, perhaps indicating that there may be more than one type of binding site on the magnetite surface.

References

1 M.C. Duff, J. Urbanik Coughlin and D.B. Hunter, *Geochimica et Cosmochimica Acta*, 2002, **66**, 3533.
2 T. Missana, M. García-Gutiérrez and V. Fernńdez, *Geochimica et Cosmochimica Acta*, 2003, **67**, 2543.
3 T. Missana, M. García-Gutiérrez and C. Maffiotte, *Journal of Colloid and Interface Science*, 2003, **260**, 291.
4 T. Missana, C. Maffiotte and M. García-Gutiérrez, *Journal of Colloid and Interface Science*, 2003, **261**, 154.
5 K. Nakata, S. Nagasaki, S. Tanaka, Y. Sakamoto, T. Tanaka and H. Ogawa, *Radiochimica Acta*, 2000, **88**, 453.
6 K. Nakata, S. Nagasaki, S. Tanaka, Y. Sakamoto, T. Tanaka and H. Ogawa, *Radiochimica Acta*, 2002, **90**, 665.
7 K. Nakata, S. Nagasaki, S. Tanaka, Y. Sakamoto, T. Tanaka and H. Ogawa, *Radiochimica Acta*, 2004, **92**, 145.
8 O. Tochiyama, S. Endo and Y. Inoue, *Radiochimica Acta*, 1995, **68**, 105.
9 J.J. Katz, G.T. Seaborg, and L.R. Morss, *The chemistry of the actinide elements*, 1986 (second edition), Chapman and Hall, London.
10 D.L. Clark, D.E. Hobart and M. P. Neu, *Chemical Reviews*, 1995, **95**, 25.
11 U. Schwertmann and R.M. Cornell, *Iron oxides in the laboratory – preparation and characterisation*, 1991, VCH, Weinheim.

THE APPLICATION OF AVALANCHE PHOTODIODES FOR THE MEASUREMENT OF ACTINIDES BY ALPHA LIQUID SCINTILLATION COUNTING

A. Reboli[1], J. Aupiais[1] and J.C. Mialocq[2]

[1] CEA/DASE/RCE, centre de Bruyères-le-Châtel, BP 12, 91680 Bruyères-le-Châtel, France
[2] CEA/DSM/DRECAM/SCM/URA 331 CNRS, centre de Saclay, 91191 Gif-Sur-Yvette, France

1 INTRODUCTION

In order to improve the resolution of alpha liquid scintillation counting, and to promote this method for the measurement of actinides in the environment, silicon avalanche photodiodes (APD) have been tested as detectors for liquid scintillation. For many years, scintillation detection has been almost exclusively based on photomultiplier tubes (PMT). PMT have low noise and high gain, but suffer from low quantum efficiency for the fluorescence emission range of many scintillators. On the contrary, avalanche photodiodes (APD) show higher quantum efficiency and a more uniform active area. Recent advances in APD device performance and in photosensitive area (up to 2 cm^2) allow APD to replace PMT for scintillation spectrometry. [1]

2 EXPERIMENTAL SET-UP

The set-up consists of a large area avalanche photodiode (16 mm diameter) coupled to a thin vial containing alpha-emitters within a liquid scintillation cocktail. A 10 mm diameter cylindrical quartz vial (THUET B., Blodelsheim, France) containing the sample in liquid scintillator solution (V = 400 μL) is placed directly on the face of the APD, using silicone grease (V 1000 Rhodorsil, Rhone Poulenc, France, refractive index n = 1.48) or immersion oil (Cargille Laboratories, Cedar Grove, USA, refractive index n = 1.51) for optical coupling. In order to minimize light loss, the vial is wrapped with several Teflon tapes except for the face coupled to the detector. The APD is biased by an ORTEC 456 (Oak Ridge, USA) regulated HT power supply. The signals from the APD are fed to an ORTEC 142IH charge-sensitive preamplifier and then to an ORTEC 572 spectroscopy amplifier (1 μs shaping time). The energy spectrum is acquired by an ORTEC 919E multichannel analyser. A Canberra 814FP (Meriden, USA) pulser is used to calibrate the spectrometry chain in the number of charges and for noise performance evaluation. The pulses are fed to the test input of the preamplifier through a capacitance of 1.15 pF. APD as well as liquid scintilators are temperature sensitive. For reliable measurements, the APD is placed in a stainless steel chamber and cooled by immersion in a cold bath (HAAKE KT 50L). The temperature during the measurements is monitored by a platinum thermoresistor. The temperature variation in the chamber does not exceed 0.5 °C.

3 RESULT AND DISCUSSION

Better resolution requires optimization of several parameters, such as bias voltage, temperature, counting geometry and composition of the scintillating cocktail.[2] Thus, new compositions of organic liquid scintillator have been tested in order to enhance energy transfer by a judicious choice of components (due to the large spectral response of APD). Components presenting the best overlap between absorption spectrum of the donor and the fluorescence spectrum of the accepting molecule (Figure 1) have been assessed. For instance, Figure 2 shows the displacement of alpha peaks (^{233}U and ^{244}Cm) to higher channel numbers, due to a higher number of photoelectrons produced by the scintillating mixture.

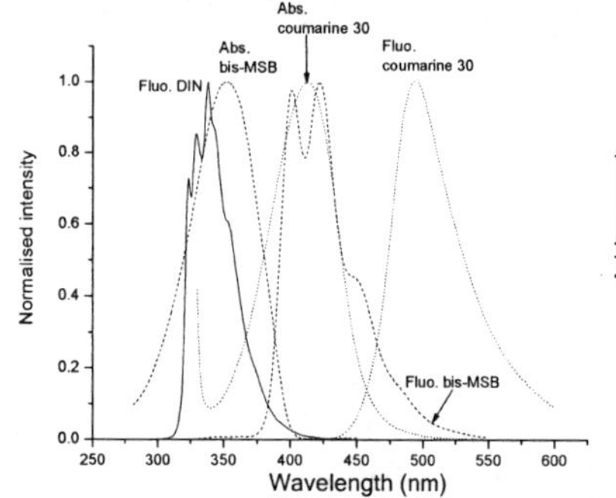

Figure 1 *Normalised absorption and fluorescence spectra of a mixture of di-isopropylnaphthalene, bis-MSB and coumarine 30.*

Figure 2 *Number of photons produced by several concentrations of bis-MSB and coumarine 30 in DIN solvent. Peaks correspond to ^{233}U and ^{244}Cm, respectively.*

Better resolution has been achieved for several actinides and α emitters (see Table 1).

Table 1 *Energy resolution (R(keV)) and relative energy resolution (R(%)) as a function of alpha particle energy for devices using APD and PMT (PERALS™ spectrometer).*

Isotope	Energy (keV)	APD(-40°C) R (keV)	APD(-40°C) R (%)	PMT R (keV)	PMT R (%)
^{232}Th	4010	200±13	(5.0)	210±13	(5.2)
^{239}Pu	5157	242±13	(4.3)	256±14	(5.0)
^{228}Th	5423	235±11	(4.3)	306±17	(5.6)
^{236}Pu	5768	243±11	(4.2)	293±14	(5.1)
^{216}Po	6778	262±14	(3.9)	448±20	(6.6)

In Figure 3, a comparison between a commercial apparatus (Perals™) and our device is given. This figure demonstrates that considerable improvement of the resolution is achieved; especially at high alpha energy for which 4 peaks are clearly visible, instead of 3 for the Perals.

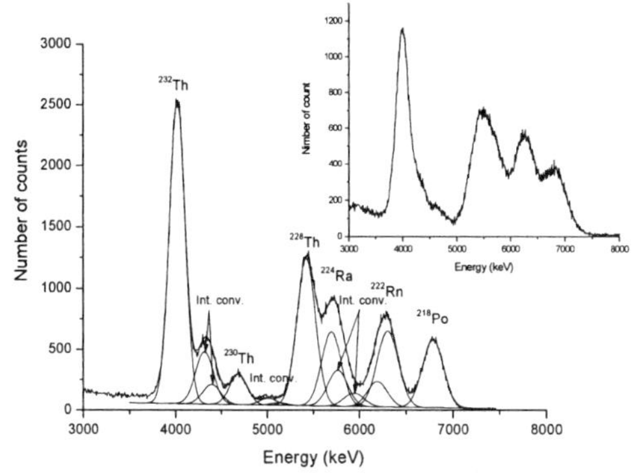

Figure 3 *Thorium isotopes spectrum obtained at -40 °C by PDA. Upper right; the same spectrum obtained at 20 °C by Perals™.*

Thus, with the APD device, it becomes possible to distinguish isotopes of americium, which is unachievable with any commercial apparatus. Finally, better resolution leads to higher reliability, because mathematical treatments are more easily performed and avoid the incorrect location of internal conversion peaks. An example of such a problem is shown in Figure 5 for plutonium isotopes.

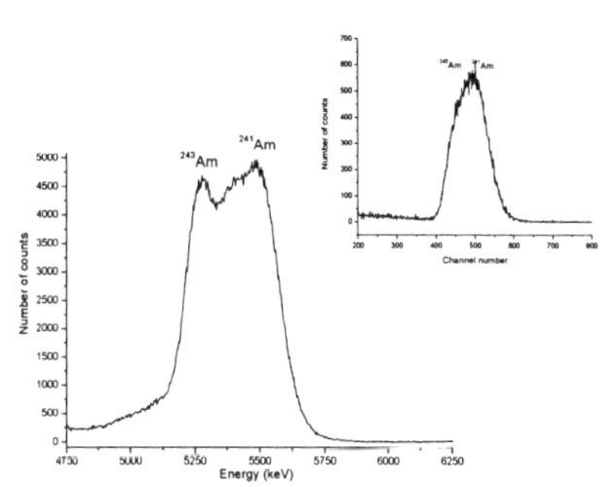

Figure 4 *Am isotopes spectrum by APD at -20°C. Upper right; the same spectrum obtained at 20 °C by Perals™*

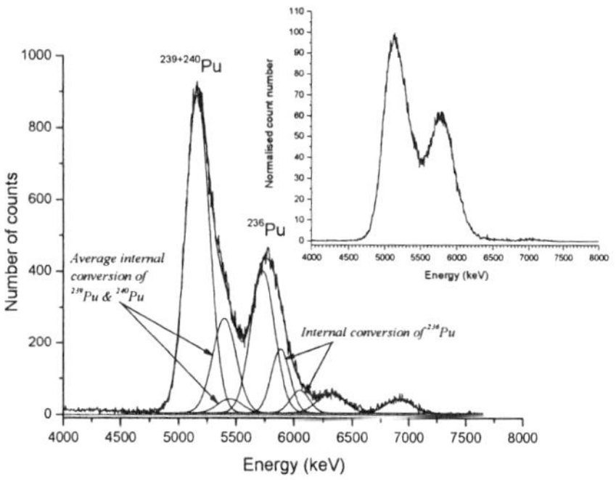

Figure 5 *Deconvolution of plutonium isotope spectrum – acquisition by APD at -40 °C, upper right; the same spectrum obtained by Perals at 20 °C.*

References

1 M. Moszynski, M. Szawlowski, M. Kapusta and M. Balcerzyk, *NIM A,* 2002, **485**, 504.

2 A. Reboli, J. Aupiais and J.C. Mialocq, *NIM A*, submitted.

THE EFFECT OF URANIUM ON THE CORROSION OF STAINLESS STEEL IN HIGH TEMPERATURE MOLTEN SALTS IN THE PRESENCE OF MOISTURE

H. Kinoshita[1*], C.A.Sharrad[1], I. May[1] and R.G. Lewin[2]

[1]Department of Chemistry, The University of Manchester, Oxford Road, Manchester M13 9PL, UK
[2]Nexia Solutions, Sellafield, Seascale, Cumbria CA20 1PG, UK
[*]Current address: Department of Engineering Materials, The University of Sheffield, Mappin Street, Sheffield S1 3JD, UK.

1 INTRODUCTION

Pyrochemical processing is one of the most feasible alternatives to the PUREX process for the safe treatment of spent nuclear fuel. This process is based on the electrochemical separation of uranium and plutonium. Due to the high operating temperature with molten chlorides, the pyrochemical process requires high corrosion resistance materials for its practical application. Various studies have addressed the corrosion of structural materials in molten chlorides focusing on corrosion under an inert atmosphere[1-4] or in the presence of oxygen[5-7]. The present study focuses on the effect of uranium (UCl_3) on the corrosion of stainless steel (316L) in the high temperature molten salt, LiCl-KCl eutectic, a strongly hygroscopic salt mixture.

2 EXPERIMENTAL

Stainless steel 316L coupons (4 mmΦ x 5 mmh) were heated in a LiCl-KCl eutectic (1.5 - 3.0) at 500 °C, which was contained in silica crucibles. Prior to heating, UCl_3 and/or moisture were added under Ar flow. The experimental conditions are summarised in Table 1. The weight loss of the coupons was measured, and the corrosion rate was calculated. Selected samples were further investigated via SEM (Phillips XL30 FEGSEM) to obtain the condition of the surface.

Table 1 *Experimental condition for corrosion tests*

	U^{3+} (Wt%)	Pre-treatment of salt			Corrosion test		
		Temp. (°C)	Treatment	Duration (hours)	Temp. (°C)	Atmosphere	Duration (hours)
Dry 0	0	300	Vacuum	1	500	Ar flow	24, 72, 168
Wet 0	0	R.T.	Exposed to air	1	500	Ar flow	24, 72, 168
Wet 1	0	R.T.	H_2O (40μl)	168	500	Ar flow	168
Wet 2	1.3	R.T.	H_2O (40μl)	168	500	Ar flow	168
Wet 3	2.5	R.T.	H_2O (40μl)	168	500	Ar flow	168
Dry 3	2.5	{ R.T. +300	{ H_2O (40μl) +Vacuum	{ 168 +1	500	Ar flow	168

3 RESULTS

Figure 1 shows back-scattering SEM micrographs of the coupons which were obtained via SEM surface analysis after the corrosion tests. Light grey parts correspond to the stainless steel that had been exposed to the salt from the direction of the dark area. The figure shows clear evidence of pitting corrosion, a type of localised corrosion that is often observed in systems where H_2O and Cl^- co-exist. The depths of the pits were less than 5 μm after 168 hours in the dry condition, but increased to 5~10 μm and more than 10 μm after 24 and 168 hours, respectively, under wet conditions.

Corrosion rates of stainless steel in molten salt without uranium (Dry 0 and Wet 0 in Table 1) are depicted in Figure 2 (a). It is clear that added moisture causes more corrosion, probably due to the conversion of the added moisture to LiOH and HCl. In the system with UCl_3 (Figure 2 (b)), the corrosion rates increased with an increase of U^{3+} concentration, both in dry and moist conditions. Since UCl_3 itself does not react with the stainless steel, because it is very stable in the molten salt, moisture in the salt must be involved in this effect. Thus, Gibbs energy changes in various possible reactions of UCl_3 with LiOH were calculated from the Gibbs energies of the reaction constituents available in the literature.[8] Figure 3 shows the calculated results. The values are normalised with respect to: (a) LiOH or (b) UCl_3, which represent the systems with a limited quantity or excess of LiOH, respectively. When the quantity of moisture, i.e. LiOH, in the system is low, UCl_3 would probably form UOCl. When the quantity of LiOH in the system is high, which is the case for our experiment, UO_2 would be formed. The reaction is thus:

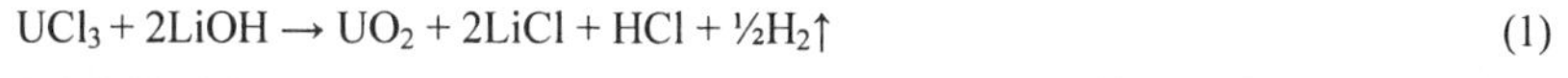

$$UCl_3 + 2LiOH \rightarrow UO_2 + 2LiCl + HCl + \tfrac{1}{2}H_2\uparrow \quad (1)$$

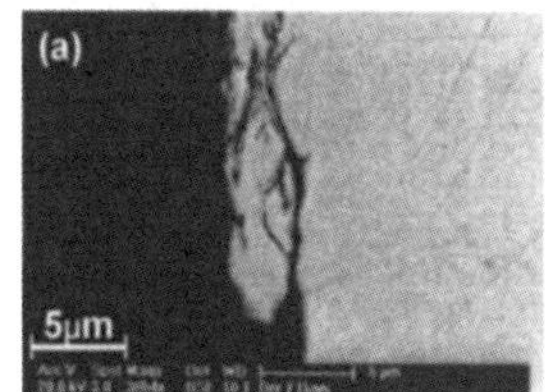

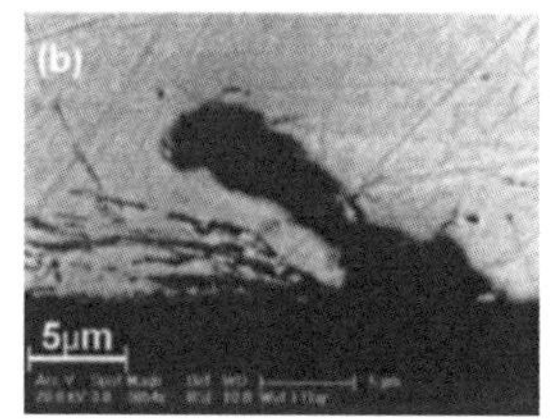

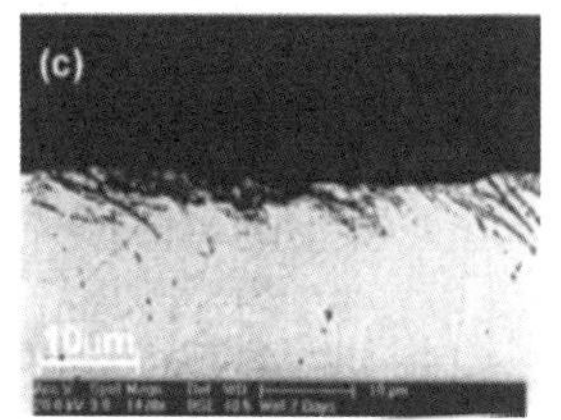

Figure 1 *Back-scattering SEM micrographs of stainless steel 316L after corrosion tests: (a) in dry salt for 168 hours, (b) in moist salt for 24 hours and (c) in moist salt for 168 hours.*

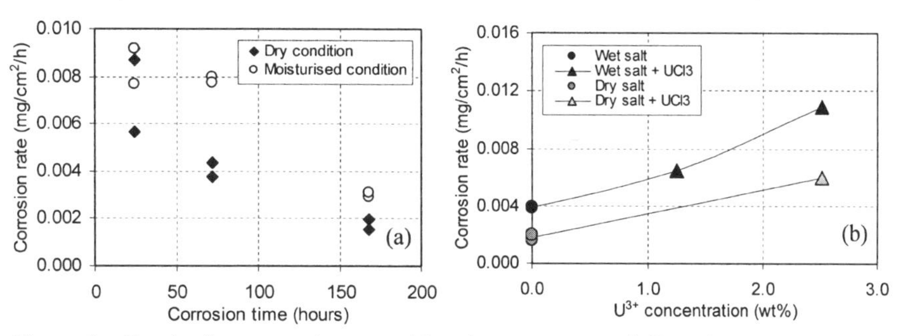

Figure 2 *Results from corrosion tests: (a) without uranium and (b) with uranium.*

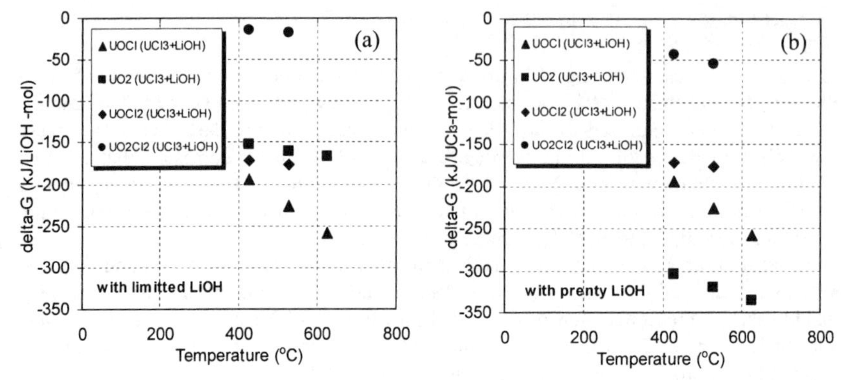

Figure 3 *Gibbs energy change in reactions of UCl_3 with LiOH. Values are normalised with respect to: (a) LiOH or (b) UCl_3, which represent the systems with limited quantity or excess LiOH, respectively.*

One of the reaction products, HCl, could be a reason for the increment in the corrosion rate. The increase of corrosion rate in dry salt with UCl_3 may perhaps be attributed to trace moisture from the added uranium halide or residual moisture after salt drying.

4 SUMMARY

The effect of U on the corrosion behaviour of stainless steel 316L in a molten LiCl-KCl eutectic was studied with added moisture. Moisture in the salt had a significant effect on the corrosion rate of the stainless steel. Surface analysis via SEM showed that the pitting corrosion occurred on the surface of the stainless steel specimen. Corrosion tests with UCl_3 in the system revealed that the presence of uranium in molten salt could increase corrosion of stainless steel. Thermodynamic studies suggested that UCl_3 would form UO_2 with significant quantities of moisture, increasing HCl in the system, thus causing the increase in corrosion rate.

References

1 N.A. Krasil'nikova, I. N. Ozeryanaya and N. D. Shamanova, *Trans. Zashchita Metallov*, 1974, **10**, 446.
2 V.P. Kochergin, O. A. Putina, V. N. Devyatkin and E. T. Kanaeva, *Trans. Zashchita Metallov*, 1975, **11**, 224.
3 Y. Hosoya, T. Terai, T. Yoneoka and S. Tanaka, *J. Nucl. Mater.*, 1997, **248**, 348.
4 N.T. Shardakov, G. N. Shardakova, V. Ya. Kudyakov, N. G. Molchanova, and V. G. Zyryanov, *Protection of Metals*, 1999, **35**, 47.
5 I.N. Ozeryanaya, N. A. Krasil'nikova, M. V. Smirnov, and N. D. Shamanova, *Trans. Zashchita Metallov*, 1978, **14**, 321.
6 X. K. Feng, and C. A. Melendres, *J. Electrochem. Soc.*, 1982, **129**, 1245.
7 Y-J. Park, *J. Kor. Nucl. Soc.*, 2000, **32**, 514.
8 Thermochemical data of pure substances, Vol. I and II, I. Barin et al., VCH, 1995.

REACTIONS OF THE FELDSPAR SURFACE: SURFACE COMPOSITION AND MICROTOPOGRAPHY AFTER CONTACT WITH RADIONUCLIDES AND HEAVY METALS.

E.S. Chardon[1,2], F. R. Livens[1,2] and D. J. Vaughan[1]

[1]Williamson Research Centre for Molecular Environmental Science, School of Earth, Atmospheric and Environmental Sciences, The University of Manchester, Manchester, M13 9PL, UK.
[2]Centre for Radiochemistry Research, School of Chemistry, The University of Manchester, Oxford Road, Manchester, M13 9PL, England.

1 INTRODUCTION

Feldspars are members of the silicate group, and are the most abundant constituents of igneous rocks, representing over 50% of the minerals in the earth's crust. The general formula of their lattice is $(Al_nSi_{4-n}O_8)^{n-}$, showing similarities with the quartz structure, where a silicon atom is in the middle of a tetrahedron formed by four oxygen atoms, except that one in two or four silicon atoms is replaced by aluminium. In order to balance the charge, a counter ion is introduced in the lattice, usually K^+, Na^+ or Ca^{2+}. Reactions involving feldspars influence geochemical cycles in many ways. Because they are the most abundant minerals in the Earth's crust, their behaviour and weathering in the environment has been extensively studied.[1] However, the chemical reactions of their surfaces, especially with radionuclides and heavy metals, are not fully understood. The aim of this project is to examine the chemical and microtopographic changes of the mineral surface after contact with solutions containing very low levels (pM to μM level) of radionuclides (U and Np) and heavy metals (Pb).

The concentrations used are comparable to those likely to be encountered in the natural environment. Lead is known to be a concern for health problems. Levels in soils are estimated to be in the range of 1-200ppm (4.8 μM-1 mM), with a mean value of 16 ppm[2] (77 μM), levels in surface sea water are of the order of 5 to 15 ppt[3] (24-72 pM), and levels in ground waters are reported within the range 1-100 ppb[4] (4.8-480 nM). Levels of uranium in the Earth's crust are 2-3 ppm[5] (8.4-12.6 μM), but can be much higher in the neighbourhood of nuclear facilities; levels in seawater are about 3.3 ppb (14 nM), but can also vary, depending on location.

2 EXPERIMENTAL METHODS

Two different feldspars have been studied: an orthoclase with an elemental composition of $KAlSi_3O_8$ and a plagioclase feldspar with an elemental composition of $(Na_{0.43}Ca_{0.58})$ $Al_{1.54}Si_{2.45}O_8$, both from Madagascar.

Sorption experiments on powder: The minerals were crushed and sieved, and the size fraction below 63 μm was used for the experiment. Liquid scintillation was used to determine uptake of lead and neptunium. In the case of lead, a ^{210}Pb radiotracer was used, mixed with stable isotope ^{208}Pb. Concentrations between 10 ppt and 10 ppm (240 pM – 48 μM) were studied. In the case of neptunium, only ^{237}Np was used and, because of the long half life, a smaller range of concentrations (10 ppb to 10 ppm, 42 nM – 42 mM) was used: ICP-MS was used in the case of uranium studies, where concentrations between 10 ppt and 10 ppm (42 pM – 42 mM) were studied. For all three elements, pH values of 2, 6 and 10 were used.

Sorption experiments on monoliths: To be able to assess the changes in both microtopography and chemistry of the surface, complementary experiments were carried out on freshly cleaved blocks of mineral. The samples were left in contact with a solution containing the required radionuclide or heavy metal for different lengths of time at the appropriate pH. The samples were then washed in DI water and air-dried before being examined by Atomic Force Microscopy (AFM).

3 RESULTS AND DISCUSSION

Sorption experiments on powder: For both minerals, sorption is strongly pH-dependent. At pH 2, there is little sorption of either Pb^{2+} or UO_2^{2} on orthoclase. Sorption of lead increases rapidly at pH 6, and remains high at pH 10, with values above 95% ± 3% sorption. Sorption of uranium on orthoclase shows much lower values, typically in the range 30 to 60% ± 10% , with a maximum at an initial concentration around 10 ppb.
The results obtained for plagioclase are significantly different. Again, sorption of lead is greatest at pH values of 6 and 10, but sorption occurs, even at pH 2, with typical values being around 40% ± 1%. At higher concentrations, saturation of the surface starts to be seen, indicated by the plateau in the isotherm at high concentration. This stronger sorption at pH 2 on plagioclase may reflect the presence of divalent Ca^{2+} in the structure of the mineral rather than the monovalent K^+, which is present in orthoclase. Ca^{2+} may be more readily exchanged for divalent Pb^{2+} in solution.

In the case of uranium, sorption is a maximum at pH 6 and for low concentrations of U, with uptake of around 20% ± 5%. As the initial U concentration increases, sorption decreases to below 10% ± 1%, but suddenly increases to above 60% ± 5% at the highest initial concentrations (10ppm). This implies precipitation on the surface at the highest concentration, confirmed by a change of topography of the surface (Figure 2). Finally, at pH 10, sorption is typically below 20% ± 5%, and no precipitation is observed, presumably reflecting the solubility of uranium at high pH.

Neptunium shows a similar behaviour on both orthoclase and plagioclase at pH 6 and 10. Sorption values around 20% ± 1% are seen at the low initial concentrations, slowly decreasing to around 5% ± 1% at the highest values. No sorption is detected at pH 2. Work is still continuing.

Sorption experiments on monoliths: The use of AFM and SEM has confirmed the formation of a precipitate on the surface, for lead at pH 10 and uranium at pH 6 (Figures 1 and 2).

PHREEQE calculations carried out considering no dissolution of the mineral predicted the precipitation of $Pb(OH)_2$ at pH 10 at the concentration considered, and precipitation of $UO_2(OH)_2$ at pH 6 for an initial U concentration of 100ppm, so it would be likely to see these species precipitating on the surface. However, it cannot be ignored that the mineral surface is reacting, especially at pH10, and thus precipitation of a silicon-containing phase is also possible. Further characterisation of the mineral phase is currently being carried out.

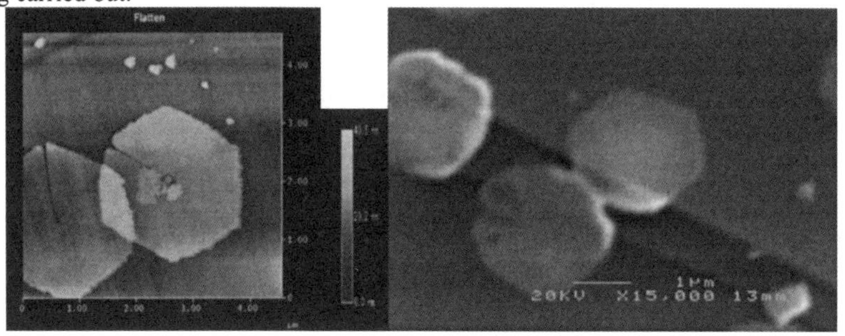

Figure 1 *AFM (left) and SEM (right) images of a plagioclase sample after 24 hours contact with Pb solution. [Pb^{2+}] = 31μM, pH = 10.*

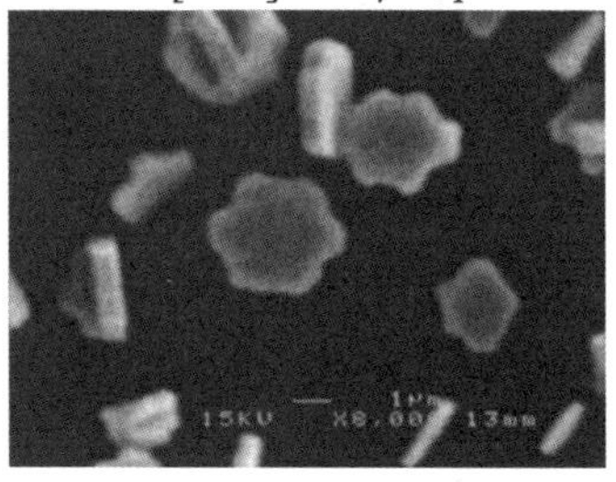

Figure 2 *SEM image of a plagioclase sample after 24 hours contact with U solution. Initial [UO_2^{2+}] = 420 μM, pH = 6.*

4 CONCLUSIONS

Evidence of the sorption of lead and uranium has been found on both plagioclase and orthoclase. Ion exchange and surface complexation mechanisms have been identified and, in some pH/metal concentration regimes, surface precipitation also occurs. Complementary experiments using NpO_2^+ are now being undertaken. Further characterisation of the precipitated phases and thermodynamic modelling are also in progress, as well as X-Ray Photoelectron Spectroscopy of the monoliths.

References

1 A.F. White and S. L. Brantley *Chemical Weathering Rates of Silicate Minerals.* A. F. White and S. L. Brantley (Editors), Mineral. Soc. Am., Washington, D. C., 1995, *Rev. Min.*, **31**.
2 V. M. Goldschmidt *Geochemistry.* A. Muir (Editor), Oxford at the Clarendon Press, London, 1954.
3 J. E. Fergusson *The heavy elements.* J. E. Fergusson (eds), Oxford, Pergamon press, 1990.
4 R. Hermann and P. Neumann-Mahlkau, *Sci. Total Environ.*, 1985, **43**, 1.
5 Plant J. A., Simpson P. R., Smith B. and Windley B. F. *Uranium: Mineralogy, Geochemistry and the Environment*, P. C. Burns and R. Finch (Editors), Mineral. Soc. Am., Washington, D. C., *Rev. Min.*, 1999, **38**, 321-432.

THE INFLUENCE OF OXALATE ON THE PARTITIONING OF TRIVALENT f-ELEMENTS TO HEMATITE

S.E. Pepper[1], L.C. Hull[2], B.N. Bottenus[1] and S.B. Clark[1]

[1]Department of Chemistry, Washington State University, Pullman, WA 99164-4630, USA
[2]Idaho National Laboratory, Idaho Falls, ID 83415, USA

1 INTRODUCTION

Adsorption to mineral surfaces has been identified as an important mechanism in the transport of actinides through the sub-surface. Iron oxides such as goethite (FeOOH) and hematite (Fe_2O_3) are widespread in nature and sorb trivalent actinide cations.[1] Organic ligands can affect the partitioning of trivalent actinides to these mineral surfaces.[2] In prior work, we have reported on the partitioning of trivalent f-elements to goethite in the presence of gluconate, which we described using a surface complexation model that considered the formation of ternary species involving the cation, ligand and surface sites.[3,4] In this study, the sorption of Eu to hematite with and without oxalate present at a single pH as a function of metal cation concentration has been investigated. A two-site Langmuir model was used to describe the experimental data.[5]

2 MATERIALS AND METHODS

Natural hematite was ground and sieved to <200 μm. XRD confirmed the presence of only hematite. Surface area was determined using BET, and found to be 23.2 ± 0.5 m^2 g^{-1}.

Partitioning was determined using a batch approach in a background electrolyte (0.1 M $NaNO_3$). Samples were prepared in triplicate for each concentration investigated. The reaction was initiated by the addition of Eu^{3+}, oxalic acid or a 1:1 molar ratio of Eu^{3+} and oxalic acid, giving a final solid to solution ratio of 5 g L^{-1}. The concentration of Eu^{3+} (and hence oxalic acid) ranged from 1×10^{-7} to 1×10^{-5} M. For those experiments carried out in the presence of oxalic acid, a radiotracer of ^{14}C-labeled oxalic acid was added. The pH of the solution was adjusted to 5.8 using 1.0 M HCl.

The suspensions were placed on a shaker for 24 hours. The supernatant was filtered and analyzed for Eu using ICP-MS, and for oxalate using liquid scintillation counting. Partitioning to the solid was determined by difference from the amount of Eu^{3+} and/or oxalic acid observed in solution.

3 RESULTS AND DISCUSSION

For the Eu-only system, speciation calculations predict that Eu^{3+} is the predominant species in solution (Table 1). In the presence of oxalate, Eu^{3+} and (Eu-oxalate)$^{+}$ are the dominant species in solution In the absence of Eu, oxalate is present in solution as (oxalate)$^{2-}$. Figure 1 shows the removal of Eu and oxalate from solution at pH 5.8. The overall shape of the uptake curves shows that, as the metal loading is increased, so does the amount sorbed to the mineral surface, suggesting a progressive filling of a limited number of sites, with the most energetically favourable being occupied first.[6] For the oxalate-only system all the oxalate was adsorbed by the hematite surface. The presence of oxalate caused a very slight reduction in the total amount of Eu adsorbed by hematite. This is similar to a previous study, which found that oxalate caused a slight reduction in the uptake of Eu at pH values above 4.2.[2] The presence of Eu did not affect the amount of oxalate adsorbed by the hematite surface.

Table 1 *Percent species present in solution under the experimental systems employed in this study. Species present at less than 2% are not shown.*

Expt system	*% species in solution*		
	Eu^{3+}	(oxalate)$^{2-}$	(Eu-oxalate)$^{+}$
Eu-only	96	-	-
Oxalate-only	-	99	-
Eu-oxalate	60	-	37

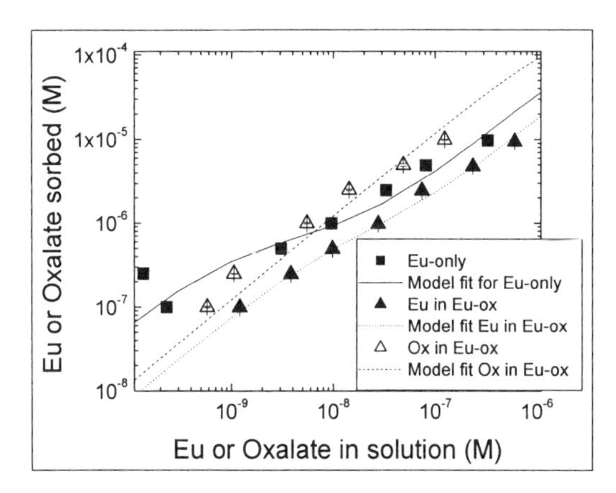

Figure 1 *Amount of Eu or oxalate sorbed for the different experimental systems investigated at pH 5.8. No data is shown for the oxalate-only system as uptake was quantitative. Lines represent Langmuir model fits applied to the data.*

The hematite surface was considered to have a fixed number of sites. The sites are divided into strong sites, which represent a small fraction of the surface, and weak sites, which comprise the rest of the surface. The ratio of weak to strong sites is approximately 40:1.[7] A two-site Langmuir model was used to describe the adsorption to hematite. The parameters are given in Table 2. The model suggests that binding occurs to both the strong and the weak sites as expected. For the Eu in the Eu-oxalate system, we were unable to fit a Langmuir isotherm for the strong site suggesting binding to the weak site only. In the case of oxalate, a Langmuir isotherm was fitted for the weak site only, regardless of the presence or absence of Eu.

Table 2 Langmuir *parameters calculated for the experimental systems investigated.*

Expt system	*pH*	*Langmuir parameter*	*Eu Langmuir parameters*		*Oxalate Langmuir parameters*	
			Strong site	*Weak site*	*Strong site*	*Weak site*
Eu only	5.8	K	1×10^9	1.08×10^5		
		N_{max}	6.26×10^{-7}	3.33×10^{-4}		
Oxalate only	5.6	K			-	No useable data
		N_{max}			-	No useable data
Eu oxalate	5.8	K	-	5.29×10^4	-	3.66×10^5
		N_{max}	6.26×10^{-7}	3.32×10^{-4}	-	3.33×10^{-4}

4 CONCLUSIONS

Speciation calculations indicate the sorption of two species, Eu^{3+} and (Eu-oxalate)$^+$. The overall shape of the uptake curves suggests a progressive filling of a limited number of sites, as the concentration of metal and complex increase. For the Eu-only system, the evidence indicates that both the strong and weak sites on the hematite surface are responsible for sorption of Eu^{3+}. Complete adsorption occurred with only oxalate present, preventing model treatment. For the Eu-oxalate, modelling suggests sorption to the weak site only, involving the two species Eu^{3+} and (Eu-oxalate)$^+$. Future work will involve investigating the effect of pH on the sorption to hematite. These data will be fitted using the diffuse double layer model, which should allow a better description of the adsorption of europium and oxalate to the hematite surface.

Acknowledgements

This work was supported by contracts from the U.S. Department of Energy's Office of Science and the Inland Northwest Research Alliance.

References

1 Th. Rabung, H. Geckeis, J.I. Kim and H.P. Beck, *J. Colloid Interface Sci.,* 1998, **208**, 153.
2 Th. Rabung, H. Geckeis, J.I. Kim and H.P. Beck, *Radiochim. Acta,* 1998**, 82**, 243.
3 S.E. Pepper, B.N. Bottenus, L.C. Hull, C.G. Shepler and S.B. Clark, The proceedings of the Separations for the Nuclear Fuel Cycle in the 21st Century symposium at ACS National Meeting 2004, Anaheim, CA (under review).
4 S.E. Pepper, L.C. Hull, B.N. Bottenus and S.B.Clark, *Radiochim. Acta,* 2005, submitted.
5 A.P. Robertson and O.J. Leckie, *J. Colloid Interface Sci.,* 1997**, 188**, 444.
6 R.H.Parkman, J.M. Charnock, N.D. Bryan, F.R. Livens and D.J. Vaughan, *Am. Mineral.,* 1999**, 84**, 407.
7 D.A. Dzombak and F.M.M. Morel, *Surface Complexation Modeling;* John Wiley and Sons: New York, 1990.

MECHANISMS FOR THE REDUCTION OF ACTINIDE IONS BY *GEOBACTER SULFURREDUCENS*

J.C. Renshaw[1,2,*], J.R. Lloyd[1], F.R. Livens[1,2], L.J.C. Butchins[2], I. May[2] and J.M. Charnock[3]

[1]Williamson Research Centre for Molecular Environmental Science and School of Earth, Atmospheric and Environmental Sciences, The University of Manchester, Oxford Road, Manchester, M13 9PL, UK
[2]Centre for Radiochemistry Research, School of Chemistry, The University of Manchester, Oxford Road, Manchester, M13 9PL, UK
[3]CLRC Daresbury Laboratory, Warrington, WA4 4AD, UK
*current address: School of Geography, Earth & Environmental Sciences, University of Birmingham, Birmingham B15 2TT

1 INTRODUCTION

The bioreduction of U(VI) by anaerobic subsurface microorganisms has been the focus of much recent interest. Both Fe(III)- and sulfate-reducing bacteria have been shown to reduce U(VI) to insoluble U(IV), with *c*-type cytochromes involved in the electron transfer to the actinide.[1-3] However, the mechanism of reduction is not clear, although it has important implications for the potential microbial reduction of transuranic species with environmentally stable lower oxidation states, such as mobile Np(V). Furthermore, there have been few studies of the microbial interactions with other transuranics, such as Pu, which exhibits an extremely complicated redox chemistry that could be controlled by microbial activity.

The model anaerobic bacterium *Geobacter sulfurreducens* is closely related to the predominant bacteria found in many contaminated subsurface environments, and so was used in this study to investigate the microbial reduction of actinide ions. This paper summarises our recent studies in this area.[4]

2 METHODS AND RESULTS

Washed cell suspensions of *G.sulfurreducens* were incubated at 30 °C in $NaHCO_3$ buffer (30 mM, pH 7) with uranyl acetate (5 mM) and acetate (10 mM) as the electron donor, all under N_2. Extended X-ray Absorption Fine Structure (EXAFS) was used to distinguish between U(VI), U(V) and U(IV) in frozen culture and supernatant samples. After 2 hours incubation, there was an indication of minor (<5%) U(V), although it could not be quantified accurately; by 4 hours, 60 ± 10% of the uranium was in oxidation state V, and by 8 hours the majority (80 ± 10%) of the uranium was transformed into $UO_2.2H_2O$. After 24 hours, all the uranium had been reduced. Analysis of the supernatant after 4 hours showed that the U(V) was not associated exclusively with the cells.

The formation of U(V) substantially before the first appearance of U(IV) suggests that biological transformation may proceed via one-electron reduction of U(VI), followed by disproportionation of the unstable U(V) intermediate (2 U(V) → U(VI) + U(IV)) and hydrolysis to generate the final U(IV) product. This mechanism would not require a second enzymatic electron transfer reaction to achieve the reductive precipitation of U(IV) in the periplasm and outside the cells of U(VI)-reducing bacteria.[2,5]

While both U(V) and Pu(V) are unstable with respect to disproportionation, Np(V) is not. If U(V) is reduced *via* disproportionation and is not reduced enzymatically, then it might be expected that *G. sulfurreducens* would not reduce the Np(V) analogue. If enzymatic reduction of U(V) does occur, then thermodynamically, reduction of Np(V) should also be possible by this organism. To test this hypothesis, washed cell suspensions of *G.sulfurreducens* in MOPS buffer (20 mM, pH 7) were challenged with 1 mM $\{Np^VO_2\}^+$ and acetate (10 mM) as the electron donor under an atmosphere of N_2 at 30°C. Control cultures contained live cells with no electron donor, or dead cells (autoclaved), whilst the abiotic control contained no cells. γ ray spectroscopy or α liquid scintillation counting were used to determine Np concentrations in solution. In all cultures, ~13% of Np was removed from solution within the first 24 hours, with little more removed after longer periods (up to 218 hours). This small loss of Np from solution is not dependent on the presence of the electron donor, nor does it require viable cells, and is therefore likely to be caused by biosorption to the cell, rather than the reduction of Np(V) to Np(IV). UV/vis/nIR spectroscopy of the culture supernatants confirmed that only $\{Np^VO_2\}^+$ was present in solution after contact with the cells for 218 hours (as determined by the characteristic 980 nm absorption band). Actively growing cells were also unable to reduce Np(V) added to the growth medium (data not shown). These results support the hypothesis that *G. sulfurreducens* is unable to reduce pentavalent actinides, despite the fact that, thermodynamically, an organism with the physiological potential to reduce Fe(III) and U(VI) should also be able to reduce Np(V) to Np(IV).[6] Therefore it is most likely that the reduction of U(V) to U(IV) proceeds *via* disproportionation.

The selective reduction of U(VI) demonstrated by *G. sulfurreducens* has important implications both for the environmental behaviour of actinides and for potential bioremediation processes. The most stable Np species in the environment is $\{Np^VO_2\}^+$, leading to high environmental mobility. These results indicate that *G. sulfurreducens* alone is unlikely to be effective in reducing and immobilising this form of Np.

Environmental Pu is present in oxidation states V or IV,[7,8] and our results suggest that bioreduction of Pu(V) by *Geobacter* species may not occur. The potential for the bioreduction of Pu(IV) to the more mobile Pu(III) is less clear, particularly since complexation can alter the redox properties substantially. In preliminary experiments with ^{239}Pu, washed cell suspensions of *G.sulfurreducens* in MOPS buffer (50 mM, pH 7) were incubated with 2.5 μM Pu(IV), with and without acetate (10 mM) as the electron donor under an atmosphere of N_2 at 30°C. Abiotic controls contained no cells. Pu concentrations in solution were determined using α liquid scintillation counting. In all cultures, ~ 11% of Pu remained in solution after 24 h, whilst in the abiotic controls, ~ 23 % of Pu was left in solution. These results suggest that *G. sulfurreducens* does not remobilize precipitated Pu(IV).

To conclude, our results with *G. sulfurreducens* suggest that it will be very difficult to predict the impact of microbial reduction on other actinides, including Pu, in the environment, and studies into the effect of microbes on actinides will require careful examination under field conditions.

References

1 R. B. Payne, D. A. Gentry, B. J. Rapp-Giles, L. Casolot and J. D. Wall, *Appl. Environ. Microbiol.*, 2002, **68**, 3129.
2 D. R. Lovley, P. K. Widman, J. C. Woodward and E. J. P. Phillips, *Appl. Environ. Microbiol.*, 1993, **59**, 3572.
3 J. R. Lloyd C. Leang, A. L. Hodges Myerson, M. V. Coppi, S. Cuifo, B. Methe, S. J. Sandler and D. R. Lovley, *Biochem. J.*, 2003, **369**, 153.
4 J. C. Renshaw, L.J.C. Butchins, F. R. Livens, I. May, J. M. Charnock and J. R. Lloyd, *Environ. Sci. Technol.*, 2005, **39**. 5657.
5 J. R. Lloyd, J. Chesnes, S. Glassauer, D. J. Bunker, F. R. Livens and D. R. Lovley, *Geomicrobiol. J.*, 2002, **19**, 103.
6 S. Ahrland, J. O. Liljenzin and J. Rydberg, *The Chemistry of the Actinides*; Pergamon Press: New York, 1975.
7 G. R. Choppin and P. J. Wong, *Aquatic Geochem.*, 1998, **4**, 77.
8 J. W. Morse and G. R. Choppin, *Reviews in Aquatic Science,* 1991, **4**, 1.

OXIDATION AND SURFACE COMPLEXATION OF ACTINYL SPECIES BY OXIDES

S.A. Stout, S.D. Reilly, J.D. Farr, P.C. Litchner and M.P. Neu

Los Alamos National Laboratory, Chemistry Division, Los Alamos, NM 87545, USA

1 INTRODUCTION

Plutonium from the production, processing and disposition of nuclear materials contaminates the environment and poses health and ecological risks. Since interactions occurring at the solid-solution interface can greatly influence the fate and transport of Pu, a thorough understanding of the reactions occurring between Pu species and vadose zone minerals is needed to enable the forecasting of Pu migration rates and pathways. Our research is aimed at characterizing and modelling the reactivity of Pu present in a geochemical system including reactions with key system components, such as birnessite, hematite, and quartz.

Most actinide-mineral adsorption studies have been conducted for U(VI). Much of that information can be used to predict the behaviour of other actinides, but the extrapolation to Pu is limited, because the nature of the contamination is generally different, and the environmental behaviour of Pu is more complicated than that of U.[1, 2] In this work, the adsorption characteristics of Pu(VI) onto δ-MnO_2, Fe_2O_3, and SiO_2 were investigated through the creation of pH dependent adsorption isotherms. A surface complexation model was fitted to the experimental isotherms. These results may improve our ability to predict the fate and transport of Pu in oxidizing environments.

2 MATERIALS AND METHODS

Manganese oxide was synthesized using the method of McKenzie.[3] Hematite and quartz were purchased from Fisher Scientific, and used without further treatment. The BET surface areas and physical properties of the mineral phases are given in Table 1. Comparison of the powder X-ray diffraction patterns with known indexed patterns indicated that the minerals were pure phases. The Pu stock solution was prepared and purified as described in the literature. A Pu(IV) stock was diluted with 2 M HCl, and then oxidized to Pu(VI) by saturating with ozone. The concentration and oxidation state purity of the Pu(VI) solution was determined using a UV-Vis spectrophotometer and the absorbances of diagnostic bands.

The Pu(VI) adsorption edge measurements utilized water-rock ratios of 1, 2, and 3 g/L for δ-MnO_2, SiO_2, and Fe_2O_3, respectively, and an ionic strength (I = 0.10 M) adjusted with NH_4NO_3. The pH of the solutions was adjusted in the range of 2 to 10 using either NH_4OH or HNO_3. To minimize the presence of CO_2, all samples were prepared using CO_2-free solutions, sparged with Ar, and tightly capped. The Pu (VI) stock was added, such that the final concentration was $1x10^{-5}$ M, and the pH was readjusted. After reacting for 24 hr, the solution pH was measured, and an aliquot removed and filtered. The Pu concentration was determined by liquid scintillation counting.

Table 1 *Physical properties of mineral phases.*

Mineral phase	Chemical formula	BET surface area (m^2/g)	Site density (sites/nm^2)	PZC
birnessite	δ-MnO_2	41.1 ± 0.2	0.67^a	2.9
hematite	Fe_2O_3	10.4 ± 0.8	2.3^b	7.2-9.3
quartz	SiO_2	1.10 ± 0.02	4.6 ± 0.15	1.8-3.4

[a] *Value calculated based on fit of adsorption data;* [b] *Value from Lenhart and Honeyman*[4].

The fit to the data was obtained using FLOTRAN combined with the parameter estimation code PEST.[5,6] A non-electrostatic approach was used to represent surface complexation reactions. The input parameters needed for the model included specific surface area, surface site density, and the fluid/rock ratio used in the experiments. The measured aqueous Pu concentrations and pH were then fitted to the measured percent Pu sorbed by adjusting the stoichiometry of the surface complexes and corresponding selectivity coefficients as well as the product of the specific surface area and surface site density. The log K values were calculated at I = 0 using the Debye-Huckel activity coefficient algorithm.

3 RESULTS AND DISCUSSION

3.1 Adsorption Isotherms and Modeling of Adsorption Curves

Adsorption edges and surface complexation fits to the adsorption of Pu(VI) by δ-MnO_2, SiO_2, and Fe_2O_3 are shown in Figure 1. The %Pu removed by δ-MnO_2 is highly pH dependent and increases from near zero at pH = 2 to almost 100% at pH = 4.5. Li *et al.*, 2004 studied the pH dependent adsorption of Zn^{2+} onto δ-MnO_2, and observed that the adsorption edge had a much lower slope and 100% adsorption was not achieved until a pH of 6.5.[7] Unlike δ-MnO_2, SiO_2 showed no Pu adsorption below pH = 4.0, but a sharp increase in %Pu removed from pH 4.0 to 5.0. At higher pH, the presence of dissolved Si may allow precipitation of a Pu silicate similar to soddyite. When studying UO_2^{2+} sorption onto SiO_2, Hongxia and TaoZuri (2002) observed a gradual increase in adsorption from pH 4 to a maximum at pH 6.5.[8] Kowal-Fouchard *et al.* also observed a gradual increase in U adsorption from pH = 2 – 5, attributing the spread of the adsorption edge over 3 pH units to the presence of numerous adsorption sites.[9] A noticeable difference in the sorption edge for hematite was observed compared to that of MnO_2 and SiO_2. The slope of the edge is lower and the %Pu removed increases steadily above pH 4.5, as the solution pH approaches the pH_{PZC}. Lenhart and Honeyman observed a much more classical isotherm shape for UO_2^{2+} adsorption onto

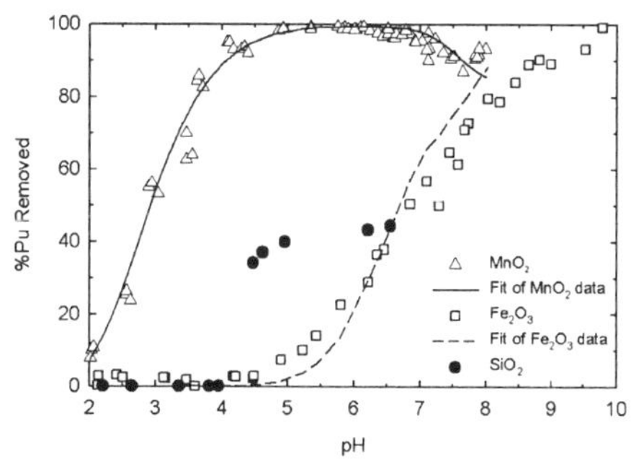

Figure 1 *Isotherms showing pH dependent adsorption Pu of onto birnessite, hematite, and quartz, initially [PuO_2^{2+}] = $1x10^{-5}$ M, following a 24 hr contact time and surface complexation fits.*

hematite, with the edge occurring between pH 3 and 5.[4] However, our adsorption isotherms are not expected to mimic those of U(VI) adsorption, since U(VI) undergoes hydrolysis at lower pH than Pu(VI), and exhibits more complex hydrolysis. The abrupt increase in the sorption edges at pH near 4.0 may also be attributed to Pu(VI) hydrolysis, which begins at approximately pH = 3.5. A comparison of the pH dependent adsorption curves with Pu hydrolysis suggests that sorption is enhanced in the pH range corresponding to the formation of hydrolyzed species.

3.2 Surface Complexation Modeling

Best fits were obtained using the sorption equilibria shown in Table 3 for MnO_2 and Fe_2O_3. Plots of these fits are shown in Figure 1. The ionic strength corrections for the surface acidity constants as well as those for the aqueous complexes were adjusted using the Debye-Huckel activity coefficient algorithm. In the case of MnO_2, the adsorption data could be fitted very well considering only four Pu(VI) surface complexes. The fit of the hematite experimental data used both Pu(IV) and Pu(VI) species.

Table 3 *Surface Reactions and log K values.*

Reactions	log K (I = 0)	
	δ-MnO_2	Fe_2O_3
$\equiv SOH + H^+ \leftrightarrow \equiv SOH_2^+$	−3.0	−7.29
$\equiv SOH - H^+ \leftrightarrow \equiv SO^-$	−5.86	−8.93
Pu(VI) Surface Species		
$\equiv SOH + PuO_2^{2+} \leftrightarrow \equiv SOHPuO_2^{2+}$	1.28	4.79
$\equiv SOH + PuO_2^{2+} - H^+ \leftrightarrow \equiv SOPuO_2^+$	1.58	-
$\equiv SOH + PuO_2^{2+} - 2H^+ + H_2O \leftrightarrow \equiv SOPuO_2OH$	−5.4	-
$\equiv SOH + PuO_2^{2+} - 3H^+ + 2H_2O \leftrightarrow \equiv SOPuO_2(OH)_2^-$	−12.8	−10.66
$\equiv SOH + Pu^{4+} - 2H^+ + H_2O \leftrightarrow \equiv SOPuOH^{2+}$	-	6.93
$\equiv SOH + Pu^{4+} - 3H^+ + 2H_2O \leftrightarrow \equiv SOPu(OH)_2^+$	-	1.29

Acknowledgements
We thank Clay Malcomber and John Rau for performing the BET surface area measurements. This research was supported by the Environmental Management Science Program, OBER-ERSD, Office of Science, U.S. DOE.

References

1 J. M. Cleveland, '*Critical Review of Plutonium Equilibria of Environmental Concern*', ed. E. A. Jenne, American Chemical Society, 1979.
2 G. Choppin, A. Bond, and P. Hromadka, *J. Radioanal. Nucl. Chem.*, 1997, **219**, 20
3 R. M. McKenzie, *Mineral. Mag.*, 1971, **38**, 493.
4 J. J. Lenhart and B. D. Honeyman, *Geochimica et Cosmochima Acta*, 1999, **63**, 2891.
5 P. C. Litchtner, '*FLOTRAN User Manual*', LA-UR-01-2349, Los Alamos National Laboratory, Los Alamos, NM, 2001.
6 J. Doherty, B. Lindsay, and P. Whyte, in '*PEST: Model Independent Parameter Estimation*', Brisbane, Australia, 1994.
7 X. Li, G. Pan, Y. Qin, T. Hu, Z. Wu, and Y. Xie, *J. Colloid Interface Sci.*, 2004, **271**, 35.
8 Z. Hongxia and T. Zuyi, *J. Radioanal. Nucl. Chem.*, 2002, **254**, 103.
9 A. Kowal-Fouchard, R. Drot, E. Simoni and J. J. Ehrhardt, *Environ. Sci. Tech.*, 2004, **38**, 1399.

Coordination and Organometallic Chemistry

STRUCTURE-PROPERTY RELATIONSHIPS IN NOVEL ACTINIDE COMPOUNDS WITH HEAVY OXOANIONS

T. Y. Shvareva,[1] T. A. Sullens,[1] T. C. Shehee,[1] P. M. Almond,[1,2] S. Skanthakumar,[2] L. Soderholm,[2] R. E. Sykora,[1,3] Z. Assefa,[3] R. G. Haire,[3] E. Jobiliong,[4] Y. Oshima,[4] J. S. Brooks,[4] T. E. Albrecht-Schmitt[1]

[1]Department of Chemistry and Biochemistry and E. C. Leach Nuclear Science Center, Auburn University, Auburn, Alabama 36849
[2]Chemistry Division, Argonne National Laboratory, Argonne, Illinois 60439
[3]Transuranium Research Laboratory, Chemical Sciences Division, Oak Ridge National Laboratory, Oak Ridge, Tennessee 37831
[4]Department of Physics/National High Magnetic Field Laboratory, Florida State University, Tallahassee, Florida 32310

1 INTRODUCTION

Over the past five years we developed the chemistry of actinide compounds containing heavy oxoanions of Se(IV), Te(IV), I(V), and Sb(III). While these anions have been selected for a number of different purposes they share a common feature in that they have a stereochemically active lone-pair of electrons that plays a dramatic role in both the structures and physico-chemical properties that these compounds display. In this report we will summarize the structure-property relationships in β-$AgNpO_2(SeO_3)$,[1,2] and $Np(NpO_2)_2(SeO_3)_3$.[3] An important feature of these compounds is the presence of so-called cation-cation interactions (CCI's) whereby an oxo atom of a neptunyl unit is shared between two Np centers, generally in the form of an apical-equatorial bridge.[4]

2 METHODS AND RESULTS

2.1 Structure of β-$AgNpO_2(SeO_3)$

The structure of β-$AgNpO_2(SeO_3)$ consists of neptunyl(V) cations that are linked to one another by both NpO_2^+–NpO_2^+ bonds and by bridging selenite anions creating the pentagonal bipyramidal NpO_7 building unit. There are neptunium oxide layers formed from NpO_2^+–NpO_2^+ cation-cation interactions. A depiction of part of the structure of β-$AgNpO_2(SeO_3)$ viewed down the *a*-axis is shown in Figure 1. As can be seen from this figure, the structure is three-dimensional, and the neptunyl oxide layers are linked together by selenite anions. When viewed down this axis it becomes apparent that there are three types of channels running through the neptunyl selenite lattice. Two of these have oxygen atoms from the selenite anions and the oxo atoms from the neptunyl(V) cations directed into the channels that are used to bind the Ag^+ cations. The third type of channel has the stereochemically active lone-pair of the selenite anions directed into it. Given the strongly nonbonding nature of these lone-pairs, it is not surprising to find these channels vacant of Ag^+ cations.

The NpO_2^+ bonds and angles of 1.867(4) and 1.881(4) Å and 178.0(2)° for Np(1) and 1.876(4) and 1.853(4) Å and 177.4(2)° for Np(2) are consistent with Np(V), and these two Np centers have bond-valence sums of 4.82 and 4.93. The selenite anions show no indication of being protonated as indicated by Se–O bonds of 1.697(4) (× 2) and 1.721(4) Å to Se(1), and 1.673(4), 1.704(4), and 1.712(5) Å to Se(2).

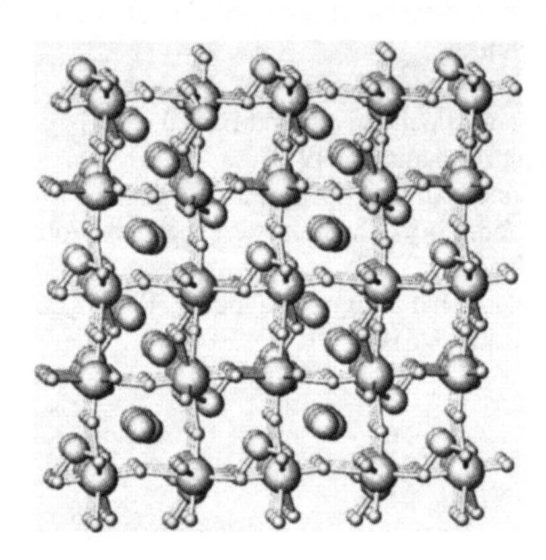

Figure 1 *A view down the a-axis of the three-dimensional structure of β-$AgNpO_2(SeO_3)$.*

2.2 Magnetic and Electronic Properties of β-$AgNpO_2(SeO_3)$

One of the perplexing features of neptunyl(V) selenites is the occurrence of mixed-valent compounds. When NpO_2 is reacted with SeO_2 in the presence of $AgNO_3$ under mild hydrothermal conditions α-$AgNpO_2(SeO_3)$ and β-$AgNpO_2(SeO_3)$ result. Crystals of these two compounds differ in morphology and coloration with the former selenite occurring as clusters of pale green acicular crystals and the latter as dark brown blocks. The pale green coloration of α-$AgNpO_2(SeO_3)$ is expected for a Np(V) compound. The presence of cation-cation interactions in Np(V) does not cause a dramatic color change although there are measurable differences in the UV-vis spectra. The brown coloration of β-$AgNpO_2(SeO_3)$ is unexpected, and has lead us to suspect that there may be partial reduction of some of the Np(V) sites to Np(IV). In order to test this we investigated the electronic properties of this compound using single crystals oriented in specific directions with respect to the magnetic field. This allows us to investigate potential anisotropy in the magnetism. The total magnetization of β-$AgNpO_2(SeO_3)$ revealed magnetic ordering below 7 K, as shown in Figure 2. Here, the field was applied at 40^0 to the [001] plane. The inverse susceptibility ($1/\chi \equiv H/M$) vs. temperature is shown in the inset of Figure 2. Based on the negative intercept, ferromagnetic ordering is indicated with two different paramagnetic states. The total magnetization versus field at 4.2 K was also measured, and indicates a saturated magnetic moment of ~ 0.8 μ_B, without showing any hysteresis. We found no difference in the zero-field cooled and field cooled magnetization, reaffirming the absence of hysteresis.

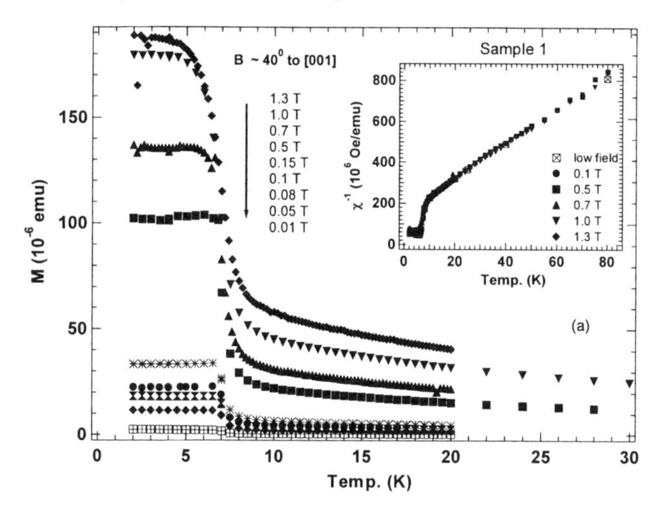

Figure 2 *The total magnetization vs. temperature for different fields of β-$AgNpO_2(SeO_3)$. Inset: Inverse susceptibility vs. temperature.*

We have investigated the electrical transport at room temperature and have found the resistivity of the material to be higher than 10^{10} Ω.cm. This implies there are only localized electronic states, so the moments must be localized. The interaction between these local moments, if sufficiently strong, will allow magnetic ordering below a certain temperature, i.e. the Curie temperature in the present case.

2.3 Structure of $Np(NpO_2)_2(SeO_3)_3$

The structure of $Np(NpO_2)_2(SeO_3)_3$ is remarkable in that it consists of three crystallographically unique Np centers with three different coordination environments in two different oxidation states. Np(1) is found in a neptunyl unit with two short Np=O bonds of 1.861(6) and 1.883(7) Å. The neptunyl unit deviates strongly from linearity with a O(10)-Np(1)-O(11) bond angle of 168.6(3)°. The Np(1) center is further ligated in the equatorial plane by three chelating SeO_3^{2-} anions with Np–O bond distances ranging from 2.409(7) to 2.614(6) Å to create a hexagonal bipyramidal NpO_8 unit. A second neptunyl cation also occurs for Np(2) with two short Np=O bonds of 1.896(7) and 1.901(6) Å. As found for the neptunyl unit containing Np(1), the Np(2) neptunyl cation is also non-linear with a O(12)-Np(2)-O(13) angle of 176.4(3)°. This cation is then bound by four bridging selenite anions and by the O(11) atom from the Np(1) neptunyl cation to form a pentagonal bipyramidal, NpO_7, unit. Therefore, there is a cation-cation, NpO_2^+–NpO_2^+, interaction between Np(1) and Np(2). The Np–O bond distances to the selenite anions range from 2.296(7) to 2.392(6) Å; whereas the Np(2)–O(11) distance of 2.514(7) Å is notably longer.

The final neptunium center, designated as Np(3), has a completely different coordination environment than found for Np(1) or Np(2). Here Np is not found to occur in the form of a neptunyl unit, but rather as a distorted NpO_8 dodecahedron. Np(3) is bound by five bridging selenite anions and by three neptunyl units via cation-cation interactions.

These cation-cation interactions occur with oxo atoms from the neptunyl units containing both Np(1) and Np(2). The Np–O distances to the selenite anions occur from 2.306(6) to 2.396(6) Å. The NpO_2^+–Np(3) distances are in the same range as those that occur with the selenite anions and occur from 2.320(6) to 2.410(7) Å. The Se–O bond distances to the three crystallographically unique selenite anions are all within normal limits and occur from 1.649(7) to 1.752(7) Å. The Np···Np distances range from 3.9951(5) to 4.1758(5) Å.

Using Np(V)–O bond-valence parameters[1] bond-valence sums were calculated for Np(1) and Np(2) to be 4.68 and 4.89, which are reasonably consistent with these sites containing Np(V). The oxidation states of Np(1) and Np(2) can also be inferred from the Np=O bond distances which are notably longer than the Np(VI)=O bonds found in $NpO_2(IO_3)_2(H_2O)$ and $NpO_2(IO_3)_2 \cdot H_2O$, which have an average distance of 1.763(8) Å.[5] Furthermore, the equatorial atoms in Np(VI) polyhedra are typically quite planar, whereas those in Np(V) units have been known to show substantial deviation from planarity, although the database of crystal structures for Np compounds is still small. For Np(2) the equatorial oxygen atoms deviate from planarity by 0.59 Å. The bond-valence sum for Np(3) can not currently be determined because the Np(IV)–O bond-valence parameter is unavailable, but using charge neutral requirements based on two NpO_2^+ cations and three SeO_3^{2-} anions, one arrives at an oxidation state of IV for Np(3), which is consistent with its coordination environment.

The NpO_7 pentagonal bipyramids and NpO_8 hexagonal bipyramids share both corners and edges. The corner-sharing occurs via the cation-cation interaction and the edge-sharing via μ_3-oxo atoms from the selenite anions. Both of these polyhedra share corners via CCI's with the NpO_8 dodecahedra. The cation-cation interactions between the Np(V) and Np(IV) units creates a puckered sheet. Two-dimensional networks formed from CCI's in neptunyl(V) compounds have been previously observed.[4] However, this network is substantially different owing to the presence Np(IV). In fact, this compound provides the first evidence for CCI's between Np(IV) and Np(V). A complete polyhedral depiction of the complex unit cell for $Np(NpO_2)_2(SeO_3)_3$ is shown in Figure 3. It can be noted from this figure that channels are created in the structure that run down the *b* axis. These channels house the stereochemically active lone-pair of electrons on the selenite anions; similar channels also occur in β-$AgNpO_2(SeO_3)$.[1]

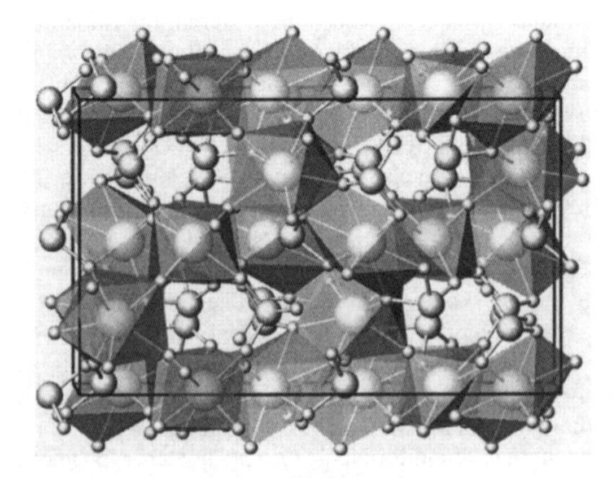

Figure 3 *A complete polyhedral depiction of the complex unit cell for* $Np(NpO_2)_2(SeO_3)_3$.

2.4 Magnetic Properties of $Np(NpO_2)_2(SeO_3)_3$

The magnetization of $Np(NpO_2)_2(SeO_3)_3$, obtained as a function of applied field at 10 K, is linear up to the highest measured field, 2.5 T. The magnetic susceptibility data, measured as a function of temperature under an applied field of 2000 G, decreases smoothly with increasing temperature. The absence of a cusp or discontinuity in the susceptibility is confirmed with low field measurements that show similar behavior and also no significant difference between data obtained from FC and ZFC measurements. Taken together, there is no evidence of either long-range or spin-glass ordering of Np moments at temperatures as low as 5 K. Instead, the Np spins behave as independent, isolated moments. The small value of the Weiss constant is consistent with single-ion magnetic properties with no evidence of magnetic ordering. The small temperature independent susceptibility indicates that the Np moments are localized and not itinerant. The inverse susceptibility data as a function of temperature are linear, showing evidence of neither a significant temperature independent susceptibility nor low-lying crystal field states. The linear fit to these data, which is sensitive to the higher temperature component, results in an effective moment $\mu_{eff} = 2.30(5)\mu_B$ and $\theta = 1.1(20)$ K.

The measured effective moment reflects an average contribution to the susceptibility of three crystallographically inequivalent neptunium atoms, Np(1) and Np(2), both of which are formally pentavalent, and Np(3), which is formally tetravalent. Free-ion effective moments $\mu^{FI}_{eff} = g[J(J+1)]^{1/2}\mu_B$ can be calculated, using Russell-Saunders coupling, for Np(V) of $3.58\mu_B$ and Np(IV) of 3.62 μ_B.[6] Averaging the contributions from the different sites the free-ion effective moment expected for this sample is 3.59 μ_B, which is significantly larger than that measured experimentally. The observation of an effective moment that is reduced from the free-ion value is not surprising because the ground term is split by a large crystal-field component in 5f systems. The overall splitting can be on the order of 600 K.[7] This splitting can significantly reduce the measured susceptibility. For example, in a cubic environment the free-ion term of Np(V) can be split such that the Γ_5 triplet is isolated as the ground state with an effective moment of 2.8 μ_B. The splitting can be sufficient to render the measured susceptibility linear over the temperature range under study here. Similarly, Np(IV) in a cubic crystal field of appropriate symmetry can have a moment ranging from 1.30-3.10 μ_B. Although there is insufficient data to yield detailed electronic information about the Np paramagnetism exhibited by $Np(NpO_2)_2(SeO_3)_3$, it is clear that the observed susceptibility is consistent with known single ion magnetic responses of Np(V) and Np(IV) systems.

There is no evidence of magnetic ordering of the Np moments down to the lowest temperature studied. There have been several published reports of Np(V)[8-10] and Np(IV)[11] compounds in which Np, coupled through oxygen superexchange pathways, order at low temperature. There is no report of ordering in a Np(V)/Np(IV) mixed-valent system. The onset temperature of long-range magnetic ordering is dependent upon the symmetry of the ground-state, the magnitude of the magnetic moment, the metal-metal distance, and the symmetry and dimensionality of the crystal lattice. Unfortunately, little direct evidence is available about the electronic and magnetic properties of the three crytallographically-independent Np to comment on the ground state moment. The two-dimensional Np(IV)-Np(V) network has metal distances and connectivity that are consistent with previous systems that have been reported to show cooperative ordering. The Np-Np distances are relatively short compared to Np compounds that have been previously reported to magnetically order.[10] In contrast, NpOSe, in which the Np(IV) moments order antiferromagnetically below 11 K, has short Np-Np bond distance of only 3.8042 and

3.8820 Å.[7] There is nothing to suggest that ordering in a mixed-valent Np(IV)/Np(V) compound might be more facile than in pure Np(IV) or Np(V) systems. Np(IV) compounds do have the advantage that nonmagnetic ground states are not obtained at low temperatures.

3 CONCLUSION

In this work we have demonstrated that heavy oxoanions can play a dramatic role in the structural chemistry of actinide compounds. In the structures of β-$AgNpO_2(SeO_3)$ and $Np(NpO_2)_2(SeO_3)_3$ we have shown the importance of cation-cation interactions, especially in mixed-valent Np(IV)/Np(V) compounds. The presence of these interactions can allow for magnetic coupling to take place that ultimately can lead to ferromagnetic ordering. These are just a few of the topics that could have covered concerning structure-property relationships in this class of compounds. We have left out key topics such as ion-exchange, selective-oxidation catalysis, and the photophysical properties of middle to late actinide iodates.

4 ACKNOWLEDGMENT

This work was supported by the U.S. Department of Energy, Office of Basic Energy Sciences, Heavy Elements Program (Grant DE-FG02-01ER15187, W-31-109-ENG-38, and DE-AC05-00OR22725), by the National Nuclear Security Administration under the Stewardship Science Academic Alliances program through DOE Research Grant #DE-FG03-03NA00066, and by NSF DMR 0203532.

References

1 T. E. Albrecht-Schmitt, P. M. Almond, and R. E. Sykora, *Inorg. Chem.* 2003, **42**, 3788.
2 E. Jobiliong, Y. Oshima, J. S. Brooks, and T. E. Albrecht-Schmitt, *Solid State Commun.*, 2004, **132**, 337.
3 P. M. Almond, R. E. Sykora, S. Skanthakumar, L. Soderholm, and T. E. Albrecht-Schmitt, *Inorg. Chem.*, 2004, **43**, 958.
4 N. N. Krot and M. S. Grigoriev, *Russ. Chem. Rev.*, 2004, **73**, 89.
5 A. C. Bean, B. L. Scott, T. E. Albrecht-Schmitt, W. Runde, *Inorg. Chem.*, 2003, **42**, 5632.
6 R. M. White, *Quantum Theory of Magnetism*, 2nd Ed. Berlin: Springer-Verlag 1983, p. 282.
7 G. Amoretti, A. Baise, M. Boge, D. Bonnisseau, P. Burlet, J. M. Collard, J. M. Fournier, S. Quezel, and J. Rossat-Mignod, *J. Magn. Magn. Mater.*, 1989, **79**, 207.
8 A. Cousson, S. Dabos, H. Abazli, F. Nectoux, M. Pagès, and G. Choppin, *J. Less-Common Met.*, 1984, **99**, 233.
9 T. Nakamoto, M. Nakada, A. Nakamura, Y. Haga, and Y. Onuki, *Solid State Commun.*, 1999, **109**, 77.
10 a) T. Nakamoto, M. Nakada, and A. Nakamura, *Solid State Commun.* 2001, **119**, 523.
b) T. Nakamoto, M. Nakada, and A. Nakamura, *J. Nucl. Sci. Tech.* 2002, **3**, 102.
11 M. Bickel and B. Kanellakopulos, *J. Solid State Chem.* 1993, **107**, 273.

ON THE ROLE OF *f*-ORBITALS IN THE BONDING IN *f*-ELEMENT COMPLEXES: THE "FEUDAL" MODEL AS APPLIED TO ORGANOACTINIDE AND ACTINIDE AQUO COMPLEXES

Bruce E. Bursten, Erick J. Palmer, and Jason L. Sonnenberg

Department of Chemistry, The Ohio State University, Columbus, Ohio 43210, USA

1 INTRODUCTION

We have been engaged for more than 20 years in the use of quantum chemical methods to study the bonding of complexes of the actinide (An) elements. The necessity to accommodate valence *f* orbitals, to treat the relativistic effects inherent to the heavier elements, and to describe the perforce large number of electrons puts severe demands on the computational aspects of actinide electronic structure; indeed, there is no other portion of the periodic table that leads to the combination of challenges with respect to the calculation of ground- and excited-state energies, bonding descriptions, and molecular properties. But there is also no place in the periodic table in which effective computational modelling of electronic structure can be more useful. The difficulties in handling many of the actinide elements provides an opportunity for computational chemistry to be an unusually important partner in developing the chemistry of these elements.

Our first contribution to theoretical actinide chemistry focussed on comparisons of some analogous transition metal and actinide organometallic complexes, namely Cp_2MX_2 (Cp = η^5-C_5H_5; M = Mo, W; X = Cl, CH_3) and $Cp^*_2UX_2$ (Cp* – η^5-C_5Me_5; X = Cl, CH_3).[1] One of our goals was to see whether an MO description could explain one of the distinctive differences between the d^2 organotransition metal complexes and the f^2 organouranium complexes: the former are diamagnetic,[2] implying that the metal-based electrons are paired, whereas the uranium complexes are paramagnetic with two unpaired electrons.[3] These early calculations used methodology that is crude by today's standards, namely the quasirelativistic Xα scattered-wave (Xα-SW) method, a precursor of the modern density functional theory (DFT) methods that dominate modern electronic structure calculations.

We used the quasirelativistic Xα-SW method to elucidate the bonding in numerous other organoactinide complexes, especially those containing Cp or Cp* ligands.[4] These studies led us to formulate the model presented here concerning the relative roles of the An 5*f* and 6*d* orbitals in the metal-ligand bonding. In this contribution, we will discuss this model and will examine some aspects of it in light of modern relativistic DFT calculations. We will look again at an organoactinide complex to provide an example of the model in practice. In addition, we will examine its suitability to a very different An-ligand system by examining nonaaqua complexes of both early and late trivalent actinide elements.

2 THE "FEUDAL" MODEL

Our early Xα-SW studies of actinide systems were dominated by complexes of Th and U because the bulk of organoactinide chemistry had been carried out on these experimentally accessible elements. For example, the chemistry of $Cp*_2AnX_2$ complexes has been and continues to be extensively developed for Th and U and a variety of anionic monodentate ligands X,[5] but there is a dearth of examples from other parts of the actinide series.

In these early studies we noted a dichotomy in the roles of the U 5*f* and 6*d* orbitals in the bonding and electronic structure of these complexes. The bonding of the ligands was dominated by interactions between the ligand orbitals and the U 6*d* orbitals. That sort of interaction parallels the metal-ligand interactions in transition metal complexes, of course, in which *f* orbitals are not generally relevant in discussions of bonding

Even though the U 5*f* orbitals are minor contributors to metal-ligand bonding, they do have an important role in these U complexes. Because the $Cp*_2UX_2$ systems are formally U(IV) complexes, they have two predominantly metal-localized electrons. Our calculations invariably show that any metal-based electrons in uranium-containing systems are >98% localized in the U 5*f* orbitals, a result that is in accord with photoelectron spectroscopic (PES) studies.[6] Further, because the ligands interact dominantly with the 6*d* orbitals, the 5*f* orbitals are scarcely split by the ligand field. As a consequence, the seven 5*f* orbitals comprise a tight band of metal-localized MOs, which explains the paramagnetism of the U complexes; the 5*f* orbitals are so close in energy that the electrons retain the same spin, and thus gain the associated exchange energy with almost no cost in orbital energy.

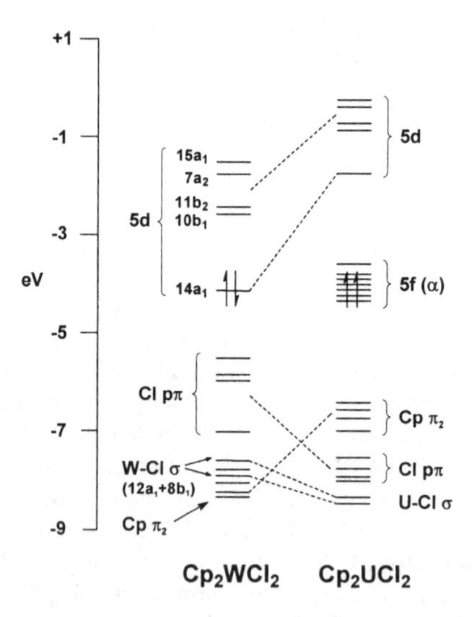

Figure 1 *Kohn-Sham orbital energies from scalar-relativistic DFT calculations on Cp_2WCl_2 and Cp_2UCl_2. The calculations on the U complex are spin-unrestricted, and only the 5f-based orbitals of α spin are shown.*

We have dubbed this description of the different roles of the An 5*f* and 6*d* orbitals as the "*f's essentially unaffected, d's accommodate ligands,*" or "*Feudal*" description of actinide bonding. In essence, the model assumes that the ligand-field splitting of the metal *d* orbitals is similar in analogous transition metal and actinide complexes, but that the An *5f* orbitals reside at lower energy than the *6d* and will therefore hold any metal-based electrons. Modern relativistic DFT calculations support these notions. For example, Figure 1 compares the Kohn-Sham orbital energies from scalar-relativistic DFT calculations on Cp_2WCl_2 and Cp_2UCl_2.[7] These results confirm those of earlier studies that the combination of two Cp and two Cl ligands destabilizes four of the W *5d* orbitals, leaving the $14a_1$ MO at lower energy than the remaining four. The two metal-localized electrons pair in this orbital, consistent with the observation that Cp_2WCl_2 is diamagnetic. In Cp_2UCl_2, the splitting pattern and energetic dispersion of the U *6d*-based MOs is very similar to that in Cp_2WCl_2, suggesting that the metal-ligand interactions are similar between the two complexes. The U *5f*-localized MOs, which are nearly pure *5f* atomic orbitals, form a tight band of orbitals below the U *6d*-based MOs. The two metal-based electrons reside in these U *5f* orbitals. The spin-parallel arrangement of the electrons is energetically preferred over the spin-paired arrangement, consistent with the paramagnetism of the U complex.

The *Feudal* description has worked well in describing the bonding in a variety of actinide complexes, especially those with strongly-bonded, formally anionic ligands (such as those in the organometallics of Th and U). Later in this paper we will examine its applicability to the case of more weakly-bound neutral ligands, such as H_2O molecules.

The low symmetry (C_{2v}) of Cp_2UX_2 systems allows the ligands to interact with either U *5f* or *6d* orbitals; all of the symmetry-adapted ligand combinations have symmetry "matches" among both sets of AOs. In other situations, however, ligand combinations are not able to interact with both *5f* and *6d* orbitals. Such is the case for centrosymmetric molecules because the *d* and *f* orbitals have different parity under inversion. Thus, in centrosymmetric molecules such as uranocene $[(\eta^8\text{-}C_8H_8)_2U]$ and the actinide hexafluorides (AnF_6), the ligand-*5f* and ligand-*6d* interactions necessarily occur in different orbitals. Nevertheless, experimental and computational results provide support for preferential interactions with the An *6d* orbitals. In the case of uranocene, for instance, the PES studies by Green *et al.* indicate *f*-orbital covalency, but to a lesser extent than that for the *d* orbitals.[8] Similarly, theoretical studies of UF_6 invariably show that the F-to-U *6d* donations are more significant than the F-to-U *5f* interactions.[9] We will see another instance in which the ligands are forced to interact with An *5f* orbitals in the next section.

3 RESULTS FROM SOME RECENT CASE STUDIES

3.1 Cp_3ThCl

The coordination of three pentahapto Cp ligands is probably the most common coordination motif in organo-f-element chemistry, with examples of Cp_3Ln (Ln = lanthanide) and Cp_3An complexes that span most of the Ln and many of the An elements.[10] In contrast, it is a very uncommon coordination motif in transition-metal chemistry, the only examples being some selected Cp_3ZrX complexes. We and others have explored a distinctive aspect of the Cp_3 ligand field that is relevant to f-element chemistry, namely the nature of the highest occupied molecular orbital (HOMO) of the $[Cp_3]^{3-}$ ligand set.[11] Under C_{3v} symmetry, the highest-energy combination of the Cp π_2 orbitals is a combination of a_2 symmetry that is ligand-ligand antibonding. By symmetry, there are no metal *s*, *p*, or *d* orbitals that are allowed to interact with this a_2 ligand combination, but one of the *f* orbitals of a central atom indeed is a basis for this representation. Thus, this ligand set is another one that forces interaction between the ligands and *f* orbitals.

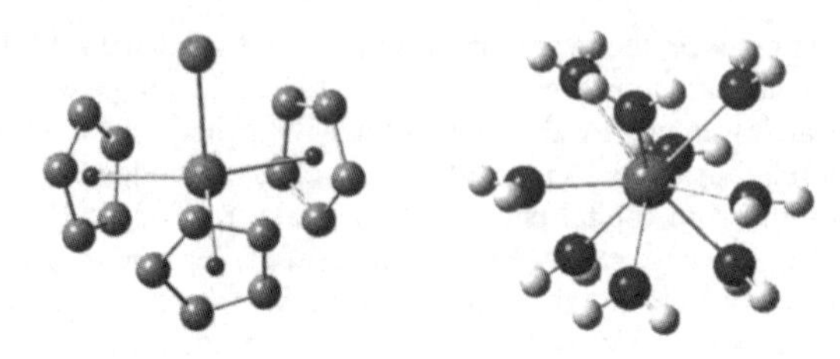

Figure 2 *Calculated structures of Cp_3ThCl (left, H atoms omitted) and D_{3h} $[An(H_2O)_9]^{3+}$ (right)*

We have recently used relativistic DFT to explore aspects of the bonding in d^0 Cp_3ZrCl and its Th analogue d^0f^0 Cp_3ThCl (Figure 2).[12] The metal-ligand bonding in Cp_3ZrCl involves the Zr 4*d*, 5*s*, and 5*p* orbitals, whereas the Th center has 5*f* orbitals available in addition to valence 6*d*, 7*s*, and 7*p* orbitals. In both complexes, the HOMO is an MO of a_2 symmetry derived from the $[Cp_3]^{3-}$ ligand set. Figure 3 presents contour representations of these HOMOs. In the Zr complex, the HOMO is rigorously metal-ligand nonbonding, as demanded by symmetry, whereas there is significant Cp-to-Th 5*f* donation in the HOMO of the Th complex, which represents a Cp-Th bonding interaction.

Figure 3 *Three-dimensional isosurfaces of the a_2 HOMOs of Cp_3ZrCl (left) and Cp_3ThCl (right)*

Table 1 *Kohn-Sham orbital energies and Mulliken percent contributions of the ligand-metal MOs of Cp_3ThCl under C_{3v} symmetry*[a]

MO symmetry and type	*Energy (eV)*	*%Th 5f*	*%Th 6d*
a_2, Cp_3-Th π_2 (HOMO)	-5.46	9.23 [$y(3x^2-y^2)$]	0 (*by symmetry*)
e, Cp_3-Th π_2	-5.83	4.10 (xz^2,yz^2),[xyz, $z(x^2-y^2)$]	2.69 (xz,yz)
a_1, Cp_3-Th π_2	-5.89	0	1.66 (z^2)
e, Cp_3-Th π_2	-6.39	0	9.07 (x^2-y^2,xy)
e, Cl-Th π	-7.09	0	10.62 (xz,yz)
a_1, Cl-Th σ	-7.47	1.93 (z^3)	9.60 (z^2)

[a]The Mulliken percent contributions from the Th 7*s* and 7*p* orbitals are nil in all of these MOs

In order to see whether the *Feudal* description is relevant to Cp_3ThCl, we can examine the composition of the principal ligand-to-metal bonding MOs, namely the Cl-to-Th σ and

π interactions and the Cp-to-Th interactions derived from the π_2 orbitals of the Cp ligands. In addition to the a_2 HOMO, there are lower π_2-based combinations of a_1 + 2e symmetry; these combinations have matches among both the Th 5*f* and 6*d* orbitals and will thus allow us to compare the relative interactions of the ligands with these Th orbitals. Table 1 presents the energies and Mulliken percent characters of the occupied Th-Cl σ and π and Th-Cp π_2-based orbitals of Cp_3ThCl. We see that the ligand-to-metal donation involves the Th 6*d* orbitals much more than the Th 5*f* orbitals, consistent with the *Feudal* description.

3.2 $[An(H_2O)_9]^{3+}$ Ions

The *Feudal* description of actinide bonding was developed largely on the basis of calculations on organometallics of Th and U, i.e. systems that contain "strong" ligands and early actinide elements. In order to check the applicability of the description with weaker ligands and other actinide elements, we examine the results of calculations on a series of nine-coordinate aquo complexes of trivalent actinide ions, $[An(H_2O)_9]^{3+}$ (An = Ac, Cm, Lr). These particular actinide elements were chosen insofar as they correspond to an empty ($5f^0$), a half-filled ($5f^7$), and a completely filled ($5f^{14}$) 5*f* subshell. The calculations were carried out under D_{3h} symmetry (Figure 2), which is actually a saddle-point on the potential-energy surface—the equatorial water ligands prefer to "gear" somewhat leading to a structure of reduced (D_3 or lower) symmetry.[13] Nevertheless, the analysis is somewhat simpler with the higher symmetry, and the qualitative aspects of the conclusions are unaffected by this constraint. We will examine only the $H_2O \rightarrow An^{3+}$ σ bonding; the H_2O-An π bonding is much weaker and can be ignored here.

Table 2 *Mulliken percent characters for selected H_2O-An σ-bonding MOs of $[An(H_2O)_9]^{3+}$ (An = Ac, Cm, Lr)*[a]

MO symmetry[b]	*An %7s*	*An %6d*	*An %5f*	*MO symmetry*[b]	*An %6d*	*An %5f*
Ac^{3+} (f^0)						
a_1' (eq)		2.45	1.61	e' (eq)	4.65	
a_1' (pr)	3.50			e" (pr)	7.68	1.14
Cm^{3+} (f^7, α spin)						
a_1' (eq)		2.85	6.08	e' (eq)	5.23	3.32
a_1' (pr)	7.72			e" (pr)	8.51	3.48
Lr^{3+} (f^{14})						
a_1' (eq)		4.21	4.16	e' (eq)	5.90	3.83
a_1' (pr)	10.51			e" (pr)	7.71	6.45

[a] Only percentages greater than 1% are listed. [b] "eq" refers to orbitals principally involving the three equatorial H_2O ligands; "pr" refers to orbitals principally involving the six prismatic H_2O ligands.

The σ-donor orbitals of the nine H_2O ligands span $2a_1' + 2e' + a_2'' + e''$ representations. The An 6*d* orbitals ($a_1' + e' + e''$) have symmetry matches with all but the a_2'' ligand combination, and the An 5*f* orbitals ($a_1' + a_2' + e' + a_2'' + e''$) have matches with all of these ligand combinations. Some of the ligand-metal overlaps might be expected to be unfavorable in these interactions. In particular, we will consider the three equatorial H_2O ligands, which lie in the horizontal mirror plane of the complex. The σ orbitals of these ligands span $a_1' + e'$. The totally symmetric a_1' combination can interact with the An $6d_{z^2}$,

$5f_{x(x^2-3y^2)}$, and 7*s* AOs. The interaction with the $6d_{z^2}$ orbital would involve overlap with the "torus" of the orbital, which is generally not favorable. Table 2 presents the Mulliken percent contributions to some of the H_2O-An σ bonding orbitals in the three complexes.

Overall, the *Feudal* description seems to be reasonably applicable to these systems. The a_1' combination of the equatorial ligands interacts with both the An $6d_{z^2}$ and $5f_{x(x^2-3y^2)}$ orbitals even though overlap considerations would seem to favor the latter. The a_1' combination of the six prismatic ligands lie very close to the nodal surface of the An $6d_{z^2}$ orbital, and the An 7s is utilized instead. The other two orbitals listed, namely the e' (eq) and e'' (pr) clearly demonstrate the preference for donation into the An 6*d* over the 5*f*.

4 CONCLUSION

We developed the *Feudal* description as a simple qualitative model of the relative roles of the An 5*f* and 6*d* orbitals in the bonding in actinide systems, especially organoactinides. It is gratifying to see that this description is supported by modern DFT methods applied to organoactinides, and apparently even to very different ligand systems, such as the aquo complexes discussed here. We hope this model continues to be useful in descriptions of actinide bonding.

Acknowledgements

We thank the Basic Energy Sciences Program of the Office of Science of the U S. Department of Energy (DE-FG02-01ER15135) and the Petroleum Research Fund (38438-AC3), administered by the American Chemical Society, for support of this research. We are also grateful to the Ohio Supercomputing Center for a generous grant of computer time.

References

1 B. E. Bursten and A. Fang, *J. Am. Chem. Soc.*, 1983, **105**, 6495.
2 R. L. Cooper and M. L. H. Green, *J. Chem. Soc. A*, 1967, 1155.
3 P. J. Fagan, J. M. Manriquez, E. A. Maatta, A. M. Seyam and T. J. Marks, *J. Am. Chem. Soc.*, 1981, **103**, 6650.
4 B. E. Bursten and R. J. Strittmatter, *Angew Chem. Int. Ed.*, 1991, **30**, 1069, and references therein.
5 See, for example: J. A. Pool, B. L. Scott and J. L. Kiplinger, *J. Am. Chem. Soc.*, 2005, **127**, 1338.
6 J. C. Green, *Structure and Bonding*, 1981, **43**, 37.
7 J. Li and B. E. Bursten, unpublished results.
8 J. G. Brennan, J. C. Green and C. M. Redfern, *J. Am. Chem. Soc.*, 1989, **111**, 2373.
9 See, for example: E. R. Batista, R. L. Martin and P. J. Hay, *J. Chem. Phys.*, 2004, **121**, 11104.
10 A. Dormond and D. Barbier-Baudry, *Science of Synthesis*, 2003, **2**, 943.
11 (a) L.T. Reynolds and G. Wilkinson, *J. Inorg. Nucl. Chem.*, 1956, **2**, 246. (b) R. Pappalardo and C. K. Jørgensen, *J. Chem. Phys.*, 1967, **46**, 632. (c) J.W. Lauher and R. Hoffmann, *J. Am. Chem. Soc.*, 1976, **98**, 1729. (d) R. J. Strittmatter and B. E. Bursten, *J. Am. Chem. Soc.*, 1991, **113**, 552.
12 E. J. Palmer, PhD Thesis, The Ohio State University, 2005.
13 (a) J. L. Sonnenberg, PhD Thesis, The Ohio State University, 2005. (b) T. Yang and B. E. Bursten, submitted for publication.

NEW CATALYTIC REACTIONS PROMOTED BY ORGANOACTINIDES

E. Barnea,[1] T. Andrea,[1] J-C Berthet,[2] M. Ephritikhine,[2] and Moris S. Eisen*[1]

[1]Department of Chemistry and Institute of Catalysis Science and Technology Technion-Israel Institute of Technology, Haifa, 32000, Israel.

[2] Service de Chimie Moléculaire, DSM/DRECAM, CNRS URA 331, CEA Saclay, Cedex 91191, France.

1 INTRODUCTION

Tailoring stoichiometric and catalytic reaction to obtain selective and if possible, regiospecific products, is a challenging goal in modern organometallic chemistry. In many cases, mild and easily controlled reactions have been obtained.[1-2] Besides the many successes, controlling the selectivity and regiospecificity of a catalytic oligomerization/polymerization, remains a challenge.[3] For the catalytic oligomerization of alkynes, the means to control the selectivity (extent of oligomerization) of the products has been successfully approached through the addition of amines.[4] We have shown that organoactinide complexes of the type $Cp^*_2AnMe_2$ (An = Th, U; Cp* = C_5Me_5) are effective precatalysts for the oligomerization of terminal alkynes.[5] Bulky alkynes, such as *t*-BuC≡CH was dimerized (eq. 1) producing the head–to–tail dimer, whereas TMSC≡CH, was partially dimerized and mainly trimerized (eq. 2) specifically towards the head–to–tail–to–head trimer, respectively.[5-6]

t-Bu-C≡C—H $\xrightarrow[\text{(An = Th, U)}]{Cp^*_2AnMe_2}$ (*t*-BuC≡C)(*t*-Bu)C=CH₂ + (*t*-BuC≡C)HC=CH(*t*-Bu) (1)

TMS-C≡C—H $\xrightarrow[\text{(An = Th, U)}]{Cp^*_2AnMe_2}$ (TMSC≡C)(TMS)C=CH₂ + (TMSC≡C)(TMS)C=CH–CH=CH(TMS) (2)

For non-bulky terminal alkynes, the oligomerization leads to a mixture of dimers to decamers with no chemo- or regio-selectivity among the different oligomers. The proposed mechanistic scenario for the oligomerization of terminal alkynes is presented in Scheme 1. This mechanism is based on kinetic experiments and trapping intermediate species such as **I** and **II**. The rate determining step, in the catalytic trimerization of TMSC≡CH was found to be the elimination of the trimer from the bis(dieneyne)thorium complex **II**. This result indicates that both rates: i) σ-bond metathesis between the actinide–carbyls and the alkyne, ii) insertion of the triple bond into the metal acetylide moiety (steps 1 and 2 and 3 in Scheme 1), are much faster than the rate for the protonolysis the metal-dialkenyl complex (step 4).

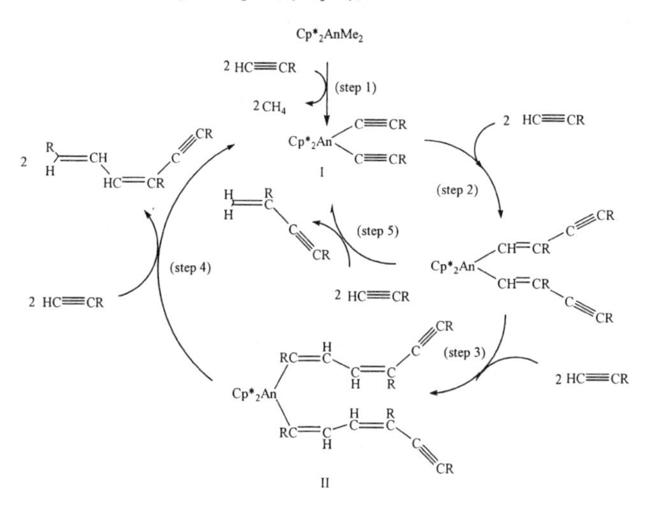

Scheme 1. *Proposed scenario for the oligomerization of terminal alkynes promoted by organoactinide complexes.*

In the past decade, neutral and cationic organoactinide complexes have been studied as catalysts for several organic transformations. Such processes comprise the hydrogenation of olefins, the intermolecular and intramolecular hydroamination of alkynes and the hydrosilylation of terminal alkynes.[7] Isonitriles, are known to undergo a 1,1-insertion into metal-acetylide bond of early or late transition metals, under stoichiometric conditions. The reactivity of the isonitrile molecules is a result of the lone pair of electrons, similar to carbenes. A basic conceptual question regards the use of organoactinides as catalysts for the coupling of terminal alkynes with isonitriles to control the oligomerization process and to induce the selective formation of substituted 1-aza-1,3-enynes, which contain conjugated acetylenic and azomethine fragments $R^1C≡C\text{-}CH=NR^2$ (α,β-acetylenic aldimines).

Herein, we report the coupling reaction of terminal alkynes and *t*-butylisonitrile (^{t}BuNC) catalyzed by the actinide complexes $[(Et_2N)_3U][BPh_4]$ (**1**) and Cp*$_2$AnMe$_2$ [Cp* = C_5Me_5, An = U (**2**), Th (**3**)].

2 RESULTS AND DISCUSSIONS

The three complexes **1**, **2** and **3** catalyzed the coupling of isonitrile and terminal alkynes (Equations 3-5) via 1,1-insertion of the isonitrile terminal carbon atom into a metal-acetylide or a metal-imine bond. The catalytic conversion of the isonitrile and alkyne to 1-aza-1,3-enynes was achieved in toluene or benzene at 90-100°C, while no reaction was observed in the absence of catalyst. The products distribution for the coupling reaction was found to depend strongly on both the catalyst and the alkyne:isonitrile ratio. The cationic catalyst **1** selectively produces the (*E*)-acetylenic imine **4** as the major product (Equation 3), from the mono-insertion reaction of ${}^tBuNC$ into the terminal alkyne.

$$R-C\equiv C-H + C\equiv N{}^tBu \xrightarrow[100^0C]{[(Et_2N)_3U][BPh_4]} R-C\equiv C-CH=N{}^tBu \quad \mathbf{4} \qquad (3)$$

R = TMS, ^{i}propyl, phenyl

$$R-C\equiv C-H + C\equiv N{}^tBu \xrightarrow[90^0C]{Cp^*{}_2UMe_2} R-C\equiv C-CH=N{}^tBu \; (\mathbf{4}) + R-C\equiv C-C(=N{}^tBu)-CH=N{}^tBu \; (\mathbf{5}) \qquad (4)$$

R = TMS, ^{i}propyl, phenyl

$$R-C\equiv C-H + C\equiv N{}^tBu \xrightarrow[90^oC]{Cp^*{}_2ThMe_2} R-C\equiv C-CH=N{}^tBu \; (\mathbf{4}) + R-C\equiv C-C(R)=CH-CH=N{}^tBu \; (\mathbf{6}) \qquad (5)$$

R = TMS, ^{t}butyl, ^{i}propyl R = TMS, ^{t}butyl, ^{i}propyl R = ^{i}propyl

Interestingly, reaction with $Cp^*{}_2UMe_2$ with the isonitrile molecule and alkynes, affords product **5** in addition to compound **4**. The former compound is obtained from the double insertion of two isonitrile molecules into one molecule of the terminal alkyne. The percentage amount of **5** can be successfully raised by increasing the amount of the isonitrile. The reaction between bulky terminal alkynes (R=TMS, ^{t}Bu) and ^{t}BuNC in the presence of the complex $Cp^*{}_2ThMe_2$, produces compound **4** (Eq. 4) as the major product. When the reaction is performed with less bulky terminal alkynes, in addition the formation of **6** was also observed. Product **6** is a result from the coupling of two alkynes and one isonitrile molecules. The variation of the molar ratio of **4** and **6** over time suggests that compound **6** was formed only after the complete formation of **4**, via the reaction of the latter with the remaining alkyne (Eq. 5). The similar behaviour of the three catalysts strongly suggests a common mechanism for the formation of compound **4**, with additional exit branches, forming products **5** and **6**, correspondingly. A plausible mechanism for this coupling reaction is presented in Scheme 2.

The organoactinide complexes **2** and **3** react with the terminal alkynes to yield the bis(acetylide) complex **A** (step 1 in Scheme 2). This complex undergoes a 1,1-insertion of the isonitrile into the metal-carbon bond to form the acetylenic imido complex **B** (step 2). Protonolysis by another terminal alkyne yields the mono-insertion product **4** and regenerates complex **A** as the active species in the catalytic cycle (step 3). For complex **2**, this protonolysis step is not as rapid as for complex **3**, permitting the complex **B** to undergo an additional 1,1-insertion of a second isonitrile molecule to yield the corresponding intermediate **C** (step 4). The double insertion product **5** is then obtained by the protonolysis with a terminal alkyne (step 5) regenerating the active bisacetylide complex **A.** With an excess of non-bulky terminal alkynes, the bisacetylide complex **A** (An = Th) can react with product **4** to yield complex **D**, by insertion of the triple bond of **4** into the Th-acetylide bond (step 6). Protonolysis of **D** by another terminal alkyne yields product **6** and the active species **A** (step 7). This is the first example of an insertion of an internal triple bond into an actinide-carbon bond. For organoactinides we have shown that the insertion of terminal alkynes into a metal acetylide bond produces dimers or higher oligomers, and when the reaction is performed in the presence of terminal and internal alkynes, only the products formed by the activation of the terminal alkyne are produced.[4-5] These results indicates that the insertion of an internal triple bond must be higher in energy in comparison with terminal alkynes. In contrast, the formation of **6** indicates that even in the presence of a terminal alkyne, the insertion of the internal triple bond of **4** was preferred, presumably due to the electronic effects of the imine fragment (^{t}Bu-N=C-), which induces the requested polarization of the internal triple bond similarly to terminal alkynes .

We have already demonstrated that the cationic complex **1** reacts with terminal alkynes to form the acetylide complex (**E**) and Et_2NH through an equilibrium process (Eq. 6).[7c] Complex **E** is an analogue of **A** and catalyzes the coupling reaction like complexes **2** and **3** by steps 1-3 in Scheme 2. The protonolysis step 3 can be performed with a terminal alkyne to yield **4** and **E** or by the free amine to regenerate **1**.

$$(Et_2N)_3U^+ + H{-}{\equiv}{-}R \rightleftharpoons (Et_2N)_2U^+{-}{\equiv}{-}R + Et_2NH \qquad (6)$$

1 **E**

3. CONCLUSIONS

In this contribution we have shown that neutral and cationic organoactinides are efficient catalysts for a number of processes. For terminal alkynes with isonitriles, these complexes promote the coupling reaction impeding the oligomerization of the terminal alkynes and producing selective compounds based on the amount ratio between alkyne:isonitrile. The implementation of organoactinides in demanding chemical transformations is under study.

4 ACKNOWLEDGMENTS

This research was supported by the Israel Science Foundation Administered by the Israel Academy of Science and Humanities under contract 83/01-1.

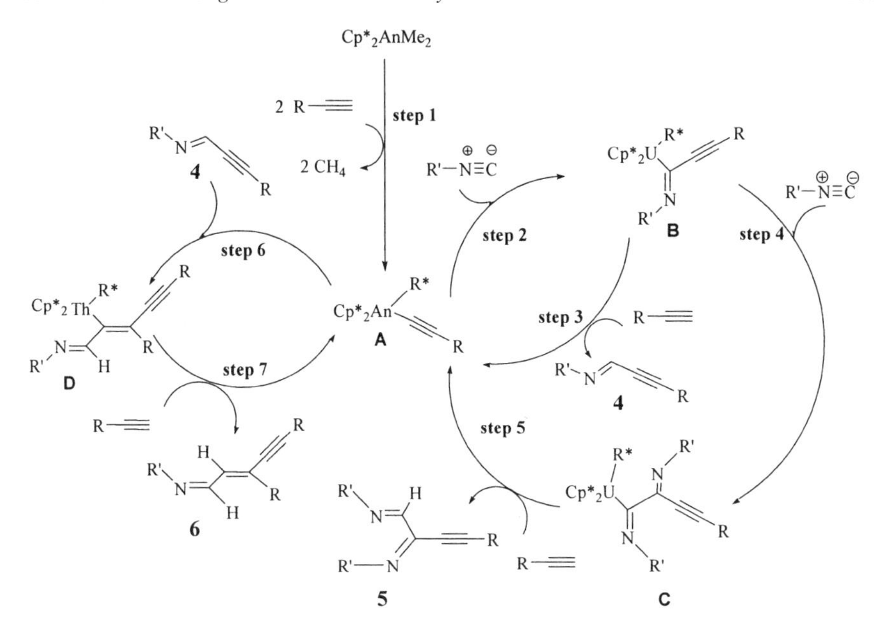

Scheme 2. *Plausible mechanism for the catalytic coupling of ^{t}BuNC and terminal alkynes mediated by $Cp*_2AnMe_2$. For clarity we use R* instead of RC≡C, and R' = ^{t}Bu.*

References

1 G. W. Parshall, S. D. Ittel, *Homogeneous Catalysis*, Wiley, New York, ed. 2, 1992.
2 M. E. Davis, S. L. Suib, *Selectivity in Catalysis*, American Chemical Society, New York, N. Y. 1991.
3 B. A. Arndtsen, R. G. Bergman, T. A. Mobley, T. H. Peterson, *Acc. Chem. Res.* 1995, **28**, 154.
4 A. Haskel, J. Q. Wang, T. Straub, T. Gueta-Neyroud, M. S. Eisen, *J. Am. Chem. Soc.* 1999, **121**, 3025, and references therein.
5 A. Haskel, T. Straub, A. Kumar Dash, M. S. Eisen, *J. Am. Chem. Soc.* 1999, **121**, 3014 and references therein.
6 T. Straub, A. Haskel, M. S. Eisen, *J. Am. Chem. Soc.* 1995, **117**, 6364.
7 For examples of catalytic activity of organoactinides in hydrogenation, see: (a) Z. Lin, T. J. Marks, *J. Am. Chem. Soc.* 1990, **112**, 5515, and references therein. For recent examples of catalytic activity of organoactinides in hydroamination, see: b) D. B. Stubbert, C. L. Stern, T. J. Marks, *Organometallics* 2003, **22**, 836; c) J. Wang, A. K. Dash, M. Kapon, J-C. Berthet, M. Ephritikhine, M. S. Eisen, *Chem. Eur. J.* 2002, **8**, 5384 d) T. Straub, A. Haskel, T. G. Neyroud, M. Kapon, M. Botoshansky, M. S. Eisen, *Organometallics* 2001, **20**, 5017, and references therein.
For examples of catalytic activity of organoactinides in hydrosilylation, see: (f) A. K. Dash, I. Gourevich, J. Q. Wang, J. Wang, M. Kapon, M. S. Eisen, *Organometallics* 2001, **20**, 5084.

OCTACOORDINATE SOFT DONOR COMPLEXES OF TRIVALENT AND TETRAVALENT LANTHANIDE AND ACTINIDE CATIONS

M.P. Jensen[1] and P.M. Almond[1,2]

[1] Chemistry Division, Argonne National Laboratory, 9700 South Cass Avenue, Argonne, IL 60439, USA
[2] Department of Civil Engineering and Geosciences, University of Notre Dame, South Bend, IN 46556, USA

1 INTRODUCTION

The 4*f*, lanthanide (Ln), and 5*f*, actinide (An), cations are considered hard Lewis acids. As such both series of elements preferentially form complexes with hard Lewis bases. However, stable complexes with ligands containing donor atoms softer than oxygen, primarily those containing nitrogen, sulphur, or halide donors other than fluoride, will form under the proper conditions. Although much is known about the thermodynamics, kinetics, and structures of *f*-element complexes with hard, oxygen donor ligands, there is little systematic information about the complexes between *f*-element cations and these so called soft donor ligands, which for our purposes will be defined as ligands that contain only donor atoms softer than oxygen.

Repeated observation has demonstrated that the complexation behaviour of Ln(III) and An(III) cations with oxygen-bearing ligands is indistinguishable when the small differences in cation size are considered. However, the An(III) cations generally from thermodynamically more stable complexes with soft donor ligands than the equivalent Ln(III) cations do. This fact is usually attributed to a slightly greater degree of covalence in the actinide-soft donor bonds as compared to the bonds in the lanthanide-soft donor ligand complexes, but an in depth understanding of the electronic origins of this enhanced covalence suggested by the thermodynamics is lacking. At various times different investigators have proposed that the actinide 7*s*, 6*d*, or 5*f* orbitals may be primarily responsible for covalent bonding in different actinide complexes.[1-5] Complicating attempts to understand this is the well-known regular change in the relative energies and spatial extents of the outermost orbitals of the actinide elements that occurs as the actinide series is traversed.[1]

Even though the origin of the enhanced thermodynamic preference of actinide cations for soft donor ligands is not known in detail, the difference in free energies of complexation can be exploited for the chemical separation of fission product Ln(III) cations from the minor actinides (Am and Cm).[6] Physically, the stronger interaction between actinides and soft donor ligands could be manifested in changes in the structures, coordination numbers, or stoichiometries of the complexes formed. For example, thermodynamically stronger interactions could result in measurably shorter bonds for the actinide complexes, or in complexes with more soft donor atoms coordinated to the actinide than are observed for the equivalent lanthanide complex. The former case has been reported for U(III) complexes with soft donor π-acceptor ligands,[3,7,8] while the latter case is represented by the chloro complexes of Am^{3+} and Pm^{3+}, where the enhanced interaction of soft donors with actinides was first recognized.[9]

To date, sulphur containing dithiophosphinate ligands, $R_2PS_2^-$, have shown the highest selectivity for An(III) cations over Ln(III) cations.[10] Previous work has shown that despite their great selectivity, the metal-sulphur bond distances in these complexes are not significantly different for An(III) and Ln(III) containing species,[11,12] and additional studies suggest that under some conditions there may be a difference in the number and type of coordinated ligands for the actinide complexes *vs.* the lanthanide complexes.[13]

Although they are more basic, and hence less useful for most *f*-element separations, another series of dithio ligands, the dithiocarbamates $R_2NCS_2^-$, provide an ideal opportunity to examine the coordination chemistry of actinide and lanthanide cations with ligands containing sulphur donors. In liquid-liquid extractions, Et_2dtc^- selectively extracts Am(III) over Eu(III) into chloroform with a separation factor of 40 (D_{Am}/D_{Eu}), which is comparable to other soft donor ligands. Moreover, the *N,N*-dialkyldithiocarbamates are easily synthesized and purified, many dithiocarbamate complexes are stable to oxidation and hydrolysis, and both lanthanide and actinide dithiocarbamate complexes have been studied by a number of different workers since the first actinide-dithiocarbamate complexes were identified in 1939.[14-20] In the present work, we compare the molecular structures and the average metal-sulphur bond distances of $Pu(Et_2dtc)_4$ (Et_2dtc^- = *N,N*-diethyldithiocarbamate anion) with those of other tetrakis dithiocarbamate complexes containing Ln(III), An(III), Ln(IV), An(IV) cations in an attempt to understand the origins of the selectivity of soft donor ligands for the actinides.

2 MATERIALS AND METHODS

The starting material for the synthesis of $Pu(Et_2dtc)_4$ was the crystalline solid $Pu(H_2O)_9(CF_3SO_3)_3$, which was prepared by a modification of the method of Matonic et al.[21] A solution containing 50 mg $^{242}Pu(IV)$ (99.97 atom% ^{242}Pu, 0.03 atom% ^{239}Pu) in 8 M HNO_3 was purified by anion exchange chromatography. The resulting solution was fumed to near dryness and the Pu(IV) was converted to the triflate salt by repeated fuming in CF_3SO_3H. The pasty, dark pink solid was dissolved in 4 mL 1.7 M CF_3SO_3H, yielding an orange-red solution containing Pu(IV) as the sole optically detectable oxidation state. The Pu was reduced to Pu(III) at a Pt electrode under an Ar atmosphere with a working voltage of +0 V vs. Ag/AgCl. The solution was concentrated to ca. 500 μL under a stream of Ar and left to stand under a normal laboratory atmosphere. Blue crystals of $Pu(H_2O)_9(CF_3SO_3)_3$, typically 3-5 mm long, formed over several days.

The crystals were isolated, and a portion of the crystals that contained 12 mg Pu was dissolved in 1 mL acetonitrile. Optical spectroscopy (300-1300 nm) indicated only Pu(III) was present in the acetonitrile solution, and the [Pu] was measured by liquid scintillation counting. The acetonitrile solution was purged with Ar for 2 hours and an excess of 2,2'-bipyridyl (Aldrich 99+%) was added to buffer the system. An acetonitrile solution of Et_2dtc^- as the diethylammonium salt (Aldrich, 98%) was added dropwise with stirring. The resulting solution immediately turned dark brown. A stream of Ar over the solution was continued and the solution was left to stand. Within 30 min. the volume of the brown solution had decreased to ca. 500 μL and very dark brown crystals had formed. The crystals were isolated and quickly washed with small volumes of acetonitrile. They were air stable for several days.

A single crystal of $Pu(Et_2dtc)_4$, 0.1 mm x 0.1 mm x 0.02 mm, was mounted on a thin glass fibre, cooled to –100 °C with a Bruker KRYO-FLEX, and optically aligned on a Bruker APEX II CCD X-ray diffractometer using a digital camera. Intensity data were measured using graphite monochromated Mo Kα radiation from a sealed tube and monocapillary collimator. APEX II (v 1.0-22, Bruker AXS) software was used for preliminary determination of the cell constants and data collection control. The determination of integral intensities, global refinement, and Lorentz polarization corrections were performed using Saint+ (v 7.09, Bruker AXS). A semiempirical absorption correction was subsequently applied using SADABS.[22] SHELXTL (v 6.14) was used for space group determination (XPREP), direct methods structure solution (XS), and least-squares refinement (XL).[23,24] The final refinements included anisotropic displacement parameters for all non-hydrogen atoms, and gave an R factor of 5.31% (R_1).

3 RESULTS AND DISCUSSION

Tetrakis *N,N*-diethyldithiocarbamate complexes with An(IV) (An = Th, U, Np, and Pu) and with Ce(IV) have been previously synthesized and studied.[15,17,19] The crystal and molecular structures of $Ce(Et_2dtc)_4$ and $Th(Et_2dtc)_4$ have been determined.[17,19] For all five of these compounds, crystallisation from benzene or toluene produced isomorphous $M(Et_2dtc)_4$ complexes assigned to the monoclinic space group *C2/c*. In our work we found that $Pu(Et_2dtc)_4$ crystallizes from acetonitrile solution in the orthorhombic space group $Pna2_1$ with a = 19.775(1) Å, b = 9.4169(7) Å, c = 3.819(2) Å. Two similar, but crystallographically unique, molecules of $Pu(Et_2dtc)_4$ are present in this crystal.

Despite this difference in the symmetry of the crystal, the general coordination environment about Pu (Figure 1) is very similar to that of the previously reported tetrakis An(IV)/Ln(IV)-Et_2dtc^- complexes and the anionic, tetrakis An(III)/Ln(III) dithiocarbamate complexes, $M(R_2dtc)_4^-$ (M = La, Pr, Nd, Sm, Eu, or Np; R = Me, Et, or R_2 = $-(CH_2)_4-$). $Pu(Et_2dtc)_4$ displays a metal coordination environment intermediate between the dodecahedral and square antiprism ideals, which is best described by two mutually perpendicular trapezoids, each composed of four sulphur atoms from two approximately co-planar Et_2dtc^- groups. The S-Pu-S and S-C-S angles within each Et_2dtc^- group vary little at 63.2 ± 0.9° and 116 ± 1°, respectively. The S-Pu-S angles formed between the co-planar Et_2dtc^- groups (i.e., the angle between S atoms of the coplanar Et_2dtc^- ligands) fall into two distinct groups, one of 164 ± 2° and one of 68.0 ± 0.2°, reflecting the obvious asymmetry in the ligand arrangement within each plane. The C-S, C-N, and C-C bond distances are all unremarkable. These observations also are true for all of the other octacoordinate *f*-ion-dithiocarbamate complexes with reported crystal structures, regardless of identity of the alkyl groups attached to the nitrogen, the metal ion oxidation state, the identity of the counter-cation (if present), or the symmetry of the crystal lattice, small differences in the angles notwithstanding.

The Pu-S bond lengths in $Pu(Et_2dtc)_4$ range from 2.749 to 2.819 Å (±0.004 Å), with an average Pu-S distance of 2.783 ± 0.011 Å. There is no significant difference in the Pu-ligand distances between the two Pu sites. This average Pu-S distance is in good agreement (±0.02 Å) with the average Ln-S distance of all structurally characterized tetrakis Ln-dithiocarbamates and the average Th-S distance in $Th(Et_2dtc)_4$ when the differences in ionic radii[25] of the various cations are considered. This correlation is summarized in Figure 2. Structurally, $Pu(Et_2dtc)_4$ appears to be a completely typical *f*-element dithiocarbamate complex. Therefore, there are no significant structural indications of a stronger interaction between Pu(IV) and Et_2dtc^- relative to the $Ln(Et_2dtc)_4^{0/-}$ complexes.

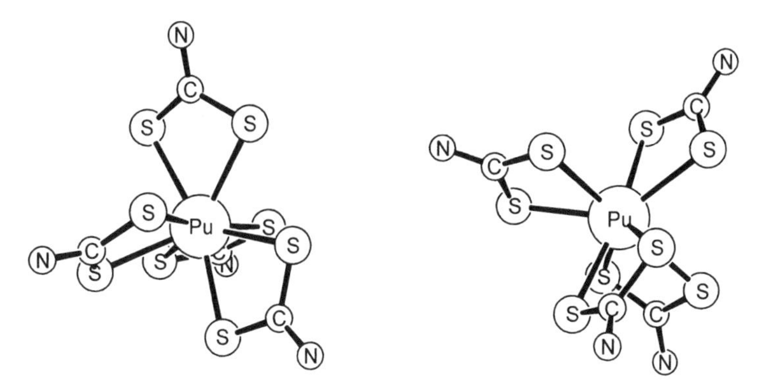

Figure 1 *Molecular structures of the two crystallographically independent $Pu(Et_2dtc)_4$ molecules as oriented relative to each other in the unit cell. The two ethyl groups attached to each nitrogen are omitted for clarity.*

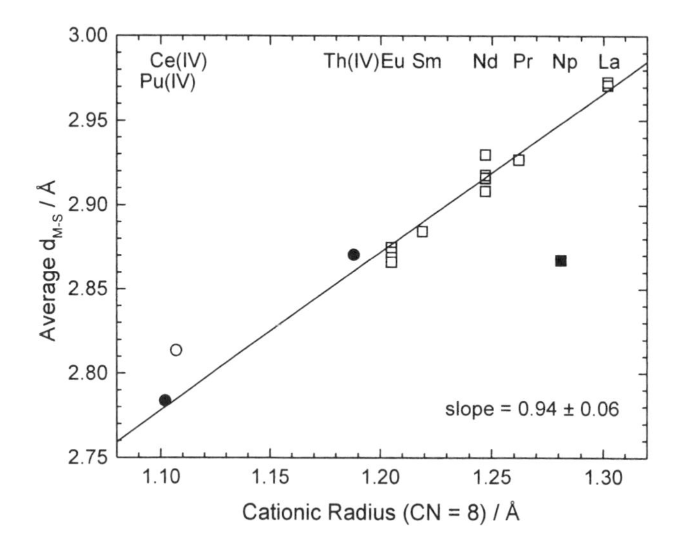

Figure 2 *Comparison of average metal-sulphur bond distances for tetrakis dithiocarbamate complexes of trivalent (squares) and tetravalent (circles) lanthanide (open symbols) and actinide (solid symbols) cations. Crystallographic radii taken from ref. 25. Data for average bond lengths are from the present work (Pu) and the Cambridge Structural Database.* [26]

One octacoordinate $An(Et_2dtc)_4^-$ complex does deviate significantly from the correlation in Figure 2, $Np(Et_2dtc)_4^-$,[20] which is represented by the closed square. The average Np-S distance in $[Et_4N^+][Np(Et_2dtc)_4^-]$ is at least 0.08 Å shorter than expected from all other reported tetrakis dithiocarbamate complexes of the trivalent and tetravalent actinides and lanthanides. Interestingly, this Np(III) complex is both isoleptic and formally isoelectronic with $Pu(Et_2dtc)_4$, which makes the reported Np results somewhat surprising. While this short average Np-S distance is consistent with the reported unit cell volumes for $[Et_4N^+][Np(Et_2dtc)_4^-]$ and $[Et_4N^+][Eu(Et_2dtc)_4^-]$, the apparatus used and amount of data collected were not ideal,[20] making additional study of Np-Et_2dtc^- and the corresponding Pu(III) complex of great interest.

4 CONCLUSIONS

Although the ligand *N,N*-diethyldithiocarbamate shows a thermodynamic preference for actinide cations over similar lanthanide cations, the crystal and molecular structure of a 1:4 Pu(IV):Et_2dtc^- complex reveals no significant structural manifestations that could be correlated to the preference of dithiocarbamate ligands for actinide cations over lanthanide cations.

Acknowledgements
This work was supported by the U. S. Department of Energy, Office of Basic Energy Science, Division of Chemical Sciences, Geosciences and Biosciences through contract number W-31-109-ENG-38.

References

1 R. J. Strittmatter and B. E. Bursten, *J. Am. Chem. Soc.*, 1991, **113**, 552.
2 G. R. Choppin, *J. Alloys Compd.*, 2002, **344**, 55.
3 M. Mazzanti, R. Wietzke, J. Pécaut, J.-M. Latour, P. Maldivi, and M. Remy, *Inorg. Chem.*, 2002, **41**, 2389.
4 M. Miguirditchian, D. Guillaneux, D. Guillaumont, P. Moisy, C. Madic, M. P. Jensen, and K. L. Nash, *Inorg. Chem.*, 2005, **44**, 1404.
5 D. Guillaumont, *J. Phys. Chem. A*, 2004, **108**, 6893.
6 K. L. Nash, *Solvent Extr. Ion Exch.*, 1993, **11**, 729.
7 J. G. Brennan, S. D. Stults, R. A. Andersen, and A. Zalkin, *Inorg. Chim. Acta*, 1987, **139**, 201.
8 J.-C. Berthet, C. Rivière, Y. Miquel, M. Nierlich, C. Madic, and M. Ephritikhine, *Eur. J. Inorg. Chem.*, 2002, 1439.
9 R. M. Diamond, K. Street, and G. T. Seaborg, *J. Am. Chem. Soc.*, 1954, **76**, 1461.
10 Y. Zhu, J. Chen, and R. Jiao, *Solvent Extr. Ion Exch.*, 1996, **14**, 61.
11 M. P. Jensen and A. H. Bond, *J. Am. Chem. Soc.*, 2002, **124**, 9870.
12 G. Tian, Y. Zhu, J. Xu, T. Hu, and Y. Xie, *J. Alloys Compd.*, 2002, **334**, 86.
13 G. Tian, T. Kimura, Z. Yoshida, Y. Zhu, and L. Rao, *Radiochim. Acta*, 2004, **92**, 495.
14 L. Malatesta, *Gazz. Chim. Ital.*, 1939, **69**, 408.
15 K. W. Bagnall, D. Brown, and D. G. Holah, *J. Chem. Soc. A*, 1968, 1149.
16 J. P. Bibler and D. G. Karraker, *Inorg. Chem.*, 1968, **7**, 982.
17 D. Brown, D. G. Holah, and C. E. F. Rickard, *J. Chem. Soc. A*, 1970, 423.
18. M. Ciampolini, N. Nardi, P. Colamarino, and P. Orioli, *J. Chem. Soc., Dalton Trans.*, 1977, 379.
19 P. B. Hitchcock, A. G. Hulkes, M. F. Lappert, and Z. Li, *Dalton Trans.*, 2004, 129.
20 D. Brown, D. G. Holah, and C. E. F. Rickard, *J. Chem. Soc. A*, 1970, 786.

21 J. H. Matonic, B. L. Scott, and M. P. Neu, *Inorg. Chem.*, 2001, **40**, 2638.
22. G. M. Sheldrick, 'SADABS', Univeristy of Göttingen, Göttingen, 2004.
23. G. M. Sheldrick, 'SHELXL97 and SHELXS97', Univeristy of Göttingen, Göttingen, 1997.
24. G. M. Sheldrick, 'XP and SHELXTL', Bruker AXS, Inc., Madison, Wisconsin, USA, 1997.
25. F. David, *J. Less-Common Met.*, 1986, **121**, 27.
26. F. H. Allen, *Acta Cryst.*, 2002, **B58**, 380.

STRUCTURE AND STABILITY OF PEROXO COMPLEXES OF URANIUM AND PLUTONIUM IN CARBONATE SOLUTIONS

Wolfgang Runde[1], Lia F. Brodnax[1], Shane M. Peper[1], Brian L. Scott[1], Gordon Jarvinen[2]

[1] Chemistry Division, Los Alamos National Laboratory, P.O. Box 1663, MS J514, Los Alamos, New Mexico 87545, USA
[2] Nuclear Materials Technology Division, Los Alamos National Laboratory, P.O. Box 1663, MS E517, Los Alamos, New Mexico 87545, USA

1 INTRODUCTION

Actinide behaviour in aqueous carbonate media is important for both environmental and industrial applications. The uranium(VI) carbonate system has been extensively studied with both solid-state and solution speciation, thermodynamics, and coordination chemistry data compiled.[1, 2] Significantly less information is available on the chemistry of plutonium in carbonate solutions. Although the solubility behaviour of plutonium(IV) in carbonate has also been reviewed,[3] information on solution and solid state complexes is limited with the coordination chemistry restricted to that of the $Pu(CO_3)_5^{6-}$ anion.[4] For decades peroxide has been used to adjust the redox state and precipitate plutonium from acidic solutions.[5] However, it is unclear to what extent peroxide complexation or precipitation of plutonium also proceed at higher pH in the presence of competing complexing ligands, i.e. carbonate or hydroxide. Our recent studies have shown intriguing results with the actinides forming mixed-ligand peroxo-carbonato complexes of extraordinary stability.

2 RESULTS AND DISCUSSION

It is well known that uranium(VI) in aqueous carbonate will form the very stable uranyl triscarbonato anionic complex, $UO_2(CO_3)_3^{4-}$.[2] The addition of minute amounts of hydrogen peroxide to a $UO_2(CO_3)_3^{4-}$ solution, even with excess carbonate present, results in the formation of mixed-ligand complexes in both the solution and solid-state. These compounds contain U(VI) in the general formula $[UO_2(O_2)_x(CO_3)_y]^{2-2x-2y}$ (x=1-3; y=3-x). Likewise, there is evidence that Pu(IV) forms similar mixed peroxo-carbonato complexes.

The speciation changes resulting from the addition of hydrogen peroxide to U(VI) triscarbonate are clearly evident in Figure 1a. An isosbestic point at ~312 nm indicates that all solutions with less than 0.93 mM H_2O_2 at 1mM U(VI) contain both $UO_2(CO_3)_3^{4-}$ and a second species that we characterized as the U(VI) monoperoxo-biscarbonato anion, $UO_2(O_2)(CO_3)_2^{4-}$. At higher peroxide/U(VI) concentration ratios, the spectra deviate from the 312 nm isosbestic point, indicating the presence of a subsequent species in solution, i.e. the uranyl bisperoxo-monocarbonate complex,$[UO_2(O_2)_2(CO_3)]^{4-}$. Under the conditions studied, we saw no indication of polymeric uranium species formation.

Using a peroxide-specific amperometric detection method, we experimentally measured the extremely low concentration of free peroxide present in solution. The experimental data points in Figure 1b correspond directly to each spectrum in Figure 1a. At a H_2O_2:U(VI) concentration ratio of less than 1:1, peroxide is quickly and nearly quantitatively complexed by U(VI), while at higher concentration ratios the amount of free peroxide in solution increases and complexation is reduced significantly. The increasing amount of free peroxide suggests a lower stability of the higher U(VI) peroxo complex.

The U(VI)-peroxide-carbonate system also shows an intriguing kinetic feature. While the addition of excess of peroxide forces the formation of higher U(VI) peroxo complexes, these degrade to the $UO_2(O_2)(CO_3)_2^{4-}$ complex over the course of several days. In fact, the $UO_2(O_2)(CO_3)_2^{4-}$ anion remains stable in solution for many months despite the degradation of free peroxide in solution. Obviously, this complex is the thermodynamically favoured product that is more stable than the uranyl triscarbonate solution species.

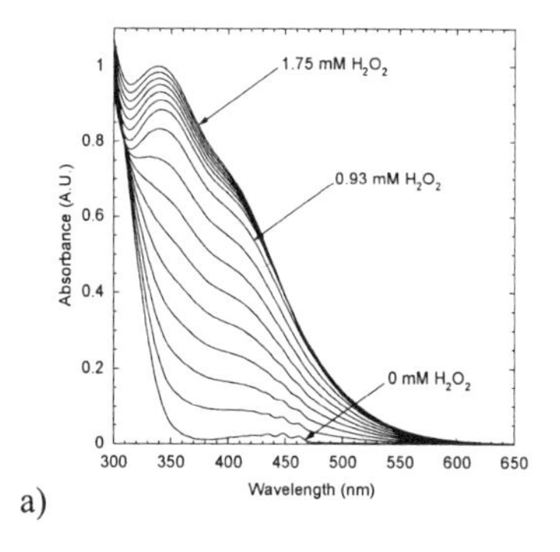

a)

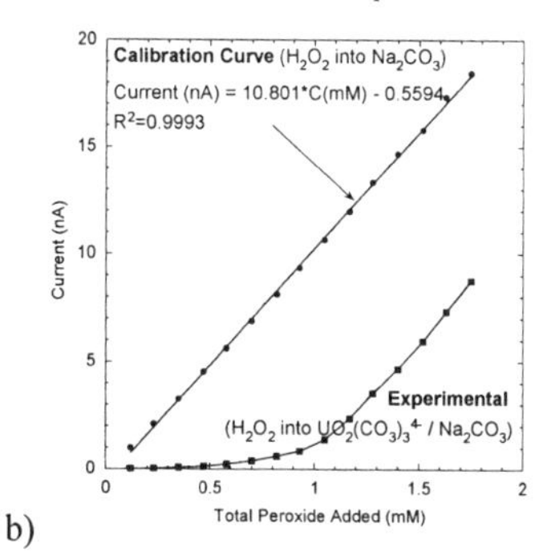

b)

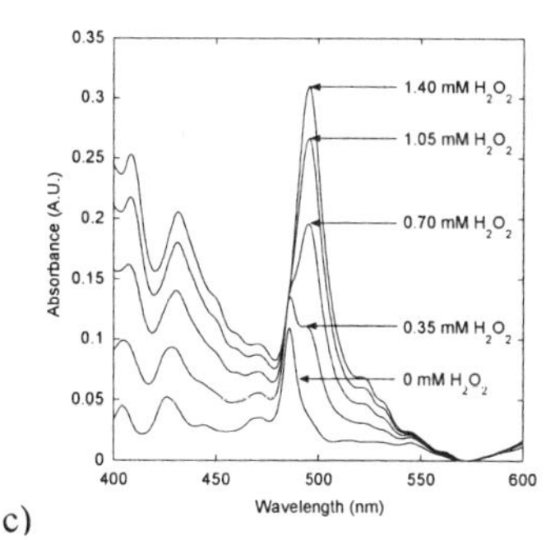

c)

Figure 1a *UV/vis absorbance spectra of 1mM uranyl triscarbonate in 0.5M sodium carbonate with the addition of H_2O_2. The spectra represent the following concentrations (mM) of total H_2O_2 added: 0, 0.12, 0.23, 0.35, 0.47, 0.58, 0.70 ,0.82, 0.93, 1.05, 1.17, 1.28, 1.40, 1.52, 1.63, 1.75.*
Figure 1b *Amperometric measurement of free H_2O_2 in solution. The calibration curve represents the measured peroxide concentration in 0.5M Na_2CO_3 (no uranium). The experimental curve shows individual peroxide measurements from a titration of 1mM $UO_2(CO_3)_3^{4-}$ with H_2O_2.*

Figure 1c *UV/vis absorbance spectra of 1.4mM plutonium(IV) in 2M sodium carbonate with the addition of hydrogen peroxide*

In contrast, plutonium(VI) is reduced to Pu(IV) in carbonate-peroxide solutions. The UV/vis spectrum of ~1mM plutonium (IV) is identical in the carbonate concentration range 0.5 M – 2 M indicating that the same Pu(IV) solution species is present. The limiting Pu(IV) complex in pure carbonate solutions, $Pu(CO_3)_5^{6-}$, is characterized by its characteristic absorbance at 485 nm. With the addition of small increments of hydrogen peroxide a second absorbance peak appears at 495 nm (Figure 1c) indicating a new solution species. An isosbestic point at about 480 nm indicates the presence of two Pu(IV) species in solution, the Pu(IV) pentacarbonate and a mixed Pu(IV) carbonate-peroxide complex. The composition of this complex is unknown but full characterization is ongoing.

Spectrophotometric and amperometric investigations at varying U(VI), carbonate, and peroxide concentrations indicate that a monomeric solution species with a 1:1 ratio of U(VI):H_2O_2 is formed. We crystallized this anionic complex and structurally characterized the solid phase as $K_4[UO_2(O_2)(CO_3)_2]\cdot 2.5H_2O$.[6] The compound contains molecular $UO_2(O_2)(CO_3)_2^{4-}$ units (Figure 2) linked by a complex network of potassium ions and interstitial water molecules.

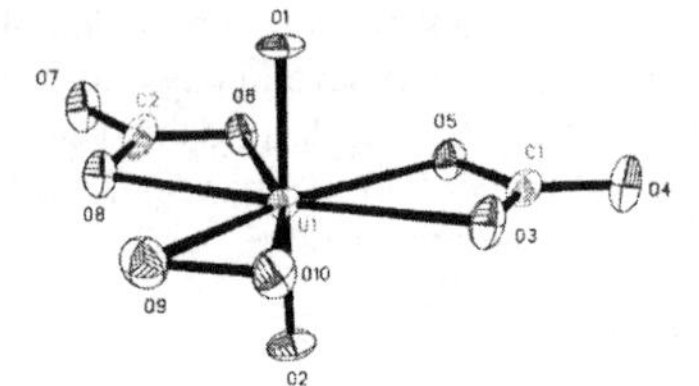

Figure 2 *A view of the anionic $[UO_2(O_2)(CO_3)_2]^{4-}$ unit in the molecular structure of $K_4[UO_2(O_2)(CO_3)_2]\cdot 2.5H_2O$. Refer to Table 1 for bond distances. Displacement ellipsoids are drawn at the 50% probability level.[6]*

Table 1 *Selected geometric parameters (Å, °) in $K_4[UO_2(O_2)(CO_3)_2]\cdot 2.5H_2O$.*[6]

U1–O2: 1.806(5)	U1–O3: 2.438(5)	U1–O6: 2.488(5)	U1–C1: 2.878(9)
U1–O1: 1.817(5)	U1–O5: 2.445(6)	U1–O8: 2.443(5)	U1–C2: 2.888(8)
U1–O10: 2.240(6)	U1–O9: 2.256(6)	O9–O10: 1.496(8)	
O2–U1–O1: 175.3(3)		O10–U1–O9: 38.9(2)	

3 CONCLUSION

In carbonate-peroxide solutions, U(VI) and Pu(IV) form mixed-ligand compounds, with carbonate and peroxide actively competing for complexation. In the U(VI) system, we have seen evidence for both the $UO_2(O_2)(CO_3)_2^{4-}$ and $UO_2(O_2)_2(CO_3)^{4-}$ complexes in solution, with the $UO_2(O_2)_2(CO_3)^{4-}$ complex being the kinetically favoured product under high peroxide conditions, but the $UO_2(O_2)(CO_3)_2^{4-}$ complex as the thermodynamically stable complex. Its presence in $K_4[UO_2(O_2)(CO_3)_2]\cdot 2.5H_2O$ also supports the high thermodynamic stability of $UO_2(O_2)(CO_3)_2^{4-}$. Small amounts of peroxide also replace carbonate ligands from the limiting carbonate complex of Pu(IV), $Pu(CO_3)_5^{6-}$, to form a mixed-ligand species. Similar to $UO_2(O_2)(CO_3)_2^{4-}$, the mixed Pu(IV) peroxo-carbonato complex is stable for months despite the degradation of free peroxide in solution.

Acknowledgements
The authors thank the Advanced Fuel Cycle Initiative (AFCI) for financial support. S.M.P is a Director's Postdoctoral Fellow, funded through LANL's LDRD Program.

References
1 D.L. Clark, D.E. Hobart, and M.P. Neu, *Chem. Rev.*, 1995. **95** 25.
2 H. Wanner, I. Forest, and O.N.E.A., eds. *Chemical Thermodynamics of Uranium.*, Chemical Thermodynamics, ed. I. Grenthe. Vol. 1. 1992, Elsevier.
3 R.J. Lemire and O.N.E.A., eds. *Chemical Thermodynamics of Neptunium and Plutonium.* Chemical Thermodynamics, Vol. 4. 2001, Elsevier.
4 D.L. Clark, S.D. Conradson, D.W. Keogh, P.D. Palmer, B.L. Scott, and C.D. Tait, *Inorganic Chemistry*, 1998. **37** 2893.
5 J.M. Cleveland, *The Chemistry of Plutonium.* Nuclear Science and Technology Series. 1979, La Grange Park, Il: American Nuclear Society.
6 R.A. Zehnder, S.M. Peper, B.L. Scott, and W. Runde, *Acta Cryst. C*, 2005. **C61** i3.

SEQUESTERED PLUTONIUM: STRUCTURAL CHARACTERISATION AND COMPLEX FORMATION OF Pu(IV) COMPLEXES WITH SIDEROPHORE-LIKE LIGANDS

A. E. V. Gorden, D. K. Shuh, B. E. F. Tiedemann, J. Xu, R. E. Wilson, and K. N. Raymond

Seaborg Center, Chemical Sciences Division, Lawrence Berkeley National Laboratory, and Department of Chemistry, University of California, Berkeley, CA 94720 USA

1 INTRODUCTION

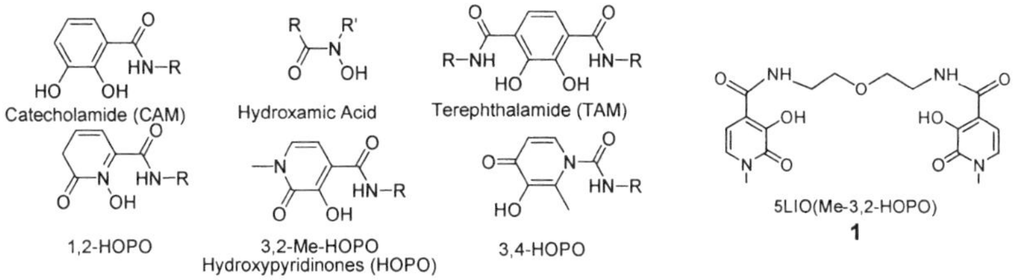

Figure 1 *At left) Structures of catecholate and hydroxamic acid derivatives like those found in the siderophores used in this series of actinide sequestering ligands. At right) The hydroxypyridonate ligand **1** at the focus of this study*

Nature has responded to the problem of the low bioavailability of Fe by developing high-affinity Fe(III) sequestering agents, called siderophores.[1] In mammals, Pu(IV) is bound to a high degree with the iron-transport protein, transferrin, or the iron-storage protein, ferritin.[2] These similarities in the chemical behaviour of Fe(III) and Pu(IV) are the foundation of a biomimetic approach to selective chelators for Pu(IV) based on catecholamide (CAM), terephthalamide (TAM), and hydroxypyridonate (HOPO) groups like those in the siderophores (Figure 1).[3] A detailed evaluation of the structure of actinide metal complexes is important to the design of new selective ligands. Anticipating the small size of crystals and safe handling considerations, we have developed procedures to use a system for structure determination of small single crystals at the ALS (Advanced Light Source) synchrotron to determine the solid-state structures of Pu complexes by X-ray diffraction (XRD). With these, Pu(IV)-Bis[5LIO(Me-3,2-HOPO)$_2$] is the first Pu(IV) hydroxypyridonate complex to be structurally characterized.

2 METHODS AND RESULTS

2.1 Experimental Methods

^{242}Pu containing samples were manipulated in a glove box designed for the safe handling of radionuclides. The ligand **1** was synthesized by previously published methods.[4] The plutonium complex was prepared through the addition of 0.05M plutonium stock solution in 1.1M $HClO_4$ (230 μl, 2.3 x 10^{-3} mmol) to a solution of 5LIO(Me3,2-HOPO) (2.08mg, 4.9 x 10^{-3} mmol) in deionized H_2O (2 ml). Crystals are obtained as purple plates. Crystals were mounted in paratone-N oil inside quartz capillaries coated with resin and sealed with epoxy.[5] Crystallographic data was collected using a Proteum300 detector at the Small-Crystal Crystallographic Beamline 11.3.1 at the Advanced Light Source (ALS). This bending magnet beamline provides an intense beam of monochromatic x-rays in the range 6-17keV and can be used with crystals as small as 15 microns on a side.[6,7,8]

2.2 Discussion

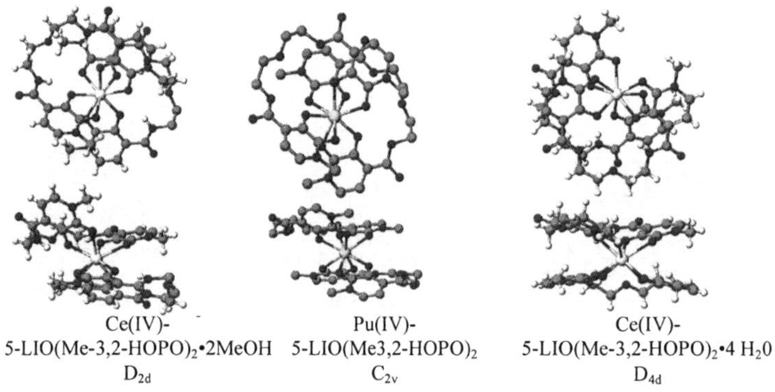

Figure 2 *Molecular structures of a) Ce(IV)[Bis(5-LIO-Me-3,2-HOPO)]$_2$·2 CH_3OH, b) Pu(IV)[Bis(5-LIO-Me-3,2-HOPO)]$_2$ and c) Ce(IV)[Bis(5-LIO-Me-3,2-HOPO)]$_2$·4 H_2O. Ce(IV) structures were generated from data from the Cambridge Crystallographic Database as previously published.*[4]

The Pu(IV) complex crystallizes in the space group Pna_{21} with Z=4. The unit cell contains two Pu molecules, with each Pu atom coordinated by eight oxygen atoms, four each from two 5LIO(Me-3,2-HOPO) ligands. Also in the unit cell are one perchlorate anion and a hydronium ion as the counter ion. Structural details are given in Table 1.

The most common coordination number encountered in *f*-element (III) and (IV) coordination complexes is 8. The ideal geometry of 8 coordinate systems can be depicted by 3 high symmetry polyhedra: trigonal dodecahedron (D_{2d}), bicapped trigonal prism (C_{2v}), and square antiprism (D_{4d}).[10] Assigning the polyhedra that best describe these systems is not straightforward. We have described shape measure (*S*) to compare the geometries of coordination systems, defining the metric *S* as a minimization comparing all possible orientations of the angles within the observed structure relative to those of the reference polyhedron allowing for the direct comparison of dissimilar metal coordination complexes.[4] We can thus compare the geometries of the Pu complex with previously prepared Ce(IV)-5-LIO(Me-3,2-HOPO) complexes from both aqueous and methanolic solutions. The geometry of the Pu molecules resembles a bicapped trigonal prism (C_{2v}, for

Pu 1 $S(C_{2v})$= 13.48°, $S(D_{4d})$= 15.43°, $S(D_{4d})$= 16.10°). The Ce(IV) complexes are different, with the aqueous system best described as square antiprism (D_{4d} , $S(D_{4d})$= 12.54°, $S(C_{2v})$= 18.56°, $S(D_{2d})$= 20.68°), and the methanol complex as trigonal dodecahedron (D_{2d}, $S(D_{2d})$= 14.33°, $S(C_{2v})$= 14.87°, $S(D_{4d})$= 17.75°). The Pu(IV) complex is more similar in shape to the methanolic Ce(IV) structure than the aqueous system. Studies are underway to determine if this is due to metal-ligand interactions or pH effects.

Table 1 *Structural Details for the Pu(IV)-[Bis(5-LIO-Me-3,2-HOPO)]$_2$ complex (goodness-of-fit on F^2, Final R indices [$I > 2\sigma(I)$])*

Orthorhombic, Z=4	Pna2$_1$	Final R indices	
Radiation source	Synchrotron	R1	0.0629
Temperature, K	298 (2)	ωR2	0.1624
Crystal size (mm)	0.50x0.50x 0.70	Pu1-$O_{phenolic}$	2.307Å
Reflections collected	63115	Pu1-O_{amide}	2.401Å
Independent reflections	16280	Pu2-$O_{phenolic}$	2.324Å
Avg. Bite Angle	67.3(4)°	Pu2-O_{amide}	2.399Å

3 CONCLUSION

The Pu(IV-[5LIO(Me-3,2-HOPO)]$_2$ complex is the first in a series of complexes to be characterized structurally. This work and the ongoing development of this method will aid in the generation of a library such complexes on which to base the design of novel ligand systems and provides a benchmark for additional structural studies with actinides.

Acknowledgements
The authors thank Dr. Allen Oliver, Brian Fairchild, and Sirine Fakra. This and the ALS are supported by the Director, Office of Basic Energy Sciences, Division of Materials Sciences and Division of Chemical Sciences of the U.S. Department of Energy at Lawrence Berkeley National Laboratory under Contract No. DE-AC03-76SF00098.

References

1 A. Stintzi, C. Barnes, J. Xu and K.N. Raymond, *Proc. Natl. Acad. Sci. USA* 2000, **97**, 10691.
2 P.W. Durbin, B. Kullgren, J. Xu and K.N. Raymond, *Radiat. Prot. Dosim.* 1998, **79**, 433.
3 A.E.V. Gorden, J. Xu, K.N. Raymond and P.W. Durbin, *Chem. Rev.* 2003, **103**, 4207.
4 J. Xu, E. Radkov, M. Ziegler and K.N. Raymond, *Inorg. Chem.* 2000, **39**, 4156.
5 A.E.V. Gorden, K.N. Raymond, and D.K. Shuh, -LBNL-53125, Lawrence Berkeley National Laboratory - Chemical Sciences Division, 2003.
6 G.M. Sheldrick, G. M.; Universität Göttingen, 1997.
7 SHELXTL, Crystal Structure Analysis Determination Package; Siemens Industrial Automation Inc.: Madison, Wisconsin 1990.
8 Crystallographic data is available with the Cambridge Crystallographic Data Center as supplementary publication no. CCDC 249885. Copies of the data can be obtained free of charge on application to CCDC 12 Union Road, Cambridge CB21EZ, UK (fax: (+44) 1223-336-033; email: deposit@ccdc.cam.ac.uk).
9 H.C. Aspinall, *Chemistry of the f-Block Elements*; Gordon and Breach Publishers: Amsterdam, 2001; Vol. 5.
10 D.L. Kepert, *Progress in Inorganic Chemistry* 1978, **24**, 179.

EXTRACTION OF ACTINIDES BY FUNCTIONALISED CALIXARENES

B. Boulet[1,2], C. Bouvier-Capely[1], C. Cossonnet[1], L. Joubert[2], C. Adamo[2], G. Cote[2]

[1] IRSN/DRPH/SDI/LRC, B.P. 17, 92262 Fontenay-aux-Roses Cedex, France
[2] ENSCP/LECA, 11 rue Pierre et Marie Curie, 75231 Paris Cedex 05, France

1 INTRODUCTION

Individual monitoring of workers exposed to a risk of internal contamination by actinides is generally achieved through *in vivo* measurements (anthroporadiametry) and *in vitro* measurements (urine and feces). The procedures currently used for actinides analysis in urine are well established and validated but are time-consuming, which limits the frequency and the flexibility of individual monitoring. The aim of this work is to propose an alternative radiochemical procedure for uranium.

Calix[n]arenes are macrocyclic molecules composed of n phenolic units linked with methylene bridges in the *ortho* positions.[1] The three main criteria to design a promising calixarene-based uranophilic extractant are cavity size, macrocycle geometry and lower rim functional groups.[2] According to these criteria, the 1,3,5-OMe-2,4,6-$OCH_2CONHOH$*p-tert*butylcalix[6]arene (L_DH_3) was identified as a possible good uranophilic extractant and compared with 1,3,5-OMe-2,4,6-OCH_2COOH*p-tert*butylcalix[6]arene (L_CH_3) [3] by theoretical calculations and by experiments.

2 EXPERIMENTAL

2.1 Apparatus and methods

The calixarenes studied in this paper are represented in Figure 1.
1,3,5-OMe-2,4,6-$CH_2CONHOH$-*p-tert*butylcalix[6]arene (L_DH_3) was designed and synthesized by Chelator S.A. (France). The extraction of UO_2^{2+} by L_DH_3 was performed by mixing equal volumes of an aqueous phase containing 10^{-9} M uranium, an acetate buffer (10^{-2} M) and 0.04 M $NaNO_3$ with an organic phase containing 10^{-3} M L_DH_3 dissolved in 1-heptanol. All the measurements of ^{238}U in aqueous phases were performed by Inductively Coupled Plasma Mass Spectrometry (ICP-MS).

2.2 Computational details

The geometry optimizations and the calculations of the interaction energies were carried out at the DFT level for all the systems, using the Gaussian code.[4] The calculations were carried out using the so-called PBE0 "parameter free" hybrid model.[5] We used a relativistic large effective core potential (ECP) of the Los Alamos group for uranium and the 6-31+G(2d,2p) basis set for the other atoms.[6] All interaction energies ΔE have been corrected for basis set superposition errors (BSSE).

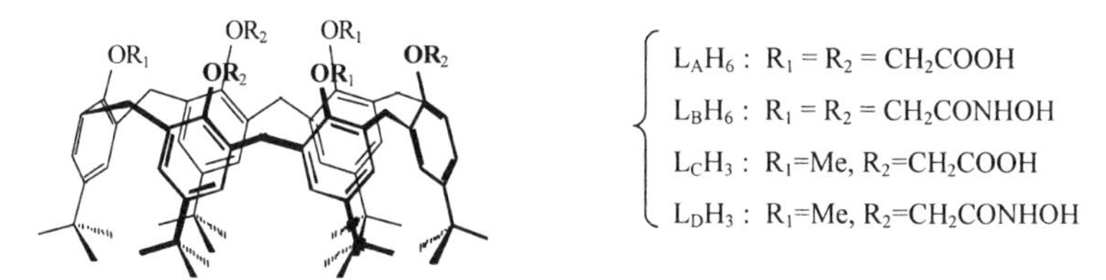

Figure 1 *Structure of 1,3,5-OR_1-2,4,6-OR_2-p-tert butylcalix[6]arene*

2.3 Results and discussion

The extraction equilibrium of UO_2^{2+} by L_CH_3 and by L_DH_3 has been studied experimentally previously.[2-3] For both calixarenes, the extraction equilibrium has been determined as follows, $UO_2^{2+} + \overline{LH_3} \longleftrightarrow \overline{(UO_2)(LH)} + 2H^+$ (1)

2.3.1 *Interaction of hydroxamic and carboxylic chelating groups with UO_2^{2+}*

To investigate the interaction of L_CH_3 and L_DH_3 with UO_2^{2+}, the interaction energies of the chelating groups of these two calixarenes with UO_2^{2+} have been calculated. In the following clause, the chelating groups are noted RH and their deprotonated form R^-, with $R_CH = CH_3OCH_2COOH$ and $R_DH = CH_3OCH_2CONHOH$.
From the equilibrium (1), it can be assumed that the uranyl ion is binding with two deprotonated chelating groups (R^-) and also probably with the third group (RH). Therefore, the first system model used to compare by theoretical calculations the binding between the calixarenes and UO_2^{2+} is ($2R^-$ + RH).

Table 1 shows the interaction energies (ΔE) between the uranyl ion and the groups ($2R_C^-$ + R_CH) and ($2R_D^-$ + R_DH). It can be observed that the interaction energy is stronger for the hydroxamic functions than for the carboxylic ones. This is in good agreement with the experimental literature data,[7] showing that L_BH_6 exhibits a higher affinity for UO_2^{2+} than L_AH_6. Figure 2 shows how the hydroxamic groups are bonded to UO_2^{2+}. It can be noticed that the equatorial coordination number (CN) of UO_2^{2+} is 5.

Table 1 *ΔE (kJ/mol) of the chelating groups of L_CH_3 and of L_DH_3 with UO_2^{2+}*

	UO_2^{2+} / [$2R_D^-$ + R_DH]	UO_2^{2+} / [$2R_C^-$ + R_CH]
ΔE	-2825	-2695

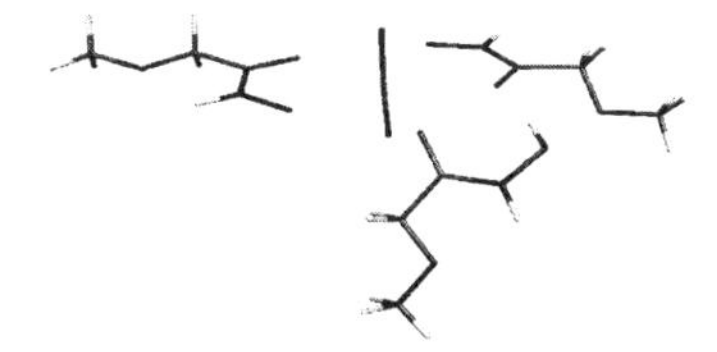

Figure 2 *UO_2^{2+} bonded to [$2R_D^-$ + R_DH]*

2.3.2 *Experimental UO_2^{2+} extraction by L_CH_3 and L_DH_3*

Figure 3 shows the percentage of extraction of U(VI) by L_CH_3 and L_DH_3 as a function of pH. It can be observed that L_DH_3 extracts quantitatively U(VI) at a slightly more basic pH than that needed for L_CH_3. This is in agreement with results obtained with L_AH_6 and L_BH_6.[7]

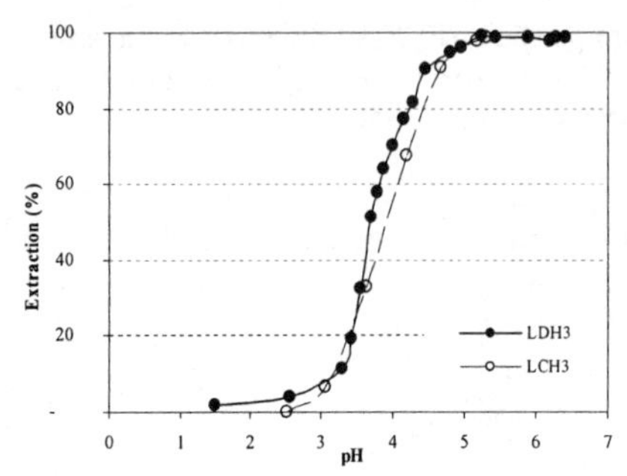

Figure 3 *Extraction (%) of U(VI) by L_CH_3 and by L_DH_3 versus pH*

3 CONCLUSION

A new calix[6]arene molecule possessing hydroxamic groups has been selected to extract U(VI): the 1,3,5-OMe-2,4,6-$OCH_2CONHOH$ *p-tert*butylcalix[6]arene (L_DH_3).
The results of theoretical calculations show that the hydroxamic function has a better affinity for UO_2^{2+} than the carboxylic function. Experimental results show that L_DH_3 extracts U(VI) with an efficiency close to that obtained with the 1,3,5-OMe-2,4,6-OCH_2COOH *p-tert*butylcalix[6]arene (L_CH_3), potentially with a higher selectivity towards cations present in urine.[2]

From the literature data and from the experimental and calculated results reported in this paper, L_DH_3 appears to be an interesting extractant for U(VI).
Further works continue on the extraction and the stripping of actinides and on the selectivity with metal cations.

Acknowledgments
The authors wish to acknowledge Dr Duval (Chelator S.A., France), Dr Pellet-Rostaing and Pr Lemaire (Claude-Bernard Lyon 1 University, France) for optimization of L_DH_3 synthesis.

References

1 C.D. Gutsche, Calixarene, Royal Society of chemistry, Thomas Graham House, Science Park : Cambridge, 1989.
2 B. Boulet, C. Bouvier-Capely, C. Cossonnet, G. Cote, ATALANTE 2004, Nîmes, 21-25 juin 2004, paper P2-09.
3 C. Dinse, N. Baglan, C. Cossonnet, J.F. Le Du, Z. Asfari, J. Vicens, *J. Alloys and Compounds*, 1998, **271-273**, 778.
4 M.J. Frisch et al., Gaussian 03, Revision B.05, Gaussian, Inc., Pittsburgh PA, 2004
5 C. Adamo and V. Barone, *J. Chem. Phys.*, 1999, **110**, 6158.
6 P.D.J. Grootenhuis, P.A. Kollman, L.C. Groenen, D.N. Reinhoudt, G.J. van Hummel, F. Ugozzoli, G.D. Andreetti, *J. Am. Chem. Soc.*, 1990, **112**, 4165.
7 T. Nagasaki, S. Shinkai, *J. Chem. Soc. Perkin Trans 2*, 1991, 1063.

5*f* ELEMENT COMPLEXES WITH 'SOFT' DONOR ATOM LIGANDS: EIGHT COORDINATE Pu(III) PYRAZINYL AND THIOETHER COMPLEXES

A. J. Gaunt, J. H. Matonic, B. L. Scott and M. P. Neu*

Actinide, Catalysis, and Separations Chemistry (C-SIC), Chemistry Division, Los Alamos National Laboratory, Mail Stop J-514, Los Alamos, New Mexico, 87545, U.S.A.

1 INTRODUCTION

Although actinide coordination chemistry is dominated by 'hard' donor atom ligands, complexes with a range of 'softer' ligands are potentially important in separations and for elucidating structure and bonding, involving 6*d* and 5*f* electrons. As a compliment to several classes of organometallic complexes, which have been prepared (in part) to maximize the covalent bonding in one or two coordination sites, and in an attempt to seek the limits of 'hard-soft' bonding, we are preparing low-valent actinide complexes with N, S, Se and Te donor ligands. Classes of anionic and neutral ligands we are currently studying include tripodal aromatic amines, thiolates, selenolates, tellurolates, dithiophosphinates, diselenophosphinates, imidodiphosphinochalcogenides and thioethers. Here we present the preparation and structural characterisation of $Pu(tpza)I_3(MeCN)$ (**1**) (tpza = tris[(2-pyrazinyl)methyl]amine) and $Pu(9S_3)I_3(MeCN)_2$ (**2**) ($9S_3$ = [$9aneS_3$]).

2 METHOD AND RESULTS

2.1 Synthesis of 1

In a dry box, plutonium metal strips (0.050 g, 0.209 mmol) were combined with I_2 (0.075 g, 0.298 mmol) and acetonitrile (5 cm^3). The mixture was stirred for 1 day, tpza added (0.061 g, 0.209 mmol) and the mixture stirred for 2 hours. A green precipitate of **1** formed and was isolated in 63 % yield (0.130 g). A second crop was isolated (0.035 g) by concentration of the supernatant. Single crystals suitable for X-ray diffraction were grown by gently heating the precipitate to dissolve in a minimum amount of acetonitrile, and allowing the solution to cool to ambient temperature.

Crystal data for **1**: $C_{14}H_{12}I_3N_8Pu$, M = 915.02, a = 11.7357(6) Å, b = 12.5099(6) Å, c = 19.1979(9) Å, α = 90, β = 95.5830(10), γ = 90, V = 2805.1(2) Å^3, monoclinic, space group $P2_1/n$, Z = 4, T = 293(2) K, reflections collected/independent = 14285/5660 [R(int) = 0.0271], $R_1(I>2\sigma(I))$ = 0.0376, and $wR_2(I>2\sigma(I))$ = 0.0967.

2.2 Synthesis of 2

In a dry box, plutonium metal strips (0.100 g, 0.40 mmol) were combined with I_2 (0.160 g, 0.60 mmol) and acetonitrile (2 cm^3). The mixture was stirred for 1 day, filtered and treated with a solution of $9S_3$ (0.075 g, 0.42 mmol) in acetonitrile (5 cm^3). A green precipitate formed within minutes. Crystals suitable for X-ray diffraction were grown, in 86 % yield (0.30 g), by heating to boiling and allowing the resulting solution to slowly cool to ambient temperature.

Crystal data for **2**: $C_6H_{12}I_3N_2S_3Pu$, $M = 831.06$, $a = 15.3412(10)$ Å, $b = 9.1175(10)$ Å, $c = 15.2098(10)$ Å, $\alpha = 90$, $\beta = 99.822(10)$, $\gamma = 90$, $V = 2096.3(3)$ Å^3, monoclinic, space group P_12_1/c_1, $Z = 4$, $T = 293(2)$ K, reflections collected/independent = 10581/4248 [R(int) = 0.0367], $R_1(I>2\sigma(I)) = 0.0338$, and $wR_2(I>2\sigma(I)) = 0.0788$.

2.3 Crystal structure and discussion of 1

Oxidation of Pu metal with iodine followed by treatment with tpza results in the isolation of the eight coordinate Pu(III) complex in a distorted square anti-prismatic geometry. One square plane is defined by N(2), N(5), I(2), I(3) and the other by N(1), N(3), N(4), I(1). The plutonium to aromatic nitrogen distances are 2.668(6) Å, 2.655(6) Å and 2.644(7) Å, for Pu-N(2), Pu-N(3) and Pu-N(4), respectively. The plutonium to aliphatic amine distance is shorter at a distance of 2.618(6) Å for Pu-N(1). The coordinated acetonitrile molecule has a Pu-N distance of 2.628(7) Å and the average Pu-I distance is 3.132(1) Å.
The closest comparable Ln(III) complex is $[Nd(tpza)(CH_3CN)_3(H_2O)_3](ClO_4)_3$ which has an Nd-N(amine) distance of 2.734(7) Å,[1] shorter than the Pu-N(amine) distance in **1** by 0.116 Å and the average Nd-N(aromatic) distance is 2.747(12) Å, shorter than the average Pu-N(aromatic) distance in **1** by 0.091 Å. Nd(III) and Pu(III) have similar ionic radii but because the complexes are not isostructural then no definitive inferences about differences in the nature of the bonding can be made. An isostructural U(III) complex, $[U(tpza)I_3(MeCN)]$, has previously been reported,[2] with a U-N(amine) distance of 2.721(8) Å and an average U-N(aromatic) distance of 2.694(14) Å, longer than in **1** by 0.103 Å and 0.038 Å, respectively. The bands in the UV/vis spectrum of **1** dissolved in acetonitrile display an approximately 15 nm red-shift relative to the Pu(III) aquo ion. The ^{1}H NMR shows three resonances in the aromatic region indicating the equivalence of the three aromatic rings. This may due to the formation of ions in solution by replacement of iodide for acetonitrile in solution[3] although fluxional processes and labilisation of the pyridyl arms are other possibilities.

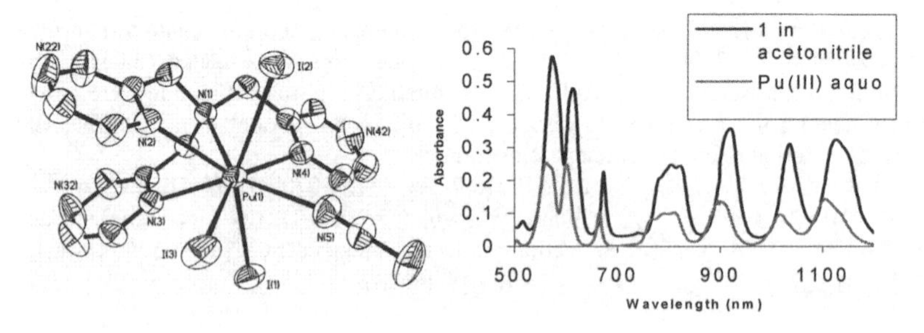

Figure 1 *ORTEP representation (left) of the crystal structure of **1*** (thermal ellipsoids at the 50 % probability level) and *UV/vis spectra of **1** in MeCN compared to the Pu(III) aquo ion (right)*

2.4 Crystal structure and discussion of 2

The structure of **2** contains an eight coordinate Pu(III) ion in a distorted square anti-prismatic geometry. The neutral crown thioether, $9S_3$, is coordinated through all three S atoms with an average Pu-S distance of 3.018(2) Å. The average Pu-N length is 2.591(10) Å and the average Pu-I length is 3.101(1) Å. Even when treating Pu with excess $9S_3$, the iodides cannot be displaced from the inner coordination sphere and only one thioether binds to the metal center. Thioethers are uncharged, neutral ligands and therefore are attractive to use for probing differences in the nature of covalent bonding between the trivalent actinides and lanthanides. Recently, the isostructural $U(9S_3)I_3(MeCN)_2$ and $La(9S_3)I_3(MeCN)_2$ complexes were reported,[4] in which the U-S distances were found to be an average of 0.047 Å shorter than the La-S lengths consistent with a greater degree of covalent interaction of U(III) than La(III) with the neutral soft sulfur donor thioether. The average U(III)-S distance is 0.028 Å longer than the average Pu(III)-S distance in **2** and can be attributed to the higher charge density of Pu(III) compared to U(III) due to the actinide contraction.

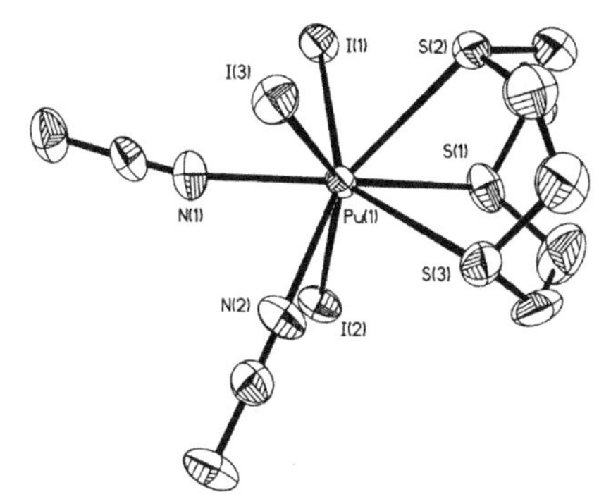

Figure 3 *ORTEP representation of the crystal structure of 2* (thermal ellipsoids at the 50 % probability level)

3 CONCLUSIONS

In pursuing our studies of actinide complexes with soft donor ligands, we have structurally characterized Pu(III) complexes with neutral nitrogen (tpza) and sulfur ($9S_3$) donor ligands, representing rare molecular Pu(III) crystal structures and the first study of transuranic tripodal amine and thioether complexes.

Acknowledgements

We thank Dr. M. Mazzanti for useful discussions and providing samples of tpza.

References

1 R. Wietzke, M. Mazzanti, J.-M. Latour, J. Pécaut, P.-Y. Cordier and C. Madic, *Inorg. Chem.*, 1998, **37**, 6690.
2 M. Mazzanti, R. Wietzke, J. Pécaut, J.-M. Latour, P. Maldivi and M. Remy, *Inorg. Chem.*, 2002, **41**, 2389.
3 A. E. Enriquez, J. H. Matonic, B. L. Scott, M. P. Neu, *Chem. Commun.*, 2003, 1892.
4 L. Karmazin, M. Mazzanti and J. Pécaut, *Chem. Commun.*, 2002, 654.

PLUTONIUM(V) CHEMISTRY IN NEUTRAL AQUEOUS COMPLEXING MEDIA: NEW CRYSTAL COMPOUNDS SYNTHESIS AND STUDY OF THEIR STRUCTURE AND PROPERTIES

A.A. Bessonov, M.S. Grigoriev, I.A. Charushnikova, N.N. Krot

Institute of Physical Chemistry, Russian Academy of Sciences
31 Leninsky prosp., Moscow, 119991, RUSSIA

1 INTRODUCTION

Transuranium elements (TRU) in the pentavalent oxidation state (both in solutions and the solid state) exist mainly as linear cations, AnO_2^+, in which the central atom holds an increased positive charge and "yl"-oxygen atoms have a negative charge. Owing to such electron density distribution, these cations show high complexing ability and may be isolated from solutions as coordination compounds with different ligands in the equatorial plane. In addition, so called cation-cation (CC) interaction, i.e. an ability of dioxocations to undertake mutual coordination through "yl"- oxygen atoms, was found in the structures of 23 neptunium(V) compounds, many of which had been synthesized and studied in the IPC RAS.[1] Thus, CC interaction is an important structure-forming factor for a lot of NpO_2^+ compounds. The role of this factor in the compounds of neighbouring PuO_2^+ is unkown, although it is known that PuO_2^+ is disposed to mutual coordination in solutions. There is an extremely small number of Pu(V) compounds that have been isolated from solution for which composition is determined reliably and there are no crystal structures of Pu(V) compounds described in literature. This can be explained by the lower stability of PuO_2^+ in comparison with NpO_2^+ as well as by the much more difficult conditions of working with higher radioactivity Pu isotopes. Thus, the scarcity of preparative chemistry of Pu(V) is a serious obstacle for revelation of regularities in changes of composition, structure and properties of similar-type AnO_2^+-compounds in the Np-Pu-Am series. An increased understanding of this system is important, both for fundamental understanding of actinide chemistry and practical application. Determining strategies for the synthesis of a number of novel Pu(V) coordination compounds, with additional investigation of structure and properties, are the goals of this work. Special attention has been paid to preparation and study of complexes where it was reasonable to anticipate the presence of CC interaction.

2 METHOD AND RESULTS

^{239}Pu was used in the work. 22 novel PuO_2^+ compounds have been obtained from neutral aqueous solutions containing oxalate-, malonate- and phthalate-ions. These are: bi-phthalate complex dihydrate $[Co(NH_3)_6]\{PuO_2[C_6H_4(COO)_2]_2\}\cdot 2H_2O$; bi-oxalate complex compounds $[Co(NH_3)_6][PuO_2(C_2O_4)_2]\cdot nH_2O$ (n = 2, 3, 5); mono-malonate complex compounds $[Co(NH_3)_6]\{PuO_2[C_3H_2O_4]\}_2A\cdot nH_2O$ (A = NO_3^-, ClO_4^-, Cl^-, Br^-, $HCOO^-$,

CH_3COO^-, $C_2H_5COO^-$, $C_3H_3O_4^-$); double mono-malonates $MPuO_2C_3H_2O_4 \cdot nH_2O$ (M = Na^+, NH_4^+, Cs^+) and derivates of Pu(V) mono-oxalate complex $MPuO_2C_2O_4 \cdot nH_2O$ (M = Na^+, NH_4^+, Cs^+). The compounds have been identified by chemical analysis, characterized by powder and single crystal XRD methods, IR-spectrometry, thermogravimetric techniques. UV/Vis spectra of related Np(V) solids have been used to find out some structure peculiarities of the complexes.

The description of $[Co(NH_3)_6]\{PuO_2[C_6H_4(COO)_2]_2\} \cdot 2H_2O$ synthesis is given below as an example of the approach to the problem, whereas details of synthesis for other Pu(V) solids are reported elsewhere in these proceedings.[2-4] Syntheses of the compound were carried out in neutral solution with phthalate-ion concentration $[L^{2-}]$ = 0.05-0.2 M, Pu(VI) was reduced to pentavalent state by a calculated quantity H_2O_2. Complex formation and stability of Pu(VI) in phthalate solutions were studied initially. It was found that plutonyl(VI) forms two complexes with the ratio $PuO_2^{2+} : L^{2-} = 1 : 1$ and 1 : 2 in solutions with concentration $[L^{2-}] < 0.3$ M. In these solutions Pu(VI) is stable even at 70 °C during several hours. The monophthalate complex has low solubility, but its crystallization is slow at room temperature and [Pu(VI)] = 0.001-0.01 M in 0.05-0.3 M Li_2L solutions. Both complexes react very quickly with H_2O_2 strictly in accordance with the reaction:

$$2Pu(VI) + H_2O_2 \rightarrow 2Pu(V) + O_2 + 2H^+ \quad (1)$$

It appeared that with increase of [Pu(VI)] > 0.15 M the H_2O_2 consumption appreciably increases over stoichiometry. So, for complete reduction of 0.025 and 0.035 M Pu(VI) the necessary H_2O_2 excess was about 5 and 20 %, respectively. The yellow colour of Pu(VI) phthalate solutions completely disappears when plutonium is reduced to Pu(V), but in time it returns back due to Pu(V) disproportionation:

$$2Pu(V) \rightarrow Pu(VI) + Pu(IV) \quad (2)$$

Experiments have shown, that this process in phthalate solutions is irreversible and is not complicated by formation of Pu(III) owing to binding tetra- and hexavalent plutonium in strong phthalate complexes. The disproportionation rate has been evaluated.

Based on the obtained information on reactions (1) and (2), synthesis of Pu(V) phthalate compounds with the $Co(NH_3)_6^{3+}$ cation was carried out. Comparison of powder XRD patterns of prepared samples with that of $Co(NH_3)_6NpO_2L_2 \cdot 2H_2O$ [5-6] has revealed their identity. Thus, $Co(NH_3)_6PuO_2L_2 \cdot 2H_2O$ is isolated under the aforementioned conditions. The crystal cell parameters have been defined as: a = 7.760(1), b = 23.152(3), c = 14.584(2) Å, β = 96.41(1)° (space group $P2_1/c$). Syntheses of other Pu(V) compounds were carried out by similar techniques. The choice of specific conditions of synthesis was based on preliminary studies of Pu(V) behaviour in malonate or oxalate solutions and on previous Np studies.

It was established that powder XRD patterns of $Co(NH_3)_6(AnO_2L)_2A \cdot nH_2O$ (An=Np, Pu; L= $C_3H_2O_4^{2-}$) are very similar, independent of the chemical nature of anion A^-, indicating similar structures. By single crystal XRD analysis it was found that infinite chains $\{AnO_2L\}_n^{n-}$, where each AnO_2^+ cation coordinates 3 L^{2-} anions, are the basic elements of the structures.[7-8] All of the compounds $Co(NH_3)_6(AnO_2L)_2A \cdot nH_2O$ are stable over an extended time period and have low solubility in water. Complex malonates $MPuO_2L \cdot xH_2O$ are found to be waterless when M = NH_4^+ or Cs^+ and the Na^+ derivate is a dihydrate. Analogues Np(V) compounds have the same structure and composition[9-10] and from the analysis of their powder XRD patterns and absorption spectra it had been concluded that Np(V) monomalonate complex forms the same infinite chains $\{NpO_2L\}_n^{n-}$ as in the structure of $Co(NH_3)_6(NpO_2L)_2A \cdot nH_2O$.[7] Therefore one can conclude that polymerization of Pu(V) monomalonate complex occurs in the same way as observed for Np(V).

Monooxalate compounds, $NaPuO_2C_2O_4{\cdot}nH_2O$ (n = 3, 1, 0), are also isostructural with analogous Np(V) compounds. For n = 3 the structure is without CC interaction.[11] For n = 1 and 0 the structure is unknown but the f-f transition bands in the near IR absorption spectra of Np(V) compounds are red-shifted relative to the $NpO_2C_2O_4^-$ complex in aqueous solution, indicating monodentate CC interaction between NpO_2^+ ions.[12] Splitting and shift of the antisymmetric vibration bands of AnO_2^+ in the IR spectra towards lower wave numbers confirm this conclusion.

$NH_4[PuO_2C_2O_4]{\cdot}2.67H_2O$ is isostructural with analogous Np(V) compound with known structure.[13] Thus, this compound contains trimer-groups of plutonoyl(V) ions that are coordinated to each other as monodentate ligands.

$CsPuO_2C_2O_4{\cdot}nH_2O$ compounds (n = 6, 4, 2) have no isostructural Np(V) analogues, but from their IR spectra we can suppose CC interaction in the cases of tetra- and di-hydrates, although further studies are required for confirmation.

In closing it should be noted, that we have succeeded in finding the conditions for preparation of single crystals of one of Pu(V) compound, $[Co(NH_3)_6][PuO_2(C_2O_4)_2]{\cdot}3H_2O$. For the first time XRD structural analysis of a Pu(V) compound has been undertaken using a single crystal sample. Dimeric complex anions $[PuO_2(C_2O_4)_2]_2^{6-}$ were found as a basis of the structure. Other important details of the structure are given in an additional conference proceedings paper[4] but here it is interesting to note practically isotropic decrease in An-O distance, which has been found both in axial and equatorial directions, at the transition from neptunium to plutonium compound. The "actinide contraction" phenomenon is the most probable reason of such decreasing.

3 CONCLUSION

It has been found that PuO_2^+ forms compounds which are close structural analogues of the similar-type NpO_2^+ solids. However, this is a non-strict rule: sometimes under the same conditions different phases are crystallized depending on the presence of Np(V) or Pu(V) in the system. CC interaction of PuO_2^+ ions in the structures of some Pu(V) compounds has been proven for the first time. In addition, the first crystal structure of a Pu(V) compound has been obtained, and indeed the first information on actinide contraction in dioxocation (AnO_2^+) compounds of TRU.

References

1 N.N. Krot, M.S. Grigoriev, *Russ. Chem. Rev.*, 2004, **73**, 94.

2 A.A. Bessonov, M.S. Grigoriev, I.A. Charushnikova, N.N. Krot, *Crystal Compounds of plutonoyl(V) monooxalate complex,* This publication.

3 I.A. Charushnikova, A.A. Bessonov, M.S. Grigoriev, N.N. Krot, *Plutonoyl(V) malonate compounds: synthesis and properties*, This publication.

4 M.S. Grigoriev, A.A. Bessonov, I.A. Charushnikova, *Novel Neptunoyl(V) and plutonoyl(V) bi-oxalate compounds: synthesis, properties, structure*, This publication.

5 M. Saeki, M. Nakada, T. Yamashita, N.N. Krot, *Radiochim. Acta,* 1998, **80**(2), 89.

6 I.A. Charushnikova, N.N. Krot, Z.A. Starikova, *Radiochemistry,* 2001, **43**(5), 438.

7 M.S. Grigoriev, I.A. Charushnikova, et al., *Radiochemistry,* 2004, **46**(3), 212.

8 I.A. Charushnikova, N.N. Krot, I.N. Poliakova, *Radiochemistry,* 2004, **46**(4), 318.

9 N.N. Krot, I.A. Charushnikova, et al., *Radiochemistry,* 1993, **35**(2), 14.

10 N.N. Krot, M.S. Grigoriev, I.A. Charushnikova, et al., *Radiochemistry,* 2004, **46**(2), 97.

11 S.V. Tomilin, Yu.F. Volkov, G.I. Visiacheva, et al., *Radiokhimiya,* 1984, **26**(6), 734.

12 N.N. Krot, A.A. Bessonov, M.S. Grigoriev, 21èmes Journées des Actinides. 1991. Montechoro, Portugal. Book of abstracts: P. 169.

13 M.S. Grigoriev, A.A. Bessonov, A.I. Yanovskii, et al., *Radiokhimiya,* 1991, **33**(5), 46.

MANAGEMENT OF TRANSURANIC ACTINIDE CONTAINING MATERIALS IN NUCLEAR CLEAN UP IN THE UK

R. J. Taylor

Nexia Solutions, BNFL Technology Centre, Sellafield, Seascale, CA20 1PG, UK

1 INTRODUCTION

Managing transuranium element (TRUe) actinide containing materials must be a priority issue for the clean up of historic nuclear legacies, current industrial operations and the development of future nuclear fuel cycles. The reprocessing of metal and oxide fuels from Magnox and light water reactors has been undertaken at the Sellafield site for many years. However, improvements to current processes and enhanced understanding of the underlying chemistry still provide benefits in processing, effluent treatment and the long term stewardship of TRU materials. This paper will provide an overview of the likely national requirements for industry in managing TRU actinide materials. Particular focus will be placed on key challenges where R&D is needed to support clean up of the UK nuclear legacy, illustrating these with projects covering, for example, actinide separations, waste processing, and stewardship issues. It is worth noting that technology development for UK clean up programmes often has common R&D issues with other programmes, *e.g.* future fuel cycle development, and there are benefits in undertaking integrated projects.

2 BACKGROUND

In the UK, national policies are to focus on legacy clean up and to keep the options open for future decisions on nuclear power generation.[1] The formation of the Nuclear Decommissioning Authority (NDA) in 2005, as a direct result of this policy, has substantially changed the structure and focus of the UK nuclear industry. Within this context, Nexia Solutions is intending to play a role in sustaining the UK's national (civil) nuclear R&D capability and laboratories, most notably the new BNFL Technology Centre at Sellafield.[2] Using its expertise and experience, Nexia Solutions will also help administer the NDA's national direct R&D programme. The NDA direct programme is focused on funding key R&D projects that accelerate clean up programmes (*i.e.* related to the site 'Life Cycle Baselines'), reduce costs, provide long term, generic or multi-site benefits and preserves key national skills. R&D relevant to shorter term and site specific issues (*i.e.* within site 'Near Term Work Plans'), continues to be funded by the Site Licence Company.

3 CHALLENGES IN TRU CHEMISTRY IN UK CLEAN UP

There are a number of key issues in clean up and remediation of UK legacy wastes and nuclear sites in which knowledge and capabilities in TRU actinide chemistry are a key component. These include,:

(a) establishing technically feasible options for Pu disposition strategies, including immobilisation of actinides, for instance within ceramic matrices for safe long term storage or disposal, and reuse in MOx fuel;
(b) asset stewardship of operating actinide processing plants and hence successful completion of planned programmes;
(c) implementation of successful strategies for dealing with a wide range of legacy fuels and residues arising from 50+ years of nuclear industry development in the UK;
(d) the long term stewardship of actinide materials (*e.g.* PuO_2 powders);
(e) the characterisation and behaviour of α (TRU) containing waste streams during clean up operations, including effects in both aqueous effluent streams and during waste processing of solids and sludges *etc.* from legacy ponds and silos;
(f) behaviour of TRU containing materials within geological repositories;
(g) long term maintenance and development of key national capabilities in experimental TRUe chemistry.

This paper briefly describes a few relevant examples of current R&D projects that address some of these UK challenges in TRU actinide science.

4 CHEMICAL SEPARATION FLOWSHEETS FOR LEGACY FUELS AND RESIDUES

Depending on the strategies ultimately adopted by the NDA, chemical processing or waste conditioning is a suitable option to deal with a range of legacy fuels and residues in the UK. There is therefore benefit in developing flexible, simplified and intensified separation flowsheets that handle a wide range of feeds, with minimised processing costs and waste generation, that could be operated from miniature to plant scales. This work builds on recent progress in developing 'advanced Purex' reprocessing flowsheets,[3] which aim to fully control U, Np and Pu within a single solvent extraction cycle, utilising annular centrifugal contactors for mixing and settling. Recent focus has been on U/Pu separation, demonstrating that hydroxamic acid complexants can strip Pu(IV) from U(VI) at high solvent Pu concentrations, such as may be encountered in Pu residues processing. A series of flowsheets ranging from 7 to 40 wt. % Pu concentrations in the initial aqueous feed have been very successfully tested at Sellafield.[4-5]. The flowsheet with a 40 wt.% Pu feed (equivalent to ~100 g/l Pu) for instance, obtained excellent U and Pu recoveries of >99.9998 and >99.997 % respectively. The Pu in U DF obtained was 1.45E6, which equates to Pu content in the U product of 0.449 mgPu/kgU. The U in Pu DF was relatively low (360), but it is proposed that this is sufficient for any likely use of the Pu product.

5 ACTINIDE SPECIATION IN WASTE STREAMS GENERATED DURING NUCLEAR CLEAN UP

There is a clear need to have speciation methods for actinides at very low concentrations, typical of effluent streams or environmental measurements. Speciation, rather than concentration, is often the key determinant of the behaviour of a particular element in a process. Hence, knowing speciation can enable the behaviour of the species to be better understood and more effective process enhancements targeted. The identity of the radionuclide, its oxidation state, charge (cationic, anionic or neutral), coordination or solvation, and phase (solution, solid or colloidal – true or pseudo colloids) will all affect its behaviour in a processing plant or in the environment.

It seems probable that the concentration of TRU elements will range from $<10^{-21}$ to $>10^{-3}$ M, across the broad spectrum of wastes that will be generated by nuclear clean up programmes. Therefore, a spectrum of techniques from chemical/ radiochemical methods to inorganic spectroscopic methods will undoubtedly be required. However, a key range is perhaps between $\sim10^{-12}$ and 10^{-6} M, which covers the majority of the feeds from legacy wastes and processing operations to Sellafield site effluent treatment plants. Advances in actinide spectroscopy using techniques such as laser induced photoacoustic spectroscopy, laser induced breakdown spectroscopy, time resolved laser induced fluorescence, coupled ICP-MS methods, surface enhanced Raman spectroscopy, all offer direct instrumental analysis of actinides at concentrations $<10^{-5}$M.[6-7]

6 THE BEHAVIOUR OF PuO_2 UNDER CONDITIONS OF LONG TERM STORAGE

Irrespective of the time-scale of any decision relating to a national strategy for the management of plutonium, the long term storage of PuO_2 in cans is a likely scenario. There are therefore significant benefits in more fully understanding the fundamental behaviour of PuO_2 under these conditions. This is required so that extrapolating to timescales beyond our current operational experience is possible using rigorously underpinned technical data. In particular, the various chemical and radiolytic mechanisms that may lead to currently unpredicted changes in the surface morphology of the PuO_2 disturbing the equilibrium environment within the can may be important. Such changes may then lead to the evolution of gases and hence can pressurisation. One such issue has been the recently raised possibility of a plutonium (V) or (VI) species formed with the liberation of H_2.[8-9]

References

1 (a) http://www.dti.gov.uk/nuclearcleanup/; (b) http://www.dti.gov.uk/energy/
2 http://www.nexiasolutions.com/
3 O.D Fox *et al.*, *Actinide Separations for the 21st Century*, ACS forthcoming.
4 J.E. Birkett *et al.*, in: Proc. ATALANTE 2004 Advances for nuclear fuel cycles, CEA, Nimes, 2004, O12-08.
5 J.E. Birkett, M.J. Carrott, O.D Fox, G. Crooks, C.J. Jones, C.J. Maher, C.V. Roube, R.J. Taylor and D. Woodhead, Proc. Actinides 2005, 2005, this volume.
6 *Actinide Separation Chemistry in Nuclear Waste Streams and Materials*, Nuclear Energy Agency Nuclear Science Committee Report, 1997, NEA/NSC/DOC(97)19, OECD.
7 *An overview of actinide redox speciation methods: their applicability, advantages and limitations. Workshop proceedings, Evaluation of Speciation Technology*, 1999, OECD/NEA.
8 J.M. Haschke *et al.*, *Science,* 2000, **287**, 285.
9 S.D. Conradson *et al.*, *Inorg. Chem.*, 2003, **42**, 3715.

EXTENDING THE CHEMISTRY OF THE URANYL ION: $\{UO_2\}^{2+}$ ACTING AS A LEWIS ACID AND A LEWIS BASE

M.J. Sarsfield,[1] and M. Helliwell,[2]

[1] Nexia Solutions, Sellafield, Seascale, Cumbria, CA20 1PG, UK. e-mail: mark.sarsfield@nexiasolutions.com
[2] School of Chemistry, The University of Manchester, Manchester M13 9PL, UK.

1 INTRODUCTION

Actinyl ions $[AnO_2]^{x+}$ are a feature of high oxidation state early *5f* element chemsitry. The trans arrangement of the oxo ligands originates from the combination of metal *5f/6d* and oxygen *2p* orbitals to form stable An=O bonds, with the stability decreasing from U to Am.[1-2] A curious characteristic of $[An^{V}O_2]^{+}$ species is their ability to act as ligands to other metal centres, bonding through one of the actinyl oxygens in a An=O—M^{x+} motif.[3] For example, at sufficient concentrations $[Np^{V}O_2]^{+}$ is capable of forming "cation-cation" complexes with $[U^{VI}O_2]^{2+}$ (through Np=O—U bonding) under 3M HNO_3 conditions.[4] Although $[An^{V}O_2]^{+}$ oxo ligands can behave "Lewis base like", $[An^{VI}O_2]^{2+}$ species do not exhibit these tendencies in solution. There is, however, a growing body of evidence that $[U^{VI}O_2]^{2+}$ oxygens have the potential to behave as Lewis bases and coordinate to electron deficient centres. In the solid state, uranyl oxygens are known to interact with cations in inorganic uranate compounds[5] and there are an increasing number of discrete molecules containing U=O-M^{x+} structural motifs (M = Li, Na, NH_4).[6]

We believed that, through tailoring the coordination environment in the equatorial plane, we could further enhance the Lewis basic properties of $[U^{VI}O_2]^{2+}$ oxygens and promote direct coordination to a Lewis acid.

2 METHOD AND RESULTS

2.1 Uranyl benzaminato complexes

The benzaminato ligand $Na[PhC(NSiMe_3)_2]$ or Na[NCN] (**I**) is easily prepared by the reaction of an equivalent of $Na[N(SiMe_3)_2]$ with benzonitrile C_6H_5CN.[7] Addition of 2 equivalents of **I** to a thf solution of $UO_2Cl_2(thf)_3$ gives the thf adduct $UO_2(NCN)_2(thf)$ (**1**) in moderate yield (Scheme 1). By adding 2 equivalents of $B(C_6F_5)_3$ to **1** the expected removal of coordinated thf and formation of a U=O-B bond takes place giving an immediate colour change from orange to deep magenta. Multinuclear NMR spectroscopy (^1H, ^{19}F, ^{11}B and ^{13}C) indicates the formation of $UO\{OB(C_6F_5)_3\}(NCN)_2$ **2** (Scheme 1).

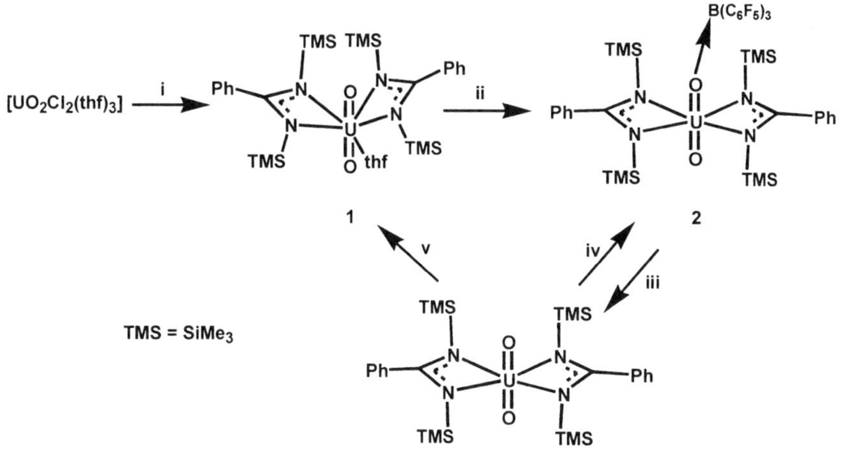

Scheme 1. *Preparation of compounds **1, 2** and **3**. i) 2 Na(NCN); ii) 2 eq $B(C_6F_5)_3$; iii) PMe_3; iv) 1 eq $B(C_6F_5)_3$, v) thf.*

The thf free derivative of **1**, $UO_2(NCN)_2$ (**3**) can be isolated and behaves as a Lewis acid and a Lewis base. It can coordinate thf to reform **1** or reacts directly with the strong Lewis acid $B(C_6F_5)_3$ to give **2** (Scheme 1).[8]

Crystals of **2** suitable for X-ray diffraction were grown from a hexane/benzene solution and confirm the structure (Figure 2). The uranyl unit remains essentially linear (O1-U1-O2, 177.45(14)°), but the coordinated oxo ligand has an elongated U=O bond (U1-O1, 1.898(3) Å) compared to the uncoordinated one (U1-O2, 1.770(3) Å) and to **3** (1.750(4) Å) the latter being typical for uranyl complexes.[9]

The U=O-B interaction appears stronger than alkali metal interactions U=O-M (M = Li, Na) judging by the changes observed in U=O bond lengths. For example, the amido complexes $[thf_2Na][UO_2(NCN)_3]$ (U=O, 1.812(3), 1.783(3) Å)[7] and $[thf_2Na][UO_2\{N(SiMe_3)_2\}_3]$ (U=O, 1.810(5), 1.781(5) Å)[10] contain one short and one elongated U=O bond, but this elongation is significantly less than in **2** (U1-O1, 1.898(3) Å).

Coordination to the boron displaces electron density away from the U=O bond leading to a reduced O=U=O stretching frequency from 818 cm^{-1}(**3**) to 780 cm^{-1} (**2**) (solid state). This is maintained in solution with Raman (C_6D_6) (O=U=O, 778 cm^{-1}) and ^{11}B NMR chemical shift (-10.6 ppm, *c.f.* 60 ppm for $B(C_6F_5)_3$) indicating that the U=O-B bonding remains intact. Toluene solutions of complex **2** show a number of intense absorptions including a broad unstructured band at λ_{max} = 600 nm that tails off at 710 nm and is not present in solutions of **3**, giving a deep magenta colour for **2**

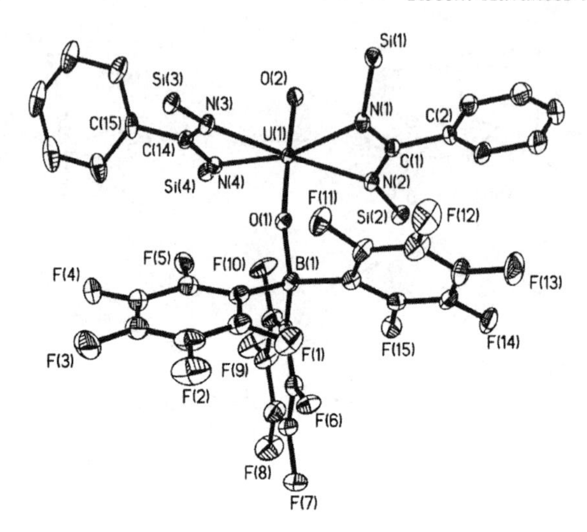

Figure 2 *ORTEP diagram of 2 at the 50% ellipsoid level. (The methyl groups on the Si atoms of 2 are omitted for clarity)*

3 CONCLUSION

We have shown that a uranyl complex can exhibit both Lewis acid and Lewis base behaviour with careful selection of equatorial ligand (NCN). Fundamental changes to the bonding in the uranyl can be probed by vibrational and electronic spectroscopy. The concept of strong equatorial ligands increasing charge on the uranyl oxygens, could be important in determining uranyl speciation in highly alkaline systems of relevance to waste storage operations. Species such as $[UO_2(OH)_4]^{2-}$ are known to lengthen the U=O bond and cause the oxo ligands to participate in a hydrogen bonding network.

We thank BNFL, the Nuclear Decommissioning Authority (NDA) and Nexia Solutions for funding (MJS).

References

1 J. J. Katz, L. R. Morss, G. T. Seaborg, in 'The Chemistry of the Actinide Elements', ed. J. J. Katz, L. R. MorssG. T. Seaborg, London, 1986.
2 R. G. Denning, *Struct. Bonding (Berlin)*, 1992, **79**, 215.
3 N.N. Krot, M.S. Grigoriev, *Russ. Chem. Rev.*, 2004, (1),89-100
4 R. Gauthier, V. Ilmstadter, K. H. Lieser, *Radiochim. Acta*, 1983, **33**, 35.
5 W. H. Zachariasen, *Acta Crystallographica*, 1948, **1**, 281.
6 D. M. Barnhart, C. J. Burns, N. N. Sauer, J. G. Watkin, *Inorg. Chem.*, 1995, **34**, 4079.
7 M. J. Sarsfield, M. Helliwell, J. Raftery, *Inorg. Chem.*, 2004, **43**, 3170.
8 M. J. Sarsfield, M. Helliwell, *J. Am. Chem. Soc.*, 2004, **126**, 1036.
9 XRD, in 'This data was collected from the Cambridge crystallographic database. From the 587 hits of uranyl structures with RFAC< 10%, the mean U=O distance is 1.762 Å with a standard deviation of 0.033 Å. This relates to an approximate range of 1.696 -1.828 Å within a 95% confidence limit', 2003.
10 C. J. Burns, D. L. Clark, R. J. Donohoe, P. B. Duval, B. L. Scott, C. D. Tait, *Inorg. Chem.*, 2000, **39**, 5464.

COVALENCY IN URANIUM COMPLEXES AND NEW *f*-ELEMENT CHEMISTRY THAT EXPLOITS SOFT ORGANOMETALLIC CARBENE LIGANDS

P.L. Arnold, S.T. Liddle, S.A. Mungur, M.R. Rodden and C. Wilson

University of Nottingham, School of Chemistry, University Park, Nottingham NG7 2RD, UK.

1 INTRODUCTION

N-heterocyclic carbene (NHC) ligands are heterocycles that bind as soft, two-electron sigma-donors through the NCN carbon atom, and are now used widely to support late transition metal homogeneous catalysts. However, very little chemistry of the electropositive metal-NHC fragment has been reported to date.

We have previously used anionic-functional groups pendant to the NHC to synthesise, and study the reactivity and hemilability of the NHC on lanthanide complexes.[1] Previously, two monodentate neutral NHC adducts have been reported on U(III) and U(IV) centres,[2] but no further chemistry of these systems has been reported, while we have explored the use of the uranyl group in observing the relative binding strength of different carbenes to U(VI).[3] Since the NHC groups should be unusually soft ligands for actinide cations, we have now extended this chemistry to lower oxidation state uranium chemistry, since a comparison of the lanthanide and actinide carbene complexes may provide a better understanding f-electron covalency, an important factor in radionuclide separation technology. For this work, we have used the alkoxide-NHC ligand $[OCMe_2CH_2(1\text{-}C\{NCHCHNPr^i\})]^-$ $[L]^-$ to support the uranium cations.[4]

2 RESULTS AND DISCUSSION

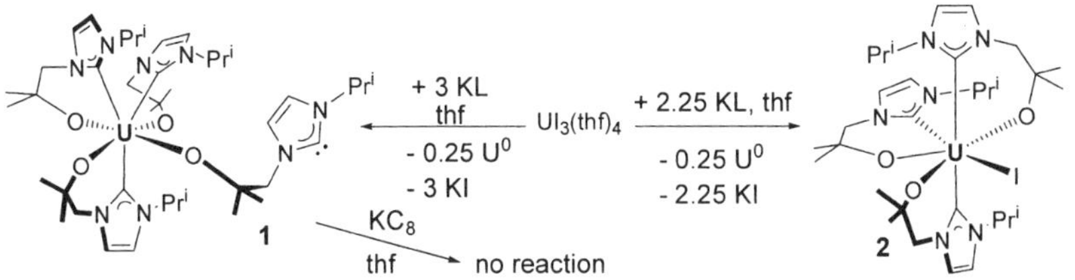

Scheme 1.

The reaction of $[U(I)_3(THF)_4]$ with the potassium alkoxide-NHC [KL] in thf affords two different U(IV) products, depending on the stoichiometry of the reaction, derived from three quarters of the U(III) reagent, and a quarter equivalent of uranium metal (Scheme 1). The complex $[UL_4]$ **1**, made from three equivalents of [KL][5], and the complex $[U(I)(L)_3]$ **2**, made from 2.25 equivalents of [KL], are both isolable from the KI and U(0) byproducts by filtration and recrystallisation from toluene, in 70 and 80 % yields respectively.[6] Complex **1** is a bright emerald green whilst **2** is a dark yellow, a more commonly observed colour for these U(IV) systems. Both are very air and moisture sensitive.

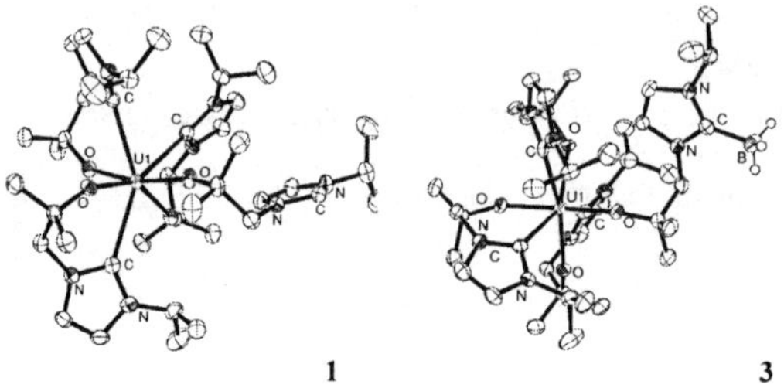

1 **3**

Figure 1. *The X-ray crystal structures of the bridged potassium carbene complex [KL] **1**, and the BH_3 adduct **3** (30 % ellipsoids). All hydrogens and methyl groups omitted for clarity.*

An ellipsoid drawing of the molecular structure of **1** is contrasted with that of **3** in Figure 1. In the solid state molecular structure of **1** one carbene remains uncoordinated. The uranium cation is seven coordinate, with an approximately pentagonal bipyramidal coordination geometry defined by four discrete alkoxide groups and only three out of the four carbene groups. There are no measurable intermolecular contacts; to the best of our knowledge, this is the first metal carbene complex in which a carbene group remains uncoordinated.

The dangling NHC group is not detected in the NMR spectrum of **1** at room temperature, since the molecule displays a fluxional process presumably involving all four carbene groups; the ^{1}H NMR spectrum contains two broad resonances at 298 K at 17 and -6 ppm (fwhm > 1000 Hz). Cooling a d_8-toluene solution of **1** to 228 K, the sharp ^{1}H NMR spectrum anticipated for a U(IV) complex is observed between 95 and -60 ppm, and assignable to four separate ligand environments, one of which appears to be less paramagnetically shifted.

The lability of the carbene has allowed us to trap small molecules with these complexes; this is demonstrated by the reaction of **1** with Lewis base-coordinated borane *eg* $BH_3.SMe_2$,[6] and triphenylborane, Scheme 2.

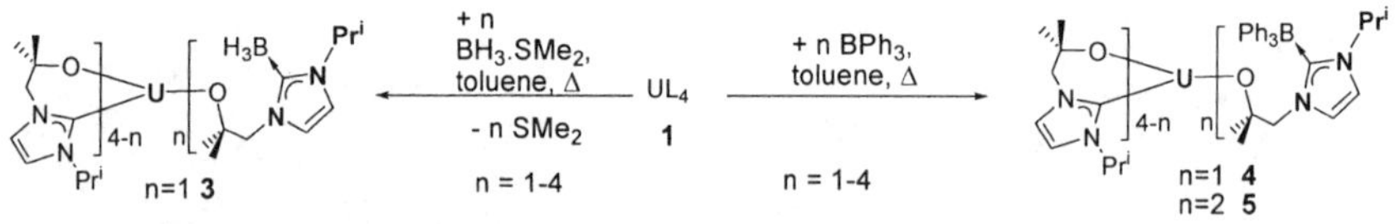

Scheme 2. *Derivatisation of the free and U(IV)-bound carbene groups with functionalised boranes*

Either borane may be titrated into the solution of **1** to yield products containing up to four insertions, according to mass spectrometry and NMR spectra data, but the increasing insolubility of the complexes with increasing numbers of insertion products has led us only to characterise the lower molecular weight adducts. Thus treatment of a dark green toluene solution of **1** with a toluene solution of $BH_3.SMe_2$ affords a paler green solution, from which a grass green powder precipitates readily upon cooling or concentrating the solution and is isolated in 80 % yield. The powder is characterised as [U(L)$_3$(OCMe$_2$CH$_2$(1-CBH$_3${NCHCHNPr^i}))] **3** by elemental analysis and ^{1}H NMR spectroscopy. The C-B distance in the molecular structure, of 1.609(7) Å, (Figure 1.) is similar to that of previously synthesised carbene borane adducts.[7] It has been suggested that such adducts may be useful starting materials for the preparation of metal NHC complexes,[8] but these data suggest that this will not hold true generally for electropositive metal NHC complexes. One equivalent of SMe$_2$ is released in the reaction, but this does not participate further. The treatment of a toluene/thf solution of **1** with a toluene solution of BPh$_3$ affords a paler, yellow-green solution, which is most readily isolated by recrystallisation from thf, to afford a solid characterised as [U(L)$_3$(OCMe$_2$CH$_2$(1-CBPh$_3${NCHCHNPr^i}))] **4**.

Interestingly, addition of a second equivalent of BPh$_3$ does not result in the formation of BPh$_3$.thf. Instead, a less soluble, pale green compound is isolated from thf that is characterised as [U(L)$_2$(OCMe$_2$CH$_2$(1-CBPh$_3${NCHCHNPr^i})$_2$] **5**.[9] Although single crystal samples of **4** were unsuitable for an X-ray diffraction experiment, elemental analysis, NMR spectroscopy and mass spectrometry suggest the formulation drawn in Scheme 2. Preliminary data from reactions of **3** and **4** with three further equivalents of BH$_3$.SMe$_2$ or BPh$_3$ respectively, in thf suggest that the relatively insoluble homoleptic complexes (n = 4 in Scheme 2) in which the U centre is no longer carbene-ligated, are stable; work is in hand to identify the structure of these potentially unusual coordination complexes.

3 CONCLUSIONS

In conclusion, new U(IV) organometallic complexes of anionic alkoxide-N-heterocyclic carbenes have been synthesised and structurally characterised, and display an example of a metal-complex with a pendant unbound carbene group. When tethered by an adjacent alkoxide group to a uranium(IV) cation, a nucleophilic N-heterocyclic carbene group can behave as a simple carbene ligand to bring in potentially reactive fragments or molecules to the primary coordination sphere of the uranium cation. Further work to study the insertion chemistry of atoms and diatomic molecules, and the use of these complexes in bifunctional catalysis is in progress.

The authors thank the EPSRC and the Royal Society for funding.

References

1. P. L. Arnold, S. A. Mungur, A. J. Blake and C. Wilson, *Angew. Chem. Int. Ed.* **2003**, *42*, 5981.
2. W. J. Evans, S. A. Kozimor, J. W. Ziller, *Polyhedron* **2004**, *23*, 2689. H. Nakai, X. L. Hu, L. N. Zakharov, A. L. Rheingold, K. Meyer, *Inorg. Chem.* **2004**, *43*, 855.
3. S. A. Mungur, S. T. Liddle, C. Wilson, M. J. Sarsfield, P. L. Arnold, *Chem. Commun.* **2004**, 2738.
4. P. L. Arnold, M. Rodden, K. M. Davis, A. C. Scarisbrick, A. J. Blake, and C. Wilson, *Chem. Commun.*, 2004, 1612.
5. P. L. Arnold, M. Rodden, and C. Wilson, *Chem. Commun.* **2005**, 1743.
6. P. L. Arnold, A. J. Blake, and C. Wilson, *Chem. Eur. J.*, **2005**, *11*, 6095.
7. N. Kuhn, G. Henkel, T. Kratz, J. Kreutzberg, R. Boese, A. H. Maulitz, *Chem. Ber.* **1993**, *126*, 2041. A. Wacker, C. G. Yan, G. Kaltenpoth, A. Ginsberg, A. M. Arif, R. D. Ernst, H. Pritzkow, and W. Siebert, *J. Organomet. Chem.*, **2002**, *641*, 195.
8. Padilla-Martinez, II, F. J. Martinez-Martinez, A. Lopez-Sandoval, K. I. Giron-Castillo, M. A. Brito, and R. Contreras, *Eur. J. Inorg. Chem.*, **1998**, 1547.
9. For **5**. ^{1}H NMR (d_5-pyridine) δ 42.8, 18.1, 15.0, 2.4, -14.6, -17.4 (s, 1H) 15.4, -12.3 (s, 2H), -6.8 (s, 3H), 41.7, 5.1, 1.2, -1.2, -4.6, -8.0 (s, 6H). Analysis (low C probably due to incomplete combustion of the metal): Calc. for $C_{76}H_{98}N_8O_4B_2U$ C 63.07, H 6.82, N 7.74. Found C 50.26, H 7.22, N 7.85.

VARIABILITY IN THE CRYSTAL STRUCTURES OF URANIUM(VI), NEPTUNIUM(VI) AND PLUTONIUM(VI) PHTHALATES

M.S. Grigoriev, I.A. Charushnikova, A.A. Bessonov and N.N. Krot

Institute of Physical Chemistry, Russian Academy of Sciences,
31 Leninsky prosp., Moscow, 119991, Russia

1 INTRODUCTION

A very limited amount of single crystal X-ray structural data is available for the compounds of the elements heavier than neptunium due to their high radioactivity. In the case of Pu(VI), an additional reason for the absence of single crystal X-ray data is the belief that Pu(VI) complexes will be analogous to known U(VI) complexes.. However, a first single crystal structural study of inorganic compound, $[PuO_2(IO_3)_2]\cdot H_2O$,[1] has shown that the composition of the coordination polyhedron (CP) of the Pu atom in this compound differs from that in its U analogue, $[UO_2(IO_3)_2H_2O]$. One more example was found from powder diffraction study of An(VI) silicates: $[(PuO_2)_2SiO_4(H_2O)_2]$ crystallizes in tetragonal space group whereas its U and Np analogues crystallise in orthorhombic space groups.[2,3] Here we present results of structural study of two series of U(VI), Np(VI) and Pu(VI) orthophthalates, showing both isostructurality and non-isostructurality of Pu(VI) compounds with their U and Np analogues.

2. METHOD AND RESULTS

It has been shown that in neutral phthalate solutions with added Np(VI) or Pu(VI) with molar ratio AnO_2^{2+}:L^{2-} = 1:1 (H_2L – orthophthalic acid) at 0°C the compounds $AnO_2L\cdot 2H_2O$ can be crystallized.[4] Np and Pu compounds have similar composition but different powder diffraction patterns. The powder diffraction pattern of the Np compound is close to that of the know U analogue.[5] The same solutions heated up to 70°C give crystalline products $AnO_2L\cdot 1\frac{1}{3}H_2O$ with the same structure for U, Np and Pu. Single crystals have been prepared and X-ray structural analysis has been carried out for rhombohedral $[NpO_2L(H_2O)]\cdot\frac{1}{3}H_2O$ (structural type **I**), triclinic $[UO_2L(H_2O)]\cdot H_2O$ (structural type **II**) and monoclinic $[PuO_2L(H_2O)]\cdot H_2O$ (structural type **III**) (the formulae given take into account the results of the structural study). Details of synthesis, crystal data, structure determination and refinement can be found in full papers.[6,7]

The actinide atoms in all 3 compounds are seven-coordinated. The CP around the An atoms can be described as distorted pentagonal bipyramids with two "yl" oxygen atoms in the apical positions. The equatorial positions are occupied by the four oxygen atoms of phthalate anions and one water molecule. The interaction between actinide dioxocations and phthalate anions in **I** (U, Np, Pu) results in formation of infinite "nanotubes" (Figure 1) filled with zeolite-type water molecules. The crystals **II** (U and Np) contain infinite ribbons (Figure 2), whereas **III** (the Pu compound) has a layer structure (Figure 3). In all three structural types phthalate anions occupy four coordination sites but the coordination types are different. In **I** each tetradentate-bridging phthalate anion is linked in monodentate way to four different An atoms. In **II** each phthalate anion forms seven-member cycle with one An atom and is linked monodentately to two other An atoms. This coordination type was found also in $[UO_2L(Urea)]\cdot\frac{1}{2}H_2O$.[8] In **III** each phthalate anion forms four-member cycle with one Pu atom and is linked in monodentate way to two other Pu atoms. The crystal structures **I** and **III** give novel coordination types for phthalate anions in actinyl(VI) compounds. It interesting to note that the principal change of phthalate anion coordination mode between structural types **II** and **III** consists in replacement of seven-member cycle by four-member one. Such change is consistent with actinide contraction because four-member cycle occupies less space in the equatorial plane of the actinyl group.

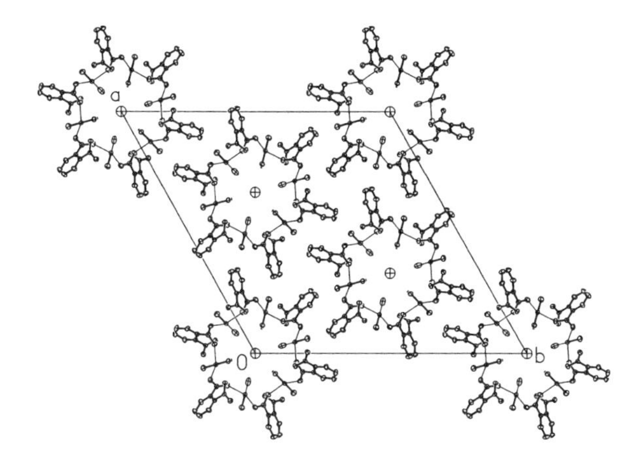

Figure 1 *Packing diagram of the structure **I***

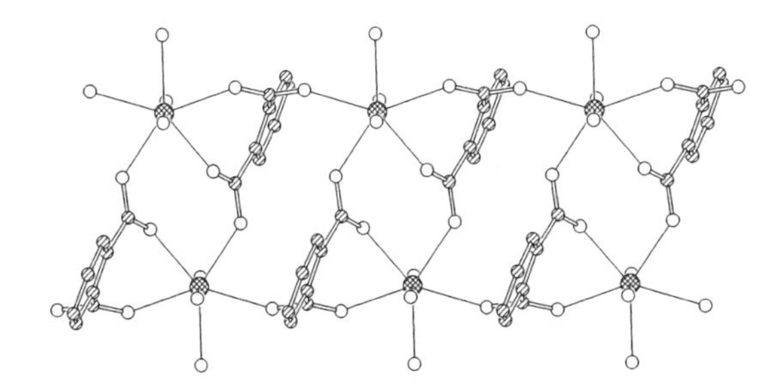

Figure 2 *Electroneutral ribbon in the structure **II***

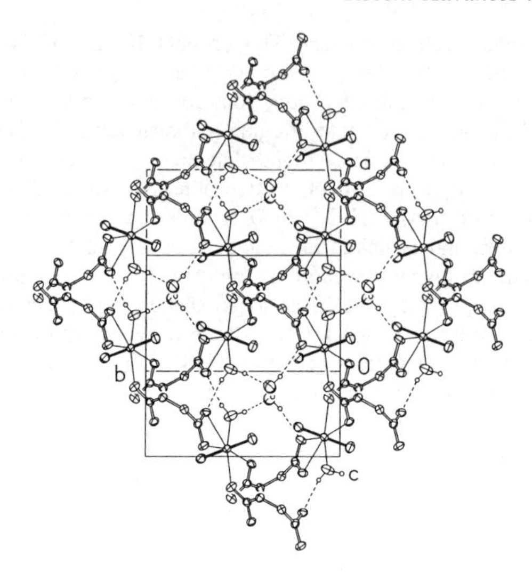

Figure 3 *Electroneutral layer in the structure **III**. Only two C atoms of each phenyl ring are shown. H-bonds are shown with dashed lines*

3. CONCLUSION

The present study has shown that the phthalate ligand can coordinate in different ways to U(VI), Np(VI) and Pu(VI) resulting in the formation of different crystal structure types: tubes, layers, ribbons. This was the case even for the same coordination number and CP composition. The fact that Pu(VI) compound studied here is not isostructural with the U(VI) and Np(VI) analogues reinforces the importance of single crystal structural investigations for similar composition actinide compounds.

References

1 W. Runde, A.C. Bean, T.E. Albrecht-Schmitt and B.L. Scott, *Chem. Comm.,* 2003, 478.
2 A.A. Bessonov, M.S. Grigoriev, A.B. Ioussov, N.A. Budantseva and A.M. Fedosseev, *Radiochim. Acta*, 2003, **91**, 339.
3 M.S. Grigor'ev, A.A. Bessonov, V.I. Makarenkov and A.M. Fedoseev, *Radiochemistry*, 2003, **45**, 257.
4 N.N. Krot, A.A. Bessonov, M.S. Grigor'ev, I.A. Charushnikova and V.I. Makarenkov, *Radiochemistry*, 2004, **46**, 421.
5 M.V. Gonzalez and J.B. Polonio, *Anales Quim*. 1975, **71**, 675.
6 M.S. Grigoriev, M.Yu. Antipin, N.N. Krot and A.A. Bessonov, *Radiochim. Acta*, 2004, **92**, 405.
7 I.A. Charushnikova, N.N. Krot and Z.A. Starikova, *Radiochemistry*, 2004, **46**, 556.
8 R.N. Shchelokov, Yu.N. Mikhailov, I.M. Orlova, A.V. Sergeev, Z.R. Ashurov, M.T. Tashev and N.A. Parpiev, *Koordinatsionnaya Khimiya*, 1985, **11**, 1144.

NOVEL NEPTUNYL(V) AND PLUTONYL(V) Bi-OXALATE COMPOUNDS: SYNTHESIS, PROPERTIES AND STRUCTURE

M.S. Grigoriev, A.A. Bessonov, I.A. Charushnikova and N.N. Krot

Institute of Physical Chemistry, Russian Academy of Sciences,
31 Leninsky prosp., Moscow, 119991, Russia

1 INTRODUCTION

Double oxalates of Np(V) with $[Co(NH_3)_6]^{3+}$ counter-cations are know and have been structurally characterized.[1] It is known also from literature[2] that Pu(V) in oxalate solutions is inclined to disproportionation but, under certain conditions, can be stabilised. This gives a possibility for attempts to synthesize Pu(V) compounds analogous to known Np(V) compounds. Here we present the results of the synthesis and characterisation of novel Pu(V) and Np(V) oxalates including first single crystal X-ray structural investigation of plutony(V) compound.

2. METHOD AND RESULTS

Novel oxalate compounds of general formula $[Co(NH_3)_6][PuO_2(C_2O_4)_2]\cdot nH_2O$ with n = 2, 3, 5 have been isolated by $[Co(NH_3)_6]Cl_3$ addition to freshly prepared PuO_2^+ neutral aqueous oxalate solutions. The number of water molecules depends on the conditions of synthesis, primarily temperature. In addition to known neptunyl(V) compounds with 3 and 4 water molecules, di- and penta- hydrates have also now been prepared. However, all attempts to prepare the tetrahydrate of the plutonyl(V) double oxalate were unsuccessful.

From powder X-ray diffraction, all neptunyl(V) and plutonyl(V) compounds of the similar composition are isostructural. Thermal behaviour of the compounds has been studied up to 800°C, and their IR spectra have been measured. Optical spectra for novel Np(V) compounds also have been obtained. Details of the synthesis, spectroscopic and thermal properties can be found in a full paper.[3] From the optical spectra of the Np complexes a hexagonal bipyramidal environment around the An atom was determined for pentahydrate compounds with pentagonal bipyramidal coordination environments proposed for all other hydrates. All Pu(V) compounds were found to be stable for at least 3 weeks in air.

A single crystal X-ray structural investigation of $[Co(NH_3)_6][PuO_2(C_2O_4)_2]\cdot 3H_2O$ has shown it to be isostructural with the Np(V) compound. The details of the structure determination are given in a full paper.[4] Dimeric complex anions, $[PuO_2(C_2O_4)_2]_2^{6-}$, (Figure 1) are a basis of the structure with the coordination polyhedra of the Pu atoms (distorted pentagonal bipyramids) sharing a common equatorial edge due to bridging O atoms of oxalate anions. A similar dimeric complex anions has also been observed in $[Co(NH_3)_6][NpO_2(C_2O_4)_2]\cdot 2H_2O$.[5]

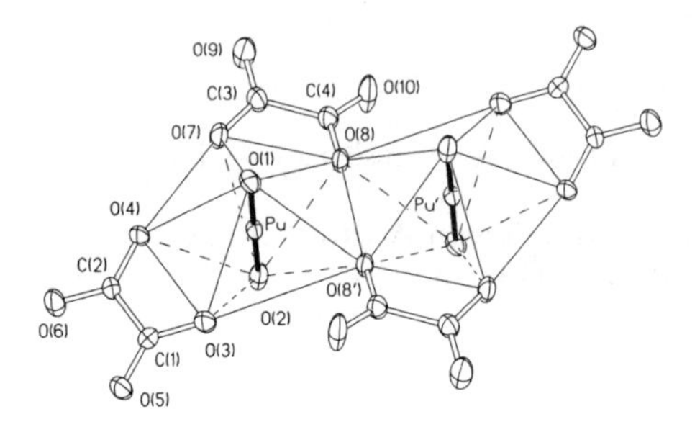

Figure 1 *The dimeric complex anion* $[PuO_2(C_2O_4)_2]_2^{6-}$ *in the structure of* $[Co(NH_3)_6][PuO_2(C_2O_4)_2]\cdot 3H_2O$

The An-O distances in the isostructural Pu(V) and Np(V) trihydrate complexes are given in Table 1. The average Pu-O distances are 1.812 Å for the axial plutonyl oxygens and 2.455 Å for the equatorial O atoms. The corresponding distances for the isostructural Np(V) complex are 1.838 and 2.476 Å respectively. While the bond lengths are comparable there appears to be a slight decrease in An-O distances both in axial and equatorial directions at the transition from Np to Pu. This is probably a result of actinide contraction.

Table 1 *Selected interatomic distances (Å) in dimeric complex anions* $[AnO_2(C_2O_4)_2]_2^{6-}$ *(atom numbering corresponds to Figure 1)*

Distance	*Pu (this work)*	*Np*[1]
An-O(1)	1.817(5)	1.841(5)
An-O(2)	1.807(5)	1.835(5)
An-O(3)	2.431(5)	2.463(5)
An-O(4)	2.436(5)	2.457(5)
An-O(7)	2.435(5)	2.447(5)
An-O(8)	2.476(4)	2.504(4)
An-O(8')	2.497(4)	2.510(4)

3. CONCLUSION

The results presented here prove the structural similarity of Np(V) and Pu(V) compounds although the conditions of the synthesis differ which sometimes makes it impossible to obtain isostructural compounds. Further single crystal X-ray studies of isostructural pairs of Np(V) and Pu(V) compounds is needed to elucidate the phenomenon of actinide contraction in a range of different complexes.

References

1 M.S. Grigor'ev, N.A. Baturin, L.L. Regel' and N.N. Krot, *Radiokhimiya*, 1991, **33**(2), 19.
2 N.P. Ermolaev, N.N. Krot and A.D. Guelman, *Radiokhimiya*, 1967, **9**, 171.

3 N.N. Krot, A.A. Bessonov, M.S. Grigor'ev, I.A. Charushnikova and V.I. Makarenkov, *Radiokhimiya*, 2005, **47**, 105.
4 M.S. Grigor'ev, M.Yu. Antipin, N.N. Krot, A.A. Bessonov, *Radiokhimiya* (in press).
5 I.A. Charushnikova, N.N. Krot and I.N. Polyakova, *Radiokhimiya* (in press).

SYNTHESIS AND CRYSTAL STRUCTURE OF A NEW Np(V) OXALATE, $Na_4(NpO_2)_2(C_2O_4)_3 \cdot 6H_2O$

I.A. Charushnikova, N.N. Krot and A.A. Bessonov

Institute of Physical Chemistry, Russian Academy of Sciences,
Leninskii pr., 31, Moscow, 119991, RUSSIA

1 INTRODUCTION

In diluted neutral Np(V) solutions, stable $[NpO_2C_2O_4]^-$ and $[NpO_2(C_2O_4)_2]^{3-}$ complexes are formed stepwise with increasing concentration of oxalate ions[1]. A series of $MNpO_2C_2O_4 \cdot nH_2O$ (where M is a singly charged cation) and $Co(NH_3)_6NpO_2(C_2O_4)_2 \cdot nH_2O$ oxalates were isolated in the solid state and characterized by various methods.[2-5] However, salts of neptunyl(V) oxalate with 2:3 NpO_2:C_2O_4 ratio have not been reported yet. In the course of our studies of the behaviour of Np(V) ions in water-ethanol solutions of sodium oxalate, we isolated the first compound containing the 2:3 complex. It was found that from solutions with high Np(V) concentration and molar ratio NpO_2:C_2O_4 from 1:1.3 to 1:3.5 the new compound can be crystallized as a mixture with $NaNpO_2C_2O_4 \cdot 3H_2O$ mono-oxalate.[3] Its composition was determined by the X-ray diffraction study as $Na_4(NpO_2)_2(C_2O_4)_3 \cdot 6H_2O$. Our attempts to prepare Np(V) oxalate solid compounds with 1:3 stoichiometry were unsuccessful.

2 METHOD AND RESULTS

2.1 Synthesis

Single crystals of $Na_4(NpO_2)_2(C_2O_4)_3 \cdot 6H_2O$ suitable for X-ray analysis were obtained as follows. A sample of 0.2 M NpO_2NO_3 solution was diluted twofold with water and treated with 0.2 M $Na_2C_2O_4$ solution in the NpO_2:C_2O_4 = 1:1.3 molar ratio. The solution was then treated with an equal volume of ethanol. After 13 days, large bright green platelike crystals were formed. One of them was chosen for the X-ray diffraction study.

2.2 X-ray diffraction and spectrophotometric study

The X-ray diffraction experiment was performed on an Enraf-Nonius CAD-4 four-circle diffractometer (λMoK_α, graphite monochromator, $\omega/2\theta$ scan mode) at room temperature. The parameters of the monoclinic unit cell are as follows: a = 14.7909(14), b = 17.656(2), c = 10.3626(10), = 126.423(10) , space group C2/c, Z= 4, d = 3.057 g/cm^3. The structure was solved by the direct method. The non-hydrogen atoms were refined by the full-matrix least-squares procedure on F^2 in the anisotropic approximation of thermal vibrations (161 parameter). The final discrepancy factors were R(F) = 0.0571 and wR(F^2) = 0.1312 for 2841 reflections with I>2σ(I).

The spectrophotometric studies were performed on a Shimadzu UV-3100 device (Japan). The IR spectra were recorded on a Specord M80 spectrometer. Thermograms were obtained on a Paulik-Paulik- Erday Q-1500D device.

2.3 Results and Discussion

The crystal structure of $Na_4(NpO_2)_2(C_2O_4)_3{\cdot}6H_2O$ consists of Np(V) dioxo cations, hydrated sodium cations, oxalate anions, and water molecules of crystallization. The structure of the dioxo cation is close to linear, the mean Np-O bond length is 1.820(5) Å and the ONpO angle is 178.8(2)°. In the equatorial plane, the dioxo cations are surrounded by five oxygen atoms of three oxalate anions (Figure 1). The mean Np-O_{Ox} distance is 2.473(5) Å.

The structure contains two crystallographically unique oxalate anions **Ox1** and **Ox2** (Figure 1). The **Ox1** anion is a tridentate bridging chelate ligand. It shows bidentate coordination through atoms O(1) and O(3), to one Np atom with monodentate coordination through the O(2) atom to a neighboring Np atom. The **Ox2** anion is a centrosymmetric tetradentate bridging chelate ligand (Fig. 1). It coordinates bidentately to both the Np and Np*b* atoms.

The structure contains two crystallographically unique cations Na(1) and Na(2). Each Na is surrounded by six oxygen atoms of oxalate anions and water molecules. Coordination polyhedra of Na are linked through common edges in the [100] direction.

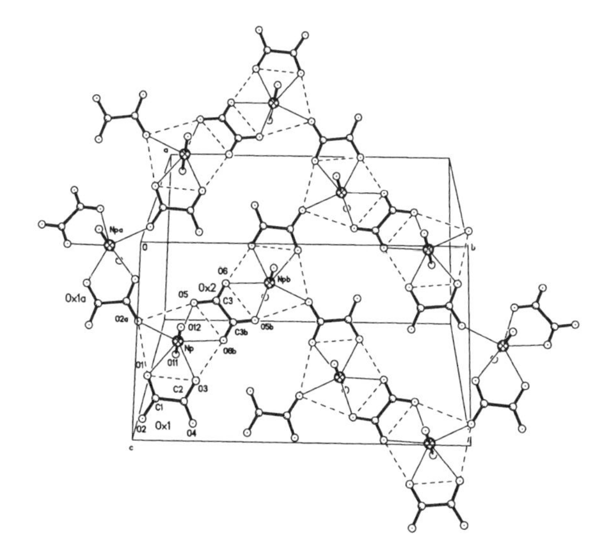

Figure 1 *Environment of Np atoms in the $[(NpO_2)_2(C_2O_4)_3]_n^{n-}$ networks. Symmetry transformations: a - (x, -y, z-1/2), b - (-x,+1/2, -y+1/2, -z+1).*

Dioxo cations and oxalate anions form loose $[(NpO_2)_2(C_2O_4)_3]_n^{n-}$ networks parallel to the *bc* plane (Fig. 1). Large cavities in the neptunyl-oxalate network accommodate Na(1) cations. They are situated at the same *x* heights as the Np atoms. As a result, corrugated mixed-cation $[Na_2(NpO_2)_2(C_2O_4)_3]_n^{2n-}$ layers are formed. The negative charge of the layer is compensated by Na(2) cations, which, together with water molecules, are situated between the layers.

There are four crystallographically unique water molecules. Two of them, O(1*w*) and O(2*w*), are included in Na environment. The O(3*w*) and O(4**w**) water molecules are located in the twofold rotation axes and add up to one crystallization water molecule per formula unit. There is also a three-dimensional hydrogen bonding network, with the O...O bond distances in the range from 2.82(1) to 3.02(1) Å.

The Np(V) has a pentagonal bipyramid oxygen environment. Two absorption bands are observed in the visible-near-IR absorption spectrum of the crystalline compound at 989 nm (D = 0.4) and 1033 nm (D = 0.07). The position of the maximum of the strong absorption band indicates the absence of the cation-cation interaction between the neptunyl groups. Note for comparison that in the spectra of solutions of the $NpO_2:C_2O_4$ = 2:3 complexes, this band appears at 989 nm. In addition, the position of the band of asymmetric stretching vibrations of neptunyl(V) in the IR spectrum range from 780 to 824 cm^{-1} which is again characteristic of groups not involved in cation-cation interactions. The participation of neptunyl groups in such interactions would shift the $\nu_{as}(NpO_2^+)$ band to smaller wavenumbers.

3 CONCLUSION

A novel neptunyl(V)oxalate complex with a 2:3 $NpO_2:C_2O_4$ ratio has been synthesised. The crystal structure is based on mixed-cation $[Na_2(NpO_2)_2(C_2O_4)_3]_n^{2n-}$ layers.

References

1 D. M. Gruen and J. J. Katz, *J. Am. Chem. Soc.*, 1953, **75**, 3772.
2 A. A. Bessonov and N. N. Krot, *Radiokhimiya (Russ)*, 1991, vol. **33**, N 3, p.p. 25-46.
3 S. B. Tomilin, Yu. F. Volkov, G. I. Visyascheva et al., *Radiokhimiya (Russ)*, 1984, vol. **26**, N 6, p.p. 734-739.
4 M. S. Grigoriev, N. A. Baturin, L. L. Regel and N. N. Krot, *Radiokhimiya (Russ)*, 1991, vol. **33**, N 3, p.p. 19-25.
5 M. S. Grigoriev, A. A. Bessonov, A. I. Yanovskii et al., *Radiokhimiya (Russ)*, 1991, vol. **33**, N 3, p.p. 46 - 53.

PLUTONYL(V) MALONATE COMPOUNDS: SYNTHESIS AND PROPERTIES

I. A. Charushnikova, A. A. Bessonov, M. S. Grigoriev and N. N. Krot

Institute of Physical Chemistry, Russian Academy of Sciences, Leninskii pr. 31, Moscow, 119991, RUSSIA

1 INTRODUCTION

Actinide (An) oxalates are widely used in radiochemical technology, preparative and analytical practice. Compounds of An(V) with the malonate ion, which is the nearest homologue of the oxalate ion, are poorly known. Earlier, we have shown that Np(V) easily forms a series of double malonates, which have a remarkable structure and properties. Np(V) compounds of general formula $MNpO_2L{\cdot}nH_2O$ ($L = CH_2(COO)_2^{2-}$, M = Li, Na, NH_4, K, Cs) and $Co(NH_3)_6[NpO_2L]_2A{\cdot}nH_2O$ (A is an inorganic or organic singly charged anion) were prepared.[1-7] We now report the synthesis of Pu(V) malonates.

2 METHOD AND RESULTS

2.1 Synthesis

Pu(V) compounds were synthesized using ^{239}Pu as a 0.15 M nitrate-perchlorate Pu(VI) solution.

2.1.1 *Synthesis of $MPuO_2L{\cdot}nH_2O$ (M = Na, NH_4, or Cs).* For Na^+ and NH_4^+ salt preparation, samples of a freshly prepared 0.15 M $PuO_2(ClO_4)_2$ solution were treated with 4 M $NaNO_3$ or 8 M NH_4NO_3 solutions and 1 M Na_2L or 0.5 M $(NH_4)_2L$ solutions. The solutions were cooled and treated with a solution of 0.8 M N_2H_4. The Pu(V) generation was established by Visible-NearIR absorption spectroscopy. In spectrum Pu(VI) band at 831 nm was absent with appearance of typical Pu(V) bands in the range 500-600 nm. Within ~2 h, the precipitation of Pu(V) malonates was observed. For Cs^+ salt preparation, a sample of a $PuO_2(ClO_4)_2$ solution was mixed with an equal volume of 5M CsCl solution and cooled in order to increase the precipitation of $CsClO_4$. The cold solution was then treated with a 0.8 M solution of N_2H_4.

2.1.2 *Synthesis of $Co(NH_3)_6[PuO_2L]_2A{\cdot}nH_2O$ where A is an inorganic anion.* Pu(V) malonates with A = NO_3^-, ClO_4^-, Cl^-, and Br^- were synthesized according to the following procedure. An aliquot of 0.1-0.2 M $PuO_2(ClO_4)_2$ at pH 3-5 was *ca.* tenfold diluted with water and treated with 0.5 M solution of LiL up to the concentration of 0.05-0.15 M. N_2H_4 was added in a 1:1 molar ratio relative to Pu(VI). Then, within ~1 min, 0.05 M solution of $Co(NH_3)_6(A)_3$ (NO_3^-, ClO_4^-, Cl^-, or Br^-) was added in the amount slightly (by 5-10%) exceeding the stoichiometric amount.

2.1.3 *Synthesis of $Co(NH_3)_6[PuO_2L]_2A{\cdot}nH_2O$ where A is an organic anion.* Compounds with organic anions A = $HCOO^-$, CH_3COO^-, $C_2H_5COO^-$, and HL^- were synthesized according to the following procedure. Samples of 0.15 M $PuO_2(ClO_4)_2$ solution at pH 4-5 were treated with equal volumes of 1 M K_2L solution. The precipitates were separated by centrifuge and resultant supernatant treated with a 0.2-0.4 M solution of an alkali formate, acetate, propionate, or hydrogen malonate. Then, a 0.8 M solution of N_2H_4 in the 1 : 1 molar ratio relative to Pu(VI) and a small excess (~10%) of a 0.125 M solution of $[Co(NH_3)_6]_2L_3$ were added.

2.2 Identification of Compounds

2.2.1 $MPuO_2L{\cdot}nH_2O$ (M = Na, NH_4, or Cs). The X-ray powder diffraction patterns of the double Pu(V) malonates are similar to those of the analogous Np(V) compounds.[1-4] All the compounds crystallise in the monoclinic crystal system, with their unit cell parameters given in Table 1. The comparison of the results of indexing of the $MPuO_2L{\cdot}nH_2O$ (M= Na, NH_4, Cs) patterns with those of Np(V) malonates reveals that the corresponding compounds are isostructural and, therefore, the plutonium compounds are described by the formulas $NaPuO_2L{\cdot}2H_2O$, NH_4PuO_2L, and $CsPuO_2L$.

Table 1 *The unit cell parameters of $MAnO_2L{\cdot}nH_2O$ (M = Na, NH_4, or Cs)*

Compound	*Cell parameters*				*Possible space group*	*V, Å³*
	a, Å	*b, Å*	*c, Å*	*β, deg.*		
$NaNpO_2L{\cdot}2H_2O$	12.935(2)	7.645(2)	7.968(4)	97.09(4)	$P2_1/n$	781.9(3)
$NaPuO_2L{\cdot}2H_2O$	12.916(4)	7.653(1)	7.916(1)	96.97(2)	$P2_1/n$	782.5(3)
NH_4NpO_2L	7.703(1)	13.020(2)	7.704(1)	111.08(1)	$P2_1/c$	720.95(17)
NH_4PuO_2L	7.650(2)	13.090(4)	7.697(2)	111.22(2)	C2/c	718.6(4)
$CsNpO_2L$	9.184(7)	13.636(8)	7.450(6)	101.97(6)	$P2_1/m$	912.7(11)
$CsPuO_2L$	9.186(2)	13.725(2)	7.459(2)	101.54(2)	$P2_1/a$	921.4(4)

Table 2 *The unit cell parameters for $Co(NH_3)_6[AnO_2L]_2A{\cdot}nH_2O$ compounds*

Compound	*Cell parameters*				
	a, Å	*b, Å*	*c, Å*	*β,deg.*	*V, Å³*
$Co(NH_3)_6[NpO_2L]_2NO_3{\cdot}H_2O$[7]	18.211(3)	7.724(5)	7.724(1)	104.18(1)	1053.4(7)
$Co(NH_3)_6[PuO_2L]_2NO_3{\cdot}3H_2O$	17.891(7)	7.718(2)	7.733(4)	100.55(1)	1049.7(7)
$Co(NH_3)_6[NpO_2L]_2Br{\cdot}H_2O$	18.140(6)	7.705(3)	7.685(3)	100.52(1)	1056.0(7)
$Co(NH_3)_6[PuO_2L]_2Br{\cdot}H_2O$	18.04(5)	7.699(10)	7.708(12)	100.7(2)	1052(3)
$Co(NH_3)_6[NpO_2L]_2HL$[6]	17.925(6)	7.736(4)	7.879(3)	103.65(3)	1061.7(8)
$Co(NH_3)_6[PuO_2L]_2HL{\cdot}2H_2O$	17.731(7)	7.723(2)	7.934(3)	102.42(1)	1061.0(6)

2.2.2 $Co(NH_3)_6[PuO_2L]_2A{\cdot}nH_2O$ (A = NO_3^-, ClO_4^-, Cl^-, and Br^-). X-ray powder diffraction patterns of all the Pu(V) compounds obtained are similar to each other and to those of the corresponding neptunium(V) salts [5-7] (Tab. 2, space group C 2/m). In the IR spectra, strong narrow bands of the asymmetric stretching vibrations (ν_{as}) of PuO_2^+ groups are observed at 824-832 cm^{-1}. The shape and positions of these bands are in good agreement with the conclusion that PuO_2^+ ions are not involved in mutual coordination (cation-cation interactions).

2.2.3 $Co(NH_3)_6[PuO_2L]_2A{\cdot}nH_2O$ (A = $HCOO^-$, CH_3COO^-, $C_2H_5COO^-$ and HL^-). The identification of the Pu(V) compounds was performed by the comparison of their X-ray diffraction patterns with those of Np(V) compounds studied earlier (Table 2).

2.3 Results and Discussion

The similarity of the X-ray powder diffraction patterns of $Co(NH_3)_6[PuO_2L]_2A{\cdot}nH_2O$ undoubtedly indicates that the structures of these compounds have the same structural motif. According analogous Np data,[6-7] the structures consist of $[AnO_2L]_n^{n-}$ chains (Fig. 1) and $Co(NH_3)_6^{3+}$ cations. In structures $Co(NH_3)_6[NpO_2L]_2 \cdot HL$[6], $Co(NH_3)_6[NpO_2L]_2NO_3 \cdot H_2O$[7], and $Co(NH_3)_6[NpO_2L]_2OH \cdot H_2O$[7], the chains and the $Co(NH_3)_6^{3+}$ cations occupy the same positions for the C 2/m space group. This motif is retained without significant changes even in the case that a relatively large Br^- or organic anion is included in the crystal. Evidently, $Co(NH_3)_6[AnO_2L]_2A \cdot nH_2O$ can be considered as salts of the AnO_2L^- anionic complex. In the course of precipitation, the AnO_2L^- complexes polymerize into structurally stable $[AnO_2L]_n^{n-}$ chains (Figure 1). Compounds of general composition $MAnO_2L \cdot nH_2O$ (where M = Li, Na, K, NH_4, or Cs) can also be regarded as salts of the AnO_2L^- complex. We can assume that the main structural elements of $MAnO_2L \cdot nH_2O$ are infinite neptunyl (plutunyl) – malonate chains, which are similar to those observed in $Co(NH_3)_6[AnO_2L]_2A \cdot nH_2O$. This assumption is supported by a series of experimental data.

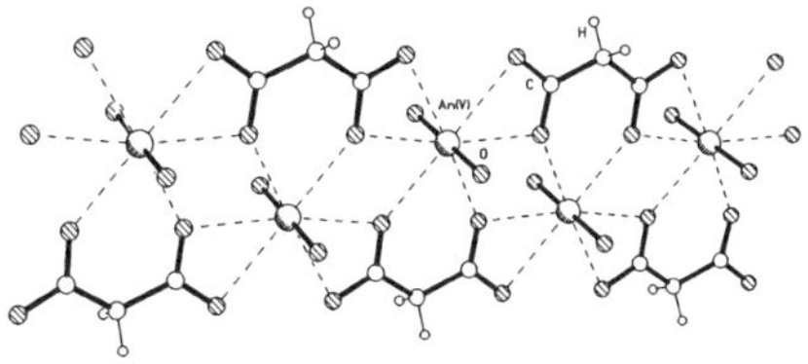

Figure 1 *The $[AnO_2L]_n^{n-}$ anionic chain in the crystal structures of $Co(NH_3)_6[AnO_2L]_2A \cdot nH_2O$, where An is Np(V) or Pu(V), A is a singly charged anion*

3 CONCLUSION

The double malonates of Np(V) and Pu(V) of general compositions $Co(NH_3)_6[AnO_2L]_2A \cdot nH_2O$ (A is a single charged anion) and $MAnO_2L \cdot nH_2O$ (M = Li^+, Na^+, K^+, NH_4^+, or Cs^+) can be considered as salts of the AnO_2L^- anionic complex. In the course of precipitation, the complexes polymerize in compact $[AnO_2L]_n^{n-}$ chains. Equal coordination capacity of dioxocations and each of malonate ligands (equal to 6) provide highly stable structures whilst all attempts to isolate bi-malonate complexes as crystal compounds were unsuccessful.

References

1 N. N. Krot, I. A. Charushnikova, T. V. Afonas'eva and M. S. Grigoriev, *Radiochemistry (Russ)*, 1993, **35**, 140.
2 M. S. Grigoriev, I. A. Charushnikova, N. N. Krot, A. I. Yanovskii, and Y. T. Struchkov, *Radiochemistry (Russ)*, 1993, **35**, 388.
3 M. S. Grigoriev, I. A. Charushnikova, N. N. Krot, A. I. Yanovskii, and Y. T. Struchkov, *Radiochemistry (Russ)*,1993, **35**, 394.
4 N. N. Krot, M. S. Grigoriev and I. A. Charushnikova, *Radiochemistry (Russ)*, 2004, **46**, 107.
5 N. N. Krot, I. A. Charushnikova and M. S. Grigoriev, *Radiochemistry (Russ)*, 2004, **46**, 111.
6 M. S. Grigor'ev, I. A. Charushnikova, Z. A. Starikova et al. *Radiochemistry (Russ)*, 2004, **46**, 232.
7 I. A. Charushnikova, N. N. Krot and I. N. Polyakova, *Radiochemistry (Russ)*, 2004, **46**, *347*.

CRYSTAL COMPOUNDS OF PLUTONYL(V) MONOOXALATE COMPLEX

A.A Bessonov, M.S.Grigoriev, I.A.Charushnikova, N.N.Krot

Institute of Physical Chemistry, Russian Academy of Sciences
31 Leninsky prosp., Moscow, 119991, RUSSIA

1 INTRODUCTION

Monooxalate complex compounds of NpO_2^+ are well known and have been studied in some detail. Structures of a number of Np(V) monooxalate compounds have been determined by single crystal XRD structural analysis.[1-3] The synthesis and some properties of AmO_2^+ oxalate compounds have also been reported.[4] Until this work there was no information in the literature on the synthesis, structure and properties for PuO_2^+ monooxalate complex derivates. We have recently studied this system,[5] with more detail concerning our results presented here.

2 METHOD AND RESULTS

2.1 Synthesis and properties of $NaPuO_2C_2O_4 \cdot nH_2O$ (n = 3, 1, 0)

The syntheses of compounds of general formula $NaPuO_2C_2O_4 \cdot nH_2O$ were carried out taking into account information obtained in the course of preparation of analogues Np(V) compounds[6.7] and the previous data on Pu(V) behaviour in oxalate solutions.[8] In the case of Np(V) tri- and monohydrate may be isolated directly from the solutions, with the anhydrous salt obtained by heating the monohydrate at 250°C. These methods, although suitable for Np(V), need to be changed in the case of plutonium, since Pu(V) in oxalate solutions is inclined to disproportionate in a second order reaction, the rate of which is sharply increased with increasing temperature and decreasing pH.[9] Taking into consideration these factors all $NaPuO_2C_2O_4 \cdot 3H_2O$ syntheses were carried out at t~0°C. Sodium oxalate was added to freshly prepared 0.1 - 0.2 M $PuO_2(ClO_4)_2$ solution at molar ratios of $C_2O_4^{2-}:PuO_2^{2+}$ between 1.3 - 1.7. $NaNO_3$ was introduced into solution up to ~0.5 M concentration, after which Pu(VI) was reduced by the addition of equimolar N_2H_4. Heating of the precipitated $NaNpO_2C_2O_4 \cdot 3H_2O$ leads to stepwise dehydration with consecutive monohydrate and anhydrous salt formation. Compounds $NaPuO_2C_2O_4 \cdot nH_2O$ with n = 3 and 1 are well crystallized and isostructural with Np(V) analogous (Table 1). The anhydrous product was practically amorphous: its XRD pattern shows only several diffuse lines, but was sufficient to compare $NaPuO_2C_2O_4$ with $NaNpO_2C_2O_4$ and to conclude their isosructurality. It is interesting to note the similar Np(V) derivates (n = 1 and 0), show cation-cation (CC) interactions between NpO_2^+ based on near IR absorption spectroscopy data.[7,9-10] and the compounds $NaPuO_2C_2O_4{\cdot}nH_2O$ (n = 1 and 0) may be described as the first Pu(V) compounds with CC coordination of PuO_2^+ ions.

Table 1. *Crystal cell parameters for* $NaAnO_2C_2O_4{\cdot}nH_2O$, *n = 3, 1, An = Np, Pu*

Compound	cell parameters					
	a, Å	*b*, Å	*c*, Å	α, degr.	β, degr.	γ, degr.
$NaNpO_2C_2O_4{\cdot}3H_2O$ [2]*	5.69(2)	8.54(2)	10.42(2)	76.2(1)	81.7(1)	71.7(1)
$NaNpO_2C_2O_4{\cdot}H_2O$	9.310(2)	10.8521(17)	9.1659(17)	101.33(2)	113.31(2)	98.07(2)
$NaPuO_2C_2O_4{\cdot}3H_2O$	5.6690(10)	8.4304(17)	10.4421(15)	76.36(1)	80.36(1)	72.36(1)
$NaPuO_2C_2O_4{\cdot}H_2O$	9.2325(12)	10.7877(13)	9.1087(11)	101.49(1)	113.33(1)	98.01(1)

*The cell calculated in [2] is transformed into given.

2.2 Synthesis, properties and structure of $NH_4PuO_2C_2O_4{\cdot}2.67H_2O$

The synthesis of $NH_4PuO_2C_2O_4{\cdot}nH_2O$ is complicated by its high solubility and very slow crystallization, even at low temperatures. Heating plutonyl(V) solutions leads to Pu(V) disproportionation in hot oxalate solutions,[8] a process promoted by an increase in the initial Pu(V) concentration. Moreover, the products of disproportionation, especially Pu(IV), are able to form precipitates in the case of a small excess of oxalate ions. A large excess of oxalate is also unacceptable, since it leads to transformation $PuO_2C_2O_4^-$ complexes into $PuO_2(C_2O_4)_2^{3-}$ and therefore prevents $NH_4PuO_2C_2O_4{\cdot}nH_2O$ precipitation. These difficulties may be overcame by using an initial Pu(V) concentration not higher 0.045M, a temperature of about 0°C and 8M NH_4NO_3 as salting-out agent. Oxalate was added as a 0.3M $(NH_4)_2C_2O_4$ solution in ~15% excess with respect to plutonium. Plutonium reduction was realized by addition of stoichiometric quantity N_2H_4. Comparison of the indexing results for the obtained precipitate with the crystal cell parameters of known $NH_4NpO_2C_2O_4{\cdot}2.67H_2O$ (space group $P6_3/m$)[3] leads to the unambiguous conclusion that these compounds have the same type of structure (Table 2). Hence the composition and crystal structure of the ammonium monooxalate plutonyl(V) complex are the same as its neptunyl(V) analogue and therefore $NH_4PuO_2C_2O_4{\cdot}2.67H_2O$ structure contains trimers of plutonyl(V) ions coordinated to each other as monodentate ligands (Figure 1). Therefore, as with the neptunyl(V) analogue, plutonyl cation:cation interactions should also be observed.

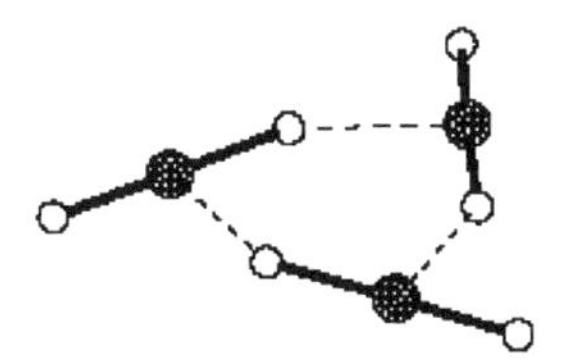

Figure 1 *The CC interaction of* PuO_2^+ *ions in the crystal structure of* $NH_4[PuO_2C_2O_4]{\cdot}2\frac{2}{3}H_2O$.

2.3 Synthesis and properties of $CsPuO_2C_2O_4{\cdot}nH_2O$ (n = 6, 4. 2)

The solubility of $CsPuO_2C_2O_4{\cdot}nH_2O$ is much lower than of the ammonium salt and its preparation is not difficult. Contrary to expectation, the powder XRD pattern of the isolated compound was found to be different from the patterns of $NH_4AnO_2C_2O_4{\cdot}2.67H_2O$

and $CsNpO_2C_2O_4{\cdot}2.67H_2O$. Indexing results are presented in Table 2: the compound is isolated from solution with 6 water molecules (space group $P2_12_12_1$), although it easily loses two water molecules at room temperature, giving the tetrahydrate (space group $P2_1/n$) and transforms to a dihydrate at ~110°C. This compound had poor crystallinity, and its powder XRD identification was impossible. Nevertheless, some specific structural peculiarities may be supposed from its IR spectra. The antisymmetric vibration bands of PuO_2^+ are noticeably shifted towards to lower frequencies in the spectra of the tetra- and dihydrate (up to 796 and 788 cm^{-1}, respectively). This is typical of mutual dioxocation coordination; essential splitting of this band in the case of dihydrate may indicate PuO_2^+-net structure, i.e. bidentate CC interaction of PuO_2^+ ions).[1] However, further structural and spectroscopic data is required for proof of cation:cation interactions.

Table 2. *Crystal cell parameters for $MAnO_2C_2O_4{\cdot}nH_2O$, M = NH_4, Cs; An = Np, Pu*

Compound	cell parameters					
	a, Å	*b*, Å	*C*, Å	A, degr.	β, degr.	γ, degr.
$NH_4PuO_2C_2O_4{\cdot}2.67H_2O$	11.309(8)	11.309(8)	11.666(2)	90	90	120
$NH_4NpO_2C_2O_4{\cdot}2.67H_2O$ [3]	11.382(2)	11.382(2)	11.734(2)	90	90	120
$CsPuO_2C_2O_4{\cdot}6H_2O$	11.9881(9)	10.7377(9)	8.2686(6)	90	90	90
$CsNpO_2C_2O_4{\cdot}2.67H_2O$	11.790(3)	11.790(3)	11.726(2)	90	90	120
$CsPuO_2C_2O_4{\cdot}4H_2O$	8.9499(17)	12.0556(12)	8.8619(12)	90	105	90

3 CONCLUSION

This work is a part of systematic study on Pu(V) chemistry in neutral complexing media. Isolation of the plutonyl(V) monooxalate complex as crystalline compounds with sodium, ammonium and caesium has been shown possible in spite of Pu(V) dismutation and reduction by the oxalate-ion. It has been shown that sodium and ammonium salts of monooxalate complex of Pu(V) have the same crystal structures as their Np(V) analogues. For the first time CC interaction of PuO_2^+ ions in the structures of crystal compounds has been experimentally observed. $CsPuO_2C_2O_4{\cdot}nH_2O$ compounds (n = 6, 4, 2) have no isostructural Np(V) analogues but CC interaction in the structures is suggested from their IR spectra.

References

1 N.N. Krot, M.S. Grigoriev, *Russ. Chem. Rev.,* 2004, **73**, 94.
2 S.V. Tomilin, Yu.F.Volkov, G.I.Visiacheva, et al., *Radiokhimiya,* 1984, **26**(6), 734.
3 M.S. Grigoriev, A.A.Bessonov, A.I.Yanovskii, et al., *Radiokhimiya,* 1991, **33**(5), 46.
4 V.G.Zubarev, N.N.Krot, *Radiokhimiya,* 1982, **24**(3), 319.
5 A.A.Bessonov, M.S.Grigoriev, I.A.Charushnikova, N.N.Krot, *Plutonium(V) chemistry in neutral aqueous complexing media: new crystal compounds synthesis and study of their structure and properties*, this publication.
6 A.D.Guelman, V.I.Blokhin, N.N.Krot, et al., *Russ. J. Inorg. Chem.,* 1970, **15**(7), 1899.
7 A.A.Bessonov, N.N.Krot, *Radiokhimiya,* 1991, **33**(3), 35.
8 N.P.Ermolaev, N.N.Krot, A.D.Guelman, *Radiokhimiya,* 1967, **9**(2), 171.
9 N.N.Krot, A.A.Bessonov, V.P.Perminov, *DAN SSSR,* 1990, **312**(6), 1402.
10 N.N.Krot, A.A.Bessonov, M.S.Grigoriev, *21èmes Journées des Actinides,* 1991, Montechoro, Portugal. Program and Abstracts. p.169.

SOLID COMPOUNDS OF Np(V) WITH 2,4,6-TRI-(2-PYRIDYL)-1,3,5-TRIAZINE: CRYSTAL STRUCTURE, SPECTRAL AND THERMAL PROPERTIES

G. B. Andreev,[1] N. A. Budantseva,[1] A. M. Fedosseev,[1] A. A. Bessonov,[1] and M. Yu. Antipin[2]

[1]Institute of Physical Chemistry of Russian Academy of Sciences, Leninsky pr. 31, Moscow, 119991, Russia
[2]A. N. Nesmeyanov Institute of Organoelement Compounds of Russian Academy of Sciences, Vavilova ul. 28, Moscow, 119991, Russia

1 INTRODUCTION

New strategies for nuclear waste management are under consideration worldwide to minimize the long-term radiotoxicity of the waste packages. The radiotoxicity can be substantially reduced if minor actinides are removed from the waste before conditioning. Separations using terdentate nitrogen planar ligands, such as 2,4,6-tri-(2-pyridyl)-1,3,5-triazine (TPT), 2,2',6',2''-terpyridine (Terpy), 4'-(4-nitrophenyl)- 2,2',6',2''-terpyridine, 4'-(4-tolyl)-2,2',6',2''-terpyridine, 2,6-bis(5,6-dialkyl-1,2,4-triazin-3-yl)pyridine, 4-amino-bis(2,6-(2-pyridyl))-1,3,5-triazine, are under investigation. Np(V) compounds with these N-donor aromatic ligands may be formed under special conditions. In this work, synthesis as well as optical properties and thermal stability of solid $[NpO_2Cl(TPT)(H_2O)]\cdot H_2O$ and $NpO_2Br(TPT)\cdot nH_2O$ were investigated. On the basis of spectroscopy data and thermal study of $NpO_2Br(TPT)\cdot nH_2O$ its composition was proposed.

2 EXPERIMENTAL DETAILS

Compounds $[NpO_2Cl(TPT)(H_2O)]\cdot H_2O$ (**1**) and $NpO_2Br(TPT)\cdot nH_2O$ (**2**) were obtained by slow evaporation of the mixture containing an aqueous solution of NpO_2Cl or NpO_2Br and TPT dissolved in methanol with the molar ratio of 1:1. NpO_2Cl and NpO_2Br were prepared by dissolution of NpO_2OH in HCl and HBr, respectively. NpO_2OH was precipitated from the neptunium stock solution by adding of an excess of concentrated NH_4OH. The precipitate was centrifugated and washed by distilled water.

X-ray diffraction data for compound **1** were collected using graphite monochromated Mo Kα (λ = 0.71073 Å) radiation on a Siemens P3/PC diffractometer. The structure was solved by direct methods. All non-hydrogen atoms were refined with anisotropic thermal parameters using full-matrix least-squares procedure on F^2. Hydrogen atoms were fixed in geometrically idealized positions. All calculations were performed using SHELXTL software. Crystallographic data for the compound **1**: space group $P2_1/c$, a=12.6240(25), b=20.9739(42), c=7.5488(15) Å, β=96.399(30)°, Z=4, R=0.022, $wR(F^2)$=0.044. Spectra in the NIR-VIS-UV and IR regions were measured using Shimadzu UV3100 and Specord M80 spectrometers, respectively. Samples for UV-Vis-NIR spectroscopy were prepared in the form of suspension in nujol coated on a quartz support. NaCl pellets containing ~2wt% of the solids were used for IR measurements.

3 RESULTS AND DISCUSSION

The crystal structure of **1** consists of molecules of the complex, $[NpO_2Cl(TPT)(H_2O)]$ (Figure 1), with water of crystallisation. The coordination polyhedron of the neptunium atom is pentagonal bipyramidal with equatorial positions occupied by three nitrogen atoms of TPT, a water molecule and a Cl atom. The NpO_2 moiety is nearly linear and symmetrical, with the Np-O distances equal to 1.7940(26) and 1.8134(26) Å and the O-Np-O angle is 178.75(12)°. The distances between Np and equatorial atoms are different: Np-N = 2.6045(27) - 2.6740(31), Np-O(H_2O) = 2.4552(25), Np-Cl = 2.8207(11) Å. Similar values of Np-N distances were found in the following structures: $[NpO_2(Dipy)(H_2O)_3](NO_3)$ (2.62(1), 2.66(1) Å)[1], $[NpO_2NCS(Dipy)(H_2O)_2]\cdot H_2O$ (2.617(5), 2.649(5) Å)[2], $[NpO_2NO_3(Terpy)(H_2O)]$ (2.580(5)-2.602(5) Å)[3], $[NpO_2(2\text{-}NC_5H_4COO)(Terpy)(H_2O)]\cdot 2H_2O$ (2.55(2)-2.65(2) Å)[4]. The mean squared displacement of atoms from the equatorial plane does not exceed 0.0915 Å.

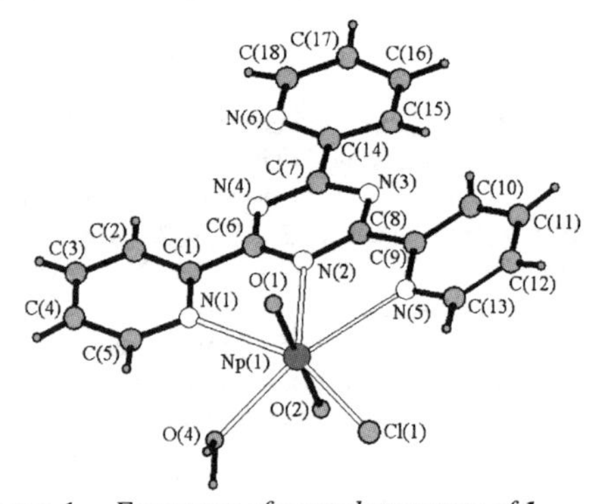

Figure 1 *Fragment of crystal structure of 1*

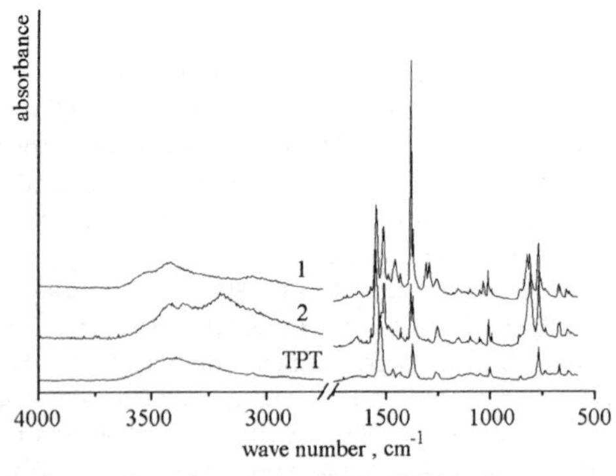

Figure 2 *Infrared spectra of pure TPT and compounds 1 and 2*

The infrared spectra of solid compounds **1**, **2** and pure TPT are shown in Figure 2. In the region 1600-500 cm^{-1}, the spectra of **1** and **2** are very similar. CC and CH vibrations of

TPT do not show substantial alteration on coordination in compounds **1** and **2**, and only the frequencies of NC bond vibrations are changed significantly due to coordination of the ligand by neptunyl. The asymmetric vibrations of the neptunyl group were found to be close – at 824 and 808 cm^{-1} for **1** and 820, 808 cm^{-1} for **2**. This fact counts in favour of suggestion that the neptunium coordination environment is similar in both compounds. In the spectrum of **2**, vibrations of water molecules are exhibited by broad band with several high resolved maxima in the range 4000-3000 cm^{-1}. This band is shifted to the lower energy in comparison with the spectrum of **1**, giving an indication of the existence of a more developed system of hydrogen bonds involving water molecules in compound **2**.

The NIR spectra of compounds **1** and **2** are identical, suggesting that the Np atom coordination spheres are similar in both cases. The maxima of bands corresponding to f-f electronic transition are both located at 984 nm, with molar absorption maxima also equivalent - 180 l mol^{-1} cm^{-1}. The effect of TPT coordination is observed in the UV region. The spectrum of pure TPT exhibits two absorption bands at 248 and 292 nm. In the spectra of compounds **1** and **2** these bands are broader and shifted to lower energy.

In the TG curves for the compounds **1** and **2**, two decomposition steps are observed. Compound **1** remains stable up to 160°C, the first step of decomposition accounts for the endothermic removal of water. The second step is the final decomposition of the compound with a slight exothermic effect at 390 °C corresponding to the reduction of Np(V) to Np(IV) with the formation of NpO_2, as proved using X-ray phase analysis. Compound **2** starts losing water molecules at 90 °C, the final decomposition is completed at 600°C. Assuming the formula $NpO_2Br(TPT){\cdot}nH_2O$ for compound **2** the water content *n* was calculated to be equal to 3.

4 CONCLUSIONS

Two new neptunyl(V) complexes with TPT, $[NpO_2Cl(TPT)(H_2O)]{\cdot}H_2O$ (**1**) and $NpO_2Br(TPT){\cdot}nH_2O$ (**2**), have been prepared. The crystal structure of compound **1** was determined using single crystal X-ray diffractometry. The neptunium atom has a pentagonal bipyramidal coordination geometry with equatorial positions occupied by three nitrogen atoms of TPT, a water molecule and a Cl atom. In the IR spectra of the compounds, frequencies of vibrations corresponding to the TPT molecule and the neptunyl group have similar values. The NIR electronic spectra of the compounds **1** and **2** are also identical, indicating that compound **2** has the structural formula $[NpO_2Br(TPT)(H_2O)]$. TG/DTA data for the compounds show their similar thermal behaviour. From these data, the water content *n* for the compound **2** was calculated. Consequently, taking into account the results of spectral and thermal analysis the formula $[NpO_2Br(TPT)(H_2O)]{\cdot}2H_2O$ is suggested for compound **2**.

References

1. G.B. Andreev, M.Yu. Antipin, A.M. Fedosseev, N.A. Budantseva, *Mendeleev Commun.*, 2001, 58
2. G.B. Andreev, M.Yu. Antipin, A.M. Fedosseev, N.A. Budantseva, *Rus. J. Coord. Chem.*, 2001, **27**, 211
3. N.A. Budantseva, G.B. Andreev, A.M. Fedosseev, M.Yu. Antipin, J.-C. Krupa, *Crystallogr. Rep.*, 2003, **48**, 58
4. A.M. Fedosseev, G.B. Andreev, N.A. Budantseva, J.-C. Krupa, *J. Nucl. Sci. Technol.*, 2002, 414.

COMPLEX FORMATION OF Np(V) WITH PICOLINIC, NICOTINIC AND ISONICOTINIC ACIDS

G. B. Andreev,[1] N. A. Budantseva,[1] A. M. Fedosseev[1] and M. Yu Antipin[2]

[1]Institute of Physical Chemistry of Russian Academy of Science, Leninsky pr. 31, Moscow, 119991, Russia
[2]A. N. Nesmeyanov Institute of Organoelement Compounds of Russian Academy of Science, Vavilova ul. 28, Moscow, 119991, Russia

1 INTRODUCTION

Neptunium in its most stable oxidation state, +V, exists as the linear dioxo cation NpO_2^+, both in aqueous solutions and in solid compounds. Similar to the uranyl ion, NpO_2^+ may be coordinated with various ligands forming coordination polyhedra like penta-, hexa- or tetragonal bipyramids with axial positions occupied by two neptunoyl oxygens, while the donor atoms of ligands are accommodated in the plane normal to the pyramidal axis. The distinctive feature of Np(V) compared to Np(VI) and U(VI) is an affinity to form cation-cation complexes. Previously, the interaction of Np(V) with N-donor ligands has not attracted considerable interest. In comparison, a large amount of uranyl(VI) complexes are known with N-donor ligands. However, structural data for U(VI) complexes with pyridinemonocarboxylic acids are rare. The dinuclear uranyl complex, $[(UO_2)_2(Pic)_4(NO_3)_2(H_2O)_2]\cdot 2H_2O$, contains picolinate groups directly bonded with two metal centers.[1] In the structure of $Na_2[(UO_2)(Pic)F_3]\cdot 4H_2O$ picolinate acts as a chelate ligand.[2] In the uranium dipicolinate $[(UO_2)(Hdpa)]\cdot HPic\cdot 6H_2O$ monopicolinic acid plays the role of solvate and in the uranium nicotinate $[(UO_2)(Nic)_2(H_2O)_2]$ the nicotinate ligands are bonded to U by oxygen atoms only.[3,4]

We present here the syntheses and structures of four novel Np(V) compounds $[NpO_2(Pic)(H_2O)_2]$ (**1**), $Cs[NpO_2(Pic)_2(HPic)]\cdot 3H_2O$ (**2**), $[NpO_2(Nic)(H_2O)]$ (**3**), and $[NpO_2(iso\text{-}Nic)(H_2O)]$ (**4**). To explore the impact of location of the carboxylic group in the isomers of pyridinemonocarboxylic acids on their interaction with Np(V) we investigated the complexation of NpO_2^+ cation with picolinic (HPic, $2\text{-}NC_5H_4COOH$), nicotinic (HNic, $3\text{-}NC_5H_4COOH$) and isonicotinic (Hiso-Nic, $4\text{-}NC_5H_4COOH$) acids in aqueous solutions.

2 RESULTS AND DISCUSSION

Compounds **1-4** represent complexes of Np(V) with all isomers of pyridinemonocarboxylic acid - picolinic, nicotinic, and isonicotinic (Table 1). In all four compounds the Np atom is found in seven coordinated pentagonal bipyramidal geometry. However, because of different connectivity of Np coordination polyhedra distinct topologies are observed. In **1** the Np atoms are connected by the picolinate-ions into infinite chains, while **2** contains isolated $[NpO_2(Pic)_2(HPic)]^-$ groups (Figure 1).

Table 1 *Crystallographic data for the structures 1-4*

Formula	$NpN_1C_6O_6H_8$ (**1**)	$NpN_3C_{18}O_{11}H_{19}Cs$ (**2**)	$NpNC_6O_5H_6$ (**3**)	$NpNC_6O_5H_6$ (**4**)
Space group	P-1	P-1	$P2_1/n$	$P2_12_12_1$
a (Å), α (deg)	6.380(2),96.49(3)	8.707(2),76.633(15)	11.901(4),90	5.742(3),90
b (Å), β (deg)	7.733(2),106.78(3)	9.777(2),78.75(2)	5.5628(15),101.05(2)	8.550(5),90
c (Å), γ (deg)	10.700(4),103.83(3)	14.640(3),86.55(2)	12.124(3),90	18.512(10),90
V ($Å^3$), Z	481.3(3),2	1189.1(4),2	788.0(4),4	908.8(9),4
R(F), wR(F^2)	0.0409,0.1010	0.0304,0.0786	0.0495,0.1189	0.0323,0.0791

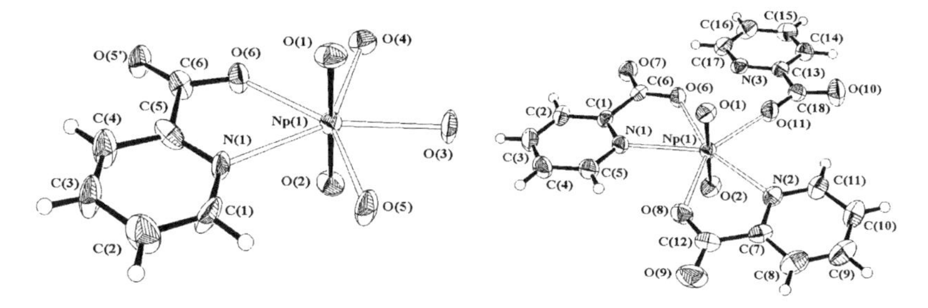

Figure 1 *ORTEP representation (50% probability ellipsoids) of the crystal structures 1 (fragment only) and 2.*

In both structures **3** and **4** the neptunoyl ions act as monodentate ligands with one oxygen atom occupying equatorial position of adjacent Np coordination polyhedron. This leads to formation of infinite neptunoyl chains in which each two neighboring Np atoms are connected through the neptunoyl oxygen and carboxylic group of the nicotinate or isonicotinate ion. Different location of the carboxylic group in nicotinic and isonicotinic acids results in different packing of these chains in **3** and **4**, the crystal structure of **3** has layered topology while **4** has the three-dimensional one structure (Figure 2).

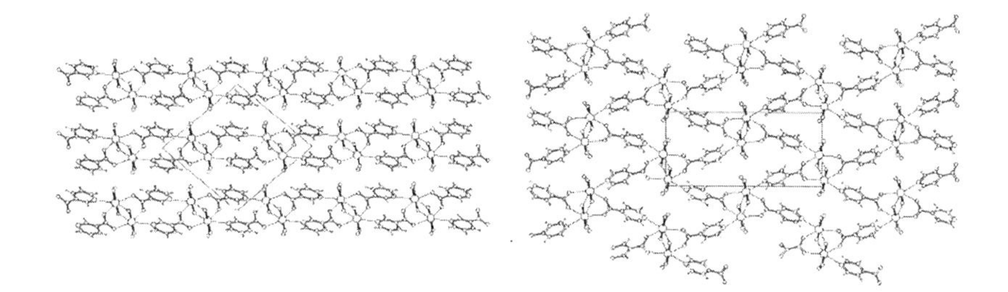

Figure 2 *Packing of infinite chains in the crystal structures 3 and 4*

In the IR spectra, the difference of Np coordination environment in **1-4** results in differences in the antisymmetric vibration region of the neptunoyl ion. In the spectra of **1** and **2**, where location of carboxylic group allows the picolinate ion to act as chelate, the

$\nu_{as}(NpO_2)$ have larger values. In the spectra of **3** and **4** where cation-cation bonds are formed and neptunoyl ions act as ligands, the shift of this band to lower energies is observed. The coordination of heterocyclic N atoms by NpO_2^+ ion leads to the shift of the ν(C-N) and ν(C-C) frequencies to higher energies relatively to the corresponding bands in the spectra of uncoordinated ligands. The frequencies of the antisymmetric modes ν_{as}(COO) for **3** and **4** are slightly less than for **1** and **2**, while the values ν_s(COO) are comparable.

The NIR spectra of solid compounds contain f-f transition bands in the range 980-1010 nm. The maxima of this bands for **1-4** are 987, 990, 1009, and 1007 nm, respectively. The intensities of band maxima are equal to 155, 120, 50, and 460 L mol^{-1} cm^{-1}. The NIR spectrum of NpO_2^+ ion is known to contain intense narrow band with maximum at 980 nm due to f-f transition from ground state $^3H_4$ to exitated state $^3\Pi_2$.[5] This transition is very sensitive to coordination environment. With complex formation, as the water molecules in coordination sphere of NpO_2^+ are substituted by ligands this f-f transition band is shifted to lower energy. The positions of the absorption bands of solid $[NpO_2(Pic)(H_2O)_2]$ and the same species dissolved in solution indicates that the picolinate ion remains coordinated to Np through a chelate ring on dissolution. Alternatively, significant shifts of absorption band to lower energies are observed in spectra of neptunium nicotinate and isonicotinate due to formation of cation-cation complex in solid **3** and **4**.

Different position of carboxylic group relative to the nitrogen atom of pyridine ring in picolinic, nicotinic and isonicotinic acids markedly affect complex formation in solution. The only complex $[NpO_2(Nic)]$ with band at 990 nm is formed in solution as determined using deconvolution of spectra recorded at various concentrations of Np(V) and nicotinic acid. The formation constant was found to be log β_1=2.94±0.05 at ionic strength I=0.2 (t=22°C). Isonicotinic acid also forms with Np(V) the unique complex of the composition $[NpO_2(isoNic)]$ but its absorption band appears at λ_{max}=992 nm. The calculated formation constant is log β_1=2.72±0.05. Picolinic acid is a stronger donor ligand with respect to Np(V) than nicotinic and isonicotinic acids. Three absorption bands 981, 987 and 994 nm were observed in the spectra indicating the presence of two complexes $[NpO_2(Pic)]$ and $[NpO_2(Pic)_2]^-$. The complexation constants are log β_1=3.68±0.04 and log β_2=6.98±0.04.

Complexation study shows that the anions of monopyridinecarboxylic acids have significant ability to form complexes with Np(V). The higher stability of the Np complexes with picolinate ion may be related to formation of the chelate ring, while in the case of nicotinate and isonicotinate ions the formation of such a stable ring structure is impossible and complexation constants appear to be similar, but significantly lower than for the picolinate.

References

1 P.R. Silverwood, D. Collison, F.R. Livens, R.L. Beddoes, R.J. Taylor, *J. Alloys Comp.*, 1998, **271-273**, 180
2 W. Aas, M.H. Johansson, *Acta Chem. Scand.*, 1999, **53**, 581
3 A. Cousson, J. Proust , E.N. Rizkalla, *Acta Crystallogr. C.*, 1991, **47**, 2065
4 N.W. Alcock, W. Errington, T.J. Kemp, J. Leceiejewicz, *Acta Crystallogr. C.*, 1996, **52**, 615
5 J.C. Eisenstein, M.H.L. Price, *J. Research NBS*, 1966, **70**, 165

THE STABILITY OF A URANYL(V) BENZOYL ACETONATE COMPLEX IN DIMETHYL SULFOXIDE

K. Shirasaki, T. Yamamura, Y. Monden and Y. Shiokawa

Institute for Materials Research, Tohoku University, Sendai, Miyagi 980-8577, Japan

1 INTRODUCTION

Light actinides such as neptunium and uranium are known to show two couples of reversible or quasi-reversible electrode reactions as seen in Np(III)/Np(IV) and Np(V)/Np(VI)[1,2]. By utilizing these two couples of fast reactions, which can minimize the overvoltage at electrodes during charge and discharge, the redox-flow battery with high efficiency has been proposed.[1,3] In the case of uranium, electrode reactions of U(III)/U(IV) and U(V)/U(VI) are regarded as reversible or quasi-reversible in the absence of proton[4, 5] which promotes the disproportionation reaction degrading U(V) to U(IV) and U(VI). The uranium redox flow battery uses aprotic solutions of U(V) and U(IV) at discharged state and U(VI) and U(III) at charged state for positive and negative electrolytes, respectively. It is well known that solutions of U(IV) and U(VI) are stable, but very few investigations have been carried out on the stability of bulk U(V) solutions[6, 7]. In our previous paper[7], the stability of the pentavalent state of the reduced uranyl(V) dimethyl sulfoxide and uranyl(V) acetylacetonate complexes were investigated in dimethyl sulfoxide solution under inert atmosphere by repeating spectral measurements by circulating solution through electrolysis cell and spectroscopic cell at 25 °C. The half-lives of the pentavalent states were found to be 1.5 - 3.7 hours. These results were contrary to our general understanding that the uranyl(V) state undergoes disproportionation reaction only in the presence of free proton cation.

In order to elucidate the decomposition mechanism of the pentavalent state of uranyl complexes in aprotic solutions, it is necessary to carry out further detailed investigation on the factors which contribute to the oxidation of the uranium(V). In this study, the pentavalent state of uranyl benzoylacetonate was studied in dimethyl sulfoxide solution by electrochemical and spectroscopic technique at various temperatures.

2 EXPERIMENTAL

Uranyl(VI) bis(benzoylacetonate) ($UO_2(ba)_2$) was synthesized by a procedure reported previously[4], and the prepared complex was identified as $UO_2(ba)_2{\bullet}(C_2H_5OH)_{0.77}$ (hereafter designated as $UO_2(ba)_2$); Calcd. for: U, 37.96; C, 41.21; H, 3.63. Found: U, 37.90; C,

41.21; H, 3.82. Dimethyl sulfoxide (DMSO) was used after distillation. Cyclic voltammetry measurements were conducted using an electrochemical measurement system (HZ-3000, Hokuto Denkou Corp., Japan) and an electrochemical cell installed inside the inert gas glove box (O_2, H_2O < 1 ppm). A platinum working electrode (1.6ϕ) and a $Ag/AgNO_3$ reference electrode were used. Each measurement was followed by the measurement of the ferrocene/ferrocenium (Fc/Fc^+) couple according to the IUPAC recommendation. Bulk electrolysis was carried out by using a flow electrolysis system as previously reported.[7] After the solutions of the pentavalent state was prepared they were immediately transferred to sample tubes and kept inside brass-made containers thermostated at 16, 25 and 50 °C. The quartz optical cell with sealing, into which small amount of solution was sampled, brought out from the glove box and absorption spectral measurements were started immediately.

3 RESULTS AND DISCUSSION

U(VI) solutions of $UO_2(ba)_2$ with [U]= 4 mM were electrolytically reduced by using flow cell at flow rate of 35.2 mL/min. at controlled potential of -2.0 V vs. Fc/Fc^+, which are little lower than the estimated formal potentials of -1.41 vs. Fc/Fc^+ (Figure 1(a), inset). Figure 1(a) summarizes the absorption spectra obtained during these electrolysis experiments. In the electrolyzed solution obtained from $UO_2(ba)_2$, strong bands appear at 1441 and 1854 nm and also a broad band at 500-1000 nm, attributed to a charge transfer band from benzoylacetone to uranium(V) (Figure 1 (a)). This is similar to the previous result for acetylacetone[7]. The colours of these solutions showed a remarkable change from yellow orange to blue. This is different form acetylacetone complex which showed a colour change from yellow to purple. These results indicate that the colours of uranium(V) solutions are greatly different according to ligands, contrary to the colour of uranium(VI) complexes which seems largely independent of ligand environment.

Absorption peaks at 1441 and 1854 nm decreases with time as shown in Figure 1(b), with an isosbestic point observed at 530 nm, but all bands attributable to the U(V) complex are still strong even after 74 hours. The ratio of the absorbance at measurement time point versus the initial time point at 1854 nm are plotted in Figure 2. Half-lives of U(V) at 16 °C and 25 °C are about 48 and 14 hours, respectively, a marked increase in stability compared with the acetylacetone system. However, this dramatic increase compared with the previous experiment may be attributed to the absolute static enviroment during storage and not to specific ligands. In Figure 2 it is apparent that there are two kinetic processes; one being a rapid decrease in stability at the initial stage and the other a slow decrease with apparent first order kinetics. The activation energy of the first order decay reaction was determined as E_a= 18.5 kJ mol^{-1} from an Arrhenius plot of ln k against $1/T$ (Figure 2(b)). In addition, for the initial stage with large decrease, a Powell plot was prepared which suggests a reaction order of three.

Acknowledgements
We would like to thank Asahi Glass Engineering Co., Ltd. and Toyobo Co., Ltd. for supplying the materials used in the flow cell. We would also like to thank Prof. I. Satoh and Mr. M. Takahashi of the Experimental Facility for Alpha-Emitters of IMR, Tohoku University, for their kind cooperation. This work was performed also at the Irradiation Experimental Facility, IMR, Tohoku University.

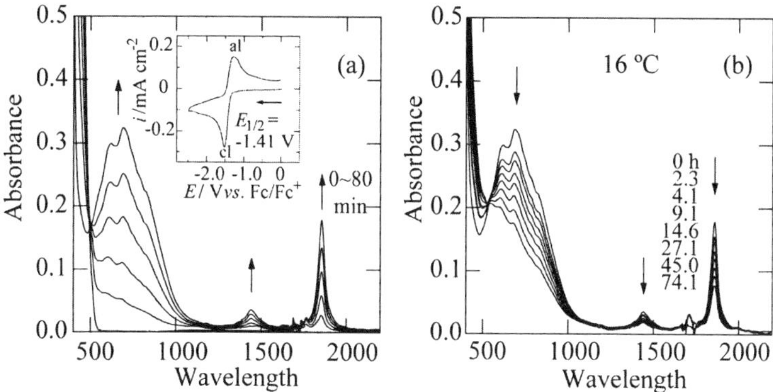

Figure 1 *(a) Abs. spectra obtained during controlled-potential electrolysis of $UO_2(ba)_2$ (5×10^{-3} mol dm^{-3}) in DMSO solution containing 0.1 mol dm^{-3} TBAP. Inset is cyclic voltammogram of $UO_2(ba)_2$ in DMSO solution containing 0.1 mol dm^{-3} TBAP. The concentration of $UO_2(ba)_2$ is 1×10^{-3} mol dm^{-3}. The initial sweep direction is cathodic and sweep velocity is 0.2 Vs^{-1}. (b) Abs. spectra obtained after electrolytic reduction corresponding to part (a) at 16 °C.*

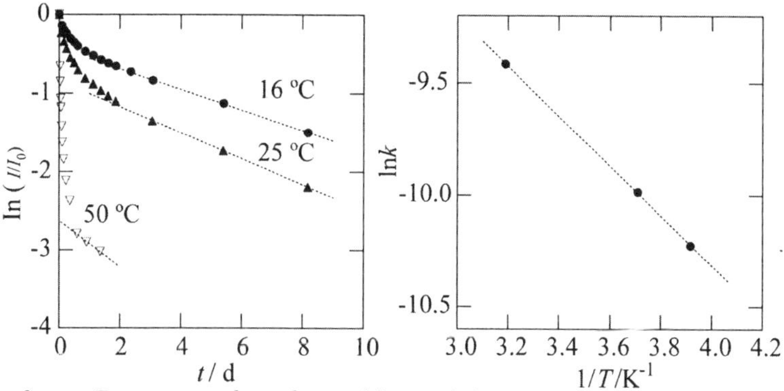

Figure 2 *Temperature dependence of decay of absorption peak at 1854 nm which is attributed to pentavalent state of $UO_2(ba)_2$ in DMSO(a) and Arrhenius plot(b).*

References

1 Y. Shiokawa, H. Yamana, H. Moriyama, *J. Nucl. Sci. Tech.* 2000, **37**, 253.
2 T. Yamamura, N. Watanabe, T. Yano, Y. Shiokawa, *J. Electrochem. Soc.* 2005, **152**, A830.
3 K. Hasegawa, A. Kimura, T. Yamamura, Y. Shiokawa, *J. Phys. Chem. Solids.* 2005, **66**, 593.
4 T. Yamamura, Y. Shiokawa, H. Yamana, H. Moriyama, *Electrochim. Acta.* 2002, **48**, 43.
5 T. Yamamura, Y. Shiokawa, Y. Nakamura, K. Shirasaki, S.-Y. Kim, *J. Alloys Compds.* 2004, **374**, 349.
6 G. Gritzner, J. Selbin, *J. Inorg. Nucl. Chem.* 1968, **30**, 1799.
7 K. Shirasaki, T. Yamamura, Y. Shiokawa, *J. Alloys Compds.* in press.

PREPARATION OF URANIUM(III) PERCHLORATE IN *N,N*-DIMETHYLFORMAMIDE AND OTHER APROTIC MEDIA

K. Shirasaki, T. Yamamura and Y. Shiokawa

Institute for Materials Research, Tohoku University, Sendai, Miyagi 980-8577, Japan

1 INTRODUCTION

Light actinide such as neptunium and uranium are known to show two couples of reversible or quasi-reversible electrode reactions as seen in Np(III)/Np(IV) and Np(V)/Np(VI).[1] By utilizing the two couples of fast reactions of uranium in aprotic solvents as active-materials of liquid rechargeable battery, a redox-flow type battery with higher efficiency than the existing vanadium redox-flow battery has been proposed.[1-2] In the charged state of the uranium redox-flow battery, trivalent uranium is used as the negative electrolyte. There is no information on preparation of U(III) by electrolysis in aprotic solvents. It is well known that uranium(III) aqueous solution is readily prepared by electrolytic reduction on a mercury cathode. Through this procedure the synthesis of various hydration sulphates and chlorides is possible.[3] Also, reduction by zinc amalgam, Zn(Hg), has been applied for preparation of aqueous solution[4] and acetonitrile, methanol and *N,N*-dimethylformamide (DMF) solutions of UCl_4, with their absorption spectra reported.[5] Sodium suspened in tetrahydrofuran was also used to reduce UCl_4 to UCl_3 and to prepare various U(III) compounds.[6] In this research, uranium(IV) dimethylsulfoxide perchlorate is used to examine U(III) bulk solution preparation by Zn(Hg) and electrolytic reduction in aprotic solvents. Absorption spectroscopy in the visible near-infrared region is applied for detection of U(III).

2 EXPERIMENTAL

Dimethyl sulfoxide (DMSO, Donor number, Index of solvation strength = 29.8), Acetonitrile (AN, 14.0), DMF (26.6) were used as aprotic solvents and tetrabutylammoniumperchlorate was used as supporting electrolyte. Uranium(IV) dimethyl sulfoxide perchlorate, prepared according to literatures, was identified as $U(dmso)_8(ClO_4)_4$; Calcd. For: C, 15.24; H, 3.84. Found: C, 15.16; H, 3.86. All experiments were carried out in inert gas glove box. A large excess of Zn(Hg)[4] and a solution of $U(dmso)_8(ClO_4)_4$ were added to a glass bottle with a screwed-cap and shaken continuously. In electrolytic reduction, Ag/Ag^+ electrode and ϕ 30mm Pt plate were used as reference and working electrode, respectively. A cell with perfluorocarbon anion-exchange membrane for separation into two compartments was used. Electrolysis by constant current of 3 mA was carried out by using a potentiostat, with the solution sampled in a quartz optical cell every 5-10 minutes.

Absorption spectra in visible near-infrared region were measured by spectrophotometer and the concentration of uranium and zinc were determined by ICP-AES spectrophotometer.

3 RESULTS AND DISCUSSION

The absorption spectrum of pure U(III) prepared by electrolytic reduction in 1M HCl solution (Fig.2(b)) is in agreement with the literature.[7] The peak positions with large molar extinction coefficient, $^4I_{11/2}$, $^4I_{13/2}$, $^4F_{7/2}$, $^2H_{11/2}$, $^2K_{13/2}$ (from low to high wavenumber), were used as typical absorption peaks of U(III), as indicated as vertical bars in Fig.1. Absorption spectra obtained during the reduction of $U(dmso)_8(ClO_4)_4$, 4.76 mmol dm^{-3}, by Zn(Hg) in DMF are shown in Fig.1(a). All the major U(III) peaks were identified in the spectrum five min after addition, and after that the peak intensity of U(III) decreased suddenly, despite that uranium concentration remaining unchanged (Fig.1, DMF-a, inset). Considering that U(IV) is stoichiometrically reduced to U(III) at 2[Zn]= [U] 10 min after addition (on the basis of reaction Zn(Hg)+ 2 $U^{4+} \rightarrow Zn^{2+} + 2\ U^{3+}$) this decrease of U(III) absorption may suggest some subsequent reaction.

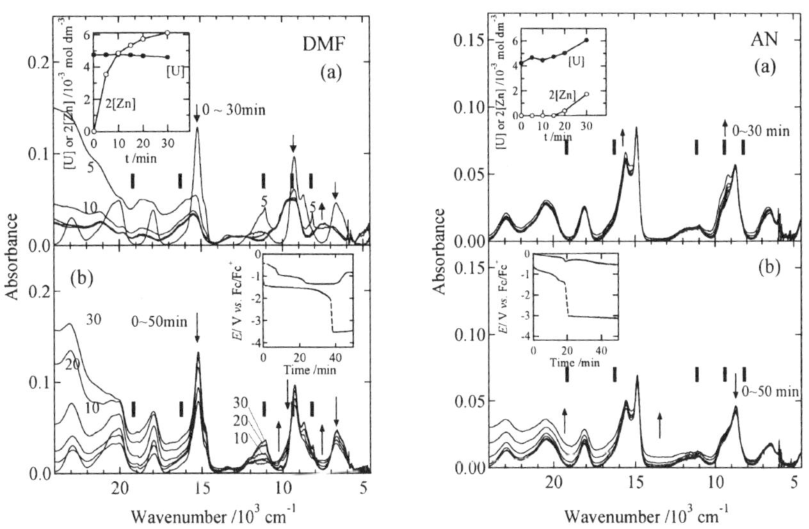

Figure 1 *Absorption spectra obtained during reduction of $U(dmso)_8(ClO_4)_4$ in DMF and AN solution by Zn(Hg) (a) and electrolysis (b). Concentrations of uranium and zinc in solution have been identified by ICP-AES (a, inset). The time course of potential of electrode (dashed curve) and solution (solid curve) during electrolysis is indicated in b, inset. See text concerning with vertical bars.*

The electrode potential remains more or less constant between 0-30 min (Fig.1, DMF-b), and the solution potential shift to lower potential slowly. Five main absorption, especially absorption of $^4F_{7/2}$, increase in intensity up to 30 min indicating generation of U(III). In DMSO the absorption change is similar to that of DMF except for exclusive suppression of $^4F_{7/2}$. In AN there are almost no spectral change both in amalgam reduction and in electrolytic reduction (Fig.1, AN-a and b) indicating that reduction from U(IV) to U(III) is difficult in AN, rather than zinc not behaving as a reducing agent in AN (Fig.1, AN-a, inset). Since the spectrum of pure U(III) is not obtained in solution including perchlorate, the spectrum of U(III) was extracted by subtracting the spectrum of existing

U(IV), as done previously. In the spectrum after 5 min (Fig.1, DMF-a), the spectra of U(III) perchlorate which was obtained by estimating the remaining U(IV) on the basis of the zinc concentration. Alternatively, since current efficiency of reducing U(IV) is evidently low, the spectrum of U(III) was deducted by assuming the appropriate amount of U(IV) which produce a similar form of spectrum with amalgam reduction in the region 8,000-10,000 cm^{-1}, *i.e.* the contribution from $^3F_3$ absorption peak of U(IV) is minimized. In the spectrum at 30 min, U(IV) concentration is assumed to 82% and the obtained spectrum of U(III) perchlorate is shown in Fig.2(c) dotted line. These spectral positions are almost identical to those obtained with chloride species in DMF. However, peak intensity, e.g. $^4F_{7/2}$, is much weaker than $1/10^{th}$ of the corresponding peak of the chloride species. This spectrum is also very similar to those obtained in 1M HCl and 1M $HClO_4$ despite the larger DMF donor number of 38.8. In contrast, the peak positions are very different from these obtained in HMPA which has a larger donor number. Since there are reports that U(III) is slowly reduced by perchloric acid,[8] it is necessary to study the stability of U(III) in system without perchlorate anion.

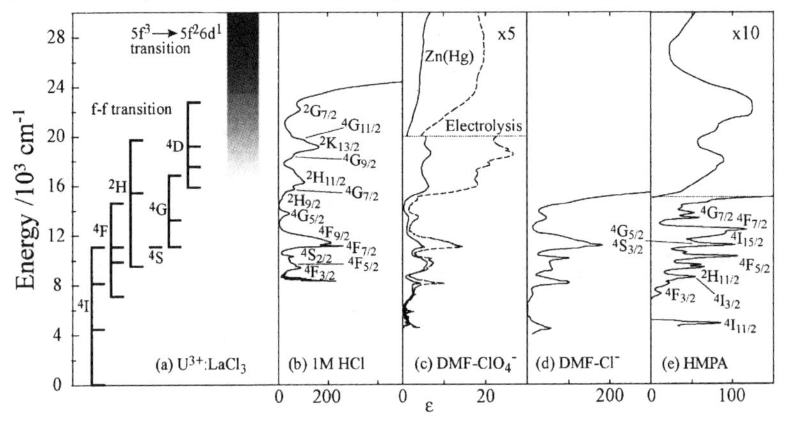

Figure 2 *Energy level assignment for obtained spectra of U(III) in this study (b, c) and found in literatures (a[9], d[5], e[10]).*

ACKNOWLEDGEMENTS

We would like to thank Prof. I. Satoh and Mr. M. Takahashi of Experimental Facility for Alpha-Emitters of IMR, Tohoku University, for their kind cooperations.

References

1 Y. Shiokawa, H. Yamana, H. Moriyama, *J. Nucl. Sci. Tech.* 2000, **37**, 253.
2 T. Yamamura, Y. Shiokawa, H. Yamana, H. Moriyama, *Electrochim. Acta.* 2002, **48**, 43.
3 R. Barnard, J. I. Bullock, B. J. Gellatly, L. F. Larkworthy, *J. Chem. Soc., Dalton Trans.* 1972, **18**, 1932.
4 A. Sato, *Bull. Chem. Soc. Jpn.* 1967, **40**, 2107.
5 J. Drozdzynski, *J. Inorg. Nucl. Chem.* 1978, **40**, 319.
7 R. A. Andersen, *Inorg. Chem.* 1979, **18**, 1507.
8 D. Cohen, W. T. Carnall, *J. Phys. Chem.* 1960, **64**, 1933.
9 H. M. Crosswhite, H. Crosswhite, W. T. Carnall, A. P. Paszek, *J. Chem. Phys.* 1980, **72**, 5103.
10 A. N. Kamenskaya, J. Drozdzynski, N. B. Mikheev, *Radiokhimiya.* 1978, **22**, 247.

BENT METAL CARBENES – ELECTROSTATIC BONDS VS COVALENCY IN ORGANOMETALLIC ACTINIDE CHEMISTRY

S.T. Liddle, S.A. Mungur, M. Rodden, A.J. Blake, C. Wilson and P.L. Arnold

University of Nottingham, School of Chemistry, University Park, Nottingham NG7 2RD, UK.

1 INTRODUCTION

N,N'-disubstituted imidazol-2-ylidenes, or N-heterocyclic carbenes (NHCs), of the form $[C\{NRCH\}_2]$, R = hydrocarbyl group, are useful as strong σ-Lewis bases, possessing a stable, singlet carbene, divalent carbon atom. NHCs bind through this carbon with negligible backbonding, to almost all metals in the periodic table.[1] As such they have already found use as LAC (ligand accelerated catalysis) additives that can improve the activity of Lewis acid metal catalyst systems, but few electropositive metal complexes of NHCs have been studied. The softness of the NHC can be considered to make it a poor donor for an electropositive metal centre. We have been studying polydentate ligands that combine the NHC group with an anionic functional group, in order to stabilise higher oxidation state, Lewis acidic metals and lanthanide catalysts.[2]

We have reported a synthesis of an amino-functionalised NHC ligand HL^1 and HL^2 Scheme 1, and showed how the coordination of the anionic amido group aids in the synthesis and manipulation of electropositive metal adducts. However, we regularly isolate Group 1 salts in which the metal carbene geometry is severely distorted despite the flexible C_2 backbone (up to 41° from the trigonal plane for a potassium alkoxy-NHC complex).[3] The extent of the deviation can be described by two angles: pitch and yaw, Fig. 1.

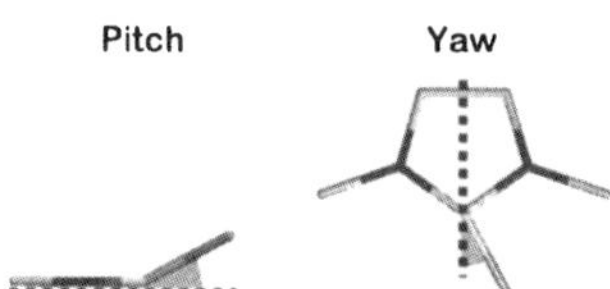

Figure 1 *Pitch and yaw in metal-NHC complexes*

We were concerned that this bend might involve a formal rehybridisation of the unusual divalent carbene carbon centre, and reduce the strength of the metal-ligand bond. The uranyl ion $[UO_2]^{2+}$, which binds ligands predominantly in the equatorial plane, and whose vibrational spectrum is very sensitive to the strength of the equatorial ligands,[4] should provide an ideal structural core at which to study this.

2 RESULTS AND DISCUSSION

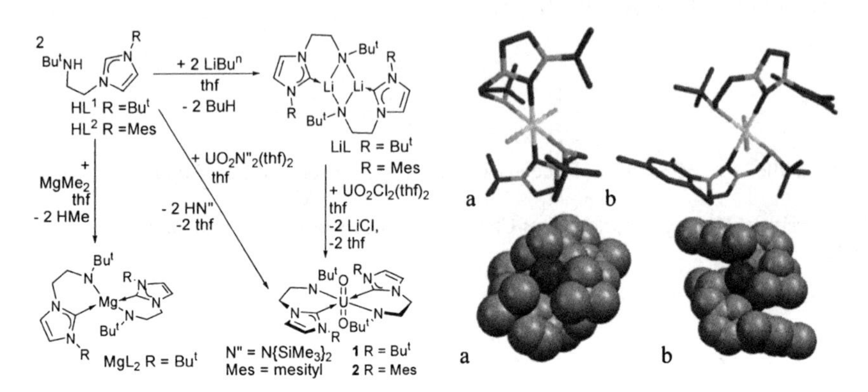

Scheme 1 *Synthesis and crystal structures of **1** (a) and **2** (b). Below: space filling projections of **1** and **2**.*

Under a dry, dinitrogen atmosphere, treatment of $[UO_2Cl_2(thf)_2]$ with two equivalents of LiL^1 affords red $[UO_2L^1{}_2]$ **1**.[5] This is the second reported example of an organometallic uranyl complex.[6] The *N*-mesityl analogue **2** (mesityl = 2,4,6-$Me_3C_6H_2$) is made similarly from LiL^2. Complex **2** is a darker red-brown than **1**. The carbene has a characteristically large chemical shift even when metal bound; 262.8 ppm in **1**, compared with 211.0 in HL.[1] Instead of a statistical mixture of *cis* and *trans* regioisomers, complex **1** forms in 90 % of the *trans*-[L] isomer (and 10 % presumed *cis*-[L], not isolated), according to ^{1}H NMR spectroscopy, while **2** forms exclusively the *trans*-[L] isomer.

In **1** and **2**, the U-C bond lengths are directly comparable to the monodentate NHC complex $[UO_2Cl_2([C\{NMesCH\}_2])_2]$.[6] In **1** the steric bulk encapsulating the metal on both sides of the ligand is similar, forming two isomers. In **2** the N-mesityl group makes it impossible to form a *cis*-NHC isomer. In **2** the ligands are flat, and can pack, but in **1** the M-CN_2 bond distorts to relieve steric congestion, showing both pitch and yaw. However, the U-C distances are the same as each other, and $[UO_2Cl_2([C\{NMesCH\}_2])_2]$, at 2.64 Å. This implies that the bidentate NHC group can bend to accommodate steric factors, without a loss of bond strength. Vibrational spectroscopy (both (ν_1) and (ν_3) stretches) is more sensitive to the electronic effects on the U=O bond.[7] The ν_3 U=O stretch is weakly visible in the FTIR spectrum of **1,** observed at 929 cm^{-1}, and of **2,** observed at 933 cm^{-1}. As the other simple uranyl NHC complex $[UO_2Cl_2([C\{NMesCH\}_2])_2]$ has a stretch of 938 cm^{-1}, we infer that the bend of the U-NHC bond observed does not affect the electron donor ability of the NHC ligand.

We have also studied uranyl alkoxy-NCH complexes, Scheme 2, but the only complex we have isolated and structurally characterised so far is $[UO_2Cl_2(HL^3)_2]$ **3**, presumably formed from a reaction with the partially deprotonated HL, an impurity in the potassium salt KL.[3] Scheme 2 shows the molecular structure of **3**. Two different molecules, and two chloride counterions form the asymmetric unit. The geometries confirm the presence of the acidic imidazolium C2-H. Interestingly, compound U(2)'s molecular geometry is *pseudo*-D_{4h} with the HL^3 ligands in a propeller-like conformation, to form an open box-shape. The imidazolium chloride ligand, now reprotonated, has the same constituents as an ionic liquid, although the alkyl groups are too short for HL^3 to melt at room temperature. The structure may have relevance to the chemistry of actinide salts in ionic liquid media.

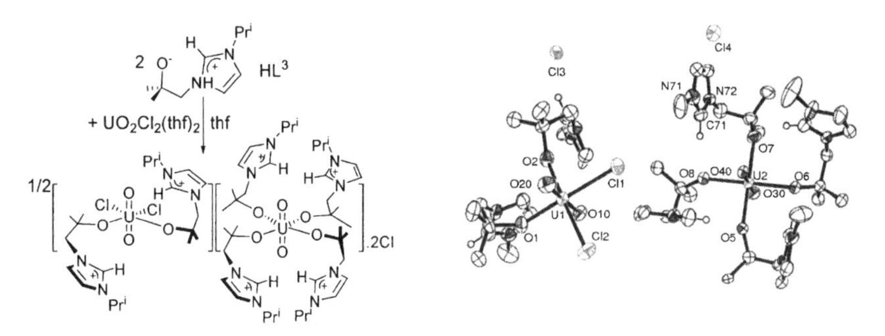

Scheme 2 *Synthesis and displacement ellipsoid drawing of **3** (30 % probability ellipsoids) non-C2 H, Pr^i methyl groups and lattice solvent omitted. Selected distances and angles: U1 O10 1.795(7), U1 Cl1 2.751(3), U1 O1 2.173(7), U2 O30 1.792(7), U2 O5 2.247(6), N71 C71 1.344(14), N72 C71 N71 107.5(10)*

3 CONCLUSIONS

The characteristic equatorial binding modes and uranyl stretching frequency in different complexes has been used to demonstrate that the N-heterocyclic carbene ligand does not have any geometrical dependence on bond strength. An alkoxide – tethered NHC adduct retains its U-O bond on partial hydrolysis, forming uranyl imidazolium salts which may have relevance to actinide chemistry and speciation in ionic liquids.

The authors thank the EPSRC and the Royal Society for funding.

4 EXPERIMENTAL DETAILS

Data for **3** were collected at the Daresbury Synchrotron in the UK. Final R_1 [8011 $F > 4\sigma(F)$] = 0.0532 and $wR(\text{all } F^2)$ was 0.119. *Crystallographic data for complex **3** are available as CCDC 279001 via www.ccdc.cam.ac.uk.*

References

1 W. A. Herrmann, *Angew. Chem.-Int. Edit.* 2002, **41**, 1291. J. C. Green, R. G. Scurr, P. L. Arnold, and F. G. N. Cloke, *Chem. Commun.*, 1997, 1963.

2 P. L. Arnold, A. C. Scarisbrick, A. J. Blake, C. Wilson, *Chem. Commun.* 2001, 2340. (b) P. L. Arnold, S. A. Mungur, A. J. Blake, C. Wilson *Angew. Chem. Int. Ed.* 2003, ***42***, 5981.

3 P. L. Arnold, M. Rodden, and C. Wilson, *Chem. Commun.*, 2005, 1743.

4 M. J. Sarsfield, H. Steele, M. Helliwell, and S. J. Teat, *Dalton Trans.*, 2003, 3443.

5 S. A. Mungur, S. T. Liddle, C. Wilson, M. J. Sarsfield, and P. L. Arnold, *Chem. Commun.*, 2004, 2738.

6 Also an NHC adduct, W. J. Oldham Jr., S. M. Oldham, B. L. Scott, K. D. Abney, W. H. Smith, and D. A. Costa, *Chem. Commun.*, 2001, 1348.

7 G. H. John, I. May, M. J. Sarsfield, H. M. Steele, D. Collison, M. Helliwell, and J. D. McKinney, *Dalton Trans.* 2004, 734. M. J. Sarsfield, M. Helliwell and J. Raftery, *Inorg. Chem.* 2004, **43**, 3170.

URANYL COMPLEXATION BY A HINGED POLYPYRROLIC MACROCYCLE; HYDROGEN BONDING WITH THE URANYL OXO ATOM AND ENFORCED π-STACKING IN THE EQUATORIAL PLANE

P.L. Arnold, J.B. Love, A.J. Blake and C. Wilson

University of Nottingham, School of Chemistry, University Park, Nottingham NG7 2RD, UK.

1 INTRODUCTION

The uranyl UO_2^{2+} ion is ubiquitous in solution and solid-state uranium chemistry. As such, its chemistry with fused pyrrolic macrocycles such as porphyrins, which display so much interesting structural and reaction chemistry with transition metals, is limited since it neither fits in the cavity of such macrocycles, nor is capable of sitting above the ligand plane. Larger macrocycles, such as expanded porphyrins have recently been shown to coordinate the uranyl ion.[1] Reactions that destroy the uranyl dicationic fragment, either by coordination or reduction, are important for understanding geological uranium chemistry, fundamental atom transfer and small-molecule activation chemistry,[2] and may provide access to the theoretically interesting, and predicted *cis*-uranyl dication.[3]

We have recently reported a potentially dinucleating, tetraanionic, hinged macrocycle **L** that contains two imino-pyrrolic binding sites. **L** is large and relatively flexible.

2 RESULTS AND DISCUSSION

The transamination reaction between the freebase macrocycle H_4**L** and the uranyl amide $[UO_2(thf)_2\{N(SiMe_3)_2\}_2]$ in thf at low temperature results in the rapid and sole formation of the khaki-green mono-uranyl complex, *trans*-$[UO_2(thf)(H_2$**L**$)]$ **1** (Scheme 1).[4] No di-uranyl complex, $[(UO_2)_2($**L**$)]$ has ever been observed. The thf coordinated between the macrocycle 'hinges' in **1** is readily displaced by donors such as pyridine, to form **2**, dark orange crystalline *trans*-$[UO_2(NC_5H_5)(H_2$**L**$)]$, Scheme 1, and Fig. 1.[5]

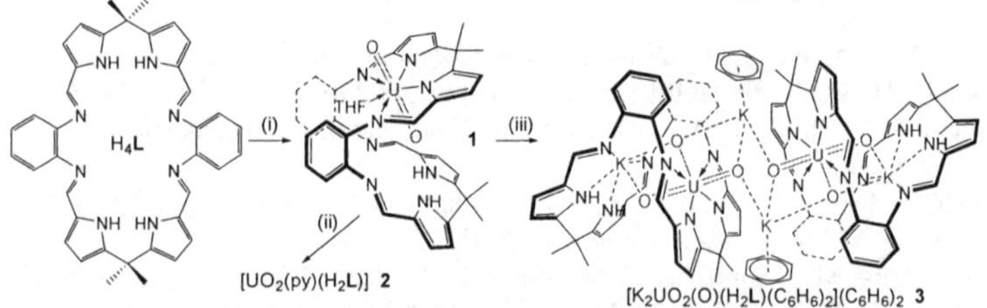

Scheme 1. *Synthesis of the macrocyclic monouranyl complexes **1**, **2** and **3**. (i) $[UO_2(thf)_2\{N(SiMe_3)_2\}_2]$, thf; (ii) pyridine (iii) K metal, benzene, reflux 48 h.*

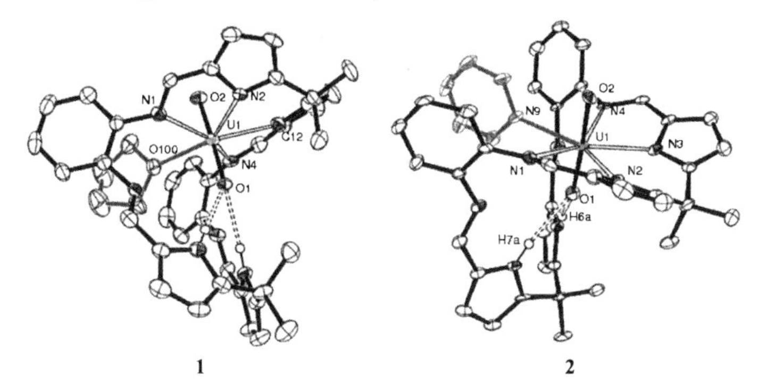

Figure 1. *The X-ray crystal structures of the mono-uranyl complexes **1** and **2** (30 % ellipsoids). All hydrogens except those on N6 and N7 have been omitted for clarity.*

In each structure the uranyl ion is five-coordinate in the equatorial plane, and the second N_4-donor compartment remains metal-free. In **1** the coordination site between the hinges is occupied by thf; in **2**, it is occupied by a molecule of pyridine. This results in an unfavourable π-stacking interaction in **2**, and a highly unusual, slightly asymmetric "sandwiched" position for the thf molecule. An appreciable H-bonding interaction occurs between the pyrrolic H6 and H7 and the uranyl oxygen O1 (N6···O1 3.111(7), N7···O1 3.146(7) Å), U1-O1 is fractionally but significantly longer (0.024 Å) than U1-O2. In **2** the hydrogen-bonding (N6···O1 3.078(3), N7···O1 3.103(3) Å) motifs are similar.

A reaction of crude sample of **1** with potassium metal afforded a mixture of products but a crop of dark orange-red single crystals was isolated from a concentrated benzene solution at room temperature in poor (< 10 %) yield. The complex is formulated as **3**, which is almost an equal mixture of *trans*-$[K_2UO_2(O)(H_2\mathbf{L})(C_6H_6)_2].2(C_6H_6)$ and *trans*-$[KUO_2(OH)(H_2\mathbf{L})(C_6H_6)_2].2(C_6H_6)$. The compound is essentially the same as **1** or **2**, with the coordinated thf or pyridine molecule replaced by an oxo atom. The (symmetry generated) dimeric structure displays an unusually high degree of coordination by potassium cations to the uranyl oxo atoms - normally the inert group in uranyl chemistry. The single crystal X-ray structure is shown in Fig. 2.

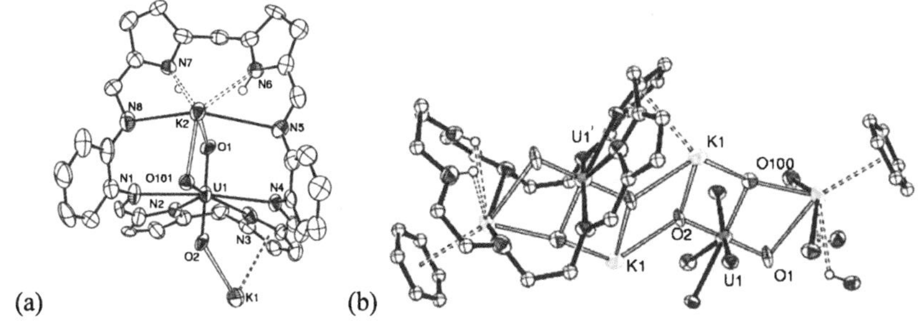

Figure 2. *Displacement ellipsoid plot of the mono-uranyl potassium complex **3** (30 % ellipsoids). (a) asymmetric unit (b) core of dimer. Selected bond lengths (Å) and angles (°): U1 O2 1.771(9), U1 O1 1.782(9), U1 O100 2.219(19), U1 N3 2.530(10), U1 N2 2.536(12), U1 N1 2.589(10), K1 O2 2.763(9), K1 O100 2.782(19), K1 O2 2.893(10), K2 O100 2.47(2), K2 O1 2.925(11).*

The complex still contains a 1:1 ratio of U:L but now the other cavity is filled by a potassium cation, designated endo. Another potassium cation is placed at the outer edge of the ligand's hinges, designated exo. The T-shaped UO_3 unit is nearly symmetrically capped in the plane by two potassium cations. Most importantly, both potassium cations coordinate to uranyl O atoms. However, a component of disorder arises from a replacement of about half of (K2) sites with a proton on the equatorial O atom.

Both uranyl U-O distances are 1.78 Å, consistent with a normal uranyl bond order of 2.5. The potassium- oxo distances are 2.8685 Å and 2.6717 Å to the uranyl and equatorial oxo groups respectively, so the metal-bridging structures are relatively symmetrical. Previously, the uranyl hydroxide systems have yielded five structurally characterised complexes with diuranium bridging hydroxides (with U-O distances of average 2.328 Å).[5] Both lithium and sodium crown ether adducts with metal-oxo interactions have been reported recently, with uranyl-M contacts of between 2.20 and 2.45 Å,[6] but the potassium coordination, and the symmetrical interactions that arise from the T-shaped UO_3 fragment are, to the best of our knowledge, unprecedented. The rest of the sample of **3** was destroyed during the structural determination. However, we will report on its spectroscopy and further reactivity of **3** in due course.

3 CONCLUSIONS

In conclusion, the polypyrrolic macrocycle L which has two cavities can only accommodate one linear uranyl moiety, but the adjacent imino-pyrrole binding site not only provides adjacent H atoms that display hydrogen-bonding reactivity with the uranyl oxo atom but also allow the incorporation of other reactive metal cations that can also display coordination chemistry with the oxo group.

The authors thank the EPSRC and the Royal Society for funding, and the EPSRC-funded synchrotron crystallography service and its director Professor W. Clegg for collection of single-crystal diffraction data at Daresbury SRS Station 9.8/16.2 SMX.

4 EXPERIMENTAL DETAILS

A crystal of **3** was mounted in silicone grease on a glass fibre on the diffractometer. $C_{62}H_{58.52}K_{1.48}N_8O_3U$, M = 1259.59, monoclinic, a = 10.630(3), b = 16.648(5), c = 31.730(8) Å, β = 99.401(4)^{O}, U = 5540(3) Å^3, T = 120(2) K, space group $P2(1)/c$, Z = 4, D_c = 1.510 g cm^{-3}, μ(synchrotron) = 6768 mm^{-1}, 9611 unique reflections (R_{int} 0.1313) used in calculations. Final R_1 [6547 F 4$\sigma(F)$] = 0.0850 and wR(all F^2) was 0.1950. *Crystallographic data for **3** CCDC 278802 available through www.ccdc.cam.ac.uk.*

References

1 J. L. Sessler, A. E. Vivian, D. Seidel, A. K. Burrell, M. Hoehner, T. D. Mody, A. Gebauer, S. J. Weghorn, and V. Lynch, *Coord. Chem. Rev.* 2001, **216-217,** 411. J. L. Sessler, A. E. V. Gorden, D. Seidel, S. Hannah, V. Lynch, P. L. Gordon, R. J.Donohoe, C. D. Tait, and D. W. Keogh, *Inorg. Chim. Acta* 2001, **341**, 54.

2 M. J. Sarsfield, M. Helliwell, and J. Raftery, *Inorg. Chem.*, 2004, **43**, 3170.

3 N. Kaltsoyannis, *Chem. Soc. Rev.*, 2003, **32**, 9. G. Schreckenbach, P. J. Hay, and R. L. Martin, *Inorg. Chem.*, 1998, **37**, 4442.

4 P. L. Arnold, A. J. Blake, C. Wilson, and J. B. Love, *Inorg. Chem.*, 2004, **43**, 8206.

5 D. L. Clark, S. D. Conradson, R. J. Donohoe, D. W. Keogh, D. E. Morris, P. D. Palmer, R. D. Rogers, and C. D. Tait, *Inorg. Chem.*, 1999, **38**, 1456.

6 C. J. Burns, D. L. Clark, R. J. Donohoe, P. B. Duval, B. L. Scott, and C. D. Tait, *Inorg.Chem.*, 2000, **39**, 5464. J. A. Danis, M. R. Lin, B. L. Scott, B. W. Eichhorn, and W. H. Runde, *Inorg.Chem.*, 2001, **40**, 3389.

STRUCTURE OF TRIVALENT AMERICIUM AND LANTHANIDE COMPLEXES WITH PHOSPHORYL ACETIC ACID AMIDES

I.L. Odinets[2], O.I. Artyushin[2], V.P. Morgalyuk[1], K.A. Lyssenko[2], E.V.Sharova[2], I.G. Tananaev[1], T.A.Mastryukova[2], and B.F.Myasoedov[1]

[1]Vernadsky Institute of Geochemistry and Analytical Chemistry, Russian Academy of Sciences, 19 Kosygina str., 119991 Moscow, Russia
[2]Nesmeyanov Institute of Organoelement Compounds, Russian Academy of Sciences, 28 Vavilova str., 119991 Moscow, Russia

1 INTRODUCTION

Neutral bidentate organophosphorus compounds are promising radiochemical extractants. Within this class of compound carbamoylmethyl phosphonates/phosphinates (CMP) and phosphine oxides $R_2P(O)CH_2C(O)NR^1R^2$ (CMPO) coordinate effectively with lanthanide and actinide ions in acidic aqueous media and selective derivatives of the latter compounds are used in liquid-liquid extraction processes[1].

Based upon equilibrium distribution measurement Siddal[2] was the first to conclude that tris bidentate chelate complexes $M(CMP)_3^{3+}$ were formed in the organic phase during solvent extraction. In contrast Horwitz *et al.*[3] concluded that monodentate tris chelates are formed via metal-phosphoryl oxygen interaction with the amide function behaving only as an internal buffer. Therefore in order to design the improved extractants it is useful to understand the fundamental coordination chemistry of such compounds in respect to *4f*- and *5f*-elements.

The structural chemistry of CMP compounds is rather well investigated with several classes of complex described including $Ln(NO_3)_3(CMP)_2$ [Ln = La, Sm, Yb, Er, Ce, Eu, Gd] [4,5], $UO_2(NO_3)_2(CMP)$[6], $Th(NO_3)_4(CMP)_2$[7], and $Ln(NO_3)_3(CMP)_2{\bullet}H_2O$ [Ln = Tb - Eu] [5]. In complexes of the first three types the ligand coordinates by O,O-bidentate mode while in the latter case CMP is bonded to the lanthanide cation via the oxygen atom of the P=O moiety and the carbonyl oxygen is H-bonded with the metal coordinated with the water molecule. For CMPO the $UO_2(NO_3)_2(CMPO)$ complexes isolated and structurally characterized[8] are similar to the complexes formed by CMP.

As part of study of the coordination characteristics of CMPO ligands we report here the similarity and difference in the coordination chemistry of N,N-dibutylcarbamoyl-diphenylphosphine oxide and a series of N-alkylcarbamoyldiphenylphosphine oxides.

2 METHODS AND RESULTS

Recently we reported the facile synthetic procedure allowing one to obtain a wide range of N-alkylcarbamoyldiphenylphosphine oxides ($Ph_2P(O)CH_2C(O)NHAlk$, known as CMPO-NHAlk), based on a direct amination by primary alkylamines of a single phosphorus substrate. The substrate used was the commercially available diphenylphosphorylacetic acid ethyl ester which gives the desired substances in practically quantitative yield.[9] In addition, we developed one-pot synthesis of N-arylcarbamoyldiphenylphosphine oxides ($Ph_2P(O)CH_2C(O)NHAr$, known as CMPO-NHAr) starting from diphenylphosphorylacetic acid and wide range of primary arylamines.

X-ray investigation carried out for $Ph_2P(O)CH_2C(O)NHPh$ and 1,2-$[Ph_2P(O)CH_2C(O)NH]C_6H_4$ have revealed that both compounds have the similar configuration and introduction of the second dentate moiety does not change the dihedral angle C(2)C(1)N(1)C(7) equal to 171.1° in both cases. Moreover conformation of phosphorylated dentate arms is also the same in both structures. The latter compound crystallizes as a solvate with water and acetonitrile molecules and analysis of crystal packing (formation of rather strong intermolecular H-bonding via NH hydrogen and phosphoryl groups) suggested that under competition bonding at extraction from acid solutions H-bonds are preferably formed via the phosphoryl oxygen atom rather than the carboxyl one.

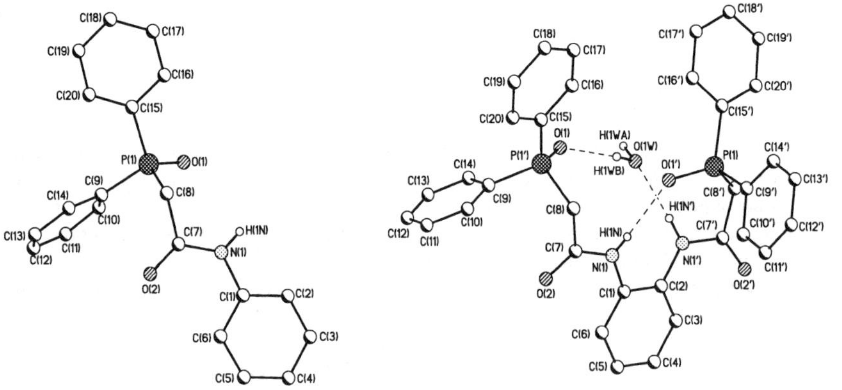

Figure 1 *Crystal structure of $Ph_2P(O)CH_2C(O)NHPh$ (left) and 1,2-$[Ph_2P(O)CH_2C(O)NH]C_6H_4.H_2O$ (right)*

The extraction experiments carried out for Am(III) CMPO-NHR have the similar lg$\mathbf{D}_{\mathbf{Am}}$ at R=n-Alk and are as good as typical N,N-dialkyl CMPO derivatives wherein the distribution coefficient is also insensitive to the structure of the substituent at the N atom[10]. One may therefore suggest that CMPO derivatives bearing either one or two alkyl groups at the nitrogen atom will lead to the similar complexes with transuranium elements. However it was found out that under neutral conditions $Ph_2P(O)CH_2C(O)NBu_2$ formed bidentate chelate complexes, $Ln(CMPO)_2(NO_3)_3$ where Ln = Pr, Gd or Eu, independently on the reactants ratio (see, e.g. fig.2). At the same time its analogue $Ph_2P(O)CH_2C(O)NHBu$-i is able to form metal-ligand complexes (Ln = Pr, Eu, Er) of different composition depending on the reactant ratio. These complexes have either two or three ligands in the inner coordination sphere. NMR and IR spectra in all cases point to a bidentate coordination mode. The same situation was observed for the ligands bearing a single aryl substituent at the nitrogen atom. Apparently such difference in coordination behavior between CMPO and CMPO-NHR ligands is connected with less steric hindrances in the later case.

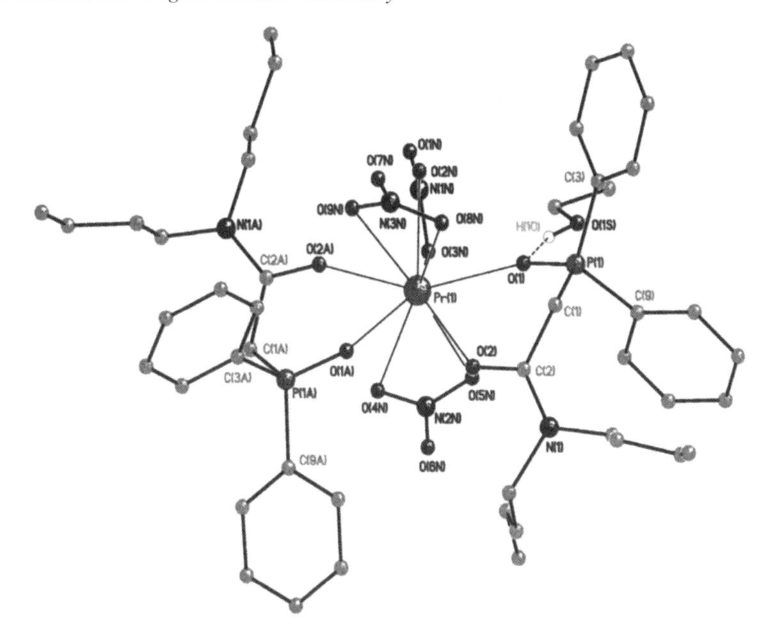

Figure 2 *Crystal structure of* $[Ph_2P(O)CH_2C(O)NBu_2]_2Pr(NO_3)_3$

Acknowledgement

Financial support by the Russian Foundation for Basic Research (grants Nos. 05-03-33094/32692) and the Grant "Leading Scientific Schools" (Project Nos. NSh-1100.2003.3 and 1693.2003.3).

References

1 W.W. Schultz and J.D.Navratil in N.N.Li (ed.). "*Recent Developments in Separation Science*", Vol.7, CRC Press, Boca Raton, Fla., 1982 (and references cited therein).
2 (a) T.H. Siddal, *J.Inorg.Nucl.Chem.*, 1963, **25**, 883; (b) ibid. 1964, **26**, 1991.
3 E.P. Horwitz, A.C.Muscatello, D.G.Kalina and L.Kaplan, *Sep.Sci.Tech.*, 1981, **16**, 417 (and references cited therein).
4 J. Petrova, S.Momchilova, and E.T.K. Haupt, *Phosphorus, Sulfur and Silicon*, 2002, **177**, 1337.
5 S.M.Bowen, E.N. Duesler, and R.T.Paine, *Inorg. Chim. Acta*, 1982, **61**, 155.
6 S.M.Bowen, E.N. Duesler, and R.T. Paine, *Inorg. Chem.*, 1983, **22**, 286.
7 S.M.Bowen, E.N. Duesler, and R.T.Paine, *Inorg. Chem.*, 1982, **21**, 261.
8 L.J.Claude, E.N.Duesler, and R.T.Paine, *Inorg. Chim. Acta*, 1985, **110**, 91.
9 O.I.Artyushin, E.V.Sharova, I.L.Odinets, S.V.Lenevich, V.P.Morgalyuk, I.G. Tananaev, G.V.Pribulova, G.V.Myasoedova, T.A. Mastryukova, and B.F.Myasoedov, *Izv. AN, Ser. Khim.*, 2004, 2395.
10 M.N. Litvina, M.K. Chmutova, N.P. Nesterova, and B.F. Myasoedov, *Radiochemistry* (rus), 1994, **36**, 325.

SURPRISING ACTIVITY OF ORGANOACTINIDE COMPLEXES IN THE POLYMERIZATION OF CYCLIC MONO- AND DIESTERS

E. Barnea,[1] D. Moradove,[1] J. C. Berthet,[2] M. Ephritikhine,[2] and M. S. Eisen[1 *]

[1] Department of Chemistry and Institute of Catalysis Science and Technology, Technion-Israel Institute of Technology, Haifa, 32000, Israel
[2] Service de Chimie Moléculaire, DSM/DRECAM, Laboratoire Claude Fréjacques, CNRS URA 331, CEA Saclay, Gif-sur-Yvette Cedex 91191, France.

1 INTRODUCTION

During the last three decades organoactinide chemistry has been flourishing, reaching a high level of sophistication. The use of organoactinide complexes in stoichiometric or catalytic conditions to promote synthetically important organic transformation has grown due to their rich, complex, and highly unique and informative chemistry. Such reactions include dimerization, oligomerization, hydrosilylation and hydroamination of terminal alkynes.[1] In many instances the regio- and chemo-selectivities displayed by organoactinide complexes are complementary to those observed with other transition-metal or lanthanide complexes. Despite the large variety of reactions catalyzed by organoactinides, the common believe was that if oxygen containing substrates will be introduced, no catalytic activity will be seen, due to the high electropositivity and high oxophilic nature of the actinides.[2] It was thought that due to these two features, oxygen containing environments will promote the deactivation of the organoactinide complex. In the last years, many main-group,[3] d-transition and lanthanide[4] metal complexes were found to be good catalysts for the polymerization of cyclic ester monomers to form polycaprolactone (PCL) and poly-L-lactide (PLLA) (Figure 1), which are two examples of polyesters which are both biodegradable and biocompatible. In contrast, organoactinide complexes were never tested for such reactions.

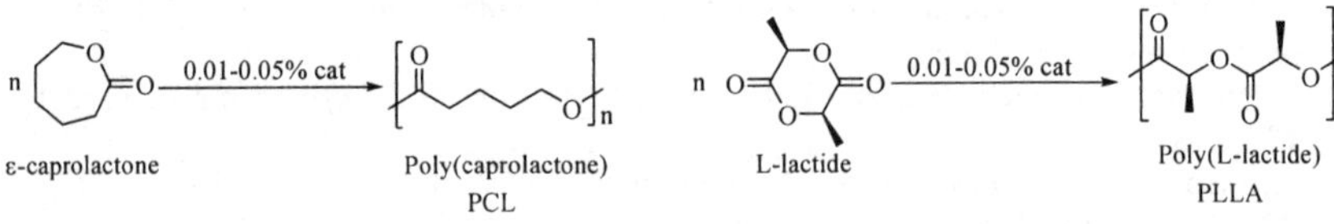

n ε-caprolactone —0.01-0.05% cat→ Poly(caprolactone) PCL

n L-lactide —0.01-0.05% cat→ Poly(L-lactide) PLLA

Figure 1 *Polymerization of cyclic mono and di-esters to polycaprolactone and poly lactide*

In this contribution we present surprising results on the polymerization of ε-caprolactone (CL) and L-lactide (LLAC) by three organoactinide complexes: the neutral complexes $Cp^*_2AnMe_2$ (**1** – An = Th, **2** – An = U, Cp* = C_5Me_5) and the cationic complex **3**, $[U(NEt_2)_3][BPh_4]$, shown in Figure 2.

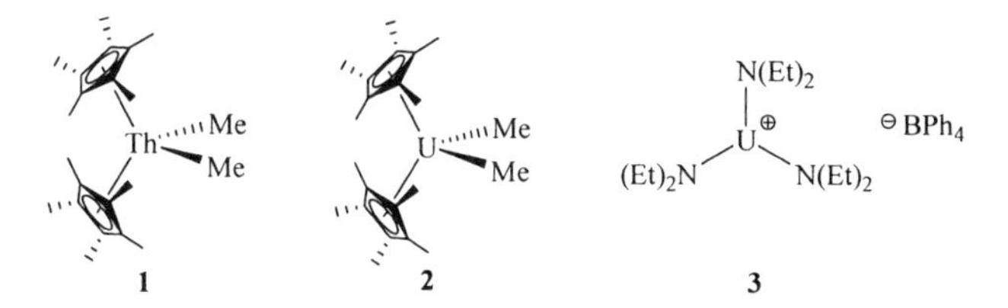

Figure 2 *Structure of the three organoactinide complexes used for the polymerization of cyclic esters.*

2 RESULTS AND DISCUSSION

2.1 ε-caprolactone polymerization

Polymerization was achieved with all the three complexes, though temperature and solvent were found to have a great effect on the complex activity, type of polymerization and the size of the polymer chains (Figure 3). Under the same conditions, the cationic complex **3** shows the highest activity, both at room and elevated temperatures. This fact may be rationalized by the easier approach of the monomer to the active site of **3** (in complexes **1** and **2** this approach might by hindered by the two Cp* ancillary ligations). In addition, and very surprisingly, the use of THF as solvent for complex **1** was found to raise the catalytic activity, despite the possible coordination of the solvent molecule to the active site.

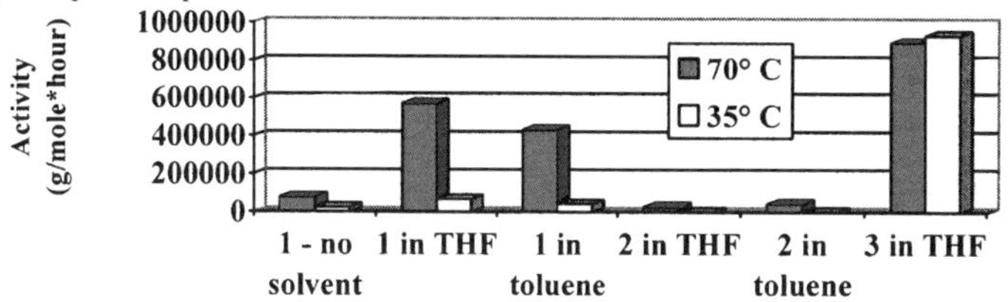

Figure 3 *Activity of the three catalysts in the polymerization of CL in various solvents and at distinct temperatures. Each point represents an average of three parallel experiments falling in a 5% error limit.*

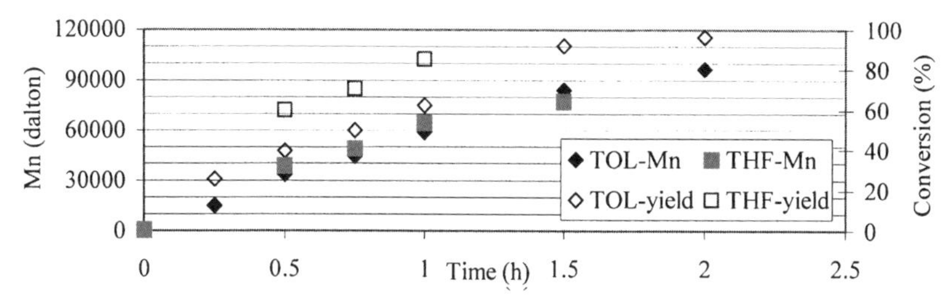

Figure 4 *Average molecular weight (Mn) and conversion (%) of PCL over time with complex **1** as catalyst (1:600 ratio) in toluene (tol) or THF at room temperature. Each point represents an average of three parallel experiments falling in a 5% error limit*

Figure 4 shows the progress of the polymerization reaction of CL with complex **1** in THF or toluene. The linear increase of the average molecular weight of the polymer suggests a living polymerization process, which is also implied from the low polydispersity values of 1.05-1.12 for toluene and 1.22-1.43 for THF. The increase in Mn and conversion is linear up to 90% conversion. At this point the viscosity of the solution slows the polymerization rate.

2.2 L-Lactide polymerization

Polymerization was performed with the three complexes, at various temperatures and in distinct solvents. As with CL, the cationic complex **3** showed the highest activity at 70°C in THF. Figure 5 shows the progress of the polymerization reaction of LLAC with complex **3** in THF. Again, the linear increase of the conversion and average molecular weight of the polymer in the early stages of the reaction suggests a living polymerization process. The reaction is stopped at 70-75% conversion by the produced polymer which blocks all movement in the reactor. The low polydispersity values of 1.04-1.23 also support a living process.

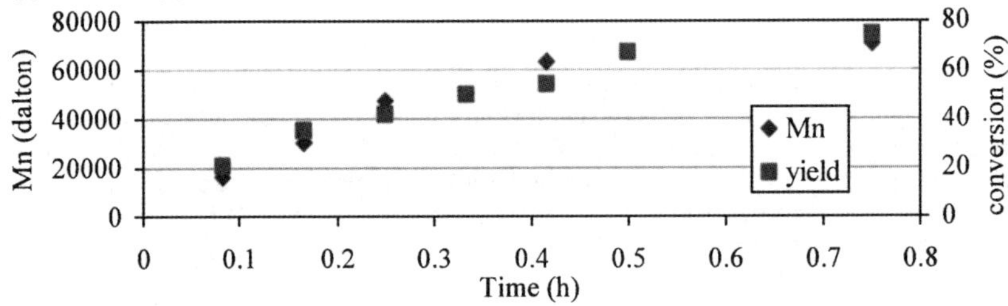

Figure 5 *Average molecular weight (Mn) and conversion (%) of PLLA over time with complex 3 as catalyst (1:600 ratio) in toluene or THF at 70°C. Each point represents an average of three parallel experiments falling in a 5% error limit.*

3 CONCLUSIONS

In this contribution we reported the novel reactivity of neutral and cationic organoactinide complexes in the polymerization of cyclic mono- and diesters. The resulting polymers are characterized by high molecular weights and yields. We have also shown the living nature of the polymerization reaction of CL with complex **1** and of LLAC with complex **3**.

References

1 For recent publications on the chemistry of organoactinide complexes see: E. Barnea, T. Andrea, M. Kapon, J. C. Berthet, M. Ephritikhine, M. S. Eisen, *J. Am. Chem. Soc.* 2004, **126**, 10860-10861 and references therein.
2 Z. Lin, T. J. Marks, *J. Am. Chem. Soc.* 1987, **109**, 7979-7985.
3 For recent reviews on polymerization of cyclic mono and diesters see: a) V. C. Gibson, E. L. Marshall, *Comprehensive Coordination Chemistry II*, 2004, **9**, 1-74. b) T. Biela, A. Duda, S. Penczek, *Macromolecular Symposia*, 2002, **183**, 1-10.
4 I. Palard, A. Soum, S. M. Guillaume, *Chem. Eur. J.* 2004, **10**, 4054-4062.

THE AFFINITY OF N DONOR LIGANDS IN AnO_2^{2+} COMPLEXES AND SYNTHETIC ROUTES TO NEW NEPTUNYL CHEMISTRY

Stéphanie M. Cornet[a], Iain May[a], Mark J. Sarsfield[b] and Madeleine Helliwell[c]

[a] Centre for Radiochemistry Research, Department of Chemistry, The University of Manchester, Manchester M13 9PL
[b] Nexia Solutions, B170, Sellafield, Seascale, Cumbria, CA20 1PG, UK
[c] Department of Chemistry, The University of Manchester, Manchester M13 9PL

1 INTRODUCTION

The linear, dioxo, actinyl cations, AnO_2^{x+} (where x = 1 or 2) dominate the +V and +VI oxidation state chemistry of the mid-actinide elements (U, Np, Pu and Am) and an increased understanding of these species will underpin the development of new actinide separation and waste treatment processes.[1] Generally between 4-6 additional ligands will coordinate in the equatorial plane of the actinyl moiety. Due to its relatively high chemical stability, coupled with ready access to long lived radioisotopes ($^{238}U/^{235}U$), the uranyl cation, $\{UO_2\}^{2+}$, has been studied most extensively.[1] Despite the continued dominance of O-donor ligands as auxilliary ligands in the equatorial plane there have been several recent studies into softer N-donor coordination. In our laboratories we have studied a range of ligand systems (e.g. $[PhC(NSiMe_3)_2]^-$) which have induced out-of-plane coordination, U-C interactions and enhanced the Lewis basicity of the uranyl oxygen, promoting coordination to the Lewis acid, $B(C_6F_5)_3$.[2] Recent additional advances include the displacement of uranyl oxygens by N-R groups (O=U=N-R, R-N=U=N-R)[3], a detailed investigation into equatorial ligand exchange reactions,[4] the first structural characterisation of a $\{UO_2\}^+$ complex[5] and incorporation of the uranyl moiety into a polyoxoanion cluster.[6]

In contrast, the high specific radioactivity of Np, Pu and Am has restricted detailed actinyl chemistry studies with the transuranic elements, with the first $\{PuO_2\}^{2+}$ complex only recently structurally characterised.[7] In contrast, the neptunyl(V) cation has been studied in greater depth, especially due to the relative ease with which the $\{NpO_2\}^+$ moiety will coordinate through its actinyl oxygen to a range of different metal cations (cation:cation complexation)[8]. Our current goal is to extend our investigations into N-donor ligand complexation to $\{NpO_2\}^{2+}$ systems, investigating how moving from an f^0 (i.e. $\{UO_2\}^{2+}$) to an f^1 system effects electronic and molecular structure.

2 RESULTS AND DISCUSSION

Our route into N-donor complexes of $\{UO_2\}^{2+}$ has been through the thf complex $[UO_2Cl_2(thf)_3]$, [9] enabling the manipulation of the cation under inert, moisture free, conditions using standard Schlenk line and dry box procedures. However, preparing a $\{NpO_2\}^{2+}$ analogue of this system has proved difficult and thus many of the more complex N-donor ligand systems are currently hard to access. We have, to an extent, circumnavigated this problem through an investigation of the $[UO_2Cl_2(R_3PX)_2]$ system, where R = Ph or Cy and X = O or NH, with all four complexes readily prepared through the reaction of $[UO_2Cl_2(thf)_3]$ and two equivalents of the required phosphine oxide or phosphinimine. ^{31}P and ^1H NMR have shown that both *cis-* and *trans-* complexes are in equilibrium in solution, as previously shown in the solid state through the structural characterisation of isomers of $UO_2Cl_2(Ph_3PO)_2$ as the α-*trans*, β-*trans* (the PR_3 groups can be situated either in a syn or anti orientation with respect to the axial oxygens) and *cis* forms.[10] The strength of the phosphinimine coordination to the $\{UO_2\}^{2+}$ moiety has been shown through the U-N bond which appeared to be considerably shorter in the $UO_2Cl_2(R_3PNH)_2$ (2.370(1)Å for R=Ph and 2.350(2)Å for R=Cy) compared to other uranyl complexes containing neutral nitrogen donor.[11] (Figure 1a). Consistent with this, there is a dramatic change in charge distribution around the coordinating HN=P moiety which is reflected in the large difference in chemical shift (Δδ=5.5 ppm) between the NH ^1H NMR signal in $[UO_2Cl_2(Ph_3PNH)_2]$ (6.51ppm) compared to the uncomplexed ligand (1.01 ppm).

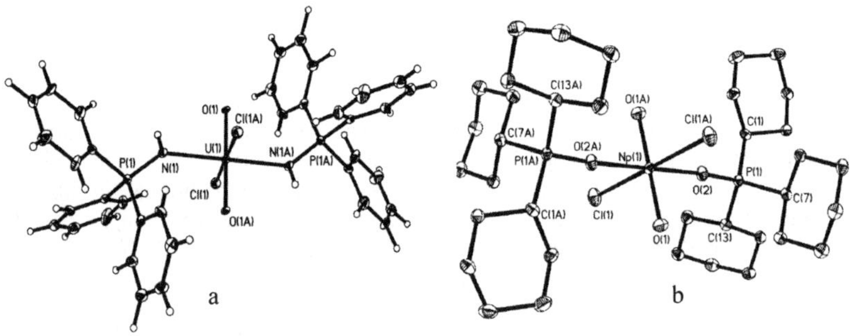

Figure 1 *Ortep representation of a) $[UO_2Cl_2(Ph_3PNH)_2]$ and b)$[Np_2O_2Cl_2(Cy_3PO)_2]$*

Given the similarities in the chemistry between R_3P=O vs R_3P=NH ligands and the strong U-N interaction in $[UO_2Cl_2(Ph_3PNH)_2]$, we investigated the hypothesis that more basic N donor ligands can effectively compete with harder O donor ligands for equatorial coordination sites. When Cy_3PO is added to dichloromethane solutions of $[UO_2Cl_2(Cy_3PNH)_2]$ there is no apparent reaction. By contrast, adding one equivalent of Cy_3PNH to a solution of $[UO_2Cl_2(Cy_3PO)_2]$ results in the release of Cy_3PO and the formation of and intermediate complex $[UO_2Cl_2(Cy_3PNH)(Cy_3PO)]$ which reacts with a further equivalent of the phosphinimine to give $[UO_2(Cy_3PNH)_2]$ and free Cy_3PO.

The preference for N *vs.* O donor ligands in the uranyl system was at the time a surprising result, perhaps in part due to the higher basicity of the phosphinimine ligands and merits more detailed experimental and computational investigation. However, this discovery has also opened the way to inert atmosphere $\{NpO_2\}^{2+}$ chemistry. As phosphine oxide ligands are not air or moisture sensitive we have been able to prepare both $[NpO_2Cl_2(Ph_3PO)_2]$ [12] and $[NpO_2Cl_2(Cy_3PO)_2]$ as pure crystalline solids, structurally characterising the *trans-* isomer of the Cy_3PO complex (figure 1 b). The reaction of both complexes with 2 equivalents of the corresponding phosphinimine ligands leads to complete displacement of coordinated phosphine oxide in solution, as shown by ^{31}P and ^1H NMR spectroscopy (figure 2) as long as the system remains completely moisture free.

This represents our first foray into the coordination chemistry of $\{NpO_2\}^{2+}$ with softer donor ligands.

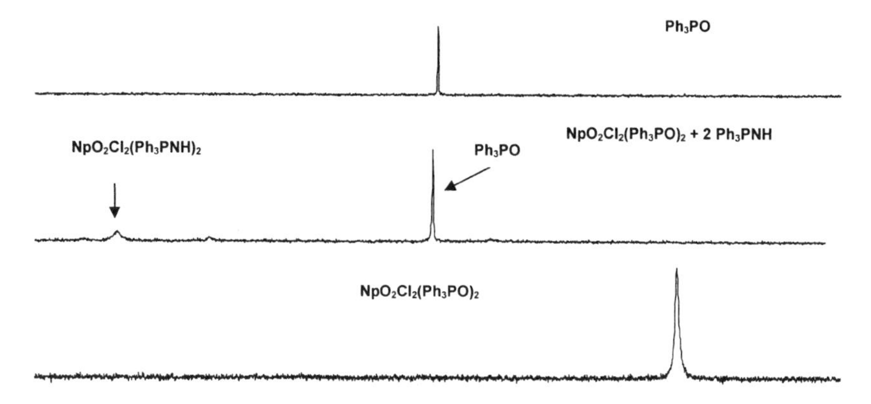

Figure 2 *^{31}P NMR spectra of the reaction between $[NpO_2Cl_2(Ph_3PO)_2]$ and Ph_3PNH*

References

1 J.J. Katz, L.R. Morss, G.T, Seaborg, "Summary and Comparative Aspects of the Actinides Elements", London, 1986
2 M.J. Sarsfield, M. Helliwell, J. Raftery, *Inorg. Chem.*, 2004, **43**, 3170; M.J. Sarsfield, M. Helliwell, *J. Am. Chem. Soc.*, 2004, **126**, 1036; M.J. Sarsfield, H. Steele, M. Helliwell, S.J. Teat, *Dalton Trans..*, 2003, 3443; M.J. Sarsfield, M. Helliwell, D. Collison, *Chem.Comm.*, 2002, 2264
3 D.R. Brown, R.G. Denning, *Inorg. Chem.*, 1996, **35**, 6158
4 V. Vallet, Z. Szabó, I. Grenthe, *Dalton Trans.*, 2004, 3799
5 J.-C Berthet, M. Nierlich, M. Ephritikhine, *Angew. Chem, Int. Ed.*, 2003, **42**, 1952
6 P.B. Duval, C.J. Burns, D.L. Clark, D.E. Morris, B.L. Scott, J.D. Thompson, E.L. Werkema, L. Jia, R.A. Andersen, *Angew. Chem., Int. Ed.*, 2001, **40**(18), 3357
7 Steven D. Conradson, Bruce D. Begg, David L. Clark et al, J.Am. Chem. Soc., 2004, **126**(41), 13443
8 I.A. Charushnikova, N.N. Krot, Z.A. Starikova, Radiochemistry, 2004, **46**(6), 565
9 M.P. Wilkerson, C.J. Burns, R.T. Paine, B.L. Scott, *Inorg. Chem.*, 1999, **38**, 4156
10 G. Bombieri, E. Forsellini, J.P. Day, W.I. Azeez, *Dalton Trans.*, 1978, 677; J. Akona, J. Fawcett, J.H. Holloway, D.R. Russel, *Acta Cryst., Sect. C, Cryst. Struct. Commun.*, 1991, 47
11 T.J. Hall, C.J. Mertz, S.M. Bachrach, W.G. Hipple, R.D. Rogers, *J. Crystallogr. Spectrosc. Res.*, 1989, **19**, 499; N.W. Alcock, D.J. Flanders, D. Brown, *Dalton Transactions*, 1985, 1001; F. Schrumpf, H.W. Roesky, M. Noltemeyer, *Z. Naturforsch., B: Chem. Sci.*, 1990, **45**, 1600; C.M.Ong, P. McKarns, D.W. Stephan, *Organometallics*, 1999, **18**, 4197
12 N.W. Alcock, M.M. Roberts, D. Brown, *Dalton Trans.*, 1982, **1**, 25

SYNTHESIS AND CRYSTAL STRUCTURE DETERMINATION OF SOME MIXED OXALATES OF ACTINIDE(S) AND/OR LANTHANIDE AND MONOVALENT CATIONS (NH_4^+ OR $N_2H_5^+$).

B. Chapelet-Arab,[1,2] S. Grandjean,[1] G. Nowogrocki,[2] and F. Abraham[2]

[1]Laboratoire de Chimie des Actinides, CEA Valrhô, bât 399, BP17171, 30207 Bagnols sur Ceze cedex,
[2]Laboratoire de Cristallochimie et Physico-chimie du Solide, UMR 8012, ENSCL, USTL, BP108, 59652 Villeneuve d'Ascq cedex.

1 INTRODUCTION

Many publications were devoted to the chemistry of actinides and lanthanides associated with oxalic acid and oxalic acid has many applications in the nuclear industry including: - as a precipitating agent in the industrial plutonium conversion process,[1] or for waste decontamination,[2,3] or as a complexing agent to adjust extracting characteristics of actinides and lanthanides or redox behavior of An[4,5]. Although the chemical properties of An and Ln oxalates were often investigated surprisingly only a few crystallographic structures of these compounds have been determined[5-14]. Recently, we have demonstrated the efficiency of crystal growth of uranyl oxalate compounds by using the controlled diffusion of the ions through silica gel impregnated with oxalic acid.[15] The use of this crystal growth method for several M^I–U^{IV}–Ln^{III} systems (M^I =NH_4, N_2H_5, Ln^{III}=Nd or Gd) enabled us to isolate crystals of a new U(IV) oxalate $(NH_4)_2U^{IV}{}_2(C_2O_4)_5.2.5H_2O$ **(1)** and of a double U(IV)–Nd(III) oxalate $(N_2H_5)_{2.6}U_{1.4}Nd_{0.6}(C_2O_4)_5.1.6H_2O$ **(2)** resulting from the substitution of U^{IV} by Nd^{III}. Furthermore, two new $N_2H_5^+$–Ln double oxalates $(N_2H_5)Ln(C_2O_4)_2.nH_2O$ were obtained for *Ln*=Nd (n=5) **(3)** and for *Ln*=Gd (n=5.5) **(4)**. Using high Ln/U ratio, two $N_2H_5^+$–Ln/U(IV) double oxalates $(N_2H_5)_{0.75}(Ln_{0.75}U_{0.25})(C_2O_4)_2.nH_2O$ were also synthesized for *Ln*=Nd (n=5.5) **(5)** and for *Ln*=Gd (n=5) **(6)**.

2 EXPERIMENTAL

Single crystals were obtained by the slow diffusion of the cations U^{4+} and Ln^{3+} (*Ln*=Nd or Gd) through silica gel impregnated with oxalic acid[16] The silica gel is prepared by pouring 1M sodium metasilicate solution into a mixture of 1M oxalic acid and 3M nitric acid. Nitric acid is used to acidify the medium and oxalic acid as a source of anions. The resulting solution is then allowed to set in tubes. To avoid formation of sodium compounds, substitution by other monovalent cations like NH_4^+ or $N_2H_5^+$ cations has been performed by ionic exchange between the silica gel and a nitric solution of oxalic acid and NH_4^+ or $N_2H_5^+$ nitrate added on the gel. When an aqueous mixture of monovalent cations ($N_2H_5^+$ or NH_4^+) and Ln^{3+} and/or U^{4+}, acidified by nitric acid is slowly added on the set $N_2H_5^+$ or NH_4^+-exchanged gel, crystals form slowly inside the gel.

3 RESULTS AND DISCUSSION

The crystal structures of the six compounds have been determined by single-crystal X-ray diffraction. The crystal structures were solved by direct methods and refined by least square method on the basis of F^2 for all unique reflections. Crystallographic data: **1**, hexagonal, space group $P6_3/mmc$, a= 19.177(3)Å, c= 12.728(4)Å, Z = 6, R1= 0.0636 for 46 parameters and 1914 reflections with $I \geq 2\sigma(I)$; **2**, hexagonal, space group $P6_3/mmc$, a= 19.243(4)Å, c= 12.760(5)Å, Z = 6, R1= 0.0714 for 57 parameters and 1774 reflections with $I \geq 2\sigma(I)$; **3**, triclinic, space group P-1, a= 8.507(3)Å, b= 9.762(4)Å, c= 10.249(4)Å, α= 62.378(5)°, β= 76.681(5)°, γ= 73.858(5)°, Z= 2, R1= 0.0335 for 172 parameters and 3430 reflections with $I \geq 2\sigma(I)$, **4**, triclinic, space group P-1, a= 8.52(3)Å, b= 9.51(3)Å, c= 10.14(3)Å, α= 62.11(4)°, β= 76.15(5)°, γ= 73.73(5)°, Z= 2, R1= 0.0325 for 172 parameters and 1742 reflections with $I \geq 2\sigma(I)$, **5**, triclinic, space group P-1, a= 8.417(2)Å, b= 9.716(3)Å, c= 10.117(3)Å, α= 61.951(4)°, β= 77.816(4)°, γ= 74.279(4)°, Z= 2, R1= 0.0494 for 170 parameters and 2911 reflections with $I \geq 2\sigma(I)$, **6**, triclinic, space group P-1, a= 8.533(4)Å, b= 9.370(4)Å, c= 10.003(4)Å, α= 62.283(6)°, β= 75.905(6)°, γ= 74.276(6)°, Z= 2, R1= 0.0352 for 164 parameters and 2824 reflections with $I \geq 2\sigma(I)$.

The honeycomb like structures of compounds **1** and **2** are built from the same three-dimensional arrangement of metallic and oxalate ions. Similar hexagonal rings of alternating metallic and oxalate ions form layers parallel to the (0 0 1) plane and pillared by another oxalate ion (Fig.1b). The monovalent cations and the water molecules occupy the hexagonal tunnels running down the [0 0 1] direction. Starting from the uranium (IV) compound $A_2U_2(C_2O_4)_5$. $2.5H_2O$ with A = NH_4^+ (**1**), the mixed U(IV)/Ln(III) oxalates are obtained by partial substitution of U(IV) by Ln(III) in a ten-coordinated site (Fig.1a), the charge deficit being compensated by intercalation of supplementary monovalent ions within the tunnels.

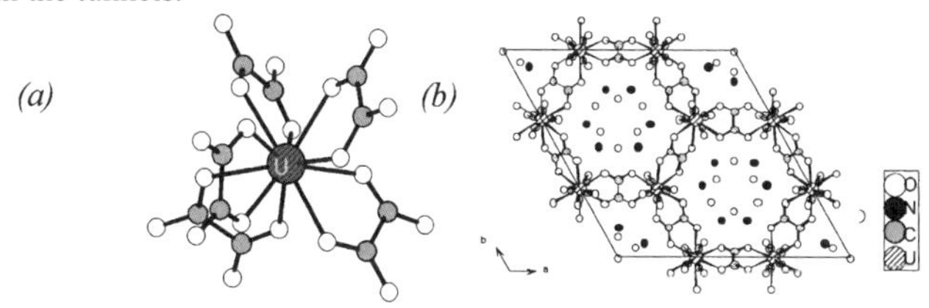

Figure 1 *(a) Environment of the U/Nd site in **1** and **2** compounds; (b) Projection of **1** and **2** structure down c axis.*

The compounds **3** to **6** are isostructural. The structures of $(N_2H_5)[Nd(C_2O_4)_2.H_2O].4H_2O$ **(3)** and $(N_2H_5)[Gd(C_2O_4)_2.H_2O].4.5H_2O$ **(4)** are built from polyhedra of rare earth atom connected by oxalate groups to form a three dimensional anionic framework $[Ln(C_2O_4)_2.H_2O]^-$.(*Ln* = Nd, Gd) which displays tunnels with elliptic cross-sections (Fig.2b). The tunnels propagate in three directions and form cavities at their intersections. In the three-dimensional network the *Ln* atoms are connected by tetradentate oxalate groups. The *Ln* atom is surrounded by nine oxygen atoms from four oxalate ions and one aqua ligand (Fig. 2a). The $N_2H_5^+$ ions and all the water molecules, excepted one, occupy the same positions in the two lanthanide oxalates structure and are disordered in the large elliptic tunnels running down the a axis (Fig. 2b). The crystal structures of **5** and **6** are similar to those of **3** and **4**. Their main characteristic is the substitution of one part of the neodymium or gadolinium by uranium compensated by reduction of the monovalent ion content without any structural modification. Ln and U atoms were fixed in the same crystallographic site with equal anisotropic displacement parameters. In the final refinement, the occupancy of the sites was fixed at 0.75 for Ln and 0.25 for U in agreement with the EDS analysis.

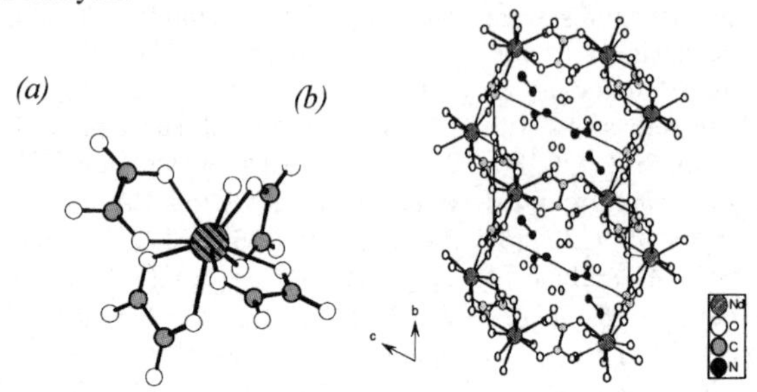

Figure 2 *(a) Environment of the Ln/U site in **3, 4, 5** and **6** compounds; (b) projection of the structure of **3** showing large elliptic tunnels running down the a axis.*

4 CONCLUSION

Single crystals of six new oxalates have been synthesised using a silica gel method. The new compounds can be separated in two families related to their crystallographic structure: hexagonal for $(NH_4)_2U_2(C_2O_4)_5$. $2.5H_2O$ **(1)** and $(N_2H_5)_{2.6}U_{1.4}Nd_{0.6}(C_2O_4)_5$. $1.6H_2O$ **(2)** and triclinic for $(N_2H_5)Nd(C_2O_4)_2$. $5H_2O$ **(3)**, $(N_2H_5)Gd(C_2O_4)_2$. $5.5H_2O$ **(4)**, $(N_2H_5)_{0.75}(Nd_{0.75}U_{0.25})(C_2O_4)_2$. $5.5H_2O$ **(5)** and $(N_2H_5)_{0.75}(Nd_{0.75}U_{0.25})(C_2O_4)_2$. $5H_2O$ **(6)**. The two families are quite different: one can be deduced from the only U(IV) containing oxalate by the substitution U(IV) $\rightarrow$ Ln(III) + M^+ whereas the second results from the reverse substitution Ln(III) + M^+ $\rightarrow$ U(IV). There are also some main differences related to the coordination sphere of the different actinide and lanthanide cations, the ratio between the number of oxalates ligands and the number of actinide and lanthanide cations and the existence of bonded water molecules. Similar studies using actinides(III) instead of lanthanides(III) are currently in progress and their preliminary results are very promising.

References

1 F. Drain, B. Gillet, G. Bertolotti, «Oxalate process: the unique way for plutonium conversion», *GLOBAL 1999 International Conference*.
2 R.D. Bhanushali, *et al., J. Radioanal. Nucl. Chem.*, 1999, **240**, N°2, 631.
3 P.P Mapara, *et al., J. Radioanal. Nucl. Chem.*, 1999, **240**, N°3, 977.
4 R. Malmbeck; O. Courson,; *et al., Radiochimica-Acta*, 2000, **88**(12) 865.
5 P. Baron; X. Heres, M. Lecomte, M. Masson, *Global 2001 international conference on*: "back-end of the fuel cycle: from research to solutions" Paris (2001).
6 W. Ollendorff and F. Weigel, *Inorg. Nucl., Chem. Letters*, 1969, **5**, 263.
7 T. Bataille, M. Louër, J-P. Auffredic, D. Louër, *J. Solid State Chem.* 2000, **150,** 81.
8 J-C. Trombe, P. Thomas and C. Brouca-Cabarrecq, *Solid State Sciences*, 2001, **3,** 309.
9 S. Romero, A. Mosset, J-C. Trombe, *Eur. J. Solid State Inorg. Chem.* 1995, **32,** 1053.
10 I. L. Jenkins and F. H. Moore, *J. Inorg. Nucl. Chem.* 1965, **27,** 81.
11 M. S. Grigor'ev, I. A. Charushnikova, *et al., Radiochemistry*, 1997, **39**, 420.
12 I. L. Jenkins and F. H. Moore, *J. Inorg. Nucl. Chem.* 1965, **27,** 77.
13 M. N. Akhtar and A. J. Smith, *Acta Crystallogr.* 1975, **B31,** 1361.
14 M. C. Favas, D. L. Kepert *et al., J. Chem. Soc. Dalton Trans.* 1983, **3**, 571.
15 B. Chapelet-Arab, *et al.*, Radiochimica Acta, (2005), in press.
16 H.K. Henisch, Crystals in gels and Liesegangs rings, The Pennsylvania State University Press, University Park and London, 1988.

COMPLEXATION OF A B-TYPE UNSATURATED POLYOXOANION, $[BiW_9O_{33}]^{9-}$, TO ACTINYL, $\{UO_2\}^{2+}$, $\{NpO_2\}^{+}$ AND $\{PuO_2\}^{2+}$, CATIONS

Roy Copping,[a] Iain May,[a] David Collison,[b] Chris J. Jones,[c] Clint A. Sharrad[a] and Mark J. Sarsfield[c]

[a] Centre for Radiochemistry Research, School of Chemistry, The University of Manchester, UK, M13 9PL
[b] School of Chemistry, The University of Manchester, UK, M13 9PL
[c] Nexia Solutions, B229 and B170, Sellafield Seascale,Cumbria, CA20 1PG

1 INTRODUCTION

Unsaturated heteropolytungstates are a highly charged subset of polyoxometalate anions. They can coordinate in the equatorial plane of the sterically demanding actinyl cations. The first structurally characterized example of such an interaction is the formation of the sandwich complex $[Na_2(UO_2)_2(PW_9O_{34})_2]^{12-}$ when the A-type tri-vacant $[PW_9O_{34}]^{9-}$ is reacted with $\{UO_2\}^{2+}$.[1] Recently the same ligand has been shown to form an analogous complex with $\{NpO_2\}^{+}$, the only difference being the overall charge of the complex, $[Na_2(NpO_2)_2(PW_9O_{34})_2]^{14-}$.[2] The B-type trivacant $[SbW_9O_{33}]^{9-}$ and $[TeW_9O_{33}]^{8-}$ ligands also form sandwich complexes with $\{UO_2\}^{2+}$, forming $[(UO_2)_2(SbW_9O_{33})_2(H_2O)_2]^{14-}$ and $[(UO_2)_2(TeW_9O_{33})_2(H_2O)_2]^{12-}$ respectively.[3] Complexes of this type may find application in nuclear waste treatment and encapsulation. In this paper we report the study of actinyl, $\{AnO_2\}^{n+}$ complexation by $[BiW_9O_{33}]^{9-}$ where An = U^{VI}, Np^{V} and Pu^{VI}.

2 RESULTS AND DISCUSSION

2.1 $[(UO_2)_2(BiW_9O_{33})(H_2O)_2]^{14-}$

The reaction of $[BiW_9O_{33}]^{9-}$ with $\{UO_2\}^{2+}$ in near neutral conditions has yielded the sandwich complex $[(UO_2)_2(BiW_9O_{33})(H_2O)_2]^{14-}$ (figure 1). Single crystal X-ray diffraction has shown the complex to be analogous to that formed by the trivacant B-type anions containing Sb^{III} and Te^{IV} heteroatoms.[3]

Each $\{UO_2\}^{2+}$ moiety is five coordinate in the equatorial plane with each uranyl moiety bonded to two terminal oxygen atoms of each $[BiW_9O_{33}]^{9-}$ with the fifth place in the coordination sphere occupied by a water molecule, the overall uranium geometry is pentagonal-bipyramidal. The average axial U=O bond distance is 1.820Å, compared to 1.775 and 1.782Å in the analogous $[SbW_9O_{33}]^{9-}$ and $[TeW_9O_{33}]^{8-}$ complexes respectively. The average bond distance for the four Uranium-POM bonds in the equatorial plane is 2.373Å and the oxygen of the water molecule occupying the fifth coordination position is at 2.485Å. The related $[SbW_9O_{33}]^{9-}$ and $[TeW_9O_{33}]^{8-}$ complexes average bond distance to uranium are 2.344 and 2.341Å respectively, with the oxygen of the water molecule at 2.50 and 2.518Å respectively.[3]

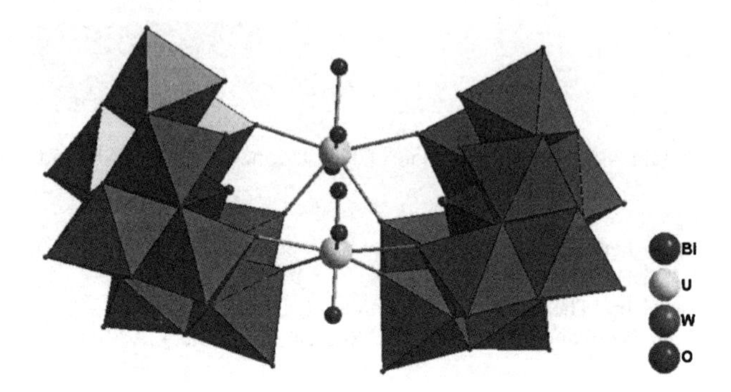

Figure 1 *Polyhedral representation of* $[(UO_2)_2(H_2O)_2(BiW_9O_{33})_2]^{14-}$

Raman spectroscopy of the complex reveals one principal band at 953cm^{-1} assigned to the symmetric W=O$_{terminal}$ stretch, which is shifted from 945cm^{-1} in 'uncoordinated' $[BiW_9O_{33}]^{9-}$ (figure 2a). The most significant feature of spectrum is the appearance of the band centered at 795cm^{-1}, which is assigned to the $\{UO_2\}^{2+}$ symmetric stretching vibration. A similar shift was observed in the related $[SbW_9O_{33}]^{9-}$ and $[TeW_9O_{33}]^{8-}$ complexes, in which the uranyl symmetric stretch was recorded at 796 and 804cm^{-1}.[3] This represents a significant weakening of the bonding in $\{UO_2\}^{2+}$ as the hydrated aquo ion has a band centered at 870cm^{-1}.[4] Note that the band at 873cm^{-1} is related to the ligand, $[BiW_9O_{33}]^{9-}$.

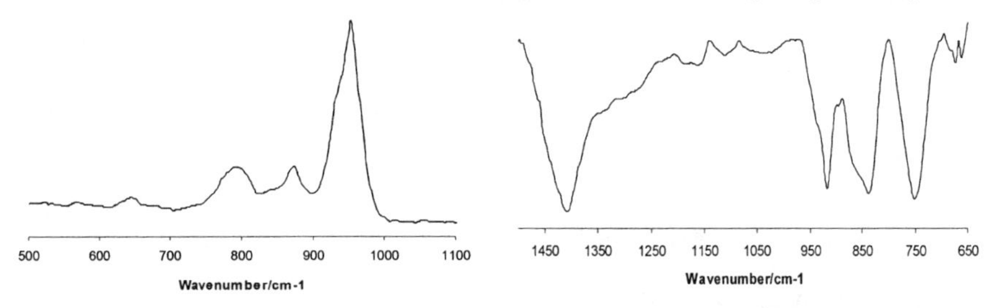

Figure 2 *Raman and ATR-IR spectrum of* $Na_{14}[(UO_2)_2(H_2O)_2(BiW_9O_{33})_2].xH_2O$

The IR spectra of the uranyl complex shows three distinctive tungsten-oxygen bands at 918, 837 and 750cm^{-1} (figure 2b) assigned to ν(W-O) terminal, edge and corner shared stretches respectively. The $\{UO_2\}^{2+}$ asymmetric stretch is almost certainly obscured by these W-O bands. The band at 1408cm^{-1} can be attributed to the NH_4^+ counter cations.

2.2 Vis/nIR spectroscopic investigation of $\{NpO_2\}^+$ and $\{PuO_2\}^{2+}$ complexation by $[BiW_9O_{33}]^{9-}$

The major $\{NpO_2\}^+$ *f-f* transition shifts from 980 to 1022 nm on coordination of $\{NpO_2\}^+$ by $[BiW_9O_{33}]^{9-}$, with coordination promoted by increasing ligand concentration and an approach to neutral pH from more acidic solution (figure 3). A similar effect is observed for isoelectronic $\{PuO_2\}^{2+}$ with the major transition centered on 831nm shifting to 851nm on complexation to $[BiW_9O_{33}]^{9-}$ (figure 4).

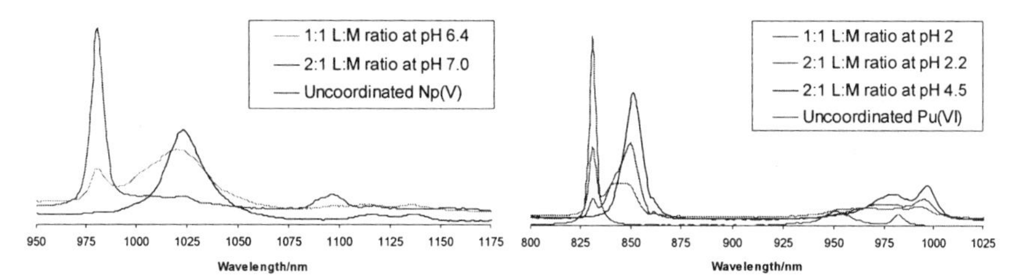

Figure 3 *nIR spectrum of* $\{NpO_2\}^{+}$ + $[BiW_9O_{33}]^{9-}$ *on increasing pH and L:M ratio*

Figure 4 *nIR spectrum of* $\{PuO_2\}^{2+}$ + $[BiW_9O_{33}]^{9-}$ *on increasing pH and L:M ratio*

These specific features signify a strong interaction between the transuranic actinyl cations and the $[BiW_9O_{33}]^{9-}$ trivacant heteropolytungstate ligand, as previously observed for $[Na_2(NpO_2)_2(PW_9O_{34})_2]^{14-}$.[2]

3 CONCLUSIONS

The reaction of $\{UO_2\}^{2+}$ with $[BiW_9O_{33}]^{9-}$ has yielded the anionic sandwich complex $[(UO_2)_2(H_2O)_2(BiW_9O_{33})_2]^{14-}$. Raman investigation has revealed a significant weakening of the bonding in $[O{=}U{=}O]^{2+}$ on coordination to $[BiW_9O_{33}]^{9-}$, indicating strong equatorial bonding. Significant shifting of the major *f-f* transition in the near IR region of the electronic absorption spectrum of $\{NpO_2\}^{+}$ and $\{PuO_2\}^{2+}$ on coordination to $[BiW_9O_{33}]^{9-}$ is also indicative of strong equatorial bonding.

References

1 K.C. Kim and M.T. Pope, *J. Am. Chem. Soc.* 1999, **121**, 8512.
2 A.J. Gaunt, I. May, M. Helliwell and S. Richardson, *J. Am. Chem. Soc*, 2002, **124**, 13350.
3 A.J. Gaunt, I. May, R. Copping, A.I. Bhatt, D.Collison, O.D. Fox, T.Holman, M.T. Pope, *Dalton Trans.*, 2003, 3009.
4 D.L. Clark, S.D. Conradson, R.J. Donohoe, D.W. Keogh, D.E. Morris, P.D. Palmer, R.D. Rogers and C.D. Tait, *Inorg. Chem.*, 1999, **38**, 1456.

URANIUM COORDINATION TO ENVIRONMENTALLY RELEVANT POLYHYDROXY CARBOXYLATE LIGANDS.

A. J. Kirkham, N. D. Bryan & I. May

Centre for Radiochemistry Research, School of Chemistry, The University of Manchester, Oxford Road, M13 9PL.

1 INTRODUCTION

Intermediate and low level nuclear waste repositories often contain cellulose based materials. Cement pore water in nuclear repositories creates a high pH environment, which is expected to remain over pH 12.5 for at least 10^5 years.[1] In these highly alkaline conditions cellulose is know to degrade to low molecular weight water soluble molecules,[1,2] with isosaccharinic acid (ISA) being the main degradation product formed.[2] It is thought that organic molecules, like ISA, will coordinate to radionuclides and hence may increase their solubility, enabling them to leach out of the repository and into the ground water.[1-5] These ISA studies were generally restricted to α–ISA, as it is less difficult to synthesise than β–ISA. The aim of this study is to relate solubility to speciation at high pH. Inital experiments were carried out with $\{UO_2\}^{2+}$, using various spectroscopic handles, and commercially available gluconic acid (Gluc). Previously, there have only been one study with $\{UO_2\}^{2+}$ and ISA[6] and only one with Gluc[7].

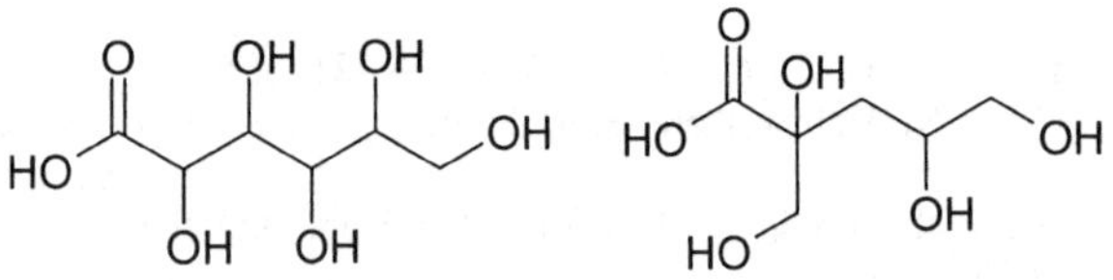

O OH OH
HO OH
OH OH

O OH
HO OH
OH
HO

Figure 1 *Skeleton structure of Gluconic acid (left) and Isosaccharinic Acid (right)*

2 EXPERIMENTAL

$Ca(ISA)_2$ was synthesised using an adapted method of Van Loon et al,[8] and then converted to NaISA or NH_4ISA *via* an Amberlite cation exchange column conditioned with NaCl or NH_4Cl.[3,9] Solution speciation was studied for the Gluc:$\{UO_2\}^{2+}$ and ISA:$\{UO_2\}^{2+}$ systems at a range of metal:ligand ratios through multinuclear NMR, vibrational and UV/vis spectroscopies as well as potentiometry and mass spectometry

3 RESULTS AND DISSCUSION

It has been shown that Gluc and ISA increases the solubility of $\{UO_2\}^{2+}$ at high pH preventing the precipitation of the actinyl cation as the hydroxide. This has been shown by both UV/vis and potentiometry. UV/vis spectroscopy shows a loss of $\{UO_2\}^{2+}$'s fine structure with an associated increase in ε with pH, in the presence of either Gluc or α–ISA. This loss of fine structure prohibits the determination of clear isosbestic points. This is indicative of strong coordination in the equatorial plane. We have also recently started to apply potentiometric methods to obtain stability constants for both systems.

Raman spectroscopy is a good spectroscopic probe for $\{UO_2\}^{2+}$, through the symmetric stretching frequency, allowing the study of the strength of ligand interactions in the equatorial plane. For the $\{UO_2\}^{2+}$:Gluc system as pH increases, the stretching frequency decreases, from 872 to 776 cm^{-1} indicating variable speciation as a function of pH with progressively stronger ligand coordination in the equatorial plane. Figure 3 shows an example for the 1:1 system.

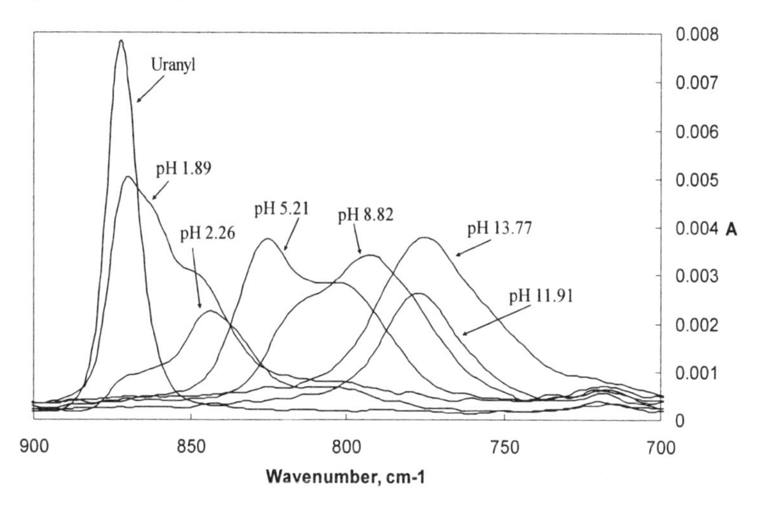

Figure 2 *Solution Raman spectra for 1:1 Gluconic acid: $\{UO_2\}^{2+}$ over a range of pH values.*

The ^{13}C NMR of $\{UO_2\}^{2+}$:Gluc at lower pH reveals many broad unresolved peaks, which cannot be resolved through variable temperature measurements, indicating the presence of several species and various rapid ligand exchange processes. As the pH is raised, the spectra simplify to show one major species. However even at 1:1 UO_2:Gluc there is still ucoordinated ligand, suggesting some minor cluster formation, perhaps with a hydroxy/oxo bridged $\{UO_2\}^{2+}$ core or mixed hydroxy-Gluc coordination. This is supported up by the loss of the minor species peaks when the concentrations of both components are decreased by a factor of 10. When the ratio of Gluc:$\{UO_2\}^{2+}$ is increased, there is no significant change in the spectra, only an increase in free ligand, indicating the formation of a 1:1 complex as the dominant species. Figure 3 shows a representative ^{13}C NMR spectrum for the 1:1 $\{UO_2\}^{2+}$:Gluc system.

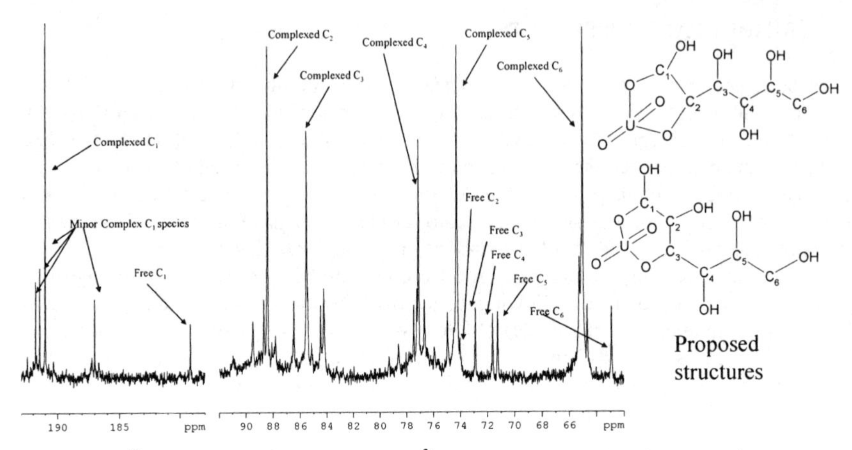

Figure 3 *^{13}C NMR Spectrum of 1:1$\{UO_2\}^{2+}$:Gluc at pH 12.25 and proposed coordinations of gluconate to $\{UO_2\}^{2+}$.*

4 CONCLUSIONS

Both Gluconate and ISA coordinate to $\{UO_2\}^{2+}$ preventing its hydrolysis and precipitation. As the pH is raised, stronger coordination in the equatorial plane is observed, indicative of strong O donor ligands *e.g.* deprotonated Gluc/ISA and perhaps also OH^-. No more than one Gluc will coordinate, suggesting a 5 or 6 membered ring is formed at high pH along with formation of a mixed hydroxyl/gluconate complex. It is probably steric hindrance that inhibits any further coordination to the metal centre. Current effort is focused on determining $\{UO_2\}^{2+}$: ISA speciation and solution thermodynamics.

References

1 L. R. Van Loon & M. A. Glaus *Journal of Environmental Polymer Degradation,* 1997, **5**, 97.

2 L. R. Van Loon; M. A. Glaus; A. Laube & S. Stallone, *Journal of Environmental Polymer Degradation,* 1999, **7**, 41.

3 M. A. Glaus; L. R. Van Loon; S. Achatz; A. Chodura & K. Fischer *Analytica Chimica Acta,* 1999, **398**, 111.

4 A. D. Moreton *Mat. Res. Soc. Symp. Proc.,* 1993, **294**, 753.

5 B. F. Greenfield; G. J. Holtom; M. H. Hurdus; N. O'Kelly; N. J. Pilkington; A. Rosevear; M. W. Spindler & S. J. Williams *Mat. Res. Soc. Symp. Proc.,* 1995, **353**, 1151.

6 L. Rao; A. Y. Garnov; D. Rai; Y. Xia & R. C. Moore *Radiochim. Acta,* 2004, **92**, 575.

7 D. T. Sawyer & R. J. Kula *Inorganic Chemistry* 1962, **1**, 303.

8 K. Vercammen; M. A. Glaus & L. R. Van Loon *Acta Chemica Scandinavica,* 1999, ***53***, 241.

9 L. R. Van Loon; M. A. Glaus; S. Stallone & A. Laube *Environmental Science & Technology,* 1997, **31**, 1243.

CO-CRYSTALLISATION PROCESSES IN A STUDY OF PHYSICOCHEMICAL PROPERTIES OF ACTINIDES AND LANTHANIDES IN LOWER OXIDATION STATES

S. A. Kulyukhin, N.B. Mikheev, A.N. Kamenskaya, N.A. Konovalova and I.A. Rumer

Institute of Physical Chemistry, Russian Academy of Sciences, 31 Leninskii prospect, Moscow, 119991 Russia

1 INTRODUCTION

The chemistry of the radioactive elements, including *f*-elements, continues to be a subject of active interest, in terms of both theoretical and experimental studies. There have recently been a large number of quantum-chemical studies on the electronic structure of actinides (An), as well as the growing interest in a study of the properties of heavy elements, broadening the field of theoretical study. In this regard, a study of the physicochemical properties of *f*-elements, including their unusual oxidation states, in particular the lower oxidation states, would appear to be an important area of investigation.

The aim of this work was to use cocrystallization processes as a method of physicochemical analysis to study the following issues:

- the possibility of the reduction of Fm, Es, and Cf to the +1 oxidation state and Pa and Th to the +2 oxidation state in the presence of divalent lanthanides (Ln);
- a comparison of the behavior of Eu^{2+}, Yb^{2+}, Es^{2+}, and an alkali-earth ion (using Sr^{2+} as an example) in hydration, solvation, and complex-formation processes;
- the behavior of An in the formation of mixed condensed clusters (MCC) with Gd_2Cl_3 and PrI_2 matrices.

2. A STUDY OF THE POSSIBILITY OF THE REDUCTION OF Fm, Es, AND Cf TO THE +1 OXIDATION STATES AND Pa AND Th TO THE +2 OXIDATION STATE IN THE PRESENCE OF DIVALENT LANTHANIDES

To study the possibility of the reduction of Fm, Es, and Cf to the +1 oxidation state we studied their cocrystallization in solution in the presence of Sm^{2+} and Tm^{2+} and in molten salts in the presence of Nd^{2+} and Pr^{2+}.[1]

A study of the experimental dependences of the changes in the cocrystallization coefficients $\boldsymbol{D}^1$ for Fm, Es, and Cf on the concentration of the halide ion in solution allowed us to conclude that in H_2O-C_2H_5OH in the presence of Sm^{2+} and in tetrahydrofurane (THF) in the presence of Tm^{2+}, Fm, Es, and Cf were only reduced to the +2 oxidation state. The reduction to the +1 oxidation state did not take place.

In molten salts in the presence of Nd^{2+} and Pr^{2+}, the behavior of Fm, Es, and Cf was similar to that of Sm, Eu, and Am in both systems and was noticeably different from that of Y. Note that the ratios between the distribution coefficients of the *f*-elements and Sr were close to 1. Thus, our research study has found that that Fm, Es, and Cf in the presence of Sm^{2+}, Tm^{2+}, Nd^{2+}, and Pr^{2+} are reduced to the +2 oxidation state and that no reduction to the +1 oxidation state takes place.

A study of the possibility of the reduction of Pa and Th to the lower oxidation states included two stages:

- a study of the possibility of the reduction of Pa^{4+} and Th^{4+} to the +3 oxidation state;
- a study of the possibility of their further reduction to the +2 oxidation state.

The study was carried out in the $LnOCl_{solid\ phase}$-$(LnCl_3$-$LnCl_2$-$SrCl_2)_{melt}$ system, which we successfully used earlier in a study of the reduction of other An to the +2 oxidation state.[1] A study of the cocrystallization of Pa and Th microquantities with LnOCl phases from $LnCl_3$-$LnCl_2$-$SrCl_2$ melts containing respective Ln^{+2} has found the following:

- Pa^{4+} is reduced to the +3 oxidation state in the presence of Tm^{2+}, with the reduction potential of the Pa^{4+}/Pa^{3+} couple ranges from -1.9 to -2.1 V *vs.* SHE;
- Th^{4+} is reduced to the +3 oxidation state in the presence of Nd^{2+}, with the reduction potential of the Th^{4+}/Th^{3+} couple ranging from -2.3 to -2.4 V *vs.* SHE;
- Pa^{3+} in the presence of Pr^{2+} is reduced to the +2 oxidation state, with the reduction potential of the Pa^{3+}/Pa^{2+} couple ranging from -2.5 to -2.6 V *vs.* SHE;
- Th^{3+} is reduced to the oxidation state +2 in the presence of Pr^{2+}, with the reduction potential of the Th^{3+}/Th^{2+} ranging from -2.8 to -2.9 V *vs.* SHE.

Thus, the cocrystallization method has allowed us to not only to identify the oxidation states of Pa and Th in the presence of Ln^{2+}, but also to estimate their reduction potentials.

3. A STUDY OF THE HYDRATE-SOLVATION AND COMPLEX-FORMATION PROPERTIES OF An^{2+} AND Ln^{2+}

A study of the cocrystallization of Eu^{2+}, Yb^{2+}, and Es^{2+} with $Sr(Sm)SO_4$ solid phase from H_2O-C_2H_5OH showed that in the range of low $[H_2O]$ in C_2H_5OH, an increase in the solubility of $EsSO_4$ relative to Eu^{2+}, Yb^{2+}, and Sr^{2+} sulfates is observed, which is related to the formation of strong solvate-hydrate complexes of Es^{2+} in solution at a molar ratio of $C_2H_5OH : H_2O = 2:1$. Increasing $[H_2O]$ in C_2H_5OH above 10 mol/l leads to the saturation of the first coordination sphere of the cations with H_2O molecules. Simultaneously, the solubility curves for all elements follow the same trend; i.e., the difference in the behavior of divalent *f*-elements with respect to one another as well as to the Sr^{2+} ion disappears.[1]

A study of the cocrystallization of Eu^{2+}, Yb^{2+}, and Es^{2+} with $SrCl_2$ and $SmCl_2$ solid phases in C_2H_5OH and CH_3CN and their mixture showed that in C_2H_5OH-CH_3CN (7.5 mol/l), ***D*** for Eu^{2+}, Yb^{2+}, and Es^{2+} with the $SrCl_2$ solid phase quickly decreases. On the other hand, ***D*** for these *f*-elements with $SmCl_2$ changes monotonically from one pure solvent to another. IR spectroscopic and thermogravimetric studies of $SrCl_2$ and $SmCl_2$ solvates, as well as a study of the absorption spectra of the *f*-*d*-transitions of Sm^{2+} solutions in these solvents have allowed us to conclude that a sharp decrease in ***D*** with $SrCl_2$ solid phase is connected with the formation of strong mixed solvate complexes of divalent *f*-elements in solution. Calculations show that an anomalous solvation effect in the mixed solvent results in a 5- to 7-fold increase in the solubility of the solvates of Eu, Yb, and Es dichlorides.[1]

A study of the solubility of $SrSO_4$ in the presence and absence of tetraphenylborate ion (BPh_4^-) in H_2O-C_2H_5OH and a study of the cocrystallization of Eu^{2+}, Yb^{2+}, and Es^{2+} with $Sr(Sm)SO_4$ solid phase have found that, in contrast to divalent *f*-elements, Sr^{2+} does not form complexes with BPh_4^-. Based on the absence of complex formation with Sr^{2+} and the data obtained by cocrystallization, conductometry and ion-exchange methods, we have calculated the stability constants for the cationic and neutral complexes of Eu^{2+}, Yb^{2+}, and Es^{2+} with BPh_4^- in H_2O-C_2H_5OH. Peaks in stability constants are observed for Es^{+2}.

A study of the solubility of MCl_2 (M = Sm^{2+}, Sr^{2+}) in the presence of ClO_4^-, BF_4^-, and BPh_4^- in THF has shown that the solubility of $SrCl_2$ does not change in the studied range of ClO_4^-, BF_4^-, and BPh_4^- in THF. On the other hand, the solubility of $SmCl_2$ in THF increases with an increase in the concentration of these ions in solution, with ClO_4^- having the strongest and BPh_4^- the weakest influence. Thus, in contrast to Sr^{2+}, divalent *f*-elements are capable of forming coordination bonds with ClO_4^-, BF_4^-, and BPh_4^-.[1] The formation of associates in the case of $NaClO_4$ and $LiBF_4$ in THF did not allow us to calculate the stability constants of respective complexes. However, based on the dependence of ***D*** for Es^{2+}, Eu^{2+}, and Yb^{2+} with [Sr(18-crown-6)]I_2 solid phase on the [$NaClO_4$] and [$LiBF_4$] concentration in THF, we calculated the ratios of the complex functions in the presence of ClO_4^- and BF_4^- for each of the studied elements. The obtained values are (2.7-2.8) for Eu^{2+} and Yb^{2+} and (2.0-2.1) for Es^{2+}, suggesting that the complexes of the studied ions with ClO_4^- are 2 to 3 times stronger than analogous complexes with BF_4^-.[1]

4. A STUDY OF THE BEHAVIOR OF An IN THE FORMATION OF MIXED CONDENSED CLUSTERS (MCC) WITH Gd_2Cl_3 AND PrI_2 MATRICES

A study of the cocrystallization of An with Gd_2Cl_3 and PrI_2 clusters from halide melts has not only found that An are capable of forming MCC with Gd_2Cl_3 and PrI_2, but also has revealed the differences in the behavior of An.[1]

Only a proportion of the selected An cocrystallizes with Gd_2Cl_3 and PrI_2 condensed phase, whereas the remainder does not participate in the cocrystallization. A comparison of some An properties, such as ionic radii, third ionization potentials, the energies of the stabilization of the *d*-level in the condensed phase, as well as electronic configurations in the +2 oxidation states has allowed us to conclude that it is only An having $f^{n-1}d^1$ electronic configuration in the +2 oxidation state are capable of forming MCC of Gd_2Cl_3 and PrI_2 composition.

In should be noted in conclusion that investigations of the cocrystallization of Ln and An in the lower oxidation states in solutions and melts have allowed us not only to obtain new data on the fundamental properties of *f*-elements and reveal previously unknown specific features in their behavior, but also to develop a number of original methods of the separation and isolation of *f*-elements.

References

1. N.B. Mikheev, I.V. Melikhov, S.A. Kulyukhin, 'Cocrystallization processes in a study of physicochemical properties of radioactive elements in different phases' in *Modern Problems of Physical Chemistry*, eds., B.F. Myasoedov, A.Yu. Tsivadze, B.G. Ershov, Yu.I. Kuznetsov and A.E. Chalykh, Granitsa, Moscow, 2005, pp. 542-564.

MONOCARBOXYLATE COMPLEXES OF URANYL: A RELATIVISTIC DENSITY FUNCTIONAL STUDY

S. Krüger, F. Schlosser, N. Rösch

Theoretische Chemie, Department Chemie, Technische Universität München, 85747 Garching, Germany

1 INTRODUCTION, METHOD AND MODELS

The interaction of actinides with humic substances, which are ubiquitous in soils and natural waters, is an important aspect of speciation, transport and immobilization of these elements in the environment.[1] Carboxyl groups are generally accepted as main complexing sites of humic substances. Therefore, carboxylic acids are frequently studied as models of humic substances. Several coordination modes have been suggested, based on crystal structures or EXAFS spectra of complexes in solution.[2] Most of this experimental work is devoted to uranyl(VI). Besides monodentate and bidentate coordination of the carboxyl group to uranyl, also chelate complexes involving a carboxyl as well as a neighboring hydroxyl group have been invoked. Moreover, depending on concentrations, pH, and other parameters, different uranyl carboxylate species are present in solution, complicating the experimental analysis.[1-4] With increasing pH, first monocarboxylate species, then bi- and tricarboxylates appear.

We studied computationally uranyl monocarboxylates, applying an accurate all-electron relativistic density functional approach.[5] Solvation effects were modelled by explicit consideration of aqua ligands of the first solvation shell in combination with an approximate treatment of long-rage electrostatic effects by means of a polarizable continuum model (PCM).[6] Geometries have been optimized in the local density approximation.[7] This approach has been proven to yield accurate results for bond lengths and vibrational frequencies of actinide complexes. Uranyl monocarboxylates of acetic and propionic acid have been modelled by complexes of the type $[UO_2(OOCR)(H_2O)_n]^+$, with R = CH_3 and CH_2CH_3. As uranyl typically is fivefold coordinated, we added four aqua ligands (n = 4) in the case of monodentate coordination of the carboxyl group, whereas we used only three aqua ligands (n = 3) for bidentate coordination. The oxygen centres of the ligands lie in the equatorial plane of the uranyl moiety. We exploited this by imposing the corresponding C_s symmetry constraints to simplify the computational effort.

2 RESULTS AND DISCUSSION

Results of geometry optimizations as well as the symmetric stretching frequencies of uranyl are collected in Table 1 together with experimental data. Because experimental results are available for monoacetate only (CH_3 sol.),[8-9] we also compare to averaged

Table 1 *Calculated structural parameters (distances in Å) and symmetric uranyl stretching frequency ν_s (in cm^{-1}) of $[UO_2(OOCR)(H_2O)_n]^+$ (R = CH_3, CH_2CH_3) with bidentate (bi, n = 3) and monodentate (mono, n = 4) carboxylate coordination. Also shown are experimental data from aqueous solution (sol.) and crystals (cryst.) as well as average changes <Δbi> for complexes in solution with monodentate coordination relative to the corresponding bidentate complexes. For the designations of atoms, see Figure 1.*

	R	$U{=}O_t$	$U\text{-}O_c$	U-C	$U\text{-}O_w$	$U\text{-}O_{eq}$	ν_s
bi	CH_3	1.787	2.371	2.769	2.360	2.364	858
	CH_2CH_3	1.786	2.369	2.768	2.369	2.369	853
Exp.	CH_3 sol.[a]	1.78(1)	2.50(2)	2.91(2)	2.38(2)	2.43(2)	861
	sol.[b]	1.78(1)	2.46(4)	2.87(4)	2.40(4)	2.42(4)	–
	cryst.[c]	1.76(3)	2.48(5)	2.86(5)	2.36(4)	2.42(6)	–
mono	CH_3	1.790	2.201	3.401	2.421	2.377	846
	CH_2CH_3	1.790	2.201	3.401	2.421	2.377	849
	<Δbi>	0.00	0.17	-0.63	-0.06	-0.01	9
Exp.	sol.[b]	1.78(1)	–	–	–	2.38(4)	–
	cryst.[c]	1.76(3)	2.39(5)	3.5(1)	2.42(6)	2.36(2)	–

[a] mono-acetate in solution,[8,9] [b] average over various carboxylates in solution,[3,8,10] [c] average over various carboxylate crystal structures.[2,8,11]

results for other carboxylic acids in solution (sol.)[3,8,10] as well as to crystal structures (cryst.)[2,8,11] of uranyl carboxylates with one or more carboxylate ligands.

The structures of monoacetate and -propionate complexes are very similar, both for bi- and monodentate coordination. The largest deviation (0.01 Å) is obtained for uranyl-water bonds $U\text{-}O_W$. Also the symmetric stretching frequencies agree very closely (Table 1). The uranyl bonds $U{=}O_t$ are in both cases 1.79 Å, in good agreement with experiment, 1.78(1). For bidentate coordination, $U{=}O_t$ is calculated a bit shorter and concomitantly the frequency slightly higher. The calculated structures differ notably in the bond lengths $U\text{-}O_c$ to the carboxyl group, 2.37 Å for bidentate and 2.20 Å for monodentate coordination, reflecting that the single $U\text{-}O_c$ bond in monodentate coordination is stronger than each of the two $U\text{-}O_c$ bonds in the bidentate case. Experimental results for these bonds are considerably longer, 2.50 Å for bidentate coordination of acetate and a little shorter values for other carboxyl groups, exceeding our result by ~0.1 Å. For monodentate coordination only crystal structures are available with $U\text{-}O_c$ = 2.39 Å, which is ~0.2 Å longer than our result (Table 1). Accordingly, the calculated U-C distances to the carboxyl C atom also underestimate the experimental values, ~0.1 Å for bi- and monodentate coordination. The difference U-C (~0.6 Å) between bi- and monodentate is calculated in agreement with ex-

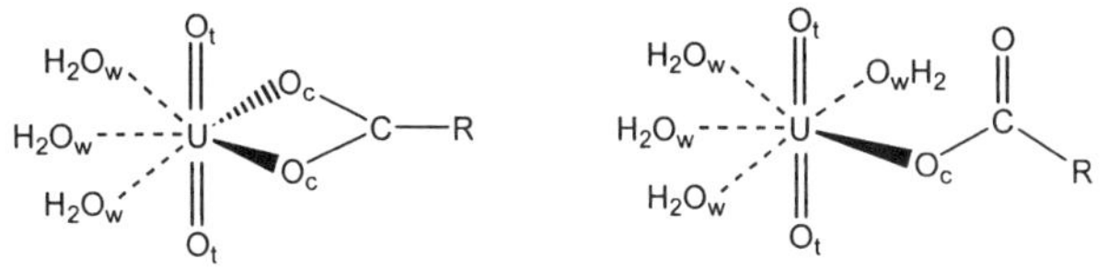

Figure 1 *Schematic structures of uranyl complexes $[UO_2(OOCR)(H_2O)_n]^+$ with the carboxylate ligand coordinated in bidentate and monodentate fashion. Three or four aqua ligands, respectively, are coordinated in the equatorial plane to achieve pentagonal coordination of the uranium centre.*

periment, where the absence of a short U-C distance is used to distinguish mono- from bidentate coordination. Also bond lengths U-O_w agree with experiment in the range of the error bars. As equatorial U-O distances are often not resolved into carboxylate and aqua ligand bonds in EXAFS evaluations, we also compare averaged equatorial oxygen bonds, U-O_{eq}. For monodentate coordination we find 2.38 Å in agreement with experiment, where as our result for the bidentate case, 2.37 Å, underestimates the experiment by ~0.05 Å.

As the differences between experiment and calculation for the uranyl-carboxylate bond clearly exceed the deviations usually obtained with an accurate density functional approach (as confirmed by our results for uranyl triacetate),[12] one needs to invoke another explanation. EXAFS averages over all species in a sample; hence, one needs to account for the effect of differently coordinated species and species with more than a single carboxyl ligand or a hydroxyl ligand (as a result of hydrolysis in the appropriate pH range). On the other hand, also approximations inherent to our model approach have to be considered. Releasing the C_s symmetry constraint may result in reorientations of the aqua ligands, and hydrogen bridges of further solvent molecules to O centers of uranyl and carboxyl ligands may slightly influence the geometry parameters. However, none of these effects is strong enough to rationalize the observed deviations. The rather short uranyl-carboxyl bonds, which are stronger than bonds to aqua ligands, point towards a case of bond competition. The experimentally determined elongation of U-O_{eq} for bidentate coordination, not reproduced in our simulation, can also be interpreted in this way. Calculated U-O_{eq} distances are found to be rather insensitive to changes in the carboxyl coordination (Table 1), but changes noticeably when the coordination number varies.[12] For sixfold coordinated uranyl triacetate, we calculated a value of 2.44 Å for U-O_{eq}, in agreement with experiment.[12] Thus, we suggest to consider a larger average coordination number of uranyl in the experimentally studied samples as a more straightforward rationalization of the larger U-O_{eq} value of bidentate complexes. A more detailed study of this interesting problem, also addressing energetic aspects and a comparison with uranyl triacetate, is in progress.[12]

References

1 R. J. Silva, H. Nitsche, *Radiochim. Acta* **1995**, *70/71*, 377.

2 M. A. Denecke, T. Reich, M. Bubner, S. Pompe, K. H. Heise, H. Nitsche, P. G. Allen, J. J. Bucher, N. M. Edelstein, D. K. Shuh, *J. All. Comp.* **1998**, *271-273*, 123.

3 H. Moll, G. Geipel, T. Reich, G. Bernhard, T. Fanghänel, I. Grenthe, *Radiochim. Acta* **2003**, *91*, 11.

4 E. H. Bayley, J. F. W. Mosselmans, P. F. Schofield, *Geochim. Cosmochim. Acta* **2004**, *68*, 1711.

5 N. Rösch, A. V. Matveev, V. A. Nasluzov, K. M. Neyman, L. Moskaleva, S. Krüger in: *Relativistic Electronic Structure Theory-Applications*, P. Schwerdtfeger, Ed.; Theor. and Comp. Chemistry Series, Vol. 14, Elsevier, Amsterdam, **2004**, p. 676.

6 M. Fuchs, A. Shor, N. Rösch, *Int. J. Quant. Chem.* **2002**, *86*, 487.

7 S. H. Vosko, L. Wilk, M. Nusair, *Can. J. Phys.* **1980**, *58*, 1200.

8 J. Jiang, L. Rao, P. Di Bernardo, P. L. Zanonato, A. Bismondo, *J. Chem. Soc., Dalton Trans.* **2002**, *8*, 1832.

9 F. Quilès, A. Burneau, *Vibr. Spectros.* **1998**, *18*, 61.

10 P. G. Allen, J. J. Bucher, D. K. Shuh, N. M. Edelstein, T. Reich, *Inorg. Chem.* **1997**, *36*, 4676.

11 M. A. Denecke, S. Pompe, T. Reich, H. Moll, M. Bubner, K. H. Heise, R. Nicolai, H. Nitsche, *Radiochim. Acta* **1997**, *79*, 151.

12 F. Schlosser, S. Krüger, N. Rösch, *Inorg. Chem., in press.*

COORDINATION CHEMISTRY OF AMERICIUM: SYNTHESIS, STRUCTURE OF Am(III) IODATES

Lia F. Brodnax[1], Amanda C. Bean[2], Brian L. Scott[1], and Wolfgang Runde[1]

[1] Chemistry Division, Los Alamos National Laboratory, P.O. Box 1663, MS J514, Los Alamos, New Mexico 87545, USA
[2] Nuclear Materials Technology Division, Los Alamos National Laboratory, P.O. Box 1663, MS E517, Los Alamos, New Mexico 87545, USA

1 INTRODUCTION

The search for new materials with nonlinear optical and ferroelectric properties led to the synthesis of over 60 lanthanide(III) iodates. The largest structural complexity belongs to the anhydrous lanthanides, $Ln(IO_3)_3$, which were found to precipitate in six different structure types[1-5]. However, crystallographic data are available for only two structure types. The structures of Type I trivalent lanthanide iodates (exceptions are Ce, Pm) were solved in the space groups $P2_1/a$ [1] or $P2_1/c$ [2]. Type II polymorphs in the space group $P2_1/c$ have a slightly increased unit cell volume and are obtained only for the anhydrous Yb and Lu iodates[1]. Four additional structure types for Ce, Pr, Nd, Sm (Type III), La, Ce, Pr (Type IV); Ce (Type V); and La (Type VI) are suggested to exist[1]. However, the polymorphs of Type II to VI are lacking structural characterization and are reported to hydrate rapidly to form the corresponding monohydrate.

Despite early interest in 5f-iodates for analytical and purification applications[6], few quantitative information on thermodynamic or structural properties of binary actinide(III) iodates are available to date. Recently, the Type I structure of $Cm(IO_3)_3$ was synthesized[7]. Americium(III) is the first element in the actinide series with solution and solid state chemistry resembling its analogous trivalent lanthanides. However, to date there are no structural information on binary americium(III) iodates reported. In this work, we attempted to synthesize hydrothermally single crystals of α-$Am(IO_3)_3$ allowing us to further discuss differences and similarities with their *4f* analogues.

2 RESULTS AND DISCUSSION

We expanded the structural variability of *f*-element iodates from the 4*f*-series to the 5*f*-series by the synthesis of a novel ternary Am(III) iodate, $KAm(IO_3)_4{\bullet}HIO_3$,[8] that has not yet been observed in the lanthanide series. Most interesting in this compound is the three-dimensional framework, in which three $[AmO_8]$ polyhedra and three $[I(5)O_3]$ groups are arranged to form irregular hexagonal channels running along the *c*-axis that are about 4.6 Å in diameter (Fig. 1a). These channels are lined with potassium cations and filled with staggered neutral HIO_3 molecules, and may be available for ionic exchange reactions. In the absence of large amounts of metal cations the binary Am(III) iodate crystallizes. We synthesized the binary anhydrous compounds $Nd(IO_3)_3$ and $Am(IO_3)_3$ by reacting KIO_4

(41.8 mg) with Nd(III) or ^{243}Am(III) in 2 mL of 0.1 M HCl in a PTFE-lined autoclave at 180 °C. After 72 hours reaction time the autoclaves were ramp-cooled at 13 °C/h to room temperature. Both compounds are isostructural, with the reported architecture of the Type I structures of the analogous 4f-element iodates[1,2]. The centrosymmetric structure of $Am(IO_3)_3$ consists of a three-dimensional network of molecular $[AmO_8]$ polyhedra that are connected by three crystallograpically distinct monodentate $[IO_3]^-$ groups (Figure 1b).

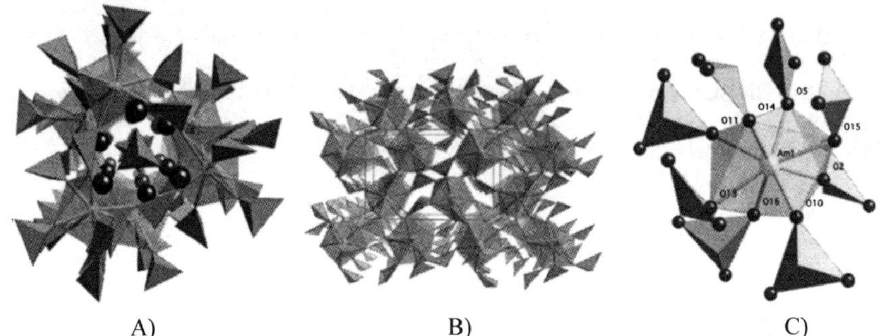

Figure 1 A) *View along the c-axis of* $K_3Am_3(IO_3)_{12}\bullet HIO_3$ *(HI(5)O*$_3$ *molecules surrounded by K atoms are staggered in the center of the channel). B) View along the a-axis of* $Am(IO_3)_3$. *C) Arrangement of eight iodate pyramids around the Am center forming the* $[AmO_8]$ *polyhedra in* $Am(IO_3)_3$.

The eight oxygen atoms around the one crystallographically unique Am atom originate from eight iodate ligands forming a distorted dodecahedral coordination sphere around the Am^{3+} ion (Figure 1c). The eight Am–O distances range between 2.34(1) Å and 2.60(1) Å, which are slightly extended compared to the Ln(III)-O distances in $Gd(IO_3)_3$ and $Tb(IO_3)_3$. The trigonal pyramidal iodate groups serve to link the americium atoms and show only slight variations in their I–O bond distances, which range between 1.77 and 1.84 Å. The μ^2-$[I(1)O_3]$ group bridges two adjacent $[AmO_8]$ polyhedra through O(13) and O(14) with I(1)-O bond distances of 1.77 Å for O(13) and 1.78 Å for O(14); a slightly shorter bond distance to the terminal O(12) atom of 1.77 Å is observed. The μ^3-$[I(2)O_3]$ group with I(2)-O bond distances between 1.77 and 1.83 Å bridges three $[AmO_8]$ polyhedra forming *zigzag* layers in the *ab* plane. The combination of $[I(1)O_3]$ and $[I(2)O_3]$ groups ensures a 3-dimensional connection of the $[AmO_8]$ polyhedra. The μ^3-$[I(3)O_3]$ group also bridges three $[AmO_8]$ polyhedra forming sheets in the *bc* plane. The I–O distances in $[I(3)O_3]$ range from 1.78(3) to 1.80(3) Å. The averaged I–O bond distances of 1.797 Å agrees well with the averaged I-O distance of 1.80 Å in $Gd(IO_3)_3$.

We used conventional UV-vis-NIR diffuse reflectance spectroscopy to verify the oxidation state of the americium ion (Figure 2). The electronic absorbance spectrum for Am^{3+}(aq) is dominated by the most intense absorbance at 503 nm. This characteristic band is shifted slightly to about 508 nm with a shoulder at 516 nm in the diffuse reflectance spectrum of $Am(IO_3)_3$ confirming the trivalent valency of the americium ion. Similar shifts are observed when Nd^{3+} ion is complexed with chloride or iodate. The Raman spectrum of the Nd(III) and Am(III) iodates exhibit three expected frequencies that are assigned to one symmetric and two asymmetric I-O stretching modes. The observed frequencies of 764, 808, and 850 cm^{-1} for $Nd(IO_3)_3$ and 761, 800, and 839 cm^{-1} for $Am(IO_3)_3$ match those found for $Cm(IO_3)_3$ (760, 804, 846 cm^{-1} [7]).

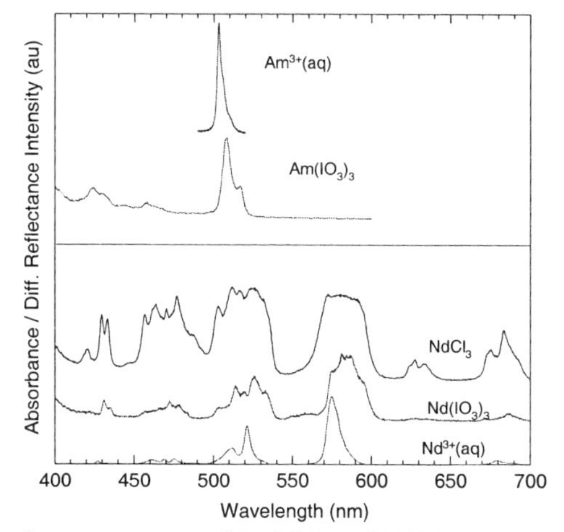

Figure 2 *Diffuse reflectance spectra of $Am(IO_3)_3$ and $Nd(IO_3)_3$ compared to absorbance spectra of the hydrated ions Am^{3+} and Nd^{3+} in solution.*

3 CONCLUSION

In this study we demonstrated that americium follows the structure and solid state chemistry displayed by its chemical analogous elements in the 4*f*-series. As curium and nearly all lanthanides, americium crystallizes in the Type I structure of binary anhydrous f-element iodates. In contrast to the lighter actinides neptunium and plutonium, which are oxidized to Np(VI) and Pu(VI) [9,10], americium remains in its most stable trivalent oxidation state.

Acknowledgements
The authors thank the G.T. Seaborg Institute at Los Alamos National Laboratory for financial support.

References

1 S.C. Abrahams, J.L. Bernstein, K. Nassau, *J. Solid State Chem.* 1976, **16**, 173.
2 P.D. Douglas, A.L. Hector, W. Levason, M.E. Light, M.L. Matthews, M.Z. Webster, *Anorg. Allg. Chem.* 2004, **630**, 479.
3 R. Liminga, S.C. Abrahams, J.L. Bernstein, *J. Chem. Phys.* 1977, **67**, 1015.
4 K. Nassau, J.W. Shiever, B.E. Prescott, A.S. Cooper, *J. Solid State Chem.* 1974, **11**, 314.
5 K. Nassau, J.W. Shiever, B.E. Prescott, *J. Solid State Chem.* 1975, **14**, 122.
6 J.M. Cleveland, *The Chemistry of Plutonium*; American Nuclear Society: La Grange Park, Il, 1979.
7 R.E. Sykora, Z. Assafa, R.G. Haire, T.E. Albrecht-Schmitt, T. E. *J. Solid State Chem.* 2004, **177**, 4413.
8 W. Runde, A.C. Bean, B.L. Scott, *Chem. Commun.* 2003, 1848.
9 A.C. Bean, B.L. Scott, T.E. Albrecht-Schmitt, W. Runde, *Inorg. Chem.* 2003, **42**, 5632.
10 W. Runde, A.C. Bean, T.E. Albrecht-Schmitt, B.L. Scott, *Chem. Comm.* 2003, **4**, 478.

THE PERFORMANCE OF TIME-DEPENDENT DENSITY FUNCTIONAL THEORY IN THE SIMULATION OF THE ELECTRONIC SPECTRA OF MOLECULAR URANIUM COMPLEXES

Kieran Ingram and Nikolas Kaltsoyannis

Department of Chemistry, University College London, Christopher Ingold Laboratories, 20 Gordon Street, London, WC1H 0AJ.

1 INTRODUCTION

Time Dependent Density Functional Theory (TD-DFT) has become a useful tool for theoretical chemists. TD-DFT enables the simulation of electronic spectra, and therefore opens another route for comparison between experiment and theory. TD-DFT has been applied in transition metal chemistry where good agreement is seen with experiment[1-2] but there is a real scarcity of TD-DFT actinide research. This work uses TD-DFT to test three functionals and six basis set combinations on three uranyl complexes with the aim of finding a reliable procedure for modelling the electronic spectra of actinide complexes.

2 COMPUTATIONAL DETAILS

All calculations were carried out using ADF2004[3-5] using the PW91 functional with a TZP STO basis set for uranium and DZP for the remaining atoms. The frozen core approximation (U 5*d*; P, Cl 2*p*; C, O, N 1*s*) was used to cut computational cost and the Zeroth Order Regular Approximation[6-7] (ZORA) was used to account for relativistic effects.

TD-DFT calculations were performed on **A**, **B**, and **C** (see fig. 1 below). Various functionals were trialled with the result that three functionals were chosen to perform the production calculations: **PW91**,[8] **LB94**[9] and **SAOP**.[10] The latter two functionals are asymptotically correct, developed specifically for use in TD-DFT calculations.

For increased accuracy the following TD-DFT parameters were specified above default values:- vectors: 400, lowest: 250, tolerance: 10^{-8} orthonormality: 10^{-10} . Six different basis set combinations were used in conjunction with the PW91 and LB94 functionals and are described in table 1. The SAOP functional can be used only in all-electron calculations and therefore only basis sets **1** and **2** are directly comparable with PW91 and LB94; for SAOP, basis sets **4**, **5** and **6** are used as in table 1 but without frozen cores.

Table 1: *Basis sets used*

1	QZ4P (all-electron) all atoms
2	TZ2P (all-electron) all atoms
3	TZ2P all atoms
4	TZ2P U & 1st coordination shell atoms, TZP rest
5	TZP U & 1st coordination shell atoms, DZP rest
6	TZP U, DZP rest

3 RESULTS AND DISCUSSION

3.1 Geometries

A variety of initial guesses for the structures of complexes **A**, **B**, and **C** were explored, both with and without symmetry constraints, and we will now discuss the lowest energy structure in each case. **A** and **C** were found to have a C_{2h} structure and **B** a D_2 structure and these are given pictorially below in fig.1, which shows that all complexes have a pseudo-octahedral geometry around the uranium metal centre. Table 2 compares selected bond length data of the three molecules. There is negligible difference between the uranyl U=O bond lengths in the three species; the U=O bond length is in good agreement with previously computed bond lengths for neutral uranyl species.[11-12]

Table 2: *Selected bond length data for complexes A, B, and C*

Bond lengths- Å	A	B	C
r (U=O)	1.802	1.799	1.798
r (U-Cl)	2.632	2.622	-
r (U-N_{NO_3})	-	-	2.96
r (U-O_{TBP})	2.450	-	2.493
r (U-O_{THF})	-	2.453	-

A difference can be seen between the U-O_{ligand} distances in **A** and **C**. This is attributed to repulsive interactions between the NO_3 groups and the TBP ligands in **C**. The similarity between the U-O_{ligand} bond lengths in **A** and **B** supports this suggestion.

Figure 1: *Ball and stick representation of the optimised geometries of complexes* ***A***, ***B***, *and* ***C***.

A: $[UO_2Cl_2(TBP)_2]$ (TBP=tributylphosphate) **B**: $[UO_2Cl_2(THF)_2]$ (THF=tetrahydrofuran) **C**: $[UO_2(NO_3)_2(TBP)_2]$

3.2 Electronic spectra

The aim of the research presented in this section is twofold; to examine the effects of increasing the quality of the basis sets on the simulated spectra, and to establish if the performance of the asymptotically correct LB94 & SAOP functionals is superior to the typical GGA functional-PW91.

Table 3: *Calculated and experimental absorption data for A*

Functional	PW91		LB94		SAOP	
Basis set	E/ cm^{-1}	Oscillator strength $x10^{-3}$	E/ cm^{-1}	Oscillator strength $x10^{-3}$	E/ cm^{-1}	Oscillator strength $x10^{-3}$
1	20330	13.6	14711	8.2	24048	13.7
2	20710	13.7	14936	8.5	24524	12.7
3	20826	13.9	15029	6.8	-	-
4	20749	13.9	14971	6.5	24458	11.6
5	20444	14.4	14983	6.6	24085	12.3
6	20516	13.4	15038	6.0	24074	12.3
Expt.	23000					

For each of the three target systems, experimental data are available for the lowest energy transition (a ligand to metal charge transfer) and we will initially focus on this. The first absorption peak for **A** is centered around 20,700cm^{-1} for most basis set combinations in conjunction with PW91, around 15,000cm^{-1} using LB94, and 24,000cm^{-1} if SAOP is employed (table 3). The experimental peak[13] is centered around 23,000cm^{-1}, i.e. SAOP agrees best with experiment in this case.

The results for complexes **B** and **C** are similar to those for **A**, LB94 gives poor agreement with experiment, PW91 adequate and SAOP good agreement. Further analysis beyond the first peak tells a similar story. Figure 2 shows an extended portion of the SAOP spectra of **A**, obtained with 5 of the 6 basis sets. The form of the spectrum changes little as the basis set are altered, a small energy shift separates them from each other. Fig. 2 is typical of the results obtained with the different functionals on **A**, **B**, and **C**, and suggests that there is little to be gained in using extended basis sets in the simulation of experimental spectra.

Figure 2: *Complex A: first 4 peaks using **SAOP***

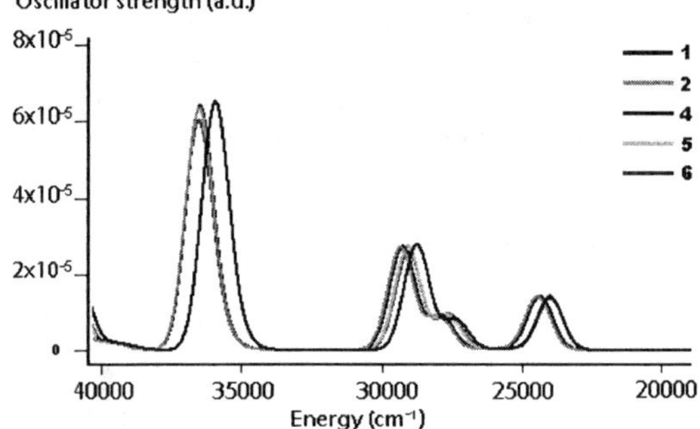

4 CONCLUSIONS

The performance of three density functionals and six basis set combinations in the simulation of the electronic spectra of three uranium compounds has been evaluated. We conclude that the use of large basis sets is not warranted in these systems, and that SAOP is the best of the functionals for reproducing the (limited) experimental data available. This latter finding is in agreement with the results of other studies on-going in our laboratory.

References

1 B. Machura, M. Jaworska and R. Kruszynski, *Polyhedron* , 2005, **24 (2)**, 267.
2 J. M. Villegas, S. R. Stoyanov, W. Huang and D. P. Rillema, *Dalton Trans.*, 2005, **6**, 1042.
3 G. te Velde, F.M. Bickelhaupt, S.J.A. van Gisbergen, C. Fonseca Guerra, E.J. Baerends, J.G. Snijders and T. Ziegler, *J. Comput. Chem.*, 2001, **22**, 931.
4 C. Fonseca Guerra, J. G. Snijders, G. te Velde, and E. J. Baerends, *Theor. Chem. Acc.*, 1998, **99**, 391.
5 ADF2004.01, SCM, Theoretical Chemistry, Vrije Universiteit, Amsterdam, The Netherlands, http://www.scm.com.
6 J. G. Snijders and A. D. Sadlej, *Chem. Phys Lett.*, 1996, **252**, 51.
7 E. van Lenthe, E. J. Baerends, J. G. Snijders: a) *J. Chem. Phys.,* 1993, **99**, 4597, b) *J. Chem. Phys.,* 1994, **101**, 9783, c) *J. Chem. Phys.,*1996, **105**, 6505.
8 J. P. Perdew et al., Physical Review B, 1992. **46**, 6671.
9 R. van Leeuwen and E.J. Baerends, *Phys. Rev. A*, 1994. **49(4)**, 2421.
10 P. R. T. Schipper et al., *J. Chem. Phys.*, 2000. **112**, 1344.
11 A. Kovacs and R. J. M. Konings, *J. Mol. Struct. –Theochem.*, 2004, **684 (1-3)**, 35.
12 Y. Oda and A. Aoshima, *J. Nucl. Sci. and Tech.*, 2002, **39 (6)**, 647-654.
13 C. Görller-Walrand, S. De Houwer, L. Fluyt and K. Binnemans, *Phys. Chem. Chem. Phys.*, 2004, **6**, 3292.

THE EFFECT OF SOLVENT ON THE MODELLING OF THE GEOMETRY AND VIBRATIONS OF $UO_2(H_2O)_5^{2+}$

Jonas Haller and Nikolas Kaltsoyannis

Department of Chemistry, University College London
20 Gordon Street, London WC1H, UK

1 INTRODUCTION

As many chemical processes take place in solution, it is important for computational chemists to have accurate and reliable models to account for solvent effects. There are several such solvent models available today, which have been tested mostly on organic molecules and have achieved very accurate results,[1] but their effects on actinides are not as well known. We are currently attempting to assess the reliability of solvent models in actinide systems, and in this contribution report some results of the geometric structure and molecular vibrations of $[UO_2(H_2O)_5]^{2+}$.

2 RESULTS AND DISCUSSION

2.1 The geometry of $UO_2(H_2O)_5^{2+}$

Geometry optimizations were performed in Gaussian 03 using DFT (PBE). A relativistic pseudopotential of the Stuttgart-Bonn variety was employed for uranium with a (14s13p10d8f6g)/[6s6p5d4f 3g] segmented valence basis set. A 6-31G** basis set was used for oxygen and hydrogen, and the conductor polarizable continuum model (CPCM) was used to account for solvent effects. As shown in Figure 1, our results indicate that the compound has the uranyl unit in an axial position and the water ligands in an equatorial plane around the uranium atom. The lowest energy structure found is with one of the water molecules mainly in the equatorial plane, while the other ligands have only their oxygen atoms in the plane and their hydrogen atoms located above and below the plane. $UO_2(H_2O)_5^{2+}$ has previously been studied both by computational[2,3] and experimental[2,4,52] methods. The calculated geometry and uranyl symmetric and antisymmetric vibrations found in this work are compared with previous theoretical studies and experiments for the gas and the solvent phase in Table 1. Both the bond lengths and the uranyl wavenumbers show excellent agreement with experiment. They are all slightly closer to the experimental values than previous work. The geometry in the solvent phase is slightly different compared with the gas phase, in

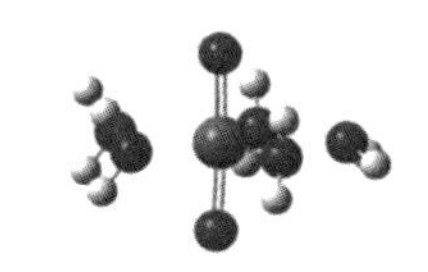

Figure 1
Optimised geometry of $UO_2(H_2O)_5^{2+}$ with water as solvent and all cavity parameters default.

particular for the water ligand lying in the equatorial plane. In the solvent phase the two hydrogens are below the equatorial plane (Figure 1), whereas in the gas phase they lie in the plane.

Table 1 *Geometric and vibrational data for $UO_2(H_2O)_5^{2+}$ in gas and aqueous phase*

	$U\text{-}O_{ax}$ distance / Å	$U\text{-}O_{eq}$ distance / Å	Uranyl sym. stretch / cm^{-1}	Uranyl asym. stretch / cm^{-1}
Solvent				
This work	1.779	2.432	862	929
Previous work	1.756[3]	2.516[3]	908[3]	1001[3]
Experiment	1.76,[2] 1.78[4]	2.41,[2] 2.41[4]	872[5]	964[5]
Gas				
This work	1.765	2.474	893	981
Previous work	1.67 (1.75)[2,*]	2.57[2]	901.7[5]	-

* corrected for correlation and error in AIMP.

2.2 Imaginary vibrations

Using only default values for the CPCM (solvent) settings gives two imaginary wavenumbers (Table 2), in contrast to the gas phase which gave no imaginary modes. Normally, an imaginary mode corresponds to a saddle point on the potential energy surface. If this is the case, it should be possible to follow the eigenvector corresponding to the imaginary mode to reach the bottom of the energy surface. This was tried using the intrinsic reaction coordinate (IRC) method, but the energy increased as a result. IRC follows the eigenvector corresponding to the lowest eigenvalue and it was tried in both directions. In addition to the IRC, single point calculations and geometry optimizations along the imaginary modes were performed; however, these also resulted in higher energy structures. HF and MP2 wavenumber calculations in solution also gave imaginary wavenumbers of similar magnitude.

2.3 Tesserae parameters

The results in section 2.2 suggest that the imaginary wavenumbers introduced by the inclusion of solvent effects are unlikely to indicate a non-minimum energy structure, but rather arise as an artefact of the solvation model. It was suggested that the imaginary wavenumbers result from small changes in the geometry altering the number of tesserae,[6] i.e. the imaginary wavenumbers are affected by the discretization of the cavity[7]. To test this hypothesis, a study of the different options for the cavity was performed. The parameters which were found to affect the wavenumbers were the average area of the tesserae (TSARE) and the overlap index between two interlocking spheres (OFAC).

When the TSARE parameter was varied, the lowest magnitude of the imaginary wavenumbers was obtained with TSARE=0.4 (Table 2), and hence this value was used in studies of the OFAC parameter. The OFAC parameter controls the smoothing of the surfaces, and it seems probable that a smoother surface should give lower or no imaginary wavenumbers, because a change in geometry will affect a smooth cavity less than a rough one. A high value of OFAC results in a smooth surface. As can be seen in Table 3, the cavity volume is almost constant while the cavity area decreases when OFAC is increased. This shows that the cavity gets smoother as OFAC is increased.

The calculation with OFAC=0.94 results in a different geometry and a magnitude of the imaginary wavenumbers that is not similar to the other calculations. If this calculation is disregarded, the magnitude of the wavenumbers generally decreases in agreement with the hypothesis. The uranyl vibrations do not change as much as the imaginary modes.

Table 2 *Cavity properties and wavenumbers for different values of the TSARE parameters (all the other cavity parameters have the default value). Default: TSARE=0.2*

TSARE	Freq. 1 / cm^{-1}	Freq. 2 / cm^{-1}	Uranyl sym stretch / cm^{-1}	Uranyl antisym stretch cm^{-1}	Number of tesserae	Average tesserae area / $Å^2$	Cavity area / $Å^2$	Cavity volume / $Å^2$
0.1	74 i	58 i	863	930	2685	0.07	180.9	189.9
0.2	73 i	56 i	862	929	1590	0.11	180.9	190.0
0.4	67 i	53 i	862	928	911	0.20	181.2	189.2
0.6	91 i	78 i	860	926	610	0.30	181.2	189.1
0.8	68 i	54 i	861	927	570	0.32	181.2	188.9

Table 3 *Cavity properties and wavenumbers for different values of the OFAC parameter (TSARE = 0.4 in all calculations). Default: OFAC=0.89*

OFAC	Freq. 1 / cm^{-1}	Freq. 2 / cm^{-1}	Uranyl sym stretch / cm^{-1}	Uranyl antisym stretch cm^{-1}	Number of tesserae	Average tesserae area / $Å^2$	Cavity area / $Å^2$	Cavity volume / $Å^3$
0.86	66 i	49 i	862	929	744	0.24	181.9	188.8
0.88	65 i	53 i	862	928	854	0.21	181.5	188.9
0.90	67 i	49 i	861	928	917	0.20	181.2	189.2
0.92	63 i	47 i	861	928	981	0.18	180.9	189.4
0.93	69 i	51 i	861	928	1051	0.17	180.6	189.4
0.94	1970 i	313 i	859	928	1250	0.14	179.8	190.5
0.96	60 i	41 i	860	927	1523	0.12	178.8	190.3
0.97	59 i	56 i	859	927	1691	0.10	177.2	189.7

4 CONCLUSIONS

The geometry and vibrations of $UO_2(H_2O)_5^{2+}$ have been studied and compared with previous theoretical studies and experiments. The new calculations show an excellent agreement in both geometry and uranyl wavenumbers. The low energy imaginary modes encountered do not seem to correspond to non-minimum structures but appear to be an artefact of the solvent model.

References

1 M. Cossi, G. Scalmani, N. Rega and V. Barone *J. Chem. Phys.*, 2002, **117** 43.
2 U. Wahlgren, et. al. *J. Phys. Chem. A,* 1999, **103** 8257.
3 P. J. Hay, R. L. Martin, G. Schreckenbach *J. Phys. Chem. A,* 2000, **104** 6259.
4 P. G. Allen et. al. *Inorg. Chem.,* 1997, **36,** 4676.
5 T. Fujii, K Fujiwara, H Yamana and H. Moriyama *J. Alloys Comp.*, 2001, **323-324** 859.
6 The charge on the surface of the cavity is calculated by solving variations of the Poisson-Boltzmann equation. For complicated cavity geometries, this equation does not have an analytical solution and numerical methods are needed. To do this, the surface of the cavity is divided into many small triangular surfaces called tesserae. Each triangle is given a point charge by solving the equation for every tessera individually.
7 V. Vallet, personal communication

Heavy Elements

CHEMISTRY OF TRANSACTINIDES

A. Türler

Institut für Radiochemie, Technische Universität München, D-85748 Garching

1 INTRODUCTION

The discovery of new elements and their correct placement in the periodic table was originally the domain of chemists, and indeed all elements up to atomic number 101 (Md, mendelevium) were discovered by chemical means.[1] Heavier elements were synthesized in heavy-ion fusion reactions at accelerators on a "one-atom-at-a-time" level. The rapidly decreasing production cross-sections, that nowadays are of the order of only picobarns or less for the heaviest known nuclei and the diminishing half-lives made manual chemical investigations of elements beyond atomic number 105 (Db, dubnium) virtually impossible. Thus, the synthesis and identification of new elements has shifted to nuclear physics and led to the discovery of more than ten new elements. However, the placement of a new element in the periodic table is not only based on its atomic number but also on its chemical properties, which are determined by its electronic structure. Due to the very high atomic charges, relativistic effects are extremely important in defining the chemical properties. Therefore, the chemistry of heavy elements is extremely interesting and rewarding to study. With atomic number 104 (Rf, rutherfordium) a new series of d-elements starts, the so-called transactinides. Since the liquid drop fission barrier also disappears around Z=104, the term superheavy elements (SHE) is often used as a synonym.

Remarkable progress was made in recent years in the experimental investigation of the chemical properties of transactinides despite the numerous difficulties involved in chemically isolating one single, relatively short-lived atom from a plethora of unwanted by-products of the nuclear production reaction. This progress can be attributed to the discovery of enhanced nuclear stability close to the vicinity of the previously predicted deformed shells at Z=108 and N=162,[2] which provided chemists with sufficiently long-lived isotopes of seaborgium (Sg, Z=106),[3] bohrium (Bh, Z=107),[4] and even hassium (Hs, Z=108).[5] The production of a heavy nucleus was detected by registering its decay, which for many superheavy nuclei consists of unique α-particle decay chains, quite often terminated by spontaneous fission (SF). The knowledge and understanding of the chemistry of SHE, especially of the early transactinides Rf, Db, and Sg and their compounds, both experimentally and theoretically, are nowadays quite extensive, as documented by the first textbook entitled "The Chemistry of Superheavy Elements".[6] While rapid chemical separations in aqueous solution were (and are) the method of choice for detailed investigations of Rf and Db,[7,8] the development of gas-phase chemical separation methods have allowed chemical studies of all transactinides up to Hs.[9–11]

The discovery of longer-lived (half-lives of several seconds) isotopes of elements 110 through 114 in ^{48}Ca-induced reactions on actinide targets at the Flerov Laboratory of Nuclear Reactions (FLNR) has opened fascinating new perspectives for studying experimentally the chemistry of even the heaviest known elements in the periodic table.[12] Currently, experiments are being conducted to elucidate the chemical properties of element 112 (E112) in its elemental state.[13–15] Recently, disagreeing decay data for E112 isotopes have rendered previous interpretations of the observed chemical properties of E112 questionable.

In the following an overview of the latest developments and accomplishments in the field of transactinide chemistry is presented.

2 CHEMICAL PROPERTIES OF TRANSACTINIDES AND THEIR COMPOUNDS

Experimental investigations of the transactinides rely strongly on predictions of their chemical properties. In order to plan an isolation scheme, the stability of compounds and their physicochemical properties have to be evaluated on the basis of predictions. Traditionally, the chemical properties of an unknown element and its compounds were predicted by exploiting the fundamental relationships of physicochemical data of elements and their compounds within the groups and the periods of the periodic table.[6] Since relativistic effects increase proportionally with Z^2, extrapolations of chemical properties from lighter homologue elements must fail at some point. The only alternative is a fully relativistic calculation. All-electron calculations of molecules containing transactinides are exceedingly difficult, due to the large number of electrons involved, and presently it is not possible to compute directly the physicochemical properties of a molecule. Nonetheless, also due to enormously increasing computing power, theoretical calculations have now progressed to the point that a number of relevant atomic and molecular properties can be predicted.[6]

Luckily chemists are sometimes facing very favourable and clear cut cases. Such a case constitutes the chemistry of Hs. If Hs has any resemblance to its lighter homolog, Os, then it should also form tetroxides. Due to the perfect tetrahedral symmetry of OsO_4, this compound only interacts very weakly with non-reducing surfaces and is therefore very volatile, similar to a noble gas. Relativistic molecular orbital calculations confirmed that indeed HsO_4 must be a very stable molecule, and that its volatility should be very similar to its lighter homologue compound OsO_4.[6] This exceptional chemical property distinguishes Os and Hs from other d-elements and allows a clear assignment to group 8 of the periodic table. Thus, experimentalists successfully took on the daunting task of building a chemical separator system that allowed the isolation of Hs as volatile HsO_4.

Another chemically very favourable case constitutes E112 as an expected homolog of Hg. E112 is expected to have an ionization potential which is more than 1.5 eV higher than that of Hg.[6] Similar to Hg, E112 should not have a bound anion.[6] Therefore, E112 is expected to be rather inert in its elemental state. It is well known that Hg is a very volatile element, but that it interacts strongly with other metals, such as Zn or Au. Single atoms of Hg can be adsorbed quantitatively on a Au surface from a stream of He.[13] Relativistic calculations of the dimmers, Hg–Au and E112–Au, indicate a lower, but still sizeable binding energy for the E112–Au dimer.[6] Therefore, current experiments focus on whether E112 also forms strong metal bonds with Au or whether E112 is so inert, that it behaves more like a heavy noble gas such as Rn. Similar considerations apply to E114.

3 EXPERIMENTAL TECHNIQUES AND DEVELOPMENTS

Chemical experiments with transactinides can be divided into four steps: synthesis, rapid transport to the chemical apparatus, fast isolation and preparation of a sample for nuclear spectroscopy, and detection of the nuclide via its characteristic decay properties.

In order to gain access to the longer-lived isotopes of transactinides, exotic actinide target nuclides are bombarded with intense heavy ion beams. Beams as intense as possible are used, but without destroying the very valuable and highly radioactive targets. A rotating ^{248}Cm target with a rotating vacuum window that can stand pressure differences of up to 1.5 atm has recently been used in heavy element chemistry experiments at the Gesellschaft für Schwerionenforschung (GSI). This rotating target carrying up to 6 mg of target material has allowed the use of high-intensity beams for heavy element chemistry experiments.[6] The rotation of the target wheel is synchronized with the time structure of the accelerator in order to spread one beam pulse over the area of one target segment.

Reaction products recoiling out of the target are stopped in a recoil chamber filled with He gas. Non-volatile products such as Sg or Bh are adsorbed on aerosol particles (KCl or C) of about 100 nm diameter and transported through a capillary over relatively long distances to the chemistry apparatus.

Four different approaches which involve direct detection of the decay of the isolated nuclides have been successful in studying chemical properties of transactinides. Two of the systems work in the liquid phase, whereas the other two are designed to investigate volatile transactinide compounds in the gaseous phase. The predominant share of today's knowledge about the chemical behaviour of Rf, Db, and Sg in aqueous solution was obtained with the Automated Rapid Chemistry Apparatus (ARCA).[6] This micro-computer controlled set-up allows fast repetitive separations on miniaturized chromatography columns. Often thousands of separations are performed. After chemical separation the fraction containing the transactinide is evaporated and the sample placed in an α–/SF spectroscopy system. This last step, which was performed manually, has now also been automated in a set-up called AIDA developed at the Japan Atomic Energy Research Institute (JAERI).[8]

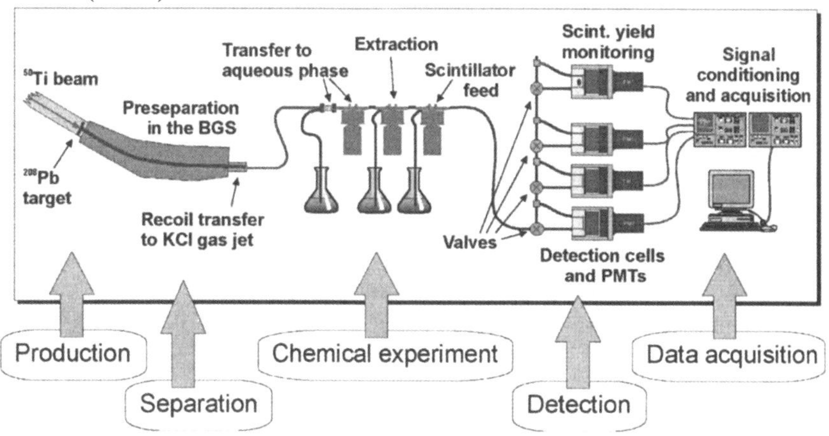

Figure 1 *The liquid-liquid extraction system SISAK coupled to the Berkeley gas-filled separator (BGS).*[6]

A continuously operating liquid-liquid extraction system using miniaturized centrifuges and liquid scintillation flow through detection cells (SISAK) was recently successful in separating and identifying 4-s ^{257}Rf.[16] The latest version of SISAK is well adapted for studying nuclides with half-lives as short as 1 s. In order to suppress the large background created by β-particles, a recoil chamber was coupled to the Berkeley Gas-filled Separator (BGS) (Figure 1). Coupling a chemistry set-up with a recoil separator opens new frontiers in direct chemical reactions with a variety of compounds, since one no longer has to deal with the harsh ionizing conditions created by the passage of the primary beam. Furthermore, contaminants that cannot be effectively removed due to their chemical similarity to SHE, such as Po or Rn isotopes, are largely removed by the separator.

In order to study elements beyond Sg, faster and much more efficient chemistry set-ups had to be used. Since transactinide nuclei are usually stopped in gas, a fast and efficient link to a gas chromatography system can be established; either by direct transport of volatile species or volatile compounds formed in-situ in the recoil chamber, or by a transport with aerosol particles. Despite the fact that the transition metals (groups 4 to 11) are very refractory, few stable inorganic compounds exist that are appreciably volatile at experimentally manageable temperatures below 1100 °C. These are the halides and oxyhalides of groups 4 to 7, the oxide hydroxides of groups 6 and 7, and the oxides of group 8. Moreover, elements 112 to 118 are expected to be rather volatile in the elemental state and thus gas-phase separations will play a crucial role in investigating their chemical properties. The On-Line Gas-chemistry Apparatus (OLGA) was developed and successfully used by the Swiss group at the Paul Scherrer Institute to study halides and oxyhalides of elements Rf,[6] Db,[6] Sg,[9] and also Bh.[10] The set-up was sensitive to cross sections of the order of ≈100 pb.

In order to assess nuclides that are produced with cross sections of a few picobarns the overall efficiency had to be improved by at least one order of magnitude. This was accomplished by the in-situ formation of volatile compounds and their condensation and detection in a thermochromatography detector, as in early experiments by Zvara.[6] Sufficiently volatile reaction products such as Hg and presumably E112 can be transported directly with the flowing gas. In experiments with Hs, recoiling atoms were converted in-situ with an admixture of O_2 gas to the volatile HsO_4. Instead of an isothermal temperature profile as in OLGA, a negative longitudinal temperature gradient is established along the column. Thus, compounds are deposited in the chromatography column according to their volatility, forming distinct deposition peaks. If the column consists of silicon detectors then the nuclear decay of the separated nuclide can be registered. The detector number indicates the temperature at which the volatile compound was deposited. Thus, every detected nuclide also reveals chemical information. A schematic of the In-situ Volatilization and On-line detection (IVO) technique is shown in Figure 2. The overall efficiency (including detection of a complete 3 member α-particle decay chain) is of the order of 30 - 50%.

Similar set-ups were used in first attempts to identify E112 chemically. If E112 is chemically similar to Hg, then it should adsorb on Au or Pd surfaces. In test experiments short-lived Hg isotopes could be isolated in the elemental form from other reaction products and transported in He quantitatively through a 30 m long Teflon™ capillary at room temperature. Hg adsorbed quantitatively on Au, Pt, and Pd surfaces at room temperature. As little as 1 cm^2 of Au or Pd surface was sufficient to adsorb Hg atoms nearly quantitatively from a stream of 1 l/min He.[13] Therefore, detector chambers containing a pair of Au or Pd coated PIPS detectors were constructed. Eight detector chambers (6 Au and 2 Pd) were connected in series by Teflon™ tubing. The detector chambers were positioned inside an assembly of 84 ^{3}He filled neutron detectors (in a polyethylene moderator), in order to detect simultaneously neutrons accompanying SF events.[14] In the latest experiments aimed at measuring the volatility of E112, one row of silicon detectors of the Cryo On-Line Detector (COLD) was removed and replaced by a Au surface,[15] or in the latest experiment by Au covered PIN-diodes.[17] A negative temperature gradient was established ranging from room temperature down to –187 °C.[15,17]

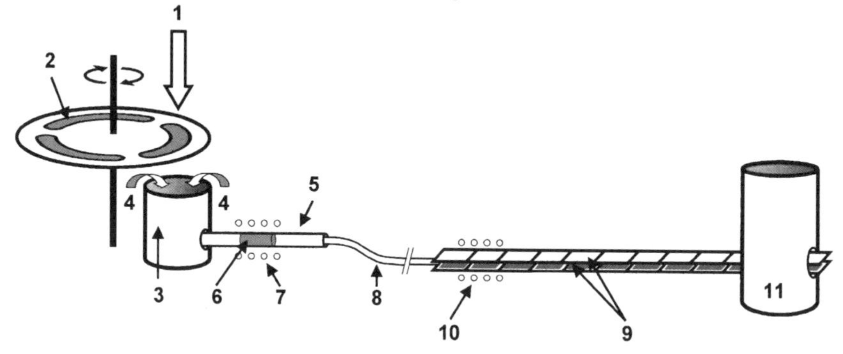

Figure 2 *The ^{26}Mg-beam (1) passed through the rotating vacuum window and ^{248}Cm-target (2) assembly. In the fusion reaction ^{269}Hs nuclei were formed which recoiled out of the target into a gas volume (3) and were flushed with a He/O_2 mixture (4) to a quartz column (5) containing a quartz wool plug (6) heated to 600 °C by an oven (7). There, Hs was converted to HsO_4 which is volatile at room temperature and transported with the gas flow through a perfluoroalkoxy (PFA) capillary (8) to the COLD detector array registering the nuclear decay (α and SF). A thermostat (10) kept the entrance of the array at 20 °C; the exit was cooled to –170 °C by means of liquid nitrogen (11). Depending on the volatility of HsO_4, the molecules adsorbed at a characteristic temperature.*[11]

4 RESULTS AND DISCUSSION

4.1 Aqueous chemistry of transactinides

The past years have focused on a detailed study of the ion-exchange and complex-formation behaviour of Rf in aqueous solution. Different adsorption behaviour for Rf, compared to its homologues Zr and Hf was observed in anion- and cation-exchange studies in 0.0001-0.1 M HF/0.1 M HNO_3 solutions.[7] Haba et al. using AIDA studied the fluoride complexation of the group 4 elements Zr, Hf, and Rf in 1.9–13.9 M HF solutions using anion-exchange chromatography.[8] A significantly different behaviour was observed for Rf. In Figure 3 the simultaneously determined K_d values for Zr, Hf, and Rf on a CA08Y anion-exchange column are shown as a function of the initial HF concentration. The slope analysis based on the law of mass action indicates that Rf is present as the hexafluoride complex RfF_6^{2-} whereas Zr and Hf are present in the forms of ZrF_7^{3-} and HfF_7^{3-}. This experimental difference was interpreted as the result of the influence of relativistic effects for Rf.

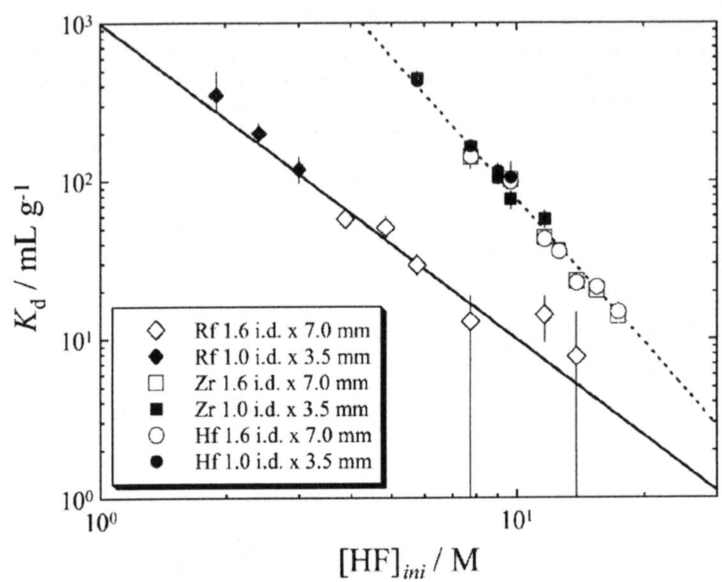

Figure 3 *Variation of the distribution coefficient, K_d, of Rf, Zr, and Hf on the anion-exchange resin CA08Y as a function of the initial HF concentration, $[HF]_{ini}$. The K_d values of Rf, Zr, and Hf are shown by diamonds, squares, and circles, respectively. Linear relationships with slopes -2.0 ± 0.3 for Rf and -3.0 ± 0.1 for Zr and Hf in the log K_d versus log$[HF]_{ini}$ plot are indicated by the solid and dotted lines, respectively.*[8]

4.2 Gas-phase chemistry of transactinides

The first chemical identification of Hs signifies the most significant accomplishment in heavy element chemistry of the past years. Even though the cross sections of the most favourable production reactions reach only a few picobarns, an unambiguous identification of the nuclide ^{269}Hs was achieved after chemical separation. In an experiment conducted at GSI by an international collaboration using the IVO technique, a total of 7 decay chains attributed to Hs isotopes were registered. In Figure 4 the distribution of decay chains along the temperature gradient is shown in comparison to the adsorption peak of OsO_4. Surprisingly, HsO_4 is slightly less volatile than OsO_4. From relativistic molecular orbital calculations a similar or even higher volatility was deduced for HsO_4 compared to OsO_4.[6] Recently, Hs was shown to form a hassate.[18] In analogy to OsO_4, which forms $Na_2[OsO_4(OH)_2]$, an osmate(VIII), with aqueous NaOH, HsO_4 was deposited under identical conditions presumably as $Na_2[HsO_4(OH)_2]$, a hassate(VIII).

Several experiments have been carried out to determine the chemical properties of E112 in the elemental state using the reaction ^{238}U(^{48}Ca, 3n)^{283}112. However, there exist conflicting reports from FLNR about the decay properties of this nuclide.[12] In the first reports ^{283}112 appeared to decay by SF with a half-live of about 5 min. In recent experiments ^{283}112 was reported to decay by α-particle emission and a half-life of about 4 s to ^{279}Ds, which in turn decayed by SF with a half-life of about 0.2 s. Also, the chemistry results are inconclusive so far. Early experiments at FLNR seemed to confirm the existence of a long-lived E112 isotope decaying by SF.[14] These experiments relied on the rather long half-life of several minutes and suggested a Rn-like behaviour of E112. Later experiments at GSI could not conclusively identify either a long-lived E112 isotope decaying by SF or a short-lived α-decaying isotope.[17]

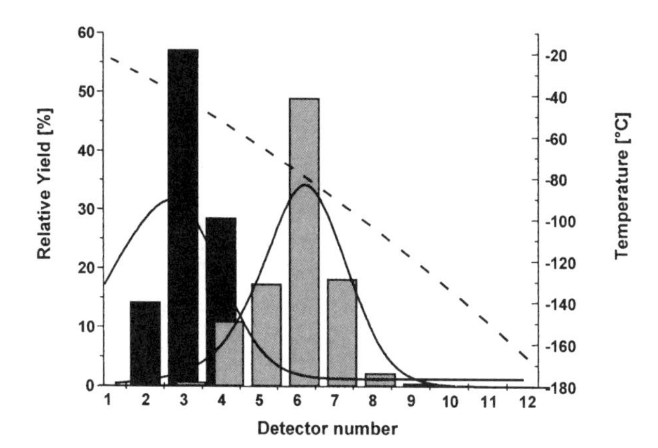

Figure 4 *Relative yields of HsO_4 and OsO_4 for each of the 12 detector pairs. Measured values are represented by bars: $^{269}HsO_4$: black; $^{172}OsO_4$: light grey. The dashed line indicates the temperature profile (right-hand scale). The maxima of the deposition distributions were evaluated as -44 ± 6 °C for HsO_4 and -82 ± 7 °C for OsO_4. Solid lines represent results of a simulation of the adsorption process with standard adsorption enthalpies of −46.0 kJ/mol for $^{269}HsO_4$ and −39.0 kJ/mol for $^{172}OsO_4$.[11]*

References

1 G.T. Seaborg and W.D. Loveland, *The Elements Beyond Uranium*, John Wiley & Sons, Inc., New York, 1990.
2 Z. Patyk, J. Skalski, A. Sobiczewski, and S. Cwiok, *Nucl. Phys. A*, 1989, **502**, 591.
3 Yu.A. Lazarev, Yu.V. Lobanov, Yu.Ts. Oganessian, V.K. Utyonkov, F Sh. Abdullin, G.V. Buklanov, B.N. Gikal, S. Iliev, A.N. Mezentsev, A.N. Polyakov, I.M. Sedykh, I.V. Shirokovsky, V.G. Subbotin, A.M. Sukhov, Yu.S. Tsyganov, V.E. Zhuchko, R.W. Lougheed, K.J. Moody, J.F. Wild, E.K. Hulet, and J.H. McQuaid, *Phys. Rev. Lett.*, 1994, **73**, 624.
4 P.A. Wilk, K.E. Gregorich, A. Türler, C.A. Laue, R. Eichler, V. Ninov, J.L. Adams, U.W. Kirbach, M.R. Lane, D.M. Lee, J.B. Patin, D.A. Shaughnessy, D.A. Strellis, H. Nitsche, and D.C. Hoffman, *Phys. Rev. Lett.*, 2000, **85**, 2697.
5 S. Hofmann, F.P. Heßberger, D. Ackermann, G. Münzenberg, S. Antalic, P. Cagarda, B. Kindler, J. Kojouharova, M. Leino, B. Lommel, R. Mann, A.G. Popeko, S. Reshitko, S. Śaro, J. Uusitalo, and A.V. Yeremin, *Eur. Phys. J. A*, 2002, **14**, 147.
6 *The Chemistry of Superheavy Elements*, ed. M. Schädel, Kluwer Academic Publisher, Dordrecht, 2003.
7 J.V. Kratz, *Pure Appl. Chem*,. 2003, **75**, 103.
8 H. Haba, K. Tsukada, M. Asai, A. Toyoshima, K. Akiyama, I. Nishinaka, M. Hirata, T. Yaita, S. Ichikawa, Y. Nagame, K. Yasuda, Y. Miyamoto, T. Kaneko, S. Goto, S.

Ono, T. Hirai, H. Kudo, M. Shigekawa, A. Shinohara, Y. Oura, H. Nakahara, K. Sueki, H. Kikunaga, N. Kinoshita, N. Tsuruga, A. Yokoyama, M. Sakama, S. Enomoto, M. Schädel, W. Brüchle and J.V. Kratz, *J. Am. Chem. Soc.*, 2004, **126**, 5219.

9 M. Schädel, W. Brüchle, R. Dressler, B. Eichler, H. W. Gäggeler, R. Günther, K.E. Gregorich, D.C. Hoffman, S. Hübener, D.T. Jost, J.V. Kratz, W. Paulus, D. Schumann, S. Timokhin, N. Trautmann, A. Türler, G. Wirth and A. Yakushev, *Nature*, 1997, **388**, 55.

10 R. Eichler, W. Brüchle, R. Dressler, Ch.E. Düllmann, B. Eichler, H.W. Gäggeler, K.E. Gregorich, D.C. Hoffman, S. Hübener, D.T. Jost, U.W. Kirbach, C.A. Laue, V.M. Lavanchy, H. Nitsche, J.B. Patin, D. Piguet, M. Schädel, D.A. Shaughnessy, D.A. Strellis, S. Taut, L. Tobler, Y.S. Tsyganov, A. Türler, A. Vahle, P.A. Wilk, and A.B. Yakushev, *Nature*, 2000, **407**, 63.

11 Ch.E. Düllmann, W. Brüchle, R. Dressler, K. Eberhardt, B. Eichler, R. Eichler, H.W. Gäggeler, T.N. Ginter, F. Glaus, K.E. Gregorich, D.C. Hoffman, E. Jäger, D.T. Jost, U.W. Kirbach, D.M. Lee, H. Nitsche, J.B. Patin, V. Pershina, D. Piguet, Z. Qin, M. Schädel, B. Schausten, E. Schimpf, H.-J. Schött, S. Soverna, R. Sudowe, P. Thörle, S.N. Timokhin, N. Trautmann, A. Türler, A. Vahle, G. Wirth, A.B. Yakushev, and P.M. Zielinski, *Nature*, 2002, **418**, 859.

12 Yu.Ts. Oganessian, V.K. Utyonkov, Yu.V. Lobanov, F. Sh. Abdullin, A.N. Polyakov, I.V. Shirokovsky, Yu.S. Tsyganov, G.G. Gulbekian, S.L. Bogomolov, B.N. Gikal, A.N. Mezentsev, S. Iliev, V.G. Subbotin, A.M. Sukhov, A.A. Voinov, G.V. Buklanov, K. Subotic, V.I. Zagrebaev, M.G. Itkis, J.B. Patin, K.J. Moody, J.F. Wild, M.A. Stoyer, N.J. Stoyer, D.A. Shaughnessy, J.M. Kenneally, and R.W. Lougheed, *Phys. Rev. C*, 2004 , **69**, 054607.

13 A.B. Yakushev, G.V. Buklanov, M.L. Chelnokov, V.I. Chepigin, S.N. Dmitriev, V.A. Gorshkov, S. Hübener, V.Ya. Lebedev, O.N. Malyshev, Yu.Ts. Oganessian, A.G. Popeko, E.A. Sokol, S.N. Timokhin, A. Türler, V.M. Vasko, A.V. Yeremin, and I. Zvara, *Radiochim. Acta*, 2001, **89**, 743.

14 A.B. Yakushev, I. Zvara, Yu. Ts. Oganessian, A.V. Belozerov, S.N. Dmitriev, B. Eichler, S. Hübener, E.A. Sokol, A. Türler, A.V. Yeremin, G.V. Buklanov, M.L. Chelnokov, V.I. Chepigin, V.A. Gorshkov, A.V. Gulyaev, V.Ya. Lebedev, O.N. Malyshev, A.G. Popeko, S. Soverna, Z. Szeglowski, S.N. Timokhin, S.P. Tretyakova, V.M. Vasko, and M.G. Itkis, *Radiochim. Acta*, 2003, **91**, 433.

15 H.W. Gäggeler, W. Brüchle, Ch.E. Düllmann, R. Dressler, K. Eberhardt, B. Eichler, R. Eichler, C.M. Folden, T.N. Ginter, F. Glaus, K.E. Gregorich, F. Haenssler, D.C. Hoffman, E. Jäger, D.T. Jost, U.W. Kirbach, J.V. Kratz, H. Nitsche, J.B. Patin, V. Pershina, D. Piguet, Z. Qin, U. Rieth, M. Schädel, E. Schimpf, B. Schausten, S. Soverna, R. Sudowe, P. Thörle, N. Trautmann, A. Türler, A. Vahle, P.A. Wilk, G. Wirth, A.B. Yakushev, and A. von Zweidorf, *Nucl. Phys. A*, 2004, **734**, 208.

16 J.P. Omtvedt, J. Alstad, H. Brevik, J.E. Dyve, K. Eberhardt, C.M. Folden III, T.N. Ginter, K.E. Gregorich, E.A. Hult, M. Johansson, U.W. Kirbach, D.M. Lee, M. Mendel, A. Nähler, V. Ninov, L.A. Omtvedt, J.B. Patin, G. Skarnemark, L. Stavsetra, R. Sudowe, N. Wiehl, B. Wierczinski, P.A. Wilk, P.A. Zielinski, J.V. Kratz, N. Trautmann, H. Nitsche, and D.C. Hoffman, *J. Nucl. Radiochem. Sci.*, 2002, **3**, 121.

17 R. Eichler et al., submitted to *Radiochim. Acta.*

18 A. von Zweidorf, W. Brüchle, S. Bürger, H. Hummrich, J.V. Kratz, B. Kuczewski, G. Langrock, U. Rieth, M. Schädel, N. Trautmann, K. Tsukada and N. Wiehl, *Radiochim. Acta*, 2004, **92**, 855.

UNUSUAL VOLATILITY OF OXIDES FORMED BY SOME ACTINIDES

V.P. Domanov

Joint Institute for Nuclear Research, 141980, Dubna, Moscow region, Russia

1 INTRODUCTION

In thermochromatographic (TC) studies of thermal oxidation of trace quantities of uranium or plutonium in a stream of He+O_2 mixture, transfer from the starting zone of the quartz TC column was observed.[1] It was found experimentally that uranium formed three volatile compounds that were adsorbed at 450±25 °C, 250±25 °C and 120±70 °C in a stream of a dry gas. The use of humid gas gave only one compound, adsorbed at 120±25 °C. The deciphering of the obtained thermochromatograms was performed by application of the results of model TC experiments, which were conducted with carrier free radioisotopes, ^{185}Os, ^{97}Ru, ^{183}Re and ^{96}Tc. It was shown that uranium deposited in the form of dioxide in the first adsorption zone, the second zone was connected with adsorption of UO_3 molecules, and the last was formed because of deposition of uranium acid (supposedly H_2UO_4). The values of adsorption enthalpy, $-\Delta H_a^0$, for UO_2 and UO_3 on quartz were equal to 172±6 kJ.mol^{-1} and 126±6 kJ.mol^{-1}, respectively. In similar conditions, plutonium formed four adsorption zones: at 450±30 °C, 250±30 °C, 130±50 °C and at −105±25 °C. As in the case of uranium, plutonium also formed dioxide, trioxide and acid, which were adsorbed in the first, the second and in the third zones accordingly. In Domonov et al[2] we assumed that the last adsorption zone could be related to the deposition of octovalent plutonium in the form of PuO_4. Later,[3] the formation of tetraoxide was confirmed. The values of $-\Delta H_a^0$ for isolated oxides on quartz were equal to: 175±7 kJ.mol^{-1} (PuO_2), 122±7 kJ.mol^{-1} (PuO_3) and 41±6 kJ.mol^{-1} (PuO_4). The mass spectrometric measurements performed by Ronchi et al.[4] soon confirmed the evidence of the formation of volatile UO_2, UO_3, PuO_2 and PuO_3 molecules. The aim of this work is to search for uncommon volatile oxides of neptunium.

2 METHOD AND RESULTS

2.1 Experimental Procedure

The experiments were based on the use of the TC method, on the study of actinides of interest in trace quantities, as well as on a fine purification of the gas mixture (He+O_2), the inside surface of the columns, connectors and all assemblies of the set-up, which consisted of an empty quartz TC column (i.d.=3mm, l=70cm), a tube furnace for heating an initial sample up to 710-730 °C, a thermogradient device (its construction was described in Eichler and Domonov[5]), helium and oxygen cylinders, a gas mixer, a tool for gas drying, involving a filter filled with P_2O_5, a fibrous filter and a cooled filter filled with purified glass powder, as well as a tool for controlling the temperature and measurement of the gas flow. At the outlet of the TC column α-spectrometric carbon filters were installed.

The typical experimental conditions were as follows: flow rate ($He+O_2$) 20 cm^3 min^{-1}, residual content of H_2O vapour $c_{H_2O} < 10^{-5}\%$, starting temperature of the thermogradient section 610±10 °C, final temperature −165 °C (cooling with liquid nitrogen), temperature gradient −19 °C cm^{-1}, duration of each experiment 30 min. After completion of each experiment, the thermogradient section of the TC column was cut into equal portions and their inner surface was treated with a suitable acid . In a similar manner, a washing from the used initial sample was obtaind and α-sources were prepared from the resulting solutions. They and the alpha spectrometric filters were measured with an α-spectrometer that consisted of a silicon surface barrier detector and related electronic units. The duration of the measurements of the low α-activity sources was as long as 2-3 days. Data handling allowed the determination of the adsorption temperature of an isolated compound, T_a. The value of T_a and other TC parameters were used for calculating the value of $-\Delta H_a^0$. The calculations were performed by using equation 15 from Eichler and Zvara.[6] The distribution of α-activity along the TC column was presented in the form of a thermochromatogram.

2.2 Neptunium

The TC properties of UO_3 and PuO_3 are similar, as exemplified by the data presented in the introduction. The formation of neptunium is not described in the literature. With allowance made for the similarity of the chemical properties of U, Np and Pu and based on the formation of volatile uranium and plutonium trioxides, we assume that formation of NpO_3 is also possible. One can suppose that given the similar properties of UO_3 and PuO_3 on the one hand and NpO_3 on the other, the centre of the adsorption zone of neptunium trioxide should be in the range 250±30 °C and the value of $-\Delta H_a^0$ for NpO_3 on quartz should be equal to 125±7 kJ mol^{-1}. In the present experiments, the initial sample was SiO_2 powder (d=150-200μm) with trace quantities of ^{237}Np deposited on its surface. Helium was used as carrier gas and oxygen as a reagent. Its concentration, c_{O_2}, was 50% and 1% (by volume). The data were expressed as a percentage of the total α-activity of ^{237}Np found on the thermogradient section. The yield of 'volatile' neptunium was symbolized by Y. The results of the experiments are shown in Figure 1. In both cases, neptunium formed three volatile compounds, which were adsorbed at 400-450 °C, 220-270 °C and 30-90 °C. One can see that deposition temperature of the second compound is actually equal to the calculated adsorption temperature of neptunium trioxide. Such a coincidence is possible, because of the similarity of the volatilities of UO_3, PuO_3 and NpO_3. The value of $-\Delta H_a^0$ for NpO_3 on quartz, based on the experimentally found adsorption temperature, was 124±6 kJ mol^{-1}. This agreed well with our estimation and with the values for UO_3 and PuO_3. The first TC peak was attributed to the deposition of a less volatile oxide NpO_2, as its adsorption temperature (425±25°C) was in a close agreement with the adsorption temperatures of UO_2 and PuO_2, and the value of $-\Delta H^0_a NpO_2$, which was equal to 167±6 kJ.mol^{-1}, agreed with the values of $-\Delta H^0_a UO_2$ and $-\Delta H^0_a PuO_2$.

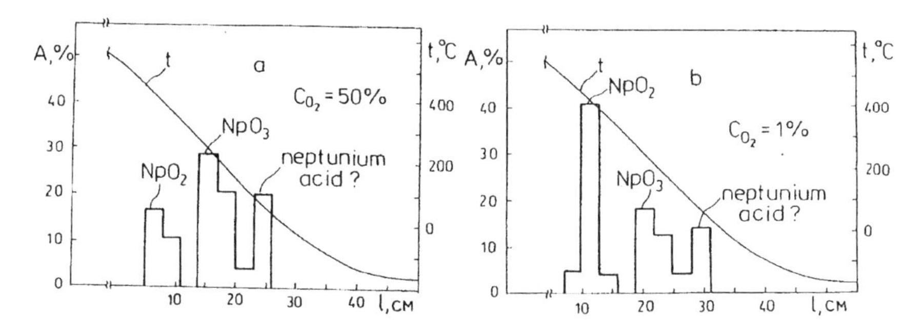

Figure 1. *Thermochrotograms of volatile neptunium compounds at different oxygen concentrations. Y=35% (a) and Y=3% (b).*

Our interpretation was supported by the chemical behaviour of the produced oxides. As can be seen from Figure 1b, with decreasing c_{O_2} the total yield of neptunium in the form of volatile compounds was also reduced from Y=35% (at c_{O_2}=50%) to 3% at c_{O_2}=1%. As this took place, the relative yield of NpO_2 increased noticeably, and the relative yield of NpO_3 reduced. A similar regularity was observed in studies of plutonium oxides.[3] The third adsorption zone, placed at 30-90°C, might be due to the formation of neptunium acid. The volatility of this compound is noticeably higher than the volatilities of H_2UO_4 and supposed H_2PuO_4.[3] Therefore, the isolated acid could have the chemical formula $HNpO_4$.

3 CONCLUSION

The formation of NpO_3 in the gas phase has been found for the first time. The volatility of NpO_2 has been observed at a relatively low temperature (<500 °C). It is believed that neptunium acid ($HNpO_4$) could be formed.

References

1 V.P. Domanov, G.V.Buklanov and Yu.V. Lobanov, Abstact Book of the 7th Inter. Conf. on Chemistry and Migration. Behavior of Actinides and Fission Products in the Geosphere, Lake Tahoe, C.E.A.Palmer (Ed.), 1999, 164.
2 V.P. Domanov, G.V. Buklanov and Yu.V. Lobanov, *Radiokhimiya*, 2002, **44**, 106.
3 V.P. Domanov, G.V.Buklanov and Yu.V. Lobanov, *J.Nucl.Sci.Technol.*, 2002, **Suppl.3**, 579.
4 C. Ronchi, F. Capone, J.V.Colle and J.P. Hiernaut, *J. Nucl. Mater.*, 2000, **280**, 111.
5 B. Eichler and V.P. Domanov, *J.Radioanal. Chem.*, 1975, **28**, 143.
6 B. Eichler and I. Zvara, *JINR Commun.*, 1975, P-12-8943, Dubna, (in Russian).

SISAK LIQUID-LIQUID EXTRACTION STUDIES OF RUTHERFORDIUM AND FUTURE PLANS TO STUDY HEAVIER TRANSACTINIDES

J.P. Omtvedt[1], J. Alstad[1], T. Bjørnstad[1], C.E. Düllmann[3,4], C. M. Folden III[3,4], K.E. Gregorich[3], D.C. Hoffman[3], H. Nitsche[3,4], D. Polakova[1], F. Samadani[1], G. Skarnemark[2], L. Stavsetra[1], R. Sudowe[3] and L. Zheng[1]

[1]Department of Chemistry, University of Oslo, P. O. Box 1033 - Blindern, N-0315 Oslo, Norway
[2]Nuclear Chemistry, Department of Materials and Surface Chemistry, Chalmers University of Technology, S-41296 Gothenburg, Sweden
[3]Nuclear Science Division, Lawrence Berkeley National Laboratory (LBNL), 1 Cyclotron Road, Berkeley, CA 94720, USA
[4]Department of Chemistry, University of California Berkeley, Berkeley, CA 94720-1460, USA

1 INTRODUCTION

During the last 10 years, the on-line liquid-liquid extraction (LLE) system SISAK[1] has been modified and enhanced to allow for studies of the chemical properties of the transactinide elements. A breakthrough was achieved in November 2000, when for the first time the SISAK system separated and detected a 4-s isotope of the transactinide element 104, rutherfordium.[1,2] This breakthrough was made possible by coupling the SISAK system to the Berkeley Gas-filled Separator (BGS) at the Lawrence Berkeley National Laboratory (LBNL). Using this technique, the rutherfordium recoil products could be separated from the beam and lighter elements produced in large quantities from undesirable reaction channels. Thus, the huge background was removed that previously prevented the liquid scintillation (LS) α-detection used by SISAK. This opened the way for a series of successful experiments.

Several experiments were performed with rutherfordium using the system, and techniques were refined to achieve accurate distribution-ratio measurements for LLE of transactinide elements. Parallel to this work an extraction chemistry for use with SISAK was developed which clearly differentiates between rutherfordium's homologues Zr and Hf.[3] A Rf experiment with this chemistry, preseparated in the Berkeley Gas-filled Separator and then transferred to the chemistry apparatus with a He/KCl gas-jet, was successfully performed in March 2003. A distribution value = $3.8^{+\infty}_{-2}$ was obtained when extracting Rf from 0.5 M H_2SO_4 with 0.02 M tri-octhyl amine in toluene.[4] An experiment with a new detector setup, enabling measurement of activity in both phases, is planned for the future.[3]

2 IMPROVED DETERMINATION OF DISTRIBUTION RATIOS

The LS detection of alpha activity with the best possible energy resolution and discrimination of beta and gamma activity prevents direct measurement of the alpha activity in the aqueous phase. This is due to the high quenching of the LS process when water and acids are mixed with the scintillation solution. Therefore, up to now the determination of activity in the aqueous phase has always been based on measurements of the total amount, R_{Tot}, of activity entering the SISAK system and the amount measured in the organic phase, i.e., $R_{Aq} = R_{Tot} - R_{Org}$. Unfortunately, the yield of the gas-jet transport system is quite variable and this leads to large uncertainties in the derived distribution ratios. A different approach is taken to determine the distribution ratio in the setup for future measurements. The activity in the aqueous phase is measured indirectly by performing a *second* extraction (Fig. 1). Depending on the second extraction agent, part of the activity in the aqueous phase is transferred to a second organic solution, where it is measured in the same way as the first organic phase.

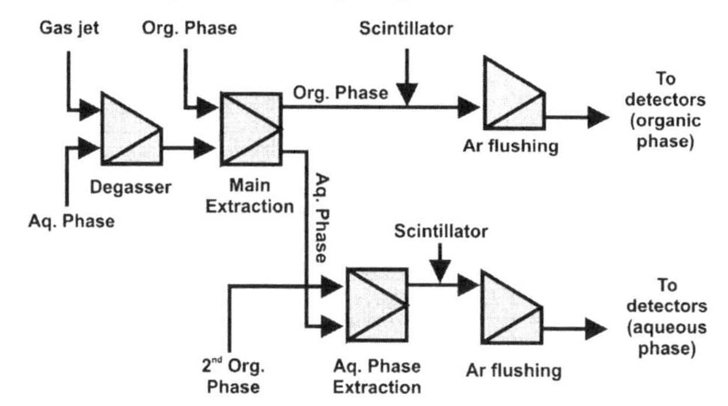

Figure 1 *Accurate measurement of distribution ratios by using a second extraction step and two sets of detectors.*

There are two ways to use this setup to measure the distribution ratio of the first extraction stage. The simplest way is to use an extractant in the second stage which has close to 100% extraction efficiency. The distribution ratio, D, will then be given by

$$D = \frac{R_{Org.1}}{R_{Aq.1}} = \frac{R_{Org.1}}{f \cdot R_{Org.2}} \tag{1}$$

where R denotes the activity in the phases indicated by the subscripts. The factor f is used to correct the $R_{Org.2}$ value if the extraction is not 100% in the second stage. A problem with this approach is that the factor f might not be very well known. In that case, an alternative is to use the same extractant in the first and second stage, i.e., both stages will have the same distribution ratio. The distribution ratio can then be calculated from the measured activity in the two organic phases according to:

$$D = \frac{R_{Org.1} - R_{Org.2}}{R_{Org.2}} \tag{2}$$

3 EXPERIMENTS WITH HEAVIER TRANSACTINIDE ELEMENTS

Based on experience gained in the series of successful rutherfordium experiments, the SISAK Oslo-LBNL-Gothenburg collaboration plans to advance to chemical studies of heavier transactinide elements in the liquid phase. Table 1 gives an overview of some of the possibilities.

Table 1 *Overview of possible SISAK transactinide experiments*

Element	Background suppression method, transport system, and SISAK setup	Experimental facility, cross section, and reaction	Transactinide half life, assumed detection rate, and usable decay chain
Rf (Z=104)	Preseparation with BGS, two stage SISAK system	LBNL 88-inch cyclotron, 12 nb, $^{208}Pb(^{50}Ti^{12+},1n)^{257}Rf$	$T_{1/2}(^{257}Rf) = 4.7$ s, one event per hour ^{257}Rf - ^{253}No
Db (Z=105)	Preseparation with BGS, two stage SISAK system	LBNL 88-inch cyclotron, 3 nb $^{208}Pb(^{51}V^{12+},1n)^{258}Db$	$T_{1/2}(^{258}Db) = 4.4$ s, one event per 4 hours ^{258}Db - ^{254}Lr - ^{250}Md
Sg (Z=106)	Preseparation with BGS, two stage SISAK system	LBNL 88-inch cyclotron, 0.1 nb (estimated), $^{244}Pu(^{26}Mg,5n)^{265}Sg$	$T_{1/2}(^{265}Sg) = 7$ s., one event per day, ^{265}Sg - ^{261}Rf - ^{257}No
Bh (Z=107)	Not currently possible with SISAK, but extraction system has been developed	LBNL 88-inch cyclotron $<$ 10 pb (estimated), $^{244}Pu(^{27}Al,4n)^{267}Bh$	$T_{1/2}(^{267}Bh) = 15$ s.,
Hs (Z=108)	Aerosol-free gas jet, transport as volatile HsO_4, SISAK back-extraction system	Rotating actinide target at GSI, 7 pb, $^{248}Cm(^{26}Mg,5n)^{269}Hs$	$T_{1/2}(^{269}Hs) = 14$ s., one event per day ^{269}Hs - ^{265}Sg - ^{261}Rf

4 CONCLUSION

The SISAK system in conjunction with the BGS has proven to be a reliable system for investigating the solution chemistry of rutherfordium. The new setup with a second extraction stage and two detector arrays enables a more accurate determination of distribution values. This, in connection with LBNL's plans to irradiate ^{244}Pu-targets in the BGS, allows SISAK to be used for the exploration of heavier transactinide elements. It is also possible to take advantage of the volatility of HsO_4 to perform a Hs experiment at the Gesellschaft für Schwerionenforschung (GSI) in Darmstadt, Germany.

References

1 J.P. Omtvedt, J. Alstad, H. Breivik, et al., *J. Nucl. Radiochem.Sci.*, 2002, **3**, 121.
2 L. Stavsetra, K.E. Gregorich, J. Alstad, H. Breivik, K. Eberhardt, C.M. Folden, T.N. Ginter, M. Johansson, U.W. Kirbach, D.M. Lee, M. Mendel, L.A. Omtvedt, J.B. Patin, G. Skarnemark, R. Sudowe, P.A. Wilk, P.M. Zielinski, H. Nitsche, D.C. Hoffman and J.P. Omtvedt, *Nucl. Instr. Meth. A*, 2005, **543**, 509.
3 D. Polakova et al., *this publication.*
4 L. Stavsetra et al., *2003 Annual Report*, Nuclear Science Division, LBNL 2004, LBNL-55057.

DEVELOPMENT OF A SYSTEM FOR CHEMICAL STUDIES OF RUTHERFORDIUM BY LIQUID-LIQUID EXTRACTION FROM SULPHURIC ACID SOLUTIONS WITH SISAK

D. Polakova[1], J. Alstad[1], T. Bjørnstad[1], G. Skarnemark[2], L. Stavsetra[1], R. Sudowe[3], L. Zheng[1], D.C. Hoffman[3], H. Nitsche[3] and J.P. Omtvedt[1]

[1]Department of Chemistry, Univ. of Oslo, P. O. Box 1033 Blindern, N0315 Oslo, Norway
[2]Nuclear Chemistry, Department of Materials and Surface Chemistry, Chalmers University of Technology, S-41296 Goteborg, Sweden
[3]Nuclear Science Division, Lawrence Berkeley National Laboratory (LBNL), 1 Cyclotron Road, Berkeley, CA 94720, USA

1 INTRODUCTION

The liquid-liquid extraction system SISAK coupled to the Berkeley Gas-filled Separator (BGS) and liquid-scintillation (LS) detectors has been successfully used for extraction of the transactinide element rutherfordium (Z=104) from nitric acid solution with dibutyl-phosphoric acid in toluene.[1,2] The aim of the work described here was to find an extraction system better suited for studying the chemical properties of Rf in more detail. The system should focus on differentiating between the Zr- and Hf-like behaviours of Rf. Since Zr and Hf, the lighter homologues of Rf, behave quite similarly due to the lanthanide contraction, comparatively small differences in the binding atomic-wave-functions will be highlighted in such an experiment.

Extraction of Zr and Hf from sulphuric acid into toluene by ion-pair formation with tri-n-octylamine (TOA) was selected as a suitable system. The extraction kinetics are quite fast and literature reports indicate that a good separation between Zr and Hf can be achieved.[3,4,5] A successful pilot experiment was performed at the Lawrence Berkeley National Laboratory (LBNL) in the USA to check the suitability of the chemical system for the extraction of Rf with SISAK.[6]

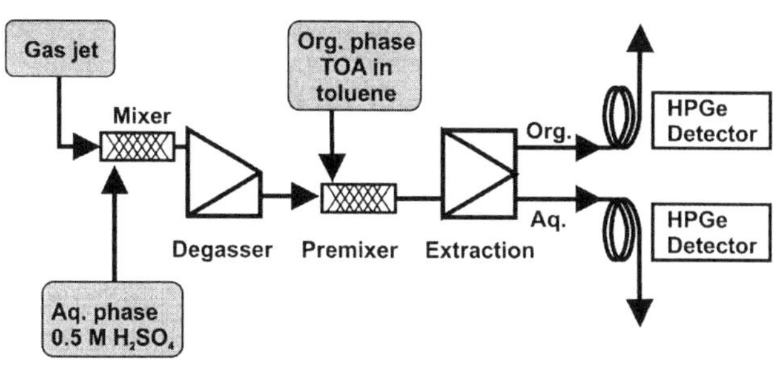

Figure 1 *SISAK set-up*

2 EXPERIMENTAL SET-UP AND PROCEDURE

On-line experiments with a He/KCl gas-jet and a SISAK setup similar to that used for Rf experiments at LBNL were performed at the Oslo Cyclotron Laboratory. 3.25-min ^{169}Hf and 14-s ^{87}Zr were produced simultaneously using a ~200 nA, 45-MeV $^{3}He^{2+}$ beam bombarding a target with strontium oxide electrodeposited onto a thin Yb-metal foil. The gas jet is fed into a mixer/degasser, which transfers the radionuclides to an acidic aqueous solution. Subsequently an organic solution containing the extraction agent (TOA) is mixed with the aqueous solution in a static mixer. The degree of extraction depends on the chemical properties of the element under study. Finally, the two phases are separated in a centrifuge. The setup is shown schematically in Figure 1. The activities of the outgoing phases were measured continuously in tube coils placed in front of HPGe detectors.

A new addition to this set-up is the small mixer in front of the centrifuge; earlier SISAK setups only used a Y-connector at this stage. The mixer ensures that the phases are mixed thoroughly enough to achieve equilibrium conditions in the liquid-liquid extraction process. As for most of the SISAK equipment, the mixer is constructed of PEEK (PolyEtherEtherKetone). The mixer, usually refereed to as the *premixer*, is constructed as a 4.3-cm long tube with an inner diameter of 0.8 cm and filled with PEEK wool. For any given chemistry it is important to determine the necessary premixer length, since this will depend on the kinetics of the selected extraction chemistry.

Distribution ratios were measured for different acid and TOA concentrations, and also for different lengths of the premixer. The flow rates for phases were kept constant at 0.4 mL/s. The organic phase was at room temperature and the aqueous phase was preheated to about 75 °C. The presented results are the mean value of between two to seven parallel measurements, and the indicated uncertainty is the standard deviation calculated from the parallels.

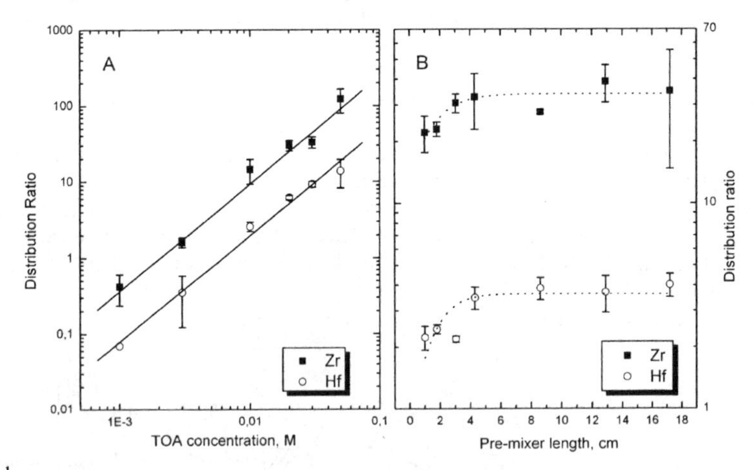

Figure 2 ***A*** *- Dependence of D value on TOA concentration for 0.5 M* H_2SO_4*:* ***B*** *- Dependence of D value on the length of PEEK premixer for 1 M* H_2SO_4 *and 0.05 M TOA at 0.4 mL/s flow for both phases.*

3 RESULTS AND DISCUSSIONS

Distribution ratios for 0.5 M H_2SO_4 increased linearly on a log-log scale with the TOA concentration (Figure 2A). As expected, Zr is extracted to a higher degree than Hf, most likely due to the more stable complexes formed with the sulphate/bisulphate by Zr.[3] The average separation factor for the measured TOA-concentration range is 6. The distribution ratio for Rf must then be measured with an uncertainty less than about 30% to be able to distinguish between Zr- or Hf-like behaviour. This can be achieved with the current SISAK system and production rates at LBNL within about 48 hours of measurement time for each distribution value, provided the distribution value is within the range 0.1 and 10.[7]

The results of the measurement with PEEK premixers of different lengths (Figure 2, panel B) show that for a premixer with a length of 4.3 cm, the distribution ratios for both Zr and Hf do not increase further with increasing length of the premixer. Thus, this is the minimum length for 0.4 mL/s flow rates. The measured hold-up time for this premixer length and flow was 0.5 s.[8] This will not lead to significant decay loss for experiments with 4-s ^{257}Rf, but ensures that the experiments will be performed at equilibrium conditions.

4 CONCLUSION

The difference in TOA extraction of Zr and Hf from sulphuric acid solutions is large enough to make SISAK experiments feasible for distinguishing between Zr- or Hf-like behaviour of element 104. From the extraction sequence, properties such as the stability of the extracted complexes can be determined.

An on-line Rf experiment is planned for the future at LBNL. Based on the presented data, this experiment will use 0.5 M sulphuric acid and 0.003 M and 0.01 M TOA solutions.

Acknowledgements

Financial support from the Norwegian Research Council, project number 148994/432 and 134882/432, is gratefully acknowledged. We wish to thank Eivind A. Olsen at the Oslo Cyclotron Laboratory for operating the cyclotron.

References

1 J.P. Omtvedt, J. Alstad, H. Breivik, et al., *J.Nucl.Radiochem.Sci.*, 2002, **3**, 121.
2 L. Stavsetra, K.E. Gregorich, J. Alstad, H. Breivik, K. Eberhardt, C.M. Folden, T.N. Ginter, M. Johansson, U.W. Kirbach, D.M. Lee, M. Mendel, L.A. Omtvedt, J.B. Patin, G. Skarnemark, R. Sudowe, P.A. Wilk, P.M. Zielinski, H. Nitsche, D.C. Hoffman and J.P. Omtvedt, *Nucl. Instr. Meth. A*, 2005, **543**, 509
3 G.A. Yagodin, O.A.Sinegribova and A.M. Chekmarev, *ISEC*, 1971, 1124.
4 T. Sato, *J.inorg.nucl.Chem.*,1974, **36**, 2585.
5 N.R. Das and S. Lahiri, *J.Radioanal.Nucl.Chem.*, 1994, **181**, 157.
6 L.Stavsetra, *2003 Annual Report*, Nuclear Science Division, LBNL 2004, LBNL-55057.
7 J. P. Omtvedt et al, *this publication.*
8 D. Polakova et al, to be published.

INFLUENCE OF KCl ON THE EXTRACTION OF RUTHERFORDIUM HOMOLOGUES HAFNIUM AND ZIRCONIUM FROM SULPHURIC ACID WITH TRI-OCTYLAMINE IN TOLUENE

L. Zheng, J. Alstad, T. Bjørnstad, D. Polakova, L. Stavsetra and J.P. Omtvedt

Department of Chemistry, Univ. of Oslo, P.O. Box 1033 Blindern, N-0315 Oslo, Norway

1 INTRODUCTION

A successful rutherfordium pilot experiment performed at Berkeley (LBLN)[1] used a new Rf solvent extraction chemistry based on aqueous sulphuric acid solution and trioctyl-amine (TOA) in toluene as the organic phase. This chemistry has currently been developed by Polakova et al.[2] for application in the on-line SISAK separation system.

In the chemical studies of Rf produced by $^{208}Pb(^{50}Ti,n)^{257}Rf$, the nuclide is separated from other reaction products in a gas separator (BGS) and recoils into a chamber (RTC) from which it is transferred to the chemistry system by a He jet. Potassium chloride aeorosol particles are used to transport the reaction products recoiling from the target to the SISAK on-line liquid-liquid extraction system. The nuclear reaction products attached to KCl aerosols in the He jet are dissolved in the aqueous phase by means of a stationary mixer. Inevitably, the potassium chloride used in carrying the reaction products enters the SISAK liquid-liquid extraction system. Literature data[3] indicate some influence of KCl on the distribution ratio for Zr and Hf in the H_2SO_4/TOA separation system. Under the conditions given for rutherfordium, the possible influence from KCl on the extraction of rutherfordium and its homologues, zirconium and hafnium, is unknown and needs to be checked for studies of the chemistry of rutherfordium.

2 EXPERIMENTAL SETUP AND PROCEDURE

Zirconium tracer, 64-d ^{95}Zr, was produced as a fission product by irradiating 1 mg UO_3, enriched (88%) in ^{235}U, with 10^{13} thermal neutrons per s for three days. The hafnium tracer, 42-d ^{181}Hf, was produced by irradiating 12 mg hafnium metal foil with thermal neutrons. The reactor at the Institute for Energy Technology (IFE) was used for these activations.

The hafnium metal foil was dissolved in 1 mL concentrated HNO_3 by adding 10 drops of hydrofluoric acid during heating in a polyethylene vial. Since Hf (and Zr) is strongly complexed by the fluoride ions in aqueous phase, fluoride was removed as HF by evaporating to dryness, the residue dissolved in concentrated nitric acid and again heated until dryness. Finally, the residue was dissolved in 5 M H_2SO_4. Preparation of the stock solution of ^{95}Zr was performed by dissolving the irradiated UO_3 in 0.5 M H_2SO_4. Then the solution was contacted with 0.05 M TOA in toluene, and the zirconium tracer was extracted into the organic phase.

The subsequent extractions were performed with equal volumes (3 mL) of organic and aqueous phase in test tubes at room temperature (20-25°C). The two phases were equilibrated in a mechanical shaker for 1 min. After phase separation, 2 mL of each phase were withdrawn and their activities were measured with an HPGe detector. Various amounts, up to 0.09 M, of KCl were added to the 0.25 M H_2SO_4 aqueous phase before extraction with 0.05 M TOA in toluene. The additions lower the measured distribution ratios of both hafnium and zirconium, as shown below. In addition to the effect of the Cl-ion on the extraction, possible interference of the K-ion was examined by performing similar extraction of hafnium and zirconium from 0.25 M H_2SO_4 containing HCl as well as LiCl in the same concentration range as that of KCl.

3 RESULTS AND DISCUSSION

According to Sato,[4] the equilibrium between TOA and sulphuric acid is expressed as:

$$H_2SO_4 + 2(R_3N)(org) \rightarrow (R_3NH)_2SO_4(org) \quad (1)$$

$$(R_3NH)_2SO_4\,(org) + H_2SO_4 \rightarrow 2(R_3NH)HSO_4\,(org) \quad (2)$$

at low acidities, the amine sulphate is first formed and at higher acidities additional acid is then extracted to form the amine bisulphate. R denotes the C_8H_{17}-branch in TOA, which is diluted in toluene to 0.05 M. The mechanism of the extraction of zirconium (IV) from sulphuric acid by TOA is similar to that for uranium (VI),[4] accordingly, the equation for zirconium extraction is

$$Zr(SO_4)_2 + 4(R_3NH)HSO_4\,(org) \rightarrow (R_3NH)_4Zr(SO_4)_4(org) + 2H_2SO_4 \quad (3)$$

at high acidities. The 0.25 M H_2SO_4 was titrated before and after equilibrium with an equal volume of 0.05 M TOA in toluene. The titration results indicate that an amount of 0.037 M H_2SO_4 was consumed by 0.05 M TOA. This implies that about half the amount of TOA is bound as sulphate and half is bound as bisulphate.

By adding KCl to the sulphuric acid solution the equilibrium between TOA and sulphuric acid is shifted. Titration of the aqueous phase after addition of 0.09 M KCl showed that only 0.029 M acid was transferred compared to 0.037 M without Cl-ions present. Thus, 0.008 mol per L less sulphuric acid was transferred. It is reasonable to assume that HSO_4^- is replaced by Cl-ions, since they have higher affinity than HSO_4^- towards TOA,[5] and since they cannot replace SO_4^{2-} ions. Furthermore, the R_3NHCl ion pair does not extract Zr and Hf under these conditions.

It is interesting to compare the measured distribution value from pure H_2SO_4 at a TOA concentration yielding similar concentrations of the amine sulphate and bisulphate as the 0.05 TOA contacted with the KCl/H_2SO_4 mixture. From the curve of the distribution ratio versus TOA concentration obtained by Polakova et al.,[6] the lowering of the distribution value corresponds to a factor of approximately 5 for both Zr and Hf. For the measured distribution ratios at zero KCl and 0.09 M KCl in aqueous phase the factor is approximately 7-8 for Zr which corresponds fairly well; for Hf, however, the factor is about 13 which shows that the effect of Cl^- is stronger than for Zr. The reduction in the distribution ratio with addition of HCl is somewhat less, and for Zr the factor is 5, i.e. similar to the estimated value, but for Hf the reduction is by a factor of 10. The effect of LiCl lies between those of HCl and KCl, as illustrated in Figure 1. The difference caused by Li^+, K^+, and H^+ ions is still under investigation.

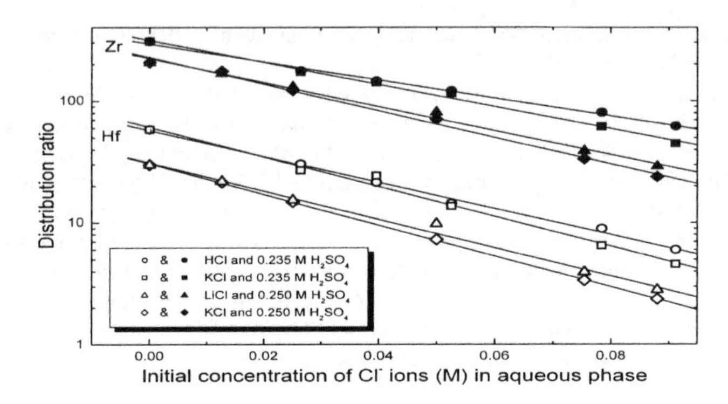

Figure 1 *Measured distribution ratios for Hf and Zr extracted into 0.05 M TOA in toluene from two different sulphuric acid solutions of 0.235 M and 0.25 M with the addition of various amounts of HCl, LiCl or KCl.*

A pure sulphuric acid solution was mixed with He/KCl under normal running SISAK conditions and the liquid was collected after passing the degasser. Samples of the solution were analysed by neutron activation. The results showed that the Cl^- content was quite low, of the order of 10^{-5} M. Based on the results obtained, less than a 1% reduction in distribution values will be observed for extraction from 0.25 M H_2SO_4 into 0.05 M TOA in toluene, provided the Cl-ion concentration is kept below 10^{-3} M.

4 CONCLUSION

The distribution ratios of both Hf and Zr decrease with the addition of KCl, Hf somewhat more than Zr. There is a small difference in the effect in the sequence: HCl < LiCl < KCl. Any distinct effect from KCl in the He jet on the extraction should not appear as long as the accumulated Cl^- concentration is no more than 10^{-3} M while the 0.25 M sulphuric acid solution is recycled over long time.

Acknowledgements

The financial support from the Norwegian Research Council is greatly appreciated.

References

1 J.P. Omtvedt, J. Alstad, H. Breivik, et al., *J. Nucl. Radiochem. Sci.*,2002, **3**, 121
2 D. Polakova et al. *SISAK Oslo Group Annual Report 2003*, University of Oslo
3 G.A. Yagodin, A.M. Chekmarev, *Ekstraktsiya. Teoriya, Sb.Statei* (1962), No.2,141-153
4 T. Sato, *J. Appl. Chem.*24,(1962);25,(1963) and T. Sato, *J. Appl. Chem.*15 (1965).
5 J. Rydberg and T. Sekine, *Solvent Extraction Equilibria. In: Principles and Practices of solvent Extraction* (J. Rydberg et al., ed.). Marcel Dekker, Inc., New York 1992.
6 D. Polakova et al., to be published.

THERMOCHROMATOGRAPHIC ADSORPTION STUDIES OF CURIUM AND BERKELIUM

S. Hübener[1], B. Eichler[2], and S. Taut[3]

[1]Institute of Radiochemistry, Forschungszentrum Rossendorf, D-01314 Dresden, Germany
[2]Laboratory for Radiochemistry and Environmental Chemistry, Paul Scherrer Institut, CH-5232 Villigen PSI, Switzerland
[3]Department Radiation Protection, Technische Universität Dresden, D-01069 Dresden, Germany

1 INTRODUCTION

The metallic state properties of the heavy actinides Fm, Md, and No, which are available only in trace amounts have been studied by thermochromatography in metal columns.[1,2,3] The thermochromatographic approach is based on analogies of the adsorption of metal atoms on metal surfaces with the desublimation of metals. The enthalpies of adsorption on metal surfaces are correlated to the enthalpies of sublimation and, moreover, related to the electronic properties of the adsorbed elements.[2] Thermochromatographic metal adsorption studies established that Fm, Md, and No are divalent metals. Isothermal gas chromatography was used to search for Lr as a p-element with a $[Rn]5f^{14}7s^2p_{1/2}$ ground state configuration, which was predicted on the base is of multidimensional relativistic Dirac-Fock calculations.[4] No evidence for Lr as a volatile p-element was found at a platinum column temperature of about 1275 K, which was far too low for studying Lr as a trivalent d-element with a $[Rn]5f^{14}6d7s^2$ ground state configuration.

In preparation for thermochromatographic Lr studies at temperatures up to 2200 K, we investigated in the present work the metal adsorption of Cm and Bk. Cm was chosen to model Lr as a trivalent d-element, and Bk as a representative of trivalent f-elements having a (divalent) $f^{n+1}s^2$ ground state configuration.

2 APPARATUS, MATERIALS, AND PROCEDURE

The thermochromatography setup, suitable for adsorption studies at temperatures up to 2200 K and the procedure are described in detail in Taut et al.[5]

$^{254}Es/^{250}Bk$ electroplated on tantalum[6] or ^{248}Bk implanted in zirconium foils were used as berkelium sources. ^{248}Bk was obtained by irradiating a ^{248}Cm target with heavy ions at the PSI Philips cyclotron. Amongst others, the nuclear reaction product ^{248}Bk recoiling from the target and curium isotopes which were sputtered from the target were caught in zirconium foils.

Sapphire tubes with an inner diameter of 3.5 mm were used to support the metal columns made of 25 μm foils. The helium carrier gas flow rate was 50 cm^3/min. The chromatography time was 12 min in case of experiments with $^{248}Bk/^{246}Cm$ and 30 min in the experiments with ^{250}Bk.

The nuclide distribution on the column was obtained from alpha and gamma spectrometric measurements of 1 cm column sections.

The standard enthalpy of adsorption, ΔH^0_{ads}, was calculated with the method presented by Taut et al.[7] Basically, this algorithm uses a third-law approach to determine both the standard adsorption entropy and enthalpy. To calculate two unknown quantities from one experiment, the method has to rely on secondary conditions. These are two empirical correlations, first between the hindrance factor, f_{hindr}, and the adsorption enthalpy, and second, between the adsorbate bonding force constant, f, and the adsorption enthalpy. f_{hindr} accounts for the non-ideal translation of the two-dimensional gas on the surface. On bcc metals the following empirical correlation exists:

$$\log(f_{hindr}) = 1.79 + 0.0062 \cdot \Delta H_{ads}[kJ/mol]$$

From experimentally determined adsorbate bonding force constants the correlation

$$\log(f[N/cm]) = -1.78 - 0.0036 \cdot \Delta H_{ads}[kJ/mol]$$

was obtained, from which the adsorption stretching frequency, ν, can be calculated:

$$\nu = \frac{1}{2\pi} \cdot \sqrt{\frac{f}{m}}$$

f_{hindr} and ν are the only unknowns in the calculation of the standard adsorption entropy:

$$\Delta S^0_{ads,mob} = \frac{1}{2}R + R \cdot \ln\left(\frac{f_{hindr} \cdot A \cdot \left(\frac{2J_{ads}+1}{2J_{gas}+1} \right)}{V \cdot \left(\frac{kT \cdot 2\pi \cdot m}{h^2} \right)^{\frac{1}{2}} \cdot \left\{ 1 - \exp\left(-\frac{h \cdot \nu}{kT} \right) \right\}} \right)$$

Here, R is the gas constant, A the molar standard surface area of the adsorption, J the electron system magnetic quantum number, V the molar standard volume of the substance to be adsorbed, k the Boltzmann constant, T the temperature, m the mass of the adsorbed atom, and h Planck's constant.

The adsorption enthalpy is calculated by solving the following equation, which is similar to that given earlier by Eichler and Zvara:[8]

$$t = \frac{T_0}{g \cdot u_0} \cdot \ln\frac{T_{ads}}{T_{start}} - \frac{4 \cdot T_0 \cdot \frac{V}{A} \cdot f_i \cdot \exp\left(\frac{\Delta S^0_{ads}}{R} \right)}{d \cdot g \cdot u_0} \cdot \int_{T_{start}}^{T_{ads}} \frac{1}{T} \cdot \exp\left(\frac{\Delta H^0_{ads}}{R \cdot T} \right) \cdot dT$$

Here, d is the inner column diameter, g the temperature gradient along the chromatography column, t the chromatography time, T_0 the ambient temperature, T_{ads} the adsorption temperature, T_{start} the temperature at starting position, and u_0 the carrier gas flow rate at T_0. The factor f_i accounts for the coverage dependence in the case of localized adsorption. In our work, mobile adsorption is assumed, so f_i equals 1. The whole calculation is carried out by the slightly modified computer program, TECRAD.[9]

3 RESULTS AND DISCUSSION

Experimental data as well as the results of calculations are listed in Table 1. An uncertainty of about 15 kJ/mol for ΔH^0_{ads} results from the fact that 1 cm column segments are counted in order to obtain the chromatograms. Taking the approximations in the evaluation of ν_{ads} and f_{hindr} into account this uncertainty should be even higher. Anyway, the trends known from the sublimation of the actinides are visible from the adsorption enthalpies too. The standard adsorption enthalpies of Bk on all metals under study are considerably lower

Table 1 *Experimental data, vibration frequencies, hindrance factors, adsorption entropies and enthalpies*

Experiment	Time [min]	T_{start} [K]	T_{ads} [K]	Gradient [K/cm]	ν_{ads} [s^{-1}]	f_{hindr}	ΔS^0_{ads} [J/(mol K)]	ΔH^0_{ads} [kJ/mol]
Cm on Nb	30	2031	1896	-24.8	2.2e12	0.08	-176	-464
Cm on Nb	12	1938	1899	-20.0	2.1e12	0.09	-175	-454
Bk on Nb	12	1889	1722	-21.3	1.6e12	0.24	-167	-388
Bk on Nb	12	1938	1671	-23.5	1.5e12	0.30	-164	-373
Bk on Nb	30	1881	1660	-21.9	1.6e12	0.25	-166	-386
Bk on Ta	30	1881	1700	-21.9	1.7e12	0.21	-168	-399
Bk on Ta	30	1881	1710	-21.9	1.7e12	0.20	-169	-403
Bk on Ti	30	1830	1700	-21.5	1.7e12	0.20	-169	-400
Bk on Ti	30	1778	1670	-21.4	1.6e12	0.23	-167	-392

(more negative) than those of the divalent heavy actinides, but not as low as that of the trivalent Cm. These results provide evidence that in analogy to its bulk properties, Bk is adsorbed as a metallic trivalent on Nb, Ta, and Ti.

The difference between the adsorption enthalpies of Bk and Cm (on Nb) of about 77 kJ/mol is lower than the $5f^9\ 7s^2 \rightarrow 5f^8\ 6d^1\ 7s^2$ promotion energy, which was spectroscopically determined to be 88 kJ/mol.[10] This discrepancy results from the uncertainties of the adsorption enthalpy values, in particular from those of Cm. Nevertheless, thermochromatographic adsorption studies are suitable for proving experimentally the predicted Lr ground state.

Acknowledgements

The authors are indebted for the use of ^{254}Es to the Office of Basic Energy Sciences, U.S. Department of Energy, through the transplutonium element production facilities at the Oak Ridge National Laboratory, managed by Lockheed Martin Energy Research corporation.

References

1 S. Hübener, *Radiochem. Radioanal. Letters,* 1980, **44**, 79.
2 S. Hübener and I. Zvara, *Radiochim. Acta*, 1982, **31**, 89.
3 S. Taut, S. Hübener, B. Eichler, A. Türler, H. W. Gäggeler, S. N. Timokhin and I. Zvara, *J. Alloys Comp.*, 1998, **271-273**, 316.
4 D. T. Jost, H. W. Gäggeler, C. Vogel, M. Schädel, E. Jäger, B. Eichler, K. E. Gregorich and D. C. Hoffman, *Inorg. Chim. Acta*, 1988, **146**, 255.
5 S. Taut, S. Hübener, B. Eichler, H. W. Gäggeler, M. Schädel and I. Zvara, *Radiochim. Acta*, 1997, **78**, 33.
6 B. Eichler, S. Hübener, N. Erdmann, K. Eberhardt, H. Funk, G. Herrmann, S. Köhler, N. Trautmann, G. Passler and F.-J. Urban, *Radiochim. Acta*, 1997, **79**, 221.
7 S.Taut, S. Hübener, B. Eichler, N. Trautmann and J.R. Peterson, *to be published.*
8 B. Eichler and I. Zvara, *Radiochim. Acta*, 1982, **30**, 233.
9 H. Funke, S. Hübener, A. Ross and B. Eichler, *Report FZR-43*, 1994, 53.
10 K.L. Vander Sluis and L.J. Nugent, *Phys. Rev. A*, 1972, **6**, 86.

SELF-DIFFUSION COEFFICIENTS AND STRUCTURE OF THE TRIVALENT *f*-ELEMENT IONS, Eu^{3+}, Am^{3+} AND Bk^{3+}, IN AQUEOUS SOLUTION

H. Latrous

Faculté des Sciences de Tunis, 2092 El-Manar, Tunisia

1 INRODUCTION

Self-diffusion coefficients, D, of the trivalent aquo ions of Bk have been determined in dilute (pH =2.5) $Nd(ClO_4)_3$-$HClO_4$ solutions at 25°C using an open end capillary method (O.E.C.M.). This method measures the transportation time of ions across a fixed distance. The variation of D versus the square root of the concentration of the solution was found to be non-linear in the concentration range studied. The limiting value, D_0, for Bk(III) ions at zero ionic strength is 5.95×10^{-6} cm^2s^{-1}. The plot of D = f ($\sqrt{C}$) for the Bk ions can be compared to data plotted for $^{241}Am^{3+}$ and $^{152}Eu^{3+}$ obtained under the same conditions. Moreover, it may be argued that the trivalent ions of Bk, Am and Eu in aqueous solutions have the same hydration number, as the 5*f* and 4*f* trivalent ions in the absence of hydrolysis, complexation or ion pairing at pH 2.5 should be the same.

A semi-quantitative extension of the Kim-Onsager[1] limiting law for metal ion concentrations up to 0.1M is proposed. The present study compares by analogy the solvation structure of Bk^{3+} ions in aqueous solution at pH 2.5 with that of the trivalent lanthanides, which provides an approach for predicting their transport properties.

2 EXPERIMENTAL

The O.E.C.M was used, and the experiments were performed at a temperature of 25 ± 0.01 °C.[2] Quartz capillaries of a fixed size (measured) were filled with an electrolyte solutions ($Nd(ClO_4)_3$-$HClO_4$) of concentration, C, that also contained the radioactive ions ($^{249}Bk^{3+}$, $^{152}Eu^{3+}$ or $^{241}Am^{3+}$). We denote by C0 (X,0) the total activity in the capillary at time, t = 0. After a diffusion period, t, the final average activity will be C(X,t). By solving Fick's equation with the proper limiting conditions, the ratio is C (X,t)/C0(X,0) and can be related to the self-diffusion coefficient, D, by the equation:

$$D_i = D_i^o \quad \frac{a\sqrt{C}}{(1+b\sqrt{C})}$$

Radioactivity measurements have been carried out separately for each f–element ion. For Eu and Am, which are gamma emitters, the measurements were obtained directly on the quartz capillary by using a hollow NaI crystal detector. For the beta emitter, Bk, the initial and final activities were measured by proportional counting, and these confirmed by alpha radiometry.

3 RESULTS AND DISCUSSION

The pH was optimized to avoid hydrolysis, pairing and/or complexation of the trivalent 4f and 5f ions.[3] Our results fit equation (1) nicely. The experimental self coefficients of diffusion were found to be in agreement with calculated values[4-6] (see Table 1).

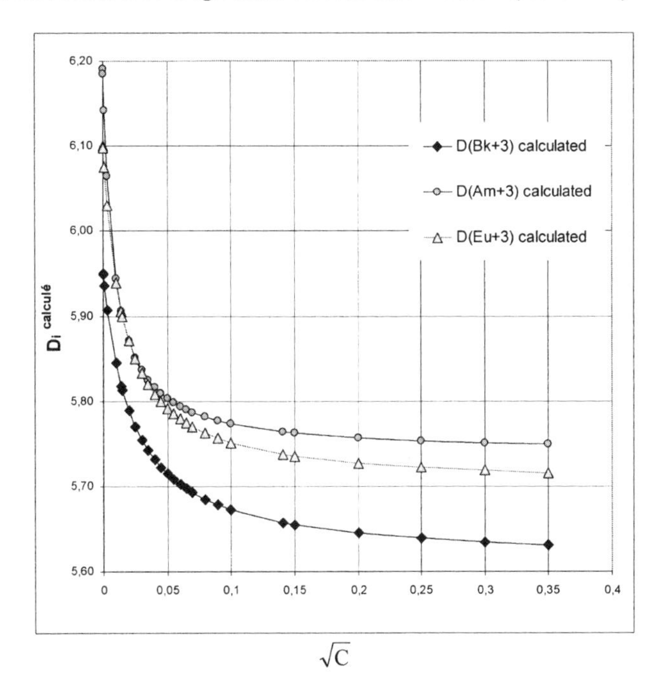

Figure 1 *Diffusion Values for the Selected f-Elements.*

Table 1 *Experimental (Exp) and Calculated (Cal) Diffusion Values.*

$\sqrt{C}$ (mol.l^{-1})	10^6 D (Bk^{+3}) Exp	10^6 (Am^{+3}) Exp	10^6 D (Eu^{+3}) Exp	10^6 D (Bk^{+3}) Cal	10^6 D (Am^{+3}) Cal	10^6 D (Eu^{+3}) Cal
0	5.95	6.19	6.10	5.95	6.19	6.10
0,0032	5.91	5.99	5.94	5.91	6.07	6.03
0,0141	5.82	5.95	5.90	5.82	5.91	5.91
0,0447	5.70	5.81	5.80	5.72	5.81	5.80
0,1000	5.69	5.78	5.78	5.67	5.77	5.75
0,1414	5.65	5.76	5.76	5.66	5.76	5.74

The variation of the self diffusion coefficient versus atomic number for the 5f elements is illustrated in Figure 2, using reported values.[4-6] The diffusion is found to decrease with Z for 5f elements, showing a strong hydration effect; the same results were demonstrated for 4f elements. The general trend of Di versus Z is very useful in predicting an unknown coefficient of self diffusion for a given Z.

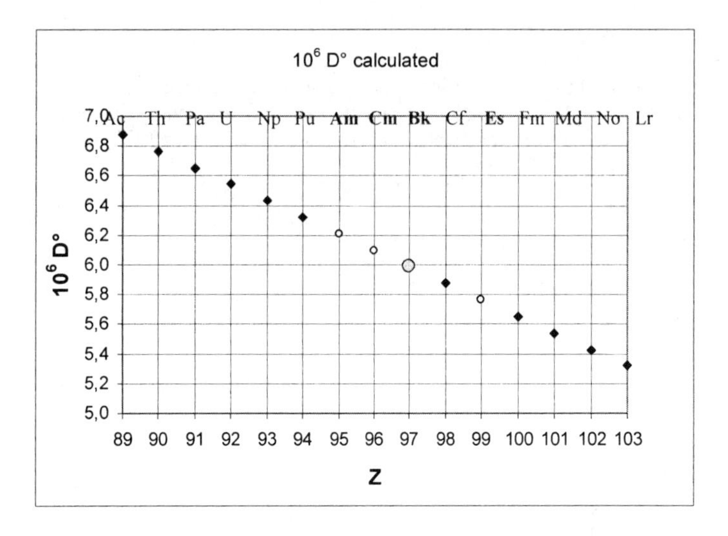

Figure 2 *Diffusion as a Function of Z (Experimental and calculated values are shown from refs. 2, 4,6).*

4 CONCLUSION

The present study contributes to our knowledge of the solvation behaviour of the trivalent f-elements in aqueous solution at pH 2.5. This information is helpful for predicting their thermodynamic properties.

Acknowledgements

This paper is dedicated to the memory of J. Oliver†.
Thanks are given to Dr R.G.Haire, member of the Chemistry Division of Oak Ridge National Laboratory, Oak Ridge, TN, USA. for his help and fruitful discussions.
Thanks are also given to Prof. M. Chemla, G. Bouissiere, R. Guillaumont, P. Turq (Paris VI P.M., Curie), H. G. Hertz (Karlsruhe, Germany) and J. Barthel (Regensburg).

References

1 S.K. Kim and L. Onsager, *J. Phys. Chem.*, 1957, **61**, 215.
2 H. Latrous, *Rev. Fac. Sci. Tunis*, 1981, **1**, 75.
3 J.P. Young, R.G. Haire, R.L. Fellows and J.R. Peterson, *J. Radioanal. Chem.*, 1978, **43**, 479.
4 H. Latrous, M. Ammar, J. M'Halla, *Radiochem. Radioanal.Lett.*, 1982, **53**, 33.
5 F. H. David and B. Fourest, *New J. Chem.*, 1997, **21**, 167.
6 H. Latrous, J. Oliver and M. Chemla, *Zeitschrift, fur Physikalishce Chemie, Neue Folge*, 1998, **159**, 195.

CERAMIC PLUTONIUM TARGET DEVELOPMENT FOR THE MASHA SEPARATOR FOR THE SYNTHESIS OF ELEMENT 114

D.A. Shaughnessy[1], P.A. Wilk[1], K.J. Moody[1], J.M. Kenneally[1], J.F. Wild[1], M.A. Stoyer[1], N.J. Stoyer[1], J.B. Patin[1], J.H. Landrum[1], R.W. Lougheed[1], Yu.Ts. Oganessian[2], A.V. Yeremin[2], S.N. Dmitriev[2], T. Hartmann[3] and K.R. Czerwinski[3]

[1]Chemistry and Materials Science, Lawrence Livermore National Laboratory, Livermore, CA 94550, USA
[2]Flerov Laboratory of Nuclear Reactions, Joint Institute for Nuclear Research, Dubna, Russia
[3]Department of Chemistry and Harry Reid Center, University of Nevada- Las Vegas, Las Vegas, Nevada 89154-4009 USA

1 INTRODUCTION

We are currently developing a Pu ceramic target for the MASHA mass separator. MASHA will use a Pu ceramic target capable of tolerating temperatures up to 2000 °C. Reaction products will diffuse out of the target into an ion source, and be transported through the separator to a position-sensitive focal-plane detector array for mass identification. Experiments on MASHA will allow us to make measurements that will cement our identification of element 114 and provide data for future experiments on the chemical properties of the heaviest elements. In this study $(Sm,Zr)O_{2-x}$ ceramics are produced and evaluated for studies on the production of Pb (homologue of element 114) by the reaction of Ca on Sm. This work will provide an initial analysis on the feasibility of using ZrO_2-PuO_2 as a target for the production of element 114.

1.1 The MASHA Separator

The MASHA (Mass Analyzer of Super Heavy Atoms) on-line mass separator is currently under development at the Flerov Laboratory of Nuclear Reactions at JINR. This separator is expected to have a number of improvements over existing recoil separators, and will provide at least a ten-fold increase in the production and detection rate for element 114. It will allow unambiguous mass identification of super heavy nuclei with a mass resolution below 1 amu.[1] An improvement in MASHA is the use of a thick, Pu ceramic target, heated to a temperature of approximately 2000 °C. The ceramic will see a large range of beam energies and a larger percentage of the excitation function will be sampled.

The target will be a combination actinide target and the ion source of the CERN-ISOLDE type.[2,3] Reaction products will diffuse out of the heated, porous target and drift to an ion source to be ionized and injected into the separator. After traveling through the separator, the products will impinge on a position-sensitive focal-plane detector array for mass measurement. Initial tests will use surrogate products, but ultimately, element 114 experiments will be performed using ceramics containing ^{244}Pu to be irradiated by ^{48}Ca ions.[4]

The Pu target for MASHA must meet several requirements; it must be physically stable and able to withstand 2000 °C without melting or undergoing large thermal expansion. The target phases must be thermodynamically stable over the temperature range without undergoing phase transitions. The Pu vapor pressure should be low and the diffusion rate of reaction products from the target must be fast enough to allow measurement of short-lived products.

2 CANDIDATE PLUTONIUM COMPOUNDS

There has been previous work on the fabrication and characterization of actinide containing ceramics including solid solutions of different actinides.[5,6] Various zirconia containing ceramics with actinides have been fabricated and studied, including ZrO_2-PuO_2.[7,8,9] The properties of ZrO_2-PuO_2 have been examined by experiment and by model, concluding that $(Pu,Zr)O_2$ based targets should have suitable properties for the production of element 114. Solids solutions of ZrO_2-PuO_2 composed the best ceramic properties, ease of synthesis, and single phase over a large range. For these reasons, ZrO_2-PuO_2 will be chosen as the final target matrix.

3 METHODS AND RESULTS

3.1 Sm Surrogate Targets

In order to begin tests of the MASHA target – ion source combination, we have synthesized surrogate targets consisting of mixtures of ZrO_2-Sm_2O_3 in several different ratios (**Table 1**). The rare earth element Sm has an electronic configuration ([Xe] $4f^5$ $6s^2$) that is analogous to that of the actinide Pu ([Rn] $5f^6$ $7s^2$), such that the two may behave similarly in a ceramic matrix. More importantly, Sm is a target homologue of Pu, the reaction between Sm and the ^{48}Ca beam results in short-lived lead isotopes, which are homologues of element 114.

Table 1 *Composition of samples. The Sm and Zr values are in mol % and the stearate and PEG in wt %*

Sample #	$SmO_{1.5}$	ZrO_2	Zn Stearate	PEG
1	65	35	3	3
2	65	35	3	6
3	80	20	3	3
4	80	20	3	6
5	50	50	3	3
6	50	50	3	6

By bombarding these targets at elevated temperatures, the diffusion of lead reaction products out of the target can be measured as a function of Sm concentration and ceramic porosity. The results from these initial experiments will aid us in the preparation of subsequent Pu ceramics. The samples were prepared with stearate as a binder and polyethylene glycol added to increase void volume and sintered at 1600 °C. The samples were analyzed by X-ray powder diffraction (PANalytical Expert Pro) using Cu $K\alpha_1$-$K\alpha_2$ emission with the addition of NIST SRM 640a, silicon as a standard. The patterns were taken between 10° and 120° 2 theta with step sizes of 0.008° 2 theta and 25 seconds count time per step. The phase analysis was performed by applying Rietveld structure refinement (Bruker AXS Topas2) (Figure 1).

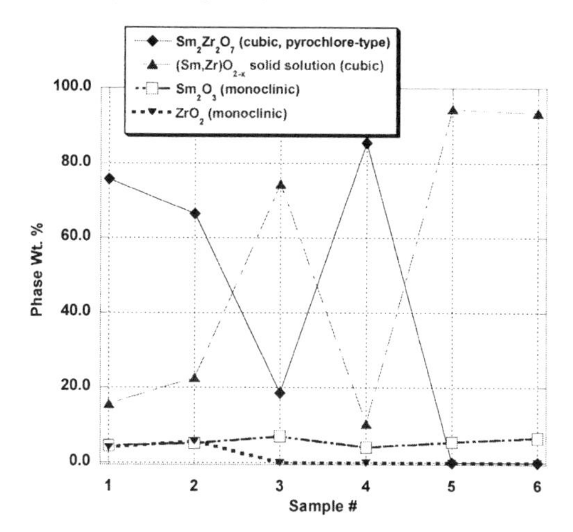

Figure 1 *Phases in the different Sm-Zr oxide samples*

4 CONCLUSIONS

Candidates for the MASHA target are currently being prepared and characterized. On-line tests with MASHA will begin with surrogate Sm targets, but subsequent irradiations with ^{242}Pu and ultimately ^{244}Pu will be performed. Once the target is prepared and tested, experiments designed to measure the mass of element 114 will begin.

Acknowledgments

This work was performed under the auspices of the U.S. Department of Energy by University of California, Lawrence Livermore National Laboratory under Contract W-7405-Eng-48.

References

1 Yu. Ts. Oganessian, V.K. Utyonkov, Yu.V. Lobanov, *et al., Nucl. Instrum. Meth. Phys. Res.*, 2003, **B204**, 606.
2 L. C. Carraz, I.R. Haldorsen, H.L. Ravn, M. Skarestad and L. Westgaard, *Nucl. Instrum. Meth.*, 1978, **148**, 217.
3 H. L. Ravn, *Phys. Reports,* 1979, **54**, 201.
4 Yu. T. Oganessian, V.K. Utyonkov, Yu.V. Lobanov, *et al., Phys. Rev. C*, 2000, **62**, 041604.
5 N.B. De Lima and K. Imakuma, *J. Nucl. Mat.*, 1985, **135**, 215.
6 V. Balek, *J. Nucl. Mat.*, 1988, **153**, 41.
7 D.F. Carroll, *J. Amer. Cer.*, 1963, **46**, 194.
8 R.B. Heimann and T.T. Vandergraaf, *J. Mat.Sci. Lett.,* 1988, **7**, 583.
9 W.L. Gong, W. Lutze and R.C. Ewing, *J. Nucl. Mat.*, 2000, **277**, 239.

Nuclear Fuels, Materials and Waste Forms

UNDERSTANDING AND PREDICTING SELF-IRRADIATION EFFECTS IN PLUTONIUM ALLOYS: A COUPLED EXPERIMENTAL AND THEORETICAL APPROACH

N. Baclet[1], P. Faure[1], G. Rosa[1], B. Ravat[1], L. Jolly[1], B. Oudot[1], L. Berlu[1], V. Klosek[1], J.L. Flament[2] and G. Jomard[2]

[1] CEA – Centre de Valduc, F-21120 Is sur Tille, France
[2] CEA – Centre Ile de France, F-91680 Bruyères le Châtel, France

1 INTRODUCTION

Among the actinides, plutonium is certainly the most complex element, due to its intermediate position in the series, where the 5f electrons lie at the edge between localization and itinerancy. Most interest is given to plutonium alloys that can be stabilized at room temperature in the face centered cubic δ-phase (fcc) by adding so-called deltagen elements such as aluminum, gallium, cerium or americium. The stabilization in the δ-phase is however limited, since a martensitic transformation can appear upon cooling or slightly increasing the pressure (polishing, machining...), leading to a mixture of (α' + δ) phases (α' having the same structure as the monoclinic α phase of pure plutonium, but containing atoms of solutes trapped in the lattice). Aging of plutonium makes this material even more complex. The decay of the different plutonium isotopes leads to the creation of displacement cascades from which vacancies, interstitials and clusters of these remain after thermalization. Stable defects created at an atomic scale can then diffuse and affect macroscopic properties. In the present study, the time scale is given in displacements per atom (dpa) to allow comparison between samples with different isotopes. The complex electronic structure of pure and even more of δ-Pu alloys adds another difficulty, since for example, modeling of displacement cascades using molecular dynamics requires specific interatomic potentials.[1] Moreover, the difficulty and restrictions in handling plutonium make data on the properties of self-irradiation defects scarce.
Understanding aging in plutonium alloys then remains a real challenge that requires the development of very ambitious models and experiments, as detailed in the following.

2 EXPERIMENTAL DETAILS

δ-Pu alloys stabilized with different deltagen elements (gallium, cerium and americium) have been synthetized and characterized. PuGa and PuCe alloys were synthetized at CEA Valduc by induction melting and quenching down to room temperature, followed by heat treatment at 733K for 200 hours under secondary vacuum to homogenize the solute distribution. PuAm alloys were prepared at the Institute for Transuranium Elements by arc melting, and were casted down to room temperature into copper moulds.

For all the δ-Pu alloys mentioned, X-ray diffraction (XRD) revealed δ-monophased alloys. Rather sharp XRD peaks indicated a good homogeneity in solute distribution, which was confirmed for the PuGa alloys through quantitative concentration profiles determined by electron microprobe analysis.

3 CHANGES IN PHYSICAL PROPERTIES DURING AGING

3.1 Swelling

Swelling is commonly the most obvious consequence of radiation damage, whatever the material (fuel, irradiated steel...) and the scale of observation. Swelling was first reported in PuGa alloys by Chebotarev[2] who indicated a saturation in swelling after about 2.5 years. Recently, microscopic swelling has been carefully followed on fully homogenized PuGa alloys with several gallium contents, and showed saturation after only 0.08 dpa (i.e. 8 months), whereas swelling of segregated PuGa alloys seemed to saturate much more slowly.[3] The change in cell parameter has also been followed versus time on a binary δ-PuAm 18 at.% alloy and ternary δ-PuAm4.3Ga2.6 at.% alloy.[4] Due to the short half-time period of ^{241}Am, self-irradiation defects appeared more rapidly in alloys containing ^{241}Am compared to binary PuGa alloys. More precisely, defects are created at a rate of 0.96 and 0.31 dpa.year^{-1}, for PuAm 18 at.% and PuAm4.3at.%Ga1at.%, respectively, compared to a rate of about 0.1 dpa.year^{-1} for classical plutonium. No swelling has been measured before 0.08 dpa (Figure 1) for binary PuAm and ternary PuAmGa alloys.[4]

Whereas both PuGa, PuAm and PuAmGa alloys have the same crystallographic structure, their swelling behaviour during aging appears significantly different; this could suggest a strong influence of dose rate on swelling, as previously observed for other irradiated materials.[5]

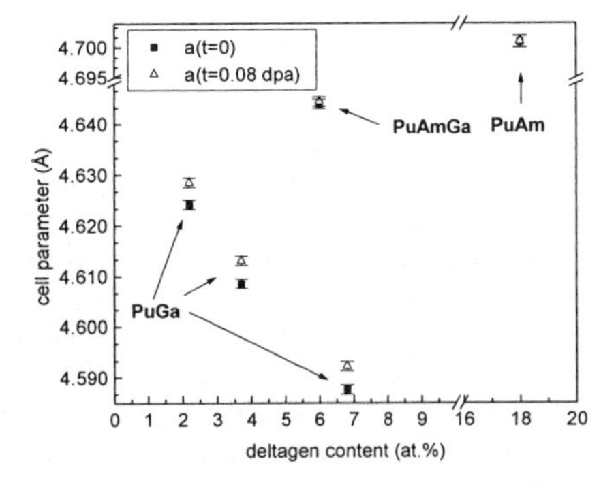

Figure 1 *Cell parameters for fresh and aged (swelling at 0.08 dpa, when saturation has been reached) δ-Pu alloys*

3.2 Phase stability

As mentioned in the introduction, the stabilization of δ-Pu alloys can be affected by cooling or applying pressure to the sample, which leads to the emergence of the martensitic α' phase, with a relative contraction of $\Delta V/V = -0.18.f_{\alpha'}$, where $f_{\alpha'}$ is the fraction of the α' phase that appears.[6] The effect of aging on phase stability in δ-Pu alloys is then of concern for technical applications.

The potential change in phase stability in δ-PuCe, PuGa and PuAm alloys has then been studied, especially for alloys with low solute content, which are the most likely to "destabilize". Whereas PuGa and PuAm alloys remain δ-monophased with time, the emergence of a monoclinic phase has been observed in three two-year old PuCe alloys (4.6, 6.1, and 8.1 at.%) by XRD (penetration depth of 2 to 3 microns). The atomic volume of the monoclinic phase did not depend on the cerium content and remained close to the atomic volume of the pure α plutonium (Table 1). Since the density of the samples remains unaffected, it seems that the monoclinic phase was localized at the surface of the sample, in a thickness of a few microns. The higher the cerium content, the lower the amount of the monoclinic phase, which is consistent with a better stability for alloys with a higher solute content (Table 1). The monoclinic phase could be reversed by heating at temperatures higher than 350°C.

Table 1 *Characteristics of the monoclinic phase that appeared with aging in δ-PuCe alloys*

Ce content (at. %)	Cell volume of monoclinic phase (Å^3)	Monoclinic phase content (volume %)
4.6	320.660	9.4
6.1	320.605	Not measured
8.1	320.240	0.7
pure Pu	320.300	-

Even if it remains difficult to identify the exact nature of the monoclininc phase (α or α'), the results show that aging can affect phase stability and must be carefully followed for other δ-Pu alloys.

3.3 Magnetic properties

Magnetic susceptibility measurements were performed on several δ-Pu alloys aged at room temperature. As illustrated in Figure 2 for the PuGa 7at.% alloy, all the alloys showed an excess magnetic susceptibility for temperatures lower than 75 K.[6] The same behaviour was observed for pure α-Pu and PuGa alloys, and more recently confirmed by McCall.[7] Annealing the sample allowed partial removal of this excess magnetic susceptibility, which suggested that self-irradiation defects might contribute to the observed phenomenon. Further annealings (longer or at higher temperatures) are planned to identify which phenomenon leads to this excess magnetic susceptibility (it must be noted here that heat treatment anneals self-irradiation defects, but the americium, uranium and helium remain in the sample).

The origin of the change in magnetic properties is not understood yet, but magnetic susceptibility results clearly indicate that the electronic properties of pure and alloyed plutonium are affected by aging.

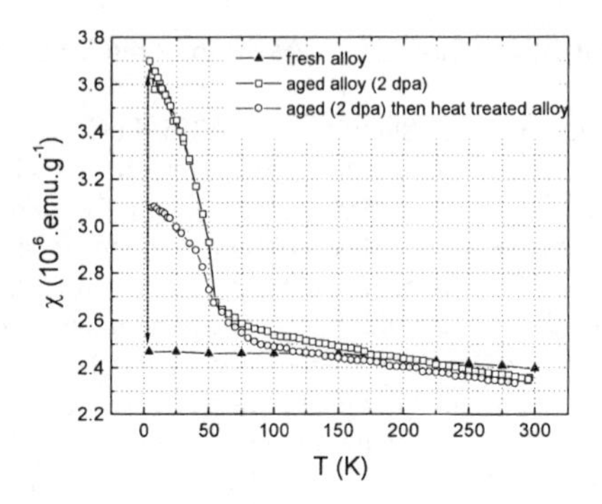

Figure 2 *Change in magnetic susceptibility versus temperature in an aged PuGa 7 at.% alloy (cooling rate of 1K.min^{-1}). Also plotted is the magnetic susceptibitity of the corresponding non aged alloy and the aged alloy after an annealing of 20 hours at 423 K. The magnetic susceptibility was calculated from measurements performed in a magnetic field of 0,6 and 1 Tesla.*

As illustrated through the different previous examples, aging of plutonium alloys affects several physical properties at very different space scales. This reveals the complexity of the mechanisms for the creation and diffusion of self-irradiation defects, and justifies the necessity to couple multi-scale modeling and experiments to better understand aging effects.

4 AGING MULTI-SCALE MODELING

Many parameters can affect changes in physical properties with aging. However, experiments on plutonium may be difficult, and only limited techniques can be used due to the handling constraints inherent to radiotoxic materials. Moreover, some physical values, cohesive energy, vacancy formation energy etc., that are necessary to describe aging phenomena cannot be measured experimentally. Multi-scale modeling might then help to identify specific defect properties that need to be obtained for a more realistic assessment of aging phenomena, from which specific experiments could be designed.

4.1 Theoretical approach

Aging effects range from the atomic scale, at which defects are created, to the macroscopic scale, which is accessed through experiments. A multi-scale approach, both in time and space, has been developed with a very strong connection between the different methods used to describe each scale. The objective was to compare modeling and experiments regarding the swelling data.

The ab-initio method allows the calculation of physical values, such as elastic constants, bulk modulus etc., that are essential input data for the interatomic potential used in molecular dynamics calculations.[8]

Displacement cascades were calculated using molecular dynamics with a progressive increase in the energy of the recoil atom. Simulations, requiring the use of a Modified Embedded Atom Model (MEAM),[1] were indeed very lengthy; as an example, a 10 keV displacement cascade termination, using 64 processors, required 18 months on the TERA

super-computer (Compaq/2560 proc). First results at 10 keV revealed that, in the first stage, displacement cascades lead to the formation of an amorphous core of 5 nm average radius with approximately 5000 Frenkel pairs. At the end of the defect recombination stage (a few nanoseconds), only a few interstitials remain out of the melting zone, with an equal number of mono-vacancies within the melting zone.
Further diffusion of the defects formed will be described with Monte-Carlo approaches (kinetic and mesoscopic), that allow the calculation of swelling induced by the remaining defects, and direct comparison to experimental swelling. In parallel, a more phenomenological approach, based on rate equations has been developed and will be compared to the Monte Carlo approaches.

Whatever the scale, modeling must be validated through comparison with experimental data. Dedicated experiments have then been designed to measure the plutonium physical properties required by the modeling approach.

4.2 Coupling with experiments

4.2.1 Single crystal elastic constants. The molecular dynamics simulation of displacement cascades depends strongly on the definition of the interatomic potential used for the calculations. This complex potential is adjusted with experimental data, such as single crystal elastic constants. These data have recently been measured for a PuGa 2at.% alloy through inelastic X-ray scattering,[9] only available at large facilities with limited experimental time. X-ray diffraction (XRD) experiments have then been designed in our laboratory to determine single crystal elastic constants for several δ-Pu alloys from XRD performed on a polycrystal.[10] This consists of applying a uniaxial stress to the polycrystalline sample through a bending device and measuring the corresponding diffracting plane strains by XRD. To determine the single crystal elastic constants, an analytical relation deduced from a self-consistent model is used. This model takes into account the mechanical compatibilities between each grain and the elastic polycrystalline matrix, and so is much more accurate than other simplified models, such as Voigt or Reuss models.[11,12] Since the experiment is performed out of a glove box, plutonium must be confined in a sealed plastic film and specific tape. The method (model, experimental device and confinement) has just been validated on a confined copper sample. The experimental results obtained are very close to those from ultrasonic techniques on a copper single crystal. The next step is a measurement on plutonium.

4.2.2 Displacement threshold energy. The displacement threshold energy E_d directly determines the number of defects induced by an α-decay. These data have however never been measured for plutonium alloys and only an empirical estimation (based on the relation $E_d \sim 175 k_B T_m$, where T_m is the melting point of the metal and k_B the Boltzmann constant) has been proposed, leading to a value of 14 eV.[13] E_d can be calculated by molecular dynamics methods, and comparing theoretical and experimental values provides a way to validate the molecular dynamics approach (code, interatomic potential).

A dedicated experiment, based on the measurement of electrical resistivity on a plutonium sample held at low temperature and irradiated by electrons has been designed. This experiment is based on the following principle: when the energy of electrons becomes sufficient to remove plutonium atoms from their stable site, isolated defects are created and an increase in electrical resistivity is observed. Further isochronal annealings should also allow the determination of the migration energy of the different defects, data that will contribute to the validation of both ab-initio and molecular dynamics calculations, and will be used as input data in Monte-Carlo methods.

Many threshold energy measurements have been made on classical metals and alloys. For plutonium, the experiment is much more difficult, since the sample needs to be confined, and the electrical resistivity is affected by self-irradiation. The experiment is under test (regarding both scientific and safety aspects) on gold and aluminum samples, before plutonium is studied.

4.2.3 Type of defects. Starting from the configuration (nature and location) of defects created at the end of a displacement cascade, Monte-Carlo simulations describe the diffusion and interactions. of these defects to determine the type of defects that remain after longer times, and then calculate the corresponding swelling. The experimental identification of the defects formed was then performed through the development of a specific technique, positron annihilation spectroscopy, which has been chosen for its sensitivity to vacancy-type defects in solids (monovacancies and larger vacancy clusters can be detected, with a detection limit of 1 atomic ppm). Immediate creation of vacancy clusters and rapid emergence of helium bubbles has then been proposed, as detailed elsewhere.[14]

5 CONCLUSION

The complexity of self-irradiation effects in plutonium alloys required the development of a multi-scale approach, both theoretical and experimental, to better understand the origin of the changes in physical properties, and even to predict aging effects at longer times. As self-irradiation involves very tiny effects, dedicated techniques have been developed, both to quantify these effects and to determine physical values as a support for the modeling approach. Preliminary results have been presented regarding self-irradiation effects and several physical values should be obtained for plutonium soon.

References

1 M.I. Baskes, *Physical Review B*, 2000, **62**, 23, 15532.
2 N.T. Chebotarev and O.N. Utkina, *Plutonium and other actinides*, North Holland Publishing Company, Amsterdam, 1976, p.559.
3 B. Oudot, PhD Thesis, Université de Franche-Comté, Besançon, France (2005).
4 M. Dormeval, PhD Thesis, Université de Bourgogne, Dijon, France (2001).
5 L.K. Mansur, *Journal of Nuclear Materials*, 1978, **78**, 156.
6 N. Baclet, M. Dormeval, P. Pochet, J.M. Fournier, F. Wastin, E. Colineau, J. Rebizant and G. Lander, *Journal of Nuclear Science and Technology*, 2002, **3**, 148.
7 S. McCall, M.J. Fluss, B. Chung, G. Chapline, M. McElfresh, D. Jackson, N. Baclet, L. Jolly and M. Dormeval, *This Volume*.
8 G. Robert, A. Pasturel and B. Siberchicot, *Physical Review B*, 2003, **68**, 075109.
9 J. Wong, M. Krisch, D.L. Farber, F. Occelli, A.J. Schwartz, T.-C. Chiang, M. Wall, C. Boro and R. Xu, *Science*, 2003, **301**, 1078.
10 N. Guibert, L. Gosmain, Ch. Valot, C. Valot, K. Inal, S. Berveiller, N. Baclet, P. Pochet, *Matériaux et Techniques*, 2004, **3-4**.
11 W. Voigt, *Lehrbuch der kristallphysik*, Berlin, 1910.
12 A. Reuss, Z. Angew, *Math. Mec.*, 1921, **2**, 49.
13 W.G. Wolfer, *Los Alamos Science*, 2000, **26**, 227.
14 B. Oudot, *This Volume*.

IDENTIFYING AND QUANTIFYING ACTINIDE RADIATION DAMAGE IN $ZrSiO_4$ MINERALS AND CERAMICS WITH NUCLEAR MAGNETIC RESONANCE

I. Farnan[1], H. Cho[2] and W.J. Weber[2]

[1]Department of Earth Sciences, University of Cambridge, CB2 3EQ, UK
[2]Pacific Northwest National Laboratory, Richland, WA 99352, USA

1 INTRODUCTION

This paper discusses how high-resolution solid-state nuclear magnetic resonance (NMR) can be used to characterise and quantify radiation damage in natural minerals and ceramic nuclear waste forms that contain actinides. The scientific goal is to identify the nature of the amorphous component of the radiation damaged material through similar approaches to those where NMR has been used to study glasses and amorphous materials.[1] NMR also allows the amount of amorphous material to be quantified as an atomic number fraction of the total. This is in contrast to traditional methods that express the damaged amorphous component as volume fractions of the total.

Very old mineral samples of $ZrSiO_4$ (zircon) containing ^{238}U and ^{232}Th with varying alpha radiation doses can be used to provide samples with differing levels of radiation damage.[2,3] Radiation damage due to the emission of an alpha particle by an actinide nucleus is believed to occur through two distinct processes.[4] The alpha particle itself (4.5 – 5.5 MeV), will mainly cause ionisations during its flight through a material, it is also thought to cause a few hundred atomic displacements (Frenkel defects) as it is stopped by collision with atomic nuclei. The recoil of the heavy actinide nucleus (70 -100 keV) is believed to cause the majority of the localised structural damage (amorphisation) as it creates a cascade of collisions with surrounding ions. The extent and nature of this 'displacement cascade' is the subject of extensive modelling by both ballistic[5,6] and increasingly molecular dynamics[7-15] methods. There is a profound need for experimental data to distinguish between these models.

2 METHODS AND RESULTS

2.1 ^{29}Si nuclear magnetic resonance to study radiation damage

High-resolution solid-state nuclear magnetic resonance requires that the sample to be investigated be reoriented macroscopically very rapidly at an angle of 54.7° to the magnetic field.[16,17] This reorientation with respect to the magnetic field removes orientation dependent line shifts from the spectra of polycrystalline or amorphous samples. The samples available to us were ancient (~570 million years old), well-characterised zircons containing up to several thousand ppm of U and Th. These were in the form of irregular monolithic pieces of several millimetres in dimensions. Originally, these would have been single crystals, but they have been subsequently transformed by radiation damage to a composite consisting of crystalline and amorphous domains within a single monolith.[18] These irregular shaped monoliths were packed together with a powder of similar density in the magic-angle spinning (MAS) rotor to provide a uniform and cylindrically symmetrical mass distribution. Experiments were carried out at spinning speeds of ~5 kHz. ^{29}Si MASNMR spectra were acquired on a Varian Chemagnetics Infinity spectrometer operating at a Larmor frequency of 79.5 MHz (B_o = 9.4 T). Typically, several hundred scans were acquired with pulse repetition delays of 300s and 1 μs (π/12) pulses with a sweep width of 20 kHz. ^{29}Si signal intensities were calibrated against a standard mass of $K_2Si_4O_9$ glass to determine the fraction of Si detected in each zircon sample. There is the possibility that some Si signals may go undetected due to paramagnetic broadening associated either with radiation induced paramagnetic defects or U^{4+} itself (other actinides will most likely be paramagnetic, too). In addition, some Si environments may exhibit very long T_1 components, with the result that they do not recover their magnetisation, for the given pulse repetition rate, to contribute to the measured signal in correct proportion to their abundance. There is evidence that signal loss is uniform between the amorphous and crystalline domains. Saturation recovery T_1 relaxation experiments with delays of 1s to 3000 s showed large increases in signal intensity but little change in relative peak areas from one delay to another.[19] It should be noted that Si signal quantification is almost never reported in the ^{29}Si NMR literature, and the detection rates found here of >80% may well be usual.

2.2 Mineral $ZrSiO_4$ samples

Figure 1(a) shows a typical ^{29}Si MASNMR spectrum of a lightly and a moderately damaged natural zircon. The narrow resonance at –81.6 ppm (lower spectrum) is expected for tetrahedral Si in a crystalline environment with 4 non-bridging oxygens from empirical correlations between structure and chemical shift. The broad resonance to low frequency

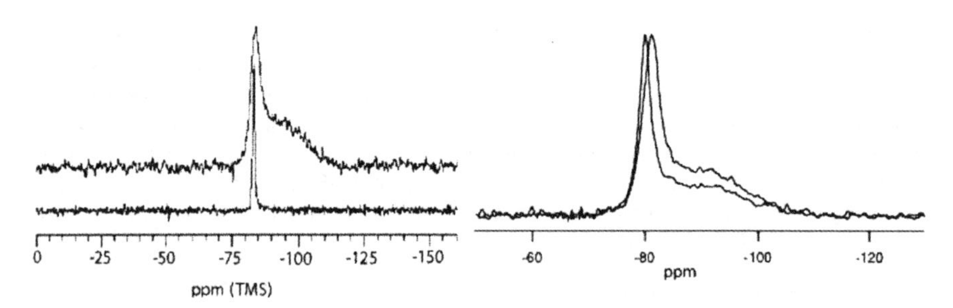

Figure 1 **(a)** *^{29}Si MASNMR spectra of a lightly damaged zircon (lower spectrum) and a moderately damaged zircon (upper spectrum).* **(b)** *^{29}Si MASNMR spectra of the zircon, Cam26 before and after annealing at 500°C for 1 hour. Integrated intensity of the narrow peak is 53 ±2 %. The peak narrows and shifts back to the position for undamaged zircon upon annealing.*

(more negative ppm) in the moderately damaged sample (upper spectrum) is from the amorphised part of the sample and contains contributions from tetrahedral Si with less than four non-bridging oxygens i.e., some of the SiO_4 tetrahedra have become polymerised. The crystalline peak in the moderately damaged sample has broadened and shifted (-83.7 ppm) due to the presence of defects in the crystalline component of the material.

The attribution of the line shape components to responses from amorphous and crystalline regions in the sample requires that the ^{29}Si NMR response from Si atoms adjacent to defects in the crystalline fraction be restricted to resonances that contribute to the broadened and shifted, but sharper peak. This peak has a maximum intensity between –81.6 ppm and –84.1 ppm in zircons of varying radiation dose. This was tested by carrying out short time scale annealing experiments. A heat treatment of 1 hour at 500 °C was applied to a zircon (Cam26) with an amorphous fraction of 47%. The resulting NMR spectra are shown in Figure 1(b). The spectra in Figure 1(b) are consistent with the idea that, during the short anneal at 500 °C, only the defects in the crystalline regions are removed and that the broad line corresponds to the response from discrete, amorphised regions. The apparent difference in the signal intensities of the amorphous components of the spectra before and after annealing is due to the normalisation to constant peak height.

A series of zircons was analysed on the basis that the amorphous and crystalline fractions were represented by the intensity of the two components of the line shape. Errors were evaluated based upon two criteria: the fraction of total Si intensity observed (determined from the mass of the sample) and in the case where all Si was observed the error in determining the areas under each component. The sharp, crystalline line was fitted to a pure Lorentzian or 20%-80% Gaussian-Lorentzian line, a broader Gaussian was used to represent the amorphous line shape in the fit. However, the fraction of amorphous material was determined by subtracting the sharper component intensity from the total line intensity. This method was used because the amorphous component was not represented very well by a single Gaussian. The fraction of silicon atoms present in the amorphous phase is plotted in Figure 2 as a function of cumulative alpha radiation dose. The damage

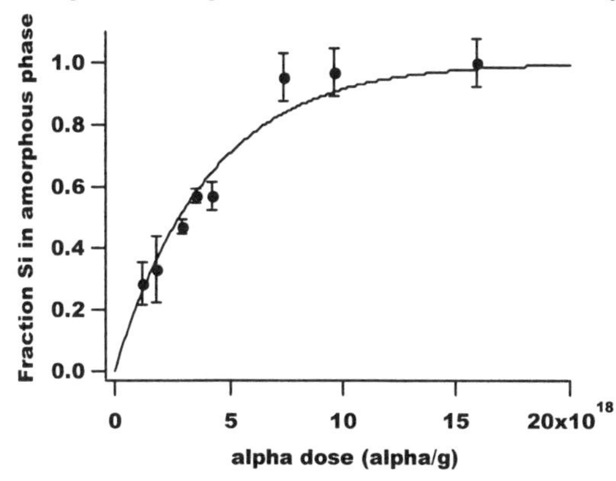

Figure 2 *The amorphous fraction (f_a) of zircon versus α-radiation dose. f_a is determined from the intensity of the amorphous component's Si NMR signal and thus represents the fraction of silicon atoms in the amorphous phase. The solid line represents a fit to direct amorphisation [$f_a = 1 - \exp(-B_d D)$, where D is the dose in alpha/g and B_d is the damage per alpha particle]*

accumulation can be fitted to a direct amorphisation mechanism with the damage fraction per alpha event, $B_d = 2.47 \pm 0.2 \times 10^{-19}$ g/α. The damaged volume may be estimated from the mass of damaged material for each event and the density of undamaged zircon (4.7 g cm^{-3}). The damaged volume is 5.225×10^{-26} m^3, which corresponds to a sphere of radius 2.32 nm. Hence a diameter of ~ 46 Å, which is in reasonable agreement with the reported observation of individual α-decay cascades as 50 Å features in transmission electron microscopy studies.[18,20]

2.3 Radiation damage: volume fraction versus number fraction.

Previous derivations of the amount of radiation damage in zircon have used a combination of x-ray diffraction measurements to determine the crystalline volume and density measurements to determine the total volume.[21,22] Data for the amorphous fraction arising from these treatments is presented in Figure 4 as a solid line. The density of zircon drops dramatically as a function of accumulated alpha dose, this manifests itself as a volume swelling of 18% for a completely amorphous sample compared with the undamaged crystal. However, this swelling does not occur in the expected exponential fashion for alpha doses up to ~ 2 x 10^{18} alpha/g. It was thus assumed that several decay events need to overlap in order for a region to be amorphised.[22] The line in Figure 3 fitted

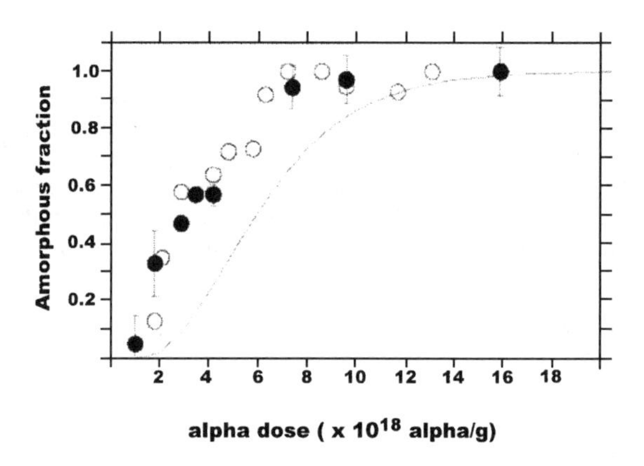

Figure 3 *Comparison of amorphous fraction vs alpha dose for zircon determined by different methods: NMR data (solid circles)[19] amorphous x-ray scattering (open circles)[23], crystalline x-ray and density determination (solid line)[3]*

best to a double overlap model of damage accumulation i.e., three events needed to overlap in order for a region to become amorphous. The NMR data and the results from measuring the x-ray scattering from the amorphous fraction[23] both appear to be consistent with the direct accumulation of damage (exponential behaviour). The interpretation we have invoked to reconcile these sets of data is that initially the radiation damaged material must be constrained by the crystalline matrix to have the same volume as the crystal. In this situation, at low doses, there will be no swelling. However, the number of atoms displaced by the alpha decay events will increase in the expected exponential fashion. This is what the NMR and the amorphous x-ray scattering experiments measure. At a certain dose there will be a percolation transition[24] and the amorphised areas will begin to connect and form a percolating cluster.[25] This point will coincide with the rapid increase in swelling that occurs over a small dose interval.

2.4 Ceramic $ZrSiO_4$

The use of mineral samples and NMR to investigate radiation damage in zircon provides a way to understand the degree and the nature of radiation damage as a function of dose. However, the range of samples is limited because uniform actinide concentrations are required and there may be uncertainties in the exact thermal history of very old samples. It would be very useful to investigate radiation damage in man-made samples of prescribed actinide content, and therefore dose, using NMR. Radioactive magic-angle spinning NMR can be achieved on highly radioactive samples if sufficient precautions are taken.[26] Figure 4 shows the ^{29}Si MASNMR spectra of two sintered ceramic zircon samples that contain 5 wt% ^{239}Pu and 10 wt% ^{238}Pu. The 5wt% ^{239}Pu sample has received 8.2 x 10^{16} α/g since it was prepared in 1981 and shows only the slightest hint of damage to low frequency of the narrow line in Figure 4(a). On the other hand, the sample containing ^{238}Pu has received an alpha dose almost two orders of magnitude greater (4.2 x 10^{19} α/g). This sample has become completely amorphous as evidenced by the broad line and absence of a crystalline peak in the region –81 to –84 ppm. In contrast to natural zircons that are completely amorphous, the centre of gravity of this amorphous line is around –103 ppm, whereas natural zircons appear to have a centre of gravity of –96 ± 1 ppm for doses from 6 x 10^{18} to 1.59 x 10^{18} α/g. There is also a shoulder to even lower frequency that may be indicative of phase separated SiO_2. This all suggests that for radiation damage by ^{238}Pu the amorphous state is different in terms of its local structure than that attained by a natural zircon damaged by ^{238}U decay. The very large difference in dose rate must be the origin of this effect. The half-life of ^{238}Pu is 87.7 years whereas for ^{238}U it is 4.47 billion years. In addition, the Pu sample contains 8 mol % actinide compared with typically ~ 1 mol % in fully amorphous natural zircons. It would seem that the massively higher dose rate (2.79 x 10^{10} α-decays/g/year versus 3.83 x 10^{18} α-decays/g/year) for the ^{238}Pu sample has led to some phase separation in the amorphous phase. The minor sharp peak in the ^{238}Pu zircon spectrum at –61.5 ppm has been attributed to Si that has been incorporated into ZrO_2. Obviously, the production of SiO_2 during phase separation would also lead to ZrO_2 being produced.

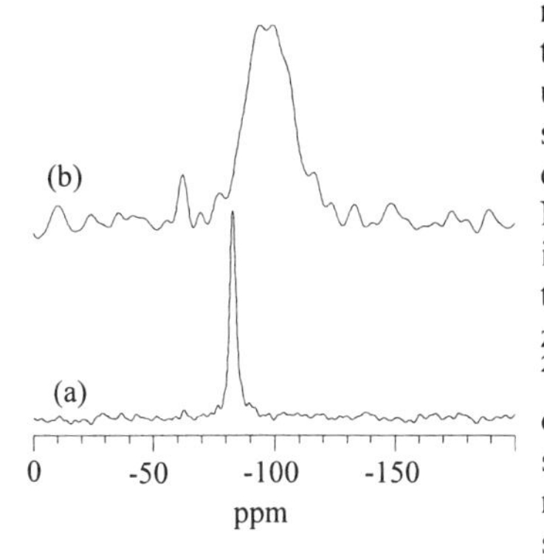

Figure 4 *^{29}Si MASNMR spectra of (a) $^{239}Pu_{0.04}Zr_{0.96}SiO_4$ and (b) $^{238}Pu_{0.08}Zr_{0.92}SiO_4$. sintered ceramic samples.*

3 CONCLUSION

High-resolution solid-state NMR is a powerful probe of the cumulative process of radiation damage in minerals and ceramics. The density and thus the local structure of the amorphous and crystalline phases do change significantly with dose. The ability of NMR to detect signal from amorphous and crystalline regions equally allows it to detect radiation damage at an early stage. It also provides an estimate of the amount of damage per alpha decay, which is a fundamental starting point for models of radiation damage accumulation.

Acknowledgements

IF acknowledges funding from UK HEFCE/EPSRC (JR98CAFA). The radioactive work was partially supported by the Environmental Management Sciences Program and the Laboratory Directed Research and Development Program at the Pacific Northwest National Laboratory, which is operated for the U.S. Department of Energy by the Battelle Memorial Institute under contract DE-AC06-76RLO-1830.

References

1 H. Eckert, *Prog. Nucl. Magn. Reson. Spectrosc.,* 1992, **24**, 159.
2 H.D. Holland and D. Gottfried, *Acta Cryst.,* 1955, **8**, 291.
3 T. Murakami, B.C. Chakoumakos, R.C. Ewing, G.R. Lumpkin and W.J.Weber, *Am. Miner.,* 1991, **76**, 1510.
4 W.J. Weber, R.C. Ewing, C.R.A. Catlow, T.D. de la Rubia, L.W. Hobbs, C. Kinoshita, H. Matzke, A.T. Motta, M. Nastasi, E.K.H. Salje, E.R. Vance and S.J. Zinkle, *J. Mater. Res.,* 1998, **13**, 1434.
5 J.F. Ziegler, *The Stopping and Range of Ions in Matter*; Pergamon Press, 1985.
6 H.L.Heinisch and W.J.Weber, *Nucl. Instrum. Methods Phys. Res. Sect. B-Beam Interact. Mater. Atoms,* 2005, **228**, 293.
7 K.O.Trachenko, M.T.Dove and E.K.H.Salje, *J. Phys.-Condes. Matter,* 2001, **13**, 1947.
8 K.Trachenko, M.T. Dove and E.K.H. Salje, *Phys. Rev. B,* 2002, **65**, art. no.-180102.
9 K. Trachenko, M.T. Dove and E.K.H. Salje, *J. Phys.-Condes. Matter,* 2003, **15**, L1-L7.
10 J.-P. Crocombette and D. Ghaleb, *J. Nucl. Mater.* 1998, **3**, 282.
11 L.R.Corrales, W.J.Weber, A.Chartier, C.Meis and J.P.Crocombette, *J. Phys.-Condes. Matter,* 2003, **15**, 6447.
12 B. Park, W.J. Weber and L.R. Corrales, *Phys. Rev. B,* 2001, **6417**, art. no.-174108.
13 B. Park, W.J. Weber and L.R. Corrales, *Phys. Rev. B,* 2002, **65**, art. no.-219902.
14 R. Devanathan, L.R. Corrales, W.J. Weber, A. Chartier and C. Meis, *Phys. Rev. B,* 2004, **69**, art. no.-064115.
15 R.Devanathan, L.R. Corrales, W.J. Weber, A. Chartier and C. Meis, *Nucl. Instrum. Methods Phys. Res. Sect. B-Beam Interact. Mater. Atoms,* 2005, **228**, 299.
16 E.R.Andrew, A. Bradbury and R.G. Eades, *Nature,* 1958, **182**, 1802.
17 I.J. Lowe, *Phys. Rev. Lett.,* 1959, **2**, 285.
18 A. Meldrum,; S.J. Zinkle,; L.A. Boatner and R.C. Ewing, *Nature,* 1998, **395**, 56.
19 I.Farnan and E.K.H. Salje, *J. Appl. Phys.,* 2001, **89**, 2084.
20 G.C. Capitani, H. Leroux, J.C.Doukhan, S. Rios, M. Zhang and E.K.H. Salje, *Phys. Chem. Miner.,* 2000, **27**, 545.
21 T. Murakami, B.C.Chakoumakos, R.C. Ewing,; G.R. Lumpkin and W. Weber, *J. Am. Miner.,* 1991, **76**, 1510.
22 W.J. Weber, R.C. Ewing and L.M.Wang, *J. Mat. Res.,* 1994, **9**, 688.
23 S. Rios, E.K.H. Salje, M. Zhang and R.C.Ewing, *J. Phys.-Condes. Matter,* 2000, **12**, 2401.
24 R. Zallen, *The Physics of amorphous solids*; 2nd ed.; Wiley: New York, 1988.
25 M.D. Rintoul and S. Torquato, *J. Phys. A-Math. Gen.,* 1997, **30**, L585.
26 I. Farnan, H.M. Cho, W.J. Weber, R.D. Scheele, N.R. Johnson and A.E. Kozelisky, *Reviews of Scientific Instruments,* 2004, **75**, 5232.

LESSONS DRAWN FROM STUDIES OF THE TRANSMUTATION OF FUELS CARRIED OUT IN THE FRAMEWORK OF THE FRENCH LAW OF DECEMBER 1991 ON THE MANAGEMENT OF LONG LIVED WASTE

S. Pillon

CEA, CE Cadarache, Bdg 151, 13108 Saint-Paul-lez Durance cedex, FRANCE

1 INTRODUCTION

For the management of high level and long lived radioactive waste (HLLW), a large and continuous research and development effort is carried out in France, in order to provide a wide range of scientific and technical alternatives. This effort has been carried out in the framework of the French Law of December 30, 1991 for 15 years.

For the demonstration of the feasibility of HLLW transmutation in nuclear reactors (essentially Np, Am and Cm), an important programme dedicated to the development of fuels containing minor actinides is underway. The transmutation of minor actinides has been envisaged to take place preferably in 4th generation fast critical power reactors (Sodium-cooled Fast Reactor: SFR or Gas-cooled Fast Reactor: GFR), or in fast sub-critical reactors specifically designed to incinerate minor actinides (Accelerator Driven System: ADS).

Concerning the critical reactors, two recycling methods can be distinguished. Minor actinides can be diluted in all or in part of the fuel, in sufficiently weak concentrations as not to affect fuel and core performance: this is known as homogeneous recycling. Minor actinides can also be concentrated into targets and managed separately from the fuel within the core or in the core periphery: this is known as heterogeneous recycling. As for sub-critical reactors, ADS employ a fuel composed of minor actinides that significantly contributes to core reactivity.[1]

Contrary to homogeneous fuel, the performance of the targets and ADS fuels is very different from that of a standard fuel, due to their composition. So a large irradiation programme is being carried out to collect the necessary data to model their behaviour under irradiation. This programme is conducted within a collaboration with industrial partners (EDF), and in the framework of various international collaborations (Europe, USA, Japan, Russia), using the main material test reactors (MTR) in operation since 1991 (Siloe and Phénix in France, HFR in Netherlands and BOR60 in Russia).

2 FUEL COMPOSITION, MICROSTRUCTURE AND DESIGN CRITERIA

While homogeneous transmutation fuels are quite similar to standard MOX fuels, targets and ADS fuels are designed as composite fuels, so called inert matrix fuels (IMF). They are composed of americium, plutonium and curium oxide for which the greatest knowledge is available (contrary to actinide nitrides, carbides or metal alloys[2]). These products result from the conversion of americium, plutonium and curium nitrate solutions produced by the advanced separation process. However, more complex compounds are being examined, such as zirconia-based solid solutions: (Am,Zr,Y)O_x or pyrochlore structures: $Am_2Zr_2O_7$.[3] These compounds are expected to keep during irradiation the fluorite or pyrochlore-type cubic phase (facilitating the incorporation of Cm and Pu elements and FP) on the one hand, and reduce chemical activity (an identified risk of cladding oxidation induced by the high oxygen potential of AmO_x) on the other hand.

Minor actinide oxides are diluted in an inert matrix to form a composite material. This inert matrix is, in fact, a material with poor neutron absorption that does not provoke the formation of activation elements. This dilution helps improve fuel thermal properties and mechanical resistance, as well as control the power density produced during transmutation. The choice of the matrix is a difficult challenge, since it has to meet a lot of requirements: compatibility with the nearest environment (coolant, clad, actinide compounds), good physico-chemical properties, good in-pile performance (resistance to irradiation damage), compatibility with fabrication and reprocessing processes, etc.

Al_2O_3, $MgAl_2O_4$, $Y_3Al_5O_{12}$ and st-ZrO_2 (stabilised zirconia) have been ranked highly, because of their double compatibility with water and sodium and their possible use in both PWR and SFR. **MgO, ZrN, TiN, W and Mo** (depleted in ^{95}Mo to avoid any Tc production) are selected only for transmutation in SFR (they are not compatible with water). For PWR only, **CeO_2** is a likely candidate.

Figure 1 gives the different IMF microstructures tested in MTR.[4] They are characterised by the type of actinide oxide dispersion: microdispersion for a direct mixing/blending of actinide oxides and inert matrix (Figure 1a) or macrodispersion in the case of dispersion of actinide macromasses or granules into the inert matrix (Figure 1b, 1c). Such microstructures are developed with MgO, $MgAl_2O_4$, Al_2O_3, and Mo. ZrO_2, CeO_2 or Y_2O_3 form a solid solution with the actinide oxides.

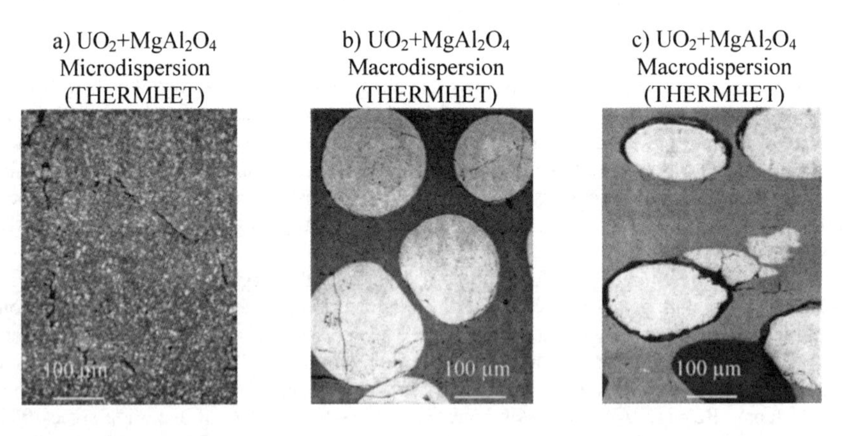

Figure 1 *Light microscope observation of different microstructures tested in MTR*

The thermal conductivity and swelling of the inert matrix – the major component of the fuel – are the main parameters governing the IMF design. Sufficient space between the fuel and cladding must be considered in order to limit chemical and mechanical interactions between the two components during irradiation, which could provoke pin failure. This space is extremely detrimental as it induces a high thermal gradient between the internal surface of the cladding and the outside of the target, and thus should be limited by choosing materials that swell very little.

To limit the operating temperature of these IMF, without restricting the minor actinide content of the fuel, choosing high thermally conductive matrices is recommended.

3 EXPERIMENTAL APPROACH

The selected inert matrices have been seriously examined based on irradiation tests performed to understand the swelling mechanisms induced either by neutrons (< 2MeV with a stopping distance of about 1 cm), fission product recoil (70 and 100 MeV, stopping distance 8 to 10 μm) or alpha decay of actinides (^{242}Cm) producing alpha particles (5 MeV, stopping distance: 20μm) and recoil atoms (0.1MeV, stopping distance: 25 nm).

Preliminary irradiation tests were conducted on sintered compacts of inert matrix (alone) in order to assess their resistance to fast neutrons. Next, experiments are conducted using a combination of matrices and actinide oxides to test resistance to fission products (FP) generated by the actinide. Uranium oxide, much easier to handle than americium oxide, is used more frequently as an actinide compound for prospective studies.

It is only in the last stage that irradiation brings into play the americium oxide. Much more complex and costly to implement, an irradiation phase is nevertheless necessary to test the effects of helium production. Helium is produced all the more in transmutation fuels, because the Am content is high. This production is essentially due to the decay of ^{242}Cm into ^{238}Pu by alpha emission, ^{242}Cm being formed by ^{242}Am decay, formed itself by neutron capture from ^{241}Am. The quantities of helium are not negligible, from 2 to 8 times higher than those of the gaseous fission products (Xe+Kr), depending on the burnup. Helium is quite insoluble in the crystal lattice and could cause, in case of retention, large swelling due to the formation of bubbles and thermal property depletion, or internal cladding pressurization, in case of release. The impact of He production on the fuel design and performance (clad-fuel gap, plenum volume, heat power, etc.) is therefore significant whatever the behaviour of the helium.

4 FUEL BEHAVIOUR UNDER IRRADIATION

4.1 Matrix resistance to neutrons and FP

Preliminary feedback concerning the behaviour of inert material under irradiation has been drawn from, among others, MATINA 1 irradiation tests performed in PHENIX, T3[6] tests in the HFR reactor, THERMHET[7] and TANOX[8,9] tests in the SILOE reactor.

MATINA 1 irradiation tests were performed using different matrices. Certain inert matrices with microdispersed uranium oxide were also irradiated: MgO, $MgAl_2O_4$ and Al_2O_3. Intermediate post-irradiation examinations did not reveal any significant swelling in the pins.

This last result, radically different from that observed in the microdispersed $MgAl_2O_4+UO_2$ target from the THERMHET experiment, where a volume swelling of approximately 20% and total macro-cracking of the composite core was observed (see Figure 2a) illustrates the importance of irradiation conditions and particularly irradiation temperatures. In contrast to the MATINA 1 maximum fuel temperature (1673K), the low irradiation temperature in the THERMHET (<1473K) experiment was not able to recover defects produced by fission products during irradiation. The in-depth analysis of the

THERMHET target revealed an amorphisation of the spinel, provoking considerable volume expansion during a long-lasting irradiation temperature drop, followed by nano-recrystallisation during a temperature rise, leading to volume reduction and the appearance of cracks. For MATINA 1, the higher temperature reached during the irradiation cycles helped avoid spinel amorphisation, even if the spinel was subjected to the beginning of an order-disorder reaction, clearly visible by X-ray diffraction.

a) Microdispersed UO_2+$MgAl_2O_4$
T_{max}<1200°C, BU=1.3 at%
(THERMHET)

b) Macrodispersed UO_2+$MgAl_2O_4$
T_{max}≈800°C, BU=0.4 at%
(THERMHET)

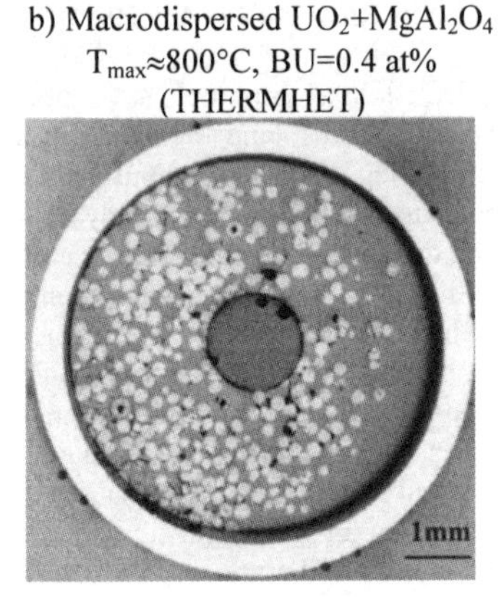

Figure 2 *Light microscope observation of irradiated IMF. The central hole has been drilled before irradiation to allowthe introduction of a thermocouple*

The $MgAl_2O_4$ matrix appears to be sensitive to swelling via amorphisation, and would not be suitable for low temperature reactor operations.[10]

In comparison to microdispersion, macrodispersion better helps to reduce extremely damaging effects upon the matrix. If the damage produced by the recoil of fission products is considered the main source of degradation to the matrix properties, such damage remains near the actinide particles and therefore is restricted to a low proportion of the target on the condition that these particles are big enough.[11] The THERMHET experiment in the Siloe reactor demonstrated the efficiency of this concept for the very first time. While the microdispersed $MgAl_2O_4$+UO_2 target underwent swelling and macro-cracking (see Figure 2a), the same macrodispersed target remained intact (see Figure 2b). This result was recently confirmed in the HFR during the OTTO experiment.[12]

4.2 Impact of the helium production on the target behaviour

The EFTTRA-T4 and T4 bis experiments performed in the HFR proved the decisive role of helium in target swelling.[13,14] A target pin composed of $MgAl_2O_4$ and AmO_2 (11% in mass) irradiated at a temperature close to 1023K with a fission rate of up to 30 at.% experienced considerable swelling of 18% in volume and 24% with a transmutation rate of 72 at.% (in some zones, even a swelling rate of 40 vol% is observed). This swelling can be partly explained by the abnormal retention of 80% of helium formed in the target, as well as the formation of a network of large-size bubbles (see Figure 3). The spinel amorphisation (as explained in § 4.1) could be the cause of this abnormal retention, when the large swelling rate would have had to favour gas release, due to pore connection, as commonly observed for standard ceramic or metal fuels. However, the mechanism of this retention is not yet clearly understood.

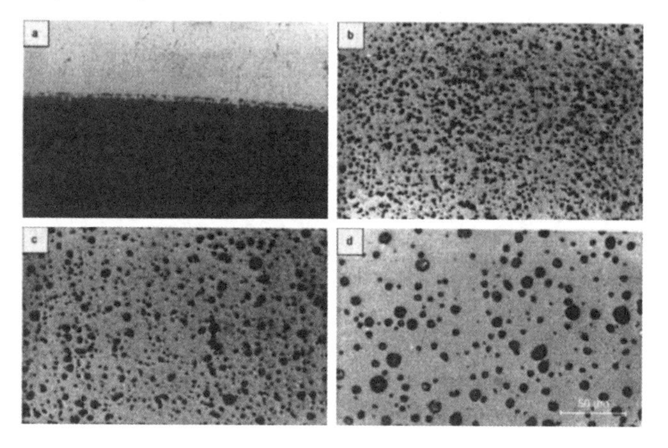

Figure 3 *Light microscope observation of AmO_2+$MgAl_2O_4$ microdispersed fuel (EFFTRA-T4, ITU)*

5 CONCLUSION

As regards the behaviour of inert matrices proposed for the transmutation of minor actinides, the 15 year programme on IMF carried out by CEA and its partners has allowed:

- The selection of ceramic and metal matrices, resistant to neutron flux and compatible with the chemical environment and fuel processing,
- The highlighting of the consequences of the FP damage for certain matrices (high volume swelling and gas retention),
- The identification of the reasons for the swelling (amorphisation), but not yet clearly those for gas retention,
- The proposal of a list of matrices resistant to FP damage (still to be validated),
- The proposal of advanced microstructures to limit damage and gas retention and to reach a high transmutation rate (up to 90% for a transmutation "once-through"),
- The optimisation of the irradiation conditions to reduce damage and helium retention,
- The preliminary design criteria for transmutation fuels and targets.

The programme still in progress at the Phénix and HTR currently addresses the next qualification and validation step, i.e.:

- the consolidation of the selected design options (inert matrices, microstructure, etc.) and operating conditions,
- the improvement of transmutation performance in terms of transmutation rate and capacity,

The development of the modelling of transmutation fuel behaviour under irradiation in normal or transient conditions, requires a relevant fuel irradiation database. This activity is supported by the European Commission in the 5th and 6th Framework Programmes.

The final stage in proving the technical feasibility of transmutation on an industrial scale will be to irradiate prototype assemblies. This last step will require the development of new facilities to separate the americium (and possibly curium) in sufficient quantities (between 1 to 10 kg) and the fabrication of transmutation fuel pins and assemblies.

References

1 S. Pillon, D. Plancq, A. Vasile, M. Mignanelli and R. Thetford, *proceedings of GLOBAL'99*, 1999, August 29-September 3, Jackson Hole, USA.
2 S. Pillon and J. Wallenius, *proceedings of ATALANTE 2004*, 2004, June 21-24, Nîmes, France.
3 P. Raison, *Prog. Nucl. Energy,* 2001, **38**, 251.
4 N. Chauvin, R. J. M. Konings and Hj. Matzke, *J. Nucl. Mater.*, 1999, **274**, 105.
5 N. Chauvin, T. Albiol, R. Mazoyer, J. Noirot, D. Lespiaux, J. C. Dumas, C. Weinberg, J.C. Ménard and J. P. Ottaviani, *J. Nucl. Mater.*, 1999, **274**, 91.
6 E.A.C Neeft, K. Bakker, R. P. C. Schram, R. Conrad and R. J. M. Konings, *J. Nucl. Mater.*, 2003, **320**, 106.
7 J. Noirot, L. Desgranges, N. Chauvin and V. Georgenthum, *J. Nucl. Mater.* 2003, **320**, 117.
8 P. Dehaudt, A. Mocellin, G. Eminet, L. Caillot, G. Delette, M. Bauer and I Viallard, *AIEA-TECDOC-970*, 1996, Technical Committee Meeting, Moscow.
9 V. Georgenthum; *Prog. Nucl. Energy*, 2001, **38**, 317.
10 N. Chauvin, C. Thiriet-Dodane, J. Noirot, Jj. Matzke, R.J.M. Konings, T. Wiss, R. Schram, K. Bakker, E. Neeft, R. Conrad, A. van Veen and T. Yamashita, *Report JRC-ITU-TN* 2002/39, 2002.
11 V. Georgenthum; *Doctoral thesis*, 2000, Université de Poitiers.
12 F.C. Klaassen, R.P.C. Schram, K. Bakker, Ch. Hellwig and T. Yamashita, *proceedings of GLOBAL'03*, 2003, New-Orleans, Louisiana, November 16-20.
13 R.J.M. Konings, R. Conrad, G. Dassel, B. J. Pijlgroms, J. Somers and E. Toscano, *J. Nucl. Mater.,* 2000, **282**, 159.
14 F.C. Klaassen, K. Bakker, R. P. C. Schram, R. Klein Meulekamp, R. Conrad, J. Somers and R. J. M. Konings, *J. Nucl. Mater.* 2003, **319**, 108.

PROPERTIES OF MINOR ACTINIDE COMPOUNDS RELEVANT TO NUCLEAR FUEL TECHNOLOGY

K. Minato, M. Takano, T. Nishi, A. Itoh and M. Akabori

Department of Materials Science, Japan Atomic Energy Research Institute, Tokai-mura, Ibaraki-ken 319-1195, Japan

1 INTRODUCTION

The long-term hazard of radioactive wastes arising from nuclear energy production is a matter of discussion and public concern. To reduce the radiotoxicity of the high-level waste and to use the repository efficiently, burning or transmutation of minor actinides (MA: Np, Am, Cm) as well as plutonium is an option for the future nuclear fuel cycle. Many concepts have been proposed for the fuels, together with reactors or accelerator-driven systems (ADS), and the research and development of MA-bearing fuels are being performed. The main chemical forms of the MA-bearing fuels studied mainly are oxides, nitrides, and metals.

The thermophysical and thermochemical properties of minor actinide compounds are essential for the design of MA-bearing fuels and analysis of their behaviour. The properties of actinide compounds of oxides and nitrides have been reviewed,[1-4] where it was recognized that the experimental data for americium and curium compounds were almost completely lacking. Some data for neptunium compounds are available in the literature, though they are insufficient.

The handling of americium and curium in general is more difficult than that of the other actinides, because of their gamma ray and neutron emissions. To support the development of MA-bearing fuels, the authors recently installed new facilities for the property measurements of minor actinide compounds. For the thermal properties of americium nitride and oxides, the lattice parameter and its thermal expansion were measured by high-temperature X-ray diffraction, the specific heat capacity by drop calorimeter, and the thermal diffusivity by a laser-flash method. The present paper gives an overview of the measurement of minor actinide nitrides and oxides properties performed at JAERI.

2 ACTINIDE NITRIDES

2.1 Miscibility of Actinide Nitrides

Actinide nitrides were synthesized from the oxides by the carbothermic reduction method, where the actinide oxides, mixed with carbon were heated in flowing N_2 gas to form the nitrides, followed by heating in N_2-H_2 mixed gas to remove the residual carbon. The preparation of the actinide nitrides AmN and (Cm,Pu)N has been reported,[5,6] along with UN, NpN, PuN, (U,Pu)N, (U,Np)N, and (Np,Pu)N.[7-11] Recently, (Pu,Am,Cm)N and (Np,Pu,Am,Cm)N were successfully synthesized.[12] The prepared nitrides were analyzed by X-ray diffraction to identify phases and determine lattice parameters. The analyses revealed that the actinide mononitrides have an NaCl-type structure, and the mixed nitrides have a single phase of NaCl-type structure forming a solid solution, as shown in Figure 1.

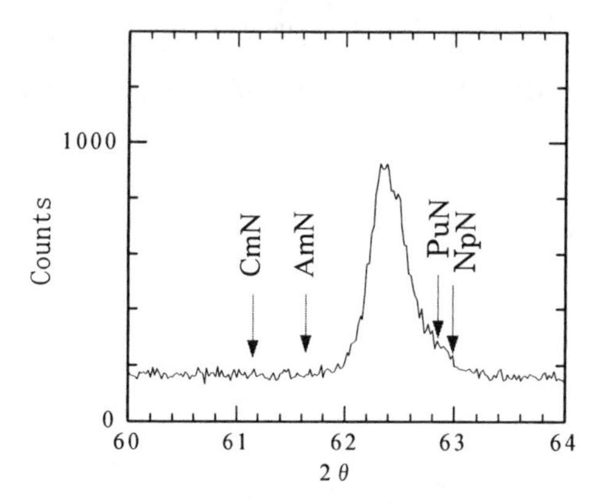

Figure 1 *X-ray diffraction analysis of (Np,Pu,Am,Cm)N, showing the formation of solid solution. Arrows indicate peak positions of respective mononitrides.*[12]

Table 1 compares the lattice parameters of the actinide mononitrides. The difference between the lattice parameters of UN and CmN is about 3%, where the two compounds are to be completely solved in each other, according to the prediction by Benedict.[13] The present result confirmed experimentally the mutual miscibility among NpN, PuN, AmN, and CmN.

Table 1 *Lattice parameters of actinide mononitrides*

Nitride	Lattice parameter; a_{MN} (nm)	Relative lattice parameter difference; $(a_{CmN}-a_{MN})/a_{CmN}$ (%)
UN	0.4888	2.765
NpN	0.4897	2.586
PuN	0.4905	2.426
AmN	0.4991	0.716
CmN	0.5027	

2.2 Thermal Expansion of Actinide Nitrides

The thermal expansions of NpN and AmN were determined for the first time from the temperature dependence of the lattice parameters measured by high-temperature X-ray diffraction.[14] The measurements were carried out in flowing N_2 in the temperature range 300-1369 K for NpN and 296-1488 K for AmN. The thermal expansions of UN and PuN have been reported. Figure 2 shows the thermal expansions of NpN and AmN, together with those of the reported values for UN and PuN.[15,16]

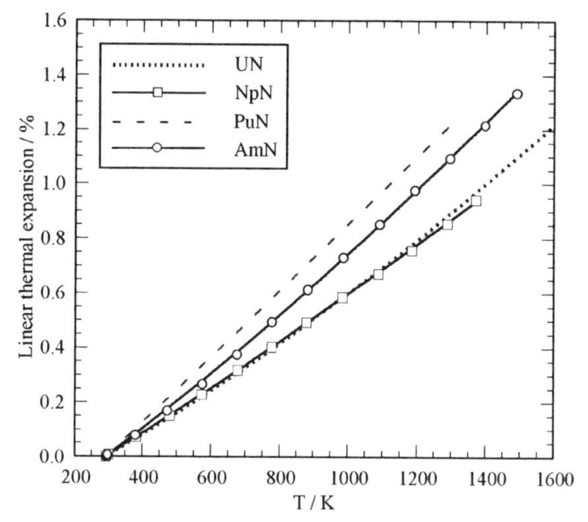

Figure 2 *Thermal expansions of NpN and AmN, together with those reported values for UN and PuN for comparison.*[14]

2.3 Specific Heat Capacity of Actinide Nitrides

Specific heat capacities of UN, PuN and NpN have been reported by Nakajima et al.,[17] where differential scanning calorimetry was used. The other data for UN and PuN are also available in literature.[18-20] To obtain the data for AmN, experiments with a drop calorimeter were initiated.

2.4 Thermal Conductivity of Actinide Nitrides

The thermal diffusivity of AmN was measured for the first time by the laser-flash method between 298 and 1473 K.[21] The specimen used was about 3 mm in diameter and 0.6 mm in thickness, whose density was about 77% of the theoretical density. Although the specific heat capacity of AmN was not available, the thermal conductivity of AmN with the theoretical density was calculated tentatively with the measured thermal diffusivity, the bulk density, and the specific heat capacity of PuN.[22] The thermal diffusivities of UN, NpN, PuN and the solid solutions of (U,Pu)N, (U,Np)N, and (Np,Pu)N were previously measured by the laser-flash method between 740 and 1630 K, and thermal conductivities have been reported as a function of temperature.[23-25] Figure 3 shows the thermal conductivity of AmN, together with the reported values for PuN, NpN, and UN.

The thermal conductivities of the actinide nitrides gradually increase with temperature over the temperature range investigated. For these nitrides, the electronic component contributes dominantly to the total thermal conductivity, and the increase in the thermal conductivity is due to the increase in the electronic component.[25] It is also seen in Figure 3 that the thermal conductivities decrease with increasing atomic number from UN to AmN. It is probable that this decrease is caused by the decrease in the electronic component to the thermal conductivity. The electrical resistivity of actinide mononitrides tends to increase with atomic number.[25]

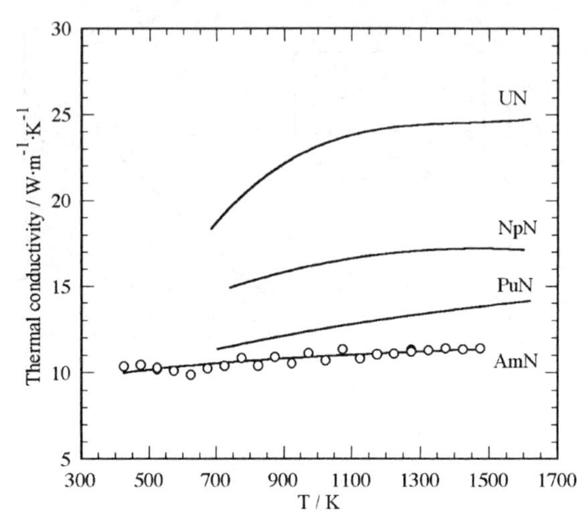

Figure 3 *Thermal conductivity of AmN, together with reported values for UN, NpN, and PuN for comparison.*[21]

3 ACTINIDE OXIDES

3.1 Thermal Expansion of Actinide Oxides

The thermal expansions of AmO_2 and $AmO_{1.5}$ were measured for the first time using high-temperature X-ray diffractometry.[26] The measurements were performed in flowing air for AmO_2 and flowing N_2 for $AmO_{1.5}$.

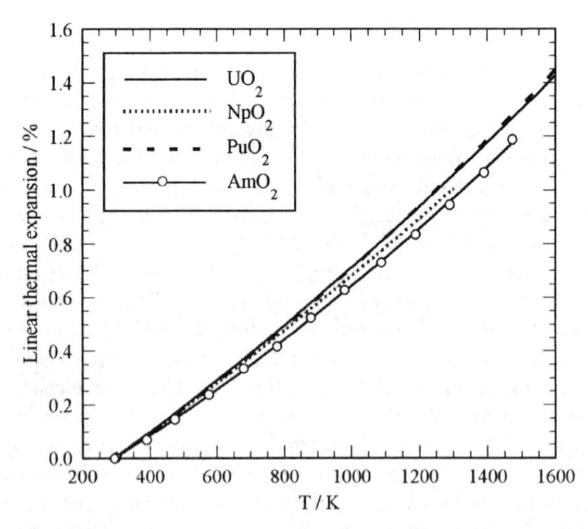

Figure 4 *Thermal expansion of AmO_2, together with those values reported for UO_2, NpO_2, and PuO_2 for comparison.*[26]

Figure 4 shows the thermal expansion of AmO_2, together with the reported values for UO_2,[16] NpO_2,[27] and PuO_2.[16] All of the actinide dioxides have the face-centered cubic crystal structure. The thermal expansion of AmO_2 is smaller than for others, though the difference is not large. Figure 5 shows the thermal expansion of $AmO_{1.5}$, which has the hexagonal close-packed structure. The volume expansion of $AmO_{1.5}$ is about 1.5 times larger than that of AmO_2.

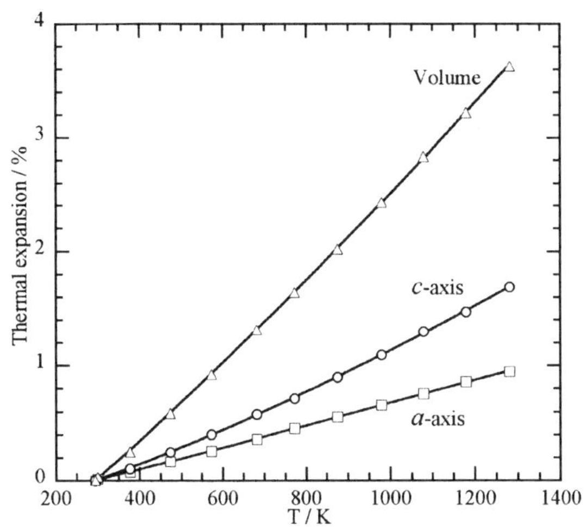

Figure 5 *Thermal expansion of $AmO_{1.5}$ with the hexagonal close-packed structure.*[26]

4 CONCLUSIONS

For MA-bearing fuel development, new facilities for property measurements of minor actinide compounds were installed at JAERI, and the thermal properties of minor actinide nitrides and oxides were measured. In the sample preparation, the mutual miscibility among NpN, PuN, AmN, and CmN was experimentally confirmed, meaning that nitride fuel could accommodate a wide range of combinations and compositions of the actinides. The thermal conductivity of americium nitride was determined, and the thermal conductivities of actinide nitrides were found to decrease with increasing atomic number, from UN to AmN. Although some new data on the thermal properties of the minor actinide compounds were presented, more measurements need to be made.

Acknowledgements

This paper contains some results obtained within the task "Technological development of a nuclear fuel cycle based on nitride fuel and pyrochemical reprocessing" entrusted from the Ministry of Education, Culture, Sports, Science and Technology of Japan. This paper also contains some results obtained within the collaborative research program of TRU behaviour in oxide fuels with Tohoku Electric Power Company, Tokyo Electric Power Company and The Japan Atomic Power Company.

References

1 M. Mignanelli and R. Thetford, 'Thermophysical and Chemical Properties of Minor-Actinide Fuels' in *Advanced Reactors with Innovative Fuels*, OECD/NEA, 2001, pp. 161-171.

2 K. Minato, Y. Arai, M. Akabori and K. Nakajima, 'Research on Nitride Fuel for Transmutation of Minor Actinides' in *Proc. AccApp/ADTTA '01*, Reno, Nevada, Nov. 11-15, 2001, CD-ROM.
3 K. Minato, M. Akabori, M. Takano, Y. Arai, K. Nakajima, A. Itoh and T. Ogawa, *J. Nucl. Mater.*, 2003, **320**, 18.
4 R. Thetford and M. Mignanelli, *J. Nucl. Mater.*, 2003, **320**, 44.
5 M. Takano, A. Itoh, M. Akabori, T. Ogawa, S. Kikkawa and H. Okamoto, 'Synthesis of Americium Mononitride by Carbothermic Reduction Method' in *Proc. Int. Conf. Global '99*, Jackson Hole, Wyoming, Aug. 29-Sept. 3, 1999, CD-ROM.
6 M. Takano, A. Itoh, M. Akabori, T. Ogawa, M. Numata and H. Okamoto, *J. Nucl. Mater.*, 2001, **294**, 24.
7 Y. Arai, S. Fukushima, K. Shiozawa and M. Handa, *J. Nucl. Mater.*, 1989, **168**, 280.
8 Y. Suzuki, Y. Arai, Y. Okamoto and T. Ohmichi, *J. Nucl. Sci, Technol.*, 1994, **31**, 677.
9 Y. Arai, Y. Suzuki and M. Handa, 'Research on Actinide Mononitride Fuel' in *Proc. Int. Conf. Global '95*, Versailles, France, Sept. 11-14, 1995, Vol. 1, p. 538.
10 Y. Arai, T. Iwai, K. Nakajima and Y. Suzuki, 'Recent Progress of Nitride Fuel Development in JAERI - Fuel Property, Irradiation Behavior and Application to Dry Reprocessing' in *Proc. Int. Conf. Global '97*, Yokohama, Japan, Oct. 5-10, 1997, Vol. 1, p. 664.
11 Y. Suzuki, T. Ogawa, Y. Arai and T. Mukaiyama, 'Recent Progress of Research on Nitride Fuel Cycle in JAERI' in *Proc. 5th OECD/NEA Information Exchange Meeting on Actinide and Fission Product P&T*, Mol, Nov. 25-27, 1998, p. 213.
12 M. Takano, et al., to be published in *J. Nucl. Sci. Technol.*
13 U. Benedict, 'Solid Solubility of Fission Product and Other Transition Elements in Carbides and Nitrides of Uranium and Plutonium' in *Thermodynamics of Nuclear Materials*, IAEA, 1980, pp. 453-470.
14 M. Takano, et al., to be published in *J. Alloy Comp.*
15 S. L. Hayes, J. K. Thomas and K. L. Peddicord, *J. Nucl. Mater.*, 1990, **171**, 262.
16 Y. S. Touloukian (Ed.), *Thermophysical Properties of Matter*, V.13, IFI/Plenum, New York, 1977.
17 K. Nakajima and Y. Arai, *J. Nucl. Sci. Technol.*, 2002, **Supplement 3**, 620.
18 F. L. Oetting and J. M. Leitnaker, *J. Chem. Thermodynam.*, 1972, **4**, 199.
19 S. L. Hayes, J. K. Thomas and K. L. Peddicord, *J. Nucl. Mater.*, 1990, **171**, 300.
20 F. L. Oetting, *J. Chem. Thermodynam.*, 1978, **10**, 941.
21 T. Nishi, et al., to be published in *J. Nucl. Mater.*
22 I. Barin (Ed.), *Thermochemical Data of Pure Substances*, Third Ed., Vol. II, VCH, Weinheim, 1995, p. 1350.
23 Y. Arai, Y. Suzuki, T. Iwai and T. Ohmichi, *J. Nucl. Mater.*, 1992, **195**, 37.
24 Y. Arai, Y. Okamoto and Y. Suzuki, *J. Nucl. Mater.*, 1994, **211**, 248.
25 Y. Arai, K. Nakajima and Y. Suzuki, *J. Alloys Comp.*, 1998, **271-273**, 602.
26 M. Takano, et al., to be published in *J. Nucl. Mater.*
27 T. Yamashita, N. Nitani, T. Tsuji and H. Inagaki, *J. Nucl. Mater.*, 1997, **245**, 72.

TRANSFORMATION CHARACTERISTICS OF ISOTHERMAL MARTENSITE IN Pu-Ga ALLOYS

K. Moore, C. Krenn, M. Wall and A. Schwartz

Lawrence Livermore National Laboratory, Livermore, CA 94550, USA.

1 INTRODUCTION

When Pu-Ga alloys are cooled to a low temperature range, an isothermal martensitic transformation occurs from the face-centered-cubic δ phase to the monoclinic α' phase. This transformation has an unusually large volume reduction of approximately 20%, causing large amounts of deformation and intriguingly unique crystal orientation relationships and particle morphologies. The time-temperature-transformation (TTT) diagram (see Figure 1) shows that transformation is most rapid at two distinct temperatures, both near -130°C and near -160°C. This unusual behaviour implies two distinct, thermally activated mechanisms must exist for this transformation.

It has been shown repeatedly[2-4] that the orientation relationship between α' and δ is $(111)_{\delta} \parallel (020)_{\alpha'}$ and $[\bar{1}10]_{\delta} \parallel [100]_{\alpha'}$. However, the habit plane has remained undetermined, mostly due to the difficulties in experimentation with the materials and the large degree of variation observed in precipitate morphology. Here, we will examine the progress of multiple experimental approaches to understanding the orientation relationship, habit plane, and transformation behaviours of this system. We show that while the α' plates exhibit what seems to be multiple habit planes, they possess a terrace and ledge structure that is faceted on 111_{δ}, which accommodates macroscopic curvature via the density of interfacial ledges.

2 METHODS AND RESULTS

2.1 Experimental procedures

TEM samples were prepared from bulk Pu-Ga material following the procedure described by Wall *et al.*[5] and Moore *et al.*[6] Special care was taken throughout the process to prevent any significant temperature rise or stress in the material and in the specimens extracted from it. Samples 31.75 mm in diameter were obtained by single point milling with trichloroethylene (TCE) steady drip lubricant. The disk was sectioned in thirds using a low speed diamond saw flooded with Dow Corning 200 cutting fluid. To insure minimal heating, the low speed diamond saw was again used to cut 6.5-mm squares from nearest the centre of the original disk. These squares were mounted on a 6.5-mm diameter aluminium rod, secured in the lathe, and turned to 2.8-mm diameter with 0.05-mm reductions per pass and TCE coolant. For final specimen preparation, the samples were moved to a dry train recirculating glove box with a base line environment of ≈ 1 ppm for O_2 and ≈ 1 ppm H_2O with an inert atmosphere of nitrogen. The cylindrical specimen was sliced to 1.0-mm thickness before lapping on 30-μm, 12-μm, then 3-μm aluminum oxide paper. Thin foils for TEM observation were prepared by electropolishing in a solution of 10% nitric acid (70% concentration), 45% methanol and 45% butoxyethanol (butylcellusove) by volume. The electropolishing parameters were: – 20°C, 35V and ≈ 40mA. The specimen was thinned continuously to perforation. After perforation, the sample was immediately re-polished at a higher voltage (50V) for 2-3 seconds to remove a thin anodic film that develops during the continuous polishing at lower voltages. The sample was then rinsed in the holder with methanol (anhydrous), followed by removal of the specimen from the holder, and placing the specimen in methanol. The sample was rinsed in several baths of methanol for approximately 1 minute, each then rinsed and stored in 200 proof ethanol until ready to transfer to the TEM. A vacuum transfer specimen holder was used for moving the specimen from the Ar environment of the glovebox to the TEM. All optical micrographs were recorded after TEM analysis, and were performed in glove boxes.

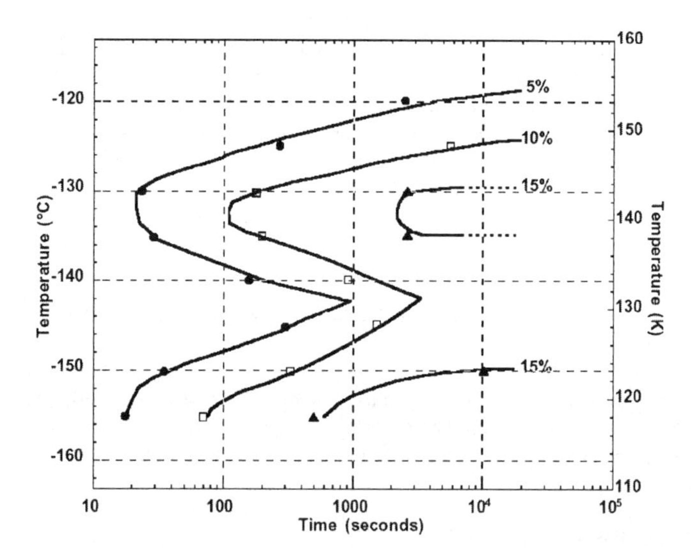

Figure 1 *The time-temperature-transformation (TTT) curve for a 0.56 wt% Pu-Ga alloy (after Orme et al. 1975[1]). Contours show a given percentage of transformation from δ to α' as a function of time and temperature.*

2.2 Combined optical and TEM analysis

One approach we have begun using to analyze the overall (macroscopic) habit plan is combining TEM with optical microscopy. This way we find an area of interest using TEM, then record a bright-field image of the area along with the corresponding electron diffraction pattern. We index the pattern via simulations, then record optical images of the area to observe and analyze the precipitate morphology. From here, we produce stereographic projections and delineate where the intersections of plates would be with the surface of the sample if they had a given habit plane. For example, Figure 2 shows a set of data from TEM to optical microscopy to simulations. In this case, we start off with a TEM image of an area of the sample with an α' plate intersecting the edge of the foil. We then

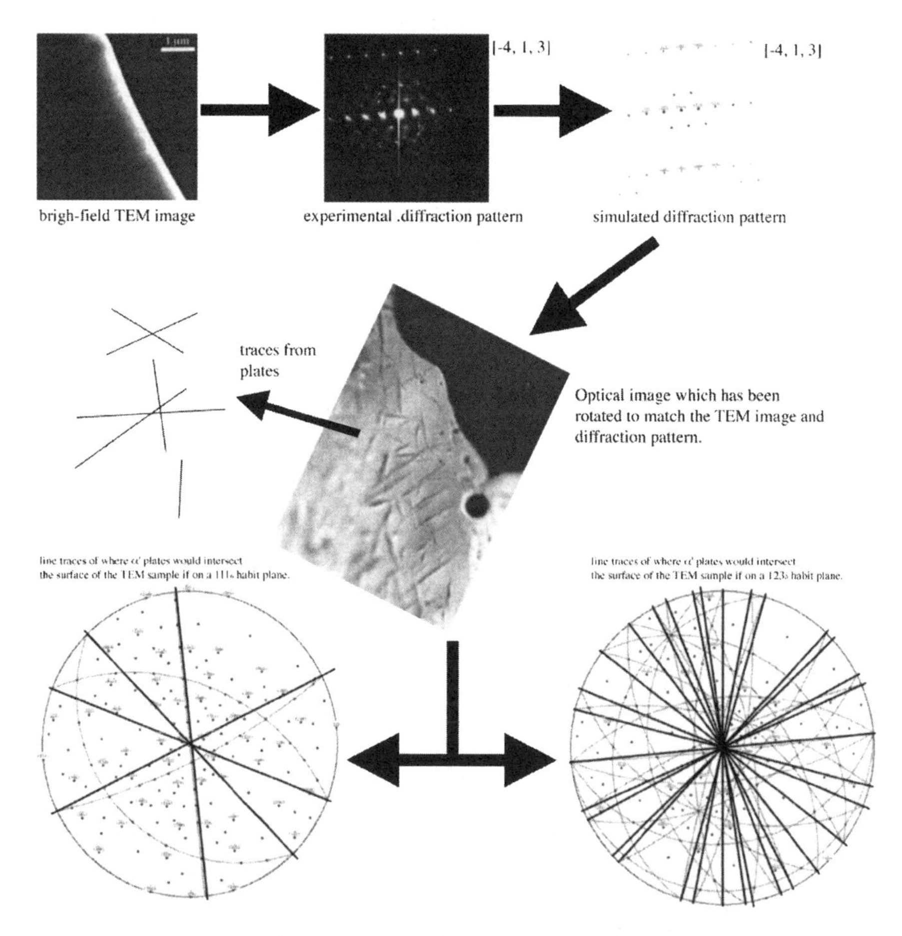

Figure 2 *A schematic diagram showing how we examine the precipitate morphology of the α' habit plane using TEM imaging and diffraction, optical microscopy, and computer simulations.*

record the corresponding electron diffraction pattern, which in this case is $[\bar{4},1,3]$ as proven by simulations of diffraction patterns. From here we remove the sample from the TEM, find the area in a light microscope, then record an image of the surface of the sample, thus allowing us to see a larger area than given in the TEM. We then compare the experimental traces from the optical images with those from simulated stereographic projections where we assume a given habit plane. In Figure 2 the lines directly to the left of the optical image are traces from the experimental image. When these are compared to the traces in the simulated stereographic projection for an assumed 111 habit plane (bottom of diagram), the match is very good, with just a few exceptions. When we then look at an assumed 123-type habit plane having 24 variants, the remaining traces can be accounted for. However, it must be considered that the 123 orientation produces so many possible line traces that it becomes quite easy to fit the traces to experimental data. This is not the case for 111, since there are only 4 possible traces. Another deficiency of this technique to keep in mind is that we are missing one variable of the precipitate orientation, namely the inclination relative to the normal of the plate face.

This represents continuing work, which needs to have a much larger number of statistical points in order to become more clear. Nonetheless, it is beginning to form a clearer picture of the general structure of the α′ precipitate morphology. By recording and examining a large number of poles, especially low-order indexes, we can begin to show trends through consistency of data. We also hope to perform synchrotron X-ray work, where we can truly observe the precipitate-matrix morphology in 3D.

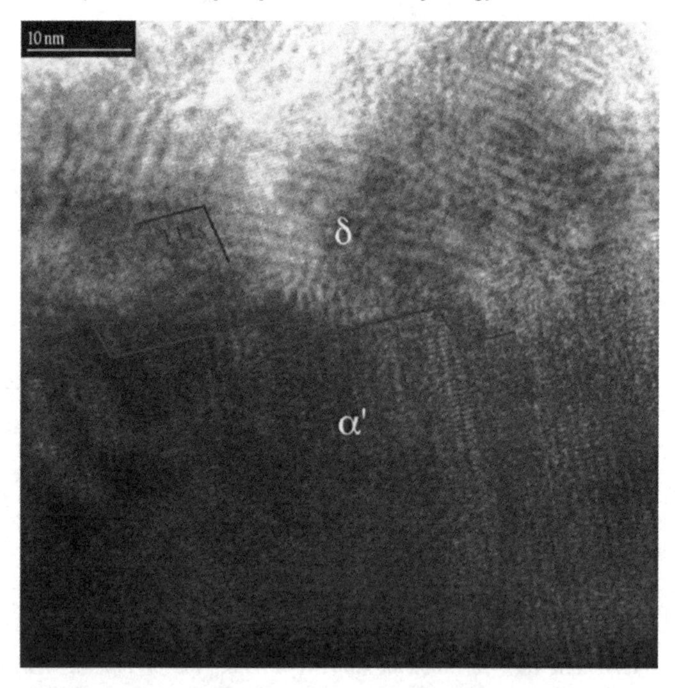

Figure 3 *High resolution TEM image of the interface showing the terrace and ledge structure faceted on $111_δ$*

2.3 high resolution TEM of the α/δ interface

The TEM yields unparalleled spatial and energy resolution, allowing for a wide range of structural and electronic investigations of actinides.[6,7] However, there has not been much high resolution TEM performed on actinides given the high atomic number, the large amount of thermal diffuse scattering produced by the material, and the rapid oxidation of the surface of the samples, all of which makes HRTEM very difficult to perform. Nonetheless, we have recorded a reasonable number of high resolution TEM images of the α'/δ interface, and in doing so have observed an important aspect of the interface. A terrace and ledge morphology is present, where both the terraces and ledges are faceted on 111_δ. This is shown in Figure 3, where the marked lines show the interface as well as where the 111 planes are in the δ matrix. Notice both the ledges and the terraces are parallel to each of the $\{111\}_\delta$ planes. Thus, while the macroscopic or overall habit plane may exhibit another orientation, at the atomic scale a $\{111\}_\delta$ habit is observed.

3 CONCLUSIONS

While we have not answered the question of what exactly the habit plane is for a given composition, lattice constant, processing condition, etc., we have shown that the α'/δ interface is faceted on 111_δ, and that interfacial curvature is therefore dictated by the number and density of these ledges.

Recent computational work[8] has pointed to a highly sensitive invariant plane strain solution for the α'-δ martensitic transformation, meaning slight changes in composition, transformation temperature, and all other factors influenced by processing of the materials will affect the precipitate morphology and habit plane. Our optical and TEM data show a wide range of precipitate behaviour, lending support to these calculations and the idea that very slight changes in the material and its processing alters the transformation product.

References

1 J.T. Orme, M.E. Faiers, and B.J. Ward. Plutonium 1975 and other actinides, ed. North Holland Publishing, 1976, p. 761.
2 T.G. Zocco, M.F. Stevens, P.H. Adler, R.I. Sheldon, and G.B. Olson, *Acta Metall. Mater.,* 1990, **38**.
3 T.G. Zocco and A.J. Schwartz, *Journal Mater.,* 2003, **55**.
4 SS. Hecker, D.R. Harbur and T.G. Zocco, *Prog. Mater. Sci.,* 2004, **49**.
5 M.A. Wall, A.J. Schwartz, and M.J. Fluss, "*Sample Preparation for Transmission Electron Microscopy Characterization of Pu Alloys*", UCRL-ID-141746.
6 K.T. Moore, M.A. Wall, and A.J. Schwartz, *J. Nucl. Mater.,* 2002, **306**.
7 K.T. Moore M.A Wall, A.J. Schwartz, B.W. Chung, D.K. Shuh R.K. Schulze, and J.G. Tobin, *Phys. Rev. Lett.,* 2003, **90**, 196404.
8 Y.M. Jin, Y.U. Wang, and A.G. Khachaturyan, *Met. Trans. A*, In Press.

REACTIVITY OF MIXED URANIUM AND PLUTONIUM NITRIDE WITH SILICON CARBIDE AS MATERIALS FOR GAS COOLED FAST REACTORS

J. Léchelle, L. Aufore and M. Reynaud

Commissariat à l'Énergie Atomique, CEN Cadarache, DEN/DEC/SPUA/LMPC, Bât 717, F-13108 Saint-Paul-lez-Durance, France.

1 INTRODUCTION

Among the different materials under consideration for Generation IV Gas Cooled Reactors, (U, Pu)N with typically 2% of the U and Pu cations substituted by minor actinides (Am, Cm or Np) as fuel, as well as SiC as an inert matrix have drawn much interest and study. Thermo-chemical compatibility and diffusionnal studies have been undertaken from foreseen normal reactor thermodynamic intensive conditions (inert gas or inert gas with low amounts of N_2 at 1200°C) to fabrication or accidental ones (1600°C and 2000°C). The main purpose is to determine the highest thermal conditions allowed by thermochemistry for the secure operation of the fuel in a reactor, as well as for fabrication. After completion of the thermal treatments, XRD was used to determine newly formed phases. Also, SEM and electron microprobe were used to investigate the occurrence of possible quenched liquid phases (eutectic or peritectic decompositions), and to estimate the material interdiffusion depth.

2 INITIAL MATERIALS

2.1 $U_{0.8}Pu_{0.2}N$ phase

This material was fabricated by the LEFCA in Cadarache for an irradiation in the PHENIX reactor named NIMPHE (**NI**trides of **M**ixed U and Pu in **PHE**nix).[1] It was made by thermal treatment under a flow of N_2 gas, during the temperature increase step, 6%H_2/N_2 gas during the plateau at 1550°C for 17h, and Ar/5%H_2 gas in order to eliminate residual oxygen and carbon from the initial mixture of $(U,Pu)O_2$ and C. The estimated structural formula from chemical analysis with the assumption N+C+1/2O=1 is $U_{0.78}Pu_{0.22}N_{0.976}C_{0.018}O_{0.012}$. XRD carried out on powder from this material gives a lattice parameter a=4.905(±0.002) Å to be compared with 4.897 Å for a Vegard's linear law between $U_{0.78}Pu_{0.22}N$ a=4.892 Å, $U_{0.78}Pu_{0.22}C$ a=4.963 Å[2] and $U_{0.78}Pu_{0.22}O_2$ a=5.454 Å. This discrepancy may be due to accumulated self-radiation damage. After annealing (Ar/H_2 2000°C – 2 hours), the lattice parameter of the starting fuel becomes: a=4.896(±0.002) Å.

In the compact material, mean grain size exhibits two modes: 3.4 μm (83 number %) and 5.9 μm (17 number %). The small grain size leads to a high contribution of grain boundary diffusion during interdiffusion phenomena.

2.2 SiC materials

Two kinds of silicon carbide materials where used:
α-SiC, from the Boostec company with about 1wt.% of B and C as impurities. XRD indicates that the 6H polytype is the most abundant, and also shows the presence of B_4C (as a sintering additive). A FEG-SEM observation of the microsctructure after a Murakami etching showed a mean grain size close to 5 μm.
β-SiC, which is made of pure silicon carbide and exhibits a cubic structure, 3C.

3 REACTIVITY AND DIFFUSION EXPERIMENTS

3.1 Reactivity experiments

Table 1 gives the results of reactivity experiments: thermal treatments of mixed SiC and (U,Pu)N powders (50% weight ratio) pellets under a flow of either Ar, Ar 96.5vol.%-N_2 3.5vol.% or under a static atmosphere of Ar at 1200°C-2000°C for 4-180 hours.[5]

Table 1 *Results of Reactivity Experiments (50wt.% of each starting material)*

	1200°C	1600°C	2000°C
Flow of gas	Experiment not realized: it would be unrealistic since during fuel manufacture the temperature under the gas flow is higher than 1200°C	*fuel manufacturing conditions* Starting 'SiC': sintered α-SiC (B_4C additive) Duration: 4h, 100h Gas: Ar, Ar/N_2 resp. Relative weight change: -9.2% Phases: - α-SiC -'(U,Pu)N' a=4.914 Å which may mean: $U_{0.78}Pu_{0.22}N_{0.7}C_{0.3}$ inferred from thermochemical calculations - small amounts of $(U,Pu)O_2$ - trace amounts of $(U,Pu)Si_2$ a=3.959 Å c=13.606 Å and $UB_xC_{(1-x)}$ a=3.588 Å b=12.011 Å c=13.606 Å	*fuel manufacturing conditions / reactor accident* Starting 'SiC': β-SiC Duration: 6h. Gas: Ar Relative weight change: -7% Phases: - β-SiC - C graphite - $U_{20}Si_{16}C_3$ (a=10.379 Å,c=8.006 Å) - '(U,Pu)N' a= 4.905 Å which may mean : $U_{0.78}Pu_{0.22}N_{0.9}C_{0.1}$ inferred from thermochemical calculations - $U_{0.78}Pu_{0.22}C_{0.9}O_{0.1}$ a= 4.953 Å - Trace amounts of UO_2 and of $(U,Pu)_3Si_2C_2$ (a=3.575 Å, c=18.867 Å)
Static Ar	*Nominal reactor conditions* Starting 'SiC' : β-SiC Duration: 180 h. Relative weight change: -19% Phases: - β-SiC - (U,Pu)N a=4.895 Å - trace amounts of $(U,Pu)O_2$ a=5.467 Å	*reactor incident* Starting 'SiC' : β-SiC Duration: 100h. Relative weight change: -10% Phases: - β-SiC and small amounts of α-SiC -'(U,Pu)N' a= 4.902 Å which may mean: $U_{0.78}Pu_{0.22}N_{0.9}C_{0.1}$ - trace amounts of tetragonal PuC_2 and $(U, Pu)Si_2$	Experiment not feasible in our HF oven which needs a gas flow

No evidence of a quenched liquid phase (broad amorphous phase peak) was found by XRD. The important weight loss at 1200°C could not be interpreted either on the basis of XRD information or on thermochemical computations with HSC Chemistry 5.1® (calculated 0.3% weight loss).

3.2 Interdiffusion experiments

SEM analysis along SiC interface of interdiffusion experiments between α-SiC and (U,Pu)N under a flow of Ar/N_2 and a static atmosphere of Ar show that U and Pu are localised as far as 20μm from the interface (depth of interdiffusion) in 4 μm large areas (in Figure 1 SiC occupies the upper part of the picture with a crack in it, its lower part rich in C is resin due to ceramographic preparation).

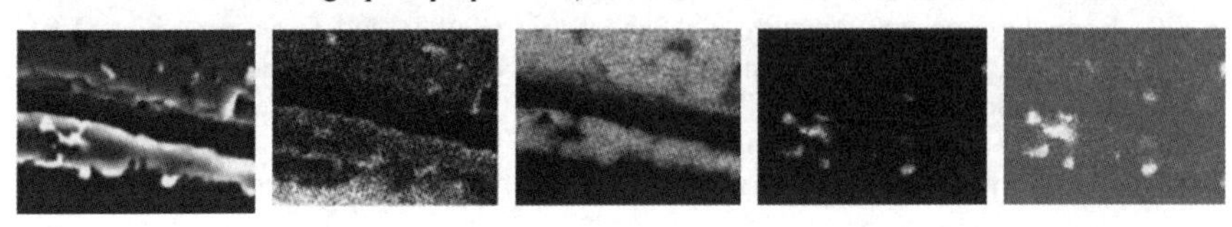

Figure 1 *Elemental mapping of a SiC interfacial area (left to right: sec. e⁻, C, Si, U, Pu)*

(U,Pu)-rich areas are not parallel to the interface: U and Pu may have diffused along grain boundaries to concentrate around small grains (higher density of grain boundaries).

4 CONCLUSION

At 1200°C for 180 hours, no new solid phase appears in a mixture of SiC and (U,Pu)N. At 1600°C, very small amounts of $(U,Pu)Si_2$[3] and $U_{20}Si_{16}C_3$[4] appear, due to diffusion of U and Pu into SiC, in agreement with the studies of Rado *et al.* with U as the actinide.[6] Carburation of the starting (U,Pu)N is suspected via lattice parameter values. At 2000°C, SiC gives abundant secondary phases: graphite and $U_{20}Si_{16}C_3$ due to elimination of N_2 by the Ar flux. No evidence of liquid phases was found. At 1600°C, grain boundary diffusion is suspected as being the main transport phenomenon responsible for a diffusion depth of 20 μm into the starting α-SiC for U and Pu.

References

1 H. Bernard, P. Bardelle and D. Warin, *Mixed nitride fuels fabrication in conventional oxide line in* IAEA Technical Committee Meeting on Advanced Fuel for FBR, 3-5 nov 1987, Vienne.

2 H.J. Matzke *Science of advanced LMFBR fuels*, A monograph on solid state physics, chemistry and technology of carbides, nitrides and carbonitrides of uranium and plutonium, , North-Holland, 1986.

3 W.H. Zachariasen, *Acta Cryst.*, 1949, **2**, 94.

4 R. Pöttgen, D. Kaczorowski and W. Jeitschko, *J. Mater. Chem.*, 1993, **3**, 253.

5 J. Léchelle, L. Aufore, V. Basini, R. Belin and S. Vaude, *Gas cooled Fast Reactor materials: compatibility and reaction kinetics of fuel/matrices couples, in* Proceedings of Atalante 2004, Nîmes, 2004.

6 O. Rapaud, C. Rado, *Fissile compound – Inert Matrix Compatibility Studies for the Development of Gas-Cooled Fast Reactor Fuels, in* Proceedings of Atalante 2004, Nîmes, 2004

CONFINEMENT OF PLUTONIUM AND THORIUM IN MONAZITE/BRABANTITE, IN VIEW OF ACTINIDE CONDITIONING

B. Glorieux[1], V. Picot[2], X. Deschanels[2], F. Jorion[2], J.M. Montel[3] and M. Matecki[1]

[1]PROMES-CNRS, Perpignan, France.
[2]DEN-CEA, Marcoule, France.
[3]LMTG-University of Toulouse, France.

1 INTRODUCTION

Monoclinic lanthanide orthophosphate is one of the ceramics potentially able to confine minor actinides. Previous research,[1] mainly conducted on natural minerals,[2] has shown its capacity to accommodate actinides, without modifying their properties in terms of structure and leaching behaviour.

The mineral form of orthophosphate is called monazite or brabantite. The monazite, MPO_4, refers to a trivalent cation, M^{3+} (M=La, Ce, Nd, Gd, Am, Pu,...), compensating the –III charge of the PO_4 entity, while the brabantite, $A_{0,5}B_{0,5}PO_4$, refers to a pair of bivalent cations, A^{2+} (A=Ca, Sr, Ba, Cd...), and tetravalent cations, B^{4+} (B=U, Th, Np,...[3]). Monazite and brabantite are isostructural (space group $P2_1/n$).

Based on previous findings,[4-7] the goal of our research is to study the conditioning of trivalent and tetravalent actinides in such orthophosphates and their influence on the structural properties and the leaching behaviour. In this communication, we present the different steps of the synthesis, the sintering and the first characterisation of a solid solution monazite-brabantite, $La_{0.73}Pu_{0.09}Ca_{0.09}Th_{0.09}PO_4$.

2 METHOD AND RESULTS

Several samples have been prepared, containing cerium and thorium. The cerium serves as a substitute for plutonium conditioning, and is used to optimise the process of synthesis and sintering. Later, the isotope ^{239}Pu replaces the cerium.

2.1 Monazite-Brabantite containing cerium and thorium

The composition $La_{0.73}Ce_{0.09}Ca_{0.09}Th_{0.09}PO_4$ is defined by the need to store 10%wt minor actinides (trivalent or tetravalent). Previous research[8] has already shown that cerium and plutonium can easily be incorporated in such a structure in a trivalent state. Among the various protocols[9] available to synthesise monazite-brabantite, glove box techniques may be used. These imply a solid reaction at high temperature in air:

$$\tfrac{1}{2}(1\text{-}3x)La_2O_3 + xCeO_2 + xCaCO_3 + xThO_2 + NH_4H_2PO_4 \quad (1)$$

$$\rightarrow$$

$$La_{1\text{-}3x}Ce_xCa_xTh_xPO_4 + {}^x\!/_4O_2 + NH_3 + {}^3\!/_2\,H_2O + x\,CO_2 \qquad x=0.0894$$

This reaction was optimised by performing TGA-DTA experiments, coupled with a mass spectrometer, and high temperature X-ray diffraction. Because thorium is not accepted inside the thermal analysis equipment, only the incorporation of cerium is analysed.

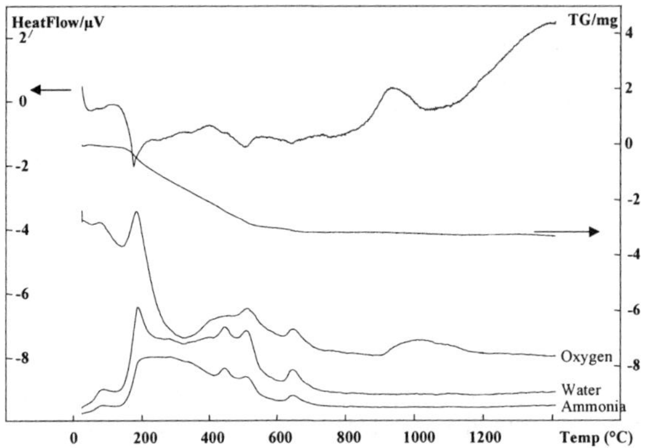

Figure 1 *Thermal analyses of the incorporation of cerium in the monazite: DTA (left), TGA (right) , O_2, H_2O and NH_3 evaporation (bottom).*

After decomposition of $NH_4H_2PO_4$ below 600°C, the reaction between the precursors occurs as a single step between 900°C and 1150°C. The reduction of cerium, shown by a slight weight lost and the oxygen evaporation, appears at the same time as an important exothermic peak, due to the reaction between precursors.

XRD refinement is performed on the final product and gives an insertion rate x (deduced from the measured cell parameters and assuming that the incorporation ratio of Ce, Ca and Th is the same) of 0.087 ± 0.02, in excellent agreement with the expected value. An XPS experiment performed on the final producects of the reaction (1) certifies that the cerium is in a trivalent state.

The sintering process appears to be different between the monazite and solid solution monazite-brabantite, depending on whether calcium and thorium are present in the structure. The pure monazite is extremely difficult to sinter and attrition equipment needs to be used to obtain a density higher than 95% of the theoretical density[10]. A solid solution monazite-brabantite easily reaches a 96% density[8] with a sintering process, defined by a compaction at 100MPa and a thermal treatment at 1450°C for 10 h. Shrinkage appears between 1200°C and 1400°C. An SEM-EDS analysis of the final sintered products reveals a homogeneous and single phase sample of $La_{0,7}Ce_{0,1}Ca_{0,1}Th_{0,1}PO_4$ composition.

2.2 Monazite-Brabantite containing plutonium and thorium

The synthesis of $La_{0,73}Pu_{0,09}Ca_{0,09}Th_{0,09}PO_4$ has been performed using the same process and the same tools described above for $La_{0,73}Ce_{0,09}Ca_{0,09}Th_{0,09}PO_4$, just replacing CeO_2 by $^{239}PuO_2$ in the reaction (1).

The sample has a geometrical density of 5.2 ± 0.1, corresponding to 96% of the theoretical density. The XRD, SEM micrograph and XANES analyses are presented in Figure 2.

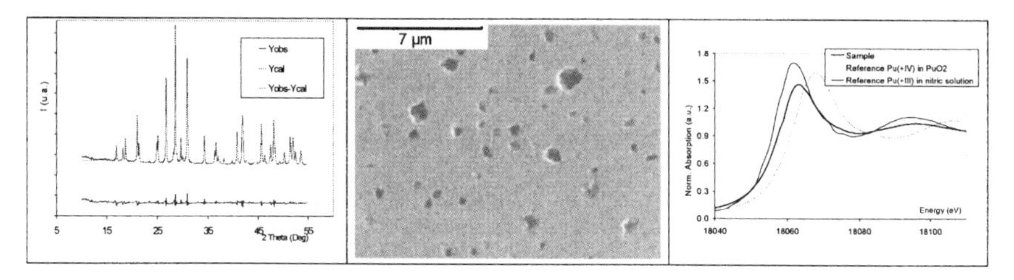

Figure 2 *XRD, SEM and XANES analyses of $La_{0,73}Pu_{0,09}Ca_{0,09}Th_{0,09}PO_4$.*

The results are in excellent agreement with the expected value. XRD revealed a homogeneous single-phase material. The insertion ratio x, deduced from the cell parameter, calculated by a Rietveld refinement, is 0.090 ± 0.05. The composition determined by EPMA analysis is $La_{0,71}Pu_{0,09}Ca_{0,1}Th_{0,1}PO_4$.

A XANES experiment, performed at ROBL-ESRF, confirms the trivalent state of the plutonium (max at 18063 eV). But that result needs to be used with caution, because of the difficulty in finding a reference sample containing only plutonium in a trivalent state.

3 CONCLUSION

The preparation of solid solution monazite-brabantite confining ^{239}Pu and thorium is successful. The sample is single-phase, dense and homogeneous. According to the XRD refinement and XANES measurement, the plutonium seems to be in a trivalent state. A leaching experiment will be performed on those samples. In order to study the effects of self-irradiation, those experiments will be repeated, using ^{238}Pu instead of ^{239}Pu.

References

1 N. Dacheux, N. Clavier, A.C. Robisson, O. Terra, F. Audubert, J.E. Lartigue and C. Guy, *Comptes Rendus Chimie,* 2004, **7**, 1141.
2 T.M. Harrison, E.J. Catlos, and J.M. Montel, *Rev. Min. & Geochem.,* 2002, **48**, 523.
3 Y. Narayana, P.K. Shetty and K. Siddappa, *Int. Cong. Ser.,* 2005, **1276**, 333.
4 K. Dias, C. V. Barros Leite and Z. Zays, *Nucl. Inst. & Met. B,* 2004, **217**, 649.
5 L. Bois, M. J. Guittet, F. Carrot and P. Trocellier, J. Nucl. Mat., 2001, **297**, 129.
6 E. H. Oelkers and F. Poitrasson, *Chem. Geol.,* 2002, **191**, 73.
7 A.M. Seydoux, R. Wirth and J.M. Montel, *Phys. Chem. Min,* 2002, **29**, 240.
8 B. Glorieux, F. Jorion, J.M. Montel, X. Deschanels and J.P. Coutures, Proceedings, *ATALANTE 2004*, **032**-04, June 21-24, 2004, Nîmes, Fr.
9 B. Glorieux, M. Matecki, F. Fayon and J.P. Coutures, *J. Nucl. Mat.,* 2004, **326**, 156.
10 Y. Hikichi and T. Ota, *Phosp. Res. Bul,* 1996, **6**, 175.

PHYSICAL AND ELECTRONIC PROPERTIES CHANGED BY AGING PLUTONIUM

B.W. Chung, J.G. Tobin, S.R. Thompson, and B.B. Ebbinghaus

Lawrence Livermore National Laboratory, P.O. Box 808, Livermore, California 94551, USA

1 INTRODUCTION

Plutonium, because of its radioactive nature, ages from the "inside out" by means of self-irradiation damage, and thus produces Frenkel-type defects and defect clusters. The defects resulting from the residual lattice damage and helium in-growth could result in microstructural, electronic, and physical property changes. This paper presents volume, density, and electronic property changes observed from both naturally and accelerated aged plutonium alloys. Accelerated alloys are plutonium alloys with a fraction of ^{238}Pu to accelerate the aging process by approximately 18 times the rate of unaged weapons-grade plutonium. After thirty-five equivalent years of aging on accelerated alloys, the samples have swelled in volume by approximately 0.12% and now exhibit a near linear volume increase due to helium in-growth. We will correlate the physical property changes to the electronic structure of plutonium, observed by the resonant photoelectron spectroscopy (RESPES).

2 METHOD AND RESULTS

We designed and installed a dilatometry system to monitor the long-term growth resulting from the residual lattice damage and helium in-growth in Pu-238 spiked alloys. The dilatometry design is based on the linear variable displacement transducer (LVDT) technology that will provide a continuous record of sample length change[1,2]. The system is periodically calibrated to maintain uncertainty in the measurement to within ±0.10 μm. During the current measurement, the sample's linear length has increased by 8.34 μm. This length change is converted to a volume change assuming expansion to be isotropic; this assumption is supported by the fact that the self-irradiation processes in plutonium are isotropic. Figure 1 shows the volume change (ΔV) normalized with the initial volume (V) of spiked alloys at 35°C. The volume has swelled by 0.12±0.01 %. The time is represented as an equivalent time (in years) obtained by multiplying the measurement time by the accelerating factor of 18. This accelerating factor is obtained by the decay rate of the spiked alloy normalized to the reference alloy. This factor will decrease as the material ages, due primarily to in-growth of Am-241 in the reference alloy. During the early stage of measurement, the sample's volume increases as a result of self-irradiation and follows an inverse exponential-type of expansion with dose (or time). This type of expansion is caused primarily by the lattice damage from both the uranium recoil and alpha particle emission. The amount of swelling in this case is approximately related to the number of Frenkel pairs that survive the radiation damage and subsequent recombination processes. The progressive accumulation of survivor vacancies provides an increasing number of alternate sites for the capture of self-interstitial plutonium atoms. As the density of these alternate sites increases, the rate of swelling thereby reduces. After the initial expansion, the volume exhibits a significantly lower rate of increase and a near linear expansion behaviour attributed to the helium in-growth mechanism. The production of radiogenic helium concentration is 41 appm (atomic parts per million) per year.

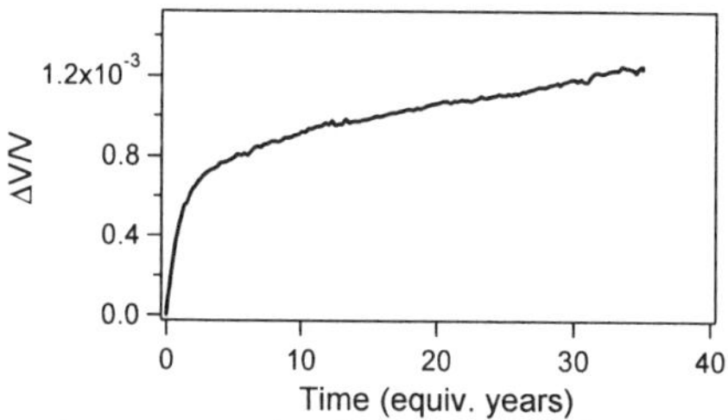

Figure 1 *The volume change normalized with the initial volume of spiked alloy at 35°C.*

Shown in Figure 2 are RESPES results from polycrystalline samples of δ-Pu (~2 atomic % Ga)[3]. A significant spectroscopic difference can be observed between the young and aged δ-Pu(Ga). Each exhibits the sigmoidal shape of RESPES: the pre-resonance at 90 eV photon energy, the minimum of the anti-resonance at 100 eV photon energy and the high intensity of the resonance, starting at a photon energy of about 120 eV. The aged sample appears to have a much more extended resonance range, with a high intensity at the Fermi Level, extending out to photon energies of 150 eV, while the young sample has a significant drop off as the photon energy moves through 140 eV, up to 160 eV.

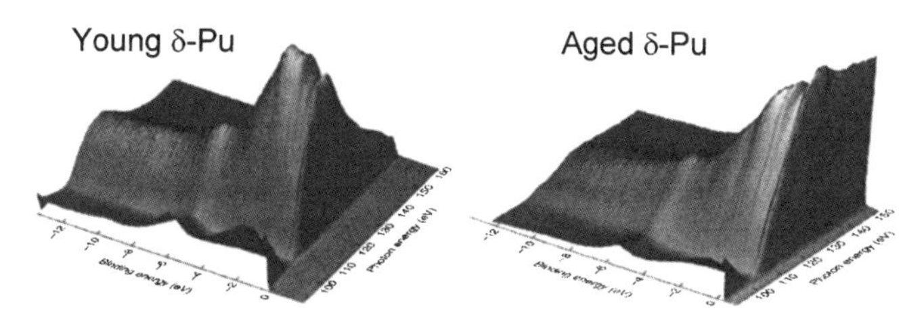

Figure 2. *RESPES comparison between young and aged δ-Pu(Ga)*[3].

The self-irradiation in plutonium alloys causes dimensional changes from the initial cascade damage (Frenkel-type defects) and helium in-growth in the plutonium alloy microstructure. These age-related phenomena will change the microstructure and influence the electronic property of plutonium alloys. Indeed, prior work has shown that the self-irradiation process (production of Frenkel-type defects and helium bubbles) increases the electrical resistance of plutonium alloys.[4,5] While the concentration of the transmutation decay products and impurities differ in the aged δ-Pu(Ga), their low concentration is unlikely to generate the observed resistance increase. Since the major difference between the young and aged δ-Pu(Ga) is their microstructure, this work suggests the valence electronic structure of Pu is dependent upon its age-related microstructure.

3 CONCLUSION

We have shown the age-dependent physical and electronic structure of plutonium. Self-irradiation damage causes both lattice damage and helium in-growth that increases volume (or decreases density) in plutonium. RESPES shows differences at the Fermi energy between young and aged δ-Pu(Ga).

Acknowledgements

This work was performed under the auspices of the U.S. Department of Energy by the University of California, Lawrence Livermore National Laboratory under Contract No. W-7405-Eng-48.

References

1 B.W. Chung, S.R. Thompson, C.H. Woods, D.J. Hopkins, W.H. Gourdin and B.B. Ebbinghaus, *Proc. Mater. Res. Soc., Symp.*, 2003, DD, Boston, MA.
2 B.W. Chung, S.R. Thompson, C.H. Conrad, D.J. Hopkins, W.H. Gourdin and B.B. Ebbinghaus, *J. Nucl. Mater.* (submitted).
3 J.G. Tobin, B.W. Chung, R.K. Schulze, J. Terry, J.D. Farr, D.K. Shuh, K. Heinzelman, E. Rotenberg, G.D. Waddill and G. Van der Laan, *Phys. Rev. B*. 2003, **68**, 1155109.
4 M.J. Fluss, B.D. Wirth, M. Wall, T.E. Felter, M.J. Caturla, A. Kubota, and T.Diaz de la Rubia, *J. Alloys Compds*, 2004, **368**, 62.
5 K.J.M. Blobaum, C.R. Krenn, J.J. Haslam, M.A. Wall, and A.J. Schwartz, *Proc. Mater. Res. Soc.*, 2004, **802**, 33.

CORROSION RESISTANCE OF THE URANIUM METAL SURFACE WITH CO AND $SCCO_2$ TREATMENT

X. Wang, G. Zhang, J. Yang and X. Fu

China Academy of Engineering Physics, P.O. Box 919-71, Mianyang 621900, Sichuan, People's Republic of China.

1 INTRODUCTION

Uranium is a highly reactive metal that can readily combine with gaseous molecules in the atmosphere. The oxidation of uranium in O_2, CO, CO_2, H_2, H_2O (v) and O_2/H_2O (v) mixtures has been the focus of many studies.[1-4] The protection of uranium against air-corrosion is of great importance in nuclear technology, especially for the long-term storage of non-irradiated components, e.g. in nuclear fuel manufacture.[5] Experimental studies indicate that the uranium surface treated with CO[6] or CO_2[7] may improve corrosion resistance, because the strong interaction between carbon and uranium would be expected to affect significantly the adsorption of oxygen and water vapour on the uranium surface and the subsequent oxidation.[8] However, the corrosion resistance of a uranium surface with CO or supercritical carbon dioxide ($SCCO_2$) treatment has not been fully investigated or well understood yet. In this work, the surface behaviour of clean uranium metal in a CO atmosphere at 25°C has been studied using XPS, and the corrosion resistance of the uranium surface with CO/$SCCO_2$ treatment is discussed.

2 METHOD AND RESULTS

2.1 XPS of Clean Uranium Metal in CO Atmosphere at 25 °C

High-purity (<120 ppm total metallic impurities) polycrystalline uranium metal coupons (Φ8 mm × 2 mm) were prepared and cleaned. The CO used in this work was of 99.99% purity. Figure 1 shows the $U4f_{7/2}$ fitting spectra and C1s spectra of clean uranium metal exposed to CO atmosphere at 25 °C. From the deconvolution spectra of Figure 1(a), it can be found that the clean metallic sample exhibited three different chemical states after exposure. The peak at 380.4 eV was attributed to UO_2 species, 378.6 eV to UC and 377.3 eV to metallic U, respectively. The peak at 282.0 eV of C1s in Figure 1(b) was attributed to UC. The surface reaction of uranium metal in a CO atmosphere is as follows,

$$3U\,(s) + 2CO\,(g) \rightarrow 2UC\,(s) + UO_2\,(s) \quad (1)$$

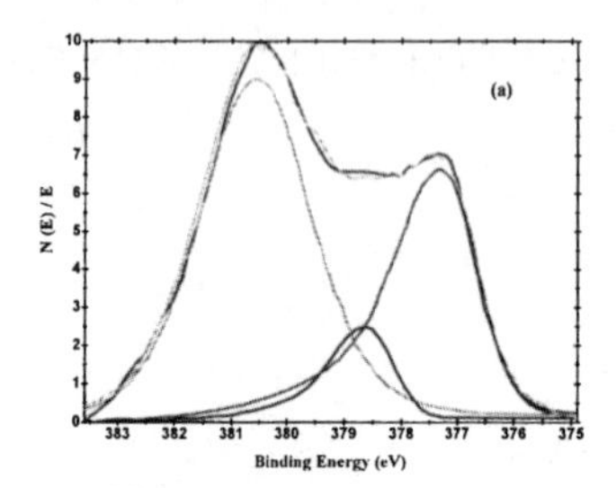

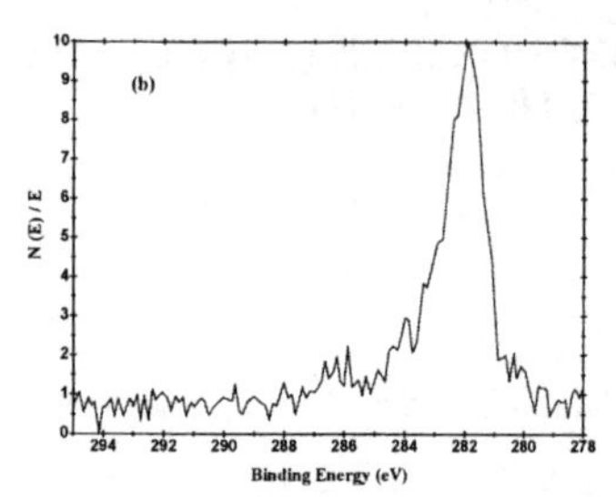

Figure 1 *XPS spectrum of uranium metal exposed to 40 L CO at 25°C (a) U4$f_{7/2}$ fitting curves; (b) C1s*

2.2 The Surface Passivation of Uranium Metal with CO Treatment

The surfaces of uranium specimens were passivated by CO under various conditions, and the corrosion resistance of those specimens in dry air at 70°C measured using the weight gain method. The experimental results for treated and untreated samples are given in Fig.2. There are two discrete stages of oxidation for every curve. In the first stage, the slopes are about 0.5, which indicates that the oxidation in the first stage seemed to fit a parabolic law. In the second stage, the rates of oxidation increase, and the slopes increase from 1 to 1.6 for different samples, and the oxidation rates of treated samples increase somewhat more rapidly than that of the untreated sample. There is a breakaway point existing for every sample. Before the breakaway point, the oxidation follows a parabolic law, however, after that, the oxidation deviates away from that parabolic law. From the break times at various doses of carbon monoxide, it can be concluded that, after uranium was treated by CO, the corrosion resistance of uranium has been enhanced and increased as the dose of CO increases in this experiment.

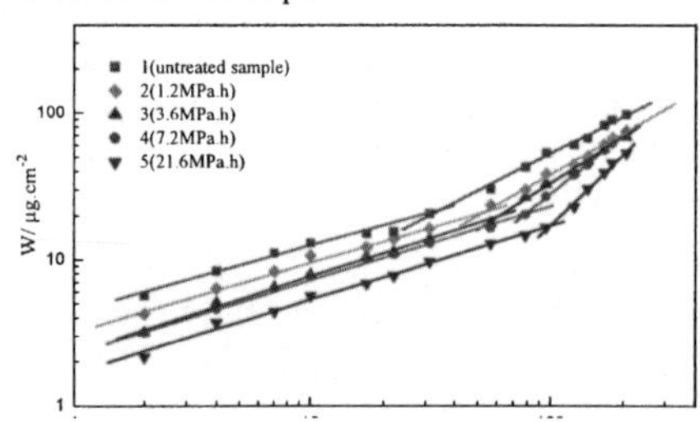

Figure 2 *Kinetic curves for the oxidation of uranium and uranium treated with CO in dry air at 70 ℃*

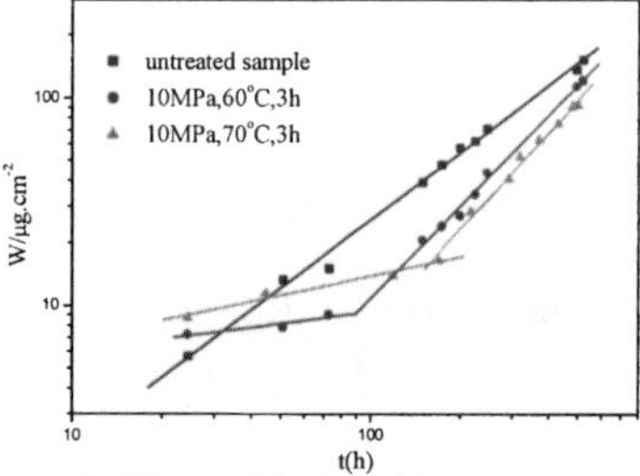

Figure 3 *The oxidation kinetic curves of uranium with little carbon impurity treated with $SCCO_2$ at different temperatures*

2.3 The Surface Passivation of Uranium Metal with $SCCO_2$ Treatment

The corrosion resistance of the uranium surface treated with $SCCO_2$ has been studied by the weight gain method. The oxidation kinetic curves of uranium have been obtained at 60°C, 70°C and 70% RH condition after the surface was treated with $SCCO_2$. The results (Fig. 3, 4, 5) show that the corrosion resistance of treated uranium (pressure, time and temperature) has been improved. The weight gain tendency of the treated sample in the first stage is very different from that of untreated sample. Comparing Fig. 3 and 4, the corrosion resistance of the treated sample in Fig. 4 is better than that in Fig.3. Though the treatment temperature of the sample in Fig.3 is higher than that in Fig. 4, the treated time in Fig. 4 is much longer. This means the formation of the passivation film is more effective under conditions of relatively low temperature and long treatment time. However, if the treatment time is longer than 24 hours at 10MPa and 50°C, i.e. when the time is about 48 hours, the passivation film may lose its protection. Because the passivation film is too thick, the film may be cracked by the high stress between the film and substrate. When the thickness of the passivation film is suitable, i.e. 10MPa, 50°C, 24 hours, the corrosion resistance of uranium surface will be much improved. Fig 5 shows that high pressure CO_2 can also do well in protecting the uranium surface, which may be because the products of the reaction of both $SCCO_2$ and high pressure CO_2 with uranium are alike.

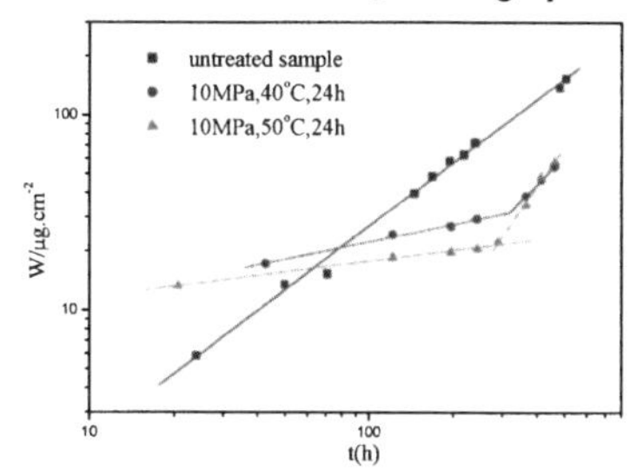

Figure 4 *The oxidation kinetic curves of uranium treated with $SCCO_2$ under different temperatures and long times*

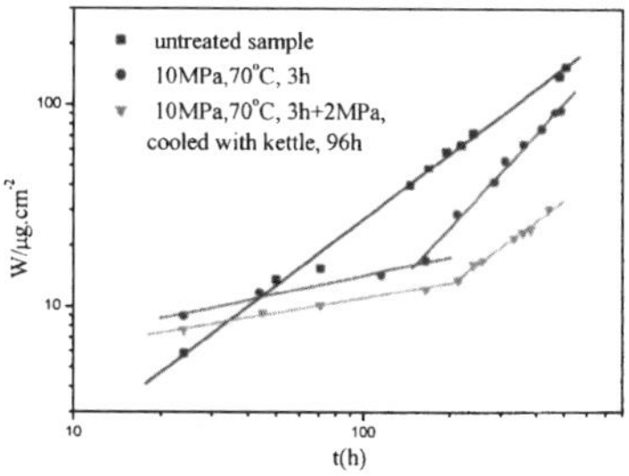

Figure 5 *The oxidation kinetic curves of uranium with small carbon impurity, treated with $SCCO_2$ and high pressure CO_2 under different temperatures*

3 CONCLUSION

Research on the interaction of uranium with CO or $SCCO_2$ is a very important subject of study for nuclear material compatibility and corrosion resistance. After the surface of uranium was treated with CO or $SCCO_2$, the corrosion-resistance of uranium was enhanced, and there is a carbon-rich layer formed at the surface layer. The thickness of the carbon-rich layer in the thin film increases as the exposed dose increases, which is mainly composed of (UO_2) and UC(UC_xO_y). The results also indicate that under the conditions used here (pressure, time and temperature), the corrosion resistance of treated uranium samples depends upon the treatment time and temperature. The composition and thickness of the passivation film may have an effect on the corrosion resistance. Further studies are required.

References

1 C.A. Colmenares, *Prog. Solid State Chem.,* 1984,**15**, 257.
2 K. Winer, C.A. Colmenares and R.L.Smith, *Surf. Sci.,* 1987,**183**, 67.
3 Y. Fu, X. Wang and Z. Zhu, *Chinese Eng. Sci.,* 2000, **2**, 59.
4 X. Wang, X. Fu and Z. Zhao, *Surf. Rev. Lett.,* 2003, **10**, 325.
5 R. Arkush, M.H. Mintiz and N. Shamir, *J. Nucl. Mater.,* 2000, **281**, 182.
6 J. Yang, J. Zou and C. Jiang, *Chinese J. Nucl. Radiochem.,* 2002, **24**, 34.
7 G. Zhang, W. Yang and X. Wang, *Chinese J. Atomic Sci. Technol.,* 2003, **37**, 233.
8 K.A. Winer, *Initial Stages of Uranium Oxidation,* PhD Thesis, UCRL-53655, 1985.

PYROCHEMICAL PROCESSING OF PLUTONIUM AT AWE

I.A. Sullivan, R.F. Watson, V.C. Freestone, A.H. Jones, T.J Paget and C.J. Bates

AWE, Aldermaston, Reading, Berkshire, RG7 4PR, UK.

1 INTRODUCTION

AWE has developed a number of pyrochemical processes to produce high purity plutonium from impure metal and oxide feeds. Details of these process are outlined below.

2 PLUTONIUM OXIDE PROCESSING

2.1 Direct Oxide Reduction

Plutonium oxide (PuO_2) is produced during the recovery of metal residues from processing equipment. The Direct Oxide Reduction (DOR) process is used to convert this impure plutonium oxide to metal.

The oxide is heated to 1100 K in a magnesia crucible. It is reduced with calcium in the presence of calcium chloride to dissolve the calcium oxide reaction product. The $CaCl_2$-CaO phase diagram[1] shows that the solubility of CaO is approximately 20 mol% at 1100K.

$$PuO_2 + 2Ca \rightarrow Pu + 2CaO \qquad \Delta G_{1100K} = -195 \text{ kJ mol}^{-1} \qquad (1)$$

About 95% of the plutonium is recovered as a metal product. However, the process does not refine the plutonium, any impurities present in the oxide will also be present in the metal. At the end of the process, the crucible is broken open and the metal removed from the salt, this metal product then undergoes casting and vacuum stripping prior to electrorefining.

3 PLUTONIUM METAL PROCESSING

3.1 Americium Extraction[2]

Impure plutonium metal requires electrorefining to produce pure plutonium. As plutonium ages the concentration of americium increases, as shown in Figure 2. This leads to an increased dose from the gamma emission (59.4 keV) associated with the α-decay of ^{241}Am. To minimise dose during electrorefining the americium is removed if the concentration in the feedstock metal is greater than 1000 ppm.

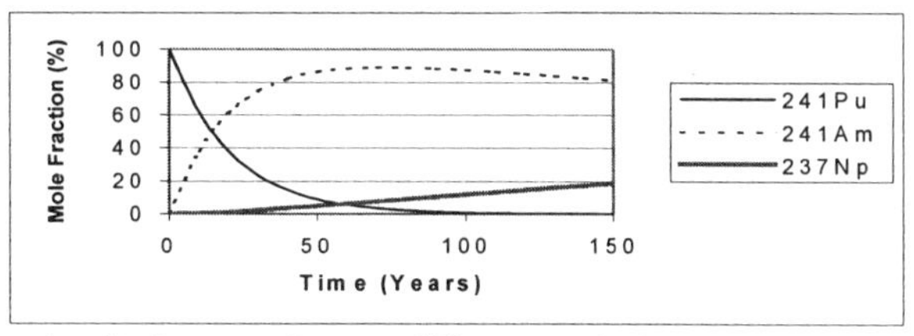

Figure 1 *Radioactive Decay of Plutonium-241*

An americium extraction process has been developed that removes approximately 90% of americium from the plutonium metal. The americium is extracted from the metal into a $CaCl_2$ salt phase by reaction with $MgCl_2$ at 1123K. The americium is concentrated in a small amount of salt which is readily recovered from the plutonium metal, the recovery of plutonium is greater than 98%.

$$2Am + 3MgCl_2 \rightarrow 2AmCl_3 + 3Mg \qquad \Delta G_{1123K} = -7 \text{ kJ mol}^{-1} \qquad (2)$$

$$2Pu + 3MgCl_2 \rightarrow 2PuCl_3 + 3Mg \qquad \Delta G_{1123K} = -79 \text{ kJ mol}^{-1} \qquad (3)$$

$$PuCl_3 + Am \rightarrow AmCl_3 + Pu \qquad \Delta G_{1123K} = -36 \text{ kJ mol}^{-1} \qquad (4)$$

3.2 Electrorefining

Impure plutonium metal is purified via an electrochemical process to give a pure metal product. Before the electrorefining, the impure plutonium is cast into a small cylindrical shape to act as the 'anode'. The anode is held in a central cup, inside a larger ceramic crucible.

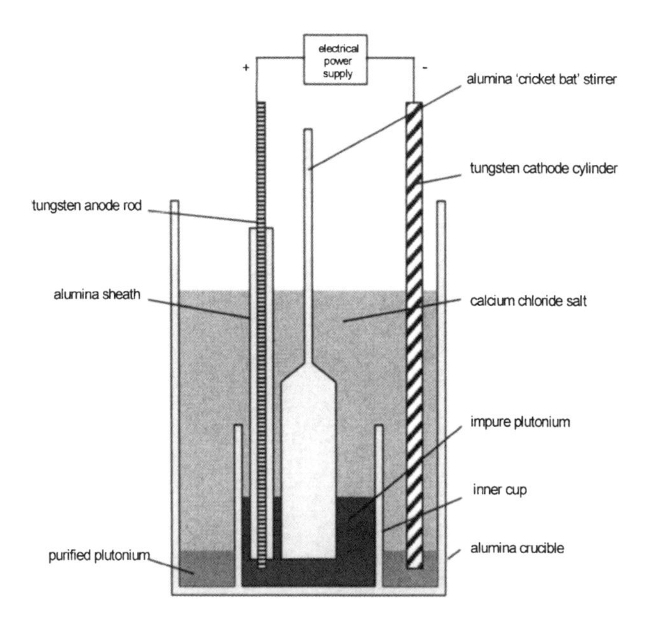

Figure 2 *Electrorefining Cell*

NaCl-KCl or $CaCl_2$ salt, which acts as the electrolyte, is loaded into the crucible and heated to 1023–1123 K, at which temperature both the salt and metal are molten. The salt and plutonium is stirred and the plutonium is oxidised at the anode, transported through the salt and reduced at the cathode. At the cathode, the plutonium collects in the bottom of the outer portion of the crucible to form a pure plutonium annulus.

During electrorefining, impurities more electropositive than plutonium concentrate in the salt phase e.g. americium. Those less electropositive remain at the anode, e.g. uranium and gallium.

The amount of plutonium that can be oxidised from the anode is limited by the formation of solid Pu-Ga phases above 20 mol% gallium. When the gallium concentration reaches 25-30 mol% the anode will solidify. If electrorefining were to continue gallium would oxidise as the solid anode surface becomes depleted in plutonium. To prevent this from happening, the open circuit potential between the anode and cathode is monitored. This rises rapidly as solid phases being to form, allowing the process to be stopped before impurities are carried over.

At the end of the process, the ceramic crucible is broken to recover the plutonium metal product ring. This process may yield up to a 90% recovery of pure plutonium, with some plutonium remaining in the anode and some dispersed throughout the salt.

The pure plutonium is then vacuum stripped to remove any volatile metals that may be present e.g sodium and potassium which have been reduced at the cathode from the salt.

4 CONCLUSIONS

The pyrochemical process developed at the AWE will continue to be developed to improve these processes by reducing operator dose, improving process efficiencies and reducing the associated arisings.

References

1 D.A. Wenz, I. Johnson and R.D. Wolsen, J. *Chem. Eng. Data*, 1969, **14**, 250.

2 R Watson and V Freestone, *Method for Separating Contaminant From Plutonium Metal*, UK Patent 2355107, 11/04/2001.

URANIUM ION VALENCES IN PEROVSKITE, ZIRCONOLITE AND PYROCHLORE FROM NEAR INFRARED DIFFUSE REFLECTANCE SPECTROSCOPY

Y. Zhang, E.R. Vance, K.S. Finnie, B.D. Begg and M.L. Carter

Australian Nuclear Science and Technology Organisation, PMB 1, Menai 2234, Australia

1 INTRODUCTION

Synroc, an assemblage of titanate minerals, is a candidate matrix for immobilisation of HLW from spent fuel reprocessing.[1] The Synroc phases, perovskite ($CaTiO_3$) and zirconolite ($CaZrTi_2O_7$) can incorporate actinide elements in their crystal structures.[2] Ceramic formulations based on the pyrochlore-structured phase ($CaATi_2O_7$, where A = actinide), in which brannerite (UTi_2O_6) was a minor phase, were also extensively studied for immobilising surplus impure weapons plutonium.[3] Previous work indicated that perovskite[4], zirconolite[5], pyrochlore[6] and brannerite[7] can incorporate U^{4+}, and of these all but perovskite can also incorporate U^{5+} by adding appropriate charge compensators or by heating in air.

Diffuse reflectance spectroscopy (DRS) is a useful tool for deriving information about actinide speciation in the solid state and has been used to probe uranium valences in polycrystalline brannerite.[7] A number of intraconfigurational *f-f* electronic absorption bands for $U^{4+}(5f^2)$ and $U^{5+}(5f^1)$ ions have been observed over the near infrared (NIR) range (4000-12000 cm^{-1}). It is of interest in the nuclear waste form application to study systematically the DRS spectra of U^{4+}/U^{5+} ions in other polycrystalline titanates, e.g. perovskite, zirconolite and pyrochlore, and compare them with those of brannerite.

2 EXPERIMENTAL

Samples doped with various amounts of U, targeted towards the Ca or Zr sites of zirconolite were made by the standard alkoxide/nitrate route[8] and sintered at 1400-1500°C for 16-20 hrs under air or argon atmospheres. The zirconolite glass-ceramic (1.26 wt% of U) was prepared by an oxide/nitrate route and hot isostatically pressed at 1220°C/ 100 MPa under an argon atmosphere for 2 hrs. X-ray diffraction and scanning electron microscopy confirmed that perovskites and zirconolites were essentially single-phase.

Roughly polished (200 grit SiC) disc samples were used for DRS measurements at ambient temperatures over the NIR range using a Cary 500 spectrophotometer equipped with a Labsphere Biconical Accessory. Spectra were converted into Kubelka-Munk units, $F(R)=(1-R)^2/2R$, by reference to the spectrum of a Labsphere certified standard (Spectralon).

3 RESULTS AND DISCUSSION

Up to 0.2 f.u. of U^{4+} can be incorporated in the Ca site of perovskite with charge compensation by Ca vacancies or adding Al^{3+} to the Ti site and sintering in argon.[4] Up to 0.4 f.u. of U^{4+} can be incorporated in the Zr site of zirconolite by sintering in argon.[5] Over 0.6 f.u. of U^{4+} in the Zr site gives rise to pyrochlore.[5] About one-third of the Ca can be replaced by U^{4+} before the zirconolite structure changes to pyrochlore.[5]

The DRS spectra of U^{4+} ions in perovskite, zirconolite, pyrochlore, a zirconolite glass-ceramic and a Th-brannerite[7] are shown in Figure 1A together with U^{4+} free ion energies[9] shown at the bottom of the graph. U^{4+} in perovskite has a few strong bands in the NIR and band broadening and splitting are obvious. Like the U^{4+} ion in $ThTi_2O_6$, U^{4+} ions in zirconolite, pyrochlore-rich ceramic and zirconolite glass-ceramic have numbers of weak bands in the NIR, and show rather similar spectral features.

The Ca site in perovskite has orthorhombic point symmetry, whilst the Ca and Zr sites in zirconolite have 8- and 7-fold coordinations, respectively, with no symmetry elements. As a result, there are additional bands and more band splittings in the case of zirconolite.

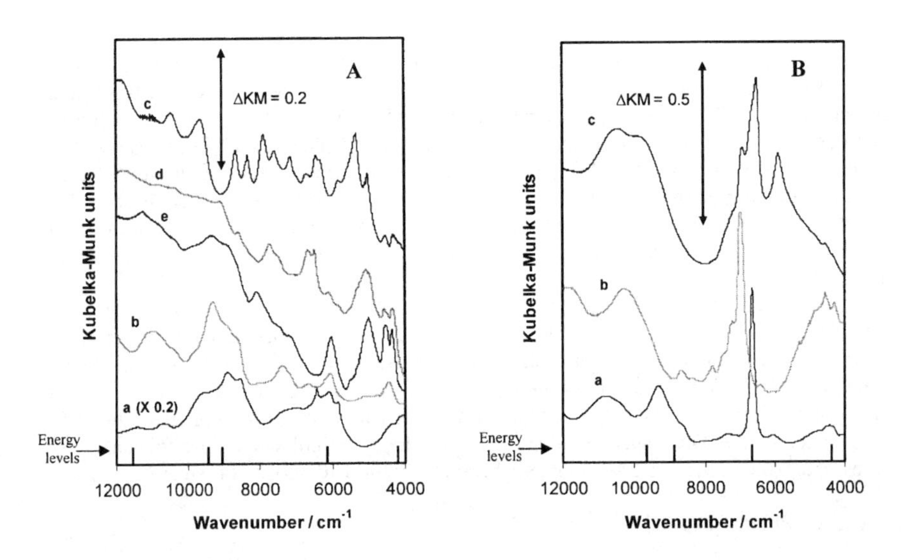

Figure 1 **A** *DRS spectra of U^{4+} ions in:* ***a****-perovskite (0.005 f.u.),* ***b****-zirconolite glass-ceramic,* ***c****-$ThTi_2O_6$ (0.008 f.u.),* ***d****-pyrochlore (0.67 f.u.) and* ***e****-zirconolite (0.01 f.u.) with free-ion energy levels of U^{4+} at the bottom*: **B** *DRS spectra of U^{5+} ions in:* ***a****-zirconolite (0.01 f.u.),* ***b****-$ThTi_2O_6$ (0.008 f.u.) and* ***c****-pyrochlore-rich ceramic (0.36 f.u.) with the energy levels of U^{5+} in $UCl_5{\cdot}AlCl_3$ at the bottom.*

No evidence of U^{5+} was observed in perovskite by trying to substitute it in either the Ca or the Ti site with appropriate charge compensation (and heating in air).[4] U^{5+} ions can be stabilised in air-heated zirconolite and pyrochlore-rich ceramic without additional charge compensators.[5]

DRS spectra of U^{5+} ions in zirconolite and pyrochlore-rich ceramic are shown in Figure 1B, together with U^{5+} bands[10] in $UCl_5{\cdot}AlCl_3$ marked at the bottom of the graph. In contrast to the DRS spectra of U^{4+}, DRS spectra of U^{5+} are much simpler with only a small number of intraconfigurational *f-f* transition bands in the NIR region, and they are fairly similar for different titanate ceramics. The U^{5+} bands shift slightly to the IR direction when U coordination numbers increase, from 6-fold for brannerite to 8 and/or 7 for pyrochlore and zirconolite.

Generally, The Kubelka-Munk relationship is valid for U^{4+}/U^{5+} ions over the low concentration range of 0.003-0.01 f.u.

4 CONCLUSION

DRS spectra have been collected for U^{4+} ions in polycrystalline perovskite, zirconolite and pyrochlore, and for U^{5+} ions in zirconolite and pyrochlore over the NIR range at room temperature. In general, the Kubelka-Munk relationship is valid for U^{4+}/U^{5+} ions over the low concentration range of 0.003-0.01 f.u. Compared to the free ion energy levels, the observed band shifts, splittings, intensities and additional bands are related to the actual U ion site symmetries in the ceramic crystal structures. DRS spectra of U^{4+}/U^{5+} in ThO_2, as well as Pu^{4+} and Np^{4+} in titanate ceramics are currently being investigated.

References

1 A.E. Ringwood, S.E. Kesson, N.G. Ware, W. Hibberson and A. Major, *Nature (London)*, 1979, **278**, 219.
2 E.R. Vance, Materials Research Society, *MRS Bulletin*, 1994, Vol. **VIX**, 28.
3 A. Jostsons, E.R. Vance and B. Ebbinghaus, Immobilisation of surplus plutonium in titanate ceramics. In: *Global'99*: *Nuclear Technology – Bridging the Millennia* (CD-ROM), 1999, American Nuclear Society, La Grange Park, IL .
4 E.R. Vance, M.L. Carter, Z. Zhang, K.S. Finnie, S.J. Thomson and B.D. Begg, *Environmental Issues and Waste Management Technologies* **IX**, 2004, American Ceramic Society, Columbus, OH, USA, Edited by J.D. Vienna and D.R. Spearing, pp. 3-10.
5 E.R. Vance, G.R. Lumpkin, M.L. Carter, D.J. Cassidy, C.J. Ball, R.A. Day and B.D. Begg, *J. Am. Ceram. Soc.*, 2002, **85**(6).
6 M.W.A. Stewart, E.R. Vance, A. Jostsons and B.B. Ebbinghaus, *Journal of the Australasian Ceramic Society*, 2003, **39**, 130.
7 K. Finnie, Z. Zhang, E. R. Vance and M. L. Carter, *J. Nucl. Mater.*, 2003, **317,** 46.
8 A.E. Ringwood, S.E. Kesson, K.D. Reeve, D.M. Levins and E.J. Ramm, in *Radioactive Waste Forms for the Future*, Eds. W. Lutze and R.C. Ewing, 1988, Elsevier, Netherlands, pp. 233-334 .
9 W.T. Carnall and H.M. Crosswhite, *ANL-84-90*, Argonne National Laboratory, USA.
10 D.M. Gruen and R.L. McBeth, *Inorg. Chem.*, 1969, **8**, 2625.

THERMAL AND MECHANICAL PROPERTIES OF SIMULATED HIGH BURNUP URANIUM OXIDE FUELS

S. Yamanaka[1], K. Kurosaki[1], M. Uno[1] and Y. Ishii[2]

[1]Department of Nuclear Engineering, Graduate School of Engineering, Osaka University, 2-1 Yamadaoka, Suita, Osaka 565-0871, Japan
[2]Tokyo Electric Power Company, 4-1 Eagsaki-cho, Tsurumi-ku, Yokohama 230-8510, Japan

1 INTRODUCTION

It is very important to understand the behaviour of fission product (FP) and transuranium (TRU) elements in an irradiated oxide nuclear fuel, because the physical and chemical properties of the fuel are affected by the FP and TRU elements.

The simulated high burnup UO_2 based fuel (SIMFUEL) has an advantage in the study of the high burnup effects of FP elements on the physical properties of the irradiated fuel. SIMFUEL is a material that replicates the chemical state and phase relation of high burnup oxide fuel. It has been studied since 1962. Recently, Lucuta et al.[1] and Matzke et al.[2] studied the effect of FP elements upon the thermophysical and thermochemical properties of fuel, using SIMFUEL with equivalent simulated burnups of 3 at.% (corresponding to 28 GWd/t) and 8 at.% (corresponding to 75 GWd/t).

In the present study, to evaluate the effect of solid state FP elements on the properties of irradiated fuel in an extremely high burnup region (up to 150 GWd/t), we prepared simulated high burnup UO_2 fuels, and measured their physical properties.

2 METHOD AND RESULTS

The composition of the simulated high burnup UO_2 fuels was determined by a burnup calculation. The compositions of the simulated high burnup UO_2 fuels with simulated burnups of 50, 100, and 150 GWd/t are summarized in Table 1. Appropriate amounts of high-purity oxides of the representative elements listed in Table 1 were mechanically mixed with UO_2 powder and calcined at 2023 K. The powder thus obtained was pressed into a pellet and sintered in a reducing atmosphere at 2023 K. In order to evaluate the effect of the oxygen content on the properties of the simulated fuels, the samples were annealed at 1400 K (40 hours), under three kinds of oxygen potentials (-390, -340, and -220 kJ/mol).

The phase identification of the simulated fuels was carried out with a powder X-ray diffraction method, using Cu Kα radiation at room temperature. The lattice parameter of the matrix phase was determined from the diffraction patterns. The total oxygen content in the samples was measured using an oxygen and nitrogen analyzer.

The specific heat capacity was measured using a differential scanning calorimeter (DSC) apparatus, in the temperature range from 300 to 1200 K in an Ar flow atmosphere.

The thermal conductivity was calculated from the thermal diffusivity, heat capacity, and density. The thermal diffusivity was measured under vacuum by the laser flash method in the temperature range from 300 to 1200 K. An ultrasonic pulse-echo measurement was carried out at room temperature in air. The Young's modulus and Shear modulus were evaluated from the measured sound velocity.

Table 1 *Compositions of the simulated high burnup fuels*

6.2 wt.%-U235	Simulated burmup					
	50.0 GWd/tM 5.24 at.%		100.0 GWd/tM 10.46 at.%		150.0 GWd/tM 15.47 at.%	
Representative elements	mol%	wt.%	mol%	wt.%	mol%	wt.%
U	92.48	96.20	85.428	92.34	78.90	88.48
Ba	0.762	0.457	1.363	0.850	1.944	1.257
Zr	1.430	0.570	2.427	1.005	3.227	1.387
Mo	1.456	0.611	2.653	1.156	3.655	1.652
Ru	0.723	0.320	1.679	0.771	2.688	1.280
Rh	0.127	0.057	0.146	0.068	0.140	0.068
Pd	0.303	0.141	1.130	0.546	2.108	1.057
Y	0.208	0.081	0.319	0.129	0.392	0.164
La	0.325	0.197	0.726	0.458	1.147	0.751
Ce	0.747	0.457	1.348	0.858	1.797	1.186
Nd	1.436	0.905	2.779	1.820	3.998	2.716
TOTAL	100.00	100.00	100.00	100.00	100.00	100.00

Figure 1 shows the X-ray diffraction patterns of the simulated burnup fuels (150 GWd/t) under three kinds of annealing conditions. There are some peaks corresponding to the matrix phase (CaF_2 type structure) and oxide precipitates in all samples. It is thought that the oxide precipitates form a perovskite-type structure, expressed as $Ba(U,Zr,Mo)O_3$ at the annealing oxygen potentials of -390 and -340 and a scheelite-type oxide structure, expressed as $Ba(U,Mo)O_4$ at the annealing oxygen potential of -220 kJ/mol. Metallic precipitates as the ε-phase composed of Mo, Ru, Rh, and Pd are clearly observed in the samples annealed under the oxygen potentials of -390 and -340 kJ/mol.

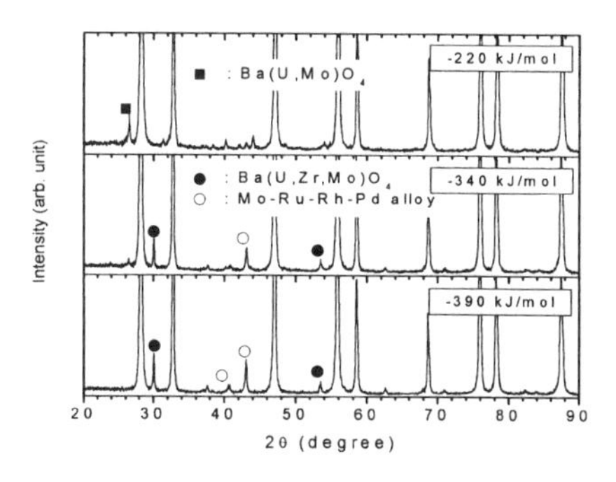

Figure 1 *X-ray diffraction patterns of the simulated burnup fuels (150 GWd/t)*

Figure 2 *Specific heat capacity of uranium dioxide*

Figure 2 shows the specific heat capacity of uranium dioxide, together with the literature data.[2] $UO_{2.001}$ and $UO_{2.029}$ were obtained from the annealing process at the oxygen potentials of -390 and -220 $kJmol^{-1}$, respectively. The specific heat capacity of $UO_{2.001}$ measured in the present study is in good agreement with that of stoichiometric

UO_2.[2] The measured specific heat capacity of $UO_{2.029}$ is slightly higher than that of stoichiometric $UO_{2.001}$, but the difference is inconspicuous.

Figure 3 shows the thermal conductivities of UO_2 and simulated fuels annealed at the oxygen potential of -390 $kJmol^{-1}$. The values of the thermal conductivity were normalized to the values for 95 % of the theoretical density. The thermal conductivities decrease with increasing temperature, indicating a phonon conduction characteristic. The thermal conductivities also decrease with increasing simulated burnup, which is due to the phonon scattering caused by the dissolved FP elements. The reduction rate of the thermal conductivity caused by increasing burnup appeared to be small. This is because of metallic precipitates, such as the Mo-Ru-Rh-Pd alloy that should increase thermal conductivity. It is expected that the thermal conductivity of the metallic precipitates is about 10 times larger than that of the matrix phase. The thermal conductivity of real irradiated fuel is smaller than those of the simulated fuels, because of irradiation effects, such as irradiation-induced point defects.[3]

Figure 4 shows the elastic moduli of the simulated fuels, as a function of burnup. It is found that the elastic moduli decrease linearly with increasing burnup. No effect of the annealing oxygen potential on the elastic modulus is observed. The reduction rates in the elastic moduli are about 6.5 % per 50 GWd/t burnup for the shear modulus and about 7 % per 50 GWd/t burnup for the Young's modulus, respectively.

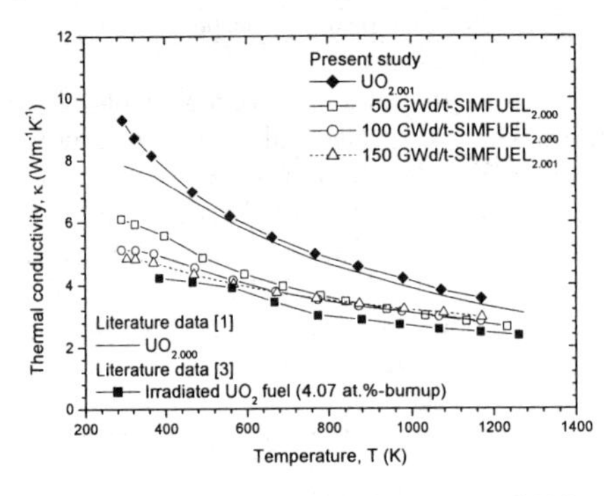

Figure 3 *Thermal conductivities of UO_2 and simulated fuels*

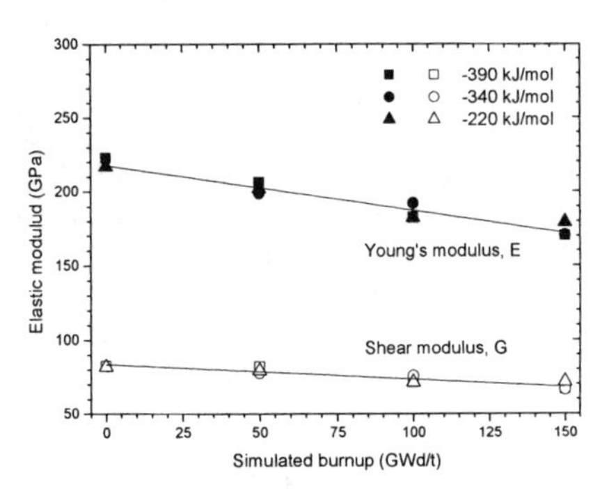

Figure 4 *Elastic moduli of UO_2 and simulated fuels*

3 CONCLUSION

Simulated high burnup UO_2 fuels with burnups of 50, 100, and 150 GWd/t were prepared and their physical properties were evaluated. Thermal conductivities decrease with increasing burnup. The reduction appears to be small with increasing burnup, due to metallic precipitates. The elastic moduli decrease with increasing burnup. The reduction rate for the Young's modulus is about 7 % per 50 GWd/t burnup.

References

1 P.G. Lucuta, H. Matzke and R.A. Verrall, *J. Nucl. Mater.*, 1995, **223**, 51.

2 H. Matzke, P.G. Lucuta, R.A. Verrall and J. Henderson, *J. Nucl. Mater.*, 1997, **247** 121.

3 K. Minato, T. Shiratori, H. Serizawa, K. Hayashi, K. Une, K. Nogita, M. Iría and M. Amaya, *J. Nucl. Mater.*, 2001, **288**, 57.

THERMODYNAMIC MODELLING AND ANALYSES OF MOLTEN SALT REACTOR FUEL

J.P.M. van der Meer[1], R.J.M. Konings[1] and H.A.J. Oonk[2]

[1]European Commission, Joint Research Centre, Institute for Transuranium Elements, P.O. box 2340, 76125 Karlsruhe, Germany
[2]Department of Geosciences, Utrecht University, Budapestlaan 4, 3584 CD Utrecht, Netherlands

1 INTRODUCTION

The Molten Salt Reactor (MSR) is one of six concepts of the Generation IV reactors. The MSR can be designed as a breeder, using the $^{232}Th/^{233}U$ cycle, or as a burner of plutonium and other actinides. Its fuel is the fissile phase dissolved in a matrix of molten fluorides. With a breeder design, a $^{7}LiF\text{-}BeF_2$ mixture is a favorable solvent, because the neutronics optimise ^{233}U breeding from ^{232}Th via ^{233}Pa. Using the MSR as a burner, a LiF-NaF eutectic composition would be preferable, because of the high solubility of actinide fluorides.

Thermodynamic models based on extrapolation from binary subsystems make it possible to predict the phase behaviour of multicomponent systems. In this study, thermodynamic assessments have been performed, making use of literature data. Also, DSC experiments have been carried out for phase diagram measurements.

Two examples are discussed here. The first is the modelling and experimental work of LiF-NaF-LaF_3, which has been used as a proxy for LiF-NaF-PuF_3. This is a possible starting mixture for the MSR burner, for which a calculated diagram is presented. The second is the modelling of the quaternary system LiF-BeF_2-ThF_4-UF_4, the selected mixture for MSR breeder fuel.

2 EXPERIMENT

Thermal analysis has been performed on binary and ternary mixtures of LiF (99.98%, met. bas.), NaF (99.99%, met. bas.) and LaF_3 (99.99%, REO), from Alfa Aesar. The fluoride salts, which are known to be hygroscopic to different extents, possibly enhancing the creation of oxyfluorides when heated, were dried under Ar 5.0 for 4 hours at 250°C and immediately brought to a Ar glovebox. There, boron nitride crucibles, grade AX05, were filled with the salt samples and transported to a Netzsch STA 449C Jupiter, where the analyses were made under a continuous flow of Ar 5.0.

3 MODELLING

Gibbs energy functions for the pure compounds were set up after careful investigations of thermodynamic tables.[1] For the unknown Gibbs energy equations of the intermediate compounds in the binary systems, the C_p was assumed to be the sum of the pure compounds, while the enthalpy and entropy of formation were optimised.

The excess Gibbs energy was described by Redlich-Kister polynomials, which are defined for binary systems A-B as in Eq. 1.

$$\Delta_{xs}G = X_A X_B \Sigma_k \, {}^kL_{A,B}(X_A - X_B)^k \quad (1)$$

The coefficients ${}^kL_{A,B}$ were optimised and they are of the form as in Eq. 2.

$$ {}^kL_{A,B} = {}^kp_{A,B} + {}^kq_{A,B}\,T \quad (2)$$

A good description with a minimum of parameters was obtained with this method. The software package of FactSage 5.3[2] was utilised to perform the optimisations. Numerical values can be found in Van der Meer et al.[3,4]

Extrapolation of the binary excess terms to ternary systems was done according to the Kohler-Toop method, which is a so-called asymmetrical method. This method is justified, because the systems concerned are chemically asymmetric, as in LiF-NaF-LaF_3, where the last compound is chemically different from the first two. The binary interactions are given a different weight in the ternary system.

4 EXAMPLES AND DISCUSSION

4.1 LiF-NaF-XF_3

The binary subsystems of LiF-NaF-LaF_3 were assessed using literature as well as our own experimental data. From the extrapolated excess Gibbs terms, the ternary diagram was calculated and checked by DSC measurements. The resulting diagram is shown in Figure 1. The overall agreement of the diagram and its experimental points is good, generally between ± 15 K.

Experimental data on the LiF-NaF-PuF_3 ternary system are not available yet, but the ternary diagram was calculated in an analogous way, which is shown in Figure 2.

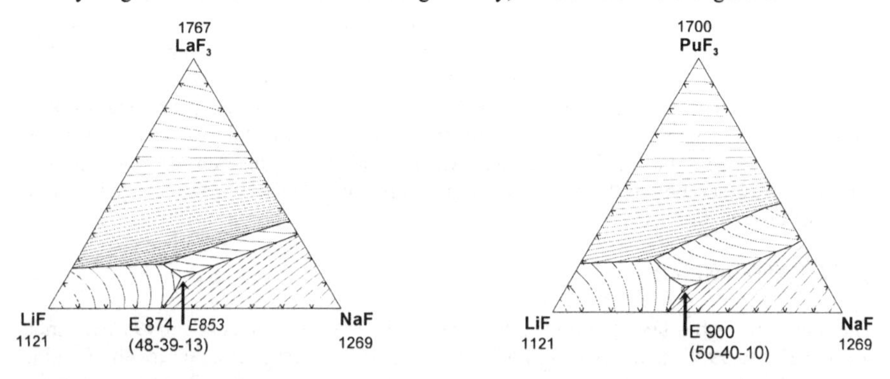

Figure 1 *Calculated liquidus surface of LiF-NaF-LaF_3, with an interval of 25 K.*

Figure 2 *Calculated liquidus surface of LiF-NaF-PuF_3, with an interval of 25 K. The temperature of melting of the pure compounds is indicated in K, as well as the calculated temperature of the ternary eutectic points. The experimental eutectic temperature is in italics. The composition is shown as percentages LiF-NaF-XF_3.*

4.2 $LiF-BeF_2-ThF_4-UF_4$

The ternary diagrams of $LiF-BeF_2-ThF_4$, $LiF-BeF_2-UF_4$, $LiF-ThF_4-UF_4$ and $BeF_2-ThF_4-UF_4$ have been calculated in a similar way. In the case of $LiF-ThF_4-UF_4$ and $BeF_2-ThF_4-UF_4$, the first component was selected as the asymmetric one. In the case of $LiF-BeF_2-ThF_4$ and $LiF-BeF_2-UF_4$, BeF_2 was selected, because molten BeF_2 behaves chemically different, as a viscous glassformer. Miscibility gaps have been found in the optimisation process in the BeF_2-rich corner of the $LiF-BeF_2-ThF_4$ and $LiF-BeF_2-UF_4$ diagrams, as has been found in the binary $LiF-BeF_2$.[4] Figures 3 and 4 show two pseudo-ternary diagrams. Figure 3 is the diagram for $LiF-BeF_2-(Th_{0.9756}U_{0.0244})F_4$, because the ThF_4/UF_4 ratio is the same as in the proposed breeder fuel composition 71.7 LiF- 16 BeF_2- 12 ThF_4- 0.3 UF_4. The diagram in Figure 4 is for $LiF-BeF_2-(Th_{0.9}U_{0.1})F_4$ to show the influence of an increase of UF_4. The differences are small, however, it can be seen that with a higher amount of UF_4, the region of demixing in the BeF_2 corner is slightly larger. There is a small decrease in the temperature of the three ternary eutectic points as well.

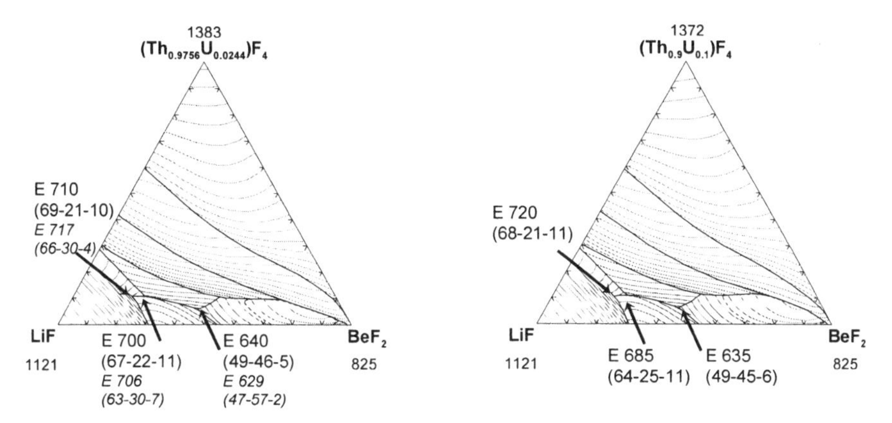

Figure 3 *Calculated liquidus surface $LiF-BeF_2-(Th_{0.9756}U_{0.0244})F_4$ with an interval of 25 K. In italics are experimental data from Thoma et al.[5] for $LiF-BeF_2-ThF_4$.*

Figure 4 *Calculated liquidus surface $LiF-BeF_2-(Th_{0.9}U_{0.1})F_4$ with an interval of 25 K. The temperature of melting of the compounds is indicated in K, as well as the temperature of the ternary eutectic points. Their composition is shown as percentages $LiF-BeF_2-(Th_{1-x}U_x)F_4$. The miscibility gap is indicated with a dashed line.*

References

1 M. Chase Jr.(ed.), NIST-JANAF, *J. Phys. Chem. Ref. Data Monograph 9*, 1994.
2 C. Bale, P. Chartrand, S. Degterov, G. Eriksson, K. Hack, R. BenMahfoud, J. Melançon, A. Pelton and S. Petersen, *CALPHAD,* 2002, **62**, 189.
3 J. van der Meer, R. Konings, K. Hack and H. Oonk, *CALPHAD, submitted*
4 J. van der Meer, R. Konings, M. Jacobs and H. Oonk, *J. Nucl. Mater.*,2005, in press.
5 R. Thoma, H. Insley, H. Friedman and C. Weaver, *J. Phys. Chem.*, 1960, **64**, 865.

STRUCTURE OF THE AMERICIUM PYROCHLORE $Am_2Zr_2O_7$ AND ITS EVOLUTION UNDER ALPHA SELF-IRRADIATION

R.C. Belin[1], P.J. Valenza[1] and P.E. Raison[2]

[1] Commissariat à l'Energie Atomique CEA-Cadarache DEN/DEC/SPUA/LMPC
13108 Saint-Paul-Lez-Durance, France.
[2] European Commission, Joint Research Center - Institute for Energy P.O. box 2,
1755 ZG Petten (N.H.), The Netherlands

1 INTRODUCTION

Nuclear waste disposal is after safety the main issue for the nuclear industry, both in terms of scientific challenge as well as public acceptance. Among the different options that have been envisioned and explored for minor actinides over the past thirty years, two alternatives currently remain: long term disposal in a safe repository or nuclear waste "burning" in a so-called transmutation process, both options being generally ultimately related. To fulfill the needs for long-term disposal, geological-time-resistant materials to be employed have to sustain both lixiviation and self-irradiation. This latter should not turn the material into an amorphous one, since lixiviation would then be dramatically enhanced. A good knowledge of radiation damage is thus a key prerequisite to selecting the suitable material. In that prospect, we have turned our attention to zirconium ceramic oxides, namely pyrochlores of formula $An_2Zr_2O_7$ (An = actinide), which have been proposed for long term radioactive waste disposal,[1] as well as for transmutation of actinides.[2]

Since Am is one of the main long-lived radioactive elements generated in reactors during the fission process, we have focused here on studying $Am_2Zr_2O_7$. Polycrystals were synthesized, and a structural analysis by the Rietveld method was followed by an X-ray diffraction study of the effect of α self-irradiation on the structure as a function of time.

2 EXPERIMENTAL

AmO_2 was handled in a special glove-box, equipped with biological protections. Only small batches of about 10mg were used for each preparation. Prior to synthesis, polycrystalline AmO_2 was calcined in air at 1173 K for 4 hours to bring the O/M ratio close to 2, and to eliminate volatile impurities. $Am_2Zr_2O_7$ was prepared with the starting polycrystalline materials AmO_2 and ZrO_2 (Cerac, 99.995% -325 mesh). Both compounds were ground/mixed with the appropriate ratios and fired in a Mo crucible under a reducing atmosphere (Ar/5% H_2) at 1673 K for 55 hours. The process was repeated twice. X-ray samples were mixed with epoxy resin according to the procedure previously described[3] and placed into a special in-house X-ray sample holder under N_2 atmosphere to prevent oxidation of the compound. Measurements were performed at RT with a Siemens D5000 X-ray diffractometer using Cu-α_1 radiation. The diffraction patterns were obtained after

72 hours by scanning with counting steps of 42 seconds, from 25° to 148° 2Θ using 0.02° step-intervals.

3 RESULTS AND DISCUSSION

The pyrochlore lattice is f.c.c with 8 formulas/unit cell (S.G.=Fd3m, N°.227). Its crystal structure can be described as a fluorite (F) superstructure with a double unit cell. Therefore, X-ray patterns for the pyrochlore (P) and the fluorite (F) structures are only differentiated by superstructure Bragg reflections, due to the cationic long-range order in the pyrochlore. Table 1 shows the atomic positions with Zr taken as origin. All positions are fixed by symmetry, except for O', having one unknown coordinate ***x=0.341*** which was determined in the present studies by Rietveld analysis (R_p=3.93, R_{wp}=4.94, R_{Bragg}=3.01). The detailed structural analysis of $Am_2Zr_2O_7$ is reported in another paper.[4]

Table 1 *Atomic coordinates for the $Am_2Zr_2O_7$ pyrochlore structure.*

Atom	Site	*x*	*y*	*z*
Am	16d	0.5	0.5	0.5
Zr	16c	0	0	0
O'	48f	***0.341***	1/8	1/8
O	8b	3/8	3/8	3/8

To follow the evolution of the structure under the effect of α self-irradiation, XRD measurements were performed on a regular basis after synthesis. Fig. 1 shows the evolution of the X-ray patterns as a function of time. One can observe that the superstructure peaks (marked *) slowly vanish and completely disappear after about 200 days, corresponding to a cumulative dose of 1.24×10^{14} α-decay events/mg. Each alpha decay event involves a He ion of 5-6 MeV energy, stopped by electronic interactions over 10-20 μm generating about 200 displacements along its trajectory and a heavy recoil atom of ^{237}Np of 91 keV energy, stopped by nuclear interactions over a much shorter distance (i.e. 20 nm), but associated with about 1500 atomic displacements.[5] The new X-ray pattern accounts for a more disordered fluorite-type structure, F (S.G.=Fm3m, N°.225). Am and Zr atoms share the same crystallographic site 4a(0,0,0), while O atoms are randomly occupying 7/8 of the 8c positions (1/4,1/4,1/4). An identical behaviour was observed for the Cf pyrochlore.[6]

Fig. 1 insert shows that the cell parameter decreases slightly, corresponding to a lattice shrinking on the order of 1% after 243 days, whereas a radiation-induced lattice expansion would be expected. This is actually due to the propensity of various complex structures, such as pyrochlores or spinels (e.g. $MgAl_2O_4$), to accommodate lattice point defects from radiation damage effects. Several mechanisms are proposed to explain this phenomenon, including cation antisite formation and intersticial-vacancy recombination. Because defect clustering is averted due to these annihilating mechanisms, the swelling generally observed in crystalline structures under irradiation is negligible in complex structures. A detailed analysis of the radiation tolerance of complex oxides is given in Sickafus et al.[7] Although our sample was kept away from air in a nitrogen glove-box, the cell contraction may be explained by a moderate oxidation due to a radiolysis phenomenon occurring in epoxy.

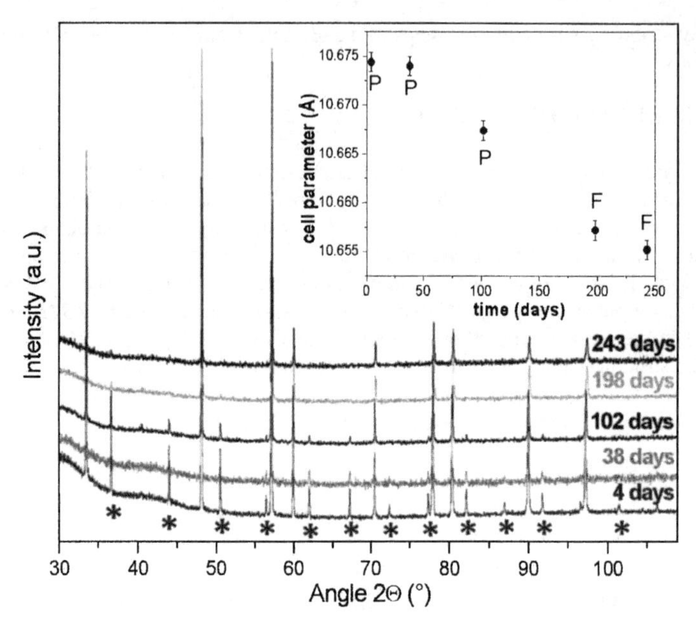

Figure 1 *Evolution of X-ray patterns and of the cell parameter for $Am_2Zr_2O_7$ as a function of time (*=superstructure peaks, P=pyrochlore, F=fluorite).*

For transmutation and, above all, long-term storage, this structural change could be of concern, as structural alteration under irradiation can be detrimental for the properties of materials. Moreover, this structural modification could end-up with an amorphization, as observed for irradiated neodymium pyrochlore.[8,9] Complementary work is in progress to determine fully the extent of these irradiation effects, and to assess the relevance of these ceramic oxides for nuclear applications.

References

1 V.S. Stubican, J.R. Hellmann, *Adv. Ceram.* 1981, **3**, 25.
2 P.E. Raison, R.G. Haire, T. Sato, T. Ogawa, *Mat. Res. Soc. Proc.*, 1999, **556**, 3.
3 R.C. Belin, P.J. Valenza, M.A. Reynaud, P.E. Raison, *J. Appl. Cryst*, 2004, **37**, 1034.
4 R.C. Belin, P.J. Valenza, P.E. Raison, to be published.
5 H. Matzke, "OECD workshop on advanced reactors with innovative fuel", PSI 1998.
6 R.E. Sykora, P.E. Raison, R.G. Haire. *J. Solid State Chem.*, 2005, **178**, 2, 578.
7 K. Sickafus, L. Minervini, R.W. Grimes, J.A. Valdez, M. Ishimaru, F. Li, K.J. McClellan, T. Hartmann, *Science*, **2000**, 289, 748.
8 S.X. Wang, B.D. Begg, L.M. Wang, R.C. Ewing, W.J. Weber, K.V. Godivan,. *J. Mat. Res.*, 1999, **14**, 12, 4470.
9 S. Lutique, Ph. D. Thesis, Université Paris XI, n°7207, May 2003.

LOCAL STRUCTURE OF $Th_{1-x}M_xO_2$ SOLID SOLUTIONS WITH M = U, Pu

G. Heisbourg[1], J. Purans[1,2], Ph. Moisy[3] and S. Hubert[1]

[1]Institut de Physique Nucléaire, Université Paris Sud, UMR8608, 91406 Orsay, France
[2]Dipartimento di Fisica del'Università di Trento, 38050 Povo, Italy
[3]CEA/VALRHO DRCP/SCPS, BP 17171, 30207 Bagnols/Cèze Cedex

1 INTRODUCTION

If the crystal structure and the lattice parameters of mixed actinide dioxides have been extensively investigated as a function of composition by using X-ray diffraction techniques,[1-3] the local structure of such important solid solutions has not yet been studied by using X–ray-absorption fine structure, except for the pure actinide dioxides. In general, the lattice parameters of solid solutions determined by XRD are found to be well approximated by the concentration weighted average of those for the constituents, usually referred to as Vegard's law. However, the most commonly used model for solid solution or alloys (zincblende) is the virtual crystal approximation (VCA).[4] Diffraction methods do not provide sufficient information on the local structure of atoms in crystals. Nevertheless, extended X-ray absorption fine structure is well suited for the determination of near-neighbour spacing, especially relative to a well-defined standard.

In the present study, the local structures were determined around Th, U, and Pu in $Th_{1-x}M_xO_2$ solid solutions with M = U, Pu by means of XAFS spectrometry over the entire range from x = 0 to 1 for both systems and compared with the data from XRD.

2 EXPERIMENTAL

Several compositions of solid solutions of $Th_{1-x}U_xO_2$ (x = 0.11, 0.24, 0.37, 0.49, 0.65, 0.81 and 0.91) and $Th_{1-x}Pu_xO_2$ (x = 0.125, 0.32, 0.66) and pure oxides were synthesized following the previously described procedure.[5] Special attention was taken with regard to oxygen. Mixed actinide oxalates were calcinated at 1300°C under a reducing atmosphere (Ar-5% H_2) for Th/U mixed oxide and at 900°C under air for Th/Pu mixed oxide. The former samples were handled in a dry glove box. All samples were characterized by using XRD. No additional peaks due to oxidized forms were observed. Moreover, XPS was used to test the tetravalent valence state of uranium at the surface.

The XAFS measurements of $Th_{1-x}U_xO_2$ and $Th_{1-x}Pu_xO_2$ were performed on the D44 (XAS4) hot beam line of the LURE DCI synchrotron radiation facility (Orsay, France). A standard transmission scheme with a Ge (400) double crystal monochromator and two ion chambers containing argon gas were used. The XAFS were treated using the "EDA" software package following the standard procedure.[6] The experimental XAFS spectra χ

(k)k[2] obtained at both Th and U(Pu) L_{III} edge were Fourier transformed (FT) in the 0.0 – 15.0 Å^{-1} range. The Th, U and Pu L_{III} edge XAFS of the solid solutions were analyzed to determine the changes in the bond lengths in the first and second coordination shells as a function of solid solution composition. As a general feature, for Th, U and Pu in the L_{III}-edge FT of both solid solutions, two main sharp peaks at 1.9 and 3.8 Å with a smaller pre-peak at 3.2 Å were observed. However, a slight and gradual shift of the first peak is observed from ThO_2 to UO_2 or PuO_2, whilst for the second peak a larger and gradual shift is observed from one member to other.

The single-shell XAFS spectra were fitted using the single-scattering curved-wave formalism with picometer accuracy.[6] The XAFS data were analyzed using two different approaches: the phases and amplitudes were either calculated or obtained experimentally from the pure oxide references. Theoretical backscattering amplitudes and phases were calculated by the FEFF8 code using different clusters that mimic the possible environment of the An^{4+} ions in solid solutions.

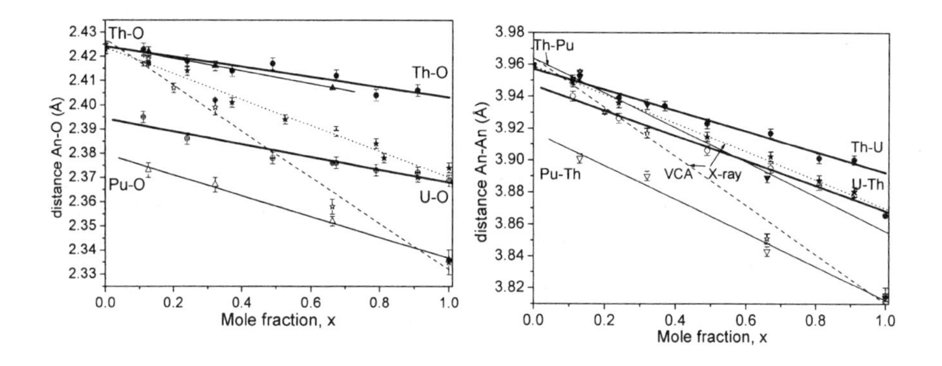

Figure 1. a) *Interatomic distances Th-O, U-O and Pu-O with composition in* $Th_{1-x}U_xO_2$ *and* $Th_{1-x}Pu_xO_2$. b) *Interatomic distances Th-Th(U,Pu), U-Th(U)) and Pu-Th(Pu) with composition. The dotted lines represent a least square fit to the VCA distance from XRD data.*

3 RESULTS AND DISCUSSION

Vegard noted that the lattice constant of solid solutions varies linearly with composition between the end members, suggesting that atomic volume was conserved regardless of the details of the local distortion of the lattice. Pauling and Huggins[7] suggest that the molar bond length of each constituent would remain constant. However one of the most commonly used models for solid solutions is the Virtual Crystal Approximation (VCA) which is an extrapolation of Vegard's law, which assumes an average value of bond length in the solid solution, ignoring any local displacement.

The first and second neighbour distances were determined with an uncertainty of 0.003 Å. Fig.1 shows the experimental values of the bond lengths, $R_{Th\text{-}O}$, $R_{U(Pu)\text{-}O}$, $R_{Th\text{-}U(Pu)}$ and $R_{U(Pu)\text{-}Th}$, as a function of composition. The VCA cation-anion (cation) distances deduced from measured XRD lattice constants are shown for comparison. A linear variation of the first and second shell distances is observed with composition for both solid solutions. However, from Fig.1a, $R_{Th(U,Pu)\text{-}O}$ and $R_{(U,Pu)\text{-}O}$ differ substantially from the VCA average cation-anion distances calculated from XRD. Nevertheless, the weighted mean cation-anion spacing is very close to the VCA distances deduced from XRD, as expected. The first shell distance has only a weak dependence on composition, approximatively one third to one half of that predicted by the VCA dependence. The total decrease in R_{Th-O} in $Th_{1-x}U_xO_2$ from ThO_2 to UO_2 is 0.019 ± 0.003 Å, and the increase in R_{U-O} from UO_2 to the lowest U concentration is 0.026 ± 0.003 Å, whereas the average near neighbour distances change by 0.055 Å for x ranging from 0 to 1. In the case of $Th_{1-x}Pu_xO_2$, the variation in R_{Th-O} between the end point compounds is 0.025 ± 0.003 Å, while the increase in R_{Pu-O} from PuO_2 to the most diluted solid solution is 0.0425 ± 0.003 Å.

The next nearest neighbour (NNN) distances (R_{Th-Th} and $R_{U(Pu)-Th}$) shown in Fig. 1b have a much greater composition dependence than the first shell distance, and the values are closer to the ones expected from the VCA model (dotted lines). The total decrease in $R_{Th-U(Th)}$ is 0.067 ± 0.003 Å or 71% of the difference between R_{Th-Th} and R_{U-U} in the end point compounds, while the increase in R_{U-Th} is 0.080 ± 0.003 Å or 86% of that predicted by the VCA dependence. It is readily apparent that the $R_{Th-U(Th)}$ and $R_{U-Th(U)}$ distances follow more closely a VCA model, shown in Fig.1b by the middle curve calculated from the XRD. However, the weighted mean cation-cation NNN calculated spacing is equal to the virtual-crystal distance, as expected. These experimental NNN distances measurements can be used to characterize qualitatively the degree of homogeneity in these solid solutions.

In $Th_{1-x}Pu_xO_2$, the same behaviour is observed. The values of NNN distance are closer to the ones expected from VCA than to their respective values in the constituent compounds. The total decrease in Th-Th(Pu) is 0.104 ± 0.010 Å and 0.103 ± 0.010 Å for Pu-Pu(Th), compared to 0.126 Å for the average NNN distance change.

4 CONCLUSION

From EXAFS measurements on Th, U and Pu L_{III}-edges of each composition of $Th_{1-x}U_xO_2$ and $Th_{1-x}Pu_xO_2$ ranging from 0 to 1, NN and NNN distances could be measured with high accuracy. We found that opposite to the lattice parameters obtained by XRD, the distances of the first and second shell given by XAFS do not follow completely either Vegard's law nor the VCA model. The NN distances vary only slightly with composition, and are closer to Vegard's law, while the NNN distances vary more strongly with composition and are closer to the VCA model. This dependency is all the more great, since the radius change between the end point compounds is larger.

References

1 I. Cohen and R. M. Berman, *J. Nucl. Mater.*, 1966, **18**, 77.
2 R. N. R Mulford and F. H. J. Ellinger, *Phys. Chem.*, 1958, **62**, 1466.
3 T.Tsuji, M. Iwatsushige,T. K.Yamashita and J. Ohuchi *Alloys and Comp.*, 1998, **271**, 391.
4 J. C. Jr Mikkelsen and J. B. Boyce, *Phys. Rev. B*, 1983, **28**, 7130.
5 G. Heisbourg, S. Hubert, N. Dacheux and J. Ritt, *J. Nucl. Mat.*, 2003, **321**, 141.
6 G. Dalba, P. Fornasini, R. Grisenti and J. Purans, *Phys. Rev. Lett.*, 1999, **82**, 4240.
7 L. Pauling and M. L. Huggins, *Z Kristallogr. Krstallgeom. Kristallphys. Kristallchem.*, 1934, **87**, 205.

FIRST-PRINCIPLES PHASE DIAGRAM OF THE Ce-Th SYSTEM

A.I. Landa and P.A. Söderlind

Physics and Advanced Technologies, Lawrence Livermore National Laboratory, University of California, P.O. Box 808, Livermore, California 94550, USA

1 INTRODUCTION

Actinide physics has seen a remarkable focus the last decade or so due to the combination of improved experimental diamond-anvil-cell techniques and the development of fast computers and more advanced theory. All *f*-electron systems are expected to have multiphase phase diagrams due to the sensitivity of the *f*-electron band to external influences such as pressure and temperature. For instance, compression of an *f*-electron metal generally causes the occupation of *f*-states to change due to the shift of these bands relative to others. This can in some cases, as in the Ce-Th system, cause the crystal to adopt a lower symmetry structure at elevated pressures. Here we study the phase stabilities of Ce, Th, and the Ce-Th system as a function of compression. Theoretically, both Ce and Th metals are rather well described by DFT,[1] although a proper treatment of the Ce-Th alloys has not yet been presented.[2] In this paper we revisit this problem by applying the modern theory of random alloys, based on the coherent potential approximation (CPA).

2 COMPUTATIONAL DETAILS

The calculations we have referred to as exact muffin-tin orbitals (EMTO) are performed using a full-relativistic Green's function technique, based on an improved screened KKR method, where the one-electron potential is represented by optimized overlapping muffin-tin (OOMP) potential spheres.[3] Within the EMTO formalism, the one-electron states are calculated *exactly* for the OOMT potentials. For the exchange/correlation approximation, we use the generalized gradient approximation. For the total energy of random substitutional alloys, the EMTO method has recently been combined with the CPA.[3]

3 RESULTS

In Figures 1 and 2 we plot the calculated *c/a* ratio for bct Ce and Th, respectively. Present results agree well with experimental data[4,5] as well as with those of previous FPLMTO calculations.[1] Figure 3 shows the calculated (EMTO) and measured[6,7] *c/a* ratios for the $Ce_{43}Th_{57}$ alloy. The results of previous FPLMTO calculations[2] are also presented.

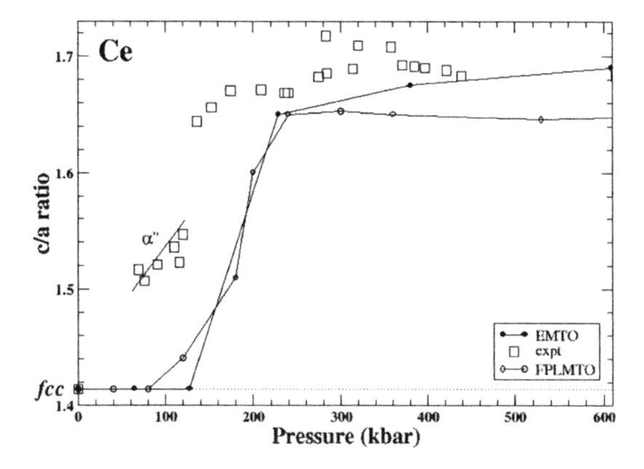

Figure 1 *The c/a axial ratio for the bct structure as a function of pressure for Ce. Experimental data (Ref. 4) are marked with open squares while theoretical results are given by a solid line and filled circles. The results of FPLMTO calculations (Ref. 1) are shown by a solid line and open circles.*

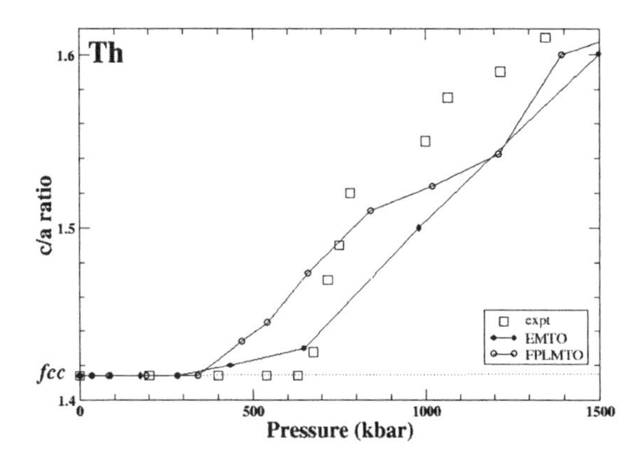

Figure 2 *The c/a axial ratio for the bct structure as a function of pressure for Th. Experimental data (Ref. 5) are marked with open squares while theoretical results are given by a solid line and filled circles. The results of FPLMTO calculations (Ref. 1) are shown by a solid line and open circles.*

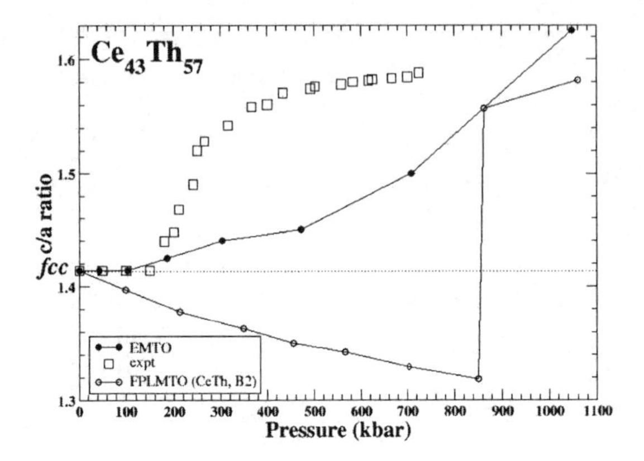

Figure 3 *The c/a axial ratio for the bct structure as a function of pressure for $Ce_{43}Th_{57}$ disordered alloy. Experimental data (Ref. 6, 7) are marked with open squares while EMTO theoretical results are given by a solid line and filled circles. Also, the results of FPLMTO calculations (Ref. 2) for Ce-Th-ordered (B2) compound are given by a solid line and open circles.*

4 CONCLUSION

We have presented accurate electronic-structure calculations for the Ce-Th system. Generally, the theory reproduced experimental data very well. For the $Ce_{43}Th_{57}$ disordered alloy, a CPA treatment is necessary to reproduce the correct structural behaviour.

Acknowledgements

This work was performed under the auspices of the U.S. Department of Energy by the University of California Lawrence Livermore National Laboratory under Contract No. W-7405-Eng. 48.

References

1 P. Söderlind, O. Eriksson, B. Johansson and J.M. Wills, *Phys. Rev. B*, 1995, **52**, 13169.
2 P. Söderlind and O. Eriksson, *Phys. Rev. B*, 1999, **60**, 9372.
3 L. Vitos, *'Quantum-Mechanical Description of Substitutional Random Alloys'* in *Recent Research and Development in Physics*, Transworld Research Network Publisher, Trivandrum, 2004, Vol. 5, pp. 103-140.
4 J. Staun Olsen, L. Gerward, U. Benedict and J.-P. Itie, *Physica B & C*, 1985, **133**, 129.
5 Y.K. Vohra and J. Akella, *Phys. Rev. Lett*, 1991, **67**, 3563.
6 G. Gu, Y.K. Vohra, U. Benedict and J.C. Spirlet, *Phys. Rev. B*, 1994, **50**, 2751.
7 G. Gu, Y.K. Vohra, J.M. Winand and J.C. Spirlet, *Scr. Metall. Mater.*, 1995, **32**, 2081.

CRYSTAL STABILITY AND EQUATION OF STATE FOR AM: THEORY

P. Söderlind and A. Landa

Physics and Advanced Technologies, Lawrence Livermore National Laboratory, University of California, P.O. Box 808, Livermore, California 94550, USA

1 INTRODUCTION

Americium metal undergoes several phase transitions under compression. Experimental studies[1,2] all agree that AmI and AmII are dhcp and fcc, respectively. AmIII and AmIV have been assigned several different structures in the past, such as monoclinic (AmIII) and face-centred orthorhombic (fco:AmIV).[1] These assignments were subsequently questioned by first-principles theory[1] and later re-interpreted.[2] The most recent experimental study suggests AmIII to be fco (very similar to γ-Pu) and AmIV primitive orthorhombic (po or pnma).[2] Here, we re-visit Am by performing density-functional electronic-structure calculations. Our motivation is to validate the new structural assignments of AmIII and AmIV, and study possible phase transitions beyond AmIV.

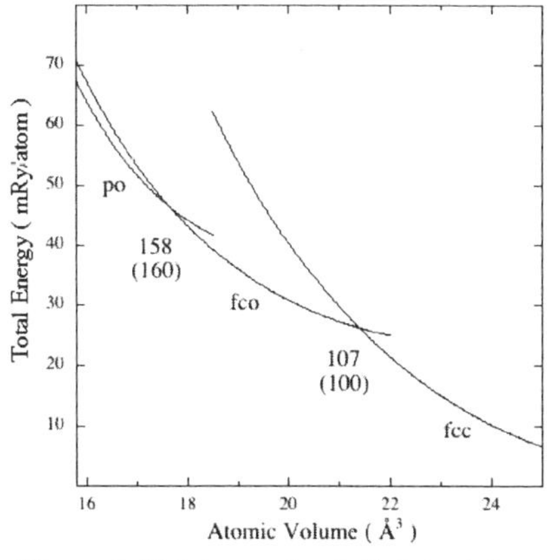

Figure 1 *Total energies (mRy/atom) for fcc (AmII), fco (AmIII), and po (AmIV). Calculated (experimental) transition pressures are given in kbar.*

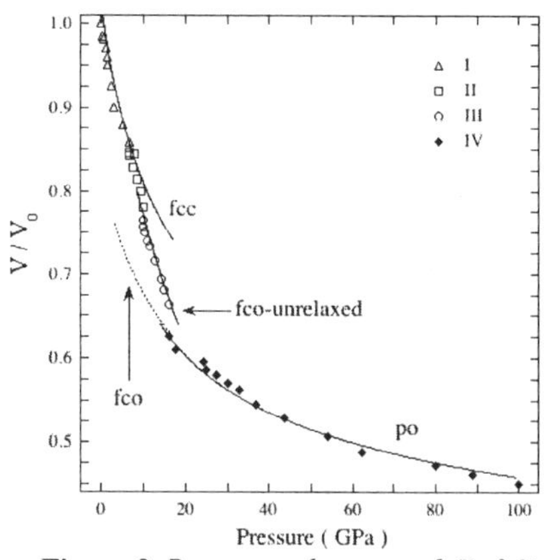

Figure 2 *Present and measured (Ref. 2) equation of state for Am. Symbols and lines refer to experiment and theory, respectively. The dashed line represents a fully relaxed AmIII calculation. V_0 is the ambient-pressure volume, 29.3 Å^3.*

2 COMPUTATIONAL DETAILS

The full-potential linear muffin-tin orbital method (FPLMTO) has been used. This all electron method treats spin-orbit coupling, spin, and orbital polarization and was recently applied for the six known phases of Pu[3] with excellent results. A more complete description of the details of the calculations can be found elsewhere.[4]

3 RESULTS

In Figure 1 we plot total energies for fcc (AmII), fco (AmIII), and po (AmIV). From these, one can calculate the free energy at zero temperature as F = E + PV, where E is the total energy, P the pressure and V the atomic volume. Transitions between phases occur when their respective free energies intersect. Theory predicts AmII-AmIII and AmIII-AmIV transitions at 107 and 158 kbar, respectively. This is in good agreement with the data[2] by Heathman et al. (100 and 160 kbar). Next, we show our calculated equation of state for Am in Figure 2. The data points are collected from Ref. 2 and the lines denote present theory. The dashed line corresponds to fully relaxed calculations of AmIII. Overall, the agreement between theory and experiment[2] is satisfactory. Only the fully relaxed AmIII phase deviates somewhat from the published data. The diamond-anvil-cell work[2] did not go beyond 1 Mbar and the AmIV phase was the highest-pressure phase. Certainly one expects higher symmetry geometries at some yet greater compression. Canonical band theory[5] predicted, years ago, that bcc is a likely high-pressure phase for any itinerant *f*-electron metal with *f* occupation between about 4-8. The 5*f* population in high pressure Am is within this interval (6-7 5*f* electrons). Therefore, we perform FPLMTO calculations of the total energy difference between bcc Am and AmIV, see Figure 3. Clearly bcc becomes stable over AmIV for atomic volumes below about 12 $Å^3$. Lastly we also plot the calculated magnetic moments for AmIII and AmIV in Figure 4. Notice that the magnetic moments of the AmIII phase are more stable than those of the AmIV phase in the transition region (between vertical dashed lines).

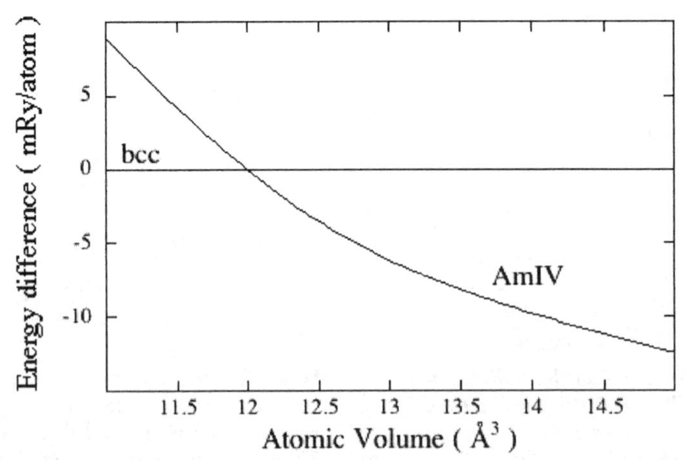

Figure 3 *AmIV – bcc energy difference in mRy/atom. Bcc becomes stable at 12* $Å^3$.

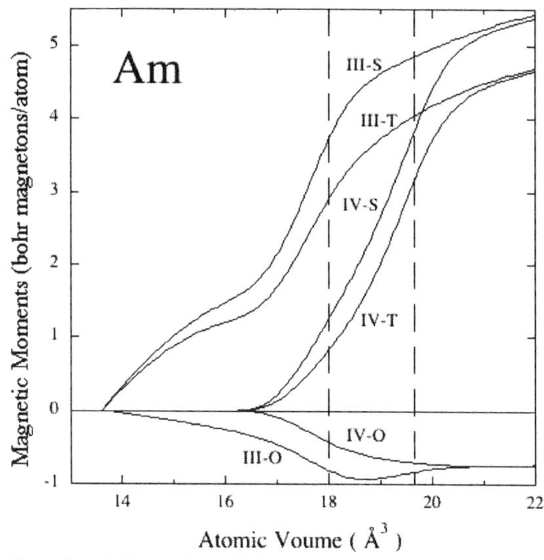

Figure 4 *Spin (S), orbital (O), and total (T) magnetic moments for AmIII (III) and AmIV (IV), as a function of volume.*

4 CONCLUSION

We have presented accurate electronic-structure calculations for Am metal. These concur with the most recent experimental studies, and inconsistencies between theory and experiment that existed in the past for Am are now fully removed. Theory also predicts a bcc phase at pressures beyond 1 Mbar.

Acknowledgements

This work was performed under the auspices of the U.S. DOE by the UC Lawrence Livermore National Laboratory under Contract No. W-7405-Eng. 48.

References

1 P. Söderlind, R. Ahuja, O. Eriksson, B. Johansson and J.M. Wills, *Phys. Rev. B*, 2000, **61**, 8119 (and references therein).
2 S. Heathman, R.G. Haire, T. Le Bihan, A. Lindbaum, K. Litfin, Y. Meresse and H. Libotte, *Phys. Rev. Lett.*, 2000, **85**, 2961.
3 P. Söderlind and B. Sadigh, *Phys. Rev. Lett.*, 2004, **92**, 185702.
4 P. Söderlind, *Europhys. Lett.*, 2001, **55**, 525.
5 P. Söderlind, *Adv. Phys.*, 1998, **47**, 959.

XPS STUDY OF HIGH TEMPERATURE OXIDATION OF A PLUTONIUM ALLOY UNDER VACUUM

F. Delaunay, B. Oudot, B. Ravat and N. Baclet

DRMN/SEMP/LSFC, CEA Valduc, 21 120 Is-sur-Tille, France

1 INTRODUCTION

Radioactivity and chemical toxicity of plutonium requires very careful handling of this kind of material. Because photoemission spectroscopy analyses a thickness of only a few nanometres, the sample cannot be confined, this means that the inner parts of the spectrometer are contaminated. As a consequence, very few laboratories in the world are able to perform photoemission spectroscopy on plutonium samples. Modifications of a photoelectron spectrometer have been recently achieved to allow experiments with radioactive material. The capability of this device is presented together with the preliminary results concerning the study of the high temperature oxidation under vacuum of a plutonium alloy stabilised in the δ-phase by addition of gallium.

2 EXPERIMENTAL

This investigation was performed on a VG Scientific ESCALAB 220i- XL electron spectrometer. XPS measurements presented in this paper were made with Al-Kα (1486.6 eV) radiation, monochromatised by two quartz crystals which also focused the x-ray beam on to the surface of the sample with a spot size 1 mm in diameter. A magnetic lens, placed under the sample, and multi-detection at the outside of the analyser (six channeltrons) allow a strong increase in the counting rate. The spectral energy scale calibration of the spectrometer was referred to the Ag $3d_{5/2}$ line (368.2 eV binding energy). A FWHM of 0.61eV was obtained with the analyser operating in CAE (Constant Analyser Energy) mode, 20 eV pass energy, 100 ms dwell time and 0.1 eV per data point. The device is also equipped with a twin anode, delivering Al-Kα or Mg-Kα radiation, a VG UVL HI differentially pumped windowless photon source to perform UPS analysis, a VG Scientific LEG 1000 electron gun to perform AES analysis and a residual gas analyser. Due to the complexity of the device, it was decided to let it out of the glove box and to connect it to a vacuum chamber used for sample preparation, which was placed inside a nitrogen filled glove box operating under a slight negative pressure. Working under UHV conditions (2.10^{-10} mbar with residual gas analysis showing trace impurities of H_2, H_2O and CO) and using metallic filters to protect the pumping system ensure a reliable confinement of the sample during experiments in the spectrometer analysis chamber.

Freshly cast δ-stabilised plutonium alloys were studied in the present work. Samples were machined to obtain a disk 10 mm in diameter and 1 mm thick, which was heated to 633 K for 6 hours in a furnace under secondary vacuum (2.10^{-7} mbar). After cooling, the samples were moved into the UHV system for subsequent analysis. Compositional depth profiles were determined by progressive Ar^+ ion bombardment, using a VG AG 5000 ion gun at a pressure of 5.10^{-6} mbar, 3-5 keV, sample current 60-80 $\mu A/cm^2$. In these conditions, the sputtering rate was estimated to be between 5 and 10 nm/mn. Sputtering was done in a special chamber, sited in the part of the device that was in the glove box to keep the level of contamination as low as possible in the analysis chamber of the spectrometer.

3 RESULTS AND DISCUSSION

A sample was analysed by grazing incident X-ray diffraction (XRD) to identify the composition of the thin oxide layer present on the surface. The results indicated that the oxide layer was mainly composed of an oxide that was PuO or PuCO, XRD being unable to distinguish these two compounds. XPS analysis was performed on a sample which was prepared under the same conditions. A strong carbon (285 eV) and oxygen (532 eV) contamination of the surface was found. Even if it was strongly attenuated by the contamination, the Pu $4f_{7/2}$ photo-ionisation spectrum exhibited a peak centred at 426.1 eV, typical of tetravalent plutonium[1,2] but also a shoulder at 424.7 eV, characteristic of trivalent plutonium oxide.[1,2] It must be noted that the sample was analysed immediately after introduction to the UHV system. Therefore, we think that the trivalent oxide observed was not due to the (self)reduction of the tetravalent oxide under UHV conditions, as previously observed by Roussel et al.,[3] and confirmed by characterisation we performed on other plutonium samples. It is assumed that the tetravalent oxide formed only a very thin layer on the surface of the sample, over a thicker layer of a trivalent oxide, mainly constituted of an oxy-carbide. This oxide carbide was probably formed on the sample during heat treatment. XRD detected only this oxide carbide, since it was the main oxide in the layer. During the transfer between the furnace and the XPS device, the surface of the oxide carbide layer was oxidised into Pu_2O_3,[4] which was oxidised into PuO_2. XPS is much more sensitive to the surface than XRD, and allowed identification of the presence of PuO_2. It was much more difficult to differentiate Pu_2O_3 from an oxide carbide, since plutonium is trivalent in both phases. Moreover, after sputtering of the PuO_2 layer, the time needed to transfer the sample before analysis was thought to be sufficient to allow oxidation of the surface of the oxide carbide layer into Pu_2O_3 which seems, considering the Pu 4f signal, to be the oxide identified by XPS. Figure 1 shows the composition depth profile obtained by measuring areas of the main peaks of the different elements, corrected by the Scofield[5] atomic sensibility factor, as a function of the sputtering time. This profile indicates a decrease in the relative oxygen concentration and a slight increase in the plutonium one between 90 and 180 minutes; this suggests that the metal-oxide interface corresponded to an abrasion time between 90 and 180 minutes. No gallium was found before 180 minutes of Ar^+ ion bombardment. This is consistent with previous results[5] reported for a δ-PuGa alloy. After 180 minutes, the gallium signal was clearly detected and it increased until a maximum was reached after 300 minutes of sputtering (after 420 mn, results not shown here, the gallium concentration was unchanged). It must be noticed that the gallium concentration found is noticeably lower than the gallium concentration in the alloy. This could be a consequence of a preferential sputtering of gallium in these experimental conditions, as mentioned by Cox[6] but contested by Roussel et al.[3] Nevertheless, starting to detect gallium between 90 and 180 minutes of sputtering was another argument to localise the interface in this region and seemed to indicate that gallium concentration in the oxide layer was much lower than in the alloy and, at least, lower than the detection limit for XPS.

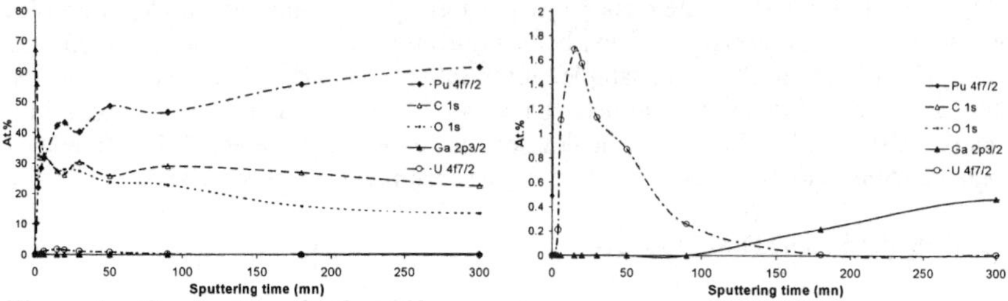

Figure 1 *Composition depth profile.*

A slight enrichment in americium was found at the surface of the sample, but it was rapidly eliminated by Ar^+ ion bombardment. This is consistent with results reported by Roussel et al.[3] A significant enrichment in uranium was also observed between 4 and 90 minutes of sputtering. A maximum of concentration was rapidly reached after 10 minutes, and then it decreased slightly until uranium was no longer detected, after 180 minutes of sputtering. This result is different from previous data published by Roussel et al. who found that uranium concentration was constant from surface to bulk, as gallium concentration. The uranium concentration in the alloy (0.08% atomic) was below the detection limit for XPS, which is consistent with the fact that uranium was not detected in the bulk. Moreover, another study,[7] coupling XPS and EPMA, performed on a different plutonium alloy, showed the same uranium and americium behaviour, and suggested that there was probably a uranium depleted zone below the surface of the sample. The same type of XPS analysis performed on a PuGa alloy that was not heat treated did not show any uranium nor americium enrichment. We conclude that uranium and americium diffused from bulk to surface during the heat treatment (6 hours at 633 K). While diffusion was also probably enhanced along grain boundaries, there was no clear evidence of uranium or americium enrichment in these regions.

4 CONCLUSION

This work has produced the first results for a plutonium alloy on a new photoemission spectrometer which is now available to perform XPS, and also UPS and AES, on trans uranium elements. Further investigations are under way, but the results presented already show a complex composition of the oxide layer and a rather easy diffusion of uranium and probably americium during heating.

References

1 D.Courteix, J. Cheyrouse, L.Heintz and R. Baptist, *Solid State Commun.*, 1981, **39**, 209.
2 D.T.Larson, *J. Vac. Sci. Technol.*, 1980, **17**, 55.
3 P.Roussel, D.S.Shaw and D.A.Geeson, *J. Nucl. Science and Technology*, 2002, **Sup.3**, 78.
4 D.T. Larson and J.M. Haschke, *Inorganic Chemistry*, 1981, **20**, 1945.
5 J.H.Scofield, *J. Electr. Spectrosc. Rel. Phen.*, 1976, **8**, 120.
6 L.E.Cox, *Phys. Rev. B*, 1988, **37**, 8480.
7 B.Oudot, *Ph.D. Thesis*, 2005.

PHASE RELATION OF MIXED OXIDE FUEL CONTAINING NP AND AM

M. Kato[1], H. Uno[2], T. Tamura[2], K. Morimoto[1], K. Konashi[3] and Y. Kihara[1]

[1]Plutonium Fuel Technology Group, Japan Nuclear Cycle Development Institute, Tokai-Mura, Ibaraki, Japan
[2]Inspection development company, Tokai-Mura, Ibaraki, Japan
[3]Institute for Materials Research, Tohoku University, Oarai-machi, Ibaraki, Japan

1 INTRODUCTION

The plutonium-and-uranium-mixed oxide fuel (MOX) containing minor actinides (MA) is expected to increase proliferation resistance and also to decrease environmental impact by removing MA from high level waste. In the development of an advanced nuclear cycle, it is important to understand the characteristics of MOX containing MA (MA-MOX) fuel. Homogeneous mixed oxide fuels containing MA have been developed for advanced fast reactors as utilized in Japan Nuclear Cycle Development Institute.[1]

It is planned to irradiate MOX fuels containing Np and/or Am in the fast reactor, *Joyo*. It is important to adjust the O/M ratio of the fuel to about 1.95 for controlling the cladding chemical interaction, and also to assure the integrity of the fuel pin at a high burn-up of 150 GWd/t.[1] The phase diagram and crystallographic data of the MOX containing Np and/or Am in the hypostoichiometric region are essentially important in understanding the irradiation and sintering behaviour. In this report, the effect of Np and Am content in MOX upon the lattice parameters, and the phase separation behaviour are evaluated.

2 EXPERIMENTAL

The samples were prepared by a mechanical blending method.[2,3] The pressed pellets were sintered at 1600-1700°C for 3h in Ar/5%H_2 mixture with added moisture. Then the O/M ratios of the samples were adjusted by annealing under appropriate conditions, which were determined in previous work.[3-6] Some samples with low O/M were annealed at 180°C for 5hr to observe microstructure. All the samples were analyzed by X-ray diffraction and ceramography. The following mixed oxide samples were prepared; $(Am_{0.005}Pu_{0.3}U_{0.695})O_{2-x}$, $(Am_{0.007}Pu_{0.29}U_{0.703})O_{2-x}$, $(Np_{0.06}Pu_{0.29}U_{0.65})O_{2-x}$, $(Np_{0.12}Pu_{0.29}U_{0.59})O_{2-x}$, $(Am_{0.024}Pu_{0.3}U_{0.676})O_{2-x}$, $(Np_{0.018}Am_{0.018}Pu_{0.3}U_{0.664})O_{2-x}$ and $(Np_{0.024}Am_{0.024}Pu_{0.3}U_{0.652})O_{2-x}$, which are described as 0.5%Am-MOX, 0.7%Am-MOX, 6%Np-MOX, 12%Np-MOX, 2.4%Am-MOX, 1.8%Np/Am-MOX and 2.4%Np/Am-MOX, respectively, in this report. The thermal gravimetry (TG) and differential thermal analysis (DTA) were employed to investigate the phase transformation temperature.

3 RESULTS AND DISCUSSION

Only a single phase of fcc structure was obtained in the samples with O/M ratio larger than 1.96. In the case of an O/M ratio smaller than 1.96, two different phases of fcc structure were found in the samples. Figure 1 shows the effect on the lattice parameters of adding MA to $(Pu_{0.3}U_{0.7})O_{2.0}$. The lattice parameter decreases with increasing MA content. Adding Am has a larger effect on the lattice parameter, compared with adding Np. The lines in Fig.1(a) show the lattice parameter estimated based on Vegard' law by following equations,

$$a_{MO2} = a_{NpO2} \times C_{Np} + a_{AmO2} \times C_{Am} + a_{PuO2} \times C_{Pu} + a_{UO2} \times C_U,$$

$$C_{Np} + C_{Am} + C_{Pu} + C_U = 1,$$

where a_i is the lattice parameter of oxide i and C_i represents the composition of element i. The lattice parameters of UO_2, PuO_2, AmO_2 and NpO_2 were 5.4702Å, 5.3960 Å, 5.3772 Å and 5.4339 Å, respectively.[7-10] It is shown that the change of the lattice parameters due to MA addition can be well explained by Vegard's law. Therefore, it is considered that MA-MOX is a substitutional solid solution. Figure 1(b) shows the lattice parameters of the samples having a fcc single phase as a function of O/M ratio. The lattice parameters of MA-free MOX are in good agreement with Sari's data,[11] and those of MA-MOX also change with the same slope as for MA-free-MOX. The lattice parameters can be described as a function of O/M ratio,

$$a = 0.248x + a_{MO2},$$

where x is deviation of O/M ratio from stoichiometric composition.

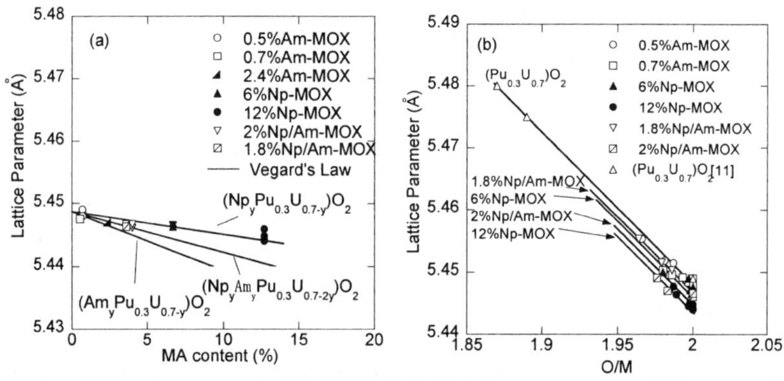

Figure 1 *lattice parameter in MA-MOX (a) Lattice parameter versus MA content in stoichiometric composition (b) Lattice parameter versus O/M ratio*

In the case of the Pu-U-O ternary system,[11] it was reported that the single phase of fcc structure found at high temperature separates to give two fcc phases with different O/M ratios during cooling to low temperature. In this study, similar separation phenomena were observed by X-ray diffraction for MA-MOX with a low O/M of about 1.92. The separation phenomenon was analysed in detail by DTA. The O/M change during the measurement was smaller than 0.001, which was estimated from the TG curve. The temperatures of phase separation, which were determined from the peaks in the DTA curves, were observed in the region from 220 to 360°C.

Figure 2(a) shows microstructures of $(Np_{0.12}Pu_{0.29}U_{0.59})O_{1.92}$ which were annealed at 180°C. The separated microstructure observed in the sample is very close to that reported by Sari[11] for MOX fuel. Figure 2(b) shows the phase separating temperature as a function of MA content. The temperature of the phase separation decreases with increasing MA content. The effect of Am addition on the separation temperature is greater than that of Np.

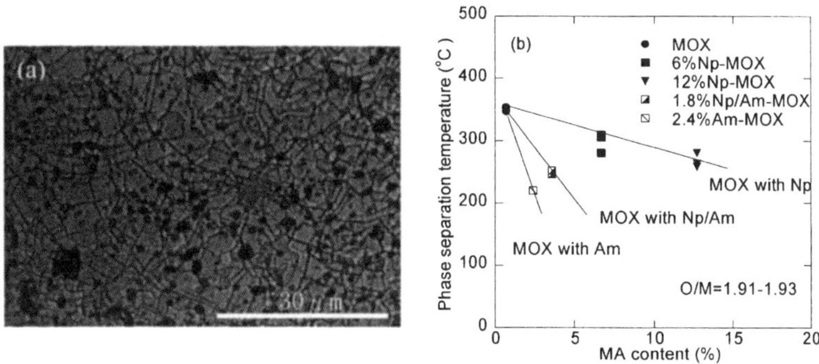

Figure 2 *valuation of phase separation (a) Microstructure of two fcc phases in $(Np_{0.12}Pu_{0.29}U_{0.59})O_{1.92}$ (b)Phase separation temperature versus MA content*

4 CONCLUSION

The lattice parameters of MOX containing Am and/or Np were evaluated as a function of MA content and O/M ratio. Since the measured lattice parameters agreed with estimations based on Vegard's law, it was concluded that the MA-MOX is a substitutional solid solution. An equation was derived from the measured lattice parameters, which describes the relationship between the lattice parameter and the composition.

Phase separation was observed in MA-MOX as well as MA-free MOX. The phase separation process may cause micro-cracking of the pellet due to local volume changes. However, phase separation of MA-MOX occurs at a temperature lower than the coolant temperature in a liquid metal cooled fast reactor. It can be concluded that the phase separation would not influence the irradiation behaviour of the fuel.

References

1 S. Nomura, A. Aoshima, T. Koyama and M. Myochin, *Global 2001*, 2001, **1**, 42.
2 K. Morimoto, M. Kato, H. Uno, A. Hanari, T. Tamura, H. Sugata, T. Sunaoshi and S. Kono, *J. of Phys. and Chem. of Solid*, 2005, **66**, 634.
3 M. Kato, K. Konashi and Y. Kihara, *Transactions*, 2004, **91**, 463.
4 K. Morimoto, M. Nishiyama, M. Kato, H. Endo and S. Kono, *Global 2001*, 2001, **1**, 38.
5 M. Kato, T. Tamura, K. Konashi and S. Aono, *STNM-11th*, 2004, Karlsruhe.
6 M. Kato, K. Konashi, S. Aono and Y. Kihara, *MMSNF-3*, Washington DC, Nov., 2004.
7 W. L.Lyon and W.E.Baily, *J.Nucl.Mater.*, 1967, **22**, 332.
8 K. Richer and C. Sari, *J.Nucl.Mater.*, 1987, **148**, 266.
9 T.D. Chikalla and L. Eyring, *J.Inorg.Nucl.Chem.*, 1968, **30**, 133.
10 T. L. Markin and R. S. Street, *J.Inorg.nucl.Chem.*, 1967, **29**, 2265.
11 C. Sari, U. Benedict and H. Blank, *J.Nucl.Mater.*, 1970, **35**, 267.

STUDY ON CORROSION RESISTANCE IMPROVEMENT OF U AND U-Nb ALLOYS BY ION IMPLANTATION WITH NIOBIUM

W. Luo, N. Ranfu, W. Sheng, L. Kezhao and L. Tianwei

China Academy of Engineering Physics, P.O. Box: 919-71, Mianyang, 621900, Sichuan, P.R. China

1 INTRODUCTION

Uranium is an active metal, and it can be corroded rapidly. Lots of methods have been adopted to hold back the corrosion of uranium, but common coatings cannot protect uranium, because of its chemical activity and radiation.[1-5] The surface optimization technology of ion implantation has some unique advantages, and it has solved a lot of traditional problems successfully.[6~14] Because implanted elements can alter natural performance, and uranium-niobium alloys are of better corrosion resistance, ion implantation technology has been used to treat uranium and uranium-niobium alloys.

2 EXPERIMENTAL AND RESULTS

2.1 Ion Implantation Apparatus

A multi-function ion implantation apparatus was used in the experiment. There are five sets of ion beam sources in the apparatus. It can be used to carry out ion implantation, ion bombardment, ion beam mixing and ion beam assisted deposit. The size of the vacuum chamber is Ø800×1000. Metal ion source and clean ion source were used in the experiment.

2.2 Corrosion Resistance Improvement of U by Nb ion implantation

2.2.1 Electrochemistry Polarization. Electrochemistry polarization tests were carried out on uranium and uranium implanted with niobium specimens. The results showed that the polarization trend of uranium and uranium implanted with niobium specimens were the same. Calculation data showed that the electrochemical corrosion resistance of uranium improved remarkably after implantation with niobium. The corrosion electric current became less than 1/25.

2.2.2 Water Vapour Corrosion Test. Water vapour corrosion tests were carried out on the Nb-implanted specimen and uranium. The results are shown in Fig.1. The figure shows that uranium can be corroded easily, and the corrosion rate varied little with corrosion time. The corrosion resistance of niobium-implanted uranium improved remarkably. After ion implantation, the corrosion rate of uranium by water vapour became less than 1/100, but in the long-term, this receded because the modified layer was too thin.

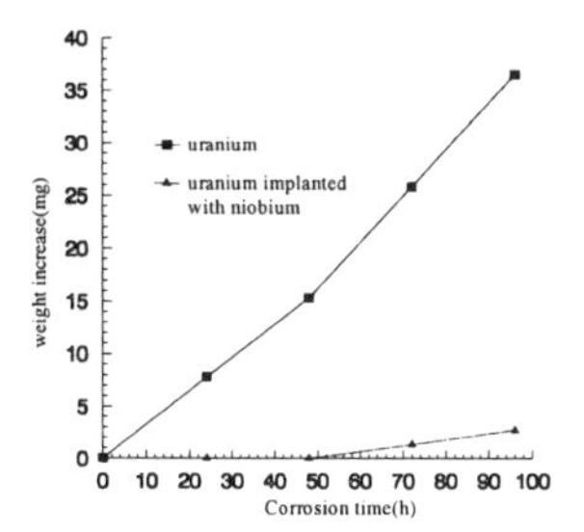

Figure 1 *weight increase by water vapour corrosion of U and U implanted with Nb*

2.2.3 Thermal Oxidation Corrosion Test. Thermal oxidation corrosion tests at 473 K were carried out on the Nb-implanted specimen and uranium. Uranium specimens corroded completely 2 hours later, but the uranium specimens implanted with niobium were not corroded, and still kept their metallic shine. Four hours later, the oxidation film of the uranium broke and fell off, but the uranium implanted with niobium corroded only at local points. After ion implantation, the corrosion rate of uranium by oxygen became less than 1/35.

2.3 Corrosion Resistance Improvement of U-Nb by Nb Ion Implantation

2.3.1 Ion Implantation of U-Nb (I). Water vapour corrosion tests were carried out on Nb-implanted specimen and U-Nb (I). The results are shown in Fig.2. The corrosion resistance of U-Nb (I) implanted with niobium improved remarkably. After ion implantation, the corrosion rate of U-Nb (I) corroded by vapour became less than 1/60. The vapor corrosion resistance improvement of U-Nb (I) is likely to be due to the formation of uranium-niobium alloys.

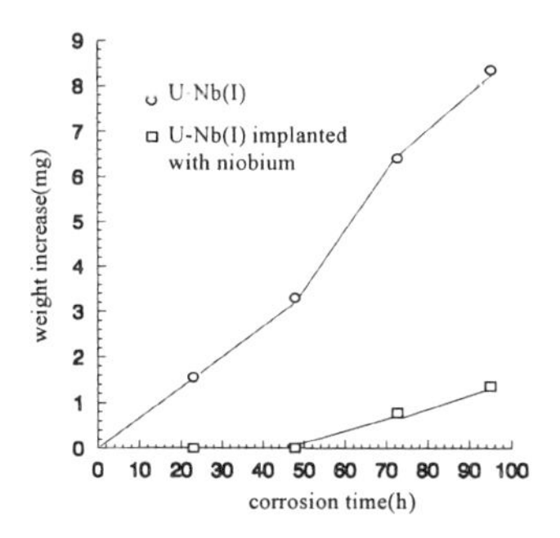

Figure 2 *Weight increase by water vapour of U-Nb(I) and Nb-implanted U-Nb(I)*

The water vapour corrosion test results of U, U-Nb (I), Nb-implanted U and Nb-implanted U-Nb (I) are shown in Table 1. The order of water vapour corrosion resistance was U<U-Nb (I) <Nb-implanted U <Nb-implanted U-Nb (I).

Table 1 *Water vapour corrosion rates*

Specimens	*U*	*U-Nb*	*Nb- implanted U*	*Nb-implanted U-Nb (I)*
Corrosion rate ($mg/cm^2.h$)	4.3×10^{-2}	5.0×10^{-3}	$<3.0\times10^{-3}$	8.4×10^{-5}

2.3.2 Ion Implantation of U-Nb (II). Thermal oxidation corrosion tests at 573 K were carried out on a Nb-implanted specimen and U-Nb (II). The results showed that the corrosion rate of U-Nb(II) specimen showed little difference after ion implantation with niobium. AES analysis was carried out on the Nb-implanted specimen before corrosion and after oxidation. The result showed that the oxidation characteristics of Nb-implanted U-Nb (II) and Nb-implanted U were the same. The peak value of niobium became even after thermal oxidation, and showed that niobium would diffuse into deeper layers if specimens were treated at 573 K or higher. Therefore, the modified layer may become thicker by this method, and it can improve the long-term corrosion resistance of uranium and uranium-niobium alloys.

3 CONCLUSION

It has been found that the corrosion resistance of uranium implanted with niobium improved, and so did uranium-niobium alloys. After ion implantation, the corrosion rates of uranium by water vapour, oxygen and electrochemical polarization became less than 1/100, 1/35 and 1/25, respectively. The corrosion rate of uranium-niobium alloys in water vapour became less than 1/60. The order of water vapour corrosion resistance was U<U-Nb (I) <Nb-implanted U <Nb-implanted U-Nb (I).

References

1 C. M. Egert and D. G. Scott, *J. Vac. Sci. Technol*, 1987, **A5**, 4.
2 E. N. Kaufmann, R. G. Musket, C.A. Colmenares, et al, *Mat. Res. Soc. Symp. Proc*, 1984, 747
3 F. Chang, M. Levy, R. Huie, et al, *Surf. Coat. Technol*, 1991, 48.
4 R.G. Musket, G.Robinson-Weis and R.G. Patterson, *Mat. Res. Soc. Symp. Proc.*, 1984, 753
5 F. Chang, M. Levy and B. Jackman, *Surf.Coat. Technol*, 1991, 49.
6 J. P. Riviere, *Materials science forum*, 1994, 163.
7 P. D. Townsend, *Contemp. Phys*, 1986, **27**, 3.
8 James K. Hirvonen, *Surf. Coat. Technol*, 1994, 65.
9 G. dearnaley, *Surf. Coat. Technol*, 1994, 65.
10 S B Ogale, *Indian Journal of Technology*, 1990, 28.
11 John T.A. Pollock, *Materials Forum*, 1986, **9**, 3.
12 T.D. Radjabov, *Vacuum*, 1988, **38**, 11
13 Ron Hutchings, *Materials science and engineering*, 1994, A184.
14 G.Dearnaley, *Nuclear Instr. Meth. Phys. Res*, 1987, B24/25.

SIMULATION BY EXTERNAL IRRADIATION OF SELF-IRRADIATION EFFECTS ON THORIUM PHOSPHATE DIPHOSPHATE (β-TPD) – CONSEQUENCES FOR ITS DISSOLUTION

C. Tamain[1], N. Dacheux[1], F. Garrido[2], A. Özgümüs[1], L. Thomé[2] and R. Drot[1]

[1] Groupe de Radiochimie, IPNO, Université Paris-Sud-11, 91406 Orsay, France.
[2] CSNSM, Université Paris-Sud-11, 91405 Orsay, France.

1 INTRODUCTION

The β-Thorium Phosphate Diphosphate (β-TPD), $Th_4(PO_4)_4P_2O_7$, can be presented as a potential actinide-bearing phase for the nuclear waste storage in geological deposits. Indeed, this ceramic can be loaded with large amounts of tetravalent actinides,[1] exhibits a high resistance to aqueous corrosion and a high thermal stability.[2] Nevertheless self-irradiation by alpha-decays due to the actinides could modify several of its physico-chemical properties. The released α-particules (~ 5 MeV) disperse their energy by ionization and electronic excitations. Recoil nuclei (~ 100 keV) lose their energy trough ballistic processes, involving elastic collisions and causing direct atomic displacements.[3-4] Thus, it is necessary to study the effects of this self-irradiation on the chemical durability of the ceramic. Self-irradiation is mainly simulated by external exposure to radiation, and the dissolution of the matrix was studied through two kinds of experiments: 1) *ex-situ* studies: samples were irradiated then submitted to dissolution tests, 2) *in-situ* studies: samples were immerged during irradiation by γ−rays and α particles, which enables the study of the effects of the radiolysis of the leaching medium on the dissolution of the sample. Complementary experiments were performed by internal irradiation through the *in-situ* loading of the samples with ^{239}Pu and leaching tests.

2 METHOD AND RESULTS

The structural effects of irradiation were first studied using highly energetic ions which interact with the matrix only by ionization (electronic stopping power), except at the end of their path where the nuclear stopping power is no longer negligible. Critical fluences of amorphization were determined from XRD (X-Ray Diffraction). The analysed thickness of β-TPD is about 10 μm, which is smaller than the irradiated zone. Ion beam irradiations were performed with high energy ions (810 MeV krypton, 450 MeV xenon, 410 MeV sulfur and 170 MeV iodine). These allowed the determination of the behaviour of β-TPD versus fluence of amorphization at several electronic stopping values (Figure 1 and Table 1). It appeared that the ceramic was completely amorphized with ions of stopping values higher than 10 MeV/μm. For stopping power lower than 4 MeV/μm, a threshold appears: the material does not reach an amorphous state, even at high fluences. Annealing studies were also performed: amorphous irradiated β-TPD can be recrystallized thanks to thermal annealing at 700°C for 10 hours. The activation energy of thermal annealing was estimated to 200 kJ/mol. Irradiations with low energy ions (5 MeV gold) were also performed to study the influence of nuclear energy loss: complete amorphization is reached for a fluence of 10^{13} cm^{-2}. For samples loaded with ^{239}Pu, the evolution of the unit cell parameters was followed by XRD analyses. It revealed that they are not modified even after several months of ageing (Table 2).

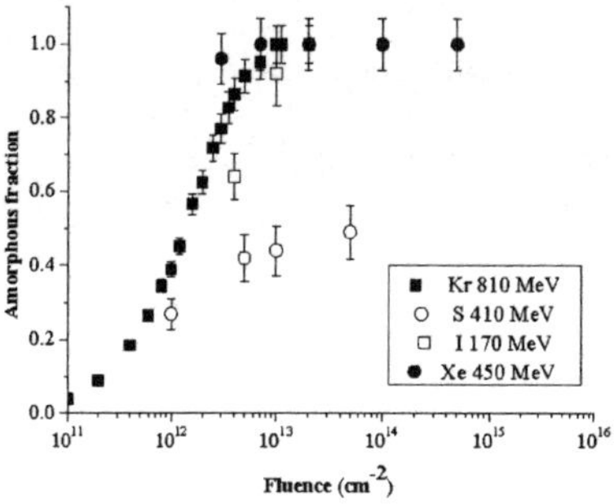

Figure 1 *Evolution of amorphous fraction of β-TPD (determined by XRD)*

Table 1 *Behaviour of β-TPD versus the electronic stopping of the ion beam.*

Ion beam (MeV)	*Electronic stopping (MeV/μm)*	*Fluence of complete amorphization*
Xe 450	20	$5{\bullet}10^{12}$ cm^{-2}
I 170	18	10^{13} cm^{-2}
Kr 810	10	10^{13} cm^{-2}
S 410	3	threshold ?
He 1.6	0.4	threshold ?

For *ex-situ* studies, the dissolution of fully and partly amorphized matrices of TUPD with 6.4 Wt.% of U was studied considering several conditions (temperature, acid concentration). From these results, the increase of the uranium release in 10^{-1}M HNO_3 is by about one order of magnitude compared to the raw material(7×10^{-4} $g.m^{-2}.d^{-1}$ at 90°C) to the fully amorphized sample (9.9×10^{-3} $g.m^{-2}.d^{-1}$at 90°C). Figure 2 shows the ratio between the normalized dissolution rate of the irradiated material compared to that of the raw material versus the amorphous fraction of the irradiated sample. It underlines the influence of the amorphous fraction, regardless of temperature. The considered normalized dissolution rate is measured at the beginning of the dissolution test, to avoid the diffusion conditions when a neoformed phase precipitates. This ratio seems to be independent of the conditions of irradiation and of the temperature of leaching: the fully amorphized material presents a kinetics of dissolution about 10 times higher than the raw material.

Table 2 *Evolution of unit cell parameters of solid solutions of β-TPuPD with 16.4 Wt.% Pu.*

Age (months)	*Cumulative dose (α.g⁻¹)*	a *(Å)*	b *(Å)*	c *(Å)*	V *(Å³)*
0	0	12.779(6)	10.365(3)	7.027(3)	930(1)
16	$1.4{\bullet}10^{16}$	12.770(1)	10.362(7)	7.013(6)	928(2)
36	$4.0{\bullet}10^{16}$	12.763(9)	10.366(7)	7.023(7)	929(2)
90	$9.9{\bullet}10^{16}$	12.778(1)	10.380(1)	7.034(1)	929(2)

In order to determine the effect of radiolysis on the dissolution of the material, *in-situ* studies were performed during γ− irradiations. They were carried out on β−TUPD pellets in aqueous and nitric media exposed to γ−irradiation at room temperature using a ^{137}Cs source. Figures 3 and 4 reveal that the only parameter which influences the release of uranium in the medium is the exposure time (not dose) due to the very short life-time radical species produced under irradiation.

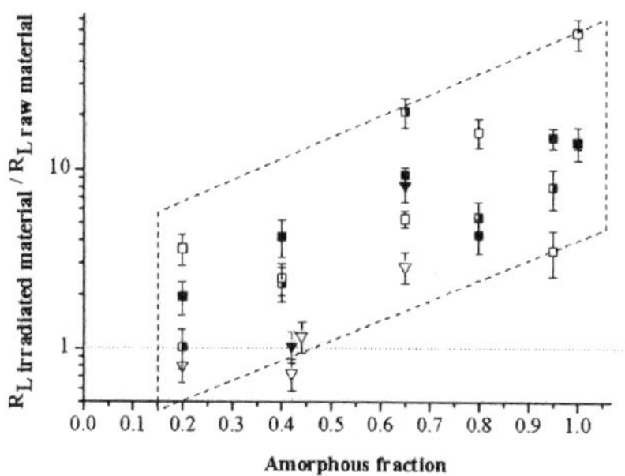

Figure 2 *Evolution of the ratio R_L irradiated material / R_L raw material with the amorphous fraction of the irradiated material (▼/▽: 410 MeV S,■ /□: 810 MeV Kr) during leaching tests (10^{-1}M HNO_3) at different temperatures (90°C : full symbols, 70°C : half full symbols, 25°C : open symbols)*

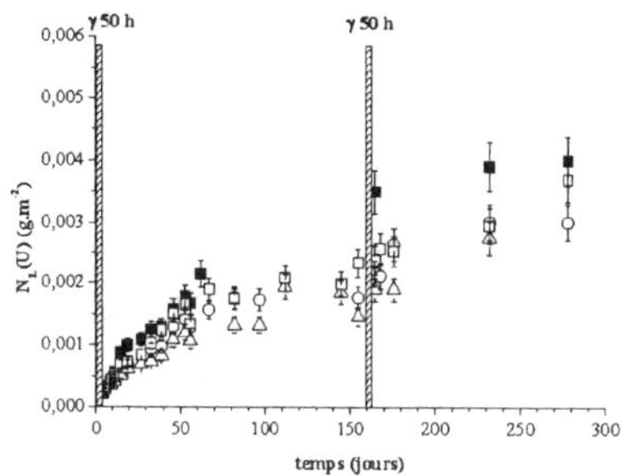

Figure 3 *Effect of dose rate on normalized leaching $N_L(U)$ (10^{-1}M HNO_3, 25°C) in-situ γ–irradiations (50 hours):■ : no irradiation ;□: 1 Gy.min^{-1} (cumulative dose:3 kGy) ;○ :10.9 Gy.min^{-1} (cumulative dose: 33 kGy) ;△: 56.6 Gy.min^{-1} (cumulative dose: 170 kGy).*

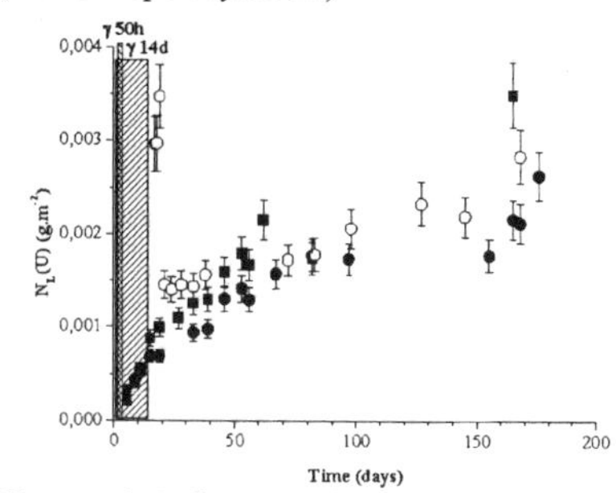

Figure 4 *Influence of irradiation time on the normalized leaching $N_L(U)$ during leaching (10^{-1}M HNO_3, 25°C) in-situ γ–irradiations (56.6 Gy.min^{-1}) : ■ : no irradiation ; ● : 50 hours (cumulative dose : 170 kGy) ; ○ : 14 days (cumulative dose : 1140 kGy).*

3 CONCLUSION

β–TPD ceramic can be fully amorphized under ion beams when the electronic energy loss is higher than 10 MeV/μm. Despite these drastic conditions of alteration, the dissolution kinetics of the irradiated material is only about 10 times higher than the dissolution kinetics of the raw material under the same conditions, which confirms the high potential of this material for tetravalent actinides storage.

References

1 N. Dacheux, A.C. Thomas, V. Brandel and M. Genet, *J. Nucl. Mater*, 1998, **257**, 179.
2 N. Dacheux, A.C. Thomas, B. Chassigneux et al., *Mat. Res. Soc.Proc.*, 1999, **556**, 85.
3 W.J. Weber, R.C. Ewing, C.R.A. Catlow et al, *J. Mater. Res.*, 1998, **13**, 1434.
4 L. Thomé and F. Garrido, *Vacuum*, 2001, **63**, 619.

FROM THORIUM PHOSPHATE-HYDROGENPHOSPHATE HYDRATE TO THORIUM PHOSPHATE-DIPHOSPHATE : A NEW WAY OF PREPARATION OF β-TPD AND ASSOCIATED SOLID SOLUTIONS WITH ACTINIDES

N. Clavier[1], N. Dacheux[1], G. Wallez[2], V. Brandel[1], J. Emery[3], M. Quarton[2] and E. Simoni[1]

[1] Groupe de Radiochimie, IPN Orsay, Université Paris-Sud-11, 91406 Orsay, France
[2] Laboratoire de Cristallochimie du Solide, Université Pierre et Marie Curie, 75252 Paris, France
[3] Laboratoire de Physique de l'Etat Condensé, Université du Maine, 72085 Le Mans, France

1 INTRODUCTION

Thorium Phosphate-Diphosphate ($Th_4(PO_4)_4P_2O_7$, β-TPD) appears as a promising material for the specific immobilization of tetravalent actinides in an underground repository. Indeed, the large replacement of thorium by tetravalent actinides in the crystal structure leads to solid solutions of general formula $Th_{4-x}An_x(PO_4)_4P_2O_7$ (β-TAnPD ; An = U, Np, Pu). The good sintering properties of these materials allow the preparation of dense pellets, which present a strong resistance to aqueous corrosion and to radiation damage.[1] Several methods of preparation have been reported for the preparation of such solids, involving both wet and dry chemical processes.[2] More recently, β-TPD was obtained through the initial precipitation of a low-temperature crystallized precursor, identified as Thorium Phosphate-Hydrogenphosphate Hydrate ($Th_2(PO_4)_2(HPO_4)\cdot H_2O$, orthorhombic, namely TPHPH), then firing at 1250°C.[3] Thus, it appears necessary to identify the different steps during the heating of TPHPH in order to provide the transformation to β-TPD, and to examine the behaviour of TUPHPH solid solutions.

2 RESULTS AND DISCUSSION

In order to follow the chemical reactions driving the transformation of TPHPH into β-TPD, the variation of the Raman spectrum of TPHPH was followed versus the temperature.[4] Below 200°C, the spectrum remained unchanged: all the vibration modes of P-O and P-O-(H) bonds were assigned, which confirmed the presence of both PO_4 and HPO_4 entities in the structure. Above this temperature, the appearance of an additional band around 770 cm^{-1} was correlated to the ν_s (P-O-P) vibration, *i.e.* to the condensation of HPO_4 groups into P_2O_7 entities. When heating above 1000°C, the spectra appeared in good agreement with that of β-TPD as a confirmation of its formation as a pure phase above this temperature.

The chemical reactions associated with the modification of the Raman spectra were identified through a TGA-DTA study.[4] The TGA curve exhibited a total weight loss between 180°C and 300°C occuring in two parts : between 180°C and 220°C, the loss of one water molecule was assigned to the dehydration of TPHPH into the anhydrous $Th_2(PO_4)_2(HPO_4)$ (TPHP), then the loss of an additional 0.5 H_2O in the 220-300°C range was correlated with the condensation of two HPO_4 groups into one P_2O_7 entity, leading to an allotropic form of TPD, namely α-TPD. These two phases act as reaction intermediates, and were previously unreported. α-TPD was found to be stable up to 950°C, where an exothermic peak on the DTA curve marks its transformation into the well known high-temperature form, β-TPD. In these conditions, the following chemical reaction scheme was proposed :

$$2\ Th_2(PO_4)_2(HPO_4)\cdot H_2O \xrightarrow{180°C \leq T \leq 220°C} 2\ Th_2(PO_4)_2(HPO_4) + 2\ H_2O\uparrow \quad (1)$$

$$2\ Th_2(PO_4)_2(HPO_4) \xrightarrow{220°C \leq T \leq 300°C} \alpha\text{-}Th_4(PO_4)_4P_2O_7 + H_2O\uparrow \quad (2)$$

$$\alpha\text{-}Th_4(PO_4)_4P_2O_7 \xrightarrow{T \geq 950°C} \beta\text{-}Th_4(PO_4)_4P_2O_7 \quad (3)$$

HT-XRD diagrams were collected at various temperatures to characterize TPHP and α-TPD intermediates.[4] as well as the initial TPHPH (Table 1). As expected from the chemical scheme, the variations of the lattice parameters remained weak below 180°C. Between 180 and 220°C, the *a* and *b* parameters decreased, inducing a net volume drop correlated with dehydration. An additional volume variation was noted during the condensation of HPO_4 groups, mainly due to the increase of the *c* parameter. All these variations remained moderate and continuous. Thus, the three forms (TPHPH, TPHP and α-TPD) were found to crystallize in closely related orthorhombic structures. Finally, only the XRD lines of β-TPD were observed above 950°C, this phase being stable up to 1350°C.

Table 1 *Unit cell parameters of TPHPH, TPHP, α-TPD and β-TPD*[4]

	T(°C)	*a* (Å)	*b* (Å)	*c* (Å)	Z	V (Å^3)
PHPTH	25	21.368(2)	6.695(1)	7.023(1)	4	1004.8(4)
PHPT	220	21.229(2)	6.661(1)	7.031(1)	4	994.2(4)
α - PDT	300	21.206(2)	6.657(1)	7.057(1)	2	996.2(4)
β - PDT	> 950	12.8646(9)	10.4374(8)	7.0676(5)	2	949.00(9)

A similar study was devoted to crystallized precursors incorporating tetravalent uranium. A complete solid solution of general formulae $Th_{2-x/2}U_{x/2}(PO_4)_2(HPO_4)\cdot H_2O$ was found between pure TPHPH and UPHPH.[5] All the analytical techniques involved, such as μ-Raman spectroscopy, led to results in good agreement with that obtained for TPHPH. From TGA-DTA, the temperature of the phase transition (960°C) did not appear significantly affected by the uranium loading in the samples. Nevertheless, as was expected from previous work, pure, homogeneous and single phase β-TUPD solid solutions were prepared only for $x \leq 2.8$ (Figure 1). For higher substitution rates, α-TUPD was first prepared, but turned into a mixture of β-TUPD, $U_{2-y}Th_yO(PO_4)_2$ and α-$Th_{1-z}U_zP_2O_7$ above 600°C. Thus, all the structures studied, except that of β-TPD, allow the complete replacement of thorium by tetravalent uranium. Moreovoer, since the incorporation of Np(IV) and Pu(IV) was observed in TPHPH up to $x = 3$ and near to $x = 4$, respectively, it appeared possible to prepare β-TNpPD or β-TPuPD solid solutions from these crystallized precursors.[6]

$$2\,Th_{2-x/2}U_{x/2}(PO_4)_2(HPO_4)\cdot H_2O \xrightarrow[-2\,H_2O]{190^{\circ}C} 2\,Th_{2-x/2}U_{x/2}(PO_4)_2(HPO_4) \xrightarrow[-H_2O]{250^{\circ}C} \alpha\text{-}Th_{4-x}U_x(PO_4)_4(P_2O_7)$$

$$x < 2.8,\ 960^{\circ}C: \alpha\text{-}Th_{4-x}U_x(PO_4)_4(P_2O_7) \rightarrow \beta\text{-}Th_{4-x}U_x(PO_4)_4(P_2O_7)$$

$$x > 2.8,\ 600^{\circ}C: \alpha\text{-}Th_{4-x}U_x(PO_4)_4(P_2O_7) \rightarrow \beta\text{-}Th_{4-x}U_x(PO_4)_4(P_2O_7) + U_{2-y}Th_yO(PO_4)_2 + \alpha\text{-}Th_{1-z}U_zP_2O_7$$

Figure 1 *Scheme of transformation of TUPHPH solid solutions versus heating temperature*

The precipitation of TUPHPH solid solutions as precursors was applied with success to the preparation of β-TUPD sintered pellets.[1] In comparison to the process previously used, based on the direct evaporation of a mixture of solutions containing actinides and phosphoric acid, this new way of preparation led to a significant improvement in the homogeneity of the solids (Figure 2). Moreover, dense pellets (95 % < $d_{\text{mes.}}/d_{\text{calc.}}$ < 98 %) were obtained after only 5 hours of heat treatment at 1250°C, instead of more than 10 hours previously.

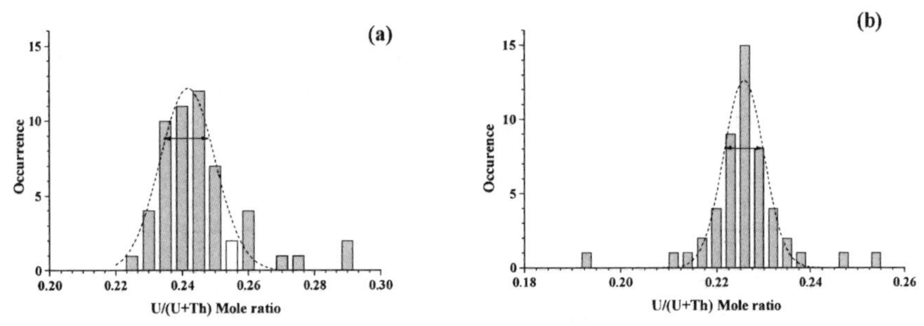

Figure 2 *Statistical EPMA study of the homogeneity of β-TUPD sintered pellets prepared by "direct evaporation" (a) or through the initial precipitation of TUPHPH (b)*

The chemical durability of β-TUPD sintered pellets obtained through the initial precipitation of TUPHPH solid solutions was tested in acidic media: both preparation method and sintering process did not influence their good resistance to alteration. Besides, TPHPH was found to precipitate as a neoformed phase in the back-end of the initial dissolution process, and thus acts as a protective layer, which could delay significantly the release of radionuclides in the biosphere.

References

1 N. Clavier, N. Dacheux, P. Martinez, E. du Fou de Kerdaniel, L. Aranda and R. Podor, *Chem. Mater.*, 2004, **16**, 3357.
2 V. Brandel, N. Dacheux and M. Genet, *Radiokhimiya*, 2001, **43**, 16.
3 V. Brandel, N. Dacheux, M. Genet and R. Podor, *J. Solid State Chem.*, 2001, **159**, 139.
4 N. Dacheux, N. Clavier, G. Wallez, V. Brandel, J. Emery, M. Quarton and M. Genet, *Mat. Res. Bull.*, 2005, in press.
5 N. Clavier, N. Dacheux, P. Martinez, V. Brandel, R. Podor and P. Le Coustumer, *J. Nucl. Mater.*, 2004, **335**, 397.
6 J. Rousselle, *PhD Thesis*, Université Paris-Sud-11 IPNO-T-04-03, 2004.

INVARIANT EQUILIBRIA IN BINARY PLUTONIUM PHASE DIAGRAMS, RELATIONSHIPS

L.F. Timofeeva

A.A. Bochvar VNIINM (All-Russian Research Institute of Inorganic Materials), Moscow Russia

1 INTRODUCTION

The universal empirical relationship[1] of invariant transformations was used to analyze eutectoid reactions in binary plutonium systems,[2] to check the parameters of eutectoid reactions of δ-solid solutions of plutonium in systems with elements of group III-B because of the differences in published diagrams of Pu–Al and Pu–Ga,[3,5,6,7] and to verify the parameters of eight invariant transformations in the diagram for Pu–Np.[9,10] The relationship establishes the relation between invariant equilibrium temperature T_n (T_E or T_P), low-melting component concentration at invariant equilibrium point C_n (C_E or C_P), and ratio of the component formation temperatures, T_A and T_B, as,

$$K_n + T_B / T_A = 1 \tag{1}$$

where $T_B/T_A = f$, which is the temperature factor; K_n is the coefficient of the invariant equilibrium, which for eutectoid reaction can be written as

$$\boldsymbol{K_n} = K_E = \frac{T_A - T_E}{T_A} C_E \tag{2}$$

and for a peritectoid reaction as

$$\boldsymbol{K_n} = \boldsymbol{K_P} = \frac{T_A - T_P}{T_P} \frac{1}{C_p} \tag{3}$$

2 RESULTS

2.1 Eutectoid transformations in binary plutonium systems

Among the known binary diagrams of plutonium,[2] there are 30 in which a total of 53 eutectoid transformations are present.[7,8] Figure 1 shows the graph of correlation between K_n and $f=T_B/T_A$ (1,2) as a straight line with intercepts of unity on both the K_n and f axes. For most eutectoid equilibria points lie on the correlation line or near it, the deviation is ≤ ± 0.05 as indicated by dotted lines. Points located far beyond the limits of the correlation band (Figure 1, points Cm-2: *δ=γ-Pu +α-Cm;* Hf: *β-Hf=L+α-Hf;* Al-5: $PuAl = Pu_3Al + PuAl_2$[5,8]) raise doubts about the validity of the parameters of reactions and should be verified. From the plot it also follows that formation of eutectoids requires a favourable temperature factor: the maximum temperatures T_A and T_B of formation of eutectoid phases do not differ by more than 60% and the absolute temperature of the eutectoid is always > $0.4T_A$.

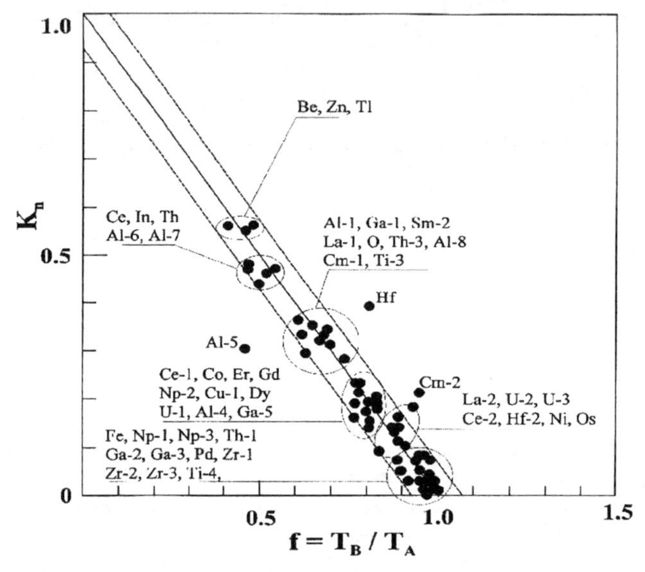

Figure 1 *The relationship between coefficients of eutectoid equilibrium $K_n = K_E = \frac{T_A - T_E}{T_A} C_E$ and temperature factor $f = T_B / T_A$ in binary Pu systems. For each point, an element is designated in the binary system with plutonium. If there are several transformations in the system, the ordinal number of the reaction is indicated next to the second element.*

2.2 Analysis of transformations in Pu systems with elements of III-B group

The parameters of the δ phase eutectoid reactions in Pu systems with Al, Ga, In, Tl[2,3,5,6] (Figure 1, points Al-1, Ga-1, In), including the Pu-Al system[5] at 0.1, 0.3 and 0.6 GPa (Figure 1, points Al-6, Al-7, Al-8), fit the correlation band with a maximum deviation of < ± 0.05. In the binary Pu systems with Ga, In, Tl, the eutectoid transformation of the δ-phase results in the formation of α(β)–Pu and the compound Pu_3(Ga, In, Tl), in contrast with the formation of the PuAl compound in the Pu-Al system.[7,8] The parameters of the PuAl= Pu_3 Al +$PuAl_2$ at 193 ^{0}C in the diagram of Ellinger et al.[4] are outside the 24% deviation range (Figure 1, Al-5). Thus, the parameters of the invariant transformations on the Pu–Al diagram, obtained experimentally,[3,5,8] are additionally confirmed by the dependency (1-3).[8,9]

2.3 Invariant transformations in Pu –Np system

The universal relationship (1-3) was used for analyses and prediction of some of the eight invariant reaction parameters in the plutonium – neptunium system.[9,10] The reaction parameters refined in the paper[9] demonstrated a better approximation to the equation of the universal dependency than the values from the literature[10] (Table 1).

Table 1 *Refined parameters of the invariant reactions in Pu-Np system*

Reaction No.	Reaction type	*T_n ^{0}C	** Pu content in phases, %(at.) Literature data [10]			Refined values[9]		
	Peritectoid reactions							
1	(γNp,εPu) +βNp = βPu	540	22	10	15	**22.8**	**7.5**	**12**
2	(γNp,εPu) + βPu =η	508	66	47	51	-	**45**	-
4	β-Np +βPu =α-Pu	300	5.0	38.5	25	**7.0**	**44**	**27.5**
	Eutectoid reactions							
7	η= βPu+γPu	288	96.5	95.5	98.3	**97.0**	-	**97.8**
8	βNp =αNp +αPu	275	1.6	1.0	4.0	-	-	**18.0**

* temperature of invariant equilibrium
**The composition for the phases is specified in the order of like in the reaction notation

3 CONCLUSION

The applicability of the universal dependency of invariant transformations for the analysis of eutectoid transformations in binary plutonium systems, and the possibility and potential of its use for the validation of previously constructed plutonium phase diagrams, and the detection of deviations and the prediction of eutectoid and peritectoid parameters have been demonstrated.

References

1 V.M.Vozdvizhensky. *Prediction of binary phase diagrams from statistical criteria.* M., *Metallurgiya*, 1975, 223.
2 *Phase diagrams of binary metallic systems*. N.P. Lyakishev, ed., in 3 volumes. Moscow, Mashinostroyeniye, 1996, **1**, 991; 1997, **2**, 1023; 2000, **3**, 448.
3 N.T. Chebotarev, V.S. Kurilo, L.F. Timofeeva, M.A. Andrianov, and V.V. Sipin, *VANT, Seriya: Materialovcdeniye i novyye materialy*, (Russian) 1990, **37**, 20.
4 F.H. Ellinger, C.C. Land, and W.N. Miner, *J. Nucl. Mater.*,1962, **5**, 165.
5 L.F. Timofeeva, *Ageing Studies and Lifetime Extension of Materials,* ed. L. Mallinson,N-Y: Kluwer Academic /Plenum, 2001,191.
6 S.S.Hecker and L.F.Timofeeva, *J.Los Alamos Science*, 2000, **26**, 196.
7 L.F. Timofeeva, *J. of Metals*, 2003, **55**, 51.
8 L.F. Timofeeva, *M I T O M*, (Russian) 2004, **11**, 41.
9 L.F. Timofeeva, N.I.Nogin and V.M.Filin, *Materials Sciences Transactions*, (Russian) 2004, **93**, 22.
10 R.I. Sheldon and D.E.Peterson, *Bull. Alloy Phase Diagrams,* 1985, **6**, 215.

PHOTOEMISSION AND SECONDARY ION MASS SPECTROMETRY STUDY OF URANIUM PASSIVATION BY C^+ IMPLANTATION

A.J. Nelson, T.E. Felter, K.J. Wu, C. Evans, J.L. Ferreira, W.J. Siekhaus and W. McLean

Lawrence Livermore National Laboratory, Livermore, CA 94550

1 INTRODUCTION

Preventing the corrosion and oxidation of uranium is important to the continued development of advanced nuclear fuel technologies. Knowledge of the surface reactions of uranium metal with various environmental and atmospheric agents, and the subsequent degradation processes, are vitally important in 21st century nuclear technology. A review of the oxidation of actinide elements and their use in catalysis[1] summarizes the present understanding of the kinetics and mechanisms of the reaction in dry and humid air.

Researchers have recently used N_2^+ and C^+ ion implantation to modify the near surface region chemistry and structure of U to affect the nucleation and growth kinetics of corrosion and to passivate the surface.[2-4] These researchers used Auger electron spectroscopy (AES) in conjunction with sputter depth profiling to show that the implanted surfaces had compositional gradients containing nitrides and carbides. In addition to chemical modification, ion implantation can create special reactive surface species that include strained defect structures that affect the initial absorption and dissociation of molecules on the surface, thus providing mechanical stability and protection against further air corrosion.

2 METHOD AND RESULTS

Oxidation of polished U in laboratory air prior to introduction into an ion implanter vacuum chamber results in ≤20 nm of oxide that was sputtered during the ion irradiation.[5] The implantation was performed on a water-cooled, lightly oxidized U sample at 10^{-7} Torr, normal incidence using CO_2 gas as the source material in a hot filament ionizer. A magnet separated the carbon +1 ions, and the beam was rastered onto the surface resulting in a pure and uniform C dose, accurate to a few percent. The TRIM calculated sputtering rates of the surface oxygen and surface uranium by the implanting carbon ions are 44% and 23%, respectively. After a year in "standard" California environment (ambient temperature, 50% relative humidity), the appearance of the implanted area remained unchanged. X-ray photoelectron spectroscopy (XPS) core-level analysis and time-of-flight secondary ion mass spectrometry (ToF-SIMS) were used to determine composition and bonding versus depth. The U sputter rate was estimated to be 2 nm/min. (SRIM-96).

Figure 1 presents the ToF-SIMS depth profile results from the 33keV, 4.3 x 10^{17} cm^{-2} C^+ implantation. We observed a thin oxide layer as indicated by the high oxygen content present in the first 50 nm of the surface. Arkush, *et al*,[2] and Musket[4] observed a similar trend in the near surface elemental composition of C^+ implanted U with AES depth profiling. The carbon profile shows a wide diffuse interface between the implanted layer (strained, defect structure) and the metal substrate. This carbon profile is comparable to previous AES results,[2,4] and in agreement with the implant depth (TRIM).

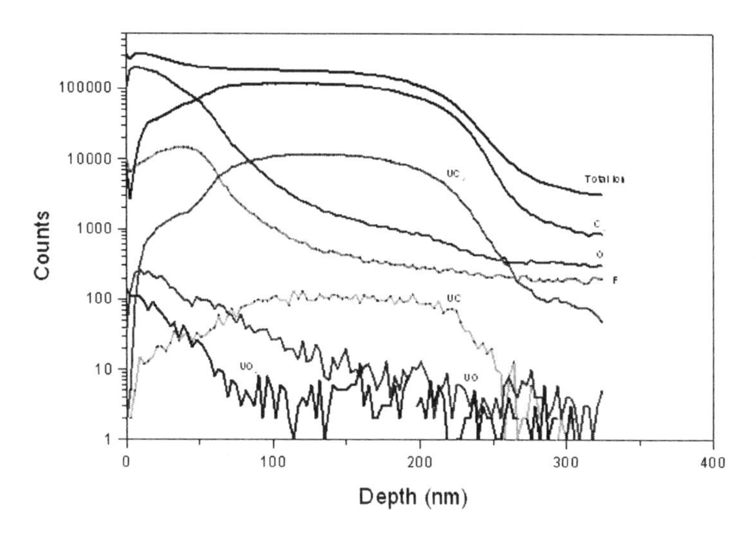

Figure 1 *ToF-SIMS depth profile of the C^+ implanted region.*

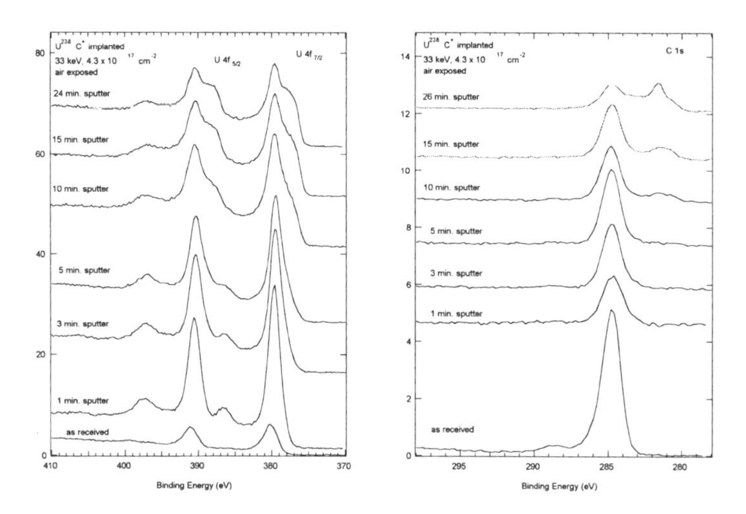

Figure 2 *High-resolution U $4f_{7/2,5/2}$ and C 1s core-level spectra versus sputtered depth for the C^+ implanted region.*

Figure 2 presents the high-resolution U $4f_{7/2,5/2}$ core-level spectra for the C^+ implanted U surface as a function of sputter etch time. The U $4f_{7/2,5/2}$ spin-orbit pair binding energies are in agreement with literature values for uranium in a U^{4+} valence state.[6-10] The initial spectra for the as received implanted surface do not exhibit the shake-up satellite feature, which is due to the excitation of an electron from the O 2*p*–U bonding orbital to a partially occupied or unoccupied U 5*f* orbital, and is indicative of oxidized U.[6]

Following further sputter depth profiling of the C^+ implanted surface, the U $4f_{7/2}$ peaks show the presence of additional components, one at ~377 eV that represents the underlying metallic uranium,[7,10,11] and the second component at 378 eV that represents UC.[12,13] These facts compliment the ToF SIMS results showing the presence of a thin oxide layer over the transitional U-carbide layer in the air exposed C^+ implanted area.

The high-resolution C 1*s* core-level spectra for the C^+ implanted U surface versus sputter etch time is presented in Figure 2. Sputter depth profiling of the C^+ implanted surface reveals a C 1*s* peak at 281.6 eV that is indicative of U-carbide.

The valence band region for the implanted U surface provided unique information about the electronic structure, and the results suggest that the UC has metallic character.

3 CONCLUSION

Core-level and valence band photoelectron spectroscopy in combination with time-of-flight secondary ion mass spectrometry (ToF-SIMS) depth profiling of C^+ implanted U revealed a buried U-carbide layer with minimal residual oxidation at the diffuse carbide/U metal graded interface. The wide defected transitional carbide layer strongly suppresses oxidation as previously described, and has metallic character.

This work was performed under the auspices of the U.S. Department of Energy by the University of California Lawrence Livermore National Laboratory under Contract No. W-7405-Eng-48.

References

1 C.A. Colmenares, *Prog. Solid State Chem.* 1984, **15**, 257.
2 R. Arkush, M.H. Mintz and N. Shamir, *J. Nucl. Mater.* 2000, **281**, 182.
3 R. Arkush, M. Brill, S. Zalkind, M.H. Mintz and N. Shamir, *J. Alloys Comp.* 2002, **330-332**, 472.
4 R.G. Musket, *Materials Research Society Conference Proceedings,* 1987, No. **93**, 49.
5 G.W. McGillivray, D.A. Geeson and R.C. Greenwood, *J. Nucl. Mat.*, 1994, **208**, 81.
6 G.C. Allen, P.M. Tucker and J.W. Tyler, *J. Phys. Chem.*, 1982, **86**, 224.
7 W. McLean, C.A. Colmenares, R.L. Smith, and G.A. Somorjai, *Phys. Rev.*, 1982, **B25**, 8.
8 K. Winer, C.A. Colmenares, R.L. Smith, and F. Wooten, *Surf. Sci.*, 1986, **177**, 484.
9 J.N. Fiedor, W.D. Bostick, R.J. Jarabek, and J. Farrell, *Environ. Sci. Tech.*, 1998, **32**, 1466.
10 S.V. Chong and H. Idriss, *Surf. Sci.*, 2002, **504**, 145.
11 T. Gouder, C.A. Colmenares and J.R. Naegele, *Surf. Sci.*, 1995, **342**, 299.
12 T. Ejima, S. Sato, S. Suzuki, Y. Saito, S. Fujimori, N. Sato, M. Kasaya, T. Komatsubara, T. Kasuya, Y. Onuki, and T. Ishii, *Phys. Rev.,* 1996, **B53**, 1806.
13 M. Eckle, R. Eloirdi, T. Gouder, M. Colarieti Tosti, F. Wastin, J. Rebizant, *J. Nucl. Mater.,* 2004, **334**, 1.

OPTICAL METALLOGRAPHY OF PLUTONIUM

R.A. Pereyra, D. Lovato, T. Baros and L. Roybal

Los Alamos National Laboratory, Los Alamos, New Mexico, 87548

1 INTRODUCTION

Plutonium (Pu) metallography has played a key role in the understanding of Pu metallurgy, and it continues to be a challenging field; one in which much scientific research remains to be conducted. The following special conditions exist when working with Pu:

(1) Work conducted with Pu must be performed in a confined space (e.g., a glovebox), because of the associated health hazards.

(2) Plutonium reacts to its environment and is sensitive to mechanical stress, which can result in the formation of a layer on the surface, masking the true microstructure.

Included in this work are examples of Pu microstructures relative to item (2) above and those of past Pu programmes at Los Alamos, as well.

2 DIFFICULTIES INVOLVED IN PLUTONIUM METALLOGRAPHY

There are several reasons why optical metallography of Pu is difficult. Firstly, because of the health hazards associated with its radioactive properties, sample preparation is conducted in gloveboxes. In the sample preparation process the sample is systematically ground with sequentially smaller abrasives that could range from hundreds of micrometers (μm) down to 0.25 μm. Therefore, cleanliness is paramount to reduce the likelihood of abrasive cross contamination, thus reducing the likelihood of ruining the sample during later stages of the process. For example, a 100-μm scratch will appear as a 5-cm-wide gouge when viewed at 500 times magnification. Since gloveboxes limit the amount of solvents that can be used, and are designed to be at negative pressure compared to the laboratory, they are by their nature counter to the philosophy of cleanliness.

Secondly, Pu will hydride or oxidize when in contact with metallographic polishing lubricants, solvents, or chemicals. Some samples are more reactive than others. For example, Figure 1 shows a delta phase alloy in which an oxide cloud has begun to spread (from the top), and if the polishing had continued, the oxide would have covered the entire sample. It is common to see Pu microstructures from the early days at Los Alamos misinterpreted because of this type of masking oxide layer. It has been noted that nucleation of the hydride/oxide begins around inclusions and samples with a higher concentration of impurities seem to be more susceptible to this reaction. Figure 2 shows a delta phase microstructure showing nucleation of a reaction layer around some inclusions. In many samples it is unavoidable, and the best one can hope for is to limit the reaction.

Lastly, strain-induced FCC delta phase Pu can transform to the monoclinic (expanded lattice) alpha′ phase. During the grinding or polishing process of delta phase samples, there is enough stress induced in the sample that the surface can transform. Figure 3 shows a cross section of a 10-μm scratch on the surface of a low-gallium alloy. The stress has transformed the dark delta phase to the light colour alpha′ to a depth of almost 20 μm. Note also that there is a continuous thin layer of alpha′ approximately 2 μm thick from the fine abrasive. This thin layer can ultimately be removed as the sample continues through the metallographic process, but the damage caused by the coarse abrasive will not. And the sample will need to be taken back to at least the size of abrasive that caused the damage.

3 OTHER PLUTONIUM MICROSTRUCTURES

The final step in the preparation is to etch the sample electrolytically in a solution of approximate 10% nitric acid in dimethylformalmide. In this process the surface of the sample is preferentially attacked revealing the microstructure. Not only are grain boundaries revealed, but also second phases and the indication of homogeneity. For example, Figures 4 – 6 show as-cast, semi-homogenized and homogenized microstructures of low-gallium alloys. In the as-cast sample (Figure 4) there is not enough gallium at the grain boundaries and the (white) alpha phase is retained. In Figure 5, although the sample has been homogenized enough to stabilize the delta phase at the grain boundaries, the shading within the grains indicates some segregation still exists. Finally, in a fully homogenized sample (Figure 6) the shading has disappeared. Note that in Figures 4 and 6, Pu_6Fe intermetallic is present at the grain boundaries.

The last two Figures (7 and 8) show the initial and final transformation from delta to alpha′ of low-gallium alloys as they are isopressed up to 10 kb. Note that in the initial stages of transformation (Figure 7), the transformation occurs along crystal orientations. This is an example of the detail that metallography, if correctly prepared, can reveal. It is also an example of the types of information that optical metallography has contributed to the understanding of plutonium metallurgy at Los Alamos.

Figure 1. *Delta phase alloy showing the spread of an oxide cloud (dark). If left unchecked it would cover the entire sample surface*

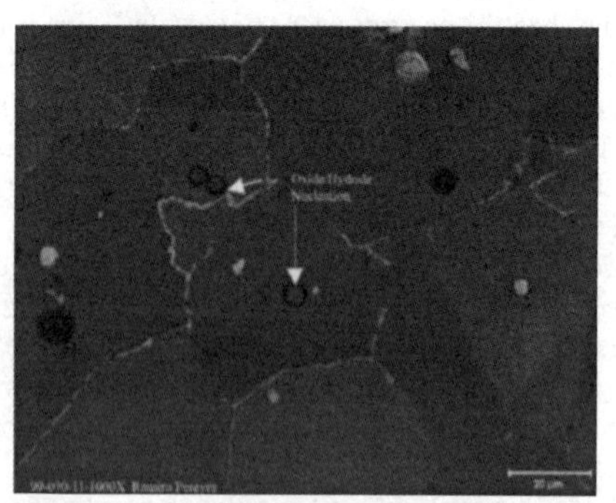

Figure 2. *Delta phase alloy showing the nucleation of hydride/oxides (dark ring) around oxide inclusions.*

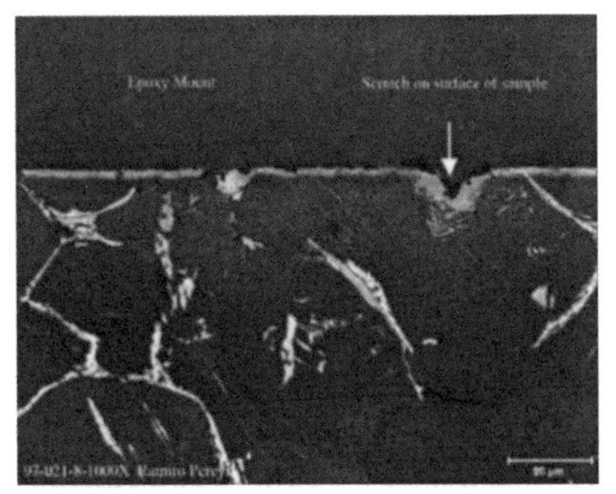

Figure 3. *The delta (dark) to alpha' (light) transformation from a 10-μm scratch on the surface of a low-gallium alloy. The scratch was due to cross contamination of a larger abrasive.*

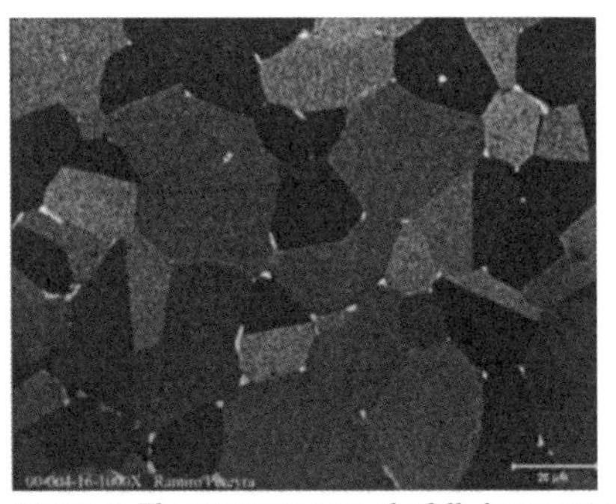

Figure 6. *The microstructure of a fully homogenized delta phase plutonium alloy. The light areas in the grain boundaries are Pu_6Fe.*

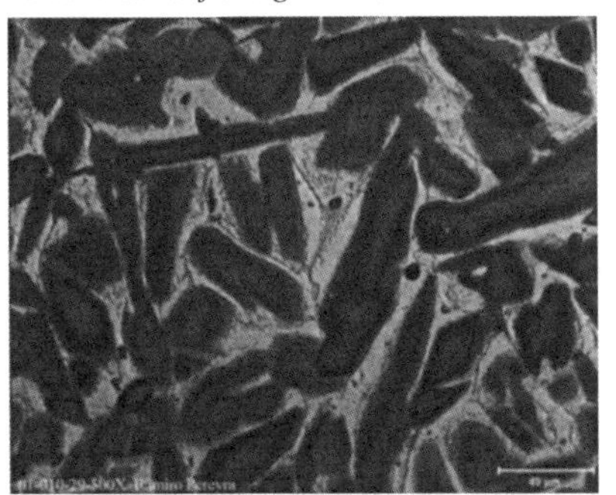

Figure 4. *Alpha (light) plus delta (dark) microstructure of an as-cast sample of a low-gallium alloy.*

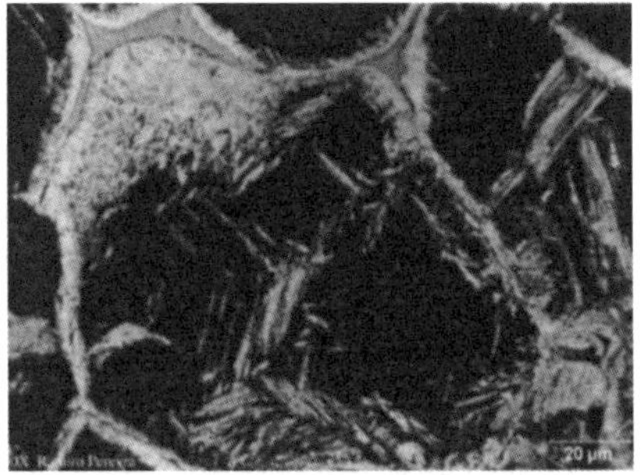

Figure 7. *The initial delta (dark) to alpha' (light) phase transformation of an isopressed low gallium alloy. Note the crystal orientation of the transformation. Pu^6Fe delineate the grain boundaries.*

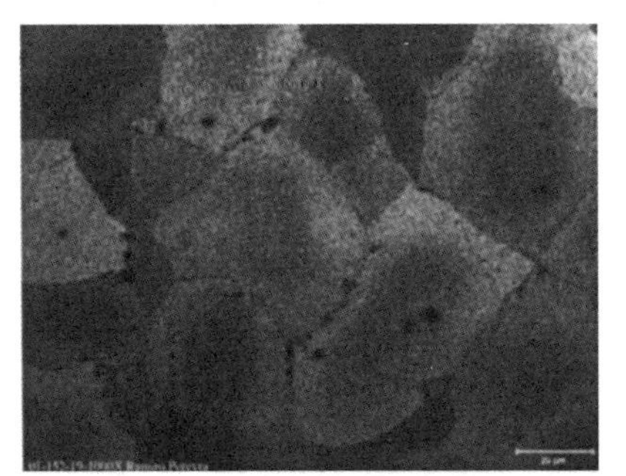

Figure 5. *Low-gallium alloy delta phase microstructure. Shading within the grains indicates that slight segregation from center of the grains to their boundary still exists.*

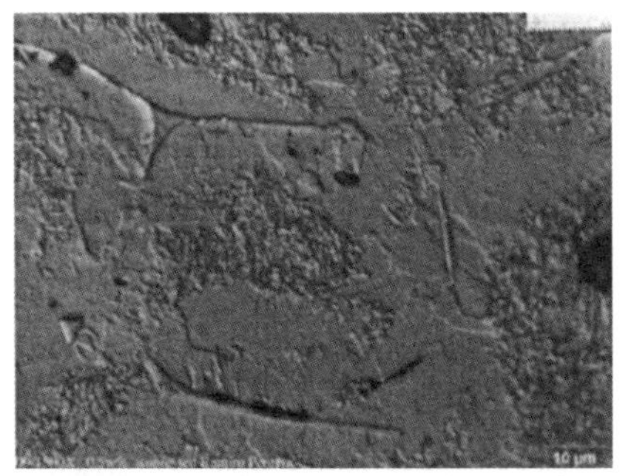

Figure 8. *Low gallium alloy, isopressed to 10k Bars. Alpha plus alpha' with Pu_6Fe delineating the grain boundaries.*

NEW FAMILY OF INTERMETALLIC HYDRIDES: $U_2T_2InH_x$ AND $U_2T_2SnH_x$

K. Miliyanchuk[1], L. Havela[1], S. Daniš[1], A.V. Kolomiets[1], L.C.J. Pereira[2] and A.P. Gonçalves[2]

[1] Charles University, Faculty of Mathematics and Physics, Department of Electronic Structures, Ke Karlovu 5, 121 16 Prague 2, The Czech Republic
[2] Departamento do Quimica, Instituto Tecnológico e Nuclear/CFMC-UL, P-2686-953 Sacavém, Portugal

1 INTRODUCTION

Hydrogen has proved to be a powerful tool for tuning the magnetic properties of actinide compounds. Hydrogen penetrates into the material expanding its crystal lattice, leading to the modification of electronic states, and therefore it affects the size of magnetic moments and the strength of exchange interactions.

U_2T_2X compounds (T – transition metal, X – In, Sn) crystallize with the tetragonal Mo_2FeB_2 structure type.[1] The magnetic order in the U_2T_2X series changes from 5*f*-antiferromagnetism to weak paramagnetism with heavy-fermion behaviour on the verge of magnetism.[2] The driving force of these changes is the increase of the hybridisation, which makes the U_2T_2X compounds very sensitive to the lattice and electronic structure modifications induced by hydrogenation.

U_2Co_2Sn was the first compound of the series found to absorb hydrogen.[3] Two hydrides are formed with different hydrogen content. The magnetism of U_2Co_2Sn, which exhibits non-Fermi liquid features and ferromagnetic spin fluctuations, changes dramatically. A weak ferromagnetism below $T_C = 33.5$ K was observed for the α-hydride, whereas the β-hydride is an antiferromagnet ($T_N = 27$ K). The results were discussed in terms of lattice expansion and 5*f*-ligand hybridization varying with the hydrogenation.

2 METHOD AND RESULTS

2.1 Experimental details

U_2T_2In and U_2T_2Sn were used as starting materials for synthesis of the respective hydrides. The synthesis was performed by a method described in Havela et. al.[4], using a hydrogen pressure of 110 bar and elevated temperatures.

The hydrogen content of the hydrides was estimated by the decomposition of the synthesized hydrides when heated in a closed volume up to $T = 1073$ K. At the end of the process, the reactor was cooled down, the pressure increase was recalibrated to ambient temperature, and the amount of released hydrogen was calculated.

The crystal structure of the parent compounds, as well as of the hydrides, was studied by X-ray diffraction. The powder patterns were obtained using a HZG3 diffractometer (Cu-$K\alpha$ radiation). The crystal structure analysis was performed using a full-profile Rietveld refinement. Positions of the hydrogen atoms could not be determined due to the very small X-ray scattering factor of hydrogen.

The magnetic susceptibilities of U_2T_2Sn and U_2T_2In hydrides wre measured in various magnetic fields using a Quantum Design PPMS extraction magnetometer. The grains of the sample were fixed in random orientation by acetone-soluble glue, which prevents rotation of individual grains under the influence of a magnetic field.

2.2 Interaction of U_2T_2In and U_2T_2Sn compounds with hydrogen

The discovery of the formation of U_2Co_2Sn hydrides[3] initiated a broader study of the U_2T_2X series. Hydrogen absorption was also registered for U_2Ni_2In, U_2Ni_2Sn, U_2Co_2In, and U_2Fe_2Sn. This contribution reports on the hydrogen absorption properties of U_2Ni_2Sn and U_2Ni_2In.

U_2Ni_2In absorbs 1.9 H atoms per formula unit. The hydride is stable up to 473 K, and the hydrogen desorption takes place completely in one step below 673 K. The penetration of hydrogen does not change the structure type and leads to weakly anisotropic lattice expansion with more pronounced expansion along the c-axis (see Table 1). Therefore, we assume that hydrogen atoms occupy interstitial positions. The shortest U-U distance is along the c-axis, and is equal to the lattice parameter, c.

Table 1 *Crystal-structure parameters of $U_2Ni_2InH_{1.9}$ and $U_2Ni_2SnH_{2.4}$, including the lattice parameters, a and c, unit cell volume, V, the shortest U-U distance, d_{U-U}, and relative lattice expansion along the a-axis, $\Delta a/a$, along the c-axis, $\Delta c/c$, and relative volume expansion, $\Delta V/V$.*

	a (Å)	c (Å)	V (Å^3)	d_{U-U} (Å)	$\Delta a/a$ (%)	$\Delta c/c$ (%)	$\Delta V/V$ (%)
U_2Ni_2In	7.390(1)	3.576(1)	195.3(1)	3.58	-	-	-
$U_2Ni_2InH_{1.9}$	7.547(1)	3.662(1)	208.6(1)	3.66	2.1	2.4	6.8
U_2Ni_2Sn[a]	7.236(3)	3.643(2)	190.8(1)	3.58	-	-	-
$U_2Ni_2SnH_{2.4}$	7.365(1)	3.754(1)	203.7(1)	3.68	1.8	3.0	6.8

[a] The lattice parameters of the initial compound are slightly smaller than those reported in the literature.[1] It may indicate a certain, non-zero, stability range.

Magnetic susceptibility exhibits a sharp cusp at 60 K, which indicates a Néel temperature much higher than in U_2Ni_2In (T_N = 15 K) (see Fig. 1).

U_2Ni_2Sn was found to absorb ~ 2.4 H atoms per formula unit (the uncertainty is due to the presence of UH_3, see below). The hydride is stable up to T = 473 K, above which one-step desorption takes place, similar to $U_2Ni_2InH_{1.9}$. This process is completely finished by T = 673 K. The hydrogen absorption leads again to lattice expansion, similar to the U_2Ni_2In case. The lattice expansion is obviously anisotropic – it is almost twice as large along the c-axis as in the basal plane. The structure parameters for $U_2Ni_2SnH_{2.4}$ are given in Table 1. Unlike U_2Ni_2In, the shortest U-U distance lies in the basal plane both for the original compound and for the hydride.

Magnetisation measurements showed that $U_2Ni_2SnH_{2.4}$ orders antiferromagnetically below T_N = 68 K, compared to 26 K for U_2Ni_2Sn. The formation of a small amount of ferromagnetic UH_3 with T_C around 180 K[5] obscures the evaluation of the data below this temperature. For both compounds, U_2Ni_2In and U_2Ni_2Sn, the value of the effective moment in the paramagnetic region (2.0-2.5 μ_B/U) indicates the itinerant character of the magnetism, and is practically unaffected by hydrogenation.

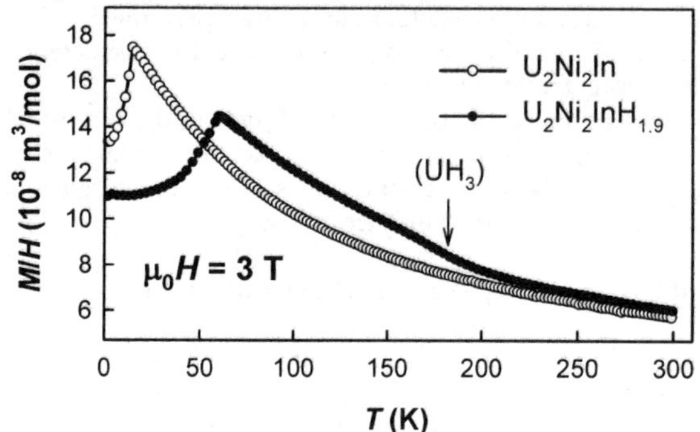

Figure 1 *Temperature dependence of magnetic susceptibility for U_2Ni_2In and $U_2Ni_2InH_{1.9}$ measured in magnetic field (3 T). (The anomaly around 180 K corresponds to UH_3)*

3 CONCLUSION

The hydrogen absorption properties of U_2T_2In and U_2T_2Sn follow the tendency found already for the U*TX* compounds,[4] namely, that the hydrogen absorption area can be framed in the upper right corner of the transition-metal series of the periodic table. This is also the area where a magnetic order can appear, due to weaker 5*f*-ligand hybridisation. In uranium compounds, hydrogen absorption leads to stronger magnetic interactions in most of the cases known. This tendency is clearly manifest also for the U_2T_2X–H systems – the Néel temperatures of $U_2Ni_2SnH_{2.4}$ and $U_2Ni_2InH_{1.9}$ considerably exceed those of all known U_2T_2Sn and U_2T_2In compounds. In general, it can be attributed to enhanced U–U spacing, reducing the 5*f*–5*f* overlap and/or reducing the 5*f*–*d* hybridisation, withdrawing partly the *d*-states due to the *d*–1*s* bonding. Which mechanism dominates in this case can be assessed only after the details of the hydrogen lattice sites are known.

Acknowledgements

This work is a part of the research program MSM 0021620834 financed by the Ministry of Education of CR and supported by the Grant Agency of CR (Grant #202/04/1103), by the Grant Agency of Charles University (Grant #224/2005) and by the exchange Program GRICES/ASCR 2004, and by FCT/POCTI, Portugal, under contract nr. QUI/46066/2002.

References

1 M.N. Peron, Y. Kergadallan, J. Rebizant et al, *J. Alloys Comp.* 1993, **201,** 203.
2 L. Havela, V. Sechovský, P. Svoboda et al., *J.Magn.Magn.Mater*, 1995, **140-144**, 1367.
3 K. Miliyanchuk, L. Havela, A.V. Kolomiets and A.V. Andreev, *Physica B*, 2005, **359-361**, 1042.
4 K. Miliyanchuk, A.V. Kolomiets, L. Havela and A.V. Andreev, *J. Alloys Comp.*, 2004, **383**, 103.
5 W. Trzebiatowski, A. Sliwa and B. Stalinski, *Rocz. Chem.*, 1952, **26**, 110.

SELF-PROPAGATING HIGH-TEMPERATURE SYNTHESIS OF ACTINIDE-CONTAINING PYROCHLORE-TYPE MATRICES

S.E. Vinokurov, Yu.M. Kulyako, S.A. Perevalov and B.F. Myasoedov

Vernadsky Institute of Geochemistry and Analytical Chemistry RAS, Kosygin str., 19, 119991, Moscow, Russia

1 INTRODUCTION

The present technology for the solidification of constituents of high-level wastes by vitrification does not ensure full immobilization of radionuclides, because of the low hydrothermal and crystallization stability of glasses.[1,2] As an alternative, isomorphous incorporation of the actinides into crystalline phases that are analogues of natural minerals and have high radiation and chemical stability (e.g. titanates with pyrochlore structure) has been studied intensively.[3,4] The processes of preparation of the mineral-like matrices are very laborious, and require complicated facilities for cold pressing and sintering,[5] hot pressing[6] or induction melting in cold crucibles.[7]

Self-propagating high-temperature synthesis (SHS)[8,9] is proposed as an alternative for these conventional methods. SHS is based on a locally induced strong exothermic chemical reaction between a metal and an oxidizer in a charge consisting of a mixture of powders of chemical elements and mineral additions. The heat released as a result of this reaction induces exothermic reaction in adjoining layers of the mixture, resulting in a rapid self-propagating process. In such a process the chemical reaction proceeds in a narrow zone, moving spontaneously with linear speed of 0.1 - 15 cm/sec at 1700-3000^{0}C.

2 METHOD AND RESULTS

The composition of the batch mixture for the SHS process is: Y_2O_3 – 43.4, Ti – 17.7, TiO_2 – 4.6, MoO_3 – 34.3 % wt. The mixing of the oxides of actinides (^{237}Np, ^{237}Pu, ^{241}Am, ^{238}U) with the batch mixture has been carried out in alcohol solution in a glove box, followed by filtration of the suspension. After filtration, the press mold was dried for 5 hours at 90^{0}C in a drying box and then pressed. The obtained tablet was placed on a layer of incendiary mixture, situated on a support of fire brick and covered with a safety quartz beaker. SHS was initiated by applying electrical firing, supplied from below, and the tablet burned out. The formation of a pyrochlore analogue occurred by a redox reaction (Scheme 1), where Ti is the reducing agent and MoO_3 is the oxidizing agent.

$$(2\text{-}x)Ti + xTiO_2 + Y_2O_3 + 2(2\text{-}x)/3\ MoO_3 = Y_2Ti_2O_7 + 2(2\text{-}x)/3\ Mo$$

Scheme 1

The actinide oxide contents in the prepared mineral-like matrices are presented in Table 1.

Table 1 *The actinide-containing matrices prepared by SHS*

Matrix number	*Content* (wt. %)	*Weight* (g)	*Density* (g/cm^3)
1	$10(PuO_2) + 0.2(Am_2O_3)$	3.2	3.0
2	$10(NpO_2)$	1.9	3.7
3	$7(UO_2) + 3(PuO_2)$	1.8	3.5
4	$9.7 (UO_2)+0.3(Am_2O_3)$	1.8	4.1

X-ray and SEM/EDS analysis were used for structural examinations of the matrices prepared by SHS. X-ray lines characteristic of the compound with a pyrochlore structure proved to be identical to those of the standard $Y_2Ti_2O_7$ phase (Figure 1). A proper actinide oxide phase is not revealed. It is seen from the SEM-photomicrographs (Figure 2) that the dominating phase of the matrices is pyrochlore, the lighter phase is molybdenum and, possibly, molybdenum dioxide and rutile.

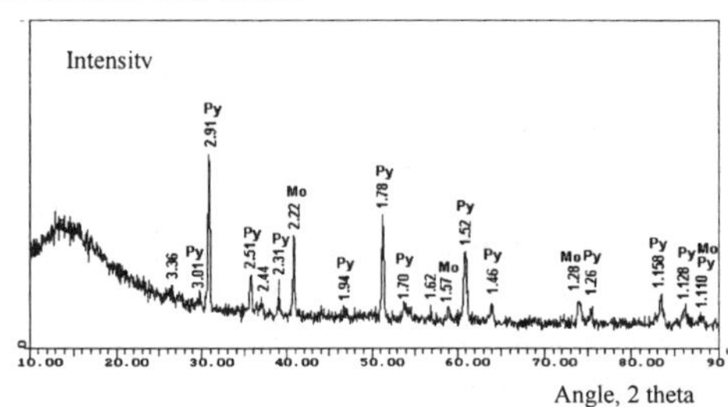

Figure 1 *X-ray diffraction patterns of matrix #1 (Py – pyrochlore, Mo – molybdenum).*

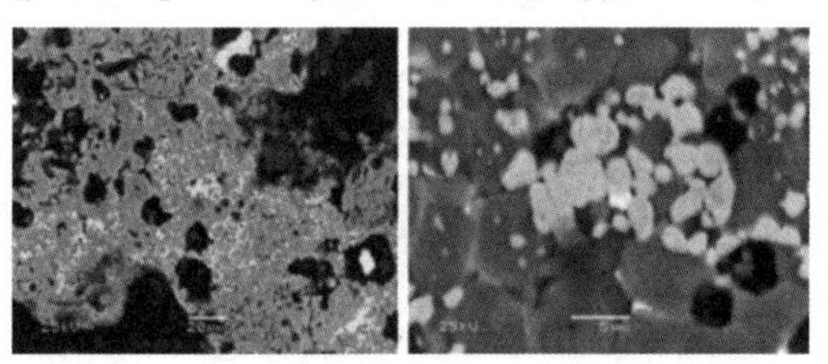

Figure 2 *SEM-photomicrograph of matrix #3 (gray – pyrochlore, light – molybdenum, black – pores)*

The study of the chemical stability of the matrices was carried out according to the MCC-1.[10] The samples under study were placed into Teflon containers. Twice-distilled water (pH=5.80) was used as the leaching agent. The containers with the suspensions were placed into a drying box at a temperature of 90 ^0C for the necessary time, and cooled to ambient temperature. Radionuclide concentrations were determined in the leaching solutions. The leaching rate of actinide i, is given by: $R_i = m_i / (f_i \cdot S \cdot t_n)$ [$g/(cm^2 \cdot day)$], were m_i is the mass of actinide, i, in the solution; f_i is the mass composition of actinide, i, in the initial sample; S is the open «geometric» surface of the sample; t_n is the duration of leaching. The data obtained in leaching tests are presented in Table 2.

Table 2 *The leaching rates of actinides contained in matrices produced by SHS*

Matrix number	*Time* (days)	*Leaching rate* (g/(cm^2·day))			
		Np	Pu	Am	U
1	28	–	7.0×10^{-8}	1.5×10^{-8}	–
2		1.1×10^{-7}	–	–	–
3		–	4.1×10^{-7}	–	5.4×10^{-7}
4		–	–	7.0×10^{-8}	8.1×10^{-7}

3 CONCLUSION

SHS produced solid final products analogous to the natural pyrochlore containing up to 10 wt.% of actinide oxides, from the initial mixtures of reagents, mineral additions and constituents of HLW. The densities of the matrices are up to 80% of the theoretical density of the natural mineral. On producing the matrices by SHS, the material balance for the actinides remains. The matrices had good hydrothermal stabilities: leaching rates of the actinides lie within the limits of 10^{-7}–10^{-8} g/(cm^2·day).

References

1 G.G. Wicks, W.C. Mosley, P.G. Whitkop and K.A. Saturday, *J. Non-Cryst. Solids*, 1982, **49**, 413.
2 H. Kamizono, *J. Amer. Chem. Soc.,* 1991, **74**, 2234.
3 P.E. Raison, R.G. Haire, T. Sato and T. Ogawa, *Sci. Bas. Nucl. Waste Manag.-XXII* (1999) MRS Symp. Proc. 556. P. 3-10.
4 R.C. Ewing, W.J. Weber and W. Lutze *Disposal of Weapon Plutonium.* eds E.R. Nerz and C.E. Walter. Kluwer Academic Publishers. 1996. P.65-83
5 B.B. Ebbinghaus, R.A. Van Konynenburg, F.J. Ryerson et. al., *Int. Conf. Waste Management.* 1998, Tucson, AZ, 1998. Proc. CD version. Rep. 65-04.
6 A.E. Ringwood, S.E. Kesson, K.D. Reeve et al. *Radioactive Waste Forms for the Future.* Eds. W. Lutze and R.C. Ewing. Amsterdam: Elsevier Science Publishers B.V., 1988. P. 233-334.
7 I.A. Sobolev, S.V. Stefanovsky, B.F. Myasoedov et. al. *Plutonium Futures – The Science*, eds K.K.S. Pillay and K.C. Kim, Santa Fe, NM, 2000, P. 122-124.
8 Yu.M. Kulyako, S.A. Perevalov, S.E. Vinokurov, B.F. Myasoedov, G.A. Petrov, M.I. Ozhovan, S.A. Dmitriev and I.A. Sobolev, *Radiochemistry*, 2001, **43**, 626.
9 T.V. Barinova, I.P. Borovinskaya, V.I. Ratnikov, T.I. Ignatjeva and V.V. Zakorzhevsky, *Int. J. Self-Prop. High-Temp. Synth.*, 2001, **10**, 77.
10 US DOE, *Nuclear Waste Materials Handbook (Test Methods)*, Technical Information Center, Washington, DC, Rep. DOE/TIC-11400. 1981.

RELATION BETWEEN SOLUBILITY AND LOCAL ENVIRONMENT OF ACTINIDES AND SURROGATES IN BOROSILICATE GLASSES

J-N. Cachia[1], X. Deschanels[1], T. Charpentier[2], D. Sakellariou[2], C. Denauwer[3] and J. Phalippou[4]

[1-3]Commissariat à l'Énergie Atomique (CEA):
[1]Valrhô/Marcoule DTCD/SECM/LMPA, BP 17171, 30207 Bagnols-sur-Cèze Cedex, France
[2]Saclay DSM/DRECAM/SCM, 91191 Gif-sur-Yvette Cedex, France
[3]Valrhô/Marcoule DRCP/SCPS/LCAM, BP 17171, 30207 Bagnols-sur-Cèze Cedex, France
[4]Université Montpellier II, Laboratoire des Verres, 34095 Montpellier Cedex 5, France

1 INTRODUCTION

The limit for plutonium loading in nuclear waste glass is uncertain. In borosilicate glass it has been specified to be less than 2 wt% PuO_2. In contrast, aluminosilicate glasses rich in rare earth elements (REE) are capable of loading up to 7 wt% PuO_2. This highlights the importance of the glass structure, which in turn depends on its composition, on the loading limit of actinides. The oxidation state (OS) of the actinide also significantly affects its solubility: the low Pu(IV) loading limit rises to 25 wt% when the plutonium is reduced to Pu(III).[1] In France, very high-level waste arising from spent fuel reprocessing is currently immobilized in R7T7 glass. The purpose of this study was to obtain basic data on actinides loadings in nuclear glasses. We first examined the solubilities of surrogate elements for trivalent and tetravalent actinides (La^{III}, Gd^{III}, Hf^{IV}), and compared them with the solubility of Pu. These elements have already been used for the same purpose in several other studies.[2-4] We then consider the structural differences in glasses containing elements of OS (III) and (IV) using spectroscopic techniques to establish correlations with the experimentally observed solubility differences.

2 EXPERIMENTAL PROTOCOL

Actinide surrogates were selected on the basis of criteria such as the oxidation state, the ionic radius or the coordination number.[5] They all have stable OS in the glass matrix. Solubility variations can thus be determined according to the material processing temperature without modifying the element OS.

2.1 Glass preparation

With the use of surrogate elements, the glass compositions were simplified to facilitate analysis and to better understand the phenomena involved. Samples were prepared from glass oxide, nitrate and carbonate precursors or from glass frit previously fabricated with the following weight composition: 58.84% SiO_2, 18.15% B_2O_3, 7.00% Na_2O, 4.28% Al_2O_3, 5.23% CaO, 2.56% Li_2O, 3.24% ZnO, 0.7% ZrO_2. The Pu glass composition was more complex (23 oxides), resembling industrial R7T7 glass. The full glass fabrication was reported in Lopez et al.[6]

2.2 Experimental techniques

The solubility is defined as the maximum concentration of an element that can be loaded in the glass, while maintaining its homogeneity (SEM analysis). The glass samples fabricated at different temperatures (1200, 1400°C) and doped to the solubility limit with La and Hf were examined by magic-angle spinning NMR spectroscopy. At the same time, we considered the local environment of lanthanum by La K-edge EXAFS spectroscopy in transmission mode at 70 K. The Pu-OS in the glass was determined at the L_{III}-edge at room temperature.

3 RESULTS

3.1 Solubility of surrogate elements and plutonium

The solubility limits of elements are plotted in Figure 1 versus the melting temperature with the solubilities of other elements (Nd, Th).[6] Above the solubility limit, the glasses contained needle-shaped inclusions of apatitic silicates $Ca_2Ln^{III}{}_8Si_6O_{26}$. Uniformly distributed heterogeneities with cubic habitus (Pu, O) formed in the Pu-doped matrix. XANES measurements showed that the Pu was in OS (IV) in the homogeneous glasses.

The solubilities can easily be subdivided in two categories. Elements with OS (III) exhibit greater solubility in borosilicate glass than those at OS (IV). A structural investigation was undertaken to determine the influence of the element oxidation states, since it was suspected that structural changes in the glass could account for the differences observed in the solubility.

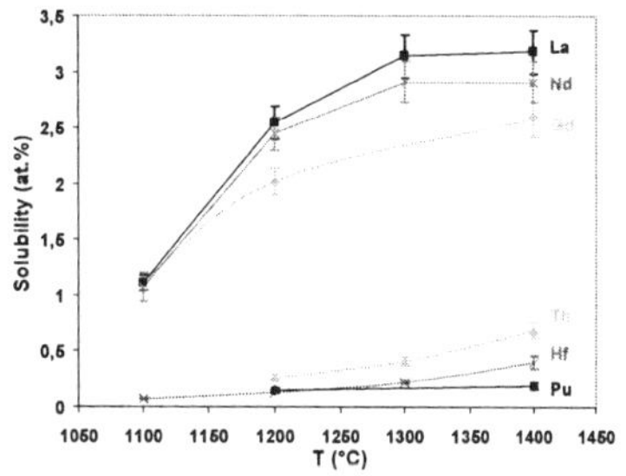

Figure 1 *Solubility behaviour of trivalent and tetravalent elements in borosilicate glass.*

3.2 Local environment and solubility

3.2.1 EXAFS analysis of the glass. Satisfactory results were obtained by refining the first coordination shell of the glass EXAFS spectra using the phases and amplitudes calculated by FEFF from the $La_2Si_2O_7$ structure. The spectrum refinement parameters confirm that the La environment is unchanged when its concentration increases in the glass. The first-neighbour shell around La consists of about 8 oxygen atoms at 2.58 Å. After processing, a second contribution was attributed to a shell of 5.6 ± 1.1 silicon atoms at a mean distance of 3.52 Å. No La–La contribution was observed up to 5 Å. Similarly, no environmental difference could be detected between 1100 and 1200°C. The La environment in the glass network resembles that of REE silicates that crystallize in the glass near the solubility limit. It is also consistent with the environments observed in different glasses containing neodymium.[7,8] In general, REE seem to impose a structure corresponding to their high coordination numbers, as very similar environments have been observed for several REE in different systems.[8]

3.2.2 NMR analysis of the glass. The results obtained with four nuclei (^{23}Na, ^{27}Al, ^{29}Si, ^{11}B) were used to follow the evolution of the glass network with increasing amounts of La and Hf. Na and Al are not subject to change as the REE concentration increases, and appear to conserve their environment. Si is relatively sensitive to the La additive concentration above 1.80 at.%. A shift to lower field values was observed, indicating the conversion of Q_4 units to Q_3 with the creation of nonbridging oxygen atoms (NBO). The creation of NBOs indicates depolymerisation of the glass network. The same behaviour was observed with boron, which was also subject to major variations. Adding La reduced the number of tetracoordinate boron atoms (B_4). The variation in the B_4/B_3 ratio with Hf doping is comparable to the behaviour observed with La, i.e. a drop in the number of tetracoordinated boron atoms as they are transformed into tricoordinated boron. The variation in the number of tetracoordinated boron entities is fully comparable to the behaviour observed in the La-doped glass over a similar composition range. Conversely, over the same range of Hf contents, the Si environment does not change. No change was observed in the NMR spectra as a function of the glass melting temperature.

4 CONCLUSION

The solubility limits of the elements increase with the glass melting temperature. Trivalent elements exhibit much greater solubilities than those at OS (IV). The EXAFS study of La-doped glass revealed a first coordination sphere with an average of 8 oxygen atoms at 2.58 Å. The results confirm that the solubility of these elements is related to the field strength parameter, which reflects their tendency to crystallize: the greater the cation field strength, the lower its solubility. The NMR study of REE doped glasses revealed depolymerisation of the glass network. The structural studies suggest that, depending on their OS, the elements have different roles within the glass: true intermediaries for tetravalent elements and network modifiers for trivalent elements. This explains to some extent the observed differences in solubility.

References

1 X. Feng, H. Li, L.L.D.L. Li, J.G. Darab, M.J. Schweiger, J.D. Vienna, B.C. Bunker, P.G. Allen, J.J. Bucher, I.M. Craig, N.M. Edelstein, D.K. Shuh, R.C. Ewing, L.M. Wang and E.R. Vance, *Ceramic Transactions*, 1999, **93**, 409.
2 W.G. Ramsey, N.E. Bibler and T.F. Meaker, *Waste management 95*, WM Symposia, (1995) Record 23828.
3 D.L. Caulder C.H. Booth, J.J. Bucher, N.M. Edelstein, P. Liu, W.W. Lukens, L. Rao, D.K. Shuh, L.L. Davis, J.G. Darab, H. Li, L. Li and D.M. Strachan, 219th American Chemical Society Meeting - Division of Nuclear Chemistry and Technology / *Symposium on Nuclear Waste Remediation and Long Term Storage,* (2000).
4 J.G. Darab, H. Li, M.J. Schweiger, J.D. Vienna, P.G. Allen, J.J. Bucher, N.M. Edelstein and D.K. Shuh, *Pu Futures – The Science – Transuranic Waste,* Santa Fe, USA, (1997) 143.
5 D.L. Lide *CRC Handbook of Chemistry and Physics*, Ed. CRC Press (1999).
6 C. Lopez, X. Deschanels, J.M. Bart, J.M. Boubals, C. Den Auwer and E. Simoni, *J. Nucl. Mater.,* 2003, **312**, 76.
7 C. Lopez, X. Deschanels, C. Den Auwer, J-N. Cachia, S. Peuget and J-M. Bart, *Physica Scripta*, 2005, **T115**, 342.
8 I. Bardez, MRS 2003, *Scientific Basis for Nuclear Waste Management* XXVII - abstracts, June 2003, Klamar, Sweden.

ELECTROCHEMICAL STUDIES OF URANIUM METAL CORROSION MECHANISM AND KINETICS IN WATER

N. Boudanova[1], A. Maslennikov[1], V. Peretroukhine[1] and C. Delegard[2]

[1]Institute of Physical Chemistry of Russian Academy of Sciences, Moscow, Russia
[2]Pacific Northwest National Laboratory, Richland, WA, USA

1 INTRODUCTION

During the long-term underwater storage of low burn-up uranium metal fuel, a corrosion product sludge forms containing uranium metal grains,[1] uranium dioxide,[1,2] uranates[1] and, in some cases, uranium peroxide.[1,3] Literature data on the corrosion of non-irradiated uranium metal and its alloys[2,4] do not allow unequivocal prediction of the paragenesis of irradiated uranium in water. The goal of the present work conducted under the programme 'Corrosion Of Irradiated Uranium Alloys Fuel In Water' is to study the corrosion of uranium and uranium alloys and the paragenesis of the corrosion products during long-term underwater storage of uranium alloy fuel irradiated at the Hanford Site. As a preliminary study to elucidate the physico-chemical nature of the corrosion of irradiated uranium alloys, the present study focuses on the corrosion of non-irradiated uranium metal. Electrochemical methods are being used to study uranium metal corrosion mechanisms and kinetics in water.

2 METHOD AND RESULTS

Uranium metal samples of reactor-grade purity were prepared for electrochemical and chemical corrosion tests. The samples have been characterized using XRD, and microstructural and chemical analysis. The electrochemical corrosion tests with a uranium metal electrode in deaerated 0.1 M $NaClO_4$ (pH=4.0-9.0) were carried out by open circuit potential (OCP) measurements, cyclic and linear voltammetry (CV and LV), chronocoulometry (CC) and potential controlled electrolysis (PCE). The electrochemical tests were carried out in a Pyrex cell with rotating disk U metal working electrode, Ag/AgCl reference electrode and Pt wire counter electrode. Before each test, the electrode was polished mechanically and etched electrochemically in a CH_3COOH/CrO_3 electrolyte. OCP values at the uranium metal electrode were measured in 0.1 M $NaClO_4$ (pH 4.0-9.0) for 60 min after first purging Ar through the electrolyte for 20 min to remove the dissolved O_2. It was found that the OCP values in all of the studied electrolytes achieved -300±50 mV steady-state values and were pH-independent for electrolytes in the range pH 3.94-7.17, shifting to more positive values (-200 mV) at pH greater than 8.0. The measured OCP values agreed well with literature data, which vary from -150 mV to -600 mV, depending on the uranium sample microstructural properties and the electrode preparation technique.[5] Comparison of OCP values at the U electrode with the formal potentials of the uranium redox reaction couples U(IV)/U(III), U(VI)/U(IV) and $2H^+/H_2$ indicated that the oxidation of U(IV) to U(VI) and cathodic water discharge were the principal electrode processes occurring at the U metal electrode in the absence of applied current. The kinetic parameters of the primary uranium corrosion reactions in 0.1 M $NaClO_4$, at pH 3.94 – 9.16, and their possible changes in time over a 90 min interval were obtained from CV measurements conducted from OCP-100 to OCP+100 mV. The CV curves were characterized by a large difference (about 150 mV) of the potentials corresponding to the maximum cathodic [I_c(max)] and anodic [I_a(max)] currents. This difference indicates the irreversibility of the observed electrode reaction (or superposition of the reactions). Analysis of the CV curves recorded at different electrode rotation rates shows that the rate of the observed electrode process (or at least of its slowest step) was determined by the electron transfer. The corrosion potentials ($E_{corr.}$), exchange current densities (i_o), transfer factors (α) and number of electrons (n) involved into the corrosion process were determined from corresponding Tafel plots. These findings are summarized in Table 1.

Table 1 *Corrosion potentials (E_{corr}), electron no. (n), transfer factors (α), exchange current densities (i_o), corrosion rates at U electrode in 0.1 M $NaClO_4$*

pH	E_{corr}, mV	N	α	i_o, $\mu A\ cm^{-2}$	Corrosion rate, $mg\ cm^{-2}h^{-1}$
3.94	-294	1.02	0.56	0.57±0.06	0.00128
4.54	-216	0.98	0.58	0.54±0.07	0.00121
6.08	-253	1.00	0.57	0.53±0.04	0.00118
7.17	-242	0.89	0.55	0.65±0.04	0.00146
8.37	-261	0.88	0.57	0.78±0.03	0.00174
9.16	-271	0.87	0.57	1.00±0.08	0.00222

Trends in $E_{corr.}$ changes with electrolyte pH and time were the same as observed for OCP values. Calculations of exchange current densities (i_o), transfer coefficients (α) and electron numbers (n) indicated that the observed reaction involved a 1-electron transfer (α close to 0.5). The i_o values, and corresponding, uranium corrosion rates were pH independent for pH between 3.94 and 6.08, and increased with further increase of electrolyte pH. The latter observation proved experimentally the theoretical expectation of the effect of OH^- ions on the uranium corrosion rate.[4] Taking into account the number of electrons participating in the electrode reaction at the uranium electrode surface, calculated from Tafel plots, and the obtained steady state potential values, H_2O reduction to form H_2 is considered to be the most probable process occurring at the U electrode. However, the possible occurrence of other reactions, for example oxidation of $UO_2{\cdot}xH_2O$ to U(V) hydroxide (which should be comparatively stable in neutral and weakly alkaline solutions) cannot be excluded. The i_o values, calculated from CV curves, did not change significantly with time during the 90-min corrosion test at a given pH, thus indicating that the corrosion rate was constant. In Table 1, the mean i_o and calculated corrosion rates values are presented. The observed uranium corrosion rates may be compared with the corrosion rates observed for non-irradiated uranium metal (0.003 $mg{\cdot}cm^{-2}h^{-1}$)[2] and for uranium metal fuel irradiated to low burn-up (0.07 $mg{\cdot}cm^{-2}h^{-1}$)[1] to illustrate the significant effect of irradiation. The presence of radioactive fission products in the uranium metal apparently shifted its E_{corr} towards more positive values.Linear voltammetry (LV) and chronocoulometry (CC) were used to determine the potential of uranium transpassivation ($E_{trans.}$) in deaerated 0.1 M $NaClO_4$ (pH=4.0-9.0). The E_{trans} for the studied uranium electrode was E=460±15 mV. This potential was weakly pH-dependent but depended strongly on the sample history (for instance, on preliminary cathodic polarization). Chronocoulonometry was performed by measuring I-t curves at 60 sec-intervals with the potential shifting by 50 mV steps starting from E_{corr} to 750 mV. The CC measurements showed the presence of "critical points", i.e., potential values at which the curve shapes and their qualitative characteristics change. We postulate that these "critical points" correspond to changes occurring at the electrode surface. Analysis of I-t curves in the potential range from $E_{corr.}$ to 450 mV show that the properties of the passive films formed at the uranium metal surface depend strongly on the applied potential and the electrolyte pH. PCE was carried out at 300 to 600 mV potentials, corresponding to transpassivation at a uranium metal electrode in 0.1 M $NaClO_4$, for times sufficient to accumulate corrosion products. Corrosion products accumulated during PCE as soluble U(VI) species in the electrolyte and as tiny solid particles of undetermined composition (presumably hydrated UO_2 and U_3O_8). The increase of U(VI) concentration in the electrolyte was linear with time at all potentials and pH. Corrosion rates calculated from PCE data were always greater than the rates estimated by I-t curve integration (Table 2). The higher rates obtained by the PCE tests are attributed to the growth of an effective electrode – electrolyte interface during the experiment. The current efficiency of the uranium electrochemical oxidation in the entire studied range of potentials and electrolyte pH was greater than 6 F/mol, proving the absence of solution species other than U(VI).

Table 2 *U corrosion rates (CR) and current efficiencies (CE) obtained from PCE data.*

E, mV	pH 3.98			pH 6.55			pH 8.46		
	CR, mg cm^{-2}h^{-1}		CE, F/mol	CR, mg cm^{-2}h^{-1}		CE, F/mol	CR, mg cm^{-2}h^{-1}		CE, F/mol
	PCE	CC		PCE	CC		PCE	CC	
300	1.21	0.295	6.84	-	-	-	-	-	-
400	2.65	0.446	7.60	7.58	0.231	6.1	6.52	0.075	5.87
500	12.1	5.95	7.88	11.5	4.964	6.3	13.6	6.842	7.15
600	40.4	21.4	5.58	-	-	-	24.9	30.71	7.96

3 CONCLUSION

Electrochemical methods were used to study uranium metal corrosion in 0.1 M $NaClO_4$ solutions (pH=4.0–9.0). The corrosion potentials, corrosion rates and transpassivation potentials of uranium metal in these solutions were determined and compared with the non-irradiated and irradiated uranium literature data. The obtained results clearly show the increase of corrosion rate for uranium metal at potentials above the transpassivation zone. It implies the shift of its corrosion potential towards the transpassivation zone when irradiated, resulting in a high corrosion rate of irradiated uranium, as reported in the literature.

References

1 C.H. Delegard, A.J. Schmidt, R. L. Sell, S.I. Sinkov, S.A. Bryan, S.R. Gano and B.M. Thornton, *Final Report - Gas Generation Testing of Uranium Metal in Simulated K Basin Sludge and in Grouted Sludge Waste Forms, PNNL-14811*, Pacific Northwest National Laboratory, Richland WA, USA, 2004, August, 55 p.

2 B.A. Hilton, *Review of Oxidation Rates of DOE Spent Nuclear Fuel Part 1: Metallic Fuel, ANL-00/24*, Argonne National Laboratory, Idaho Falls ID, USA, 2000, 85 p.

3 M. Amme, *Radiochim. Acta*, 2002, **90,** 399.

4 V.V. Gerasimov, *Corrosion of Uranium Metal and Alloys*, "Atomizdat" Publ. Co., 1965, 96 p. (in Russian).

5 E.T. Shapovalov and V.V. Gerasimov, *Soviet Atomic Energy*, 1969, **27**, 289.

THE EFFECT OF OXYGEN POTENTIAL ON THE SINTERING BEHAVIOUR OF MOX FUEL CONTAINING AMERICIUM

S. Miwa[1], M. Osaka[1], H. Yoshimochi[1], K. Tanaka[1], T. Seki[2] and S. Sekine[3]

[1]O-arai Engineering Center, Japan Nuclear Cycle Development Institute, O-arai-machi, Higashi-ibaraki-gun, Ibaraki 311-1393, Japan
[2]Inspection Development Co. Ltd., O-arai-machi, Higashi-ibaraki-gun, Ibaraki 311-1393, Japan
[3]Nuclear Technology and Engineering Co. Ltd., O-arai-machi, Higashi-ibaraki-gun, Ibaraki 311-1393, Japan

1 INTRODUCTION

The Japan Nuclear Cycle Development Institute promotes R&D programmes for future nuclear cycle technology based on fast reactors. The development of mixed oxide (MOX) fuel containing Am (Am-MOX) is one programme that is underway. Am-MOX fuel is viewed as a promising candidate fuel, because it offers a reduced environmental burden by transmutation of Am into less radiotoxic nuclides. So far, a remote process fabrication technology for Am-MOX pellets, which have high radioactivity, has been established.[1]

Recently, the oxygen potential of Am-MOX, which is an important property from the viewpoint of fuel design, was found to be extremely high.[2] Oxygen potential is known to affect the sintering behaviour of nuclear fuel. Therefore, it is important to investigate the effect of oxygen potential on the sintering behaviour of Am-MOX. In this study, Am-MOX pellets sintered at various oxygen potentials are examined by ceramography and metrology. The mechanism of sintering is discussed in terms of the cationic diffusivity of each metallic element.

2 EXPERIMENTAL

Green pellets of Am-MOX were prepared by a conventional powder metallurgical technique. Known amounts of UO_2, PuO_2 and $(Pu,^{241}Am)O_2$ powders were weighed for the preparation of $(U_{0.68}Pu_{0.27}Am_{0.05})O_2$, and mixed for 8 h using tungsten balls in a ball mill. Organic binder was added to the mixed powder and this was compacted at 40 MPa to get green pellets. The green pellets were heated at 800 °C for 2.5 h in a reducing atmosphere to remove the organic binder.[1]

Sintering tests for the green pellets were performed at 1700 °C for 3 h under an Ar atmosphere, containing 5% H_2. Both the heating and cooling rates were 200°C/h. The oxygen potentials of the sintering atmosphere were adjusted by adding moisture to get values of -520 kJ/mol, -390 kJ/mol and -340 kJ/mol.

The density was obtained from metrological results for the pellets. The microstructural analyses of the sintered Am-MOX pellets were done by cross-sectional ceramography.

3 RESULTS AND DISCUSSION

Figure 1 shows the (percentage of theoretical) density of the sintered Am-MOX pellets as a function of the oxygen potential. The densities of the Am-MOX pellets sintered at -520 kJ/mol, -390 kJ/mol and -340 kJ/mol are 92.3 %, 94.0 % and 91.1 % of the theoretical density, respectively. Clearly, the sintered densities are affected by the oxygen potential. The sintered density increases with increasing oxygen potential up to -390 kJ/mol (threshold oxygen potential), then decreases above the threshold. This tendency is similar to that observed in the $(U,Gd)O_2$ system.[3]

Figure 2 shows ceramographic images of as-polished surfaces of Am-MOX pellets. For the pellet sintered at -520 kJ/mol, both small and large pores with rough inner surfaces are observed and its grain boundaries are wider than those of the other two. Relatively large pores with smooth inner surfaces are distributed in the pellets sintered at -390 kJ/mol and -340 kJ/mol. The microstructure observed for the pellet sintered at -520 kJ/mol is like that observed with low O/M ratio (O/M < 1.975).[4]

For a detailed discussion of the microstructure, the total area of pores, pore radius and pore number density are obtained by image analysis (Figure 3). The "pore radius" is that of an equivalent one having the same area as the original pore. While the pellet sintered at -340 kJ/mol has almost the same number density of pores as that sintered at -390 kJ/mol, it has a larger pore radius, and thereby a larger total area of pores. This difference in the pore structures can explain the difference in the sintering behaviour of Am-MOX pellets shown in Figure 1. A mechanism for pore evolution is discussed below by adopting the hypothesis of Yuda and Une[3] for $(U,Gd)O_2$, based on the difference in cation diffusivities for different sintering atmospheres, i.e. oxidizing and reducing.

Reports have been made on the dependences of the sintering behaviour of UO_2[5] and $(U,Pu)O_2$[6] on the oxygen potential. These studies noted that the diffusion coefficients of U and Pu in $UO_{2\pm x}$ and $(U,Pu)O_{2\pm x}$ were higher in the oxidizing atmosphere by 4-5 orders of magnitude at a constant temperature, compared to those in the reducing atmosphere. In the present case, U and Pu have relatively high diffusivities at -340 kJ/mol (oxidizing atmosphere), compared to that at -390 kJ/mol (reducing atmosphere).[3,7] These results were attributed to the variable valences of U and Pu. It was also said that densification began at a much lower temperature in the oxidizing atmosphere.[5,6]

On the other hand, Since the oxygen potentials of AmO_{2-x} and Am-MOX are high,[8] all Am is considered to be in a trivalent state under the experimental conditions.[4] Therefore, the diffusivity of Am is unlikely to be dependent on the oxygen potential. The diffusion of Am is, thus, mainly dependent on temperature, the same as for Gd, which has only a trivalent state in $(U,Gd)O_2$ solid.

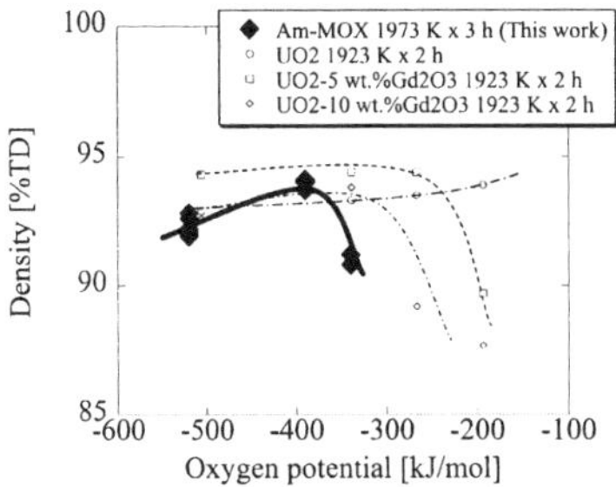

Figure 1 *Dependence of sintered theoretical density of 5% Am-MOX and $(U,Gd)O_2$[3] on the oxygen potential.*

In the oxidizing atmosphere, the interdiffusion between U or Pu and Am is considered to take place at relatively high temperature, compared to interdiffusion of the U/U and U/Pu systems. Accordingly, in the final sintering stage, larger pores can be generated by the diffusion of Am into $(U,Pu)O_2$ from the region where $(Pu,Am)O_2$ particles were originally present. On the other hand, in the reducing atmosphere, since interdiffusions of U/U and U/Pu were not enhanced, simultaneous interdiffusion between U, Pu and Am may take place. Consequently, the pores become small due to the further densification in the intermediate and final sintering stages.

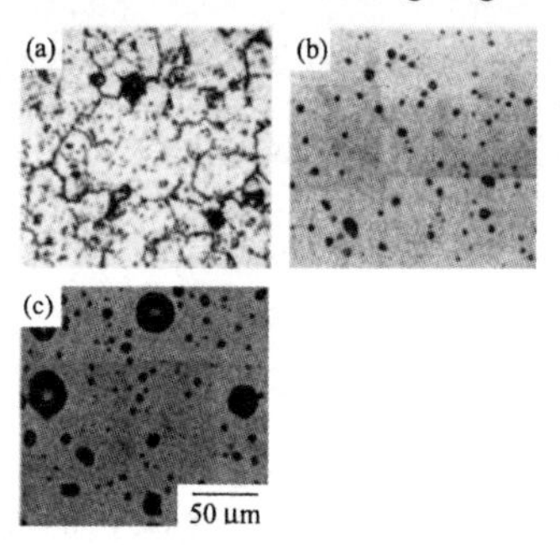

Figure 2 *Ceramographic images for as-polished surface of Am-MOX pellets sintered at (a) -520 kJ/mol, (b) -390 kJ/mol and (c) -340 kJ/mol.*

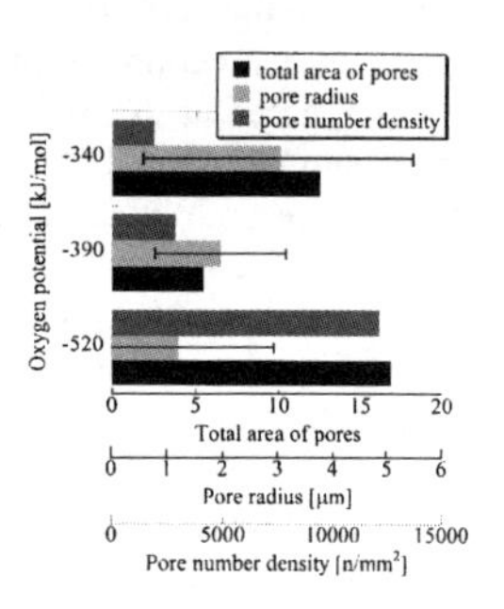

Figure 3 *Pore analysis, total pore areas, pore radius and pore number density*

4 CONCLUSION

The effect of oxygen potential on the sintering behaviour of Am-MOX was investigated. The differences in sintering behaviour for Am-MOX pellets that were observed by changing the oxygen potential were attributable to differences of pore structure, which was caused by the valence state of Am in the oxides.

References

1 H. Yoshimochi, M. Nemoto, S. Koyama and T. Namekawa, *J. Nucl. Sci. Technol.*, 2004, **41**, 850.
2 M. Osaka, I. Sato, T. Namekawa, K. Kurosaki and S. Yamanaka, *J. Alloys and Comp.* 2005, **397**, 110.
3 R. Yuda and K. Une, *J. Nucl. Mater.*, 1991, **178**, 195.
4 M. Osaka, S. Miwa, H. Yoshimochi, K. Tanaka, K. Kurosaki and S. Yamanaka, *This publication.*
5 T. R. G. Kutty, P. V. Hegde, K. B. Khan, U. Basak, S. N. Pillai, A. K. Sengupta, G. C. Jain, S. Majumdar, H. S. Kamath and D. S. C. Purushothem, *J. Nucl. Mater.*, 2002, **305**, 159.
6 T. R. G. Kutty, P. V. Hegde, R. Keswani, K. B. Khan and D. S. C. Purushothem, *J. Nucl. Mater.*, 1999, **264**, 10.
7 Hj. Matzke, *J. Nucl. Mater.*, 1983, **114**, 121.
8 T. D. Chikalla and L. Eyring, *J. Inorg. Nucl. Chem.*, 1967, **29**, 2281.

VARIATION OF LATTICE PARAMETER AND THERMAL EXPANSION COEFFICIENT OF $(U,Er)O_2$ AS A FUNCTION OF $ErO_{1.5}$ CONTENT

S.H. Kim[1], Y.K. Kim[1], H.S. Kim[1], S.H. Na[1], Y.W. Lee[1], D.S. Sohn[1] and D.J. Kim[2]

[1]Department of Advanced Nuclear Fuel Fabrication Development, Korea Atomic Energy Research Institute, Daejeon 305-600, South Korea.
[2]Department of Nuclear Engineering, Hanyang University, Seoul 133-791, South Korea

1 INTRODUCTION

Thermal expansion data for sintered UO_2 is precisely reported in the literature.[1] $ErO_{1.5}$ can be used as a burnable poison[2,3] for PWR reactors for the purpose of high burnup and extended cycle operation. However, there are no reports of the basic properties of $(U,Er)O_2$. In this work, we report the variation of the room temperature lattice parameter and thermal expansion as a function of $ErO_{1.5}$ in $(U,Er)O_2$.

2 METHOD AND RESULTS

The UO_2 powder was mixed with weighed amounts of $ErO_{1.5}$, at concentrations of 5, 10 and 20 mol%, by a Turbula® mixer for 1 hour, and then successively milled by a dynamic ball mill for 1~6 hours. The milled oxide powders were compacted with a compaction pressure of 300 MPa, and then sintered at 2023 K in H_2 for 6 hours. The XRD patterns were recorded in the range of $20° < 2\theta < 120°$ using an X-ray diffractometer. The bulk thermal expansion of the pellets is investigated in the temperature range of 298-1673 K in an Ar atmosphere using a thermo-mechanical analyzer.

2.1 Lattice Parameter

Une and Oguma[4] analyzed the O/M ratio of $(U_{1-y}Gd_y)O_2$ solid solutions with $0.04 \leq y \leq 0.27$, sintered in H_2 at 1973 K. Their results showed that the O/M ratios of all the solid solutions within the range of y were in the range of 1.995 to 2.000, independent of the $GdO_{1.5}$ content, and were averaged to 1.997. According to Tagawa and Fujino,[5] hypo-stoichiometric $U_{1-y}LaO_{2-x}$ has been reported to oxidize easily in air, even at room temperature, to a near-stoichiometric composition. Although the O/M ratio of the samples was not measured chemically in this study, the deviation from the stoichiometry is assumed to be very small up to 20 mol% $ErO_{1.5}$, based on the near stoichiometric behaviour of other substitutional impurities as discussed above. Therefore, we will indicate the chemical formulae of the Er-doped UO_2 solid solutions as approximately $(U_{1-y}Er_y)O_2$.

Figure 1 shows the room temperature XRD patterns of all the sample $ErO_{1.5}$ peaks are not observed in the $(U,Er)O_2$ pellets, indicating the formation of a solid solution between UO_2 and $ErO_{1.5}$. The diffraction peaks around 94° are magnified in Figure 1(b) in order to show the peak shift trend. The peak positions shift to a higher angle as the Er content increases, showing that the lattice contracts when the Er^{3+} ions are incorporated into the UO_2 lattice.

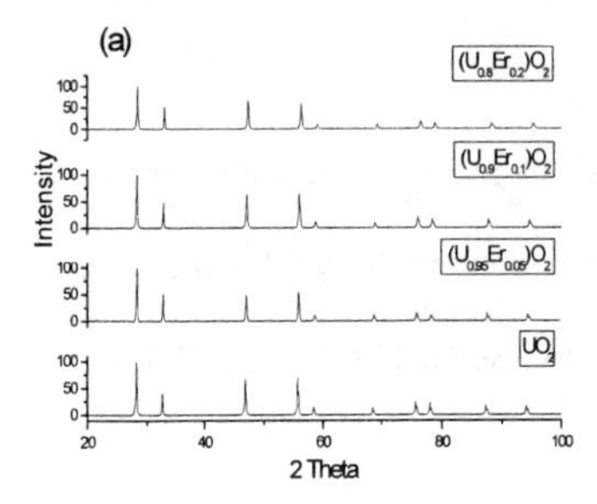

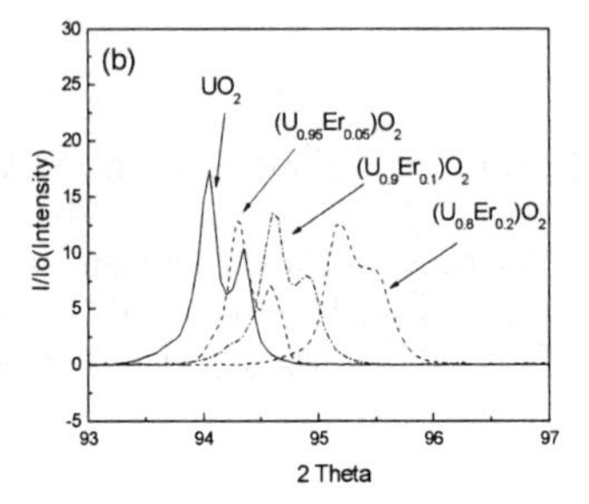

Figure 1 *XRD patterns of UO_2 and $(U,Er)O_2$ pellets at room temperature*

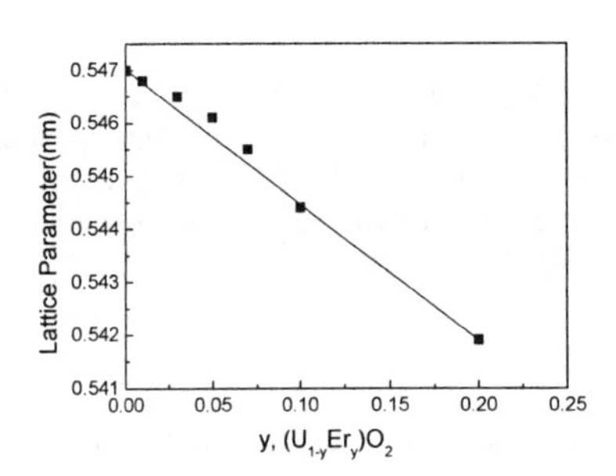

Figure 2 *Lattice parameters of the $(U_{1-y}Er_y)O_2$ solid solutions as a function of the $ErO_{1.5}$ content*

Table 1 *Sintered densities of the samples*

$ErO_{1.5}$ (mol%)	Theoretical density (g/cm^3)	Sintered density (g/cm^3)
0	10.96	10.66
5	10.87	10.41
10	10.82	10.44
20	10.68	10.39

Figure 2 shows that the lattice parameters of the $(U_{1-y}Er_y)O_2$ decrease linearly as a function of $ErO_{1.5}$ content, and follow Vegard's law. The lattice parameters of $(U_{1-y}Er_y)O_2$ can be expressed as :

$$\text{Lattice parameter(nm)} = 0.5471-0.0264y \ (0 \leq y \leq 0.2) \quad (1)$$

Table 1 shows the measured densities of UO_2 and $(U,Er)O_2$ pellets. It was confirmed by optical microscopy that all the samples prepared in this work had relatively homogeneous microstructures.

2.2 Thermal Expansion

The linear thermal expansion (LTE) data as a function of the temperature for the $(U,Er)O_2$ pellets are shown in Figure 3. The lattice parameter of the $(U,Er)O_2$ pellets is lower than that of UO_2, but the thermal expansion of the $(U,Er)O_2$ pellets is higher than that of pure UO_2. The expansion data of all the samples were fitted using a polynomial regression as shown below. For pure UO_2,

$$LTE(\%) = -0.29784 + 0.00102T - 1.02657 \times 10^{-7} T^2 + 8.19047 \times 10^{-11} T^3 \quad (2)$$

For $(U_{0.95}Er_{0.05})O_2$,

$$LTE(\%) = -0.29135 + 9.88399\times10^{-4}T - 3.54189\times10^{-8}T^2 + 5.58482\times10^{-11}T^3 \quad (3)$$

For $(U_{0.9}Er_{0.1})O_2$,

$$LTE(\%) = -0.29835 + 0.00102T - 4.38551\times10^{-8}T^2 + 5.70305\times10^{-11}T^3 \quad (4)$$

For $(U_{0.8}Er_{0.2})O_2$,

$$LTE(\%) = -0.30173 + 0.00103T - 3.75175\times10^{-8}T^2 + 5.41373\times10^{-11}T^3 \quad (5)$$

At 1473 K, around the reactor centreline temperature, the expansion data as a function of Er content are linearly fitted by,

$$LTE(\%) = 1.2525 + 0.0029y \quad (6)$$

where y denotes the Er content.

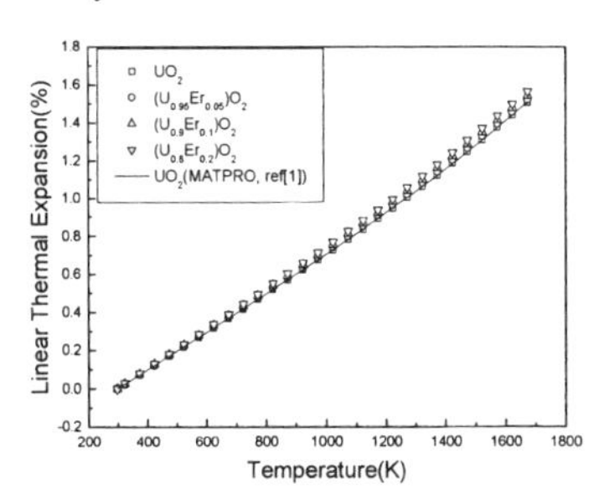

Figure 3 *Comparison between the thermal expansions of UO_2 and $(U,Er)O_2$ pellets*

Table 2 *Average linear thermal expansion coefficients ($\bar{\alpha}$) for UO_2 and $(U,Er)O_2$ in the temperature range of 298-1673K*

Composition	average cte ($\times10^{-6}$ K^{-1})
Pure UO_2	10.97
$(U_{0.95}Er_{0.05})O_2$	11.08
$(U_{0.9}Er_{0.1})O_2$	11.23
$(U_{0.8}Er_{0.2})O_2$	11.39

The values of the average linear thermal expansion coefficients ($\bar{\alpha}$) in the temperature range of 298-1673 K are given in Table 2. $\bar{\alpha}$ values of UO_2 in the same temperature range were found to be 10.97×10^{-6} K^{-1} and increased to 11.39×10^{-6} K^{-1} in the $(U_{0.8}Er_{0.2})O_2$ pellet.

3 SUMMARY

Lattice parameters of $(U,Er)O_2$ pellets are lower than that of UO_2, and decrease as the Er content increases. The linear thermal expansion and average thermal expansion coefficient of $(U,Er)O_2$ are higher than that of UO_2. As the Er content increases in the $(U,Er)O_2$ pellets, the average thermal expansion coefficient increases continuously.

Acknowledgements

We acknowledge that this project has been carried out under the Nuclear R&D Programme by the Ministry of Science and Technology in South Korea.

References

1 World Wide Web, *INSC Materials Properties Database*, http://www.insc.anl.gov/matprop/.
2 K.W. Song, K.S. Kim, H.S. Yoo and Y.H. Jung, *J. Kor. Nucl. Soc.*, 1998, **30**, 128.
3 J.H. Yang, K.W. Kang, K.S. Kim, K.W. Song and J.H. Kim, *J. Kor. Nucl. Soc,.* 2001, **33**, 307.
4 K. Une and M. Oguma, *J. Nucl. Mater.* 1985, **110**, 215.
5 H. Tagawa and T. Fujino, *J. At. Energ. Soc. Japan*, 1980, **22**, 871.

AN EXPERIMENTAL INVESTIGATION OF THE EFFECTS OF AMERICIUM ADDITION TO (U, Pu)O_{2-x} ON PHASE RELATION

M. Osaka[1], S. Miwa[1], H. Yoshimochi[1], K. Tanaka[1], K. Kurosaki[2] and S. Yamanaka[2]

[1]O-arai Engineering Center, Japan Nuclear Cycle Development Institute, O-arai-machi, Higashi-ibaraki-gun, Ibaraki, 311-1393, Japan.
[2]Department of Nuclear Engineering, Graduate School of Engineering, Osaka University, Yamadaoka 2-1, Suita-shi, Osaka 565-0871, Japan

1 INTRODUCTION

Recovery of minor actinides (MAs) from spent nuclear fuel, and their recycling into reactors represent key technologies in a future nuclear cycle based on fast reactors; the two are especially critical to a reduced environmental burden. Specific R&D for an Am containing mixed oxide of U and Pu (MOX), (U,Pu,Am)O_2, which is considered a promising candidate for a fuel of a future fast reactor, is now underway. So far, the fabrication technique with remote handling has been established,[1] and several properties such as oxygen potential[2] have been determined experimentally. It was revealed through these studies that only a small amount of Am addition to MOX had significant effects on various properties.

In this study, the effects on the phase relation of Am addition up to 5% to MOX are investigated experimentally. The (U,Pu,Am)O_2 solid solutions are prepared and their oxygen contents (O/M ratios) are adjusted. Samples are examined by XRD, ceramography and DTA. Finally, the valence state of Am in (U,Pu,Am)O_{2-x} is discussed.

2 EXPERIMENTAL

The (U,Pu,Am)O_2 solid solutions were prepared by a conventional powder metallurgical route; i.e. ball-milling of UO_2, PuO_2 and (^{241}Am,Pu)O_2 powders, uni-axial pressing into a pellet, followed by sintering in an Ar atmosphere containing 5% H_2 (Ar/H_2) for 3 h at 1973 K.[1] The density of the sintered (U,Pu,Am)O_{2-x} pellets was about 90 %T.D. The homogeneity of each solid solution was confirmed by ceramography, XRD and electron probe micro analysis results. These pellets were then thermally treated under Ar/H_2 at different oxygen potentials to adjust the O/M ratio from 1.94 to 2.00. A series of examinations, namely XRD, ceramography, and DTA analyses, were performed for pellets of different O/M ratios.

3 RESULTS AND DISCUSSION

3.1 Stoichiometric Solid Solution (U,Pu,Am)O_2

Figure 1(a) shows the lattice parameter dependence of stoichiometric ($U_{1-y-z}Pu_yAm_z$)O_2 solid solutions on Am content. The lattice parameter decreases with increasing Am content.

Theoretical lines showing the lattice parameter versus Am content obtained by using the method of Ohmichi et al.[3] with the effective ionic radii of Shannon[4] are also shown. The slope of the experimentally obtained line agrees with that of the theoretical one, assuming Am as trivalent, $(U^{5+}{}_{z}U^{4+}{}_{1-y-2z}Pu^{4+}{}_{y}Am^{3+}{}_{z})O_2$. This implies that the valence state of Am in $(U,Pu,Am)O_2$ is trivalent, like lanthanide elements in mixed oxides with U, e.g. $(U,Gd)O_2$[3].

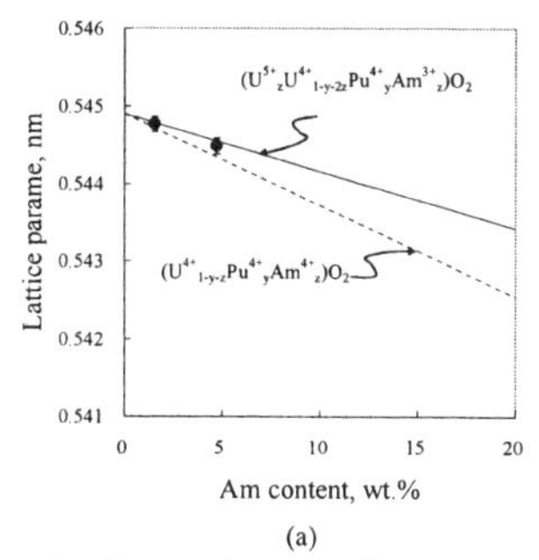

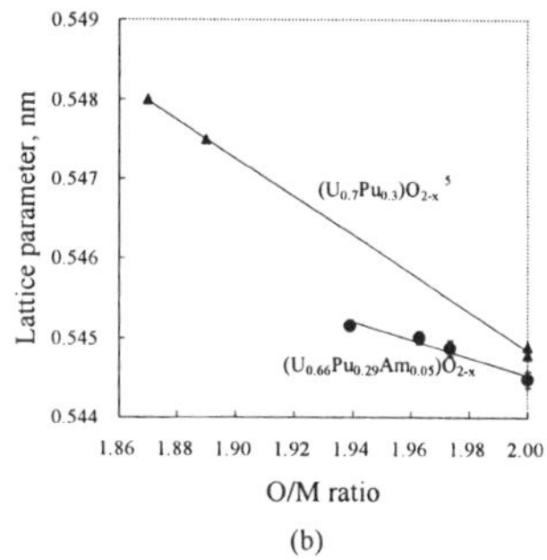

Figure 1 *Dependences of lattice parameter of $(U,Pu,Am)O_{2-x}$ on: (a) americium content, (b) O/M ratio.*

3.2 Hypostoichiometric Solid Solution $(U,Pu,Am)O_{2-x}$

Figure 1(b) shows the lattice parameter dependence of hypostoichiometric $(U_{0.66}Pu_{0.29}Am_{0.05})O_{2-x}$ on O/M ratio. The lattice parameter decreases with decreasing O/M ratio. It is noted that the slope of the line for the lattice parameter versus O/M ratio of $(U_{0.66}Pu_{0.29}Am_{0.05})O_{2-x}$ is smaller than that of non Am-containing $(U_{0.7}Pu_{0.3})O_{2-x}$.[5]

Figure 2 shows ceramographies of stoichiometric and hypostoichiometric $(U_{0.66}Pu_{0.29}Am_{0.05})O_{2-x}$. The structures of $(U_{0.66}Pu_{0.29}Am_{0.05})O_{2-x}$ with O/M=2.00 and 1.99 are homogeneous, while the structure of hypostoichiometric $(U_{0.66}Pu_{0.29}Am_{0.05})O_{1.97}$ is not; in particular, grain boundaries are wider, and their surfaces are rougher.

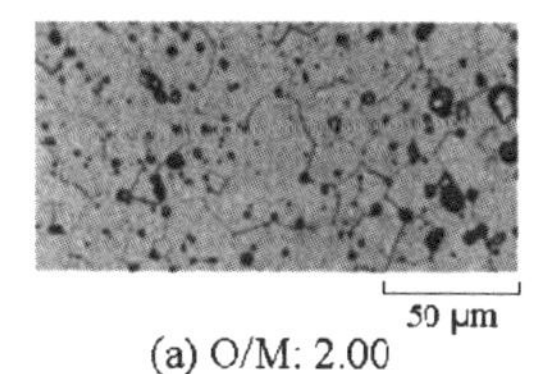

(a) O/M: 2.00

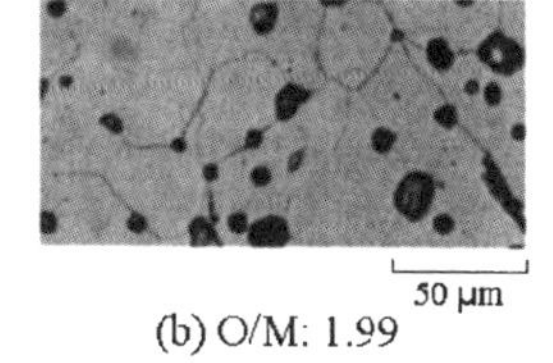

(b) O/M: 1.99

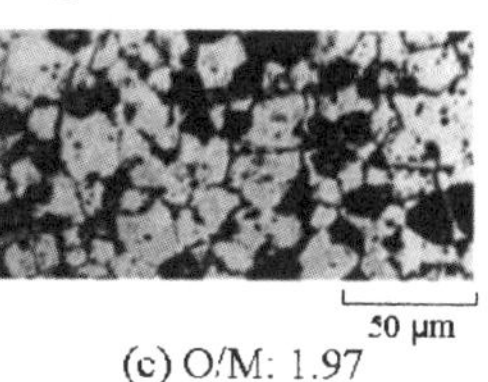

(c) O/M: 1.97

Figure 2 *Ceramographies of: (a) stoichiometric $(U_{0.66}Pu_{0.29}Am_{0.05})O_{2.00}$, (b) $(U_{0.66}Pu_{0.29}Am_{0.05})O_{1.99}$ (c) $(U_{0.66}Pu_{0.29}Am_{0.05})O_{1.97}$.*

Figure 3 shows the peak locations observed in DTA, plotted on the phase diagram of temperature versus O/M ratio of $(U_{0.66}Pu_{0.29}Am_{0.05})O_{2-x}$. Many DTA peaks were observed on both raising and lowering the temperature at constant O/M ratios. The DTA peaks were so small that they cannot be due to stable phase transitions, but rather be derived from quasi-stable ones. Since such DTA peaks are not observed for $(U,Pu)O_{2-x}$, these are considered to be caused by the Am addition.

To discuss the different aspects of $(U_{0.66}Pu_{0.29}Am_{0.05})O_{2-x}$ from MOX as mentioned above, the following hypothesis is adopted: all Am atoms in $(U,Pu,Am)O_{2-x}$ are trivalent, and a corresponding number of U atoms are pentavalent to compensate for positive charges from oxygen deficiencies. This hypothesis has been proved to be correct in the $(U_{0.5}Am_{0.5})O_2$ system.[6] In addition, Pu atoms become trivalent below the O/M ratio, where all U atoms become tetravalent (boundary O/M ratio). This is expressed by the following formulae: $(U^{5+}{}_{z-2x}U^{4+}{}_{1-y-2z+2x}Pu^{4+}{}_{y}Am^{3+}{}_{z})O_{2-x}$ above O/M=2-*z*/2, $(U^{4+}{}_{1-y-z}Pu^{4+}{}_{y-2x+z}Pu^{3+}{}_{2x-z}Am^{3+}{}_{z})O_{2-x}$ below O/M=2-*z*/2. In the case of $(U_{0.66}Pu_{0.29}Am_{0.05})O_{2-x}$, the boundary O/M ratio is 1.975.

From this hypothesis, the smaller slope for the line of $(U_{0.66}Pu_{0.29}Am_{0.05})O_{2-x}$ (Figure 1 (b)) can be caused by Am valence, which may form a different type of oxide, such as pervoskite, $AmUO_4$.[6] Also, the structural change around O/M=1.98 as shown in Figure 2 can be explained by the fact that this O/M is close to the boundary O/M ratio, 1.975. Regarding the many DTA peaks shown in Figure 3, the corresponding U-O phase diagram above O/M=2 is plotted according to the hypothesis. It seems that the plotted phase boundary of the U-O system agrees with the experimentally observed DTA peaks. This also supports the above hypothesis.

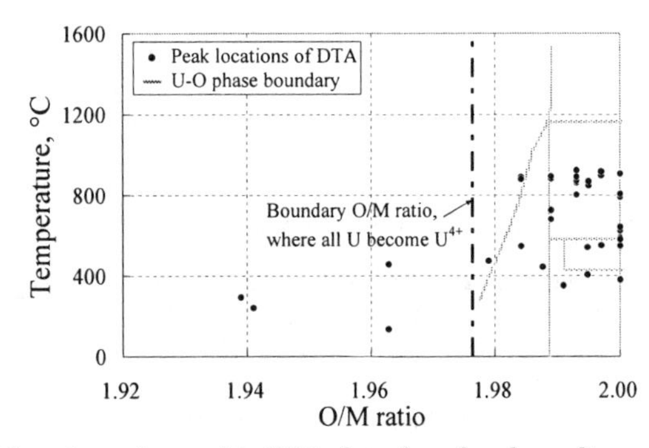

Figure 3 *Peak locations observed in DTA plotted on the phase diagram of temperature versus O/M ratio of* $(U_{0.66}Pu_{0.29}Am_{0.05})O_{2-x}$.

4 CONCLUSION

The phase relation of $(U,Pu,Am)O_{2-x}$ solid solution was experimentally investigated by using XRD, ceramograhy and DTA. It was concluded that Am was likely to exist as Am^{3+} and a corresponding number of U^{4+} ions were substituted by U^{5+} to preserve electrical neutrality.

References

1 H. Yoshimochi, M. Nemoto, S. Koyama and T. Namekawa, *J. Nucl. Sci. Technol.*, 2004, **41**, 850.
2 M. Osaka, I. Sato, T. Namekawa, K. Kurosaki and S. Yamanaka, *J. Alloys Comp.*, 2005, **397**, 110.
3 T. Ohmichi, S. Fukushima, A. Maeda and H. Watanabe, *J. Nucl. Mater.*, 1981, **102**, 40.
4 R. D. Shannon, *Acta Cryst.*, 1976, **A32**, 751.
5 T. L. Markin and R. S. Street, *J. Inorg. Nucl. Chem.*, 1967, **29**, 2265.
6 K. Mayer, B. Kanellakopoulos, J. Naegele and L. Koch, *J. Alloys Comp.*, 1994, **213/214**, 456.

MOLECULAR DYNAMICS STUDIES OF URANIUM-PLUTONIUM MIXED OXIDE FUELS

K. Kurosaki[1], J. Adachi[1], M. Osaka[2], K. Tanaka[2], M. Uno[1] and S. Yamanaka[1]

[1]Department of Nuclear Engineering, Graduate School of Engineering, Osaka University, 2-1 Yamadaoka, Suita, Osaka 565-0871, Japan
[2]Alpha-Gamma Section, Fuels and Materials Division, Irradiation Center, O-arai Engineering Center, Japan Nuclear Cycle Development Institute, Narita-cho 4002, O-arai-machi, Ibaraki 311-1393, Japan

1 INTRODUCTION

The uranium-plutonium mixed oxide (MOX) is a candidate fuel for not only fast breeder reactors (FBR), but also liquid water reactors (LWR).[1] Although the thermophysical properties of the MOX fuel are very important to evaluate the fuel performance, there is limited information on these properties, due to the difficulties associated with the high radiation fields. Therefore, it is necessary to develop a new technique to evaluate the thermophysical properties of MOX fuel.

In recent years, we have performed extensive molecular dynamics (MD) studies on actinide compounds, and succeeded in evaluating their thermophysical properties.[2,3] In the present study, the MD calculation for the MOX fuel in the whole plutonium composition range, viz. $(U_{1-x}Pu_x)O_2$ (x: 0, 0.1, 0.2, 0.3, 0.4, 0.5, 0.6, 0.7, 0.8, 0.9, 1.0), is performed to evaluate the thermophysical properties, such as the thermal expansion coefficient and thermal conductivity. The effect of the PuO_2 content on the properties of $(U,Pu)O_2$ is studied.

2 METHOD AND RESULTS

The MD calculation for $(U,Pu)O_2$ was performed for a system of 324 ions (108 cations and 216 anions) initially arranged in a CaF_2 type crystal structure. The desired number of plutonium ions was substituted randomly onto the uranium ion sites. We used a molecular dynamics program based on MXDRTO.[4] The calculations were made in a temperature range from 300 K to 2500 K, and in a pressure range from 0.1 MPa to 1.5 GPa. The calculation procedure and conditions are the same as those reported previously.[2,3]

We employed the semi-empirical two-body potential function proposed by Ida for cation-anion interactions.[5] The potential is a partially ionic model including a covalent contribution:

$$U_{ij}(r_{ij}) = \frac{z_i z_j e^2}{r_{ij}} + f_0(b_i + b_j)\exp\left(\frac{a_i + a_j - r_{ij}}{b_i + b_j}\right) - \frac{c_i c_j}{r_{ij}^6} + D_{ij}\left\{\exp[-2\beta_{ij}(r_{ij} - r_{ij}^*)] - 2\exp[-\beta_{ij}(r_{ij} - r_{ij}^*)]\right\} \quad (1)$$

where f_0 equals 4.186, z_i and z_j are the effective partial electronic charges on the ith and jth ions, r is the atom distance, r_{ij} is the interatomic distance, r_{ij}^* is the bond length of the cation-anion pair in vacuum, and a, b, and c are the characteristic parameters depending on the ion species. In this potential function, D_{ij} and β_{ij} describe the depth and shape of this potential, respectively. The potential parameters are determined by trial and error using the experimental values of the lattice parameters. The potential parameters used in the present study are summarized in Table 1.

Table 1 *Values of the interatomic potential function parameters for $(U,Pu)O_2$*

Ions	*z*	*a*	*b*	*c*	D_{ij}	β_{ij}	r_{ij}
O	-1.2	1.926	0.160	20			
					(for U-O pairs)		
U	2.4	1.659	0.160	0	18	1.25	2.369
					(for Pu-O pairs)		
Pu	2.4	1.229	0.080	0	13	1.56	2.339

The thermal conductivity of the system was calculated using the Green-Kubo relation.[6] Details of the relation have been described in our previous papers.[2,3]

Figure 1 shows the temperature dependence of the lattice parameters of $(U,Pu)O_2$ calculated by the MD method, together with the experimental data.[7,8] The calculated values of the lattice parameter at room temperature of UO_2 and PuO_2 are 0.5474 and 0.5398 nm, respectively. These values agree well with the literature data. The lattice parameter of $(U,Pu)O_2$ decreases with PuO_2 content, which is consistent with Vegard's law. The lattice parameters increase with increasing temperature, showing thermal expansion behaviour.

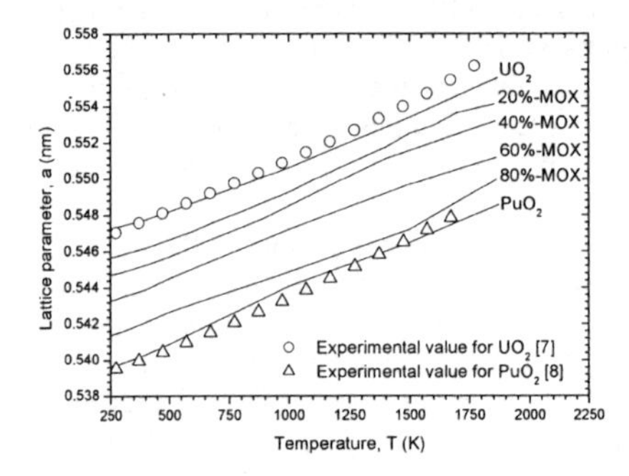

Figure 1 *Temperature dependence of the lattice parameters of $(U,Pu)O_2$*

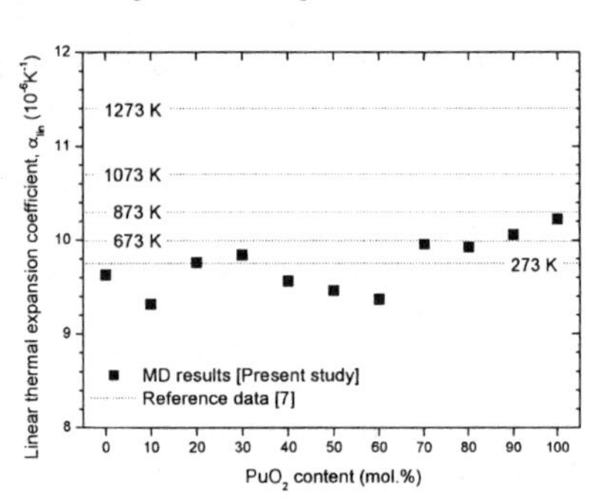

Figure 2 *Average linear thermal expansion coefficients (300-2000 K) of $(U,Pu)O_2$*

We can calculate the linear thermal expansion coefficient from the slope of the lattice parameter versus temperature plot. Figure 2 shows the calculated results of the average linear thermal expansion coefficient from 300 to 2000 K for $(U,Pu)O_2$ as a function of PuO_2 content. In this figure, the literature data are also plotted for comparison, which are obtained from the recommended equation developed by Martin.[7] Although UO_2, PuO_2, and MOX fuels have very similar thermal expansions according to Martin's work, our calculated data increase slightly with PuO_2 content. In the temperature range from 300 K to 2500 K, the thermal conductivity of $(U,Pu)O_2$ were evaluated by the MD calculation. Only the lattice contributions to the thermal conductivity can be evaluated in the present case. Figure 3 shows the calculated thermal conductivities of UO_2, $(U_{0.8}Pu_{0.2})O_2$, and PuO_2, together with the literature data.[9-12] The calculated thermal conductivities decrease with increasing temperature. Our calculated values well agree with the literature data, so it can be said that the MD method is very useful for evaluating the thermal conductivity of MOX fuels. Figure 4 shows the calculated results of the thermal conductivities at 700 K of

$(U,Pu)O_2$, as a function of PuO_2 content. The thermal conductivity of $(U,Pu)O_2$ decreases with PuO_2 content up to about 50 mol.%, and then it increases gradually. This behaviour is characteristic of the phonon conduction, as observed in real materials.

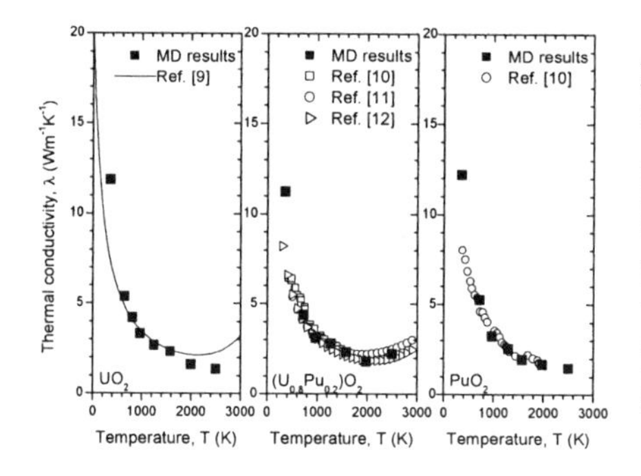

Figure 3 *Temperature dependence of the thermal conductivities for UO_2 $(U_{0.8}Pu_{0.2})O_2$, and PuO_2*

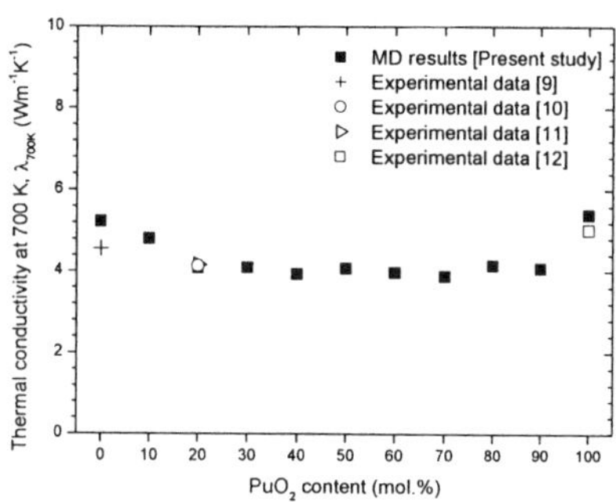

Figure 4 *Thermal conductivities at 700 K of $(U,Pu)O_2$*

3 CONCLUSION

The MD calculation was performed for $(U_{1-x}Pu_x)O_2$ (x: 0, 0.1, 0.2, 0.3, 0.4, 0.5, 0.6, 0.7, 0.8, 0.9, 1.0). Although information about the physical properties of $(U,Pu)O_2$ is very limited, several thermophysical properties can be evaluated by MD calculations. The calculated lattice parameters, linear thermal expansion coefficients, and thermal conductivities are almost identical with the literature data. The present study shows that the MD method can be usefully applied to determine the thermophysical properties of MOX fuels.

References

1 L.C. Walters, D.L. Porter and D.C. Crawford, *Progress in Nuclear Energy*, 2002, **40**, 513.
2 K. Kurosaki, K. Yamada, M. Uno, S. Yamanaka, K. Yamamoto and T. Namekawa, *J. Nucl. Mater.*, 2001, **294**, 160.
3 K. Kurosaki, M. Imamura, I. Sato, T. Namekawa, M. Uno, S. Yamanaka, *J. Nucl. Sci. Tech.*, 2004, **41**, 827.
4 K. Kawamura and K. Hirao, "*Material Design using Personal Computer*", Shokabo, Tokyo, (1994).
5 Y. Ida, *Phys. Earth Planet Interiors*, 1976, **13**, 97.
6 R. Zwanzig, *Ann. Rev. Phys. Chem.*, 1965, **16**, 67.
7 D.G. Martin, *J. Nucl. Mater.*, 1988, **152**, 94.
8 *Thermophysical properties of matter*, The TPRC data series vol. 13. IFI/Plenum Data, New York, 1970.
9 MATPRO-Version 11, NUREG/CR-0497, TREE-1280, Rev. 2, August (1981).
10 R.L. Gibby, *J. Nucl. Mater.*, 1971, **38**, 163.
11 Y. Philipponneau, *J. Nucl. Mater.*, 1992, **188**, 194.
12 C.S. Olsen and G.A. Reymann, "THREE-NUREG-1005 (1976)", eds. P.E. MacDonald and L.B. Thompson.

THERMODYNAMIC MODELLING OF THE URANIUM-ZIRCONIUM-IRON-OXYGEN SYSTEM

M. Ito, K. Kurosaki, M. Uno and S. Yamanaka

Department of Nuclear Engineering, Graduate School of Engineering, Osaka University, 2-1 Yamadaoka, Suita, Osaka 565-0871, Japan

1 INTRODUCTION

This study intends to establish a technique for Inner Vessel Reactor - Accident Management (IVR-AM), which is expected to be applied to light water nuclear reactors in the future. Although an understanding of the oxidation behaviour of molten fuels under conditions of high-temperature steam is required, it is difficult to investigate experimentally the phase relation of the molten fuels in the actual complex system. Therefore, it is necessary to establish a thermodynamic database that would help in the prediction of the thermochemical behaviour of molten fuels.

In recent years, we have performed thermodynamic modelling for O-Pu-Zr, ZrO_2-MO_2 (M=Th, U, Pu and Ce) and PuO_2-ZrO_2-UO_2,[1-3] and succeeded in evaluating their multiple phase diagrams. In the present study, thermodynamic modelling is performed for the U-Zr-Fe-O quaternary system, which is the main system of the molten fuels. High temperature oxidation behaviour of a representative model alloy of the molten fuels is also investigated. The modelling is carried out by means of the CALculation of PHAse Diagrams (CALPHAD) technique.[4]

2 METHOD AND RESULTS

In the present study, the program Thermo-Calc version L. was applied for the assessment of the thermodynamics of the system. The thermodynamic modelling of the binary systems, U-Zr, U-Fe, U-O, Zr-Fe, Zr-O and Fe-O, were carried out and these data can then be extended into the ternary and quaternary systems. Sets of the thermodynamic data for each binary system are available in the literature.[5-12] These thermodynamic data were optimized using a software which is originally developed by the author's group. The calculation was performed in the temperature range from room temperature to about 3273 K.

Figure 1 shows the calculated U-Fe-Zr, U-Zr-O, Fe-U-O and Fe-Zr-O ternary phase diagrams at 1273 K. It has been confirmed that the calculated ternary or pseudo-binary phase diagrams are quite similar to those reported in the literature. The free energy data used in the present study are appropriate. The U-Fe-O and Zr-Fe-O phase diagrams are assessed in the present study.

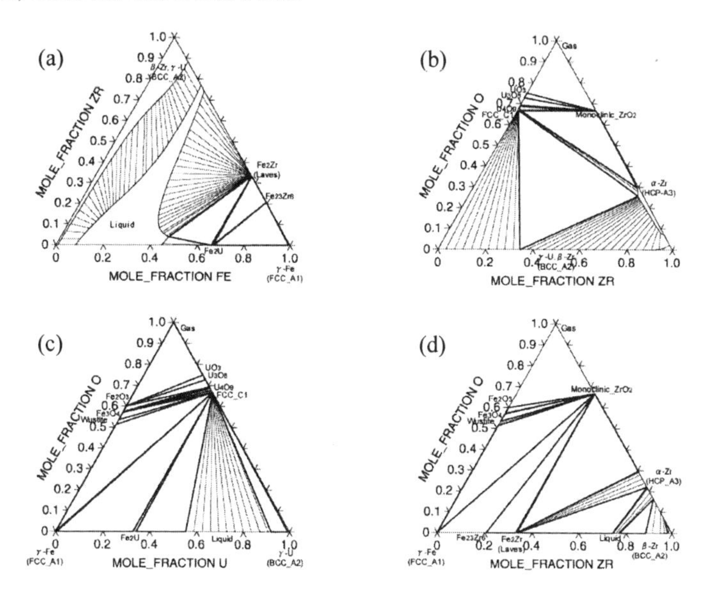

Figure 1 *Isothermal section of calculated ternary phase diagrams; (a) U-Fe-Zr, (b) U-Zr-O, (c) Fe-U-O and (d) Fe-Zr-O*

The high temperature oxidation behaviour of the representative model alloy of the molten fuels was investigated. Figure 2 shows the U-Zr-Fe-O quaternary phase diagram which is discussed in this study. The representative composition of the model alloy was chosen as U: 38.0%, Fe: 28.7% and Zr: 33.3%. The oxidation behaviour of the molten fuels was evaluated by observing the change of the phases in the U-Fe-Zr isothermal diagram with increasing oxygen content. The oxygen content range is from 0 to 50 at. %-O.

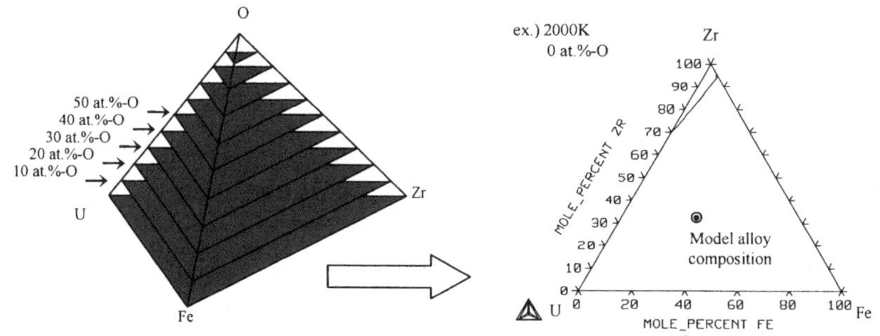

Figure 2 *U-Zr-Fe-O quaternary phase diagram*

Figure 3 shows the change of the mol fraction of existing phases in the model alloy with oxygen content at (a) 2000 K and (b) 1500 K. It is confirmed from Figure 3 (a) that the model alloy shows a liquid single phase at 0 at. %-O and that the FCC_C1 phase appears and increases with oxygen content. On the other hand, it is confirmed from Figure 3 (b) that the same behaviour occurs in the oxygen content range from 0 to 33 at. %-O. However, at higher oxygen content, the Fe_2Zr intermetallic compound and tetragonal ZrO_2 appear and increase with oxygen content at 1500 K.

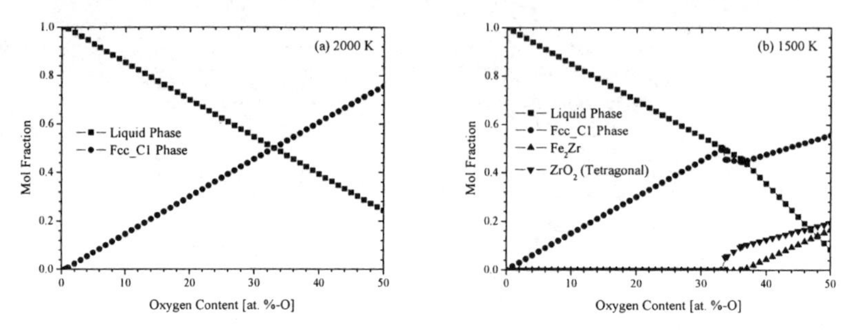

Figure 3 *High temperature oxidation behaviour at (a) 2000 K and (b) 1500 K*

3 CONCLUSION

Thermodynamic modelling was performed to assess the phase relation in the U-Zr-Fe-O system. We have evaluated the oxidation behaviour of a model alloy which had a representative composition at 2000 K and 1500 K. It is confirmed that the FCC_C1 phase appears and increases with oxygen content at both temperatures. Additionally it is confirmed in the case of 1500 K that the Fe_2Zr intermetallic compound and tetragonal ZrO_2 appear and increase with oxygen content above 33 at. %-O.

References

1 H. Kinoshita, M. Uno and S. Yamanaka, *J. Alloys Comp.*, 2003, **354**, 129.
2 H. Kinoshita, M. Uno and S. Yamanaka, *J. Alloys Comp.*, 2004, **370**, 25.
3 H. Kinoshita, M. Uno and S. Yamanaka, *J. Nucl. Mater.*, 2004, **334**, 90.
4 N. Saunders and A.P. Miodownik, In: *CALPHAD (Calculation of Phase Diagrams): A Comprehensive GuidePergamon Materials Series* vol. 1, Pergamon, Oxford (1998).
5 M. Kurata, T. Ogata, K. Nakamura and T. Ogawa, *J. Alloys Comp.*, 1998, **271/273**, 636.
6 P.Y. Chevalier and E. Fischer, *J. Nucl. Mater.*, 1998, **257**, 213.
7 M. Selleby and B. Sundman, *Calphad*, 1996, **20**, 381.
8 V.W.A. Fischer and A. Hoffmann, *Arch. Eisenhuttenw.*, 1957, **28**, 743.
9 T.S. Jones, S. Kimura and A. Muan, *J. Am. Ceram. Soc.*, 1967, **50**, 139.
10 T. Katsura, M. Wakihara, S. Hara and T. Sugihara, *J. Solid State Chem.*, 1975, **13**, 107.
11 C. Servant ,C. Gueneau and I. Ansara, *J. Alloys Comp.*, 1995, **220**, 19.
12 SGTE Pure Substance Database (Edit. 1998), Provided by GTT Technol., Herzogenrath, Germany 1998.

NANOINDENTATION STUDIES OF URANIUM INTERMETALLIC COMPOUNDS

J. Adachi, K. Kurosaki, M. Uno and S. Yamanaka

Department of Nuclear Engineering, Graduate School of Engineering, Osaka University, 2-1 Yamadaoka, Suita, Osaka 565-0871, Japan

1 INTRODUCTION

Uranium-plutonium mixed nitride, (U,Pu)N, is a promising candidate as a fast breeder reactor fuel because of its many desirable properties, such as its high melting point, high fuel density, and high thermal conductivity.[1] The intermetallic compounds consisting of uranium and the platinum family metals are generated in the irradiated nitride fuel and may affect the mechanical properties of the fuel. The mechanical properties of the intermetallic compounds are very important in evaluating the safety of fuel under irradiation.

In our previous study,[2] the mechanical properties of URu_3, URh_3, and UPd_3 were measured by ultrasonic pulse echo measurement and a micro Vickers hardness tester. In the present study, we have evaluated the nanoscale mechanical properties of uranium intermetallic compounds, as well as the bulk mechanical properties.

The nanoscale mechanical properties, which are the Young's modulus and nano-hardness, can be measured using a diamond nanoindenter instead of a cantilever of a atomic force microscope (AFM). The nanoindentaion technique[3-7] has been developed over some decades, and the mechanical properties within a sub-micro or nano scale have been widely discussed.

2 METHOD AND RESULTS

Polycrystalline samples of URu_3, URh_3, and UPd_3 were prepared from uranium and platinum family metal ingots by an arc melting under high vacuum. The purities of the starting materials were above 99.9 %. The prepared bulk samples were annealed at 1073 K for 10 h under a vacuum below 10^{-5} Pa.

The crystallographic properties, such as the lattice parameter and structure of UPd_3, URh_3, and URu_3 were measured by powder X-ray diffraction. The crystallographic properties and other properties are summarized in Table 1.

The nanoindetation tests were performed at room temperature in air using an AFM supplied by JEOL with a TriboScope (Hystron Inc.) nanoindenter. The nanoindenter is attached to the SPM head. A Berkovich type diamond indenter was used in the present study. The nanoscale mechanical properties, such as the reduced modulus and nanohardness can be evaluated from the load-displacement curves. The calculation method is described in our previous study.[8] We can calculate the Young's modulus from the reduced modulus and Poisson's ratio using the following equation:

$$\frac{1}{E_r} = \frac{1-\nu_s^2}{E_s} + \frac{1-\nu_i^2}{E_i},$$

where E_r is the reduced modulus, E_s and E_i are the Young's moduli of a sample and the indenter, and ν_s and ν_i are the Poisson's ratio of the sample and the indenter, respectively. Assuming that E_i = 1140 GPa and ν_i = 0.07 for a diamond tip, the Young's modulus of the sample was calculated. The values of the Young's modulus and nano-hardness of URu_3, URh_3, and UPd_3 obtained at an indentation load of 2000 μN are listed in Table 1. In the case of URh_3, the nanohardness was very much higher than the Vickers hardness.

Table 1 *Crystallographic and mechanical properties of URu_3, URh_3, and UPd_3.*

Compounds		URu_3	URh_3	UPd_3
Crystal structure		$AuCu_3$ type	$AuCu_3$ type	$TiNi_3$ type
Lattice parameter	a (nm)	0.3979	0.3991	0.5778
	c (nm)			0.9721
Bulk density	(gcm^{-3})	13.7	13.6	12.5
	(%T.D.)	96.1	95.2	95
Poisson's ratio		0.343	0.361	0.340
Young's modulus	(GPa)#1	248	218	161
	(GPa)#2	213±20	238±15	126±14
Vickers hardness#3[2]	(GPa)	11.6	5.93	5.09
Nanohardness #4	(GPa)	9.8±1.1	11.7±0.4	7.2±0.5

#1: Evaluated from the ultrasonic pulse echo measurement[2] ; #2: Evaluated from the nanoindentation test; #3: Load: P = 4.9 N and Load time: T = 30 s; #4: Load: P = 2000 μN

Figure 1 shows the load dependence of the nano-hardness of UPd_3, URh_3, and URu_3. The values of the hardness decrease with increasing load, indicating a typical load dependence of hardness. In addition, it can be confirmed from Table 1 that the nano-hardness is larger than the micro Vickers hardness.

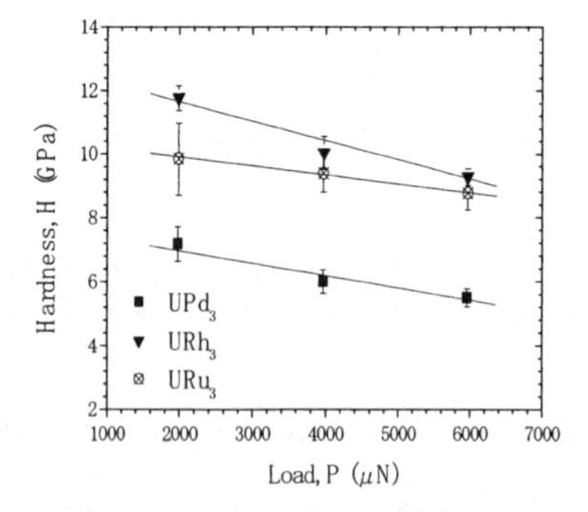

Figure 1 *Load dependence of the hardness for UPd_3, URh_3, and URu_3.*

UPd_3 gives the lowest Young's modulus and nanohardness among UPd_3, URh_3, and URu_3, in both bulk and nano mechanical properties. This characteristic could be due to its crystal structure, which for UPd_3 is the hexagonal, $TiNi_3$ type, but those of URu_3 and URh_3 are the cubic, $AuCu_3$ type.

Figure 2 shows the indentation images of UPd_3, (a): micro Vickers indentation image, (b): nanoindentation image. The figure shows that the Vickers hardness is affected by grain boundaries and/or pores, but the nanohardness is not affected by those microstructures, because the indentation size of the micro Vickers indentation is vastly larger than that of the nanoindentation. For example, cracking can be observed in Fig.2, and the measured Vickers hardness values could be lower than the true values if the specimens cracked during the Vickers hardness testing. In addition, if there are oxide layers on the sample surface, they could affect the nanohardness values. These are the reasons for the nanohardness being higher than the Vickers hardness, especially for URh_3. These results show that the measurement of nanoscale mechanical properties is better than the measurement of microscale mechanical properties when evaluating the mechanical properties of materials.

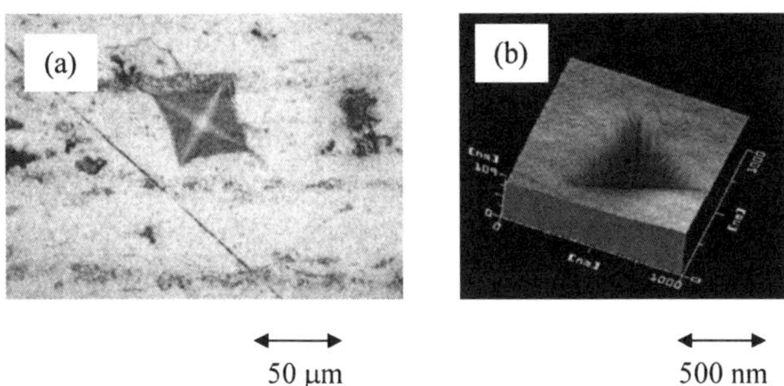

Figure 2 *Indentation images of UPd_3, (a): micro Vickers indentation image (optical microscope image), (b): nanoindentation image (AFM image).*

3 CONCLUSION

Nanoindentation tests were performed for UPd_3, URh_3, and URu_3, and their nanoscale mechanical properties were measured. Their hardness decreased with increasing indentation load. UPd_3 has low values of the mechanical property in comparison with those of URh_3 and URu_3, which is due to the crystal structure. The nanoindetation tests are very useful measurement in evaluating the mechanical properties of materials.

References

1 Hj. Matzke, in *Scinece of Advanced LMFBR Fuels*, Amsterdam, North-Holland, 1986.
2 S. Yamanaka, K. Yamada, T. Tsuzuki, T. Iguchi, M. Katsura, Y. Hoshino and W. Saiki, *J. Alloys Compd.*, 1998, **271-273**, 549.
3 W. C. Oliver and G. M. Pharr, *J. Mater. Res.*, 1992, **7**, 1564.
4 M. Nishibori and K. Kinosita, *Thin solid Films*, 1978, **48**, 325.
5 D. Newey, M.A. Wilkins and H.M. Pollock, *J. Phs. E: Sci. Intrum*, 1982, **15**, 119.
6 J. B. Pethica, in; V. Ashworth, W. Grant, R. Procter (Eds.), *Ion implantation into Metals*, Pergamon Press, Oxford, 1982, p. 147.
7 J.B. Pethica, R. Hutchings and W.C. Oliver, *Phil. Mag.*, 1983, **A48**, 593.
8 K. Kurosaki, Y. Saito, H. Muta, M. Uno and S. Yamanaka, *J. Alloys Compd.*, 2004, **381**, 240.

SELF-IRRADIATION OF CERAMICS AND SINGLE CRYSTALS DOPED WITH PLUTONIUM-238

B.E. Burakov, M.A. Yagovkina, M.V. Zamoryanskaya, V.M. Garbuzov, V.A. Zirlin and A.A. Kitsay

Laboratory of Applied Mineralogy and Radiogeochemistry, V. G. Khlopin Radium Institute, 28, 2-nd Murinskiy Ave., St. Petersburg, 194021, Russia

1 INTRODUCTION

Crystalline ceramics are the prospective materials most suggested for the immobilization of long-lived radionulcides, in particular, weapons grade Pu and other actinides. Methods of immobilization might include: 1) transmutation (burning) followed by geological disposal of the irradiated materials or 2) direct geological disposal of actinide matrices. Different durable host phases have been suggested for actinide (An) incorporation in the form of solid solutions. These are: different polymorphs of zirconia, $(Zr,An,...)O_2$, in particular, one of cubic fluorite-type structure;[1,2] zircon, $(Zr,Hf,An,...)SiO_4$;[3,4] monazite, $(La,An,...)PO_4$;[5,6] Ti-pyrochlore, $(Ca,Gd,Hf,Pu,U)_2Ti_2O_7$[7] etc. To investigate the resistance of actinide host phases to accelerated radiation damage, which simulates effects of long term storage, ^{238}Pu-doped samples have been studied using X-ray diffraction analysis (XRD) and other methods.

2 METHOD AND RESULTS

2.1 Sample features

Polycrystalline ^{238}Pu-doped samples of gadolinia-stabilized cubic zirconia, $Zr_{0.79}Gd_{0.14}Pu_{0.07}O_{1.99}$;[8-10] Ti-pyrochlore, $(Ca,Gd,Hf,Pu,U)_2Ti_2O_7$;[9,12] zircon/zirconia based ceramic, $(Zr,Pu)SiO_4/(Zr,Pu)O_2$;[11] La-monazite, $(La,Pu)PO_4$,[13] and Pu-phosphate, $PuPO_4$, with a monazite structure,[13] were obtained from previous research. Single crystal ^{238}Pu-doped zircon, $(Zr,Pu)SiO_4$, has been synthesized by the flux method.[14] Single crystal ^{238}Pu-doped Eu-monazite, $(Eu,Pu)PO_4$, has been recently obtained in our laboratory using a similar approach. The main features of the samples are summarized in Table 1. The XRD measurements were carried out at ambient temperature after different cumulative doses using a special technique developed at the V.G. Khlopin Radium Institute.[15]

2.2 Ti-pyroclore

The Ti-pyrochlore became nearly amorphous at a cumulative dose of $(1.1\text{-}1.3)x10^{25}$ alpha decays/m^3. The radiation damage was accompanied by a decrease of the ceramic density of approximately 10 % in comparison with initial sample (Table 1). Many cracks have been observed by optical microscope in the ceramic matrix after a dose 5.7x10^{24} alpha decays/m^3. Inclusions of $(U,Pu)O_x$ and $(Hf,Ti,Ca)O_x$ were identified in the ceramic matrix by SEM.[16]

Table 1 *Principal features of* ^{238}*Pu-doped samples*

Sample, formula of main phase	*Bulk Pu content and distribution wt.% el.*	^{238}*Pu content wt.% el.*	*Geometric density (g/cm*^{3}*)*	
			Initial	*at highest cumulative dose*
Ti-pyrochlore ceramic $(Ca,Gd,Hf,Pu,U)_2Ti_2O_7$	10.5 inhomogeneous from 3.4 to 26.8	8.7	4.8	4.3
Zircon/zirconia ceramic $(Zr,Pu)SiO_4$+15% $(Zr,Pu)O_2$	5.7 homogeneous	4.7	4.4	No data no cracks visually observed
Cubic zirconia ceramic $Zr_{0.79}Gd_{0.14}Pu_{0.07}O_{1.99}$	12.2 homogeneous	9.9	5.8	5.8
La-monazite ceramic $(La,Pu)PO_4$	9.9	8.1	4.7	4.7
Pu-monazite ceramic $PuPO_4$	65.2 homogeneous	7.2	4.9	No data visually observed cracks
Zircon single crystal $(Zr,Pu)SiO_4$	3.3 inhomogeneous from 1.9-4.7	2.7	No data	No data visually observed cracks
Eu-monazite single crystal $(Eu,Pu)PO_4$	6.0 homogeneous	4.9	No data	No data

2.3 Zircon and tetragonal zirconia

Self-irradiation of zircon/zirconia based ceramics caused amorphization of the zircon phase at a cumulative dose of (1.3-1.5)x10^{25} alpha decays/m^3. Tetragonal zirconia, $(Zr,Pu)O_2$, which was a minor phase (15 wt. %) in the zircon based ceramic, has demonstrated significantly higher resistance to radiation damage. No evidence of ceramic matrix swelling or cracking was found by optical microscopy after zircon amorphization.

Self-irradiation changed the colour of the ^{238}Pu-doped single crystal zircon from the initial pink-brown to yellow-gray, and caused crack formation in the crystal.

2.4 Gadolinia stabilized cubic zirconia

Cubic zirconia, $Zr_{0.79}Gd_{0.14}Pu_{0.07}O_{1.99}$, retained a crystalline structure after extremely high self-irradiation doses of 2.77x10^{25} alpha decay/m^3. No changes in the measured ceramic density, matrix swelling, or cracking were observed. No inclusions of separate Pu phases have so far been found in the ceramic matrix.

2.5 Monazite and Pu-phosphate with monazite structure

La-monazite, $(La,Pu)PO_4$, remained crystalline at a cumulative dose of 1.19x10^{25} alpha decay/m^3. No swelling or crack formation in the ceramic matrix has so far been observed. Under radiation damage, La-monazite changed colour from the initial light blue to gray. Self-irradiation of a single crystal of Eu-monazite, doped with 4.9 wt.% ^{238}Pu for 14 months did not cause crack formation or change the initial strong crimson colour of the crystals. However, the accumulation of tiny particles around the monazite crystals has been observed after 14 months of crystal storage inside hermetically sealed glass cassettes. Pu-phosphate, $PuPO_4$, with a monazite structure became amorphous at a relatively low dose of 4.2x10^{24} alpha decay/m^3. Swelling and crack formation as a result of self-irradiation damage were observed in this ceramic. Also, under self-irradiation, this sample completely changed colour from an initial dark blue to black.

3 CONCLUSION

Actinide host phases demonstrate different resistance to self-irradiation. It is assumed on the basis of the XRD data that self-irradiation of cubic zirconia and La-monazite is accompanied by two processes: the accumulation of defects in crystalline structures and repeated self-annealing of those defects at ambient temperature. Increasing Pu content in the monazite structured solid solutions, $(TR,Pu)PO_4$, decreases the resistance of monazite to self-irradiation.

References

1 D. Carroll, *J. Am. Ceram. Soc.*, 1963, **46**, 194.
2 R. Heimann and T. Vandergraaf, *J. Mater. Sci. Lett.*, 1988, **7**, 583.
3 B. Burakov, *SAFE WASTE'93 Conf. Proc.*, 1993, **2**, 19.
4 R. Ewing, W. Lutze and W. Weber, *J. Mat. Res.*, 1995, **10**, 243.
5 L.A. Boatner, G.W. Beall, M.M. Abraham, C.B. Finch, P.G. Huray and M. Rappaz, *Sci. Basis Nucl. Waste Manag.*, 1980, Vol. 2, 289.
6 L.A. Boatner and B.C. Sales, in *Radioactive Waste Forms for the Future*, 1988, eds. W. Lutze and R.C. Ewing, 495.
7 B. Ebbinghaus, R. VanKonynenburg, F. Ryerson, E. Vance, M. Stewart, A. Jostsons, J. Allender, T. Rankin and J. Gordon, *WM-98 Conf. Proc. CD-ROM*, 1998, Rep. 65-04.
8 B.E. Burakov, E.B. Anderson, M.V. Zamoryanskaya, M.A. Yagovkina and E.V. Nikolaeva, *Mat. Res. Soc. Symp. Proc. Sci. Basis Nucl. Waste Manag. XXV*, 2002, Vol. **713**, 333.
9 B.E.. Burakov, E. Anderson, M. Yagovkina, M. Zamoryanskaya and E. Nikolaeva, *J. Nucl. Sci. and Tech., Suppl.*, 2002**, 3**, 733.
10 B.E. Burakov, M.A. Yagovkina, M.V. Zamoryanskaya, A.A. Kitsay, V.M. Garbuzov, E.B. Anderson and A.S. Pankov, *Mat. Res. Soc. Symp. Proc. Sci. Basis Nucl. Waste Manag. XXVII*, 2004, Vol. **807**, 213.
11 B.E. Burakov, M.A. Yagovkina and A.S. Pankov, *Pu Future–The Science Conf. Proc.*, 2003, 274.
12 B. Burakov and E. Anderson, *in Review of Excess Weapons Disposition: LLNL Contract Work in Russia,* 2002, eds. L.J. Jardine, G.B. Borisov, UCRL-ID-149341, 265.
13 B.E. Burakov, M.A. Yagovkina, V.M. Garbuzov, A.A. Kitsay and V.A. Zirlin, *Mat. Res. Soc. Symp. Proc. Sci. Basis Nucl. Waste Manag. XXVIII*, 2004, Vol. **824**, 219.
14 J.M. Hanchar, B.E. Burakov, E.B. Anderson and M.V. Zamoryanskaya, *Mat. Res. Soc. Symp. Proc. Sci. Basis Nucl. Waste Manag. XXVI*, 2003, Vol. **757**, 215.
15 B. Burakov, *in Excess Weapons Plutonium Immobilization in Russia*, 2000, eds. L.J. Jardine, G.B. Borisov, UCRL-ID-138361, 251.
16 M.V. Zamoryanskaya and B.E. Burakov, *Mat. Res. Soc. Symp. Proc. Sci. Basis Nucl. Waste Manag. XXVIII*, 2004, Vol. **824**, 231.
17 T. Geisler, B. Burakov, M. Yagovkina, V. Garbuzov, M. Zamoryanskaya, V. Zirlin and L. Nikolaeva, *J. Nucl. Mater.*, 2005, **336**, 22.

ACTINOID PHOSPHATES OF KOSNARITE STRUCTURE

S.V. Tomilin[1], A.I. Orlova[2], A.N. Lukinykh[1] and A.A. Lizin[1]

[1]Federal State Unitary Enterprise
"State Scientific Center–Research Institute of Atomic reactors", Dimitrovgrad-10, Ulyanovsk region, Russia.
[2]Nizhny Novgorod State University, Nizhny Novgorod, Russia

1 INTRODUCTION

Synthetic orthophosphates are analogues of the natural mineral kosnarite[1] (ideal formula $KZr_2(PO_4)_3$), which possess many unique physical and chemical properties, in particular, a tendency to wide isomorphism.[2] In combination with high chemical, thermal and radiation stabilities this property makes phosphates of this family promising crystalline matrices for radioactive waste immobilization,[3] especially alkali and alkali-earth nuclides. The prospective of immobilization of actinide elements within the kosnarite-structured orthophosphates (this structural type is mostly known as NZP, from $NaZr_2(PO_4)_3$ orthophosphate) is of interest too. It is obvious that such immobilization is possible, if the crystalline structure of actinide orthophosphates is similar or related to the structure of NZP. However, the previously known group of alkali actinide orthophosphates of $A^IM_2^{IV}(PO_4)_3$ structure, where M^{IV}=Th, A^I=Li, Na, K, Rb, Cs; M^{IV}=U, A^I=Li, Na, K; M^{IV}=Np, A^I=Na has another structural type,[4-5] i.e. $NaTh_2(PO_4)_3$ (NThP) with a monoclinic lattice and an actinide coordination number of 9. Not long ago, actinides (IV) were found to form NZP-structured phases with an octahedral arrangement around the central atom in an anion skeleton. A tendency for a typical polymorphous transition at low-temperature to a monoclinic form (α) and to a high-temperature rhombohedral one of NZP type (β)[6,7] in isoformular orthophosphates $A^IM_2^{IV}(PO_4)_3$, where A^I=Na, K, Rb, M^{IV}=U, Np, Pu was revealed. As a continuation of the work[6,7] the first X-ray characteristics of seven double actinide and alkali orthophosphates of $A^IM_2^{IV}(PO_4)_3$ type, where M^{IV}=U, Np, Pu, A^I=Na, K, Rb (β-forms) were obtained. The results of crystalline structure refinement of $KU_2(PO_4)_3$ by the Rietveld method using X-ray diffractometry data are presented.

2 METHODS AND RESULTS

2.1 Synthesis and identification of crystalline forms

The compounds were synthesized by the method described previously[6]. The products underwent X-ray diffraction analysis (Debye chamber, copper irradiation) after rapid cooling (50^0C/min). Analysis of the diffraction data showed that after cooling from low temperatures (500-1300°C), crystalline modifications similar to monoclinic ones

$NaTh_2(PO_4)_3$ in their structure are mostly formed (α-form). The rhombohedral structures ($NaZr_2(PO_4)_3$- β-forms) were crystallized after rapid cooling from high temperatures (1300-1600°C). For $KU_2(PO_4)_3$, this modification did not form after slow cooling (initial temperature of cooling 1300°C, 250°C/hour). For other compounds slow cooling did not work. Evidently, the α→β transition is reversible. Temperature intervals of the rhombohedral orthophosphates are presented in Table 1, while Table 2 provides crystalline lattice parameters (CLP).

2.2 Changes of lattice parameters

Table 2 shows that CLP and lattice cell volumes change regularly with changing alkali and actinide element radii.

Table 1 *Temperature intervals for forming rhombohedral orthophosphates*

Compound	*Temperature, °C*	*Compound*	*Temperature, °C*
$KU_2(PO_4)_3$	1200-1450 [a]	$RbNp_2(PO_4)_3$	650-1500 [a]
$RbU_2(PO_4)_3$	950-1400 [a]	$NaPu_2(PO_4)_3$	>1100
$NaNp_2(PO_4)_3$	1300-1650 [a]	$KPu_2(PO_4)_3$	>1000
$KNp_2(PO_4)_3$	1100-1650 [a]	$RbPu_2(PO_4)_3$	>900

[a] - upper temperature is the melting point

Table 2 *Crystallographic characteristics of rhombohedral orthophosphates (hexagonal arrangement)*

Compound	*Synthesis temperature, °C*	*a, nm*	*c, nm*	*V, nm³*	*Space group*
$NaZr_2(PO_4)_3$	1200	0.8816(2)	2.281(1)	153.53	$R\bar{3}c$
$KU_2(PO_4)_3$	1200	0.9111(6)	2.499(4)	179.65	$R\bar{3}c$
$RbU_2(PO_4)_3$	1100	0.9059(4)	2.560(3)	181.94	$R\bar{3}c$
$NaNp_2(PO_4)_3$	1450	0.9137(5)	2.424(3)	175.75	Rhombohedral[a]
$KNp_2(PO_4)_3$	1200	0.9081(5)	2.514(3)	179.58	Rhombohedral[a]
$RbNp_2(PO_4)_3$	1200	0.9113(7)	2.568(3)	184.68	$R\bar{3}c$
$NaPu_2(PO_4)_3$	1200	0.9109(8)	2.384(3)	171.31	Rhombohedral[a]
$KPu_2(PO_4)_3$	1000	0.9050(2)	2.481(1)	175.99	$R\bar{3}c$
$RbPu_2(PO_4)_3$	1200	0.9030(7)	2.540(4)	179.38	Rhombohedral[a]

[a] Possible space groups R3, $R\bar{3}$, R32, $R\bar{3}m$, R3m;
Parenthesized – errors of lattice parameter determination

With increasing ionic radii of both the alkali and actinide elements, the cell volumes increase (one exception: in Rb-phosphate Np substitution by U decreases the volume). Moreover, for the same actinide the orthophosphates demonstrate consistent changes in parameters **a** and **c** (**a** decreases, **c** increases) with the increasing alkali cation radius. A similar dependence was noted for the first time in the NZP isoformular and isostructural d-element orthophosphates (Ti, Zr, Hf).[4,8-9] For these d-element compounds, the CLP changes regularly, i.e. parameters **a** and **c** increase with the increasing of f- and d-element ion radius.

2.3 Refining of $KU_2(PO_4)_3$ crystalline structure

The structure was refined by Rietveld full-profile analysis using X-ray diffraction data (diffractometer DRON-3M, 2θ=10-150 degrees, number of reflexes 862, space group $R\bar{3}c$, $R_F = 3,52$). β- $KU_2(PO_4)_3$ was shown to belong to the NZP type of complex skeleton $[U_2(PO_4)_3]^-_\infty$, which has K^+ cations in its vacancies. The uranium atom is located on the triple axis in a slightly distorted octahedral arrangement, forming two groups of slightly different interatomic distances $[U\text{-}O(1)]_3$=0.2269(2) nm and $[U\text{-}O(2)]_3$=0.2320(2) nm. The potassium atom is in the middle of a trigonal antiprism and forms six equal bonds K-O=0.2907(3) nm. As expected, the U-O interatomic distances in β- $KU_2(PO_4)_3$ turned out to be shorter, than those in α-$KU_2(PO_4)_3$, where the average value of distances is K-O=0.246 nm[10] at a uranium coordination number of 9. The α-β transition is accompanied not only by re-arrangement of the inner coordination sphere of the uranium atom and reduction of the U-O bond lengths, but also by an increase of the shortest distances U-U from 0.411 to 0.420 nm. As a result, the arrangement of the polyhedral skeleton becomes looser (but more symmetric), and there is a 29% increase in V/Z volume per one formula element: from 23.15 nm^3 in the α-modification up to 29.97 nm^3 in the β-modification.

3 CONCLUSION

The presence of U, Np and Pu actinides (IV) of the total NZP crystalline modifications formulae $A^IM_2^{IV}(PO_4)_3$ brings together the crystal chemistry of the d- and f-elements to some extent. In addition to the purely scientific interest, these phenomena have a practical application. In particular it suggests the possibility of concentrating and immobilizing the actinide fraction of radioactive phases in homogeneous NZP phases.

References

1 M.E. Brownfield, E.E. Foord and S.J. Sutley, *Am. Mineralogist*, 1993, **73**, 653.
2 A.I.Orlova, *Radiochemistry*, 2002, **44**, 385.
3 B.E. Scheetz, D.K. Agrawal, E. Breval and R. Roy, *Waste Management*, 1994, **14**, 489.
4 B. Matkovic, B. Prodic and M. Sljukic, *Bull. Soc. Chem. France*, num. Spec. 1968, 1777.
5 A.A. Burnayeva, Yu.F. Volkov, A.I. Kryukova, O.V. Skiba, V.I. Spiryakov, I.A.Korshunov and T.K. Samoilova, *Radiochemistry*, 1987, **29**, 3.
6 Yu.F. Volkov, R.F. Melkaya, V.I. Spiryakov and G.A. Timofeyev, *Radiochemistry*, 1994, **36**, 205.
7 A.A. Burnayeva, Yu.F. Volkov, A.I. Kryukova, I.A. Korshunov and O.V. Skiba, *Radiochemistry*, 1992, **34**, 12.
8 A.I. Kryukova, *Journal of Inorganic Chemistry*, 1991, **36**, 1962.
9 J. Alamo and R. Roy, *J. Mat. Sci.*, 1986, **21**, 444.
10 H.T. Hawkins, D.R. Spearing, D.K. Veirs, J.A. Danis, D.M. Smith, C.D. Tait and W.H. Runde, *Chem. Mater.*, 1999, **11**, 2851.

MONAZITE AND KOSNARITE-BASED CERAMICS CONTAINING URANIUM AND PLUTONIUM

S.V. Tomilin[1], A.I. Orlova[2], A.N. Lukinykh[1] and A.A. Lizin[1]

[1]Federal State Unitary Enterprise
"State Scientific Center–Research Institute of Atomic reactors", Dimitrovgrad-10, Ulyanovsk region, Russia.
[2]Nizhny Novgorod State University, Nizhny Novgorod, Russia

1 INTRODUCTION

Phosphate minerals draw more and more attention as potential matrices for the immobilization of radioactive waste. Compound cation orthophosphates of kosnarite (ideal formulae $KZr_2(PO_4)_3$)[1] and monazite (Ce, La, Y, RE, Ca, Th)(P, Si)O_4[2] structure, having the most known synthetic analogues as orthophosphates $NaZr_2(PO_4)_3$ (NZP) and $CePO_4$ are of the greatest interest. The advantage of these skeletal structures is their ability to accumulate various combinations of different cations without changing their structural type. This property of kosnarite and monazite-like structures is typical of f-elements, too, in particular, the rare earths (RE) and actinides (An), as indicated in Volkov and Orlova.[3] Thus, in phosphates $R_{1/3}Zr_2(PO_4)_3$ (R=La, Pr – Tb, Er, Tm, Yb) and $Na_{1-3x}Eu_xTi_2(PO_4)_3$ ($0 \leq x \leq 1$), lanthanide cations occupy vacancies in the three-dimensional anion skeleton $[M_2(PO_4)_3]^-_\infty$ (M=Ti, Zr). At the same time, there are phosphate compounds with lanthanides occupying octahedral positions in the skeleton statistically sharing them with zirconium: $Na_{1+x}Zr_{2-x}R_x(PO_4)_3$, where R=Tm, Yb, Lu. Formation of kosnarite-like phases was observed for high-temperature phosphate modifications with actinides (IV) $AM_2(PO_4)_3$, where A and M are certain combinations of alkali elements and actinides (A=Li, Na, K, Rb, Cs; M=U, Np, Pu). Solid NZP solutions were found to form in $NaZr_{2-x}M_x(PO_4)_3$ (M=Np, Pu) and $KZr_{2-x}U_x(PO_4)_3$, too.

There are a wide variety of monazite-structured f-element phosphates. For an oxidation state of 3+, it is RPO_4 phosphates, where R=Ce – Gd, Pu, Am, Cm, and for an oxidation state of 4+ it is $AM_2(PO_4)_3$ (A=Na – Cs, M=Ce); (A=Na, M=Np) and $B_{0,5}M_2(PO_4)_3$ (B=Mg, Ca, Sr, Gd, M=Ce); (B=Mg, Ca, Sr, M=Np); (B=Ca, M=Pu) phosphates.

Both structural types have a similar cation/anion ratio, equal to 1:1, and despite essential differences in structure and coordination number (CN) of the skeletal elements (in kosnarite CN=6, in monazite it is 9) we can speak of the polymorphous similarity of these structural types under certain conditions.

This work is aimed at the synthesis of phosphate ceramics containing uranium and plutonium. Known titanate ceramic compounds of pyrochlorine structure, intended for plutonium immobilization were taken into account in formulating the composition.[4] These data can be helpful for the subsequent comparison of the properties of titanate and phosphate ceramics of similar cation composition.

2 METHODS AND RESULTS

2.1 Synthesis and identification of phase composition

Four phosphate compositions (specimens 1-4) containing the cations Ca^{2+}, Cd^{2+}, Gd^{3+}, Hf^{4+}, Ce^{4+}, U^{4+} and Pu^{4+} were synthesized and studied. Compositions (Ca, Gd, Hf, Ti, U, Ce)$(PO_4)_3$ (specimen 1) and (Ca, Gd, Hf, Ti, U, ^{239}Pu)$(PO_4)_3$ (specimen 2) are proposed for phosphate specimens of the expected kosnarite structure, while for the expected structural analogues of monazite the compositions are as follows: (Ca, Cd, Gd, U, Ce)$(PO_4)_3$ (specimen 3) and (Ca, Cd, Gd, U, Ce, ^{239}Pu)$(PO_4)_3$ (specimen 4). Cerium was used as a plutonium simulator in specimens 1 and 3. Reactions between oxides and phosphoric acid with denitration, dispersion and sintering of the intermediate products, then compacting and high-temperature sintering in the final stage were applied to synthesize the specimens. Table 1 presents mass fractions of the components in the initial compositions. The gross chemical composition of the produced compounds rests on stoichiometric calculations (Table 2). The stoichiometric cation-anion ratios were equal to 0.85:1 and 1:1 in specimens 1, 2 and 3, 4, respectively, which corresponds to electric neutrality of all the final compositions.

Table 1 *Mass fraction of components in initial compositions*

Component	Mass fraction, %			
	Specimen 1	*Specimen 2*	*Specimen 3*	*Specimen 4*
CaO	5.877	5.771	8.036	7.840
Gd_2O_3	4.747	4.611	3.768	3.664
CdO	-	-	10.534	10.244
CeO_2	6.861	-	29.462	24.037
TiO_2	21.180	20.799	-	-
UO_2	14.725	13.361	14.755	14.349
PuO_2	-	7.324	-	7.262
Hf_2O	6.291	6.178	-	-
P_2O_5	42.388	41.701	33.418	32.604

Table 2 *Molar fraction of chemical components in obtained orthophosphates*

Specimen	Ca^{2+}	Gd^{3+}	Cd^{2+}	Hf^{4+}	Ti^{4+}	U^{4+}	Ce^{4+}	Pu^{4+}	PO_4^{3-}
1	0.525	0.130	-	0.150	1.329	0.273	0.138	-	3
2	0.525	0.130	-	0.150	1.329	0.273	-	0.138	3
3	0.304	0.044	0.174	-	-	0.116	0.362	-	1
4	0.304	0.044	0.174	-	-	0.116	0.304	0.058	1

The obtained products were identified by X-ray in a Debye chamber, as well as by copper radiation in a diffractometer. Specimens 1 and 2 turned out be two-phase, while specimens 3 and 4 were monophase (Table 3). A kosnarite phase, similar to $Ca_{0,5}Ti_2(PO_4)_3$[5] in interplanar distances (d) and reflex intensities of powder roentgenograms (I) is dominant in specimen 1. Monazite, whose seven lines were registered, is an additional minor phase. Conversely, in specimen 2 monazite dominates, and only five weak lines of a kosnarite-like phase were observed. In specimens 3 and 4, I and d were similar to the corresponding $CePO_4$ monazite phase.[6] Table 3 presents X-ray characteristics of the dominating phases and corresponding data for $Ca_{0,5}Ti_2(PO_4)_3$ and $CePO_4$.

Table 3 *Crystallographic characteristics of produced orthophosphates*

Spec. No.	Dominant phase	Addit. phase	Lattice parameters of dominating phase a, Å	b, Å	c, Å	β, град	V, $Å^3$
1	kosnarite	monazite	8.427(3)		22.20(1)		1365.6
2	monazite	kosnarite	6.667(5)	6.828(4)	6.347(4)	103.64(5)	280.8
3	monazite	no	6.775(8)	7.031(8)	6.46(1)	103.6(1)	299.2
4	monazite	no	6.72(1)	6.965(7)	6.42(1)	103.5(1)	292.3
$Ca_{0,5}Ti_2(PO_4)_3$, ICPDS 35-740			8.361		21.997		1331.7
$CePO_4$,	ICPDS 26-355		6.777	6.993	6.445	103.54	296.9

2.2 Peculiarities of lattice parameters

In specimen 1, the crystalline lattice parameters a, c and V of kosnarite type are slightly higher than those of $Ca_{0,5}Ti_2(PO_4)_3$ (Table 3). This is because larger structure-forming cations Hf^{4+} (r=0,71Å), Ce^{4+} (r=0,87Å), Gd^{3+} (r=1,10Å), U^{4+} (r=0,89Å) are a part of the structure, along with Ti^{4+} (ion radius r=0,61Å).[7] Hence, in this case formation of a complicated monophase cation composition takes place. The volumes of the unit cells of the monazite-like phases are comparable or slightly less, than those of $CePO_4$, which testifies to the presence of cations smaller than Ce^{3+} (r=1.20Å) in the structure. It suggests formation of a monophase containing various cations, too. The average ion radius of cations in specimens 1 and 2 per one PO_4^{3-} group is equal to 0,66Å, while for specimens 3 and 4 this value is 1,11Å, i.e. much higher. This fact contributes to the formation of monazite-structured phases in specimens 3 and 4, with a higher coordination number (CN 9) than that of kosnarite phases, where cations occupy positions in the anion skeleton have CN 6.

3 CONCLUSION

Monophase ceramics are more preferable for property reproducibility and the control of actinide inclusion. At the same time, if both structural types have high chemical, thermal and radiation stabilities, the presence of two phases with high strength characteristics in the ceramics does not decrease the high protection parameters.

References

1 M.E. Brownfield, E.E. Foord and S.J. Sutley, *Am. Mineralogist*, 1993, **78**, 653.
2 V.Y. Terekhov, N.I. Yegorov, I.M. Batyushkin and D.A. Mineyev. *Mineralogy and Geochemistry of Rare and Radioactive Metals,* M.: Energoatomizdat, 1987, 358.
3 Yu.F. Volkov and A.I. Orlova, *Systematization and Crystallochemical Aspects of Inorganic Compounds with Uninuclear Tetrahedral Oxoanions.* Dimitrovgrad: FSUE "SSC RIAR", 286.
4 L.J. Jardine and G.B. Borisov, *Excess Weapons Plutonium Immobilization in Russia.* Proc. Meet. Coord. Review of Work Held in St. Petersburg, Russia. St. Petersburg, 1999.
5 S. Senbhagaraman, T.N. Guru Row and A.M. Umarji, *J. Mater. Chem.*, 1993, **3**, 309.
6 Y.-X. Ni, J.M. Hughes and A.N. Mariano, *Am. Mineralogist.*, 1995, **80**, 21.
7 R.D. Shannon, *Acta Crystallogr.*, 1976, **A32**, 751.

APPLICATION OF CERAMICRETE MATRICES FOR LOW-TEMPERATURE SOLIDIFICATION OF LIQUID ACTINIDES-CONTAINING WASTES

Yu.M. Kulyako[1], S.E. Vinokurov[1], B.F. Myasoedov[1], A.S. Wagh[2], S.I. Rovny[3] and M.D. Maloney[4]

[1] Vernadsky Institute of Geochemistry and Analytical Chemistry RAS, Kosygin Str., 19, 119991, Moscow, Russia
[2] Argonne National Laboratory, 9700 South Cass Avenue, Argonne, IL 60439, USA
[3] PA Mayak, 456780, Ozyorsk, Russia
[4] CH2MHill, 9191 South Jamaica Str. Englewood, CO 80112, USA

1 INTRODUCTION

The development of atomic power engineering is inseparably linked with the problem of the environmentally non-hazardous treatment and storage of accumulated liquid radioactive wastes, which have a complex chemical composition, including alkaline wastes. The use of Ceramicrete technology[1,2] for cementation of such wastes by their low-temperature solidification in immobilizing matrices, based on chemically bonded phosphate ceramics (CBPCs) is proposed as one of the methods of solving this problem. Such matrices are formed as a result of the chemical interaction between magnesium oxide and potassium dihydrogenphopsphate.

This method, developed at Argonne National Laboratory, appears to be promising, since it exhibits several undoubted merits: the possibility of waste solidification at room temperature; preparation of matrices analogous to natural minerals (monazite, apatite), having high stability under geological conditions; the possibility of waste solidification within a wide pH range; simplicity of facilities, low power inputs and minimum quantities of secondary radioactive wastes.

A conclusion about possibility of using CBPCs for immobilizing radioactive waste cannot be made without an integrated study of the physicochemical properties of CBPCs, including the study of the behaviour of incorporated radionuclides and hazardous elements, determining the environmental impact of the solidified wastes over a long period of time. This paper is devoted to this study, the results of which will allow the creation of a scientific basis for the development of a new technology of solidification of liquid radioactive wastes.

2 METHOD AND RESULTS

The experiments on the solidification of two types of simulated high-level radioactive wastes with chemical compositions corresponding to those of actual wastes were performed: Stream-1 – liquid stream containing the components of the supernatant in HLW tanks at Hanford and Stream-2 – sludge stream containing the components of sludge in

HLW tanks plus the highly radioactive materials that are separated from HLW liquids. The radiochemical compositions of the prepared waste simulants are presented in Table 1.

Table 1 *Nuclear and physical characteristics of radionuclides belonging to Stream-1 and Stream-2*

Radionuclide	*Half-life*	*Specific activity* (Bq/g)	*Energy* (MeV)	*Radionuclide conc.* (Bq/L)	
				Stream-1	*Stream-2*
^{85}Sr	64.8 days	8.76×10^{11}	γ (1.57)	–	1.2×10^{7}
^{237}Np	2.14×10^{6} yrs	2.61×10^{7}	α(4.787); γ	1.2×10^{8}	2.4×10^{6}
^{239}Pu	24060 yrs	2.30×10^{9}	α(5.150); γ	1.2×10^{8}	2.9×10^{8}
^{241}Am	432 yrs	1.27×10^{11}	α(5.476); γ	–	8.0×10^{8}

The experiments on the solidification of two types of radioactive liquid waste, the compositions of which corresponded to the supernatant from HLW storage tanks at a radiochemical plant (Mayak_Stream 1) and radioactive iron hydroxide pulp of a chemical metallurgic plant (Mayak_Stream 2), were carried out at PA Mayak. The choice of these wastes was determined by their closeness of their chemical and radiochemical compositions to Hanford wastes.

CBPCs were prepared by solidification of simulated Streams 1&2 in PTFE containers by adding stoichiometric quantities of MgO and KH_2PO_4. It has been demonstrated that there are minor releases of nitrogen, water vapour, carbon dioxide and nitrogen oxides during solidification of CBPCs. No hydrogen release was detected.

After thorough mixing the reagent masses were sucked in to syringes (d=2 cm), and after 3-week storage, their density (1.6-1.8 g/cm^3) and compressive strength (25-55 MPa) were determined. Scanning electron microscopy of the matrices showed that the samples consist mainly of granules of potassium-magnesium phosphate. Autoradiography of the CBPCs showed that the distribution of Np and Sr is rather uniform, whereas Pu is nonuniformly concentrated in certain zones, formed apparently by phosphates.

Radiation resistance was estimated after γ-irradiation of the matrices (^{60}Co emitter, total absorbed dose 2.8×10^{8} rad). It was shown that irradiation of the matrices does not result in a noticeable change in their compressive strength.

The degree of radionuclide retention by CBPCs was determined in accordance with the ANS-16.1 test[3] during long-term leaching of the samples with distilled water at room temperature. For this test, CBPCs samples with diameter and height of 2 cm were prepared. Leaching was performed in PTFE containers with a leachant volume of 190 mL. The data obtained are presented in Table 2.

Table 2 *Radionuclide leachability indexes L_i* for CBPCs prepared by solidification of Stream-1&2*

Radionuclide	*Leachability indexes*	
	Stream-1	Stream-2
^{85}Sr	–	12.0
^{237}Np	12.6	13.5
^{239}Pu	13.2	14.2
^{241}Am	–	14.5

* $L_i = \frac{1}{n}\sum_{1}^{n}(-\log D_i)$; $D = \pi\cdot\left[\frac{a_n/A_0}{(\Delta t)_n}\right]^2\cdot\left[\frac{V_{sp}}{S_{sp}}\right]^2\cdot T$,

where D is the effective diffusivity, cm/s; V_{sp} and S_{sp} are the volume and geometric surface area of the specimen, cm^3 and cm^2; a_n and A_0 are the activity of a nuclide released from the specimen during the leaching interval and total activity of a given nuclide in the specimen at the beginning of the first leaching interval, Bq; $(\Delta t_n) = t_n - t_{n-1}$, the duration of the nth leaching interval, s; $T = \{0.5\ [t_n^{1/2} + t_{n-1}^{1/2}]\}^2$, the leaching time representing the "mean time" of the leaching interval, s.

The resistance of CBPCs to the leaching of environmentally hazardous elements was determined in accordance with the TCLP[4] test using extraction fluid #1 (pH=4.93) as a leachant. The results obtained are presented in Table 3. Comparison of these data with UTS limits shows that the prepared CBPCs retain hazardous elements rather strongly.

Table 3 *Content of environmentally hazardous elements in the solutions after leaching according to TCLP test*

Element	*Stream-1*		*Stream-2*		*UTS limits*
	pH	Content (mg/L)	pH	Content (mg/L)	(mg/L)
Pb		0.003		0.008	0.75
Cr	11.0	0.04	7.1	0.016	0.85
Cd		0.00005		0.0004	0.2

3 CONCLUSION

Solidification of simulated alkaline high-level supernatant and sludge in matrices based on chemically bonded phosphate ceramics (CBPCs) was performed. The densities (1.6-1.8 g/cm^3) and compressive strengths (no less than 25 MPa) of the CBPC matrices were determined. The compressive strength of the matrices was found to be unchanged at a radiation doses less than 10^8 rad.

High chemical stability and resistance of CBPCs to long-term leaching of the actinides (Np, Pu, Am) and Sr were established: their leachability indexes are 12.6-14.5 and 12.0, respectively. The matrices immobilized hazardous elements (Pb, Cr, Cd) very strongly. It was found that the contents of these elements in the solutions after long-term leaching are considerably smaller than the UTS limits.

The advantages of the Ceramicrete method and the established physicochemical properties of CBPC matrices prepared by this method, indicate the possibility of application of these matrices to the immobilization of actual liquid actinide-containing radioactive wastes.

Acknowledgement

This work was founded by US DOE, Project RCO-10116-MO-03 (ANL), Contract ANL-T2-0208-RU.

References

1 A. Wagh, R. Strain, S. Jeong, D. Reed, T. Krouse and D. Singh, *J. Nucl. Mater.*, 1999, **265**, 295.
2 D. Singh, A. Wagh, J. Cunnane and J. Mayberry, *J. Environ. Sci. Health.*, 1997, **32**, 527.
3 *Measurement of the Leachability of Solidified Low-Level Radiactive Wastes by a Short-Term Test Procedure*. ANSI/ANS-16.1-1986.
4 Method 1311.Toxicity Characteristic Leaching Procedure. Revision 0, July 1992.

AB INITIO STUDY OF URANIUM AND PLUTONIUM CARBIDES AND NITRIDES

M. Freyss[1] and I. Sato[2]

[1] CEA-Cadarache, DEN/DEC/SESC/LLCC, Bâtiment 151, 13108 Saint-Paul lez Durance, France
[2] JNC, Japan Nuclear Cycle Development Institute, 4002, Narita-cho, O-arai-machi, Ibaraki 311-1393, Japan

1 INTRODUCTION

We present the results of an *ab initio* study of the bulk properties of actinide carbides and nitrides. The structural and cohesive properties of uranium and plutonium carbides and nitrides UC, PuC, UN, PuN, and of the mixed (U,Pu)C and (U,Pu)N compounds are calculated and compared. The comparison between (U,Pu)C and (U,Pu)N is of special interest, since these compounds are envisioned as possible future nuclear fuels. This study is a first step in the subsequent study of the stability of point defects and of the behaviour of fission gases in these materials.

2 METHOD OF CALCULATION

We use an *ab initio* pseudopotential plane-wave method[1] based on the Density Functional Theory and implemented in the ABINIT code.[2] The pseudopotentials are of the norm-conserving Troullier-Martins type.[3] Exchange-correlation interactions are taken into account in the Generalized Gradient Approximation (GGA).[4] Most of the previous studies of actinide carbides and nitrides were done using LDA.[5,6,7] The cut-off energy for the electron plane-wave basis is fixed at 160 Ryd. Spin-polarization is not taken into account, and the calculations are done in the scalar relativistic approximation. The mixed (U,Pu)C and (U,Pu)N compounds are modeled using a 8 atom supercell in the NaCl structure. The use of a supercell, as required here, makes the pseudopotential approach particularly appropriate for this study and for the subsequent study of the stability of point defects in these materials.

3 RESULTS FOR U AND Pu MONOCARBIDES AND MONONITRIDES

The calculated equilibrium lattice parameters, bulk moduli and cohesive energies for UC, UN, PuC and PuN are given in Table 1. The lattice parameters are in agreement with the experimental data within 1%. On the other hand, the bulk modulus of UN is in agreement within only 15%. Taking into account the antiferromagnetic order of UN might improve this result. The stability of an antiferromagnetic order in PuN should also be investigated, since it is not definitely established experimentally. As to the cohesive energies, they are larger for uranium nitride and uranium carbide than for the corresponding plutonium compounds.

Table 1 *Calculated lattice parameter a, bulk modulus B, cohesive energy E_{co} and comparison to experimental data (Exp.)*

	GGA			Exp.	
	a (bohr)	*B (GPa)*	*E_{co} (eV)*	*a (bohr)*	*B (GPa)*
UC	9.36	145	14.7	9.38	158
UN	9.15	230	15.7	9.24	200
PuC	9.35	127	10.2	/	/
PuN	9.26	133	11.6	9.27	/

4 RESULTS FOR MIXED U AND Pu CARBIDES AND NITRIDES

In the mixed $(U_x Pu_{1-x})C$ and $(U_x Pu_{1-x})N$ compounds, the 8 atom supercell allows us to consider different atomic concentrations, *x*, of uranium relative to plutonium: 0.25, 0.50 and 0.75. The calculated lattice parameters and cohesive energies as a function of *x* are given in Figure 1 and Figure 2, respectively.

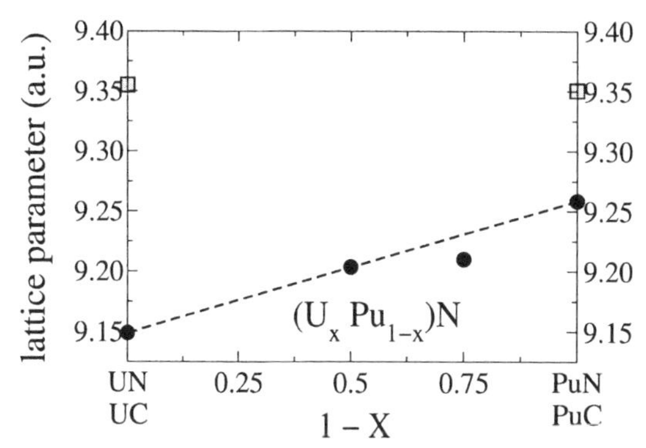

Figure 1 *Calculated lattice parameter (in bohr) of UC, PuC (gray squares) and of (U,Pu)N (black circles) as a function of the Pu atomic concentration. The dashed line corresponds to Vegard's law.*

Figure 1 shows that the lattice parameter of $(U_x Pu_{1-x})N$ as a function of the plutonium atomic concentration (*1-x*) does not follow a linear behaviour (Vegard's law). This was also found experimentally.[8] On the other hand, the cohesive energy of $(U_x Pu_{1-x})N$ appears to vary linearly with the plutonium atomic concentration. For $(U_x Pu_{1-x})C$, results for intermediary values of *x* are still needed.

The formation energy of the compounds is also calculated. It is defined as:

$$E_{form} = E_{(Ux\,Pu1-x)N} - [x \,.\, E_{UN} + (1-x) \,.\, E_{PuN}]$$

where $E_{(Ux\,Pu1-x)N}$, E_{UN} and E_{PuN} the energy per molecule of $(U_xPu_{1-x})N$, UN and PuN, respectively. The formation energy gives an insight on the stability of the mixed compound in a monophasic state, relative to UN and PuN rich regions (respectively UC and PuC rich regions in the case of the carbides). A negative value is indicative of a stable monophasic state. For $(U_{0.5}Pu_{0.5})N$ and $(U_{0.25}Pu_{0.75})N$, the calculated formation energies are -0.14eV and -0.12eV, respectively. Thus, the mixed nitride compounds are found to be stable in a monophasic solid solution state, as expected from experimental data.

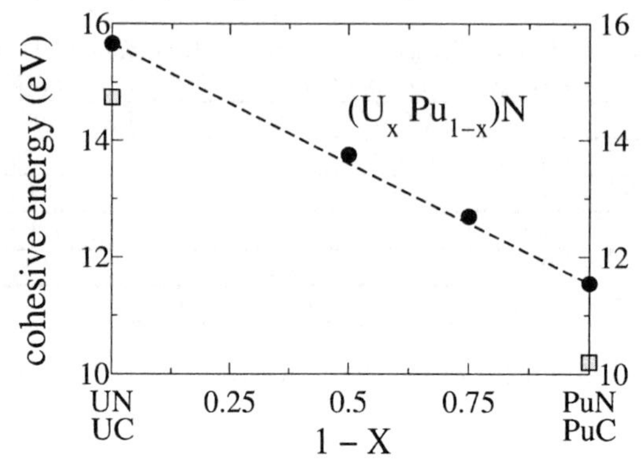

Figure 2 *Calculated cohesive energy (in eV) of UC, PuC (gray squares) and of (U,Pu)N (black circles) as a function of the Pu atomic concentration. The dashed line is a guide to the eye.*

5 CONCLUSION

Prior to the study of point defects and of the behaviour of fission gases in uranium and plutonium carbides and nitrides, we have investigated some cohesive properties of these materials in an *ab initio* pseudopotential plane-wave approach with the GGA approximation. This approach enables the satisfactory description of the structural properties of the carbides and nitrides. Uranium nitride and carbide are found to have a larger cohesive energy than the corresponding plutonium compounds. The lattice parameter of the mixed $(U_x\ Pu_{1-x})N$ nitrides does not follow Vegard's law, whereas the cohesive energy follows a linear behaviour as a function of the plutonium atomic concentration. A negative formation energy is obtained for $(U_{0.5}Pu_{0.5})N$ and $(U_{0.25}Pu_{0.75})N$, indicative of their stability in a monophasic state. Similar calculations for the mixed carbides are in progress. The influence of the antiferromagnetic order of these compounds on their bulk properties should also be investigated.

References

1 M.C. Payne, M.P. Teter, D.C. Allan, T.A. Arias and J.D. Joannopoulos, *Rev. Mod. Phys.,* 1992, **64**, 1045.

2 X. Gonze, J.M. Beuken, R. Caracas, F. Detraux, M. Fuchs, G.M. Rignanese, L. Sindic, M. Verstraete, G. Zerah, F. Jollet, M. Torrent, A. Roy, M. Mikami, P. Ghosez, J.Y. Raty and D.C. Allan, *Comp. Mat. Science,* 2002, **25**, 478. (http://www.abinit.org)

3 N. Troullier and J.L. Martins, *Phys. Rev. B*, 1991, **43**, 1993.

4 J.P. Perdew, K. Burke and M. Ernzerhof, *Phys. Rev. Lett.,* 1996, **77**, 3865.

5 M.S.S. Brooks, *J. Magn. Magn. Mat.*, 1982, **29**, 257.

6 M.S.S. Brooks, *J. Phys. F: Met. Phys.*, 1984, **14**, 639.

7 J. Trygg, J.M. Wills, M.S.S. Brooks, B. Johansson and O. Eriksson, *Phys. Rev. B,* 1995, **52**, 2496.

8 K. Minato, M. Akabori, M. Takano, Y. Arai, K. Nakajima, A. Itoh and T. Ogawa, *J. Nucl. Mat.*, 2003, **320**, 18.

EFFECT OF HIGH STATIC PRESSURE ON CRYSTALINE AND ELECTRONIC STRUCTURES OF TRANSITION, LANTHANIDE AND ACTINIDE METALS

B.A. Nadykto and O.B. Nadykto

RFNC-VNIIEF, Arzamas-16 (Sarov), Nizhni Novgorod region, 607190, Russia.

1 INTRODUCTION

When the crystalline structure of a material remains unchanged under pressure, kinks or discontinuities in the *P(ρ)* curve[1-11] can be explained by changes in the electronic structure (i.e. redistribution of electrons between inner and outer shells) only. Here we analyze experimental compressibility data, and present values of the equilibrium energy, E_n, and density, ρ_n, for a number of phases of different elements. We present data both for the equilibrium states at standard pressure and temperature and various high pressure (HP) phases. The *P(ρ)* dependency for each phase has been calculated using an equation of state,[1,2] and the obtained results have been compared with the experimental data.[3-11]

2 RESULTS AND DISCUSSION

The state of the solid phase under pressure can be described in terms of the energy of outer electrons (denoted by *OE*) in the unit atomic cell, which is calculated accounting for the interaction of *OE* with electrons from inner shells and the atomic nucleus. This approach allows the equation of state for solids to have the form[1,2],

$$E(\sigma) = (2E_n N_A / A)(0.5\,\sigma^{2/3} - \sigma^{1/3}), \qquad P(\sigma) = (2E_n N_A \rho_n / 3A)(\sigma^{5/3} - \sigma^{4/3}) \quad (1)$$

Here, $\sigma=\rho/\rho_n$ is the compression ratio, ρ_n is the equilibrium density at P=0, T=0, N_A is the Avogadro number, A is the atomic mass, and E_n the equilibrium energy of *OE* of the atomic cell. The elastic pressure is determined as $P = -\partial E/\partial V$. In Eq.1, $2E_nN_A\rho_n/3A=3B_0$, where B_0 is the bulk modulus at $\sigma=1$. When the bulk modulus, B_0, is known, the energy of *OE* in the atomic cell can be calculated as $E_n = (9AB_0/2N_A\rho_n)$.

2.1 Transition metals

The changes in the compressibility of 3d metals with nucleus charge and under HP can be explained by the competition between the occupation of the inner 3d shell and outer 4s-4p shell. At standard pressure, the number of outer 4s-4p electrons for elements from Ti to Ni does not exceed three. Our analysis suggests that Cu, which is normally considered as univalent, has states with nearly equal numbers of atoms with 2 and 3 electrons in the outer shell. The 3d shell in metals is completed in Zn, for which the conventional bivalent

$3d^{10}4s^2$ state is corroborated by the analysis of the compressibility data. At HP the number of 4s-4p *OE* in 3d metals increases, while the number of 3d electrons decreases. States with four *OE* were found in Ti-Cu elements under pressure, whereas states with five 4s-4p *OE* were found in Fe, Co, Ni in experiments at ultra-high pressure. Zn at HP has states with three and four *OE*, which shows that the 3d shell in Zn under compression is not completely occupied. Our analysis shows that compression of Ti and Fe leads to the appearance of the states (1-electron states in Ti and 2-electron states in Fe), in which the number of *OE* is smaller than that under standard conditions. In standard conditions, the number of *OE* in the 4d metals from Zr to Rh is either 4 (Zr, Nb, Rh) or in the range from 4 to 5 (Mo, Tc, Ru). At HP, most of them transform to a state with five *OE*. Pd under standard conditions has three *OE*; however, the number of *OE* increases with pressure, first to 4 and then to 5. In Ag, as for Cu, the numbers of atoms with 2 and 3 electrons in the outer shell are nearly equal. The data on compressibility in standard conditions shows that the 4d shell in Cd ($4d^{10}5s^2$ configuration) is fully occupied. However, Cd under compression transforms to states with three and four *OE*. This means that the 4d shell in Cd is not fully occupied. The filling of the 5d shell is similar to that of the 3d and 4d shells.

2.2 Lanthanides

Changes in the electronic and crystalline structures of lanthanides have been observed in a number of experiments at HP. The experimental points on the normal isotherm of Ce have been reported in Vohra et al.[3] The experimental behaviour of Ce in the normal isotherm at P > 10 GPa corresponds to a phase of B_o = 70 GPa and ρ_o = 9.21 g/cm^3, while that at P > 60 GPa corresponds to another phase of B_o = 250 GPa and ρ_o = 11.93 g/cm^3. These transitions can be considered as electronic phase transitions, because they are accompanied by dramatic enhancement (by factors of more than 2 and 5, respectively) of the energy of the OE. An unusual γ-α transition, which is accompanied by +14.5 % jump in the specific volume with no changes in the *fcc* lattice, has been observed in Ce at P=0.67 GPa. The enhancement in the density of the α-phase is often explained by the valence increase, due to the transfer of 4f electrons to the conduction band. On the other hand, our calculations show that the energies of the atomic cells for the γ and α phases obtained from the compressibility data are quite close, and the energy of electrons in the α-phase is even a bit lower than that in the γ-phase. This means that the numbers of electrons in α-Ce and γ-Ce are equal; however, the OE in α-Ce and γ-Ce are in different quantum states. Recent measurements of the crystalline structures in Nd,[4] Pr,[5] and Sm[5] show that at HP the low-symmetry (i.e. monoclinic and orthorhombic) structures are dominant. In the region of the highest accessible pressures, HP phases of these elements have an equilibrium density, which is higher by a factor of ~ 2 than the parent phase density, a high bulk modulus (320-430 GPa) and a high energy (160-200 eV) of the atomic cell, corresponding to 5 or 6 electrons in the outer shell. The volume per atom in HP phases of Ce, Pr, Nd and Sm is close to that in HP phases of Th, U and Am.

2.3 Actinides

There are experimental data for compressibility of Th,[7] Pa,[8] U[9] and Am.[10] As may be seen from theoretical analysis, there are phases with a close bulk modulus, B_0=340-400 GPa, in Th, U and Am at P>50-150 GPa. Lindbaum et al[10] reported experimental data on Am compression under HP, up to 100 GPa. The transition from phase AmI to AmII and AmIII leads to a significant change in the slope of the $P(\rho)$ curve, which corresponds to a severe decrease in the bulk modulus (by a factor of ~2) and increase in equilibrium volume.

Phases AmII and AmIII have different crystalline structures; however, the $P(\rho)$ dependency for both phases can be described with the single curve, corresponding to a bulk modulus, B_0, of 13.0 GPa and equilibrium density, ρ_n, of 11.8 g/cm^3. At $P > 150$ GPa, Th experiences a transition to a phase with an equilibrium density of 19.1 g/cm^3, bulk modulus of 400 GPa, and OE energy of 227 eV per atom. This OE energy corresponds to 6 electrons in the outer shell, and means that completed shells are involved in the rearrangement of electrons. Compressibility of Pa at HP up to 129 GPa has been reported in Haire et al.[8] Our analysis suggests that at P<95 GPa, the experimental points correspond to a phase with ρ_0 = 15.37 g/cm^3 and B_0 = 115 GPa. At HP, the experimental points deviate significantly from the theoretical curve and they correspond to another electronic phase of Pa with ρ_0 = 19.827 g/cm^3 and B_0 = 400 GPa. It has been suggested[8] that at P=77 GPa the tetragonal structure of Pa gets transformed to the low-symmetry orthorhombic structure of α-U. There is no noticeable change in the slope of $P(\rho)$ curve at 77 GPa, which shows that the initial electronic structure of Pa remains unchanged. An abrupt change in the slope of $P(\rho)$ curve and electronic structure occurs at P= 95 GPa, and this transition is not accompanied by changes in the orthorhombic crystalline structure of α-U. The structure of uranium has been studied at pressures up to 100 GPa in Le Bihan et al.[9] The experimental P(ρ) data agree with the theoretical curves for the normal isotherm of U obtained from shock-wave measurements.[11] The original electronic structure of U changes first at P=13 GPa and then at $P \approx 80$ GPa. A phase transition at $P \approx 80$ GPa may be seen clearly from the shock-wave data. The crystalline structure of U[9] at P<100 GPa remains unchanged. Metal behaviours under pressure are analyzed in Nadykto and Nadykto.[12]

3 CONCLUSION

This study suggests that the experimental data on the compressibility of materials is an essential source of information on electronic structure. We found that the energies of different phases in simple solids are surprisingly close to the sum of the ionization potentials of free atoms. This suggests that quantum states typical of the free atoms may also be found in solids.

Acknowledgements
The work was supported by the ISTC (Project #1662).

References

1 B.A Nadykto, *Physics-Uspekhi*, 1993, **36**, 794.
2 B.A Nadykto, *VANT. Ser. Teor. i Prikl. Fizika*, **3**, 1996, 58.
3 Y.K.Vohra, S.L.Beaver, J. Akella et al., *J. Appl. Phys*., 1999, **85**, 2451.
4 G. N. Chesnut and Y. K Vohra., *Phys.Rev B*., 2000, **61**, 3768.
5 G. N. Chesnut and Y. K.Vohra, *Phys.Rev. B*., 2000, **62**, 2965.
6 G. N. Chesnut and Y. K. Vohra, *Proceedings of International Conference on High Pressure Science and Tecnology (AIRAPT-17)*, 1999, eds., Manghnani M.H., Nellis W.J., Nicol M.F, pp. 483-486.
7 Y.K. Vohra and J.Akella, *Phys. Rev. Lett*., 1991, **67**, 3563.
8 R.G.Haire, S. Heathman, M. Idiri at al., *Phys. Rev.B*, 2003, **67**, 134101.
9 T. Le Bihan., S. Heathman, M. Idiri et al., *Phys. Rev. B*, 2003, **67**, 134102.
10 A. Lindbaum, S. Heathman., K. Litfin at al., *Phys. Rev. B*, 2001, **63**, 214101.
11 S. P. Marsh, *LASL Shock Hugoniot Data,* 1980, University of California, Berkley.
12 B.A Nadykto and O.B. Nadykto, In *Plutonium Futures – Science 2003*. Conference transactions. Edit. by G.D. Jarvinen. P. 184-186.

POSITRON ANNIHILATION SPECTROSCOPY: A POWERFUL TOOL TO INVESTIGATE THE EARLY STAGES OF SELF-IRRADIATION EFFECTS IN δ-Pu ALLOYS

B. Oudot[1], L. Jolly[1], N. Baclet[1], B. Ravat[1], P. Julia[1] and R. I. Grynszpan[2]

[1]CEA-Centre de Valduc, 21120 Is-sur-Tille, France.
[2]Dept. Lasers, Optics and Thermo-Optics, DGA/CEP, Dept. of Defence, Arcueil, SERAM, Ecole Nat. Sup. Arts et Métiers, Paris & LCMTR, CNRS-ISCA, Thiais, France.

1 INTRODUCTION

Plutonium remains the most peculiar element of the periodic table. It lies at the boundary between the light actinides (with delocalised 5*f* electrons) and the heavy actinides (with localised 5*f* electrons). This induces special behaviours, such as, for instance, the occurrence of six allotropic phase transitions at ambient pressure associated with large volume changes. At room temperature, plutonium crystallises in the monoclinic phase (α), but can be stabilised in the fcc phase (δ) by adding "deltagen" elements such as aluminium, gallium, americium or cerium. Plutonium decay leads to the creation and recoil of uranium and helium ions, which produce defects through displacement cascades. These self-irradiation defects tend to change plutonium properties. As already observed,[1-3] radiation damage induces swelling, which may depend on the deltagen doping element, or on its segregation within the δ-stabilised alloy. Since possible differences may appear between macroscopic and microscopic swelling, as revealed by dilatometry and X-ray Diffraction (XRD), respectively, it is very important to study the basic mechanisms of radiation damage for a better understanding of their effects on pure and alloyed plutonium. Positron annihilation spectroscopy (PAS) is an appropriate method to tackle this problem, owing to its high sensitivity to vacancy-type defects in solids, and to the possibility of remotely acquiring data. The goal of this work was to monitor, after melting, the early stages of formation of defects in a δ-stabilised PuGa alloy, and to correlate the results obtained by dilatometry, XRD and PAS.

2 SWELLING MEASUREMENTS

Both types of swelling were investigated versus time, which is expressed hereafter in units of displacements per atom (dpa). The first type, macroscopic swelling ($\Delta L/L_{t=0}$), was measured by optical fibre Bragg grating sensor.[4] The time dependence of the sample elongation was followed under isothermal conditions. In order to be able to observe a significant change in swelling through the Bragg grating sensor, the fibre has to be carefully stuck in a groove of the sample.[4,5] The sample equipped with the optical fibre is then introduced to a furnace. This dilatometer is in a glove box.

The second type of swelling (microscopic) was observed by XRD, through the measurement of the change in variation of the lattice parameter versus time ($\Delta a/a_{t=0}$). Since the diffractometer is out of the glove box, the sample is confined between two sealed polyamide sheets.

3 THE POSITRON ANNIHILATION SPECTROSCOPY TECHNIQUE

Depending on the sample depth to be probed, numerous techniques are available to study point defects in materials. However, due to radiotoxicity, rapid oxidation or nuclear stability of ^{239}Pu, some techniques had to be dismissed, including: i) Transmission Electron Microscopy, requiring very thin samples (sensitive to oxidation), which precludes confinement; ii) Small Angle Neutron Scattering, only applicable to ^{242}Pu isotope, which is less fissile, but not readily available; and iii) Small Angle X-Ray Scattering, as the defect contribution in the confinement can not easily be separated from that of the Pu sample.

In contrast, PAS allows remote detection (i.e., outside a glove box) of gamma photons resulting from positron annihilation, and hence the use of an easy sample confinement. The basic principles of the related techniques have already been described long ago, for instance by West.[6] A positron implanted in the material is sensitive to the electronic density, and hence to the presence of vacancy-type defects, which trap the thermalised positive probe and strongly affect its lifetime. In this study, we mainly used the positron lifetime technique, which, in principle, can associate a specific lifetime component (τ_i) with a specific defect (trap) size or type i, with a contribution (I_i) proportional to the defect concentration. For several types of traps, the mean lifetime can therefore be defined as :

$$\tau_{mean} = \Sigma \tau_i \, I_i \qquad (1)$$

A standard "sandwich" configuration consisting of a ^{22}Na positron source inserted between two plutonium samples was confined in an aluminium box, which in turn was enclosed in a plexiglas box. These operations were done in a glove box. The sample holder was then sealed between two polymer sheets, ready for measurements on the lifetime spectrometer.

4 RESULTS AND DISCUSSION

Only the mean lifetime is presented in Figure 1a, since analyses of lifetime spectra could not allow a reliable deconvolution into several possible components. We may assume that displacement cascades yield similar size distributions of defects, as long as no overlap occurs. While enough vacancy type defects are produced (even at low dpa values) to ensure that all positrons are trapped, those with the largest trapping efficiency will provide the dominant contribution to the mean lifetime.

The increase of τ_{mean} may therefore be associated with an increase in the concentration of vacancy clusters, rather than with an increase in the defect average size, according to the increase in macroscopic and, especially, in microscopic swelling, observed by dilatometry and XRD, respectively (Figure 1b). Indeed, an increase in defect size tends to deform the lattice, resulting in X-ray line broadening, whereas an increase in concentration leads to an increase in lattice parameters (line shift).

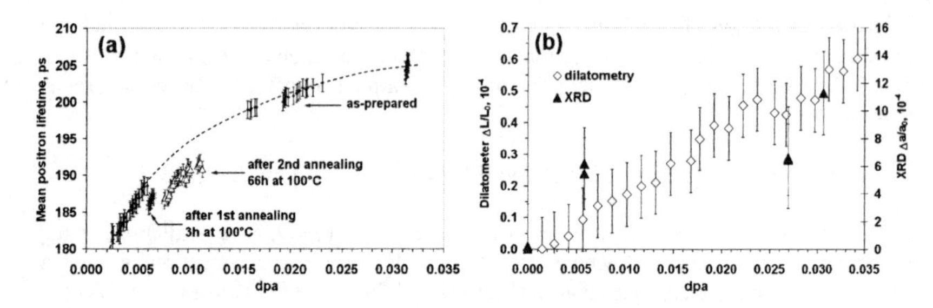

Figure 1 *Evolution versus dpa (time), for a δ-PuGa alloy:*
a) mean positron lifetime of two samples, respectively as-prepared (dashed line: guide for the eye), and annealed;
b) macroscopic(dilatometry) and microscopic(XRD) swelling.

The effect of thermal treatment on mean positron lifetime was also studied. After a treatment at 100°C for 3 hours, the positron lifetime is reduced but does not reach its initial value. This implies that defect annealing is not complete, as confirmed by the result of an additional annealing (66 hours at 100°C - Figure 1a). We assume that such an annealing temperature ($\approx 0.4\ T_{melting}$) is high enough to affect the smallest defects produced by self-irradiation. Hence, the mean lifetime value measured after each heat treatment most likely corresponds to large voids resulting from the fraction of defects produced at room temperature, which either withstood annealing or clustered into more thermally stable open-volumes.

In this investigation, using the positron lifetime technique, we monitored structural changes versus time in a δ-stabilised PuGa alloy. The time dependence unveiled by PAS is corroborated by XRD and dilatometry results, which show that swelling appears very rapidly.

References

1 J. Jacquemin and R. Lallement, *Plutonium 1970 and other actinides*, edited by W.N. Miner, pp. 616-622.
2 J. A. C. Marple, A. Hough, M. J. Mortimer, A. Smith and J. A. Lee, *Plutonium 1970 and other actinides*, edited by W.N. Miner, pp. 623-634.
3 B. Oudot, PhD Thesis, University of Franche-Comté, 2005.
4 V. Dewynter-Marty, P. Julia, A. Jacob, S. Rougeault, J. Boussoir and P. Ferdinand, *16th International Conference on Optical Fiber Sensors OFS 16*, October 2003, Nara Japan, pp. 236-239.
5 P. Julia, *Plutonium Futures – The Science*, Jarvinen Am. Instit. Phys. edn., 2003, pp. 109-110.
6 R.N. West, in *Positrons in Solids*, P. Hautijärvi, Springer-Verlag edn., 1979, p. 89.

HIGH PRESSURE X-RAY DIFFRACTION STUDY OF A $Pu_{0.92}Am_{0.08}$ BINARY ALLOY

V. Klosek, P. Faure, C. Genestier and N. Baclet

CEA – Centre de Valduc, F-21120 Is-sur-Tille, France

1 INTRODUCTION

Plutonium lies at the borderline between the light actinides (Pa-Np) with itinerant *5f* electrons, and the heavy actinides (Am-Cf) with localised *5f* states. Due to this particular position, Pu exhibits six allotropic solid phases depending upon temperature. The face centered cubic δ phase, stable between 315 and 457 °C, is by far the most studied, especially because of its intermediate behaviour between delocalisation and localisation of the *5f* electrons, which still remains a great challenge for physicists to describe. Alloying Pu with gallium, aluminium, cerium or americium allows the stabilisation of this δ phase at room temperature. The existence of the δ-stabilised $Pu_{1-x}Am_x$ solid solution for an Am content ranging from 5 to 75 at. % is an exciting opportunity for investigating localisation of *5f* electrons between Pu and Am, and especially the effect of high pressures. Whereas pure Pu is known to exhibit one single transition at about 40 GPa from the monoclinic α phase to a phase reported to be either hexagonal[1] or orthorhombic[2,3], pure Am exhibits four phase transitions when applying pressure up to 100 GPa, traducing the progressive delocalisation of its *5f* orbitals.[4] No experimental data are reported concerning the pressure phase diagram of the Pu - Am system that only consists, up to now, of extrapolations from pure Pu and Am data.

In the present paper we report the first known experimental determination of phase transitions under pressure (up to 20 GPa) of a $Pu_{0.92}Am_{0.08}$ binary alloy.

2 EXPERIMENTAL PROCEDURE

High pressure X-ray diffraction (XRD) experiments were carried out on a polycrystalline foil of $Pu_{0.92}Am_{0.08}$ loaded in a large aperture (80 degrees) diamond anvil cell (DAC) equipped with a boron seat and a gas membrane. We used 520 μm culet diamonds and a 290 μm thickness rhenium gasket preindented to 90 μm and drilled with a hole of 270 μm diameter. A piece of $Pu_{0.92}Am_{0.08}$ alloy was first rolled at room temperature then annealed (3 hours at 450°C) to restore the metallic structure (elimination of both structural defects caused by self-irradiation and stress induced by rolling). Then a micro-sized sample (about 100*100 μm²) was cut from the resulting foil and put in the gasket hole together with four little ruby spheres ($Al_2O_3:Cr^{3+}$) used as pressure gauges. A pressure transmitting medium, argon, was loaded at room temperature using a high pressure gas-loading system.

XRD experiments were carried out on a Mo rotating anode, Nonius FR591 (λ Kα = 0.7102 Å), in transmission geometry and angular dispersive mode. Pressure was increased *in situ* by tuning the gaseous pressure inside the membrane of the DAC. Sample pressure was measured before and after each diffraction measurement using a PRL (Pressure Ruby Luminescence) system. XRD exposure times ranged from ~ 4 to ~ 17 hours to ensure good statistics. The sample was oscillating by ± 7° around the two rotation axes of the goniometer perpendicular to the incoming X-ray beam. The diffraction patterns, recorded on an image plate, were integrated with the FIT2D software.[5] The data were then analysed by the Rietveld method using the FULLPROF program.[6]

3 RESULTS AND DISCUSSION

When increasing pressure, the δ-Pu fcc form (S.G. *Fm3m*) first converts at P = 0.7 GPa to a face-centred orthorhombic structure (S.G. *Fddd*), similar to γ-Pu and Am III structures, and which consists of a slightly distorted hexagonal close-packed arrangement (Fig. 1). The corresponding phase will be referred to in the following as the γ' phase, the prime standing for *expanded*, since substitional Am atoms are trapped in the structure. Traces of a similar expanded γ' phase were already found to appear under pressure in Pu-Ga alloys with low Ga contents (below 2 at. %) during the δ → α' transformation,[7,8] but these were reported to disappear very rapidly at room temperature.

The stability domain of the γ' phase appears to be very narrow. Pure γ'-$Pu_{0.92}Am_{0.08}$ was not observed, and the orthorhombic structure rapidly transformed to a monoclinic form (S.G. *P2₁/m*), thus adopting the same structure as the one known for pure α-Pu (Fig. 1). This phase will be referred to in the following as the expanded α' phase. Thus, between P = 0.7 GPa and P = 1.3 GPa, two or three phases coexist in the sample (Fig. 2). The phase transformation is complete above P = 1.3 GPa, where the monoclinic α' phase is the only observed phase. Moreover, after complete downloading (P = 0), the structural sequence could not be reversed and the alloy remained single-phase α'-$Pu_{0.92}Am_{0.08}$.

The structural transition δ → γ' involves a contraction of atomic volume of about 7.5 %, whereas the γ' → α' transition is accompanied by a contraction of about 11 % (Fig. 3). *5f* electrons in γ-Pu are generally considered as mainly involved in bonding, although no theoretical work has really been aimed at precisely determining to what degree. In α-Pu, *5f* states are known to be fully itinerant, and the electronic structure is dominated by narrow *5f* bands with a large density of states, responsible of the stabilisation of the low symmetry structure.[9] Conversely, the *5f* electrons in δ-Pu are neither fully itinerant nor localised (some recent theoretical models had to consider that about 20 % of *5f* electrons should be itinerant in δ-Pu[10]). Insertion in α-Pu of americium atoms, whose *5f* electrons are fully localised, leads to an increase in Pu - Pu interatomic distances, and hence to a decrease in overlap between Pu *5f* states that thus become more localised. This

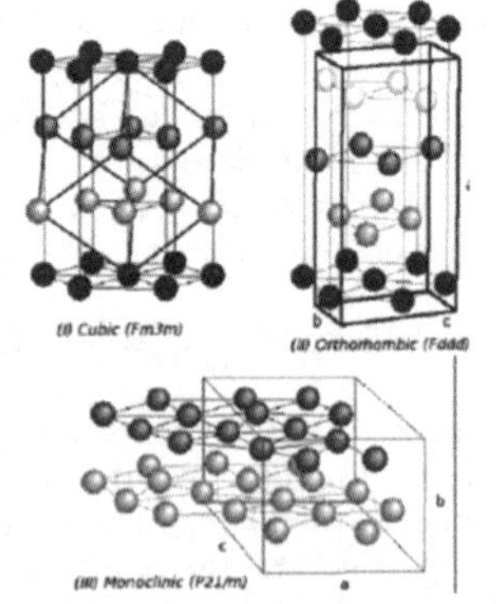

Figure 1 *3D views of the three structures observed for $Pu_{0.92}Am_{0.08}$ under pressure.*

process results in the stabilisation of the fcc δ-Pu structure at room temperature. Application of pressure on this δ phase stabilised Pu-Am alloy decreases the interatomic distances, and thus tends to force the *5f* states to overlap and the *5f* electrons to delocalise. It clearly appears from the present study that this delocalisation occurs in two steps. The first step corresponds to a sudden partial increase in the involvement of *5f* states in bonding, leading to the stabilisation of the γ' phase. In the second step, the *5f* electrons become finally fully itinerant, and the α' phase becomes stable. Electronic structure and total energy calculations are planned to shed more light on the phase stability under pressure in our $Pu_{0.92}Am_{0.08}$ alloy.

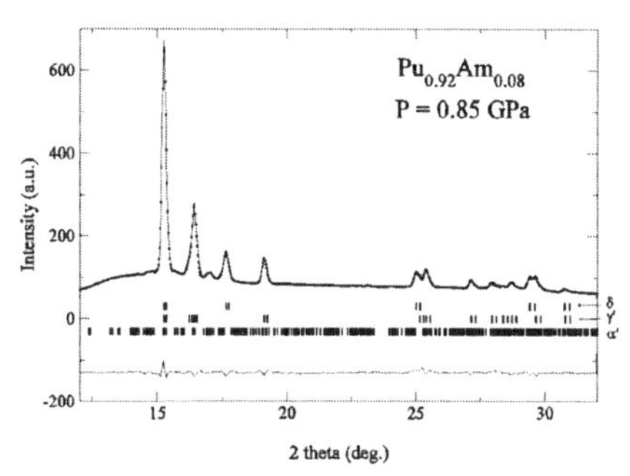

Figure 2 *Rietveld refinement of the diffraction pattern recorded at P = 0.85 GPa. Lattice parameters (Å): δ-phase a = 4.6332(4); γ'-phase a = 3.1701(2), b = 5.7224(5), c = 10.065(2); α'-phase a = 6.197(6), b = 4.826(3), c = 10.978(12), β = 102.6(1)°.*

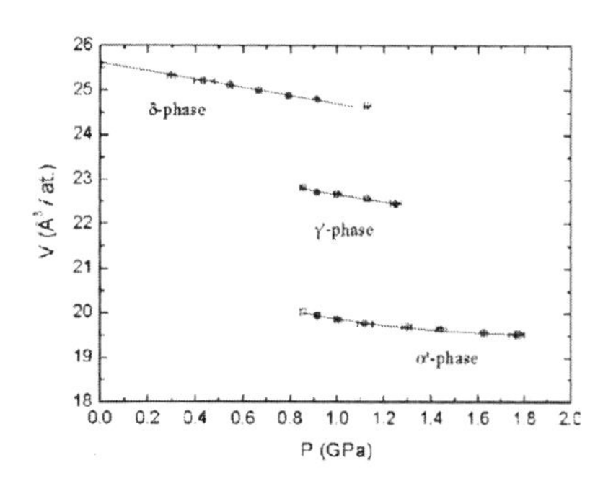

Figure 3 *Atomic volume vs. pressure for $Pu_{0.92}Am_{0.08}$ up to P = 2 GPa*

References

1. S. Dabos-Seignon, J.P. Dancausse, E. Gering, S. Heathman and U. Benedict, *J. Alloys Compd.,* 1993, **190**, 237.
2. M. Pénicaud, *J. Phys.: Condens. Matter,* 2002, **14,** 3575.
3. S.K. Sikka, *Solid State Comm.,* 2005, **133**, 169.
4. A. Lindbaum, S. Heathman, K. Litfin, Y. Mresse, R.G. Haire, T. Le Bihan and H. Libotte, *Phys. Rev. B,* 2001, **63,** 214101.
5. A.P. Hammersley, S.O. Svensson, M. Hanfland, A.N. Fitch and D. Hausermann, *High Pressure Research,* 1996, **14,** 235.
6. J. Rodriguez-Carvajal, *Physica B,* 1993, **192**, 55.
7. S.S. Hecker, D.R. Harbur and T.G. Zocco, *Prog. In Mat. Sci.*, 2004, **49,** 429 & refs therein.
8. P. Faure, private comm.
9. P. Soderlind, J.M. Wills, B. Johansson and O. Eriksson, *Phys. Rev. B,* 1997, **55**, 1997.
10. O. Eriksson, J.D. Becker, A.V. Balatsky and J.M. Wills, *J. Alloys Compd.,* 1999, **287**, 1.

CALCULATION OF THE CRITICAL CONCENTRATION OF HELIUM RELEASE FROM URANIUM TRITIDE

B.Y. Ao and X.L. Wang

China Academy of Engineering Physics, P.O. Box 919-71, Mianyang 621900, Sichuan, People's Republic of China.

1 INTRODUCTION

With the development of fusion research, a safe and simple method of tritium storage must be developed. At present, the method using metal hydride is considered as the most suitable, because such hydrides have many favourable characteristics, such as low decomposition pressure at room temperature, large hydrogen capacity, high pumping speed of hydrogen without or with an easy activating process, *etc*. Among these characteristics, low decomposition pressure is the most important one for safety. This is the reason why materials that form stable hydrides, such as zirconium, titanium and uranium, are considered to be suitable. Especially for the temporary storage, uranium is regarded as the most promising material and, in fact, is in use in many tritium handling systems.[1,2]

However, like all other metal hydrides used for tritium processing, uranium tritide accumulates helium (^{3}He), produced by the β decay of tritium in the metal tritide lattice. The decayed helium stays in the solid and does not release until a critical helium concentration has been reached. The presence of helium in the matrix induces structural changes, which in turn considerably influence the absorption-desorption characteristics of the metal tritide. Therefore, we have been investigating aging effects in uranium tritide. We report in this paper the first results obtained from theoretical calculations for the critical concentration of helium release from uranium tritide.

2 METHOD AND MODEL

Although helium is inert and extremely insoluble in metals, the decayed helium stays in the metal tritide, and does not release until a critical helium concentration has been reached. The critical helium-to-metal atomic ratio, η_c, appears to determine uniquely when accelerated release begins, independent of the initial amount of tritium.

As pointed out by Spulak,[3] this critical behaviour is relevant to percolation, where the percolation probability that a site randomly occupied on a regular lattice is connected to an infinite set of other randomly occupied sites is zero until a critical site occupancy is reached. The critical occupancy, or concentration, is near 0.3 for three dimensional lattices, and above this concentration, the percolation probability rapidly approaches unity. This similarity has led to a qualitative description, where the percolation of individual helium atoms was proposed to determine when accelerated release occurs.

Helium bubbles have been observed in many metal tritides. Percolation theory can be used to determine the volume of bubbles at the onset of accelerated release. When percolation occurs, the helium has a flow path to the surface and escapes. The volume fraction of the bubbles when the critical site occupancy is achieved is a dimensional invariant. That is, for a regular lattice, the fraction of the total volume occupied when the critical occupation fraction is reached by spheres of a radius, such that adjacent ones touch, centered on lattice sites, does not depend on the type of lattice, but only on the dimensionality of the lattice. For a three-dimensional lattice this critical volume fraction is 0.15±0.01. For a randomly-placed collection of bubbles, this still applies, because the bubbles can be placed on various lattices locally.

The helium-to-metal atomic ratio at the onset of accelerated release is then

$$\eta_c = \frac{fn}{n_m} \tag{1}$$

where n is the number density of helium atoms in the bubbles, n_m is the initial number density of metal atoms in the lattice and f is the volume fraction of the bubbles when accelerated release begins. Generally, for high-early-release metal tritides, $f = 0.15$, and for low-early-release metal tritides, $f = 0.225$.

The pressure, p, in a helium bubble is given by

$$p \approx \frac{2\gamma}{r} \tag{2}$$

where r is the bubble radius and γ is the specific free energy; for a typical value $\gamma \approx 2$ N m^{-1}, $p \approx 4$ GPa. At this high pressure, the ideal gas equation of state is inappropriate. Trinkaus[4] has derived equations of state for helium in small bubbles in metals. Near $r = 1$ nm, $p \propto n^3$. Since $p \propto r^{-1}$, scale n as

$$n = n_0\left(\frac{r_0}{r}\right)^{1/3} \tag{3}$$

where n_0 is the density for radius r_0. The total number density of helium atoms macroscopically in the material from decay is

$$Nn_b = xn_m(1 - e^{-\lambda t}) \tag{4}$$

where N is the number of helium atoms per bubble, n_b is the number density of bubbles, x is the initial concentration of tritium (composition MT_x, M ≡ metal), and the decay constant for tritium is $\lambda = 5.62\times10^{-2}$ year^{-1}. Then the number density of helium atoms in a bubble is

$$n = \frac{3N}{4\pi r^3} \text{ or } n = n_0\left\{\frac{4\pi r_0^3 n_0 n_b}{3xn_m(1 - e^{-\lambda t})}\right\}^{1/8} \tag{5}$$

Note that n is weakly dependent on all variables, except the equation-of-state scaling parameters. Even that dependence is not strong; for a given n_0 different materials will scale to a different r_0 but $n \propto r_0^{3/8}$. At the onset of accelerated release, $f = n_b 4\pi r^3/3$. Then from eqn. (3),

$$n = n_0\left(\frac{4\pi r_0^3 n_b}{3f}\right)^{1/9} \tag{6}$$

where f is used implicitly to eliminate the explicit time dependence in eqn. (5)

Equations (5) or (6) can be used with eqn. (1) to predict the critical helium-to-metal atomic ratio, η_c. Using data from different studies on helium bubbles in metal tritide,

Spulak[3] obtained calculated values for η_c in good agreement with experimental values measured for Er, Sc, Ti and Zr tritides.

3 CALCULATION

This qualitative model was used to predict the critical helium-to-metal atomic ratio for uranium tritide, η_c, at the onset of accelerated release for comparison with experimental observation. Some experimental study of helium release from uranium tritide has shown that the early helium release rate is relatively fast. The release fraction is about 10% during about the first 200 days, after which the release increases abruptly. So f in eqn. (1) is chosen as 0.15. It takes about 3 years for the helium concentration in uranium titide to attain the critical helium-to-metal atomic ratio. So, t in eqn. (5) is chosen as 3 years. According to Bowman's NMR observations,[5] the radius of helium bubbles in uranium tritide ranged from 1 nm to 100 nm. So, r_0 in eqn.(5) and eqn. (6) is chosen as 1 nm. For most metal tritides, the typical values of n_0 and n_b are chosen as 1.0×10^{29} m^{-3} and 2.6×10^{24} m^{-3}, respectively. The density of stoichiometric uranium tritide UT_3 is 11.2 $g.cm^{-3}$ from which n_m is calculated as 2.84×10^{28} m^{-3}. Using the percolation model and these data, the calculated value of critical helium-to-metal atomic ratio is 0.38 if the aging time of the uranium tritide is considered, or the calculated value is 0.40 if eliminating the explicit time dependence. The two calculated values are in reasonably good agreement with the generally accepted experimental values in the literature (0.38 ~ 0.42).[6]

4 CONCLUSION

Using percolation theory, we have calculated the critical helium-to-metal atomic ratio for uranium tritide. The calculated results show that the critical values are in good agreement with experimental values. We hope that this research plays an important role in understanding the critical behaviour of helium in metal tritides. However, this calculation cannot be used to obtain a deep understanding of helium effects in metal tritide. In order to do this, an atomic level computer simulation method and experiments are in progress in our laboratory.

Acknowledgements

This research was supported by the Science and Technology Foundation of China Academy of Engineering Physics under contract No. 20040546.

References

1 T. Yamamoto, Supardjo, T. Terai, F. Ono, S. Tanaka and M. Yamawaki, *Fusion Technol.*, 1988, **14**, 764.
2 L.K. Heung, *Fusion Sci. Technol.*, 2002, **41**, 753.
3 R.G. Spulak, *J. Less-Common Met.*, 1987, **132**, L17.
4 H. Trinkaus, *Radiat. Eff.*, 1983, **78**, 189.
5 R.C. Bowman and A. Attalla, *Phys. Rev.*, 1977, **16**, 1828.
6 R. Lässer, *Tritium and Helium-3 in Metals*, Springer-Verlag, Berlin, 1989, Chapter 6, p. 112.

RECYCLING OF CALCIUM IN THE INCINERATION OF LOW LEVEL RADIOACTIVE WASTES

K. Sawada[1], M.Suzuki[2], Y. Enokida[1] and I.Yamamoto[2]

[1]EcoTopia Science Institute.
[2]Department of Materials, Physics and Energy Engineering; Nagoya University, Furo-cho, Chikusa-ku, Nagoya 464-8603.

1 INTRODUCTION

Low-level radioactive wastes (LLW), such as contaminated paper, clothes, and gloves, discharged from nuclear facilities should be reduced in volume, because repository space is limited. Incineration is one method, which enables a marked reduction in the volume of burnable wastes. When the wastes are incinerated, chloride and sulphur are released from polybynylchloride and rubber, and changed into acidic gases such as HCl and SO_2. These chemicals are harmful to human beings and the environment; especially HCl, which causes not only corrosion of the facilities, but also formation of dioxins with other thermal decomposition products. Calcium hydroxide may be injected into the flue to absorb the acid gases.[1,2] Calcium hydroxide reacts with HCl and SO_2, and converts them into $CaCl_2$ and $CaSO_4$, which prevents corrosion of the facilities. The injection of calcium hydroxide, however, increases the amount of fly ash, because a large excess amount of calcium hydroxide is necessary due to the gas-solid reaction.[3] If calcium in the fly ash were recycled as an acid gas sorbent, it would lead to a promising incineration process for LLW treatment. To recycle calcium in fly ash, we propose a new method of treatment by oxalic acid. It is known that calcium compounds convert into insoluble calcium oxalate with oxalic acid.[4] Calcium oxalate will thermally convert into calcium oxide which may be used as an acid gas sorbent. Moreover, if the fly ash contains actinide compounds, this method might contribute to the removal of actinides from fly ash, since the solubility of actinide oxalates in water is controllable by pH and or Eh. In this work, we have showed the possibility of the recycling of calcium in fly ash, by converting calcium compounds to the oxide.

2 EXPERIMENTAL

2.1 Materials

Two samples of fly ash were employed. They were real samples discharged from the melting process of municipal solid wastes in Japan, which simulates fly ash from LLW incineration very well, except that no actinides are present. Table 1 shows the elemental contents of the fly ash samples. Aqueous solutions of oxalic acid whose concentrations were 0 (i.e. water), 0.5 and 1 mol dm^{-3} were prepared by diluting oxalic acid (Wako Pure

Chemical Ind., Ltd.) with pure water. The chemicals used in this work were of special grade.

2.2 Procedure

0.25 g of fly ash and 10 cm^3 of oxalic acid solution were mixed in a test-tube, and then shaken for 3 h. This shaking time (3 h) was enough for the fly ash to react with the oxalic acid. The residue was collected with a filter (mean pore diameter was 0.2 μm), and the effluent was used for chemical analyses. After decomposing oxalic acid in the filtrate by a microwave digester (MARS5 XP1500Plus, CEM Co., USA), the concentrations of elements, such as calcium, silicon and lead, were measured by ICP-AES (ICPS-7000, Shimadzu, Japan). The filter cake was dried, and analyzed by thermogravimetry (TG, TGA-50, Shimadzu, Japan) under the following conditions: heating rate, 5 K min^{-1}; gas flow rate, 80 cm^3 min^{-1} for N_2 and 20 cm^3 min^{-1} for O_2. The cakes, heated in a muffle furnace (KDF-S80, Denken Co., Ltd., Japan) under the atmospheric pressure at 823 and 1173 K, were analysed by X-ray diffraction (XRD, RINT-2500TTR, Japan).

Table 1 Elemental contents in fly ash

Sample	Elemental contents [mg g^{-1}]							
	Ca	K	Si	Al	Cu	Zn	Pb	Fe
Sample 1	218	56	73	21	2.0	24	5.0	8.9
Sample 2	280	-	7.5	8.8	12.6	80.1	48.3	2.0

3 RESULTS AND DISCUSSION

As shown in Table 2, when the samples were contacted with pure water, the concentration of calcium in the filtrate reached approximately 1.9-2.5 g dm^{-3}, i.e. 40% of calcium in the samples was leached into the liquid phase. By mixing with 0.5 and 1 mol dm^{-3} oxalic acid solution, the concentration of calcium in the filtrate decreased to 38-51 mg dm^{-3}. Less than 1% of the total calcium was retained by 0.5 or 1 mol dm^{-3} oxalic acid solutions. These results indicate that calcium can convert into insoluble compounds. Lead showed similar behaviour to calcium. On the other hand, aluminium, silicon, zinc and iron were leached into the oxalic acid solutions. This means that calcium and lead can be enriched in the precipitate, compared with to the other elements. As they are similar to some of those elements, actinides could be fractionated into the insoluble solid or aqueous solution. The TG analysis for the cake from Sample 1 treated with 1 mol dm^{-3} oxalic acid showed that decomposition of the cake proceeded through three steps. The decomposition temperature of the cake was almost same as for calcium oxalate hydrate reagent. The major component was considered to be calcium oxalate hydrate. From the weight change of the reagent, calcium oxalate hydrate was converted into calcium oxalate at 450-700 K, calcium carbonate at 750-850 K, and calcium oxide at temperatures higher than 970 K. In the XRD analysis of the dried

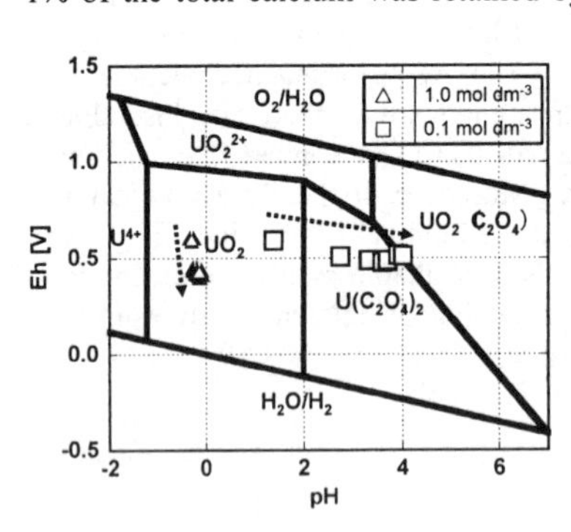

Figure 1 *Eh-pH diagram for a system of UO_2 and oxalic acid*

cake, the peaks of calcium oxalate hydrate were observed; this indicates that calcium in fly ash can convert to calcium oxalate with oxalic acid. The analysis also shows that calcium oxalate decomposes to calcium carbonate by heating at 823 K. Although XRD analysis was attempted for the sample heated at 1073 K, we observed no peak for calcium carbonate or calcium oxide. It could be that calcium carbonate decomposed into calcium oxide at 1073 K, which then changed into the other compounds of low crystallinity by reaction with moisture in the air. Although lead may co-exist in the cake, it will evaporate at temperatures higher than 1273 K; lead, from shielding materials, waste gloves *etc.* in LLW, could be removed from the recycled acid gas sorbent.

Figure 1 shows Eh-pH diagram for a system of UO_2 and oxalic acid based on reported data in the literature.[5, 6] The two lines in the figure are loci of Eh-pH after 3 h for the leaching experiment of the Sample 1. This implies that actinide behaviour can be controlled when Eh and pH are selected properly by adjusting the initial concentration of oxalic acid and or solid-liquid ratio. HERE

Table 2 *Effect of the oxalic acid concentration on the solubility of the elements contained in fly ash*

Sample	Oxalic acid concentration [mol dm^{-3}]	Concentration [mg dm^{-3}]							
		Ca	K	Si	Al	Cu	Zn	Pb	Fe
Sample 1	0	1910	790	N.D.	2	2	14	54	N.D.
	0.5	38	740	510	370	N.D.	33	2	110
	1	51	560	710	460	17	36	N.D.	150
Sample 2	0	2500	560	N.D.	21	18	N.D.	420	1
	0.5	50	900	170	95	43	140	N.D.	19
	1	48	520	170	100	44	56	4	24

N.D. : < 0.1 mg dm^{-3}

4 CONCLUSIONS

Calcium compounds in fly ash were converted into calcium oxalate with oxalic acid, and precipitated. The calcium oxalate was successfully converted into calcium carbonate at 773 K. Although confirmation of formation of calcium oxide was not obtained by XRD analysis, the cake was expected to convert into calcium hydroxide by heating at 1073 K. From these results, it seems feasible to recycle calcium compounds in fly ash as acid gas sorbents in the incineration process for LLW to contribute to a reduction in the final disposal volume of LLW. Actinide behaviour contacted with oxalic acid can be controlled with by adjusting the initial concentration of oxalic acid and or solid-liquid ratio.

References

1 F. Bodenan and Ph. Deniard, *Chemosphere*, 2003, **51**, 335.
2 W. Duo, N. F. Kirkby, J. P. K. Seville and R. Clift, *Chem. Eng. Sci.*, 1995, **50**, 2017.
3 T. Tanosaki, M. Matsumoto, K. Nozaki, K. Izumi, K. Nakamura, M. Nambu and T. Maruta, *Inorganic Materials*, 1998, **5,** 149.
4 U. Palaniswamy, B. B. Bible and R. J. McAvoy, *Scientia Horticulturae*, 2004, **102**, 267.
5 I. Casas, J. Pablo, J. Giménez, M. E. Torrero, J. Bruno, E. Cera, R. J. Finch and R. C. Ewing, *Geochemica et Cosmochimica Acta*, 1998, **62,** 2223.
6 M. Nakahara, *Mukikagoubutu Sakutai Jiten, Maruzen*, 1997 (in Japanese)

VAPORIZATION BEHAVIOUR OF Pu-Cd INTERMETALLIC COMPOUNDS

K. Nakajima, Y. Nakazono and Y. Arai

Department of Nuclear Energy system, Japan Atomic Energy Research Institute, Oarai-machi, Higashiibaraki-gun, Ibaraki-ken, 311-1394, Japan.

1 INTRODUCTION

A pyrochemical process for the metallurgical treatment of spent nuclear fuels, such as metallic and nitride fuels, has been developed for advanced reprocessing in the nuclear fuel cycle.[1,2] This process contains the Cd-distillation step to recover the plutonium as a residue. Therefore, the vaporization behaviour of the Pu-Cd system is a key issue to be addressed for understanding this Cd-distillation behaviour. In a previous paper[3] the authors determined the vapour pressures of Cd(g) over $PuCd_2$+Pu and evaluated the standard Gibbs energy of formation of $PuCd_2$. In the present study, Knudsen-cell mass-spectrometric measurements of $PuCd_6$+$PuCd_4$ and $PuCd_4$+$PuCd_2$ were carried out to determine the vapour pressures of Cd(g) over these samples and to obtain the thermodynamic quantities of $PuCd_4$ and $PuCd_2$ from these vapour pressures.

2 EXPERIMENTAL

The diphasic $PuCd_2$+$PuCd_4$ and $PuCd_4$+$PuCd_6$ samples are prepared from pure Cd and plutonium hydride powders. These powders are obtained from plutonium metal by heating at about 600K in an Ar-8%H_2 mixed gas stream. The Pu metal is obtained from Pu_2C_3, by tantalothermic reduction and selective evaporation methods.[3] These intermetallic samples are prepared by weighing the filed Cd and Pu hydride, and compacting the mixture into a pellet, and heating in an Ar gas stream in an Y_2O_3 Knudsen cell. The samples are identified by means of powder X-ray diffractometry, using Ni-filtered Cu-Kα radiation. In the X-ray diffraction (XRD) pattern there is an unidentified peak in the vicinity of 35 degrees. The relative intensity of this peak is found to increase with the mass-spectrometric measurement of $PuCd_4$+$PuCd_6$. So, this peak is thought to be the peak corresponding to $PuCd_4$. The XRD pattern also shows Cd metal, $PuCd_6$, and a large amount of PuO_2 and hex-Pu_2O_3 in the $PuCd_2$+$PuCd_4$ sample. A previous paper[3] reported that $PuCd_2$ easily decomposed into PuO_2 and Cd, so the Cd peaks probably appear during the XRD measurement. Further, it is presumed that $PuCd_6$ also appears due to the oxidation of $PuCd_4$ during the XRD measurement. The samples prepared in this study have porous or powder shapes unlike those of Pu+$PuCd_2$, which is prepared not from the Pu hydride, but Pu metal.[3] The $PuCd_2$+$PuCd_4$ sample could have a large amount of plutonium oxides, due to its larger surface area.

The Knudsen-cell mass-spectrometric measurements of diphasic $PuCd_2$+$PuCd_4$ and $PuCd_4$+$PuCd_6$ are carried out in the temperature range of 525-641 K and 490-602 K, respectively. A quadrupole mass spectrometer (MEXM-1200, ABB EXTREL, USA), combined with an Y_2O_3 Knudsen cell, which has a lid with an effusion orifice of 0.5 mm in diameter, is employed. An impact electron energy of 28eV is selected.

The absolute vapour pressure of Cd(g) is determined from its ion currents by the modified integral method.[3] In this study, the total lost amount of Cd(g) is evaluated from the weight change of the Knudsen cell containing the sample before and after the mass-spectrometric measurement, ΔW. The conversion of the ion currents to the absolute vapour pressures of Cd(g), p, is performed by using the following equation:

$$p(t)=\frac{\sqrt{2\pi R}\Delta W I(t)T(t)}{aL\int_0^{t_{end}} I(t)\sqrt{T(t)}dt\sum_i \gamma_i\sqrt{M_i}} \tag{1}$$

where these symbols have the same meaning as those used in the previous paper.[3] The product of the orifice area and the Clausing factor, or aL, is determined from the mass-spectrometric measurement of pure Cd metal by comparison with reference data.[4]

3 RESULTS AND DISCUSSION

The vapour pressures of Cd(g) over $PuCd_2$+$PuCd_4$ and $PuCd_4$+$PuCd_6$ determined by a least-squares treatment are expressed, respectively, by the following equations:

$$\log p(\mathrm{Pa}) = -7096\pm79/T+11.31\pm0.14 \tag{2}$$

$$\log p(\mathrm{Pa}) = -6084\pm77/T+10.46\pm0.14. \tag{3}$$

Since only the vapour species of Cd(g) over both these two samples is detected the following reactions can be assumed:

$$PuCd_6(s)=PuCd_4(s)+2Cd(g) \tag{4}$$

$$PuCd_4(s)=PuCd_2(s)+2Cd(g). \tag{5}$$

Then, the standard Gibbs energies of formation of $PuCd_2$(s) and $PuCd_4$(s), or $\Delta_f G^\circ(PuCd_2,s)$ and $\Delta_f G^\circ(PuCd_4,s)$, can be calculated by using the reference data of $\Delta_f G^\circ(PuCd_6,s)_{ref}$[5] and the vapour pressures over pure Cd, p_{ref},[4] from the following equations:

$$\Delta_f G^\circ(PuCd_4,s) = -2RT\ln p + 2RT\ln p_{ref} + \Delta_f G^\circ(PuCd_6,s)_{ref} \tag{6}$$

$$\Delta_f G^\circ(PuCd_2,s) = -2RT\ln p + 2RT\ln p_{ref} + \Delta_f G^\circ(PuCd_4,s) \tag{7}$$

assuming that the activities of these Pu-Cd intermetallic compounds are unity. The evaluated $\Delta_f G^\circ(PuCd_2,s)$ and $\Delta_f G^\circ(PuCd_4,s)$ are obtained from a least-squares fit to $\Delta_f G^\circ$=A+BT and expressed as the following equations:

$$\Delta_f G^\circ(PuCd_2,s)(\mathrm{J/mol}) = -106100\pm3100+111.0\pm5.4T \quad (525\text{-}641\mathrm{K}) \tag{8}$$

$$\Delta_f G^\circ(PuCd_4,s)\ (\mathrm{J/mol}) = -154000\pm2900+125.4\pm5.4T \quad (490\text{-}602\mathrm{K}). \tag{9}$$

The values of $\Delta_f G^\circ(PuCd_2,s)$ and $\Delta_f G^\circ(PuCd_4,s)$ evaluated in this study almost agree with those extrapolated from the data given by Shirai.[2] However, it is found that $\Delta_f G^\circ(PuCd_2,s)$ previously evaluated by the authors[3] deviates significantly. If it is assumed that Eqs.(2) and (3) can be extrapolated to room temperature, then it is possible to calculate the standard enthalpy and entropy of formation of these intermetallic compounds at room temperature, or $\Delta_f H^\circ_{298}$ and $\Delta_f S^\circ_{298}$. The results are presented in Table 1, which also includes the estimated values of $\Delta_f H^\circ_{298}$ given by Miedema.[6] The standard deviations, σ, tabulated in this table are derived from those for the constants in Eqs.(2) and (3), and do not include possible errors in p_{ref},[3] $\Delta_f G^\circ(PuCd_6,s)_{ref}$,[5] and the measured temperatures.

Table 1 *The standard enthalpy and entropy of formation of $PuCd_2$ and $PuCd_4$ in units of kJ/mol and J/K/mol: the values in the last column are given by Miedema*[6]

	$\Delta_f H^\circ_{298}$	$\sigma(\Delta_f H^\circ_{298})$	$\Delta_f S^\circ_{298}$	$\sigma(\Delta_f S^\circ_{298})$	$\Delta_f H^\circ_{298}$ [6]
$PuCd_2$	- 106.4	4.2	- 129.1	7.6	- 108
$PuCd_4$	- 155.1	3.0	- 113.5	5.4	- 109

The experimental values of $\Delta_f H^\circ_{298}$ in Table 1 are found to lie within the scatter of the data in the figure from Miedema's original paper, which shows the comparison between the experimental and the expected values.[6] However, the values of $\Delta_f H^\circ_{298}(PuCd_2)$ and $\Delta_f S^\circ_{298}(PuCd_2)$, calculated in a similar manner from the previous result[3] are - 337.4 kJ/mol and - 319 J/mol/K, respectively, and much different from those presented in Table 1. Indeed, the following relation must hold:

$$\Delta_f S^\circ_{298}(PuCd_2) = S^\circ_{298}(PuCd_2) - S^\circ_{298}(Pu) - 2S^\circ_{298}(Cd) > - S^\circ_{298}(Pu) - 2S^\circ_{298}(Cd) \quad (10)$$

and the value for $S^\circ_{298}(Pu) + 2S^\circ_{298}(Cd)$ given by Barin[7] is 155 J/mol/K. Then, $\Delta_f S^\circ_{298}(PuCd_2)$ derived from the previous result[3] is contradictory to the above relation.

One of the reasons for these disagreements is that the effective vaporization area of the sample in the previous Knudsen-effusion experiment[3] might become small, due to surface oxidation, which could lead to diminishing measured vapour pressures.[8] On the other hand, the $PuCd_2$+$PuCd_4$ and $PuCd_4$+$PuCd_6$ samples used in this study have porous or powder shapes, so these samples could have enough surface area to realize their intrinsic vapour pressures, although the $PuCd_2$+$PuCd_4$ sample also seems to be easily oxidized, as in the case of the Pu+$PuCd_2$ sample.

References

1 M.A. Vest, E.F. Lewandowski, R. D. Pierce and J.L. Smith, *Nucl. Tech.*, 1997, **120**, 232.
2 O. Shirai, M. Iizuka, T. Iwai, Y. Suzuki and Y. Arai, *J. Electroanal. Chem.*, 2000, **490**, 31.
3. K. Nakajima, Y. Arai and T. Yamashita, *J. Phys. Chem. Solids*, 2005, **66**, 639.
4 Kubaschewski, E.L. Evans and C.B. Alcock, *Metallurgical Thermochemistry*, 5th ed., Pergamon Press, 1979, p. 360
5 P. Chiotti, V.V. Akhachinskij, I. Ansara, M.H. Rand, in: V. Medvedev, M.H. Rand, E.F. Westrum, Jr. (Eds.), *The Chemical Thermodynamics of Actinide Elements and Compounds, Part 5 The Actinide Binary Alloys*, International Atomic Energy Agency, Vienna, 1981, p.228.
6 A.R. Miedema, in: H. Blank and R. Lindner (Eds.), *Plutonium 1975 and other Actinides*, North Holland Publishing Co., Amsterdam, 1976, p.3.
7 I. Barin, *Thermochemical Data of Pure Substances*, 3rd Ed., VCH, Weinheim, 1995.
8 G. Rosenblatt, *J. Electrochem. Soc.*, 1963, **110**, 563.

DEVELOPMENT OF THE SCIENTIFIC CONCEPT OF THE PHOSPHATE METHODS FOR ACTINIDE-CONTAINING WASTE HANDLING (PYROCHEMICAL FUEL REPROCESSING)

A.I. Orlova[1], O.V. Skiba[2], A.V. Bychkov[2], Yu.F. Volkov[2], A.N. Lukinykh[2], S.V. Tomilin[2], A.A. Lizin[2] and V.A. Orlova[1]

[1]Department of Chemistry, Nizhny Novgorod State University, 23, Gagarin Ave., Nizhny Novgorod, 603950, Russia.
[2]Laboratoty of radwaste immobilization, Federal State Unitary Enterprise State Scientific Center, Research Institute of Atomic Reactors, Dimitrovgrad, Ulyanovsk, 433510, Russia.

1 INTRODUCTION

The phosphate reprocessing method for actinide-containing wastes, involving the conversion of hazardous components into mineral like compounds are of interest. The scientific concept of these methods is based on the knowledge of the crystal chemistry of actinide compounds having complex cationic compositions and containing cations of different nature in various combinations and proportions. The concept in question has been successfully developed using pyroelectrochemical irradiated fuel reprocessing in molten alkaline element chlorides at the Research Institute of Atomic Reactors. Irradiated fuel is dissolved in molten chlorides of alkaline elements by means of treating a melt with a chlorinating agent. Uranium and plutonium dioxides are then removed electrochemically. The molten salt solvent-electrolyte, when used many times, is contaminated by residual actinides, fission products and so called “process” elements. This melt is unsuitable for future use.

Phosphate methods can be applied for solving the following tasks: *a* – molten salt refining and regeneration; *b* - converting the used alkaline chloride melt into a stable product for safe storage and disposal. Moreover, in melt refining (task *a*) phosphate methods could be used in two ways: *a1* – only phosphate refining the melt by means of a reaction (precipitation, coprecipitation, ionic exchange, other chemical interactions) and *a2* – not only refining the salt melt, but also incorporating the impurities in the structures of stable crystalline phosphates.

To choose the most effective ways of carrying out the above tasks, crystal chemistry knowledge of actinide (III) and (IV) phosphates and the phosphates of other waste elements is used as a basis. The chemical resistance data for phosphates (in molten salts, including the presence of soluble chlorine, in hydrothermal conditions) and their behaviour at high temperatures and in radiation fields are also important.

2 RESULTS AND DISCUSSION

The present paper summarizes the results obtained by the authors for the complex phosphates of uranium, plutonium, americium, curium with 1-4 valent elements, including the synthesis methods and crystallographic properties.

The compounds were prepared by precipitation and coprecipitation if soluble (for example $A^I_3PO_4$ or insoluble (for example Zr or Ti-containing phosphates) reagents were added to the melt. Methods of formation for phosphates of types: $A^IM_2^{IV}(PO_4)_3$ (A^I = Na, K, M^{IV} = U; A^I = Na, M^{IV} = Pu), $Na_3Pu_2(PO_4)_3$, $Na(K)_3R^{III}{}_2(PO_4)_3$ with R^{III} = Am, Cm, phosphates of similar type to the lanthanides, $Na(K)UO_2PO_4$, Zr phosphates such as $A^IZr_2(PO_4)_3$, $B^{II}{}_{0.5}Zr_2(PO_4)_3$, A^I = Li – Cs, B^{II} = Ca, Sr, Ba, and rhodium and ruthenium phosphates were established. Sometimes we added thermally treated phosphates, $Zr(HPO_4)_2$, ZrP_2O_7, $Na(K)M^{IV}{}_2(PO_4)_3$ (M^{IV}=Ti, Zr, Hf), to the melt and heated these mixtures at 700-800 ^{0}C for 1 to 4-6 hours in the presence of Ar, Cl_2 or air atmosphere above the melt. In the last case, the Ti and Zr compounds are responsible for conversion of the radionuclide into complex phosphates, due to the chemical reactions between these Ti and Zr phosphates and the alkaline element chlorides. Also, we used the well-known sol-gel method and high-temperature reactions between oxides. Among the prepared phosphates there are some possessing the monazite structure, $NaZr_2(PO_4)_3$ (NZP), apatite and other known or unknown structures.

We have considered several, generally mineral-like structures, when developing the waste immobilization matrix for the phosphate concentrating of the products containing actinides (task *a*, methods *1* and *2*) and for the reprocessing of spent alkaline chloride solvents into environmentally "friendly" insoluble chemical forms suitable for long-term storage (task *b*). Preference was given to structures capable of accumulating actinides (and lanthanides) and capable of wide cationic isomorphism. Much prominence is given to the studies of phosphates having crystal structures of the kosnarite type (analogue $NaZr_2(PO_4)_3$, NZP), monazite (analogue $CePO_4$) and langbeinite (analogue $K_2Mg_2(SO_4)_3$).

The list of the phosphates obtained with such structures contains the compounds of uranium, plutonium, americium, curium, lanthanides, cesium and other alkaline elements, barium and other elements. They have been characterized by X-ray and neutron powder diffraction methods, including the high-temperature variations, infrared spectroscopy, thermal analysis and radiometry.

Phosphates of a framework structure with rhombohedral and cubic framework modifications are:

- $A^IM^{IV}{}_2(PO_4)_3$, A^I= Na, K, Rb, M^{IV} = U, Np, Pu
  high-temperature modifications, the NZP structural type, the trigonal system,
  low-temperature modifications, the $NaTh_2(PO_4)_3)$ structural type, the monoclinic system,
- $Am_{1/3}Zr_2(PO_4)_3$, the NZP type,
- $R^{III}{}_{1/3}Zr_2(PO_4)_3$, R^{III} = Ce - Lu, the NZP type, sp.gr. $P\bar{3}c$,
- $M^{IV}{}_{1/4}Zr_2(PO_4)_3$, M^{IV} = Th, U, the NZP type, sp.gr. $R\bar{3}c$,
- $A^I{}_2R^{III}Zr(PO_4)_3$, A^I = K, Rb, Cs, R^{III} = Ce-Lu, the langbeinite type, sp.gr. $P2_13$,
- $A^IBaR^{III}{}_2(PO_4)_3$, A^I = K, Rb, Cs, R^{III} = Er-Lu, the langbeinite type, sp.gr. $P2_13$.

New phosphates of the monazite structural type, sp. gr. $P2_1/n$: $Ca_{0.5}Pu_2(PO_4)_3$, $Cd_{0.5}Pu_2(PO_4)_3$, $CaAmPu(PO_4)_3$, thorium and uranium phosphates seires of the $B^{II}R^{III}Th(U)(PO_4)_3$ type, where B^{II} = Mg - Ba, R^{III} = Sm, Gd were also synthesized and investigated. Phosphates with kosnarite (NZP) and monazite structures are considered as the forms for concentrating impurities when decontaminating the melts (task *a*). The phosphates with the langbeinite structure were studied as chemically and thermally stable forms for alkaline elements, including cesium, other large alkaline and also alkaline earth elements (task *b*). The phosphates of langbeinite structure were synthesized through reactions in molten chlorides of alkaline elements (Na, K, Cs). The possibilities for the immobilization of alkaline elements in this structure were established. We also established that the phosphates of the langbeinite type can incorporate lanthanides (III). The compounds $A^{I}_{2}R^{III}Zr(PO_4)_3$ (A^I = K, Rb, Cs, R^{III} = Ce – Lu, Y) and $A^{I}BaFe_{2-x}R^{III}_{x}(PO_4)_3$ (A^I = K, Rb, Cs, R^{III} = Tb – Lu: x = 2 and R^{III} = Pr – Gd: x < 2) possess the structural modification of langbeinite, sp.gr. $P2_13$.

The NZP phosphates being studed do not undergo any chemical or phase transformations up to 1100 – 1600 °C. They are stable in hydrothermal conditions (ceramic samples of the NZP structural type were tested in water at temperatures of 200 – 400 °C and under a pressure of 6×10^7 Pa for up to 2 years). Investigation of the behaviour of langbeinite phosphates in hydrothermal conditions (90 °C, 28 days) has been carried out. The phase and chemical compositions of the samples remain unchanged.

It was found that the internal α-radiation affects the phosphate stability and their amorphisation. For this, we used the phosphate samples $NaPu_2(PO_4)_3$. The plutonium isotopic composition in this phosphate was 79 % ^{238}Pu + mixed ^{239}Pu, ^{240}Pu. Metamictization (amorphisation under radiation) occurs after 1 month if 9.3×10^{18} α-particles/g are emitted along with the same quantity of the recoil atoms (^{238}Pu content in the sample was near 48 % wt.). The changes in the X-ray diffraction patterns were noticed when 1.3×10^{18} α-particles/g had been emitted. These data on the phosphate radiation tests were compared with the similar ones for the Synrock ceramics, containing 2-5 % wt. of plutonium-238 (metamictization after emission of 4.7×10^{18} α-particles/g, perovskite and zyrkonolite phases) and the US Titanate Ceramisc, containing near 9 % wt. of ^{238}Pu (metamictization after emission of 8.8×10^{17} α-particles/g, changes on the X-ray diffraction patterns were observed after emission of 0.5 – 0.25 $\times10^{17}$ α-particles/g).

3 CONCLUSIONS

The results allow us to make the following conclusions:

1. New data in the field of chemistry and structural chemistry of actinide phosphates and phosphates of many 1-4 valent elements are the basis for solving the problems of radioactive waste using pyroelectrochemical technology, by means of immobilization into ecologically stable crystalline forms.
2. We have established that phosphate precipitation and coprecipitation, also in combination with ionic exchange and other chemical interactions are useful methods for removing dangerous products from used chloride melts.
3. In our opinion these phosphate methods could be more effective if the solid phases formed are converted into mineral-like crystalline forms, such as kosnarite, monazite or other structure types that have a high isomorphous capacity.
4. We have shown that tetrahedron-octahedron framework structures (NZP, langbeinite) are suitable for the reprocessing of used decontaminated alkaline chloride melts into stable chemical formes for long-term storage and deposition.

References

1 A.I. Orlova, Yu.F. Volkov, R.F. Melkaya, L.Yu. Masterova, I.A. Kulikov and V.A. Alferov, *Radiokhimiya*, 1994, **36**, 295.
2 A.I. Orlova, *Radiokhimiya*, 2002, **44**, 385.
3 Yu.F. Volkov, S.V. Tomilin, A.I. Orlova, A.A. Lizin, V.I. Spiryakov and A.N. Lukinykh, *Radiokhimiya*, 2003, **45**, 289.
4 A.I. Orlova, V.A. Orlova, A.V. Buchirin, K.K. Korchenkin, A.I. Beskrovniy and V.T. Demarin, *Radiokhimiya*, 2005, **47**, 213.

SYNTHESIS AND INVESTIGATIONS OF THE PHOSPHATES OF Th, U, Np, Pu, Am AND LANTHANIDES WITH MONAZITE, ZIRCON, KOSNARITE AND LANGBENITE MINERAL-LIKE STRUCTURES.

A.I. Orlova,[1] D.B. Kitaev,[1] M.P. Orlova,[1] D.M. Bykov,[1] V.A. Orlova,[1] S.V. Tomilin,[2] A.A. Lizin,[2] A.N. Lukinich[2] and S.V. Stefanovskiy[3]

[1]Department of Chemistry, Nizhny Novgorod State University, 23, Gagarin Ave., Nizhny Novgorod, 603950, Russia.
[2]Laboratoty of radwaste immobilization, Federal State Unitary Enterprise State Scientific Center, Research Institute of Atomic Reactors, Dimitrovgrad, Ulyanovsk region, 433510, Russia.
[3] SPS "Radon", 7-th Rostovsky st., 2/14, Moscow, 119121, Russia.

1 INTRODUCTION

The results of structural investigations of actinide (Th, U, Np, Pu, Am)- and lanthanide- (Ce-Lu) containing phosphates are presented. The known simple and complex possible phosphates of actinides represent only a very small fraction of all possible phosphates $Me_m(PO_4)_n$ ($Me_m = A^I_xB^{II}_y R^{III}_zM^{IV}_pC^V_q$; A, B, R, M, C = cations possessing oxidation state 1+, 2+, 3+, 4+, 5+), calculated in accordance with the classification in Volkov et al.[1] Several new crystalline compounds of f-elements were synthesised, and their new crystal chemical data are discussed here. This knowledge provides a theoretical basis for development of phosphate crystalline materials (ceramics) for radioactive nuclide immobilization.

2 RESULTS AND DISCUSSION

The known phosphates of actinides (III) and (IV) are represented by the following phases: eulitine, zircon, monazite, kosnarite (NZP), NThP, rabdafanite, SbBi-phohsphate, whitlockite, arkanite, glazerite, langbeinite, sheelite. The majority of these structural types are known minerals.

Phosphates with structures of mineral types, such as monazite, zircon, kosnarite and langbeinite have been studied here. The newly synthesized and characterized compounds contain Th, U (IV), Np (IV), Pu (IV), Am, (Ce – Lu) and other 1-4 –valent elements.

Crystal chemical design of new phosphates with the expected structures and desired cation composition was based on the evaluation of the results of our own investigations and literature data analysis.[2]

Synthesis of the phosphates was performed using methods based on sol-gel technology and solid state reactions. The process comprised several alternate stages of heating and grinding (in some instances without grinding). Upon each stage the samples were estimated by the X-Ray method. Structural X-ray analysis (Rithveld technique), IR spectroscopy, DTA, radiometry were also applied for the characterization of the samples.

2.1 Phosphates with the monazite and zircon structures. Structures having a single type of cation sides and random cation positions.

The monazite structure was adopted in the phosphates of the following formula types: $B^{II}M^{IV}(PO_4)_2$ (M^{IV} = Th, B^{II} =Mg, Ca, Sr); $B^{II}_{0.5}M^{IV}_2(PO_4)_3$ (M^{IV} = Np, B^{II} = Mg, Ca, Sr and M^{IV} = Pu, B^{II} = Ca, Cd); $B^{II}GdTh(PO_4)_3$, $B^{II}Gd(Nd)U(PO_4)_3$ (B^{II} = Mg, Ca, Sr, Ba, Cd) and $CaAmPu(PO_4)_3$. Cell parameters were determined (space group $P2_1/n$), and the cell changes related to cationic radii (r_{eff}). A Rietveld structural refinement was carried out for the compound $CaGdTh(PO_4)_3$.

Some phosphates of family $B^{II}_{1.5(1-x)}R^{III}_{3x}M^{IV}_{1.5(1-x)}(PO_4)_3$, containing erbium (R), zirconium (M) and divalent elements (B = Mg, Ca, Sr, Ba, Mn, Co, Ni, Cu, Zn, Cd) crystallized in the zircon structural type (space group $I4_1/amd$). For a certain value of x, the analogue is $ZrSiO_4$, with a tetragonal cell. A transition between monazite and zircon in this family, with an associated change of coordination environment ($MeO_9 \leftrightarrow MeO_8$) for the structure-forming cations was observed. The environment cation substitution ranges in the monazite-like and zircon-like phosphates were determined, and the relationship between the nature of the tetravalent elements (*f- or d-element*) and effective radii with the structure type and cell size were ascertained.

After generalizing our results and known data,[3] it is evident that the phosphates with monazite and zircon structures can accomodate many kinds of 1-4 –valent cations, along with actinides and lanthanides(III) and (IV): Monazite (c.n. = 9) - Na, K, Rb, Cs, Mg, Ca, Sr, Cd, Ba, La, Ce – Lu, Y, Pu (III), Am, Cm, Bk, Cf, Es, Zr, Th, U (IV), Np (IV), Pu (IV); Zircon (c.n. = 8) - Mg, Ca, Sr, Cd, Ba, Mn, Co, Ni, Cu, Zn,, Sc, Y, Gd, Tb, Dy, Ho, Er, Tm, Yb, Lu, Zr. In these phases, many cations possess unusual coordination numbers.

2.2 Phosphates with kosnarite (structural type $NaZr_2(PO_4)_3$, NZP) and langbeinite structures. Structures having several cation sides and ordered cation positions.

The kosnarite (NZP)- and langbeinite-type structures are related. They both comprise a mixed octahedron-tetrahedron framework, but have connectivities within the framework and hence different numbers, shapes and sizes of cavities. Different types of cation positions are realized in the both structures. Therefore, the variations of cation composition are available in both structures.

The NZP structure is attained in the actinide phosphates $A^IM_2(PO_4)_3$, (A^I = Na, K, Rb; M^{IV} = U, Np, Pu) and can also exist in the β-modification above a temperature around 1270-1470K. The *a* and *c* parameters of the trigonal cell and their changes under the influence of M^{4+} and A^+ cations were determined. The temperature stability fields and transition temperature fields of the low-temperature, α-modification ($NaTh_2(PO_4)_3$-type) and the high-temperature, β-form (NZP type) were defined. In addition, it was shown that actinides M^{IV} = Th, U, Pu and R^{III} = Am can be included in the NZP phosphates, $M^{IV}_{1/4}Zr_2(PO_4)_3$ and $R^{III}_{1/3}Zr_2(PO_4)_3$, presumably occupying non-framework positions in the structure (c.n. = 6). The existence and structure of these compounds had been predicted on the basis of "crystal chemical" design.

We also carried out systematic crystal chemical investigations of cubic type phosphates (langbeinite type) for the first time: $A^I_2R^{III}Zr(PO_4)_3$, A^I = K, Rb, Cs, R^{III} - all lanthanides; $A^IBaR_2^{III}(PO_4)_3$, A^I = Cs, R^{III} = Dy, Er, Tm, Yb, Ho. In these, the *f*- element cations fully occupy the framework positions. This is interesting, because cations in this structure, as well as in NZP structure, have octahedral Met-O_6 surroundings. The results of crystal chemical analysis of a new cubic lanthanide phosphate family were used for the prediction of the chemical composition of isostructural actinide (III) phosphates.

Thus, we have been able to extend the list of cations contained in the framework of NZP- and langbeinite-like phosphates. NZP- Li, Na, K, Rb, Cs; H, Cu (I), Ag; Mg, Ca, Sr, Ba, Mn, Co, Ni, Cu, Zn, Cd, Hg; Al, Sc, V, Cr, Fe, Ga, Y, In, Sb, La, Ce– Lu; Bi; Ge, Sn,Ti, Zr, Mo, Hf, Ce, Th, U, Np, Pu; Nb, Sb, Ta; Langbeinite - Na, K, Rb, Cs; Mg, Ca, Sr, Ba; Sc, Cr, Fe, Ti, Ga, Y, La, Ce – Lu, In, Bi; Ti, Zr, Hf; Nb.

There are more than 50 elements here. We have established from the empirical evidence that a wide range of cation combinations can form such phosphates, and that in some examples, the actinide and lanthanide cations possess an unusual coordination: c.n. = 6.

3 CONCLUSIONS

The general pattern of structure formation in the anhydrous orthophosphate system shows the following:

1. The crystal chemical behaviour of the lanthanides and actinides depends on cation compositions of the phosphates and their formation conditions. As a result,

   a) Polymorphism in the series "NThP – NZP", "monazite – zircon", "NZP – langbeinite" exists. Cation coordination number may be changed from 9 to 6; from 9 to 8 or not change depending upon the circumstances. For example, $A^{I}M^{IV}{}_2(PO_4)_3$ (M^{IV} = U, Np, Pu) "NThP – NZP"; $R^{III}PO_4$ (R^{III} = Gd, Tb, Dy) "monazite – zircon", $K_2GaZr(PO_4)_3$ "NZP – langbeinite".

   b) Structural transitions occur in these families. For example, there is a "monazite – zircon" transition in the rows $R^{III}PO_4$, $B^{II}R^{III}Zr(PO_4)_3$ (B^{II} – 2-valent element, R^{III} – lanthanide); "NZP – langbeinite" in the row $A^{I}YbZr(PO_4)_3$ (A^{I} = Na - K); "langbeinite - unknown structure" in the rows $A^{I}BaR^{III}{}_2(PO_4)_3$ (A^{I} = K, Rb, Cs, R^{III} – lanthanide).

2. Extensive cation isomorphism is typical for these structures, with participation of the lanthanides and actinides.
3. Now we can use crystal chemical data to "construct" crystalline (ceramic) materials for actinide immobilisation, and also for the immobilisation of fission products containing diverse cation compositions.
4. Transuranium element phosphates have not yet been investigated sufficiently. Therefore, it is very convenient to use crystal chemical considerations to predict compositions and structures of new and unknown actinide phosphates.

References

1. Yu. Volkov and A. Orlova, *Radiohimiya*, 1996, 38, №1, 3.
2. Yu. Volkov and A. Orlova, *Systematization and crystal chemical aspect of inorganic compounds containing one-nuclear tetrahedral oxoanions*, Dimitrovgrad, 2004, 286.
3. W. Lutze and R. Ewing, *Monazite, in Radioactive waste forms for the future*, Elseveir, 1988, 497.

BASIC CONCEPT OF HYDRO-PYRO HYBRID TYPE SULFIDE PROCESS FOR SPENT FUEL

N. Sato and O. Tochiyama

Institute of Multidisciplinary Research for Advanced Materials, Tohoku University, 2-1-1 Katahira, Aoba-ku, Sendai 980-8577, JAPAN

1 INTRODUCTION

In recent years, dry processes are drawing much attention in nuclear engineering field, since they can attain the crude separation of radioactive materials producing less waste volume, compared with conventional wet processes. As one such dry processes, we propose a pyro-reprocessing of spent nuclear fuel via sulfurization, called the "Hydro-Pyro Hybrid Type Sulfide Process". This paper presents the basic concept of this new process, and discusses its applicability using thermodynamical considerations and experimental results.

2 CONCEPT OF THE PROCESS

The flow sheet for the proposed hydro-pyro hybrid type sulfide process is given in Fig. 1. After the spent fuel voloxidation step, in which the UO_2 is oxidized to U_3O_8 by heat treatment in oxidising conditions, there are three main steps, i.e. 1) selective sulfurization of rare-earth oxides, 2) magnetic separation of rare-earth sulfides, and 3) acid leaching of sulfides.

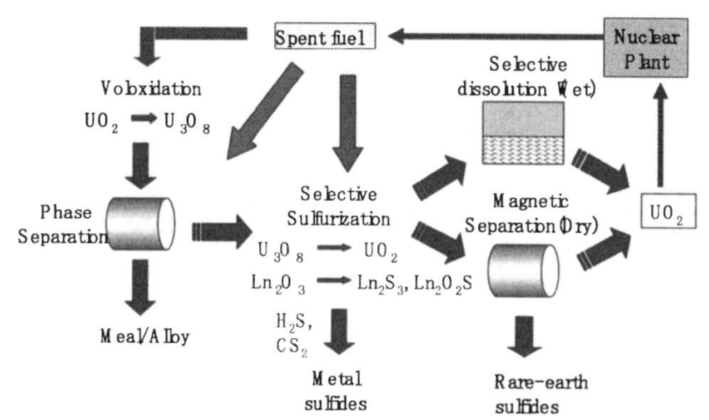

Figure 1 *Flowsheet for the proposed hydro-pyro hybrid type sulfide process.*

In the selective sulfurization process with H_2S and CS_2, the rare-earth oxides R_2O_3, which are poison elements in the reactor, are to be sulfurized to form oxysulfides R_2O_2S, sesqui sulfides R_2S_3 and other sulfides, while the U_3O_8 is to remain inert, except for a possible reduction to UO_2. Separation of rare-earth sulfides from the uranium oxides would be possible by magnetic field, as these sulfides possess higher magnetic susceptibilities than oxides. A separation of sulfides from uranium oxides by acid leaching would be also possible, as the sulfides tend to react with moisture in the air or water forming H_2S gas, while the uranium oxides might not dissolve in the solution and be recovered. The recovered UO_2 would be recycled to the power plant. The applicability of each step is discussed in the following sections.

3 SELECTIVE SULFURIZATION

Gibbs free energy changes were obtained using the database MALT2[1] for the sulfuization reactions with H_2S. As shown in Fig. 2, the formation of rare-earth oxysulfides R_2O_2S have the lowest values (eqs. (3) and (5)), and thus these compounds could be formed even at low temperatures. Then a metal such as Pd would be sulfiurized, although PdS would decompose to Pd and S at temperatures higher than 600 °C (eq.(7). After this, rare-earth sulfides would be formed from their oxysulfides (eqs.(4) and (5)). Although the sulfurization of UO_2 would not occur at low temperature, UOS might be formed at higher temperatures, since the values for eq.(1) change from positive to negative at 600 °C. The formation of uranium sulfides by H_2S is difficult in the temperature range considered (eq.(2)). Next, the sulfurization of the major components of spent fuel with CS_2 was considered in the same manner, and the results are shown in Fig. 3. The sulfurization of the metal component is omitted in the figure, since sulfurization with CS_2 is simply easier than that with H_2S. The sulfurization of rare-earth oxides by CS_2 is similar to that by H_2S, though the reaction might occur at lower temperatures than the H_2S case. Since the lowest value is for the reaction of U_3O_8, the reduction of U_3O_8 to UO_2 would occur prior to the formation of R_2O_2S. The formation of UOS from UO_2 by CS_2 would require a higher temperature compared with that by H_2S, as given in Fig. 2.

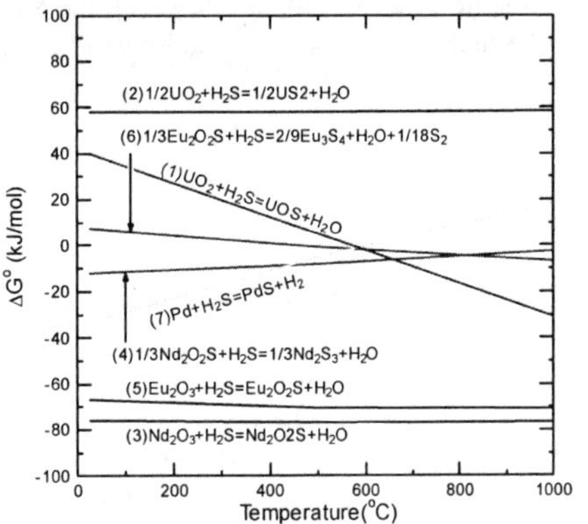

Figure 2 *Gibbs free energy change for the related sulfuization reactions with H_2S.*

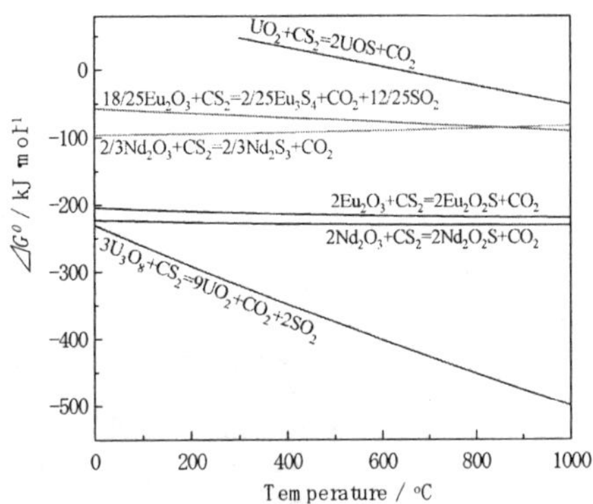

Figure 3 *Gibbs free energy change for the related sulfuization reactions with CS_2.*

4 MAGNETIC SEPARATION

Among the major components of the spent fuel, the rare-earth oxides group has a higher magnetic susceptibility than UO_2 as seen in Fig. 4.[2-4] Other components such as metals and oxides have smaller susceptibilities than UO_2. Therefore, there is a possibility of the magnetic separation of rare-earth sulfides from UO_2. In our preliminary experiments with a mixture of UO_2 and rare-earth sulfides, when the magnetic field was applied, the rare-earth powders tended to be attracted to the magnet, while UO_2 did not. Although the separation effect was not high in a once through operation, it was improved by a circulation system, using an organic suspension with the powders.

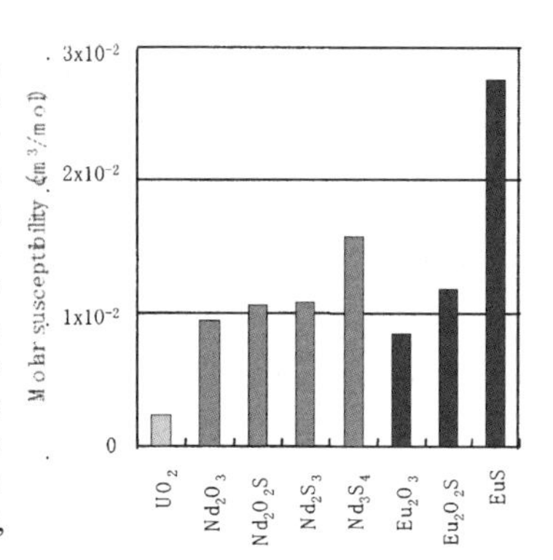

Figure 4 *Magnetic susceptibility of neodymium and europium oxides and sulfides compared with UO_2.*

5 SELECTIVE DISSOLUTION

Since the sulfides dissolve easily in acidic solution forming H_2S, a selective dissolution step in which only sulfides selectively dissolve into acid, would be also applicable. Figure 5 shows the dissolution ratio of UO_2, Eu_2O_2S and Nd_3S_4 in 1M-HNO_3 at room temperature obtained in the present study. Sulfide (Nd_3S_4) dissolves quite fast, then oxysulfide (Eu_2O_2S) more gradually. However, the dissolution ratio of UO_2 is very low, i.e. a large selectivity can be obtained in this step. When CS_2 is used as the sulfurizing agent, soluble rare-earth sulfides are formed. However, sulfurization of UO_2 with CS_2, which leads to the dissolution of UO_2 may also occur. To decrease the dissolution of UO_2, sulfurization by H_2S would be favourable, compared to that by CS_2.

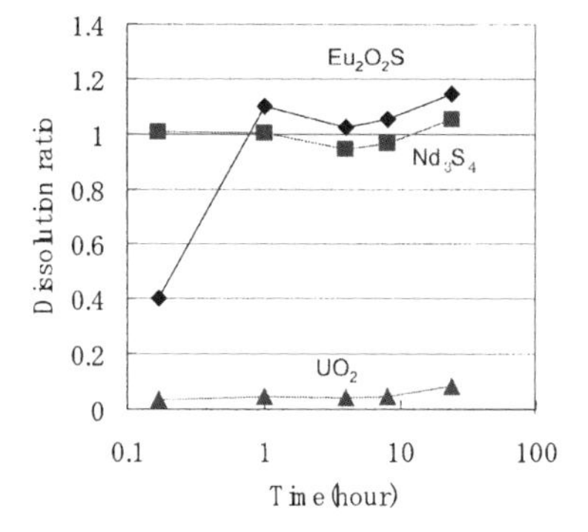

Figure 5 *Dissolution ratio of UO_2, Eu_2O_2S and Nd_3S_4 in 1M-HNO_3.*

6 CONCLUSION

In this paper, the basic concept of the hydro-pyro hybrid type sulfide process for spent fuel was introduced, based on thermodynamic considerations and experimental results. The potential of the process, which consists of three major steps (selective sulfurization, magnetic separation and selective dissolution), was clearly shown in this preliminary study.

References

1 Thermodynamic Database MALT for Windows, *Kagaku Gijutsu-Sha*, 2002, Japan.
2 M.A.Taher and J.B. Gruber, *Phys. Rev.*, 1977, **B16**, 1624.
3 P. Wachter, *Crit. Rev. Solid State Sci.*, 1972, **3**, 189.
4 N. Elliot, *Phys. Rev.*, 1949, **76**, 498.

LOW TEMPERATURE SELECTIVE REDUCTION OF U_3O_8 IN THE PRESENCE OF SULFURIZING AGENTS

G. Shinohara, N. Sato and O. Tochiyama

Institute of Multidisciplinary Research for Advanced Materials, Tohoku University, 2-1-1 Katahira, Aoba-ku, Sendai 980, JAPAN

1 INTRODUCTION

In nuclear engineering, dry processes have been attracting much attention in recent years since the crude separation of radioactive materials, leading to waste volume reduction compared with conventional wet processes, would be achievable. As one such dry processes, we propose a sulfide process. This process is composed of three main steps, i.e. 1) selective sulfurization of rare-earth oxides, 2) magnetic separation of rare-earth sulfides, and/or 3) acid leaching of sulfides, for the separation of rare-earth elements and recycling of UO_2. In a mixture of uranium and rare-earth oxides, R_2O_3 (R=Nd, Eu) would be selectively sulfurized with CS_2. Reduction of U_3O_8 to UO_2 could also occur in the presence of the sulfurizing agent. The uranium and rare-earths would be separated by selective acid leaching, since soluble rare-earth oxysulfides and sulfides are formed, whereas uranium remains as insoluble oxides. Since the sulfides have higher magnetic susceptibilities than the oxides, the separation of uranium oxide and rare-earth sulfides could also be attained with a magnetic field. In the present paper, the reduction behaviour of uranium oxide (U_3O_8) in the presence of sulfurizing agents such as CS_2 and FeS_2 was studied using XRD and TG methods. In the scheme of magnetic separation, the addition of iron in the system could accelerate the separation effect of the rare-earth. If iron sulfides are used as a solid sulfurizing agent for R_2O_3, they could also serve as a reducing agent for U_3O_8.

2 EXPERIMENTAL

U_3O_8 was obtained by the oxidation of U metal turnings in air at 800 °C. Stoichiometric UO_2 was prepared by the H_2 reduction of U_3O_8 at 1000 °C. Analytical grade CS_2 with a boiling point of 46-47 °C and maximum water content of 0.02% (Wako Pure Chemicals Co., Ltd.) and nitrogen gas of 99.99% purity (Nippon Sanso Co., Ltd.) were used as received. Iron disulfide was also obtained from Wako Pure Chemicals Co., Ltd. The weighed amount of U_3O_8 or the mixture of U_3O_8 and FeS_2 powder were set in a quartz boat. The boat was heated in a flow of CS_2 and N_2 gas mixture, which was obtained by bubbling N_2 gas through a bubbler containing liquid CS_2, at temperatures between 300 to 500 °C. The sulfurization experiments were carried out at a heating rate of 1 °C/min with a CS_2/N_2 gas flow rate of 5/50 ml min^{-1}. The N_2 gas flow rate was measured using a digital mass flow meter (Kofloc Model DPM-2A). For the phase analysis of the products, the X-ray powder diffraction was carried out with a Rigaku Type RAD-IC diffractometer using CuKα radiation (40kV, 20mA) monochromatized by curved pyrolytic graphite. The apparatus for the thermogravimetric experiments was reported in elsewhere.[1] TG profiles were obtained in a vertical resistance tube furnace, equipped with a quartz spring. A quartz basket containing the weighed sample, approximately 100 mg, was suspended from a quartz spring at the centre of the uniform (±1 °C) temperature zone of the furnace. The sample was heated at constant rates from room temperature to 1000 °C. Before heating, the reaction tube was evacuated to approximately 100 Pa for 30 min and then refilled with N_2 up to ambient pressure. After that, CS_2 and N_2 gas mixture was introduced. The change in length of the quartz spring caused by the weight change during heating was measured by a level meter (Mitsutoyo) with an accuracy of 0.001 mm. The change in length was calibrated by using a standard weight. The sensitivity of the spring was 7.04 mg/mm.

3 RESULTS AND DISCUSSION

3.1 Reduction of U_3O_8 with CS_2

The XRD patterns for the products (U_3O_8 powder was heated in a flow of CS_2/N_2 with rates of 5/50 ml/min at temperatures between 300 and 500 °C for 1 hour) are shown in Fig.1. The reported patterns of UO_2 (JCPDS41-1422) and UOS (JCPDS4-207) are also given in the figure as diamond and reverse triangle marks, respectively. U_3O_8 was reduced to UO_2 with a single phase at low temperatures, such as 300 and 400 °C. At 500 °C, small peaks for the UOS phase were observed with those of the UO_2 phase. For temperatures higher than 500 °C we have already reported that β-US_2, UOS and UO_2 with a single phase were formed by the reaction of UO_2SO_4 with CS_2 at 800°C, H_2S at 1000°C and H_2 at 1000°C, respectively[2], and that the decomposition of UO_2SO_4 to UO_2 appeared at around 350 °C, and that the formation of UOS from UO_2 started at 480 °C. Our results were consistent with those reported. The results of the TG experiments for the reaction of U_3O_8 and CS_2 are shown in Fig. 2 as two different curves for the different heating rates. The calculated weight decrease for the formation of U_4O_9 and UO_2, and the weight increase for those of UOS and US_2 are given in the figure as solid lines. When the sample was heated at a rate of 5 °C/min, the TG curve suddenly decreases at 300 °C and becomes flat up to 600 °C. The ΔM value of 4.0 wt.% agrees well with that of the reduction

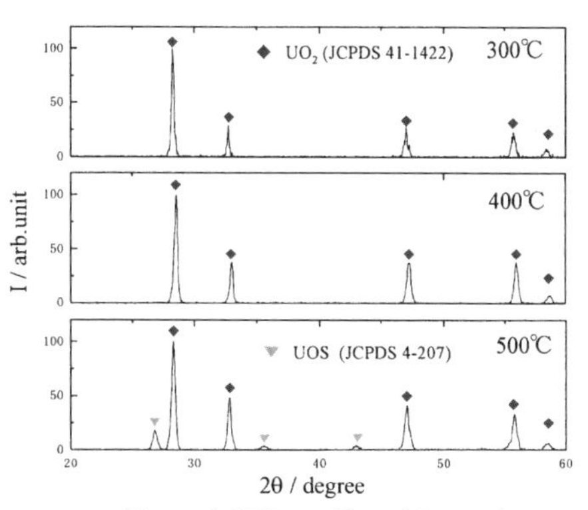

Figure 1 *XRD profiles of the products obtained by the reaction of U_3O_8 with CS_2 at different temperatures.*

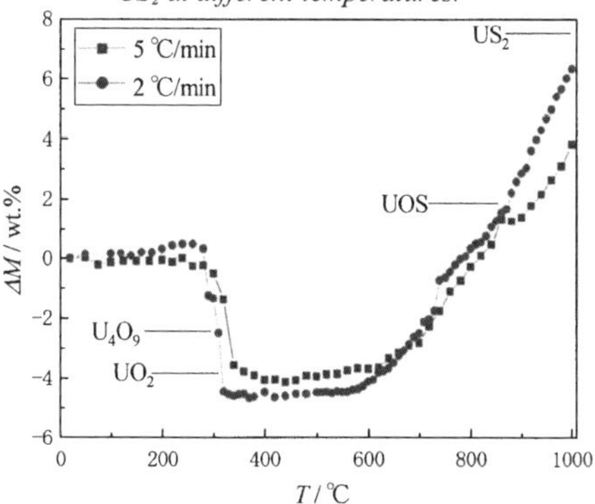

Figure 2 *TG curves obtained by reaction of U_3O_8 with CS_2 at heating rates of 2 and 5 °C/min.*

of U_3O_8 to UO_2 (-3.8 wt.%). After this the curve changes to show a weight increase, forming UOS and US_2. A similar TG curve was obtained with a heating rate of 2 °C/min. The products at 1000 °C were identified as a mixture of UOS and β-US_2 by XRD analysis[2], which agrees with the TG results.

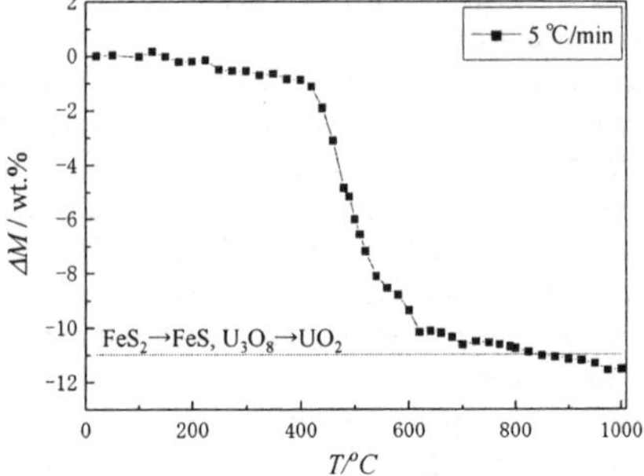

Figure 3 *TG curves obtained by reaction of U_3O_8 with FeS_2 at a heating rate of 5 °C/min.*

3.2 Reaction of U_3O_8 with FeS_2

The TG curve for the mixture of U_3O_8 and FeS_2 powder heated with a heating rate of 5 °C /min in a flow of N_2 is shown in Fig. 3. There seems to be three stages of weight decrease, i.e. a gradual one up to 400 °C, a rapid one between 400 and 600 °C and a gradual one over 600 °C. The ΔM value of 10.2 wt.% for the rapid decrease agrees well with that caused by the reduction of U_3O_8 to UO_2 and decomposition of FeS_2 to FeS. The gradual part may be caused by the decomposition of sulfides. The XRD pattern for the product after the TG experiment at 1000 °C is shown in Fig.4 with reported patterns of UO_2(JCPDS41-1422), FeS(JCPDS37-477) and the synthesized FeS_2. The product seems to be a UO_2 phase with a small amount of FeS. The peaks for FeS_2 were not observed. Thus, the reduction of U_3O_8 could be caused by the oxidation of sulfur, which was generated from the decomposition of FeS_2,

$$U_3O_8 + FeS_2 \rightarrow 3UO_2 + FeS + SO_2 \quad (1)$$

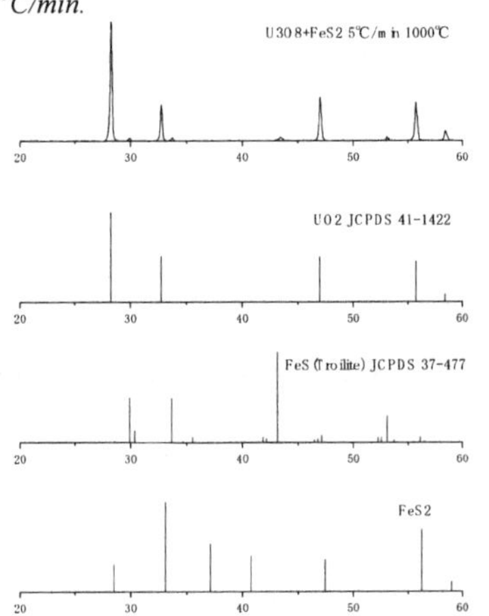

Figure 4 *XRD profiles of the products obtained by the reaction of U_3O_8 with FeS_2.*

Furthermore, the ΔG^0 values for the above reaction are negative for all the temperatures considered, the oxidation of sulphur is enough for the reduction of U_3O_8 to UO_2, but not enough for the sulfurization of these oxides.

4 CONCLUSION

In the present study, the low temperature selective reduction of uranium oxide (U_3O_8) in the presence of sulfurizing agents such as CS_2 and FeS_2 was examined using XRD and TG methods. The U_3O_8 was found to be reduced to UO_2 at around 300 °C in the presence of CS_2 vapour. Such reduction was also observed at around 600 °C in the presence of FeS_2 solid.

References

1 M. Skrobian, N. Sato and T. Fujino, *Thermochim. Acta*, 1994, **249**, 211.

2 N. Sato, H. Masuda, M. Wakeshima, K. Yamada and T. Fujino, *J. Alloy Compd.*,1998, **265**, 115.

Separation and Solution Chemistry

ACTINIDE SOLUTION CHEMISTRY AND CHEMICAL SEPARATIONS: STRUCTURE-FUNCTION RELATIONSHIPS IN THE GRAND SCHEME OF ACTINIDE SEPARATIONS SCIENCE

K.L. Nash

Department of Chemistry, Washington State University, PO Box 644630, Pullman, WA, 99164-4630, USA

1 INTRODUCTION

Actinide partitioning has received global attention during the last two decades to support the development of advanced fuel cycles that will secure future supplies of nuclear fuels, optimise repository performance and hinder the spread of nuclear weapons (through transmutation). Interest in actinide partitioning has supported the development of new reagents and processes for the recovery and isolation of these species, though solutions to some issues remain elusive. Both pyrometallurgical and improved conventional aqueous methods show promise. Concern about weapons proliferation, terrorism and increased emphasis on protection of the environment has also elevated the importance of fast and efficient analytical scale separations methods and remediation strategies for actinides.

There are both important similarities and differences between separations for processing and analysis. In analytical-scale separations: there is a need for highly efficient single-stage separation reactions, thus highly selective receptors; volatile solvents like chloroform are often preferable; solid phase separations materials offer the advantage of compactness and large numbers of separation stages if chromatographic operations are possible; and analyte concentrations are typically low. For spent fuel reprocessing by hydrometallurgy: reversibility of extraction reactions is essential (hence less efficient single stage efficiency is often acceptable); low volatility, non-flammable solvents are essential; continuous operation is readily achieved; and concentrations of solutes are high. Complex feed solutions are the rule for hydrometallurgical separations (less common for analytical separations). Preconcentration is often needed for analytical separations; dilution is sometimes necessary for spent fuel processing. Waste minimization is a much higher priority for hydrometallurgical separations, as are concerns about the degradation of reagents through hydrolysis or radiolysis. Each is best served by rapid reactions.

The same sorts of metal ion receptors and solvent compatibility considerations apply for both types of separations, though the relative importance of these features shift with priorities. Ultimately, our understanding of the solution chemistry of actinide ions has both permitted and limited advances. In the following, selected aspects of the interface between fundamental actinide solution chemistry and the development and application of separations processes will be discussed.

2 FUNDAMENTAL ACTINIDE SOLUTION CHEMISTRY IN SEPARATIONS

Some features of the basic chemistry of actinides in the acidic solutions of relevance to most separations are well understood. The diverse redox chemistry of U and Pu is essential to the operation of the PUREX process, and at the root of the inconsistent partitioning of Np in conventional PUREX. The trivalent oxidation state dominates the chemistry of transplutonium actinides. The kinetics of actinide redox reactions in acid solutions has been well studied for many important systems in acidic solutions; redox kinetics in neutral-basic solutions and complexation and phase transfer kinetics less so. In a given oxidation state, actinides behave similarly (but not identically). Aside from their stronger interactions with donor atoms softer than O (N, S, Cl^-), trivalent actinides have nearly identical chemistry with lanthanides. The cations are strongly hydrated in each oxidation state and suffer different degrees of susceptibility to hydrolysis in each oxidation state. The oxidized species (An(V,VI)) are substitution-inert dioxo cations in acidic solutions.

Actinide chemistry in strongly alkaline solutions is possibly foremost among features of actinide solution chemistry that have been under-investigated. The energetics of solvation of metal complexes in aqueous and organic media remains unpredictable. Because of the general absence of directed valence in actinide complexes, structure-function relationships for the design of actinide-specific reagents are not as well developed as they are for d-transition elements. Radiation-stable materials for actinide-lanthanide separations remain an elusive goal. Most fundamental experimental investigations have been conducted under well-defined and controlled experimental conditions. The conditions most relevant to process conditions are often far more complex. Therefore, studies of actinide chemistry under non-ideal conditions (i.e., complex mixtures, high concentrations) are needed. Actinide solution chemistry in unconventional media (room temperature ionic liquids, molten salts, micelles and third phases, hydrolytic polymers, supercritical fluids) has been only minimally explored.

3 ENERGETICS OF SEPARATIONS REACTIONS

In aqueous media, the high dielectric constant, significant dipole moment and hydrogen bonding ability of H_2O support the existence of free ions as independently mobile species. In organic solutions, the corresponding low dielectric constant and general absence of important hydrogen bonding interactions (long chain alcohols aside) demand neutral species be present at equilibrium, i.e., significant charge separation is poorly tolerated. London dispersion forces and van der Waals interactions are most important in nonpolar organic solutions. Thus, the most useful extractant molecules saturate the coordination sphere of the metal ion, masking its polar nature, while presenting to the solvent an appropriately arranged lipophilic shell of alkyl groups. The species present in the organic phase, usually created by self-assembly at the aqueous-organic interface, is larger than the hydrated ion present in the aqueous phase. One would expect that molecular diffusion should occur at a much slower rate.

While the dynamic features of actinide complexation and redox reactions in aqueous media have received some attention, the chemical dynamics of ligand exchange processes in organic media have not. In one example, the rate of exchange of TBP molecules on $UO_2(NO_3)_2(TBP)_2$ solutions in *o*-xylene, investigated by ^{31}P NMR, is defined by the kinetic parameters $\Delta G^* = 63.4$ kJ mol^{-1} ($k = 48\ M^{-1}s^{-1}$), $\Delta H^* = 30.8$ kJ mol^{-1}, and $\Delta S^* = -109$ J $K^{-1}mol^{-1}$. The negative value for ΔS^* indicates that the exchange reaction proceeds by an associative mechanism, presumably involving an intermediate in which free and bound TBP molecules exchange places. The lability of NO_3^- in this complex is unknown.

In addition to NMR relaxation methods, stop-flow spectrophotometry techniques also probe millisecond-second time regime and can yield information complementary to that obtained by NMR. Studies have been conducted on both UO_2^{2+} and NpO_2^+ complex formation and dissociation reactions with bidentate and tridentate complexing agents including a series of structurally similar diphosphonic acids, diglycolic and dipicolinic acid using this technique.[2] Isokinetic plots establish that the structural features of the ligands determine common mechanistic pathways for UO_2^{2+} and NpO_2^+ species, i.e., the hydration and charge density of the free cations are of secondary importance. It has been noted that Np(V) complex formation reactions are often more rapid than U(VI) despite greater thermodynamic driving force for U(VI). These observations suggest that the rate of exchange of water molecules on a precursor complex controls the rate of complexation.

Thermodynamic data can provide other insights into the arrangement of molecules in condensed liquid phases. For example, in the solid state, complexes of lanthanide nitrates clearly indicate that a tridentate coordination mode predominates for the 2,6-bis[(dialkylphosphino)methyl]pyridine-N,P,P'-trioxide (R_4NOPOPO) complexes.[3] For this coordination mode to continue in the organic solution relevant to solvent extraction, it becomes necessary to accommodate up to 12 oxygen donor atoms in the inner coordination sphere, two from each bidentate nitrate, three from each of two NOPOPO ligands. If we consider the thermodynamic parameters for the extraction of Am^{3+} from 0.5 M nitrate by NOPOPO and octyl(phenyl)-N,N-diisobutylcarbamoylmethylphosphine oxide (CMPO) at pH 2.6 to 3, we find that the parameters are nearly identical ($Am(NO_3)_3NOPOPO_2$: $\Delta G° = -31.0$ kJ/mol, $\Delta H° = -42.6$ kJ/mol, $\Delta S° = -39$ J/mol K, $Am(NO_3)_3CMPO_3$: $\Delta G° = -32.2$ kJ/mol, $\Delta H° = -49.4$ kJ/mol, $\Delta S° = -58$ J/mol K), suggesting a lower denticity than the maximum is prevalent.[4] The similarity of the thermodynamic parameters for extraction of UO_2^{2+} by NOPOPO with the series of monodentate organophosphorus extractant molecules in Table 1 suggests low denticity for the NOPOPO ligand in the uranyl system as well. Two important conclusions can be drawn from these observations 1) solid state structures do not necessarily describe solution phase species of actinide complexes, and 2) the relationship between the structural features of a ligand and its ability to bring all possible donor atoms to bear on an actinide ion in solution is a complex function of overlapping effects.

Table 1 *Thermodynamic parameters for extraction of UO_2^{2+} by selected organophosphorus reagents*

System	ΔG° kJ/mol	ΔH° kJ/mol	ΔS° J/mol K
TBP	-51.7	-60	-28
debebp	-48.2	-58	-33
NOPOPO	-43.4	-56	-41
4mepep	-36.3	-46	-32
db[bp]	-34.5	-44	-32
B[DBP]	-15.5	-25	-32

4 ORGANIZATION OF SOLUTES IN ORGANIC SOLUTIONS

The low dielectric constant of organic solvents combined with the amphiphilic nature of solvent extraction reagents leads to the spontaneous organization of solute species in the organic media of importance to solvent extraction. Some examples of the inherent organization of extractant molecules include the dimerization of dialkylphosphoric acid or carboxylic acid extractants, polymerization of dialkyldiphosphonic acid extractants and the micellization of strongly acidic sulfonic acid extractants.[5] The process of extraction of metal nitrate salts by solvation with neutral donor ligands is a self-assembly process that generally causes a significant increase in order in solvent extraction reactions of this type.

It is also well-known that high concentrations of polar solutes can cause secondary organization of solute molecules in nonpolar organic solutions, leading in the extreme to the formation of multiple organic phases. This phenomenon, known as third phase formation in solvent extraction, is an undesirable (and potentially dangerous) outcome for a solvent extraction reaction and is at present understood only in the most rudimentary fashion. Third phase formation has been most extensively investigated in connection with PUREX processing of spent nuclear fuels.[6] Most investigations have addressed Th(IV), Pu(IV), U(IV), or U(VI) extraction by TBP. Detailed investigations of third phase formation in developing processes (like DIAMEX) have become more common in recent years. Third phase formation is a complex function of concentrations of extractant, mineral acid, metal salt, nature of the diluent, ionic strength and temperature. The phenomenon most commonly occurs in normal alkane solutions. "Phase modifiers" (secondary solvents - alcohols, branched or aromatic hydrocarbons) or highly branched diluents are employed to reduce third phase formation. More sophisticated experimental methods and elaborate models have been developed in recent years attempting to develop a fundamental understanding of the factors that drive and limit third phase formation.[7]

5 NEPTUNIUM CHEMISTRY IN SEPARATIONS AND WASTE DISPOSITION

Neptunium is produced in spent nuclear fuel with a yield of about 500 g Np/ton of fuel for typical light water reactor fuels and average burnup times. The complex redox chemistry and 2.1×10^6 yr half-life of ^{237}Np combine to make Np one of the more challenging isotopes in the nuclear fuel cycle. It has value as feedstock for the production of ^{238}Pu, but is otherwise problematic. From the refluxing HNO_3 used to dissolve fuel, a mix of Np(V) and Np(VI) emerges. Np(V) is weakly extracted from HNO_3 by TBP while Np(VI) or Np(IV) would co-extract with U and Pu. In historic PUREX processing, it was mainly preferred the Np be maintained in the pentavalent oxidation state and so left with fission products, Am and Cm for direct disposal as waste. Several modern actinide partitioning schemes take advantage of the more extractable Np(IV) or Np(VI) forms to allow its controlled co-extraction with U or Pu in PUREX-like processing schemes. Notably, butyraldehydes, hydroxamic acids and electrochemical adjustments have all been demonstrated as useful approaches to Np partitioning control.[8]

In the single pass (i.e., no recycle) nuclear fuel cycle, or if Np is left in the PUREX raffinate, the long-lived ^{237}Np becomes the most important contributor to the alpha radiation dose at 250,000 years after disposal. The predominance of NpO_2^+ under neutral pH conditions will allow comparatively high mobility for Np in the geo/hydrosphere. In addition, the operation of spent fuel recycling for Pu recovery at production facilities like the Hanford site has left a legacy of alkaline wastes to be transitioned to a more stable form for final disposition. Comparatively few experimental

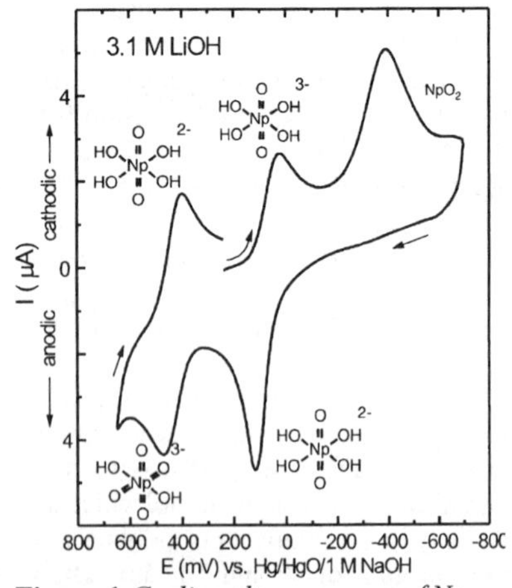

Figure 1 *Cyclic voltammogram of Np Species in 3.1M LiOH*

observations of the chemistry of Np in strongly alkaline solutions have been made. In Figure 1 is shown a cyclic voltammogram of Np in 3.1 M LiOH. The traces clearly indicate that the Np(VI)/Np(V) transition (E ≈ +100 mV, relative to Hg/HgO reference electrode) is electrochemically reversible, that the Np(VII)/Np(VI) transition (E ≈ +450 mV) is quasi-reversible, and the Np(V)/Np(IV) transition, (E ≈ -350 mV) is irreversible.[9] The most probable soluble species to account for this behavior are shown in the figure. Electrochemical reversibility becomes less common as the carbonate concentration increases. It has also been reported that Np(V)-Np(VI) couples are quasi-reversible in 1 M NaOH solutions containing sodium oxalate, and that more complex speciation is indicated in basic solutions of EDTA. Thermodynamic data in the literature do not explain these observations, emphasizing the essential fact that much of the chemistry of actinide ions in alkaline solutions remains undiscovered.

6 STRUCTURE-FUNCTION RELATIONSHIPS AND LIGAND DESIGN

Ion exchange separations of the actinides beyond Pu from other actinides and from fission product lanthanides played a central role in the discovery and characterization of these elements. However, ion exchange resins alone exhibit limited inherent ability to accomplish the needed separations. Introduction of appropriate eluants into the mobile phase proved essential. The work of Diamond et al.[10] demonstrated the clear importance of chloride donor atoms for the separation of trivalent actinides from lanthanides, and provided the first clear evidence for a covalent contribution in the bonding of actinides to ligand donor atoms softer than oxygen.

To accomplish the even more demanding task of mutually separating adjacent trivalent actinide cations, reagents capable of size recognition were needed. The separations supporting the first characterization of Am and Cm could be readily accomplished using buffered lactate or citrate as eluant. These complexants proved less useful for the heavier and more transient actinides. The synthesis of Md was accomplished as the first example of "one-atom-at-a-time" chemistry, demanding reliance on the parallel chemistries of trivalent actinides with the corresponding lanthanide ions for chemical identification. Ultimate predictability of elution behavior was achieved with the introduction of α-hydroxyisobutyric acid eluant for cation exchange separations.[11] The hindrance to free rotation of ligand functional groups provided by the introduction of a second methyl group onto lactic acid provided the perfect balance between rapid kinetics and size selectivity to make the separation sufficiently predictable. This is perhaps the first example of intentional ligand "design" to tailor specificity in actinide separations.

In the subsequent years of the late 1950's through at least the 1970's, enormous efforts were expended world-wide to apply similar rational design to the creation of increasingly efficient solvent extraction reagents, one particularly notable example being the development of the highly size-selective bis(2-ethyl(hexyl))phosphoric acid extractant (HDEHP).[12] Subsequently, the emphasis of waste management in spent fuel treatment provided incentive for creation of the purpose-designed extractant CMPO, cornerstone of the TRUEX process.[8] This reagent was the product of more than a decade of targeted ligand design, including considerable fine-tuning of ligand donor strength and phase compatibility characteristics. Its development is a landmark as the first dedicated reagent designed for actinide partitioning and waste management.

To overcome some of the limitations of CMPO for actinide partitioning applications, French researchers have developed an alternative complexant/separations system based on lipophilic analogs of malonic acid, the malonamide extractants, and a process for their use, DIAMEX.[8] Unlike the CMPO reagent for which no water-soluble analog exists, the performance of malonamide extractants can be predicted based on the thermodynamic properties of water-soluble malonic acid complexes. Thermodynamic data for aqueous malonate complexes are consistent with the solvent extraction results. To illustrate the positive impact of preorientation of ligand donor atoms on complex stability, Lumetta and coworkers have noted a 10^7 stronger extraction of lanthanides in a structurally hindered analog of malonamide extractants.[13]

A clear demonstration of the connection between thermodynamic parameters for aqueous complexation reactions and the same cation receptor as an extraction reagent is seen in the diglycolic acid system. The aqueous chemistry of this system has been investigated in great detail.[14] Diglycolate has been adapted for use as a solvent extraction reagent by Sasaki and coworkers in the form of a tetraoctyldiglycolamide (TODGA) extractant.[15] The two sets of data in Figure 2 have been arranged to emphasize the overlap of the "free energies" for the two systems in the mid-lanthanide region. The tridentate coordination geometry of DGA fully displaces water molecules from the inner coordination sphere of the lanthanides in the tris-complex. In the SX system, both DGA extractants and nitrates must be accommodated, hence the overlap between the data sets is not perfect. In fact, the slight net flattening of the curve in the solvent extraction system may reflect principally the effect of coordinated nitrate anions in the TOGDA system. Both the overlapping region and the discrepancies contain essential information relating the coordination chemistry in aqueous and organic media. Additional work (both experimental and computational) is needed to increase understanding of the differences. Sasaki and co-workers also have reported that at [HNO_3] above 1 M, trivalent actinide nitrates are more strongly extracted than hexavalent actinides, a circumstance rarely encountered in solvent extraction. The reduced extractability of U(VI) is consistent with the relative stability of the aqueous complexes of UO_2^{2+} and the trivalent lanthanides (e.g., log β_2 (U(VI)-L) = 7.54 , log β_2 (Eu(III)-L) = 11.28). A steric conflict between the planar geometry of the uranyl axial cooordination band and the ability of DGA to adopt a planar arrangement of donor atoms is implicit in this observation.

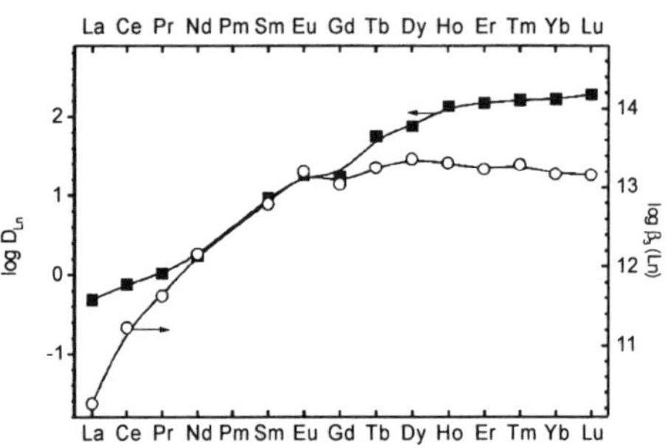

Figure 2 *Lanthanide complexes with DGA_{aq} (log β_3 [16]) and distribution ratios (log D [17]) into TODGA from 1 M HNO_3.*

7 LANTHANIDE-ACTINIDE SEPARATIONS

Among the most challenging tasks in actinide separations is the mutual separation of trivalent transplutonium actinides from fission product lanthanides. Diamond's work has been followed by the development of a number of complementary approaches to accomplishing this separation. The application of chloride as the agent of An(III)/Ln(III) separation has seen its most effective application in the TRAMEX process[16], developed at ORNL in mid 1960's. In this process, ternary or quaternary amine extractants achieve An(III)/Ln(III) separation factor of about 100 from 11 M LiCl solutions. A similar approach has been used for analytical-scale separations.[17]

Another significant development in trivalent actinide/lanthanide separations was the observation that aminopolycarboxylic acids could affect this separation (besides being useful reagents for size-selective lanthanide separation in ion exchange). The TALSPEAK (**T**rivalent **A**ctinide/**L**anthanide **S**eparation by **P**hosphorus reagent **E**xtraction from **A**queous **K**omplexes) Process developed by Weaver and Kappelmann[18] relied on the increased strength of the interaction of trivalent actinides with the amine donor atoms of diethylenetriamine-N,N,N',N",N"-pentaacetic acid (DTPA). As a "soft" donor atom, amine nitrogen is somewhat more effective than Cl^-, as its comparatively small size (thus higher charge density) makes it hard enough to interact more strongly with the actinide ion than Cl^-. The general features of TALSPEAK have been applied in recent years for trivalent actinide/lanthanide separation in several actinide partitioning schemes.[8]

Aromatic amine donor atoms (pyridines, pyrazines to name two) have also been demonstrated as potential contributors to efficient separations. The overall reduction in the strength of bonding between the cation and the ligand donor atom that occurs as N donors replace O donors requires that such reagents be preferentially composed of numerous N-donors in a suitably pre-arranged geometric fashion.[8] Thiophosphinic acids also have been demonstrated in recent years to be quite capable reagents for this separation. Unfortunately, both classes are susceptible to radio- and hydrolytic degradation that renders them less effective under typical process conditions.[8]

8 UNCONVENTIONAL APPROACHES TO ACTINIDE SEPARATIONS

The extreme demands of process chemistry, particularly radiolytic degradation and the need to process moderately concentrated solutions of diverse mixtures, have fostered a number of alternative separations methods, those most developed relying on pyrometallurgy or volatility. In each of these systems, the coordination chemistry of the actinide ions is generally simple, involving up to six halide ions (F^- for volatility and F^- or Cl^- for pyrometallurgy). Solvation is either irrelevant or inconsequential in each case. Separations in these categories either work or they don't, i.e., there is little opportunity to significantly alter the separation performance by modifying the conditions.

Extraction using an HNO_3-TBP complex in supercritical CO_2 has been demonstrated as an alternative approach to dissolving UO_2, the matrix of spent nuclear fuel.[8] Because of the high pressures and elevated temperatures required to generate supercritical fluids, applicability of this technique to the dissolution of spent nuclear fuels remains an open question. The development in recent years of room temperature ionic liquids (RTILs), organic fluids of low volatility, generally high viscosity and wide electrochemical windows, has raised expectations that these materials might create new avenues to accomplishing important actinide separations. One would expect to see unique coordination environments around actinide ions dissolved in RTILs, thus the opportunity to do separations not readily achievable in more conventional media. To date, significant intrinsic solubility of actinide salts in RTILs has not been seen without the introduction of carrier ligands, as in conventional solvent extraction. Because RTILs can be made either hydrophilic or lipophilic depending on the choice of anion, it might be possible to develop separation systems based on the combination of a conventional organic solution with an RTIL. Other potential obstacles are that these materials tend to exhibit chemical instability in contact with water and prices are high enough to be prohibitive for large-scale applications (unless they provide a breakthrough improvement in separation efficiency).

9 CONCLUSIONS

The world's growing need for energy supplies that do not adversely impact the global climate can be well served by expansion of the use of nuclear power for electricity production. However, this resource will ultimately have a limited impact without intelligent use of the fuel available, the breeding of additional fuel and close attention to waste management and weapons proliferation issues. Aqueous-based technologies will likely continue to dominate the field for the next 20 years. "Dry" concepts (pyrometallurgy and volatility in particular, but also less conventional techniques like those based on supercritical fluids or RTILs) will likely assume a greater role in the future. Continued studies of actinide solution chemistry have great potential for advancing this future through improved understanding of structure-function relationships, and of detailed descriptions of solvation phenomena in particular. An impending major impediment to growth in this field is the virtual disappearance of educational opportunities for a next generation of experts in this demanding science and technology.

References

1 A. K. Mohammed, J. C. Sullivan, K. L. Nash, *Solvent Extr. Ion Exch.,* 2000, **18**, 809
2 H. Hall, J.C. Sullivan, P.G. Rickert, K.L. Nash, *Dalton Trans.,* 2005, 2011
3 R. T. Paine, Design of Ligands for f Element Separations. in *Separations of f Elements*. K. L. Nash, G. R. Choppin. Eds., New York, Plenum, 1995, pp. 63-75
4 K. L. Nash, C. Lavallette, M. Borkowski, R. T. Paine, X. Gan, *Inorg. Chem.,* 2002, **41**, 5849
5 J. Rydberg, M. Cox, C. Musikas, G. R. Choppin, Solvent Extraction Principles and Practice, Second Edition, Marcel Dekker, NY, 2004, pp. 134, 658-667
6 P. R. V. Rao, Z. Kolarik, *Solvent Extr. Ion Exch.,* 1996, **14**, 955
7 R. Chiarizia, K. L. Nash, M. P. Jensen, P. Thiagarajan, K. C. Littrell, *Langmuir,* 2003, **19**, 9592
8 J. N. Mathur, M. S. Murali, K. L. Nash, *Solvent Extr. Ion Exch.,* 2001, **19**, 357
9 A. V. Gelis, P. Vanysek, M. P. Jensen, K. L. Nash, *Radiochim. Acta,* 2001, **89**, 565
10 R. M. Diamond, J. K. Street, G. T. Seaborg, *J. Am. Chem. Soc.,* 1954, **76**, 1461
11 G. R. Choppin, R. J. Silva, *J. Inorg. Nucl. Chem.,* 1956, **3,** 153
12 D. F. Peppard, G. W. Mason, J. L. Maier, W. J. Driscoll, *J. Inorg. Nucl. Chem.,* 1957, **4,** 334-343
13 G. J. Lumetta, B. M. Rapko, B. P. Hay, P. A. Garza, J. E. Hutchison and R. D. Gilbertson, *Solvent Extr. Ion Exch.,* 2003, **21,** 29-39
14 A. E. Martell, R. M. Smith, *Critically Selected Stability Constants of Metal Complexes Database Database 46.* Gaithersburg, MD 20899, 1998, NIST
15 Y. Sasaki, Y. Sugo, S. Suzuki, S. Tachimori, *Solvent Extr. Ion Exch.,* 2001, **19**, 91
16 W. D. Bond, R. E. Leuze, Removal of Americium and Curium from High-Level Wastes. in *Actinide Separations*. J. D. Navratil, W. W. Schulz, Eds. Washington, D. C., American Chemical Society, 1980, **ACS Symposium Series Vol. 117,** 441
17 E. P. Horwitz, M. L. Dietz, R. Chiarizia, H. Diamond, I. S. C. Maxwell, D. R. Nelson, *Anal. Chem. Acta,* 1995, **310,** 63
18 B. Weaver, F. A. Kappelmann, *J. Inorg. Nucl. Chem.,* 1968, **30,** 263

X RAY ABSORPTION SPECTROSCOPY OF ACTINIDES INVOLVED IN TOXICOLOGICAL PROCESSES

C. Den Auwer[1], E. Ansoborlo[1], P. Moisy[1], E. Simoni[2] and C. Vidaud[3]

[1]CEA Marcoule, DEN/DRCP/SCPS, 30207 Bagnols sur Cèze Cedex, France
[2]IPN Orsay, 91405 Orsay Cedex, France
[3]CEA Marcoule, DSV/DIEP/SBTN, 30207 Bagnols sur Cèze Cedex, France

1 INTRODUCTION

Actinide chemistry is surprisingly rich because of their large atomic numbers (from $Z = 89$ for Ac) and the relative availability of their valence electrons. For industrial, environmental and public health purposes, this chemistry has been the subject of considerable efforts since the 50's. Aqueous redox chemistry, ionic selective recognition, uptake by specific biomolecules or compartments of the geosphere are some of the major fields of investigation. In the field of human toxicology, internal contamination with actinides under either acute or chronic conditions has the potential to induce both radiological and chemical toxicity. Whatever the route of contamination (inhalation, ingestion or wound), the radionuclide is absorbed into, and then transported by the blood fluid (the absorption rate depends on the dissolution properties of the initial physico-chemical form) prior to deposition in the target organs (e.g. bone, kidney, liver) in which it is stored and then slowly eliminated through urines and faeces. These different biological steps are dominated by kinetics but underpinned, and to a significant extent controlled by the thermodynamic equilibrium underlying the speciation of compounds. Speciation studies,[1] which refer to the distribution of species in a particular medium, are necessary to improve the description, understanding and prediction of actinide trace element behaviour and toxicity. It may also have an important input to decorporation studies or treatment of the contamination by providing guidance on the structure, affinity and design of potential specific chelating agents synthesized and used for the elimination of an incorporated radionuclide. The geochemical and microbial behaviour of metal cations has been the subject of extensive research and it is well known that sorption, transport, interaction with minerals of the geosphere, natural degradation products or micro organisms can affect the cation speciation. Although these processes have long been investigated for transition metals, far fewer studies have been devoted to interaction mechanisms with actinides. At the molecular level, the general understanding of the interactions engaged in actinide aducts, in other words physical chemical mechanisms that drive the affinity of possible coordination sites for actinide cations still needs to be increased. As a result, the intramolecular interactions of actinide elements with either smart chelates designed for coordination and bioinorganic chemistry or naturally occurring chelating agents are relatively unknown.

Many traditional spectroscopic or analytical techniques have been used as speciation and molecular probes. Thermodynamic stability, structural data, formal oxidation state of the cation are some of the variables of the system. Among the spectroscopic techniques, X-ray Absorption Spectroscopy (XAS) is an element specific structural probe.[2] While the edge part of the absorption spectrum (XANES) is particularly sensitive to the properties of the central atom (often the metal cation) valence orbitals, the modulations of the absorption spectrum at higher energy (EXAFS) is a fingerprint of the central atom environment. Its identity is then available (coordination type and number, bond distances) using suitable structural models. The technique has been largely applied to environmental sciences.[3] It is particularly suited when the central atom is surrounded by a large number of atoms and the contrast in electronic density is poor. This is the case for instance when focus on the cation first coordination sphere is needed as for specific uptake by metaloprotein chelation sites or sorption studies onto mineral surfaces where a site by site structural (and electronic) analysis is needed. Since the 1980's, the considerable theoretical developments in XAS data analysis has made the technique available to large intricate systems and for instance BioXAS[4] is now considered has one of the major technique for structural biology and in particular metallogenomics. A large number of reviews have been covering the subject and the reader can refer to the issue of *Coordination Chemistry Reviews* (2005, **249**, issue 1-2) on synchrotron radiation in inorganic and bioinorganic chemistry, the issue of the *Journal of Synchrotron Radiation* (2005, **12**, part 1) on bioXAS and metallogenomics or the article of G. E. Brown *et al.*[5] on metal oxide surfaces.

The biokinetics and bioinorganic chemistry of all the actinides have been reviewed by the International Commission on Radiological Protection (ICRP Publications 67, 69), and a review and synthesis of these data have been previously published by Taylor.[6] The actinide chemistry is particularly rich because of the large number of available valence electrons. Stable or metastable oxidation states from III to VII are expected for the earlier actinide elements in aqueous chemistry. Oxidation states V and VI are often characteristic of the transdioxo cation unit in the form of AnO_2^{n+} (An = U, Np, Pu; n = 1, 2). For instance, the stability of the uranyl cation (UO_2^{2+}) under atmospheric conditions and its ubiquity in the geosphere defines it as a major potential contaminant.

This paper addresses some of the issues related to the "molecular speciation" of actinides with geochemical or microbial systems using the XAS probe. An extensive review of the actinide behaviour (and in particular Pu) in biological or natural systems is by far beyond the scope of this paper and is well referenced in the literature.

2 INTERACTION WITH THE GEOSPHERE

Migration of actinides in the aqueous environment, living organisms and plants is largely influenced by sorption mechanisms onto mineral surfaces. This sorption behaviour of actinides is currently described using surface complexation thermodynamical models and site by site interpretation of the surface complexes is often lacking. Among the questions that XAS has addressed are inner *versus* outer-sphere complexation, formation of polynuclear precipitates, and mono *versus* polydentate surface binding modes. Such data are the prerequisite to further model the interactions between the heavy metal ions and the mineral surfaces in aqueous medium although one of the drawbacks of the technique is to average the signal over all the possible species. Some of the more recent studies[7] exemplify this method, considering a topological approach between the cation coordination sphere and the surface. Such data have to be compared to the coordination processes to natural substances as humic acids described in the next chapter. The uranium cation has been the most extensively studied because of its ubiquity in the geosphere and the absence of radiological concerns. However, recent work has also been devoted to thorium,[8] neptunium,[9-11] plutonium,[12] and americium.[13] Many mineral substrates have been investigated, based on their occurrence in the geosphere, surface reactivity and potential redox properties : goethite (FeOOH),[14] muscovite ($KAl_2(AlSi_3O_{10})(F,OH)_2$), mackinawite (FeS),[10.15] hematite (Fe_2O_3),[16-17] ferrihydrite ($Fe_2O_3H_2O$),[18] montmorillonite ($(Na,Ca)(Al,Mg)_6(Si_4O_{10})_3(OH)_6$-$nH_2O$),[19-20] zircon ($ZrSiO_4$),[21] zirconium diphosphate (ZrP_2O_7),[22] titania (TiO_2),[23] silica[24-25] and alumina or feldspar minerals[14.26] etc. The physico-chemical parameters at the solid-solution interface such as pH, Eh, structure of the water molecules, may be different than those in the bulk solution, which could change drastically the radionuclides chemical properties at the interface. Therefore, careful investigation of the microscopic properties of this interface should be carried out. The redox chemistry of uranium in aqueous solution is predominantly governed by the +VI oxidation state (the +VI/+IV potential with respect to the standard electrode is equal to 0.27 V). As a consequence, no reduction to the +IV oxidation state is expected upon sorption in atmospheric conditions as confirmed by the invariance of the uranium L_{III} edges observed in most of the above references. Similarly, the +V oxidation state is conserved when NpO_2^+ is sorbed onto goethite (FeOOH).[9] Conversely, when mackinawite (FeS) is used as a substrate, reduction of Np(V) to Np(IV) has been observed.[10] Similarly, reduction of Pu(VI) to Pu(IV) has been reported upon sorption onto manganite and hausmannite.[12] When complexing counter-ions are present, such as carbonate anions, ternary inner-sphere complexes can be formed with the surface and the counter-ion. The final stoichiometry often depends on the sorption conditions (pH, concentration in solution). In presence of perchlorate ions, no inner-sphere complex with the counter-ion has been published to date. On average, the equatorial uranium coordination number ranges from 5 to 6 as in aqueous uranyl. When a two-shell fit of the equatorial oxygens is performed, a shorter bond distance is often observed between the uranium atom and the surface than between the uranium atom and the outer oxygens.[23] In addition, to provide a theoretical support to these experimental investigations, the interactions between uranyl ions and the TiO_2 (110) surface are currently studied by the authors using density functional calculations. The aim of such studies is to explore the bonding properties (covalency, sorption energy) between the actinide ion and the mineral surface.

3 INTERACTION WITH THE BIOSPHERE

Humic acids (HA) are natural, polyelectrolytic organic macromolecules that are present in soils, sediment and water. They play an important role in the migration behaviour that is intimately linked to the transport properties of mineral surfaces. Many studies have been devoted to the thermodynamics of actinide/HA systems, *i.e.* formation constants. However the HA-cation interaction at the molecular level is once again a difficult problem given the intricacy of the possible coordination functions. In the case of uranyl complexation, donor sites are therefore difficult to identify and alternate strategies as complexation by synthetic identified functional groups as xylose, phenylalanine or glycine are needed in order to understand the complex formation.[27] Comparison between the natural and synthetic humic acids has revealed no severe discrepancies, showing that the carboxylate groups act predominantly as monodentate ligands.[28-29] Similarly, natural wood-degradation processes lead to possible complexation by either hydroxybenzoic acids or phenols.[30] In acidic conditions, the carboxylic group coordinates in a bidentate mode to the uranyl cation, whereas in basic conditions the phenolic OH functions are the coordinating ones. Other actinides such as thorium,[31] neptunium,[32] plutonium,[33] and even curium[34] have been investigated.

Microbial effects related to actinide contamination are also a critical issue of public health and safety. How does the actinide cation interact with the micro organisms in terms of migration behaviour, chemical speciation and further transport to the superior organisms (*e.g.* human) are some of the questions to be addressed.[35] Although the aqueous geochemistry of actinides is an area of extensive research as described above, fewer studies have focussed on the interaction of actinides with micro organisms and microbial products[36] and their effect on redox and speciation data. For instance upon uptake by *bacillus sphaericus*, an aerobic soil bacteria, U(VI) forms a stable phosphate complex[37] On the contrary, Pu(VI) (which also forms a phosphate complex) is reduced with increasing contact time.[38] Uranium EXAFS measurements performed after uptake by various bacterial systems or plants indicates that in many cases, uranium is present at oxidation state VI and is predominantly bound to phosphoryl groups.[39-41]

In superior organisms, the intricacy of the biological medium challenges the spectroscopic investigation. To illustrate this difficulty, a selection (including major component concentrations in terms of ligand family such as minerals, organics, proteins, amino acids and metals) of the different media present in the human body (saliva, blood serum, gastric juice, mothers' milk) is presented in Table 1. In addition, the composition of a cell culture medium, used for *in vitro* experiments in order to simulate actinide behaviour is also included.

Table 1. Composition of some human biological media

Ligands (M)	Blood serum	Saliva	Gastric juice	Cultur medium	Mother milk (g/L)
pH	7.4	5.6 - 7.6	1.5 - 5.5	7.3	4.5 - 7.5
Mineral					
Carbonate	$2.5\ 10^{-2}$	$6.6\ 10^{-3}$		$2.6\ 10^{-2}$	
Phosphate	$1.1\ 10^{-3}$	$3.7\ 10^{-3}$	$2.8\ 10^{-3}$	$5.6\ 10^{-3}$	0.1
SCN^-		$2.1\ 10^{-3}$			
F^-		$7.3\ 10^{-6}$			
SO_4^{2-}	$3.3\ 10^{-4}$	$3.3\ 10^{-4}$		$0.4\ 10^{-3}$	0.1
Cl^-	$9.0\ 10^{-2}$	$2.9\ 10^{-2}$	$1.2\ 10^{-1}$	$1.3\ 10^{-1}$	0.4
NH_4^+		$2.6\ 10^{-3}$	$6.0\ 10^{-3}$		
NO_3^-				$0.4\ 10^{-3}$	
Organic					
Ascorbate		$8.0\ 10^{-6}$	$4.7\ 10^{-5}$		
Citrate	$1.6\ 10^{-4}$	$5.4\ 10^{-5}$			0.8
Lactate	$1.5\ 10^{-3}$	$1.4\ 10^{-4}$	$3.3\ 10^{-4}$		
Oxalate	$9.2\ 10^{-6}$				
Aspartate				$0.15\ 10^{-3}$	
Glutamate				$0.15\ 10^{-3}$	
Glucose				$11.0\ 10^{-3}$	70
Proteins + Amino-acids	10^{-4} à 10^{-5}	10^{-4} à 10^{-5}	10^{-4} à 10^{-5}		10
HEPES Buffer				$17.5\ 10^{-3}$	
Albumin	$6.3\ 10^{-4}$				
Transferrin	$3.7\ 10^{-5}$				
Urea		$3.0\ 10^{-3}$	$3.3\ 10^{-4}$		
Cystein				$0.1\ 10^{-3}$	
Glycin				$0.25\ 10^{-3}$	
Metals					
Ca^{2+}	$1.4\ 10^{-3}$	$1.5\ 10^{-3}$	$1.9\ 10^{-3}$	$1.0\ 10^{-3}$	0.3
Mg^{2+}	$5.6\ 10^{-4}$	$2.7\ 10^{-4}$	$3.1\ 10^{-4}$	$0.4\ 10^{-3}$	0.5
Na^+	$9.0\ 10^{-2}$	$1.0\ 10^{-2}$	$4.9\ 10^{-2}$	$1.4\ 10^{-1}$	0.2
K^+	$4.9\ 10^{-4}$	$2.0\ 10^{-2}$	$1.2\ 10^{-2}$	$5.0\ 10^{-3}$	$3.0\ 10^{-1}$
Fe^{3+}	$3.0\ 10^{-5}$				$5.0\ 10^{-4}$

Most studies related to biokinetics of actinides focus on blood, since the overall behaviour of radionuclides is dependent on their time residence in this compartment. In the frame of the French Human and environmental toxicology program (CEA, CNRS, INRA, INSERM, MRT) the authors have undertaken a project based on spectroscopic measurements (spectrophotometry and XAS) involving transferrin, one of the major carrier of iron in serum and potential metal ion mediator.[42] First XAS involving Np(IV) uptake by apotransferrin[43] have suggested that it is mainly coordinated by carboxylate functions. However a definite analysis of the chelation site in terms of protein donor groups is currently under investigation.

The increasing need in "molecular speciation" of actinide cations involved with biosystems defines XAS as a major structural and electronic probe as already demonstrated by the BioXAS application to metallogenomics. Combining theoretical chemistry as molecular dynamics or quantum chemistry with EXAFS and XANES data analysis should lead to a better understanding of the metal protein binding sites.

References

1 Definition of IUPAC (Templeton et al, 2000) : Speciation is the distribution of an element amongst defined chemical species in a system.
2 See for instance A. Levina, R. S. Armstong and P. A. Lay, *Coord. Chem. Rev.*, 2005, **249**, 141.
3 R. J. Silva and H. Nitsche, *Radiochim. Acta*, 1995, **70/71**, 377.
4 I. Ascone, R. Fourme, S. Hasnain and K. Hodgson, *J. Synch. Rad.*, 2005, **12**, 1.
5 G. E. Brown, Jr., V. E. Henrich, W. H. Casey, D. L. Clark, C. Eggleston, A. Felmy, D. W. Goodman, M. Grätzel, G. Maciel, M. I. McCarthy, K. H. Nealson, D. A. Sverjensky, M. F. Toney and J. M. Zachara, *Chem. Rev.*, 1999, **99**, 77.
6 D. M. Taylor, *Journal of Alloys and compounds*, 1998, **271**, 6.
7 M. Walter, T. Arnold, G. Geipel, A. Scheinost and G. Bernhard, *J. of Colloid and Interface. Science*, 2005, **282**, 293.
8 P. Dähn, A. M. Scheidegger, A. Manceau, E. Curti, B. Baeyens, M. H. Bradbury and D. Chateigner, *J. of Colloid and Interface Science*, 2002, **249**, 8.
9 J-M Combes, C. J. Chisholm-Brause, G. E. Jr. Brown, G. A. Parks, S. Conradson, P. G. Eller, I. R. Triay, D. E. Hobart and A. Meijer, *Environ. Sci. Technol.*, 1992, **26**, 376.
10 L. N. Moyes, M. J. Jones, W. A. Reed, F. R. Livens, J. M. Charnock, J. F. W. Mosselmans, C. Hennig, D. J. Vaughan and R. A. D. Pattrick, *Environ. Sci. Technol.*, 2002, **36**, 179.
11 E. Östhols, A. Manceau, F. Farges and L. Charlet, *J. of Colloid and Interface Science*, 1997, **194**, 10.
12 D. A. Schaughnessy, H. Nitsche, C. H. Booth, D. K. Shuh, G. A. Waychunas, R. E. Wilson, H. Gill, K. J. Cantrell and R. J. Serne, *Environ. Sci. Technol.*, 2003, **37**, 3367.
13 C. Degueldre, D. Reed, A. J. Kropf and C. Mertz, *J. Synchrotron Rad.*, 2004, **11**, 198.
14 M. Walter, T. Arnold, T. Reich and G. Bernhard, *Environ. Sci. Technol.*, 2003, **37**, 2898.
15 L. N. Moyes, R. H. Parkman, J. M. Charnock, D. J. Vaughan, F. R. Livens, C. R. Hugues and A. Braithwaite, *Environ. Sci. Technol.*, 2000, **34**, 1062.
16 J. R. Bargar, R. Reitmeyer, J. J. Lenhart and J. A. Davis, *Geochim. Cosmoschim. Acta.*, 2000, **64**, 2737.
17 J. R. Bargar, R. Reitmeyer, J. A. Davis, *Environ. Sci. Technol.*, 1999, **33**, 2481.
18 T. D. Waite, J. A. Davis, T. E. Payne, G. A. Waychunas and N. Xu, *Geochim. Cosmochim. Acta*, 1994, **58**, 5465.

19 C. Chisholm-Brause, S. D. Conradson, C. T. Buscher, P. G. Eller and D. E. Morris, *Geochim Cosmoschim. Acta*, 1994, **58**, 3625.
20 C. Hennig, T. Reich, R. Dähn and A. M. Scheidegger, *Radiochim. Acta* 2002, **90**, 653.
21 C. Lomenech, E. Simoni, R. Drot, J-J. Ehrardt and J. Mielczarski, *J. of Colloid and Interface Sci.*, 2003, **261/2,** 221.
22 R. Drot, E. Simoni and C. Den Auwer, *C. R. Acad. Sci. Paris, Série II c*, 1999, **2**, 111.
23 C. Den Auwer, E. Simoni, R. Drot, M. Gailhanou, S. D. Conradson and J. Mustre de Leon, *New J. Chem.*, 2003, **27**, 648.
24 T. Reich, H. Moll, T. Arnold, M. A. Denecke, C. Hennig, G. Geipel, G. Bernhard, H. Nitsche, P. G. Allen, J. J. Bucher, N. M. Edelstein and D. K. Shuh, *J. Elect. Spect. Related Phenom.*, 1998, **96**, 237.
25 E. R. Sylwester, E. A. Hudson and P. G. Allen, *Geochim. Cosmochim. Acta*, 2000, **64**, 2431.
26 M. A. Denecke, J. Rothe, K. Dardenne and P. Lindqvist-Reis, *Phys. Chem. Chem. Phys.*, 2003, **5**, 939.
27 S. Pompe, M. Bubner, M. A. Denecke, T. Reich, A. Brachmann, G. Geipel, R. Nicolai, K. H. Heise and H. Nitsche, *Radiochim. Acta*, 1996, **74**, 135.
28 M. A. Denecke, S. Pompe, T. Reich, H. Moll, M. Bubner, K. H. Heise, R. Nicolai and H. Nitsche, *Radiochim. Acta*, 1997, **79**, 151.
29 M. A. Denecke, T. Reich, S. Pompe, M. Bubner, K. H. Heise, H. Nitsche, P. G. Allen, J. J. Bucher, N. M. Edelstein, D. K. Shuh and K. R. Czerwinski, *Radiochim. Acta*, 1998, **82**, 103.
30 A. Roβberg, L. Baraniak, T. Reich, C. Hennig, G. Bernhard and H. Nitsche, *Radiochim. Acta*, 2000, **88**, 593.
31 M. A. Denecke, D. Bublitz, J. I. Kim, H. Moll and I. Farkes, *J. Synch. Rad.*, 1999, **6**, 394.
32 S. Sachs, K. Schmeide, T. Reich, V. Brendler, K. H. Heise and G. Bernhard, *Radiochim. Acta*, 2005, **93**, 17.
33 C. M. Marquardt, A. Seibert, R. Artinger, M. A. Denecke, B. Kuczewski, D. Schild and Th. Fanghänel, *Radiochim. Acta*, 2004, **92**, 617.
34 J-M. Monsallier, R. Artinger, M. A. Denecke, F. J. Scherbaum, G. Buckau and J-I. Kim, *Radiochim. Acta*, 2003, **91**, 567.
35 A. J. Francis, C. J. Dodge, J. B. Gillow and H. W. Papenguth, *Environ. Sci. Technol.*, 2000, **34**, 2311.
36 A. E. V. Gorden, J. Xu, K. N. Raymond and P. Durbin, *Chem. Rev.*, 2003, **103**, 4207.
37 C. Hennig, P. J. Panak, T. Reich, A. Roβberg, J. Raff, S. Selenska-Pobell, W. Matz, J. J. Bucher, G. Bernhard and H. Nitsche, *Radiochim. Acta*, 2001, **89**, 625.
38 P. J. Panak, C. H. Booth, D. L. Caulder, J. J. Bucher, D. K. Shuh and H. Nitsche, *Radiochim. Acta*, 2002, **90**, 315.
39 A. Günther, G. Bernhard, G. Geipel, T. Reich, A. Roβberg and H. Nitsche, *Radiochim. Acta*, 2003, **91**, 319.
40 M. Merroun, C. Hennig, A. Roβberg, T. Reich and S. Selenska-Pobell, *Radiochim. Acta*, 2003, **91**, 583.
41 A. J. Francis, J. B. Gillow, C. J. Dodge, R. Harris, T. J. Beveridge and H. W. Papenguth, *Radiochim. Acta*, 2004, **92**, 481.
42 See for instance H. Sun, H. Li and P. J. Sadler, *Chem. Rev.*, 1999, **99**, 2817 and references herein.
43 I. Llorens, C. Den Auwer, P. Moisy, E. Ansoborlo, C. Vidaud and H. Funke, *FEBS J.*, 2005, **272**, 1739.

SOLUBILITY AND REDOX REACTIONS OF PLUTONIUM(IV) HYDROUS OXIDE IN THE PRESENCE OF OXYGEN

V. Neck,[1] M. Altmaier,[1] and Th. Fanghänel[1,2]

[1] Forschungszentrum Karlsruhe, Institut für Nukleare Entsorgung,
PO Box 3640, D-76021 Karlsruhe, Germany
[2] Ruprecht-Karls Universität Heidelberg, Physikalisch-Chemisches Institut,
Im Neuenheimer Feld 253, D-69120 Heidelberg, Germany

1 INTRODUCTION

The solubility of Pu(IV) hydrous oxide, PuO_2(s, hyd), has been critically discussed in recent OECD/NEA reviews[1,2] which provide a well ascertained set of thermodynamic data and ion interaction coefficients for the system $Pu/e^-/H^+/OH^-/NaClO_4$ or $NaCl/H_2O$. The selected data are consistent with the solubility of PuO_2(s, hyd) in equilibrium with Pu^{3+} (under reducing conditions),[3,4] with PuO_2^+ and PuO_2^{2+} or Pu^{3+}, PuO_2^+ and PuO_2^{2+}, the predominant aqueous species at pH 1 - 3 under air,[5-8] and also with the equilibrium Pu(IV) concentrations ascribed to the species $Pu(OH)_n^{4-n}$.[2,9] However, particularly the low redox potentials (pe) at pH > 4 in the presence of O_2 are not yet understood.[7] Haschke et al.[10-13] claimed that PuO_{2+x}(s), mixed valent $(Pu^V)_{2x}(Pu^{IV})_{1-2x}O_{2+x}$(s, hyd),[14,15] is more stable but also more soluble than PuO_2(s), so that the measured Pu concentrations and oxidation state distributions do not represent equilibrium thermodynamics but kinetically controlled steady-state concentrations.[12] In order to clarify these questions we have analyzed the solubility of PuO_2(s, hyd) in solutions where radiolysis effects and complexing ligands like carbonate are negligible. The total Pu concentrations and the aqueous redox speciation are discussed with regard to the amount of oxygen present in the system and the redox potentials measured in the solubility studies.

2 DISCUSSION OF SOLUBILITY DATA

2.1 The effect of oxygen on the solubility of PuO_2(s, hyd)

Fig. 1 shows a comparison of total Pu and Pu(IV) equilibrium concentrations measured in solubility studies with PuO_2(s, hyd). The studies of Rai *et al.*[5-7] Kasha,[16] Kim and Kanellakopulos[17] at pH 1 - 8 were performed under air, those of Rai *et al.*[18] at pH 8 - 13, Lierse and Kim[19] at pH 1 - 12 and Altmaier et al.[20] at pH 2 - 13 under Ar. The initial Pu(IV) stock solution used in the study at our laboratory[20] contained 0.5 % Pu(VI) and the Ar glove box used had a certain O_2 contamination (ca. 10 ppm).

The total Pu concentration (log $[Pu]_{tot}$) measured at pH < 1.5 in samples under air (Fig. 1a) is slightly increased compared to log $[Pu(IV)]_{aq}$. It passes through a plateau at pH 1 - 3 and decreases with a slope of -1 at pH > 3. Aqueous speciation by spectroscopy and solvent extraction showed that the dissolved Pu at pH 1 - 3 consists mainly of PuO_2^{2+} and PuO_2^{+}, while PuO_2^{+} predominates at pH 3 - 9.[5-7,20] It should be noted that the oxygen concentration in solutions exposed to air and then kept in closed vials is not equal to $[O_2]_{aq}$ in equilibrium with the O_2 partial pressure of air. The available oxygen in these closed systems, i.e., the sum of $[O_2]_{aq} = 2.5\times10^{-4}$ M at $pO_2(g) = 0.2$ bar and $O_2(g)$ in the gas phase above the solution, corresponds to the sum [Pu(VI)] + [Pu(V)] at pH 1 - 3. This level (plateau of $[Pu]_{tot}$) is considerably lower in the present study [20] where only 0.5 % of the Pu is oxidised (Fig. 1a). Lierse and Kim[19] titrated PuO_2(s, hyd) solutions in 1 M $NaClO_4$ under a continuous Ar stream from pH 12 to pH 1 (Fig. 1b). The similar solubility curve, but with log $[Pu]_{tot}$ somewhat lower than in studies under air, can be explained if we assume that the Ar stream included ca. 1 - 3 ppm O_2 and that the amount of oxidised Pu accumulated during the titration experiment (2 - 3 months). Batch samples prepared in an Ar glove box (with negligible total amounts of O_2 in the closed vials) did not show this increase of $[Pu]_{tot}$.[19] The concentration of oxidised Pu species in equilibrium with PuO_2(s, hyd) is obviously correlated to the amount of oxygen present in the system.

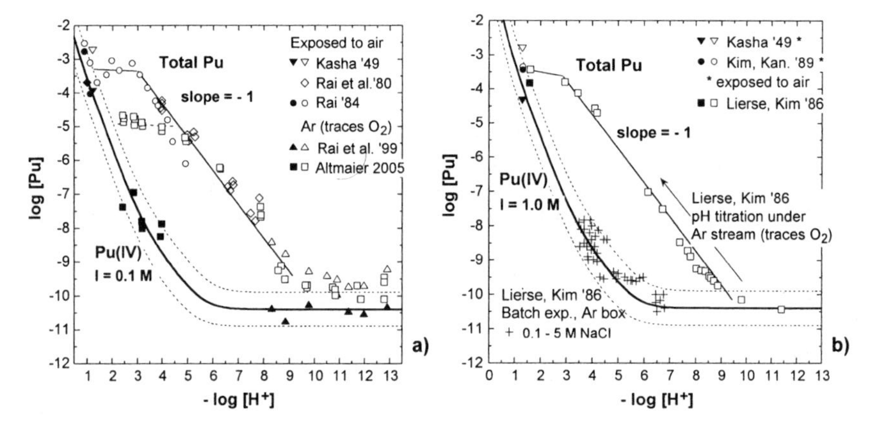

Figure 1 *Solubility of PuO_2(s, hyd) at 20-25°C [5,6,16-20] at $I \leq 0.1$ M (a) and $I = 1$ M (b) in the presence of different amounts of oxygen; total Pu (open symbols) and Pu(IV) concentrations (filled symbols) determined after ultrafiltration.*

The Pu concentrations shown in Fig. 1 were measured after ultrafiltration (usual pore size: 1.5 - 2 nm).[5-7,18-20] It is well known that tetravalent actinides have a high tendency towards polynucleation and colloid formation. In contact with solid AnO_2(am, hyd), these oxyhydroxide colloids $An_mO_{mx}(OH)_{m(4-2x)}(H_2O)_n$, in the following designated as AnO_2(coll, hyd), remain stable in solution as recently shown for Th(IV).[21] On the one hand they have properties of small solid particles, on the other hand they must be considered as large polynuclear aqueous species in equilibrium with both solid AnO_2(am, hyd) and aqueous species $An_m(OH)_n^{4m-n}$. In a recent study we quantified the pH-independent concentration of polymeric/colloidal Th(IV) in neutral and alkaline solutions by comparing Th(IV) concentrations in aliquots from the supernatant with those after ultracentrifugation or ultrafiltration.[21] In an analogous study with PuO_2(am, hyd) [20] the concentration $[Pu(IV)]_{coll}$ was determined after equilibration for 6 - 77 days:

$$AnO_2(am, hyd) \Leftrightarrow AnO_2(coll, hyd) \qquad (1)$$

with log K°_{coll} = log $[An(IV)]_{coll}$ = - 6.3 ± 0.5 for Th(IV)[21] and - 8.3 ± 1.0 for Pu(IV).[20] Using $\Delta_fG^\circ_m(PuO_2(am, hyd))$ = - 965.5 ± 4.0 kJ/mol,[2] a mean molar standard Gibbs energy of formation, $\Delta_fG^\circ_m(PuO_2(coll, hyd))$ = - 918.1 ± 7.0 kJ/mol, can be calculated for these Pu(IV) colloids or polymers. Their size, estimated from a correlation between standard Gibbs energy of formation and particle size, is in the range 1.5 – 2 nm.[20]

2.2 Solubility and pe controlling reactions

The major reactions in solubility studies performed with PuO_2(s, hyd) in the presence of oxygen cannot be understood by regarding only the measured Pu concentrations (Fig. 2a). The simultaneously measured redox potentials (Fig. 2b) must be considered as well.

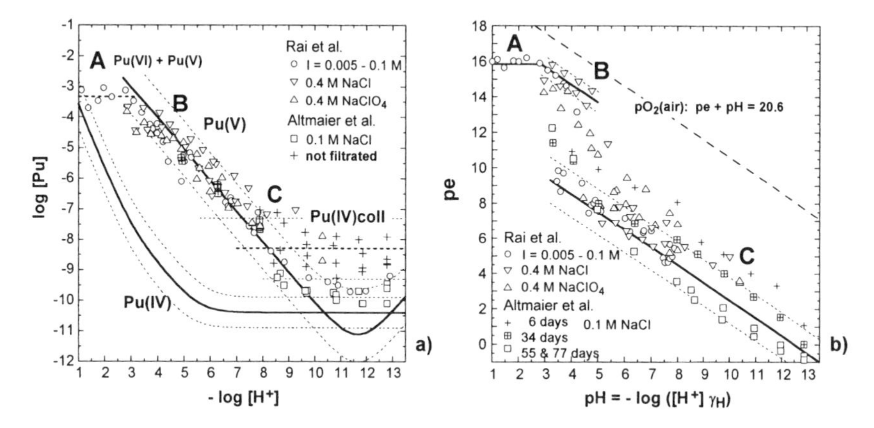

Figure 2 *a) Solubility of PuO_{2+x}(s, hyd) at 20-25°C; calculated Pu(IV) concentration,*[9] *Pu(V) concentration after ultrafiltration*[5-7,18,20] *(open symbols) and $[Pu]_{tot}$ including colloidal Pu(IV) polymer*[20] *(crosses); b) corresponding pe values.*

It is helpful to divide the experimental data into three pH regions with a certain overlap. In region A (pH 1 - 3) the solubility of PuO_2(s, hyd) in equilibrium with Pu(IV) aqueous species is increased by the formation of Pu(V) and Pu(VI) and the measured pe values are consistent with the PuO_2^+/PuO_2^{2+} redox couple. As shown by Rai[6] the solubility and pe values under air at pH < 4 (regions A and B) are controlled by the reactions (2) and (3):

$$PuO_2(s, hyd) \Leftrightarrow PuO_2^+ + e^- \quad (\log K^\circ_{IVs/V} = -19.45 \pm 0.23\,^{6} \text{ or } -19.78 \pm 0.86\,^{2}) \qquad (2)$$

$$PuO_2^+ \Leftrightarrow PuO_2^{2+} + e^- \quad (\log K^\circ_{V/VI} = -16.16 \pm 0.43\,^{6} \text{ or } -15.82 \pm 0.09\,^{2}). \qquad (3)$$

At pH > 3 (regions B and C) the solubility (log $[Pu]_{tot}$ ≈ log $[PuO_2^+]$) decreases with a slope of -1. The Pu(V) concentration alone does not require to distinguish between regions B and C, but the pe values drop to drastically (ca. 7 log-units) lower values (Fig. 2b). The solubility constant log $K^\circ_{IVs/V}$ for reaction (2), derived from the experimental values of pe and $[PuO_2^+]$, would differ by 7 orders of magnitude from the value derived in regions A and B.[5,6] Accordingly, the solubility limiting solid phase cannot be PuO_2(s, hyd).

The pe values in regions B and C (pe + pH = const., slope -1), can neither be explained by equilibria between aqueous Pu species nor by the O_2 partial pressure.[6,7] Aqueous Pu concentrations decrease to values below 10^{-5} M and the values of (pe + pH) in regions B and C correspond to $pO_2(g)$ = 10^{-8} and 10^{-33} bar, respectively. As the Pu(V) concentration at pH > 4 is much lower than that of initially dissolved oxygen, aqueous Pu species alone cannot have consumed the oxygen present in the system. Combining all this information allows only one interpretation: O_2 must be scavenged by the abundantly present solid phase PuO_2(s, hyd) yielding PuO_{2+x}(s, hyd) according to the net reaction of the water-catalyzed oxidation reported by Haschke et al.[11,12]:

$$PuO_2(s, hyd) + {}^{x}/_{2}\, O_2 \rightarrow PuO_{2+x}(s, hyd) \qquad (4)$$

As a consequence, the maximum concentration of PuO_2^+ at pH > 3 (slope -1 in regions B and C) is limited to the solubility of PuO_{2+x}(s, hyd). At pH < 3 (region A), the oxidised fractions of PuO_{2+x}(s, hyd) are below this solubility line (slope -1) and hence completely soluble.

2.3 Pu(V) concentration in equilibrium with PuO_{2+x}(s, hyd)

Recent EXAFS studies[14,15] indicate that PuO_{2+x}(s, hyd) is a mixed valent oxyhydroxide $(Pu^{V})_{2x}(Pu^{IV})_{1-2x}O_{2+x-n}(OH)_{2n}$(s, hyd). It may be considered as solid solution with Pu(IV) and Pu(V) end-members: PuO_{2+x}(s, hyd) = $(PuO_{2.5})_{2x}(PuO_2)_{1-2x}$(s, hyd). Haschke et al.[10-13] observed a maximum value of x = 0.27 ($PuO_{2.25}$(s) = $^{1}/_{4}$ Pu_4O_9(s)) but they supposed that $PuO_{2.5}$(s) = $^{1}/_{2}$ Pu_2O_5(s) should be most stable.[12-13] The values of x in the solubility studies can be estimated from the maximum of oxidised Pu (plateau at pH < 3, Fig. 1a), which is 10 % of the total amount of Pu in the samples of Rai *et al.*[6,7] under air and 0.5 % in the study of Altmaier *et al.*[20], i.e., x = 0.05 and 0.003, respectively. In the following simplifying treatment the PuO_2^+ concentrations at pH 3 - 9 (slope -1) are ascribed to reactions (5) or (6)

$$PuO_{2.5}(s, hyd) + H^+ \Leftrightarrow PuO_2^+ + {}^{1}/_{2}\, H_2O \qquad (\log{}^*K_{s,0}) \qquad (5)$$

$$PuO_{2.5}(s, hyd) + {}^{1}/_{2}\, H_2O \Leftrightarrow PuO_2^+ + OH^- \qquad (\log K_{sp}) \qquad (6)$$

The different sets of solubility data determined by Rai *et al.*[5-7] in 0.0015 M $CaCl_2$, dilute media (I < 0.1 M), 0.4 M $NaClO_4$, 0.4 M NaCl, 4.0 M $NaClO_4$, and 4.0 M NaCl (in the latter case radiolysis led to higher pe values and Pu(VI) formation after 100 days[7]) under air but also those of Lierse and Kim[19] and Altmaier *et al.*[20] under Ar (in the presence of only traces of oxygen) yield consistent equilibrium constants at I = 0, with a mean value ($\pm 2\sigma$) of $\log{}^*K°_{s,0}(PuO_{2.5}(s, hyd)) = 0.0 \pm 0.8$ or $\log K°_{sp}(PuO_{2.5}(s, hyd)) = -14.0 \pm 0.8$.

This solubility product calculated for $PuO_{2.5}$(s, hyd) as solid solution in PuO_{2+x}(s, hyd) = $(PuO_{2.5})_{2x}(PuO_2)_{1-2x}$(s, hyd) is 5 orders of magnitude lower than that of Pu(V) hydroxide ($\log K°_{sp}(PuO_2OH(am, hyd)) = -9.0 \pm 0.5$ [1,2]). It compares well with known values for Np(V) pentoxide: $\log K°_{sp}(NpO_{2.5}(cr)) = -12.2 \pm 0.8$ (calculated from thermochemical data)[1,2] and $\log K°_{sp}(NpO_{2.5}(s, hyd)) = -11.4 \pm 0.4$ (solubility study of Efurd *et al.*[22]), in particular if one takes into account that the known solubility constants for (PuO_2OH(am), PuO_2(am, hyd), and PuO_2(cr)) are also 0.3 - 1.7 log-units lower than those of the analogous Np(V) and Np(IV) hydroxides and oxides.[2,9]

2.4 The effect of $PuO_{2+x}(s)$ and Pu(IV) colloids on the redox potentials at pH > 3

The pe values at pH < 3 (region A in Fig. 2) are consistent with the redox equilibrium between PuO_2^+ and PuO_2^{2+},[6] but the redox potentials at pH > 3, where the total dissolved Pu is dominated by Pu(V) and Pu(IV) colloids or polymers,[6,7,20] could not yet be explained. The understanding of the pe values in regions B and C requires the equilibrium constants derived above for $PuO_{2.5}$(s, hyd) and PuO_2(coll, hyd). Combining the solubility controlling equilibria (2) and (5), with $\log K°_{IVs/V} = -19.8 \pm 0.9$[2] and $\log{}^*K°_{s,0} = 0.0 \pm 0.8$, yields

$$PuO_2(s, hyd) + {}^1/_2\, H_2O \Leftrightarrow PuO_{2.5}(s, hyd) + e^- + H^+ \quad (7)$$

with $\log K° = -pe - pH = -19.8 \pm 1.2$. This explains the slope of -1 in region B (Fig. 2b) and is reasonably consistent with the experimental values of (pe + pH) = 18.7 ± 0.6.[6,7] The value of (pe + pH) is buffered by $PuO_{2+x}(s)$ containing both PuO_2(s, hyd) and $PuO_{2.5}$(s, hyd). At pH > 4, Pu(IV) colloids/polymers contribute significantly to the total Pu in solution (Fig. 2a). Combining the equilibria (7) and (1), with $\log K°_{Pu(IV)coll} = -8.3 \pm 1.0$, the redox control is given by:

$$PuO_2(coll, hyd) + {}^1/_2\, H_2O \Leftrightarrow PuO_{2.5}(s, hyd) + e^- + H^+ \quad (8)$$

with $\log K° = -pe - pH = -11.5 \pm 1.6$. This is in accord with the slope of -1 in region C (Fig. 2b) and, within the uncertainties, with the experimental values of (pe + pH) = 12.5 ± 1.2.[5-7,20] Of course, Pu(IV) colloids/polymers can also be oxidised by oxygen, but the oxidised fraction is highly soluble (like PuO_2OH(am)), so that the remaining colloids consist of Pu(IV) and the Pu(V) concentration remains controlled by solid PuO_{2+x}(s, hyd).

3 DISCUSSION OF THERMODYNAMIC DATA FOR $PuO_{2+x}(s)$

If the solubility constant of $PuO_{2.5}$(s, hyd), i.e., the standard Gibbs energy of reaction (6), is combined with $\Delta_fG°_m(PuO_2^+) = -852.6 \pm 2.9$ kJ/mol, $\Delta_fG°_m(H_2O(l)) = -237.14 \pm 0.04$ kJ/mol and $\Delta_fG°_m(OH^-) = -157.22 \pm 0.07$ kJ/mol[1,2] the molar standard Gibbs energy of formation of $PuO_{2.5}$(s, hyd) in PuO_{2+x}(s, hyd) = $(PuO_{2.5})_{2x}(PuO_2)_{1-2x}$(s, hyd) is calculated to be -971.2 ± 5.4 kJ/mol. Hence the molar standard Gibbs energy of formation of PuO_{2+x}(s, hyd), with x < 0.1 in the evaluated solubility studies, is given by

$$\begin{aligned}\Delta_fG°_m(PuO_{2+x}(s,hyd)) &= 2x\, \Delta_fG°_m(PuO_{2.5}(s,hyd)) + (1-2x)\, \Delta_fG°_m(PuO_2(s,hyd)) \\ &= \{2x\,(-971.2 \pm 5.4) + (1-2x)(-965.5 \pm 4.0)\}\ \text{kJ/mol}\end{aligned} \quad (9)$$

Accordingly $\Delta_fG°_m(PuO_{2+x}$(s, hyd)) is slightly lower than $\Delta_fG°_m(PuO_2(s,hyd)) = -965.5 \pm 4.0$ kJ/mol. This is consistent with the observation that PuO_2(s, hyd)) can be oxidised by $O_2(g)$, whereas experimental attempts to oxidise dry $PuO_2(cr)$ by $O_2(g)$ in absence of water failed ($\Delta_fG°_m(PuO_2(cr)) = -998.1 \pm 1.0$ kJ/mol, $\Delta_rG°_m > 0$). The $\Delta_fG°_m$ values reported by Haschke et al.,[12,13] - 1080 kJ/mol for $PuO_{2.25}(s)$ and - 1146 ± 25 kJ/mol for $PuO_{2.5}(s)$, are much too negative. They refer to the oxidation of $PuO_2(s)$ by water according to

$$PuO_2(s) + xH_2O \rightarrow PuO_{2+x}(s) + xH_2(g) \quad (10)$$

The observed formation of $H_2(g)$ is most likely caused by experimental artefacts, e.g., radiolysis. For thermodynamic reasons reaction (10) is not possible ($\Delta_rG°_m > 200$ kJ/mol).

4 CONCLUSIONS

The critical evaluation of solubility data for PuO_2(am, hyd) at pH 1 - 13, including the aqueous redox speciation, the amount of oxygen present in the system and the measured redox potentials, shows that O_2 is scavenged by PuO_2(s, hyd) yielding PuO_{2+x}(s, hyd). For the first time all experimental data are explained consistently in terms of equilibrium thermodynamics. The presently available database for Pu is not sufficient, in particular the understanding of the equilibrium Pu(V) concentration and pe values requires additional thermodynamic data for PuO_{2+x}(s, hyd) and small Pu(IV) colloids or polymers present in neutral to alkaline solutions. The equilibrium constants derived in the present paper, $\log K°_{Pu(IV)coll} = -8.3 \pm 1.0$ and $\log K°_{sp}(PuO_{2.5}$ in PuO_{2+x}(s, hyd)) $= -14.0 \pm 0.8$, are comparable with known values for analogous Th(IV) oxyhydroxide colloids and $NpO_{2.5}$(s), respectively. Further efforts are necessary to quantify the maximum value of x in PuO_{2+x}(s, hyd) and to obtain more accurate thermodynamic data as a function of x.

References

1 R.J. Lemire, J. Fuger, H. Nitsche, P. Potter, M.H. Rand, J. Rydberg, K. Spahiu, J.C. Sullivan, W.J. Ullman, P. Vitorge, H. Wanner (OECD, NEA-TDB). *Chemical Thermodynamics of Neptunium and Plutonium*. Elsevier, North-Holland, 2001.
2 R. Guillaumont, Th. Fanghänel, J. Fuger, I. Grenthe, V. Neck, D.A. Palmer, M.H. Rand (OECD, NEA-TDB). *Update on the Chemical Thermodynamics of Uranium, Neptunium, Plutonium, Americium and Technetium*. Elsevier, North-Holland, 2003.
3 D. Rai, Y.A. Gorbi, J.K. Fredrickson et al., *J. Solution Chem.*, 2002, **31**, 433.
4 K. Fujiwara, H. Yamana, T. Fujii and H. Moriyama, *Radiochim. Acta*, 2002, **90**, 857.
5 D. Rai, R.J.Serne and D.A. Moore, *Soil Sci. Am. J.*, 1980, **44**, 490.
6 D. Rai, *Radiochim. Acta*, 1984, **35**, 97.
7 D. Rai, D.A. Moore, A.R. Felmy et al., *Radiochim. Acta*, 2001, **89**, 491.
8 H. Capdevila and P. Vitorge, *Radiochim. Acta*, 1998, **82**, 11.
9 V. Neck and J.I. Kim, *Radiochim. Acta*, 2001, **89**, 1.
10 J.M. Haschke, T.H. Allen and L.A. Morales, *Science*, 2000, **287**, 285.
11 J.M. Haschke, T.H. Allen and L.A. Morales, L.A., *J. Alloys Comp.*, 2001, **314**, 78.
12 J.M. Haschke and V.M. Oversby, *J. Nucl. Mat.*, 2002, **305**, 187.
13 J.M. Haschke and T.H. Allen, *J. Alloys Comp.*, 2002, **336**, 124.
14 D.J. Farr, R.K. Schulze and M.P. Neu, *J. Nucl. Mat.*, 2004, **328**, 124.
15 S.D. Conradson, B.D. Begg, D.L. Clark et al., *J. Am. Chem. Soc.*, 2004, **126**, 13443.
16 M. Kasha, in: *The Transuranium Elements, Research Papers*, eds. G.T. Seaborg, J.J. Katz, W.M. Manning), p. 295, McGraw-Hill, New York, 1949.
17 J.I. Kim and B. Kanellakopulos, *Radiochim. Acta*, 989, **48**, 145.
18 D. Rai, N.J. Hess, A.R. Felmy et al., *Radiochim. Acta*, 1999, **86**, 89.
19 Ch. Lierse and J.I. Kim, Report RCM 02286, Technische Universität München, 1986.
20 M. Altmaier, V. Neck, A. Seibert and Th. Fanghänel, unpublished results.
21 M. Altmaier, V. Neck and Th. Fanghänel, *Radiochim. Acta*, 2004, **92**, 537.
22 D.W. Efurd, W. Runde, J.C. Banar et al., *Environ. Sci. Technol.*, 1998, **32**, 3893.

URANIUM SPECIATION IN MOLTEN SALTS FROM X-RAY ABSORPTION AND ELECTRONIC ABSORPTION SPECTROSCOPY MEASUREMENTS

V. A. Volkovich,[1] I. May,[2] C. A. Sharrad,[2] H. Kinoshita,[2] I. B. Polovov,[1] A. I. Bhatt,[2] J. M. Charnock,[3] T. R. Griffiths[4] and R. G. Lewin[5]

[1]Department of Rare Metals, Ural State Technical University – UPI, Ekaterinburg, 620002, Russia
[2]Centre for Radiochemistry Research, Department of Chemistry, the University of Manchester, Manchester, M13 9PL, UK
[3]CLRC Daresbury Laboratory, Warrington, WA4 4AD, UK
[4]Redston Trevor Consulting Ltd., Leeds, LS17 8RF, UK
[5]Nexia Solutions, Sellafield, Seascale, CA20 1PG, UK

1 INTRODUCTION

Understanding the speciation of uranium and other actinides in molten salts is important for developing pyrochemical reprocessing of spent nuclear fuels. Since 5*f*-orbitals are localised inside the atom and shielded from surrounding fields by 6*s* and 6*p* electrons, the 5*f*-levels and Laporte forbidden *f-f* transitions are relatively weakly dependent on the coordination environment. Electronic spectroscopy thus provides rather limited information concerning uranium speciation. Increasing temperature results in shifting the bands arising from *f-f* transitions towards lower energies, with simultaneous intensity lowering and broadening but with little oscillator strength change. Nevertheless, the electronic absorption spectra of uranium(III), (IV), (V) and (IV) in molten salts have been studied over the past fifty years and their distinctive profiles allow the oxidation state of uranium species to be determined. X-ray absorption spectroscopy (XAS), in its application to molten salt systems, is a more recent technique. It has the advantage of being element specific. Uranium oxidation states can be judged from XANES (X-ray Absorption Near Edge Structure) spectra and the coordination environment, surrounding ligands, bond lengths and coordination numbers, determined from an analysis of their EXAFS (Extended X-ray Absorption Fine Structure) spectra.

2 EXPERIMENTAL

Experiments were performed in LiCl, LiCl-KCl, NaCl-KCl, NaCl-CsCl, CsCl and LiCl-$BeCl_2$ based melts between 380 and 750 °C. Uranium was introduced into the melts by dissolving anhydrous uranium chlorides (UCl_3, UCl_4 or UO_2Cl_2), chlorinating uranium metal or oxides (UO_2 or UO_3) with Cl_2 or HCl in the melt or by anodic dissolution of uranium metal. The experimental set-up for *in situ* electronic spectroscopy employed an Ocean Optics SD2000 fibre optic spectrometer, described elsewhere.[1] Average oxidation state of uranium was determined by oxidimetric analysis of melt samples rapidly quenched under an inert atmosphere.

Uranium $L_{(III)}$-edge XAS measurements were performed in transmission mode (for LiCl, LiCl-$BeCl_2$, LiCl-KCl) and in fluorescence mode (for LiCl-KCl) on stations 9.3 and 16.5, respectively, using the CLRC Daresbury Radiation Source operating at a typical beam current of 150 mA and an energy of 2 GeV. Monochromatic radiation (Si 220 double crystal), was detuned to 50% maximum intensity to minimise harmonic generation. Molten salt samples were contained in specially designed silica cells, (inverted T-shape) with melt thickness of 2 or 3 mm. Slightly modified cells were used for on-line *in situ* electrochemical experiments. Up to 16 scans per sample were collected. A 30-element germanium detector was used to collect the fluorescence data. The spectra were summed, calibrated and had their background subtracted using the Daresbury Laboratory programs EXCALIB and EXBACK and were simulated using the EXCURV98 program; the effects of multiple scattering were included in the final analysis where necessary. Due to the limited size of this communication we concentrate here on the results obtained by XAS.

3 RESULTS AND DISCUSSION

Melts containing U(III) and U(IV) ions were obtained by direct dissolution of UCl_3 and UCl_4 in molten LiCl at 750 °C. Their electronic spectra corresponded to the established literature data. The U $L_{(III)}$-edge energy for U(III) was 5.1 eV less than for U(IV), consistent with the decrease in Z_{eff}. Analysis of EXAFS spectra showed the presence of octahedrally coordinated UCl_6 groups, Table 1. Collecting data up to 18 keV enabled more information on long-range interactions to be obtained.

Table 1 *Structural parameters from EXAFS curve fitting of U(IV) and U(III)-containing melts*

Shell	CN	Distance, Å	Debye-Waller factor ($2\sigma^2$), Å^2	R
		UCl_4 in LiCl, 750 °C, k_{max} = 10 Å^{-1}, 4.54 wt.% U		
U-O	0.5	1.79	0.018	34.65
U-Cl	6	2.63	0.040	
		UCl_4 in LiCl, quenched melt, k_{max} = 10 Å^{-1}, 4.54 wt.% U		
U-O	0.5	1.77	0.010	31.97
U-Cl	6	2.69	0.032	
		UCl_4 in LiCl-KCl, 450 °C, k_{max} = 11 Å^{-1}, 4.2 wt.% U		
U-Cl	6	2.67	0.031	35.25
U-U	1	3.16	0.019	
		UCl_3 in LiCl, 750 °C, k_{max} = 10 Å^{-1}, 10.31 wt.% U		
U-O	0.5	1.69	0.011	27.88
U-Cl	6	2.72	0.050	
U-Li	12	3.26	0.057	
		UCl_3 in LiCl-KCl, 450 °C, k_{max} = 8 Å^{-1}		
U-Cl	6	2.84	0.038	37.59
		UCl_3 in LiCl, quenched melt, k_{max} = 11 Å^{-1}, 10.31 wt.% U		
U-Cl	6	2.91	0.014	28.88

Complete EXAFS spectra (Fig. 1) show high frequency oscillations at high k values that at present can only be attributed to long-range (3.5-5.5 Å) uranium-uranium interactions, but a reliable fit with a statistically justified number of parameters could not here be obtained. Interestingly, this high energy ordering seems to disappear once the melt is quenched, Fig. 1. Although there is no evidence of uranium-uranium bridging in electronic spectra, such interactions were proposed on the basis of electrochemical studies in uranium-containing melts and from electronic spectra of chloride melts containing uranium and neptunium.[2]

The electronic absorption spectra of uranium-containing melts do not show noticeable profile differences when measured at various U concentrations,[3] , for example, upon chlorinating the metal in the melt, Fig. 2.

We attempted to investigate the effect of lowering uranium concentration in the melt on the structures of uranium complex ions but could not obtain reasonable quality spectra at uranium content below 2 wt.% (0.5 mol.%). At 2.15-2.86 wt % (0.5-0.7 mol.%) uranium in the melt uranium atoms were always found at around 3.2 Å, indicating that some kind of oligomerisation was taking place.[2]

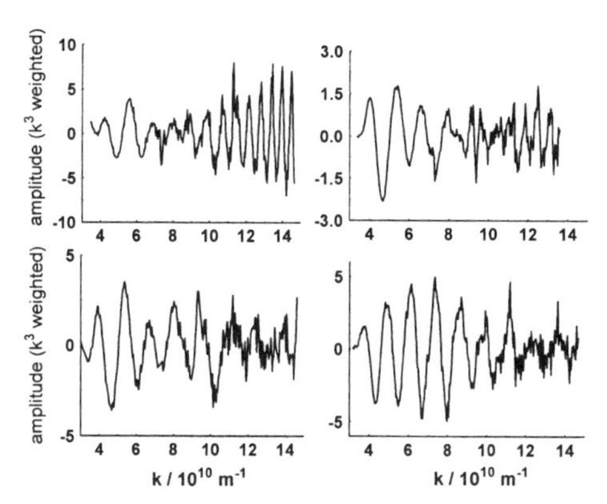

Figure 1 *EXAFS spectra of U(IV), left, and U(III), right, in LiCl at 750 °C, top, and in quenched state, bottom*

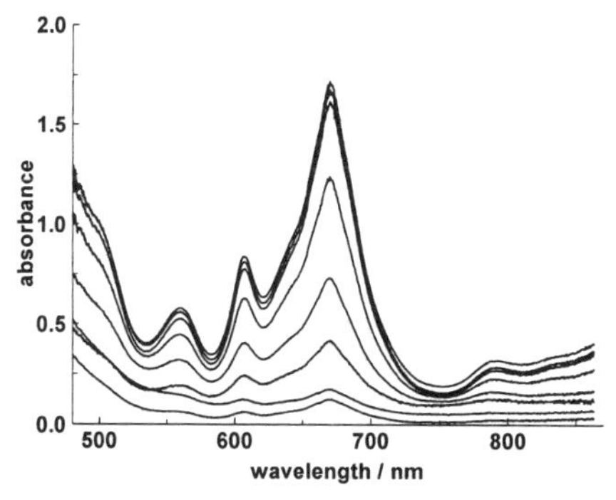

Figure 2 *Electronic absorption spectra showing a steady increase in the concentration of U(IV) as it is formed during the reaction of uranium metal with excess chlorine in LiCl-KCl melt at 650 °C*

Uranium(V) and (VI) in chloride melts form oxygen-containing chloro species. Above 600 °C, under an inert atmosphere, uranyl chloride, UO_2Cl_2, decomposes to UO_2Cl. The results for U(VI) and U(V) in chloride melts are given in Table 2. For U(V), both U-O and U-Cl distances were slightly longer than in melts containing pure U(VI), due to the longer uranium-ligand bonds expected for U(V) than with U(VI). Using a simple linear extrapolation of the various bond lengths *vs.* oxidation state we found that in melts containing only a U(V) species, a U-O distance of *ca.* 1.82 Å and a U-Cl distance of *ca.* 2.73 Å are expected.

Tables 1 and 2 show that, contrary to normal expectations based on temperature effects, U-Cl and U-O distances are *greater* for quenched samples of LiCl based melts than for their corresponding melts. The increase upon quenching 'UO_2Cl_4' and 'UCl_6' units is most likely the result of distortion of the uranium moieties from the tetragonal bipyramids (D_{4h}) or octahedra (O_h), respectively, present in the melts. Shannon[4] has shown that distortion of octahedral sites leads to an overall increase in interatomic distances. Since previous XAS studies of molten salt solutions have not reported interatomic data for their quenched samples this is the first confirmation of Shannon's conclusion in relation to molten and quenched systems.

Table 2 *Structural parameters of U(VI) and U(V)-containing species in LiCl and LiCl-KCl melts and after quenching from EXAFS curve fitting*

Shell	CN	Distance, Å	Debye-Waller factor ($2\sigma^2$), Å^2	R
		$UO_3 + Cl_2$ in LiCl, 750 °C, k_{max} = 10 Å^{-1}, 9.66 wt.% U, n_U=5.96		
U-O	2	1.77	0.008	24.17
U-Cl	4	2.67	0.036	
		$UO_2 + Cl_2$ in LiCl, 750 °C, k_{max} = 10 Å^{-1}, 13.64 wt.% U, n_U=5.98		
U-O	2	1.75	0.014	40.43
U-Cl	4	2.63	0.037	
		$UO_2 + Cl_2$ in LiCl, quenched melt, k_{max} = 13 Å^{-1}, 13.64 wt.% U, n_U=5.98		
U-O	2	1.77	0.010	41.43
U-Cl	4	2.75	0.027	
		UO_2Cl_2 in LiCl, 750 °C, k_{max} = 10 Å^{-1}, 7.90 wt.% U, n_U=5.68		
U-O	2	1.78	0.010	26.84
U-Cl	4	2.67	0.043	
		UO_3 + HCl in LiCl-KCl, 700 °C, k_{max} = 10 Å^{-1}		
U-O	2	1.78	0.004	32.90
U-Cl	4	2.69	0.030	
		UO_3 + HCl in LiCl-KCl, quenched melt, k_{max} = 12 Å^{-1}, 9.3 wt.% U		
U-O	2	1.81	0.012	35.56
U-Cl	4	2.69	0.009	
		$UO_2 + Cl_2$ in LiCl-KCl, 700 °C, k_{max} = 10 Å^{-1}		
U-O	2	1.77	0.005	34.50
U-Cl	4	2.67	0.030	
		UO_2 + HCl in LiCl, 750 °C, k_{max} = 11 Å^{-1}, 4.88 wt.% U, n_U=5.27		
U-O	2	1.80	0.011	41.35
U-Cl	4	2.71	0.046	
		UO_2 + HCl in LiCl, quenched melt, k_{max} = 10 Å^{-1}, 4.88 wt.% U, n_U=5.27		
U-O	2	1.76	0.011	33.76
U-Cl	4	2.76	0.024	
U-U	2	4.04	0.015	

n_U – average oxidation state of uranium in the sample.

To study possible temperature effects on U-Cl and U-O bond lengths in a given solvent melt, XAS spectra of LiCl-KCl-UO_2Cl_2 melt were measured between 450 and 700 °C, Fig. 3. Since a temperature increase of 250 degrees showed no bond length changes the pronounced difference in U-Cl and U-O interatomic distances observed for molten and solidified samples is thus due to symmetry changes in "UCl_6" and "UO_2Cl_4" groups upon quenching and not a temperature effect.

Electrochemical processes (anodic dissolution and cathodic reduction) are important in pyrochemical processing of nuclear materials. We studied for the first time the anodic dissolution of uranium in LiCl-KCl eutectic melt at 450 °C by *in situ* X-ray absorption spectroscopy to see if any other species besides UCl_6^{3-} could be detected. Measuring the spectra of melts during anodic dissolution was also the most sure way of monitoring U(III) since the influence of external factors, *e.g.*, possible oxygen impurities in the atmosphere and the effect of silica, were minimised. Data were collected using fast EXAFS spectroscopy over a smaller energy range and were best fitted with a "UCl_6" moiety. The U–Cl distance obtained, 2.84 Å, correlated well with the 2.82 Å found by Okamoto *et al.*[5] in a LiCl-KCl-UCl_3 (15%) melt at 600 °C.

Reduction of U(IV) on a platinum cathode in LiCl-KCl eutectic at 450 °C was also tried as a method of producing U(III) containing melt. Increasing the oxidation state of the central metal ion leads to shortening the metal-ligand distance and the value of the U–Cl distance was used here to monitor the reduction progress, Fig. 4. EXAFS spectroscopy is an averaging technique and for a mixture of UCl_6^{3-} and UCl_6^{2-} thus gives an intermediate value of the interatomic distance. During the reduction the green colour of the melt changed to dark reddish-purple but even after 10 h complete conversion of U(IV) to U(III) was still not achieved (the complex shape of the EXAFS cell probably slowed the diffusion rate).

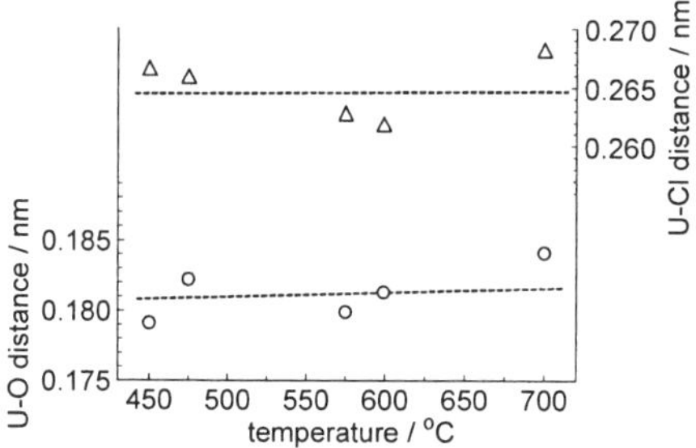

Figure 3 *Effect of temperature on U-Cl and U-O interatomic distances in $UO_2Cl_4^{2-}$ complex ion in LiCl-KCl melt (from XAS measurements)*

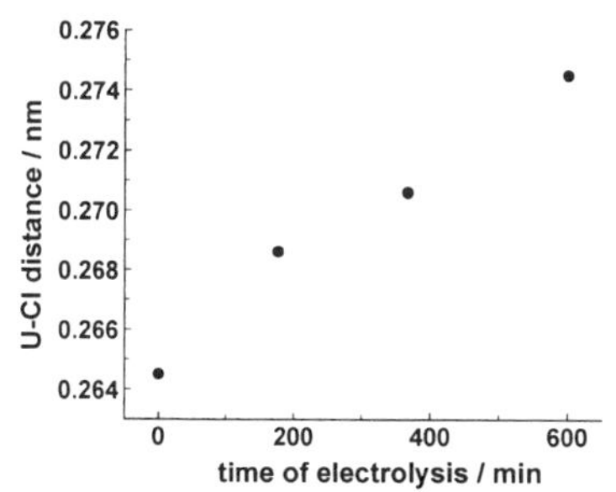

Figure 4 *Changes in average U-Cl distance during electrochemical reduction of U(IV) to U(III) on Pt cathode in LiCl-KCl melt at 450 °C*

Electronic absorption spectra of U(IV) chloro complexes in chloride melts containing strong complexing agents ($ZnCl_2$, $AlCl_3$, *etc.*) have often been interpreted in terms of tetrahedral coordination about uranium or an octahedral-tetrahedral equilibrium.[6] To check this assumption we conducted XAS experiments in the LiCl-$BeCl_2$ eutectic melt (Table 3), where Be(II) ions would compete with uranium for Cl^- ligands. Uranium in this melt forms predominantly U(IV) and U(III) complex ions. After dissolving UCl_4 and UCl_3

in LiCl-$BeCl_2$ eutectic most of the uranium remained in the original oxidation state. Dissolution of UO_2Cl_2 under an HCl atmosphere resulted in the reduction of uranium to the tetravalent state, and only in the presence of chlorine in the atmosphere was the mean oxidation state of uranium in the melt above four; and in this case uranium was present as a mixture of U(IV) and U(VI) since the electronic spectra of the melt contained none of the bands that could be attributed to U(V). In the presence of Be(II) ions, which have a high affinity for oxygen, $UO_2Cl_4^{2-}$ is reduced to UCl_6^{2-}. Analysis of the EXAFS spectra showed that in all cases uranium formed six-coordinated complex ions.

Table 3 *Structural parameters of uranium-containing complexes formed in LiCl-$BeCl_2$ melts determined from EXAFS curve fitting*

Shell	CN	Distance, Å	Debye-Waller factor ($2\sigma^2$), $Å^2$	R
UO_2Cl_2 in LiCl-$BeCl_2$ (prepared under Cl_2), 380 °C, k_{max} = 9 $Å^{-1}$, 7.30 wt.% U, n_U=4.23				
U-O	0.5	1.72	0.062	27.59
U-Cl	6	2.56	0.034	
UO_2Cl_2 in LiCl-$BeCl_2$ (prepared under Cl_2), quenched melt, k_{max} = 13 $Å^{-1}$, 7.30 wt.% U, n_U=4.23				
U-O	0.5	1.74	0.021	25.41
U-Cl	2	2.54	0.012	
U-Cl	4	2.74	0.019	
U-U	2	4.48	0.020	
UO_2Cl_2 in LiCl-$BeCl_2$ (prepared under HCl), 380 °C, k_{max} = 9 $Å^{-1}$, 7.71 wt.% U, n_U=3.87				
U-Cl	6	2.59	0.026	38.08
UO_2Cl_2 in LiCl-$BeCl_2$ (prepared under Cl_2), quenched melt, k_{max} = 11 $Å^{-1}$, 7.71 wt.% U, n_U=3.87				
U-Cl	6	2.71	0.024	26.63
UCl_4 in LiCl-$BeCl_2$, 380 °C, k_{max} = 10 $Å^{-1}$, 5.75 wt.% U, n_U=3.76				
U-Cl	6	2.58	0.029	35.93
UCl_4 in LiCl-$BeCl_2$, quenched melt, k_{max} = 15 $Å^{-1}$, 5.75 wt.% U, n_U=3.76				
U-Cl	6	2.73	0.016	31.34
UCl_3 in LiCl-$BeCl_2$, quenched melt, k_{max} = 11 $Å^{-1}$, 11.66 wt.% U, n_U=3.01				
U-Cl	6	2.92	0.015	27.46
U-U	6	5.19	0.016	

4 CONCLUSION

Uranium speciation determined in LiCl, LiCl-KCl and LiCl-$BeCl_2$ melts by XAS has shown that uranium forms six-coordinated complex ions (in the first coordination sphere) and indicated that at uranium concentrations above 2 wt.% (*ca.* 0.5 mol.%) oligomerisation or chloride bridging is taking place.

References

1 V.A. Volkovich, I. May, J.M. Charnock and R.G.. Lewin, *Phys. Chem. Chem. Phys.*, 2002, **4**, 5753.
2 A.I. Bhatt, E. du Fou de Kerdaniel, H. Kinoshita, F.R. Livens, I. May, I.B. Polovov, C.A. Sharrad, V.A. Volkovich, J.M. Charnock and R.G. Lewin, *Inorg. Chem.*, 2005, **44**, 2, and references cited therein.
3 V.A. Volkovich, A.I. Bhatt, I. May, T.R. Griffiths and R.C. Thied, *J. Nucl. Sci. Technol.*, 2002, **Suppl. 3**, 595.
4 R.D. Shannon, *Acta Cryst.*, 1976, **A32**, 751.
5 Y. Okamoto, M. Akabori, A. Itoh and T. Ogawa, *J. Nucl. Sci. Technol.*, 2002, **Suppl. 3**, 638.
6 S.V. Volkov and K.B. Yatsimirskii, Spectroscopy of Molten Salts, Kiev, Naukova Dumka, 1977, p. 129, and references cited therein.

THE SEPARATION OF PLUTONIUM FROM URANIUM BY SOLVENT EXTRACTION: RECENT DEVELOPMENTS FOR ADVANCED FUEL CYCLES REPROCESSING

J. E. Birkett, M. J. Carrott, O. D. Fox,* G. Crooks, C. J. Jones, C. J. Maher, C. Roube, R. J. Taylor and D. Woodhead

B229 and BTC, Nexia Solutions, Sellafield, CA20 1PG, UK
*Presenting author, danny.fox@nexiasolutions.com

1 INTRODUCTION

1.1 Advanced Fuel Cycles

There are numerous possible Advanced Fuel Cycle (AFC) scenarios including open, and partially or fully-closed variants.[1] Inevitably, these AFC scenarios give rise to numerous reprocessing options. It is sufficient for the purposes of this paper that, based on recent worldwide developments, if closed fuel cycles are adopted an 'Evolutionary Concept' would appear the most likely variant to find favour. In this scenario, the fuel from the current generation of Light Water Reactors (using enriched U) is recycled and returned in the form of Mixed Oxide Fuel (MOX). The fuel cycle is closed at this point with respect to Pu only (and presumably U if this is reused). Eventually, recycle of Pu will be accompanied by that of the TRansUranic elements (TRUs) necessitating the use of fast reactors (or accelerator driven systems) to achieve total actinide burning. The fuel cycle would then be closed with respect to all TRUs; this maximises the energy efficiency of the cycle with respect to energy extracted per gram of U and significantly reduces the losses of the long-term radiotoxic actinide isotopes to the High Level Waste (HLW) and thereby reduces timescales of care.

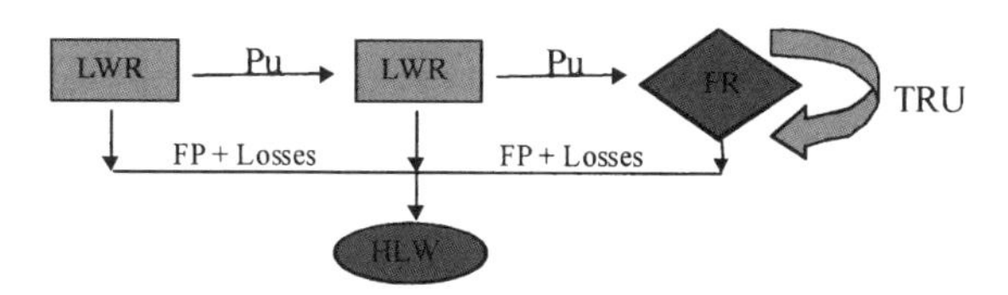

Figure 1 *Schematic of an 'Evolutionary Concept' in AFC development*

This paper discusses features of flowsheets for AFCs with particular reference to the separation of Pu from U. An important feature of AFCs is the presence of much higher Pu-loadings which make much greater demands on the PUREX process. PUREX technology manipulates the redox chemistry of Pu to achieve separation from the bulk U. There are, however, drawbacks associated with the use of reductive separations, many of which are exacerbated with increased Pu-loading of the solvent. An alternative option of complexant backwashing of Pu is presented and comparison made with reductive backwashing. Recently, a series of U/Pu separation flowsheets have been demonstrated in our laboratories and an example flowsheet is reported and discussed herein. Besides the relevance of these flowsheets to Advanced Fuel Cycles, the development of technologies designed to deal with particularly high Pu loadings is pertinent to the requirements for processing a number of Pu residues and experimental fuels. These legacy fuels and residues reflect sixty years' plus pioneering nuclear industry development in the UK that witnessed operation of thermal and fast reactors.

2 RESULTS

2.1 The Separation of Pu from U in Advanced Fuel Cycles

Plutonium will readily exist in the tetravalent state under PUREX conditions and it is in this oxidation state that extraction occurs in all primary extraction cycles. Early processes used ferrous sulphamate to achieve reduction of Pu(IV) to the almost inextractable Pu(III) state, and thereby achieve separation from the bulk U, present as U(VI). Modern plants, such as Thorp, use a 'salt free' flowsheet based on U(IV) stabilised with hydrazine.

The rates of reduction of Pu(IV) by ferrous and uranous ions are fast: the half-lives for both reactions are of the order of a few seconds under mixed phase conditions even at quite high acidity. The reductive stripping of Pu from TBP-diluent phases would be relatively straightforward were it not for the oxidising properties of nitric acid. Nitrous acid catalyses, through an autocatalytic mechanism, the re-oxidation of Pu(III) to Pu(IV) and for this reason a nitrous acid scavenger is added. In process terms, the consequence of Pu(III) reoxidation is the superstoichiometric consumption of the reductant; the extent being largely dependent on the Pu (and in some cases other species) content of the fuel, and the acidity and residence time of the solvent extraction separation.

$$Pu^{3+} + HNO_2 + H^+ \longrightarrow Pu^{4+} + NO + H_2O$$
$$2NO + HNO_3 + H_2O \longrightarrow 3HNO_2$$

Modern reprocessing plants use U(IV) stabilised with hydrazine for the reductive back-washing of Pu(III) as the reaction is both strongly thermodynamically favoured and rapid even at quite high acidity. The reagent is also 'salt free' as the resultant U(VI) accompanies the U solvent product. There are, however, drawbacks associated with the use of U(IV)/hydrazine, many of which are exacerbated by increased Pu-loading of the solvent. Firstly, the use of U(IV) results in isotopic dilution of enriched U. Significant U(IV) is oxidized rapidly by nitrous acid in the solvent phase and because the distribution of U(IV) is lower than U(VI) the flowsheet is operated at a reasonably high acidity to minimise losses of U(IV) to the Pu product. These factors mean there is a trade-off between the DF U vs. DF Pu. Finally, U(IV) is consumed in catalytic cycles involving hydrazine and technetium.

To achieve efficient U/Pu separations in AFC flowsheets one option is to replace the reductive stripping of Pu by complexation: Pu(IV) is selectively complexed by a hydrophilic ligand and stripped in to the aqueous phase. The advantages of complexation include fast kinetics, relative temperature insensitivity compared to redox reactions and no reoxidation of Pu(III), hence no need for a stabiliser and likely improved criticality control.

2.2 Hydroxamic acids as Pu stripping reagents

Hydroxamic acids (*e.g.* AHA, Acetohydroxamic acid) are small, organic ligands that selectively complex tetravalent [e.g. Np(IV), Pu(IV)] over hexavalent actinides [U(VI)] in nitric acid. The resultant complexes are hydrophilic allowing the selective back-extraction of tetravalent actinides from uranium-loaded solvent.[2]

A considerable volume of spectroscopic data has established actinide(IV) (U, Np, Pu) coordination by hydroxamic acids in nitric acid and shown that the complexes formed can be considered stable on the timescale anticipated for a process separation.[2] In addition, both Np(VI) and Pu(VI) are reduced by hydroxamic acids to inextractable states. Therefore, hydroxamic acids provide complete control over all available oxidation states of Np and Pu encountered in PUREX separations.

To support flowsheet development, distribution data for the Pu:AHA system as function of a series of process parameters e.g. acidity, Pu:AHA ratio, Pu-loading, U-loading, has been collected at macro and trace Pu levels;[3] for example, see Figure 2. The data show that Pu is effectively retained in the aqueous phase by the complexant acetohydroxamic acid (AHA), D is less than or equal to 0.05 at 1M HNO_3,

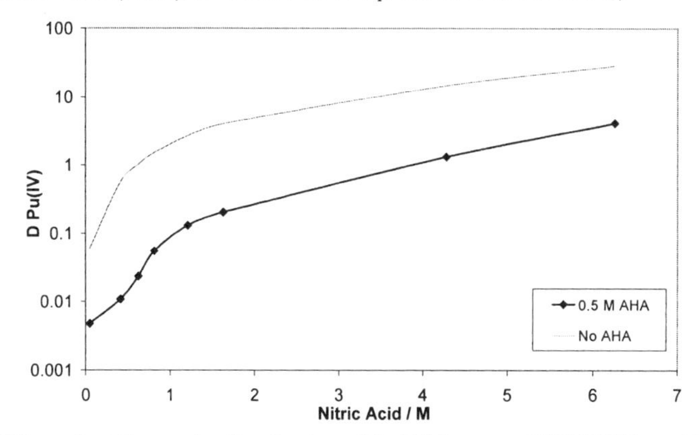

Figure 2 *Effect of acidity on the distribution of Pu(IV) between 30%TBP/Exxsol D80 and 0.5 M AHA at 25 °C*

A comparison with distribution data for Pu(III) stabilised by U(IV)-hydrazine [not shown] shows that reduction to Pu(III) gives marginally lower Pu distribution coefficients at probable flowsheet acidities, <1.5 M, and shows much better performance at higher acidities. Based on the distribution data it would appear that a hydroxamic acid flowsheet would not require overly tight acidity control, although a low level of acidity (ca. 1 M HNO_3) should ideally be maintained.

2.3 Flowsheet demonstration of U/Pu separations for an AFC Scenario

To determine the effectiveness of hydroxamic acids at separating Pu from U under conditions relevant to AFCs a series of flowsheet trials were performed on a glovebox-housed 28-stage centrifugal contactor rig using U/Pu/Tc-active simulant solutions. The first of these trials – relevant to recycle of MOX fuel of approximately 7% wt Pu content - was reported recently.[4] This trial demonstrated that, a high specification can be achieved on both the U and Pu streams and a concentrated Pu product can be obtained which maximises solvent throughput and minimises Pu product volumes.

Recently, a second trial was run at approximately 20% wt. Pu to demonstrate performance at much higher Pu concentrations relevant to an AFC processing FR fuel. This second trial allowed further optimisation of the flowsheet in terms of process parameters such as acidity, number of SX stages, product volumes and decontamination factors of products (U, Pu). The flowsheet tested, is shown in Figure 3, this included 12–stages of extraction, designed to simulate the HA cycle of a reprocessing facility and to provide a suitable feed for 16-stages of U/Pu separation.

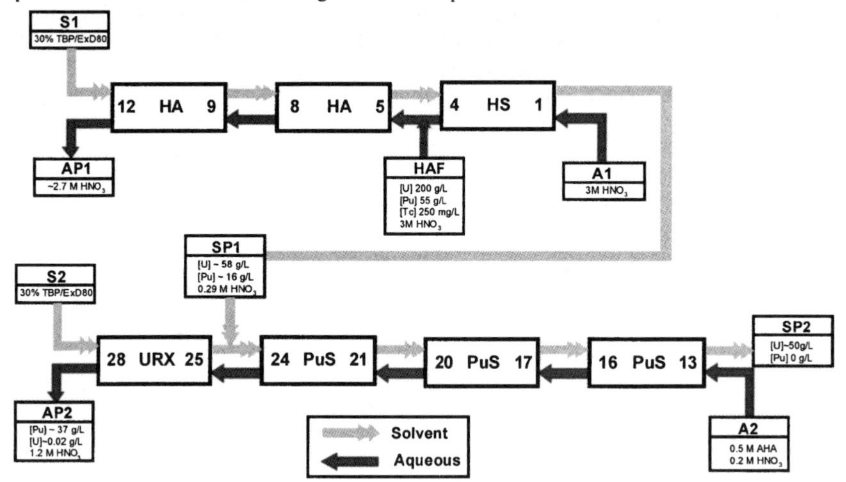

Figure 3 *A U/Pu separations flowsheet tested using AHA for complexant backwashing of Pu; the Pu was 20 wt% of the HM content.*

The designed flowsheet was tested U/Pu/Tc active and run for over four hours with this active feed, ~202g/L U, ~55 g/L Pu. In-line spectrophotometric analyses were used to monitor product streams and thereby run-up of the flowsheet to steady state, and confirm satisfactory operation of the flowsheet on the day of the trial, Figure 4. Analyses by in-line and off-line methods are in excellent agreement with a solvent U-loading of ~48 g/L observed at steady state. A summary of U and Pu concentrations in the feed and products streams is given in Table 1. The HAR was sampled hourly and results for the fourth samples showed only a trace of U and Pu present in the raffinate giving recoveries of 99.999 and 99.995 % respectively. The mass balances for U, Pu and acid across the flowsheet were good.

The concentration of elements in the solvent feed, SP1, to the Pu split section was [U] = 58 g/L, [Pu] = 16.3 g/L. The concentration of U in SP2 was in excess of 47 g/L, once the flowsheet reached steady state after approximately two hours. Decontamination factors for the Pu in U product, were in excess of 60000. The concentration of Pu in the AP2 product stream was determined by both alpha spectrometry and by spectrophotometry and averaged 37 g/L in the final two hours of the trial. The level of U in the Pu product is consistently low which corresponds to a U DF (i.e. measure U contamination of Pu product) in excess of 5000.

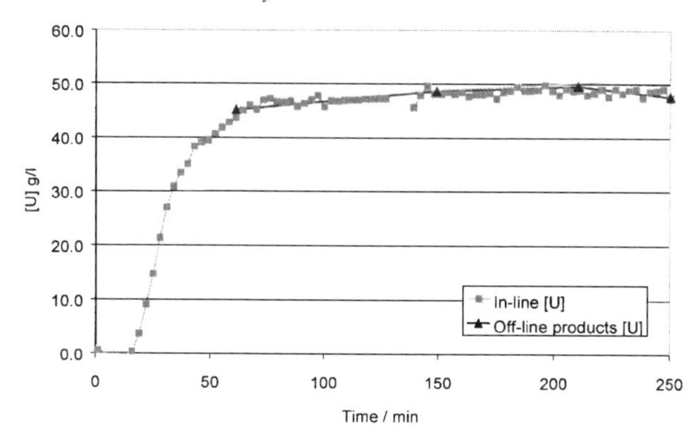

Figure 4 *In-line spectrophotometric data of SP2 U-loaded solvent product showing approach of flowsheet to steady state*

Table 1 *Feed and Product streams after 4 hours of flowsheet operation*

Sample	*Hrs:mins*	*[U] (g/l)*	*[Pu] (g/l)*	*[H^+](M)*
HAF	0	202	55	3.1
AP1_4	4:10	0.7E-3	1.9E-3	2.5
AP2_4	4:10	2.3E-2	36.0	1.11
SP1	4.10	58	16.3	0.29
SP2_4	4:10	47.5	2.23E-04	0.01
Mass Balance (%)	4:10	101	97	94

Concentration profiles for Pu, U and H^+ across the split contactor are shown in Figure 5. The gradual stripping of Pu as the solvent moves across the contactor is evident and the most pertinent feature of the flowsheet. The Pu concentrations of both the solvent and aqueous phases fall steadily across stages 23 –13. The aqueous phase acidities show a marked increase on approach to the feed plate from around stage 20. This is expected due to the back-washing of significant mass of acid carried through with the solvent from the extraction contactor. Through the whole of the split contactor the acidity is maintained above approximately 0.4 M.

4 EXPERIMENTAL

Three blocks of four-stage centrifugal contactors were used to emulate a typical PUREX HA extraction section. Four blocks of four-stage centrifugal contactors were used to achieve back-washing of Pu into the aqueous phase. Feeds to the rig were by automated reciprocating syringe pumps. Product streams and profile stage samples were analysed for a range of metals (U, Pu, Tc) and acidity. A range of methods were employed depending on concentration of analyte, concentrations of interfering species and sample matrices present. The methods employed included: off-line spectrophotometry, titrations,

Inductively-coupled plasma mass spectrometry (ICP-MS), X Ray fluorescence (XRF), Alpha spectrometry, Gamma and low energy photon spectroscopy (LEPS), Liquid scintillation counting (LSC).

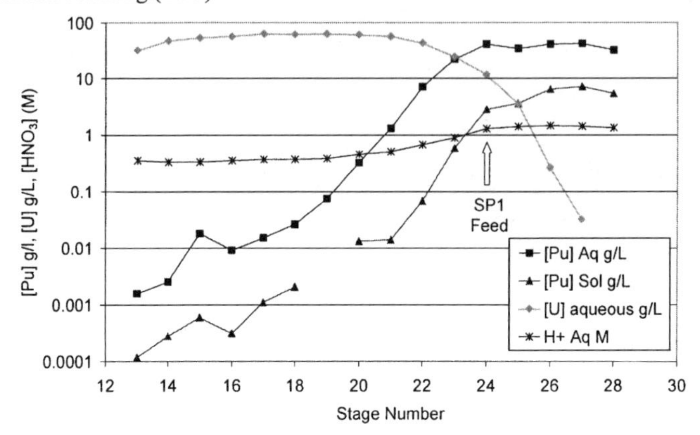

Figure 5 *Concentration profiles for Pu, U and H across the split contactor*

5 CONCLUSIONS

A specific key challenge identified in the development of flowsheets for AFCs is the selective complexant stripping of Pu(IV) at much higher concentrations than those currently encountered within reprocessing plants. An experimental flowsheet trial to evaluate the performance of the reagent, acetohydroxamic acid (AHA), for the separation of Pu under such high loadings (~ 20% wt).

Results showed that a high specification can be achieved on both the U and Pu streams, even at such high Pu concentrations in the feed. Based on the final samples at 4 h flowsheet operation, the DFs obtained for the U (SP2) and Pu (AP2) streams were >60000 and >5000 respectively. The highly concentrated nature of the Pu product demonstrates both the efficiency of AHA at stripping large masses of Pu and its ability to maintain Pu in the aqueous phase. These results are promising for the use of AHA for the recovery of Pu under conditions relevant to AFCs.

References

1 Accelerator–driven Systems (ADS) and Fast Reactors (FR) in Advanced Nuclear Fuel Cycles, A Comparative Study, NEA, OECD Report, 2002.
2 R. J. Taylor, Progress Towards Understanding The Interactions Between Hydroxamic Acids And Actinide Ions, *J. Nucl. Sci. Tech., Supp.* 3, 2002, 886.
3 O. D. Fox and R. J. Taylor, Plutonium Distribution in the Presence of Hydroxamic Acids, *AIP Conf. Proc. Plutonium Futures*, 2003, 673(1), 282.
4 J. E. Birkett, M. J. Carrott, O. D. Fox, C. J. Maher, C. V. Roube, R. J. Taylor, Plutonium and neptunium stripping in single cycle solvent extraction flowsheets: recent progress in flowsheet testing, *Proc. Atalante* 2004, O12-8.

TERNARY COMPLEX FORMATION AND COORDINATION MODES OF Am, Cm AND RARE EARTHS WITH EDTA+NTA: STABILITY CONSTANTS, TRLFS AND ^{13}C NMR STUDIES.

J.N. Mathur, P. Thakur and G.R. Choppin

Department of Chemistry and Biochemistry, Florida State University,
Tallahassee, FL, 32306, USA. Email: choppin@chem.fsu.edu

1 INTRODUCTION

There are numerous reports on the aqueous complexation and thermodynamics of trivalent Am, Cm and Eu with aminopolycarboxylate ligands (e.g., ethylenendiaminetetraacetic acid (EDTA) and nitrilitriacetic acid (NTA) at low ionic strength (<1.0 M) [1]. However, few reports are available describing the complexation behavior of these metal ions at high ionic strength (I= 6.60 mol kg^{-1}), where the thermodynamics of complexation may be different due to reduced water activity. Herein, we report the thermodynamic parameters describing binary and ternary complexation of Am, Cm, and Eu with NTA and with EDTA+NTA. The coordination modes of the complexes have been established by Time Resolved Laser Fluorescence (TRLFS) and ^{13}C NMR spectroscopy.

2 METHOD AND RESULTS

2.1 Stability constant and Thermodynamics

The stability constants of trivalent Am Cm and Eu with NTA and with EDTA+ NTA have been measured by a solvent extraction technique using di-(2-ethylhexyl)phosphoric acid, HDEHP, in heptane as the extractant at pcH of 3.60, I= 6.60mol kg^{-1} and the temperature range of 0-60.0 (±0.1) °C. The details of the experimental techniques, purity of radioactive tracers and the sampling have been described elsewhere[2].

The complexation of the metal ions can be written as:

$$M^{3+} + NTA^{3-} \overset{\beta_{101}}{\Leftrightarrow} MNTA \qquad (1)$$

and

$$M^{3+} + NTA^{3-} + EDTA^{4-} \overset{\beta_{111}}{\Leftrightarrow} M(EDTA)(NTA)^{4-} \qquad (2)$$

From the values of the stability constants at 0, 25, 45 and 60± 0.1°C, the enthalpies of complexation ΔH_{101} and ΔH_{111} were computed by Van't Hoff equation and the entropy changes $\Delta S_{101,\,m}$ and $\Delta S_{111,\,m}$ by the Gibbs-Helmoltz equation. The protonation constants and the ΔH associated with the deprotonation of NTA and EDTA at 6.60 mol kg^{-1} $NaClO_4$ at temperatures between 0- 60 °C were determined, and used in the calculation of stability constants. The stability constants and thermodynamic parameters have been reported in the molality scale and the necessary corrections applied for that of the heat of deprotonation of ligands were described previously.[2] The values of $\log\beta_{101}$ for the M-EDTA complexes were taken from an earlier study.[2]

The stability constants for the formation of M-NTA ($\log\beta_{101}$) and M-EDTA-NTA($\log\beta_{111}$) in Table 1 shows an increasing trend with increased in temperature. The positive ΔH (after applying corrections for the heat of deprotonation of EDTA and NTA) (Table 2) as compared with the negative values reported in literature for these metal ions with polyaminocarboxylate ligands at lower ionic strengths is likely due to the reduced activity of water at such high ionic strengths. The secondary hydration of M^{3+} decreases, resulting in the increased interaction of H_2O molecules in the primary coordination of M^{3+}. The heat energy required to break the M^{3+}-H_2O bonds is larger than the heat released during the formation of M^{3+}-ligand bonds, the net effect is positive enthalpy.

Table 1 *The stability constants for the formation of binary and ternary complexes of Am^{3+}, Cm^{3+} and Eu^{3+} with NTA and NTA+EDTA at I= 6.60mol kg^{-1} and the pcH of 3.60 at different temperatures.*

Temp		log β_{101}			log β_{111}	
(^{0}C)	Am	Eu	Cm	Am	Eu	Cm
0	9.71±0.08	9.62±0.07	9.85±0.08	23.73±0.08	24.40±0.10	23.80±0.09
25	10.48±0.09	10.52±0.08	10.82±0.08	25.73±0.09	25.96±0.09	25.91±0.08
45	10.83±0.08	10.91±0.09	11.08±0.09	26.82±0.11	27.44±0.11	27.39±0.10
60	11.33±0.10	11.38±0.08	11.66±0.10	28.06±0.11	28.71±0.11	28.80±0.10

Table 2 *Thermodynamic parameters for the formation of binary and ternary complexes of Am^{3+}, Cm^{3+} and Eu^{3+} with NTA and NTA+EDTA at I= 6.60mol kg^{-1} and the pcH of 3.60 at 25°C*

M	*ΔG_{101}* *kJ/mol*	*ΔH_{101}* *kJ/mol*	*ΔS_{101}* *J/mol K*	*ΔG_{111}* *kJ/mol*	*ΔH_{111}* *kJ/mol*	*ΔS_{111}* *J/mol K*
Am	- 59.58±0.50	23.3±2.7	278±9.0	-146.29±0.46	40.0±4.2	625±65
Cm	- 61.52±0.48	24.7±2.5	289±8.0	-147.31±0.48	41.3±3.7	633±58
Eu	- 59.81±0.52	24.5±2.5	283±10	-147. 59±0.51	41.4±5.5	634±68

2.2. TRLFS of Eu(III) with NTA and EDTA and their mixture

The details of the instrumental setup, data accumulation and calculations were given earlier[2]. The excitation spectra of Eu:EDTA:NTA of 1:1:2 , 1:1:20 and 1:10:10 in 6.60 mol kg^{-1} $NaClO_4$ at pcH ~ 9.0 were recorded. The two peaks at 579.66 and 580.16 nm at ratio of 1:1:2, had lifetime of 488 and 729 μs which corresponds to N_{H2O} (hydration numbers) of 1.4 and 0.7, respectively at two wavelengths indicating the formation of ternary complex $Eu(EDTA)(NTA)(H_2O)^{4-}$ and $Eu(EDTA)(NTA)^{4-}$.

At the ratio of 1:1:20, a single peak at 580.15 nm, could be deconvoluted into two peaks at 580.01 and 580.21 nm with lifetime = 937±24 μs and N_{H2O}= 0.4. Similarly at the ratio of 1:10:10, a single unsymmetrical peak at 580.21 nm was observed which on deconvolution gave two peaks at 579.93 and 580.21 nm with average lifetime of 990±30 μs which gave an N_{H2O} of 0.3, indicating the formation of a ternary complex $Eu(EDTA)(NTA)^{4-}$ at both these ratios. However, in both the cases NTA binds in two modes (1) via three carboxylates and (2) via two carboxylates and one nitrogen.

2.3 ^{13}C NMR spectroscopy

The ^{13}C NMR spectra of La-EDTA-NTA, 1:1:1 at pcH 9.64, had three peaks (only carboxylates resonance is discussed) at 185.8, 185.4 and 185.1 ppm indicating that in the ternary complexes $La(EDTA)(NTA)(H_2O)^{4-}$ and $La(EDTA)(NTA)^{4-}$, NTA binds in two modes, via two carboxylates in the former and via three in the later.

The above results were further confirmed by taking the spectra of the solution at pcH 11.12, where NTA^{3-} is the dominant anion (~98%). Two peaks at 186.1 and 185.7 ppm confirmed the formation of only one ternary complex $La(EDTA)(NTA)^{4-}$, where NTA binds with three carboxylates.

The spectra of Y-EDTA-NTA at pcH=9.60 had two peaks at 183.4 and 183.2 ppm, suggesting that NTA binds to Y(III) via two of its three carboxylates for a total CN of 8. The 9 coordinated cations (e.g., La^{3+}, Eu^{3+} and Am^{3+}, Cm^{3+}) bind with NTA via two or three carboxylates.

3 CONCLUSION

The stability constants ($\log\beta_{101}$ and $\log\beta_{111}$) of the trivalent Am Cm and Eu with NTA and with NTA+EDTA at I=6.60 mol kg^{-1} $NaClO_4$ were found to increase with increased temperature from 0-60^{0}C. The positive ΔH calculated for both the binary and the ternary complexes in the present investigation is in contrast to the negative values of ΔH reported for complexation of trivalent Am, Cm and Eu with NTA or EDTA at lower ionic strengths in the literature. The TRLFS study of the Eu-EDTA-NTA system at pcH ~9.0 at the ratio of 1:1:2, confirmed the formation of the ternary complexes $Eu(EDTA)(NTA)(H_2O)^{4-}$ and $Eu(EDTA)(NTA)^{4-}$ where NTA is binding with two and three carboxylates respectively. The ternary complex $Eu(EDTA)(NTA)^{4-}$ is formed at 1:1:20, and 1:10:10, in which NTA binds in two modes, via two carboxylates and one nitrogen and via three carboxylates. ^{13}C NMR spectra of La-EDTA-NTA system support the findings of TRLFS studies and confirmed the formation of ternary complex.

References

1 R.M. Smith, A.E. Martell and R.J. Motekaitis, NIST Critically Selected Stability Constants of Metal Complexes Database, Version 6.0. Users guide, U.S. Department of Commerce. Technology Administrative, National Institute of Standards and Technology, Standards Reference Data Program, Gaithersburg, MD (1999).
2 K. Cernochova, J.N. Mathur and G.R. Choppin, Radiochim Acta (In Press).

Acknowledgement

This research was supported by the US-DOE- Environment Management Science Program.

INFLUENCE OF TEMPERATURE AND IONIC STRENGTH ON THE HYDRATION OF CURIUM(III) IN AQUEOUS SOLUTION STUDIED WITH TRLFS

P. Lindqvist-Reis,[1] R. Klenze[1] and Th. Fanghänel[1,2]

[1]Institut für Nukleare Entsorgung, Forschungszentrum Karlsruhe, P.O. Box 3640, 76021 Karlsruhe, Germany
[2]Ruprecht-Karls-Universität Heidelberg, Physikalisch-Chemisches Institut, Im Neuenheimer Feld 253, 69120 Heidelberg

1 INTRODUCTION

It is well-established that the inner-sphere hydration numbers for the early and the late trivalent lanthanide aqua ions are 9 and 8, respectively.[1] However, detailed knowledge of the structure and dynamic properties of the aqua ions in the middle of the lanthanide series is lacking.[2] A similar trend is expected for the trivalent actinide aqua ions, although the transition between 9- and 8-coordination may occur later in this series due to the larger ionic radius.[3] Previously, the An^{3+} hydration numbers have been estimated by interpolation using the values obtained for Ln^{3+} ions with similar ionic radii.[1] From those results the hydration number for Cm^{3+} was estimated to be 8.9, for the heavier ions, Bk^{3+}, Cf^{3+} and Es^{3+}, 8.7, 8.2, and 8.0, respectively, while the lighter ions, Ac^{3+}–Am^{3+}, were all found to be 9-coordinate. More recently, EXAFS has been used to derive hydration numbers and bond distances for the U^{3+}–Cf^{3+} aqua ions.[3] The results showed a steady decrease in the An–O bond distance with increasing atomic number, while the reported hydration numbers were somewhat scattered between about 7 and 10. A re-evaluation of the EXAFS data for Cm^{3+}(aq) showed that the first hydration shell is clearly asymmetric, indicating a tricapped trigonal prism coordination geometry with six shorter and three longer Cm–O distances.[4a]

We have recently studied the temperature-dependency of Cm^{3+}(aq) with time-resolved laser fluorescence spectroscopy (TRLFS) from 20 to 200 °C.[5] An equilibrium between 9- and 8-hydrated Cm^{3+} was established, with a dominant moiety of the 9-coordinate species at room temperature. In this paper, we provide further evidence for the coexistence of such species. The emission spectra of the aqueous species are compared with those of the corresponding species in crystalline hosts with known crystal structures. Reference is made to recent quantum chemical calculations on the structure and thermodynamic data of nona- and octahydrated Cm^{3+} in aqueous solution.[4b] Finally, we report on dehydration effects of Cm^{3+}(aq) in perchlorate and trifluoromethanesulfonate solutions at low water activities. All details for the sample preparation and the TRLFS set-up have been given elsewhere.[4a,5]

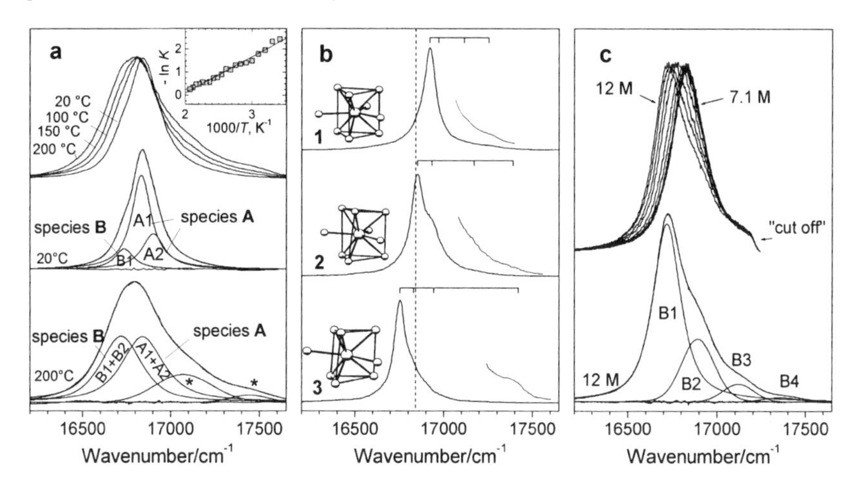

Figure 1 *$^6D'_{7/2} \rightarrow {}^8S'_{7/2}$ emission spectra of Cm^{3+} in: (a) 0.1 M $HClO_4$ aqueous solution at 20, 100, 150, and 200 °C. Species **A** and **B** refer to nona- and octahydrated species, respectively; (b) **1–3** at 20 °C, in comparison with the Cm^{3+} aqua ion spectrum, which is represented with a vertical dashed line at 16840 cm^{-1} (the X-ray structures of the hydrated cation entities are also shown); (c) 7.1 to 12 M $HClO_4$ at 20 °C. The curve-fitted component bands correspond to the emission contributions from the B_1–B_4 crystal field levels of the exited $^6D'_{7/2}$ state. The Excitation wavelength is 375 nm.*

2 RESULTS AND DISCUSSION

Figure 1a shows the normalized emission spectra of 0.5 μM Cm^{3+}(aq) at 20, 100, 150, and 200 °C. Emission is observed from the first excited $^6D'_{7/2}$ state to the $^8S'_{7/2}$ ground state, both of which are split by the ligand field into four Kramers doublets. The ground state splitting is not resolved, while the thermal population of the higher excited crystal field levels of the emitting $^6D'_{7/2}$ state gives rise to three 'hot bands', A_2–A_4, situated at the blue side of the main transition, A_1. The spectra are broadened asymmetrically and shifted to lower energies with increasing temperature, whilst the emission intensity and lifetime decrease notably.[5] Two spectral components, **A** and **B**, assigned to nona- and octahydrated Cm^{3+}, respectively, can be deduced from the spectra. The **B**/**A** intensity ratio follow the Van't Hoff equation with $\Delta H = 13.1 \pm 0.4$ kJ mol^{-1} and $\Delta S = 25.4 \pm 1.2$ kJ mol^{-1} K^{-1} (insert to Figure 1a); the ratio between **A** and **B** are roughly 90:10 at 20 °C and 60:40 at 200 °C.[5]

These findings are in good agreement with recent quantum chemical calculations, showing the transition to be entropy driven. The use of a solvent model, in this case a continuum model, to describe the interaction between the first coordination shell and the bulk water was found to be mandatory to obtain a reasonable ΔH.[4b] Moreover, the assignment of species **A** and **B** is supported by direct comparison with the spectra of the crystalline host compounds $[La(H_2O)_9](CF_3SO_3)_3$ (**1**), $[La(H_2O)_9]Cl_3 \cdot 15\text{-crown-5} \cdot H_2O$ (**2**), and $[Y(H_2O)_8]Cl_3 \cdot 15\text{-crown-5}$ (**3**). Figure 1b shows the spectra of $[Cm(H_2O)_9]^{3+}$ in **1** and **2** and $[Cm(H_2O)_8]^{3+}$ in **3**. The asymmetry/shoulder seen at the high-energy flanks of **2** and **3** are mainly due to transitions from the A_2 levels, whereas in **1** the A_1 and A_2 levels are rather narrowly spaced and can only be discerned at low temperature. The fact that the energy at peak maximum of component **A** of Cm^{3+}(aq) and **2** are similar, while that of **1** and **3** are blue- and red-shifted, respectively, suggests that the main Cm^{3+} aqua species have a TTP geometry room temperature, probably with a lower symmetry than C_{3h}.

Due to their low coordination affinities to hard Lewis acids such as Ln^{3+} and An^{3+}, perchlorate and trifluoromethanesulfonate are often preferred as counter ions over, e.g., chloride, nitrate, and sulphate. Until now, there has been no direct evidence for inner-sphere Ln–ClO_4 or An–ClO_4 complexes in aqueous solution, even with an excess of ClO_4^-. We observe a decrease in the Cm^{3+} emission lifetime from 64 to 58 μs with increasing $HClO_4$ concentration from 0.01 to 8.8 M; at higher concentration the lifetime increases sharply and reaches ~80 μs at 12 M $HClO_4$, which is accompanied by a redshift of the $^6D'_{7/2} \rightarrow {}^8S'_{7/2}$ emission band (Figure 1c). Similar observations were made with triflic acid, where the onset for the increase in the lifetime and the redshift occurs at ~7.5 M HCF_3SO_3. These concentrations correspond to water to acid ratios of approximately 4 and 3 for the perchloric and triflic acid solutions, respectively. We note that the emission spectrum of Cm^{3+} in 12 M $HClO_4$ is similar to that of $[Cm(H_2O)_8]^{3+}$ in **3**, but also to that of species **B** of Cm^{3+}(aq) at elevated temperature (cf. Figure 1a, b, c). Therefore, it is reasonable to interpret the redshift as a change in the coordination number from 9 to 8 occurring at high ionic strength where the water activity is very low. On a molecular level, this may be understood as a competition for water of hydration between Cm^{3+} and H^+ ions.

In summary, we have shown that the $^6D'_{7/2} \rightarrow {}^8S'_{7/2}$ emission spectrum of Cm^{3+}(aq) shifts to lower energy with increasing temperature. The redshift, which is due to an increase of the ligand field, is interpreted as a change in the inner-sphere hydration number of the Cm^{3+} ion from 9 to 8. Spectra of $[Cm(H_2O)_9]^{3+}$ and $[Cm(H_2O)_8]^{3+}$ in crystals are consistent with these findings. The hydrated Cm^{3+} ion persists as a stable nonahydrate species in aqueous solution up to very high ionic strength and over a wide concentration range of $HClO_4$ and HCF_3SO_3. However, its coordination geometry changes, most likely to eight-coordination, when the water/acid concentration ratio exceeds a certain critical value; these are about 4 and 3 for perchloric and triflic acid, respectively.

References

1 E. N. Rizkalla and G. R. Choppin, 'Lanthanides and Actinides Hydration and Hydrolysis' in *Handbook on the Physics and Chemistry of Rare Earths. Lanthanides/Actinides: Chemistry*, eds., K. A., Jr. Gschneidner, L. Eyring, G. R. Choppin and G. H. Lander, Elsevier Science, Amsterdam, 1994, Vol. 18, Chapter 127, pp 529–558.

2 L.Tilkens, K. Randall, J. Sun, M. T. Berry, P. S. May and T. Yamase, *J Phys. Chem. A*, 2004, **108**, 6624.

3 F. H. David and V. Vokhmin, *New J. Chem.*, 2003, **27** 1627 (Table 3 and refs therein).

4 (a) P. Lindqvist-Reis, R. Klenze, M. A. Denecke, T. Fanghänel and A. Eichhöfer, to be submitted. (b) B. Schimmelpfennig, to be submitted.

5 P. Lindqvist-Reis, R. Klenze, G. Schubert and T. Fanghänel, *J. Phys. Chem. B*, 2005, **109**, 3077.

INSIGHTS INTO THIRD PHASE FORMATION USING NEPTUNIUM

J.W. Plaue[1, 3] A.V. Gelis[2] and K.R. Czerwinski[3]

[1] Defense Nuclear Facilities Safety Board*, 625 Indiana Ave, NW, Suite 700, Washington DC, 20004-2901. Email: jplaue@alum.mit.edu
[2] Chemical Engineering Division, Argonne National Laboratory, 9700 S. Cass Ave, Argonne, IL 60439. Email: guelis@cmt.anl.gov
[3] Department of Chemistry, University of Nevada, Las Vegas, 4505 Maryland Parkway, Box 454003, Las Vegas, Nevada. 89154-4003 Email: czerwin2@unlv.nevada.edu
*The views expressed are solely those of the author and no official support or endorsement of this presentation by the Defense Nuclear Facilities Safety Board or the United States government is intended or should be inferred.

1 INTRODUCTION

Third phase formation occurs under certain conditions in solvent extraction systems involving polyvalent metal ions, like the actinides. As a third phase forms, the organic phase splits into a heavy phase rich in tri-*n*-butyl phosphate (TBP), acid, and metal; and a light phase consisting mostly of diluent. A fundamental understanding of third phase formation is helpful for supporting process safety. In this investigation, the Np ion was used to investigate third phase formation in the 1.1 M TBP/HNO_3/dodecane system. Only limited previous work on Np third phase formation in PUREX systems has been published. However, basic data on third phase formation of Th(IV), U(IV), and Pu(IV) has been published[1]. Researchers have focused on the conditions at the boundary for organic phase separation. A common metric, known as the limiting organic concentration (LOC), is used to describe this boundary as the highest organic phase metal concentration prior to phase separation. As one compares the LOC behaviour, there is an inconsistency between the hexavalent species of U and Pu. For U, the trend found by numerous authors is for the LOC of the U(IV) to be substantially lower than U(VI) [1]. Because of its industrial relevance, Pu(IV) has been extensively studied, but there are few reports on the influence of Pu(VI)[2]. However, Pu(VI) has shown a much stronger tendency to induce third phase formation, opposite of the U trend. Clearly, third phase formation is a complex phenomenon with respect to metal ion speciation. Recent investigations have examined third phase formation from the perspective of surfactant and colloid sciences[3]. Small angle neutron scattering studies have revealed small reverse micelle-like aggregates of 3 to 4 TBP molecules per metal–nitrate–acid center. While the formation of reverse micelles seems plausible, there is currently no explanation for why radical variations should be observed for the same oxidation states of similar metal ions such as U and Pu. It also contrasts with the view that both tetravalent and hexavalent nitrate species form disolvate complexes with TBP ($An^{m+}(NO_3)_m \cdot 2TBP$) at low metal to TBP concentration ratios[4]. With these observations in mind, examination of the third phase behaviour for Np became an obvious choice. Attempts at correlating the behaviour of U, Np, and Pu third phase formation to the fundamental properties of the metals could then be made. Thus conditions for the formation of third phase with Np(IV) and Np(VI) have been examined.

2 METHODS AND RESULTS

2.1 Preparation of Neptunium Third Phases

Work was performed at the Chemical Engineering Division at Argonne National Laboratory. Neptunium nitrate solutions were prepared from the HNO_3 dissolution of NpO_2 stocks followed by anion exchange purification. Pure Np(IV) and Np(VI) states were respectively prepared by reduction of the stock solution with hydrogen peroxide and oxidation with hot concentrated HNO_3. Valence adjustments were monitored by UV-Visible spectrophotometry at the characteristic absorptions for Np(IV) and Np(VI) [5] before and after extraction. Organic phases of 30 vol. % (1.1 M) TBP (Aldrich 99+ %) and *n*-dodecane (Aldrich) were prepared without further purification. Prior to extraction experiments, all organic phases were pre-equilibriated with equal volumes of the appropriate HNO_3 concentration. Consistent with literature practice, the LOC was determined by visual observation as the highest metal concentration in the organic phase prior to any obvious phase splitting or cloudiness. Neptunium concentrations were determined by gamma ray spectroscopy using a high purity germanium well detector.

2.2 Third Phase Boundary

The third phase formation boundaries for both extractable Np oxidation states were determined (Figure 1). LOC for Np(VI) follows a nearly linear trend, while Np(IV) appears to be curved similarly to Pu(IV). As with the LOC curve for Pu(IV), the shape of the Np(IV) curve also strongly resembles a plot of distribution values of the metal nitrate species in the TBP solvent extraction system as a function of initial aqueous acid concentration. The similarity in curves potentially suggests a link between the aqueous phase speciation and third phase formation.

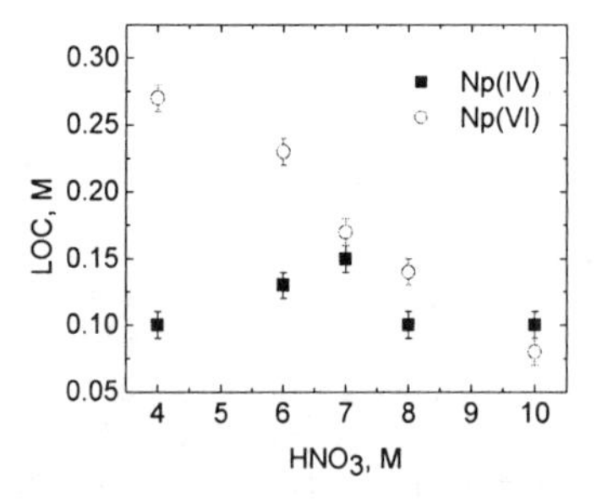

Figure 1 *Third phase boundary for Np in 1.1 M TBP/dodecane/HNO_3 at 22 °C*

In terms of magnitudes of the LOCs, Np behaves similarly to U in lower molarity acid systems, with the hexavalent state generally less susceptible to third phase formation than the tetravalent. However, in systems with acid concentrations above approximately 7 M, the two valence states behave similarly with respect to third phase formation, indicating HNO_3 is possibly dominating the third phase chemistry. This behaviour is unique to Np, and appears to support a trend in the increased tendency for hexavalent actinides to form third phase as one moves from U to Pu.

2.2 Comparison with Other Actinides

The LOC values for U, Np, and Pu at constant conditions are presented in Table 1. For the tetravalent species, U(IV) has the strongest propensity to form a third phase, while the order is reversed for the hexavalent actinides. In fact, U(VI) is known not to form third phases in dodecane in HNO_3 concentrations below 10 M.A comparison of these trends with the ionic radii for each metal reveals that for the hexavalent actinides, the third phase tendency increases with decreasing radii [6]. The opposite trend of an increasing tendency toward third phase formation with increasing radii is found with the tetravalent actinides, including Th(IV). The observed trend with ionic radii for the differing oxidation states can be related to charge density of the cation charge in the ionic volume and differences in speciation due to the formation of oxo species for the hexavalent state. U(IV), with the largest ionic radius and hence lowest charge density, has the lowest LOC value. Based on a study of effective cationic charge of the hexavalent actinides [7], the lowest charge on the metal center is associated with Pu(VI), with a charge of 2.9. The effective charges for U(VI) and Np(VI) are 3.2 and 3.0 respectively. Using literature data for 6-coordinate hexavalent U, Np, Pu radii[6], the charge density of Np is the lowest, followed by Pu then U. The relative differences are slight and within the error of the effective charge values. Compared to the tetravalent oxidation state, there is no evidence of a relationship between LOC and charge density. However, the oxo groups will have a consequence on the overall effective charge of the hexavalent species and need to be considered. Future efforts will explore the influence of charge density on third phase formation.

Table 2 Actinide LOC values in 7M HNO_3 / 1.1M TBP / dodecane, M at T=20-25 °C

	U	Np	Pu
An(IV)	0.08 [8]	0.15	0.27[9]
An(VI)	no 3Φ [10]	0.17	0.10[11]

3 CONCLUSION

The third phase formation boundary for Np in 1.1 M TBP/dodecane was determined over a range of HNO_3 concentrations from 4 to 10 M. For acid concentrations less than 7 M, the Np(IV) system is more prone to third phase than Np(VI). Above 7 M, the tendencies are similar. In comparison with U and Pu, Np behaves as expected with intermediate LOCs in both systems.

Acknowledgements. The authors wish to thank the Professional Development Program of the Defense Nuclear Facilities Safety Board and the Transmutation Research Program at UNLV.

References

1 P.R. Vasudeva Rao, Z. Kolarik, Solvent Extr. Ion Exch. 1996, **14**, 955.
2 A.L. Mills, W.R. Logan, In *Solvent Extraction Chemistry*, D. Dyrssen, J.O. Liljenzin, J. Rydberg, Eds.; North-Holland: Amsterdam, 1967, 322.
3 R. Chiarizia, K.L. Nash, M.P. Jensen, P. Thiyagarajan, K.C. Littrell, *Langmuir*, 2003, **19**, 9592.
4 T.V. Healy, H.A.C. McKay, *Trans Faraday Soc,* 1956, ***52***, 633.
5 V.A. Mikhailov, *Analytical Chemistry of Neptunium*, Halsted Press, NY, 1973.
6 R.D. Shannon, *Acta Cyrst.* 1976, **A32**, 751.
7 G.R. Choppin, L.F. Rao, L. F Radiochim Acta, 1984, **37**, 143.
8 P.D. Wilson, J.K. Smith, *Inst. Chem. Eng. Symp. Ser.* 1987, **103**, 67.
9 Z. Kolarik, CIM Spec. Vol. 21, Canad. Inst. of Mining and Metallurgy: Montreal, 1979, 178.
10 R. Chiarizia, M.P. Jensen, M. Borkowski, J.R. Ferraro, P. Thiyagarajan, K.C. Littrell, *Solvent Extr. Ion Exch.*, 2003, **21(3)**, 423.
11 J.W. Plaue, A.V. Gelis, K.R. Czerwinski, *in-progress*.

INVESTIGATION OF THE BEHAVIOR OF Am (IV), (V) AND (VI) IN ALKALINE MEDIA

M.V. Nikonov[1], I.G. Tananaev[1], B.F. Myasoedov[1], and D.L Clark[2]

[1]Vernadsky Institute of Geochemical and Analytical Chemistry RAS, 19, Kosygina str. 119991 Moscow, Russia
[2]G.T. Seaborg Institute for Transactinium Science, Los Alamos National Laboratory, NM, U.S.A.

1 INTRODUCTION

Knowledge of actinide chemical properties and their chemical forms during the long term aging of highly alkaline radioactive wastes (HARW) in Hanford, Savannah River, INEL, West Valley and Oak Ridge[1-3] is essential to optimize the regime of storage and to estimate the possibility of element separation. In addition, studying the behavior of actinide elements in the highest oxidation states in alkaline solutions is of considerable scientific interest. The present work is directed to study the composition and properties of Am(III-VII) solutions and compounds formed in alkaline media.

2 METHOD AND RESULTS

The most interesting results, which were obtained in these studies are briefly presented below.

2.1 $Am(OH)_3$ oxidation by the ozone.

Oxidation of Am(III) hydroxide by ozone was studied in sodium bicarbonate solutions, alkaline carbonate solutions and in other alkalines. It was shown that ozone oxidation of $Am(OH)_3$ pulp in 0.1M $NaHCO_3$ leads to consequent oxidation of Am(III) into higher oxidation states and causes the appearance of sudden oscillations [Am(V)]↔[Am(VI)]. The rates of $Am(OH)_3$ oxidation by the ozone in 0.1M Na_2CO_3 + (0.1-2.0)M NaOH mixture, or (0.1-1.0)M NaOH/LiOH do not differ greatly. The final product of oxidation was Am(VI). In contrast under similar conditions in both KOH or CsOH, and in concentrated RbF solutions no noticeable concentration of Am(VI) was recorded.

2.2 Interaction of Am(VI) with water

It was shown that the reduction rate of ^{243}Am(VI) in basic aqueous solution (~20% per hour) was found to be of the same order as the rate for ^{241}Am(VI). The values of the apparent rate constant (k_a) of the first order reaction in 3M and 1M NaOH are the same ~$(5.3\pm0.3)\cdot10^{-4}$ s^{-1} and in 0.1M NaOH k_a is increased to $1.7\cdot10^{-3}$ s^{-1}. On the basis of the data obtained we can conclude that the reduction of Am(VI) by water molecules is the main contributor to the process above and any considerable contribution from Am α-radiolysis effects to americium reduction under these conditions can be excluded.

2.3 Synthesis of Am(VII) solid state species

Ozonisation of moist precipitates of Am(VI) with perxenate ions, obtained by means of the mixing of ^{243}Am(VI) solution in 1M NaOH at 20°C with solid state sodium perxenate, leads to formation of green Am species with a specific electronic absorbtion spectrum (EAS) much like the spectra of Am(VII) in alkaline media (Figure 1).[4] It would therefore appear that an americium (VII) compound had been synthesized.

2.4 Interamericium hydroxide interactions

It was shown that the mixing of equivalent amounts of ^{243}Am(III) and ^{243}Am(V) hydroxides in 1M NaOH at 50-70^{0}C, or in 7M NaOH at 25°C during a few minutes caused the appearance of pure Am(IV) species via reproportionation reaction: Am(III) + Am(V) $\rightarrow$ 2Am(IV). The EAS of the initial mixture and the interaction product in the different conditions are shown in Figure 2. The reaction between Am(VI) and Am(III) hydroxide in equivalent amounts produced the same result.

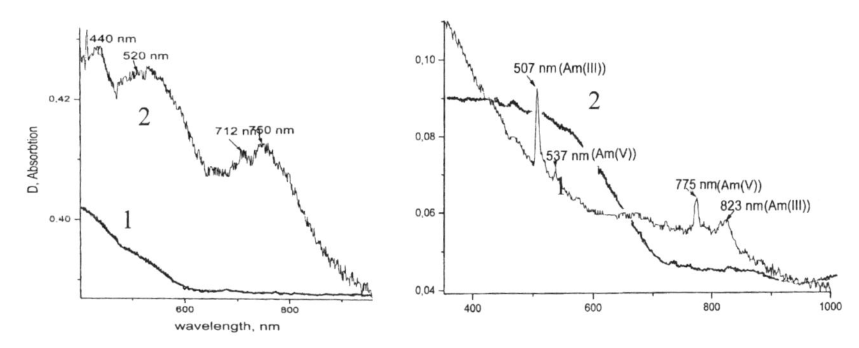

Figure 1
The absorption spectra of (1) product of interaction of Am(VI) and Na_4XeO_6 in 1M NaOH; (2) solid Am(VI) perxenate treated by ozone.

Figure 2
The absorption spectrum of the mixture of Am(III) and Am(V) hydroxides in 1M NaOH (1) at 25°C; and (2) at >60°C, or in 7M NaOH at 25°C.

2.5. Reversible Am redox-reaction in alkaline media

It was shown that the redox reaction of Am(VI) and RuO_4^{2-} ions is reversible in basic solutions. Based on the literature data we obtained an approximate value of redox potential of Am(VI)/Am(V) pair. In 1 M NaOH E^f(Am(VI)/Am(V)) was determined as ≈ 0,7 V (vs NHE). $Fe(CN)_6^{3-}$ ions were shown to oxidize Am(III) hydroxide into insoluble Am(IV) in diluted alkaline solutions ($[OH^-]<0.1M$) or into Am(V) at higher $[OH^-]$. The yield of Am(V) rose with increase of NaOH concentration.

3 CONCLUSION

i. The reduction of Am(VI) by water in alkaline solutions was studied separately in 0.1M-3M NaOH solutions, using ^{241}Am and ^{243}Am radionuclides, which have specific radioactivities differing by a factor of 15. It was shown that the contribution of radiolysis factor is insufficient to be taken into account. The reduction of Am(VI) in alkaline solutions takes place because of the interaction with water molecules.
ii. A solid state Am(VII) compound was stabilized for the first time.
iii. The results obtained have identified a number of important reactions affecting the oxidation state and stability of Am in alkaline waste environments. Such reactions as the interaction of Am(III) and Am(V,VI) compounds, the interaction of Am(III) with $Fe(CN)_6^{3-}$ (present in tank environs) and the interaction of Am(VI) with water molecules in alkaline media are important for a better understanding of processes which can take place in alkaline waste solutions.

Acknowledgments

The work was supported by the U.S.DOE-OBES, Project RC0-20004-SC14.

References

1 R.E. Gephart, and R.E. Lundgren. *Hanford Tank Clean up: A Guide to Understanding the Technical Issues*, Pacific Northwest Laboratory. PNL-10773. 1995.
2 S.F. Agnew. *Hanford Defined Wastes: Chemical and Radionuclide Compositions*. Los Alamos National Laboratory, LAUR-94-2657 Rev.2. 1995.
3 Congress. U.S. "Long-Lived Legacy: Managing High-Level and Transuranic Waste at the DOE; N.N. Egorov. *Voprosy radiatsionnoy bezopasnosti*. (In Russian), 1997, **4**, 3.
4 N.N. Krot, A.D. Gelman, M.P. Mefod'eva, V.P. Shilov, V.F. Peretrukhin, and V.I. Spitsyn. Available in English as *"Heptavalent State of Neptunium, Plutonium and Americium"*, UCRL-Trans-11798, Lawrence Livermole National Laboratory, Livermore, California. 1977.

A COMPARATIVE STUDY OF THE COMPLEXATION OF URANIUM(VI) WITH OXYDIACETIC ACID AND ITS AMIDE DERIVATIVES

Linfeng Rao and Guoxin Tian

Lawrence Berkeley National Laboratory, One Cyclotron Road, Berkeley, CA 94720, USA.
E-mail: lrao@lbl.gov

1 INTRODUCTION

There has been significant interest in recent years in the studies of alkyl-substituted amides as extractants for actinide separations because the products of radiolytic and hydrolytic degradation of amides are less detrimental to separation processes than those of organophosphorus compounds traditionally used in actinide separations. Stripping of actinides from the amide-containing organic solvents is relatively easy. In addition, the amide ligands are completely incinerable so that the amount of secondary wastes generated in nuclear waste treatment could be significantly reduced.

One group of alkyl-substituted oxa-diamides have been shown to be promising in the separation of actinides from nuclear wastes.[1-3] For example, tetraoctyl-3-oxa-glutaramide[1] and tetraisobutyl-oxa-glutaramide[3] form actinide complexes that can be effectively extracted from nitric acid solutions. To understand the thermodynamic principles governing the complexation of actinides with oxa-diamides, we have studied the complexation of U(VI) with dimethyl-3-oxa-glutaramic acid (DMOGA) and tetramethyl-3-oxa-glutaramide (TMOGA) in aqueous solutions, in comparison with oxydiacetic acid (ODA) (Figure 1). Previous studies have indicated that the complexation of U(VI) with ODA is strong and entropy-driven.[4-6] Comparing the results for DMOGA and TMOGA with those for ODA could provide insight into the energetics of amide complexation with U(VI) and the relationship between the thermodynamic properties and the ligand structure.

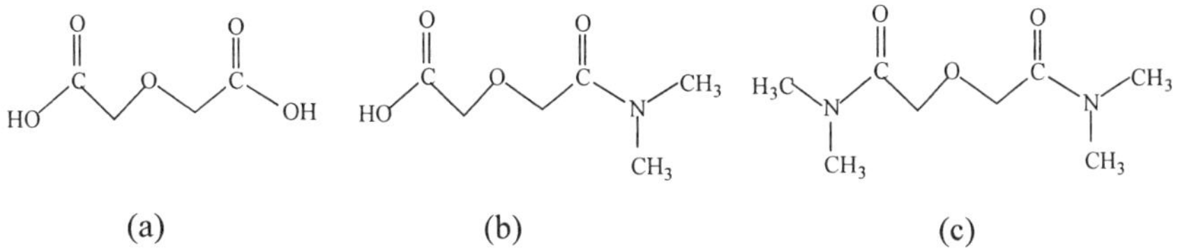

(a) (b) (c)

Figure 1 (a) Oxydiacetic acid (ODA); (b) dimethyl-3-oxa-glutaramic acid (DMOGA); (c) tetramethyl-3-oxa-glutaramide (TMOGA).

2 METHOD AND RESULTS

2.1 Stability Constants of the U(VI) Complexes with DMOGA and TMOGA

All thermodynamic measurements were conducted at $I = 1.0$ mol·dm^{-3} ($NaClO_4$) and $t = 25$ °C. Potentiometry was used to determine the protonation constant of DMOGA and the stability constants of the U(VI) complexes with DMOGA. Details of potentiometry are provided elsewhere.[6] Prior to each complexation titration, the electrode was calibrated with a standard acid/base titration in order to calculate the hydrogen ion concentrations from the electrode potential in the subsequent titration. Spectrophotometry was used to determine the stability constants of the U(VI) complexes with DMOGA and TMOGA. Sets of absorption spectra of U(VI) in solutions of different concentrations of DMOGA or TMOGA were collected on a Cary-5G spectrophotometer.

The program Hyperquad[7] was used to calculated the stability constants from potentiometry and spectrophotometry. The results are summarized in Table 1.

Table 1 *Thermodynamic parameters of U(VI) complexation with DMOGA and TMOGA*

Ligand	Reaction	log β	ΔH, kJ·mol^{-1}	ΔS, J·K^{-1}·mol^{-1}
DMOGA	$H^+ + L^- = HL(aq)$	3.49 ± 0.03	0.83 ± 0.01	70 ± 1
	$UO_2^{2+} + L^- = UO_2L^+$	3.84 ± 0.02[a]	15.0 ± 0.1	124 ± 1
	$UO_2^{2+} + 2L^- = UO_2L_2(aq)$	5.88 ± 0.02[a]	21.6 ± 0.2	185 ± 1
TMOGA	$UO_2^{2+} + L = UO_2L^{2+}$	1.71 ± 0.03	5.72 ± 0.09	52 ± 1
	$UO_2^{2+} + 2L = UO_2L_2^{2+}$	2.94 ± 0.01	19.0 ± 0.2	120 ± 1
ODA[b]	$UO_2^{2+} + L^{2-} = UO_2L(aq)$	5.01 ± 0.04	16.4 ± 0.2	152 ± 1
	$UO_2^{2+} + 2L^{2-} = UO_2L_2^{2-}$	7.64 ± 0.07	23.8 ± 0.1	227 ± 2

[a] The average value from potentiometry and spectrophotometry.
[b] Data for ODA are from the literature.[6]

2.2 Enthalpy of Complexation

The enthalpy of complexation of U(VI) with DMOGA and TMOGA was determined by calorimetry with an isothermal microcalorimeter (Model ITC 4200, Calorimetry Sciences Corp.). Detailed descriptions of the microcalorimeter and its calibration are provided elsewhere.[8] The enthalpy of complexation was calculated from the reaction heat by the computer program Letagrop.[9] The results are summarized in Table 1.

3 CONCLUSION

From ODA to DMOGA and TMOGA, the enthalpy of complexation (ΔH) becomes less endothermic and more favourable to the complexation while the entropy of complexation (ΔS) becomes smaller and less favourable to the complexation (Figure 2). Such trends imply that the amide group is less hydrated than the carboxylate group. When forming complexes with U(VI), less energy is required to dehydrate the amide group and fewer water molecules are released from the hydration sphere of the amide group than the carboxylate group. As a result, the enthalpy and the entropy of complexation both decrease in the order: ODA > DMOGA > TMOGA, making opposite contributions to the stability of the complexes. The complexation of U(VI) with DMOGA and TMOGA is weaker than that with ODA, but is still entropy driven.

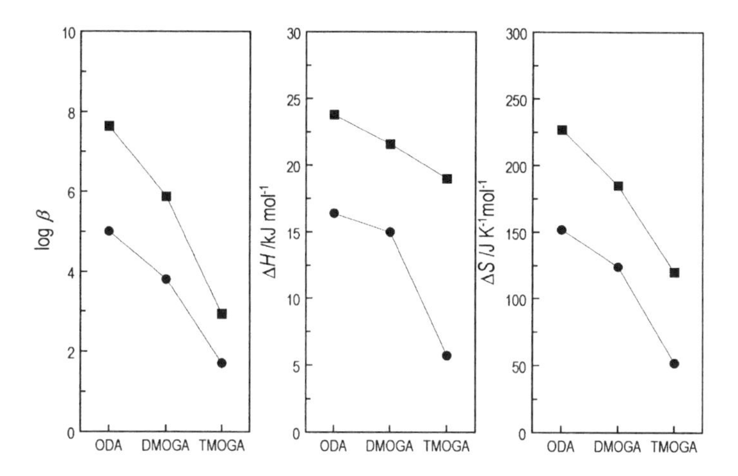

Figure 2 *Comparison of thermodynamic parameters of U(VI) complexation with ODA and its amide derivatives.* (●) *ML;* (■) *ML_2 complexes.*

Acknowledgements

This work was supported by the Director, Office of Science, Office of Basic Energy Sciences of the U. S. Department of Energy under Contract No. DE-AC03-76SF0098 at the Lawrence Berkeley National Laboratory.

References

1 H. Suzuki, Y. Sasaki, Y. Sugo, A. Apichaibukol and T. Kimura, *Radiochim. Acta*, 2004, **92**, 463.
2 H. Narita, T. Yaita and S. Tachimori, *Solv. Extr. Ion Exch.*, 2004, **22**, 135.
3 G. Tian, P. Zhang, J. Wang and L. Rao, Extraction of actinide(III, IV, V, VI) ions and Tc(VII) by N,N,N',N'-tetraisobutyl-3-oxa-glutaramide, *Solv. Extr. Ion Exch.*, 2005, **23**, 631.
4 P. Di Bernardo, G. Tomat, A. Bismondo, O. Traverso and L. Magon, *J. Chem. Res.* (M), 1980, 3144.
5 J. Jiang, J. C. Renshaw, M. J. Sarsfield, F. R. Livens, D. Collison, J. M. Charnock and H. Eccles, *Inorg. Chem.*, 2003, **42**, 1233.
6 L. Rao, A. Y. Garnov, J. Jiang, P. Di Bernardo, P.L. Zanonato and A. Bismondo, *Inorg. Chem.*, 2003, **42**, 3685.
7 P. Gans, A. Sabatini and A. Vacca, *Talanta*, 1996, **43**, 1739.
8 P. Zanonato, P. Di Bernardo, A. Bismondo, G. Liu, X. Chen and L. Rao, *J. Am. Chem. Soc.*, 2004, **126**, 5515.
9 R. Arnek, *Arkiv Kemi*, 1970, **32**, 81.

ACTINIDE ELECTROCHEMISTRY IN LOW TEMPERATURE IONIC LIQUIDS

G. Lallemand,[1] D. Bottomley,[1] R. Malmbeck,[1] P. Masset,[1] J. Serp,[1] J-P. Glatz,[1] J. Rebizant[1] and C. Madic

[1]European Commission, Joint Research Centre, Institute for Transuranium Elements
P.O. Box 2340, 76125 Karlsruhe, Germany
[2]Commissariat à l'Energie Atomique, DEN/SAC/DIR, 91191 Gif-Sur-Yvette Cedex, France

1 INTRODUCTION

Low Temperature or Room Temperature Ionic Liquids (RTILs) are viscous media entirely made up of ionic species but in contrast to traditional ionic media such as high temperature molten salts (LiCl/KCl, LiF/AlF_3...), the melting point is typically under 373K instead of over 600K. RTILs also have specific features so that they can be considered as a separate class of solvents. The interest of ITU in these media lies in their application to reprocessing of spent nuclear fuels. The objective is to deposit electrochemically actinides, lanthanides and fission products in a metallic form according to their reducing ability. Many of these metals are quite electronegative, making their deposition in aqueous media impossible. Molten salts have been the only media studied this far to undertake this process. The drawback is that they are liquids only at high temperatures: the experimental equipment must withstand the extremely hot and corrosive operating conditions (chloride and fluoride-based media). Ionic liquids have a wide electrochemical window (4V up to 6V), are liquid at lower temperatures (< 470K), and would be less corrosive. One of ITU's research programmes is to examine in a purpose-built glove-box the actinide behaviour in RTILs and the potential of these solvents to achieve the electrolytic deposition. This paper briefly describes the set-up for performing the electro-deposition of metals as well as the first results obtained in low temperature ionic media.

2 METHODS AND RESULTS

2.1 Experimental set-up

A purpose-built glove box has been constructed at ITU to investigate low temperature ionic liquids by electrochemical techniques. It is kept under purified argon atmosphere, oxygen and moisture-free conditions (< 10 ppm) and allows handling of actinide metals. Experiments can be conduced from ambient temperature up to 473 K. It is also equipped with a UV-Vis spectro-photometer that makes possible in-situ analysis. Electrochemical studies are carried out through a three-electrode set-up positioned in a electrochemical cell (Metrohm). Vitreous carbon and gold are used as working electrodes (WE) - respectively 0.23 and 0.08 cm^2 surface area. The reference electrode (RE) is made of a silver wire immersed in a separated compartment (with a ceramic frit of 1.6 μm porosity) filled up with 0.01M $AgCF_3SO_3$-containing RTIL. The counter electrode (CE) is a spiral of a 1 mm diameter platinum wire so as to expose a large surface area in comparison to the WE. The electrolytic bath consists of 6 ml [N,N-butyltrimethyl ammonium][TFSI], TFSI being the (bis(trifluoromethyl)sulfonyl)imide , $(CF_3SO_2)_2N^-$. Cations of interest are introduced in the RTIL at approximately 10^{-2} M by anodic dissolution. The main condition is to introduce them in a free form and to keep the medium clean of any other ions that are likely to interact. Anodic dissolution consists of applying a constant potential or a constant current between an anode of the chosen metal and a cathode which gives to the system a potential value superior to the oxidation potential of the metal (M^0), in order to release the metallic cations (M^{n+}) into the melt. The cathodic compartment is isolated by a large surface glass filter (6 cm^2, 6-40 μm dia. pores) so as to prevent any reduction of the metallic cations produced at the anode and electro-transported to the cathode and also prevent any contamination of the bulk medium by the solvent degradation products. The recovery of cations introduced by anodic dissolution is conducted by electro-refining. In this case, the anode is made of the same metal as the cations to be deposited and dissolved in the melt and the cathode is a metallic foil, for instance copper, on which the cations are reduced. A potential inferior to the reduction potential of the couple cation/metal is imposed on the cathode. On the anodic side, the metal is oxidised into cationic species which are electro-transported to the cathode to be reduced.

2.2 Results

The first experiments were undertaken with easily-oxidizable metals such as silver in order to set up the experimental system. A silver spiral (0.8 cm^2 surface area) was anodically dissolved in the RTIL by applying a constant positive current (3.5 mA/cm^2). Cyclic voltammetry was then carried out to observe the Ag^+/Ag^0 couple in solution (Fig.1). A single peak of reduction is seen (peak Ic) and is associated with a sharp anodic peak (peak Ia); its shape is consistent with the dissolution of the silver metal formed during the cathodic sweep. This silver solution is later used to study the reduction of Ag^+ ions to metal. The anode is a silver spiral and the cathode is a copper foil. A constant negative current (– 0.13 mA/cm^2) is imposed and a stable white silver deposit is formed (Fig.1). These results confirmed by those obtained with nickel and copper validate our experimental set-up and demonstrate that anodic dissolution can be used to produce metallic cations in RTILs .

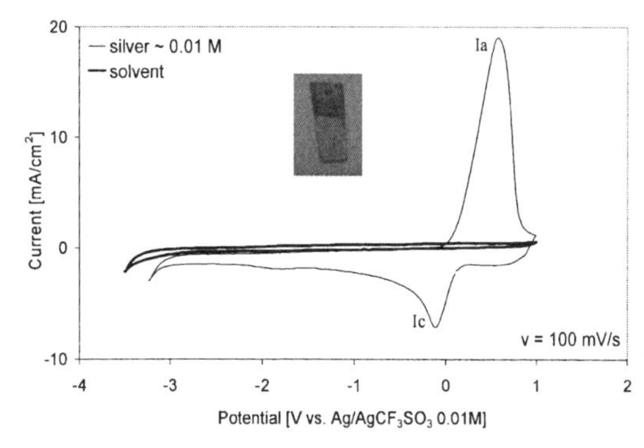

Figure 1. *Cyclic voltammetry of Ag^+ (~10^{-2} M) in [BuMe₃N] [TFSI] after an electrodissolution of an hour at 3.5 mA (10C charge, T = 30°C ; WE =Au, CE = Pt, RE = $Ag/AgCF_3SO_3$ 0.01M*

Inset. Silver deposit on a copper foil (1 cm^2) after an electro-refining of 4 hours at –0,13 mA/cm^2 and 30°C

The same set of experiments is carried out with uranium. A metallic uranium plate (~ 0.3 cm^2) was used for the anodic dissolution. The observed equilibrium potential in $[BuMe_3N]$ [TFSI] was –1.6V (vs. $Ag/AgCF_3SO_3$ 10^{-2} M) and was in line with values obtained in molten salts. A first cyclic voltammogram determines the potential at which anodic dissolution should be done. An anodic current was detected only from +0.2V (vs. $Ag/AgCF_3SO_3$ 10^{-2}M), which is a particularly high value compared to the equilibrium potential. A dissolution at 3 mA/cm^2 during 36000s was set (the resulting potential ~ +0.5V). Cyclic voltammetry of this solution (~10^{-2} M U) shows a high reduction peak at –1.8 V that could be attributed to the U^{3+}/U^0 transition even if no typical reoxidation peak is seen (Fig. 2). We agree the explanation suggested by May et *al.* [1] that the formed U metal is immediately converted to uranium oxide - probably due to the presence of moisture - and that this prevents its re-oxidation. No other uranium couples are evident in Fig. 2. In future work, it is intended to associate the UV-spectrometry to characterise species and also modify the coordination environment of uranium cations by adding complexing agents such as chlorides. After the anodic dissolution, the recovery of cations dissolved in the melt was attempted by setting a constant reduction potential (–1.6V) at 160°C (S_{copper} ~1 cm^2, $S_{uranium}$ ~0.3 cm^2). A black deposit was gradually formed (~7 C in total) and an analysis of this deposit by energy dispersive X-ray (EDX) confirms that it was uranium (Fig. 3).

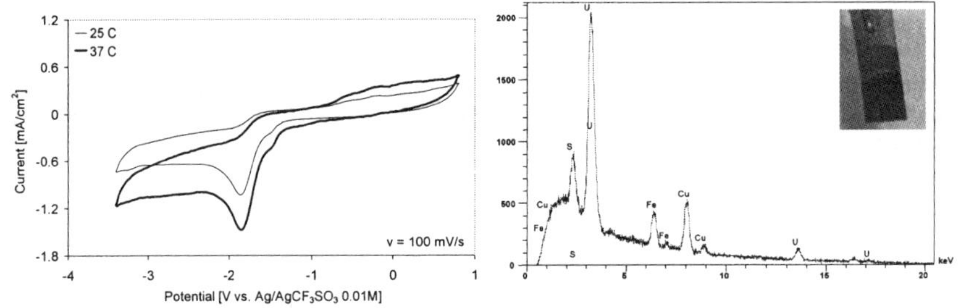

Figure 2. Cyclic voltammetry of U in $[BuMe_3N]$ [TFSI] as a function of the coulometry ; T = 100 °C; WE = vitreous carbon, CE = Pt, RE = $Ag/AgCF_3SO_3$ 0.01M

Figure 3. EDX analysis of the uranium deposit (inset) after an electro-deposition of 2 hours at –1.6V (~7C), 160°C on a copper foil (1 cm^2)*
**Copper not totally pure (iron present)*

3 CONCLUSIONS

Anodic dissolution and metal electro-deposition have been shown to be applicable in ionic liquids. First tests with silver has verified the performance of the equipment and provided a check of the glove box conditions. Initial testing using U has enabled us to perform electro-dissolution and cathodic plating on copper. However other U redox peaks are not observed. Further testing will concentrate on characterization of species, better understanding of cation behaviour, work at higher temperature and finally plutonium chemistry in RTILs.

References

1. A. Bhatt, I. May, Actinide, Lanthanide and Fission Product Speciation and Electrochemistry in High and Low Temperature Ionic Melts, Atalante 2004, Nîmes (France), June 21-25 2004.

NEW DATA ON EXTRACTION OF NITRIC ACID WITH 30% TBP FROM SOLUTIONS CONTAINING URANYL NITRATE

E.A. Puzikov, B.Ya. Zilberman, I.V. Blazheva, E.N. Mishin and N.V. Ryabkova

RPA "V.G.Khlopin Radium Institute", 2nd Murinsky av. 28, 197021 St-Petersburg, Russia

1 INTRODUCTION

Rig trials on Purex process improvement have indicated that the real distribution of nitric acid through counter-current units in the presence of U(VI) substantially differs from that calculated using published data on extraction equilibria. The databases on joint extraction of HNO_3 and U(VI) by 30% TBP in alkanes [1-3] demonstrate that the data on U(VI) distribution is in good agreement, though the data on HNO_3 distribution differs substantially. So, additional experiments using refined procedures to obtain equilibrium data on HNO_3 concentration in U(VI) extract have been required[4].

2 METHOD AND RESULTS

2.1 Experimental

Alkanes (Isopar L), CCl_4 and hexachlorobutadiene (C_4Cl_6) were used as TBP diluents. Solvent and aqueous phases were contacted for 10 min. and then centrifuged, initial and final phase volumes being accurately measured in a calibrated test tube to check material balance.

Direct potentiometric titration of the extract sample in NaF media by NaOH aqueous solution with intense stirring has been used to determine HNO_3 concentration in solvent, containing U(VI). The procedure has been verified by addition of HNO_3 sample in 30% TBP. The full quantity of HNO_3 has been determined by the second titration peak in the aqueous phase, $Na[HF_2]$ being the reacting complex in the presence of UO_2F_2 and $NaNO_3$ formed in this media. It has been confirmed also by direct titration of HF in NaF solution.

2.2 Extraction of nitric acid

The full data on joint U and HNO_3 extraction by 30% TBP with Isopar L as function of solvent loading with uranium for 0.05-11.5 mole/L of HNO_3 is shown in Figure 1. The maximum solvent loading at acidity higher than 5.6 mole/L is limited by third phase formation. The obtained data is compared with the reference data for 0.6 and 4 mole/L HNO_3 in Figure 2. It shows that our data is in good agreement with the reference data on U distribution, but HNO_3 distribution coefficients are much higher than well-known[1,2] values and correspond only to the more recent data[3]. It should be mentioned, that U distribution coefficients at 0.05 mole/L HNO_3 appeared to be lower than the major part of reference data (except[1]) because of additional nitric acid introduced with U(VI) salt.

The effect of diluents on joint U(VI) and HNO_3 extraction by 30% TBP is compared in Figure 3. It shows, that at acidity of 0.05 mole/L U(VI) distribution coefficients are higher for Isopar than for CCl_4 (as well as for C_4Cl_6). The difference in diluents disappears at 0.6 mole/L HNO_3, and at higher acidity U distribution coefficients are higher for heavy diluents. This fact disagrees with the common idea of independent diluent interaction with free TBP, it's disolvate with $UO_2(NO_3)_2$ and HNO_3 monosolvate. Probably, interaction of diluent with U and HNO_3 hydrated solvates, which are formed at low acidity and/or low solvent loading should be taken into consideration[5]. At the same time, the data shows that using alkanes or heavy diluents (Figure 4) does not affect HNO_3 distribution in 30% TBP in the presence of U(VI).

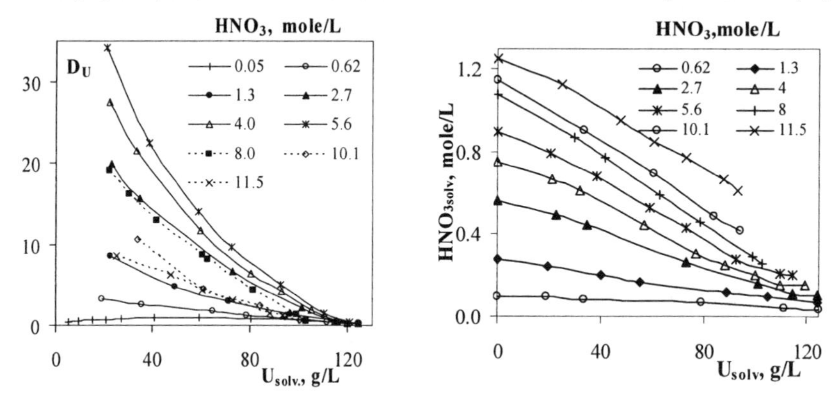

Figure 1. *Extraction of U (left) and HNO_3 (right) by 30% TBP with Isopar L at various uranium concentration in extract. Temperature 20°C.*

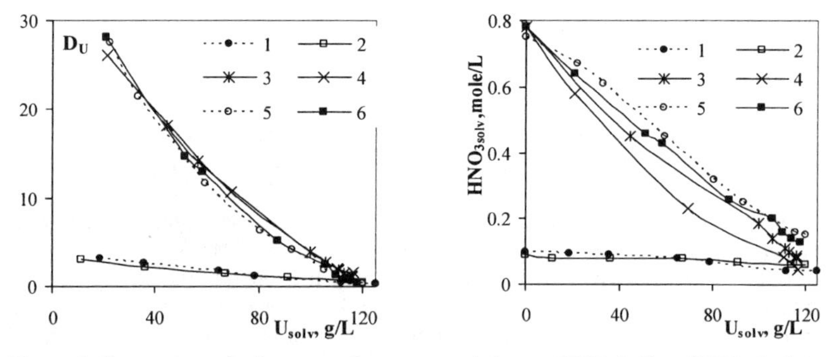

Figure 2. *Comparison of reference and experimental data on U(VI) (left) and HNO_3 (right) extraction. Nitric acid aqueous concentration: 0,6 (1,2) and 4,0 (3-6) mole/L. 1, 5 – experimental data; 2, 6 – [3]; 3 – [2]; 4 – [1]*

It is well-known that U(VI) distribution coefficients decrease with an increase of temperature (Figure 5), however the position of the peak on the curve of U(VI) distribution coefficients as a function of acidity shifts to the lower acidity region in the aqueous phase at all solvent loadings with uranium. Distribution coefficients of HNO_3, which are almost unaffected by temperature in the absence of uranium, noticeably increase at higher solvent loading with uranium (Figure 6) especially at high acidity of the aqueous phase. It demonstrates that interaction of HNO_3 with $UO_2(NO_3)_2(TBP)_2$ is more enforced with an increase of temperature than we supposed earlier.[6] The nature of this interaction requires deeper investigation with the use of independent physical and chemical methods.

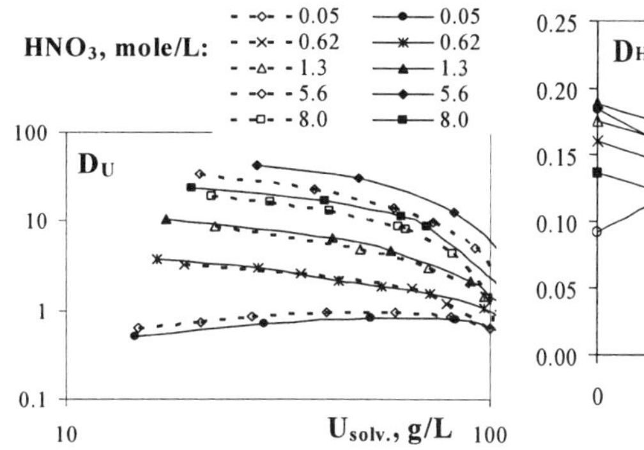

Figure 3. *The effect of diluents on U extraction by 30% TBP at various solvent loading (20°C).* —— CCl_4; ---- *Isopar L..*

Figure 4. *Extraction of HNO_3 by 30% TBP with CCl_4 in the presence of uranium (20°C).*

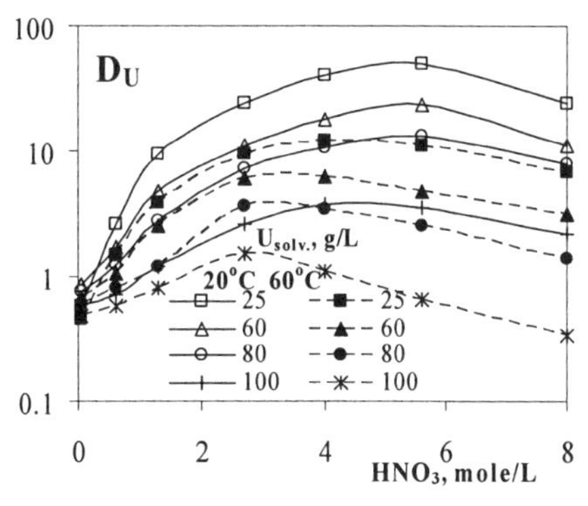

Figure 5. *U distribution coefficients as a function of acidity at 20 and 60°C for various solvent loadings (30% TBP with C_4Cl_6)*

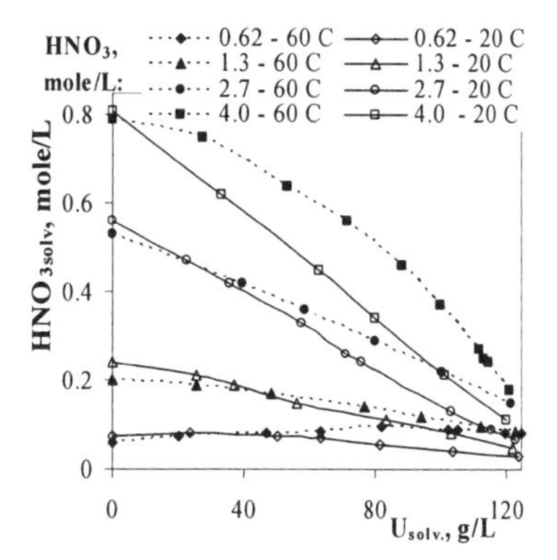

Figure 6. *Extraction of HNO_3 at 20 and 60°C for various solvent loadings (30% TBP with C_4Cl_6).*

3 CONCLUSION

Improved experimental procedures allowed more accurate data to be obtained showing higher values on HNO_3 distribution into 30% TBP with various diluents in the presence of U(VI). These data make it possible to correct significantly the mathematical model of HNO_3 extraction and obtain more accurate simulation of stripping stages in a Modified Purex process for NPP spent fuel. The corrected simulation model is given in the next paper.

References

1 J.W. Codding, W.O. Haas and F.K. Heumann. *Ind. Eng. Chem.*, 1958, **50**, No. 12, 145.
2 G. Petrich and Z. Kolarik. *KfK-2536*, 1977; *KfK-3080*, 1983.
3 V.E. Vereschagin and E.V. Renard. *Atomic Energy (rus),* 1978, 44, No 5, 422; *ibid.* 1978, **45,** No 1, 45.
4 I.V. Blazheva, B.Ya Zilberman and N.V. Ryabkova. *13 Rus. Solv. Extr. Conf. (Moscow, 19-24 Sept. 2004). Exten. abstr.(rus).* v.1, p. 92. M. RAS & Mend. St. Ac. Chem. Tech. 2004.
5 Yu.S. Fedorov and B.Ya. Zilberman. *Radiochemistry*, 2000, 42, No 3, 223.
6 B.Ya. Zilberman and Yu.S. Fedorov. *Radiochemistry (rus)*, 1986, 28, No 1, 37.

NEW DATA ON EXTRACTION OF TETRAVALENT URANIUM WITH 30% TBP FROM SOLUTIONS CONTAINING URANYL NITRATE

E.A. Puzikov, B.Ya. Zilberman, I.V. Blazheva, E.N. Mishin and N.V. Ryabkova

RPA "V.G.Khlopin Radium Institute", 2nd Murinsky av. 28, 197021 St-Petersburg, Russia

1 INTRODUCTION

Rig trials on Purex process improvement have indicated that real distribution of U(IV) in the presence of U(VI) through counter-current units substantially differ from values calculated with the use of published data on extraction equilibria[1-4]. Therefore, additional experiments using refined methods of obtaining accurate equilibrium data both on U(IV) and U(VI) content in extract has been required, which is the aim of this work.

2 METHOD AND RESULTS

2.1 Experimental

Alkanes (Isopar L or dodecane) and hexachlorobutadiene (C_4Cl_6) were used as TBP diluents. The solvent was consequently rinsed by soda and hydrazine nitrate solutions, the latter being introduced during preliminary acidifying of solvent by HNO_3 in order to prevent U(IV) oxidation and obtain more accurate data on U(IV) distribution in the presence of U(VI). The solvent and aqueous phases were mixed in closed centrifugal tubes, which were centrifuged immediately after the rough phase separation. Concentration of U(IV) was determined by titration with ammonium vanadate, first in solvent in less than 5 min. after centrifugation and then in the aqueous phase. This procedure allows to obtain much higher U(IV) distribution coefficients at low nitric acid concentration.

2.2 Extraction of U(IV)

The data on U(IV) extraction by 30% TBP in Isopar L and C_4Cl_6 at low solvent loading is compared to the reference data for paraffin diluent in Figure 1. It demonstrates that the obtained distribution coefficients of U(IV) with Isopar in the range of acidity $\leq$2 mole/L HNO_3 are higher than the reference data,[1-3] including our data obtained earlier.[4] The difference is more significant the lower the acidity, confirming the ability of U(IV) partially hydrolyzed species to oxidise in contact with air. The obtained values for Isopar coincide with that for C_4Cl_6 at acidity higher than 4 mole/L, but at low acidity they are significantly higher. It could be clearly seen also from solvent loading curves with U(IV) for 1 and 5 mole/L HNO_3 for both diluents (Figure 2). The data for dodecane includes the range of third phase formation (dash lines), where the average concentration in both solvent phases is plotted, as calculated through balance. The difference in our experimental and reference data on U(IV) extraction was also found in the presence of U(VI) both at low acidity and/or at low solvent loading with U (Table 1). The obtained data are in satisfactory agreement with U(IV) distribution profile across the stages of counter-current rig.

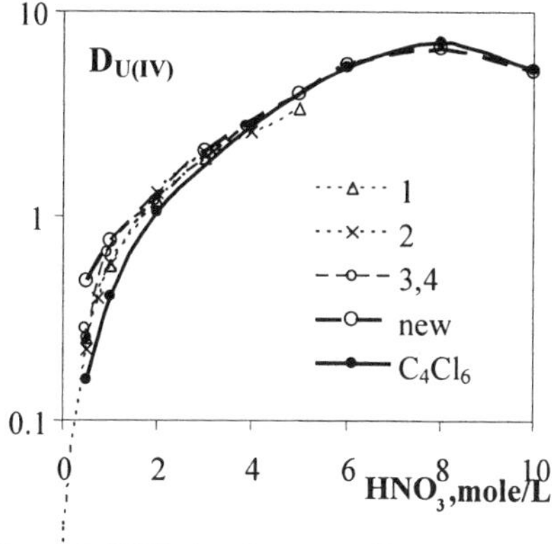

Figure 1. *U(IV) distribution coefficients as a function of HNO_3 concentration*

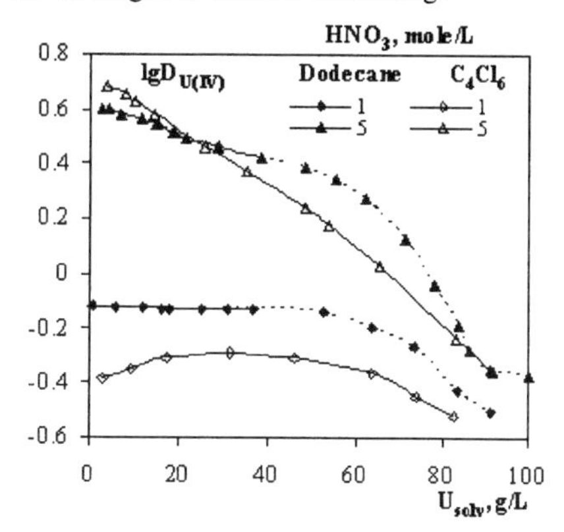

Figure 2. *Solvent saturation with U(IV) for 30% TBP in dodecane and C_4Cl_6 at aqueousHNO_3 concentration of 1 and 5 mole/L . 3-rd phase formation region is shown with dotted lines.*

Table 1. *U(IV) extraction by 30% TBP with Isopar L at various solvent loading with U(VI) at 20°C and comparison of experimental and reference data[6].*

HNO_3 aq., mole/L	U(IV) distribution coefficients at solvent loading with U(VI), g/L									
	Experimental					Interpolation according to reference data [6]				
	1	25	50	75	100	1	25	50	75	100
0,5	0,48	0,26	0,13	0,071	0,024	0,18	0,11	0,072	0,042	0,018
1,0	0,76	0,45	0,22	0,088	0,035	0,45	0,24	0,13	0,058	0,022
3,0	2,1	1,2	0,57	0,25	0,055	1,83	0,82	0,39	0,14	0,046
5,0	3,8	2,5	1,35	0,52	0,15	4,3	1,85	0,85	0,30	0,068

2.3 Mathematical description of extraction

U(IV) extraction by TBP is commonly represented by the following reaction:

$$\mathbf{U^{4+}_{(aq)} + 4NO_3^-{}_{(aq)} + 2TBP_{(s)} = [U(NO_3)_4(TBP)_2]_{(s)}}, \qquad (1).$$

Equilibrium concentration constant **K** depends on solution composition, which is characterized by ionic strength $\mathbf{I = 0.5\,\Sigma\,X_i\,Z_i^2}$, where X_i is cation concentration in aqueous solution, Z_i - its charge.[5] The values of **K** could be approximated by corrected polynomial:

$$\ln K_{U(IV)} = 11.85 - 19.17I^{1/2} + 8.67I - 1.25I^{3/2} \qquad (2)$$

where values of **I** for U(IV) are calculated according to a new empirical equation

$$I_{U(IV)} = X_{HNO_3} + 5X_{U(VI)} + 6X_{U(IV)}. \qquad (3)$$

The values of **K** calculated with the use of old [5] and new model are shown in Figure 3.

U(VI) extraction is described with the similar approach by the well-known equation,[5] with our corrections for decrease of its extraction at low acidity and/or solvent loading because of hydrated species formation in this range.[6] Equations describing HNO_3 extraction were also given previously[6] as a sum of four processes: HNO_3 extraction by free TBP ($Y_{H/S}$), formation of higher solvates ($Y_{H/H}$), formation of hydrated solvates ($Y_{H/H2O}$) and extraction by $UO_2(NO_3)_2(TBP)_2$ ($Y_{H/U}$), where W_{H2O} is free H_2O in the solvent, A_W - H_2O activity, S –concentration of free TBP. However, the data on HNO_3 extraction in the presence of U(VI) given in the above paper (5P01), leads to correction of the coefficients in equations marked * (Table2). Satisfactory coincidence of experimental and calculated data on HNO_3 extraction demonstrated in Fig. 4, was confirmed by additional rig trials with U(IV).

Table 2. *Mathematical description of nitric acid extraction in the presence of U(VI)*

$Y_H = Y_{H/S} + Y_{H/H_2O} + Y_{H/H} + Y_{H/U}$	$Y_{H_2O} = 0,24 A_W^2 S^2; F_d = 3,6 X_H^2;$
$Y_{H/S} = f_H S;$ $f_{H/S} = 0,18 X_H [NO_3]$	$W_{H_2O} = \dfrac{Y_{H_2O}}{1 + \sqrt{1 + 8F_d Y_{H_2O}}}; Y_{H/H_2O} = F_d W_{H_2O}^2$
$Z = 0,018 f_{H/S}(1 + f_{H/S})$	$Y_{H/H} = 0,005 Y_H X_H^2$ *
$A_W = 1 - 0.0719 X_H$, if $X_H < 10$ $A_W = 0.28$, if $X_H \geq 10$	$Y_{H/U} = 0.004 Y_U X_H^2 + Y_H \begin{bmatrix} 0.11 - 2.2(2Y_U - 0.2)^2 + \\ 0.03(11 - X_H) 4Y_U^2 \end{bmatrix}$ *

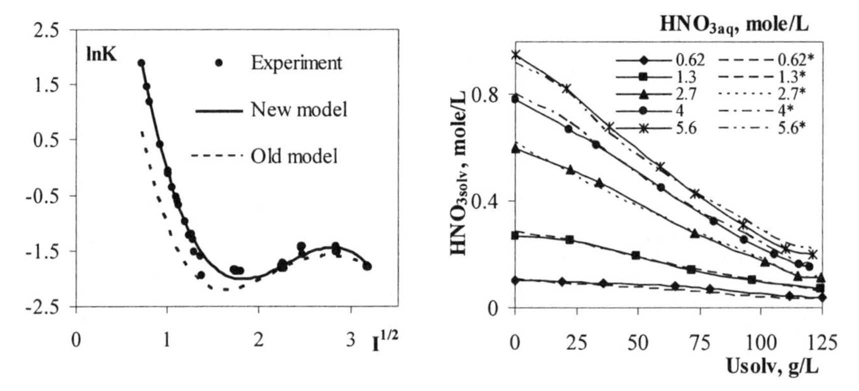

Figure 3. *Ln $K_{U(IV)}$ as a function of ionic strength of solution*

Figure 4. *Comparison of experimental and calculated data on HNO_3 extraction in the presence of $UO_2(NO_3)_2$.*

3 CONCLUSION

Improved experimental procedures generate more accurate data on U(IV) extraction by 30% TBP in various diluents, also in the presence of U(VI). It is possible to significantly improve the model of U(IV) extraction, obtaining more accurate simulation of stripping stages in a modified Purex process for NPP spent fuel. The model of HNO_3 extraction in the presence of U(VI) should be corrected additionally due to its higher extractability than calculated through free TBP concentration in combination with its co-extraction with $UO_2(NO_3)_2(TBP)_2$, although the chemical nature of the effect is not yet clear.

References

1. H.A.C. McKay and T.Y.W. Streeton. *J. Inorg. Nucl. Chem.* 1965, **27**, No 4, 879.
2. Z. Kolaric, S. Tachimori and T. Nakashima, *Solv. Extraction & Ion Exch.*, 1984, **2**, No 4&5, 607.
3. A.S. Solovkin, V.N. Rubisov, V.I. Druzherukov *et al.*, *Atomic Energy (rus),* 1987, **62**, No 5, 334.
4. B.Ya. Zilberman, Yu.S. Fedorov, A.A. Kopyrin, S.A. Arhipov, I.V. Blazeva and R.G. Glekov. *Radiochemistry(rus)*, 2001, **43**, No 2, 155.
5. A.M. Rosen and M.Ya. Zelvensky. *Radiochemistry(rus)*, 1976. **18**, No 4, 572.
6. E.A. Puzikov, B.Ya. Zilberman, Yu.S. Fedorov, E.N. Mishin, O.V. Shmidt, N.D. Goletsky and L.V. Sytnik. *Radiochemistry,* 2004, **46**, No 2, 136

BEHAVIOUR OF SOME ANHYDROUS U(IV) COMPLEXES IN HYDROPHOBIC ROOM TEMPERATURE IONIC LIQUIDS

S.I. Nikitenko,[1] C. Cannes,[2] C. Le Naour,*[2] D. Trubert,[2] J.C. Berthet,[3] and P. Moisy [4]

[1]Institute of Physical Chemistry, RAS, Leninskii Pr.31, 117915 Moscow, Russia
[2]Institut de Physique Nucléaire, Radiochimie, CNRS-IN2P3-UPS, BP1-91406 Orsay cedex, France. E-mail : lenaour@ipno.in2p3.fr
[3]Service de Chimie Moléculaire, DSM, DRECAM, CNRS URA 331, CEA Saclay, France
[4] DEN, DRCP, SCPS, LCA, Bat.399, CEA, Valrho/Marcoule, BP17171 30207 Bagnols sur Cèze, France

1 INTRODUCTION

Room temperature ionic liquids (RTIL) have a considerable potential for the recovery and purification of actinides due to their negligible vapour pressure, good electrical conductivity, wide electrochemical window, high thermal stability, and the ability to dissolve both organic and inorganic compounds.[1-3] Most of the published studies have been focused on spectroscopic and electrochemical investigations of actinides in water sensitive ionic liquids made from $AlCl_3$ and organic chloride salts.[4-6] Water stable RTILs are composed of bulky organic cations, like dialkylimidazolium, alkylpyridinium, tetraalkylammonium *etc.*, and various inorganic anions, like PF_6^-, $CF_3SO_3^-$, $(CF_3SO_2)_2N^-$ (Tf_2N^-), *etc.*[7] The behaviour of hydrated complexes of actinides has been recently studied in $[Me_3BuN][Tf_2N]$.[8] XANES and EXAFS study indicated coordination of Tf_2N^- ligands to UO_2^{2+} and U(IV) in the solutions of hydrated complexes $UO_2(Tf_2N)_2{\cdot}xH_2O$ and $U(Tf_2N)_4{\cdot}xH_2O$ respectively.[8] It was shown also that Np(IV) is oxidized spontaneously to NpO_2^+ in the solutions of $Np(Tf_2N)_4{\cdot}xH_2O$.[8] In this communication we are presenting the spectroscopic and voltammetric studies of anhydrous U(IV) complexes $[Cation]_2UCl_6$ in RTILs [Cation][Tf_2N], where the $Cation^+$ is 1-butyl-3-methyl imidazolium ($BuMeIm^+$) and methyl-tri-*n*-butyl ammonium ($MeBu_3N^+$), as well as the preliminary data on behaviour of $U(TfO)_4$, and $UI_4(MeCN)_4$ in $[MeBu_3N][Tf_2N]$.

2 METHOD AND RESULTS

2.1 Preparation of U(IV) complexes and RTILs

Ionic liquids [BuMeIm][Tf_2N] and $[MeBu_3N][Tf_2N]$ were prepared by a metathesis reactions from [BuMeIm]Cl and $[MeBu_3N][OH]$ respectively with HTf_2N in aqueous medium. [BuMeIm]Cl was synthesized *via* quaternisation of 1-methylimidazole by 1-chlorobutane as described recently.[7] Both prepared RTILs were washed with deionised water. Organic impurities were removed with activated carbon for 12 h. Drying of RTILs was performed under reduced pressure (~5 mbar) at 70-80°C overnight. Concentration of water in dried RTILs measured with coulometric Karl-Fisher titration were found to be equal to 125±20 and 174±30 ppm for [BuMeIm][Tf_2N] and $[MeBu_3N][Tf_2N]$ respectively.

Complexes of general formula $[Cation]_2[UCl_6]$ have been prepared by precipitation from U(IV) solutions in 10M HCl in the presence of a small excess of corresponding [Cation]Cl. Complexes $U(TfO)_4$ and $UI_4(MeCN)_4$ have been obtained by published procedures.[9] All uranium(IV) solids, their solutions, as well as the samples for spectroscopic and voltammetric measurements were prepared and stored in an argon filled drybox.

2.2 Spectroscopic studies

The absorption spectra of $[Cation]_2[UCl_6]$ complexes in RTILs are similar to the diffuse solid-state reflectance spectra of corresponding solid complexes. The solid-state reflectance spectra of prepared complexes are also very close to those of other hexachlorides of U(IV) with different cations.[10] The strong similarity of the solid and solution spectra allows us to conclude that the octahedral complex UCl_6^{2-} is a predominant chemical form of U(IV) in RTILs solutions.

UV/VIS spectra of $U(TfO)_4$ and $UI_4(MeCN)_4$ in RTILs are quite different from those of $[Cation]_2[UCl_6]$. The spectrum of $U(TfO)_4$ in both ionic liquids exhibits the absorption band at 630-650 nm known to be characteristic for nonoctahedral U(IV) complexes.[10] The UV/VIS spectrum of $UI_4(MeCN)_4$ in $[MeBu_3N][Tf_2N]$ shows broad intensive band with the maximum less than 250 nm, which is most probably related to the first electron-transfer band of uranium.

Addition of water until 0.5M has no significant effect on the absorption spectrum of UCl_6^{2-} in RTILs indicating its stability with respect to hydrolysis in hydrophobic ionic liquids. The similar phenomenon observed recently for $NpCl_6^{2-}$ and $PuCl_6^{2-}$ in $[Bumim][Tf_2N]$ was explained by strong solvation of $AnCl_6^{2-}$ anion in ionic liquids and by H-bonding of water molecules with Tf_2N^- anions.[11] By contrast, complexes $U(TfO)_4$ and $UI_4(MeCN)_4$ was found to be highly sensitive to hydrolysis in RTILs solutions.

2.3 Voltammetric studies

The electrochemical measurements were performed under argon at glassy carbon working electrode (d=3 mm) *vs* Ag/Ag^+ (Ag wire/0.01M AgTfO in RTIL) reference electrode and Pt wire counter electrode. The electrochemical window of RTILs spans in the potential range of -2.5 - 1.0 V and -3.5 - 1.0V for $[Bumim][Tf_2N]$ and $[MeBu_3N][Tf_2N]$ respectively.

Voltammograms of UCl_6^{2-} in both RTILs at the potential range of -2.5 - 1.0 V reveal several electrochemical systems, which can be attributed to the following processes :

Oxidation: $UCl_6^{2-} - e^- \rightarrow UCl_6^-$

Reduction I: $UCl_6^{2-} + e^- \rightarrow UCl_6^{3-}$

Reduction II: $UCl_6^{2-} + e^- + xTf_2N^- \rightarrow UCl_6(Tf_2N)_x^{3-(6-y)-x}$

The peak currents for quazi-reversible ***Oxidation*** and ***Reduction I*** processes have comparable intensities. By contrast, the peak intensity of irreversible ***Reduction II*** process is much lower. It is assumed, that the ***Reduction II*** involves adsorption of the complex $UCl_{(6-y)}(Tf_2N)_x^{3-(6-y)-x}$ at the electrode followed by its passivation. The half-wave potentials, $E_{1/2}$, for observed redox couples are summarised in Table 1.

The $E_{1/2}$ potential of U(V)/U(IV) couple is at 80 mV and those of U(IV)/U(III) at 250 mV more positive in $[Bumim][Tf_2N]$ than that in $[MeBu_3N][Tf_2N]$. This observation could be explained by the difference in solvation of uranium complexes in the two ionic liquids.

Table 1. *The values of half-wave potential for UCl_6^{2-} redox processes in the RTILs at the potential range of -2.5 - 1.0 V vs. Ag/Ag$^+$, v=100 mV·sec^{-1}, T=25°C.*

Process	*$E_{1/2}$, V vs. Ag/Ag$^+$*	
	[Bumim][Tf$_2$N]	*[MeBu$_3$N][Tf$_2$N]*
Oxidation	+ 0.27	+0.19
Reduction I	-1.96	-2.20
Reduction II*	-1.41	~-1.6

*E_{pc} – cathodic peak potential

Complexes $U(TfO)_4$ and $UI_4(MeCN)_4$ exhibit broad cathodic irreversible waves U(IV)→U(III) at -0.80V and -1.16V respectively. Complex $U(TfO)_4$ is stable to electrochemical oxidation : any anodic process is observed during the potential scan to positive direction until +1V. Complex $UI_4(MeCN)_4$ demonstrates quazi-reversible U(IV)→U(V) oxidation at $E_{1/2}$=-0.11V.

Scan to negative potential direction until -3.5V in [MeBu$_3$N][Tf$_2$N] solutions reveals the presence of irreversible cathodic processes at the half-wave potentials equal to -3.0V for UCl_6^{2-}, -2.8V and -3.1V for $U(TfO)_4$, and -3.1V for $UI_4(MeCN)_4$. In principle, these cathodic waves could be assigned to the reduction of U(III) to U(0). However, the additional investigation is needed to establish the nature of these processes.

References

1 W.J. Oldham, D.A. Costa and W.H. Smith, Ionic Liquids. Industrial Applications for Green Chemistry, Rogers, R.D.; Seddon, K.R. Eds., ACS Symposium Series, 818, Washington, DC, 2002, *Ch.15*, 188.

2 A.E. Visser and R.D. Rogers, *Journal of Solid State Chemistry*, 2003, **109,** 113.

3 S. Dai, Y.H. Ju and C.E. Barnes, *J.Chem.Soc. Dalton Trans.*, 1999, 1201.

4 J.P. Schoebrechts and B.Gilbert, *Inorg .Chem.*, 1985, **24**, 2105.

5 C.J. Anderson, M.R. Deakin, G.R. Choppin, W. D'Olieslager, L. Heerman and D.J. Pruett, *Inorg. Chem.*, 1991, **30**, 4013.

6 D.A. Costa, W.H. Smith and H.J. Dewey, *Electrochem.Soc. Proceedings*, 2000, **99-41**, 80.

7 Ionic Liquids in Synthesis, P. Wasserscheid and T. Welton, Eds., Wiley: Weinheim, Germany, 2003.

8 A.I. Bhatt, H. Kinoshita, A.L. Koster, I. May, C.A. Sharrad, V.A. Volkovich, O.D. Fox, C.J. Jones, B.G. Lewin, J.M. Charnock and C. Hennig, *Proceedings ATALANTE 2004, June 21-25, Nimes, France,* 2004, O13-04.

9 J.C. Berthet, M. Lance, M. Nierlich, M. Ephritikhine, *Eur. J. Inorg. Chem.* 1999, 2005.

10 J.L. Ryan, C.K. Jørgensen, *Molec.Phys.* 1963, **7**, 17.

11 S.I. Nikitenko, P. Moisy, C. Berthon, I. Bisel, *Proceedings ATALANTE 2004, June 21-25, Nimes, France,* 2004, O25-05.

THE HYDROLYSIS OF PROTACTINIUM(V): EQUILIBRIUM CONSTANTS AND DETERMINATION OF THE STANDARD THERMODYNAMIC DATA

C. Le Naour,* D. Trubert and C. Jaussaud

Institut de Physique Nucléaire, Radiochimie, CNRS-IN2P3-UPS, BP1-91406 Orsay cedex, France

1 INTRODUCTION

Protactinium is now experiencing a renewal of interest, because the isotope ^{233}Pa is involved in the nuclear reaction $^{232}Th \rightarrow {}^{233}U$ that occurs in thorium fuel reactors, e.g. in Accelerator-Driven System (ADS).[1] However, the strong tendency of protactinium(V) toward hydrolysis, polymerisation, sorption on any solid support (leading to an erratic behaviour of this element in absence of complexing agent), and its "unique" chemical properties (little resemblance with Nb and Ta, but any with pentavalent actinides), are reflected in controversial literature.[2-4] Therefore, scarce thermodynamic data relative to hydrolysis and complexation of Pa(V) are available.[5] Concerning hydrolysis, data were determined at high ionic strength (3 and 5 M),[6,7] but only at 20-25°C.
In non-complexing (acidic) media, Pa(V) is known to exist under the following forms: $PaO(OH)^{2+}$, $PaO(OH)_2^+$ and $Pa(OH)_5$.[3,6] In previous papers,[8-11] the two hydrolysis constants relative to the following equilibria,

$$PaO(OH)^{2+} + H_2O \rightleftarrows PaO(OH)_2^+ + H^+ \quad K_2 \quad (1)$$

$$PaO(OH)_2^+ + 2\,H_2O \rightleftarrows Pa(OH)_5 + H^+ \quad K_3 \quad (2)$$

were determined as a function of ionic strength at 10, 25, 40 and 60 °C. The thermodynamic constants were obtained from systematic studies of the variations of the partition coefficient of Pa(V) in the system TTA/toluene/Pa(V)/H_2O/H^+/Na^+/ClO_4^- as a function of proton and TTA concentrations, ionic strength and temperature.

2 EXPERIMENTAL AND DETERMINATION OF HYDROLYSIS CONSTANTS

Solvent extraction experiments were performed between 10 and 60 °C with protactinium (isotope ^{233}Pa - 27d) at tracer scale (~ 10^{-12} M). The preparation of the organic (TTA/toluene) and the aqueous ($HClO_4$/$NaClO_4$) phases as well as the overall experimental protocol used in this work were described previously.[11] The hydrolysis constants, at fixed ionic strength and temperature, were deduced from the variations of the distribution coefficient as a function of the proton concentration, for a given TTA concentration. This systematic study of the variations of the partition coefficient of Pa(V) as a function of proton and TTA concentrations, ionic strength and temperature represents more than 2000 experimental points[11]. Hydrolysis constants at zero ionic strength were calculated using SIT modelling. Interaction coefficients were also derived. Figure 1 depicts the 3D-variations of the constant K_2 as a function of ionic strength and temperature.

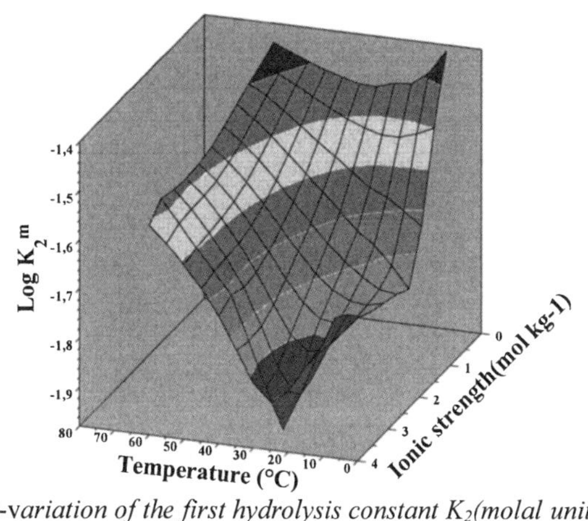

Figure 1 3D-*variation of the first hydrolysis constant K_2(molal unit) at tracer scale, as a function of ionic strength and temperature.*

3 STANDARD THERMODYNAMIC DATA

3.1 First hydrolysis equilibrium [K_2]

Since only four values are available, the change of heat capacity has been considered to be constant throughout the interval of temperature investigated. The variations of extrapolated K_2^0 are well fitted, leading to the determination of Δ_rH^0, $\Delta_rC_p^0$, Δ_rG^0 and Δ_rS^0 with acceptable uncertainties. Standard thermodynamic data related to the first hydrolysis equilibrium of Pa(V) are listed in table 1. The endothermic character of the hydrolysis reaction of Pa(V) is also observed for U(IV), Pu(IV) and Pu(VI) ($\Delta_rH^0 > 0$). For actinides IV and VI, the standard Gibbs energy change lies between 30 - 60 kJ mol^{-1}.[12-14] The Δ_rG^0 value associated to Pa(V) is smaller and of the same order of magnitude of Δ_rG^0 for tetravalent actinides. Concerning entropy change, high positive values (~+ 100 J K^{-1} mol^{-1}) are obtained when bare ions are involved [U^{4+}, Pu^{4+}]. The Δ_rS^0 value is close to those of PuO_2^{2+}, which confirm that Pa(V) is never under the form of a bare ion.

3.2 Second hydrolysis equilibrium [K_3]

The high error bars assigned to the constants as well as the dispersion of the extrapolated values do not allow the same fitting as for the K_2^0 variations. Therefore, the Van't Hoff expression used do not include the heat capacity change, (i.e. considering that Δ_rH^0 does not vary with temperature). Although this equation is not rigorous, it was used to give estimations of thermodynamic data related to the second hydrolysis equilibrium of Pa(V). Fitting the log(K_3^0) variations lead to the estimated value of Δ_rH^0 listed in Table 1 together with Δ_rG^0 and Δ_rS^0.
Since these values are only estimations and there are no thermodynamic data for species with both charge and structure analogous to the protactinium hydroxide complexes, Pa(V) can not be compared with other actinides.

Table 1 *Standard thermodynamic data (at 298.15 K) related to the hydrolysis equilibria (K_2 and K_3) of Pa(V) and deduced from the temperature dependence of extrapolated constants to zero ionic strength*

Equilibrium (K_2): $PaO(OH)^{2+} + H_2O \rightleftarrows PaO(OH)_2^+ + H^+$			
Δ_rH^0=5.7 ± 1.3kJ mol^{-1}	$\Delta_rC_p^0$=-200±89 JK^{-1} mol^{-1}	Δ_rG^0=7.1± 0.1 kJ mol^{-1}	Δ_rS^0=-4.5±4.7 J K^{-1} mol^{-1}
Equilibrium (K_3): $PaO(OH)_2^+ + 2\,H_2O \rightleftarrows Pa(OH)_5 + H^+$			
Δ_rH^0 = 61 ± 31 kJ mol^{-1}	Δ_rH^0 = 36.3 ± 4 kJ mol^{-1}	Δ_rS^0 = 81 ± 118 J K^{-1} mol^{-1}	

4 CONCLUSION

The hydrolysis constants of Pa(V) at tracer scale were determined at different temperature and ionic strength using the solvent extraction technique. The constants were extrapolated to I=0 using SIT modelling.

Acknowledgments

This work was fully supported by ANDRA-France under contract No 00875 lot 3.

References

1. M. Lung, and O. Gremm, *Nucl. Eng. Design*, 1998, **180**, 133.
2. R. Guillaumont, G. Bouissières, and R. Muxart, *Actinides Rev.*, 1968, **1**, 135.
3. R. Muxart, and R. Guillaumont, Complément au Nouveau Traité de Chimie Minérale-2-Protactinium ,Masson, Paris, 1974.
4. H. W. Kirby 'Protactinium' in *The Chemistry of Actinide Elements* 2nd ed. J.J. Katz, G.T. Seaborg and L.R. Morss, eds., Chapman & Hall, London, 1986 pp. 102-168.
5. J. Fuger, I. L. Khodakovsky, E. I. Sergeyeva and J. D. Navratil, *The Chemical Thermodynamics of Actinide Elements and Compounds: Part 12 . The Actinide Aqueous Inorganic Complexes* IAEA, Vienna, 1992.
6. R. Guillaumont, Ph.D. Thesis (No. 5528), Université de Paris, France,1966.
7. J.O. Liljenzin, and J. Rydberg, in Colloque International sur la Physico-Chimie du Protactinium, Editions du CNRS, Paris, 1966 pp.255-271.
8. D. Trubert, C. Le Naour, and C. Jaussaud, *J. Sol. Chem.* 2002, **31**, 261.
9 C. Le Naour, D. Trubert, and C. Jaussaud, *J. Sol. Chem.* 2003, **32**,489.
10 D. Trubert, C. Le Naour, and C. Jaussaud, *J. Sol. Chem.* 2003,**32**,505.
11. C. Jaussaud, Ph.D. Thesis (N° 7149), Université Paris XI-Orsay, France 2003.
12. R. J. Lemire, J. Fuger, H. Nitsche, P. Potter, M. H. Rand, J. Rydberg, K. Spahiu, J. C. Sullivan, W. J. Ullman, P. Vitorge and H. Wanner, *Chemical Thermodynamics of Neptunium and Plutonium,*OECD/NEA, Paris, 2001.
13. I. Grenthe, J. Fuger, R. J. Konings, R. J. Lemire, A.B. Muller, C. Nguyen-Trung, and H. Wanner, *Chemical Thermodynamics of Uranium* OECD/NEA, Paris, 1992.
14. R.J. Silva, G. Bidoglio, M. H. Rand, P. B. Robouch, H. Wanner, and I. Puigdomenech, *Chemical Thermodynamics of Americium* OECD/NEA, Paris, 1995.

THE STUDY OF POLYMERIZATION OF $Pu(OH)_4$ AND ITS SOLUBILITY IN 0.01M NaCl SOLUTIONS AT VARIOUS pH VALUES

Yu.M. Kulyako,[1] S.A. Perevalov,[1] T.I. Trofimov,[1] D.A. Malikov[1], B.F. Myasoedov[1], A. Fujiwara, [2], and O. Tochiyama [3]

[1] Vernadsky Institute of Geochemistry and Analytical Chemistry RAS, Kosygin Str., 19, 119991, Moscow, Russia
[2] Radioactive Waste Management Funding and Research Centre, 15, Mori Bldg., 2-8-10, Toranomon, Minatoku, Tokyo, 105-0001, Japan
[3] IMRAM, Tohoku University, 1,1 Katahira, 2-Chome, Aobaku, Sendai 980-8577 Japan

1 INTRODUCTION

A rapid drop in acidity of solutions containing Pu(IV) causes deep hydrolysis of Pu and precipitation of monomeric $Pu(OH)_4$ at pH>3. The slow lowering of the acidity at $[H^+]<0.3M$ promotes an appearance of Pu polymers with subsequent formation of highly polymerized $Pu(OH)_4$ on achieving high pH values.[1] Polymeric $Pu(OH)_4$ is also formed as a result of oxidation of Pu(III) to Pu(IV) in solid $Pu(OH)_3$. This behavioural variety of Pu(IV) at the formation of its hydroxide as well as further polymerization of the formed $Pu(OH)_4$ determine its solubility in the aqueous solutions. Available literature data on $Pu(OH)_4$ solubility often vary by several orders of magnitude.[3-4] Information on the compositions of Pu(IV) hydroxo-compounds and Pu occurrence in natural waters is urgent for a prediction of environmental behaviour of these compounds during long-term Pu storage. We have studied the properties of $Pu(OH)_4$ samples, prepared in different ways and kept for a long period of time. Concentrations of Pu(IV) and its speciation in 0.01M NaCl solutions in contact with these samples at various pH values were also determined.

2 METHOD AND RESULTS

The study on solubility of various $Pu(OH)_4$ samples in mineral acid solutions was conducted. Hydroxides of Pu(IV) were prepared, using ^{239}Pu isotope with ^{241}Am impurity (~9% by α–activity). Two $Pu(OH)_4$ samples were obtained from $Pu(OH)_3$ as a result of oxidation of Pu(III) to Pu(IV) in solid $Pu(OH)_3$. One of these samples was aged for more than 9 months, and another one was prepared just before the study. Two other $Pu(OH)_4$ samples were prepared by rapid alkaline precipitation (one portion of 8 M concentrated alkali was added in excess relative to Pu(IV)) and slow alkaline precipitation (portions of 0.1 M diluted alkali were added at about 2 min intervals) from acid solution of Pu(IV). In the process of storage of $Pu(OH)_4$ samples they were periodically analysed for their solubility in mineral acid solutions. The Pu oxidation states and its concentration in the acid solutions were determined by spectrophotometry and alpha-radiometry, respectively.

It was shown that Pu(IV) exists in all the prepared hydroxides in two forms. The slightly polymerized form of Pu(IV), which is soluble in 4M HCl solution, predominates in fresh-prepared hydroxides. With time the ratio of this form of Pu(IV) in the hydroxide precipitates decreases, whereas the portion of $Pu(OH)_4$ which is insoluble in 4M HCl but soluble in the mixture of 6M HNO_3 and 0.1M HF, increases. After 4-month storage, regardless of the preparation method of $Pu(OH)_4$, samples consists of ~90% of the highly polymerized form and of ~10% of the slightly polymerized one. Further, the samples containing both highly and slightly polymerized Pu(IV) are denoted as "$Pu(OH)_4$–untreated" and the samples treated with 4M HCl are denoted as "$Pu(OH)_4$–washed with 4M HCl". The presence of polymeric Pu(IV) in $Pu(OH)_4$ is proved by spectrophotometry after adding "$Pu(OH)_4$–washed with 4M HCl" into aqueous solution with pH~2. In the spectrum of this solution are present only maxima of absorption bands at 510, 579, 618, 690 and 736 nm corresponding completely to those of polymeric Pu(IV)[5,6].

To study speciation of Pu(IV) there were prepared 4 suspensions of $Pu(OH)_4$ in 0.01M NaCl solutions with pH values of 4, 6, 8 and 10 for each of 4 "$Pu(OH)_4$–untreated" and 4 "$Pu(OH)_4$–washed with 4M HCl" samples. After settling the suspensions Pu concentration was determined in the respective supernatants and centrifugates (centrifugal force 8000 rpm or 6500g). In all 16 determinations of Pu concentration in the settled and centrifuged solutions for all of "$Pu(OH)_4$–untreated" and "$Pu(OH)_4$–washed with 4M HCl" samples were undertaken. The data obtained on Pu concentration had wide scatter, ranging from 10^{-4} to 10^{-7} M and it was impossible to reveal the dependence of Pu concentrarion on the type of the $Pu(OH)_4$ samples as well as of the solutions pH. Ultra-filtration of the centrifugates through the filters cutting off the particles with nominal molecular weight limit (NMWL)>100 kDa (particle size >5 nm) leaves behind more than 99.9% of Pu on the filters. Therefore, Pu in the settled and centrifuged 0.01M NaCl solutions exists practically completely as the highly polymerized colloidal particles with various NMWL>100 kDa. This fact explains the wide scatter of the data obtained on Pu concentration.

Solubility values of the actinide hydroxides are determined as concentrations of actinides passed into solutions equilibrated with these compounds (achievement of equilibrium "from below") after ultra-filtration of these solutions through the filters with pore size < 0.45μm.[4,7] To determine solubility values of the prepared $Pu(OH)_4$ samples, we used 5-fold sequential ultra-filtration of the same 0.01M NaCl solutions with pH 4-10 through the same precipitates of "$Pu(OH)_4$–untreated" and "$Pu(OH)_4$–washed with 4M HCl" which left on the filters cutting off NMWL >30 kDa (particle size >1 nm). This method allowed an achievement of the equilibrium "from below".

Table 1 *Average equilibrium Pu concentrations as a function of pH value of 0.01M NaCl solutions 5 times passed through for the respective "$Pu(OH)_4$–untreated" and "$Pu(OH)_4$–washed with 4M HCl" samples*

	"$Pu(OH)_4$–untreated"			"$Pu(OH)_4$–washed with 4M HCl"		
pH	Pu concentration/M	%Pu[a]	%Am[a]	Pu concentration/M	%Pu[a]	%Am[a]
4	7.6×10^{-7}	44	56	8.5×10^{-8}	60	40
6	2.9×10^{-8}	75	25	3.2×10^{-9}	95	5
8	1.7×10^{-8}	58	42	2.5×10^{-9}	98	2
10	6.2×10^{-9}	72	28	2.2×10^{-9}	98	2

[a] percentages of Pu and Am relative to a total alpha-activity of the sample

As seen from the data presented in Table 1 the Pu(IV) concentrations in ultra-filtrates equilibrated with the "$Pu(OH)_4$–untreated" are about an order of magnitude higher than those in the respective ultra-filtrates equilibrated with "$Pu(OH)_4$–washed with 4M HCl". Enrichment of Pu species in the filtrates passed through the "$Pu(OH)_4$–untreated" with Am(III), which is not subject to polymerization, pointed to the presence of slightly polymerized Pu(IV) in this hydroxide type. The concentration of the slightly polymerized Pu(IV) species in the ultra-filtrates equilibrated both with "$Pu(OH)_4$–untreated" and "$Pu(OH)_4$–washed with 4M HCl" depends on the solutions pH within the pH range of 4-6. On further increase of pH this dependence is revealed not so clearly, especially for the solutions equilibrated with "$Pu(OH)_4$–washed with 4M HCl". These solubility data are of the same order as the published values.[3-4]

3 CONCLUSION

Regardless of the preparation method of $Pu(OH)_4$, most of the samples (~90%) convert to the highly polymerized form with storage time. The 0.01M NaCl solutions (pH~4–10) equilibrated with aged $Pu(OH)_4$ contain practically all Pu in the form of highly polymerized colloidal particles with NMWL>100 kDa. Therefore, the wide scatter of Pu concentrations (10^{-4} - 10^{-7}M) is observed in these solutions. Concentrations of the slightly polymerized Pu species in the same solutions range from 10^{-7} to 10^{-9} M and depend on the solution pH within the pH range of 4-6. Within the pH range of 6-10 this dependence is practically absent. It is clear that content of slightly polymerized Pu species in the solutions in contact with $Pu(OH)_4$ can not be used for determination of solubility product of $Pu(OH)_4$ because this hydroxide is formed as a result of both chemical interaction between Pu^{4+} cations and OH^- anions and of stages of polymerization and colloid formation.[8,9]

Acknowledgements

This work was performed under financial support of the Ministry of Economy, Trade and Industry(METI), Japan and Russian grant 1693.2003.3.

References

1 M. Kasha, G.E. Sheline, *Transuranium Elements*- Nat. Nucl. Energy Ser., Div. IV, 1949, v.14B, 3.100
2 M.S. Milukova, N.I. Gusev, I.G. Sentyurin, I.S. Sklyarenko. "Analytical chemistry of plutonium". Publisher "Science", Moscow, 1965, 30.
3 V. Neck, J. I. Kim, *Radiochim. Acta,* 2001, **89**, 1
4 K. Fujiwara, H. Yamana, T. Fudjii, H. Moriyama, *Radiochim. Acta*, 2002, **90**, 857
5 D.W. Ockenden, G.A. Welch, *J. Chem. Soc.,* 1956, 3358
6 M.H. Lloid, R.G. Haire, *Radiochim. Acta,* 1978, **25**, 139
7 D. E. Hobart "Actinides in the Environment" p. 394. "The Robert A. Welch Foundation Conference on Chemical Research, XXXIV FIFTY YEARS WITH TRANSURANIUM ELEMENTS, October 22-23, 1990, Houston, Texas M.
8 Y.P. Davydov, *Radiokhimiya,* 1967, **9**, 1, 52 (see p. 50 in the English translation of the Russian original)
9 Haissinsky, *Acta Phys.*, 1935, **3**, 4, 517

DISSOLUTION OF ACTINIDE OXIDES IN SOME ORGANIC REAGENTS SATURATED WITH NITRIC ACID

Yu.M. Kulyako, T.I. Trofimov, M.D. Samsonov and B.F. Myasoedov

Vernadsky Institute of Geochemistry and Analytical Chemistry RAS, Kosygin Str., 19, 119991, Moscow, Russia E-mail: kulyako@geokhi.ru

1 INTRODUCTION

Extraction, separation and purification of U and Pu are the key goals of the currently widely used PUREX process. The main drawback of the PUREX process is the generation of large volumes of aqueous and organic high-level radioactive wastes (HLW) [1,2]. Application of supercritical fluid extraction (SFE) technique was recently shown[3] to be very promising for oxide nuclear fuel reprocessing, with supercritical CO_2 (sc-CO_2) being a non-toxic and environmentally benign solvent which allow minimizing hazardous liquid HLW generation. It was shown that uranium oxide fuel can be directly dissolved in sc-CO_2 containing the adduct of tri-n-butyl phosphate (TBP) with HNO_3.[4-5] It was of interest to investigate other adducts of organic reagents with HNO_3 suitable for direct dissolution of actinide oxides followed by the investigation of this process under SFE conditions.

2 METHOD AND RESULTS

Solid solutions of the samples of $^{239}PuO_2$ in the matrix of $^{233+238}UO_2$ (MOX) were prepared (Table 1) by calcination of mixed oxalates of U and Pu in the atmosphere of the mixture of Ar with 10% H_2 at 850^{0}C for 8 hours.

Table 1 *Composition of the MOX samples prepared*

Sample	Actinide oxide content/wt%	
	$^{233+238}UO_2$	$^{239}PuO_2$
MOX-1[a]	95.0 (4.53[b])	5.0
MOX-2	89.7 (3.53[b])	10.3
MOX-3	92.6 (4.14[b])	7.4

[a] see ref.[5]
[b] in brackets the $^{233}UO_2$ content in $^{233+238}UO_2$ is given

The MOX samples prepared were analyzed by X-ray diffraction. In all cases the matrix of the solid solutions were shown to be uranium dioxide. The methyl isobutyl ketone $(CH_3)(C_4H_9)CO$ (hexone) was used in our work. Like TBP, this reagent can extract U and Pu and separate them from fission products in highly acidic HNO_3 solutions. In addition, hexone is highly soluble in sc-CO_2.[6]

Considerable recent attention, especially in France, has been focused on investigation of new extractants such as malonamides as alternative to TBP and carbamoyls being used in the TRUEX-process.[7] These reagents are easily synthesized in a pure form and do not generate new solid wastes, since they are completely burned. In our work N,N'-dimethyl-N,N'-dioctyl-hexylethoximalonamide (DMDOHEMA) was the chosen amide.

The hexone-HNO_3 and DMDOHEMA-HNO_3 adducts were prepared by intense mixing of equal volumes of the reagent and 8 M HNO_3 in 15-mL centrifuged test-tubes with screw caps for 15 min. It was determined that HNO_3 concentrations in the adducts hexone-HNO_3 and DMDOHEMA-HNO_3 comprise ~4M and ~4.6M, respectively.

Known volume of obtained adducts were introduced into 15-mL centrifuged test-tubes with the MOX samples. The test-tubes were placed into an ultrasonic bath and held at 60 °C for about 3 hours for dissolution of the samples. The obtained solutions of the complexes of U and Pu with hexone-HNO_3 and DMDOHEMA-HNO_3 adducts were analyzed for $^{233+238}U$ and ^{239}Pu content by radiometry using alpha-spectrometer Alpha Analyst ("Canberra"), with the data shown in Table 2.

Table 2 *Dissolution of the MOX samples in the systems "TBP-HNO_3", "Hexone-HNO_3" and "DMDOHEMA-HNO_3"*

Adduct	Sample	Actinide added/mg		Actinide found /mg in solution		Actinide found /mg in solid residue	
		U	Pu	U	Pu	U	Pu
TBP-HNO_3[a]	MOX-1	14.6	0.67	13.62	0.60	–	–
Hexone-HNO_3	MOX-2	40.9	4.7	39.2	0.15	–	4.5
DMDOHEMA-HNO_3	MOX-3	25.6	2.1	24.6	2.1	–	–

[a] see ref.[5]

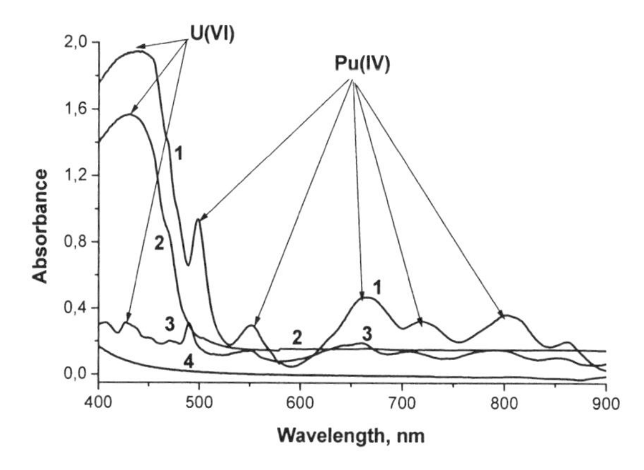

Figure 1 *Spectra of the organic phase obtained on dissolution of the MOX samples in the systems "DMDOHEMA-HNO_3" (spectrum 1), "Hexone-HNO_3" (spectrum 2), "TBP-HNO_3" (spectrum 3, see ref.[5]) and of neat adduct "DMDOHEMA-HNO_3" (spectrum 4)*

As seen from the data obtained the MOX samples which is the solid solution of PuO_2 in the matrix of UO_2, are completely dissolved only in the adducts of TBP-HNO_3 and DMDOHEMA-HNO_3. The mechanism of dissolution has been presented previously.[5] When dissolving the MOX sample in the adduct “hexone-HNO_3”, extraction of only uranium occurs, with a consequent separation of U(VI) from Pu(IV), which remains in the solid phase, probably as PuO_2.

Oxidation states of U and Pu in the obtained solutions of the actinides complexes with the adducts involved were identified by absorption spectroscopy. As evident from the spectra (Figure 1) on dissolution of the MOX sample in the system "Hexone-HNO_3" uranium exists in this solution in oxidation state +6, whereas in the systems TBP-HNO_3 and DMDOHEMA-HNO_3 uranium and plutonium exist in oxidation state +6 and +4, respectively.

3 CONCLUSION

Direct dissolution of actinide dioxides in hexone and DMDOHEMA, saturated with nitric acid was studied. It was shown that MOX completely dissolved only in the DMDOHEMA-HNO_3 adducts, as previously found for the TBP-HNO_3 adduct.[5] When dissolving the MOX sample in the adduct “hexone-HNO_3”, extraction of only uranium occurs, resulting in separation of U(VI) from Pu(IV), which remains in the solid phase, probably as PuO_2. Hence, the “hexone-HNO_3” and “DMDOHEMA-HNO_3” systems, as well as the “TBP-HNO_3” one can find application in the future technology of oxide nuclear fuel reprocessing under SFE conditions

Acknowledgements

The work was performed under financial support Russian grant 1693.2003.3.

References

1 J.J. Katz, G.T. Seaborg and L.R. Morss, The Chemistry of Actinide Elements. // Vol. 1. 2nd Ed. Chapman and Hall. London. 1986. p. 525.
2 I.S. Denniss and A.P. Jeapes in P.D. Wilson (Ed.) The Nuclear Fuel Cycle: From Ore to Waste. // Oxford University Press. Oxford. UK. 1996. Ch. 7. p. 123.
3 N. Smart, C. Phelps and C.M. Wai, *Chemistry in Britain*, 1998, **34**, 34.
4 M.D. Samsonov, T.I. Trofimov, S.E. Vinokurov, S.C. Lee, C.M. Wai and B.F. Myasoedov, “Dissolution of Actinides Oxides in Supercritical CO_2 containing Various Organic Ligands.” Journal of NUCLEAR SCIENCE and TECHNOLOGY, Supplement 3, p. 263-266 (November 2002)
5 Y.M. Kulyako, T.I. Trofimov, M.D. Samsonov and B.F. Myasoedov B. F. ‘Dissolution of U, Np, Pu, and Am oxides in tri-n-butyl phosphate saturated with nitric acid’ *Mendeleev Communication*, 2003, №6, 15-16
6 L. T. Taylor, Supercritical Fluid Extraction, John Wiley and Sons, Inc. New York, 1996.
7 C.Cuillerdier, *et al.* Malonamides as New Extractants. Commissariat a l’Energie Atomique – IRDI/DERDCA/DGR-BP N 6, 92265, France.

SEPARATION OF URANIUM AND PLUTONIUM BY THE METHOD OF COUNTERCURRENT CHROMATOGRAPHY

M.N. Litvina, D.A Malikov, T.A Maryutina, Yu.M Kulyako and B.F Myasoedov

Vernadsky Institute of Geochemistry and Analytical Chemistry RAS, Kosygin Str., 19, 119991, Moscow, Russia

1 INTRODUCTION

Tributylphosphate (TBP) diluted by organic solvent (kerosene) and nitric acid solutions are used in the PUREX-process for reprocessing of spent nuclear fuel (SNF) as extracting and salt-forming agents, respectively. Earlier it was shown that the TBP-HNO_3 complex is an effective solvent for solid UO_2 and PuO_2, these being basic constituents of SNF,[1] and the use of this solvent may considerably decrease volumes of aqueous and organic high-level wastes. Further processing of TBP-HNO_3-U(Pu) complexes may be performed using environmentally appropriate supercritical (liquid) CO_2 followed by transport into a planetary centrifuge for separation of uranium and plutonium by the method of countercurrent chromatography (CCC) under dynamic conditions. The CCC method of uranium (VI) and plutonium (IV) separation using the system "TBP-white spirit (WS)-HNO_3" proposed and developed in this work permits the separation of uranium and plutonium rather effectively.

2 METHOD AND RESULTS

2.1 Batch extraction

The uranium (^{233}U) and plutonium (^{239}Pu) contents in aqueous and organic phases as well as in eluates were determined by radiometry. The preliminary data obtained on simultaneous extraction of ^{233}U and ^{239}Pu by 100% TBP from nitric acid solutions as well as by TBP solutions in WS from 0.5M HNO_3 have shown that in all the cases distribution ratios of U(VI) were higher than those of Pu(IV), and distribution ratios of both U and Pu depended on TBP and HNO_3 concentrations. On the basis of the data obtained appropriate concentrations of mobile (HNO_3 solution) and stationary (TBP solution in WS) phases were chosen for U and Pu separation by the CCC method under dynamic conditions.

2.2 Separation of U(VI) and Pu(IV) under dynamic conditions

The possibility of U and Pu separation by the CCC method in the system "30% TBP in WS - 0.5M HNO_3" is demonstrated (Figure 1). The following working characteristics of the process were used for separation of U and Pu: F=0.7 mL/min; ω=660 rpm.

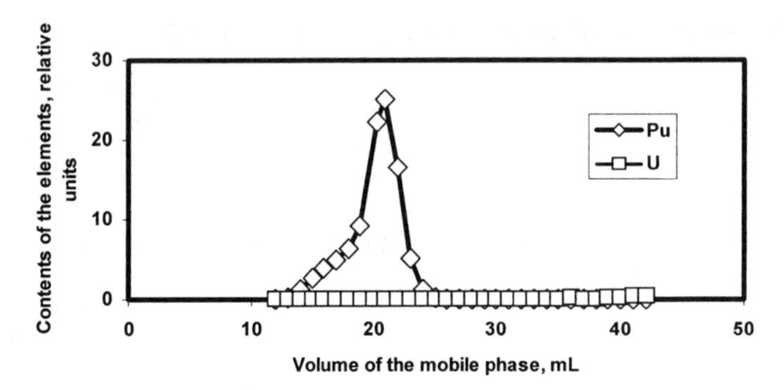

Figure 1 *Separation of Pu and U by the CCC method in the system "30% TBP solution in WS - 0.5M HNO_3. Working characteristics of the process: ω=660 r.p.m., V_{test}=0.5 mL, F=0.7 mL/min. Test sample: 0.5 mL of 30% TBP - WS↔8M HNO_3-U-Pu.*

It is seen from the data presented in Fig. 1 that Pu is eluted from the column with flow of mobile phase, while U remains in the stationary phase. At that Pu fraction (99.7%) no U is present.

A decrease of TBP concentration in the stationary phase by mixing 5 mL of 30% TBP-WS with 5 mL of white spirit in the column during the chromatographic experiment (creation of concentration gradient of TBP in the stationary phase) results in eluting U from the column. At that 8 mL plutonium fraction (99.7% of all Pu) contains 2.2% of U, and the following 24 mL uranium fraction (97.8% of all U) contains 0.3% of Pu.

A decrease of nitric acid concentration in a mobile phase to 0.2M results in decreasing distribution ratios of both elements. In this case U and Pu are separated by the method of isocratic elution. However, there is insufficient completeness of separation at this concentration of nitric acid: 10 mL plutonium fraction contains 99% of plutonium and 8.5% of uranium, and 17 mL uranium fraction contains 91.5% of uranium and 1% of plutonium.

The data obtained permitted the method of step elution for separation of U and Pu. For that the test sample (0.5 mL) containing MOX-fuel, dissolved in 100-% TBP↔8M HNO_3, is introduced into the rotating column with a flow of 0.3M HNO_3 solution, and then the eluent is substituted by 0.1M HNO_3 solution (Figure 2).

It is seen from the data presented in Figure 2 that the method of step CCC permits the efficient separation of uranium and plutonium in the TBP extract of MOX-fuel: the first 10 mL plutonium fraction contains 98.9% of all Pu and 0.07% of U, and the second 18 mL uranium fraction contains 99.93% of all U and 1.1% of Pu. It should be noted that analogous results are obtained on using test samples consisting of nitric acid solutions of U(VI) and Pu(IV) instead of organic extracts containing these elements.

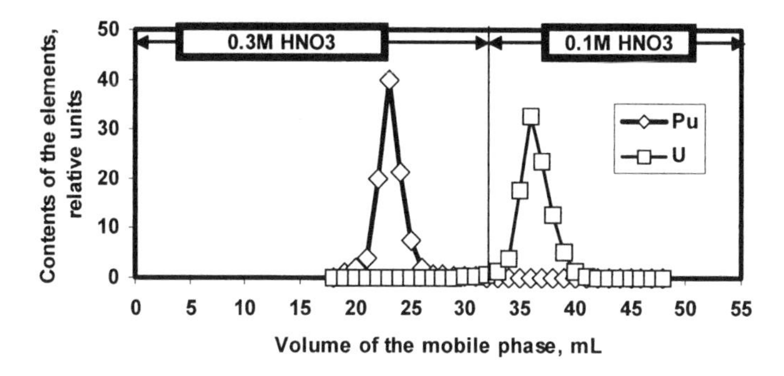

Figure 2 *Step separation of U and Pu in the system "30% TBP –WS - HNO_3". Mobile phase: 0.3M HNO_3 solution and 0.1M HNO_3 solution. Parameters of the column: L=15.6 m, S_f= 0.36 (the factor of the stationary phase retention) V_{column}= 30.5 mL. Working characteristics of the process: ω=660 r.p.m., F=0.76 mL/min. Test: 0.5 mL of 100% TBP↔8M HNO_3-U-Pu.*

3 CONCLUSION

The possibility of U(VI) and Pu(IV) separation in the nitric acid solutions or in the organic extracts (TBP) by the CCC method using two-phase liquid systems, based on TBP, under dynamic conditions at one chromatographic cycle was shown. It was found that U and Pu can be separated under the conditions of concentration gradients of both TBP in stationary phase and nitric acid in mobile phase.

The chromatographic systems and optimum operation characteristics of planetary centrifuge as well as parameters of chromatographic column were proposed for an effective separation of U and Pu. The system "30% TBP -WS - 0.5M HNO_3" allowed the concentration of uranium in the stationary phase, while plutonium was eluted with a flow of the mobile phase. The step elution permitted practically complete separation of uranium and plutonium. In this case first plutonium fraction (10 mL of 0.3M HNO_3 solution) contained 98.9% of all Pu and 0.07% of U, and the second uranium fraction (18 mL of 0,1M HNO_3) contained 99.93% of all U and 1.1% of Pu.

Acknowledgements

The work was performed under financial support of Russian foundation for basic research (grant 03-03-32765).

References

1 Yu. M. Kulyako, T. I. Trofimov, M. D. Samsonov, B. F. Myasoedov, *Radiokhimiya*, 2003, **45**, 453 (in Russian).
2 T.A. Maryutina, M.N. Litvina, D.A. Malikov, B.Ya. Spivakov, B.F. Myasoedov, M. Lecomte, C. Hill, C. Madic, *Radiokhimiya*, 2004, **46**, 549 (in Russian).

ALKALINE SEQUESTRANTS IN HANFORD TANK WASTE PROCESSING

L. R. Martin, R. L. P. Witty and K. L. Nash

Department of Chemistry, Washington State University, Pullman, WA 99164-4630, USA

1 INTRODUCTION

To complete the remediation of the Hanford tank sludges, alkaline leaching strategies are being developed. Recently, the suggestion has been made that organic complexants could be employed for selective leaching of problematic species in these tanks to enhance removal of matrix metal ions. One reagent identified, gluconate, is of interest because it is known to strongly complex metal ions in alkaline solutions. Gluconate is believed to be present in some wastes as a by-product of process application or of cellulose degradation. Though it is well known that polyvalent metal ions interact with gluconate,[1-4] there is little literature on the interactions of actinides with this ligand at high pH, ionic strength and temperature. The work presented here describes initial results of an investigation into the coordination chemistry UO_2^{2+} in similar media.

2 EXPERIMENTAL

All chemicals were of reagent grade and used as received. Sodium gluconate was used exclusively to avoid uncertainties related to the presence of lactones, which are know to exist in gluconic acid solutions. Lactones are formed in acidic solutions at comparatively slow rates, hence their influence can be minimized. All solutions were prepared in deionized water. Uranyl solutions were prepared from a 0.47 M laboratory stock of $UO_2(ClO_4)_2$. UV-visible spectrophotometric data were obtained on a Cary 14 recording spectrophotometer with OLIS upgrade. A ^{13}C NMR investigation of free gluconate was conducted by adding equivalents of NaOD into a 0.5 mM solution of sodium gluconate in D_2O. Spectra were recorded on a Varian DX 300 spectrometer at 75.419 MHz and externally referenced to TMSP. All titrations were carried out at 25°C under a N_2 atmosphere, using a Mettler Toledo DL 50 Graphix titrator and a Ross combination pH electrode. Standardized NaOH (0.1 M) was titrated into a 50 mL solution containing uranyl and gluconate at 0.1 M ionic strength ($NaClO_4$). The experiments were designed to exclude/minimize the effects of CO_2. Titrations were carried out in triplicate and analyzed using PSEQUAD.[5] Uranyl hydrolysis constants were taken into account in the data fitting.[2]

3 RESULTS AND DISCUSSION

In the ^{13}C NMR investigation, 0.5, 1.5, 2.5, 3.5, 4.5, 5.5, 6.5 and 10 equivalents of NaOH were sequentially added to a 0.5 mM solution of sodium gluconate seeking evidence for the spontaneous ionization of alcoholic protons. Over this broad range (pHs *ca.* 7-12), no shift in any of the 6 carbon peaks was noted as increasing amounts of base were added. A small downfield shift in the C2 peak (δ *ca.* 2 ppm) was observed upon addition of 1 equivalent of yttrium to a separate sample of the starting solution. Such a shift indicates an interaction with the metal centre at C2, but we believe that the shift is too small to represent the loss of a hydrogen ion from the alcohol upon complexation.

The electronic absorption spectra of 0.5 mM UO_2^{2+} in 0.5 mM sodium gluconate with varying equivalents of OH^- added are shown in Figure 1. No precipitation was observed at any base loading. The absorption maximum is seen near 436 nm, this is red shifted from that of uranyl in acidic perchlorate solution.[6] The absence of an isosbestic point, on the addition of base, indicates that there are multiple species present. The absorbance and pH of these solutions are seen to decrease with time and exposure to air, signifying CO_2 absorption into solution.

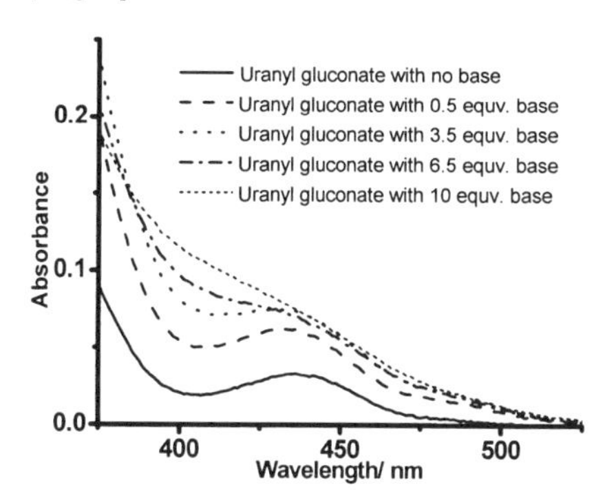

Figure 1 *Absorption spectra of uranyl-gluconate solutions 1:1 M ratio U:Glu with various equivalents of base (pH ca. 7-12).*

The results of our potentiometric titrations are shown in Figure 2. The pK_a for gluconate was found to be 3.61 (σ = 0.03), in reasonable agreement with previous reports.[2] In a previous study of this system using this same technique, Sawyer reported the existence of only one complexed species $[(UO_2)(Glu)(OH)_2]^-$.[1] The data for the titrations in Figure 2 cannot be fit using any single species. We found the best fit of the potentiometric titration results required six species of the general stoichiometry $(UO_2)_{1\text{-}2}(OH)_{2\text{-}5}(glu)_1^{-2\text{-}0}$, each species representing at least 20% of the total U species at some point in the titration. Our preliminary values for the equilibrium constants for UO_2^{2+} complexes with gluconate are given in Table 1. The standard deviations ($\pm 1\ \sigma$) are derived from the minimized sum of squares of residuals for three replicate titrations at each of the two sets of conditions using the PSEQUAD program. Figure 3 depicts the distribution of major species as a function of pH for 1:1 molar ratio of Glu:U and 0.001 M total concentrations. These data indicate that at low concentrations of UO_2^{2+} and gluconate, and at pH > 10 there are two important species in solution, $[(UO_2)_2(Glu)(OH)_5]^{2-}$ and $[UO_2(Glu)(OH)_3]^{2-}$. The latter species becomes increasingly important at pH 12 and above. These thermodynamic data provide no explicit information regarding the structure or connectivity of the complexes. Several literature reports assume the direct coordination of the alcoholic oxygens of gluconate to metal ions in alkaline solutions. However, it is difficult to envision a reasonable coordination mode for gluconate that will accommodate two uranyl cations. The alternative structure of a single gluconate coordinated to the well-known uranyl hydrolytic dimer $(UO_2)_2(OH)_2^{2+}$ could account for the observed stoichiometry of the dinuclear uranyl-gluconate complex and allows simpler coordination modes for gluconate.

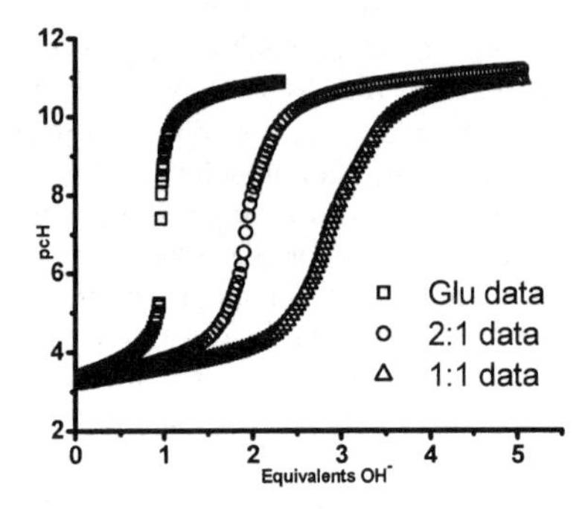

Figure 2 *pH titrations of sodium gluconate, a 2:1 mixture of Glu: U and a 1:1 mixture of Glu:U ($[Glu]_{init}$ = 0.001 M in each).*

Figure 3 *Distribution of major species as a function of pH for a 1:1 M ratio of Glu:U (0.001 M each).*

Additional experiments (multinuclear NMR studies, calorimetry and attempts to grow crystalline solids) are either planned or underway to further substantiate the existence of these species, and to allow more detailed characterization of the nature of this interaction.

4 CONCLUSION

Mono and dinuclear complexes between uranyl and gluconate involving either hydroxide ions or the loss of hydrogen ions from alcohol groups appear to account for the unusually high solubility of uranyl ions in strongly alkaline solutions. The structural features of these complexes are under investigation.

Table 1 *Stability constants of UO_2^{2+} with Gluconate for the complex $(UO_2^{2+})_m(OH)_h(L)_l$ (I=0.1 ($NaClO_4$), 25 C).*

Stoichiometry *Mhl*	$Log\beta_{mhl}$	$\pm 1\ \sigma$
1, -1, 1	-0.63	0.07
1, -2, 1	-5.08	0.03
1, -3, 1	-14.36	0.03
2, -3, 1	-4.35	0.03
2, -4, 1	-10.24	0.05
2, -5, 1	-18.27	0.04

Acknowledgements

The authors acknowledge the support of the US Department of Energy, Environmental Management Science Program.

References

1 Sawyer, D. T.; Kula, R. J. *Inorg. Chem.* 1962, **1**, 303.
2 Motekaitis, R. J.; Martell, A. E. *Inor. Chem.* 1984, **23**, 18.
3 Escandar, G. M.; Olivieri, A. C.; Gonzalez-Sierra, M.; Frutos, A. A.; Sala, L. F. *J. Chem. Soc., Dalton Trans.* 1995, 1393.
4 Pecsok, R. L.; Sandera, J. *J. Am. Chem. Soc.* 1955, **77**, 1489.
5 Zékány, L.; Nagypál, I. In *Computational Methods for the Determination of Stability Constants*; Leggett, D., Ed.; Plenum Press: New York, 1985, pp 291-299.
6 Clark, D. L.; Conradson, S. D.; Donohoe, R. J.; Keogh, D. W.; Morris, D. E.; Palmer, P. D.; Rogers, R. D.; Tait, C. D. *Inorg. Chem.* 1999, **38**, 1456.

ANION CONTROL IN HANFORD LEACHATES BY SOLID-LIQUID SEPARATION

L.R. Martin, A.R. Routt, R.M. Olsen and K.L. Nash

Department of Chemistry, Washington State University, Pullman, WA 99164-4630, USA

1 INTRODUCTION

Plutonium manufacture at the Hanford site in the USA created large volumes of complex wastes that have been stored in the 177 underground tanks.[1] Under current plans these wastes are to be retrieved, pre-treated and immobilized for final disposal. Dependent on the classification of the immobilized waste the disposal site will either be a geologic repository for High Level Waste (HLW) or sub surface burial for Low Level Waste (LLW).[2] Immobilization is to be carried out by vitrifying the waste in a borosilicate glass matrix. However, for this approach to be successful the removal of anionic components like chromate, phosphate and sulphate is highly desirable. Each considered to be a glass limiting species.[3-5]

One option for the removal of these species from aqueous solution is to precipitate them from solution using polyvalent heavy metal ions. Barium is known to form insoluble salts with CrO_4^{2-}, SO_4^{2-} and PO_4^{3-}. This paper presents the results of our initial studies on the viability of removing these species from tank leachates by precipitating them as heavy metal solids. The effects of an organic complexant that may be present in the sludge leachates will also be discussed. $^{152,154}Eu$ has been used to model the behaviour of the trivalent actinides in this process.

2 EXPERIMENTAL

All chemicals were of reagent grade and used as received. Deionized water was used in the preparation of all solutions. Gravimetric analysis was performed in the initial investigations to determine the yield of precipitated solid. The pH of the $BaCl_2$ (75 mM) and Na_2SO_4 (50 mM) solutions were adjusted before (and after mixing if required) using 0.1 M HNO_3 and 0.1 M NaOH. The sorption and co-precipitation behaviour of Eu in the $BaSO_4$ system was investigated using radiotracer Eu. The $^{152,154}Eu$ solutions were prepared from laboratory stocks. For the co-precipitation experiments, a solution containing $6x10^{-7}$ moles $^{152,154}Eu$ tracer was added to 1 mL Na_2SO_4 this was then mixed with 1 mL $BaCl_2$. In the sorption experiments, the sulfate and Ba solutions were mixed before the Eu was added. After 24 hours these solutions were centrifuged and the supernatant separated from the solid. The activities of both phases were determined using a Packard Cobra II Auto-gamma counter with a modified NaI well detector.

3 RESULTS

Initial investigations of $BaCrO_4$, $BaSO_4$ and $Ba_3(PO_4)_2$ precipitation at various pHs indicated that $BaCrO_4$ precipitation is complete by pH 2, $Ba_3(PO_4)_2$ is complete by pH 10 and the precipitation of $BaSO_4$ is unaffected by pH, Figure 1. Prior to the Eu precipitation/sorption investigations control experiments were run to determine if there was any formation of colloidal $Eu(OH)_3$. The experiments indicated that only at pH 13 was there any evidence for the formation of colloidal $Eu(OH)_3$. In the presence of citrate there was no evidence of $Eu(OH)_3$ formation at any pH studied. The Eu precipitation/sorption experiments with $BaSO_4$ were carried out at pHs of 5, 8, 11 and 13 in the presence and absence of citrate. Figures 2 and 3 depict the results of the Eu sorption and precipitation experiments. The precipitation experiment identified significant quantities of Eu associated with the resulting solid both in the presence and absence of citrate at pH 5, 8 and 11. Interestingly there was found to be more Eu in the precipitate when citrate was present. However, at pH 13 this trend was reversed with 85 % of the Eu associated with the precipitate when no citrate was present. Only 11% of the Eu was found in the solid phase when citrate was present.

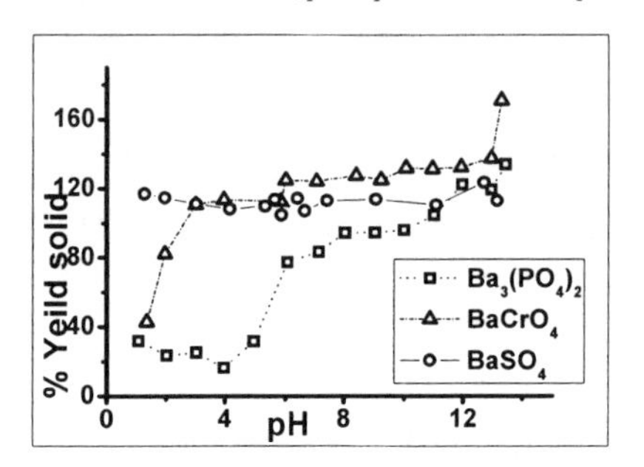

Figure 1 *Results from the investigation into the effects of pH on the precipitation of the barium solids.*

The results of the sorption experiments in the absence of citrate found that at pH 5, 8 and 11 only about 4% of the Eu present was associated with the $BaSO_4$. However, in the presence of citrate increasing the pH was found to increase the amount of Eu associated with the solid phase. At pH 13 though, there was found to be no Eu associated with the precipitate.

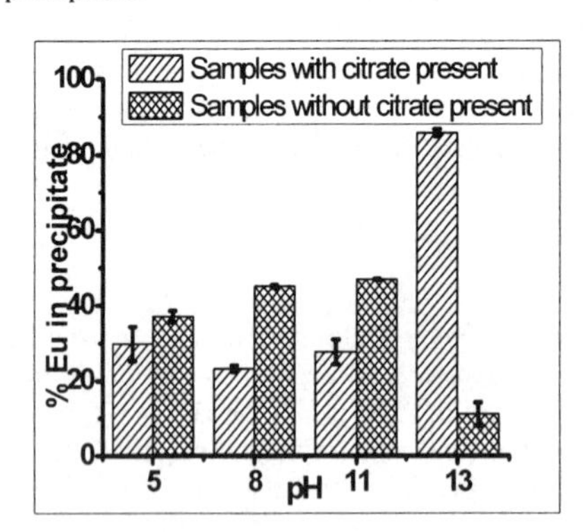

Figure 2 *Percentage of total Eu associated with the $BaSO_4$ precipitate 24 hours after co-precipitation.*

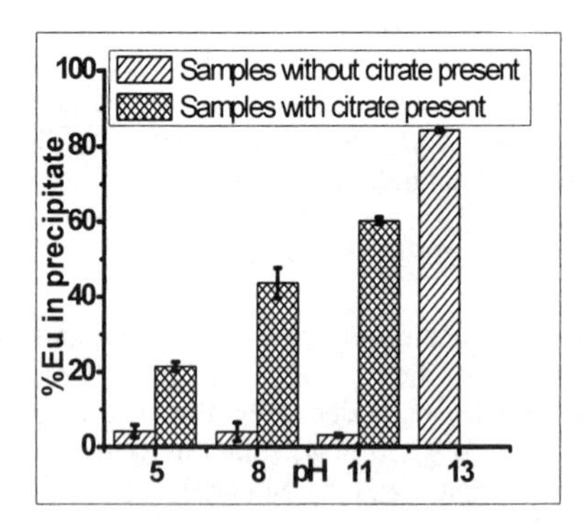

Figure 3 *Percentage of Eu associated with the$BaSO_4$ precipitate 24 hours after initiation of the sorption experiments.*

4 DISCUSSION AND CONCLUSION

In the absence of citrate, it appears that *ca.* 30% of the Eu is co-precipitated with $BaSO_4$, and when also considering the results of the sorption experiments it would appear that only a small amount of additional Eu would be adsorbed. In the presence of citrate, increasing the pH significantly increased the amount of Eu precipitated. It is interesting that in both co-precipitation and sorption experiments the presence of the organic complexant, citrate, has facilitated Eu co-precipitation at pH 5, 8 and 11. It would appear that either citrate forms a complex with Eu that is more susceptible to co-precipitation at pHs up to 11 or citrate may sorb to the $BaSO_4$ surface and removes Eu from solution. At pH 13 there are one of two explanations for the observed behaviour, the first being that at around pH 13 the alcohol groups on the citrate molecule may start to hydrolyse.[6] This in turn may change the characteristics of the interaction between Eu and the ligand allowing it to stay in solution. The second explanation is that citrate could be interacting with $BaSO_4$ at these high pHs and may inhibit the precipitation of the solid.[7]

For our application it is preferable that the radioactive elements do not associate with the precipitate. Initial experiments have been carried out investigating the removal of the Eu from the solid. The preliminary results have indicated that 0.1 M oxalic acid and 0.1 M HNO_3 scrub of the precipitate may be sufficient to remove the Eu from the solid. Further investigations are planned to validate these results. The sorption/co-precipitation characteristics of other radionuclides must also be investigated. We anticipate that the removal of most elements other than Sr^{2+} should occur readily. Separate studies will also address the other anionic species of concern, their mixtures and additional cationic precipitants.

References

1 K.L. Nash, K. L. A.V. Gelis, M.P. Jensen, A.H. Bond, J.C. Sullivan, L. Rao and A. Garnov, *J. Nucl. Sci. and Technol.* 2002, 512.
2 M.J. Kupfer, *Chemical Pretreatment of Nuclear Waste for Disposal, [Proceedings of an American Chemical Society Symposium on Chemical Pretreatment of Nuclear Waste for Disposal], Washington, D. C., Aug., 1992,* 1994, 25.
3 G.J. Lumetta, B.M. Rapko, J. Liu, and D.J. Temer, *Science and Technology for Disposal of Radioactive Tank Wastes, [Proceedings of the American Chemical Society Symposium on Science and Technology for Disposal of Radioactive Tank Wastes], Las Vegas, Nev., Sept. 7-11, 1997,* 1998, 203.
4 G.J. Lumetta and B.M. Rapko, *Separation Science and Technology,* 1999, **34**, 1495.
5 B.M. Rapko and J.D. Vienna, *Sep. Sci. and Technol.* 2003, **38**, 3145.
6 R.M. Izatt, J.H. Rytting, L.D. Hansen and J.J. Christensen, *J. Am. Chem.l Soc.* 1966, **88**, 2641.
7 K. Dunn and T.F. Yen, *Environ. Sci. Technol.* 1999, **33**, 2821.

URANIUM REMOVAL FROM PROCESS EFFLUENTS CONTAINING CHROMATE AND ALUMINUM NITRATE

L. R. Martin, R. C. Harrington, J. J. Neeway and K. L. Nash

Department of Chemistry, Washington State University, Pullman, WA 99164-4630, USA

1 INTRODUCTION

Remediation of the 177 waste tanks at the Hanford site remaining from four decades of plutonium production is a daunting task. There are some 54 million gallons of waste from various processes that were used in plutonium production, considered to have an overall activity of almost 1.7×10^8 Curies.[1] To complicate the issue the alkaline wastes, over years of storage, have stratified into a mix of a clay-like sludge phase, a solution/slurry phase and a surface salt cake. Currently all of the waste is considered high level waste (HLW). At present the remediation plans call for the vitrification of the radioactive materials so that they can be sent to a geologic repository.[2] As vitrification is an expensive process, it is most cost effective to separate the radionuclides with short half lives and non-radioactive materials from the long-lived radioactive species. The less radioactive/non-radioactive material can then be disposed of as low level waste and buried near surface.

Dependent on the tank composition, the radioactive materials are generally perceived to be distributed between the three phases, although the transuranics and Sr reside within the sludge phase in the absence of chelating agents.[3] Prior to the vitrification process the sludges are to be leached under alkaline conditions to remove some of the elements problematic to the vitrification process, including Al, Cr, S and P. It is important to remove the majority of these elements as they are considered glass limiting species. However, it appears that to adequately remove Al and Cr from the sludges dilute acid washing is required. An undesirable side effect is that this method of leaching also facilitates the mobilization of actinides (An).[3-,5]

In present work we asses whether U(VI) can be removed from dilute acid solutions containing high concentrations of Al and Cr using conventional solvent extraction by tributyl phosphate (TBP). The assets and liabilities of this approach will be discussed.

2 EXPERIMENTAL

All chemicals were of reagent grade and used as received. All solutions were prepared in deionized water. Uranyl solutions were prepared from 0.5 M $UO_2(NO_3)_2$ stock solution in 0.2 M HNO_3. The aqueous solutions containing various concentrations of Al and Cr used in this investigation were prepared from reagent grade $Al(NO_3)_3$ and K_2CrO_4 respectively. The 30% TBP/dodecane solution was prepared as described elsewhere.[6,7]

The extractant solution was pre-equilibrated by contact with 0.1 M HNO_3. The aqueous solutions used for extraction were prepared by combination of 1 mL of uranyl solution with 1mL of Al/Cr solution of interest. This solution was then contacted with 2 mL of the pre-acidified 30% TBP/dodecane phase for 10 minutes using a vortex mixer. All extractions were carried out in triplicate. Back extractions were accomplished using 2 mL 0.2 M Na_2CO_3. This step was repeated in an attempt to scrub all U(VI) and Cr from the organic phase. Metal concentrations were determined using a Perkin Elmer Optima ICP-OES. The distribution (D) ratios were calculated using $D_m = (C^{\circ}_m - C_m)/C_m$, where D_m is the distribution of metal ions between the aqueous and organic phase, C°_m is the metal concentration in the aqueous phase before extraction, and C_m is the aqueous phase metal concentration after reaching partition equilibrium.

3 RESULTS

We have observed efficient partitioning of U(VI) into and back from TBP/dodecane from $Al(NO_3)_3$ solutions at low acidity. This work will be described in more detail in a separate publication. Aluminum is not seen to extract into 30% TBP/dodecane in the absence of U(VI). However, chromium is observed to have some affinity for the organic phase. Extraction of U(VI) from solutions containing variable concentrations of $Al(NO_3)_3$ indicated that two NO_3^- ions are extracted with each uranyl ion. Increasing the initial aqueous phase concentration of U(VI) had only a slight effect on the amount of chromate extracted although the increase could be partially due to media effects across the concentration range studied, Figure 1. Addition of increasing amounts of chromate into uranyl nitrate solutions had no significant effect on U(VI) partitioning, as indicated in Figure 2.

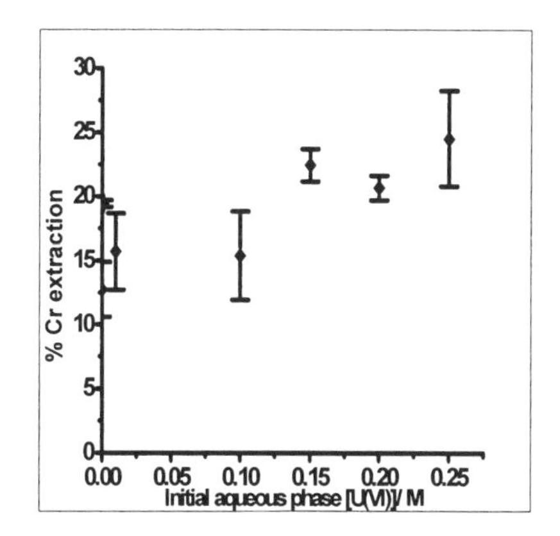

Figure 1 *Initial $[U(VI)]_{aq}$ vs. % Cr extraction for the extraction of 1.0 mM Cr(VI) from 0.5 M $Al(NO_3)_3$ in 0.1 M HNO_3 into 30% TBP/dodecane.*

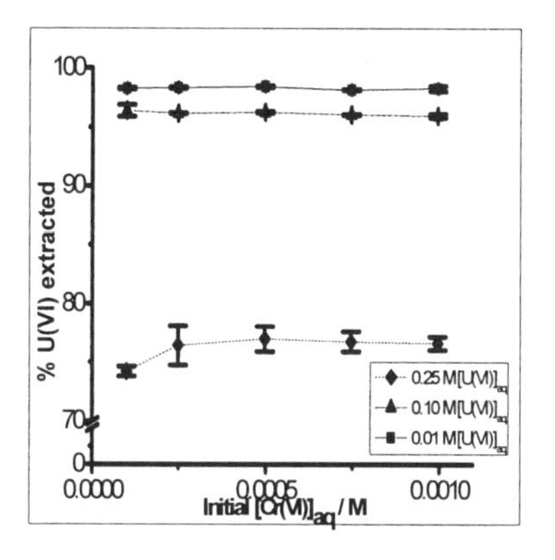

Figure 2 *Initial $[Cr(VI)]_{aq}$ vs. % U(VI) extraction for three concentrations of U(VI) from Cr(VI) in 0.5 M $Al(NO_3)_3$ aqueous phases into 30% TBP/dodecane.*

In back extraction experiments using 0.2 M Na_2CO_3, U(VI) was quantitatively recovered from the TBP phase in 2 equal volume contacts. In the same experiments it was established that chromium removal from the organic phase was incomplete despite repeated scrubbing. No third phase formation was observed when scrubbing with Na_2CO_3.

4 DISCUSSION AND CONCLUSION

To be used effectively for waste remediation, solvent extraction demands the ability to recycle reagents, therefore the removal of U and Cr from the TBP phase is essential. Our work to date has demonstrated that 30% TBP/dodecane can be used to remove U(VI) from $Al(NO_3)_3$ solutions containing Cr(VI). We have also established quantitative back extraction of U(VI) with Na_2CO_3. Chromium(VI) extraction is observed to increase with increasing [U(VI)] in the aqueous phase. This is presumably due to the extraction of uranyl chromate, although we cannot completely rule out the extraction of chromic acid. There appears to be no significant impact on the extraction of U(VI) into TBP/dodecane by increasing the initial aqueous phase Cr(VI) concentration.

The retention of chromium in the organic phase after scrubbing with Na_2CO_3 is somewhat perplexing. The observed behaviour could be the result of a number of processes. One possibility is that there may be oxidizable species present in the organic phase that are subject to attack by the Cr(VI). Through this process the Cr(VI) would be reduced to Cr(III) which could then be retained in the organic phase by solvent degradation products. Another suggestion, again based on the reduction of Cr(VI) to Cr(III), is that the kinetics of scrubbing Cr(III) from the organic phase may be much slower than those of Cr(VI) and so the solutions were not contacted with the Na_2CO_3 for long enough. There is also the prospect there may be no Cr(VI) reduction and the chromium retained in the organic phase is a complex between chromic acid and TBP.

From these initial results it is apparent that this system requires further study, not only to determine the reaction pathways that lead to the retention of Cr in the organic phase but also to investigate the behaviour of transuranic elements under similar conditions.

Acknowledgements

The authors acknowledge the support of the US Department of Energy, Environmental Management Science Program.

References

1 K.L. Nash, A.V. Gelis, M.P. Jensen, A.H. Bond, J.C. Sullivan, L. Rao and A. Garnov, *J. Nucl. Sci. Technol.* 2002, 512.
2 M.J. Kupfer, *Chemical Pretreatment of Nuclear Waste for Disposal, [Proceedings of an American Chemical Society Symposium on Chemical Pretreatment of Nuclear Waste for Disposal], Washington, D. C., Aug., 1992,* 1994, 25.
3 W.A. Reed, A. Garnov, L. Rao, K.L. Nash, A.H. Bond, *Sep. Sci. Technol.* 2005, i*n Press*.
4 A.H. Bond, K.L. Nash, A.V. Gelis, J.C. Sullivan, M.P. Jensen and L. Rao, *Sep. Sci. Technol.* 2001, **36**, 1241.
5 A.Y. Garnov, L. Rao, K.L. Nash and A.H. Bond, *Sep. Science Technol.* 2003, **38**, 359.
6 L.R. Martin, Ph D. Thesis; The University of Manchester: Manchester, 2003.
7 W.W. Schulz, J.D. Navratil and A.E. Talbot, *Eds. Science and Technology of Tributyl Phosphate, Vol. 1: Synthesis, Properties, Reactions, and Analysis*, CRC Press, Boca Raton, Florida 1984.

DIFFUSION COEFFICIENTS OF ACTINIDE AND LANTHINDE ELEMENTS IN LIQUID METAL FOR REDUCTIVE EXTRACTION

H. Moriyama[1], K. Moritani[1], T. Sasaki[1], I. Takagi[1], K. Kinoshita[2] and H. Yamana[3]

[1] Department of Nuclear Engineering, Kyoto University, Yoshida, Sakyo-ku, Kyoto 606-8501, Japan, moriyama@nucleng.kyoto-u.ac.jp
[2] Central Research Institute of Electric Power Industry, Tokyo 201-8511, Japan
[3] Research Reactor Institute, Kyoto University, Osaka 590-0494, Japan

1 INTRODUCTION

In recent years, there has been a renewed interest in new nuclear fuel cycles or new nuclear waste management strategies, especially for efficient transmutation of long-lived radionuclides in order to minimize the radiological toxicity of nuclear wastes. The reductive extraction process, which has been developed for reprocessing molten salt reactor fuels, is expected to be more useful for separation and recovery of transuranic elements. Recently, extensive studies have been performed concerning the thermodynamics of reductive extraction.[1-2] However little is still known about the kinetics. For the development of reductive extraction process the rate of extraction of actinide and lanthanide elements in a two-phase system of molten LiCl-KCl eutectic salt and liquid cadmium was measured at 723-873K in our previous study.[3] The mass transfer coefficients were found to be as high as expected. In some cases, however, it was found that the rate of reductive extraction was possibly affected by the solubility limits of the solute elements in the metal phase.

For comparison, in the present study, the diffusion coefficients of actinide and lanthanide elements were measured as a function of temperature in liquid cadmium for reductive extraction. By combining the present results with the previous ones on the rate of extraction, the extraction mechanism and the system performance will be discussed for further development of the extraction system.

2 METHOD

Diffusion coefficients were measured by a capillary method.[4] All the chemicals were of reagent grade obtained from Nacalai Tesque, Inc., and the radioactive tracers were produced by neutron irradiation of metal specimens. All experiments used high-purity reagents and were performed in a glove box under argon atmosphere containing <0.5ppm of O_2 and <0.2ppm of H_2O.

In the measurement, a quartz capillary tube of an i.d. of 0.2 or 0.3mm and a length of about 50mm sealed from one end was used. The open end of the evacuated capillary was immersed into liquid Cd containing the radioactive tracers in order to allow the liquid to rise up. The capillary filled with liquid Cd was then immersed into pure liquid Cd. The temperature of the system was controlled within ±0.5K. After a known diffusion time, the capillary was removed, allowed to cool, and cut into several sections. The concentration profile of each radioactive tracer was determined by direct γ-ray spectrometry. No significant differences were observed for the capillaries of 0.2 and 0.3mm, and it was concluded that thermal convection was not an important source of uncertainty.

3 RESULTS

The concentration profile relative to the distance from the capillary edge was fitted by

$$C_x = C_0 erf\left[\frac{x}{2(Dt)^{1/2}}\right] \tag{1}$$

where C_x denotes the concentration at a distance x from the open edge of the capillary, C_0 the initial concentration, D diffusion coefficient, and t the diffusion time. The diffusion coefficient of each ion was determined by a least-squares fit of the data to eq 1 with standard deviations of 20-30%.

In Figure 1, the diffusion coefficients of solute elements in liquid Cd are plotted as a function of inverse temperature and compared with literature values for U in liquid Cd.[5] Similar temperature dependences are found for all the elements, although some differences are observed possibly due to the different interactions of solute elements with the solvent and due to different sizes of the diffusing species. For comparison, the diffusion coefficients in liquid Bi[6] are also plotted in Figure 2. By comparing these figures it is recognized that the diffusion coefficients are strongly dependent on the solvent viscosity and that different interactions of solute elements may be present in different metals. The temperature dependence is thus analyzed using the Stokes-Einstein equation, in Table 1. For further details, it is suggested to study possible deviations of the data from the equation due to formation of non-ideal solutions or intermetallic compounds.

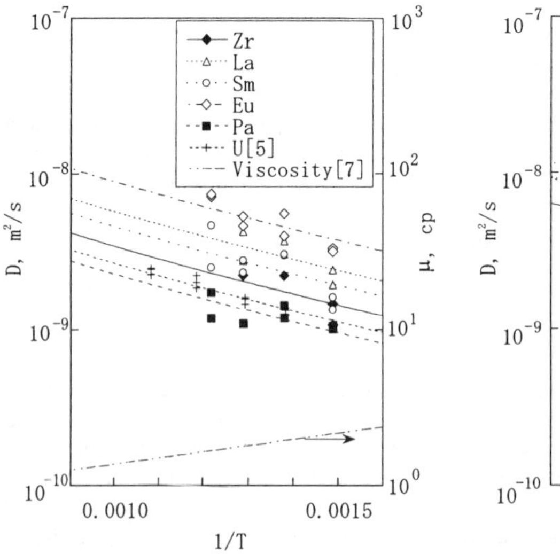

Figure 1 *Diffusion coefficients in liquid Cd*

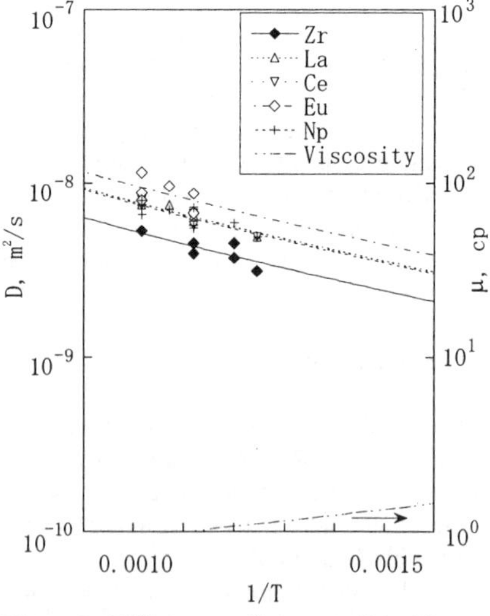

Figure 2 *Diffusion coefficients in liquid Bi*

Table 1 *Diffusion coefficient in liquid Cd and Bi*

Element	$D = kT/(\alpha\mu r) = AT\exp[-E/(RT)]$*, m^2/s	
	log A for Cd	log A for Bi[6]
Zr	-11.06±0.10	-10.94±0.05
La	-10.84±0.10	-10.77±0.02
Ce	-	-10.78±0.03
Sm	-10.94±0.12	-
Eu	-10.65±0.07	-10.68±0.07
Pa	-11.24±0.08	-
U	-11.17±0.03	-
Np	-	-10.78±0.04

* k is the Boltzmann constant, T the temperature, α a constant, μ the solvent viscosity, r the effective radius of the diffusing species, and E the activation energy obtained by the solvent viscosity measurements: 7.71 kJ/mol for Cd,[7] 6.39 kJ/mol for Bi.[8] The errors in the log A values are of standard deviations.

4 CONCLUSION

It was found that the diffusion coefficients of solute elements in liquid metals are much dependent not only on the solvent viscosity but also on the interactions of solute elements with solvent metals (Cd and Bi). For further development of the reductive extraction process using these solvent metals, it is important to consider the effects of such interactions on the kinetics of extraction.

We thank Messrs. D. Yamada and T. Murai for their cooperation in the experiment. This work was supported by MEXT (Ministry of Education, Culture, Sports, Science, and Technology of Japan), Development of Innovative Nuclear Technologies Program.

References

1 T. Inoue, M. Sakata, H. Miyasiro, A. Sasahara and T. Matsumura, *J. Nucl. Technol.* 1991, **93**, 206.
2 K. Kinoshita, T. Inoue, S. P. Fusselman, D. L. Grimmett, J. J. Roy, R. L. Gay, C. L. Krueger, C. R. Nabelek and T. S. Storvick, *J. Nucl. Sci. Technol.* 1999, **36**, 189.
3 H. Moriyama, D. Yamada, K. Moritani, T. Sasaki, I. Takagi, K. Kinoshita, and H. Yamana, *J. Alloys and Compounds*, in press.
4 H. Moriyama, K. Moritani and Y. Ito, *J. Chem. Eng. Data* 1994, **39**, 147.
5 J. C. Hesson, L. Burris, *Trans. Met. Soc. AIME* 1963, **227**, 571.
6 H. Moriyama, Y. Asaoka, K. Moritani and Y. Ito, Extraction Rate of Metal Elements in Molten Salt and Liquid Metal Binary System, (III) Diffusion Coefficient, presented at the annual Mtg. of Atomic Energy Society of Japan, Osaka, March 28-30, 1991.
7 *Liquid Metals Handbook*, 2nd ed. (revised), Government Printing Office, Washington, D.C., 1954.
8 *CRC Handbook of Chemistry and Physics*, 64th ed., CRC Press, Boca Raton, Florida, 1983-1984.

ELECTROCHEMICAL AND SPECTROSCOPIC CHARACTERISTICS OF URANIUM IONS IN A HYDRATE MELT

A. Uehara,[1] T. Fujii,[1] T. Nagai,[1] O. Shirai,[1] N. Sato[2] and H. Yamana[1]

[1] Division of Nuclear Engineering Science, Research Reactor Institute, Kyoto University, Asashironishi, Kumatori, Osaka, 590-0494, Japan
[2] Institute for Multidisciplinary Research for Advanced Materials, Tohoku University, 2-1-1, Katahira, Sendai, 980-8577, Japan

1. INTRODUCTION

Calcium chloride hydrate melt, $CaCl_2 \cdot nH_2O$ with n less than *ca.* 10, can be recognized as an extremely highly concentrated electrolyte medium with limited amount of water, and it gives quite characteristic chemical behavior to the solute in it. The chemical behavior of actinides in hydrate melts merits further investigation, because it may lead to establishing a new chemical separation method for actinides and fission products. For studying the mechanism of the characteristic variation of cations in hydrate melts, spectroscopic technique is effective, because the change of the coordination environment of cations in the hydrate melt can be observed directly[1]. Electrochemical technique is also effective for this purpose, because it can provide information about the stability of the valence states through the variation of redox potentials.[2] In this context, in this study, U^{4+} and UO_2^{2+} in $CaCl_2 \cdot nH_2O$ hydrate melt were studied by means of absorption spectrophotometry and electrochemical measurement.

2. EXPERIMENTAL

Calcium chloride hexahydrate of analytical grade (Fluka Chemie GmbH) was used after filtration with a filter of 0.45 μm pore size. The melt samples of $n = 7 - 10$ were prepared by adding weighed amount of water to $CaCl_2 \cdot 6H_2O$. Anhydrous UCl_4 was synthesized from the reaction of U_3O_8 with carbon tetrachloride under an inert atomosphere, which was followed by XRD analysis for confirmation. Anhydrous UO_2Cl_2 was prepared by heating UCl_4 under atmosphere at 623 K. Melt samples of $CaCl_2 \cdot nH_2O$ ($n = 6 - 10$) containing uranium species at [Ca]/[U] > 100 were prepared by heating slightly, and each melt was transferred into a quartz cell. It was sealed with stopcock and kept at 298 K, and the absorption spectrum between 5000 to 25000 cm^{-1} was measured by an absorption spectrophotometer, SV-570 (JASCO Co.). Cyclic voltammetry was undertaken using a HZ-3000 (Hokuto Denko.Co.). Ag/AgCl electrode containing 1 M NaCl was used as the reference electrode and pyro-carbon electrode rod (ϕ4mm) was used as a working electrode. A Platinum mesh electrode was used as the counter electrode.

3. RESULTS AND DISCUSSION

3.1. Absorption spectra of UCl_4 in hydrate melts

The absorption spectra for U^{4+} in $CaCl_2{\bullet}nH_2O$ is shown in Fig. 1, with the spectrum in 1 M hydrochloric acid also shown (curve A) for comparison. It is clear that the water content influences the spectrum of U^{4+} to a great extent. The influence is characterized with the observations, (i) drastic changes of the ratio of the intensities of two peaks in two doublets around 15000 and 9000 cm^{-1}, (ii) decrease of the intensity of the peaks at 8660, 14800, 15300, 17800 and 22600 cm^{-1} with increasing of water content, (iii) increase of the intensity of the peak of 9340 cm^{-1} with increasing water content, and (iv) blue shift for every peak with increasing water content. It is clear that the absorption spectrum in $CaCl_2{\bullet}nH_2O$ gradually changes with increasing water to approach to the spectrum of aqueous solution. The electronic *f-f* transition is considered to be influenced by the gradual change of the chemical circumstance from the balance of water and chloride ions.

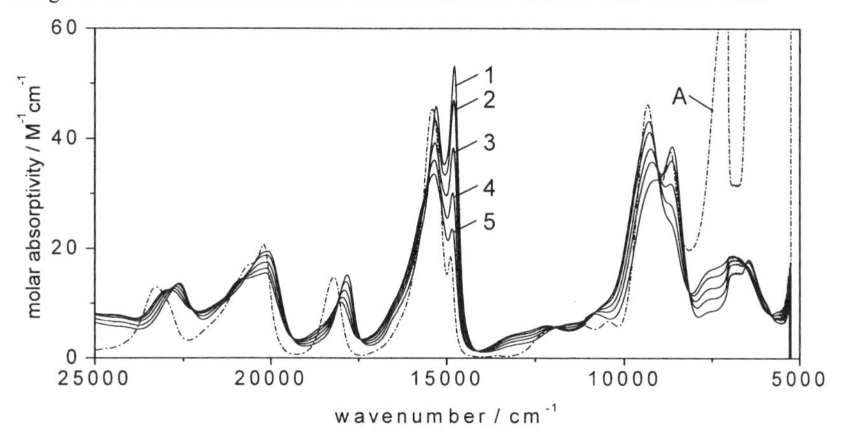

Figure 1 *The absorption spectra of U^{4+} in various $CaCl_2{\bullet}nH_2O$ melts (curves 1 to 5 are n = 6 to 10, respectively). Curve A indicates the absorption spectrum of U^{4+} in 1 M HCl.*

To help understand the spectral changes with the change of water/Cl^- ratio in the more water-abundant region, absorption spectra of U^{4+} in 1 M HCl solution with various concentration of Cl^- were studied. The concentration of Cl^- was controlled by adding $CaCl_2$ to 1M HCl solution. In solutions with 2.5 – 5.6 M Cl^-, the peak at 8660 cm^{-1} showed similar intensity to that in 1M HCl solution without $CaCl_2$. In contrast, in the solution with 5.6 – 11.9 M Cl^-, it changed with increasing water amount to approach to the spectrum that was observed in $CaCl_2{\bullet}6H_2O$. This suggests that the primary coordination sphere of chloro complex of U^{4+} in 5.6 – 11.9 M Cl^- solution is rather similar to that in $CaCl_2{\bullet}nH_2O$ with n from 6 to 10, but that in 2.5 – 5.6 M Cl^- solution it is closer to that in aqueous solution. Consequently, we can suggest that there is a change of the chloro/aquo complex at 5.6 M Cl^-, beyond which U^{4+} starts behaving as in hydrate melt.

3.2. Electrochemical behavior of UO_2Cl_2 in hydrate melt

The redox chemistry of UO_2^{2+} in $CaCl_2{\bullet}6H_2O$ was studied by cyclic voltammetry using pyro-carbon rod electrode. Curve 1 in Fig. 2-(a) shows the result of the cyclic voltammetry starting from the rest-potential (0.1 V vs. SSE) toward the negative direction. A pair of cathodic and anodic peaks were observed, and a particular continuous reduction current was observed in the region lower than –0.3 V. When the potential was reversed at –0.2 V and 0.2 V, the peak potentials of the two peaks were found to be almost independent of the scan rate, and the potential gap between two peaks was found to be constantly 0.06 V as shown in Fig. 2-(b). This suggests that the corresponding redox reaction is reversible and it has one-electron exchange, and thus this is attributable to the redox pair of UO_2^{2+}/UO_2^{+}. The mid-potential of anodic and cathodic peaks was constantly –0.019 V vs. SSE, which is approximately the formal redox potential of the pair of UO_2^{2+}/UO_2^{+}. By changing the water content from $n = 6$ to 10, two peaks slightly shifted to negative potential. This suggests that increasing water content in hydrate melt possibly changes the chemical activity of chloro complex of UO_2^{2+}.

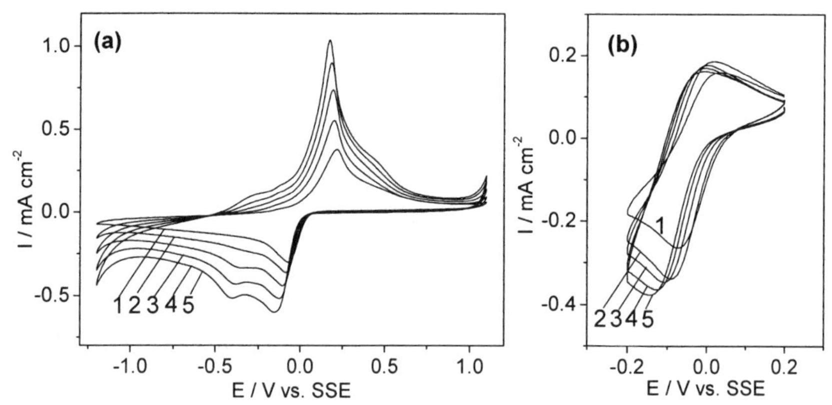

Figure 2-(a) and (b) *Curves 1 to 5; cyclic voltammograms for the redox reaction of 0.0237 M UO_2^{2+} in various $CaCl_2{\bullet}nH_2O$ melt. Scan speed : 0.02 V/s.($n = 6 – 10$).*

The continuous negative current observed in the range lower than –0.3 V seems to be the result of a slow reduction of uranyl ion that starts from –0.3 V. It is clear that U^{4+} was not formed by this reaction because the reduction current of U^{4+}, that occurs at –0.8 V, was not observed. Here, redox couple of U^{4+}/U^{3+} in $CaCl_2{\bullet}nH_2O$ could be observed when UCl_4 was dissolved in the melt. These observations suggest that this reduction reaction is a deposition reaction from uranyl to solid UO_2. Similarly, the deposition of UO_2 by the electrolytic reduction of UO_2^{2+} has been reported in various non-protonic solvents such as organic solvents[3] and weak acid aqueous solutions[4]. As shown in Fig. 2-(a), within the hydrate melt region, the reduction current from UO_2^{2+} to UO_2 increased with the water content. This means that, with the increase of the diffusion coefficient of uranyl ion, the deposition of UO_2 proceeds.

References

1 C.A. Angell and D.M. Gruen, *J. Am. Chem. Soc.,* 1966, **88**, 5192.
2 M. Zielinska-Ignaciuk and Z. Galus, *J.Electroanal. Chem.,* 1974, **50**, 41.
3 L. Martinot, L. Lopes, J. Marien and C. Jerome, *J. Radioanal. Nucl. Chem*., 2002, **253**, 407.
4 W.E. Harris and I.M. Kolthoff, *J. Am. Chem. Soc*., 1946, **69**, 446.

SPECTROSCOPIC STUDY FOR THE OXYCHLORIDE PRECIPITATION OF NEODYMIUM AND URANIUM IN CHLORIDE MELTS

T. Fujii,[1] A. Uehara,[1] T. Nagai,[1] O. Shirai,[1] N. Sato,[2] and H. Yamana[1]

[1] Research Reactor Institute, Kyoto University, 2-1010, Asashiro Nishi, Kumatori, Sennan, Osaka 590-0494, Japan
[2] Institute for Multidisciplinary Research for Advanced Materials, Tohoku University, 2-1-1, Katahira, Sendai 980-8577, Japan

1 INTRODUCTION

The pyrochemical reprocessing technique using molten salts is considered to be one of the candidates for the future reprocessing of spent nuclear fuels. Calcium dichloride attracts a particular interest for the electrolytic reduction of actinide dioxide fuels into metal. Lanthanide cations easily form precipitates of oxychlorides and oxides in chloride melt in the presence of oxygen ion, O^{2-}. This may become a problem for controlling the electrolytic refining system of actinide oxides. Higher solubility[1] of oxychloride in $CaCl_2$ mitigates the difficulty of the process control against the increase of O^{2-} ion content in the melt.

UV/Vis spectrophotometry is a strong analytical technique for characterizing chemical species and their behavior in molten salts. We have studied absorption characteristics of neodymium in various molten chlorides.[2-3] Possible higher solubility of oxychloride in $CaCl_2$ enables us to study the spectrophotometric characteristics of Nd^{3+} and U^{4+} in the presence of O^{2-}. In this context, the change of the absorption spectrum of Nd^{3+} and U^{4+} in $CaCl_2$ at 1073 K was studied with the stepwise addition of O^{2-}. For comparison, experiments were also performed in NaCl-2CsCl at 923 K.

2 EXPERIMENTAL

Anhydrous uranium tetrachloride was synthesized by the reacting U_3O_8 with carbon tetrachloride under an inert atmosphere. The CCl_4/N_2 mixed gas was introduced to the reaction tube under N_2 atmosphere, and the product identified as UCl_4 by XRD analysis. Anhydrous NaCl-2CsCl eutectic mixture (mole ratio of sodium to caesium = 1/2), $CaCl_2$, and $NdCl_3$ were products of Anderson Physics Laboratory Engineered Materials.

All the experiments were carried out in a glove box system filled with dry Ar continuously purified to remove oxygen and moisture. The content of O_2 and H_2O in the atmosphere was continuously maintained at less than 1 ppm. Absorbance measurements were performed with molten chlorides containing *ca.* 0.03 mol dm^{-3} (M) $NdCl_3$ or UCl_4. CaO and Na_2CO_3 were used as the O^{2-} suppliers for $CaCl_2$ and NaCl-2CsCl, respectively. The details of our experimental apparatus and the procedure for the absorbance measurements were presented elsewhere.[2-3] The temperature of the sample during the experiment was kept at the desired value within ±3 degrees.

3 RESULTS AND DISCUSSION

Figure 1 shows the variation of the spectrum of Nd^{3+} or U^{4+} in the chloride melts with the addition of O^{2-} ion. A rise of the baseline was here observed, similar to our previous study[3]. This is due to the formation of oxychloride precipitate that could be confirmed visually. Each spectrum was measured after the steady state was attained (~6 hours). For neodymium, the absorbances at 562 and 625 nm were set to zero, and the molar absorption at 589 nm was used for the quantitative analysis. For uranium, the absorbances at 917 and 1355 nm were set to zero, and the molar absorption at 1167 nm was used for the quantitative analysis. In $U^{4+}/CaCl_2$ system, a small portion of UCl_4 evaporated and a viscosity increase was observed with the addition of CaO, thus making the quantitative discussion difficult.

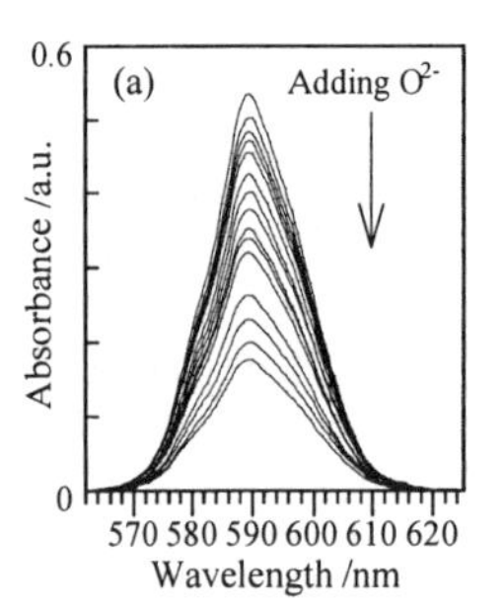

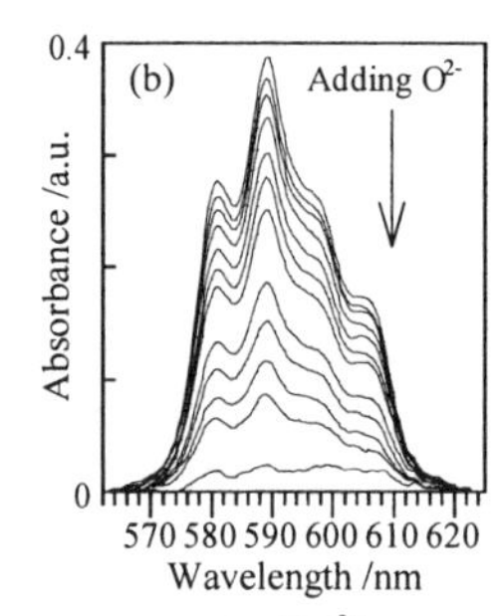

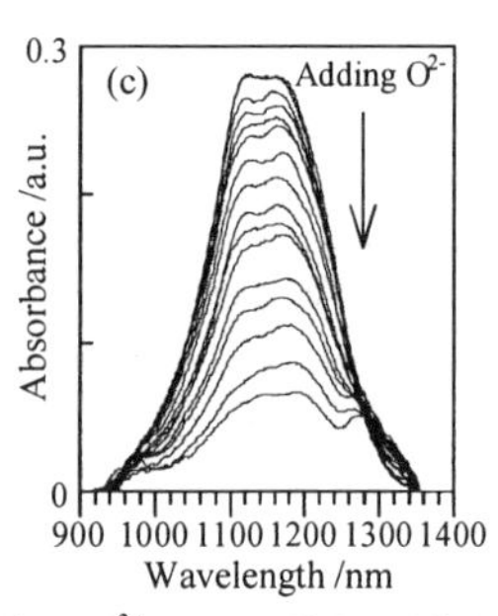

Figure 1 *Electronic absorption spectra; (a) Nd^{3+} in $CaCl_2$; (b) Nd^{3+} in NaCl-2CsCl; (c) U^{4+} in NaCl-2CsCl*

The absorption spectrum (b) is very different from (a). This is attributable to the structural symmetry of the Nd complex. Nd^{3+} forms $NdCl_3^{6-}$ in the chloride melts and its distortion from the octahedral symmetry is smaller[2] in NaCl-2CsCl than that in $CaCl_2$. Since the formation of the oxychloride should cause a distortion of the symmetry, appreciable spectral change should be observed in the NaCl-2CsCl system. The shape of spectrum (b) does not change upon the addition of O^{2-} ion. This suggests a small solubility product of NdOCl. On the other hand, the shape of spectrum (c) changes with increasing O^{2-} ion content, and a distinguishable new absorption peak was found at 1280 nm. According to the reported data,[4,5] this peak is assigned to $UOCl_2$. Hence, varied spectrum (c) at higher $[O^{2-}]$ suggests the formation of soluble oxychloride in the melt.

At each step of the addition of O^{2-}, $[Nd^{3+}]$ or $[U^{4+}]$ (mole/L) in the chloride melt can be roughly estimated by their absorbances, as shown in Fig. 2 as titration curves.
The oxychloride formation reaction can be written as,

$$Nd^{3+} + O^{2-} + Cl^- \leftrightarrow NdOCl \qquad K_{sp} = [Nd^{3+}][O^{2-}] \qquad (1)$$

and,

$$U^{4+} + O^{2-} + 2Cl^- \leftrightarrow UOCl_2 \qquad K_{sp} = [U^{4+}][O^{2-}] \qquad (2)$$

where the unit of the concentration in Eqs. (1) and (2) is mole/kg solvent. $[O^{2-}]$ can be calculated from the balance between the added amount of O^{2-} and the decrease of $[Nd^{3+}]$ or $[U^{4+}]$, and correlation between $[Nd^{3+}]$ and $[O^{2-}]$ or $[U^{4+}]$ and $[O^{2-}]$ was obtained as plotted in Figure 3. By using the data that satisfy linear relations, which are shown with open circles, the solubility products were determined.

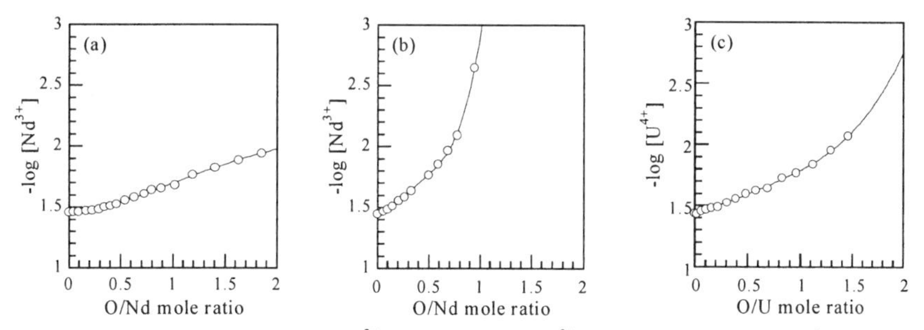

Figure 2 *Titration curves; (a)* Nd^{3+} *in* $CaCl_2$*; (b)* Nd^{3+} *in NaCl-2CsCl; (c)* U^{4+} *in NaCl-2CsCl*

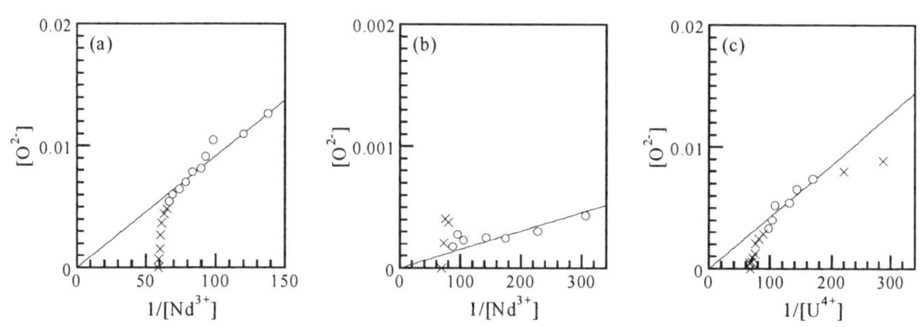

Figure 3 *Correlation between* $[Nd^{3+}]$ *and* $[O^{2-}]$ *or* $[U^{4+}]$ *and* $[O^{2-}]$*; (a)* Nd^{3+} *in* $CaCl_2$*; (b)* Nd^{3+} *in NaCl-2CsCl; (c)* U^{4+} *in NaCl-2CsCl*

The determined solubility products are, K_{sp} = 9.2×10^{-5} (pK_{sp}=4.04) for $Nd^{3+}/CaCl_2$ system at 1073 K, K_{sp} = 1.5×10^{-6} (pK_{sp}=5.82) for Nd^{3+}/NaCl-2CsCl system at 923 K, and K_{sp} = 4.2×10^{-5} (pK_{sp}=4.37) for U^{4+}/NaCl-2CsCl system at 923 K.

Acknowledgment
The authors wish to thank Mr. Roy Jacobus for his help in improving the English expressions of this paper.

References

1 D. Lambertin, J. Lacquement, S. Sanchez and G. Picard, *Electrochem. Comm.*, 2002, **4**, 447.
2 T. Fujii, T. Nagai, N. Sato, O. Shirai and H. Yamana, *J. Alloys Comp.*, 2005, **393**, L1.
3 H. Yamana, T. Fujii, and O. Shirai, Proc. International Symposium on Ionic Liquids in Honour of Marcelle Gaune-Escard, Carry le Rouet (2003) pp.123-135.
4 R. E. Ewing, *J. Inorg. Nucl. Chem.*, 1961, **17**, 177.
5 S. Dai, L. M. Toth, G. D. Del Cul and D. H. Metcalf, *Inorg Chem.*, 1995, **34**, 412.

EXTRACTION OF TRANSURANIUM ELEMENTS AND Tc(VII) FROM ACID SOLUTIONS BY NEUTRAL PHOSPHOROUS COMPOUNDS

V.P. Morgalyuk,[1] G.A. Pribylova,[1] I.G. Tananaev,[1] O.I. Artyushin,[2] E.V. Sharova,[2] I.L. Odinets,[2] T.A. Mastryukova,[2] and B.F. Myasoedov[1]

[1]Vernadsky Institute of Geochemistry and Analytical Chemistry, Russian Academy of Sciences, 19 Kosygina str., 119991 Moscow, Russia
[2]Nesmeyanov Institute of Organoelement Compounds, Russian Academy of Sciences, 28 Vavilova str., 119991 Moscow, Russia

1 INTRODUCTION

The development of the nuclear industry and processing of accumulated liquid radioactive wastes necessitates selective extraction of long-lived isotopes (primarily, actinides). For this purpose, liquid-liquid extraction procedures using neutral organophosphorus compounds, *e.g.*, carbamoylmethylphosphine oxides (CMPO) such as (N,N-diisobutylcarbamoyl)octylphenylphosphine oxide[1] or N,N-dibutylcarbamoyl-diphenylphosphine oxide[2], are used in current practice. As we reported earlier[3], the extraction ability of N-dialkylcarbamoyl-(O-alkyl)phenylphosphinates (CMP) is highly competitive with that of CMPO but the former ones are easier and cheaper to obtain.

Throughout our search for reasonably priced and active organophosphorus extractants we obtained via simple synthetic procedure two novel series of CMP(O) derivatives, namely N,N-dibutylcarbamoylmethyl(methyl)phosphinates **1** and N-alkylcarbamoyldiphenylphosphine oxides **2** (so called CMPO-NHR). Now we report our results concerning their extraction ability in respect to transuranium elements from nitric acid solutions.

2 METHODS AND RESULTS

2.1. Synthesis

CMP based ligands **1** were synthesized by Arbuzov rearrangement starting from di(O-alkyl)methylphosphonites and chloroacetic acid dibutylamides. The synthesized compounds are viscous oils which can be purified by column chromatography.

For CMPO derivatives **2** there was developed the facile and economical procedure based on a direct amination by primary alkylamines of a single phosphorus substrate, the commercially available diphenylphosphorylacetic acid ethyl ester, yielding the desired substances in practically quantitative yield.[4] All compounds were structurally characterized by ^{1}H, ^{31}P, ^{13}C NMR and IR spectroscopies, their purity was additionally confirmed by elemental analysis.

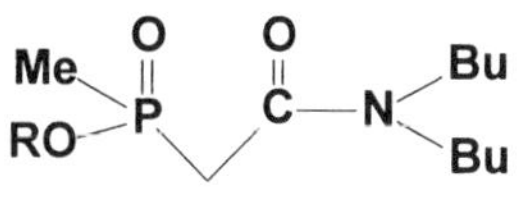

1

R=*n*-C_6-C_{10}-alkyl

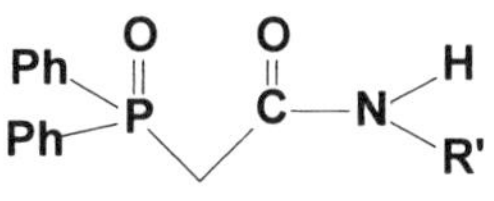

2

R'=*n*- and *i*-C_2-C_{10}-alkyl

Figure 1 *Structure of CMP(O) derivatives under investigation*

2.2 Extraction procedure

An aqueous phase of the selected elements were prepared at varying concentrations of nitric acid (1-5 M). A 3 ml aliquot was then shaken at room temperature for 3 minutes (time required for achievement of equilibrium) with equal volume of organic phase (solution of the reagent in 1,2-dichloroethane). After centrifugation the concentrations of Am(III)/Eu(III), U(VI) and Tc(VII) in the aqueous phase were determined using a Bekman scintillation counter LSC-4900 by γ-, α- and β-activity respectively. The distribution coefficient (*D*) was calculated from metal concentrations determined before and after extraction.

2.3 Extraction properties of N,N-dibutylcarbamoylmethyl(methyl)phosphinates (1)

To select the preferential ligand **1**, the extraction of Am(III) from a 3M nitric acid solution into dichloroethane was investigated. According to the results, (O-n-heptyl)(N,N-dibutylcarbamoyl)methylphosphinate **1b** is the most efficient and further elongation of alkyl chain in alkoxy group at the phosphorus atom leads to decreasing of the extraction ability. Thus $\mathbf{lgD_{Am}}$ decreases from 0.43 for n-heptyl CMP derivative **1b** up to –1.05 for its n-nonyl analogue **1c**. The extractive properties of compounds **1a** and **1b** bearing either n-hexyl or n-heptyl group correspondingly are similar to each other and therefore (O-n-hexyl)(N,N-dibutylcarbamoyl)methylphosphinate **1a** was used in further testing.

For the extraction of Am(III) and Eu(III) by 0.1M solution of **1a** in DCE the maximum ***D*** value was achieved at $[\mathbf{HNO_3}]$ = 3 M and $\mathrm{lg}\mathbf{D_{Am}}$ = -0.49 while $\mathrm{lg}\mathbf{D_{Eu}}$ is equal to -0.79. Under these conditions the selectivity factor for the couple americium/europium ($f_{Am/Eu}$) is *ca.*2 for one stage and this fact may be used for multistep extraction procedures. For the extraction of **U(VI)** by 0.1 M solutions of **1a** in DCE the maximum was achieved at $[HNO_3]$ = 4 M, where $\mathbf{lgD_{U(VI)}}$ is equal to 1.38. In the case of Tc(VII) extraction by 0,1 M solutions of **1a** in DCE, $\mathbf{lgD_{Tc(VII)}}$ has the maximum value equal to 0.97 at $[HNO_3]$ = 0.001 M and Tc(VII) extraction decreases with the increase of $[\mathbf{HNO_3}]$. The selectivity factor for the couple technetium/europium $f_{Tc/Eu}$ is *ca.*90 for a single step using 0.1 M solutions of **1a** in DCE from 0.1M HNO_3 and $f_{U/Tc}$ averages 5 from 1M HNO_3.

The investigation of **D** for Am(III), Eu(III), U(VI) and Tc(VII) versus the concentration of **1a** in logarithmical coordinates demonstrated that complexes having the composition Am•2L, Eu•2L, U•L and 2Tc•L were extracted into the organic phase.

2.4 Extraction properties of N-alkylcarbamoyldiphenylphosphine oxides 2

According to quantum-chemical calculations all CMPO-NHR derivatives **2** bearing straight alkyl substituents at the nitrogen atom have the similar electron-statistic potentials

(ESP) and dihedral angles between P=O and C=O moieties. Therefore their extraction ability must also be very similar. The experimental estimation of the extraction efficiency proved this suggestion.

Thus all CMPO-NHR compounds, **2,** with R' = n-Alk have the similar $\lg \mathbf{D}_{Am}$ and are as good as typical N,N-dialkyl CMPO derivatives wherein the distribution coefficient also is insensitive to the structure of the substituent at the N atom.[5] The highest distribution coefficient of Am^{III} (D_{Am}) was attained at $[HNO_3]$ = 3 M for all the compounds e.g. 0.68 for n-octyl derivative. However, ligands **2** with branched substituents are noticeably inferior to their analogues *n*-alkyl substituents in extraction ability toward Am^{III}. Earlier,[5] the negative effect on the extraction of actinides was reported for acidic and neutral organophosphorus extractants having branched alkyl substituents in the alkoxy group at the P atom.

2.5 Re-extraction

The metals extracted into the organic phase by CMP(O) ligands **1** and **2** were shown to be washed out quantitatively into the re-extract under the action of aqueous solutions of complexing agents, e.g. OEDPA.

3 CONCLUSION

N,N-dibutylcarbamoylmethyl(methyl)phosphinates **1** and N-alkylcarbamoyldiphenyl phosphine oxides **2** proved to be effective extractants allowing one to transfer quantitatively into the organic phase americium(III), europium(III), uranium(VI) and technetium(VII). Although the extraction ability of the ligands **1** is about two-fold lower than that for N-dialkylcarbamoyl-(O-alkyl)phenylphosphinates, the cost of the latter is much higher. Thus compounds based on **1** are more attractive for extractive concentration of lanthanoids from an economical viewpoint. Moreover, CMPO-NHR derivatives *n*-alkyl substituents are as effective as *N,N*-dialkylcarbamoylmethyl(diphenyl)phosphine oxides but the simplicity of the synthesis of CMPO-NHR makes them promising for extraction of actinides from nitric acid media.

Financial support by the Russian Foundation for Basic Research (grants Nos. 05-03-33094/32692) and the Grant "Leading Scientific Schools" (Project Nos. NSh-1100.2003.3 and 1693.2003.3).

References

1 W.W. Schulz, and E.P Horwitz, *Sep. Sci. Technol.* 1988, **23**, 1191.
2 V.N Romanovskiy, I.V Smirnov, A.Y Shadrin, and B.F Myasoedov, *Spectrum '98. Proc. Int. Top. Meet. Nuclear and Hazardous Waste Management.* La Grange Park, Illinois. Amer. Nucl. Soc. 1998, 576.
3 I. Tananaev, L.Ivanova, V. Morgalyuk, N. Molochnicova, G. Pribylova, D. Drozhko, T.A.Mastrykova, G. Myasoedova, and B. Myasoedov *J. Nucl. Sci. Tech.* 2002, Suppl. **3**, 422.
4 O.I. Artyushin, E.V. Sharova, I.L. Odinets, S.V. Lenevich, V.P. Morgalyuk, I.G. Tananaev, G.V. Pribulova, G.V. Myasoedova, T.A. Mastryukova, and B.F. Myasoedov, *Izv. AN, Ser. Khim.*, 2004, 2395.
5 M.N. Litvina, M.K. Chmutova, N.P. Nesterova, and B.F. Myasoedov, *Radiochemistry* (rus), 1994, **36**, 325.

TRANSURANIUM, RARE EARTH ELEMENTS AND TECHNETIUM (VII) EXTRACTION BY CALIX[4]ARENES FROM ACIDIC AND ALKALINE MEDIA

I.S. Antipin,[1] G.A. Pribylova,[2] S.E. Solovyeva,[1] I.G. Tananaev,[2]
A.I. Konovalov,[1] and B.F. Myasoedov[2]

[1]Arbuzov Institute of Organic and Physical Chemistry Kazan SC RAS, Russia, 420088, Kazan, Arbuzov str., 8.
[2]Vernadsky Instituite of Geochemistry and Analytical Chemistry RAS, Russia, 119991, Moscow, Kosygin str., 19

During the last two decades calixarenes and crown ethers have attracted a great deal of interest for the molecular design of preorganized three-dimensional receptors for different ionic species,[1-3] including radionuclides.[1-2,4-5] Presently, some information is available about the thermodynamics and control/design of radionuclide recognition by macrocycles.[1,5] but a detailed mechanistic and structural understanding remains elusive. The goal of our work is to apply new technological methods, allowing us to carry out isolation and separation of actinides, lanthanides, and technetium effectively using the calixarenes matrixes.

The influence of pH and the ionic composition of aqueous phase in relation to extraction of TPE and Rare Earth elements was studied in order to estimate the complex forming ability of dimetylamine-metylized calyx [4] resorcinearene (I) in dichlorethane (DCE) for affectivity and selectivity of extraction. It was shown that separation factor for Eu/Am pair under optimal conditions increases from 0.63 up to 4.53 with increase of extractant concentration, that was stipulated by the different stoiochiometry of extracting complexes: 1:2 for Eu and 1:3 for Am (Table 1).

Table 1 *The Influence of concentration of extractant (I) in DCE and $NaNO_3$ on the effectivity of extraction of Am(III) and Eu(III) at pH=4-6*

C(**I**), mol/L	C($NaNO_3$), mol/L	Am	Eu	$f_{Am/Eu}$
		D	D	
0.0013	2	0.017	0.027	0.63
0.005	2	0.82	0.37	2.22
0.01	2	7.24	1.6	4.53
0.005	4	2.91	0.39	7.46

It was found that amino-phosphonate derivatives of calix [4] resorcinearene extracted ions of Ln and Lu from aqueous solutions into chloroform much more effectively in comparison with the o,o-diethyl[(4-nitrophenyl)aminobenzyl]phosphonate. The optimal concentrations of calixarenes in conventional organic solvents were found on the base of the obtained data.

It is shown that during Tc(VII) extraction by 0.025 M solutions of calix[4]arene with 4 ester in dichlorethane from solutions of [HNO_3] in the range from 10^{-3} to 5M distribution coefficients decrease from 3.2 to 0.19 accordingly. Extraction of Tc (VII) by calix[4]arene with 4 methylketone groups is more effective in identical conditions and in the range of nitric acid concentrations from 0.001 to 5M values of D_{Tc} vary from 17.3 to 0.19. It was determined that increase of [HNO_3] during Tc extraction by the aforementioned calixarenes leads to decrease of D_{Tc}. This fact can be explained that Tc(VII) exists as technetium acid, its dissociation is lower at increase of H^+ in the solution and equilibrium $R + HTcO_4 = RTcO_4 + H^+$ (where R – function group of compound) shifts to the left. Dependence of D_{Tc} as a function of reagent concentration was investigated at Tc(VII) extraction from 0.01 M HNO_3. Calix[4]arene with R = -CH_2-C(O)-CH_3 was used in the experiment. It was shown that in 0.01M HNO_3 in the range of reagent concentrations from 0.0025 to 0.025 M D_{Tc} changes from 0.63 to 16.6. Using logarithm coordinates "lgD_{Tc} – lg[R]" dependence mentioned above is expressed as straight line with angular coefficient of 1. On the basis of obtained results one can assume that Tc (VII) forms extractable complex with calix[4]arene of 1:1 composition.

It was determined that in alkaline media both investigated reagents have extremely high extraction ability for pertechnetate ions. Thus for calix[4]arene containing R = -CH_2-C(O)-O-Et, increase of [NaOH] from 0.001 to 1M during Tc extraction leads to D_{Tc} increase from 2.4 to 360 and in the case of R = -CH_2-C(O)-CH_3 D_{Tc} increases from 4.6 to 400. Increase of salt concentration leads to a drop in distribution coefficients values during extraction of Tc by 0.025 M solutions of calix[4]arenes with –CH_2-C(O)-Et and –CH_2-C(O)-CH_3 groups in dichlorethane from alkaline solutions. Thus D_{Tc} = 142 during extraction of Tc (VII) by calix[4]arene solutions with ester group from 1M NaOH +1M $NaNO_3$ and D_{Tc} = 102 in 1M NaOH + 2M NaNO solutions. For reagent with methylketone group at the same conditions D_{Tc} is equal to 118 and 108 respectively.

As distinct from simple calixarenes during Tc extraction by tio derivatives nonlinear dependence of D_{Tc} from [HNO_3] is observed with maximum at 0.01-0.1M. Further increase of nitric acid concentration leads to a drop of Tc distribution coefficients. The extraction efficiency of tio derivatives is higher at other comparable conditions comparing to simple calixarenes. The compound structure is of vital importance during extraction. Thus D_{Tc}=2.3 at extraction of Tc by 0.025M solutions of tiocalixarenes with ester group in the cone form in dichlorethane from 0.01M solutions of HNO_3 and as 1,3-alternate form distribution coefficient of technetium is equal only to 0.038. It was discovered that tiocalixarenes could be used for Tc(VII) and Pu(IV) separation by extraction from acidic solutions. Thus distribution coefficients values D_{Tc} changes from 167 to 127 during Tc extraction by 0.025 M solutions of tiocalixarene in dichlorethane in the range of [HNO_3] from 10^{-2} to 0.1M while D_{Pu} is equal to 0.007 at the same conditions. The data obtained are presented at the Table 2.

Table 2 *Logarithms of Tc distribution constants lgD_{Tc} in the system of water – 0.025M solution of calix[4]arenes with groups $–CH_2-C(O)-O-Et$ (2) and $–CH_2-C(O)-CH_3$ (3) in 1,2 – dichlorethane at different concentrations of acid and alkali*

R	$[HNO_3]$, M							[NaOH], M			
	0.001	0.01	0.1	0.5	1	3	5	0.001	0.028	0.1	1
2	0.51	0.26	0.02	0.004	-0.07	-0.32	-0.72	0.39	0.4	0.97	2.6
3	-	1.22	-	0.928	0.66	-0.11	-0.7	0.67	1.29	2.31	>4

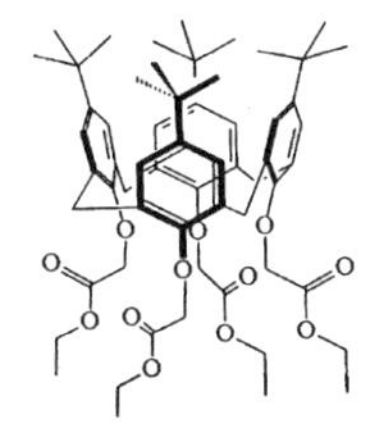

2

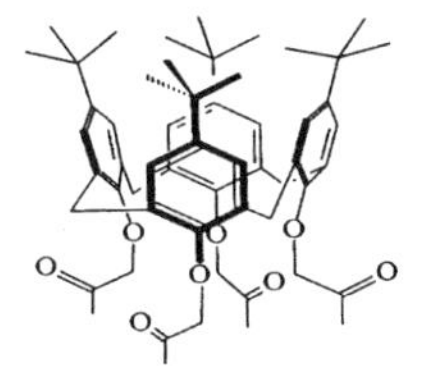

3

It is seen from the obtained data that the extraction ability of ligand **3** is considerably higher in comparison with ligand **2** in slightly acid media that can be explained by differences in the values of Lewis basicity of ligands **2** and **3**. Solvatechroum parameters of basicity of Camlet-Taft β,[6] and calorimetric scales ΔH^0_{BF3} [7] are 0.48 and 0.45 and -76.03 and -75.55 kJ/mole for **3** and **2**, respectively. These data explain the fact that ethoxycarbonyl group interacts with Lewis acids slighter than with methylcarbonyl one.

The work was supported by the U.S. DOE-OBES, Project RUC2-20010-MO-04.

References

1 M.A. McKervey, M.-J. Schwing-Weill, and F. Arnaud-Neu. Cation binding by calixarenes. Comperehensive supramolecular chemistry; Ed. G.Gokel, Pergamon, Oxford, 1996, **1**, 537.
2 R. Ludwig. Calixarenes in analytical and separation chemistry. *Fres. J. Anal. Chem.*, 2000, **367**, 103.
3 C. Wiether, C.B. Dieleman, and D. Matt. Calixarene and resorcinarene ligands in transition metal chemistry. *Coord. Chem rev.*, 1997, **165**, 93.
4 M.-J. Schwing-Weill, and F. Arnaud-Neu. Calixarenes for radioactive waste management. *Gazz. Chim. Ital.*, 1997, **127**, 687.
5 F. Arnaud-Neu, M.-J. Schwing-Weill, and J.F. Dozol. Calixarenes for nuclear waste treatment // in: *Calixarenes 2001*, Eds. Z. Asfari, V. Bohmer, J. Harrowfield, J. Vicens, Kluver Academic publishers, Netherlands, 2001. 642.
6 M.J.Kamlet, J.L.M.Abboud, M.H.Abraham, and R.W.Taft. *J.Org.Chem.*, 1983, **48**, 2877.
7 V. Gutman. *Chemistry of coordination compounds in nonaqueous solutions*, Mir, Moscow, 1971, 220.

ADSORPTION PROPERTIES OF AMIDOXIME RESINS FOR URANIUM(VI) AND SOME FISSION PRODUCTS SOLUBLE IN AQUEOUS CARBONATE SOLUTIONS

M. Nogami,[1] I. Goto,[1] N. Asanuma,[2] Y. Ikeda[2] and K. Suzuki[1]

[1]Institute of Research and Innovation, 1201, Takada, Kashiwa, Chiba 277-0861, Japan. E-mail: nogami@iri.or.jp
[2]Research Laboratory for Nuclear Reactors, Tokyo Institute of Technology, 2-12-1-N1-34, O-okayama, Meguro-ku 152-8550, Tokyo, Japan

1 INTRODUCTION

Uranium and Plutonium, which are important elements in the field of reprocessing and radioactive waste management, have a specific property in aqueous carbonate solutions from the viewpoint of solution chemistry of actinide elements, i.e. U(VI), Pu(IV), and Pu(VI) ions form stable anionic carbonate complexes of $UO_2(CO_3)_3^{4-}$, $Pu(CO_3)_4^{4-}$, and $PuO_2(CO_3)_3^{4-}$ and $(PuO_2)_2(CO_3)_5^{6-}$, respectively.[1,2] On the other hand, it is well known that resins or adsorbents containing amidoxime groups ($-C(NH_2)=NOH$) as a functional group adsorb U(VI) species in seawater.[3,4] U(VI) in seawater exists mainly in the form of $UO_2(CO_3)_3^{4-}$.

Based on the knowledge, we have investigated the possibility to separate U and Pu from most of the fission products (FPs) by using amidoxime resins in aqueous carbonate solutions, where only a few FPs are dissolved in these solutions, and in low concentrations. Two ways have been investigated to obtain carbonate solutions containing these elements; one is adding basic carbonate solutions to acidic nitrate solutions[5] and the other is dissolving simulated spent fuels directly to basic carbonate media.[6]

Our previous works for amidoxime resins using the former solutions have shown that U(VI), Pu(IV), and Pu(VI) were adsorbed from carbonate solutions at pH9 which were prepared by adding a Na_2CO_3-$NaHCO_3$ solution to acidic nitrate solutions. It has been also shown that U(VI) was separated from some FPs by a column experiment using a similar medium.[7] In this study, adsorption properties of amidoxime resins for U(VI) and some FPs in the latter solutions were investigated using Na_2CO_3 and $(NH_4)_2CO_3$ media.

2 METHOD AND RESULTS

2.1 Adsorption properties of amidoxime resin in various concentrations of carbonate

Dependence of adsorption properties of amidoxime resins for U(VI) and soluble FPs on carbonate concentrations was examined by a batch method using a commercial resin (Mitsubishi Kasei : particle diameter 149 to 297μm). Zr(IV), Mo(VI), Cs(I) were selected as soluble FP ions from our previous study[5] and we had not measured the solubility for these FPs in Na_2CO_3 and $(NH_4)_2CO_3$ media. In operations, the conditioned resin (wet 1g) was immersed in the sample solution of the metal ion ($10cm^3$, 1mM) dissolved in Na_2CO_3

or $(NH_4)_2CO_3$ at room temperature followed by shaking for 24h at 323K in a water bath. 1mM of the initial concentration was determined based on the solubility of U(VI) salts in both media.[8] In the case of Zr(IV) in Na_2CO_3, however, the solutions after filtration were used for the adsorption experiments due to the low solubility. The concentrations of each metal ion in the sample solutions before and after contact with the resin were determined by AA for Cs and ICP-AES for the other metal cations. The adsorption properties were evaluated by distribution ratio (K_d), where the dry form was used for the weight of the resin.

The results are shown in Figure 1. The concentration of Zr(IV) in each sample solution for Na_2CO_3 (mM) is described in the quotation. For Cs(I) in Na_2CO_3, no adsorptions have been observed by the experiments using the resin synthesized in this study (See below.). K_d for U(VI) has a maximum around 2 to 3% of Na_2CO_3. In general, there is small dependence of K_d on the concentration of Na_2CO_3. K_d for U(VI) is more than ca. 5 times higher than those for FPs where the concentrations of Na_2CO_3 are less than 5%. Lower initial concentrations of Zr(IV) means that the direct comparison of K_d for Zr(IV) with those of other ions is impossible. It is, however, expected that U(VI) is separated from the FPs under these solution conditions. On the contrary, K_d for U(VI) in $(NH_4)_2CO_3$ are more than 100 times greater in lower concentrations, and decrease remarkably with carbonate concentration. Zr(IV) and Mo(VI) are also strongly adsorbed at lower concentration of $(NH_4)_2CO_3$, and are more adsorbed in higher concentrations.

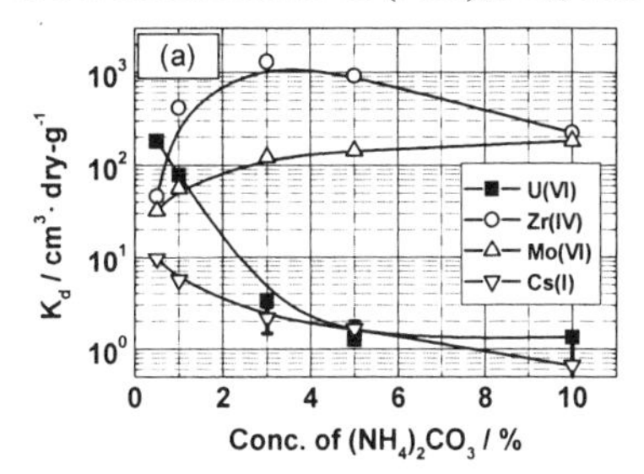

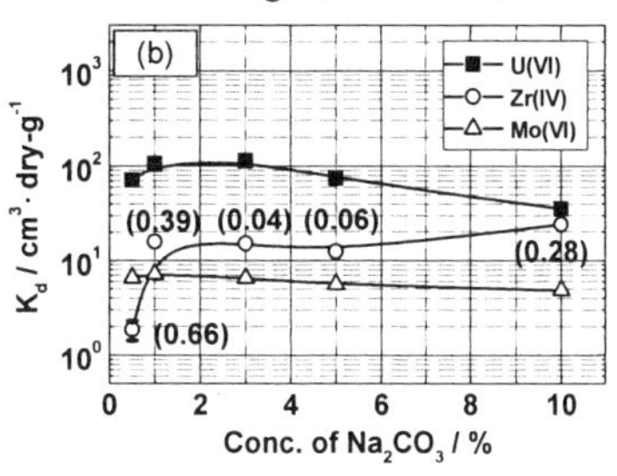

Figure 1 *Dependence of adsorption properties of amidoxime resin on carbonate concentration (a) Na_2CO_3, (b) $(NH_4)_2CO_3$*

The adsorption mechanism of U(VI) by amidoxime resin has been revealed as a complex formation between one UO_2^{2+} cation and two amidoxime functional groups whose oxime groups are dissociated ($-C(NH_2)=NO^-$).[4] Thus, the remarkable decrease in K_d for U(VI) with increasing carbonate concentrations in $(NH_4)_2CO_3$ means that uranyl carbonato complexes become more dominant species in the solution with increasing carbonate concentrations. It is also expected that the complex formation of U(VI) with CO_3^{2-} is stronger in $(NH_4)_2CO_3$ than in Na_2CO_3 where the concentrations of carbonate are more than 1%. The clarification of the adsorption mechanisms of Zr(IV) and Mo(VI), particularly of large adsorptions in $(NH_4)_2CO_3$, needs further studies because both elements seem to exist as oxo anion complexes without carbonate in basic carbonate media.[8]

2.2 Adsorption rates of modified amidoxime resin

The preliminary experiments showed that the commercial amidoxime resin had slow adsorption rates for U(VI), namely 24h or more time were necessary to attain the equilibrium. Amidoxime resins with micropores (less than 100nm) have been mainly studied for the adsorption of a very low concentration of U in seawater (ca. 3ppb).

However, treatment of much higher concentrations of U in much shorter time is necessary for our system, and a modified amidoxime resin with macropores (several hundred nanometers) was newly synthesized.

A conventional suspension polymerization of acrylonitrile and the subsequent amidoximation of acrylonitrile were applied.[7] For producing macropores in the resin, methyl benzoate and m-xylene were added to the monomers during polymerization. The measurement of porosities using a mercury porosimeter showed that the commercial and the modified resins had main pores *of ca.* 40nm and 350 to 400nm, respectively.

Adsorption rates of both resins for U(VI) were examined in a similar manner to that described in 2.1. The results are shown in Figure 2. For 1% $(NH_4)_2CO_3$, an increase in the adsorption rate is observed because the adsorption equilibrium is attained in *ca.* 2h in the modified resin. Moreover, it can be seen that the modified resin has a significantly faster adsorption rate in 1% Na_2CO_3 where the resin reaches the equilibrium in 5 to 10min. One of the reasons of such a significant difference in the adsorption rates in the two media may be the different complex formation with U(VI) as discussed in 2.1. There may be a lot of difference in salting-out effect between Na^+ and NH_4^+ in carbonate media.

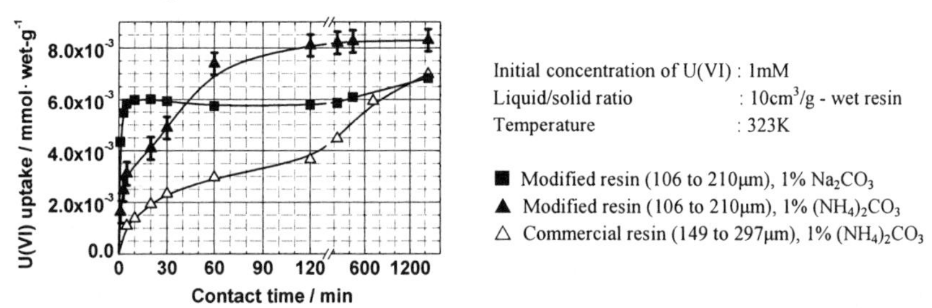

Figure 2 *Relationship between U(VI) uptake versus time by two amidoxime resins*

3 CONCLUSION

This study has clarified the fundamental adsorption properties of amidoxime resins for U(VI) and some FPs from Na_2CO_3 and $(NH_4)_2CO_3$ media. It has been shown that Na_2CO_3 is preferable to $(NH_4)_2CO_3$ for the separation of U from the FPs and for the adsorption rate of U(VI). The modified amidoxime resin has enhanced the adsorption rate of U(VI).

References

1 D.L. Clark, D.E. Hobart and M.P. Neu, *Chem. Rev.*, 1995, **95**, 25.
2 J.J.Katz, G.T.Seaborg and L.R.Morss, *The Chemistry of the Actinide Elements*, 2nd edn., Champman and Hall, 1986, Vol. 1, Chapter 7, p. 765.
3 H. Egawa, H. Harada and T. Nonaka, *Nippon Kagaku Kaishi*, 1979, 1767.
4 K. Saito, T. Hori, S. Furusaki, T. Sugo and J. Okamoto, *Ind. Eng. Chem. Res.*, 1987, **26**, 1977.
5 N. Asanuma, M. Harada, Y. Ikeda and H. Tomiyasu, *J. Nucl. Sci. Technol.*, 2001, **38** 866.
6 N. Asanuma, M. Harada, Y. Ikeda, M. Nogami, K. Suzuki, T. Kikuchi and H. Tomiyasu, *Proc. GLOBAL 2003*, New Orleans, USA, 2003, 696-701
7 M. Nogami, S.-Y. Kim, N. Asanuma and Y. Ikeda, *J. Alloys Comp.*, 2004, **374**, 269.
8 N.P. Galkin, A.A. Maiorov and U.P.Veryatin, *International Series of Monographs on Nuclear Energy, Division IX: Chemical Engineering*, Pergamon Press, Oxford, 1963, Vol. 1.

SOLUBILITY OF UO_2(β-DIKETONATO)$_2$DMSO COMPLEXES (β-DIKETONATE = ACETYLACETONATE, TRIFLUOROACETYLACETONATE, HEXAFLUORO-ACETYLACETONATE; DMSO = DIMETHYL SULFOXIDE) IN SUPERCRITICAL CO_2

T. Tsukahara,[1] Y. Kachi,[3] Y. Kayaki,[2] T. Ikariya[2] and Y. Ikeda[*3]

[1] Department of Applied Chemistry, The University of Tokyo, 7-3-1, Hongo, Bunkyo-ku, Tokyo, 113-8656, Japan
[2] Department of Chemical Technology, Tokyo Institute of Technology, 2-12-1 O-okayama, Meguro-ku, Tokyo 152-8550, Japan
[3] Research Laboratory for Nuclear Reactors, Tokyo Institute of Technology, 2-12-1-N1-34 O-okayama, Meguro-ku, Tokyo 152-8550, Japan. E-mail: yikeda@nr.titech.ac.jp

1 INTRODUCTION

Supercritical carbon dioxide (scCO_2) has become of interest as promising alternatives of organic solvents for the extraction of metal ions from solid and liquid wastes.[1] Previous studies have demonstrated that formations of neutral metal complexes with chelating ligands (MCC) lead to an increase in solubility of metal species in scCO_2.[1-3] The most effective ligands would be *β*-diketones, which have been applied to the extraction of uranyl and lanthanide ions from the radioactive wastes generated from the reprocessing processes of spent nuclear fuels.[2-4] The extractability can be enhanced by adding tributyl phosphate (TBP) to scCO_2 containing *β*-diketones through a so-called synergistic effect.[4] To clarify the extraction mechanism and to optimize extraction conditions, it is essential to determine the solubility of MCC in scCO_2. However, limited information is available concerning solubility of uranyl complexes in scCO_2.

Hence, we measured the solubility of uranyl complexes with *β*-diketonate, UO_2(*β*-diketonato)$_2$dmso [*β*-diketonate = acetylacetonate (acac), trifluoroacetylacetonate (tfacac), hexafluoroacetylacetonate (hfacac), dmso = dimethyl sulfoxide], in scCO_2 by using UV-visible spectrophotometer, and discussed the relationship between solubility and ligand-CO_2 interactions.

2 EXPERIMENTAL

2.1 Apparatus for solubility measurements:

The solubility was measured by UV-visible spectrophotometer (SHIMADZU UV-2400PC) equipped with high-pressure cell having two sapphire windows (TAIATSU Glass Co., Ltd.). The high-pressure cell body with 216 cm^3 is made from hastelloy. The optical path length of cell and the volume of sample were 1.7 cm and 0.54 cm^3, respectively. The temperature was controlled with electric heating rods put into cell body and detected by a thermocouple inserting inside the cell. The pressure was controlled with the syringe pump (ISCO Model-260D) connected to a 1/16 inch stainless steel tube. Pressure to the cell was controlled by exerting CO_2 gas via the syringe pump from the head of cell. The samples were stirred by directly injecting CO_2 gas to the cell.

2.2 Measurements of solubility

The molar extinction coefficients (ε) of $UO_2(acac)_2$dmso, $UO_2(tfacac)_2$dmso, and $UO_2(hfacac)_2$dmso [abbreviated as UO_2(ACAC), UO_2(TFA), and UO_2(HFA)] were determined from the plots of absorbance *vs.* their concentrations. Then, the powders of above $UO_2(\beta$-diketonato$)_2$dmso in larger excess than their expected solubility in $scCO_2$ were placed in the high-pressure cell, and the UV-visible absorption spectra of samples measured at 40°C in the pressure range of 10 to 25MPa (density; $0.6 < \rho$ (g cm^{-3}) < 0.9).

3 RESULTS AND DISCUSSION

3.1 Solubility

We found that the ε values (M^{-1}cm^{-1}, M = mol dm^{-3}) for UO_2(ACAC), UO_2(TFA), and UO_2(HFA) in $scCO_2$ were estimated as 105.65 (at 360nm), 95.94 (at 440.5 nm), and 7.95 (at 488.0 nm), respectively. The absorbance of $scCO_2$ samples containing excess amount of UO_2(ACAC), UO_2(TFA), and UO_2(HFA) was measured. Their spectra were found to increase with an increase in pressure. The solubility at each pressure was obtained from the corresponding absorbance using ε value. The plots for the solubility of UO_2(ACAC), UO_2(TFA), and UO_2(HFA) in $scCO_2$ *vs.* pressure are shown in Figure 1 (a). This figure shows that the solubility of UO_2(ACAC), UO_2(TFA), and UO_2(HFA) in $scCO_2$ increases linearly with increasing pressure and that the solubility of $UO_2(\beta$-diketonato$)_2$dmso in $scCO_2$ is enhanced with the fluorination of coordinated β-diketonates.

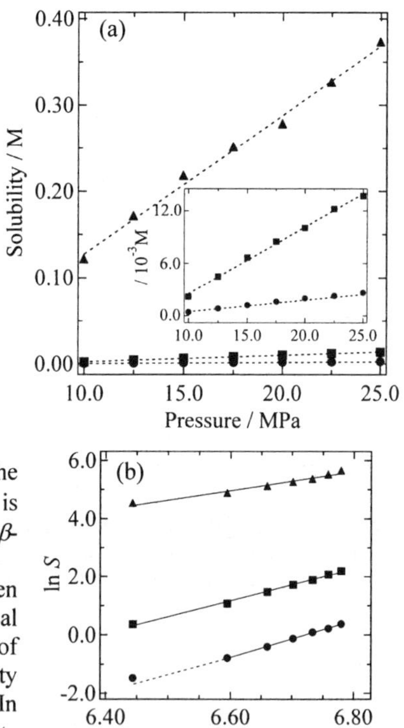

Figure 1 (a) *Plots of solubility of UO_2(ACAC)(●), UO_2(TFA)(■), and UO_2(HFA)(▲) in $scCO_2$ as a function of pressure at 40 °C. The inset shows pressure dependences of solubility for UO_2(ACAC), UO_2(TFA). (b) Plots of ln S vs. ln ρ for UO_2(ACAC)(●), UO_2(TFA) (■), and UO_2(HFA) (▲) in $scCO_2$.*

The solubility of solutes in $scCO_2$ have been mainly explained by two different theoretical approaches, *i.e.*, one is Peng-Robinson equation of state[5] and another is Hildebrand solubility parameter based on regular solution theory.[4] In fact, the solubility of some metal β-diketonate complexes, M(β-diketonato)$_n$ [M = metal ions, n = coordination number], were measured and discussed on the basis of two theories mentioned above. Unfortunately, it is difficult to discuss the solubility of $UO_2(\beta$-diketonato$)_2$dmso in $scCO_2$ in a similar manner as those of M(β-diketonato)$_n$, because there is no useful information concerning the thermodynamic properties of $UO_2(\beta$-

diketonato)$_2$dmso, and the solubility for $UO_2(\beta$-diketonato$)_2L$ with two kinds of ligands should be affected by changing L group.[3] On the other hand, Chrastil has proposed that the experimental solubility data of solutes in $scCO_2$ are simply related to the density of $scCO_2$ as follows:[6]

$$\ln S = k \ln \rho + C \tag{1}$$

where S (g dm^{-3}) is the solubility of MCC in $scCO_2$, ρ (g dm^{-3}) is density of $scCO_2$ at each measurement condition, k corresponds to the number of CO_2 molecules associated with MCC, and C is temperature-dependent constant related to the volatility of solutes. This model has been proved to be useful for evaluating the relationship between solubility and ligand-CO_2 interactions. Therefore, we examined whether *Eq*. (1) is applicable to the solubility of UO_2(ACAC), UO_2(TFA), and UO_2(HFA) in $scCO_2$. The plots of $\ln S$ *vs*. $\ln \rho$ in the present systems show the linear relationships as shown in Figure 1(b). The k and C values of UO_2(ACAC), UO_2(TFA), and UO_2(HFA) are 6.3 and -42.3, 5.5 and -35.1, and 3.3 and -16.6, respectively. The absolute values of k and C were found to become large in order of UO_2(ACAC) > UO_2(TFA) > UO_2(HFA), which is opposite to the tendency of solubility. In spite of large solubility of UO_2(HFA) in $scCO_2$, the k value, *i.e*., the number of $scCO_2$ molecules associated with UO_2(HFA), is the smallest in three complexes. These phenomena strongly suggest that the solubility of $UO_2(\beta$-diketonato$)_2$dmso depends on strength of interactions between $scCO_2$ and particular sites (CF_3, CH_3) of ligands in $UO_2(\beta$-diketonato$)_2$dmso rather than the number of solvation. A similar tendency, that the k values decrease with an increase in molecular weight has been observed in the solubility of organophosphorus compounds.[7] However, Wai *et al*. have showed that in the case of uranyl complexes with mixed ligands of β-diketone and phosphine oxides, the k values become large and increase with increasing solubility.[2] Hence, to examine the solubility mechanism, we have focused on the NMR measurements for characterization of structures and dynamics of uranyl complexes in $scCO_2$, and such studies are in progress.[8]

4 SUMMARY

The solubility of $UO_2(\beta$-diketonato$)_2$dmso in $scCO_2$ has been investigated at 40°C in the pressure range of 10 to 25 MPa (density; $0.6 < \rho$ (g cm^{-3}) < 0.9) by using UV-visible spectrophotometer equipped with high-pressure cell. It was verified that the solubility of $UO_2(\beta$-diketonato$)_2$dmso under $scCO_2$ increases with increasing pressure or fluorination of β-diketonato ligands, and suggested that the interactions between $scCO_2$ and particular sites of ligands in $UO_2(\beta$-diketonato$)_2$dmso plays an important roles for determining solubility of $UO_2(\beta$-diketonato$)_2$dmso in $scCO_2$.

References

1 J.A. Darr and M. Poliakoff, *Chem. Rev*., 1999, **99**, 495.
2 Addleman, R.S.; Carrot, M.J.; Wai, C.M. *Anal. Chem*., 2000, 72, 4015.
3 Wai, C.M.; Waller, B. *Ind. Eng. Chem. Res*., 2000, 39, 4837.
4 C.M. Wai, S. Wang, and J.J. Yu, *Anal. Chem*., 1996, **68**, 3516.
5 W.C. Cross, A. Akgerman and C. Erkey, *Ind. Eng. Chem. Res*., 1996, **35**, 1765.
6 J. Chrastill, *J. Phys. Chem*., 1992, **86**, 3016.
7 Y. Meguro, S. Iso, T. Sasaki and Z.Yoshida, *Anal. Chem*., 1998, **70**, 774.
8 T. Tsukahara, Y. Kayaki, T. Ikariya and Y. Ikeda, *Angew. Chem. Int. Ed.,* 2004, **43**, 3719.

KINETIC STUDY ON EXCHANGE REACTION OF HEXAFLUOROACETYL-ACETONATE IN BIS(HEXAFLUOROACETYLACETONATO)(DIMETHYL SULFOXIDE)DIOXOURANIUM(VI) IN SUPERCRITICAL CO_2 BY ^{19}F NMR

Y. Kachi,[1] Y. Kayaki,[2] T. Tsukahara,[3] T. Ikariya[2] and Y. Ikeda[1*]

[1]Research Laboratory for Nuclear Reactors, Tokyo Institute of Technology, 2-12-1-N1-34 O-okayama, Meguro-ku, Tokyo 152-8550, Japan. E-mail: yikeda@nr.titech.ac.jp
[2]Department of Chemical Technology, Tokyo Institute of Technology, 2-12-1 O-okayama, Meguro-ku, Tokyo 152-8550, Japan
[3]Department of Applied Chemistry, The University of Tokyo, 7-3-1, Hongo, Bunkyo-ku, Tokyo, 113-8656, Japan

1 INTRODUCTION

In the extraction field, supercritical CO_2 (Sc-CO_2) has become of interest as alternative to organic solvents,[1] *e.g.*, application of Sc-CO_2 containing extractants such as tributhyl phosphate (TBP) and β-diketone to the separation media of metal ions in the reprocessing spent nuclear fuel, the radioactive waste treatments etc.. However, the complex formation reactions between metal ions and extractants, which are the most fundamental data for understanding extraction mechanisms of metal ions in Sc-CO_2 medium, have not been studied sufficiently.[2]

In this present study we have examined the ligand exchange reactions in $UO_2(hfacac)_2DMSO$ (hfacac = hexafluoroacetylacetonate, DMSO = dimethyl sulfoxide) in Sc-CO_2 based on our previous studies for the interaction of Sc-CO_2 with β-diketone and its complexes, and for the structures of UO_2(β-diketonato)$_2$DMSO complexes in Sc-CO_2.

2 EXPERIMENTAL

2.1 Materials

The $UO_2(hfacac)_2DMSO$ complex was synthesized by the same method as reported previously.[2,3] Hexafluoroacetylacetone (Hhfacac, Wako Pure Chemical Ind. Ltd., 99%) was purified by distillation. Dimethyl sulfoxide (Wako, 99%) and DMSO-d_6 (Kanto Kagaku, 98 atom% D) were stored over 4A molecular sieves.

2.2 Measurements of NMR Spectra of Sc-CO_2 Containing Solutes

^1H and ^{19}F NMR spectra of Sc-CO_2 containing solutes were measured using a JEOL JNM LA 300 WB FT-NMR at 25 MPa in the temperature range from 40 to 120 °C without spinning. A zirconia cell (ZC) was used as a high-pressure NMR sample tube. CO_2 gas was charged into ZC by using a syringe pump (ISCO Model-260D). Pressure and temperature of samples were controlled by using back-pressure regulator (JASCO 880-81) and by providing heated air from the lower part of probe.

3 RESULTS AND DISCUSSION

3.1 Structure of $UO_2(hfacac)_2DMSO$ in Sc-CO_2 Containing Free Hhfacac

We have measured ^1H and ^{19}F NMR spectra of Sc-CO_2 containing $UO_2(hfacac)_2DMSO$ and confirmed that the structure of this complex in Sc-CO_2 is pentagonal bipyramidal as observed in non-aqueous solvents.[3] In order to examine the structure of $UO_2(hfacac)_2DMSO$ in Sc-CO_2 containing free Hhfacac, we measured ^1H and ^{19}F NMR spectra of Sc-CO_2 containing $UO_2(hfacac)_2DMSO$ (1.03×10^{-2} M) and free Hhfacac (2.96×10^{-2} M). The results are shown in Figure 1. Signals a, b, and c in Figure 1(a) are assigned as the –CH group of coordinated hfacac, the –CH_3 one of coordinated DMSO, and the –CH one of free Hhfacac, respectively. Signals d, e, and f in Figure 1(b) correspond to the –CF_3 groups of coordinated hfacac, and of the enol (Henol) and keto (Keto) isomers of free Hhfacac, respectively. From the area ratios of signals a to b to c and of signals d to e at 40 °C, it was found that two hfacac and one DMSO coordinate to the uranyl ion. This indicates that even in the presence of free Hhfacac in Sc-CO_2, the $UO_2(hfacac)_2DMSO$ complex maintains the pentagonal bipyramidal structure and the following equilibrium does not exist to any appreciable extent in Sc-CO_2:

$$UO_2(hfacac)_2DMSO + Hhfacac = UO_2(hfacac)_2cacafhH + DMSO \quad (1)$$

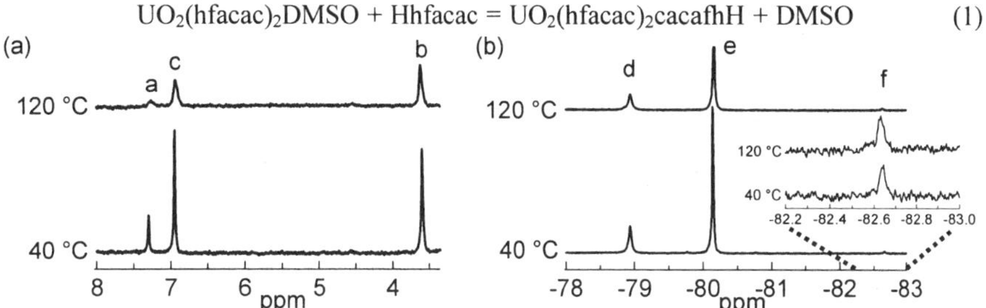

Figure 1 *^1H (a) and ^{19}F (b) NMR spectra of Sc-CO_2 containing $UO_2(hfacac)_2DMSO$ and free Hhfacac at 40 and 120 °C.*

3.2 Exchange Reaction between Free Hhfacac and Coordinated hfacac in $UO_2(hfacac)_2DMSO$ in Sc-CO_2

As seen from Figure 1, the signals d and e become broad with increasing temperature, while the signal f does not show such a broadening. These results indicate that the free Henol exchanges with the coordinated hfacac in $UO_2(hfacac)_2DMSO$ as follows:

$$UO_2(hfacac)_2DMSO + Henol^* \leftrightarrow UO_2(hfacac)(hfacac^*)DMSO + Henol \quad (2)$$

where the asterisk denotes the exchanging species. Hence, the first-order rate constants (k_{ex}) for the exchange reaction (2) were obtained by the conventional NMR line-broadening method, *i.e.*, by using the following equations:[4]

$$\tau_c P_f = \tau_f P_c \quad (3)$$

$$k_{ex} = 1/\tau_c = (\text{rate})/2[UO_2(hfacac)_2DMSO] \quad (4)$$

where the τ and P with the subscripts of c and f express the mean lifetimes and the mole fractions of the coordinated hfacac and the free Henol, respectively.

The k_{ex} values were measured for Sc-CO_2 containing $UO_2(hfacac)_2DMSO$ (1.0×10^{-2} M) at various free Hhfacac concentrations in the range from 9.07×10^{-3} to 7.50×10^{-2} M. The k_{ex} values were found to increase with an increase in [Henol]. Plots of k_{ex} values *vs.* [Henol] give straight lines with intercepts and result in the following equation.

$$k_{ex} = k_a + k_b[\text{Henol}] \quad (5)$$

3.3 Mechanism

The [Henol] dependence of k_{ex} indicates that the exchange reaction of hfacac in $UO_2(hfacac)_2DMSO$ proceeds through two processes, *i.e.*, one is independent of [Henol] and another is dependent on [Henol]. Two mechanisms can be proposed (see Scheme), *i.e.*, mechanism 1 (I→II→III→III'→II'→I') corresponds to the path dependent on [Henol] and mechanism 2 (I→IV→V→V'→I') is the path independent of [Henol].

In the mechanism 1, the rate-determining step is the pathway II→III, because the dissociation of coordinated DMSO in pathway I→II is expected to occur rapidly on the basis of previous papers concerning the intra- and intermolecular exchange reaction of DMSO.[5] The III species is formed by the bond cleavage of coordinated hfacac in the intermediate II. The proton transfer takes place in the III→III' pathway, followed by the ring closure of unidentate hfacac*. Applying the steady-state approximation to the intermediate II, the concentration of II should be given as $[II] = k_1[Henol]/(k_{-1}[DMSO] + k_2)$. In mechanism 2, the intermediate IV is formed by the bond cleavage of coordinated hfacac in $UO_2(hfacac)_2DMSO$, which is the rate-determining step, followed by the rapid coordination of the incoming Hhfacac*(IV→V) and the proton transfer(V→V').
From these mechanisms, the exchange rate is expressed as follows:

$$\text{Rate} = k_{ex}[UO_2(hfacac)_2DMSO] = (k_3 + k_2[II])\,[UO_2(hfacac)_2DMSO] \quad (6)$$

$$k_{ex} = k_3 + k_1k_2[Henol]/(k_{-1}[DMSO] + k_2) \quad (7)$$

In the present system without adding free DMSO, $k_{-1}[DMSO]$ term should be negligible and hence the following equation can be derived.

$$k_{ex} = k_3 + k_1[Henol] \quad (8)$$

Equation (8) is consistent with Eq. (5), that is, k_3 and k_1 correspond to k_a and k_b, respectively. This supports that the proposed mechanisms are reasonable.

Scheme

The kinetic parameters for k_1 and k_3 were obtained from their temperature dependence, and are $k_1(60°C) = 1.97 \times 10^2\ M^{-1}s^{-1}$, $\Delta H^{\neq} = 24.5$ kJ mol^{-1}, $\Delta S^{\neq} = -130$ J mol^{-1} K^{-1}; $k_3(60°C) = 5.72 \times 10^{-1}\ s^{-1}$, $\Delta H^{\neq} = 62.3$ kJ mol^{-1}, $\Delta S^{\neq} = -52.5$ J mol^{-1} K^{-1}.

References

1 *Supercritical Fluids*, eds., R. Noyori, *Chem. Rev.*, 1999, **99**.
2 T. Tsukahara, Y. Kayaki, T. Ikariya and Y. Ikeda, *Angew. Chem. Int. Ed.*, 2004, **43**, 3719.
3 G.M. Kramer, M.B. Dines, R. Kastrup, M.T. Melchior and T. Maas, JR., *Inrog. Chem.*, 1981, **20**, 3.
4 Y. Ikeda, H. Tomiyasu and H. Fukutomi, *Inorg. Chem.*, 1984, **23**, 3197.
5 Y. Ikeda, H. Tomiyasu and H. Fukutomi, *Inorg. Chem.*, 1984, **23**, 1356.

HEAT OSCILLATING EXTRACTION OF RARE EARTH ELEMENTS

A.A. Kopyrin, M.A. Afonin and A.A. Fomichev

Department of Rare Earth, Saint-Petersburg State Institute of Technology, Saint-Petersburg, Moskovsky av. 26, 190013, Russia. E-mail: fomich@lti-gti.ru

1 INTRODUCTION

Almost all modern processes for element and isotope separation are based on achievement by system stationary conditions. In these conditions the separation of elements with similar properties is difficult. The separation and isolation of lanthanides and other elements with similar properties requires new methods for element separation. There is a novel method for undertaking element separation using non-stationary non-equilibrium processes, based on differences in the kinetics of chemical reactions. Realisation of the process in non-stationary non-equilibrium conditions using phenomena of oscillating extraction allows the exploitation of differences in kinetics of ion complexation and extraction rates, using their transport through the boundary area in both directions for element separation.[1] However, the influence of oscillatory temperature changing in the extractor(s) is not described in the literature. To achieve high elements separation factors using oscillating extraction it is possible to use multiple repetition of processes similar to those described in literature as "parametric pumping"[2]. This can change, and essentially enrich, existing methods of isolation and separation of elements and isotopes.

2 METHOD AND RESULTS

2.1 Experimental Setup and Description of Procedures

The setup flowsheet (fig. 1.) is described as following: the emulsion is pumped from thermostatic extractors 16, 17 to centrifugal separators 5, 6 through the turbidimeters 11, 12 by peristaltic pump 10. After separation from the separator 5 the organic phase flows to the extractor 17 and from the separator 6 the organic phase flows to spectrophotometer 8 for analysis and it flows to the extractor 16, passing through both aqueous phases organic phase forms bulk liquid membrane. After spectrophotometers 7, 9 the aqueous phases come back to their corresponding extractors. Data acquisition of pH, red/ox potential of the aqua phase, temperature and emulsion turbidity are performed with frequency above 1 Hz using the DAQ board in computer 18 and specially designed software. Diode-array spectrophotometers are used to obtained spectra every 6-15 seconds. By deconvolution of each spectrum, using specially designed software concentration of several elements can be obtained.

2.2 Heat Oscillating Extraction of Neodymium and Praseodymium

The behavior of the extraction system based on liquid membrane 0.5 M TBP in kerosene, moving between two aqueous phase solutions (6 M $NaNO_3$ and 0.1 M HNO_3) under periodical temperature oscillations were investigated (fig. 2). All phase solutions are initially free from REE. In the phase - 6 M $NaNO_3$ (fig. 2), where extraction occurs, the periodic temperature oscillations was initiated within the range 20–35 °C. The temperature of the solution where the stripping process occurs, 0.1 M HNO_3, is maintained constant – 25 °C. The stirrers of both extractors support low stirring of phases to reduce mass transfer. As can be seen from figure 2 oscillations of concentrations of Nd and Pr in organic phase (3) respond inversely to oscillations of temperature (1) in extractor 16. Separation coefficient (6) is increased up to 5 while the metals are in macro-concentration in the 1st aqueous phase (5). The high value of separation coefficient obtained is because the speed of Nd transport from 1st to 2nd aqueous phase is higher than the speed of Pr transport. Processing error (6, fig. 2) is increased when the metals concentration is decreased to zero.

3 CONCLUSION

Fission products of spent fuel contain REE, therefore the results on separation of elements in non-stationary conditions using heat oscillating extraction can be used in radiochemical practice.

Acknowledgements
This work was supported by the U.S. Department of Energy, Office of Basic Energy Sciences, under grant RC0-20000-SC14 administered by the Civilian Research and Development Foundation.

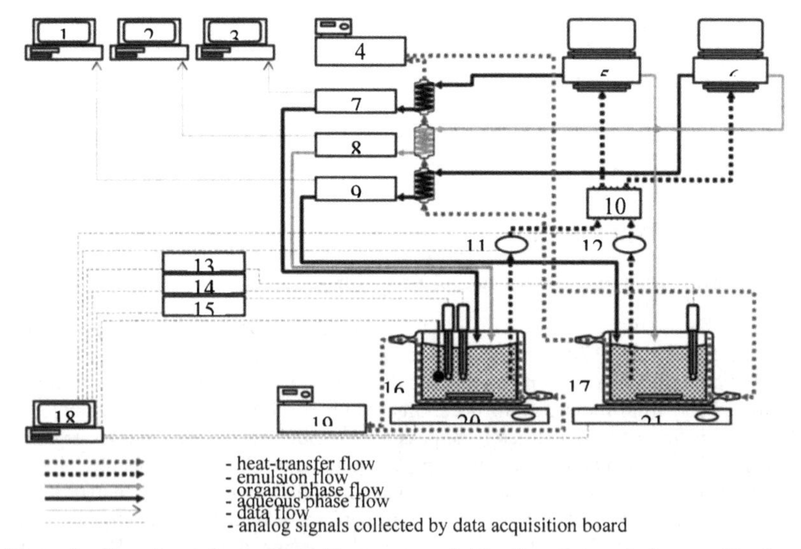

Figure 1 *Experimental setup: 1-3, 18 computers; 4, 19 – thermostats; 5, 6 – centrifugal separators EC-33 NIKIMT; 7, 8, 9 – spectrophotometers SF-2000 OKB Spectrum; 10 – peristaltic pump; 11, 12 – turbidimeters; 13 –pH-meter; 14, 15 - potentiometers; 16, 17 – extractors; 20, 21 – magnetic stirrers*

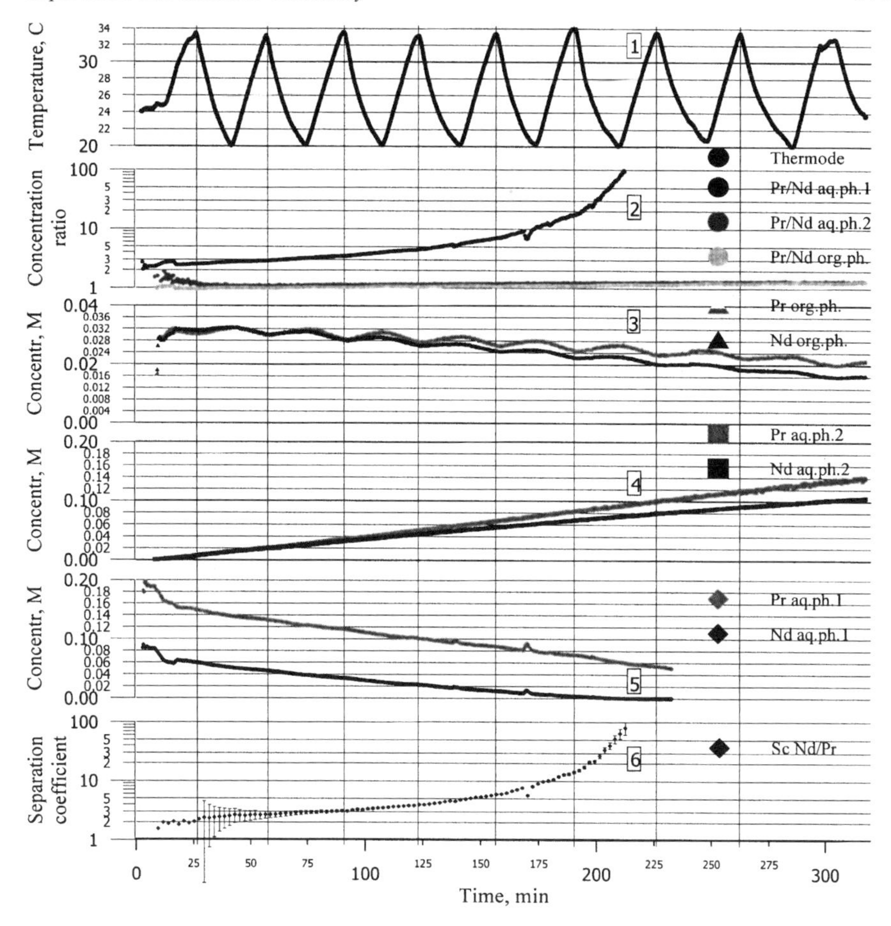

Figure 2 *Temporal dependence of extraction system behaviour under periodical temperature oscillations. 1 –temperature of the extractor 16; 2- concentration ratio Pr/Nd in three phases; 3 - concentration of Nd and Pr in bulk liquid membrane; 4 – concentration* C^{aq2} *of Nd and Pr in 2-nd aqueous phase; 5 - concentration* C^{aq1} *of Nd and Pr in 1-st aqueous phase;*

6 - separation coefficient calculating as $S_c = C^{aq2}_{Nd} / C^{aq2}_{Pr} \cdot C^{aq1}_{Pr} / C^{aq1}_{Nd}$

References

1 M. A. Afonin, A. A. Kopyrin, A. I. Shvalbe, A. A. Baulin, A. S. Petrov, *Journal of Alloys and Compounds*, 2004, **374**, 426.

2 C. Stöhr, C. Bartosch, R. Kiefer, W. H. Höll, *Chemical Engineering & Technology*, 2001, **24**, 879.

MATHEMATICAL MODEL OF EXTRACTION SEPARATION OF F-ELEMENTS IN NON-STATIONARY CONDITIONS

A.A. Kopyrin, M.A. Afonin and A.A. Baulin

Saint-Petersburg Institute of Technology; Moskovsky pr., 26, saint-Petersburg, Russia, 190013

1 INTRODUCTION

Liquid-liquid extraction driven by oscillatory oxidation-reduction reaction in aqueous phase could give an opportunity to separate similar elements using repetition of extraction/stripping in one extractor. The multiple extraction scenario should improve the separation by exaggerating the small kinetic differences between similar metal ions. The same effect could possibly be applied to isotopes separation.

The investigated system consists of two extractors connected by bulk liquid membrane. The aqueous phases of extractors consist of 6M $NaNO_3$ with $CeNO_3$ and hypothetical elements – I1 and I2 (it could be either different elements – neodymium and praseodymium, for example, or different isotopes of an element – for example, neodymium). The organic phase extractant consists of 0.5 TBP in diluent (kerosene or tetradecane). Electrochemical oxidation/reduction of cerium in aqueous phases for initialization of the oscillatory extraction process were used in both extractors.

2 RESULTS AND DISCUSSIONS

The special mathematical model based on the system of ordinary differential equations (of extraction/stripping) – ODE with 100% (electrochemical) conversion of cerium for the investigation of the system described above was originated. The change of the ratio of Ce^{4+} and Ce^{3+} in aqueous phase of extraction system is in accordance with the following equations:

$$[Ce^{4+}] = (Ce_{tot})*Sin(OMEGA*T-TIMESHIFT) \quad (1)$$
$$[Ce^{3+}] = Ce_{tot} - [Ce^{4+}] \quad (2)$$

where Ce_{tot} is analytical concentration of cerium in aqueous phase, M; OMEGA - is circular frequency, Hz; T- time, sec; TIMESHIFT – time shift. Solutions of the system of equations – time dependences of elements concentrations were calculated by the Gear method. In both extractors the concentration of Ce^{3+} and Ce^{4+} was changing periodically according to the formulae 1 and 2. The calculation of temporal dependencies of substances concentration in aqueous and organic phases was performed by the program oscmem2, using constants obtained from processing the experimental data. The results of calculating oscillating curves of enrichment rate of organic phase, Ce(III), Ce(IV) concentrations in aqueous and organic phases, free TBP concentration in organic phase in both extractors and I1 and I2 concentrations in the aqueous phases of first and second extractor and organic phase of first extractor are shown on the Figure 1.

The fluctuations of free TBP concentration force the I1 and I2 to change concentrations in the organic phase (in liquid membrane). The transport from and into each extractor was calculated continuously on each integration step by Gear algorithm during the calculation of the ODE. The extraction rate constants are attributed to hypothetical elements with identical equilibrium partition coefficients: - element I1 and the element I2. The equilibrium extraction constant of I1 is equal to $K_{I1}=k_{extraction\ I1}/k_{stripping\ I1}=0.026/0.0039=6.59$. The equilibrium extraction constant of I2 is equal to $K_{I2}=k_{extraction\ I2}/k_{stripping\ I2}=0.0099/0.0015=6.59$. The extraction rate constant of element I1 is 2.6 times greater than the extraction rate constant of element I2. From the graphs on the Figure 2 it is clear that in the oscillatory conditions the model predicts the non-stationary temporal dependencies of all metals concentration in both aqueous phases and in organic phase (liquid membrane) during the simulated experiment. The ratio of average concentrations of element I1 and I2 in aqueous phase of 1 extractor is equal to 0.99 and this ratio for the second extractor is equal to 1.023. The calculation was interrupted after the system became "stable". This happened after 1000 min of simulated experiment.

The influence of frequency and time shift (phase shift between the oscillatory concentration profiles of Ce^{4+} in aqueous phases in both extractors) were investigated. The resonance dependence of separation factor against frequency (very small difference in frequency resulted in the big difference in separation factor) was established. In some cases new value of frequency taken from the interval of two adjacent frequencies resulted in the drop in value of the separation factor. The periodical dependence of separation factor against time shift was established. The data about dependence separation factor against time shift is given on the Figure 2.

3 CONCLUSIONS

The mathematical model of oscillation extraction process taking place in two extractors coupled by bulk liquid membrane with 100% (electrochemical) conversion of Ce was generated. Numeric data of experiments with kinetics constants ratio equal to 2.6 were given. The periodical dependence of separation factor against shift was established. The resonance dependence of separation factor against frequency (small difference in frequency resulted to the big difference in the separation factor) was established.

Acknowledgements

This work was supported by the U.S. Department of Energy, Office of Basic Energy Sciences, under grant RC0-20000-SC14 administered by the Civilian Research and Development Foundation.

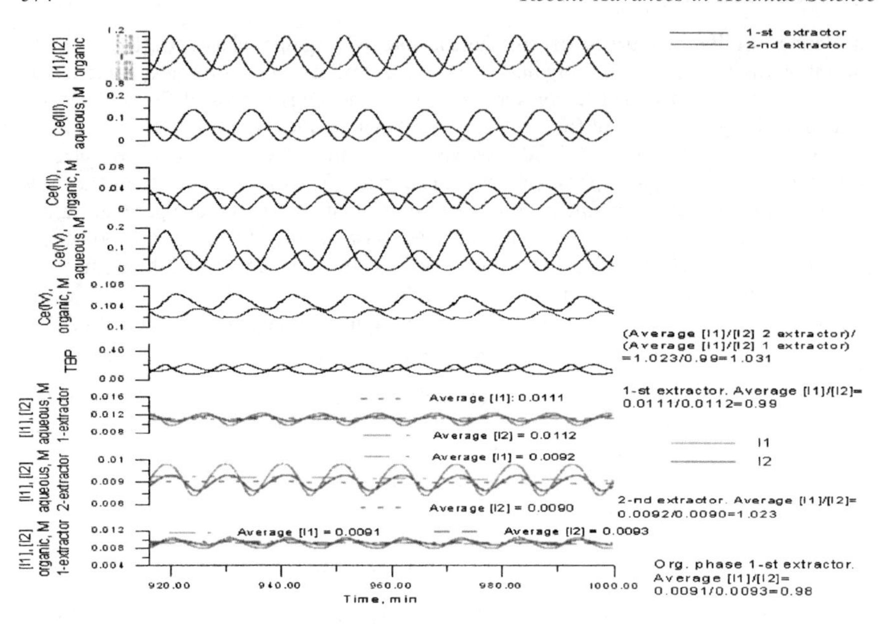

Figure 1. *Results of the simulation of the temporal dependence of the substances concentration. V organic in 1 extractor = V organic in 2 extractor = V aqueous in 1 extractor = V aqueous in 2 extractor = 150 ml. Iinitial concentrations: Ce – 0.25 M, I1 – 0.02 M, I2 – 0.02 M. Parameters of formulate 1 - OMEGA = 0.01 Hz, TIMESHIFT = 0 sec (first extractor), 240 sec (second extractor). Flowrate of liquid membrane = 4.69 ml/min.*

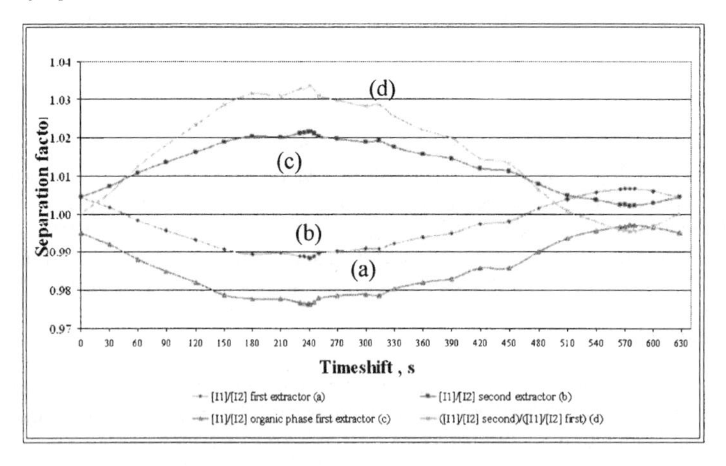

Figure 3. *Results of the simulation of the dependence of the concentration ratios (a,b,c) and separation factor (d) of two hypothetical elements - I1 and I2 in aqueous (a,b) and organic phases(c) against time shift (in second extractor).*

FINAL TESTS USING THE BNFL ELECTROREFINER AT HARWELL

S.A. Beetham[1], B. Hanson[2], D.J. Hebditch[3], P. Hopkins[2], R.G. Lewin[4] and H.E. Sims[5]

[1]UKAEA B220, Harwell, Didcot, Oxfordshire, [2]NexiaSolutions, Hinton House, Risley, Warrington, WA3 6AS, [3]British Nuclear Group, Berkeley Centre, Berkeley, GL13 9PB, [3]Nexia Solutions, B170, Sellafield, Seascale, Cumbria, CA20 1PG, [5]NexiaSolutions, B168, Harwell, Didcot, Oxfordshire, OX11 0QJ.

1 INTRODUCTION

Molten salt technologies have received increased attention as an alternative to aqueous nuclear fuel reprocessing.[1] In order to investigate this technology an electrorefiner was constructed and operated at Harwell and initial tests with solid and liquid cathodes have been described previously.[2] In summary the electrorefiner vessel contained 1,400g LiCl/KCl eutectic, 3,200g Cd, 65g $PuCl_3$ and smaller quantities of UCl_3, $AmCl_3$, $NdCl_3$ and $CeCl_3$. In the final part of the programme the following tests were carried out:

1) A small unstirred liquid cadmium cathode was operated at high current density ~200mA cm^{-2}. This was carried out to investigate the effect of a liquid cadmium cathode becoming fully saturated.
2) A standard liquid cadmium cathode was operated at an intermediate current density and with more frequent mixing (0.1Hz plunger) compared with previous tests (almost static) to investigate the effect of mixing on separation factors.
3) Partial decontamination of the salt using Cd/Li prior to disposal.

2 RESULTS AND DISCUSSION

2.1 Small cathode at high current density

This test was carried out with a small mass of Cd (7.9g) in a cut down alumina crucible. The crucible supports had been plasma sprayed with alumina to prevent them becoming conductors if they made contact with uranium dendrites. The original crucible and growth of dendrites are shown Figure 1. There were two immediately obvious features of this test: (i) the current at a potential of -200mV[#] vs. a U quasi reference electrode (QRE) was –450mA compared with that in the normal configuration (taller crucible) where a potential of -250mV vs. a U QRE gave a current of -220mA. Cutting down the crucible and improving the mass transport characteristics probably explains this difference i.e. a much lower IR loss in the salt.

[#] note voltages are as measured and not compensated for IR drop.

(ii) Initially the Fe wire was immersed in the Cd button. After the first deposition, dendrite growth was obvious but so also was the end of the wire, the Cd appeared to have migrated away. The dendrite growth continued up over the rim of the crucible and onto the outer sides so that it became difficult to extract the crucible through the penetration in the electrorefiner top plate. It appeared that the Cd was associated with the dendrites.

Figure 1 Small cathode before and after operation

The U/Pu ratio in the sample was 15.4 (corrected for salt) compared with their ratio in the salt of 0.58, clearly demonstrating that the sample was concentrated in U above that expected from its equilibrium separation factor in liquid Cd which is indicative of a solid electrode. The sample definitely contained Pu because the sample was enriched in Pu over Am compared with salt. If the Pu was just from the associated salt, the Pu/Am ratio would be the same as the salt. The U:Cd ratio was 1.3/1.

2.2 Effect of mixing on separation factor

In previous work with liquid cadmium cathodes, the cathode was unmixed apart from occasionally "plunging" the surface to push down deposits. It was noted previously[2] that one method of increasing the effective separation factor would be to improve the mixing at the salt cadmium interface. CRIEPI prefer a stirred liquid cadmium cathode[3] and ANL operated their plunger at about 1Hz compared with the much lower frequency at Harwell. In this test the plunger was operated at ~0.1Hz. It should be noted that in these tests the plunger area was ~20% of the liquid cadmium cathode surface area whereas ANL have operated theirs with a fractional area of 80%. The following results were obtained:

1) The average current rose from ~150mA expected at a cathode voltage of –175mV in the normal almost static conditions to an average of 195mA at the same potential (ranging from 175mA to ~240mA) an average increase of 30%. This implied that the current was mass transfer limited.

2) The Pu/Am separation factor also increased to 1.14 from 1.07 expected for that average current under near static conditions.

3) The U/Nd separation factor was 22 compared with a literature equilibrium value of 43.

4) The U/Pu separation factor at 0.1Hz plunging was 1.5 compared with the previous high current run averages of 1.3 (150mA) and ~1 (230mA) for several previous static runs. The literature equilibrium value is ~1.9.

2.3 Salt decontamination

500g Cd containing 8.6g Li in an alumina crucible was contacted with the salt in the electrorefiner at 450°C for ~120 hours. This was an attempt partially to clean the salt by chemical reduction before dissolution and treatment. There was originally 43.5g Pu dissolved in the salt (assuming 1,400g salt), this has been reduced to 14.2 g corresponding to 67% reduction of Pu. Similarly 1.11g Am originally in the salt was reduced to 0.564g or ~ 50% reduction of Am. About 50% of the Ce was reduced and probably a corresponding amount of Nd. No analysis for U was carried out. On this basis it was calculated that the Cd lump contained 29.3g Pu, 0.54g Am and ~7g each of Nd and Ce. The difference between Pu and Am could be associated with their equilibrium separation factor, but a larger difference between them and Ce would be expected. On breaking open the crucible a large mass of metal above the bulk of the Cd could be clearly seen in Figure 2 below.

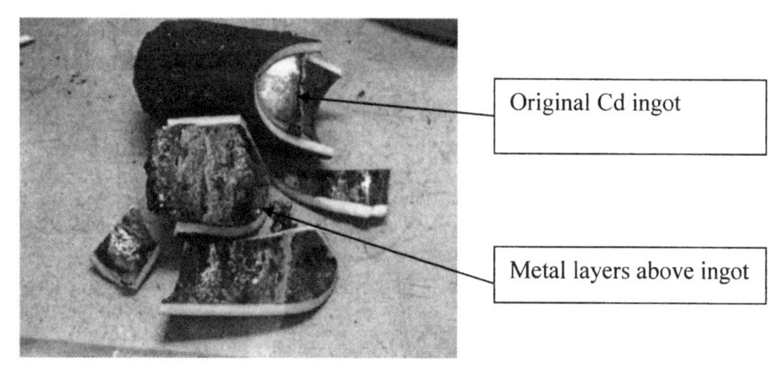

Figure 2 Crucible contents after salt clean-up test

These metallic layers were probably U, Pu and RE which had been reduced by Li but had not dissolved in the liquid Cd.

3 CONCLUSIONS

The second phase of the work showed that:

- Operation of liquid cadmium cathodes at high current densities transforms them into solid cathodes, the Cd appears to migrate out of the pool to the U dendrites.
- Mixing increases the mass transfer at the surface of the electrode and appears to improve the effective separation factors for a given current density.
- Partial salt decontamination using Li/Cd is effective, but the product would require considerable processing either for recycle or disposal.

References

1 See for example I. Johnson, *J. Nucl. Mat.* 1988, **154**, 169; C.C. McPheeters, R.D. Pierce and T.P. Mulcahey, *Prog. Nucl. Energy.*,1997, **31**, 175.

2 D. Hebditch, B. Hanson, R. Lewin, S. Beetham, J. Jenkins and H. Sims, "Electrorefining of uranium and electropartition of U, Pu, Am, Nd and Ce., *Proc. Global 2003*, 1574.

3 T. Koyama, M. Iizuka, N. Kondo, R. Fujita and H. Tanaka, *J. Nuc. Mat.*,1997, **247**, 227.

BEHAVIOUR OF FPs AND TRU IN A FLUORIDE VOLATILITY PROCESS

I.Amamoto[1,2], K.Sato[1] and T.Terai[2]

[1]Reprocessing System Engineering Group, System Engineering Technology Div. O-Arai Engineering Centre, Japan Nuclear Cycle Development Institute
[2]Dept.of Quantum Engineering and Systems Science, School of Engineering, The Univ. of Tokyo

1 INTRODUCTION

For the development of an advanced non-aqueous reprocessing system by the fluoride volatility process as part of the feasibility study on commercialisation of the fast reactor cycle system in Japan[1], it is necessary to determine the physicochemical properties of fission products (FPs) and transuranic elements (TRU) in spent mixed oxide (MOX) fuel. However experiment using such high radioactive substances, i.e., FPs and TRU, is not easy to carry out due to radiological restriction. In this paper, the separation possibility of the principal fluorides which would be formed by fluorination of FPs and TRU are estimated from their vapour pressures as well as the effects of chemical adsorbents which are quoted from publicised data of relevant literature.[2] The obtained data are then compared with known data from the recycled uranium (RU) conversion tests[3] to determine their validity.

The concept of the fluoride volatility process assumed in this study, comprises of the fluorination step and purification step as shown in Figure 1[4]. The separation between volatile and involatile substances is to be carried out at the fluorination step. Subsequently, gaseous target substance(s) are purified at the purification step with chemical traps which are filled by some adsorbents such as LiF, NaF, MgF_2, or other suitable materials.

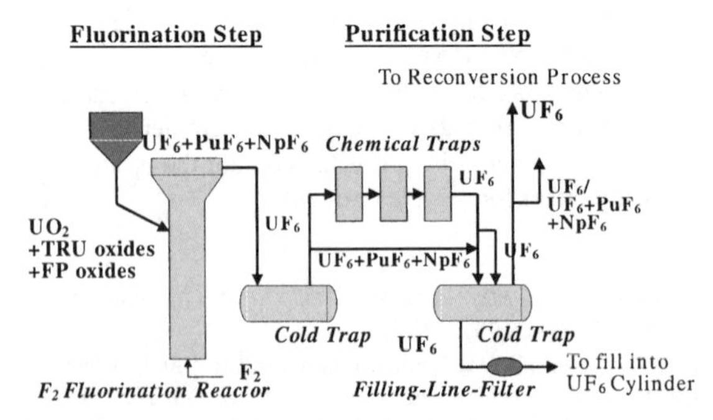

Figure 1 *Schematic Diagram of Fluoride Volatility Process*

2 METHOD AND RESULTS

2.1 Objective

The objective of this study is to evaluate the possibility of separation and purification of uranium and TRU by the fluoride volatility process. With this objective in mind, the following were undertaken.

2.2 Selection of Elements to Consider

To determine the nuclide composition in the spent MOX fast reactor fuel for reprocessing, ORIGEN2[5] were applied and 22 kinds of elements among more than 0.1wt % of whole with burnup 150GWd/t and neptunium (0.03 wt%) were selected for the comparison of vapour pressure.

2.3 Results from RU Conversion Test

The RU conversion test was carried out in Japan from 1994 to 2000 for the practical establishment on the effective use of RU. Despite the presence of very small amount of impurities in RU, some useful results were obtained eg., the behaviour of impurities in RU after 300t-U of conversion (see Table 1[3]).

Table 1 *Accumulating Point*

Element \ Investigating Point / Temperature	Fluidised-bed reactor (Fluidised medium)	CT (MgF_2)	CT (NaF)	CT (Al_2O_3)	Filling line-filter (Ni)
	703K	393 ~ 473K	363K	423K	353K
F (Fluorine)	-	-	-	◎	-
Tc (Technetium)	-	◎	-	-	-
Ru (Rutenium)	○	○	○	◎	-
U (Uranium)	◎	-	◎	-	-
Np (Neptunium)	◎	○	○	-	◎
Pu (Pultonium)	◎	○	-	-	◎
Am (Americium)	◎	-	-	-	-

CT:Chemical Trap
◎ :Main Accumulating Point
○ :Detected Point of Accumulation

2.4 Vapour Pressure Consideration

At the fluorination step, whilst each component in the spent MOX fuel is fluorinated by fluorine gas and forms a variety of fluorides, separation takes place between gaseous fluorides and other phases. It is necessary to understand the different vapour pressures of each fluoride at specific temperatures for proper separation to take place.The vapour pressures of the relevant fluorides from literature,[6-8] 20 different vapour pressures for fluorides in 13 species among selected elements, are summarised in Figure 2. Based on these figures the removal of MoF_6, RuF_5, TeF_4 and TeF_6 from UF_6 and higher valent TRU fluorides would be difficult. In the case of TeF_6, the data could not be presented here because of its higher vapour pressure in lower temperature ranges. Substances entraining UF_6 and TRU fluorides are removed at the purification step, for example, TcF_6 can be removed using MgF_2.[2] Other volatile substances can be separated by adsorbent mentioned in section 1. On the other hand, involatile fluorides would remain in the fluidised-bed reactor with the fluidised medium (Al_2O_3) or would become ash when a flame-tower type reactor is used. Most of FPs and lower valent TRU fluorides would form involatile substances.

The comparison of Figure 2 with Table 1, found most of elements' behaviour could be explained theoretically except for ruthenium compounds, presuming activities of all elements are considered as 1 for convenience.

· At the fluorination step,

* a small amount of uranium remains as involatile compounds (mainly UO_2) because of its lower reaction rate;
* some of PuF_6 and NpF_6 are decomposed or reduced easily and form involatile fluorides;
* AmF_3 is accumulated as expected.

· At the purification step,

* a small amount of UF_6 forms involatile compounds such as Na_2UF_8[9] (or $2NaF \cdot UF_6$);
* nickel, the filling line-filter material, reduces liquid PuF_6 and NpF_6 to tetrafluorides;
* TcF_6 can be removed using MgF_2.

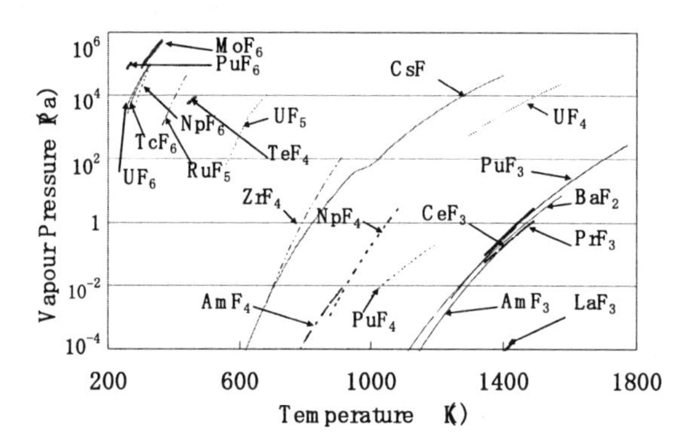

Figure 2 *Vapour Pressure of Relevent Fluorides*

3. CONCLUSION

Results based on literature investigation and RU conversion test, showed the behaviour of FPs and TRU in spent MOX fuel conformed to the theoretical estimate, *eg.* most of FPs and lower valent TRU would remain in the reactor at the fluorination step whilst some fluorides such as Ru and Tc would accompany UF_6 and volatile TRU fluorides. Measures should be taken to reduce the waste quantity of such accumlating high radioactive substances.

References

1 H.Tanaka, F.Kawamura, *et al.*, *Proc. of Global 2001 Int. Conf.*, Paris,France, 2001,054, 10-13
2 A.A.Jonke, *Reprocessing of Nuclear Reactor Fuels by Processes Based on Volatilization, Fractional Distillation, and Selective Adsorption*, Atomic Energy Rev.3(1),1965
3 I.Amamoto, T.Terai, *et al.*, *J.Nucl.Sci.Technol.*, 2002, **Suppl.3**, .769-771
4 I.Amamoto, K.Sato, Japan Pat. Appl. No.2004-224444; Fig3 [in Japanese]
5 A.G.Croff, *Nucl.Technol.*, 1983, **62**, 335,
6 Gmelin-Institut, et al (Eds.), *Gmelin Handbuch der Anorganischen Chemie (Uran Erganzungsband C8)*, 1980, Springer-Verlag, Berlin [in German]
7 O.J.Wick (Ed), *Plutonium Handbook*, Gordon & Breach, Science Publishers, Inc., N.Y.,1967
8 N.P.Galkina(Ed), *Databook on Properties of Principal Inorganic Fluorides*, Moscow Atom, Moscow 1976 [in Russian]
9 I.Amamoto, T.Terai, *et al.*, *J.Nucl.Sci.Technol.*, 2002, **Suppl.3**, .870-873

SPECIFIC FEATURES OF KINETICS AND MECHANISM OF Np(IV) OXIDATION WITH NITRIC ACID IN AQUEOUS AND TBP SOLUTIONS

V.S. Koltunov[1], R.J. Taylor[2], V.I. Marchenko[1], O.A. Savilova[1], K.N. Dvoeglazov[1], G.I. Zhuravleva[1], I.S. Denniss[2]

[1] A.A.Bochvar All-Russia Research Institute of Inorganic Materials, Moscow, Russia.
[2] Nexia Solutions, BNFL Technology Centre, Sellafield, CA20 1PG, UK.

1 INTRODUCTION

Tetravalent neptunium is known to be stable in solutions of HNO_3 at average concentrations, also in the presence of nitrous acid. However, our experiments have demonstrated the oxidation of Np(IV) proceeding at a very low concentration of H^+ ions ($<\sim$0.1M) and a relatively high temperature; this reaction being catalyzed with UO_2^{2+} ions. The final product of the reaction in aqueous solution was found to be Np(V), however, no nitric acid reduction to HNO_2 or nitrogen oxides was observed. Nitrous acid was also not found to influence the reaction as oxidation proceeds in the presence of hydrazine, i.e. in the absence of HNO_2. These data indicate that nitric and nitrous acids are not oxidants of Np(IV) which is also corroborated by experiments in hydrochloric acid solutions where under comparable conditions Np(IV) oxidizes at about the same rate as in nitric acid solutions.

2 METHOD AND RESULTS

Kinetic studies carried out by the spectrophotometric method to trace optical density variations in the Np(V) maximum range at 980 nm have revealed that the rate of the Np(IV) oxidation is adequately described by the equation:

$$d[Np(V)] / dt = k_1'[Np(IV)] + k_2'[Np(IV)]^2 \quad (1)$$

This suggests two parallel routes of the reaction which correspondingly show first and second order dependencies relative to Np(IV).

The analysis of the rate constants k_1' and k_2', determined via numerically solving equation (1), in experiments under different conditions (see Table 1), shows that the base contribution (>80%) of the observed rate of Np(IV) oxidation is made by the reaction route with the rate constant k_2'. It is evident from the table that orders of the reaction proceeding via this route are equal to 1 with respect to U(VI) concentration and –3 with respect to H^+ ions. The overall rate equation of Np(IV) oxidation therefore has the form:

$$\frac{d[Np(V)]}{dt} = k_1 \frac{[Np(IV)]}{[H^+]^2} + k_2 \frac{[Np(IV)]^2[U(VI)]}{[H^+]^3} \quad (2)$$

where $k_1 = (2.0 \pm 0.3)\times10^{-5}$ M^2min^{-1} and $k_2 = (5.50 \pm 0.47)\times10^{-2}$ $Mmin^{-1}$ at 50°C and a solution ionic strength $\mu = 0.5$. The activation energies for these two routes of the reaction were determined over the temperature range 30-50 °C and are equal to: E_1 = (148 ± 31) and E_2 = (122 ± 12) $kJmol^{-1}$, respectively.

The second order Np(IV) dependence of the reaction proceeding via its major route leads to an assumption that the limiting stage of the reaction is Np^{4+} ion disproportionation. This assumption is indirectly confirmed by the kinetic equation:

$$\frac{d[Np(V)]}{dt} = k_2'' \frac{[Np(IV)]^2}{[H^+]^3} \quad (3)$$

Which is derived on the basis of the equilibrium constant equation for the Np(IV) disproportionation reaction as well as the equation of the rate of the reverse reaction between Np(V) and Np(III)[1].

The fact that H^+ ions inhibit the reaction (with an order of -3) while UO_2^{2+} ions accelerate it indicates the participation in its slow stage of hydrolyzed Np(IV) ions, i.e. $NpOH^{3+}$ ions and complexed Np(IV)·U(VI) ions, i.e. $Np(OH)_2UO_2^{4+}$ ions. The apparent lack of Np(III) ions among the final reaction products might be explained by its high rate oxidation with UO_2^{2+} ions[2]. Thus, the reaction of Np(IV) oxidation in an aqueous HNO_3 solution might be schematically presented by the following consecutive stages:

$$Np^{4+} + H_2O \rightleftharpoons NpOH^{3+} + H^+ \quad (4)$$
$$Np^{4+} + 2H_2O \rightleftharpoons Np(OH)_2^{2+} + 2H^+ \quad (5)$$
$$Np(OH)_2^{2+} + UO_2^{2+} \rightleftharpoons Np(OH)_2UO_2^{4+} \quad (6)$$
$$NpOH^{3+} + Np(OH)_2UO_2^{4+} \rightarrow Np^{3+} + NpO_2H^{2+} + UO_2^{2+} + H_2O \quad (7)$$
$$2Np^{3+} + UO_2^{2+} + 4H^+ \rightarrow 2Np^{4+} + U^{4+} + 2H_2O \quad (8)$$
$$NpO_2H^{2+} \rightarrow NpO_2^+ + H^+ \quad (9)$$

Here stage (7) is relatively slow and all the other stages are quick.

Table 1 *Experimental Kinetic Determinations Of Second Order Rate Constants k_2' ($T=50^oC$, $\mu = 0.5$)*

$[H^+]$ M	[U(VI)] M	$[Np(IV)]_0$ 10^3, M	k_2' $M^{-1}min^{-1}$	$k_2 \cdot 10^2$ * $M \cdot min^{-1}$
0.033	0	1.0	2.0	-
0.027	0.0105	1.0	37.6	7.05
0.029	0.021	1.0	37.0	4.30
0.033	0.042	1.1	62.6	5.35
0.023	0.063	1.0	300	5.79
0.019	0.084	1.0	830	6.78
0.023	0.113	1.0	630	6.78
0.018	0.042	1.0	320	4.44
0.033	0.042	1.1	62.0	5.30
0.043	0.042	1.0	32.0	6.06
0.050	0.042	0.95	20.0	5.95
0.065	0.042	1.1	7.00	4.58
0.084	0.042	1.1	2.75	3.88
0.103	0.042	1.0	2.00	5.20
0.019	0.042	0.48	370	6.04
0.018	0.042	0.51	400	5.55
0.019	0.042	0.87	275	4.49
0.021	0.042	1.24	265	5.84
			Average	5.50 ± 0.47

* $k_2 = k_2'[H^+]^3 / [U(VI)]$

The mechanism of the minor route of the reaction with the rate constant k_1' is not clear. One might assume that hydrolyzed ions $Np(OH)_2^{2+}$ participate in the slow stage of this route of the reaction. However, the nature of the Np(IV) oxidant in the reaction remains difficult to understand.

As distinct from aqueous solution, in a solution of TBP Np(IV) oxidizes to the hexavalent state and the reaction stoichiometry is expressed via equation (10) (N.B. this is without accounting for complexation by TBP molecules)[3]:

$$Np(NO_3)_4 + HNO_3 + H_2O = NpO_2(NO_3)_2 + HNO_2 + 2HNO_3 \qquad (10)$$

Both the reaction products, viz., Np(VI) and HNO_2, play the part of autocatalysts for this reaction. Taking this fact into account the oxidation of Np(IV) is controlled by the rate equation:

$$-d[Np(IV)] / dt = k_1'[Np(IV)][HNO_3] + k_2'[Np(IV)][HNO_2] + k_3'[Np(IV)][Np(VI)] \qquad (11)$$

Or alternatively this can be expressed as:

$$Dx/dt = k_1(a-x) + k_2'(a-x)x + k_3'(a-x)x \qquad (12)$$

Which after integration leads to:

$$\ln\frac{a-x}{K+x} = \ln\frac{a}{K} - k_4'(a+K)t \qquad (13)$$

where $k_1 = k_1'[HNO_3]$, $a = [Np(IV)]_0$ is original concentration of Np(IV); $K = k_1'/k_4'$, $k_4' = k_2' + k_3'$.

The processing of the experimental kinetic data acquired under different conditions using linear equation (13) shows complete agreement with this form. The contribution of the route of the reaction that has the rate constant k_1 was found to be significant only at the initial stages of the reaction when the concentrations of the Np(VI) and HNO_2 products were low. The rate of the autocatalytic reaction route is then described by the equation[3]:

$$-\frac{d[Np(IV)]}{dt} = k_4\frac{[Np(IV)][Np(VI)][H_2O]^3}{[HNO_3]^4} \qquad (14)$$

where $k_4 = (9.5 \pm 0.8)\cdot 10^{-4}$ min^{-1} at 60.2 °C and [TBP] = 100 %. The activation energy of the reaction in this route was E = (90 ± 2) kJmol^{-1}. It has been established in this case that the major contribution to the Np(IV) oxidation rate is made by the route of the reaction that has a limiting stage of Np(V) reproportionation:

$$Np(NO_3)_4 + NpO_2(NO_3)_2 + 2H_2O \rightarrow 2NpO_2NO_3 + 4HNO_3 \qquad (15)$$

The dilution of TBP with n-dodecane results in a significant deceleration of the Np(IV) oxidation reaction due to the low rate of reaction (15) under these conditions. In this case equation (14) is observed to convert to an equation of the first order relative to Np(IV).

References

1 J.C. Hindman, J.C. Sullivan and D. Cohen, *J.Amer. Chem. Soc.*, 1958, **80**, 1812.

2 T.W. Newton, *J.Phys.Chem.*, 1970, **74**, 1655

3 V.S. Koltunov, R.J. Taylor, O.A. Savilova, G.I. Zhuravleva, I.S. Denniss and A.L. Wallwork, *Radiochimica Acta*, 1997, **76**, 45.

LANTHANIDE SALTS SOLUTIONS: REPRESENTATION OF OSMOTIC COEFFICIENTS WITHIN THE BINDING MEAN SPHERICAL APPROXIMATION

Alexandre Ruas,[1] Philippe Moisy,[1] Jean-Pierre Simonin,[2] Olivier Bernard,[2] Jean-François Dufrêche[2] and Pierre Turq[2]

[1]CEA-Valrhô Marcoule, DEN/DRCP/SCPS/LCA, Bât 399, BP 17171, 30207 Bagnols-sur-Cèze Cedex, France
[2]Laboratoire LI2C (UMR 7612), Université P.M. Curie, Boîte n° 51, 4 Place Jussieu, 75252 Paris Cedex 05, France

1 INTRODUCTION

It is generally admitted that any theory suitable for the lanthanide(III) salts may be expected to give predictions for the actinide(III) salt (such as $Am(NO_3)_3$ and $Cm(NO_3)_3$) properties, and to be a complement to their direct experimental acquisition. On the basis of this simple idea, osmotic coefficients were described by the BIMSA for aqueous binary lanthanide(III) salts solutions at 25°C (Binding Mean Spherical Approximation). Departures from ideality of different lanthanide salts binary solutions are represented.

2 THEORY

The microscopic model solved by the BIMSA consists of considering cations (respectively anions) as hard spheres of charge z+ (respectively z-) and diameter σ_+, (respectively σ_-). The model takes into account ionic association: the cations and the anions can form a pair, the proportion of associated and non-associated electrolytes being deduced from the equilibrium thermodynamic constant at zero ionic strength K.
In this work, the anion size is kept constant whereas ε^{-1}, the inverse of the relative permittivity of solution, and the diameter of the cation σ_+, were chosen as linear functions of the salt concentration: $\sigma_+^{(0)}$ being the cation diameter at infinite dilution and ε_W the relative permittivity of pure solvent, we have $\sigma_+ = \sigma_+^{(0)} + \sigma^{(1)} C_S$ and $\varepsilon^{-1} = \varepsilon_W^{-1}(1 + \alpha C_S)$
Therefore, the BIMSA theory allows, with the use of a set of data which have a microscopic meaning ($\sigma_+^{(0)}$, σ_-, z+, z-, K, $\sigma^{(1)}$, α), the computation of osmotic coefficient variation with concentration of a given binary system.
In this work we focus on aqueous binary salts solutions of lanthanide perchlorates, nitrates and chlorides. Therefore z+ = 3, z- = 1. The diameters of the anions, σ_-, were taken from previous data.[1] Then, the remaining parameters $\sigma_+^{(0)}$, K, $\sigma^{(1)}$ and α, were optimized such as to best reproduce experimental osmotic coefficient variation with concentration that can be found from previous studies.[2] This could be done with help of a least-square algorithm.[3] One of the difficulties of the procedure is that the diameter of a given cation, at infinite dilution, $\sigma_+^{(0)}$, must have a common value for all (chloride, nitrate and perchlorate) salts

containing this cation. In the next part we discuss about the used refined parameters $\sigma_+^{(0)}$, K, $\sigma^{(1)}$ and α and the calculated osmotic coefficients for all the studied binary salts.

3 RESULTS AN DISCUSSION

Since association of lanthanide perchlorates is weak compared to lanthanide chloride or nitrate,[4-5] our calculations were made by assuming that lanthanides and perchlorates do not associate (i.e., K=0). In practice a very weak association does not impact significantly the calculated osmotic coefficient variation with concentration. Figure 1 compares the experimental and calculated osmotic coefficients for two binary aqueous salts solutions, $Lu(ClO_4)_3$ which has the lowest deviation with experimental results among the studied salts, and $LaCl_3$ which has the highest deviation. Even for $LaCl_3$, the discrepancy is acceptable and comparable with experimental uncertainty.

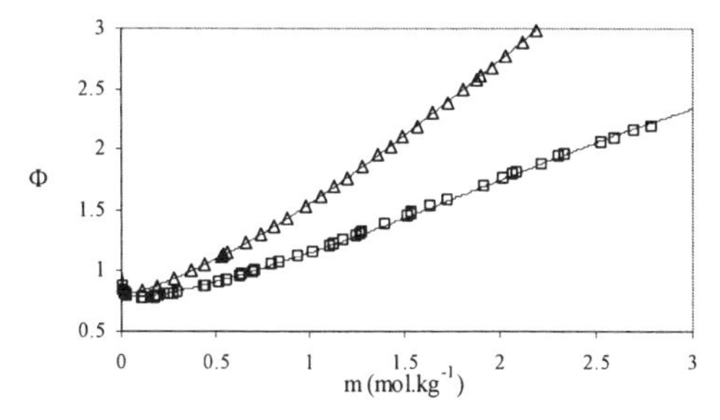

Figure 1. *Calculated (solid line) and experimental $LaCl_3$ (□) and $Lu(ClO_4)_3$ (△) osmotic coefficient variation with concentration of the binary lanthanide salt.*

The BIMSA has shown to be able to accurately reproduce the osmotic coefficient variation with concentration for all the studied lanthanide salts. Next, we discuss about the reliability of the parameters used in the theory to obtain the calculated osmotic coefficient variations. Figure 2 shows $\sigma_+^{(0)}$ values along the lanthanide series. Our $\sigma_+^{(0)}$ values are close to the values proposed by David and Fourest.[6] A significant gap is observed in the middle of the series, which may be attributed to the change in the coordination number of lanthanide +III occurring around the gadolinium element.

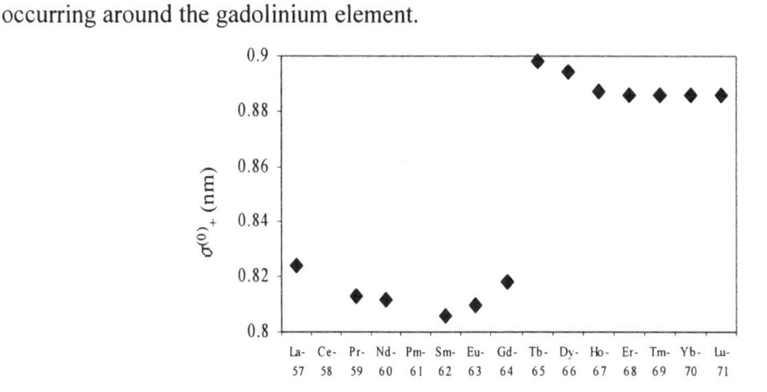

Figure 2. *Hydrated cation diameter at infinite dilution along the lanthanide series.*

In all cases, the expected conditions $\sigma^{(1)} \leq 0$ and $\alpha \geq 0$ are satisfied (the solvation number and the local solution permittivity are expected to decrease with concentration).
Figure 3 shows the dependence of our calculated association constant for lanthanide nitrates, compared to experimental values obtained by Bonal *et al*[7] from microcalorimetric measurements. Above samarium, the two curves exhibit noticeable parallel variation. Regarding lanthanide-chloride association, our calculated values can be compared with the data of Mironov *et a,l*[8] both sets of data being surprisingly close.[3]

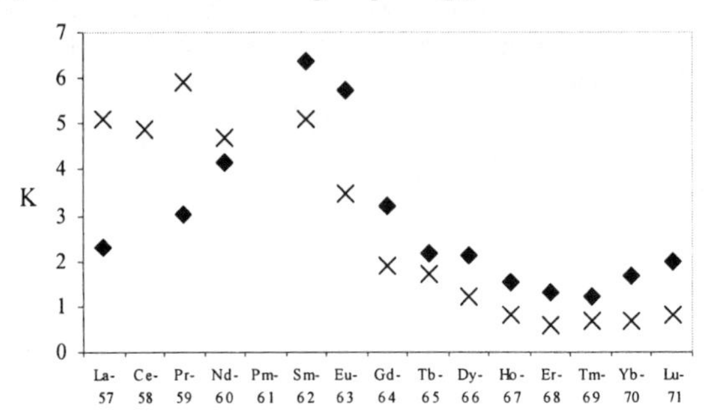

Figure 3. Variation of the lanthanide nitrate association constant at zero ionic strength along the lanthanide series. (×): experimental values,[7] (♦): our results.

4 CONCLUSION

The use of the BIMSA theory allows a good representation of a macroscopic thermodynamic property of the lanthanide salts solutions, the osmotic coefficient. All the studied salts solutions (lanthanide perchlorate, chloride and nitrate) could be treated successfully up to high concentration with the use of parameters that have microscopic physical meaning. These parameters are found to have reasonable adjusted values.
Also, in future work, we will try to use this model as a predictive tool for properties of concentrated actinide solutions that are difficult to study experimentally, because of their radiotoxicity. We hope also to modify the theory in order to take into account other microscopic phenomena, such as distinguish between inner-sphere complexes or outer-sphere ones, take into account the existence of complexes higher than 1-1 complexes or charge modification in associated species.

References

1 J.-P. Simonin, O. Bernard and L.Blum, *J. Phys. Chem. B*, 1998, **102**, 4411.
2 J.A. Rard, *J.Chem Eng. Data*, 1987, **32**, 334 and literature therein.
3 A. Ruas, Ph. Moisy, J.-P. Simonin, O. Bernard, J.F. Dufrêche and P. Turq, *J. Phys. Chem. B*, 2005, **109**, 5243.
4 J.A. Rard, *Chem. Rev.*, 1985, **85**, 555.
5 Z. Chen and C. Detellier, *J. Sol. Chem.*, 1992, **21**, 941.
6 F. H. David and B. Fourest, *New J. Chem,*. 1997, *21*, 167.
7 C. Bonal, J.-P. Morel and N. Morel-Desrosiers, *J. Chem. Soc., Faraday Trans.*, 1998, **94**, 1431.
8 V.E. Mironov, N.I. Avramenko, A.A. Koperin, V.V. Blokin, M. Yu Eike and I.D. Isayev, *Koord. Khim.*, 1982, **8**, 636.

THE REDUCTION OF Np(VI) BY ACETOHYDROXAMIC ACID IN NITRIC ACID SOLUTION

D.Y. Chung and E.H. Lee

Department of Nuclear Chemical Engineering, Korea Atomic Energy Research Institute, P.O. Box 105, Yuseong, Daejeon, 305-600, Korea

1. INTRODUCTION

BNFL is currently undertaking extensive research and development work both to enhance the current reprocessing technology and to develop further Advanced Purex processes. One major goal of this processes is the development of a flowsheet which will reduce both the cost and environmental impact of reprocessing in the future.[1] Several different methods for Np routing in the process and several new salt-free reagents to separate Pu(IV) and Np(IV) from U(VI) have been reported in the literature.[2,3,4] Formohydroxamic acid(FHA) and acetohydroxamic acid(AHA) are especially suited to the separation of Np(IV) from U(VI) by a selective formation of a hydrophilic complex with Np(IV). U(VI) extraction in to 30%TBP is unaffected. Additionally these hydroxamic acids will reduce Np(VI) to inextractable Np(V), thus allowing for the separation of Np from U.[5,6,7] The reduction of Np(VI) by FHA in 2 M HNO_3 has been reported by Colston et al.[8] A rapid reduction, taking less than a few seconds to complete, was demonstrated. The kinetics, as determined by stopped-flow spectrophotometry, have been shown to be first order with respect to [FHA] and [Np(VI)]. This paper will determine the rate constant for the reduction reaction of Np(VI) to Np(V) by AHA.

2. EXPERIMENTAL

$^{237}Np(t_{1/2}=2.14x10^6$ y) was supplied by AEA Technology. UV-Vis-NIR spectrophotometry showed that the purchased Np was present as Np(V) and Np(VI). Np(V) can be oxidized to Np(VI) by $K_2Cr_2O_7$ and, after increasing the HNO_3 concentration, extracted into 30% TBP/n-dodecane. If an aqueous stock solution is required, the extracted Np(VI) can then be stripped back from the solvent phase into 0.01M HNO_3 and the acidity adjusted. Variation in the absorbance as a function of the time were recorded in a 1 cm optical cell with a cell stirring module installed in a model 8453 Hewlett-Packard UV-VIS spectrophotometer. During the kinetic experiments the rate of the decrease of the Np(VI) concentration was followed by monitoring the increase in the concentration of Np(V), which was determined from the intensity of the absorption band at 980nm.

3. RESULTS AND DISCUSSION

Hydroxamic acids, as hydrophilic complexants for tetravalent actinides, are also reducing agents, which are thermodynamically capable of reducing Np(VI).[9-10] Further reduction of Np(V) to Np(IV) by FHA and AHA, as determined by the onset potentials,[5-6] is not expected thermodynamically. Colston *et al.* reported the reduction kinetics of Np(VI) to Np(V) by FHA in a nitric acid solution.[8] Their results indicate that the reduction of Np(VI) by FHA is particularly fast. The stoichiometry of the Np(VI)-AHA reaction was not directly studied but, by an analogy to the stoichiometric equations of the reaction between Np(VI) and FHA, it is perhaps likely that nitrogen and acetic acid are the products of an AHA oxidation. In this case, the stoichiometry of the reduction routes might be expressed by the following equations:

$$2NpO_2^{2+} + 2CH_3CONHOH \rightarrow 2NpO_2^{+} + 2CH_3COOH + N_2 + 2H^{+} \qquad (1)$$

For the case of a constant acidity, the kinetic equation for the reduction of Np(VI) by AHA, are described by:

$$-\frac{d[NpO_2^{2+}]}{dt} = k[NpO_2^{2+}][AHA] \qquad (2)$$

As seen in Figure 1, the dependency of $\log([Np(VI)]/[Np(VI)]_o)$ on the time is a linear function, which indicates that the reaction is a first-order with respect to Np(VI).

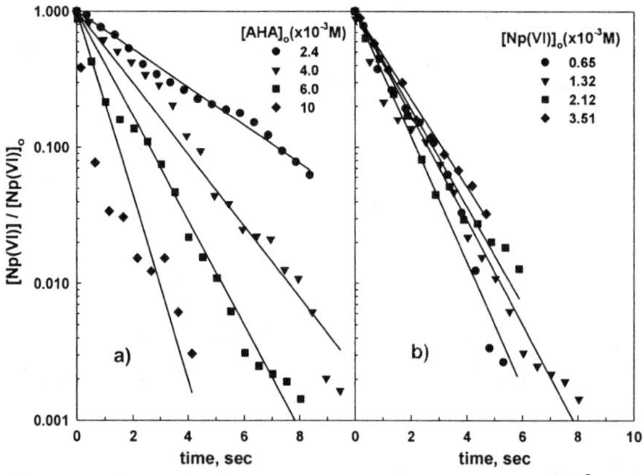

Figure 1 *Plot of $[Np(VI)/Np(VI)]_o$ vs. time.(a) at $[Np(VI)]=1.32x10^{-3}$ M, $[HNO_3]=1.0$ M, 25°C, b) at $[AHA]=6.0x10^{-3}$ M, $[HNO_3]=1.0$ M, 25°C)*

Noting that the amounts of N(VI) and AHA which have reacted at any time are equal and given by $[Np(VI)]=[Np(VI)]_o-[Np(V)]$, $[AHA]=[AHA]_o-[Np(V)]$. After an integration and rearrangement of equation (3), the final result in a number of different forms is

$$\frac{1}{[Np(VI)]_O - [AHA]_O} \ln\left(\frac{[AHA]_O}{[Np(VI)]_O} \cdot \frac{[Np(VI)]_O - [Np(V)]}{[AHA]_O - [Np(V)]} \right) = kt \qquad (3)$$

Hence, a plot of the left-hand side (2nd order function) versus t will give a straight line of slop k through the origin. Figure 2 shows that the experimental data is consistent with equation (3). Consequently, the values of k were determined under various $[Np(VI)]_o$ and $[AHA]_o$ conditions and are listed in Table 1. Also, since all the experiments were run at 25°C±1°C, k has a value of $k=191.2\pm11.2\ M^{-1}s^{-1}$ at 25°C and $[HNO_3]=1.0$ M.

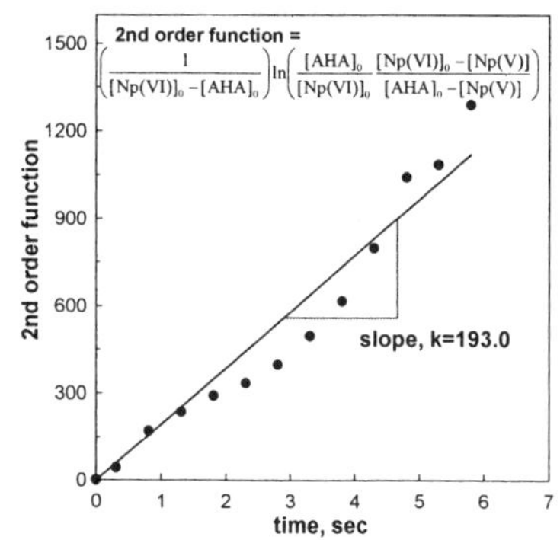

Figure 2 *The 2nd order function against time for the reduction of Np(VI) by AHA at 25℃.($[Np(VI)]_o$=6.5x10^{-4} M and $[AHA]_o$=6.0x10^{-3} M)*

Table 1 *Rate constants of Np(VI)-AHA reaction at $[HNO_3]$=1.0M and 25□.*

$[Np(VI)]_o$ (M)	$[AHA]_o$ (M)	k ($M^{-1}s^{-1}$)
1.32x10^{-3}	2.4x10^{-3}	201.8
1.32x10^{-3}	4.0x10^{-3}	191.2
1.32x10^{-3}	6.0x10^{-3}	182.4
1.32x10^{-3}	1.0x10^{-2}	173.3
6.5x10^{-4}	6.0x10^{-3}	193.0
2.12x10^{-3}	6.0x10^{-3}	187.2
3.51x10^{-3}	6.0x10^{-3}	209.3
Average		191.2±11.2

4. CONCLUSIONS

The rate equation for the reduction of Np(VI) ions by acetohydroxamic acid in nitric acid has been determined. The reduction of Np(VI) to Np(V) by AHA in nitric acid is rapid and first order with respect to both [Np(VI)] and [AHA]. The reduction of Np(V) to Np(IV) by AHA in a nitric acid media was not observed.

References

1 R.J. Taylor and I. May, *Separation Science and Technology,* 2001, **35**, 1225.
2 V.S. Koltunov, R.J. Taylor, S.M. Baranov, E.A. Mezhov and I. May, *Radiochim. Acta.,* 1999, **86**, 115.
3 V.S. Koltunov, S.M. Baranov, *Inorganica Chimica Acta***,** 1987, **140,** 31.
4 R.J. Taylor, I. May, V.S. Koltunov, S.M. Baranov, V.I. Marchenko, E.A. Mezhov, V.G. Pastuschak and O.A. Saviliva, *Radiochim. Acta*, 1998, **81**, 149.
5 R.J. Taylor and I. May, *Czech. J. Phys.*, 1999, **49/S1**, 617.
6 R.J. Taylor, *J. of Nucl. Sci. and Tech.*, 2002, **Supp.3**, 886.
7 I. May, R.J. Taylor, I.S. Denniss, G. Brown, A.L. Wallwork, N.J. Hill, J.M. Rawson and R. Less, *J. of Alloys and Compounds*, 1998, **275-277**, 769.
8 B.J. Colston, G.R.Choppin and R.J. Taylor, *Radiochim. Acta*, 2000, **88**, 329.
9 I. May, R.J. Taylor and G. Brown, *J. of Alloys and Compounds*, 1998, **271-273**, 650.
10 I. May, R.J. Taylor, I.S. Denniss, G. Brown, A.L. Wallwork, N.J. Hill, J.M. Rawson and R. Less, *J. of Alloys and Compounds*, 1998, **275-277**, 769.

PHOTOREDUCTION OF UO_2F_2 ENRICHED WITH ^{18}O IN ISOPROPANOL SOLUTION

N.G. Gorshkov, I.N. Izosimov, S.V. Kolichev, V.V. Smirnov, and N.G. Firsin

Khlopin Radium Institute, 194021 St.Petersburg, Russia. E-mail:izosimov@atom.nw.ru

1 INTRODUCTION

In this work the behavior of isotopic species $[^{16}O\text{-}U\text{-}^{16}O]^{2+}$, $[^{16}O\text{-}U\text{-}^{18}O]^{2+}$ and $[^{18}O\text{-}U\text{-}^{18}O]^{2+}$ of uranyl fluoride complexes in isopropanol solutions under UV irradiation was studied. The choice of isopropanol is explained by its ability to form complexes with UO_2F_2 contrary to methanol and ethanol.[1-2] The presence of isopropanol molecules in the first coordination sphere of the uranyl ion can affect the photochemical behavior of that ion. We compared the rates of decreasing content of various isotopic species of uranyl ion in isopropanol solution of UO_2F_2.

2 EXPERIMENTAL AND DATA ANALYSIS

The solutions were irradiated by a pulse xenon lamp with the use of a UV filter in the range 300-400 nm. The solution composition was as follows: isopropanol 19 ml, water 0.9 ml, hydrofluoric acid 0.1 ml, and uranyl fluoride, concentration approximately 1mg/ml. The absorption spectra were recorded on a Shimadzu UV-3101 PC spectrophotometer. The IR spectra were recorded on a Shimadzu FTIR-8700 spectrometer. Introduction of ^{18}O into the uranyl slightly affects the solution extinction coefficients: there are only isotopic shifts of the electronic vibration bands corresponding to stretching modes of the uranyl group. These shifts do not exceed 1 nm. The spectra are shown in Fig. 1.

The fraction of the uranyl groups containing one or two ^{18}O atoms was determined from the area of the asymmetric stretching vibrations band (as) corresponding to the isotopic species of uranyl ion transferred into the form of $NaUO_2(CH_3COO)_3$ and placed into a KBr pellet [3]. The IR spectra are shown in Fig. 2. Example of variation in the fractions of all three isotopic species of uranyl as a result of irradiation of the solution is presented in Table 1. Results of the analysis of the IR spectra of $NaUO_2(CH_3COO)_3$ samples are presented in the Table 2.

Data listed in Table 2 indicate that the photoreduction yield for uranyl complexes containing asymmetric $[^{16}O\text{-}U\text{-}^{18}O]^{2+}$ isotopic species is higher than that for uranyl containing symmetric $[^{18}O\text{-}U\text{-}^{18}O]^{2+}$ species. The difference in S_{a1}/S_{a2} ratios at 3σ level (Table 2) before and after irradiation was observed in 3 runs, at 2σ level, in 3 runs.

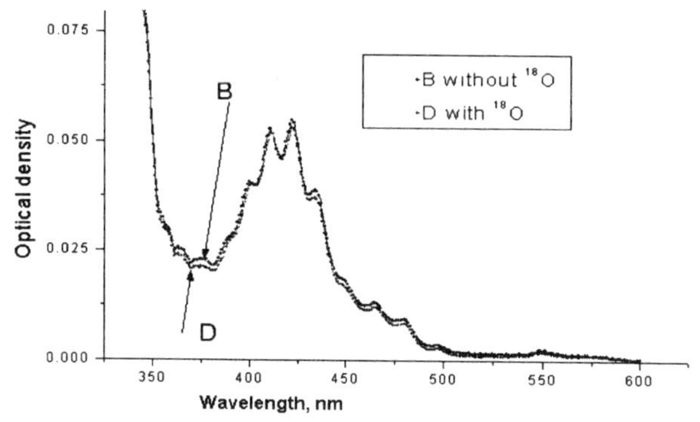

Figure 1. *Absorption spectra of UO_2F_2 in isopropanol. (B) Nonlabeled sample, (D) ^{18}O content in the uranyl group 42 at%. Uranium concentration, μg·ml^{-1}: (B) 762 and (D) 753.*

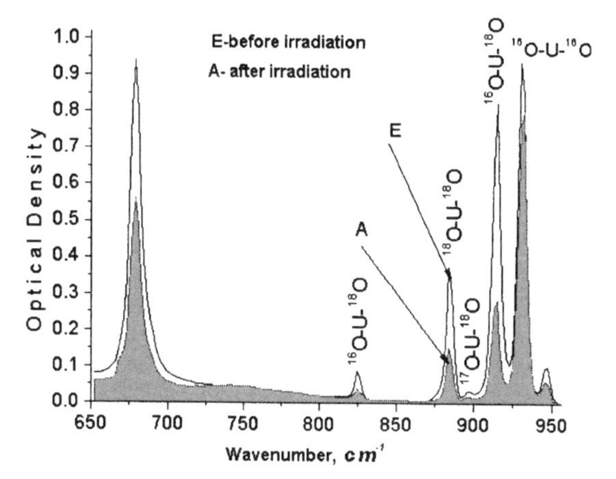

Figure 2. *The IR spectrum of $NaUO_2(CH_3COO)_3$ recovered from a solution of UO_2F_2 in isopropanol. The content of ^{18}O in the initial UO_2F_2 is 42 % (E).*

Table 1. *Concentrations (mg/l) of uranyl isotopic species as a result of UV irradiation*

Exposition	*[A+B+C]**	*[A]***	*[B]***	*[C]***
0 Xe lamp flash	635.7±0.5	214±2	310±3	112±1
45 Xe lamp flashes	531.9±0.5	251±2	195±2	86±1

* Total uranium concentration determined with the use of arsenazo III.

** Concentrations evaluated from the IR spectra. A, B, and C are concentrations of [^{16}O-U-^{16}O], [^{16}O-U-^{18}O] and [^{18}O-U-^{18}O] species, respectively.

3 DISCUSSION AND CONCLUSION

The content of [^{16}O-U-^{18}O]$^{2+}$ and [^{18}O-U-^{18}O]$^{2+}$ species in the solution can decrease as a result of both reduction of uranyl and oxygen exchange between uranyl ion and water. At the chosen concentration of H^+ we can expect exchange of both one and two oxygen atoms of the uranyl group with water.[4] In the case of simultaneous exchange of two oxygen atoms the concentrations of both isotopic species should decrease with the same rate. In the case of exchange of one oxygen atom the concentration of [^{16}O-U-^{18}O] species should decrease at a lower rate than that of [^{18}O-U-^{18}O] species. Thus, the existing conceptions on oxygen exchange in the uranyl ion do not explain the observed effect. The second process resulting in decreasing content of ^{18}O in UO_2F_2 is reduction of UO_2^{2+} to U(IV) with precipitation of insoluble UF_4. Our preliminary experiments on photo induced reduction of uranyl fluoride enriched with ^{18}O and uranyl fluoride with natural isotopic composition of oxygen did not elucidate a noticeable difference in the rate of these processes. Previously, the photochemical behavior of uranyl fluoride enriched with ^{18}O in methanol solution was studied.[5] No differences were found in the quantum yield of photo induced reduction for different isotopic species of uranyl in light irradiation of the solutions by xenon pulse lamp and laser with various light wavelengths, while a decrease in the total content of ^{18}O in UO_2F_2 was observed.[5] It is the authors' opinion that under conditions of complete suppression of oxygen exchange the decreasing content of ^{18}O in UO_2F_2 in photo induced reduction is caused by the difference in the extinction coefficients of UO_2F_2 solutions with different isotopic composition of oxygen. As seen from Fig. 1, the absorption of a UO_2F_2 solution insignificantly varies in enrichment of uranyl groups with ^{18}O; therefore, the difference in the reduction rates cannot be explained by the difference in the coefficients of absorption of isotopic species of uranyl and we must conclude that there are other reasons for accelerated decay of the [^{16}O-U-^{18}O]$^{2+}$ groups in a solution. It is well known that uranyl fluoride forms dimers in both aqueous and nonaqueous solutions and the constant of dimerization is fairly high.[6-7] It would appear reasonable that the light quanta are absorbed by dimers. Since as a result of excitation only one UO_2^{2+} group is reduced, the excitation energy is localized on one moiety forming a dimer. Our experiments suggest that when the dimer contains the [^{16}O-U-^{18}O]$^{2+}$ group, this group is mainly involved in photo induced reduction. This can be one of the possible explanations of the observed phenomenon.

Table 2. *Intensities (errors at 1σ level) of the bands in the IR spectrum of $NaUO_2(CH_3COO)_3$.*

Sample No	*content of ^{18}O(%)*	*S_{a1} ($\nu_{as}^{16,18}$) Before irradiation*	*S_{a2} ($\nu_{as}^{18,18}$) Before irradiation*	*S_{a1} ($\nu_{as}^{16,18}$) After irradiation*	*S_{a2} ($\nu_{as}^{18,18}$) After irradiation*	*S_{a1}/ S_{a2} Before irradiation*	*S_{a1}/ S_{a2} After irradiation*
1	42%	2.67±0.02	1.22±0.02	1.38±0.02	0.76±0.02	2.19±0.04	1.82±0.06
2	17%	0.63±0.02	0.31±0.02	0.24±0.02	0.16±0.02	2.0±0.1	1.5±0.2
3	42%	2.21±0.02	1.01±0.02	1.74±0.02	0.97±0.02	2.19±0.05	1.80±0.04
4	38%	4.15.±0.02	1.64±0.02	1.65±0.02	0.71±0.02	2.53±0.03	2.32±0.07
5*	38%	4.15±0.02	1.64±0.02	1.86±0.02	0.76±0.02	2.53±0.03	2.45±0.07
6*	38%	4.15±0.02	1.64±0.02	0.65±0.02	0.27±0.02	2.53±0.03	2.4±0.2
7	38%	4.15±0.02	1.64±0.02	1.28±0.02	0.59±0.02	2.53±0.03	2.17±0.08
8	38%	4.15±0.02	1.64±0.02	2.26±0.02	0.96±0.02	2.53±0.03	2.35±0.05
9	38%	4.15±0.02	1.64±0.02	2.56±0.02	1.08±0.02	2.53±0.03	2.37±0.05

*dilute solutions of uranyl fluoride in isopropanol

References

1 V.A.Mikhalev, D.N. Suglobov, Radiokhimiya, 1992, **34**, 1.
2 V.A.Mikhalev, D.N.Suglobov, Radiokhimiya, 1993, **35**, 1.
3 S.A.Gaziev, et.al., Inorg. Chim. Acta,1987, **139**, 345.
4 S.A Gaziev, et.al., Radiokhimiya, 1984, **27**, 316.
5 G.L.DePoorter, C.K. Rofer-DePoorter, J. Inorg. Nucl. Chem., 1978, **40**, 2049.
6 J.S. Johnson, K.A.Kraus, T.E. Young, J. Am. Chem. Soc., 1954, **74**, 1436.
7 V.M. Vdovenko, et al., Radiohimiya, 1967, **12**, 2863.

ELECTROREFINING OF PLUTONIUM IN CALCIUM CHLORIDE

A. Jones, T. Paget, I. Sullivan and R. Watson

AWE, Aldermaston, Reading, RG7 4PR, UK

1 INTRODUCTION

Electrorefining (ER) is a long established technique for the purification of plutonium.[1] In this process a potential is applied between a liquid plutonium anode and a tungsten cathode with equimolar NaCl-KCl as the electrolyte. Plutonium is oxidised at the anode, transported through the salt and reduced and collected at the cathode.

The product metal can contain Na and K as well as NaCl and KCl and needs to be vacuum cast to remove these volatile salts and metals before it can be used. It was realised that by changing the salt to $CaCl_2$, the need to vacuum cast the ER product could be eliminated. $CaCl_2$ has a low vapour pressure at 850°C and calcium metal is unlikely to be formed as the reduction potentials of Pu^{3+} and Ca^{2+} are well separated. Changing from NaCl-KCl to $CaCl_2$ necessitated increasing the process temperature from 750 to 850°C because of the higher melting point of $CaCl_2$ (782°C).

2 EXPERIMENTAL

The process is carried out at approximately 850 °C in an alumina ceramic crucible with an alumina stirrer and tungsten electrodes. Prior to electrorefining, the tungsten cylinder cathode was made anodic and a current of 15A was passed through the cell for 5 minutes with stirring. This reversed polarity mode was used to clean the cathode surface and strip impurities from the salt. During electrorefining, polarity was returned to convention and a current of 25 amps (j_{anode} = 0.7 A.cm^{-2}, $j_{cathode}$ = 0.06 A.cm^{-2}) was passed for 40 to 50 hours at a potential of 1.5 to 4V, with stirring. During this time the applied potential was interrupted periodically and the open circuit potential or back emf (bemf) of the cell measured and recorded. When the bemf exceeded a predetermined value, the process was terminated.

Micro sections of product plutonium were viewed on a Reichert Polyvar microscope. The micro-hardness measurements were taken with a Vickers Micro Hardness Indenter at 100 g load.

3 RESULTS

3.1 Successful Runs

Fourteen trial runs have been completed using $CaCl_2$ as the solvent salt. Of these, 10 were successful (Table 1), producing well formed high density metal with no loose material. The average yield of the successful runs was 89.9% compared with 74.7% in NaCl-KCl. The high current efficiencies at the anode and cathode indicate there are no significant side reactions taking place. Vacuum casting of the product showed there was no volatile material to be removed and no salt entrained in the metal.

Table 1 *Results of successful electrorefining runs in $CaCl_2$*

Run No	ER Yield %	Current efficiency (% of theoretical)	
		Anode	Cathode
1	89.6	93.3	87.5
2	90.9	96.3	89.6
3	90.4	99.0	91.2
4	88.9	101.0	90.9
5	89.9	98.3	93.6
6	88.8	92.6	87.5
7	94.6	91.9	85.5
8	85.1	85.2	77.4
9	88.7	92.9	83.2
10	91.8	96.6	88.9

The process operates with a low bemf throughout; there is a rapid increase at the end, with the process being terminated on high bemf (>0.4V). This is similar to the behaviour of the process in NaCl-KCl. Assay of the spent anodes show the composition is approximately 44mol% Ga. At this composition the anode has solidified at 850°C. However, there is no evidence of Ga transfer to the product.

Chemical analysis shows the ER product is high purity, and vacuum casting has no effect. With the exception of Am and Np, the degree of impurity removal is broadly similar to that found in NaCl-KCl. Removal of Am and Np is reported to be 93% and 55% respectively in $CaCl_2$.[2] This compares with 70% and 87% in NaCl-KCl. The difference is caused by the relative position of the reduction potentials of Pu, Am and Np differing in NaCl-KCl and in $CaCl_2$. Phase diagram studies suggest the formation of complexes of the type K_3PuCl_6 and K_2PuCl_5.[3] We are not aware of evidence of analogus Np and Am compounds. In LiCl, NaCl and $CaCl_2$, phased diagram studies do not indicate compound formation with $PuCl_3$.[4] This suggests the change in the relative values of the reduction potentials is due to the stabilisation of plutonium chloro-complexes by K^+.[5] A further factor may be the change in temperature.

3.2 Low Product Density Runs

Four runs using $CaCl_2$ have resulted in reduced process yields and low product metal density. Both the anode and cathode current efficiencies are lower than when good metal is produced, showing that significant side reactions are taking place. The back emf was high throughout the run. This is indicative of depletion of plutonium at the anode surface, possibly caused by the presence of solid phases. All these poor runs occurred when the feed metal had dross on its upper surface.

Table 3 *Results of less successful electrorefining runs in $CaCl_2$*

Run No	ER Yield %	Current Efficiency (% of theoretical)	
		Anode	Cathode
11	67.3	76.4	62.0
12	66.9	82.2	68.0
13	87.0	89.9	78.5
14	81.2	83.5	71.7

Typically the product in these runs is poorly formed, in particular on the outer surface, where it covers a layer of better formed metal. This product was vacuum cast and showed a poorly formed surface layer. Analysis showed the bulk metal was high purity plutonium but that the surface contained significant amounts of calcium and chlorine, presumably as $CaCl_2$ and possibly Ca metal.

It is proposed the presence of dross on the anode surface disrupts the refining process. At some point oxidation of plutonium at the anode is inhibited and the salt becomes depleted in Pu^{3+}. Reduction of Ca^{2+} starts at the cathode, and the presence of calcium results in poorly formed plutonium. In two of the four runs, assay of the used salt showed it was severely depleted in plutonium; physically the salt was a fine crystalline powder, usually associated with pure $CaCl_2$, rather than the hard salt mass produced in other runs.

Similar feed metal to that used in these poor runs has been recast to remove the dross and successfully processed at high yield, producing high density product.

4 CONCLUSION

Electrorefining in $CaCl_2$ results in a high purity product at high yield. No volatiles are present, the vacuum casting process can be eliminated. Dross on the surface of the feed metal results in a reduced yield of low density metal. The degree of americium and neptunium removal changes with the solvent salt. This is believed to be caused by the greater stabilisation of plutonium chloro-species in NaCl-KCl.

References

1 J.L. Willit, W.E. Miller and J.E. Battles, *J. Nucl. Mater.*, 1992, **195**, 229.

2 S. Owens, K. Axler, G. Bird, M. Reimus and G. DePoorter, *Proceedings of the Eigth International Symposium on Molten Salts* eds., RJ Gale, G Blomgren and H Kojima, The Electrochemical Society, Pennington, 1992, pp. 204-214.

3 R. Benz, M. Kahn and J.A. Leary, *J. Phys. Chem.*, 1959, **63**, 1983.

4 C.W. Bjorklund, J.G. Reavis, J.A. Leary and K.A. Walsh, *J. Phys. Chem.*, 1959, **63**, 1774.
KWR Johnson, M Khan and JA Leary, *J. Phys. Chem.*, 1961, **65**, 2226.

5 A.H. Jones, T.J. Paget and R.F. Watson, *Proceedings of the Fourteenth International Symposium on Molten Salts*, The Electrochemical Society, Pennington, *in press.*

SEPARATION OF MINOR ACTINIDES AND LANTHANIDES FROM NITRIC ACID SOLUTION BY R-BTP EXTRACTION RESIN

H. Hoshi,[1] Y.-Z. Wei,[1] M. Kumagai,[1] T. Asakura[2] and Y. Morita[2]

[1]Nuclear Chemistry and Chemical Engineering Center, Institute of Research and Innovation, Kashiwa, Chiba, Japan. E-mail: hoshi@iri.or.jp
[2]Process Safety Lab., Dep. of Fuel Cycle Safety Research, Safety Research Center, Tokai

1 INTRODUCTION

For nuclear fuel cycle development one of the most important tasks is to improve reprocessing processes, making them both more economical and efficient.[1] For the establishment of a Fast Breeder Reactor (FBR) cycle system for the future, it is strongly desirable to develop a new reprocessing process which uses more compact equipment and produces less radioactive wastes compared to the present PUREX process. For this purpose, we have proposed a novel aqueous reprocessing system, the ERIX Process (The Electrolytic Reduction and Ion Exchange Process for Reprocessing Spent FBR-MOX Fuel) to treat spent FBR-MOX fuels. This process consists of (1) Pd removal by selective adsorption using a specific anion exchanger; (2) electrolytic reduction for the valence adjustment of the major actinides including U, Pu, Np and some fission products (FP) such as Tc and Ru; (3) anion exchange separation for the recovery of U, Pu and Np using a new type of anion exchanger, AR-01; and (4) selective separation of long-lived minor actinides (MA = Am and Cm) by extraction chromatography.[2-3]

Recently, excellent selectivity for MA(III) over Ln(III) has been found for some extractants containing soft donors, such as S or N.[4-6] Kolarik *et al.* reported that a new N-donor ligand, 2,6-bis(5,6-dialkyl-1,2,4-triazine-3-yl)-pyridine (R-BTP), shows high selectivity for MA (III) over Ln(III).[7]

This work is focused on the recent study on mutual separation of MA(III) and Ln(III) from nitric acid solution using the silica-based extraction resin containing the newly developed chelating extractant, R-BTP.

2 METHOD AND RESULTS

2.1 Preparation of silica-based extraction resin

Spherical silica particles with a diameter of 40-60 μm, a mean pore size of 600 nm and a pore fraction of 0.69 were used. As the extractant support, an inert copolymer of formylstyrene and divinylbenzene was synthesized and embedded in the pores of the silica particles. The preparation method of the polymer-immobilized silica particles (SiO_2-P) was reported previously.[8] The extractant, 2,6-bis(5,6-diisobutyl-1,2,4-triazine-3-yl)-pyridine (*iso*-Bu-BTP), was impregnated into the SiO_2-P particles. Figure 1 shows the chemical structure of extractant. The impregnation procedures of the extractant into the SiO_2-P were described in a previous report.[9]

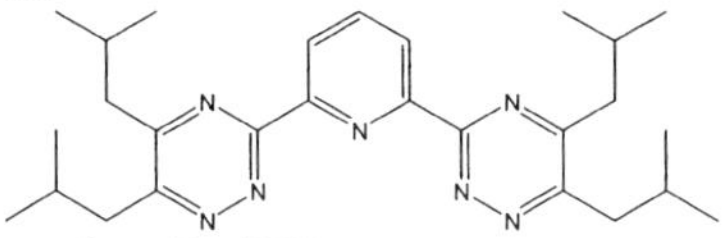

Figure 1 *Chemical structure of iso-Bu-BTP*

2.2 Adsorption experiment

All distribution coefficients (K_d) were measured by batch experiments. A definite amount of extraction resin (0.25 g) was combined in a glass vial with screw cap with a measured volume (5 cm^3) of an aqueous solution. The aqueous solution contained 1 mM Ce(III), Nd(III) and Gd(III) and was spiked with ^{243}Am. The glass vial was packed in a vinyl bag and was shaken mechanically for 3 hours in a water bath at 25°C. The aqueous phase was filtrated through a membrane filter with 0.45 μm pore. The gamma activity was measured by gamma spectrometry (GMX-20190-P; ORTEC). The concentration of metal was determined by ICP-AES (ICPS-1000; Shimadzu) in a glove box. The distribution coefficient was calculated by:

$$K_d = \frac{A_0 - A_S}{A_S} \times \frac{V}{W} \text{ or } \frac{C_0 - C_S}{C_S} \times \frac{V}{W} \qquad (1)$$

where A and C denote activity and the concentration of metal in the aqueous phase, respectively. The subscript 0 and S indicate before and after adsorption, respectively. V indicates the volume of aqueous phase and W is the weight of the dry extraction resin.

2.3 Adsorption from sodium nitrate solution

Adsorption kinetics of Am and Ln from sodium nitrate solution (3 M $NaNO_3$ + 0.1 M HNO_3) was examined. The distribution coefficients of Am were over 10^4 after 1 hour and reached equilibrium in about 2 hours. On the other hand, the distribution coefficients of Ln slightly increased after 4 hours. When using the *n*-butyl substituted compound, both Am and Ln reached equilibrium within 2 hours.[9] Thus, the *iso*butyl substituted BTP-resin has slower kinetics than the *n*-butyl analogue.

2.4 Adsorption from nitric acid solution

As mentioned above, *iso*-Bu-BTP/SiO_2-P has high affinity for Am from sodium nitrate solution. However, it is desired that MA are removed by a salt-free process. Thus, adsorption of Am and Ln from nitric acid solution was examined. The distribution coefficients of Am and Ln are plotted as a function of nitric acid concentration (Figure 2). The *iso*-Bu-BTP/SiO_2-P indicated quite high affinity for Am from nitric acid solution. The distribution coefficients of Am were over 10^4 cm^3/g from 1 - 4 M nitric acid solution and reached equilibrium after 2 hours. Both Ce(III) and Nd(III) also reached equilibrium at 2 hours, while Gd(III) was slightly slow and reached equilibrium at 3 hours. On the other hand, the distribution coefficients of Gd(III) from nitric acid solution were higher than that from sodium nitrate solution. Therefore, separation factor between Am and Gd were less than 100. However, in this ERIX Process, MA are separated by packed column and this separation factor is considered high enough for mutual separation of MA and Ln. This impregnated resin has great potential in the development of a salt-free separation process.

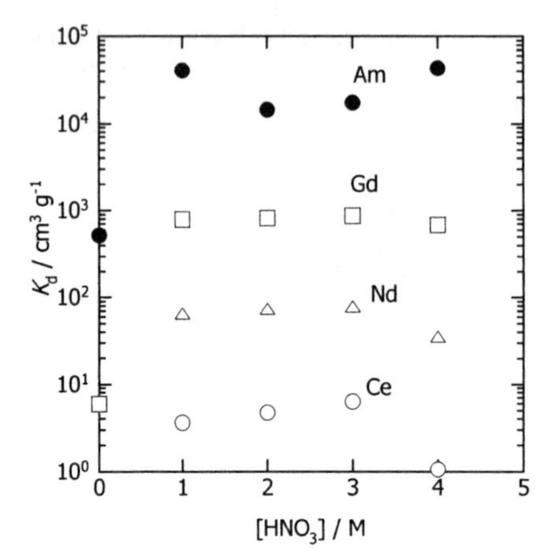

Figure 2 *Adsorption onto iso-Bu-BTP/SiO$_2$-P from nitric acid solution.*

3 CONCLUSION

A novel silica-based extraction resin impregnated with *iso*-Bu-BTP was prepared. It has high adsorption affinity for Am(III) not only in sodium nitrate solution but also in concentrated nitric acid medium. The distribution coefficient of Am(III) was over 10^4 cm^3/g from nitric acid solution. This impregnated resin looks to be a promising adsorbent for the develop of a salt-free separation process.

4 ACNOWLEDGEMENT

This work was financed by the Ministry of Education, Culture, Sports, Science and Technology of Japan (MEXT) under the framework of "The Development of Innovative Nuclear Technologies".

References

1 Y.-Z. Wei, T. Arai, M. Kumagai, A. Bruggeman, L. Vos, *Proc. of 13th Pacific Basin Nuclear Conference (PBNC-2002)*, Shenzen, China, October 21 - 25, 2002.
2 H. Hoshi, Y.-Z. Wei, M. Kumagai, T. Asakura, Y. Morita, *J. Alloy. Compd.* 2004, **374**, 451.
3 Y.-Z. Wei, H. Hoshi, M. Kumagai, T. Asakura, Y. Morita, *J. Alloy. Compd.*, 2004, **374**, 447.
4 Y. Zhu, *Radiochimica Acta*, 1995, **68**, 95.
5 C. Madic, M. J. Hudson, *Nuclear Science and Technology*, EUR 18038, 1998, 140.
6 M. J. Hudson, M. R. Foreman, C. Hill, N. Huet, C. Madic, *Solv. Extr. Ion Exch.*, 2003, **21**, 637.
7 Z. Kolarik, U. Müllich, F. Gassner, *Solv. Extr. Ion Exch.* 1999, **17**, 23.
8 Y.-Z. Wei, M. Kumagai, Y. Takashima, G. Modolo, R. Odoj, *Nucl. Technol.*, 2000, **132,** 413.
9 H. Hoshi, Y.-Z. Wei, M. Kumagai, T. Asakura, Y. Morita, *Proc. of Rare Earth'04*, Nara, Japan, November 7 - 12, 2004.

UC ELECTROCHEMICAL BEHAVIOUR IN AQUEOUS SOLUTION

A. Maslennikov,[1] N. Boudanova,[1] B. Fourest,[2] Ph. Moisy[3] and M. Lecomte[3]

[1]Institute of Physical Chemistry of Russian Academy of Sciences, Moscow, Russia
[2]Institute of Nuclear Physics, Orsay, France
[3]Research Center VALRHO, France

1 INTRODUCTION

Uranium monocarbide (UC) or mixed carbide, (U,Pu)C, may find application as high temperature gas cooled fuel. If aqueous flowsheet will be accepted for the carbide fuel reprocessing, the development of the process of irradiated UC (or (U,Pu)C) dissolution would become of great importance. The recent analysis of the data on UC dissolution in different aqueous electrolytes[1] indicated that even in strong oxidative conditions (15 M HNO_3) UC dissolution was not quantitative. Oxidation of UC to U(VI) was also accompanied by the accumulation of a number of organic species.[1-3] The possibility of increasing the yield of UC dissolution and destruction of accumulated organic species by application of electrochemical techniques in presence of mediators Ag(II) and Ce(IV) was previously demonstrated.[4-5] However, the development of a reliable technique for UC dissolution requires fundamental knowledge on the electrochemical properties of this compound.

The present study was aimed at an electrochemical investigation of UC dissolution in 0.5-6.0 M HNO_3, and 0.1-4.0 M NaOH. The essential parameters of the dissolution processes – dissolution potentials, estimates for the dissolution rate constants and current efficiency (CE) of the process were determined.

2 METHOD AND RESULTS

UC was fabricated by induction melting followed by quenching to form a microsphere of 4 mm diameter. The characterization of the sample was carried out using XRD and SEM. The microsphere was pressed into a Teflon tip, connected to EDI101T rotating disk electrode (RDE) assembly (Radiometer). Before the electrochemical tests the UC electrode was polished electrochemically in 1 M Na_2CO_3 solution. A UC working semi-spherical electrode, Hg/Hg_2Cl_2 reference electrode and Pt wire counter electrode comprised the conventional three electrode cell for the electrochemical measurements. Cyclic voltammetry (CV) and multistep potential sweep coulometry (MPSC) were used to determine the principal UC electrochemical characteristics. All the potentials in the present study are referred to a Hg/Hg_2Cl_2 electrode.

CV curves indicate three ranges of potentials, corresponding to different processes, occurring at the UC electrode in 0.5-6.0 HNO_3. In the range from 0 to 400 mV the UC electrode was found to be in a passive state. This state corresponds to the UC surface oxidation with formation of the uranium oxycarbide ($UC_{1-x}O_x$) protective layer. Between 500 and 1300 mV the pseudopassivation of the UC electrode was observed. The latter state was associated with simultaneous occurrence of two electrode reactions: $UC_{1-x}O_x$ accumulation at the electrode surface and its further oxidation with formation of soluble products. The increase of current density in the pseudopassivation range of potentials with the increase of the HNO_3 concentration in the electrolyte, indicates the simultaneous acceleration of both ($UC_{1-x}O_x$) formation and its further oxidation. At potentials exceeding 1300 mV intensive dissolution of UC was observed. The anodic reaction was accompanied with a significant increase of the interface electrode-electrolyte due to the pitting dissolution mechanism. Therefore, during reverse potential scan the current densities at a fixed potential value exceeded significantly those, observed during direct potential scan. The peculiar electrochemical UC behavior in HNO_3 solutions with concentrations more than 4.0 M was associated with the NO_3^- ions and HNO_2 reduction at the potentials $E<700$ mV. However, the increase of HNO_3 concentration to 6.0 M did not shift the potential of UC dissolution to more positive values. The addition of F^- ions able to form stable complexes with the products of UC oxidation, U(IV) and U(VI), resulted in the increase of UC dissolution rate in the narrow range of potentials from 800 to 1100 mV. The further increase of potential was accompanied by the decrease of anodic current, apparently due to the formation of the U(VI) oxyfluoride of comparatively low solubility at the UC electrode surface (so-called secondary passivation).

In 0.1-4.0 M NaOH the UC electrode stays in passive state at the potentials up to 400 mV. The rapid increase of anodic current density was observed at the potentials exceeding 400 mV up to the peak value at 1200 mV. The peak potential shifted slightly to more positive potentials with the increase of NaOH concentration. The further potential increase to 1500 mV resulted in the decrease of current density, more pronounced in diluted NaOH. A gas evolution at the electrode was observed at the potentials exceeding 800 mV. The current density increase may be associated either with UC dissolution accompanied with CO_2 formation or with O_2 evolution at the electrode. Addition of 0.02 M F^- to the alkaline electrolyte caused the decrease of the anodic current densities in the range from 500 to 1500 mV. This decrease may be associated with the increase of the passive layer stability in presence of F^- ions.

The estimates of UC electrochemical dissolution rates using CV data were made on the assumption that U(VI) and CO_2 had been the final product of UC oxidation in aqueous solutions. The calculations provided the values of 55-75 $mg \cdot cm^{-2}h^{-1}$ in 2.0-4.0 M HNO_3 and 140-190 $mg \cdot cm^{-2}h^{-1}$ in 2.0-4.0 M NaOH.

MPSC measurements were carried out in 0.5-6.0 M HNO_3 and in 0.1-4.0 M NaOH in the range of potentials from $E_{i=0}$ to 1500 mV. They included registration of I-t curves at a constant potential E during 1 min in the range from the open curcuit potential $E_{I=0}$ to 1500 mV, integration of I-t curve to obtain the charge passed through the cell, followed by the positive shift of the applied potential with an increment of 50 mV. The analysis of MPSC data showed, that in the range of potentials corresponding to the UC anodic dissolution, I-t curves in both HNO_3 and NaOH were characterized by a minimum, observed at t=5-10 s, followed by the increase of the current density to a constant value. The observed current density drop may be associated with the immediate increase of the surface film resistance after application of anodic potential pulse, followed by its dissolution with the rate proportional to the applied potential.

The charge values (Q, mC), obtained by integration of corresponding I-t curves, were summarized for all the range of potentials of MPSC measurements and the uranium concentrations in the resulting electrolyte were calculated on assumption that all the uranium present on electrochemical dissolution is oxidized to U(VI). These data were compared to the U(VI) concentrations, determined using laser induced fluorescence technique (LIF). The CE values in HNO_3 solutions were greater than 100% in all the studied interval of the acid concentration and exceeded 200% in 4.0-6.0 M HNO_3. This overestimation may be accounted for through the incomplete oxidation of UC through to U(VI) during the measurement at low (<1 M) HNO_3 concentration and for the increasing role of chemical oxidation in HNO_3 solutions with concentration exceeding 4.0 M. The CE values for the UC electrochemical dissolution in NaOH did not exceed 15%, indicating the formation of low-soluble UC oxidation products, apparently mono- or polyuranates. Therefore, the significant current densities observed at UC electrode in NaOH solutions at the potentials between 800 and 1500 mV are associated with the water decomposition at the electrode surface rather than UC anodic dissolution.

3 CONCLUSIONS.

The investigation of UC electrochemical dissolution in HNO_3 and NaOH showed, that the mechanism of the electrode reactions, including effect of complex forming ions (F^-) in the studied electrolyte, was completely different. The effective dissolution of UC in HNO_3 was observed at the UC potentials exceeding 1500 mV (dissolution rates 55-75 $mg \cdot cm^{-2}h^{-1}$) with formation of U(VI). The contribution of UC chemical oxidation with HNO_3 in 4.0-6.0 M HNO_3 provided the current efficiency more than 200%. The oxidation of UC in NaOH resulted in formation of low-solubility products and the efficiency of the dissolution (about 15%) could not be achieved by the introduction of complex forming agent (F^-) to the electrolyte. The suggested mechanism of UC electrochemical dissolution requires further experimental validation through calculation of material balance, perhaps through potential controlled electrolysis. This study now is being in progress.

References

1. C. Terrasier, Etude de la formation d'acides organiques à partir de carbone à l'état de traces en milieu acide et oxydant. *Ph.D. Thesis, Université Paris VI*, France March 2003.
2. G. R. Choppin, H. Bokelund, M. S. Caceci and S. Valkiers. *Radiochim. Acta*, 1983, **34,** 151.
3. V. Chandramouli, N. L. Sreenivasan and R. B. Yadav, *Radiochim. Acta,* 1990, **51**, 23.
4. H. Von Bildstein and K. Knotik, *Kerntechnik,* 1966, **8**, 110.
5. A.Palamalai, S.K. Rajan, Chinnusamy, M. Sampath, P.K. Varghese, T.N. Ravi, V.R. Raman and G.R. Balasubramanian. *Radiochim. Acta,* 1991, **55**, 29.

REDOX BEHAVIOUR OF PLUTONIUM(IV) IN ACIDIC SOLUTIONS

H.R. Cho,[1] C.M. Marquardt,[1] V. Neck,[1] A. Seibert,[1] C. Walther,[1] J.I. Yun[1]
and Th. Fanghänel[1,2]

[1]Forschungszentrum Karlsruhe, Institut für Nukleare Entsorgung,
PO Box 3640, D-76021 Karlsruhe, Germany
[2]Physikalisch-Chemisches Institut, Ruprecht-Karls Universität,
Im Neuenheimer Feld 253, D-69120 Heidelberg, Germany

1 INTRODUCTION

The redox behaviour of plutonium in acidic solutions has been studied for many decades. The formation of Pu(III), Pu(V) and Pu(VI) in Pu(IV) solutions exposed to air is usually ascribed to the disproportionation of Pu(IV) into Pu(III) and Pu(V), followed by the reaction of Pu(V) with Pu(IV) or the disproportionation of Pu(V) into Pu(III) and Pu(VI).[1-4] As the measured oxidation state distributions led to doubts on this reaction path,[3,5] we have revisited this topic and studied 10^{-5} to $5 \cdot 10^{-4}$ M Pu(IV) solutions ($pH_c = 0.3 - 2.1$ in 0.5 M HCl/NaCl, 22°C) as a function of time. Most of the solutions, obtained by dilution of electrochemically prepared Pu(IV) stock solutions, were kept under air in closed vials or cuvettes. Part of the experiments were performed in an Ar glove box. The concentrations of $Pu^{IV}(aq)$, Pu^{3+}, PuO_2^+ and PuO_2^{2+} were determined by UV/Vis/NIR absorption spectroscopy, using a capillary cell (path length 1 m) for low concentrations. Partly the solutions were colloid-free, partly they included Pu(IV) colloids, in particular at Pu(IV) and H^+ concentrations above the reported solubility of PuO_2(am, hyd).[6] The presence or absence of Pu(IV) oxyhydroxide colloids > 5 nm was confirmed by laser-induced breakdown detection (LIBD) as described previously.[5,6] As the fraction of polymeric or colloidal Pu(IV) cannot be quantified by spectroscopy it was calculated from the difference $[Pu]_{tot} - \{[Pu^{IV}(aq)] + [Pu^{3+}] + [PuO_2^+] + [PuO_2^{2+}]\}$.

2 DISCUSSION OF EXPERIMENTAL RESULTS

2.1 Disproportionation of Pu(IV) ?

The disproportionation of Pu(IV) solutions at pH 0 - 2 leads to Pu(III) and, depending on pH, to Pu(V) (1), Pu(VI) (2) or both:[1-4]

$$2\,Pu^{4+} + 2\,H_2O \Leftrightarrow Pu^{3+} + PuO_2^+ + 4\,H^+ \quad (1)$$

$$3\,Pu^{4+} + 2\,H_2O \Leftrightarrow 2\,Pu^{3+} + PuO_2^{2+} + 4\,H^+ \quad (2)$$

Accordingly and independent of whether an equilibrium state is reached or not, the following balance must be valid for Pu(III), Pu(V) and Pu(VI) formed from Pu(IV):[3]

$$[Pu(III)] = [Pu(V)] + 2\,[Pu(VI)] \tag{3}$$

However, none of the solutions investigated in the present study (four examples are shown in Figs. 1a-d) fulfils this balance at reaction times < 10 days. Instead, the formation of Pu(III) is always approximately equal to the simultaneous decrease of Pu(IV)aq:

$$D[Pu(III])/dt = - d[Pu(IV)aq]/dt \tag{4}$$

i.e., {[Pu(IV)aq] + [Pu(III)]} = constant and, as a consequence:

$$D\{[Pu(V)] + [Pu(VI)]\}/dt = - d[Pu(IV)coll]/dt. \tag{5}$$

2.2 A mechanism accounting for the observed oxidation state distributions

The present results indicate that the so-called "disproportionation of Pu(IV)" is a two-step process. The initial step is the formation of PuO_2^+, either by the redox equilibrium with PuO_2(am, hyd)[7,8] which is equal to $PuO_n(OH)_{4-2n} \cdot xH_2O$ colloids > 5 nm:

$$PuO_2(coll, hyd) \Leftrightarrow PuO_2^+ + e^- \tag{6}$$

or by the oxidation of colloidal or smaller polynuclear Pu(IV) species by O_2, analogous to the water-catalysed oxidation of solid PuO_2(s, hyd) to PuO_{2+x}(s, hyd),[9] followed by the dissolution of the oxidised Pu(V) fractions:

$$PuO_2(coll, hyd) + x/2\ O_2 + x\ H_2O \rightarrow \{PuO_{2+x}(coll, hyd)\} + x\ H_2O \rightarrow 2x\ (PuO_2^+ + OH^-) + (1-2x)\ PuO_2(coll, hyd) \tag{7}$$

The second step is the simultaneous equilibration of the redox couples Pu(V)/Pu(VI) and Pu(IV)/Pu(III) which are related by pe (and pH because of Pu(IV) hydrolysis equilibria):

$$PuO_2^+ \Leftrightarrow PuO_2^{2+} + e^- \tag{8}$$

$$Pu^{4+} + e^- \Leftrightarrow Pu^{3+} \qquad (Pu(OH)_n^{4-n} + n\,H^+ + e^- \Leftrightarrow Pu^{3+} + n\,H_2O) \tag{9}$$

This mechanism explains that the sum {[Pu(III)] + [Pu(IV)aq]} always remains constant. It also explains the different behaviour of a colloid-free 3.8×10^{-4} M Pu(IV) solution at pH = 0.3 (Fig. 1a), i.e., far below the solubility of PuO_2(am, hyd),[6] and at pH = 1.0 (Fig. 1b) where polynuclear and colloidal Pu(IV) is formed immediately. In the experiment at $[Pu]_{tot} = 1.0 \times 10^{-5}$ M and $pH_c = 2.1$ (Fig. 1c), considerably above the solubility of PuO_2(am, hyd),[6] Pu(V) is evidently formed faster than Pu(III), not simultaneously as required by the disproportionation reaction (1).

The presence of O_2 is not necessary for reaction (6), but a prerequisite for reaction (7). Analogous experiments under air and in an Ar glove box gave very similar results (*c.f.*, Figs. 1b and d). This favours reaction (6) as the initial step. However, the Ar box used had a certain O_2 contamination (ca. 10 ppm), possibly sufficient to produce Pu(V) and Pu(VI) from colloidal or polynuclear Pu(IV), so that a final decision is not yet possible.

The oxidation state distributions and pe values measured after equilibration times of more than 20 days are consistent with the known redox equilibria (6), (8) and (9).[7,8,10] The "disproportionation" reactions (1) and (2) also describe correctly equilibrium state thermodynamics, but not the underlying reaction mechanism.

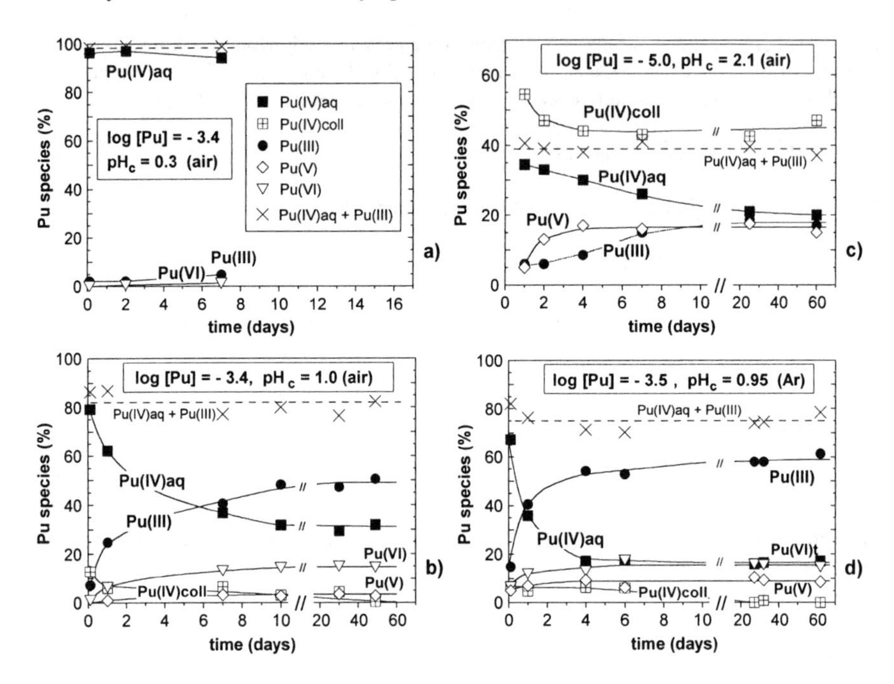

Figure 1 *Oxidation state distributions of initially Pu(IV) solutions as a function of time; a - c) solutions exposed to air, d) solution kept in an Ar glove box containing ca. 10 ppm $O_2(g)$*

References

1 R.E. Connick and W.H. McVey, *J. Am. Chem. Soc.*, 1953, **75**, 474.

2 S.W. Rabideau, *J. Am. Chem. Soc.*, 1953, **75**, 798 and 1957, **79**, 6350.

3 D.A. Costanzo, R.E. Biggers and J.T. Bell, *J. Inorg. Nucl. Chem.*, 1973, **35**, 609.

4 H. Capdevila, P. Vitorge and E. Giffaut, *Radiochim. Acta*, 1992, **58/59**, 45.

5 C. Walther, C. Bitea, J.I. Yun, J.I. Kim, Th. Fanghänel, C.M. Marquardt, V. Neck and A. Seibert, *Actinides Research Quarterly*, Los Alamos Natl. Lab., 2003, **11**, 12.

6 R. Knopp, V. Neck and J.I. Kim, *Radiochim. Acta*, 1999, **86**, 101.

7 D. Rai, *Radiochim. Acta*, 1984, **35**, 97.

8 H. Capdevila and P. Vitorge, *Radiochim. Acta*, 1998, **82**, 11.

9 J.M. Haschke and V.M. Oversby, *J. Nucl. Mat.*, 2002, **305**, 187.

10 R.J. Lemire, J. Fuger, H. Nitsche, P. Potter, M.H. Rand, J. Rydberg, K. Spahiu, J.C. Sullivan, W.J. Ullman, P. Vitorge, H. Wanner (OECD, NEA-TDB). *Chemical Thermodynamics of Neptunium and Plutonium*. Elsevier, North-Holland, 2001.

THE EFFECT OF FISSION PRODUCT ELEMENTS ON CHLORINATION OF URANIUM METAL IN MOLTEN SALTS

V.A. Volkovich,[1] B.D. Vasin,[1] I.B. Polovov,[1] S.A. Kazakov[1] and T.R. Griffiths[2]

[1]Department of Rare Metals, Ural State Technical University – UPI, Ekaterinburg, 620002, Russia
[2]Redston Trevor Consulting Ltd., Leeds, LS17 8RF, UK

1 INTRODUCTION

There is now growing interest towards developing new safer methods for reprocessing spent nuclear fuels. One possible alternative to the existing PUREX process is pyrochemical reprocessing in molten salts. Two processes currently in existence (ANL in the USA and RIAR in Russia) employ two different approaches for dissolving the fuel in the head end of reprocessing – metallic fuel is dissolved anodically (ANL) and ceramic is treated with Cl_2 or Cl_2/O_2 gas (RIAR). A possible alternative for anodic dissolution of uranium metal is reacting it with chlorine in molten alkali chlorides. This could also be of interest for developing a pyrochemical process for converting highly enriched uranium to reactor-grade for energy production.

The chlorination of metals in molten salts has mainly involved dissolution and speciation of platinum group metals and only one report for uranium.[1] The effect of other elements present on the chlorination of the metal of interest is unreported, and spent fuels contain a variety of fission product elements (but at low levels). We now report the effect of temperature and melt composition on the chlorination of uranium metal and the influence of one other metal from molybdenum, niobium, zirconium, neodymium and tellurium. It was expected that addition of U metal to a melt containing Mo, Nb or Zr chloro-species would reduce them to the metal and uranium would be chlorinated, but a different reaction was expected for Nd. The absence of electrode potentials for tellurium meant no prediction in this case.

2 EXPERIMENTAL

The experiments were performed in LiCl-KCl, NaCl-KCl, NaCl-CsCl and CsCl melts between 450 and 750 °C. The progress of the reaction was followed using *in situ* fibre optic electronic spectroscopy.[2] The experiments were conducted under chlorine atmosphere with the pressure inside the cell exceeding ambient, Cl_2 was obtained by electrolysis of molten $PbCl_2$. Uranium was added as small pieces, total surface area of *ca.* 25 mm^2. When the effect of another metal on uranium chlorination was studied, that metal had been previously added and reacted with Cl_2 for 10-30 minutes to have both the metal and chloro species in the melt before uranium addition.

3 RESULTS AND DISCUSSION

Reaction of uranium with excess chlorine results in the formation of UCl_6^{2-} ions, easily identifiable from their electronic spectra. No reaction between uranium metal and chlorine was observed in LiCl-KCl melt at 450 °C; but on rising the temperature uranium dissolution commenced and increased with temperature increase, Table 1. Similar behaviour was observed in the NaCl-CsCl melt. Solubility of chlorine in alkali chlorides increases with increasing temperature and with increasing radius of alkali metal cation. The rate of uranium chlorination at 750 °C, however, decreased in the order LiCl-KCl > NaCl-KCl > NaCl-CsCl > CsCl, Table 1. Since in molten salts the diffusion coefficients for both dissolved chlorine and UCl_6^{2-} decrease with increasing cation radius of the solvent the obtained data indicate that uranium chlorination is controlled by diffusion.

Table 1 *Chlorination of U metal in molten alkali metal chlorides*

Melt	T, °C	Time, min	Final U conc., wt. %	Chlorination rate, mg h^{-1}
LiCl-KCl	550	325	0.19	2
LiCl-KCl	600	146	1.05	31
LiCl-KCl	650	53	2.24	127
LiCl-KCl	750	96	3.75	142
NaCl-KCl	750	92	2.37	109
NaCl-CsCl	550	214	2.59	44
NaCl-CsCl	750	174	3.81	78
CsCl	750	101	1.94	69

The rate of dissolution of uranium in the presence of molybdenum or niobium was higher than in pure alkali chloride melts (Table 2). Reaction of Mo and Nb with Cl_2 resulted in the formation of $MoCl_6^-$ and $NbCl_6^{2-}$ ions, respectively, identified from their spectra. Addition of U to such melts (with chlorine continuously bubbled through) resulted in additional uranium dissolution, *e.g.*:

$$4\ MoCl_6^- + 5\ U + 6\ Cl^- \rightarrow 4\ Mo + 5\ UCl_6^{2-} \qquad (1)$$

Finely divided molybdenum formed then reacted with Cl_2 producing Mo(III), (IV) and (V) species that again oxidised uranium. Thus in the presence of Mo and Nb uranium was dissolving by reaction with Cl_2 and also due to oxidation by molybdenum or niobium chlorides that acted as chlorine-transferring species. Even after 1-2 hours of reaction there were still Mo or Nb chlorides present in the melt. Chemical analysis of the quenched melt showed that molybdenum was mainly present in the form of $MoCl_6^{2-}$ species.

Table 2 *Chlorination of U metal in molten alkali metal chlorides in the presence of other elements*

Melt	T, °C	Metal added	Chlorination rate, mg h^{-1}	
			uranium	added metal
LiCl-KCl	550	Mo	112	108
NaCl-KCl	750	Nb	90	3
LiCl-KCl	550	Nb	111	11
LiCl-KCl	550	Nd	20	49
LiCl-KCl	550	Nd	14	23
LiCl-KCl	550	Zr	4	---
LiCl-KCl	550	Te	605	403

Zirconium very slowly reacts with chlorine in molten salts and thus had essentially no effect on uranium chlorination, Table 2. In contrast, Tellurium, reacts efficiently with Cl_2 very efficiently, yielding $TeCl_6^{2-}$. Addition of uranium metal to such melt resulted in a very vigorous exothermic reaction:

$$U + [TeCl_6]^{2-} \rightarrow [UCl_6]^{2-} + Te \tag{2}$$

and the tellurium produced was again converted to $[TeCl_6]^{2-}$ by reaction with Cl_2. The rate of U dissolution in this system was by far the fastest of all the systems investigated.

An unusual and unexpected (but reproducible) behaviour was observed when uranium was chlorinated in the presence of neodymium metal. The concentration of neodymium in the melt increased but that of uranium reached a maximum and declined, Fig. 1. Measured spectra showed only the presence of $NdCl_6^{3-}$ and UCl_6^{2-}, and no U(III) formation. After experiments some unreacted U and Nd metals were always found in the cells, but since Nd is aggressive towards silica cells, experiments had to be completed in under 3 hours. The initial stages of the process can be described by the following reactions:

$$Nd + 3\,MeCl + 3\,Cl^- \rightarrow [NdCl_6]^{3-} + 3\,Me \text{ (Me = alkali metal)} \tag{3}$$

$$2\,Me + Cl_2 \rightarrow 2\,MeCl \tag{4}$$

$$2\,Nd + 3\,Cl_2 + 6\,Cl^- \rightarrow 2\,[NdCl_6]^{3-} \tag{5}$$

$$U + 2\,Cl_2 + 2\,Cl^- \rightarrow [UCl_6]^{2-} \tag{6}$$

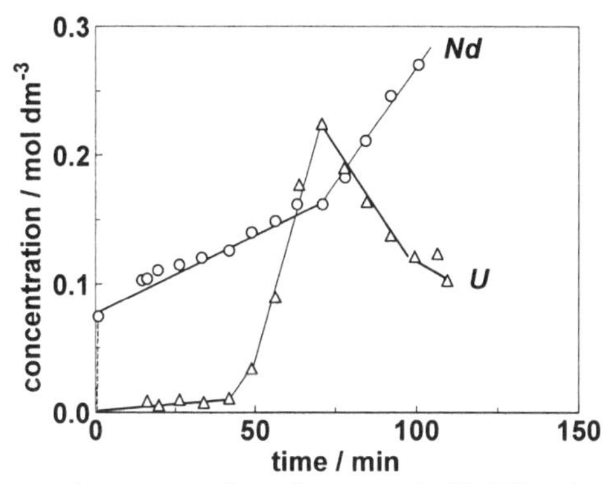

Figure 1 *Chlorination of uranium and neodymium in LiCl-KCl melt at 550 °C*

There is no definite explanation for diminishing U(IV) concentration in the melt. Upon reaching a certain concentration of Nd(III) it can react with Nd(0) to form Nd(II) species. The more Nd(III) is in the melt the more Nd(II) can be formed and the latter can act as a reductant for U(IV) converting it to uranium metal.

References

1 V.A. Volkovich, A.I. Bhatt, I. May, T.R. Griffiths and R.C. Thied, *J. Nucl. Sci. Technol.*, 2002, **Suppl. 3**, 595.

2 V.A. Volkovich, I. May, J.M. Charnock and B. Lewin, *Phys. Chem. Chem. Phys.*, 2002, **4**, 5753.

SPECTROELECTROCHEMICAL STUDIES OF URANIUM BEHAVIOR IN $(Li\text{-}K)Cl_{eut}$ BASED MELTS AT 450 °C

I.B.Polovov[1], C.A. Sharrad[2], V.A. Volkovich[1], H. Kinoshita[2], S.A. Kazakov [1], I. May [2], B.D. Vasin [1]

[1]Department of Rare Metals, Ural State Technical University – UPI, Ekaterinburg, 620002, Russia
[2]Centre for Radiochemistry Research, School of Chemistry, The University of Manchester, Oxford Road, Manchester, M13 9PL, United Kingdom

1 INTRODUCTION

Development of safe modern technologies for reprocessing spent nuclear fuels and conversion of highly enriched metallic uranium into reactor grade material requires knowledge of uranium species in molten chloride mixtures. The most useful and direct information concerning oxidation state of uranium in such systems can be obtained from *in situ* electronic spectroscopy. On the other hand the investigation of electrochemical properties of uranium in fused halides allows us to determine standard potentials and kinetic parameters of electrode reactions. The combination of these two techniques in one spectroelectrochemical cell provides unambiguous information concerning speciation in studied systems. Previous investigations were mostly performed in aqueous or organic low-temperature solutions.

There are two reports dealing with high temperature spectroelectrochemistry.[1-2] Smirnov and Potapov[1] investigated speciation of chromium and nickel in molten equimolar mixture of sodium and potassium chlorides and Nagai *et al.*[2] studied UO_2^{2+}/UO_2^{+} equilibrium in NaCl-CsCl melt. In the present work we report the design of spectroelectrochemical cell and the set-up for conducting experiments in fused salts. Behaviour of uranium in a lithium-potassium chloride eutectic mixture was studied at 450 ^{0}C and the information obtained concerning coordination and electrochemical properties of uranium (III) and (IV) species is presented.

2 EXPERIMENTAL

The experiments were carried out in a standard 1 cm silica optical cell attached to a silica tube (20 mm i.d.) with two side-arms. All electrochemical measurements were performed *vs.* chlorine or silver/silver chloride reference electrodes. When studying anodic dissolution of metallic uranium, electrooxidation of uranium(III) and electroreduction of uranium(IV), the chlorine reference electrode was also used as a counter electrode. Platinum wire in a silica sheath acted as a working electrode with the exception of the experiments on anodic dissolution of uranium metal. A uranium rod suspended on a molybdenum current conductor wire was used to study the interaction of uranium(IV) containing electrolyte with uranium metal.

Details of the high temperature spectroscopy set-up were described elsewhere.[3] Electrochemical measurements were performed using a Solartron 1287 potentiostat which was also used to control the oxidation-reduction potential of the system.

All experiments were performed in LiCl-KCl eutectic melt at 450 °C. The progress of electrode reactions was followed by *in situ* electronic spectroscopy and voltammetry. Uranium was introduced into the melt by anodic dissolution of the metal or by dissolving anhydrous uranium tetrachloride. The oxidation state of uranium was also controlled by oxidimetry.[4]

3 RESULTS AND DISCUSSION

Absorption spectra of the melts obtained after dissolution of uranium tetrachloride were typical for the uranium(IV) species UCl_6^{2-}, with major bands at 505, 565 and 670 nm, Fig. 1. The melt was light green and the average oxidation state of uranium was *ca* 4.0. The spectra didn't change during HCl bubbling through the melt.

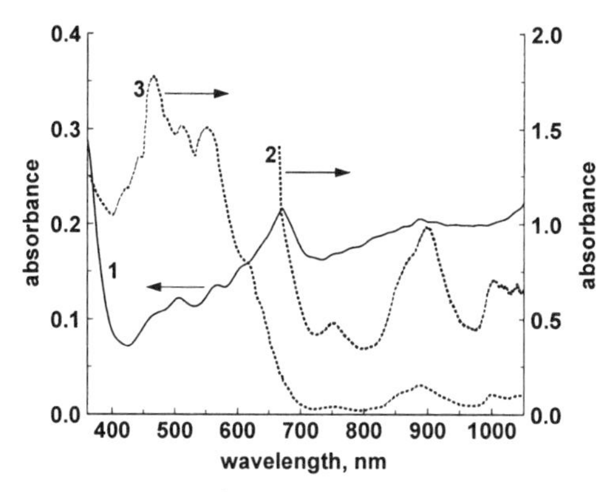

Figure 1 *Electronic spectra of uranium species in LiCl-KCl melt at 450 °C (1- uranium(VI), 0.01 wt.% U; 2 – uranium(III), 0.012 wt. % U; 3 – uranium(III) after anodic dissolution of U metal, 0.0018 wt. % U)*

When uranium metal was introduced into the melt or a reducing potential was applied to a platinum working electrode, the colour of the melt changed to deep purple. Absorbance increased considerably in the entire wavelength range (the extinction coefficient of U(III) is about two orders of magnitude higher than that of U(IV)). When uranium was used as a reductant, the potential between the platinum working electrode and the reference electrode didn't change. Despite the fact that the melt was purple, the fraction of uranium(III) ions was relatively small and the potential was close to the initial value. A considerable length of time was necessary to convert all uranium into trivalent state, Fig. 2. After contacting LiCl-KCl-UCl_4 melt with U metal for 4 hours the oxidation state of uranium (determined oxidimetrically in a quenched melt sample) was close to 3.0.

When reduction of uranium tetrachloride was studied, we were not able to obtain a good spectral picture for the U(III) species in the entire wavelength range even when the initial concentration of UCl_4 was very low. The best data for U(III) were obtained during anodic dissolution of uranium metal, Fig. 1. The spectra of uranium(III) contain peaks at around 465, 510, 550, 750 and 900 nm. The last two have lower intensity (by about an order of magnitude) compare to the others. According to oxidimetry uranium anodically dissolves forming U^{3+} ions. Coulometry didn't give reasonable results, possibly due to high corrosion rate of uranium in $(Li\text{-}K)Cl_{eut}$ or/and interaction of U(III) ions with quartz.

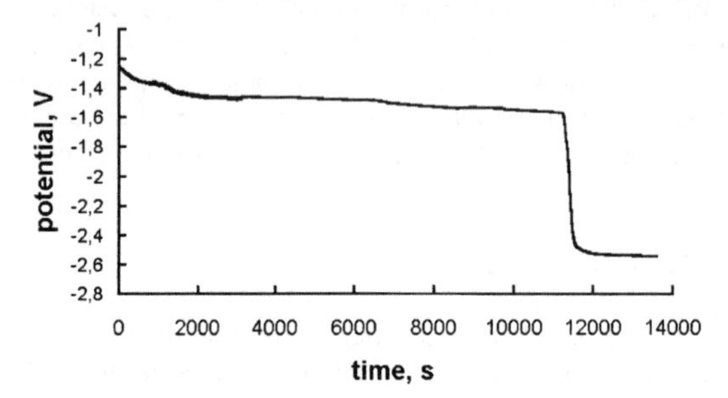

Figure 2 *Potential change during contact of uranium with uranium(IV) containing melt (uranium content, wt. %: initial – 0.01, final – 0.012)*

Additionally, some electrochemical studies were carried out in UCl_3-$(Li\text{-}K)Cl_{eut}$ melt by chronoamperometry, chronopotentiometry, and linear and cyclic voltammetry. A typical cyclic voltamogramm is presented in Fig. 3. Because of interaction of U(III) species with silica and platinum-uranium alloy formation on the working electrode, the obtained electrochemical results can only be used for estimation of characteristic potentials and kinetic parameters.

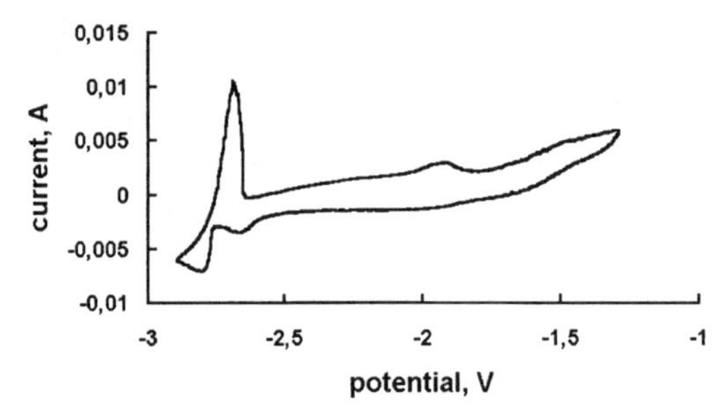

Figure 3 *Cyclic voltamogramm of UCl_3-$(Li\text{-}K)Cl_{eut}$ melt on platinum electrode at 450 ^0C (scan rate – 20 mV/sec, uranium content – 0.08 wt. %)*

Acknowledgements

IBP thanks INTAS (Grant No. 03-55-1453) for financial support.

References

1 M.V. Smirnov and A.M. Potapov, *Electrochem. Acta*, 1994, **39**, 143.
2 T. Nagai, T. Fujii, O. Shirai and H. Yamana, *J. Nucl. Sci. Technol.*, 2004, **41**, 690.
3 V.A. Volkovich, A.I. Bhatt, I. May, T.R. Griffiths and R.C. Thied, *J. Nucl. Sci. Technol.*, 2002, **Suppl. 3**, 595.
4 V.A. Volkovich, I. May, A.I. Bhatt, T.R. Griffiths, J.M. Charnock and B. Lewin, In: *Int. symp. ionic liquids*, Carry le Rouet, France, 2003, p. 253.

EVALUATION OF NUCLEAR FUEL CYCLE SYSTEMS FOR ACTINIDE RECYCLE IN FEASIBILITY STUDY ON COMMERCIALIZED FAST REACTOR CYCLE SYSTEMS

K. Nomura, Y. Nakajima, T. Ogata, Y. Nagaoki and T. Namba

FBR Cycle System Development Office, Japan Nuclear Cycle Development Institute, O-Arai Engineering Center, Ibaraki, 311-1393, Japan

1 INTRODUCTION

Japan Nuclear Cycle Development Institute (JNC) started the Feasibility Study on Commercialized FBR Cycle Systems (FS) in the Japanese fiscal year (JFY) 1999 in cooperation with electric power companies and related organizations, establishing five development targets (ensuring safety, economical competitiveness, reduction of environmental burden, efficient utilization of natural resources and enhancement of nuclear non-proliferation). In the FBR cycle system being pursued in the FS, minor actinides (MA; Am, Cm and Np) will be recycled with U and Pu from the viewpoint of reduction of environmental impact.[1]

In Phase I (JFY 1999 - 2000), the feasibility of various concepts of nuclear fuel cycle systems, which combined reprocessing and fuel fabrication, were extensively examined through literature survey and experimental studies. Several concepts were selected for further study; advanced aqueous process, metal electrorefining process and oxide electrolysis process as reprocessing systems and simplified pelletizing process, vibro-packing process and casting process as fuel fabrication systems, respectively.

In Phase II (JFY 2001 - 2005), the conceptual design study on the four nuclear fuel cycle systems, namely (1) advanced aqueous process + simplified pelletizing process or (2) vibro-packing process, (3) metal electrorefining process + casting process and (4) oxide electrolysis process + vibro-packing process has been continued and experimental studies on the key technologies such as MA recovery process have been conducted in order to estimate their technological performances and the feasibility of the respective systems. Attractive nuclear fuel cycle systems will be identified by the end of JFY2005.

An outline of the selected reprocessing system concepts and the state of MA recovery technologies to be applied in respective reprocessing systems are mainly described in this paper.

2 OUTLINE OF THE SELECTED REPROCESSING SYSTEM CONCEPTS

Figure 1 shows the block flow-sheet of the proposed reprocessing system. Each system treats MA as a product. Decreasing MA contents in the radioactive waste by recovering MA as products leads to the reduction of environmental impact.

2.1 Advanced Aqueous Process

The advanced aqueous reprocessing system, consisting of U crystallization, single cycle extraction and MA separation, has been newly established. This system is based on the solvent extraction method, using tri-butyl phosphate (TBP), which is well established in the conventional aqueous system. Nuclear proliferation resistance of the system is enhanced by recycling U, Pu and MA slightly contaminated with fission products, a so-called low DF product. Np is recovered with U and Pu in the single cycle extraction by TBP. Am and Cm are separated in an MA separation process.

2.2 Metal Electrorefining Process

This system is based on the pyrometallurgical process for metal fuel developed at the Argonne National Laboratory (ANL) in the United States. MA, in this process, are deposited together with U and Pu from spent fuel anodically dissolved into the KCl-LiCl molten salt, with fission products remaining in the molten salt. The fact that the metal electrorefining process cannot separate pure Pu guarantees high proliferation resistance.

When this reprocessing option is applied to oxide fuel a head-end process to reduce oxide fuel to metal form is needed as well as a process to oxidize metallic products to mixed oxide (MOX).

2.3 Oxide Electrolysis Process

The oxide electrolysis process is what has been originally developed at RIAR in Russia. The system is expected to be simple because the reprocessing products, UO_2 and MOX granules, can be directly used as fuel materials for vibro-packed fuels. This leads to the potential to achieve high economic competitiveness even for a small-scale fuel cycle facility. In addition, the oxide electrolysis process, which is intrinsically low DF product system, has high proliferation resistance. Several new processes such as simultaneous electrolysis for MOX fuel dissolution and partial UO_2 deposition, MOX co-deposition and drawdown electrolysis for MA recovery are incorporated into the system considered in FS.

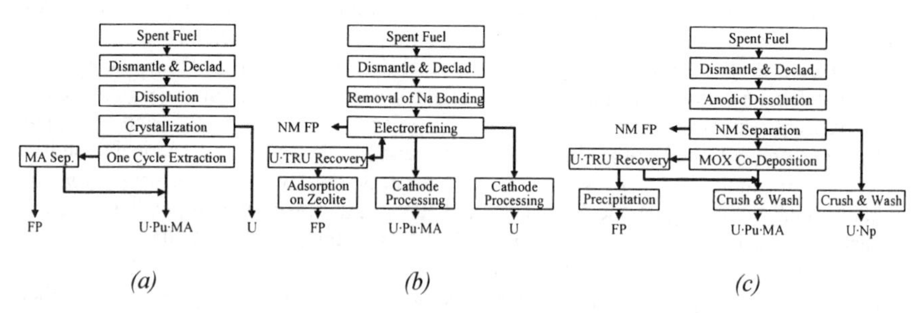

Figure 1 *Proposed reprocessing system: (a) advanced aqueous process, (b) metal electrorefining process, (c) oxide electrolysis process*

3 STATUS OF STUDIES ON MA RECOVERY

3.1 "SETFICS" Process and Extraction Chromatography Method

The SETFICS (Solvent Extraction for Trivalent f-elements Intra-group Separation in CMPO-Complexant System) process has been developed by JNC.[2] It is based on solvent extraction with so-called TRUEX solvent, which is mixture of CMPO, TBP and normal paraffinic hydrocarbon diluent. The TRUEX solvent is an outstanding extractant for trivalent Am, Cm and lanthanide. The separation step of Am and Cm from lanthanide is applied to the SETFICS process. Am and Cm are stripped from the TRUEX solvent phase to the aqueous phase containing di-ethylene tri-amine penta-acetic acid (DTPA), which is selectively complexed with Am, Cm and heavy lanthanides that are quantitatively rather minor component in the spent fuel than light lanthanides, while major part of light lanthanides remains in the organic phase.

Since the amount of secondly waste from the SETFICS process is so large that a significant economical impact will be incurred, the amount of solvent and aqueous solutions should be minimized. In the present design of the advanced aqueous process proposed, organic solvent waste is minimized by replacing the liquid solvent with silica based adsorbent impregnated with CMPO. In general, chromatographic separation is suitable for quantitative recovery of minor components such as MA in the highly active liquid waste from the PUREX process. Extraction chromatography does not need diluent, therefore its application has some effects concerning reducing the amount of organic waste. An adsorbent based on macroporous silica particles coated with styrene-divinylbenzene polymer, SiO2-P, provides a variety of extraction capabilities, because SiO2-P can hold many types of extractant molecules, including CMPO. Results of some cold tests by using lanthanide elements show that the extraction chromatographic system with CMPO will also have a superior performance as MA recovery method.

3.2 MA Recovery in Metal Electrorefining Process

In the metal electrorefining process, as described above, MA is deposited and recovered together with U and Pu from spent fuel anodically dissolved into the KCl-LiCl molten salt. The metal electrorefining process is potentially favorable in terms of enhancement of nuclear non-proliferation because it is, in principle, difficult to extract directly high-purity Pu. U, Pu and MA are deposited into the liquid Cd phase at almost the same potential at around 500 °C. Verification by experiments with TRU and irradiated fuel will be required for the future experimental plan.

3.3 MA Recovery in the Oxide Electrolysis Process

A few technologies for MA recovery to be applied to the oxide electrolysis process have been studied so far. The drawdown electrolysis is, however, difficult to be controlled because the electrochemical potentials of Am and Cm deposition are close to that of the molten salt decomposition. Although the Am and Cm precipitation method by carbonate is also investigated, the separation of lanthanide from MA seems difficult. An alternative method is required for MA recovery in the oxide electrolysis process.

References

1 T. Namba, H. Funasaka, Y. Nagaoki and Y. Sagayama, *ATALANTE 2004*, 2004, **O11-03.**

2 H. Hirano, Y. Koma and T. Koyama, *The 7th Information Exchange Meeting on Actinide and Fission Product Partitioning and Transmutation*, 2002, **S2-5**, 339-346.

EXTRACTION OF TRANSPLUTONIUM AND RARE-EARTH ELEMENTS WITH THE USE OF ZIRCONIUM SALT OF DIBUTYL PHOSPHORIC ACID

B.Ya. Zilberman[1], Yu.S. Fedorov[1], O.V. Shmidt[1], N.D. Goletskiy[1], E.N. Mishin[1], Yu.V. Palenik [1] and G.R Choppin [2]

[1] RPA "V.G.Khlopin Radium Institute", 2nd Murinsky av. 28, 197021 St-Petersburg, Russia
[2] Florida State University, 32306-4390 Talahasse, Florida, USA

1 INTRODUCTION

The zirconium salt of dibutylphosphoric acid (ZS HDBP) dissolved in TBP diluted with hydrocarbons extracts effectively TPE, RE and Mo from nitric acid media[1,2]. Separation of TPE and cerium subgroup of RE from yttrium subgroup of RE and Mo is possible with this reagent at 5 mole/L HNO_3. TBP acts as a solvating agent in the formation of a Ln complex with ZS HDBP, releasing 3 molecules of HDBP. The HDBP can additionally bind RE, doubling their maximum concentration in solvent phase. However, it is still not clear in the case of TBP absence, whether molecules of HDBP taken in excess to $ZrA_4(HA)_4$ and it's compound with RE participate in solvation and what the solvate structure is.

2 METHOD AND RESULTS

Distribution coefficients of Ce, Eu, Am and Mo as a function of the HDBP:Zr ratio are plotted in Figure 1. The maximum values for TPE and RE are observed at HDBP:Zr=9, in the absence of TBP, decreasing from Eu to Ce and then to Am. The optimum HDBP:Zr ratio for Mo recovery is 12. Distribution coefficients for Ce, which are similar to those of Am, are plotted as function of aqueous acidity in Figure 2, the difference of free H^+ and bulk acidity being insignificant in the scale of the plot. It shows that the slope of the acidity curve is around -6 and slightly increases at HDBP:Zr=6 up to -4 indicating that dimers $[Zr(DBP)_4]_2[Ce(DBP)_3(HDBP)_3]_2$ or $[Zr(DBP)_4]_2[CeNO_3(DBP)_2(HDBP)_3]_2$ are extracted correspondingly, while in the presence of TBP a monomer $Zr(DBP)_4[CeNO_3(DBP)_2(TBP)_3]$ is extracted.[2] Ce(III) extraction from HNO_3 media by ZS HDBP with xylene at various HDBP:Zr ratio starting from Zr free solutions (Figures 3 and 4) was studied in detail. It was found that a decrease in Zr concentration in the solvent phase at constant HDBP content results in decrease in the maximum solvent loading both in the presence and absence of TBP. At the same time Figure 5 shows that in the absence of TBP at 0.2 mole/L HDBP and HDBP:Zr=9 and at 0.4 mole/L HDBP and HDBP:Zr=18 the solvent loading is the same. However, the Ce distribution coefficients are different, increasing as the "free" HDBP share increase at the same concentration of Zr (Figure 6), independent of TBP presence. It is possible that HDBP could become a solvating agent in the coordination sphere of Ln, analogous to TBP, increasing the extraction ability of ZS HDBP even more than TBP, but without increasing the maximum solvent loading. The study of HNO_3 extraction by ZS HDBP (Figure 7) revealed that the HDBP:Zr ratio (also in the presence of TBP) affects HNO_3 distribution significantly. The breakpoint on the curves is at HDBP:Zr=12

(Figure 8), when the second coordination sphere of the Zr atom is filled with HDBP anions (Figure 9), whereas at HDBP:Zr≤6 inner sphere nitrate coordination is observed.

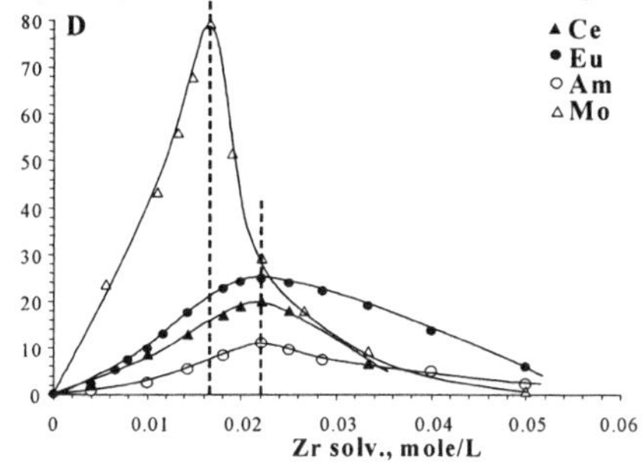

Figure 1. *Distribution coefficients of Ce, Am and Mo as a function of Zr conc. in solvent phase at extraction by 0.2 mole/L ZS HDBP with xylene (absence of TBP). Aq. acidity: 1,5 mole/L HNO_3.*

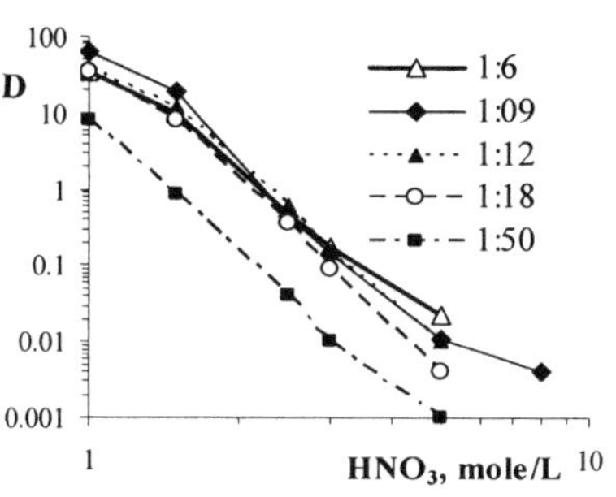

Figure 2. *Ce distribution coefficients as a function of acidity at extraction by 0.2 mole/L HDBP with xylene in the absence of TBP at various molar ratios HDBP:Zr in solvent phase.*

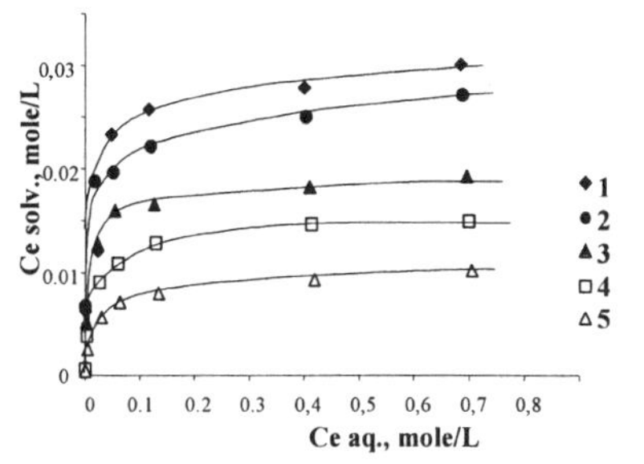

Figure 3. *Ce(III) extraction from 1.5 mole/L HNO_3 by 0.2 mole/L HDBP with xylene in the absence of TBP at various ratios HDBP:Zr in the solvent phase:* ***1*** *– 6;* ***2*** *– 9;* ***3*** *– 12;* ***4*** *– 18;* ***5*** *– 30.*

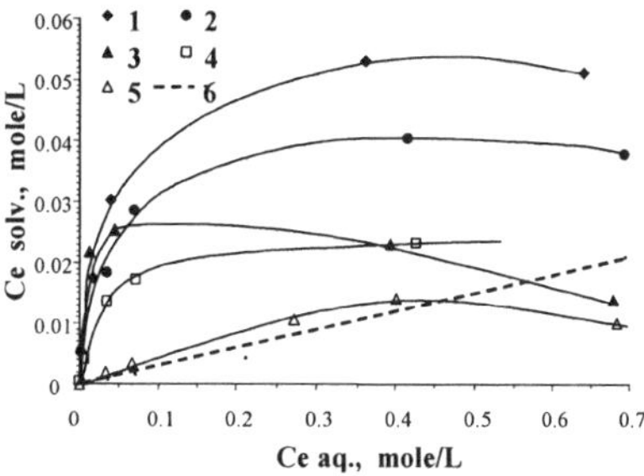

Figure 4. *Ce(III) extraction from 1.5 mole/L HNO_3 by 0.2 mole/L HDBP with 30% TBP in xylene minus Ce(III) extraction by TBP itself at various ratio HDBP:Zr in the solvent phase:* *1 – 6; 2 – 9; 3 – 12;* ***4*** *– 18;* ***5*** *– 30;* ***6*** *–30% TBP*

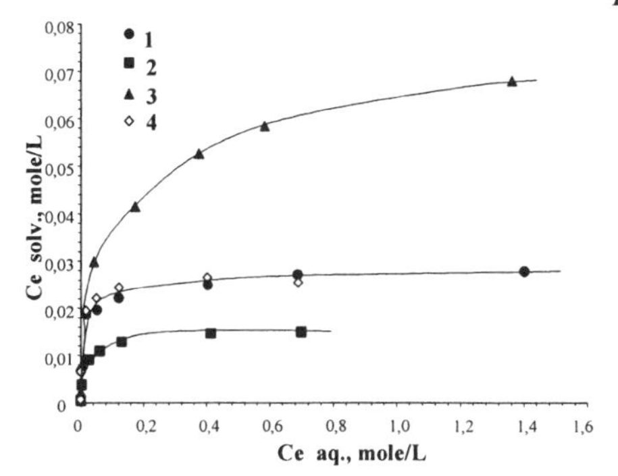

Figure 5. *Ce(III) extraction from 1.5 mole/L HNO_3 by 0.2 mole/L HDBP with xylene. HDBP concentration, mole/L:* ***1, 2, 3*** *– 0,2;* ***4*** *- 0,4. HDBP:Zr ratio:* ***1, 3*** *- 9;* ***2, 4*** *- 18*

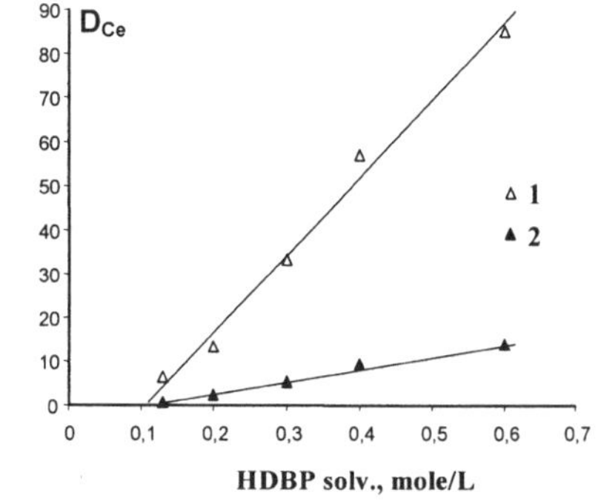

Figure 6. *D_{Ce} as a function of HDBP concentration at constant Zr content (2 g/L) in the solvent phase, absence of TBP (**1**) and in 30% TBP with xylene (**2**). Aqueous acidity: 1.5 mole/L HNO_3.*

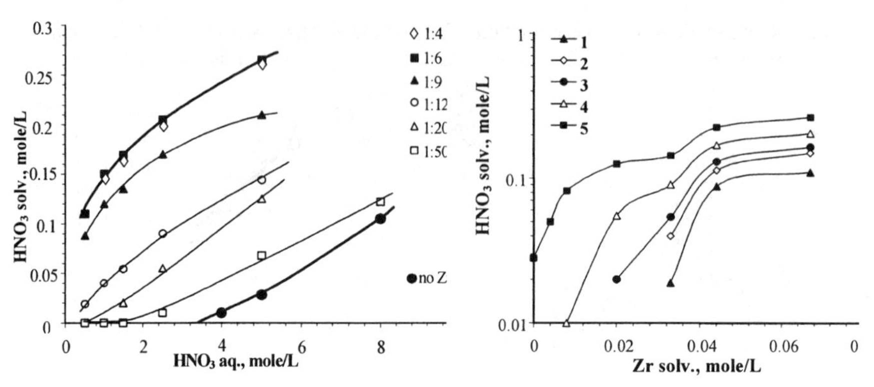

Figure 7. *HNO_3 extraction by ZS HDBP (0.4 mole/L HDBP) with xylene at various Zr:HDBP ratio.*

Figure 8. *HNO_3 extraction by ZS HDBP (0.4 mole/L HDBP) with xylene as a function of Zr concentration in solvent. Aqueous acidity, mole/L:* ***1*** *– 0.5;* ***2*** *– 1.0;* ***3*** *– 1.5;* ***4*** *– 2.5;* ***5*** *– 5.0.*

Figure 9. The structure of ZS HDBP-HNO_3 complex at HDBP:Zr $\leq$ 6 (**A**) and 9 (**B**)

3. CONCLUSION

It was found that the full substitution of nitrate anions in the inner sphere of Zr^{4+} ions occurs at an excess of HDBP above the stoichiometry $ZrA_4(HA)_4$. Simultaneously a competition of HNO_3 and HDBP begins in the outer sphere of ZS HDBP. The excess of HDBP acts as solvation agent in the coordination sphere and promotes Am and RE extraction, in contrast to that of TBP, but without enlarging the maximum solvent loading with RE. The structure of the extracted complex at maximum loading in the absence of TBP might look in this case as $[(HDBP)_3LnNO_3(DBP)_2][Zr(DBP)_4][(DBP)_2LnNO_3(HDBP)_3]$ and/or $[(HDBP)_3Ln(DBP)_3][Zr(DBP)_4][(DBP)_3Ln(HDBP)_3]$ depending on HNO_3 concentration.

References

1 B.Ya. Zilberman, Yu.S Fedorov, O.V. Shmidt, E.N. Inkova, A.A. Akhmatov and N.D. Goletskiy *Radiochemistry*, 2000, **42**, No 4, 338.

2 O.V. Shmidt, B.Ya. Zilberman, Yu.S Fedorov, D.N. Suglobov, E.A. Puzikov, L.G. Mashirov, Yu.V. Palenik and R.G. Glekov. *Radiochemistry*, 2002,.**44**, No 5, 428.

THE EFFECT OF DIBUTYLPHOSPHORIC ACID ON ACTINIDE (IV) EXTRACTION BY 30% TBP FROM SOLUTIONS CONTAINING URANYL NITRATE

I.V. Blazheva, B.Ya. Zilberman, Yu.S. Fedorov, E.N. Mishin, E.A. Puzikov and N.V. Ryabkova

RPA "V.G.Khlopin Radium Institute", 2nd Murinsky av. 28, 197021 St-Petersburg, Russia

1 INTRODUCTION

The effect of dibutylphosphoric acid (HDBP) on the extraction of several tetravalent actinides (An(IV)) by 30% TBP in the presence of $UO_2(NO_3)_2$, as well as on uranyl nitrate itself was investigated by our group previously.[1-2] Spectrophotometric studies of U(IV) and Np(IV) extracts showed that An(IV) could be extracted in the form of a mixed complex $An^{IV}(NO_3)_3(TBP)_2$-R_2OPO-$UO_2(NO_3)_2(TBP)$ (Figure 1). However, it seems necessary to develop a procedure for comparison of tetravalent elements behavior, even for those without bands in visible range of spectrum in the same conditions. We introduced a factor of influence: $\mathbf{F}=\mathbf{R}_{An(IV)}/\mathbf{R}_{0An(IV)}$ to evaluate the influence of solvent loading with U(VI) on tetravalent actinide interaction with HDBP in 30% TBP, where $\mathbf{R}_{An(IV)}=\mathbf{D}_{AAn(IV)}/\mathbf{D}_{An(IV)}$ is relative enhancement of An(IV) extraction as a result of HDBP addition to 30% TBP, $\mathbf{D}_{AAn(IV)}$ and $\mathbf{D}_{An(IV)}$ are distribution coefficients of An(IV) extracted by (HDBP + 30% TBP) and 30% TBP correspondingly at the fixed U(IV) concentration, $\mathbf{R}_{0An(IV)}$ is the value of $\mathbf{R}_{An(IV)}$ for solution without U(VI). F is equal to 1 if U(VI) does not affect An(IV) extraction by TBP in the presence of HDBP, $F<1$ if U(VI) opposes extraction and $F>1$ if U(VI) enhances extraction. Results of the investigation of HDBP influence on Th extraction by 30% TBP in the presence of $UO_2(NO_3)_2$ and comparison of the obtained values of F factor with that for other actinides are presented below.

2 METHOD AND RESULTS

Concentration of Th has been determined by specrtrophotometry with thoron.[3] It was preliminary separated from HDBP by alkaline precipitation and from U by selective extraction. The treatment[1] of data on Np extraction (Figure 2) shows that solvent loading with uranium results in relative enhancement of Np extraction, however the change in HDPB:Np(IV) molar ration from 0.57 to 2.15 results in the change of F factor from 4 in the maximum to less than 1 indicating the change in extraction mechanism. The data on Th extraction in the same conditions is shown in Figures 3 and 4. Figure 3 demonstrates that increase in HDBP concentration results in significant increase of Th distribution coefficient (D_{Th}), especially at molar ratio HDBP:Th>1. The curves for 1 mole/L HNO_3 and 3 and 5 mole/L are different. The influence of solvent loading with uranium on Th extraction in the presence of HDBP is shown in Figure 5. The values of F factor are always less than 1 at microconcentrations of Th and high HDBP:Th ratio, whereas at high Th concentration and HDBP:Th = = 1.1 solvent loading with uranium does not affect Th extraction from 3 and 5 mole/L HNO_3 (F~1). It increases Th extraction from 1 mole/L HNO_3 with maximum of F values lying in the range of ~60 g/L of U in the solvent phase. It seems that a weak mixed Th complex with HDBP and uranyl nitrate with probable composition $Th(NO_3)_3(TBP)_2$-R_2OPO-$UO_2(NO_3)_2(TBP)$ is formed at low acidity.[1] However, even in this range at the ratio HDBP:Th greater than 1 the enhancing effect gradually decreases and at HDBP:Th≈ 2 it completely disappears (F<<1), confirming destruction of the mixed complex.

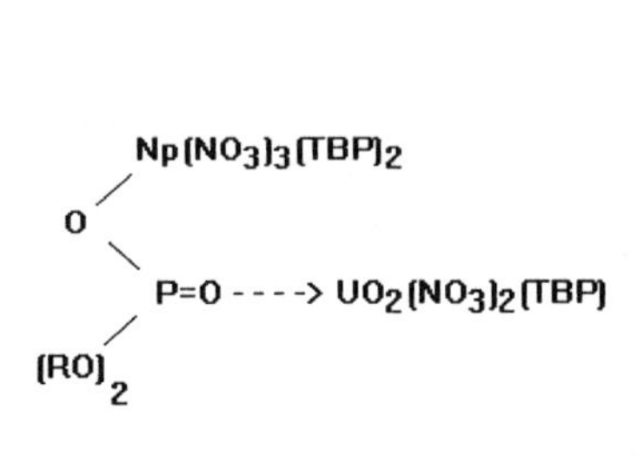

Figure 1. The structure of the mixed complex of Np(IV) with U(VI) and HDBP (at low HDBP concentration) in diluted TBP as a solvent.

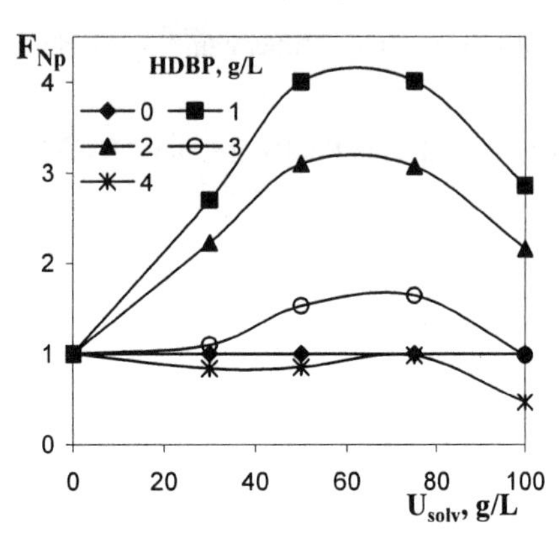

Figure 2. *F factor as function of solvent loading with U(VI) for Np(IV) extraction by HDBP dissolved in 30% TBP.*
Aqueous acidity - 3 mole/L HNO_3
Initial concentration of Np- 2 g/L.

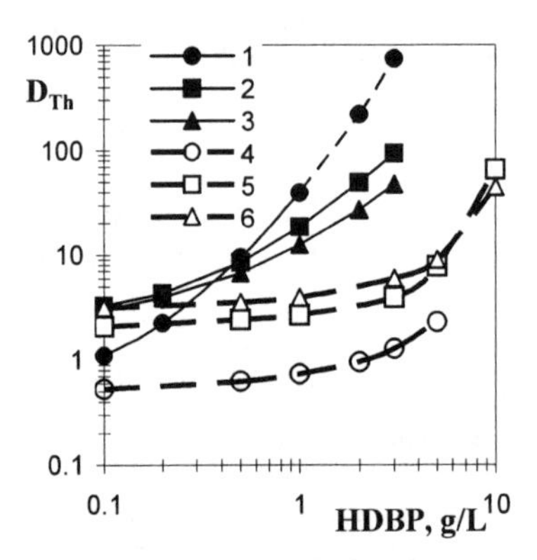

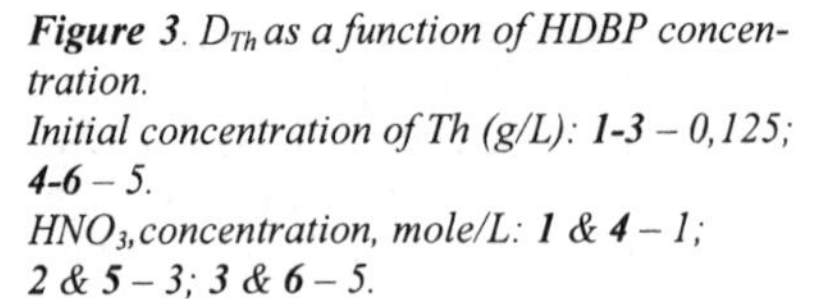

Figure 3. D_{Th} as a function of HDBP concentration.
Initial concentration of Th (g/L): ***1-3*** *– 0,125;* ***4-6*** *– 5.*
HNO_3 concentration, mole/L: ***1*** *&* ***4*** *– 1;* ***2*** *&* ***5*** *– 3;* ***3*** *&* ***6*** *– 5.*

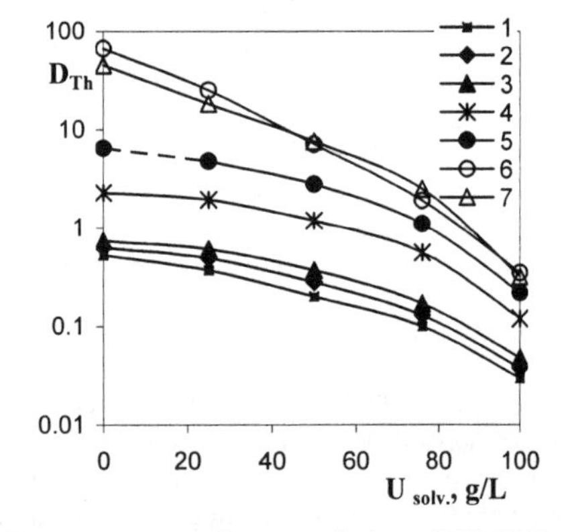

Figure 4. *D_{Th} as a function of solvent (30% TBP) loading with U(VI) at various concentrations of HDBP (the range of precipitation is shown by dashed line).*
Initial concentration of Th – 5 g/L. HNO_3, concentration, mole/L: ***1-5*** *–1,* ***6***-*3,* ***7*** *– 5*
HDBP:Th: ***1*** *– 0,* ***2*** *– 0,11;* ***3*** *– 0,22;* ***4*** *– 1,1,* ***5*** *-* ***7*** *– 2,2.*

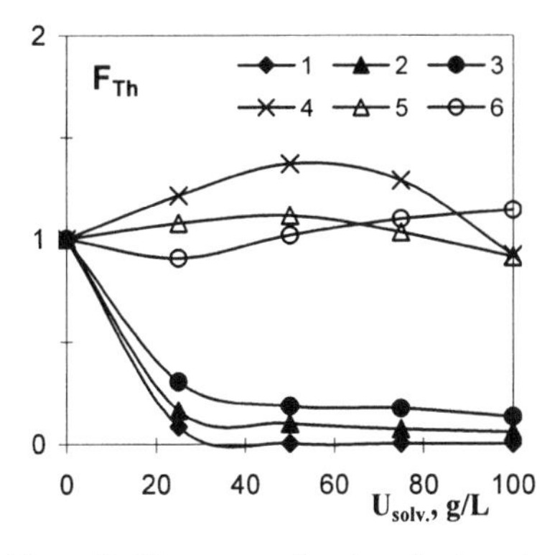

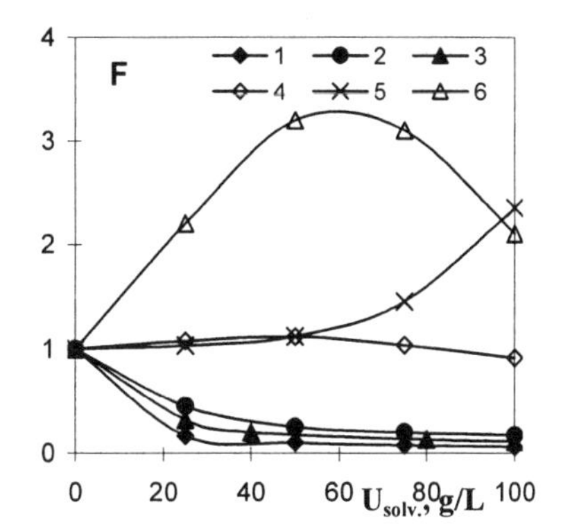

Figure 5. *F factor as a function of solvent loading with U(VI) for Th extraction by 30% TBP in the presence of HDBP.*
Initial concentration of Th (g/L) and the ratio HDBP:Th, correspondingly: ***1-3*** *– 0,125 & 30;* ***4-6*** *– 5 & 1,1. Acidity, HNO_3, mole/L:* ***1*** *&* ***4*** *– 1;* ***2*** *&* ***5*** *– 3;* ***3*** *&* ***6*** *– 5.*

Figure 6. *F factor as function of solvent loading with U(VI) for An(IV) extraction by HDBP dissolved in 30% TBP. Aqueous acidity 3 mole/L HNO_3.*
Initial concentration of An (g/L) and the ratio HDBP:An, correspondingly: ***1-3*** *– 0,125 & 30;* ***4-6*** *– 5 & 1,1.*
1 *&* ***4*** *– Th,* ***3*** *&* ***6*** *– Np(IV),* ***2*** *– Pu(IV),* ***5*** *– U(IV)*

In terms of F factor the behavior of Th at microconcentrations is similar to that of other tetravalent actinides, Pu & Np. However at higher concentrations the behavior of teravalent Th, Np and U all differ markedly (Figure 6). For example, the F factor for Np(IV) is much higher than 1 in high acidity range, particularly at 3 mole/L, whereas the curve for U(IV) coincides with that for Th up to 50 g/L U(VI) loading, then rising, reaching Np(IV) values. The data on Zr behavior will be published later.

3 CONCLUSION

HDBP addition to diluted TBP results in increase of tetravalent actinide distribution coefficients both in the presence and in the absence of U(VI). At low acidity and in the presence of U(VI) a mixed complex, $An_{IV}(NO_3)_3(TBP)_2$-R_2OPO-$UO_2(NO_3)_2(TBP)$ is formed.

The use of the **F** factor, an expression derived from ratios of actinide distribution data in the presence and absence of U(VI) and DBP, allows a quantitative estimation of the extent of this interaction. The values of $F<1$ were obtained at microconcentrations of An(IV) and high HDBP:An(IV) ratio. However, solvent loading with U(VI) affects An(IV) behavior in different ways at their high concentration and HDBP:An(IV) = 1. The maximum values of $F>>1$ for Np(IV) were obtained at 3 mole/L HNO_3 and a solvent loading of 60 g/L of U(VI), whereas the maximum values of F for Th(IV) were found at 1 mole/L of HNO_3, the values for U(IV) lying in between.

References

1 I. May, R.J. Taylor, A.L. Wallwork, J.J. Hastings, Yu.S. Fedorov, B.Ya. Zilberman, E.N. Mi-shin, S.A. Arkhipov, I.V. Blazheva, L.Ya. Poverkova, F.R. Levins and J.M. Charnok. *Radiochimica Acta*, 2000, **88**, No 2, 283.

2 Yu. S. Fedorov, B. Ya. Zilberman, S. M. Kulikov, I. V. Blazheva, E. N. Mishin, A.L. Wallwork, I.S. Denniss, I. May and N. J. Hill. *Solv. Extr. & Ion. Ecxch.* 1999, **17**, No 2, 243.

3. V.I. Kuznetsov. *J. of Organic Chemistry (rus)*, 1944, **14**, p.914.

SORPTION PROPERTIES OF NEW SOLID EXTRACTANTS BASED ON MALONAMIDE COMPOUNDS

J. Sulakova, J. John and F. Sebesta

Czech Technical University in Prague, Department of Nuclear Chemistry and Centre for Radiochemistry and Radiation Chemistry, Brehova 7, 115 19 Prague 1, Czech Republic

1 INTRODUCTION

The DIAMEX process, with malonamide compounds as extractants, is one of the processes proposed for chemical separation of minor actinides and lanthanides from high level liquid nuclear wastes arising from the reprocessing of spent nuclear fuels. For similar type of applications, granular solid materials with malonamide extractants impregnated onto a suitable support, were recently proposed.[1] At the Czech Technical University in Prague, novel composite solid extractants (SEX) / extraction chromatographic systems, with N,N'-dimethyl-N,N'-dibutyltetradecylmalonamide (DMDBTDMA–PAN) or N,N'-dimethyl-N,N'-dioctyl-hexyloxyethylmalonamide (DMDOHEMA–PAN) as extractants and modified polyacrylonitrile (PAN) binding matrix, have been developed. The main advantage of the new SEX-based procedures is a combination of the selectivity of common extractants used in liquid-liquid extraction with the simplicity of the column arrangement. In this study, sorption properties of these materials were characterised in detail.

2 EXPERIMENTAL

The DMDBTDMA–PAN and DMDOHEMA–PAN composite solid extractants were prepared from the respective extractants produced by PANCHIM, France (batch 4/93 or 02/02, respectively). The procedure used was a modification of the method developed for the preparation of composite ion exchangers,[2] the beads were produced by coagulation (in an aqueous bath) of the solution of the extractant in a solution of polyacrylonitrile (PAN). Two solvents were used to prepare PAN solution – (a) concentrated (65 %) nitric acid or (b) dimethylsulfoxide (DMSO). All the SEXs prepared contained 33 % (w/w, dry weight) of the extractants. Fractions with grain size 0.31–0.85 mm were used for the experiments.

Weight distribution ratios, D_g, were determined by batch adsorption experiments performed at volume of solution to the mass of the solid extractant ratio V/m = 250 ml/g, 20 hours contact time, and at room temperature. The solution of 0.1 M $NaNO_3$ in HNO_3 with the desired concentration was labelled with $^{152+154}Eu$, ^{241}Am, ^{233}U or ^{239}Pu radionuclides for the experiments. For the case of europium, its concentration was adjusted to 10^{-5} mol/l, for ^{233}U, the respective uranium concentration in the solution was $2.1.10^{-6}$ mol/l. The tracer concentrations of americium and plutonium used for these experiments were 0.2 kBq ^{241}Am/ml or 80 Bq ^{239}Pu/ml. Glass microfibre filtration was used to separate the SEX from the solution at the end of the experiment. High resolution gamma-ray spectrometry or liquid scintillation counting were used for counting the aliquots of the initial labelled solution and the solution after equilibration with the SEX.

Kinetic experiments were performed by contacting 0.1 g of SEX material with 50 ml of 10^{-5}M Eu + 0.1M $NaNO_3$ + 0.001 M HNO_3 solution in a stirred glass reactor. For the measurement, 0.2 ml samples were periodically taken.

Practical dynamic extraction capacities were determined from the break-through curves of europium through a column of solid extractant. They were expressed per ml or per gram of the bed of the absorber as volume or mass capacities – *Q(V)* or *Q(m)*, respectively. The experiments were performed in polypropylene Resorian™ A161 Cartridge (Supelco, USA) with ID = 9 mm and absorber bed volume BV = 1.28 ml; 10^{-3}M Eu + 0.1M $NaNO_3$ + 0.01M (3M) HNO_3 solution labelled with $^{152+154}$Eu was passed through the column at a flow rate of 4–4.6 BV/hr.

3 RESULTS AND DISCUSSION

In the first phase, dependences of Eu, Am, Pu, and U weight distribution coefficients, D_g, were determined for all the new solid extractants. The results obtained are summarised in Figure 1. From these data, it can be seen that the behaviour of the solid extractants prepared from solution of PAN in DMSO or from its solution in cc HNO_3 is almost identical for all the nuclides studied.

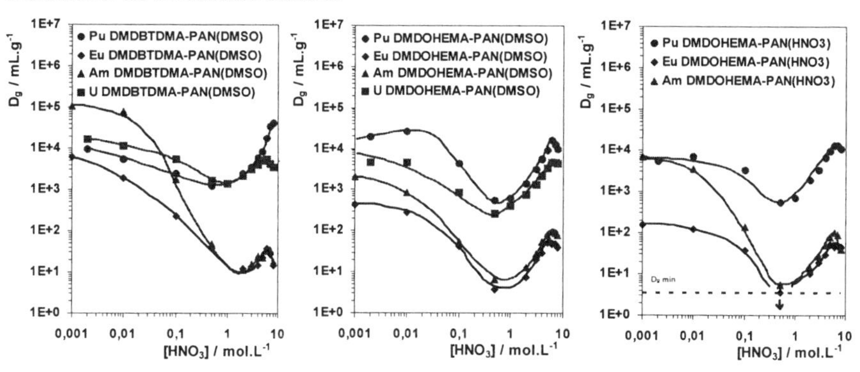

Figure 1 *Comparison of experimental values of weight distribution ratios D_g of Eu, Am, U and Pu on the new solid extractants (0.1 M $NaNO_3$ + HNO_3, V/m = 250 ml/g, 20 hrs contact time)*

When compared[3] with the published data, it can be seen that in the nitric acid concentration range 2–8 mol/l the behaviour of the solid extractants closely follows the behaviour of DMDBTDMA or DMDOHEMA in liquid-liquid extraction – for all the nuclides a sharp peak was observed on the dependence of D_g on nitric acid concentration at the HNO_3 concentration 6 mol/l. However, the measured D_g values are significantly higher than the values that can be calculated from the published values of distribution ratios, *D*, for liquid-liquid extraction which is another argument in favour of the application of the new solid extractants. As expected, for Eu and Am, the D_g values are generally higher for the DMDOHEMA–PAN material and an Am/Eu separation factor $SF_{Am/Eu} \sim 2$ is achieved with this material. For U and Pu, the performance of the DMDBTDMA–PAN and DMDOHEMA–PAN materials is almost identical.

When correlating the behaviour of the solid extractants with that of DMDBTDMA or DMDOHEMA proper in liquid-liquid extraction at low nitric acid concentrations (below 1 mol/l), the D_g values of all the nuclides were unexpectedly found to increase sharply with decreasing nitric acid concentrations, while the respective D values steadily decrease. The reason for this difference is currently under investigation.

The kinetics of uptake of europium from 0.1M $NaNO_3$ + 0.001 M HNO_3 solution was found to be relatively fast. It is somewhat faster for the materials prepared from PAN solution in DMSO – the equilibrium has been reached in approximately in 1 hour for both the DMDBTDMA–PAN(DMSO) and DMDOHEMA–PAN(DMSO) materials. For the materials prepared from PAN solution in cc HNO_3 the kinetics is ~ 50 % slower.

Practical dynamic extraction capacities for europium were determined for solid extractants prepared from the solution of PAN in cc HNO_3. All the break-through curves measured were almost symmetric, practical capacities *Q(V)* and *Q(m)* could therefore be calculated from the mass of europium contained in the volume of feed solution that passed through the column till 50% break-through. In Table 1 these data are compared with the theoretical capacities calculated from the known contents of malonamides in SEXs and assuming 1 : 2 ratio of Eu and malonamides[4] in the extracted complex.

Table 1 *Comparison of experimental practical capacities and theoretical capacities of the SEX materials prepared from HNO_3 solutions of PAN (0.001M Eu + 0.1 M $NaNO_3$ + 0.001M or 3M HNO_3, ~ 4 BV/hr)*

SEX	*Theoretical capacity (mmol/g)*	*HNO_3 (mol/l)*	*Practical capacity*	
			Q(m) (mmol/g)	*Q(V) (mmol/ml)*
DMDBTDMA-PAN(HNO_3)	0.379	0.01	0.070	0.008
		3	0.029	0.004
DMDOHEMA-PAN(HNO_3)	0.344	0.01	0.034	0.003
		3	0.017	0.002

4 CONCLUSION

The results of a detailed study of new solid extractants comprising two extractants from malonamides family incorporated into a binding matrix of polyacrylonitrile revealed that their properties are promising for their potential application in the partitioning of lanthanides and minor actinides from high active waste. More research will be needed to explain the behaviour of the new materials at low nitric acid concentration.

Acknowledgements

This research was supported by EC FP6 F16W-CT-2003-508854 project EUROPART. Drs. Ch. Madic and M. Lecomte of CEA, France, provided the malonamide extractants.

References

1 K.V. Hecke and G. Modolo, *J. Radioanal. Nucl. Chem.*, 2004, **261(2)**, 269.
2 F. Sebesta, *J. Radioanal. Nucl. Chem.*, 1997, **220(1)**, 77.
3 J. Sulakova, F. Sebesta and J. John, *Chem. listy*, 2004, **98(8)**, 752 (in Slovak).
4 G.R. Choppin, M.K. Khankhasayev and H.S. Plendl, *Chemical Separation in Nuclear Waste Managment*, Florida State University, USA, 2002, p. 40.

DIRECT EXTRACTION OF URANIUM AND PLUTONIUM WITH SUPERCRITICAL CARBON DIOXIDE FROM MOX FUEL

S. Ogumo,[1] T. Shimada,[2] Y. Mori,[2] Y. Kosaka,[3] M. Mizuno,[3] S. Miura,[4] M. Kamiya,[4] T. Koyama[4] and Y. Enokida[5]

[1] Advanced Reactor Technology Co., LTD: kounan2-16-5, minato-ku, Tokyo, Japan
[2] Mitsubishi Heavy Industries, Ltd.: Nishi-ku, Yokohama 2208401 Japan
[3] Nuclear Development Corporation: Tokai-Mura, Naka-Gun 3191111 Japan
[4] Japan Nuclear Cycle Development Institute: Tokai-Mura, Naka-Gun 3191194 Japan
[5] Nagoya University: Chikusa-ku, Nagoya 4648603 Japan

1 INTRODUCTION

The Super-Direx process is a reprocessing method based on supercritical fluid extraction by a mixture of carbon dioxide (CO_2) and a complex of tri-*n*-butyl phosphate with concentrated nitric acid (TBP-HNO_3). In this process, uranium (U) and plutonium (Pu) are extracted directly from the spent fuel at 313-333 K and 18-26 MPa without the acid dissolution process. The Super-Direx process needs only a single column where dissolution of the fuel and extraction of heavy metals proceed simultaneously and efficiently. This simplicity leads to drastic reduction of construction cost as compared with the conventional reprocessing plant.[1-3] The Super-Direx process was experimentally applied to MOX fuel processing in Japan.

2 EXPERIMENTS

2.1 Methods

The experimental apparatus is shown in Fig.1 and 2. Non-irradiated UO_2 or MOX (Pu: 22 wt%) powder was placed in a direct extraction column. The amount of fuel was 6 g. The diameter of fuel powder was under 0.1 mm. The pressure of the column was increased by CO_2 and kept at 18 MPa or 26 MPa. The temperature was kept at 313 K or 333 K. Extraction was started by supplying the mixture of TBP-HNO_3 complex (TBP: 2.8 M, HNO_3; 4.8 M) and CO_2. The concentration was 0.52 mol-TBP/L-scf. The extracted complex was separated from CO_2 by reducing pressure. The concentration of U and Pu extracted was calculated on the basis of analysis by absorption spectroscopy or alpha spectrometer.

2.2 Results

The concentration of U and Pu extracted in SCF is shown in Fig.3. The mole fraction of U + Pu and TBP in the extracted complex is shown in Fig.4. Contacted time means time in which CO_2 passed fuel in the column. In two cases of No.1 and No.2 using UO_2, the concentration was saturated in a contacted time of 2 minutes and the maximum was more than 65 g-(U+Pu)/L-scf. The mole fraction of U + Pu and TBP was about 0.5. The equivalent mole fraction of U or Pu and TBP is supposed to be 0.5 in the chemical forms $UO_2(NO_3)_2(TBP)_2$ or $Pu(NO_3)_4(TBP)_2$. Therefore, almost all the TBP was used for complexation reaction with U or Pu. In the case of No.3 (MOX, 313 K, 18 MPa), the maximum concentration was about 31 g-(U+Pu)/L-scf. The mole fraction was about 0.25. Therefore, about half of the TBP was used for complexation reaction with U or Pu. In the case of No.4 (MOX, 333 K, 26 MPa), the maximum was 41 g-(U+Pu)/L-scf. The mole fraction was about 0.25-0.45. Increasing contacted time increased the mole fraction of. Pu that was extracted in almost the same recovery ratio as U in all cases.

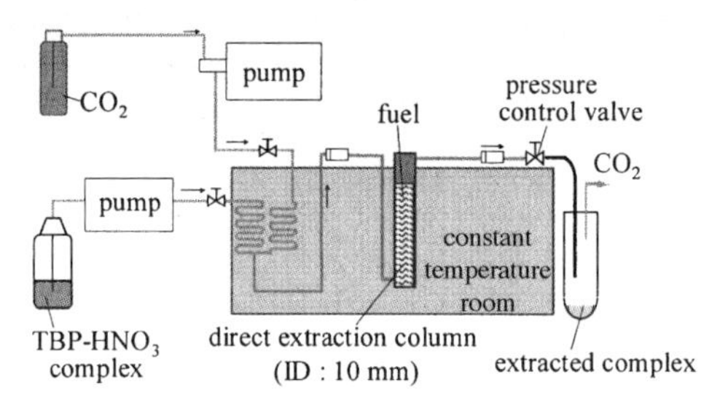

Figure 1 *Experimental apparatus*

Figure 2 *Appearance of experimental apparatus (at the Chemical Processing Facility in the Japan Nuclear Cycle Development Institute)*

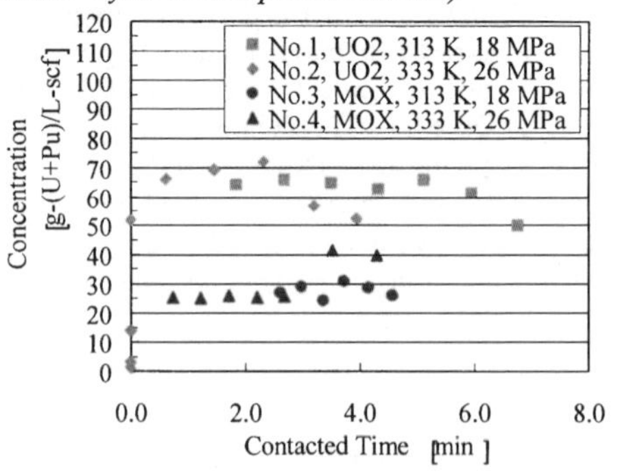

Figure 3 *Concentration of U and Pu extracted in SCF*

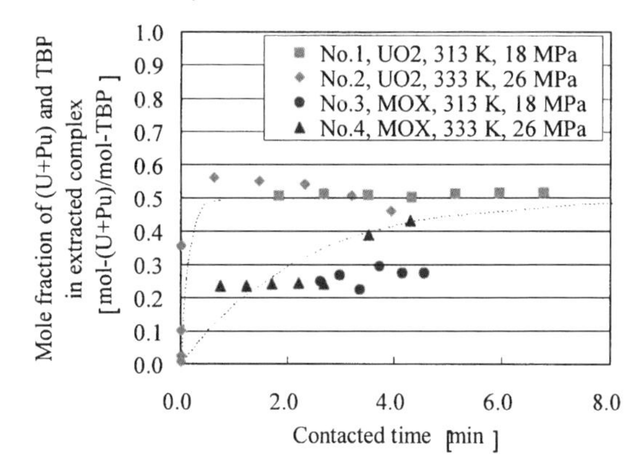

Figure 4 *Mole fraction of U + Pu and TBP in extracted complex*

3 DISCUSSION

The direct extraction process is operated continuously by supplying fuel to the column semi-continuously to give enough contacted time. The capacity is the product of concentration of U and Pu in SCF and flow rate of SCF. If the flow rate is 6 L/min, the annual reprocessing capacity is more than 50 t-U+Pu. The diameter of the column is constrained under the critical dimension e.g. 130 mm for FBR enriched ~30 %-Pu. In this case, the rate of streamline flow is 7.5 mm/s, 8 times the previous experimental condition.[2]

The data demonstrated that direct extraction to SCF has the same tendency as follows: generally, the rate of dissolution of MOX to nitric acid is lower than that of UO_2 and that of non-irradiated MOX is lower than that of irradiated MOX. To make a direct extraction column compact, it is desirable that the mole fraction of U + Pu and TBP reaches a saturation point in a short contacted time. If this contact time is 4 minutes, the column is about 6m high.

Hereafter, basic experiments investigating the decontamination factor of minor actinides and fission products are needed for optimization of the process, and experiments under the real flow rate with a column of industrial scale are needed.

4 CONCLUSION

The experiments on direct extraction of U and Pu with supercritical CO_2 from MOX or UO_2 fuel were conducted and the data demonstrated that the direct extraction column could have enough capacity for practical application.

Acknowledgements

This work was partially supported by the Institute of Applied Energy, Japan.

References

1 Y. Enokida, *Proc. Super Green 2002,* 2002, pp. 35-41.
2 T. Shimada *et al.*, *Proc. Super Green 2003*, 2004, pp. 169-171.
3 Y. Enokida *et al.*, American Chemical Society, 2003, pp. 10-22.

MODELLING THE HYDROLYSIS OF ACTINIDE COMPLEXED HYDROXAMIC ACID LIGANDS

R.J.Taylor,[1] C. Boxall,[2] F. Andrieux[2] and C. Mason[1]

[1]Nexia Solutions, BNFL Sellafield, Seascale CA20 1PG UK
[2]Centre for Materials Science, University of Central Lancashire, Preston PR1 2HE UK

1 INTRODUCTION

The separation of U from Np and Pu are major stages in the Purex process.[1-2] Simple hydroxamic acids (XHA) are salt free, hydrophilic organic compounds, RCONHOH, with affinities for cations such as Fe^{3+}, Np^{4+} and Pu^{4+} [3-6]. They are also redox active, capable of reducing a range of metal ions e.g. Np(VI) to Np(V).[6] These two properties have led to them being identified as useful reagents for the control of Pu and Np in an Advanced Purex process.[6-9] The kinetics of the acid hydrolysis of free formo- (FHA) and acetohydroxamic (AHA) acids to hydroxylamine and the parent carboxylic acid are well known[10]. Hydrolysis of hydroxamates bound to metal ions also occurs and preliminary studies have shown that the Pu(IV)-FHA and AHA complexes are slowly reduced to free Pu(III) ions.[9] An understanding of these processes is vital if they are to be controlled within the design of an Advanced Purex process. To this end, we have used UV-visible & nIR spectrophotometry to study and theoretically model the kinetics of the hydrolysis of metal-HA systems in nitrate media – Fe^{3+} / AHA (as a non-active analogue); Np^{4+} / FHA, Np^{4+}/AHA – wherein the metal ion complexes with, but does not oxidise the ligand.

2 RESULTS & DISCUSSION

The ferric-AHA and Np(IV)-AHA systems exhibit coloured complexes with AHA:M ratios of 1:1 to 3:1 and 1:1 to 5:1 respectively.[10] The Np(IV)-FHA system is expected to behave similarly to the AHA system. Conversions between the free metal ion and the complexes can be described by the following equilibria,[10] at 298 K in ClO_4^- media:

$$HL + M^{n+} \xrightleftharpoons{K_1} ML^{(n-1)+} + H^+ \qquad K_1(Np) = 2754;\ K_1(Fe) = 5.83$$

$$HL + ML^{(n-1)+} \xrightleftharpoons{K_2} ML_2^{(n-2)+} + H^+ \qquad K_2(Np) = 5.49;\ K_2(Fe) = 2.04$$

$$HL + ML_2^{(n-2)+} \xrightleftharpoons{K_3} ML_3^{(n-3)+} + H^+ \qquad K_3(Np) = 4.5 \times 10^{-2};\ K_3(Fe) = 7.4 \times 10^{-3}$$

$$HL + ML_3^{(n-3)+} \xrightleftharpoons{K_4} ML_4^{(n-4)+} + H^+ \qquad K_4(Np) = 6.02 \times 10^{-5}$$

$$HL + ML_4^{(n-4)+} \xrightleftharpoons{K_5} ML_5^{(n-5)+} + H^+ \qquad K_5(Np) = 5.88 \times 10^{-7}$$

Speciation diagrams (omitted) were calculated and used to identify solution conditions under which the mono-XHA complex dominates (using the Np(IV)-AHA system as a guide for the as yet uncharacterised Np(IV)-FHA system). Use of such conditions then allows assumption of the following mechanism in modelling the hydrolysis of HA ligands in the presence of non-oxidising metal ions and competitive complexation by nitrate ions.

$$M(NO_3) \xleftrightarrow{K_D} M + NO_3^- \tag{1}$$

$$HL + M^{n+} \xleftrightarrow{K_1} C + H^+ \tag{2}$$

$$HL + H^+ \xrightarrow{k'_0} RCOOH + NH_3OH^+ \tag{3}$$

$$C + mH^+ \xrightarrow{k'_1} M + RCOOH + NH_3OH^+ \tag{4}$$

Where M_t = $[Fe^{3+}]$ or [Np(IV)] at time t; HL_t = [XHA] at time t; (5a,b)

C_t = [metal-HA complex] at time t; $k_1 = k'_1[H^+]^m$ in analogy to (5c,d)

$k_0 = k'_0[H^+]$ in accordance with equation 2. (5e)

Using this mechanism, it can be shown that

$$C_t = \frac{K'_1 \, HL_t M_T}{[H^+] + K_1 HL_t} \tag{6}$$

where HL_t can be calculated using

$$t = \frac{1}{k_1}\ln\left(\frac{\frac{K'_1 HL_t}{[H^+]}+1}{\frac{K'_1 HL_0}{[H^+]}+1}\right) - \left(\frac{\frac{K'_1 M_T}{[H^+]}+1}{\frac{k_1 K'_1 M_T}{[H^+]}+k_0}\right)\ln\left(\frac{HL_t}{HL_0}\right) - \left(\frac{\frac{k_1^2 K'_1 M_T}{[H^+]}+k_0^2}{\frac{k_0 k_1^2 K'_1 M_T}{[H^+]}+k_0^2 k_1}\right)\ln\left(\frac{\frac{K'_1 HL_t}{[H^+]}+1+\frac{k_1 K'_1 M_T}{k_0[H^+]}}{\frac{K' HL_0}{[H^+]}+1+\frac{k_1 K'_1 M_T}{k_0[H^+]}}\right) \tag{7}$$

where, for the Np and Fe systems, K'_1 is given respectively by

$$K'_1 = \frac{K_1}{1 + ([NO_3^-]_T / K_D)} \quad \text{and} \quad K'_1 = K_1 \tag{8a,b}$$

Results of kinetic absorbance experiments are shown in Figures 3 & 4. Equation 7 is insoluble with respect to HL_t. However, in conjunction with equation 6, it can be used to fit theoretical C_t vs t curves to experimental data, a process facilitated by only two parameters, K'_1 and k_1 (= $k'_1[H^+]^m$, equation 5d), being unknown for the Np(IV)-FHA system, and only one, k_1 being unknown for the Fe(III)-AHA system.

Parameter k'_1 affects the overall duration of the decay in C_t, an effect that vanishes for very small values of k'_1. This is because, under conditions where the process associated with k'_1 is much slower than the parallel process associated with k'_0, the latter determines the overall rate. In contrast, K'_1 only affects the degree of sigmoid character in C_t vs t – the larger K'_1, the longer the observed induction period before C_t starts to decay *e.g.* $[Np(FHA)^{3+}]$ vs t data at $[HNO_3]$ = 0.79 M, Figure 2. Again, this may be readily understood in that the larger the value of K'_1, the longer a significant/measurable concentration of complex will maintain as a fraction of the overall decay time.

For the Fe(III)-AHA system, curve-fitting indicates that m = 2.01 and $k'_1 = 2.2 \times 10^{-3}$ $dm^6\ mol^{-2}\ s^{-1}$. Use of these parameters to calculate values of k_1 and to compare them with k_0 as a function of pH indicates that, at pH < 2, $k_1 > k_0$ i.e. Fe(III)-complexed AHA is more susceptible to acid catalysed hydrolysis than the free AHA. This trend is reversed at pH > 2. For the Np(IV)-FHA system, $k'_1 < 1 \times 10^{-5}$ $dm^6\ mol^{-2}\ s^{-1}$ and K_1 = 2029, the latter indicating that complexation of Np(IV) with FHA is slightly weaker than that with AHA.

3 CONCLUSIONS

Kinetic analysis allows for the calculation of a value of K_1 for the Np(IV)-FHA system of 2029 at 293 K. This same analysis indicates that complexation with Fe(III) protects AHA against hydrolysis at pH > 2. This effect is reversed at pH < 2. Importantly for reprocessing applications, at 0.1 > pH > -0.11, complexation by Np(IV) protects FHA against hydrolysis, thus enhancing the robustness of the flow sheet. Further study of the hydrolysis of FHA in the Np(IV)-FHA system over a wider pH range would allow for determination of whether, in analogy to the Fe(III)-AHA system, FHA hydrolysis is enhanced below some threshold pH value. Extension of these modelling approaches to Pu(IV) will provide a sound basis to fully understand the stability of hydroxamic acids in process flow sheets.

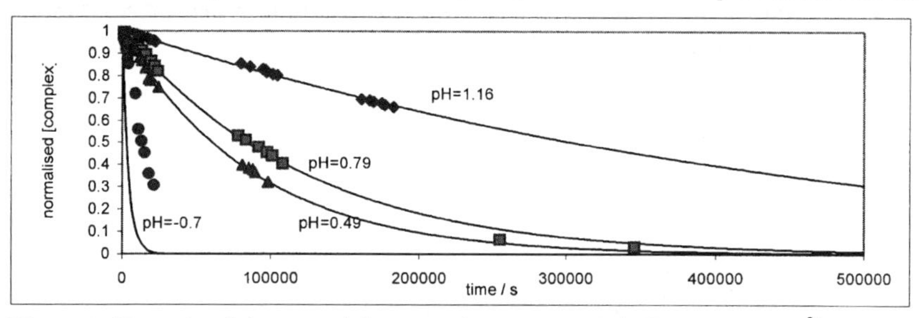

Figure 1 *Normalised theoretical (lines) and experimental (bullets) $[Fe(AHA)^{2+}]$ vs. t $[Fe^{3+}]$=2.5 mM and [AHA]=4 mM at a range of pH values T=293K.*

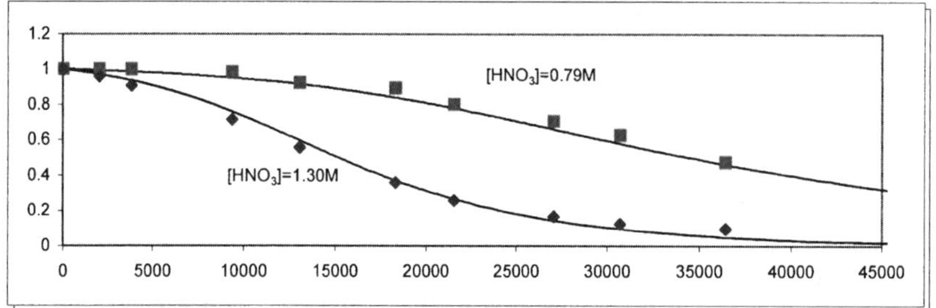

Figure 2 *Normalised experimental (bullets) and theoretical (lines) $[Np(FHA)^{3+}]$ vs t for $[Np^{4+}]$=5 mM and [FHA]=0.1 M in 0.79 & 1.30 M HNO_3, (K'_1 = 2029).*

References

1 P.D.Wilson (ed.), I.S.Dennis and A.P.Jeapes, *The Nuclear Fuel Cycle*, Oxford Science Publications, 1996, Chapter 7, p. 116.
2 P.Baron, B.Boulis, M.Germain, J.P.Gué, P.Miquel, F.J.Poncelet, J.Dormant and F.Dutertre, Proceedings GLOBAL '93, 1993, p. 63.
3 P.Desaraju and A.Winston, *J.Co-ord.Chem.*, 1986, **14**, 241.
4 A.Barocas, F.Baroncelli, G.B.Biondi and G.Grossi, *J.Inorg.Nucl.Chem.*, 1966, **28**, 2961.
5 F.Baroncelli and G.Grossi, *J.Inorg.Nucl.Chem.*, 1965, **27**, 1085.
6 R.J.Taylor, I.May, A.L.Wallwork, I.S.Dennis, N.J.Hill, B.Y.Galkin, B.Y.Zilberman, Y.S.Fedorov, *J.Alloys Comp.*, 1998, **271-273**, 534.
7 I.May, R.J.Taylor, I.S.Denniss, G.Brown, A.L.Wallwork, N.J.Hill, J.M.Rawson and R.Less, *J.Alloys Comp.*, 1998, **275-277**, 769.
8 I.May, R.J.Taylor and G.Brown, *J.Alloys Comp.*, 1998, **271-273**, 650.
9 R.J.Taylor, C.Mason, R.Cooke and C.Boxall, *J.Nucl.Sci.Tech. Supplement*, 2002, **3**, 278.
10 R.J.Taylor, *J.Nucl.Sci.Tech.*, Supplement, 2002, **3**, 886.

A NEW APPROACH FOR ESTIMATING STABILITY CONSTANTS BETWEEN ACTINIDES AND ORGANIC LIGANDS BY SEMI-EMPIRICAL GIBBS ENERGY FRAGMENTAL METHOD

T. Sasaki, S. Kubo, T. Kobayashi, I. Takagi, and H. Moriyama

Department of Nuclear Engineering, Kyoto University, Yoshida, Sakyo-ku, Kyoto 606-8501, JAPAN, sasaki@nucleng.kyoto-u.ac.jp

1 INTRODUCTION

In the separation chemistry of trivalent f-elements, it is still hard to reach the principle for the stability and selectivity of ligand with a variety of donor atoms, because the complex formation reaction is controlled by many kinds of factors; the electrostatic force, the ligand conformation energy, and the intra- and inter-ligand steric hindrance, *etc*. For a theoretical approach to interpret the stabilization, though computational approach has been applied for decades, the knowledge is not complete enough to understand complex stability. An empirical approach is thus still useful for a better understanding of complex stability on the basis of systematic experimental data. In the present study, the formation constants of Am and Cm with a series of carboxylates containing neutral donors such as amine and alcohol oxygen were determined by solvent extraction. In order to analyze the trend of formation constants, a simple model was developed in which these ligands were considered to be a number of fragment components of functional groups and chemical bridges for their different contributions to the formation constants.

2 METHOD AND RESULTS

The experimental procedure for determining formation constants by a solvent extraction was described elsewhere.[1] Several carboxylic acids, L, were used in this study as shown in Fig. 1.

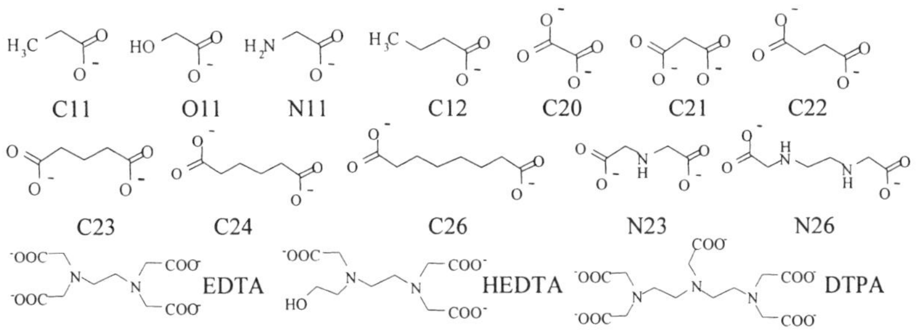

C11 O11 N11 C12 C20 C21 C22

C23 C24 C26 N23 N26

EDTA HEDTA DTPA

Figure 1 *Scheme of chemical structures of several carboxylic acids as anion forms.*

The formation constant of 1:*n* complex can be described as $\beta_n = [ML_n]/[[M][L]^n$. A simple semi-empirical model to analyze the formation constant values has been proposed.[2] In short; the free energy change ΔG_l is evaluated by considering a hypothetical process. The ΔG_l can be divided into three energy factors, ΔG_{frg}, ΔG_{ele} and ΔG_{str}, for the fragmentation process, the electrostatic and steric effects, respectively. At the first step of the process, the ligand molecule is fragmented into functional groups and chemical bridges with the change of energy, ΔG_{frg}. The fragment components are of carboxyl, amine and methylene chain groups, and so forth. Following this, the energy changes by the electrostatic and steric effects are taken into consideration. The coordination power of a functional carboxylic group as an enthalpic contribution can be described such as $-\Delta G_{ele,(-COO-)} = RT \ln K_{ele,(-COO-)}$. In the case of multiple anionic donors, it is needed to take into consideration an energy term ($\Delta G_{ele,rep}$) for the Coulomb repulsion between anionic donors and Neck *et al.* have introduced the repulsion energy term E_L.[3] For simplicity the charge q of carboxyl group and neutral donors such as water molecule and amine group is set at -1 and 0 respectively. The ΔG_{str} of a ligand is defined as the steric effect in a sense of entropic energy concept. All chelating bridges between the nearest two donor atoms are considered. The extent of the effect is categorized by the position of the two donor atoms (Table 1) as reported previously.[2] The total strain energy of the ligand can be described as the summation of the strain effect of bridges. Parentheses denote the unit number of methylene group $-CH_2-$. All ligands can be fragmented in a similar manner. The experimental data of log β_ns are used in the least squares fitting analysis to obtain the parameter values for the electrostatic and steric effects of fragment components of the ligands. The formation constants (25°C, I=0.1) are summarized in Table 2. Also, the parameter values fitted are summarized in Table 1.

Table 1
Assignment of parameters for electrostatic and steric effects to ligands.

Parameters			Values fitted	
log K_{ele}			Am(III)	Cm(III)
$-COO^-$			2.17	2.23
	-ROH		-0.67±0.19	-0.59±0.18
		$-^{\#1}N$	0.12±0.44	0.20±0.42
		$-^{\#2}N$, $-^{\#3}N$	0.69±0.19	0.88±0.18
log K_{str}				
Type I		Type II		
$^-$OOC-C(0)-COO$^-$	HOC-C(0)-COO$^-$ H_2NC-C(0)-COO$^-$	$^{\#2}$NC-C(0)-COO$^-$	1.21±0.15	1.07±0.14
$^-$OOC-C(1)-COO$^-$	HOC-C(1)-COO$^-$ H_2NC-C(1)-COO$^-$		0.27±0.04	0.13±0.04
$^-$OOC-C(2)-COO$^-$			-0.60±0.05	-0.74±0.05
$^-$OOC-C(3)-COO$^-$	HOC-C(3)-COO$^-$		-0.96±0.07	-1.10±0.07
$^-$OOC-C(4)-COO$^-$			-0.93±0.08	-1.07±0.07
$^-$OOC-C(6)-COO$^-$			-1.12±0.10	-1.26±0.09
		$^{\#3}$NC-C(0)-COO$^{-\#3}$ NC-C(0)-COH	1.97±0.18	1.68±0.17
Type III				
$^{\#2}$NC-C(0)-C$^{\#2}$N			3.07±0.41	3.15±0.40

For instance, the $\log K_{ele}$ values of "-ROH" are found to be negative, suggesting the strong hydration energy to the trivalent actinide ions compared to alcohol OH. On the other hand, the values of amine groups are positive, which may suggest the stronger coordination energy and some additional non-electrostatic interactions between 5f-elements and the nitrogen donor. The summation of the obtained fitting parameter values (calc.) is consistent with the experimental formation constants with small standard errors as shown in Table 2.

Table 2 *Stability constants calculated (β_1; upper, β_2; lower) with the difference from experimental values*; $\Delta\beta$ = $\log\beta$(calc)-$\log\beta$(exp). *Ref.4 for lanthanides.*

	Am(III)		Cm(III)	
	$\log\beta$	$\Delta\beta$	$\log\beta$	$\Delta\beta$
C11	2.17±0.13 3.02±0.44	0.0 1.0	2.23±0.13 3.22±0.44	0.0 1.0
O11[4]	2.75±0.14 5.05±0.11	0.0 0.1	2.75±0.14 5.05±0.11	0.0 0.1
N11[4]	3.50 -	0.0	3.50 -	0.0
C12[4]	2.04 -	0.1	2.04 -	0.2
N23	7.04±0.04 12.42±0.05	0.1 0.3	7.12±0.05 12.58±0.06	0.1 0.3
N26	10.21±0.58 19.59±0.23	0.7 0.6	10.76±0.71 20.09±0.38	0.5 0.9
C20	5.25 8.85	0.0 0.1	5.25 8.85	0.0 0.1
C21[4]	4.29 6.99	0.0 0.0	4.29 6.99	0.0 0.1
C22[4]	3.42 -	0.0	3.42 -	0.0
C23	3.08±0.04 4.69±0.08	0.0 -0.1	3.11±0.06 5.00±0.08	0.0 -0.4
C24[4]	3.09 -	0.0	3.09 -	0.0
C26	2.92±0.12 4.49±0.27	0.0 -0.2	3.00±0.08 4.66±0.14	-0.1 -0.3
EDTA	17.80	-0.5	18.10	-0.9
HEDTA	15.70	-0.1	15.70	-0.2
DTPA	22.92	0.7	22.99	0.7

3 CONCLUSION

The formation constants of Am and Cm with a series of carboxylates were analyzed by a simple fragment model in which the ligand molecules were fragmented to functional groups and chemical bridges. It was found that the negatively charged oxygen and the amine enhance the β of trivalent actinides by the electrostatic interaction while the steric effect of longer bridge length reduces it.

References

1 T. Sasaki, S. Kubo, T. Kubota, I. Takagi, H. Moriyama, *J. Alloys Comp., in press.*
2 T. Sasaki, S. Kubo, T. Kobayashi, I. Takagi, H. Moriyama, *J. Nucl. Sci. Technol.* 42 (2005) 724-731.
3 V. Neck, J.I. Kim, *Radiochim. Acta* 88 (2000) 815-822.
4 R.M. Smith, A.E. Martell (Eds.): *Critical Stability Constants Vol.6, Second Supplement,* Plenum Press. New York and London (1989).

MEASUREMENT OF SOME ACTINIDES REDOX POTENTIALS IN TBP MEDIUM WITH ULTRAMICROELECTRODE

H. Mokhtari and S. Picart

CEA, DEN/DRCP/SCPS, Actinides Chemistry Laboratory, BP 17171, 30207 Bagnols-sur-Cèze FRANCE

1 INTRODUCTION

The determination of the redox potentials of actinides, in both aqueous and organic media,[1] is of prime importance if one wants to control the process of liquid-liquid extraction. Indeed, the affinity of the extracting phase for the actinide depends essentially on its redox state which leads to the formation of coordination complex in the organic medium. Our aim was to measure the redox potential of different actinides solvated in tributylphosphate (TBP) by the technique of ultra-microelectrodes (UME) which are well adapted to electrochemical measurements in poorly conducting solutions.[2]

2 RESULTS

2.1 Qualification of the equipment and of the measurement

A first test of the UME was carried out with reference compounds of the family of ferrocene (Fc) for both aqueous and TBP solutions at room temperature under argon. Those electroactive species are characterised by a rapid electron transfer reaction and their redox potential are considered independent of the solvent according to the Strehlow hypothesis.[3] The voltammetry of ferrocene-methanol (FcMeOH) 2.5 mmol/L in KCl 0.1 mol/L aqueous solution is reported in Figure 1 for both micro (5mm diam.) and ultramicro-electrodes (10μm diam.). The value of the apparent potential of the ferrocinium-methanol/ferrocene-methanol redox pair and the diffusion coefficient of FcMeOH were measured and compared for both electrodes: those values are equivalent (Table 1) within error. Thus the ultramicroelectrodes performed well in water.

Table 1 *Potential and diffusion coefficient values of ferrocene-methanol measured in KCl 0.1 mol/L aqueous solution on micro and ultramicro electrodes.*

FcMeOH/KCl 0.1M	*Potential*/mV/SHE	*Diffusion coefficient*/$cm^2\ s^{-1}$
macroelectrode	424 ±5	0.64 ±0.05
UME	419 ±5	0.67 ±0.05

The behaviour of ferrocene-methanol in HNO_3 5 mol/L aqueous solution is quite different because it is readily oxidised in ferricinium-methanol by nitrous acid contained in concentrated nitric acid. The voltametry of ferrocinium-methanol is then a reduction wave practically symmetric by the potential axis to the response of FcMeOH in KCl 0.1 mol/L.

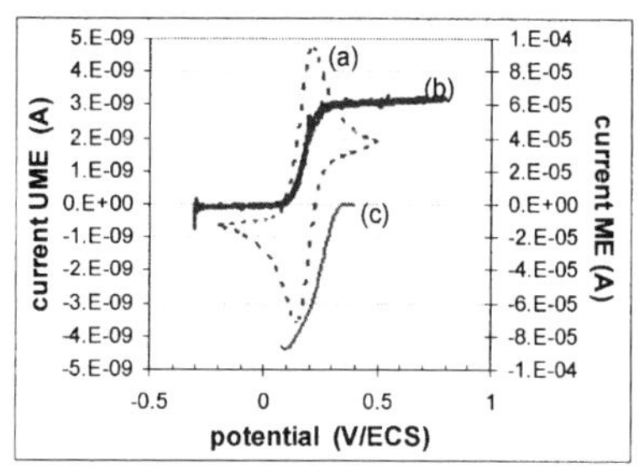

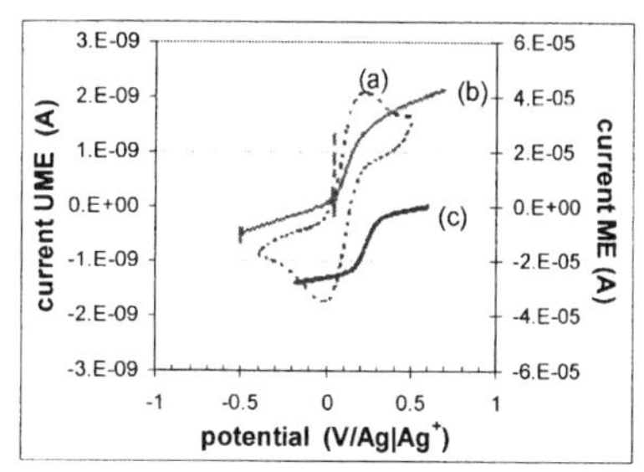

Figure 1 *On the left : Voltammetry of ferrocene-methanol in KCl 0.1M aqueous solution on microelectrode (a) ; on UME (b) and in* HNO_3 *5M on UME (c).* $V=20mV\,s^{-1}$*. On the right : Voltammetry of ferrocene in TBATf 0.1M TBP solution on microelectrode (a) ; on UME (b) and in* HNO_3 *2M TBP solution on UME (c).* $V=10mV\,s^{-1}$*.*

The voltammetry of Fc (2.5 mmol/L) in 0.1 mol/L TBATf /TBP solution is given in Figure 1 for both micro and ultramicro-electrodes. It is similar to that of FcMeOH in aqueous solution except that ohmic resistance of the solution is far stronger for TBP than for water. The measurements of potentials and diffusion coefficients on micro and ultramicro electrodes are collected in Table 2. They are in good agreement which confirms the qualification of the use of UME in organic medium like TBP.

Table 2 *Potential and diffusion coefficient values of ferrocene measured in TBATf 0.1 mol/L TBP solution on micro and ultramicro electrodes.*

Fc/TBP-TBATf 0.1M	*Potential*/mV/Ag+|Ag	*Diffusion coefficient*/$cm^2\,s^{-1}$
macroelectrode	110	0.18
UME	130	0.21

The diffusion coefficient of Fc is lower in TBP than in H_2O which is a consequence of the difference of viscosity between those two media (3,32 and 1,00 cP at 25°C respectively).[4] If we look at the behaviour of Fc in TBP contacted with an HNO_3 5 mol/L aqueous solution (TBP 2 mol/L HNO_3), the same phenomenon as in H_2O is observed : the nitrous acid co-extracted with nitric acid oxidises Fc in Fc^+. The half wave potential of ferrocene is close to 130 mV/Ag|Ag^+ and the diffusion coefficient is about 0.21 $cm^2\,s^{-1}$.

2.2 Measurement of the redox potential of Pu(IV)/Pu(III) in TBP

The transfer of plutonium in TBP was achieved by contacting an HNO_3 5 mol/L aqueous solution of Pu(IV) (3.8 mmol/L) with TBP saturated with aqueous nitric acid (2.8 mol/L). The voltammetry was recorded on the UME for the Pu(IV) TBP solution under argon (see Figure 2) and is compared to the response of Fc (3.2 mmol/L) in the same condition. The half wave potential corresponding to the formal potential of the Pu(IV)/Pu(III) redox pair in a HNO_3 2.8 mol/L TBP solution is then estimated to be close to 0 V/Fc .

2.3 Measurement of the redox potential of Am(VI)/Am(V) in TBP

The first step was to prepare the aqueous solution of Am(VI) by the anodic oxidation of Am(III) in a 5 mol/L HNO_3 solution in the presence of lacunary phosphotungstate ligand and silver nitrate redox mediator. This oxidation was completed in 1 h 30 and led to a HNO_3 5 mol/L aqueous solution of Am(VI) (6.3 mmol/L). The transfer of Am(VI) in TBP was performed by contacting the aqueous solution of Am(VI) with TBP.

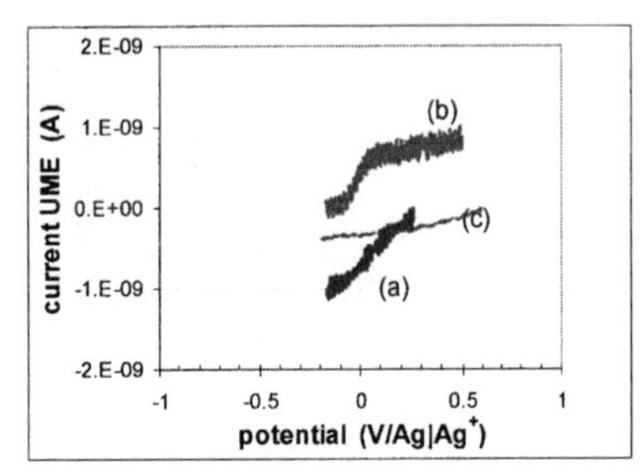

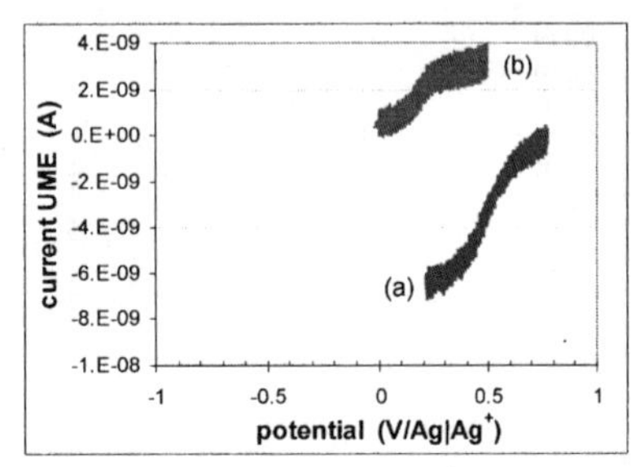

Figure 2 *Left : voltammetry on UME in TBP : (a) Pu(IV) 3.8mM. $V=100mV\,s^{-1}$; (b) Fc 3.2mM $V=20mV\,s^{-1}$; (c) blank solution $V=10mV\,s^{-1}$. Right: voltammetry on UME in TBP solution of : (a) Am(VI) 5.8mM $V=100mV\,s^{-1}$; (b) Fc3.8mM. $V=100mV\,s^{-1}$.*

The voltammetry was recorded on the UME for the Am(VI) TBP solution (see Figure 2) and shows a wave corresponding to the reduction of Am(VI) into Am(V). The value of the formal potential of the Am(VI)/Am(V) redox pair versus Fc, in a HNO_3 2.4 mol/L TBP solution, is estimated to be close to 330 mV/Fc.

3 INTERPRETATION

Those measurements enable us to make a comparison between the potentials of Pu(IV) and Am(VI) in water and TBP considering that the potential of Fc is constant between the two solvents. According to Table 3, one observes a negative shift of potential from aqueous to TBP solvent for both actinides which indicates a stabilisation of the higher oxidation state of the actinides (Pu^{4+} and AmO_2^{2+}) by TBP through a strong electronic donor effect.

Table 3 *Potentials of Pu(IV)/Pu(III) and Am(VI)/Am(V) redox pairs in water and TBP.*

E°'/Fc (mV)	*H_2O/HNO_3* [5]	*TBP/HNO_3*
Pu^{4+}/Pu^{3+}	400	0
AmO_2^{2+}/AmO_2^{+}	1100	330

4 CONCLUSION

This study was initiated to characterise the redox properties of actinides in organic solvent as they play a great role in the process of liquid-liquid extraction. UME were the tools selected for the electrochemical measurements because they are perfectly adapted to poorly electric conducting medium like organic solvents. The TBP solvent was first looked at because it can easily dissolve actinides in different oxidation states. Pu(IV) and Am(VI) were also chosen because of their affinity for TBP. The UME allowed us to measure the potentials of the Pu(IV)/Pu(III) and Am(VI)/Am(V) redox couples in acidic TBP solvent and those results compared to values obtained in acidic water, showing clearly a stabilisation of the actinides in their (IV) or (VI) oxidation states by TBP.

References

1 L. Martinot, C. Licour, L. Lopes, J. Alloys and Compounds, 1995, **228**, 6.
2 D.Hauchard et al, J. Electroanal. Chem., 1993, **347**, 399.
3 B. Trémillon, "Electrochimie analytique", Masson, Paris, 1993, 271.
4 L.L. Burger, "Physical properties of TBP" in "Science and technology of tributyl phosphate",eds., W.S. Schulz and J.D. Navratil, CRC Press, 1984, vol. I, Ch. 3, 31.
5 J.J. Katz et al, "The chemistry of the actinides elements", 1986, Chapman, 818.

DEVELOPMENT OF HIGH THROUGHPUT ELECTROREFINING OF URANIUM IN METALLIC FUEL CYCLE

K. Uozumi, M. Iizuka and T. Ogata

Nuclear Technology Research Laboratory, Central Research Institute of Electric Power Industry, Komae, Tokyo 201-8511, Japan

1 INTRODUCTION

In the electrorefining step of pyro-reprocessing the spent metallic fuel used in FBRs is dissolved anodically in molten LiCl-KCl eutectic salt. In this step, most of uranium is solely collected on a solid cathode and plutonium is collected on a liquid cadmium cathode together with residual uranium.[1] It is important to enhance the processing rate of uranium while collecting on the solid cathode because uranium occupies 70-80% in the spent fuel. For this purpose, Argonne National Laboratory (ANL) in USA has developed a high throughput electrorefiner (HTER) for processing the blanket metallic fuel, which was made of uranium, irradiated at EBR-II reactor.[2]

For adopting the HTER for processing both the blanket and the driver metallic fuels, the difference of the molten salt composition should be considered. Since the driver fuel contains much more plutonium than the blanket fuel, the molten salt composition should be controlled to contain about 6wt% of plutonium and 2wt% of uranium in order to recover plutonium together with some of the uranium into the liquid cadmium cathode.[3] Accordingly, the concentration of uranium in the salt should be reduced by half from the value of around 4wt% in the ANL's HTER.

The objective of this study is to determine the processing rate of uranium at low concentration of uranium in the salt. Furthermore, some ideas were tested for improving the performance of the electrorefiner.

2 EXPERIMENTAL

A laboratory scale high throughput electrorefining apparatus, in which about 500g of uranium can be used, was fabricated. The apparatus is shown in Figure 1. This apparatus employs an anode-cathode module, which consists of a cylindrical cathode and a rotating anode located in the centre of the cathode. The anode is equipped with anode baskets that hold simulative spent fuel and also has blades to scrape off the dendritic uranium deposit growing on the inner surface of the cathode. The scraped uranium is collected in a mesh basket placed beneath the module. Though the basic concept of this apparatus is similar to the ANL's HTER, some original ideas, such as angled split blades, were adopted. The dimensions of the apparatus were the following; cathode inner diameter was 70mm or

90mm, cathode effective height was 60mm. The electrode volume, which is the outer volume of the electrode module immersed in the molten salt phase, was 1.5 x 10^{-3} m^3.

The experiments were conducted in an argon atmosphere glove box that contained less than 10ppm of moisture and oxygen. The electrorefining apparatus was immersed in LiCl-KCl eutectic salt bath that is fused at 773K. The molten salt contained 2-4wt% of uranium as UCl_3. The maximum anode rotation rate was 100rpm. The performance of the apparatus was examined under several operational conditions, such as anode rotation rate, concentration of uranium in the salt phase, and shape of the blades.

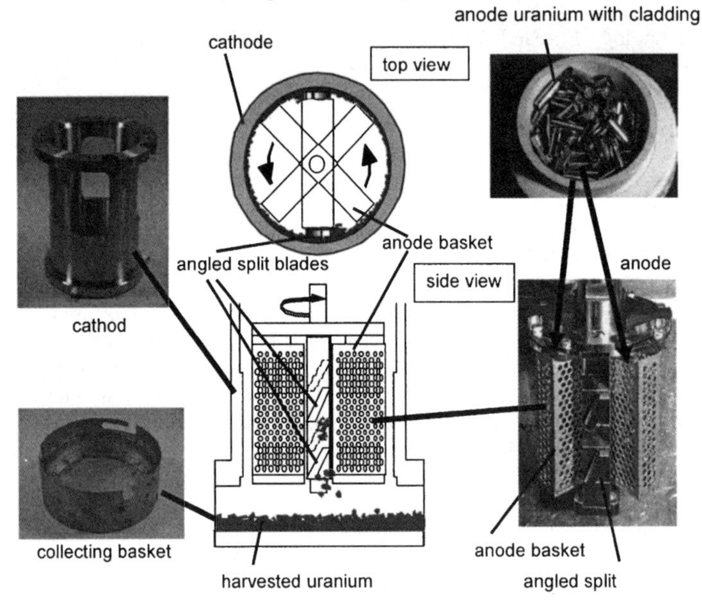

Figure 1 *High throughput electrorefining apparatus equipped with angled split blades.*

3 RESULTS

The collection rate of uranium at 2wt% of uranium in the salt similar to that of obtained at 4wt% of uranium in the salt. Since several electrode modules will be installed in an electrorefining vessel in our plant design,[1] we evaluate the performance of the electrode module as a processing rate of uranium versus its electrode volume immersed in the salt, assuming that a linear relationship between them can be extrapolated. In the present experiments, maximum precessing rate of 9.6 x 10^{-6} kg-U/second was achieved. This value corresponds to processing 10kg of uranium in 18 hours using an engineering scale electrode module, whose electrode volume is 2.4 x 10^{-2} m^3. As shown in Figure 2, the result is more than twice as high as the result obtained using a conventional mandrel type cathode, which is just a rod made of iron. [4]

Prevention of anode jamming due to the cathode deposit was achieved by using angled split blades. Additionally, electrolysis proceeded successfully until 510g of anode uranium rod, clad with stainless steel, were completely dissolved by adopting intermittent reverse rotation of anode and reverse current electrolysis during the electrorefining. The purpose of the reverse current electrolysis was to dissolve the surface of the cathode deposit slightly in order to prolong the electrolysis. The processing rate of the last experiment, in which uranium rod with clad was used, was 8.4 x 10^{-6} kg-U/second until 90% of the anode uranium was dissolved. It means that reduction of the processing ratio due to the reduction of thc surface area by the clad was not crucial.

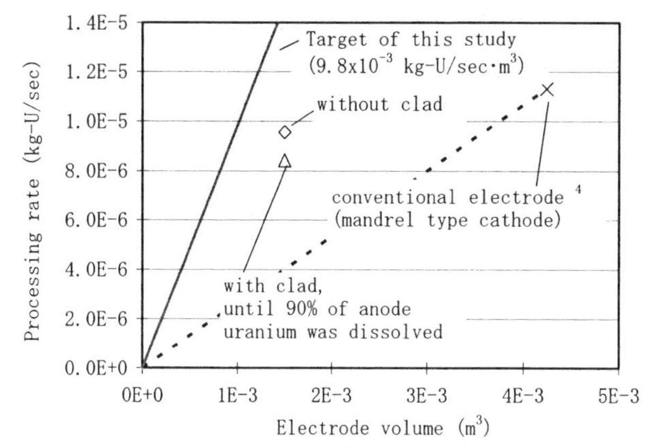

Figure 2 *Processing rate of uranium using high throughput electrorefining apparatus.*

4 CONCLUSION

To investigate the processing rate of uranium at low concentration of uranium in the salt, several experiments were conducted using a laboratory scale high throughput electrorefining apparatus. It was found that high processing rate of uranium can be achieved even at 2wt% of uranium in the salt. Angled split blades that scrape off uranium deposit were effective for avoidance of jamming, and intermittent current and rotation reversals were employed. Based on these results, engineering scale electrorefining tests that use up to 10kg of uranium will be conducted.

Acknowledgements

This work was conducted under "Development of Innovative Nuclear Technologies Program", sponsored by MEXT (Ministry of Education, Culture, Sports, Science, and Technology) of Japan. Deep appreciation is further expressed for unfailing support extended to the authors by Messrs. H. Nakamura, K. Utsunomiya, and Dr. R. Fujita of Toshiba Corporation.

References

1 T. Inoue, Y. Sakamura, M. Sakamura, M. Iizuka, K. Kinoshita, T. Usami, M. Kurata, and T. Yokoo, 201st meeting of the electrochemical society, 13th international symposium on molten salts, Philadelphia, PA, USA, May (2002).
2 D. Vaden, S. X. Li, and T. A. Jonson, 5th topical meeting, DOE owned Spent Nuclear Fuel and Fissile Materials Management, Charleston, SC, USA, Sept. (2002).
3 K. Uozumi, M. Iizuka, T. Kato, T. Inoue, O. Shirai, T. Iwai, and Y. Arai, *J. Nucl. Mater.*, 2004, **325**, 34.
4 T. Koyama, M. Iizuka, Y. Shoji, R. Fujita, H. Tanaka, T. Kobayashi, and M. Tokiwai, *J. Nucl. Mater.*, 1997, **34**, 384.

EXPERIMENTAL DATA POINTS TO EXISTENCE OF PLUTONIUM (VIII) IN ALKALINE SOLUTIONS

M.V. Nikonov, A.V. Gogolev, I.G. Tananaev and B.F. Myasoedov

Vernadsky Institute of Geochemistry and Analytical Chemistry of RAS, 19 Kosygin str., 119991 Moscow, Russia

1 INTRODUCTION

The discovery and studies of 5f elements in unknown oxidation states are of a noticeable interest for theoretical inorganic chemistry and enrich the chemistry of the transuranium elements. The outstanding attainments, such as the discovery of Np(VII), Pu(VII),[1] Am(II),[2] Am(VII)[3], stabilization and studies of properties of Am(IV),[4] Bk(IV),[5] Pu(V),[6] had served as the base for the new phase of development of fundamental and applied chemistry of the actinides. Before our experiments it was well known that the highest oxidation state for Pu was (VII). Hence the electron structure of Pu(VII) as $5f^1$ permits us to hypothesise that this actinide may exists in the oxidation state +8 with the electron configuration of an inert gas.

During 40+ years several attempts to prepare Pu(VIII) have been made. The main aim of these experiments was to oxidize Pu(VII) by different agents in the various media. Firstly it was shown that extended bubbling of 50 mg/l O_3 in O_2 mixture through ~10^{-3}M Pu(VII) in 4-15M NaOH did not transform the oxidation state. Electrochemical oxidation of Pu(VII) at 20-40°C in 4-15 M NaOH at [Pu(VII)]=10^{-4}-10^{-2}M and high values of the anode potential and current density didn't produce positive results either. The long-time γ-irradiating of Pu(VII) solutions in 4M NaOH, saturated with N_2O had led to the reduction of Pu(VII) into Pu(VI). The interaction between dispersed mixture of $Na_2Pu_2O_7$ and Li_2O_2 at 350-450°C in O_2 was also studied. It was shown that Li_2O_2 oxidised Pu(VI) only to Pu(VII).[7] The attempted oxidation of Pu(VII) by XeF_4, XeF_2 and F_2 were also unsuccessful.[7-9] Despite the fact that Pu(VIII) synthesis looked to be impossible our work indicates that Pu(VIII) species can indeed be prepared.

2 METHODS AND RESULTS

The ozonization of suspensions of hydroxide or solution of Pu(VI) in 1-3M NaOH is the most known and widely used method of Pu(VII) solutions preparation [1,10]. First of all we prepared pure 2.9 mM Pu(VI) solution in 1.5M NaOH. During 30 min interaction of this solution of Pu(VI) with 3.5% vol. O_3/O_2 mixture (gas flow 5-7 dm^3/h, 20°C) we obtain the blue-black ozonized solution with the characteristic absorption spectrum having the main maximum at 635 nm (Figure 1, spectrum 1). After than we added a small (0.05 ml) drops of a non ozonized Pu(VI) step by step to the 3 ml portion of ozonized (colored) solution. It was detected that the optical density of the blue-black solution instantly decreased proportionally to the growth of amount of the added initial solution of Pu(VI) (Figure 1, spectra 2-7). Therefore, it was supposed that the ozonizing of alkaline Pu(VI) solution in 1-3M NaOH leads to formation of a mixture of Pu(VII), and Pu(VIII). During the experiments above Pu(VIII) interacted with the initial non ozonized Pu(VI) solution by the following scheme:

$$\text{Pu(VI)} + \text{Pu(VIII)} \rightarrow 2\text{Pu(VII)} \quad (1).$$

When the Pu(VIII) amount was spent, the reaction (1) is finished and the optical density in the solution cease to change (Figure 1, spectra 7 and 8). Hence, the spectra 7, and 8 in Figure 1 could be considered as the characteristic spectrum of Pu (VII) in 1,5M NaOH.

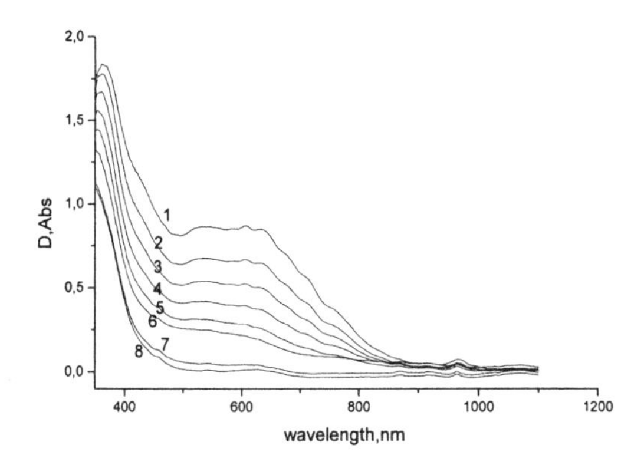

Figure 1 *The absorption spectra (1) of a solution of 2.9 мM Pu(VI) in 1.5M NaOH (1 ml) ozonizing within 60 min, and subsequent addition of 1.3 мM solution of Pu(VI) in 1.5M NaOH (ml): 2.- 0.05; 3.- 0.1; 4.- 0.15; 5.- 0.2; 6.- 0.25; 7 and 8 – after subsequent storage for 30 and 90 min, respectively*

Knowing the amount of Pu(VI) in the first aliquote (Figure 1, spectra 1 and 2), added to the Pu solution treated with ozone (0.05 ml $\times$ 1.3 mM = 6.5×10^{-5} mmol) and decrease of an optical density of the mixture at 635 nm (~ 0.18), the probably value of molar extinction coefficient of Pu(VIII) was estimated to be equal to 2600 $\pm$ 400 $M^{-1}\cdot cm^{-1}$ approximately. The value of $\varepsilon^{Pu(VII)}$ at the chosen wavelength 500-550 nm was found to be ~100 $M^{1}\cdot cm^{-1}$. It was supposed, that the extinction coefficient of Pu(VII) detected from previous work, *ca.* 530 $M^{-1}\cdot cm^{-1}$,[1] is the average value of ε for a mixture of Pu(VII) and Pu(VIII). Using the parameters: $\varepsilon^{Pu(VII)} + \varepsilon^{Pu(VIII)} = 530\ M^{-1}\cdot cm^{-1}$; $\varepsilon^{Pu(VII)} = 100\ M^{-1}\cdot cm^{-1}$; and $\varepsilon^{Pu(VIII)} = 2600\ M^{-1}\cdot cm^{-1}$, the $\varepsilon^{Pu(VII)}/\varepsilon^{Pu(VIII)}$ ratio must be about 0.25, and from these calculations the yield of Pu(VIII) in the experimental solution was estimated to be ~(15±5)%.

The reaction (1) takes place also if Pu(VI) is applied in the solid state as $Pu^{VI}O_3{\cdot}H_2O$ hydroxide. The scheme of the experiment is as follows: ~5 mg of Pu(IV) hydroxide, precipitated by means of ammonia from nitric acid solution and washed twice by distillated water was synthesized. An aqueous suspension of Pu(IV) hydroxide (pH~7.5) was ozonized for 60 min. The completeness of oxidation of Pu(IV) hydroxide into $PuO_3{\cdot}H_2O$ was controlled by well known spectrophotometric techniques. A portion of the obtained $PuO_3{\cdot}H_2O$ precipitate was dissolved completely in 3M NaOH. This solution of Pu(VI) was ozonized for 40 min. The $PuO_3{\cdot}H_2O$ precipitate was then treated by the just ozonized Pu(VI) solution. After ~10 sec mixing, the solution was centrifugated to record the absorbtion spectrum. It was detected that the absorbance in the wavelength range of 500-700 nn decreased by ~10 times on result of the interaction between the ozonized solution an $PuO_3 \cdot H_2O$ precipitate. However, a noticeable amount of the precipitate remained unreactec For confirmation of our supposition the spectrophotometric titration of the product o reaction (1) by hydrogen peroxide was carried out. The addition of H_2O_2 cause characteristic changes in the spectra of the solution. The stoichiometric ratio of [Pu]:[H_2O_2] determined after the completing of the titration was found to be 2.2:1. This ratio correspon the following equation of the reaction:

$$2Pu(VII) + H_2O_2 + 2OH^- \rightarrow 2Pu(VI) + 2H_2O + O_2 \quad (2)$$

The data obtained can confirm that Pu in the tested solution was mainly in the +7 oxidatio state. The slightly increased experimental value of the ratio [Pu]:[H_2O_2], can be explaine by the errors in the determination of reactant concentration and by the presence of proportion of Pu(VI) in the initial solution. The total amount of Pu in the tested solutio after titration (2) was determined after the treatment of the alkaline solution by 1M HNO_3 b the magnitude of absorbance at 830 nm. The spectrum did not show the presence of an oxidation forms of Pu, besides Pu(VI).

For supplementary determination of the highest oxidation state of Pu in ou experiments two redox-reactions had been studied: (1) interaction of Pu(VI) with Fe(VI) i 1M NaOH solution; and (2) contact of ozonizing Pu(VI) alkaline solution with Fe(III). As results of these experiments it was found that Pu(VI) solutions are found to be oxidized b ferrate-ions in 1.5 M NaOH. On the other hand, the preliminary ozonized plutoniun solutions oxidize Fe(III) to Fe(VI). These data contradict the known literature about th behavior of Pu ions in the highest oxidation states in alkaline solutions. As ferrate-ion oxidize Pu(VI) ions we conclude that the value of the redox potential of the pai Pu(VII)/Pu(VI) is less than 0.72 V (vs NHE). As the redox potential of the pai Am(VI)/Am(V) in 1 M NaOH is equal 0.65V and Am(VI) does not oxidize Pu(VI) unde these conditions.[10] These results confirm the existence of Pu(VIII) in the mixture wit Pu(VII) during ozonizing of Pu(VI) alkaline solution.

Acknowledgments

We are especially grateful to Dr. David L. Clark (G.T. Seaborg Institute for Transactiniun Science) for unceasing attention to our work. The work was supported by the U.S.DOE OBES, Project RC0-20004-SC14.

References

1 N.N. Krot, and A.D. Gelman. *Doklady AN SSSR (Rus)*, 1967, **177**, 124.
2 B.F. Myasoedov, and C. Musikas. *Radiochem. Radionalyt. Lett.* 1969, **2**, 21.
3 N.N. Krot, V.P. Shilov, A.D. Gelman, et al. *Doklady ANSSSR (Rus)*. 1974, **217**, 589.
4 V.F. Peretrukhin, E.A. Erin, N.N. Krot et al. *Doklady ANSSSR (Rus)*. 1978, **242**, 1359
5 V.I. Chepovoy, I.A. Lebedev, B.F. Myasoedov. *Rad. Rad. Lett.* 1973, **15**, 39.
6 N.A. Boudantseva, I.G. Tananaev, A.M. Fedosseev, C. Delegard. *Journal of Alloys an Compounds*. 1998, **271-273**, 813.
7 N.N. Krot, A.D. Gelman, F.A. Zakharova et al. *Radiokhimiya (Rus)*. 1972, **14**, 890.
8 Yu.K. Gusev, V.F. Peretrukhin, V.P. Shilov et al. *Radiokhimiya (Rus)*. 1972, **14**, 888.
9 V.F. Peretrukhin, N.N. Krot, A.D. Gelman. *Radiokhimiya (Rus)*. 1972, **14**, 628.
10 N.N. Krot, A.D. Gel'man, M.P. Mefod'eva et al *Heptavalent State of Neptunium, Plutonium an Americium.* UCRL-Trans-11798, LLNL, Livermore, California. 1977.

CONTROL OF NEPTUNIUM ROUTING DURING THE REPROCESSING OF SPENT NUCLEAR FUEL USING PUREX

J.E. Birkett, M.J. Carrott, G. Crooks, C.J. Maher, O.D. Fox, C.J. Jones, C.V. Roube, R.J. Taylor and D.A. Woodhead

Nexia Solutions, B229 and B170, Sellafield, CA20 1PG, UK

1 INTRODUCTION

The routing of neptunium, either to High Level Waste (HLW) or with actinide product streams, is an important issue in the reprocessing of spent nuclear fuel using the PUREX process. Depending upon the prevailing conditions Np may be present in any of three oxidation states, Np(IV), Np(V) and Np(VI), making control of Np routing difficult.[1] In the 1st solvent extraction cycle of the PUREX process, where the solvent and aqueous phases are continuously mixed and separated, any Np(V) present will be rejected to the aqueous raffinate and hence to the High Level Waste (HLW). In contrast any Np(IV) or Np(VI) present will extract into the solvent stream and continue through the process with the extracted U(VI) and Pu(IV). In processes where it is intended that Np is directed to the HLW, it is necessary to maintain Np as inextractable Np(V) if the need for further decontamination of the U or Pu products from Np is to be avoided. However, in some Advanced Fuel Cycles (AFCs), where mixed Np/Pu or U/Np/Pu products are required, it is necessary to extract the Np in the first solvent extraction cycle. In these cases the Np oxidation state needs to be maintained as Np(IV) or, more probably, Np(VI). Chemical methods of controlling the Np oxidation state which exploit the effects of acidity and nitrous concentration are attractive in PUREX as they add no new reagents or salts to the process streams. However, in mixed phase systems it is difficult to gain complete control over Np oxidation states in this way. In particular the oxidation of Np(V) to Np(VI) by nitric acid is catalysed by nitrous acid yet nitrous acid can act as a reductant for Np(VI).[2,3] The situation is further complicated by the extraction of nitrous acid and Np(VI) into the solvent. In a single cycle advanced PUREX process[4] control of the Np oxidation state will be crucial if essentially all the Np is to be directed, either to follow uranium and plutonium to a mixed actinide product, or to follow the fission products to the HLW. In the former case conditioning to Np(VI) before the first solvent extraction would send Np forward with the U/Pu stream. Thereafter reagents such as hydroxamic acids, which can function both as a complexant for Np(IV) and a reductant for Np(VI), provide a means of returning Np to the aqueous phase after the first extraction.[5-7] In the latter case it is necessary to prevent the oxidation of Np(V) to Np(VI) to retain Np in the aqueous raffinate. To assist the design of flowsheets a model has been developed to describe Np(V) oxidation in mixed phase systems, and its predictions have been compared with experimental data.

2 RESULTS & DISCUSSION

In order to better understand the speciation of Np in mixed phase systems the stability of Np(V) in nitric acid mixed with 30% tri-*n*-butylphosphate (TBP) in Exxon D80 (OK) at S/A 1:1 has been studied. In 1.8 M HNO_3, with no added nitrite, no oxidation of Np(V) was observed over several hours. The addition of 1.1 x 10^{-3} M $NaNO_2$ resulted in the gradual oxidation of Np(V) at a rate increasing with $[HNO_3]$ following an induction period of *ca.* 20 mins (Figure 1). Oxidation was also promoted by the presence of uranyl and by elevated temperature both of which eliminate the induction period (Figure 2).

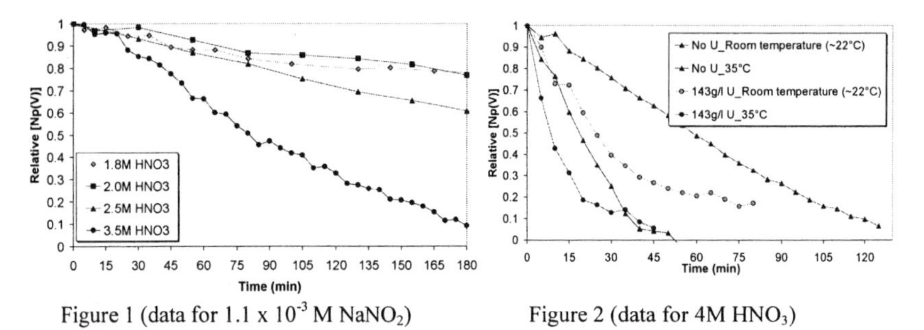

Figure 1 (data for 1.1 x 10^{-3} M $NaNO_2$) Figure 2 (data for 4M HNO_3)

An algorithm incorporating Koltunov's published kinetic equations (1) to (3)[8,9] was used to model the two phase system (Figure 3).[10]

Equation (1) the aqueous phase oxidation of Np(V) in nitric acid:

$$-\frac{d[NpO_2^+]}{dt} = k_{1A}\,[NpO_2^+][HNO_2]^{0.5}\,[H^+]^{z}\,[NO_3^-]^{0.5} + k_{1B}\,[NpO_2^+][H^+][HNO_2]$$

Equation (2) the aqueous phase reduction of Np(VI) by nitrous acid:

$$-\frac{d[NpO_2^{2+}]}{dt} = 2k_{2A}\,[NpO_2^{2+}][HNO_2][H^+]^{-1} + 2k_{2B}\,[NpO_2^{2+}][HNO_2]^{1.5}[H^+]^{-0.5}\,[NO_3^-]^{-0.5}$$

Equation (3) the solvent phase oxidation of Np(V):

$$-\frac{d[NpO_2^+]}{dt} = k_3\,\frac{[NpO_2^+][HNO_2]^{0.5}\,[HNO_3]^{0.5}}{[H_2O]^{0.2}}$$

The change in [Np(V)] with time was simulated using the algorithm with different values of z in equation (1) and the result found to be highly sensitive to the value of z used. Values for z of both 0.5 and 2 were quoted by Koltunov. The model did not give a good fit to the experimental results obtained from the mixed phase system at the higher nitrous acid concentrations of 10^{-2} to 1 M. However, much better fits were obtained at nitrous acid concentrations of *ca.* 10^{-3} M, these being more similar to the concentrations originally used to derive equation (1). An example is shown in Figure 3 where Simulation A used z = 2 in equation (1) for solutions of 103.5 mg/l Np in 4 M HNO_3 containing 143 g/l U(VI) and 1.1 x 10^{-3} M HNO_2 in contact with 30%TBP/OK with an S/A ratio of 2:1 at 22°C. In comparison Simulation B used z = 1 and did not provide a good model of the system.

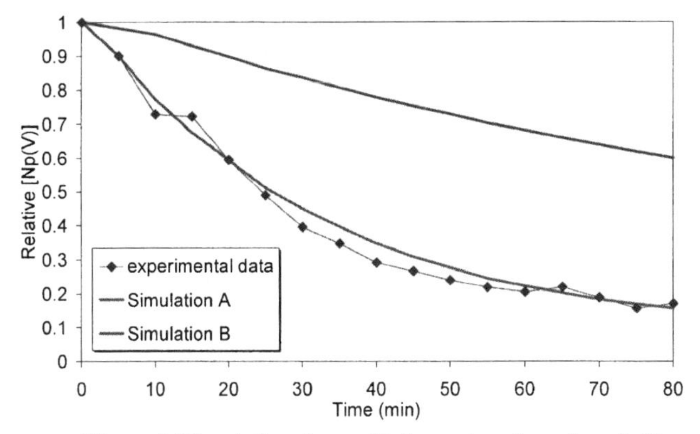

Figure 3 (Simulation A uses Koltunov's value of z = 2, B uses z = 1)

A U/Np active flowsheet trial has been carried out using a 22 stage 1 cm centrifugal contactor rig to represent the Highly Active extraction (HA 16 stages) and scrub (HS 6 stages) sections of a PUREX process with $NaNO_2$ added to the feed to initiate Np(V) oxidation. The results are being used to validate, under multistage conditions, a process simulation incorporating the kinetic model. This includes mass transfer rates but these have not yet been optimised. An analysis of the findings will be presented at a later date.

3 CONCLUSIONS

A model is being developed which can describe the change in concentration with time of Np(V) in the mixed phase Np(V)/U(VI)/HNO_2/HNO_3/30% TBP/OK system for nitrous acid concentrations of *ca.* 0.001M. This is being incorporated into process simulations.

Acknowledgement: This work was funded in part by the Nuclear Decommissioning Authority through British Nuclear Group and Nexia Solutions.

References

1. R.J. Taylor, I. S. Denniss and A. L. Wallwork, *Nuclear Energy,* 1997, **36**, 39-46.
2. J.P. Moulin, *Contribution à l'étude de la réaction d'oxydation du neptunium pentavalent par l'acide nitrique catalysée par l'acide nitreux*, Docteur-Ingénieur, PARIS VI, PARIS, 1974.
3. O. Tochiyama, Y. Nakamura, M. Hirota, Y, Inoue, *J. Nucl. Sci. Tech.*, 1995, **32**, 118-124 .
4. R.J. Taylor, I. May, I.S. Denniss, A.L. Wallwork, G. Hunt, S. Hutchison, P. Bothwell, V. Richards, N.J. Hill, RECOD 98 5th International Conference on Recycling, Conditioning and Disposal, October 25-28, Nice, France, 1998, pp. 417-424.
5. R. J. Taylor, I. May, A. L. Wallwork, I. S. Denniss, N. J. Hill, B. Ya. Galkin, B. Ya. Zilberman, Yu. S. Federov, *J. Alloy and Compounds*, 1998, **535,** 271-273.
6. R. J. Taylor, *Progress Towards Understanding The Interactions Between Hydroxamic Acids And Actinide Ions, J. Nucl. Sci. Tech., Supp. 3,* 2002, 886-889.
7. O. D. Fox, C. J. Jones, J. E. Birkett, M. J. Carrott, C. J. Maher, C. V. Roube, R. J. Taylor, *Advanced PUREX flowsheets for future Np and Pu fuel cycle demands*, *ACS Symposium Series*, 2005, *in press.*
8. V.S. Koltunov, translation from *The kinetics of the actinide reactions, Atomizdat,* Moscow, 1974.
9. V.S. Koltunov, K.M. Frolov, M.Yu. Sinez and Yu.V. Isaev, *Radiokhimiya*, 1992, **34**, 28-34, and 118-123.
10. $k_{1A} = 2.884x10^{11}\ e^{(-9922/T)}$, $k_{1B} = 5.405x10^{12}\ e^{(-10031/T)}$, $k_{2A} = 6.928x10^{10}\ e^{(-7507/T)}$, $k_{2B} = 2.497x10^{12}\ e^{(-7806/T)}$, $k_3 = 1.952x10^{11}\ e^{(-9008/T)}$; T = temperature (K).

CRYSTALLIZATION BEHAVIOR OF URANIUM AND PLUTONIUM IN NITRIC ACID SOLUTION

K. Yano, A. Shibata, K. Nomura, T. Koizumi and T. Koyama

Recycle Process Technology Group, Advanced Fuel Recycle Technology Division, Waste Management And Fuel Recycle Research Center, Japan Nuclear Cycle Development Institute Tokai Works, 4-33 Muramatsu, Tokai-mura, Naka-gun, Ibaraki, 319-1194, Japan

1 INTRODUCTION

The Japan Nuclear Cycle Development Institute (JNC) has been developing a U crystallization process as a part of "New Extraction System for TRU Recovery (NEXT)"[1] to separate the majority of the U from the dissolver solution of FBR spent fuel. The crystallization process for NEXT yields some advantages over the conventional PUREX process with regard to cost performance, waste volume, and nuclear proliferation.

A crystallization process as a part of reprocessing had been developed in Germany for purification of U product from LWR reprocessing.[2] Compared with their process, U crystallization for NEXT process is different on dealing with highly concentrated dissolver solution containing more Pu and FPs. Therefore, it is important to know the behaviour of Pu under the operating condition of U crystallization from highly concentrated dissolver solution to prevent Pu from accompanying U. In previous work,[3] it was clarified that Pu(VI) co-crystallized with U but Pu(IV) did not, although Pu concentration was lower than its solubility.

This report gives the results of test (i) and (ii) as follows;

(i) crystallizing out of U/Pu(IV)/Pu(VI) mixed solution
(ii) crystallizing out of U/Pu(IV) mixed solution under the condition of various Pu/(U+Pu) ratio.

2 EXPERIMENTAL

U/Pu nitric solutions were prepared by dissolution of UO_2 and MOX powder with nitric acid solutions, and concentrated by evaporation. Pu was then adjusted to the tetravalent state with a few drops of hydrogen peroxide solution. Pu(VI) was obtained by boiling the solution over 3 hours, followed by confirmed that Pu(IV) was no longer present. Table 1 shows the composition of the solutions prepared for the test (i) and (ii).

50mL of the feed solution was initially kept at over 40°C in the crystallization vessel and cooled down to about 15°C to crystallize uranyl nitrate hexahydrate. After it was confirmed that the temperature of crystallization vessel didn't change during over 30 minutes, the mixture of crystals and mother solution was separated by a glass filter. Separated crystals were washed out by 20mL of 180gU/L uranyl nitrate solution at 4N in order to remove the mother solution stuck on their surface. The crystals washed only once to decrease an amount of the U crystals dissolved into washing solutions. The contents of U and Pu in solutions and crystals were measured by UV/vis spectrometry and acidity was by titration.

Table 1 *Composition of Solutions for the Test (i) and (ii)*

No.	Concentration		Acidity (N)	Pu valence
	U (g/L)	Pu (g/L)		
(i) –1	540	51	2.7	IV
(i) –2	610	47	5.6	VI
(i) –3	550	40	5.6	IV and VI[a]
(ii)–1	420	26	3.7	IV
(ii)–2	590	28	6.7	IV
(ii)–3	560	12	5.0	IV

[a] The proportion of Pu(VI) to whole Pu contained in the solution is 38% calculated on the premise that the Pu(VI) absorbance at 830nm is proportional to its concentration regardless of U and nitrate concentration.

3 RESULTS AND DISCUSSION

3.1 Pu Behavior under the Condition that Pu(IV) and Pu(VI) Co-exist.

Decontamination factor of Pu (DF_{Pu}) to U in crystals were obtained in test (i) to investigate the effects on Pu behavior of its valence. Table 2 shows DF_{Pu} in test (i).Under the condition that Pu is tetravalent, the colour of the obtained crystals was lemon yellow and DF_{Pu} after washing was 29. When Pu valence is hexavalent, however, orange crystals were obtained and DF_{Pu} was very low. Therefore Pu(VI) co-crystallized with U. In the case of (i)-3, that Pu(IV) and Pu(VI) co-exist, the color of the obtained crystals was yellowish-orange compared with (i)-2 and DF_{Pu} was intermediate value between (i)-1 and 2.

On the premise that the Pu(VI) absorbance at 830nm is proportional to its concentration and regardless of U and nitrate concentration, it is possible to evaluate the proportion of Pu(VI) to whole Pu contained in solutions. From the results of the test (i)-1 and 2, it is thought that the valence of almost all of the Pu in the crystals was hexavalent in test (i)-3. So, the decontamination factor of Pu(VI) ($DF_{Pu(VI)}$) under the condition of test (i)-3 can be calculated on the premise that all the Pu valence in the crystal was hexavalent. The evaluated value of $DF_{Pu(VI)}$ in test (i)-3 are 1.3 before washing and 1.8 after washing. Therefore it implies that the behavior of Pu(IV) and Pu(VI) in the mixed system is similar to that in the solo system.

Table 2 DF_{Pu} to U in crystals

No.	DF_{Pu}[b]	
	Before washing	After washing
(i) –1	4.9	29
(i) –2	1.3	1.5
(i) –3	3.6	4.9

[b] DF_{Pu} was calculated by

$$DF_{Pu} = \frac{C_{f,Pu}/C_{f,U}}{C_{c,Pu}/C_{c,U}}$$

$C_{x,y}$: Conc. of element Y in X
X; f: feed solution, c: crystal
Y; U: uranium, Pu: plutonium

3.2 The Influence of Pu on the Crystallization Behavior of U

Table 3 shows the composition of mother solutions in test (i)-1 and (ii) compared with the solubility of U in U-HNO_3-H_2O system.

Table 3 *Composition of Mother Solutions with U Solubility in U-HNO_3-H_2O system*

No.	Final Temp.	Composition of Mother Solution			U Solubility in Nitric Acid[c]
		U	Pu/(U+Pu) ratio	Acidity	
(i)–1	11.1 °C	290 g/L	19.9 %	4.1 N	240 g/L
(ii)–1	13.8 °C	240 g/L	13.7 %	5.5 N	230 g/L
(ii)–2	12.3 °C	290 g/L	13.7 %	9.6 N	200 g/L
(ii)–3	13.1 °C	230 g/L	8.4 %	8.0 N	190 g/L

[c] calculated by interpolation of acidity and temperature on data by Hart *et al.*[4]

In every case, the U concentration of the mother solution was different from its solubility in nitric acid and slightly higher than its solubility. Therefore it is suggested that Pu(IV) has some influences for the solubility of U in dissolver solution. Generally, the solubility of a metal nitrate decreases when other nitrates are introduced into the solution. In highly ionic concentrated solution, however, activity of solutes may go down to increase its solubility. In addition, as phenomena of raising solubility, there are the salting-in effect and formations of another chemical structure e.g. micelle or another complex. To consider whether the phenomenon that the U solubility increase was caused by Pu or not, it is essential to understand how U and Pu exist in the solution. However, there is limited data on the chemical states of U and Pu in highly concentrated solution. To evaluate the influence of Pu(IV) on the solubility of U, it is necessary to investigate chemical states of U and Pu in highly concentrated dissolver solution by additional analytical techniques.

4 CONCLUSION

Pu(IV) was not crystallized with U and DF_{Pu} was 29. On the other hand, Pu(VI) was co-crystallized with U and DF_{Pu} was about 1~2. When Pu(IV) and Pu(VI) co-exist, their behaviour were similar to those in the solo system, respectively.

It is suggested that there are some influences of Pu(IV) on U solubility in highly concentrated solution. When Pu(IV) exists, it tends to increase the solubility of U compared with that in the U-H_2O-HNO_3 systems in this test, although further experimental and theoretical study are still required to understand the phenomenon..

References

1 T. Takata, Y. Koma, K. Sato, M. Kamiya, A. Shibata, K. Nomura, H. Ogino, T. Koyama and S. Aose, *J. Nucl. Sci. Technol.*, 2004, **41**, 3, 307
2 E. Henrich, U Bauder, R. Marquart, W.G. Druckenbrodt and K. Wittmann, *Atomkernenergie Kerntechnik*, 1986, **48**, 4, 241
3 K. Yano, A. Shibata, K. Nomura, T. Koizumi and T. Koyama, *Proc. ATALATE2004.* Nimes, France, June 21-24, 2004, P1-66.
4 R.G. Hart and G.O. Morris, *Progr. Nucl. Energy III*, 1958, **2**, 544

A HOT TEST ON MINOR ACTINIDES SEPARATION FROM HIGH-LEVEL-WASTE BY CMPO/SiO_2-P EXTRACTION RESIN

Y.-Z. Wei[1], H Hoshi[1], M. Kumagai[1], P. Goethals[2] and A. Bruggeman[2]

[1]Institute of Research and Innovation (IRI), 1201 Takada, Kashiwa, 277-0861 Japan
[2]Belgian Nuclear Research Center (SCK•CEN), B-2400 Mol, Belgium

1 INTRODUCTION

To separate the long-lived minor actinides (MA=Am, Cm) from High-Level-Waste (HLW) we have proposed an advanced separation process based on extraction chromatography that uses a minimal organic solvent and compact equipment.[1-2] This process consists of two separation columns packed with CMPO (octyl(phenyl)-N,N-diisobutylcarbamoylmethylphosphine oxide) and R-BTP (2,6-bis-(5,6-dialkyl-1,2,4-triazine-3-yl)pyridine) extraction resins, respectively. In the CMPO-column, the elements contained in HLW can be separated into (a) non-adsorptive fission products (FP), (b) MA-RE(rare earths), and (3) Zr-Mo, depending on their different adsorption-elution behaviors with CMPO and the eluents. The effluent of MA-RE is then applied to the second column packed with R-BTP extraction resin, where the MA are separated from the RE, since R-BTP shows high adsorption selectivity for MA over RE.[2-3] For this purpose we prepared the novel silica-based extraction resins by impregnating CMPO, or R-BTP into a styrene-divinylbenzene copolymer, which was immobilized in porous silica particles with a diameter of 50 μm. Compared to conventional polymeric matrix resins, these new type of silica-based extraction resins have rapid adsorption kinetics and a notably low pressure drop in a packed column.[1,4]

In our previous work,[4] it was found that Am(III) together with RE(III) can be successfully separated from a simulated HLW solution by using a column packed with the CMPO extraction resin. In this work, a hot test for a practical HLW solution was undertaken to examine the separation behaviour of Am and some typical fission products.

2 METHODS AND RESULTS

Chemical structure and SEM image of the CMPO extraction resin are illustrated in Figure 1. As the support, the silica/polymer composite (SiO_2-P) developed in our previous work was utilized.[1] This support contains a macroreticular styrene-divinylbenzene copolymer immobilized in porous silica particles with a pore size of 0.6 μm and mean diameter of 50 μm. CMPO was impregnated into the support at 33.3 wt% CMPO.

CMPO

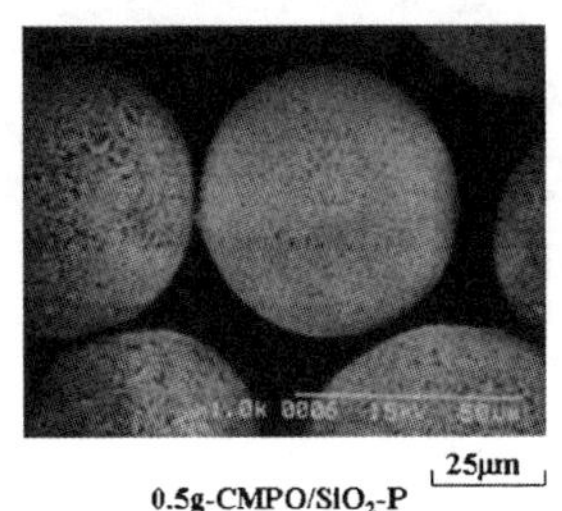

Figure 1 *Structure and photograph of the CMPO extraction resin*

Table 1 *Analytic results of the sample solution and recoveries in the separation test*

Nuclide	Concentration (μg/g-soln)	Recovery (%)	Analysis method
^{241}Am	3.90	102.5	γ-spec
^{240}Pu	0	-	ICP-MS
^{239}Pu	0	-	ICP-MS
^{237}Np	0	-	ICP-MS
^{235}U	0	-	ICP-MS
^{154}Eu	0.095	101.7	γ-spec
^{153}Eu	1.30	105.9	ICP-MS
^{145}Nd	9.01	102.3	ICP-MS
^{140}Ce	14.60	111.8	ICP-MS
^{137}Cs	11.74	102.7	γ-spec
^{133}Cs	14.90	107.1	ICP-MS
^{105}Pd	0	-	ICP-MS
^{101}Ru	0	-	ICP-MS
^{99}Tc	0	-	ICP-MS
^{95}Mo	7.04	105.3	ICP-MS
^{93}Zr	6.28	103.6	ICP-MS
^{89}Y	6.46	109.9	ICP-MS
^{88}Sr	19.60	107.8	ICP-MS

The HLW sample solution containing 6 M HNO_3 was obtained from an anion exchange process that had separated the U, Pu, Pd, Ru and Tc from a spent LWR-fuel solution.[5] Table 1 shows the analytic results of the representative nuclides contained in the sample solution. The separation experiment was carried out using a Pyrex-glass column with a 10-mm inner diameter and 500-mm length. The column was packed with the CMPO/SiO_2-P extraction resin to about 480-mm (resin height) and the resin bed volume was 38 cm^3. The packed column was installed in a hot cell and the temperatures of the column and the throughput solutions were kept at 323 K. Prior to the chromatographic operation, the extraction resin was conditioned by passing 200 cm^3 of 6 M HNO_3 through the resin bed. 50 cm^3 of the HLW solution was loaded onto the column at a constant flow velocity of 1.0 cm^3/min (0.76 m/h) by a metering pump. Subsequently, 50 cm^3 of 6 M HNO_3, 150 cm^3 of H_2O, 150 cm^3 of oxalic acid ($H_2C_2O_4$) and 150 cm^3 of H_2O were fed to the column, successively, as eluents. The effluents from the column were collected by an auto-fractional collector in 10-cm^3 aliquots. The concentrations of nuclides in each fraction were determined by ICP-MS or γ-spectrometry.

Figure 2 shows the results of separation experiment for the HLW solution with the CMPO/SiO_2-P column. As can be seen, Cs and Sr showed no adsorption. These elements leaked out with the sample solution and the 6 M HNO_3 rinse solution. Am and the RE elements, i.e. Y, Ce, Nd and Eu were completely adsorbed by the CMPO/SiO_2-P extraction resin. The adsorbed Am and RE were eluted off efficiently by supplying only water to the column. According to previous studies,[6-7] the trivalent metal ions such as Am(III) and RE(III) are adsorbed by CMPO extraction resin as neutral nitrato-complexes.

$$M^{3+} + 3\,NO_3^- + 3(\text{CMPO/SiO}_2\text{-P}) = M(NO_3)_3(\text{CMPO/SiO}_2\text{-P})_3 \quad (1)$$

Therefore, the elution effect of Am and RE is considered to result from the decomposition of the complexes with the decrease of NO_3^- concentration in the resin bed by supplying water to the column. Note that the Am and RE showed sharp elution peaks with almost no tailing. This indicates that the silica-based CMPO extraction resin has no hydrodynamic or chemical kinetics restriction to hinder elution. The rapid kinetics is considered to result from the fine particle size. On the other hand, the Zr and Mo contained in the HLW solution were strongly retained by the CMPO/SiO_2-P. The adsorbed Mo and most of the Zr were eluted effectively by 0.5 M $H_2C_2O_4$. A small portion of Zr is finally eluted by water and the reason is unknown. Zr and Mo are known to form complexes with oxalic acid and the complexes are not adsorbed by CMPO extraction resin.[4] From the recovery shown in Table 1, all of the detected nuclides were quantitatively recovered from the extraction resin. The results of this hot test demonstrate that successful separation and recovery of Am together with some RE from HLW can be achieved. Moreover, in our previous study,[8] it was found that the CMPO/SiO_2-P has relatively excellent radiation stability in strong nitric acid medium. These results indicate that the proposed MA-RE separation process is essentially feasible, though further investigations such as pilot-scale testing and evaluation of the process regarding practical use conditions are necessary.

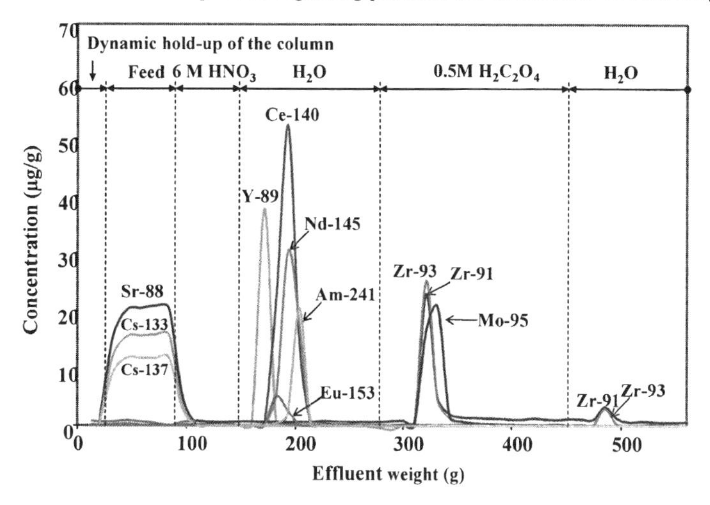

Figure 2 *Results of column separation test for a HLW with CMPO extraction resin*

3 CONCLUSIONS

A hot separation test for a high-level-waste solution was conducted using the column packed with CMPO/SiO_2-P extraction resin. Am and RE were strongly adsorbed on the resin and were then effectively eluted off by H_2O. Am together with some RE was successfully separated from the other FPs such as Cs, Sr, Zr and Mo. The test results indicate that the first step of the MA-RE separation process is feasible.

References

1 Y.-Z. Wei, M. Kumagai, Y. Takashima, *et al.*, *Nucl. Technol.*, 2000, **132**, 413.
2 Y.-Z. Wei, H. Hoshi, M. Kumagai, *et al.*, *J. Alloys Comp.*, 2004, **374**, 447.
3 Z. Kolarik, U. Müllich and F. Gassner, *Solvent Extr. Ion Exch.*, 1999, **17,** 23.
4 H. Hoshi, Y.-Z. Wei, M. Kumagai, *et al.*, *J. Nucl. Sci. Technol.* 2002, Suppl.3, 874.
5 Y.-Z. Wei, T. Arai, H. Hoshi, *et al.*, *Nucl. Technol.*, 2005, **149**, 217.
6 K. L. Nash, *Solvent Extr. Ion Exch.*, 1993, **11**, 729.
7 K. Tekeshita, Y. Takashima, S. Matsumoto *et al.*, *J. Chem. Eng. Japan*, 1995, **28**, 91.
8 A. Zhang, Y.-Z. Wei, M. Kumagai, *et al.*, *Radiation Phys. Chem.*, 2005, **72**, 455.

RECOVERY OF URANIUM AND PLUTONIUM METALS THROUGH CATHODE PROCESSING OF ELECTRODEPOSITS FROM REDUCED OXIDE FUEL ANODES

T. Hijikata,[1] T. Koyama,[1] T. Usami,[1]
S. Kitawaki,[2] T. Shinozaki,[2] and M. Fukushima[2]

[1]Nuclear Technology Research Laboratory, Central Institute of Electric Power Industry (CRIEPI), 2-11-1 Iwadokita Komae-shi Tokyo 201-8511, Japan. E-mail: hizikata@criepi.denken.or.jp
[2]Waste management and Fuel Cycle Research Center, Tokai Works, Japan Nuclear Cycle Development Institute (JNC), 4-33 Muramatsu Tokai-mura Naka-gun Ibaragi 319-1194, Japan

1 INTRODUCTION

Electrometallurgical pyroprocessing is one of the most promising technologies for the advanced fuel cycle, with favourable economic potential and intrinsic proliferation-resistance. Though Pu data is still lacking for this process, the feasibility of each process step has almost been demonstrated. CRIEPI and JNC have jointly started "*the integrated experiments*" of electrometallurgical reprocessing of metal and oxide Pu-containing fuels at the chemical processing facility (CPF) in the JNC-Tokai Works.[1-2] The integrated experiment was proposed to evaluate the applicability of the electrometallurgical pyroprocess for both metal and oxide fuel.[1-2] The integrated experiment consists of (1) reduction of oxide fuel into metal form by lithium reductant, (2) molten salt electrorefining to recover U and U-Pu, (3) cathode processing to remove salt and Cd from actinide by distillation, and (4) injection casting of actinide to form metal fuel. After the success of the integrated experiments of UO_2,[3] several tens of grams of PuO_2 were treated in each continuous operation for the first three processes.

2 METHOD AND RESULTS

2.1 Electrorefining of the Reduced Actinide Oxide

In the electrorefining test, reduced UO_2 or PuO_2 are anodically dissolved, and U-Pu is recovered by a liquid Cd cathode while U is recovered on an iron solid cathode. The electrorefiner was reported previously.[2] As for the reduced UO_2 anode (RU1 to RU4), the current was passed at 1A (area: *ca.* 9cm^2) in galvanic mode, and was decreased gradually to 50mA because the anode potential became noble. During electrorefining, the solid cathode potential was kept around −1.5 V. Theoretical deposits were calculated by electrical quantity. As seen in Table 1, the weights of cathode deposits with adhering salt were measured from 15g to 21g. The deposit (see Fig.1 (a)) was removed mechanically, and was heated in the cathode processor. As for the reduced PuO_2 anode (RPu), the potential of the cadmium cathode was kept at around -1.3 V while the anode potential was increased from -1.3V to -0.7V. 10300 coulombs (8.5 g Pu equivalents) of current (500 to

50mA, area: around 8.5cm^2) were passed, and the concentration of U and Pu in the Cd cathode rose to 0.8 wt% and 6.1wt%, respectively. The high Pu/U ratio in the Cd cathode against the high Pu/U ratio (0.8 wt%/0.05 wt%) in the salt phase agreed with equilibrium distribution between salt and Cd as discussed in elsewhere.[4] The Cd-Pu-U alloy in aluminum nitride (AlN) crucible (see Fig.1 (b)) was heated at 800K, and poured into the tungsten (W) crucible of the cathode processor.

(a) Solid cathode deposit (RU1) (b) Cd cathode deposit (RPu)

Figure 1 Photographs of cathode deposit: (a) solid cathode deposit, (b) Cd cathode deposit

Table 1 Results of the electrorefining and distillation test for the reduced UO_2 and PuO_2

RUN NO.	Loaded oxides (g)	Reduced metals (g)	Electrical quantity (Coulombs)	Theory deposits (g)	Measured deposits (g)	Distillation residue (g)	Recovery yield of reduced metals (%)
RU1(UO_2)	10.3	9.1	14700	12.1	15.5	8.6	95
RU2(UO_2)	10.3	9.1	10500	8.6	16.6	7.1	78
RU3(UO_2)	10.0	8.9	12400	10.2	17.7	6.9	78
RU4(UO_2)	10.4	7.4*	11800	9.7	20.9	9.1	123
RU5(UO_2)	20.7	18.3	-	-	31.7	19.5	106
RPu(PuO_2)	18.4	16.2	10300	8.5	124.4	6.5	76**

*: A part of reduced metal was taken for the sample

**: Recovery yield is calculated by the denominator of theory deposits

2.2 Cathodic process of electrodeposition

In cathodic processing, the cathode deposit of the electrorefiner was heated for distillation of adhering salt and Cd. The whole experimental apparatus was shown elsewhere,[2] while the main vessel is shown schematically in Fig. 2. The typical temperature profiles of crucible and nozzle are shown in Fig. 3. The pressure in the distillation vessel was kept below 1 kPa. For the solid cathode deposit, the temperature profile of the nozzle had a peak at around 1250K, as shown in Fig.3 (a). The peak can ascribe to heat transportation due to salt vaporization. As summarized in Table 1, high recovery yield of actinides from the reduction product was demonstrated for continuous operation of electrorefining and cathode processing. The form of recovered U metal varied, however, with increasing RUN number, as seen in Fig. 4. The metal in latter experiments seemed to keep the form of dendritic deposit, suggesting that the metal was not molten up to 1673 K. Because Li_2O was carried over from reduction step to the electrorefiner, oxygen was dissolved in U deposit to increase liquidus temperature.[5] In the case of the Cd cathode (RPu), two peaks were found around 750K and 1250K in the temperature profile of the nozzle as shown in Fig.3 (b). The first peak corresponds to the evaporation of Cd, while the second peak is due to the evaporation of salt. Though mass balance of the experiment (RPu) was satisfactory (more than 99%), the obtained metal (6.5g) was separated in the crucible (5g) and in the bottom of nozzle (1.5g). Another 2g of U and Pu metal was left in the AlN crucible.

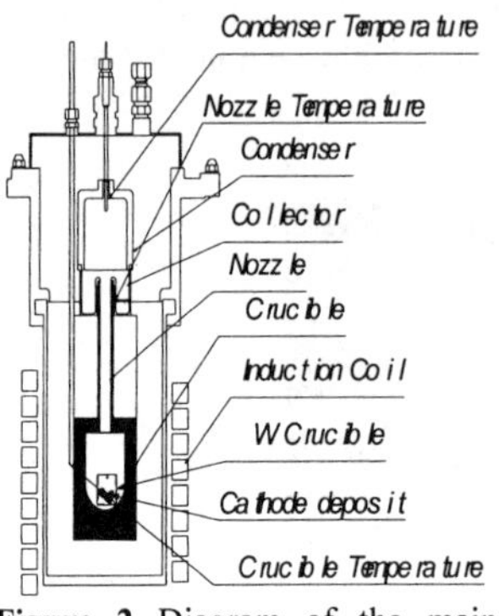

Figure 2 Diagram of the main vessel in the cathode processor

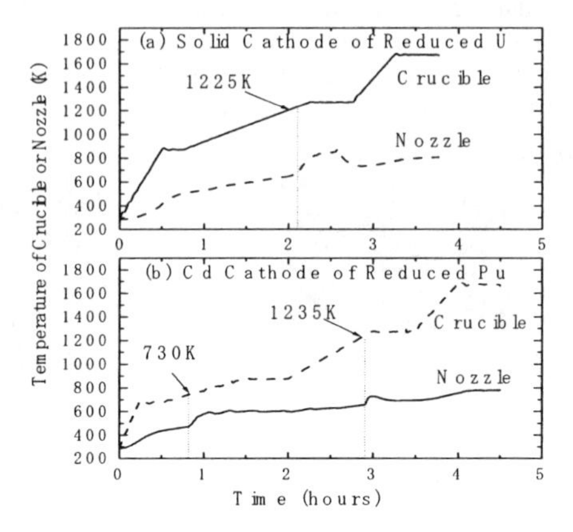

Figure 3 Temperature profiles of crucible and nozzle for (a) solid cathode of reduced UO_2: (b) Cd cathode of reduced PuO_2

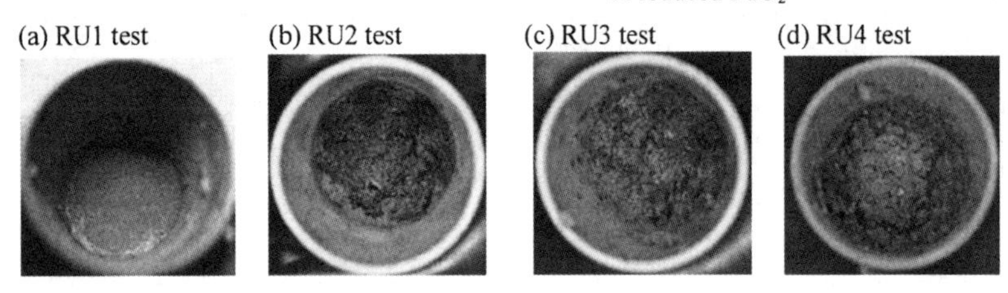

Figure 4 Photographs of U metal in W crucible after distillation: (a) RU1 test at 1673K, (b) RU2 test at 1573K, (c) RU3 test at 1673K, (d) RU4 test at 1673K

3 CONCLUSION

On average 94% of the reduced U metal was recovered as U metal through the electrorefining and distillation. Most of the salt adhering on U dendrites was found to be distilled up to 1273K, and U metal was obtained by heating to 1673K. Most of the Cd was found to be distilled from U-Pu-Cd alloy at around 823K and most of salt on the Cd cathode was distilled below 1273K. A metal U-Pu product was obtained after distillation.

References

1 T.Koyama, T.Hijikata, T.Inoue, S.Kitawaki, T.Shinozaki, and M.Myochin, Proceeding of International Conference on Innovative Technologies for Nuclear Fuel Cycles and Nuclear Power, 23-26 June 2003 Vienna, Austria.

2 T.Hijikata, T.Koyama, T.Usami, S.Kitawaki, T.Shinozaki, and T.Kobayashi, Proceeding of GLOBAL2003, 763-772 November 2003, New Orleans.

3 T.Koyama, T.Hijikata, T.Usami, S.Kitawaki, T.Shinozaki, and T.Kobayashi, Proc. of 11th Symposium on The Thermodynamics of Nuclear Materials, Sept. 2004, Karlsruhe, Germany.

4 K. Kinoshita, T.Koyama, T.Inoue, M.Ougier, and J.P.Glatz, *J.Phys.Chem. Solids*, 2005, **66**, 619.

5 A.E.Martin and R.K. Edwards, *J.Phys.Chem.*, 1965, **69**, 1788.

KINETICS OF ELECTROCHEMICAL GENERATION OF SILVER(II)

K.L. Berg, P.J.W. Rance and R.J. Taylor

Nexia Solutions, Sellafield, B170, Seascale, Cumbria CA20 1PG, UK

1 INTRODUCTION

Electrochemical mediated oxidation (EMO) is a means of oxidising scarcely oxidisable materials in aqueous solution, by using electric current to generate a very oxidative species electrolytically. It has been particularly interesting for the nuclear industry due to its ability to rapidly dissolve plutonium(IV) oxide into nitric acid solutions, by oxidising it to plutonyl ions, thus avoiding the use of hydrogen fluoride. Ag(II) is generally preferred as the EMO oxidant, although there are alternatives such as Ce(IV) and Co(III). Work began at BNFL, Sellafield in the nineties to examine the use of silver(II) as an oxidant for PuO_2 in fuels and residues. Prior to this, a substantial body of research had been carried out by American and French workers,[1-2] and the process had come into commercial use in the French nuclear industry. In Great Britain, the main body of work on EMO has concentrated on the use of silver(II) to treat industrial organic wastes, leading to the invention of the SilverII™ process, patented by AEA Technology, which is essentially a divided flow cell reactor with parallel plate electrodes. The cell must be divided in order to prevent catholyte reduction products from reaching the anode. However, optimum conditions for the mineralisation of organics are not necessarily favourable for the oxidisation of PuO_2. In particular, the temperature seems to be most beneficial around room temperature for PuO_2 dissolution, whereas 60-90°C is common for organics oxidation.

Early reports on silver(II) catalysed plutonium(IV) oxide dissolution[1-2] found that the rate of dissolution is controlled mainly by the generation rate of silver(II), not the kinetics of the dissolution reaction. Hence, for engineering purposes it is important to have good knowledge of the kinetics of the anodic silver(II) generation. There are a few papers describing the kinetics of this reaction in nitric acid solutions,[3-4] but they do not altogether provide a very clear understanding of the system, thus more fundamental experimental work was felt to be needed.

2 METHOD AND RESULTS

Our work concentrated on using platinum as the anode material, since this had been the first choice in previous works. However, there may be alternatives which are more favourable, in terms of price, durability etc.

A sketch of the rotating disk electrode apparatus used is shown in Fig. 1. Polarisation curves at various rotation rates of the anode were obtained by stepping the potential in small increments from the open circuit potential up to potentials where the oxygen evolution became significant, holding it for a few seconds at each potential. Silver(II) generation rates as a function of the potential were obtained by subtracting oxygen evolution data from the overall polarisation data, assuming additivity of the two reactions. An example of silver(II) generation rate data thus obtained are shown in Fig. 2.

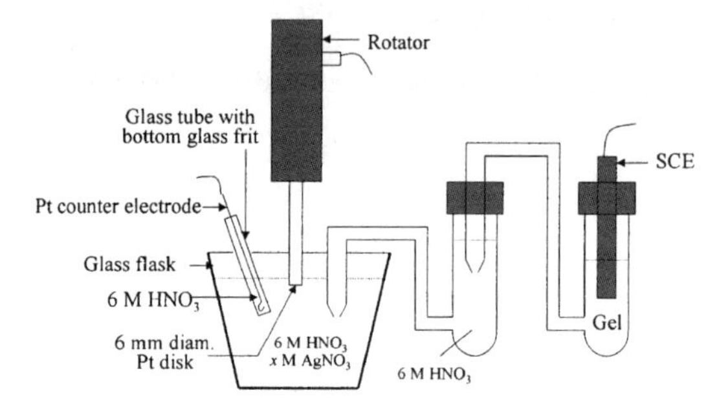

Figure 1 *Sketch of apparatus used for fundamental studies*

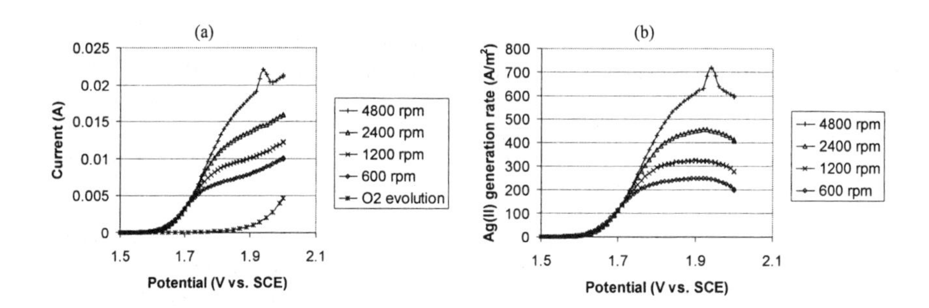

Figure 2 *(a) Steady state polarisation data for a Pt rotating disk at various rotation rates in 6 M HNO_3 and 50 mM $AgNO_3$ (20°C) + O_2 evolution data in 6 M HNO_3 only. (b) Silver(II) generation rates calculated by subtracting O_2 evolution data from overall data in (a).*

The curves in Fig. 2(a) indicate that the efficiency of silver(II) generation is very good up to about 1.9 V vs. SCE, where the oxygen evolution rate starts to increase rapidly. As seen in Fig. 2(b), increasing the potential above that level will just lead to increased oxygen generation, since the limiting current of silver(II) generation is then exceeded. In an applied case of PuO_2 dissolution or organics treatment by silver(II), one would normally prefer to work with as high a current as possible (giving a maximum oxidation rate), but not beyond the point where oxygen evolution becomes significant. How does one know what current a given reactor should be set to operate at?

One option is obviously to use a reference electrode to measure the anode potential, and adjusting the current until the desired potential is reached (1.9 V_{sce} for the case above). However, this could have practical difficulties, depending on the design of the reactor.

Also, there may be some uncertainty as to whether the additivity of the oxygen evolution rate and silver(II) generation rate strictly applies. A better option is to measure the rate of build-up of silver(II) at various currents, without any materials to be treated present, and determine the current efficiency in terms of silver(II) generation. An example is shown below, where a small batch reactor (called 'H-cell', after its shape), with a bottom magnet stirrer in the anolyte compartment, was used. The current efficiency was determined by measuring the silver(II) concentration as function of time, using spectrophotometry, and comparing the experimental data with the theoretical line of 100% faradaic efficiency. It can be seen that the limiting silver(II) current is reached at a current of about 0.3 A, or 400 A/m^2. Comparing this with the data in Fig. 2(b), it seems that the mass transfer conditions for this particular cell correspond to those with the RDE rotating at about 2000 rpm.

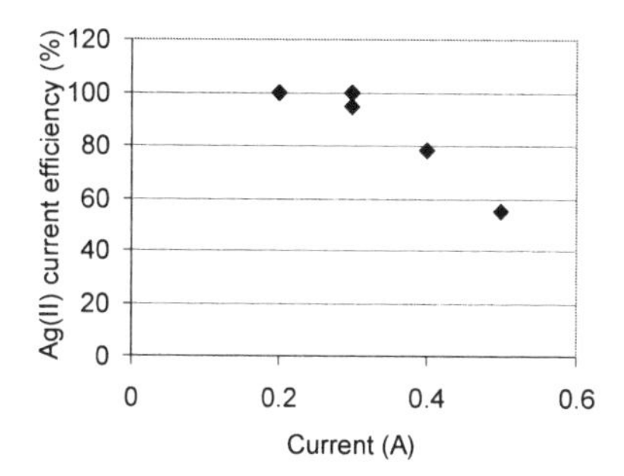

Figure 3 *Current efficiency of silver(II) generation in a H-cell (batch) reactor, at various levels of current, using a coil shaped platinum anode (area approx. 7.5 cm^2) and 150 mL of anolyte solution (6 M HNO_3 and 50 mM $AgNO_3$, 20°C).*

Probably, the most important question to be considered for a silver(II) process is what type of reactor design offers the best conditions of mass transfer or generation rate of silver(II) and, hence, materials oxidation rate. We are currently investigating a flow cell system, based on the use of Ineoschlor's FM01-LC laboratory electrolyser as a generator of silver(II), similar to the SilverII™ process. Also we will be testing a larger H-cell.

Acknowledgements

KLB wishes to thank the European Commission for financial support through a Marie Curie Fellowship Scheme.

References

1 L.A. Bray, and J.L. Ryan, in *Actinide Recovery from Waste and Low-Grade Sources*, J.D. Navratil and W.W. Schultz, (eds.), Harwood, London, 1982, 129.
2 J. Bourges, C. Madic, G. Koehly, and M. and Lecomte, *Journal of the Less-Common Metals*, 1986, **122**, 303.
3 M. Fleischmann, D. Pletcher and A. Rafinski, *Journal of Applied Electrochemistry*, 1971, **1**, 1.
4 O. Arnaud, C. Eysseric and A. Savall, ICHEME Symp. Series, 1999, **145**, 229.

IONIC LIQUIDS FOR ACTINIDES AND LANTHANIDES CHEMISTRY

I. Billard[1], C. Gaillard[1], S. Mekki[1], A. Ouadi[1], S. Stumpf[1], P. Hesemann[2], G. Moutiers[3], D. Trubert[4] and C. Le Naour[4]

[1] Institut de Recherches Subatomiques, Chimie Nucléaire, CNRS-IN2P3-ULP, BP 28, 67037 Strasbourg cedex 2, France.
[2] Laboratoire de Chimie Organométallique, CNRS UMR 5076, 8, rue de l'Ecole Normale, 34296 Montpellier Cedex, France.
[3] CEA Saclay, DEN/DPC/SCP/DIR, 91191 Gif-sur-Yvette cedex, France.
[4] Institut de Physique nucléaire, Radiochimie, CNRS-IN2P3-UPS, BP1, 91406 Orsay cedex, France.

1 INTRODUCTION

RTILs (Room Temperature Ionic Liquids) are salts composed of a cation, (the most popular ones being imidazolium, alkylammonium or pyridinium), and an anion (for example, PF_6^-, Cl^-, $CF_3SO_3^-$, BF_4^-, $(CF_3SO_2)_2N^-$, further denoted as Tf_2N^-). They are currently attracting an increasing interest in organic synthesis, catalysis, electrochemistry and material science.[1] This is due, in part, to their "green properties" (non flammable, non volatile) but these may be considered more as "fashionable introductory arguments" than really "scientifically ascertained facts". A deeper reason is that they display unusual solvating properties, large electrochemical windows and thermal stability ranges, opening a wide field of investigations. Furthermore, they are "designer solvents" because slightly changing their chemical structure changes their physico-chemical properties, such as miscibility to water or other usual solvents, although their green aspects remain. RTILs display the additional advantage of being radiation-resistant and to allow for a higher criticality limit than dodecane. These considerations have been the basis on which various radiochemistry groups are currently studying them, both for fundamental research or in view of applied goals. In this paper, a brief overview of the studies currently underway in Strasbourg will be given and the way they insert in the worldwide trends will be presented.

2 METHODS AND RESULTS

2.1 Fundamental aspects

Although RTILs are said to be "ionic", various experiments indicate they are neither fully associated nor dissociated. The water they may contain strongly favours dissociation.[2] Thus, it is expected that ion solvation would imply alternated cationic and anionic shells, which is rather unusual. Consequently, in RTILs, complexation might be hampered, as compared to the situation in molecular solvents, owing to charge screening. In this respect, molecular dynamics (MD) is a way to get insights into this question. EXAFS can also be used successfully and other spectroscopic techniques, such as TRLFS (Time Resolved Laser Fluorescence Spectroscopy) or UV-vis, might prove useful. In particular, we have shown that a Eu-hexachloro-complex forms in 1-methyl-3-butylimidazoliumTf_2N (C_1C_4-imTf_2N).[3] However, although water appears to be a strong Eu complexant, it does not compete favourably with Cl^- for the Eu inner sphere. Our most recent studies, based both on experimental data and MD simulations,[4] indicate that although dissolved, Eu^{III} salts might well not be dissociated, depending on the nature of the RTIL anionic counterpart. Thus, complexation in RTILs should be considered as a three-body competition between RTIL anions, solutes and metallic counter-anions. We also have evidence that PF_6^- is, unexpectedly, a stronger complexant than Cl^-. These peculiarities may favour unusual compounds, by competitive solubilisation of various salts in RTILs.

2.2 Applied studies: extraction processes

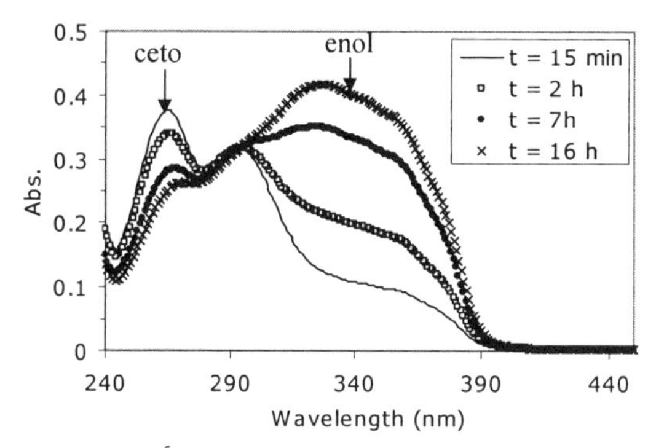

Figure 1 *TTA kinetics (4.9 10^{-5} M) in C_1C_4-imTf_2N containing 0.126 M of water.*

RTILs could replace other solvents so that one will benefit from their green properties but this is insufficient to undertake such an important industrial change. RTILs should prove, at the very least, as useful an extracting solvents as dodecane is, in terms of radiolysis stability, ease of use and cost. However, this does not prejudge other uses of RTIL for actinide/lanthanide partitioning. Therefore, an increasing number of studies worldwide[5-7] examine the extraction efficiencies of usual extracting agents (Tri-butyl-phosphate (TBP), CMPO, crown-ethers...) dissolved in RTILs. Although data are somehow limited, general trends can already be identified: extraction is feasible, in some cases with even better efficiencies than in usual solvents but, depending on the RTILs structure, the extraction can be tuned from an ion-exchange process to the extraction of neutral species.[5] Clearly, ion exchange, although interesting on fundamental grounds, is not suitable for industrial applications as it implies pollution of the aqueous solution. We have shown by EXAFS spectroscopy that the TBP-uranyl complex extracted in C_1C_4-imTf_2N from a nitric solution is different from the $UO_2(NO_3)_2$.TBP_2 moiety obtained with dodecane, thus confirming the change in mechanism. In order to better understand the behaviour of extracting agents in RTILs, we have undertaken a UV-vis study of TTA (4,4,4-trifluoro-1-(2-thienyl)-1,3-butanedione) dissolved in C_1C_4-imTf_2N (cf. fig.1), as compared to usual solvents (water, dioxane and cyclohexane). In RTILs, the kinetics of the enol/keto-hydrate equilibrium is

roughly two orders of magnitude faster than in cyclohexane, while the viscosity is one order of magnitude larger in RTILs. There is thus a need for exploratory studies in various RTILs, in order to understand the parameters responsible for such changes.
Task-Specific Ionic Liquids (TSILs) is another way to tackle the question of actinide/lanthanide partitioning, by grafting the complexing entity of interest onto the ionic liquid. The studies are very scarce at the moment.[8-9] We have synthesised one TSIL,[2] allowing the Am(III) pH-dependent extraction from basic water solutions together with quantitative back extraction. Basic conditions are not fully adapted to industrial needs but we could understand what a TSIL should *not* be in view of An extraction. We are now applying these considerations to the synthesis of a better suited TSIL.
The last approach is the use of RTILs in conjunction of supercritical CO_2 (Sc-CO_2). Sc-CO_2 is highly soluble in RTILs.[10-11] We first demonstrated the extraction of Ln and Cu[12] from C_1C_4-imTf_2N into Sc-CO_2. High extraction efficiencies (up to 90 %) can be obtained, without the adjunction of the usual Sc-CO_2 modifiers (TBP, methanol *etc.*). These experiments, performed in collaboration with Prof. C. Wai, will be pursued in the future.

3 CONCLUSION

RTILs are a fascinating new solvent media in which exciting fundamental and applied science can be performed. Electrochemical studies are also promising, in view of the electrodeposition of metals, such as Pu or Np. Recent studies seem, for these applications, to favour the use of asymmetrical alkylammonium based ionic liquids instead of alkylimidazolium, because of a more reductive electroactivity range of potentials.

References

1 T. Welton, *Chem. Rev.*, 1999, **99**, 2071.
2 C. Gaillard, G. Moutiers, C. Mariet, T. Antoun, B. Gadenne, P. Hesemann, J. J. E. Moreau, A. Ouadi, A. Labet, and I. Billard, in 'Potentialities of RTILs for the nuclear fuel cycle: electrodeposition and extraction', ed. R. D. Rogers and K. R. Seddon, 2005.
3 I. Billard, S. Mekki, C. Gaillard, P. Hesemann, G. Moutiers, C. Mariet, A. Labet, and J. C. Bünzli, *Eur. J. Inorg. Chem.*, 2004, 1190.
4 C. Gaillard, I. Billard, A. Chaumont, G. Wipff, S. Mekki, A. Ouadi, M. A. Denecke, G. Moutiers, and P. Hesemann, submitted.
5 M. P. Jensen, J. Neuefeind, J. V. Beitz, S. Skanthakumar, and L. Soderholm, *J. Am. Chem. Soc.*, 2003, **125**, 15466.
6 M. L. Dietz, J. A. Dzielawa, I. Laszak, B. A. young, and M. P. Jensen, *Green Chem.*, 2003, **5**, 682.
7 A. E. Visser and R. D. Rogers, *J. Solid State Chem.*, 2003, **171**, 109.
8 A. E. Visser, R. P. Swatloski, W. M. Reichert, R. Mayton, S. Sheff, A. Wierzbicki, and R. D. Rogers, *Environ. Sci. Technol.*, 2002, **36**, 2523.
9 A. E. Visser, R. P. Swatloski, W. M. Reichert, R. Mayton, S. Sheff, A. Wierzbicki, J. H. Davis, and R. D. Rogers, *Chem. Commun.*, 2001, 135.
10 J. S. Wang, Y. Lin, and C. M. Wai, *Sep. Sci. Technol.*, 2003, **38**, 2279.
11 C. Cadena, J. Anthony, J. Shah, T. Morrow, J. Brennecke, and E. Maginn, *J. Am. Chem. Soc.*, 2004, **126**, 5300.
12 S. Mekki, C. M. Wai, I. Billard, G. Moutiers, C. H. Yen, J. S. Wang, A. Ouadi, C. Gaillard, and P. Hesemann, *Green Chem.*, 2005, **7**, 421.

SOLVENT EXTRACTION OF ACTINIDES BY TETRAALKYLPYRIDINE-N-OXIDE-BIS PHOSPHINE OXIDE COMPLEXANTS

K.L. Nash[1], B. Gannaz[2], C. Lavallette[2], G. Cote[2], M. Borkowski[3] and R. T. Paine[4]

[1]Chemistry Division, Argonne National Lab, Argonne, Illinois – present address - Department of Chemistry, Washington State University, PO Box 644630, Pullman, WA 99164-4630, USA,
[2] Ecole Nationale Superieure de Chimie de Paris, France,
[3] Los Alamos National Laboratory, Carlsbad, NM
[4] University of New Mexico, Albuquerque, NM

1 INTRODUCTION

Polydentate solvent extraction reagents remain an important component of actinide separations from the ultra-low detection level to industrial scale production facilities. Organophosphorus compounds have played a particularly important role in actinide separations, largely due to the donor ability of the functional groups, and the unique coordination geometries that can be established around them. In this investigation, we describe selected features of solvent extraction studies of Am^{3+}, Eu^{3+} and UO_2^{2+} extraction by 2,6-bis[(dialkylphosphino)methyl]pyridine-N,P,P'-trioxide (R_4NOPOPO where R is either 2-ethyl(hexyl) (TEH(NOPOPO) or *n*-octyl TO(NOPOPO)) extractants in dodecane solutions.

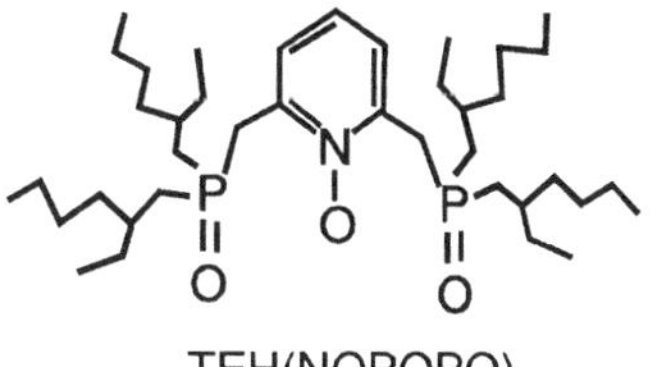

TEH(NOPOPO) 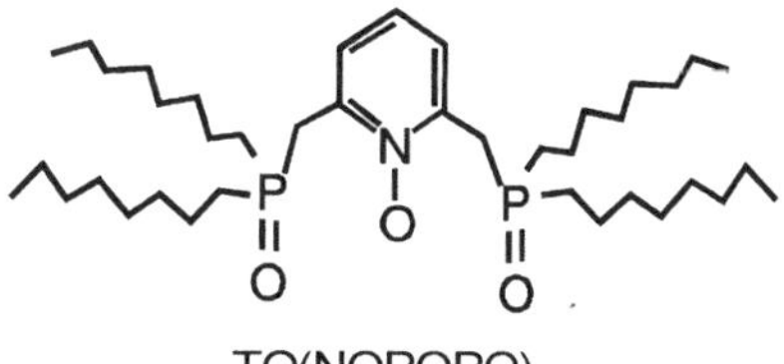 TO(NOPOPO)

2 EXPERIMENTAL

The methods used for synthesis of the compounds have been described previously.[1-2] Sodium nitrate, nitric acid, *n*-dodecane, and toluene used in the solvent extraction studies were of reagent grade and used as received. Aqueous solutions were prepared with deionized water. The tracer stock solutions were prepared from the purified radionuclides ^{233}U, ^{241}Am and ^{152,154} Eu (ANL stock), and were assayed radiometrically on a Packard Cobra II automatic gamma counter (^{241}Am, ^{152,154}Eu) or by liquid scintillation (^{233}U)

The metal nitrate distribution ratios were determined by equilibrating equal volumes of aqueous and organic phases. The organic phases were pre-equilibrated three times with equal volumes of fresh $NaNO_3$ or HNO_3 solutions of the appropriate concentration. Fifteen minutes contact with vortex mixing was adequate to reach the biphasic equilibrium. To determine the stoichiometry of the metal complexes in the organic phase, the phases were mixed for 30 minutes using a vortex mixer at room temperature. The phases were centrifuged, separated, and duplicate 200 μL aliquots were withdrawn from each phase for the radiometric analysis. Under all conditions, mass balance was greater than 95%.

The metal ion distribution experiments as a function of temperature were conducted at five temperatures between 10.0°C and 45.0°C. Chemical equilibrium was attained by vortex mixing of the two phases (3.5 mL of each) for one minute followed by 10 minutes for thermal equilibration in a water bath. At each temperature the process of vigorous mixing/thermal equilibration was repeated eight times prior to sampling. At this stage, the two-phase sample was centrifuged and duplicate 200 μL aliquots of each phase were taken for radiometric assay. Equal volume samples of both organic and aqueous phases were taken to maintain a constant ratio of the phases throughout data acquisition. The residual solutions were then equilibrated at the next temperature. The distribution data were collected in three independent experiments.

3 RESULTS AND DISCUSSION

The stoichiometry of the phase transfer reaction of Am(III), Eu(III) and UO_2^{2+} into TEH(NOPOPO) /*n*-dodecane or TO(NOPOPO)/toluene can be represented as follows:

$$M^{3+}_{aq} + 3\ NO_3^-{}_{aq} + 2\ \text{TEH(NOPOPO)}_{org} \rightarrow M(NO_3)_3(\text{TEH(NOPOPO)})_{2,\ org} \qquad (1)$$

$$UO_2^{2+}{}_{aq} + 2\ NO_3^-{}_{aq} + 2\ \text{TO(NOPOPO)}_{org} \rightarrow UO_2(NO_3)_2\ (\text{TO(NOPOPO)})_{2,\ org} \qquad (2)$$

Dependence of the extraction reaction on $[NO_3^-]$ or $[R_4NOPOPO]$ were determined by conventional slope analysis at room temperature. Besides the actinide nitrate complexes, the extractants were observed to extract up to two molecules of HNO_3, necessitating that a correction be made to determine the actual concentration of free $R_4NOPOPO$ under each set of conditions. In the Am system, an additional series of experiments were conducted in $NaNO_3$ solution. In the temperature variation experiments, electroneutrality of the extracted complex was assumed, implying that the nitrate dependence of the equilibrium constant for this reaction remained as written in equations 1 and 2. The dependence of the phase transfer reaction on $[R_4(NOPOPO)]_{cor}$ was determined experimentally at each temperature. In the range of temperature from 10°C to 45°C, the stoichiometry of the extracted species remained constant. Thermodynamic parameters resolved from the temperature dependent data are presented in Table 1. The observed negative entropies and enthalpies are consistent with those reported previously for extraction reactions of this type.[3]

In the reported crystal structures of lanthanide-NOPOPO complexes,[4-5] a tridentate configuration of ligand donor atoms is most typically observed, incorporating one or fewer bidentate or monodentate nitrate counterions in the inner coordination sphere. Denticity of the extractant and the ordering of donor atoms around the metal ions are not directly addressed by thermodynamic studies of this sort. However, comparing these data with reports from the literature can provide insights. In the thoroughly investigated CMPO-nitrate system, the extracted trivalent lanthanide or actinide nitrates contain three bidentate nitrates and three monodentate CMPO's, with the latter coordinated through the P=O group. Two bidentate nitrates and two monodentate CMPO molecules are indicated as the predominant coordination mode for the uranyl complex. Little is known about the dynamics of nitrate coordination with these metal ions in organic solutions. The low dielectric constant would seem to demand close association between the cation and the anions, a requirement probably best served by bidentate coordination of nitrate molecules in these media. The lower denticity of extractant molecules in such systems as the NOPOPO or CMPO media leave highly nucleophilic oxygen donor atoms in position to interact with organic solvent molecules. A more direct investigation of the dynamics of ligation in such systems appears to be a useful territory for exploration.

Table 1. *Thermodynamic parameters for extraction of Am^{3+}, Eu^{3+} and UO_2^{2+} by R_4NOPOPO extractants from nitrate media.*

System	$[HNO_3]_{aq}$	ΔG (kJ/mol)	ΔH (kJ/mol)	ΔS (J/mol K)
$Am(NO_3)_3(TEHNOPOPO)_2$	0.001 M	-31.0 ± 0.3	-42.6 ± 2.6	-39 ± 9
$Am(NO_3)_3(TEHNOPOPO)_2$	0.5 M	-32.0 ± 0.4	-47.2 ± 2.0	-51 ± 7
$Eu(NO_3)_3(TEHNOPOPO)_2$	0.5 M	-30.8 ± 0.6	-35.0 ± 2.4	-14 ± 8
$UO_2(NO_3)_2(TONOPOPO)_2$	0.25 M	-43.3 ± 0.6	-56 ± 2.8	-42.7 ± 9

4 CONCLUSIONS

The extraction systems behave predictably, partitioning electroneutral nitrate complexes of each metal ion with two extractant molecules associated with the metal complex in the organic phase. Thermodynamic parameters for the phase transfer reactions are interpreted to indicate that the NOPOPO extractant molecules probably do not achieve full tridentate coordination in either Am^{3+}/Eu^{3+} or UO_2^{2+} complexes. As extractants, the NOPOPO ligands perform similarly to CMPO-based extractants for nitric acid systems. The availability of an apparent excess of oxygen donor atoms (beyond the capacity of the inner sphere coordination ability of Am^{3+}, Eu^{3+}, or UO_2^{2+}) implies that additional reactive donor atoms are available for interactions with the medium.

References

1 K. L. Nash, C. Lavallette, M. Borkowski, R. T. Paine, X. Gan, *Inorg. Chem.*, 2002, **41**, 5849.

2 X.-M. Gan, E. N. Duesler, R.T. Paine, *Inorg. Chem.*, 2001, **40**, 4420.

3 K. L. Nash, Studies of the Thermodynamics of Extraction f-Elements in *Solvent Extraction for the 21st Century, Proceedings of the International Solvent Extraction Conference, 1999*, M. Cox, M. Hidalgo, M. Valiente, Eds., Society of Chemical Industry, London, 2001, pp 555-559.

4 R. T. Paine, Design of Ligands for f Element Separations. in *Separations of f Elements*. K. L. Nash and G. R. Choppin. Eds., 1995, New York, Plenum Press: pp. 63-75.

5 D. J. McCabe, A. A. Russell, S. Karthikeyan, R. T. Paine, R. R. Ryan, B. F. Smith, *Inorg. Chem.*, 1987, **26**, 1230.

KINETICS AND MECHANISM OF ACTINIDE COMPLEXATION BY POLYDENTATE LIGANDS

K.L. Nash[1], J.C. Sullivan[2], M. P. Jensen[2], H. Hall[2] and J.I. Friese[3]

[1] Washington State University, Department of Chemistry, PO Box 644630, Pullman, WA, 99164-4630, USA,
[2] Argonne National Laboratory, Chemistry Division, 9700 S. Cass Ave., Argonne, IL 60439,
[3] Pacific Northwest National Laboratory, PO Box 999, Richland, WA 99183

1 INTRODUCTION

As compared with the amount of information available on the thermodynamic properties of actinide complexes, far less is known about the kinetics of actinide complexation reactions. It is generally recognized that the reactions occur with millisecond lifetimes under most conditions in aqueous solutions, and so are most amenable to investigation using either relaxation techniques (stopped flow, temperature jump, pressure jump) or NMR spectroscopy. Water exchange rates have been measured for some trivalent lanthanides (generally good analogs for the trivalent actinides in most aspects of their coordination chemistry) and hexavalent species, but have not been reported for pentavalent or tetravalent actinide ions. Some of the earliest studies of actinide complexation kinetics[1] were of the rates of exchange of simple inorganic ions like sulfate, acetate, and carbonate. Later investigations established that the rates of complex formation and dissociation of hexadentate aminopolycarboxylate complexes of Ln(III) and An(III).[2]

During the past decade, we have investigated the kinetics of actinide complexation reactions with ligands having structural complexity somewhere between these extremes. We have observed using stopped-flow spectrophotometry kinetic parameters describing the interactions of dioxocations (UO_2^{2+} and NpO_2^{+}) with bidentate diphosphonate and tridentate diglycolate and dipicolinate ligands. Solvation of precursor complexes has been suggested as a key feature of the kinetics in these systems. Herein we compare the kinetic features of UO_2^{2+} and NpO_2^{+} complexation reactions in 0.1 M acid media.

2 EXPERIMENTAL

The details of solution and complexant preparation have been described previously.[3-5] All reactions were run under pseudo-first order conditions in which an excess of the complexant of interest reacts with 10^{-5} to 10^{-4} M solutions of UO_2^{2+} or NpO_2^{+}. Either Arsenazo III (2,7-bis(2,2'-arsonophenylazo)-1,8-dihydroxynaphthalene-3,6-disulfonic acid) or Chlorophosphonazo III (2,7-bis(4-chloro-2-phosphonobenzylazo)-1,8-dihydroxynaphthalene-3,6-disulfonic acid) were employed at a limiting concentration to monitor the progress of the reaction. It was demonstrated experimentally that the indicator does not participate in the rate law for formation of the complex of interest. An OLIS

RSM-16 stopped flow spectrophotometer was used for data acquisition and analysis of the traces completed using the OLIS RSM Robust Global Fitting software package. Data were collected over a wavelength range of 500 nm to 800 nm. All experimental observations were made in at least triplicate and the individual rate constants averaged for subsequent weighted regression analysis. Reactions were repeated as a function of temperature to allow calculation of activation parameters. For the Uranyl-BzDPA system, the reaction rate was determined in deionized water, 50:50 methanol-water, and 50:50 *t*-butanol-water.

3 RESULTS AND DISCUSSION

In each of the systems studied, a plot of k_{obs} vs. $[L]_t$ is linear with a finite positive intercept. We have interpreted these plots to indicate a first order approach to equilibrium, in which the slope defines the rate constant for complex formation, the intercept defines the corresponding dissociation rate constant. The resolved activation parameters for complex formation and dissociation reactions at 0.1 M total acid are shown in Table 1. Some systems were investigated at higher pH conditions as well. In the dipicolinate system, we noted a more complex reaction mechanism overall. In addition to the straight-forward second-order metal-ligand association reaction seen in other systems, a parallel pathway whose concentration dependence indicates a Michaelis-Menten reaction pathway involving a preformed thermodynamically stable complex. Substituting methanol or *tert*-butanol for portions of the water gives a change in the solvation properties and a corresponding change in the rate of reaction in the uranyl-BzDPA system. A correlation of the second-order rate constants with the dielectric constant of the medium indicates that in this system the activated complex features the coordinated release of a hydrogen cation from the precursor complex, apparently $UO_2(H_4BzDP)^{2+}$ in which the uranyl cation is coordinated to BzDP via the two phosphoryl oxygen atoms of BzDP. Activation entropies indicate a predominance of associative reaction mechanism in the each system.

Table 1. *Activation parameters for complex formation and dissociation of NpO_2^+ and UO_2^{2+} complexes with selected ligands at pH 1 (±1σ uncertainties in parentheses).*

System	ΔH^*_{form} kJ/mol	ΔS^*_{form} J/mol-K	ΔG^*_{form} kJ/mol	ΔH^*_{dis} kJ/mol	ΔS^*_{dis} J/mol-K	ΔG^*_{dis} kJ/mol
U(VI)-BzDPA, pH 1	42.0(1.1)	-70(5)	62.9(4.0)	45.0(1.3)	-70(5)	65.8(5.1)
U(VI)-BzDPA, pH 1, 50:50 methanol-H_2O	34.2(1.8)	-105(8)	65.5(6.0)	30.9(1.7)	-128(6)	69.1(5.0)
U(VI)-Dipic, pH 1	47.3(2.6)	-39(9)	58.9(5.3)	62.4(1.6)	-41(6)	74.6(3.4)
Np(V)-Dipic, pH 1	42.0(1.9)	-58(7)	59.3(4.0)	51.9(0.7)	-54(3)	68.0(1.6)
U(VI)-HEDPA, pH1	38.8(0.6)	-40(2)	50.7(2.7)	57.8(20.9)	-22(62)	64.6(23.4)
U(VI)-MDPA, pH 1	38.0(1.4)	-44(4)	51.1(5.0)	44.8(2.4)	-66(8)	64.5(8.5)
U(VI)-E12DPA, pH 1	46.8(0.1)	-43(1)	59.6(1.4)	44.0(4.1)	-65(13)	63.6(14.0)

The relative rates of complex formation reactions (determined by ΔG^*_{form}) are UO_2^{2+} - BzDPA (MeOH-H_2O) < UO_2^{2+} - BzDPA (H_2O) < UO_2^{2+} - E12DPA ≤ NpO_2^+ - dipic ≤ UO_2^{2+} - dipic < UO_2^{2+} - MDPA ≤ UO_2^{2+} - HEDPA. Aside from UO_2^{2+} - dipic, the dissociation rate constants are seen to lie within the rather narrow range of 5-45 s^{-1} (in 0.1 M acid) including the single observation on NpO_2^+ rates. Activation enthalpies and entropies demonstrate a wider range of values and the usual evidence for enthalpy-entropy compensation. The comparatively elevated ΔH^* for UO_2^{2+} - HEDPA and depressed ΔH^*

for UO_2^{2+} - BzDPA (MeOH-H_2O) (relative to the corresponding values for the other diphosphonate data) are consistent with the general thesis that solvation of the precursor complex is an important determinant of reaction rates.

Microscopic reversibility requires that forward and reverse reactions proceed through the same intermediate, hence the activation parameters in Table 1 can be used to establish the relative thermodynamic stability of each species formed ($\Delta G° = \Delta G^*_{form} - \Delta G^*_{dis}$, $\Delta H° = \Delta H^*_{form} - \Delta H^*_{dis}$, $\Delta S° = \Delta S^*_{form} - \Delta S^*_{dis}$, Table 2). Complex stability ($\Delta G°$) increases in the order UO_2^{2+} - BzDPA (H_2O) < UO_2^{2+} - BzDPA (MeOH-H_2O) < UO_2^{2+} - E12DPA ≤ NpO_2^+ - dipic ≤ UO_2^{2+} - MDPA ≤ UO_2^{2+} - HEDPA < UO_2^{2+} - dipic. Complex enthalpies follow a similar pattern, though the most exothermic reaction is that between UO_2^{2+} and HEDPA rather than UO_2^{2+} - dipic. Substitution of 50% methanol for H_2O in the UO_2^{2+} - BzDPA system shifts the net reaction from slightly exothermic to slightly endothermic, with a compensating increase in the net entropy of reaction. In each system the net entropy change is small. We have demonstrated in previous reports that metal-ligand systems react by a common mechanism (as indicated by linear isokinetic plots) based more strongly on the structure of the ligand, that is, NpO_2^+ and UO_2^{2+} data are mutually correlated based on ligand type.

Table 2 *Thermodynamic parameters for U(VI) and Np(V) complexation reactions.*

System	ΔG° kJ/mol	ΔH° kJ/mol	ΔS° J/mol K
U(VI)-BzDPA, pH 1	-2.9	-3.0	0
U(VI)-BzDPA, pH 1, 50:50 methanol-H_2O	-3.6	+3.3	+23
U(VI)-Dipic, pH 1	-15.7	-15.1	+2
Np(V)-Dipic, pH 1	-8.7	-9.9	-4
U(VI)-HEDPA, pH 1	-13.9	-19.0	-18
U(VI)-MDPA, pH 1	-13.4	-6.8	+22
U(VI)-E12DPA, pH 1	-4.0	+2.8	+22

2 CONCLUSIONS

Investigations of the rates and mechanisms of NpO_2^+ and UO_2^{2+} complexation by dipic, BzDPA, MDPA, HEDPA and E12DPA in 0.1 M acid establish that the reaction is associative in nature. The patterns of reaction rates as ligand structure and medium properties change are consistent with the rate of exchange of water molecules on the activated complex as the rate limiting process under these conditions.

References

1 K. L. Nash and J. C. Sullivan, Kinetics and Mechanisms of Actinide Redox and Complexation Reactions in *Advances in Inorganic and Bioinorganic Reaction Mechanisms, V. 5* (ed. A. G. Sykes) Academic Press, New York (1986) pp. 185-213.
2 K. L. Nash and J. C. Sullivan J. *Alloys and Compounds,* 1998, **271-273**, 712.
3 J. I. Friese, K. L. Nash, M. P. Jensen and J. C. Sullivan, *Radiochim. Acta,* 2001, **89**, 35.
4 H. Hall, J.C. Sullivan, P.G. Rickert and K.L. Nash, *Dalton Trans.* 2005 2011.
5 M. A. Hines, J. C. Sullivan and K. L. Nash, *Inorg. Chem.,* 1993, **32,** 1820.

THE SIGNIFICANCE OF MINOR ACTINIDE BASED FUEL CYCLES IN ADVANCED FUEL CYCLES

H.P. Nawada and C. Ganguly

Nuclear Fuel Cycle and Materials Section, Division of Nuclear Fuel Cycle and Waste Technology, Department of Nuclear Energy, International Atomic Energy Agency, IAEA, Vienna, AUSTRIA E-mail h.nawada@iaea.org

1 INTRODUCTION

The long-term hazard of spent fuel or high level liquid waste is associated with the radiotoxicity of some actinides particularly from plutonium and minor actinides (MA) as well as some long-lived fission products (such as ^{99}Tc and ^{129}I). There is considerable interest in many Member States of the IAEA in the partitioning and transmutation (P&T) of long-lived nuclides as a potential complement to the reference concept fuel cycle.[1] Efficient P&T schemes could shorten the time needed for isolation of nuclear waste to less than 500 years. In the context of transmutation of these transuranic elements, the development of recovering MA elements from spent fuel and fabrication of MA-based fuel or target are the vital components of P&T approach. Apart from reducing the radio-toxicity, the MA-based fuel cycle enhances the proliferation resistance of any fuel cycle with in-built facilitation of high resource utilization of actinides. Keeping in view of recent research on MAs, the Nuclear Fuel Cycle and Materials Section of the Agency is cooperating in the following subject-areas: fuel cycle approaches to minimize MAs viz., by thorium fuel cycle, liquid metal cooled reactor fuel cycle and inert matrix fuels, minimization of MA-losses in pyro-chemical processes, MA-fuel and target, coated particle fuel, MA-property data-bank and reprocessed uranium management. This report describes the global developments in some of the above topics.

2 MINIMIZATION OF MA-LOSSES IN PYRO-CHEMICAL PROCESSES

The IAEA activities on P&T especially pyro-chemical processes has been recently reviewed and presented.[2] The development of pyro-chemical processes calls for significant R&D initiatives to develop materials and process control equipment owing to the use of highly radioactive molten salts and metals at high temperature in a pyro-chemical separation and MA-based fuel / target manufacturing processes. The IAEA has initiated a Coordinated Research Program (CRP) with emphasis on the development of an Advanced Fuel Cycle in its Member States. The overall objective of this CRP is to develop methods of minimize losses in partitioning steps with a view to minimizing environmental impact.[2] The salient first outcomes of this CRP meeting are: i) A definition of sources of possible losses of long-lived radio-toxicity and their chemical form as waste from pyro-chemical processing, ii) an explanation of target value for the reduction of ratio of radiotoxicity iii) establishment of a quantitative relationship between environmental impact and waste reduction ratio considering separation losses, leading subsequently to a model being developed.

3 DEVELOPMENT IN MA-FUEL AND MA-PROPERTY DATA BANK

Current status of fabrication and processing of MA-based fuel has been critically analysed by JRC/EC[3] and ANL.[4] The technologies of minor actinide-bearing-fuel materials are not well established. To enhance the capacity of the interested Member States in developing innovative nuclear fuel cycle technologies for sustainability, the IAEA is coordinating and preparing a technical document to review the current status and future trends in the processing of MAs and their pertinent properties for the fabrication of nuclear fuels (targets) to incinerate in thermal as well as fast neutron spectrum. The latest directions in the subject of minor actinide fuel (target) development namely measurement of properties, processing, fabrication, irradiation behaviour in-pile and back-end issues would be covered. Properties included in the discussion are thermal, mechanical, chemical, physical and irradiation. Development of liquid metal-cooled reactors (LMRs) and its fuel cycle is of paramount importance in the context of improving nuclear fuel cycle technologies for sustainability. In addition, several Member States are evaluating various fuel cycle scenarios in which LMR fuel cycles would play a critical dual role of accomplishing transmutation of long-lived actinides to reduce the radiological risk of the nuclear waste and conserving effective utilization of natural resources. The Agency is conducting a technical meeting on LMFR fuel cycle to clarify the status on: a) LMFRs fuel cycle issues which includes safety, waste management, environment, economics and proliferation resistance as well as P&T; b) Status and programme of conventional, advanced and directions for innovations in LMFR fuels including non-oxide and metallic fuels and their methods of fabrication, out-of-pile properties, in-pile performance, irradiation-testing and post-irradiation examination.

A large number of physico-chemical and thermo-mechanical data are required for the design and fabrication of an MA-based fuel or specialized targets. The effectiveness of separation of MAs from other elements such as rare earths in the specified pyro-chemical separation method is determined by the relative difference in the physico-chemical properties of MAs from that of the rare earth elements in that particular molten-salt and liquid alloy medium. Generation of experimental data for the materials containing MAs would involve enormous impediments such as i) requirement of shielded facilities (in some cases highly purified argon atmosphere) in dealing a with MAs and ii) decay heat, decay products and self-radioloysis limiting the applicability of many of the experimental measurement techniques. Hence, the information is very limited, and the available information is scattered in documents and other sources some of which are proprietary. In this context the Agency is developing a database on minor actinides (MA) used for advanced nuclear fuel cycles.

4 COATED PARTICLE FUEL AND INERT MATRIX FUEL DEVELOPMENT

This coated particle fuel technology is primarily of interest in future nuclear scenarios that employ high-temperature gas reactors. The deep burn capability of this fuel makes it attractive for once through cycles for plutonium and minor actinides destruction. In the Deep-Burn concept, destruction of the transuranic component of light water reactor waste could be achieved in one burn-up cycle, accomplishing the virtually complete destruction of ^{239}Pu and up to 90% of all transuranic waste, including the near totality of ^{237}Np and its precursor, ^{241}Am. Novel Pu based kernels have been fabricated at ITU using an innovative fabrication route based on actinide infiltration of porous yttria stabilized zirconia kernels[5]. In this context, IAEA has recently conducted a technical meeting on "Current status and future prospects of gas cooled reactor fuels" to review the progress of coated particle fuel development and a document is being prepared.

The reduction of the accumulated plutonium inventories by the use of inert matrix fuels in current nuclear power reactors is a subject of imminent interest in several Member States. The recent progress made in the synthesis of crystalline inert matrices for the incineration of excess plutonium has been summarized in a recent IAEA technical document.[6] Some of these materials could be proposed for the transmutation of separated MAs. However, specific environmental studies are also required to understand the long-term behaviour of the inert matrix forms with regard to the conditions of a geological disposal. Partitioning of MAs followed by waste conditioning into a very stable matrix could also be considered for the near future.

5 CONCLUSIONS

National and International collaborative efforts have clarified during the last decade most of the scientific issues and associated challenges with P&T. Innovations in the scientific and engineering R&D on key MA-processing technologies and scenario-studies could be judiciously utilized to assist future decisions. The development of MA- processing technologies calls for a remarkable R&D initiatives and international collaborations to develop materials and process control equipments. This calls for the setting up of several international collaborations to accommodate various aspects of advanced fuel cycle development. In this regard, a "regional approach", can help to build-up synergies, and that different countries with different objectives, could develop and operate "regional" facilities in common, sharing costs and resources.[7]

References

1 *Emerging nuclear energy and transmutation systems: Core physics and engineering aspects,* INTERNATIONAL ATOMIC ENERGY AGENCY, IAEA-TECDOC-1356, IAEA, Vienna (2003)

2 H.P. Nawada and K. Fukuda, J. Phys. Chem. Solids, 2005, **66,** 647

3 R.J.M. Konnings, *Advanced fuel cycles for ADS: fuel fabrication and reprocessing,* European Commission Report, EUR 1928 En, 2001

4 D.C. Crawford et. al., *Current US plans for development of fuels for accelerator transmutation of waste,* in IAEA-TECDOC-1356, IAEA, Vienna (2003)

5 J. Somers and A. Fernandez, *Novel Inert Matrix Fuel Kernels for Pu and Minor Actinide Incineration in High Temperature Reactors,* in Proc. of 2nd International Topical Meeting on High Temperature Reactor Technology, (HTR-2004), Beijing, CHINA, Sep 22-24, 2004

6 *Viability of inert matrix fuels in reducing plutonium inventories in the reactors,* INTERNATIONAL ATOMIC ENERGY AGENCY, IAEA, Vienna (to be published)

7 M. Salvatores, *P/T Potential for Waste Minimization in a Regional Context,* from Proc. 8th Information Exchange Meeting on Actinide and Fission Product Partitioning and Transmutation 9-11 November 2004 Las Vegas, Nevada, USA

SEPARATION OF URANIUM FROM RADIOACTIVE WASTE USING SUPERCRITICAL CARBON DIOXIDE WITH NITRIC ACID – TRI-N-BUTYLPHOSPHATE COMPLEX

Y. Meguro, O. Tomioka, T. Imai, S. Fujimoto and M. Nakashima

Department of Decommissioning & Waste Management,
Japan Atomic Energy Research Institute, Tokai, Ibraki 319-1195, Japan.
E-mail: meguro@popsvr.tokai.jaeri.go.jp

1 INTRODUCTION

Removal of actinides such as uranium and plutonium from apparatus and wastes contaminated by actinides leads to reduction in cost of waste management. Less generation of secondary radioactive wastes is required in the development of a removal method of actinides. Recently, much attention has been paid to direct separation of metals from solid samples using supercritical CO_2 as a medium by utilizing its large penetration force and high diffusivity into the solids.[1 - 11] The authors have developed a supercritical fluid leaching (SFL) method to remove metals from solid samples and this process generates less acid solution and organic solvent waste. The SFL method using a HNO_3-TBP complex as a reactant was applied to remove uranium from a sea sand sample and some simulated wastes such as a HEPA filter, a firebrick and incinerated ash,[3] and several techniques to accelerate mass transfer rate in supercritical CO_2 were developed.[11]

The wastes generated from nuclear fuel facilities has an uncertain history and properties and therefore analysis of the removal behaviour of uranium from the wastes becomes difficult. In the present study, samples systematically simulating the firebrick waste were prepared and the removal behaviour of uranium from them was investigated by the SFL method using the HNO_3-TBP complex as a reactant.

2 EXPERIMENTAL

2.1 Apparatus

An apparatus for the SFL was identical to that reported previously,[2] the main part consisting of a reaction vessel (50 cm^3), a reactant mixing vessel, a collection vessel, a syringe pump, a pressure-reducing regulator, a thermostatic water bath and a pre-heating coil.

2.2 Materials

A circular cylinder of 10 mm in diameter and 20 mm in length made of porous alumina brick of 40% porosity, 1.3 g/cm^3 density and *ca.* 1 μm pore size was prepared as a matrix material for a sample. The brick was soaked in 1 mol/dm^3(M) HNO_3 solution containing

0.023 M uranyl ion and kept in a vacuum desiccator until generation of air bubbles from pores in the brick was finished. The pressure in the desiccator was returned to normal and the brick was kept standing for one night. The brick was dried for a week and heated first at 500 °C for 4 hours with air and then at 900 – 1800 °C for 6 hours with reducing gas (H_2:N_2 = 3:1). The prepared sample contained *ca.* 1500 ppm uranium as UO_2. A stock solution of the HNO_3-TBP complex solution was prepared by shaking a mixture of the same volume of anhydrous TBP and 70% HNO_3 for 30 min.

2.3 Procedure

The solid waste sample was placed in the reaction vessel and the temperature of the vessel was kept at 60 °C using the thermostatic water bath. The HNO_3-TBP complex solution (2 cm^3) was taken into the reactant-mixing vessel. A mixture of CO_2 and the HNO_3-TBP complex was introduced into the reaction vessel through the pre-heating coil using the syringe pump at 15 MPa, the valves before and after the reaction vessel were closed and then the system was allowed to stand for 60 min (a complexation process). CO_2 of 15 cm^3/min was allowed to flow for 60 min by the syringe pump at 25 MPa by means of the pressure-reducing regulator (a dissolution process). The U(VI)-TBP complex and the HNO_3-TBP complex eluted with CO_2 were recovered in the collection vessel at normal pressure. One SFL operation consisted of one complexation-dissolution processes cycle.

The activity (A) of γ-ray at 186 keV of ^{235}U in the sample was measured before and after the SFL operation and the removal efficiency was determined as {(A before SFL) – (A after SFL)}/(A before SFL) x 100.

3 RESULTS AND DISCUSSION

Removal efficiencies of uranium from the simulated firebrick samples prepared at 900, 1200, 1500 and 1800 °C were determined. The sample was decontaminated several times by the SFL operation and then accumulated removal efficiencies were calculated from the efficiency of each decontamination. The relationships between the accumulated efficiency and number of decontaminations for the every sample are shown in Fig. 1. Uranium could be removed from the samples prepared at different temperatures and the accumulated efficiency increased with an increase of the number of decontaminations. From the samples prepared at 900 °C and 1200 °C, *ca.* 70% of uranium was removed by one SFL operation and more than 80% was removed after decontaminating 3 or 4 times. For the samples prepared at 1500 °C and 1800 °C, the accumulated removal efficiency also increased with an increase of the number of SFL, however removal rate of uranium from these samples was slow and only 40% of uranium was removed by four SFL operations.

The relationship between the removal efficiency and a period of the complexation process in the SFL operation was investigated using the sample prepared at 900 °C. The removal efficiency increased with an increase of the period and it was 68% when the period was one hour, 80% for 2 hours, 86% for 4 hours and 92% for 6 hours. The efficiency however did not go up more when the period became more than 6 hours. Only once did a SFL operation of 6 hours complexation gave higher efficiency than the 6 times SFL of one hour complexation did. From these results it is extrapolated that there are at least three types of uranium compounds on the alumina surface of the sample, those are an easily removed uranium compound (70 - 80%), a slowly removed uranium compound (*ca.* 10%) and a hard-to-remove uranium compound (*ca.* 10%). The slowly removed and hard-to-remove compounds were estimated to be formed by heating at high temperature under the reduction atmosphere in the sample preparation. This is supported by the results that the samples prepared at high temperature required much more SFL operations to remove uranium.

Real firebrick wastes generated from sintering furnaces for uranium fuel pellets have been used under different temperature depending on the position of the furnace. Uncomplicated removal behaviour is not expected to be observed when uranium is removed from the real firebrick wastes.

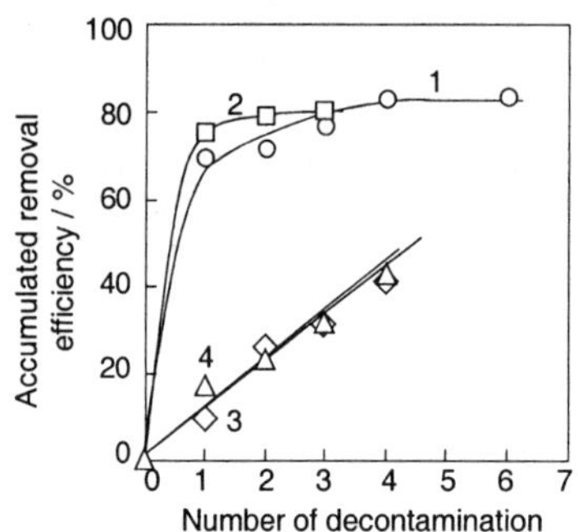

Figure 1 *Relationship between accumulated removal efficiency and number of decontamination.*
Sample preparation temperature
1; 900 °C, 2; 1200 °C, 3; 1500 °C, 4; 1800 °C.

References

1 O. Tomioka, Y. Meguro, S. Iso, Z. Yoshida, Y. Enokida, I. Yamamoto, *J. Nucl. Sci. Technol.*, 2001, **38**, 461.

2 Y. Meguro, S. Iso, Z. Yoshida, O. Tomioka, Y. Enokida, I. Yamamoto, *J. Supercrit. Fluids*, 2004, **31**, 141.

3 Y. Meguro, O. Tomioka, T. Imai, S. Fujimoto, M. Nakashima, Z. Yoshida, T. Honda, F. Kouya, N. Kitamura, R. Wada, I. Yamamoto, S. Tsushima, *Proc. Waste Manage. 2004 Symp.*, 2004, WM-4216.

4 A. A. Murzin, V. A. Babain, A. YU. Shadrin, I. V. Smirnov, V. N. Romanovskii, M. Z. Muradymov, *Radiochemistry*, 1998, **40**, 47.

5 M. Shamsipur, A. R. Ghiasvand, Y. Yamini, *J. Supercrit. Fluids*, 2001, **20**, 163.

6 R. Kumar, N. Sivaraman, E. Senthil Vadivu, T. G. Srinivasa, P. R. Vasudeva Rao, *Radiochim. Acta*, 2003, **91**, 197.

7 K. Park, M. Koh, C. Yoon, H. Kim, H. Kim, *Proc. 2nd Internat. Symp. Supercrit. Fluid Technol. Energ. Environ. Appl.*, 2003, 88.

8 C. Perre, S. Sarrade, *Proc. 2nd Internat. Symp. Supercrit. Fluid Technol. Energ. Environ. Appl.*, 2003, 172.

9 Y. Nagase, K. Masuda, R. Wada, I. Yamamoto, O. Tomioka, Y. Meguro, R. Fukuzato, *Proc. 2nd Internat. Symp. Supercrit. Fluid Technol. Energ. Environ. Appl.*, 2003, 254.

10 A. Shadrin, A. Murzin, V. Kamachev, V. Romanovskyi, S. Podoinitsyn, S. Bychkov, I. Efremov, T. Koyama, M. Kamiya, *Proc. 2nd Internat. Symp. Supercrit. Fluid Technol. Energ. Environ. Appl.*, 2003, 385.

11 Y. Meguro, O. Tomioka, T. Imai, S. Fujimoto, M. Nakashima, R. Wada, S. Tsushima, I. Yamamoto, *Proc. 3rd Internat. Symp. Supercrit. Fluid Technol. Energ. Environ. Appl.*, 2004, PTh-1.

Spectroscopy and Magnetism

SPECTROSCOPIC ACTINIDE SPECIATION FOR NUCLEAR WASTE DISPOSAL

M.A. Denecke, P.J. Panak, M. Plaschke, J. Rothe and M. Weigl

Forschungszentrum Karlsruhe, Institut für Nukleare Entsorgung, P.O. Box 3640, 76021 Karlsruhe, Germany

1 INTRODUCTION

Spectroscopic actinide speciation, providing fundamental molecular scale information under varying conditions, is requisite to performance assessment of nuclear repositories and useful for the development of partitioning strategies and transmutation fuels. Actinide speciation aids in the understanding of processes which determine the behaviour of actinides released into the environment. This, in turn, determines transport properties (mobilization/immobilization), reactivity, bio-availability and hence, potential risk. Partitioning may be used to isolate long-lived radionuclides from nuclear fuel reprocessing waste for subsequent transmutation to shorter lived nuclides, or for their conditioning and immobilization in special matrices for disposal. In this work we present examples of the application of X-ray spectroscopic techniques for actinide speciation in the aforementioned fields of research. Examples include the determination and comparison of coordination structures for trivalent curium complexed with the partitioning ligand 2,6-di(5,6-dipropyl-1,2,4-triazin-3-yl)pyridine ($(n\text{-}C_3H_7)_4$-BTP, Figure 1) to that of the europium counterpart and the identification of functional groups responsible for complexation of uranyl cations with humic substances by comparison of electronic structure changes induced by complexation to those of model compounds.

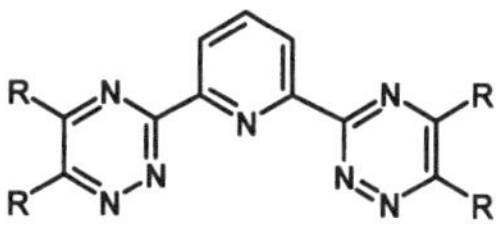

Figure 1 *2,6-di(5,6-dipropyl-1,2,4-triazin-3-yl)pyridine*

2 EXPERIMENTAL

Cm L3 edge extended X-ray absorption fine structure (EXAFS) measurements were performed at the Argonne National Laboratory (ANL) Advanced Photon Source (APS) in the same manner as reported in Weigl *et al.*[1] Analysis of EXAFS data was made by performing theoretical least squares fits in R-space from R-Δ = 1.3 to 4.7 Å using the "feffit" software,[2] thereby obtaining coordination numbers (N), interatomic distances (R), mean square radial displacements, or EXAFS Debye-Waller factors (σ^2), and relative shifts in ionization potential (ΔE_0) describing the coordination structure. Phase shift and backscattering amplitude functions used in the fits were calculated with "feff8"[3] using a cluster of 55 atoms having Cartesian coordinates calculated for the crystal structure of Ce[$(CH_3)_4$-BTP]$_3$.[4]

Ligand-to-metal concentration ratios ([L]/[M]) were varied from about 8 to 630 in the titration experiments by adding (n-$C_3H_7)_4$-BTP dissolved in organic solvent (TPH/n-octan-1-ol) to the same solvent containing metal cation and recording time-resolved laser-induced fluorescence spectroscopic (TRLFS) spectra after each addition. TRLFS measurements were performed using an excimer pumped dye laser system (Lambda Physics, EMG 201 and FL 3002). Excitation wavelengths of 375 nm and 394 nm were used for Cm(III) and for Eu(III), respectively. Details on the experimental set-up are given elsewhere.[5]

3 RESULTS

3.1 Coordination Structures for Cm(III) Complexed with (n-$C_3H_7)_4$-BTP

EXAFS spectra and theoretical fit curves are shown in Figure 2 for Cm(III) and Eu(III) complexed with (n-$C_3H_7)_4$-BTP. Metal complexes were prepared by extracting ^{248}Cm(III) or Eu(III) from acidic (HNO_3) aqueous solutions into an organic phase containing (n-$C_3H_7)_4$-BTP. The spectra of the two complexes are qualitatively similar.

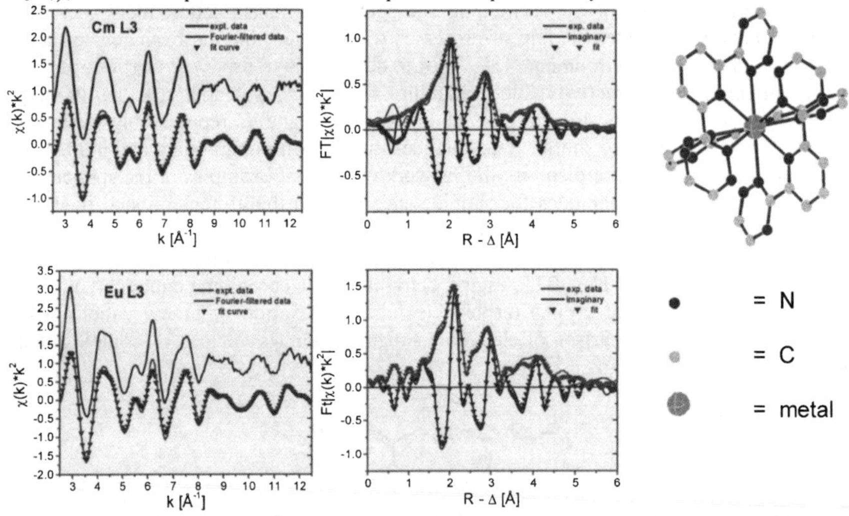

Figure 2 *Cm and Eu L3 k^2- weighted EXAFS for the extracted (n-$C_3H_7)_4$-BTP complexes together with their corresponding Fourier-filtered data and back-transformed fit (top left) and FT EXAFS magnitude, FT imaginary part together with corresponding R-space fit results (top right). Far right: structure of the Cm/Eu-[(n-$C_3H_7)_4$-BTP]$_3$ complex (H atoms and n-propyl groups not shown)*

Initial fits to the data, using a model of two coordination shells, reveal that the number of coordinating N atoms in the first shell was found to be ~9 and two times as many C/N atoms were found in the second shell. This indicates that three ligands were bound to both Cm(III) and Eu(III). For this reason the data was modeled with four coordination shells, with 9 nearest N atoms directly bound to the metal cations, the second and third shells of 18 C/N atoms each (designated C/N and C′/N′, respectively), and a fourth, most distant shell of 9 C/N atoms, corresponding to the C atom located *para* to the N atom in the pyridine ring and the N atoms *para* to the ligating N atoms in the triazin rings (see Figure 1). Three significant three-legged multiple scattering paths (Cm/Eu→N→C/N→Cm/Eu, Cm/Eu→C/N→C′/N′→Cm/Eu, and (Cm/Eu→N→ C′/N′→Cm/Eu paths) were included in the fits with their effective path lengths correlated to single scattering distances involved and Debye-Waller factors varied. The n-propyl groups of the triazin rings were not included. The metrical results are listed in Table 1.

Comparison of results in Table 1 shows that those for Cm are not significantly different than those for Eu, indicating that the coordination structure of Cm-[(n-C_3H_7)$_4$-BTP]$_3$ and Eu-[(n-C_3H_7)$_4$-BTP]$_3$ are the nearly same. The observed selectivity of (n-C_3H_7)$_4$-BTP for Cm(III), over Eu(III), is not structural in origin; the metal cation complexes, coordinated with three ligands, have the same coordination structure.

Table 1 *Metrical parameters from fits of Cm/Eu L3 edge R-space data in Figure 2 and compared to those reported from XRD studies of Ce[(CH$_3$)$_4$-BTP]$_3$. Estimated standard deviations are listed in parentheses and do not include systematic errors. S_0^2 is fixed at 1*

Sample	Shell	N [a]	R [Å]	σ^2 [Å^2]	ΔE_0 [eV]	*r-factor* [b]
Cm[(n-C_3H_7)$_4$-BTP]$_3$	N	9	2.568 (0.007)	0.0051 (0.0007)	1.1 (0.8)	0.025
	C/N	18	3.431 (0.009)	0.005 (0.001)		
	C/N	18	4.81 (0.03)	0.004 (0.003)		
	C	9	5.30 (0.04)	0.001 (0.006)		
Eu[(n-C_3H_7)$_4$-BTP]$_3$	N	9	2.559 (0.008)	0.0044 (0.0008)	2.5 (0.9)	0.028
	C/N	18	3.42 (0.01)	0.006 (0.001)		
	C/N	18	4.82 (0.02)	0.004 (0.003)		
	C	9	5.30 (0.03)	-0.002 (0.003)		
Ce[(CH_3)$_4$-BTP]$_3$ [c]	N	9	2.618			
	C/N	18	3.485			
	C/N	18	4.770			
	C	9	5.282			

[a] held constant at given values; [b] parameter describing goodness of fit = weighted sum of squares of residuals divided by the degrees of freedom; [c] from Iveson *et al.*[4]

We also investigated Cm(III) and Eu(III) complexation with varying amounts of (n-C_3H_7)$_4$-BTP using TRLFS. The fluorescence spectra are depicted in Figure 3 and compared to that of the Cm^{3+} and Eu^{3+} aquo species. A strong bathochrome shift of 19.3 nm is observed for the Cm-complex spectra compared to that of the Cm^{3+} aquo species. However, the Cm fluorescence is independent of [L]/[M]; Cm(III) is observed to form only one species, Cm-[(n-C_3H_7)$_4$-BTP]$_3$. The formation of 1:1 or 1:2 complexes is not observed for [L]/[M] > 6.

In contrast, Eu(III) forms at least two different species for the varying [L]/[M] studied. Comparison with the literature shows that our fluorescence spectrum, at high [L]/[M] (> 300), is identical to the Eu-[(i-C_3H_7)$_4$-BTP]$_3$ measured in water/ethanol,[6] meaning that at [L]/[M] > 300 the 1:3 complex was formed. The species lifetimes in the titration series (not shown) exhibited a linear increase with increasing number of coordinating ligands. We found Cm-[(n-C_3H_7)$_4$-BTP]$_3$ formed at much lower [L]/[M] than Eu-[(n-C_3H_7)$_4$-BTP]$_3$. This demonstrates that the higher affinity (corresponding to a larger formation constant) of (n-C_3H_7)$_4$-BTP toward trivalent actinide cations over lathanide cations, is the basis for the observed selectivity of BTP ligands in liquid/liquid extraction. Together EXAFS and

TRLFS results show that the ligand selectivity for Cm(III) over Eu(III) was associated with a higher affinity and not simple structural differences between complexes, as there is no substantial difference between the coordination structures of Cm-[(n-C_3H_7)$_4$-BTP]$_3$ and Eu-[(n-C_3H_7)$_4$-BTP]$_3$.

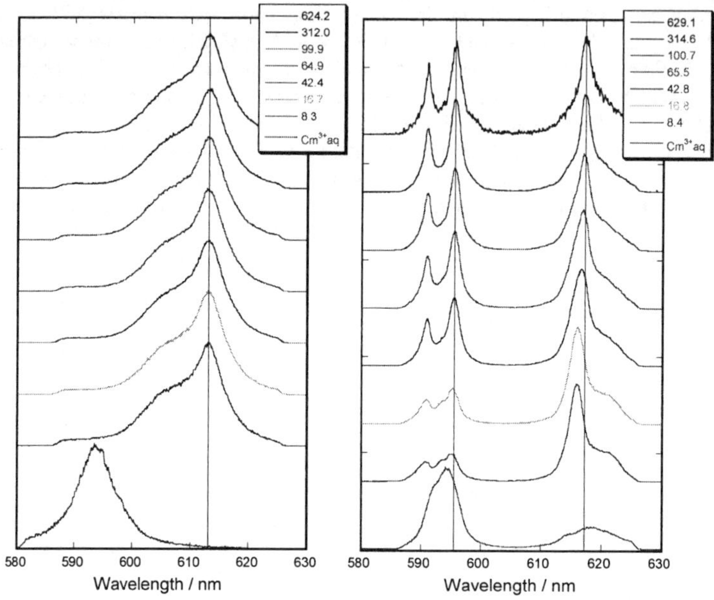

Figure 3 *Fluorescence spectra of Cm(III) (left) and Eu(III) (right) as a function of ligand-to-metal concentration ratio indicated in the legend (inverted series) compared to that of the Cm^{3+} and Eu^{3+} aquo species (bottom)*

3.2 STXM Investigation of Actinide Complexation by Humic Acid

We applied scanning transmission X-ray microscopy (STXM) for characterizing the humic acid (HA) chemical functionality, with simultaneous visualization of HA morphology on a sub-µm scale.[7,8,9,10] To aid in our understanding of complex, heterogeneous HA systems we compared results to those from model substances in a "bottom-up" approach. By means of STXM, both morphological and microchemical information can be obtained at the same time *in situ* for aqueous colloid species within the 'water window' (i.e. between the C 1s and O 1s absorption edges at 284 eV and 537 eV, respectively). Microscopic images of colloidal morphologies were obtained by scanning selected sample areas. By recording images at varying incident radiation energies near the C 1s ionization energy, we could adjust the absorption contrast. We could extract high resolution C 1s near edge X-ray absorption fine structure (NEXAFS) spectra from selected sample regions in resulting stacks of images, recorded at discrete energies. C 1s NEXAFS provides direct speciation of carbon-containing macromolecules from their characteristic 1s $\rightarrow \pi^*$, σ^* resonances.

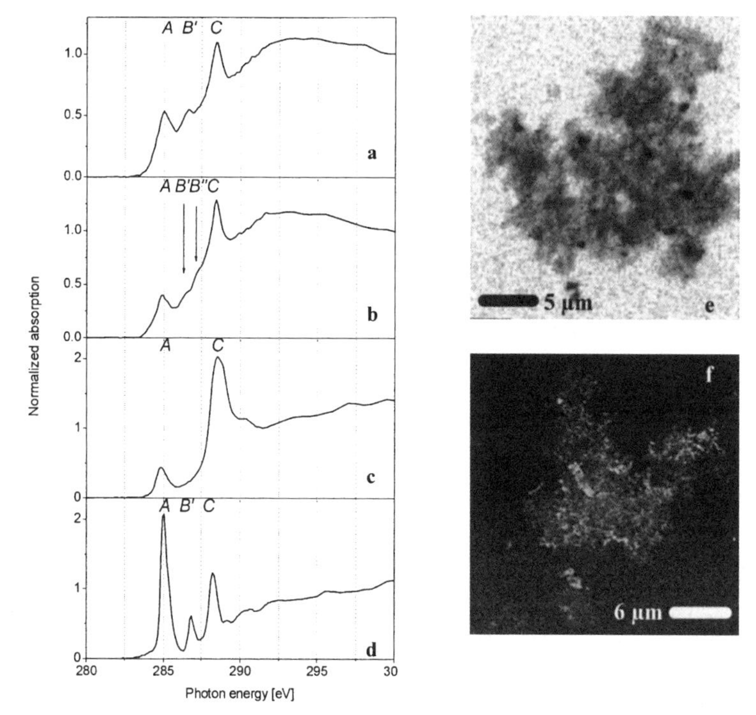

Figure 4 *C 1s NEXAFS of (a) U(VI)-HA (signature 1), (b) U(VI)-HA (signature 2, see text for details), (c) HA without metal loading, (d) 4-hydroxy benzoic acid, (e) STXM micrograph at 288.5 eV of U(VI)-HA aggregate and (f) STXM micrograph (contrast enhancing negative image) at 290 eV of HA aggregate at pH 4.3 (without metal loading)*

The normalized C 1s-NEXAFS spectra of purified Aldrich HA, reacted with U(VI), are depicted in Figure 4a-b. The spectrum of HA before metal loading is plotted in Figure 4c. Similar to results in Rothe *et al.*,[11] the uncomplexed HA NEXAFS could not be reproduced from a superposition of spectra extracted from image stacks of the complexes. This indicates that the spectral changes observed upon metal complexation with HA resulted from distinct changes in the HA carbon molecular states. The STXM micrograph of a typical HA aggregate, formed upon HA-metal interaction, is depicted in Figure 4e. The U(VI)-HA aggregates were similar to those formed for Eu(III)-HA, showing segregation of carbonaceous material into zones with different morphologies and optical densities.[8] Patches with higher optical densities were found embedded in a less optically dense matrix. The STXM micrograph of unreacted HA (Figure 4f) at pH 4.3, reveals a less dense, network-like structure with particles and larger sheets of carbonaceous material embedded in a fibrous network. At least two different C 1s-NEXAFS signatures characterized the U(VI)-HA (Figure 4a-b): a signature with a relatively strong phenolic resonance (C 1s (C_{ar}-OH) $\rightarrow \pi^*_{C=C}$; position *B'*) and the second signature showing two shoulders *B'* and *B"*. Peak positions and shapes could be assigned in accordance with model compound spectra, such as substituted benzoic acids.[12] The energy of peak *B'* (286.6 eV) coincided with the phenolic transition for 4-hydroxy benzoic acid (Figure 4d).[13] Transitions observed in the energy range near *B′* are generally indicative of substituted aromatic carbon atoms, its intensity a measure of the number. The position of peak *B′* correlates to the overall electronegativity of aromatic substituents.[12] This correlation is the basis for identification of the origin of this peak in U(VI)-HA and HA spectra, as phenolic groups. However, here we observe that the energy position of *B′* is not affected by metal complexation. Resonance *A* represents the aromatic carbon (C_{ar}) in these compounds. Feature *B″* is due to transitions from molecular states formed following metal complexation.[8] The intensity of peak *C* in the HA spectrum, resulting from carboxyl C transitions (C 1s (COOH) $\rightarrow \pi^*_{C=O}$), is observed to have decreased upon complexation. We interpret the observed spectral changes and the similarity between the U(VI)-HA NEXAFS signatures and those for Eu(III)-HA, reported by Plaschke *et al.*[13] as evidence for a similar complexation mechanism, which involves HA carboxyl functional groups, for both metal cations.

To corroborate this interpretation, C 1s-NEXAFS spectral changes induced by U(IV) complexation in polyacrylic acid (PAA), as a HA model,[14] were also investigated. The C 1s-NEXAFS of Na(I)-PAA and U(VI)-PAA are displayed in Figure 5a-b, respectively. The morphology of the U(VI)-PAA aggregates is shown in the image in Figure 5c. The carboxyl peak *(C)* at ~288.4 eV dominates the Na(I)-PAA spectrum. As observed for U(VI)-HA, this peak's intensity is strongly decreased in the U(VI)-PAA sample spectrum. U(VI)-PAA also exhibits a broad absorption feature at ~287.5 eV adjacent to peak *C* (*B″* in Figure 5). Again, we interpret the simultaneous decrease in carboxyl peak intensity and appearance of a new peak *B″* as a result of metal ion complexation. Comparison of PAA to HA results allows us to conclude that C 1s-NEXAFS peaks, *B″* and *C,* are sensitive to metal ion complexation by carboxyl functional groups. Theoretical calculations are underway to help identify the exact nature of the source of the *B″* transition.

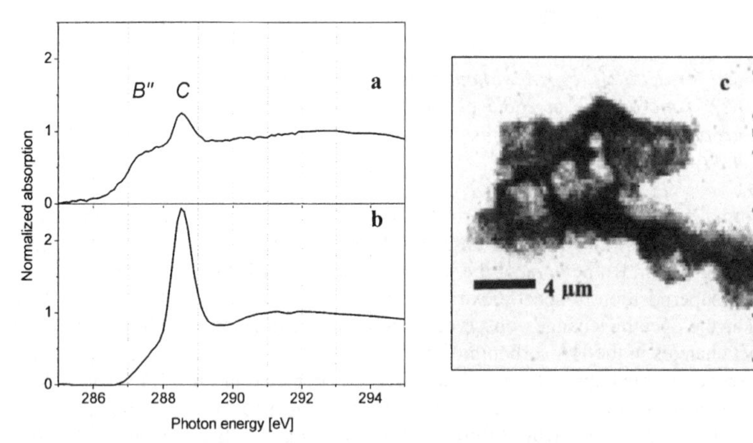

Figure 5 *C 1s NEXAFS of (a) U(VI)-PAA, (b) PAA without metal loading and (c) STXM micrograph at 288.5 eV of U(VI)-PAA aggregate*

4 CONCLUSION

We have presented two examples of spectroscopic actinide speciation in the field of nuclear waste disposal research: the first being useful for the development of partitioning strategies and the second for understanding the interaction of uranium with natural organic colloids, as potential vehicles for actinide mobilization. The first example aimed at providing insight into the potential that combining different spectroscopies generally has for understanding processes and reactions involving actinides. The information gained from the partitioning case may be used to optimize ligand design and hence, extraction performance. The second example is a case where a "bottom-up" approach of investigating model systems was used to understand a complex, heterogeneous, natural system. In our case example of humic acid-uranium interaction, comparison with model compounds as building blocks and a more complex, but simple polymer allowed identification of a distinct metal-complexation effect on the carbon molecular states.

Acknowledgements

We are grateful for beamtime allocations at Brookhaven National Laboratory (BNL), National Synchrotron Light Source (NSLS) and at the APS BESSRC BM12 station, support by the BESSRC beamline staff and for experimental assistance and use of the infrastructure of the Actinide Facility. Use of the APS was supported by the U.S. Department of Energy, Office of Science, Office of Basic Energy Sciences, under Contract No. W-31-109-ENG-38. C 1s-NEXAFS/STXM data was recorded by the X-1A STXM developed by the group of Janos Kirz and Chris Jacobsen at SUNY Stony Brook, with support from the Office of Biological and Environmental Research, U.S. DoE under contract DE-FG02-89ER60858, and the NSF under grant DBI-9605045.

References

1 M. Weigl, M.A. Denecke, P.J. Panak, A. Geist and K. Gompper, *Dalton Trans.*, 2005, 1281.
2 E.A. Stern, M. Newville, B. Ravel, Y. Yacoby and D. Haskel, *Physica B*, 1995, **208-209**, 117.
3 A.L. Ankudinov, B. Ravel, J.J Rehr and S.D. Conradson, *Phys. Rev. B,* 1998, **58**, 7565.
4 P.B. Iveson, C. Riviere, D. Guillaneux, M. Nierlich, P. Thuéry, M. Ephritikhine and C. Madic, *Chem. Commun.*, 2001, 1512.
5 K.H. Chung, R. Klenze, K.K. Park, P. Paviet-Hartmann and J.I. Kim, *Radiochim. Acta,* 1998, **82**, 215.
6 S. Colette, B. Amekraz, C. Madic, L. Berthon, G. Cote and C. Moulin, *Inorg. Chem.*, 2004, **43**, 6745.
7 J. Rothe, M.A. Denecke and K. Dardenne, *J. Colloid Interface Sci.*, 2000, **231**, 91.
8 M. Plaschke, J. Rothe, T. Schäfer, M.A. Denecke, K. Dardenne, S. Pompe and K.-H. Heise, *Colloids Surf. A*, 2002, **197**, 245.
9 A.C. Scheinost, R. Kretzschmar, I. Christl and C. Jacobsen, *Spec. Pub. – Roy. Soc. Chem.*, 2002, **273**, 39.
10 J. Thieme, C. Schmidt, G. Abbt-Braun, C. Specht and F.H. Frimmel, in: *Refractory Organic Substances in the Environment*, eds., F.H. Frimmel and G. Abbt-Braun, Wiley, New York, 2000.
11 J. Rothe, M. Plaschke and M.A. Denecke, *J. Phys. IV France*, 2003, **104**, 421.
12 M. Plaschke, J. Rothe, M.A. Denecke and Th. Fanghänel, *J. Electron Spectrosc. Relat. Phenom.*, 2005 (submitted).
13 M. Plaschke, J. Rothe, M.A. Denecke and Th. Fanghänel, *J. Electron Spectrosc. Relat. Phenom.*, 2004, **135**, 53.
14 O. Tochiyama, H. Yoshino, T. Kubota, M. Sato, K. Tanaka, Y. Niibori and T. Mitsugashira, *Radiochim. Acta*, 2000, **88**, 547.

SUPERCONDUCTIVITY IN ACTINIDE MATERIALS

J.D. Thompson,[1] J.L. Sarrao,[1] N.J. Curro,[1] E.D. Bauer,[1] L.A. Morales,[1] F. Wastin,[2] J. Rebizant,[2] J.C. Griveau,[2] P. Boulet,[2] E. Colineau[2] and G.H. Lander[2]

[1]Los Alamos National Laboratory, Los Alamos, NM 87545 USA
[2]European Commission, JRC, Institute for Transuranium Elements, Post Box 2340, 76125 Karlsruhe, Germany

1 INTRODUCTION

More than four decades after the first discovery of superconductivity in elemental mercury, Bardeen, Cooper and Schrieffer showed theoretically that superconductivity arises when conduction-band electrons form pairs with precisely opposite spin and momentum due to an attractive interaction mediated by lattice vibrations (phonons).[1] With zero net spin and angular momentum, these Cooper pairs condense into a macroscopic quantum state that is separated energetically from all unpaired electrons by a finite gap Δ, which is proportional to the superconducting transition temperature, T_c. Though unable to predict what materials might be superconducting, this theory established conditions favourable for superconductivity, namely that there be a high density of electronic states at the Fermi energy ($N(E_F)$) and phonons should be sufficiently soft to create Cooper pairs effectively. Relative to *s*, *p* metals, transition metals have a large density of electronic states, and, consequently, these concepts provided a natural explanation for why superconductivity prevailed in *d*-electron metals.[2]

The same conditions for the appearance of superconductivity apply to actinides that are *d*-electron-like. Tetravalent Th is one such example and is superconducting below 1.39 K.[3] As with *d*-electron metals, certain crystal structures also tend to favour superconductivity in actinide-based metals. A simple rationale is based on Hill's criterion for the overlap of 5*f* wave functions.[4] For actinide-actinide spacings less than ~0.35 (~0.34) nm for U (Pu), wave functions of 5*f* electrons from adjacent actinides overlap and create a narrow band of itinerant electrons, with predominantly 5*f* symmetry, that crosses the Fermi energy. Qualitatively, the width of this electronic band is inversely proportional to $N(E_F)$, i.e., narrower bands correspond to a higher density of electronic states at E_F. Crystal structures that impose nearest actinide distances below Hill's limit, then, are a necessary, but not sufficient condition, for phonon-mediated superconductivity. For sufficiently narrow bands, Coulomb interactions between electrons become increasingly important. In the BCS theory, these repulsive Coulomb interactions compete with the attractive interaction provided by phonons and reduce the superconducting transition temperature that otherwise would have been possible.[2]

In metals with actinide-actinide spacings beyond Hill's limit, 5*f* wave functions do not overlap directly; the 5*f* electrons are quasi-localized in states below E_F, and consequently, they should not participate in superconductivity. Nevertheless, there are several examples of materials that, by this reasoning, should not superconduct, but they do; furthermore, the 5*f* electrons play an essential role in the superconductivity.[5] These 'anomalous' superconductors belong to a family called strongly correlated or heavy-fermion materials in which a very narrow band, of order 1-10 meV wide, forms at low temperatures through hybridisation of 5*f* and ligand wave functions. Precisely how the physics of these heavy-fermion compounds should be understood theoretically remains a major challenge for condensed matter physics. At least qualitatively though, some things are known.[6] Above a few to tens of Kelvins, the magnetic susceptibility of these materials follows a Curie or Curie-Weiss temperature dependence with an effective moment close to that expected if the *f*-electrons were fully localized. At much lower temperatures, the magnetic susceptibility tends toward a large constant value that is typical of a massively enhanced Pauli-type susceptibility, which is proportional to $N(E_F)$. In this same low temperature regime, the electronic coefficient of specific heat, γ, also proportional to $N(E_F)$, grows correspondingly. At extremely low temperatures, direct measurements of the effective mass, m*, of itinerant electrons at E_F, find that m* reaches 100-1000 times the mass of a free electron, consistent with values of m* implied by the massively enhanced values of γ. These same heavy charge carriers participate in superconductivity as evidenced by a jump in specific heat at T_c that follows the BSC prediction $\Delta C/\gamma T_c \approx 1.5$.[5,6] Excluding their superconductivity, properties of heavy-fermion materials are qualitatively similar to behaviours exhibited by an isolated local moment that interacts antiferromagnetically with a broad band of conduction electrons through the Kondo effect to create a resonance in the density of states near E_F.[6] Associated with this resonance is a large $N(E_F)$ with character dominated by the symmetry of the local moment, i.e. 5*f* character in the case of actinides. This simple picture, however, presents a dilemma. Introduction of Kondo impurities into conventional superconductors rapidly suppresses T_c to zero because of the pair-breaking effect of the local moments,[7] but heavy-fermion superconductors are built from a periodic array of approximately 10^{23} Kondo impurities, e.g. Ce, U and Pu. The mere existence of heavy-fermion superconductivity suggests that the conventional mechanism of phonon-mediated superconductivity is inappropriate and that alternative mechanisms should be considered.

One alternative is that Cooper pairing is mediated by spin fluctuations,[5,8] which, like phonons, are bosonic excitations. Cooper pairs formed by the exchange of spin fluctuations do not necessarily have zero net spin and momentum that is required in conventional superconductivity.[8] Because paired states with spin, $S \geq 0$, and angular momentum, $L \geq 0$, are allowed, the superconducting energy gap, Δ, goes to zero on certain parts of the Fermi surface. Figure 1 compares an unconventional superconducting gap with $S = 0$ and $L = 2$ to a conventional gap, which is finite over the entire Fermi surface. One consequence of gap nodes, where $\Delta=0$, is that $N(E_F)$ is non-zero as temperature goes to zero, and physical properties that depend on $N(E_F)$ will grow as a power-law in temperature for $T \ll T_c$, in stark contrast to a thermally activated temperature dependence in a conventional superconductor.[9] In the following, we take a working definition of an unconventional superconductor as one in which the uniform magnetic susceptibility is local-moment-like, the actinide-actinide spacing is close to or beyond the Hill limit, and power-law temperature variations appear below T_c, in physical properties that depend on $N(E_F)$.

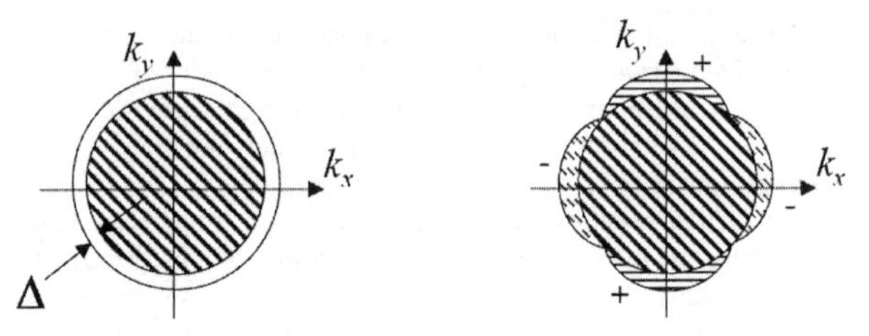

Figure 1 *Schematic representation of the energy gap of a conventional superconductor with finite gap Δ over the entire Fermi surface (left) and of an unconventional superconductor with gap nodes (right)*

2 DISCUSSION

Table 1 gives, to our knowledge, a complete list of all actinide superconductors. Uranium-based superconductors known prior to 1989 (those listed before UGe_2 in Table 1) have been discussed already in excellent reviews in which arguments are presented that some of

Table 1 *Actinide superconductors, with their transition temperature, T_c, and nearest f-f spacing d.*

Material	T_c (K)	*d* (nm)	*Material*	T_c (K)	*d* (nm)
α-U	<0.1[10,a]	0.31	U_2PtC_2	1.47[20]	0.35
β-U	0.75-0.85[11,b]	0.31	UAl_2Si_2	1.34[16]	0.41
γ-U	1.85-2.07[12,c]	0.29	UAl_2Ge_2	1.60[16]	0.42
UCo	1.22[13]	0.27/0.36[c]	UGa_2Ge_2	0.87[16]	0.42
U_6Fe	3.78[13]	0.32	URu_2Si_2	1.5[21,f]	0.41
U_6Mn	2.31[13]	0.32	UGe_2	0.4[22,g]	0.38
U_6Co	2.33[13]	0.32	UIr	0.14[23,h]	0.33-0.38[g]
U_6Ni	0.33[14]	0.32	UPd_2Al_3	1.9[24,i]	0.40
UPt_3	0.54[15,d]	0.41	UNi_2Al_3	1.0[25,j]	0.40
URu_3	0.145[16]	0.40	URhGe	0.25[26,kj]	0.35
UBe_{13}	0.9[17]	0.51	$PuCoGa_5$	18.5[27]	0.42
U_3Ir	1.3[18]	0.40	$PuRhGa_5$	8.7[28]	0.43
U_5Ge_3	0.99[19,e]	0.29/0.36[c]	Am	0.79[29,l]	0.30

[a]T_c exceeds 2 K under pressure; [b]T_c depends on elements included to stabilize crystal structure; [c]Two inequivalent U sites; [d]Superconductivity co-exists with weak magnetism (T_N = 5 K); [e]Superconductivity controversial; [f]Superconductivity co-exists with weak magnetism (T_N = 17.5 K); [g]Superconductivity pressure induced at $1 < P_c < 1.6$ GPa and co-exists with weak itinerant ferromagnetism; [h]Four inequivalent U sites and superconductivity pressure induced at $P_c \approx 2.6$ GPa, near the boundary of weak itinerant ferromagnetism; [i]Superconductivity co-exists with strong antiferromagnetism (T_N = 14 K); [j]Superconductivity co-exists with weak magnetism (T_N = 4.6 K); [k]Superconductivity co-exists with weak itinerant ferromagnetism (T_C = 9.5 K); [l]T_c exceeds 2 K under pressure, as discussed in the text.

these superconductors, e.g. UPt_3, UBe_{13}, are unconventional.[5,30] Since that time, several new actinide-based superconductors have been discovered, including those based on Pu, and these are subjects of brief discussion in the next three sections.

2.1 Uranium Superconductors

In each of the U-based superconductors, UGe_2, UIr, URhGe, UPd_2Al_3 and UNi_2Al_3, unconventional superconductivity, by our working definition, exists simultaneously with some form of magnetism that derives from their 5*f* electrons. The first three compounds are 5*f*-band ferromagnets; whereas, the hexagonal U123 materials are antiferromagnetic with Néel temperatures well above T_c. The normal and superconducting state properties of UGe_2, URhGe and UPd_2Al_3 are summarized by Flouquet,[31] which includes an extensive bibliography of the literature on these materials.

There is little doubt that U's 5*f* electrons are responsible for the magnetism in these materials, but a crucial question is whether these same 5*f* electrons participate in superconductivity. Even in conventional superconductors, local moment antiferromagnetism and superconductivity can co-exist, provided that the antiferromagnetic order of localized *f* electrons does not couple to *d*-electrons responsible for superconductivity.[32] Like UPt_3[33] and URu_2Si_2,[34] UPd_2Al_3 is an example of superconductivity co-existing with antiferromagnetism, but in UPd_2Al_3 the relationship between superconductivity and magnetic order has been made especially clear and may bear more broadly on understanding other actinide superconductors. Neutron-diffraction studies of UPd_2Al_3[35] find an atomic-like staggered moment (0.84 μ_B) below T_N, as expected for localized 5*f* orbitals; however, the jump in specific heat at T_c, $\Delta C/\gamma T_c \approx 1.48$ with $\gamma = 115$ mJ/mol-K^2, is consistent with superconductivity developing out of a band of delocalized heavy-mass electrons.[36] This apparent dichotomy has led to the suggestion[36] that 5*f* electrons in UPd_2Al_3 assume dual roles, with two of the three 5*f* electrons being localized (magnetic)[37] and the other being itinerant (and superconducting) even though all U sites are crystallographically equivalent. An approximately 1% decrease in the ordered moment[38] and pronounced changes in the spin-excitation spectra below T_c confirm that these 5*f* electrons, though with very different characters, are intimately coupled to each other and to superconductivity.[39] These experiments are among the most definitive in showing that antiferromagnetic spin fluctuations are a viable mechanism for unconventional superconductivity.

The isostructural compound UNi_2Al_3 is similar to, but also distinctly different from, UPd_2Al_3. Though antiferromagnetism and superconductivity also coexist in UNi_2Al_3 and these two orders are coupled, the ordered moment is much smaller (~ 0.2 μ_B) and is an incommensurate spin-density-wave type.[40] In further contrast to UPd_2Al_3, so far there is no clear evidence that the 5*f* electrons in UNi_2Al_3 assume dual roles, possibly because all of the *f* electrons are more nearly itinerant.[41] The most striking difference, however, is in the nature of superconductivity. Knight shift studies[42] of UNi_2Al_3 are consistent with an odd-parity superconducting gap, i.e. S = 1 and L = 1; whereas, similar measurements on UPd_2Al_2 imply an even-parity gap (S = 0, L = 2) with line nodes.[43] A spin-triplet state (S = 1), also found in hexagonal UPt_3,[44] is typically expected if Cooper pairing is mediated by ferromagnetic spin fluctuations, as opposed to antiferromagnetic fluctuations that favour spin-singlet (S = 0) pairing.[9]

Hybridisation of 5*f* and ligand electrons creates a narrow, highly correlated electronic band at E_F out of which 5*f*-band ferromagnetism emerges in UGe_2, UIr and URhGe with Curie temperatures (ordered moments) of 54 K (1.48 μ_B),[23] 46 K (0.5 μ_B),[24] and 9.5 (0.42 μ_B),[27] respectively. Superconductivity appears at atmospheric pressure in URhGe[27] and is induced by pressure in UGe_2[23] and UIr.[24] Because their superconductivity develops in proximity to ferromagnetism, it is plausible that superconductivity in these compounds might be unconventional, spin-triplet. Very little is known about the superconductivity in UIr, except that it exists in a very narrow pressure window slightly above 2.5 GPa where a second ferromagnetic transition in UIr appears to extrapolate to T=0.[24] On the other hand, UGe_2 and URhGe, which form in similar orthorhombic crystal structures, have been studied somewhat more extensively. Their large normal state specific heat coefficients, $\gamma \approx$120-160mJ/mol-K^2, clearly imply that 5*f* electrons participate in band ferromagnetism, but $\Delta C/\gamma T_c$ in both compounds is only about one-third that expected by BCS theory.[27] This second property suggests either that the sample volume may not be completely superconducting, e.g. superconductivity might exist only at the interface of ferromagnetic domains, or that a reduced $\Delta C/\gamma T_c$ could the manifestation of an unavoidable vortex state created by the internal magnetic field associated with ferromagnetic order. Whether superconductivity and ferromagnetism coexist or compete in these compounds is an important issue that requires further investigation, but initial spin-lattice relaxation experiments on polycrystalline UGe_2 are consistent with bulk, spin-triplet superconductivity.[45] Finally, we note that pressure studies reveal strikingly different relationships between superconductivity and ferromagnetism in UGe_2 and URhGe. The Curie temperature of UGe_2 decreases monotonically to T_C=0 at a critical pressure of ~1.6 GPa, and superconductivity exists between ~0.9 and ~1.6 GPa.[23] In URhGe, however, superconductivity at atmospheric pressure is completely suppressed with about 3-3.5 GPa applied pressure, even though its T_C increases linearly to over 20 K at 13 GPa.[31]

2.2 Plutonium Superconductors

Perhaps the most surprising recent development in actinide superconductivity has been the discovery of superconductivity in isostructural compounds $PuCoGa_5$[28] and $PuRhGa_5$.[29] Not only are these the first Pu-based superconductors, their T_c's are nearly an order of magnitude higher than those for other actinides, and are comparable to many of the highest T_c's of transition metal compounds. With nearest *f-f* spacings well beyond the Hill limit and Curie-Weiss-like uniform magnetic susceptibilities, these Pu-based superconductors appear to have localized 5*f* electrons. Their enhanced Sommerfeld coefficients, $\gamma \approx$70-100 mJ/mol-K^2, and BCS-like jump $\Delta C/\gamma T_c$, however, are consistent with bulk superconductivity developing out of a relatively narrow, correlated band of conduction electrons.[28,29,46,47] In these respects, $PuCoGa_5$ and $PuRhGa_5$ are reminiscent of UPd_2Al_3. A further similarity is revealed in photoemission spectra of $PuCoGa_5$ that are described best by a model in which four of Pu's five 5*f* electrons are localized and one 5*f* electron is itinerant, as deduced as well for δ-Pu.[48] This picture of Pu's 5*f* configuration still leaves open the possibility that superconductivity is conventional, especially given the surprisingly high T_c's. Because Pu is much more radioactive than U, self-heating makes it difficult to study properties of these Pu compounds at temperatures much less than 4 K << T_c, where ideally experiments should be made to differentiate between conventional and unconventional superconductivity. Nevertheless, power-laws in specific heat[46] and, most convincingly, in the spin-lattice relaxation rate $1/T_1$ well below T_c of $PuCoGa_5$[49] and $PuRhGa_5$,[50] argue strongly for a superconducting gap with nodes. Combined with Knight-shift measurements, that establish spin-singlet pairing in $PuCoGa_5$,[49] these power-law dependences are consistent with Cooper pairs having net angular momentum, L = 2. This pairing state is the same as that in the isostructural heavy-fermion superconductor $CeCoIn_5$, whose γ is about ten times larger and T_c about ten times smaller than that found in $PuCoGa_5$.[51] Further, above T_c, $1/T_1$ of both $CeCoIn_5$ and $PuCoGa_5$ exhibits a temperature dependence that is dominated by low-energy antiferromagnetic spin fluctuations,[49] which are favourable to formation of an unconventional superconducting state.

2.3 Am Superconductivity

Elemental Am is the only known trans-Pu superconductor. With six well-localized *5f* electrons, Hund's rules require that the ground state of Am^{3+} be non-magnetic. Consequently, it is probable that, at atmospheric pressure, phonon-mediated superconductivity arises[30] in a band of weakly correlated non-*f* electrons, which is inferred from Am's transition-metal-like $\gamma \approx 3$ mJ/mol-K^2.[52] Applying pressure to Am induces a cascade of structures with progressively smaller unit-cell volumes and a remarkable increase of T_c from 0.79 K at P=0 to a maximum of 2.3 K at the Am-I/Am-II border near 6 GPa.[53,54] T_c decreases monotonically across the Am-II and Am-III phases, reaches a minimum of ~1.1 K near the Am-III/Am-IV border (~16 GPa) and varies non-monotonically with increasing pressure in Am-IV. Pronounced changes in the resistivity accompany these changes in structure and T_c, leading to the suggestion of a pressure-induced localized to delocalised transition of the 5f electrons that plausibly accounts for the non-monotonic $T_c(P)$ in Am-IV.[54] Though much work remains to be done, it seems likely that superconductivity, in at least one of these high-pressure phases of Am, could be unconventional.

3 PERSPECTIVE

For unconventional superconductivity mediated by spin fluctuations, $T_c \sim T_{sf}e^{-1/\lambda}$, where T_{sf} is the characteristic energy scale of spin fluctuations and λ is the parameter of order unity that characterizes the coupling of Cooper pairs by spin fluctuations. Provided λ does not vary significantly among different materials, this analogue of the BCS expression (where T_{sf} is replaced by a characteristic phonon energy) implies an approximately linear relationship between T_c and T_{sf}, which is shown in Figure 2. This correlation argues for a continuum in the pairing mechanism among Ce- and U-based heavy-fermion and the other major family of unconventional superconductors, high-T_c cuprates.[49] Significantly, $PuCoGa_5$ bridges the gulf between these extremes. Qualitatively, this can be understood because T_{sf} reflects the degree to which nearly localized and ligand electrons hybridise, i.e. the bandwidth of strongly correlated itinerant electrons.

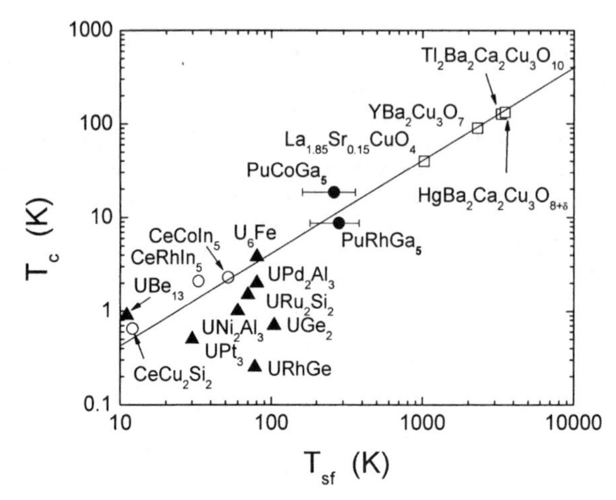

Figure 2 *T_c versus T_{sf} on logarithmic scales for several actinide, rare-earth and cuprate superconductors. The solid line has unity slope. Adapted after Moriya and Ueda*[9]

The relatively large radial extent of 3*d*-electron wave functions allows greater mixing with ligand wave functions relative to ligand mixing with 4*f* wave functions, which are confined more closely to the ionic core. Plutonium's 5*f* electrons are intermediate between 3*d* and 4*f* behaviours. To a first approximation, then, T_c's of these superconductors are determined by the relative widths of their highly correlated conduction bands. Though there is scatter among points included in Figure 2, due in part to different methods used to estimate T_{sf},[9] it is remarkable that the correlation is as good as it is, given the diversity of ground states and crystal structures, but it is interesting to note that those believed to be spin-triplet superconductors lie relatively further away from the straight line. The rich spectrum of physical properties exhibited by actinide superconductors has challenged our understanding of 5*f* electrons and will continue to exert significant influence on the study of strongly correlated electron phenomena and materials. As with transition-metal physics and conventional superconductivity, the discovery of new actinide superconductors will be important for guiding the development of appropriate theory for both the condensed matter physics of 5*f* electrons and for unconventional superconductivity. In this regard, exploration of ternary compounds with a tetragonal structure that provides *f-f* spacing beyond Hill's limits may prove beneficial.

References

1 J. Bardeen, L. N. Cooper and J. R. Schrieffer, *Phys. Rev.*, 1957, **108**, 1175.
2 R.D. Parks, *Superconductivity*, Marcel-Dekker, New York, 1969.
3 W.R. Decker, D.T. Peterson and D.K. Finnemore, *Phys. Rev. Lett.*, 1967, **18**, 899.
4 H.H. Hill, in *Plutonium 1970 and Other Actinides*, ed., W.N. Miner, The Metallurgical Society of AIME, New York, 1970, pp. 2-19.

5 H.R. Ott and Z. Fisk, in *Handbook on the Physical and Chemistry of Actinides*, eds., A.J. Freeman and G.H. Lander, North Holland, Amsterdam, 1987, Vol. 5, pp. 85-147.
6 Z. Fisk, D.W. Hess, C.J. Pethick, D. Pines, J.L. Smith, J.D. Thompson and J.O. Willis, *Science*, 1988, **239**, 33.
7 M.B. Maple, L.E. DeLong and B.C. Sales, in *Handbook on the Physical and Chemistry of Rare Earths*, eds., K.A. Gschneidner and L. Eyering, North Holland, Amsterdam, 1978, Vol. 1, pp. 797-846.
8 P.W. Anderson, *Phys. Rev. B*, 1985, **32**, 499.
9 T. Moriya and K. Ueda, *Rep. Prog. Phys.*, 2003, **66**, 1299.
10 J.C. Ho, N.E. Phillips and T.F. Smith, *Phys. Rev. Lett.*, 1966, **17**, 694.
11 B.T. Matthias, T.H. Geballe, E. Corenzwit, K. Anders, G.W. Hull, J.C. Ho, N.E. Phillips and D.K. Wohlleben, *Science*, 1966, **151**, 985.
12 B.B. Goodman, J. Hillairet, J.J. Veyssie and L. Weil, in Proc. VII Int. Conf. on Low Temp. Phys., eds., G.M. Graham and A.C. Hollis Hallet, Univ. Toronto Press, Toronto, 1961, p. 350.
13 B.S. Chandraesekar and J.K. Hulm, Jr., *J. Phys. Chem. Solids*, 1958, **7**, 259.
14 H.H. Hill and B.T. Matthias, *Phys. Rev.*, 1968, **168**, 464.
15 G.R. Stewart, Z. Fisk, J.O. Willis and J.L. Smith, *Phys. Rev. Lett.*, 1984, **52**, 679.
16 H.R. Ott, F. Hulliger, H. Rudiger and Z. Fisk, *Phys. Rev. B*, 1985, **31**, 1329.
17 H.R. Ott, H. Rudiger, Z. Fisk and J.L. Smith, *Phys. Rev. Lett.*, 1983, **50**, 1595.
18 M.B. Maple, M.S. Torikachvili, C. Rossel and J.W. Chen, *Physica*, 1985, **135B**, 430.
19 Y. Onuki, I. Ukon, T. Komatsubara, S. Takayanagi, N. Wada and T. Watanabe, *J. Phys. Soc. Jpn.*, 1989, **58**, 795; P. Boulet, M. Potel, J. C. Levet and H. Noel, *J. Alloys Compds.*, 1997, **262-263**, 229.
20 B.T. Matthias, C.W. Chu, E. Corenzwit and D. Wohlleben, *Proc. Nat. Acad. Sci.*, 1969, **64**, 459.
21 W. Schlabitz, J. Baumann, B. Pollit, U. Rauchschwalbe, H.M. Mayer, U. Ahlheim and C.D. Bredl, *Z. Phys. B*, 1986, **62**, 177.
22 S.S. Saxena, P. Agarwal, K. Ahilan, F. M. Grosche, R.K.W. Haselwimmer, M.J. Steiner, E. Pugh, I.R. Walker, S.R. Julian, P. Monthoux, G.G. Lonzarich, A. Huxley, I. Sheikein, D. Braithwaite and J. Flouquet, *Nature*, 2000, **406**, 587.
23 T. Akazawa, H. Hidaka, T. Fujiwara, T.C. Kobayashi, E. Yamamoto, Y. Haga, R. Settai and Y. Onuki, *J. Phys. Condens. Mat.*, 2004, **16**, L29.
24 C. Geibel, C. Schank, S. Thies, H. Kitazawa, C.D. Bredl, A. Bohn, M. Raus, A. Grauel, R. Gaspari, R. Helfrich, U. Ahlheim, G. Weber and F. Steglich, *Z. Phys. B*, 1991, **84**, 1.
25 C. Geibel, S. Thies, D. Kaczorowski, A. Mehner, A. Grauel, B. Seidel, U. Ahlheim, R. Helfrich, K. Petersen, C.D. Bredl and F. Steglich, *Z. Phys. B*, 1991, **83**, 305.
26 D. Aoki, A. Huxley, E. Ressouche, D. Braithwaite, J. Flouquet, J.B. Brison, E. Lhotel and C. Paulsen, *Nature* 2001, **413**, 613.
27 J.L. Sarrao, L.A. Morales, J.D. Thompson, B.L. Scott, G.R. Stewart, F. Wastin, J. Rebizant, P. Boulet, E. Colineau and G.H. Lander, *Nature*, 2002, **420**, 297.
28 F. Wastin, P. Boulet, J. Rebizant, E. Colineau and G.H. Lander, *J. Phys.: Condens. Matter*, 2003, **15**, S2279.
29 J.L. Smith and R.G. Haire, *Science*, 1978, **200**, 535.
30 Z. Fisk and H.R. Ott, *Int. J. Mod. Phys.*, 1989, **3**, 535.
31 J. Flouquet, http://xxx.lanl.gov/abs/cond-mat/0501602.

32 M. Sigrist and K. Ueda, *Rev. Mod. Phys.*, 1991, **63**, 239.
33 G. Aeppli, E. Bucher, C. Broholm, J. K. Kjems, J. Baumann and J. Hufnagl, *Phys. Rev. Lett.*, 1988, **60**, 615.
34 M.B. Maple, J.W. Chen, Y. Dalichaouch, T. Kohara, C. Rossel, M.S. Torikachvili, M.W. McElfresh and J.D. Thompson, *Phys. Rev. Lett.*, 1986, **56**, 185.
35 A. Krimmel, P. Fischer, B. Roessli, H. Maletta, C. Geibel, C. Schank, A. Grauel, A. Loidl and F. Steglich, *Z. Phys. B*, 1992, **86**, 161.
36 R. Caspary, P. Hellmann, M. Keller, G. Sparn, C. Wassilew, R. Kohler, C. Geibel, C. Schank, F. Steglich and N.E. Phillips, *Phys. Rev. Lett.*, 1993, **71**, 2146.
37 G. Zwicknagl and P. Fulde, *J. Phys.: Condens. Matter*, 2003, **15**, S1911.
38 N. Mekoki, Y. Haga, Y. Koike and Y. Onuki, *Phys. Rev. Lett.*, 1998, **80**, 5417.
39 N. Bernhoeft, N. Sato, B. Roessli, N. Aso, A. Hiess, G. H. Lander, Y. Endoh and T. Komatsubara, *Phys. Rev. Lett.*, 1998, **81**, 4244; N. K. Sato, N. Aso, K. Miyake, R. Shiina, P. Thalmeier, G. Varelogiannis, C. Geibel, F. Steglich, P. Fulde and T. Komatsubara, *Nature*, 2001, **410**, 340.
40 J.G. Lussier, M. Mao, A. Schroder, J.D. Garrett, B.D. Gaulin, S.M. Shapiro and W.J.L. Buyers, *Phys. Rev. B*, 1997-II, **56**, 11749.
41 N.K. Sato, *J. Phys.: Condens. Matter*, 2003, **15**, S1937.
42 K. Ishida, D. Ozaki, T. Kamatsuka, H. Tou, M. Kyogaku, Y. Kitaoka, N. Tateiwa, N. K. Sato, N. Aso, C. Geibel and F. Steglich, *Phys. Rev. Lett.*, 2002, **89**, 037002.
43 R. Feyerherm, A. Amato, F.N. Gygax, A. Schenck, C. Geibel, F. Steglich, N. Sato and T. Komatsubara, *Phys. Rev. Lett.*, 1994, **73**, 1849.
44 R. Joynt and L. Taillefer, *Rev. Mod. Phys.*, 2002, **74**, 235.
45 H. Kotegawa, S. Kawasaki, A. Harada, Y. Kawasaki, K. Okamoto, G.Q. Zheng, Y. Kitaoka, E. Yamamoto, Y. Haga, Y. Onuki, K.M. Itoh and E.E. Haller, *J. Phys. Condens. Matter*, 2003, **15**, S2043.
46 E.D. Bauer, J.D. Thompson, J.L. Sarrao, L.A. Morales, F. Wastin, J. Rebizant, J.C. Griveau, P. Javorsky, P. Boulet, E. Colineau, G.H. Lander and G.R. Stewart, *Phys. Rev. Lett.*, 2004, **93**, 147005.
47 A.B. Shick, V. Janis and P. Oppeneer, *Phys. Rev. Lett.*, 2005, **94**, 016401.
48 J.J. Joyce, J.M. Wills, T. Durakiewicz, M.T. Butterfield, E. Guziewicz, J.L. Sarrao, L.A. Morales, A.J. Arko and O. Eriksson, *Phys. Rev. Lett.*, 2003, **91**, 176401.
49 N.J. Curro, T. Caldwell, E.D. Bauer, L.A. Morales, M.J. Graf, Y. Bang, A.V. Balatsky, J.D. Thompson and J.L. Sarrao, *Nature*, 2005, **434**, 622.
50 H. Sakai, Y. Tokunaga, T. Fujimoto, S. Kambe, R. E. Walstedt, H. Yasuoka, D. Aoki, Y. Hommi, E. Yamamoto, A. Nakamura, Y. Shiokawa, K. Nakajima, Y. Arai, T.D. Matsuda, Y. Haga and Y. Onuki, *J. Phys. Soc. Jpn.*, 2005, **74**, 1710.
51 J.D. Thompson, J.L. Sarrao, L.A. Morales, F. Wastin and P. Boulet, *Phys. C*, 2004, **412-414**, 10.
52 W. Mueller, R. Schenkel, H.E. Schmidt, J.C. Spirlet, D.L. McFlroy, R.A. O. Hall and M.J. Mortimer, *J. Low Temp. Phys.*, 1978, **30**, 561.
53 P. Link, D. Braithwaite, J. Wittig, U. Benedict and R.G. Haire, *J. Alloys Compds.*, 1994, **213-214**, 148.
54 J.C. Griveau, J. Rebizant, G.H. Lander and G. Kotliar, *Phys. Rev. Lett.*, 2005, **94**, 097002.

NEUTRON SCATTERING STUDIES OF $NpTGa_5$ (T = Fe, Co, Ni AND Rh)

N. Metoki

Advanced Science Research Center, Japan Atomic Energy Research Institute
Department of Physics, Tohoku University

1 INTRODUCTION

The 115 rare earth and actinide compounds with a $HoCoGa_5$ structure have attracted much attention after the discovery of superconductivity in Ce-[1] and Pu-based[2] systems. We can systematically grow high quality and large single crystals of 115 compounds with different numbers and characteristics of *f* electrons (e.g. Ce, U, Np, Pu, Am) and transition metal *d*-electrons, which show a variety of electronic and magnetic properties. This variety allows a systematic understanding of these highly correlated *f* electron systems.

Prior to $PuTGa_5$, $UTGa_5$ has been studied. The strong itinerant character of 5*f* electrons, with Pauli-paramagnetic behavior and/or itinerant antiferromagnetic order, was revealed by systematic studies.[3-9] Very recently, $NpTGa_5$ (T = Fe, Co, Ni, Rh, Pt) have been studied intensively.[10-19] The macroscopic measurements revealed the antiferromagnetic nature for $NpTGa_5$. No superconductivity has been reported so far in $UTGa_5$ and $NpTGa_5$ compounds. This difference is probably due to the character of the 5*f* electrons. In this work, our recent neutron scattering studies on the magnetic structure of $UTGa_5$ and $NpTGa_5$ will be reviewed and the results are discussed in terms of the 5*f* electronic structure.

2 EXPERIMENTAL

Neutron scattering experiments have been carried out on the thermal and cold triple-axis spectrometers TAS-1, TAS-2, and LTAS, installed at JRR-3 in the Japan Atomic Energy Research Institute, JAERI. The majority of the experiments were carried out with an unpolarized, monochromatic neutron beam, and analysed using vertically bent pyrolytic graphite (PG) crystals. Neutron spin polarization was analyzed using a Heusler analyzer. Magnetic fields were applied to study the *H-T* phase diagram and the magnetic structure under an external field of up to 10 T. The single crystalline samples were grown by the Ga self-flux method at the Oarai facility of Institute for Materials Research in Tohoku University (IMR). The detail of the sample preparation technique has been published elsewhere.[10-13]

3 RESULTS

3.1 $NpFeGa_5$

In $NpFeGa_5$, AFM reflections with $q = (^1/_2\ ^1/_2\ 0)$ have been observed.[17] The ferromagnetic structures of Np and Fe magnetic sublattices have been revealed by neutron diffraction. The magnetic moment is perpendicular to the tetragonal *c*-axis and coupled anti-parallel to the nearest neighbors along the [100] and [010] directions in the basal plane. The moment is ferromagnetically ordered along the *c*-axis. The Np and Fe sublattices are antiferromagnetically coupled. It should be noted, however, that the Np and Fe spin moments align ferromagnetically along the *c*-axis, because the Np total moment is dominated by orbital moment, thus the spin moment is opposite to the total moment. The magnetic moment was derived as 0.24 μ_B/Fe and 0.86 μ_B/Np, respectively. The presence of the magnetic moment at Fe was also confirmed by Mössbauer spectroscopy measurements.

$NpFeGa_5$ exhibits successive transition at T_{N1} = 118 K and T_{N2} = 77 K, where there are clear anomalies in resistivity.[13] We observed the antiferromagnetic ordering below T_{N1}, and a slight change in the reflection intensities was found at T_{N2}. No significant change in magnetic transition can be expected from our unpolarized neutron scattering experiments. Thus the mechanism for the successive transition remains an open question.

3.2 $NpCoGa_5$

The magnetic structure of $NpCoGa_5$ was studied as an international joint research project for Japan Atomic Energy Research Institute (JAERI) with the Institute for Transuranium (ITU) and CEA-Grenoble.[11,12,14,16] The magnetic structure and *H-T* phase diagram are shown in Figure 1. Antiferromagnetic reflections with the propagation vector $q = (0\ 0\ ^1/_2)$ have been observed below T_N. Thus, the Np moments ordered ferromagnetically in the basal plane and show a simple + - stacking along the *c*-axis. Co has no magnetic moment. The absence of the $(0\ 0\ ^l/_2)$ (l = odd integer) peak indicates the moment direction parallel to the *c*-axis. The magnetic structure is the same as that for $UPdGa_5$ and $UPtGa_5$. It is a consequence of the electronic structure, especially, Fermi surface topology of $NpCoGa_5$, which is very similar to $UPdGa_5$ and $UPtGa_5$ because of the identical number of valence electrons.

$NpCoGa_5$ exhibits a metamagnetic transition with relatively small critical field $\mu_0 H_c$ = 4.5 T at T = 3 K for the field direction $H \parallel c$. The phase boundary observed by neutron, magnetization, and specific heat measurements on single crystals grown in the different institutes (Tohoku University and ITU) shows excellent agreement. The weak critical field can be understood in terms of a weak out-of-plane antiferromagnetic coupling broken by the applied field, while the strong in-plane ferromagnetic interaction plays a dominant role in the magnetic ordering at T_N = 47 K, where the ferromagnetic basal plane ordering remains unchanged throughout the metamagnetic transition. This two-dimensional magnetic interaction is consistent with the large positive value of the paramagnetic Curie temperature despite the antiferromagnetic ordering. The dashed line in Figure 1 denotes a cross-over region from the paramagnetic to the field-induced ferromagnetic phase.

3.3 $NpNiGa_5$

The magnetic structure and *H-T* phase diagram of $NpNiGa_5$ (Figure 2) have been studied by neutron scattering, including spin polarization analysis and high field experiments.[10,18] There is the successive transition from simple ferromagnetic order below T_c = 30 K to a canted antiferromagnetic structure below $T_N = 18$ K. The magnetic moment of Np was 0.8 μ_B and the angle between two adjacent Np moments was estimated as ~80 degrees at 3 K from the ferromagnetic (// c) and the antiferromagnetic component (// [110]) with $q = (^1/_2\ ^1/_2\ ^1/_2)$. Neutron spin polarization analysis clarified, unambiguously, that the magnetism is carried by Np, whereas Ni is non-magnetic.

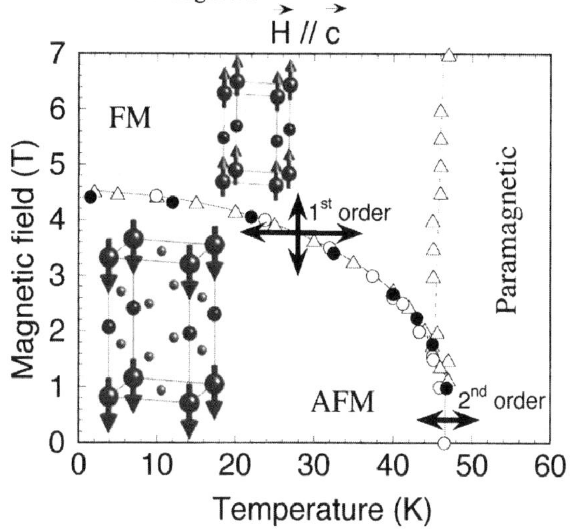

Figure 1 *H-T magnetic phase diagram of NpCoGa₅. Determined magnetic structures in each phase are shown in the insets*

A remarkable increase of the Np magnetic moment was observed below $T_N = 18$ K. This strong enhancement of the Np moment indicates that the magnetic transition at T_N is associated with a change in 5*f* electronic state. $NpNiGa_5$ exhibits a metamagnetic transition at 8 T with application of the magnetic field perpendicular to the *c*-axis.

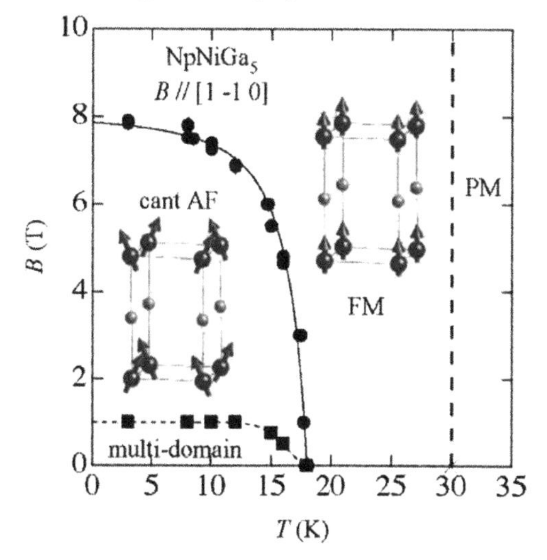

Figure 2 *H-T magnetic phase diagram of NpNiGa₅*

A dramatic reduction of the Np moment, from 0.8 to 0.5 μ_B at the metamagnetic critical field, H_c, was observed, due to a sudden disappearance of the antiferromagnetic component and a discontinuous reduction of the ferromagnetic moment with a magnetic field above H_c. From these observations we concluded that the ground state, with a canted antiferromagnetic structure, is a large moment state, while a high temperature and a high field ferromagnetic phase is the low moment state of Np 5*f* electrons. Thus the metamagnetism is associated with the transition of the 5*f* electronic state between the low and high moment states.

3.4 $NpRhGa_5$

$NpRhGa_5$ orders antiferromagnetically with the propagation of q = (0 0 $^1/_2$).[19] We found that the direction of the magnetic moment is parallel to the *c*-axis for $T_{N2}<T<T_{N1}$, whereas the Np moment is along the [110] below T_{N2}, as shown in Figure 3. This is clearly indicated by the temperature dependence of the antiferromagnetic Bragg intensities. The (1 1 $^1/_2$) reflection intensity increases below T_{N1} and exhibits a discontinuous change at T_{N2}. On the other hand the (0 0 $^1/_2$) reflection is not observed for $T_{N2}<T<T_{N1}$, but the intensity abruptly appeared below T_{N2}. These temperature dependencies can be understood by the magnetic structure shown in Figure 3. Note that the in-plane moment direction was also determined by neutron scattering on a single domain sample under a magnetic field. The unusual temperature dependence of the magnetic susceptibility[13] can be well explained with the switching of the moment direction[15] reveled by the present neutron scattering study.

The antiferromagnetic moment exhibits a continuous increase below T_{N1}, indicating the second order magnetic transition. On the other hand, a large discontinuous change of the magnetic moment from 0.3 to 0.6 μ_B was observed at T_{N2}. Therefore the first order transition at T_{N2} is not a simple re-orientation of the moment direction, but a significant change of the 5*f* electronic state should be accompanied by this magnetic transition.

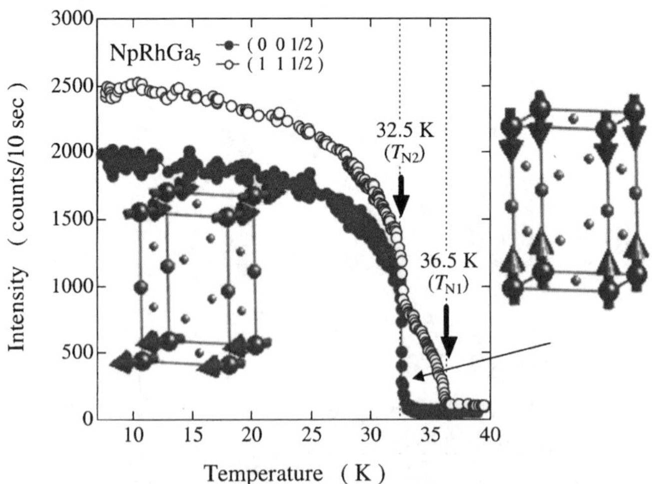

Figure 3 *Temperature dependences of the (0 0 $^1/_2$) and (1 0 $^1/_2$) antiferromagnetic scattering intensity in $NpRhGa_5$*

4 DISCUSSION

The variety of the magnetic structure in $NpTGa_5$ is most probably ascribed to the unique electronic structure. $NpTGa_5$ shows antiferromagnetic ground state for T = Fe, Co and Ni, belonging to the different groups in the periodic table. According to the band structure calculations,[20] a sharp $j = {}^5/_2$ branch of the Np 5*f* band is just at the Fermi level for T = Fe, Co, and Ni, which is the origin of the antiferromagnetic instability of $NpTGa_5$. In contrast the 5*f* electrons in $UFeGa_5$ and $UCoGa_5$ are strongly itinerant with the main part of 5*f* bands above E_f,[21] which is the reason for the non-magnetic ground state with small effective mass enhancement. With increasing valence electron number, E_f shifts towards high energy and cross the 5*f* bands in $UNiGa_5$ and $UPtGa_5$,[22] resulting in a similar electronic structure with the magnetic ground state as in $NpTGa_5$. Thus a rigid band model is roughly correct in $UTGa_5$, but it is not the case for $NpTGa_5$. The large paramagnetic density of state originating from 5*f* bands involves the itinerant magnetism in $NpTGa_5$ and $UTGa_5$.

The transition metal *d* bands would be highly affected by 5*f* electrons *via* hybridization,[17] because of the preferable magnetic nature of 5*f* bands in $NpTGa_5$. The existence of Fe moment in $NpFeGa_5$ indicates the unoccupied 5*f*-3*d* hybridization band contributing to the Fermi surface, whereas the Co and Ni 3*d* band in $NpCoGa_5$ and $NpNiGa_5$ are deep and fully occupied. The magnetic structures of $NpTGa_5$ changes significantly with T. $NpFeGa_5$ with $q = ({}^1/_2\ {}^1/_2\ 0)$, called C-type, exhibits in-plane antiferromagnetic and inter-plane feromagnetic interaction, whereas the magnetic interactions in $NpCoGa_5$ with $q = (0\ 0\ {}^1/_2)$, A-type, are opposite to $NpFeGa_5$ both in-plane and out-of-plane direction. $NpNiGa_5$ exhibits a ferromagnetic ordering and canted structure with in-plane antiferromagnetic component $q = ({}^1/_2\ {}^1/_2\ {}^1/_2)$, namely G-type. The $NpRhGa_5$ has the same ordering vector $q = (0\ 0\ {}^1/_2)$ as in $NpCoGa_5$. It is reasonable because these iso-electronic compounds have a similar two-dimensional electronic structure. The variety of the magnetic structure is theoretically calculated based on the *j-j* coupling scheme.[23]

The origin for the canted antiferromagnetic structure in $NpNiGa_5$[18] and the mechanism of the unusual successive transition in $NpNiGa_5$[18] and $NpRhGa_5$[19] remain open questions. We showed that a notable change in 5*f* electronic states is associated with the successive magnetic ordering and the metamagnetic transition between the low and high moment states. The *J* multiplet of Np^{3+} splits into doublet and singlet states in tetragonal symmetry. When a doublet ground state is assumed, as in the case of $NpCoGa_5$,[11] an in-plane magnetic component can be allowed only by the mixing of the excited singlet. Therefore the competing magnetic interactions for the in-plane and out-of-plane component, and quadrupolar interactions should be taken into consideration to account for the magnetic structure and the phase diagram of $NpTGa_5$. A theoretical study according to this scenario qualitatively reproduces the unusual behavior of $NpTGa_5$.[24] Further microscopic studies such as NMR, μSR, Mössbauer spectroscopy, and resonant X-ray scattering studies are planned to shed light on the nature of 5*f* electrons in $NpTGa_5$.

5 CONCLUSIONS

The magnetic structure and the *H-T* phase diagram of $NpTGa_5$ (T = Fe, Co, No, and Rh) have been studied by means of neutron scattering. The variety of the itinerant antiferromagnet is discussed in terms of the 5*f* electronic structure. The successive magnetic transitions in $NpNiGa_5$, $NpRhGa_5$, and possibly $NpFeGa_5$ may be ascribed to the notable change in $5f$ electronic state due to the competing magnetic and quadrupolar interactions within different $5f$ orbital.

References

1 H. Hegger, C. Petrovic, E.G. Moshopoulou, M.G. Hundley, J.L. Sarrao, Z. Fisk and J.D. Thompson, *Phys. Rev. Lett.*, 2000, **84**, 4986.
2 J.L. Sarrao, L.A. Morales, J.D. Thompson, B.L. Scott, G.R. Stewart, F. Wastin, J. Rebizant, P. Boulet, E. Colineau and G.H. Lander, *Nature*, 2002, **420**, 297.
3 Y. Tokiwa, Y. Haga, E. Yamamoto, D. Aoki, N. Watanabe, R. Settai, T. Inoue, K. Kindo, H. Harima and Y. Onuki, *J. Phys. Soc. Jpn.*, 2001, **70**, 1744.
4 Y. Tokiwa, T. Maehira, S. Ikeda, Y. Haga, E. Yamamoto, A. Nakamura, Y. Onuki. M. Higuchi and A. Hasegawa, *J. Phys. Soc. Jpn.*, 2001, **70**, 2982.
5 S. Ikeda, Y. Tokiwa, Y. Haga, E. Yamamoto, T. Okubo, M. Yamada, N. Nakamura, K. Kindo, Y. Inada, H. Yamagami and Y Onuki, *J. Phys. Soc. Jpn.*, 2003, **72**, 576.
6 S. Ikeda, Y. Tokiwa, T. Okubo, Y. Haga, E. Yamamoto, Y. Inada, R. Settai, and Y Onuki, *J. Nucl. Sci. Tech.*, 2002, **Suppl 3**, 206.
7 Y. Tokiwa, Y. Haga, N. Metoki, Y. Ishii and Y Onuki, *J. Phys. Soc. Jpn.*, 2002, **71**, 725.
8 K. Kaneko, N. Metoki, N. Bernhoeft, G. H. Lander, Y. Ishii, S. Ikeda, Y. Tokiwa, Y. Haga and Y Onuki, *Phys. Rev. B*, 2004, **68**, 214419.
9 S. Ikeda, N. Metoki, Y. Haga, K. Kaneko, T. D. Matsuda and Y Onuki, *J. Phys. Soc. Jpn.*, 2003, **72**, 2622.
10 D. Aoki, E. Yamamoto, Y. Homma, Y. Shiokawa, A. Nakamura, Y. Haga, R. Settai and Y Onuki, *J. Phys. Soc. Jpn.*, 2004, **73**, 519.
11 D. Aoki, Y. Homma, Y. Shiokawa, E. Yamamoto, A. Nakamura, Y. Haga, R. Settai, T. Takeuchi and Y Onuki, *J. Phys. Soc. Jpn.*, 2004, **73**, 1665.
12 D. Aoki, Y. Homma, Y. Shiokawa, E. Yamamoto, A. Nakamura, Y. Haga, R. Settai and Y Onuki, *J. Phys. Soc. Jpn.*, 2004, **73**, 2608.
13 D. Aoki, Y. Homma, Y. Shiokawa, H. Sakai, E. Yamamoto, A. Nakamura, Y. Haga, R. Settai and Y Onuki, *J. Phys. Soc. Jpn.*, accepted.
14 E. Colineau, P. Javorsky, P. Boulet, F. Wastin, J.–C. Griveau, J. Rebizant, J.P. Sanchez and G.R. Stewart, *Phys. Rev. B*, 2004, **69**, 184411.
15 E. Colineau, F. Wastin, P. Boulet, P. Javorsky, J. Rebizant and J.P. Sanchez, *J. Alloys Compd.*, 2005, **386**, 57.
16 N. Metoki, K. Kaneko, E. Colineau, P. Javorsky, D. Aoki, Y. Homma, P. Boulet, F. Wastin, Y. Shiokawa, N. Bernhoeft, E. Yamamoto, Y. Onuki, J. Rebizant and G.H. Lander, *Phys. Rev. B*, 2005 **72**, 014460.
17 N. Metoki, K. Kaneko, H. Yamagami E. Yamamoto, Y. Haga, D. Aoki, Y. Homma, Y. Shiokawa and Y. Onuki, unpublished results.
18 F. Honda, N. Metoki, K. Kaneko, S. Jonen, E. Yamamoto, D. Aoki, Y. Homma, Y. Shiokawa and Y. Onuki, unpublished results.
19 S. Jonen, N. Metoki, F. Honda, K. Kaneko, E. Yamamoto, Y. Haga, D. Aoki, Y. Homma, Y. Shiokawa and Y. Onuki, *Phys. Rev. B*, submitted.
20 H. Yamagami, unpublished results.
21 T. Maehira, M. Higuchi and A. Hasegawa, *Phys. B*, 2003, **329-333**, 574.
22 H. Yamagami, *Acta Phys. Pol. B*, 2003, **34**, 1201.
23 H. Onishi and T. Hotta, *New Journal Phys.*, 2004, **6**, 193.
24 A. Kiss and Y. Kuramoto, unpublished results.

TUNING THE SUPERCONDUCTING BEHAVIOUR OF THE $PuTGa_5$ COMPOUNDS

F. Wastin, F. Jutier, E. Colineau, J. Rebizant, P. Boulet, J.C. Griveau and G.H. Lander

European Commission, Joint Research Centre, Institute for Transuranium Elements, Postfach 2340, D-76125 Karlsruhe, Germany

1 INTRODUCTION

The discovery of a relatively high superconducting $T_c \approx 18$ K in $PuCoGa_5$,[1] followed by a $T_c \approx 9$ K in $PuRhGa_5$[2] has attracted much interest onto the actinide-based '115' compounds *AnT*Ga_5 (*An* = U, Np, Pu, Am and *T* = a *d*-transition metal). These materials crystallize in the tetragonal $HoCoGa_5$-type structure with the space group P4/mmm. The structure consists of a sequential stacking along the tetragonal axis of *An*Ga_3 layers with $AuCu_3$ structure and *T*Ga_2 layers. Isostructural Co-based compounds form with the rare earths, $Ce(4f^1)$[3-5] and $Pr(4f^2)$ on substituting In for Ga, whereas the actinide elements, $U(5f^3)$,[6] $Np(5f^4)$[7-9], $Pu(5f^5)$[1,2] and $Am(5f^6)$[10] form Ga-based compounds with various transition metal elements *T* = Fe, Co, Ni, Ru, Rh, Pd, Os, Ir, Pt. These compounds are isotructural to an extensively studied series of Heavy-Fermion superconductors that form in the Ce*T*In_5 system (*T* = Co, Rh, Ir).[3,4,5]

A variety of electronic properties is found in these isostructural series of compounds and is summarized in Figure 1 for the *f*-systems known so far. $CeRhIn_5$[3] is a Heavy-Fermion superconducting under pressure, $CeCoIn_5$ and $CeIrIn_5$ are Heavy-Fermion superconductors at ambient pressure.[4,5] In each of the compounds, superconductivity emerges in proximity to magnetic order, allowing the possibility of spin fluctuations to mediate Cooper pairing. In the case of the 5*f*-elements, no superconductivity has been reported either at ambient or under high pressures for uranium or neptunium compounds. The uranium series displays either Pauli paramagnetism (PP) for the Fe series, possible mixed valence state behaviour for the Co series or is magnetically ordered for the Ni series.[6,11-14] So far, all neptunium compounds exhibit magnetically ordered ground states[7-9,15] and a Np^{3+} valence state as unambiguously established by Mössbauer spectroscopy.[7,9] Pu-materials with Fe and Ni display paramagnetic behaviour[16] and superconductivity is only observed for Co and Rh compounds. These two Pu-based superconducting compounds appear to be close to a magnetic phase instability[17-19] and their magnetic susceptibility in the normal state indicates local-moment (Curie-Weiss) behaviour close to that expected for Pu^{3+}. This magnetic behaviour is unexpected in the case of "conventional" superconductivity.[20]

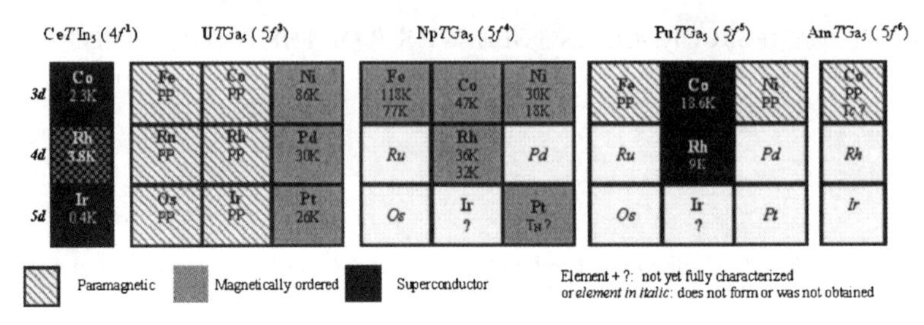

Figure 1 *Electronic properties of the $AnTGa_5$ systems compared to their $CeTIn_5$ homologues.*

Moreover, although their electronic coefficient of specific heat γ is about ten times smaller and the superconducting transition temperature, T_c, about ten times higher than that found in $CeTIn_5$, both systems display strong correlation suggesting that $PuTGa_5$ superconducting compounds display an unconventional superconducting state. Finally, the only known Am compound ($AmCoGa_5$) seems to display temperature-independent paramagnetism that, assuming by analogy to other actinide compounds with a 3+ valence state, can be understood to be due to the $5f^6$ atomic configuration resulting in a total moment $J = 0$. The possible appearance of superconductivity below 2 K remains unclear.[10]

In this paper, we report several approaches to tune the superconducting behaviour in the $AnTGa_5$ series of compounds.

2 RESULTS AND DISCUSSION

Two types of systems have been studied. The first one consists in doping (or partially substituting) Pu with another 5*f* element, i.e. $Pu_{1-x}An_xCoGa_5$. The following compounds were synthesized and investigated: $Pu_{0.9}U_{0.1}CoGa_5$, $Pu_{0.9}Np_{0.1}CoGa_5$, $Pu_{0.95}Am_{0.05}CoGa_5$, $Pu_{0.9}Am_{0.1}CoGa_5$, $Pu_{0.88}Am_{0.12}CoGa_5$, $Pu_{0.8}Am_{0.2}CoGa_5$. The second one consists in substituting the *d*-transition metal, i.e. $Pu(Co_{1-x}T_x)Ga_5$, and $PuCo_{0.9}Fe_{0.1}Ga_5$, $PuCo_{0.8}Fe_{0.2}Ga_5$, $PuCo_{0.9}Rh_{0.1}Ga_5$, $PuCo_{0.5}Rh_{0.5}Ga_5$, $PuCo_{0.5}Ir_{0.5}Ga_5$ and $PuCo_{0.9}Ni_{0.1}Ga_5$ were studied. It may be worth to note that this last series corresponds to two different sub-types of doping, one with isoelectronic *d*-transition metal (same count of outer electrons; i.e. Co, Rh, Ir) and the other one bringing elements with a different count of outer electrons (i.e. Fe, Co, Ni).

2.1 "Structural" Tuning

All doped systems crystallize in the same structure type as their parent pure compounds. The first effects evident of partially substituting one element by another are slight modifications of the crystallographic lattice parameters and to decrease the superconducting transition temperature (Figure 2). However, as reported by Boulet *et al.*,[21] these two effects do not follow Vegard's law, indicating that a steric effect is not the only parameter involved. An interesting correlation of the *c/a* ratio of the tetragonal lattice parameters and T_c was observed in the $CeMIn_5$ heavy fermion superconductors[22] and is amazingly reproduced for the $PuCo_{1-x}Rh_xGa_5$ series[23] which also exhibits a linear relation of *c/a* and T_c with nearly identical slopes $d\ln T_c/d(c/a) \approx 100$ (Figure 2). This correlation seems to apply also for $PuCo_{0.5}Ir_{0.5}Ga_5$ (isoelectronic) but does not at all fit for non-isoelectronic element substitution (*An* and *d*-metals) which provokes a stronger decrease in T_c. The superconducting transition temperature decrease is even faster with *An*- than with *T*-substitution. Even, in the case of Am-doping, despite the *c/a* ratio remaining almost unchanged, one observes a dramatic decrease of T_c as Am content increases.

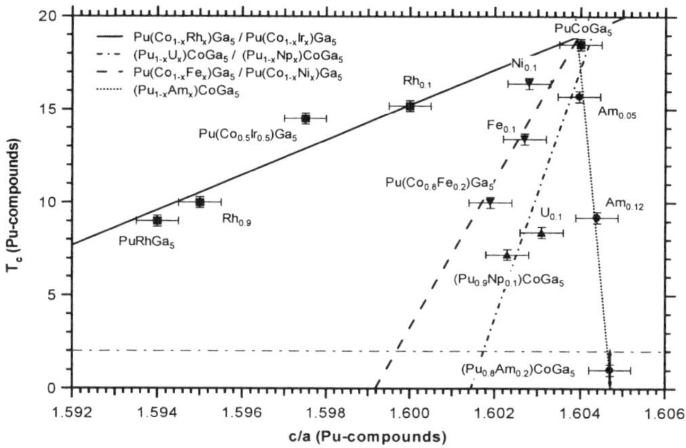

Figure 2 *c/a ratio of tetragonal lattice parameters vs superconducting transition temperature T_c of $Pu(Co_{1-x}T_x)Ga_5$ (T = Co, Rh, Fe, Ni) and $Pu_{1-x}An_xCoGa_5$ (An = U, Np, Am). The full line represents the linear fit of the correlation observed in the Co-Rh series (see text) whereas dashed lines are guide for the eyes. The horizontal dashed-dot line represents the lowest temperature experimental limit (2K) due to self-heating effect.*

Another way of fine tuning the structural parameters is to apply pressure on the systems. The temperature-pressure phase diagrams of $PuTGa_5$ (T = Co, Rh, Ir) have been investigated.[24,25] $PuCoGa_5$ and $PuRhGa_5$ phase diagrams show similar characteristics to that of $CeCoIn_5$.[22] In all three compounds, the superconducting transition temperature first increases with increasing pressure and passes through a maximum before decreasing again at higher *P*, resulting in a "dome-like" shape of *Tc(P)* and providing new hints of the common mechanism of superconductivity between the Ce- and Pu-systems. It would be also interesting to corroborate this behaviour by the evolution of *c/a* under pressure in order to comment on the influence of the structural tuning.

2.2 "Electrons Count" Tuning

Figure 3 provides a new perspective on the influence of doping the systems investigated with different elements. As already stressed, it is important to notice that Pu(Co, Rh, Ir)Ga_5 compounds are isoelectronic whereas Pu(Co, Fe/Ni)Ga_5 and (Pu, U/Np)$CoGa_5$ have different electron counts than $PuCoGa_5$. When plotting the variation of T_c as a function of the electron count (here only outer-shell electrons are counted), Figure 3 shows a spectacular concentration of all $AnTGa_5$ superconducting compounds in a narrow band at approximately 32 ± 0.2 electrons, whereas all compounds outside of this band (i.e. with different electron count) are not superconducting.

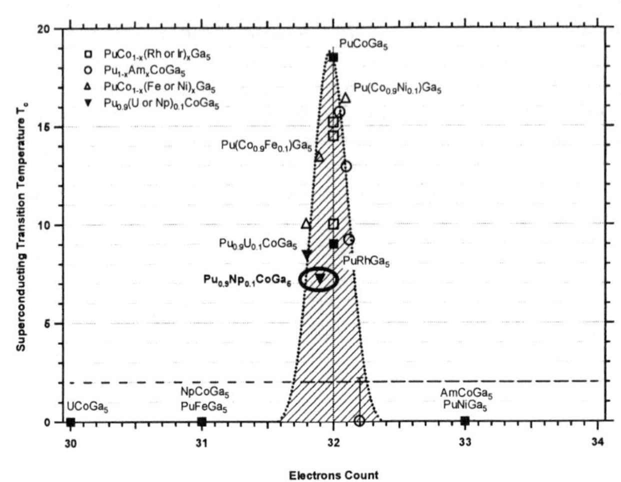

Figure 3 *Critical temperature T_c in $Pu(Co_{1-x}T_x)CoGa_5$ and $(Pu_{1-x}An_x)CoGa_5$ systems, as a function of the "valence" electrons count (taken as the total s, p, d and f electrons count from the outer shells, e.g. for $PuCoGa_5$, a total of 32 electrons is obtained by summing 6f and 2s for Pu, 7d and 2s for Co and 5×(2s and 1p) for Ga). The dashed area illustrates the electrons count interval in which materials display superconductivity. The vertical full line indicates isoelectronic compounds and the c/a ratio dependence (see 2.1). The horizontal dashed-dot line represents the lowest temperature experimental limit (2K) due to self-heating effect. The anomalous position of $(Pu_{0.9}Np_{0.1})CoGa_5$ is highlighted (see text).*

With closer inspection, it is even noticeable how regular the distribution of T_c is as a function of the electron count, with two exceptions: one for the isoelectronic compounds for which the *c/a* dependence within the series was already discussed; the second for $Pu_{0.9}Np_{0.1}CoGa_5$. However, in the latter, the anomalously low T_c may be related to the magnetic character of Np (all Np '115' compounds display magnetic ordering). It is interesting to point out that a similar direct empirical relation between superconductivity and the number of valence electrons per atom was already observed by Matthias[26] and Cooper *et al.*[27] showing both optimum conditions for different counts of valence electrons/atom. The correlation between those earlier observations and ours, would certainly deserve such a potential relation to be re-addressed carefully.

2.3 Tuning with Time

Sarrao *et al.*[1] reported that, in $PuCoGa_5$, T_c was decreasing at a rate of about -0.2 K/month due to the alpha decay of the ^{239}Pu isotope used for the compound's synthesis. Indeed through its decay, ^{239}Pu creates lattice damage that causes a decrease of T_c together with an increase of the critical current (J_c) and of the upper critical field (H_{c2}) with ageing of the compound. A careful examination of these phenomena has been undertaken by Jutier *et al.*[28] who reported a T_c decrease of -0.24 K/month and -0.39 K/month in $^{239}PuCoGa_5$ and $^{239}PuRhGa_5$ respectively, and after about two years it leads to the disappearance of the superconducting transition in the latter. Within the limit of time investigated, the authors did not report a strong variation in the lattice parameters due to the lattice swelling and the c/a ratio remaining almost unchanged. Moreover, substituting ^{239}Pu by ^{242}Pu shows that T_c remains unchanged on a long time scale, whereas the T_c decrease is even enhanced when substituting ^{239}Pu by more unstable isotopes like ^{241}Am or ^{243}Am (e.g. $^{239}Pu_{0.88}{}^{241}Am_{0.12}CoGa_5$ shows a T_c decrease of -1.13 K/month). These simple experiments unambiguously prove that, in this case, the tuning of the superconducting behaviour results exclusively from the self-irradiation process, inducing the creation of defects and the "growth of impurities". The apparent discrepancy observed in the T_c decrease of the Co and Rh compounds, could be explained by the same authors considering the classical clean/dirty limit model, based on the coherence length and mean free path of the Cooper pairs in these materials, these parameters being almost twice as large as the Rh compound.[29]

3 SUMMARY AND CONCLUSIONS

The study of solid solutions of the $PuCoGa_5$ and $PuRhGa_5$ compounds, combined with the investigation of $PuCoGa_5$ doped either on the actinide or transition metal sites, provides a first comprehensive experimental study of several structural and electronic parameters involved in the superconducting behaviour of these compounds. It shows that isoelectronic Rh substitution has only a small effect on T_c, as pure $PuRhGa_5$ is also a superconductor. These studies also suggest that integrity of the Pu sublattice is more important for superconductivity than is periodicity of the *d*-electron elements. The striking similarity of the properties between $PuTGa_5$ and $CeTIn_5$ that has been recently demonstrated and illustrated by the linear correlation in both systems between the critical temperature and the ratio of the tetragonal lattice parameters c/a, suggests that structural aspects may play a key role and is necessary but NOT sufficient to allow the appearance of superconductivity in these systems. This study also evidences a remarkable correlation between T_c and the total electron counts of the compounds, recalling similar earlier empirical relations between superconductivity and the number of valence electrons/atom. The stronger decrease of T_c upon *An* compared to *T*-substitution suggests that superconductivity is effectively of *5f*-electrons origin. The whole emerging picture is in very good agreement with the present theoretical approaches but would also deserve further investigations. Within the different tuning approaches presented here none has led to the appearance of magnetic ordering. Further investigations in this frame, as well as Ga-site substitution, are in progress.

Acknowledgements

The authors would like to thank J.D. Thompson, E.D. Bauer, J.L. Sarrao for fruitful discussions and close exchanges of information. F.J. acknowledges the European Commission for support in the frame of the "Training and Mobility of Researchers" programme.

References

1 J.L. Sarrao, L.A. Morales, J.D. Thompson, B.L. Scott, G.R. Stewart, F. Wastin, J. Rebizant, P. Boulet, E. Colineau and G.H. Lander, *Nature*, 2002, **420**, 297.
2 F. Wastin, P. Boulet, J. Rebizant, E. Colineau and G.H. Lander, *J. Phys.: Condens. Matter*, 2003, **15**, S2279.
3 H. Hegger, C. Petrovic, E.G. Moshopoulou, M.G. Hundley, J.L. Sarrao, Z. Fisk and J.D. Thompson, *Phys. Rev. Lett.*, 2000, **84**, 4986.

4 R. Movshovich, M. Jaime, J.D. Thompson, C. Petrovic, Z. Fisk, P.G. Pagliuso, and J.L. Sarrao, *Phys. Rev. Lett.*, 2001, **86**, 5152.

5 C. Petrovic, R. Movshovich, M. Jaime, P.G. Pagliuso, M.F. Hundley, J.L. Sarrao, Z. Fisk and J.D. Thompson, *Europhys. Lett.*, 2001, **53**, 354.

6 K. Kaneko, N. Metoki, N. Bernhoeft, G.H. Lander, Y. Ishii, S. Ikeda, Y. Tokiwa, Y. Haga and Y. Onuki, *Phys. Rev. B*, 2004, **68**, 214419 and references therein.

7 E. Colineau, P. Javorsky, P. Boulet, F. Wastin, J.–C. Griveau, J. Rebizant, J.P. Sanchez and G.R. Stewart, *Phys. Rev. B*, 2004, **69**, 184411.

8 D. Aoki, Y. Homma, Y. Shiokawa, E. Yamamoto, A. Nakamura, Y. Haga, R. Settai, T. Takeuchi and Y Onuki, *J. Phys. Soc. Jpn.*, 2004, **73**, 1665.

9 E. Colineau, F. Wastin, P. Boulet, P. Javorsky, J. Rebizant and J. P. Sanchez, *J. Alloys Compd.*, 2005, **386**, 57.

10 J. Rebizant *et al.*, unpublished.

11 Y. Tokiwa, Y. Haga, E. Yamamoto, D. Aoki, N. Watanabe, R. Settai, T. Inoue, K. Kindo, H. Harima and Y. Onuki, *J. Phys. Soc. Jpn.*, 2001, **70**, 1744.

12 Y. Tokiwa, T. Maehira, S. Ikeda, Y. Haga, E. Yamamoto, A. Nakamura, Y. Onuki. M. Higuchi and A. Hasegawa, *J. Phys. Soc. Jpn.*, 2001, **70**, 2982.

13 S. Ikeda, N. Metoki, Y. Haga, K. Kaneko, T.D. Matsuda and Y. Onuki, *J. Phys. Soc. Jpn.*, 2003, **72**, 2622.

14 R. Troć, Z. Bukowski, C. Sułkowski, H. Misiorek, J.A. Morkowski, A. Szajek and G. Chełkowska, *Phys. Rev. B*, 2004, **70**, 184443.

15 D. Aoki *et al.*, private communication.

16 P. Boulet, E. Colineau, F. Wastin, P. Javorský and J. Rebizant, *Phys. B: Condens. Matter,* 2005, **359-361**, 1081-1083.

17 I. Opahle and P.M. Oppeneer, *Phys. Rev. Lett.*, 2003, **90**, 157001.

18 T. Maehira, T. Hotta, K. Ueda and A. Hasegawa, *Phys. Rev. Lett.*, 2003, **90**, 207007.

19 I. Opahle, S. Elgazzar, K. Koepernik, and P.M. Oppeneer, *Phys. Rev. B*, 2004, **70**, 104504.

20 J.D. Thompson, J.L. Sarrao, N.J. Curro, E.D. Bauer, L.A. Morales, F. Wastin, J. Rebizant, J.C. Griveau, P. Boulet, E. Colineau and G.H. Lander, *Superconductivity in actinide materials*, this publication.

21 P. Boulet, E. Colineau, F. Wastin, J. Rebizant, P. Javorský, G.H. Lander and J.D. Thompson, , *Phys. Rev. B*, in press.

22 P.G. Pagliuso, C. Petrovic, R. Movshovich, D. Hall, M.F. Hundley, J.L. Sarrao, J.D. Thompson and Z.P. Fisk, *Phys. Rev. B*, 2001, **64**, 100503.

23 E.D. Bauer, J.D. Thompson, J.L. Sarrao, L.A. Morales, F. Wastin, J. Rebizant, J.C. Griveau, P. Javorský, P. Boulet, E. Colineau, G.H. Lander and G.R. Stewart, *Phys. Rev. Lett.*, 2004, **93** 147005.

24 J.C. Griveau, C. Pfleiderer, P. Boulet, J. Rebizant and F. Wastin, *J. Magn. Magn. Mater.*, 2004, **272-276**, 154.

25 J.C. Griveau, P. Boulet, E. Colineau, F. Wastin and J. Rebizant, *Phys. B: Condens. Matter*, 2005, **359-361**, 1093-1095.

26 B.T. Matthias, *Phys. Rev.*, 1955, **97**, 74

27 A.S. Cooper, E. Corenzwit, L.D. Longinotti, B.T. Matthias and W. Zachariasen, *Proc. Nat. Acad. Sci.*, 1970, **67**, 313.

28 F. Jutier, J.C. Griveau, E. Colineau, J. Rebizant, P. Boulet, F. Wastin and E. Simoni, *Physica B: Condens. Matter*, 2005, **359-361**, 1078-1080.

29 F. Jutier, J.-C. Griveau, E. Colineau, J. Rebizant, P. Boulet, F. Wastin and E. Simoni, *Influence of self-irradiation damage on the superconducting behaviour of plutonium-based compounds*, this publication.

THE ELECTRONIC STRUCTURE OF THE Pu-BASED SUPERCONDUCTOR $PuCoGa_5$: LSDA AND LSDA+U INVESTIGATIONS

P.M. Oppeneer,[1] A.B. Shick,[2] I. Opahle,[3] S. Elgazzar[3] and V. Janiš[2]

[1]Department of Physics, Uppsala University, Box 530, S-751 21 Uppsala, Sweden
[2]Institute of Physics, Academy of Sciences of the Czech Republic, Na Slovance 2, CZ-182 21, Prague, Czech Republic
[3]Leibniz-Institute of Solid State and Materials Research, P.O. Box 270016, D-01171 Dresden, Germany

1 INTRODUCTION

The recent discovery of superconductivity[1] at a high transition temperature, T_c= 18.5 K, in the plutonium-based compound $PuCoGa_5$ has ignited a tremendous interest in this new material.[2-10] Prior to the discovery of superconductivity in $PuCoGa_5$ not a single Pu-based compound was known to exhibit superconductivity. Moreover, among the known *f*-electron (actinide or lanthanide) superconductors one rarely finds a material with a T_c over 2 K. From this perspective, $PuCoGa_5$ can be regarded to have an astonishingly high transition temperature. The nature of the superconducting pair formation has been a point of special focus. From its initial discovery it was speculated[1] that possibly an unconventional pairing mechanism might be responsible for the anomalously high T_c (i.e., a pairing mechanism *distinct* from the common phonon-mediated pairing). The magnetic properties of $PuCoGa_5$ consequently received particular attention. Susceptibility measurements provided no evidence for long-range magnetic order, but at elevated temperatures the susceptibility obeys a modified Curie-Weiss behaviour with an effective moment of 0.7 μ_B. Such moment corresponds approximately to that of a Pu^{3+} ion ($5f^5$ configuration). The temperature dependence of the resistivity exhibits a characteristic *S*-like shape, which indicates Bloch electron scattering from spin-fluctuations.[1,3,8] Furthermore, the linear-temperature specific heat coefficient, with $\gamma \sim 77$ mJ/mol K^2, appears to be significantly enhanced, which could be due to the presence of strong dynamic spin fluctuations.

Recently an affirmative answer to the question whether the superconductivity in $PuCoGa_5$ is unconventional could be given. Experimental studies, employing nuclear magnetic resonance (NMR) and nuclear quadrupolar resonance (NQR) showed that indeed $PuCoGa_5$ is an unconventional, spin-singlet, *d*-wave type superconductor.[10] Very recently, similar NMR/NQR studies proved that the isoelectronic compound $PuRhGa_5$ is also an unconventional *d*-wave superconductor.[11] Such type of unconventional superconducting gap symmetry could possibly correspond to pairing mediated by antiferromagnetic spin fluctuations.[10] Thus, although no long-range antiferromagnetic order could be detected in $PuCoGa_5$ or $PuRhGa_5$, short-range antiferromagnetic fluctuations are not only expected to be present but to constitute the essential bosons responsible for the pairing.

While these latest findings definitely categorise $PuCoGa_5$ as a most exceptional and intriguing material, many important questions are still open. In particular, the electronic structure of $PuCoGa_5$ and the role played by the Pu $5f$ electrons are not understood. *Ab initio* calculations are an ideal tool to investigate the underlying electronic structure and unravel its relationship to the superconductivity. It is well-known, however, that in the actinide series precisely at Pu, a transition in the behaviour of the $5f$ electrons occurs. The $5f$ electrons of the early actinides (up to Pu) are relatively delocalized and bonding. The $5f$ electrons in the heavy actinides, starting from americium, are inert and retracted from the bonding. The cross-over from delocalised to localised behaviour occurs for Pu, particularly with the electronic structure of fcc δ-Pu being a topic of ongoing debate.[12-17] The delocalised $5f$ electrons of the early actinides are quite well described by density functional calculations employing the local spin-density approximation (LSDA). So far not much is known about the electronic structure of Am and the heavier actinides, but first investigations indicate that the electronic structure cannot be sufficiently well described with the LSDA, delocalised $5f$ electron approach.[18] The $5f$ electrons in Am are modestly localised and intra-atomic $5f$ Coulomb correlation effects are to be taken into account. For δ-Pu various electronic structure models have been put forward: the LSDA model, in which the $5f$'s are treated as delocalised,[12,13] the LSDA+U or LSDA+DMFT model,[14-16] in which the $5f$ electrons are modestly localised through additional intra-atomic Coulomb correlations, and the mixed level model (MLM), in which some of the $5f$ electrons are fully localised and some other $5f$'s are delocalised.[17] A similar issue as for δ-Pu occurs for $PuCoGa_5$. The appropriate description of the $5f$'s in $PuCoGa_5$ can ultimately only be established through comparison of computed and experimental electronic data.

In the following we shall discuss electronic structure results obtained for $PuCoGa_5$ focusing mainly on the LSDA and LSDA+U approaches. We shall discuss to what extent the respective electronic structure approaches can explain the available experimental data and consider the implications for the unconventional superconductivity.

2 ELECTRONIC STRUCTURE METHODOLOGIES

Employing the LSDA approach, the electronic structure of $PuCoGa_5$ has been calculated in several papers.[3,4,7] Also, the LSDA has been used by Opahle *et al.*[7] to compute the electronic structures of the related actinide 115 compounds, $UCoGa_5$, $NpCoGa_5$, $PuRhGa_5$ and $PuIrGa_5$. The effect of Coulomb correlations in the $5f$ shell on the electronic structure were recently investigated through LSDA+U calculations by Shick *et al.*[9] The supplementary Coulomb interaction of the $5f$ electrons is modelled through the Coulomb U and exchange J parameters.[9] The values of U and J which are reasonable for actinide materials are known.[9] Recently, a variant of the LSDA, the GGA+OP (orbital polarisation) was applied to $PuCoGa_5$ by Söderlind.[19] Investigations on the basis of the MLM model were recently reported by Joyce e*t al.*[5]

The above mentioned studies can be regarded as state-of-the-art investigations, using electronic structure models valid for specific material classes. Each of these electronic structure models for $PuCoGa_5$ has to be tested against existing experimental data. Unfortunately, for $PuCoGa_5$ not many experimental data have been reported so far. Apart from the reported lattice parameters, susceptibility, resistivity, and specific heat,[1,6] crucial information on the electronic structure of $PuCoGa_5$ was recently provided by photoemission (PE) measurements.[5] The PE showed that Pu $5f$ states are present at the Fermi energy (E_F) which strongly suggests that the f electrons participate in the pairing. The PE revealed also second $5f$-related response which needs yet to be explained. Other recent data are the behaviour of the superconducting transition temperature under pressure and the peculiar linear scaling of T_c with the crystallographic c/a ratio.[6]

3 RESULTS

3.1 Results of LSDA Calculations

The recent LSDA calculations for $PuCoGa_5$ and for the related actinide 115 compounds provided detailed information on the electronic structures as well as on the trends within the series.[3,7] The obtained results can be summarised as follows: the lowest total energy of the Pu-115 compounds (see Figure 1) is computed to be an antiferromagnetically ordered state, which appears to be in agreement with experiment for $PuIrGa_5$. The superconductivity in $PuCoGa_5$ and $PuRhGa_5$ occurs thus in materials on the verge of magnetic ordering. The density of states (DOS) at E_F is largely dominated by the Pu 5*f* states. Due to a peculiarity of the band filling in these materials, the Co 3*d* states (and Rh 4*d*, Ir 5*d* states, respectively) are practically filled and contribute only a small portion to the DOS at E_F. The adjacent Pu layers are magnetically weakly coupled. On account of the tetragonal structure, with its large c/a ratio (~1.6), the electronic structure becomes quasi-two-dimensional. This feature is particularly reflected in the highly anisotropic Fermi surface, which consists of three two-dimensional "tubes" as well as two small, three-dimensional hole pockets.[3,4] The LSDA calculations, furthermore, predict lattice parameters for the Pu-115 compounds in reasonably good agreement with experiment.[3,7] The LSDA calculations provide, in addition, an accurate description of the electronic structures of $UCoGa_5$ and $NpCoGa_5$: $UCoGa_5$ is calculated to be a paramagnet with a very small DOS at E_F, in agreement with experiment. Also, the Fermi surface of $UCoGa_5$ agrees very well with data inferred from de Haas-van Alphen measurements. $NpCoGa_5$, furthermore, is calculated to be an antiferromagnet in agreement with experiment. Some of the properties of $NpCoGa_5$, however, as, e.g. the spin and orbital moments appear to demand closer inspection. Altogether, the LSDA approach explains very well various properties of the actinide-115 compounds.

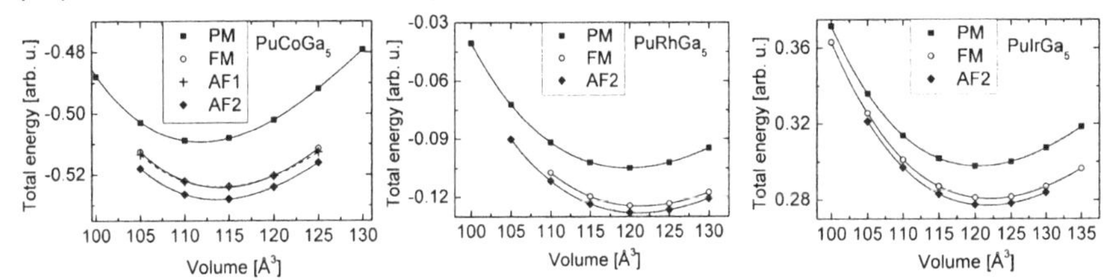

Figure 1 *Calculated LSDA total energies for different magnetic phases of $PuCoGa_5$, $PuRhGa_5$ and $PuIrGa_5$.[7] For all three iso-electronic Pu-compounds an antiferromagnetically ordered configuration has the lowest total energy*

The recent photoemission experiment[5] puts restrictions on the electronic structure models for $PuCoGa_5$. The PE spectrum revealed a 5*f* response at E_F, important for models of the superconductivity, but also a second 5*f*-related response at a binding energy of 1.2 eV. The second response might be due to partially localised 5*f* electrons. The PE spectrum obviously provides a critical test for electronic structure models of $PuCoGa_5$. The LSDA calculations show that there is a 5*f*-related peak in the DOS at about 0.7 eV below E_F (see Figure 2). This energy position appears too small compared to the PE result. This indicates that the electronic structure of $PuCoGa_5$ requires an account of electron correlation effects beyond those already incorporated in the LSDA. We remark, however, that a recent investigation employing the GGA+OP approach, in which the 5*f*'s are delocalized, fairly well reproduced the PE spectrum.[19] It was also well reproduced by the mixed level model, in which four Pu 5*f* electrons are, conversely, treated as *localized* and their energy position is adjusted to the experimental PE position.[5]

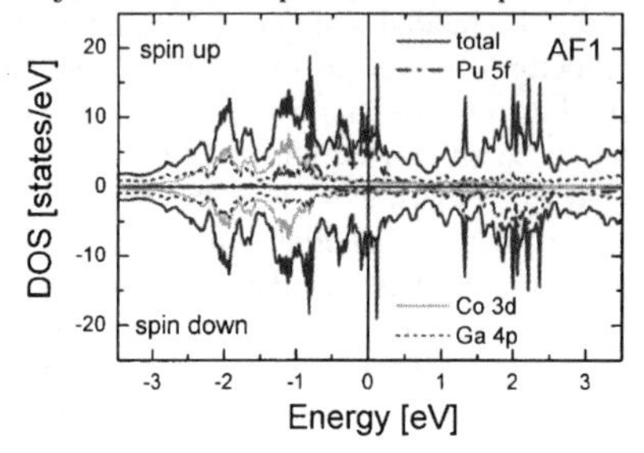

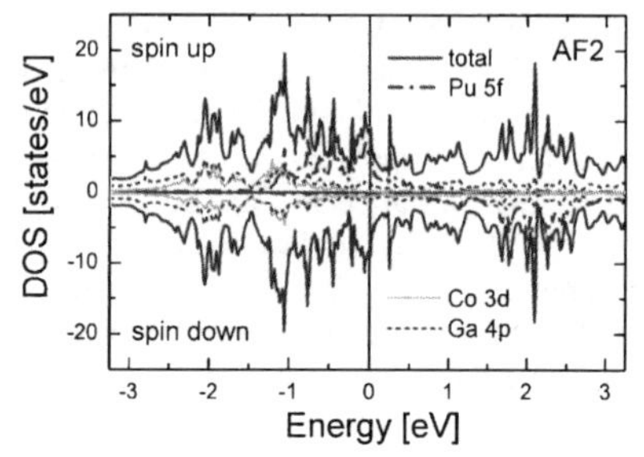

Figure 2 *Calculated[7] LSDA total and partial DOS for $PuCoGa_5$ in the AF1 (Q=(0 0 ½), left) and AF2 (Q=(½ ½ 0), right) antiferromagnetic configurations (with Q the antiferromagnetic wave-vector)*

3.2 LSDA+*U* Results for $PuCoGa_5$

A third possible route to explain the PE spectrum would be through LSDA+*U* calculations. In this approach the intra-atomic Coulomb correlation *U* causes a rearrangement of the 5*f* levels, leading to a moderate increase of the binding energy of the occupied 5*f* states. The LSDA+*U* scheme in the so-called atomic limit (for details, see Ref. 20) was applied recently.[9] The *U* value was varied in the accepted range of 2–5 eV,[21] while for the exchange constant *J* the value *J*= 0.7 eV was used.[22] The inclusion of the Coulomb *U* and exchange *J* leads to an interesting reconstruction of the 5*f* manifold. First, the Pu 5*f* states are shifted down to a binding energy below that obtained from the LSDA. Their final energy position depends slightly on the *U* value used, but for reasonable *U* values of about 3 eV the main weight of the 5*f* DOS occurs at binding energies of 1–2 eV, in good agreement with experiment.[5] A further interesting modification occurs in the Pu spin and orbital magnetic moments. The Coulomb *U* strongly reduces the Pu spin moment M_s from 4.75 μ_B (in the LSDA) to 2.56 μ_B in the LSDA+*U* (*U*= 3 eV), while the orbital moment M_l is slightly decreased from -1.87 μ_B to -1.60 μ_B. Normally, the LSDA+*U* approach does not drastically modify the spin moment, but tends to enlarge substantially the orbital moment. The exceptional result for $PuCoGa_5$ can be traced back to an enhancement of the effective spin-orbit coupling due to modifications in the spin and orbital dependent contributions to the LSDA+*U* effective potential (including spin off-diagonal terms). The enhanced effective spin-orbit coupling leads to a rearrangement of the occupation of the Pu j_z-states ($j_z= m_z+s_z$). While the LSDA fills the *f* states from $j_z= -^5/_2$ up to $j_z= ^7/_2$, the LSDA+*U* occupies the states from $j_z= -^5/_2$ to $^5/_2$ (see Figure 3). The main difference appears in the $j_z= ^7/_2$ subset (i.e., $m_z= 3$, $s_z= +½$), which is excluded in the LSDA+*U* (in the LSDA+*U* a spin flip to $s_z= -½$ is forced and j_z therefore becomes equal to $^5/_2$). As a result, the spin moment becomes considerably reduced and the ratio $C_2 = M_l/(M_l+M_s)$ is much enhanced, yet still smaller than what is expected for a trivalent Pu ion. This extraordinary LSDA+*U* result can be interpreted as a strengthening of the effective spin-orbit coupling leading to the formation of a Pu ground state formed solely out of the $j= ^5/_2$ manifold. While neither the LSDA nor LSDA+*U* are based on any kind of atomic coupling scheme (e.g., *LS* or *jj* coupling), the LSDA+*U* suggest (more than the LSDA) a *jj*-atomic-like coupled, Pu ground-state configuration in $PuCoGa_5$. The importance of a *jj*-coupling scheme for the electronic structure of elemental Pu as well as for $PuCoGa_5$ has recently been suggested.[4,23] Aspects of the angular momentum coupling scheme are particularly important for Pu, because Pu is close to relativistic half-filling and the competition of exchange splitting and spin-orbit interaction will determine the eventual magnetic state.

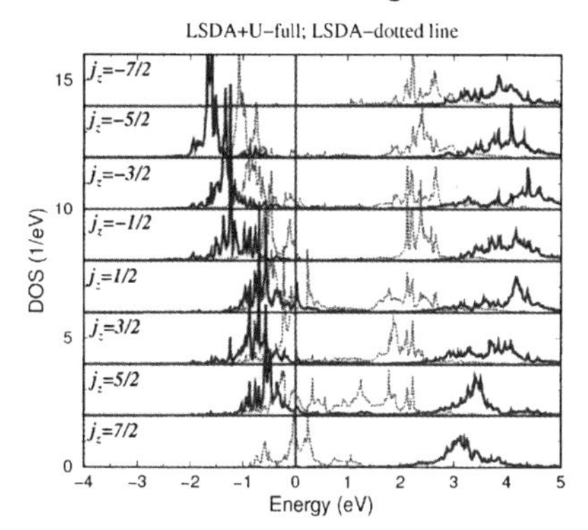

Figure 3 *The j_z resolved Pu 5f DOS as obtained with the LSDA (dotted lines) and LSDA+U (U= 3 eV) (full lines) schemes*

4 CONCLUSIONS

The LSDA calculations[3,4,7] explain various electronic structure properties of $PuCoGa_5$. The LSDA+*U* method,[9] however, improves the energetic position of the 5*f* states as compared to the PE spectrum.[5] In accordance with the PE spectrum, a picture of the Pu 5*f* states emerges, in which the 5*f*'s appear to be fairly hybridized and yet retain some atomiclike features. To explore further this electronic structure picture, more experiments would be desirable, for example, de Haas-van Alphen measurements. Such experiments would also provide valuable information on the Fermi surface, which is predicted by the LSDA calculations to be two–dimensional.[3,7] Concerning the origin of the superconductivity, it has been conjectured[3,7] that the physics underlying the $PuCoGa_5$ and $PuRhGa_5$ superconductors is analogous to that of the heavy-fermion superconductors $CeCoIn_5$ and $CeIrIn_5$. The recent discovery of an unconventional, d-wave order parameter in $PuCoGa_5$ supports this analogy, because, earlier a similar unconventional order parameter was discovered for $CeCoIn_5$.[24] The scaling of the transition temperature T_c with the c/a ratio is identical in the Pu and Ce-115 compounds.[6] Also, in both $PuCoGa_5$ and $CeCoIn_5$ no long-range antiferromagnetic order has been detected, but both materials are on the verge of antiferromagnetic ordering, as is, e.g. apparent from the long-range antiferromagnetism occurring in the isoelectronic materials $CeRhIn_5$ and possibly $PuIrGa_5$.[25,26] These observations, again, support the idea that antiferromagnetic spin fluctuations[3,6,10] constitute the essential bosons responsible for the unconventional pair formation.

Acknowledgements

We are grateful for valuable discussions and scientific exchange with J.L. Sarrao, J.D. Thompson, J.J. Joyce, T. Durakiewicz, N.J. Curro, F. Wastin, G.H. Lander, E. Coulineau, P. Boulet, O. Eriksson, L. Havela, N. Metoki and K. Kaneko. This work was supported financially through Grant No. 202/04/1055 of the Grant Agency of the Czech Republic.

References

1 J.L. Sarrao, L.A. Morales, J.D. Thompson, B.L. Scott, G.R. Stewart, F. Wastin, J. Rebizant, P. Boulet, E. Colineau and G.H. Lander, *Nature*, 2002, **420**, 297.
2 F. Wastin, P. Boulet, J. Rebizant, E. Colineau and G.H. Lander, *J. Phys. Condens. Matter*, 2003, **15**, S2279.
3 I. Opahle and P.M. Oppeneer, *Phys. Rev. Lett.*, 2003, **90**, 157001.
4 T. Maehira, T. Hotta, K. Ueda and A. Hasegawa, *Phys. Rev. Lett.*, 2003, **90**, 157001.
5 J.J. Joyce, J.M. Wills, T. Durakiewicz, M.T. Butterfield, E. Guziewicz, J.L. Sarrao, L.A. Morales, A.J. Arko and O. Eriksson, *Phys. Rev. Lett.*, 2003, **91**, 176401.
6 E.D. Bauer, J.D. Thompson, J.L. Sarrao, L.A. Morales. F. Wastin, J. Rebizant, J.C. Griveau, P. Javorsky, P. Boulet, E. Colineau, G.H. Lander and G.R. Stewart, *Phys. Rev. Lett.*, 2004, **93**, 147005.
7 I. Opahle, S. Elgazzar, K. Koepernik and P.M. Oppeneer, *Phys. Rev. B*, 2004, **70**, 104504.
8 Y. Bang, A.V. Balatsky, F. Wastin and J.D. Thompson, *Phys. Rev. B*, 2004, **70**, 104512.
9 A.B. Shick, V. Janiš and P.M. Oppeneer, *Phys. Rev. Lett.*, 2005, **94**, 016401.
10 N.J. Curro, T. Caldwell, E.D. Bauer, L.A. Morales, M.J. Graf, Y. Bang, A.V. Balatsky, J.D. Thompson and J.L. Sarrao, *Nature*, 2005, **434**, 622.
11 H. Sakai, Y. Tokunaga, T. Fujimoto, S. Kambe, R.E. Walstedt, H. Yasuoka, D. Aoki, Y. Homma, E. Yamamoto, A. Nakamura, Y. Shiokawa, K. Nakajima, Y. Arai, T.D. Matsuda, Y. Haga and Y. Ōnuki, *J. Phys. Soc. Jpn.*, 2005, **74**, 1710.
12 P. Söderlind, *Europhys. Lett.,* 2001, **55**, 525.
13 P. Söderlind, A. Landa and B. Sadigh, *Phys. Rev. B*, 2002, **66**, 205109.
14 J. Bouchet, B. Siberchicot, F. Jollet and A. Pasturel, *J. Phys. Condens. Matter*, 2000, **12**, 1723.
15 S.Y. Savrasov, G. Kotliar and E. Abrahams, *Nature*, 2001, **410**, 793.
16 A.B. Shick, V. Drchal and L. Havela, *Europhys. Lett.*, 2005, **69**, 588.
17 J.M. Wills, O. Eriksson, A. Delin, P.H. Andersson, J.J. Joyce, T. Durakiewicz, M.T. Butterfield, A.J. Arko, D.P. Moore and L. Morales, *J. Electron Spectrosc. Relat. Phenom.*, 2004, **135**, 163.
18 D.B. Ghosh, S.K. De, P.M. Oppeneer and M.S.S. Brooks, 2005, unpublished.
19 P. Söderlind, *Phys. Rev. B*, 2004, **70**, 094515.
20 A.B. Shick and W.E. Pickett, *Phys. Rev. Lett.*, 2001, **86**, 300.
21 J.F. Herbst, R.E. Watson and I. Lindgren, *Phys. Rev. B*, 1976, **14**, 3265.
22 D. van der Marel and G.A. Sawatsky, *Phys. Rev. B*, 1988, **37**, 10674.
23 K.T. Moore, M.A. Wall, A.J. Schwarz, B.W. Chung, D.K. Shuh, R.K. Schulze and J.G. Tobin, *Phys. Rev. Lett.*, 2003, **90**, 196404.
24 K. Izawa, H. Yamaguchi, Y. Matsuda, H. Shishido, R. Settai and Y. Ōnuki, *Phys. Rev. Lett.*, 2001, **87**, 057002.
25 W. Bao, G. Aeppli, J.W. Lynn, P.G. Pagliuso, J.L. Sarrao, M.F. Hundley, J.D. Thompson and Z. Fisk, *Phys. Rev. B*, 2002, **65**, 100505R.
26 F. Wastin, D. Bouexière, P. Boulet, E. Colineau, J.C. Griveau and J. Rebizant, 2003, unpublished. (Antiferromagnetic order in $PuIrGa_5$ was initially reported, but later investigations could not confirm its existence).

THE ELECTRONIC STRUCTURE OF THE Pu-BASED SUPERCONDUCTOR $PuCoGa_5$: LSDA AND LSDA+*U* INVESTIGATIONS

P.M. Oppeneer,[1] A.B. Shick,[2] I. Opahle,[3] S. Elgazzar[3] and V. Janiš[2]

[1]Department of Physics, Uppsala University, Box 530, S-751 21 Uppsala, Sweden
[2]Institute of Physics, Academy of Sciences of the Czech Republic, Na Slovance 2, CZ-182 21, Prague, Czech Republic
[3]Leibniz-Institute of Solid State and Materials Research, P.O. Box 270016, D-01171 Dresden, Germany

1 INTRODUCTION

The recent discovery of superconductivity[1] at a high transition temperature, T_c= 18.5 K, in the plutonium-based compound $PuCoGa_5$ has ignited a tremendous interest in this new material.[2-10] Prior to the discovery of superconductivity in $PuCoGa_5$ not a single Pu-based compound was known to exhibit superconductivity. Moreover, among the known *f*-electron (actinide or lanthanide) superconductors one rarely finds a material with a T_c over 2 K. From this perspective, $PuCoGa_5$ can be regarded to have an astonishingly high transition temperature. The nature of the superconducting pair formation has been a point of special focus. From its initial discovery it was speculated[1] that possibly an unconventional pairing mechanism might be responsible for the anomalously high T_c (i.e., a pairing mechanism *distinct* from the common phonon-mediated pairing). The magnetic properties of $PuCoGa_5$ consequently received particular attention. Susceptibility measurements provided no evidence for long-range magnetic order, but at elevated temperatures the susceptibility obeys a modified Curie-Weiss behaviour with an effective moment of 0.7 μ_B. Such moment corresponds approximately to that of a Pu^{3+} ion ($5f^5$ configuration). The temperature dependence of the resistivity exhibits a characteristic *S*-like shape, which indicates Bloch electron scattering from spin-fluctuations.[1,3,8] Furthermore, the linear-temperature specific heat coefficient, with $\gamma \sim 77$ mJ/mol K^2, appears to be significantly enhanced, which could be due to the presence of strong dynamic spin fluctuations.

Recently an affirmative answer to the question whether the superconductivity in $PuCoGa_5$ is unconventional could be given. Experimental studies, employing nuclear magnetic resonance (NMR) and nuclear quadrupolar resonance (NQR) showed that indeed $PuCoGa_5$ is an unconventional, spin-singlet, *d*-wave type superconductor.[10] Very recently, similar NMR/NQR studies proved that the isoelectronic compound $PuRhGa_5$ is also an unconventional *d*-wave superconductor.[11] Such type of unconventional superconducting gap symmetry could possibly correspond to pairing mediated by antiferromagnetic spin fluctuations.[10] Thus, although no long-range antiferromagnetic order could be detected in $PuCoGa_5$ or $PuRhGa_5$, short-range antiferromagnetic fluctuations are not only expected to be present but to constitute the essential bosons responsible for the pairing.

While these latest findings definitely categorise $PuCoGa_5$ as a most exceptional and intriguing material, many important questions are still open. In particular, the electronic structure of $PuCoGa_5$ and the role played by the Pu $5f$ electrons are not understood. *Ab initio* calculations are an ideal tool to investigate the underlying electronic structure and unravel its relationship to the superconductivity. It is well-known, however, that in the actinide series precisely at Pu, a transition in the behaviour of the $5f$ electrons occurs. The $5f$ electrons of the early actinides (up to Pu) are relatively delocalized and bonding. The $5f$ electrons in the heavy actinides, starting from americium, are inert and retracted from the bonding. The cross-over from delocalised to localised behaviour occurs for Pu, particularly with the electronic structure of fcc δ-Pu being a topic of ongoing debate.[12-17] The delocalised $5f$ electrons of the early actinides are quite well described by density functional calculations employing the local spin-density approximation (LSDA). So far not much is known about the electronic structure of Am and the heavier actinides, but first investigations indicate that the electronic structure cannot be sufficiently well described with the LSDA, delocalised $5f$ electron approach.[18] The $5f$ electrons in Am are modestly localised and intra-atomic $5f$ Coulomb correlation effects are to be taken into account. For δ-Pu various electronic structure models have been put forward: the LSDA model, in which the $5f$'s are treated as delocalised,[12,13] the LSDA+U or LSDA+DMFT model,[14-16] in which the $5f$ electrons are modestly localised through additional intra-atomic Coulomb correlations, and the mixed level model (MLM), in which some of the $5f$ electrons are fully localised and some other $5f$'s are delocalised.[17] A similar issue as for δ-Pu occurs for $PuCoGa_5$. The appropriate description of the $5f$'s in $PuCoGa_5$ can ultimately only be established through comparison of computed and experimental electronic data.

In the following we shall discuss electronic structure results obtained for $PuCoGa_5$ focusing mainly on the LSDA and LSDA+U approaches. We shall discuss to what extent the respective electronic structure approaches can explain the available experimental data and consider the implications for the unconventional superconductivity.

2 ELECTRONIC STRUCTURE METHODOLOGIES

Employing the LSDA approach, the electronic structure of $PuCoGa_5$ has been calculated in several papers.[3,4,7] Also, the LSDA has been used by Opahle *et al.*[7] to compute the electronic structures of the related actinide 115 compounds, $UCoGa_5$, $NpCoGa_5$, $PuRhGa_5$ and $PuIrGa_5$. The effect of Coulomb correlations in the $5f$ shell on the electronic structure were recently investigated through LSDA+U calculations by Shick *et al.*[9] The supplementary Coulomb interaction of the $5f$ electrons is modelled through the Coulomb U and exchange J parameters.[9] The values of U and J which are reasonable for actinide materials are known.[9] Recently, a variant of the LSDA, the GGA+OP (orbital polarisation) was applied to $PuCoGa_5$ by Söderlind.[19] Investigations on the basis of the MLM model were recently reported by Joyce e*t al.*[5]

The above mentioned studies can be regarded as state-of-the-art investigations, using electronic structure models valid for specific material classes. Each of these electronic structure models for $PuCoGa_5$ has to be tested against existing experimental data. Unfortunately, for $PuCoGa_5$ not many experimental data have been reported so far. Apart from the reported lattice parameters, susceptibility, resistivity, and specific heat,[1,6] crucial information on the electronic structure of $PuCoGa_5$ was recently provided by photoemission (PE) measurements.[5] The PE showed that Pu $5f$ states are present at the Fermi energy (E_F) which strongly suggests that the f electrons participate in the pairing. The PE revealed also second $5f$-related response which needs yet to be explained. Other recent data are the behaviour of the superconducting transition temperature under pressure and the peculiar linear scaling of T_c with the crystallographic c/a ratio.[6]

3 RESULTS

3.1 Results of LSDA Calculations

The recent LSDA calculations for $PuCoGa_5$ and for the related actinide 115 compounds provided detailed information on the electronic structures as well as on the trends within the series.[3,7] The obtained results can be summarised as follows: the lowest total energy of the Pu-115 compounds (see Figure 1) is computed to be an antiferromagnetically ordered state, which appears to be in agreement with experiment for $PuIrGa_5$. The superconductivity in $PuCoGa_5$ and $PuRhGa_5$ occurs thus in materials on the verge of magnetic ordering. The density of states (DOS) at E_F is largely dominated by the Pu 5*f* states. Due to a peculiarity of the band filling in these materials, the Co 3*d* states (and Rh 4*d*, Ir 5*d* states, respectively) are practically filled and contribute only a small portion to the DOS at E_F. The adjacent Pu layers are magnetically weakly coupled. On account of the tetragonal structure, with its large c/a ratio (~1.6), the electronic structure becomes quasi-two-dimensional. This feature is particularly reflected in the highly anisotropic Fermi surface, which consists of three two-dimensional "tubes" as well as two small, three-dimensional hole pockets.[3,4] The LSDA calculations, furthermore, predict lattice parameters for the Pu-115 compounds in reasonably good agreement with experiment.[3,7] The LSDA calculations provide, in addition, an accurate description of the electronic structures of $UCoGa_5$ and $NpCoGa_5$: $UCoGa_5$ is calculated to be a paramagnet with a very small DOS at E_F, in agreement with experiment. Also, the Fermi surface of $UCoGa_5$ agrees very well with data inferred from de Haas-van Alphen measurements. $NpCoGa_5$, furthermore, is calculated to be an antiferromagnet in agreement with experiment. Some of the properties of $NpCoGa_5$, however, as, e.g. the spin and orbital moments appear to demand closer inspection. Altogether, the LSDA approach explains very well various properties of the actinide-115 compounds.

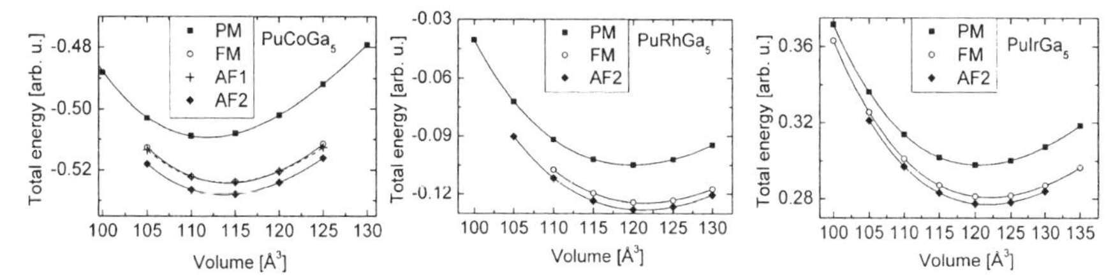

Figure 1 *Calculated LSDA total energies for different magnetic phases of $PuCoGa_5$, $PuRhGa_5$ and $PuIrGa_5$.[7] For all three iso-electronic Pu-compounds an antiferromagnetically ordered configuration has the lowest total energy*

The recent photoemission experiment[5] puts restrictions on the electronic structure models for $PuCoGa_5$. The PE spectrum revealed a 5*f* response at E_F, important for models of the superconductivity, but also a second 5*f*-related response at a binding energy of 1.2 eV. The second response might be due to partially localised 5*f* electrons. The PE spectrum obviously provides a critical test for electronic structure models of $PuCoGa_5$. The LSDA calculations show that there is a 5*f*-related peak in the DOS at about 0.7 eV below E_F (see Figure 2). This energy position appears too small compared to the PE result. This indicates that the electronic structure of $PuCoGa_5$ requires an account of electron correlation effects beyond those already incorporated in the LSDA. We remark, however, that a recent investigation employing the GGA+OP approach, in which the 5*f*'s are delocalized, fairly well reproduced the PE spectrum.[19] It was also well reproduced by the mixed level model, in which four Pu 5*f* electrons are, conversely, treated as *localized* and their energy position is adjusted to the experimental PE position.[5]

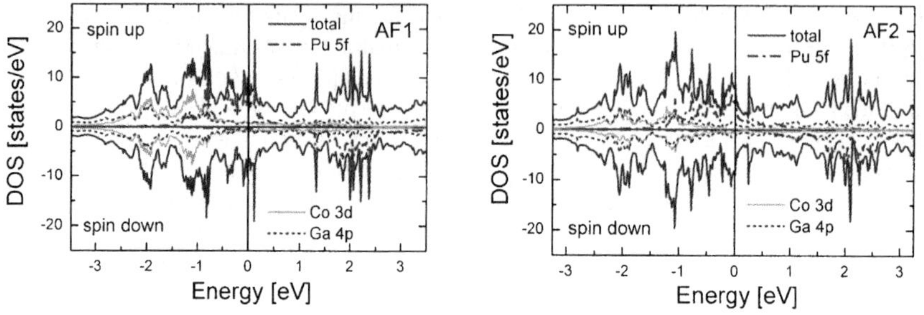

Figure 2 *Calculated[7] LSDA total and partial DOS for $PuCoGa_5$ in the AF1 (Q=(0 0 ½), left) and AF2 (Q=(½ ½ 0), right) antiferromagnetic configurations (with Q the antiferromagnetic wave-vector)*

3.2 LSDA+*U* Results for $PuCoGa_5$

A third possible route to explain the PE spectrum would be through LSDA+*U* calculations. In this approach the intra-atomic Coulomb correlation *U* causes a rearrangement of the 5*f* levels, leading to a moderate increase of the binding energy of the occupied 5*f* states. The LSDA+*U* scheme in the so-called atomic limit (for details, see Ref. 20) was applied recently.[9] The *U* value was varied in the accepted range of 2–5 eV,[21] while for the exchange constant *J* the value *J*= 0.7 eV was used.[22] The inclusion of the Coulomb *U* and exchange *J* leads to an interesting reconstruction of the 5*f* manifold. First, the Pu 5*f* states are shifted down to a binding energy below that obtained from the LSDA. Their final energy position depends slightly on the *U* value used, but for reasonable *U* values of about 3 eV the main weight of the 5*f* DOS occurs at binding energies of 1–2 eV, in good agreement with experiment.[5] A further interesting modification occurs in the Pu spin and orbital magnetic moments. The Coulomb *U* strongly reduces the Pu spin moment M_s from 4.75 μ_B (in the LSDA) to 2.56 μ_B in the LSDA+*U* (*U*= 3 eV), while the orbital moment M_l is slightly decreased from -1.87 μ_B to -1.60 μ_B. Normally, the LSDA+*U* approach does not drastically modify the spin moment, but tends to enlarge substantially the orbital moment. The exceptional result for $PuCoGa_5$ can be traced back to an enhancement of the effective spin-orbit coupling due to modifications in the spin and orbital dependent contributions to the LSDA+*U* effective potential (including spin off-diagonal terms). The enhanced effective spin-orbit coupling leads to a rearrangement of the occupation of the Pu j_z-states ($j_z= m_z+s_z$). While the LSDA fills the *f* states from $j_z= -^5/_2$ up to $j_z= {}^7/_2$, the LSDA+*U* occupies the states from $j_z= -^5/_2$ to $^5/_2$ (see Figure 3). The main difference appears in the $j_z= {}^7/_2$ subset (i.e., $m_z= 3$, $s_z= +½$), which is excluded in the LSDA+*U* (in the LSDA+*U* a spin flip to $s_z= -½$ is forced and j_z therefore becomes equal to $^5/_2$). As a result, the spin moment becomes considerably reduced and the ratio $C_2 = M_l/(M_l+M_s)$ is much enhanced, yet still smaller than what is expected for a trivalent Pu ion. This extraordinary LSDA+*U* result can be interpreted as a strengthening of the effective spin-orbit coupling leading to the formation of a Pu ground state formed solely out of the $j= {}^5/_2$ manifold. While neither the LSDA nor LSDA+*U* are based on any kind of atomic coupling scheme (e.g., *LS* or *jj* coupling), the LSDA+*U* suggest (more than the LSDA) a *jj*-atomic-like coupled, Pu ground-state configuration in $PuCoGa_5$. The importance of a *jj*-coupling scheme for the electronic structure of elemental Pu as well as for $PuCoGa_5$ has recently been suggested.[4,23] Aspects of the angular momentum coupling scheme are particularly important for Pu, because Pu is close to relativistic half-filling and the competition of exchange splitting and spin-orbit interaction will determine the eventual magnetic state.

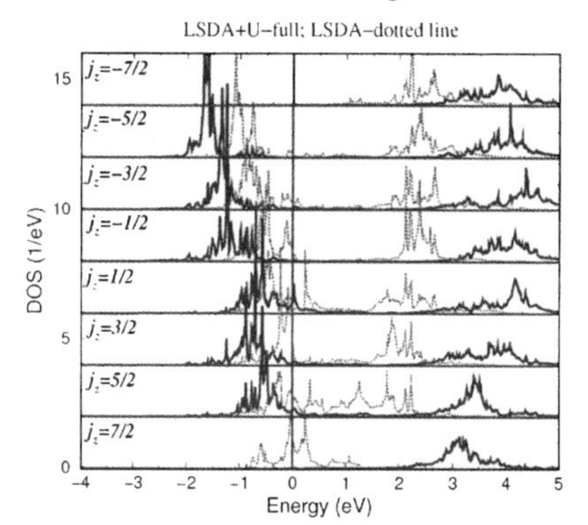

Figure 3 *The j_z resolved Pu 5f DOS as obtained with the LSDA (dotted lines) and LSDA+U (U= 3 eV) (full lines) schemes*

4 CONCLUSIONS

The LSDA calculations[3,4,7] explain various electronic structure properties of $PuCoGa_5$. The LSDA+*U* method,[9] however, improves the energetic position of the 5*f* states as compared to the PE spectrum.[5] In accordance with the PE spectrum, a picture of the Pu 5*f* states emerges, in which the 5*f*'s appear to be fairly hybridized and yet retain some atomiclike features. To explore further this electronic structure picture, more experiments would be desirable, for example, de Haas-van Alphen measurements. Such experiments would also provide valuable information on the Fermi surface, which is predicted by the LSDA calculations to be two–dimensional.[3,7] Concerning the origin of the superconductivity, it has been conjectured[3,7] that the physics underlying the $PuCoGa_5$ and $PuRhGa_5$ superconductors is analogous to that of the heavy-fermion superconductors $CeCoIn_5$ and $CeIrIn_5$. The recent discovery of an unconventional, d-wave order parameter in $PuCoGa_5$ supports this analogy, because, earlier a similar unconventional order parameter was discovered for $CeCoIn_5$.[24] The scaling of the transition temperature T_c with the c/a ratio is identical in the Pu and Ce-115 compounds.[6] Also, in both $PuCoGa_5$ and $CeCoIn_5$ no long-range antiferromagnetic order has been detected, but both materials are on the verge of antiferromagnetic ordering, as is, e.g. apparent from the long-range antiferromagnetism occurring in the isoelectronic materials $CeRhIn_5$ and possibly $PuIrGa_5$.[25,26] These observations, again, support the idea that antiferromagnetic spin fluctuations[3,6,10] constitute the essential bosons responsible for the unconventional pair formation.

Acknowledgements

We are grateful for valuable discussions and scientific exchange with J.L. Sarrao, J.D. Thompson, J.J. Joyce, T. Durakiewicz, N.J. Curro, F. Wastin, G.H. Lander, E. Coulineau, P. Boulet, O. Eriksson, L. Havela, N. Metoki and K. Kaneko. This work was supported financially through Grant No. 202/04/1055 of the Grant Agency of the Czech Republic.

References

1 J.L. Sarrao, L.A. Morales, J.D. Thompson, B.L. Scott, G.R. Stewart, F. Wastin, J. Rebizant, P. Boulet, E. Colineau and G.H. Lander, *Nature*, 2002, **420**, 297.
2 F. Wastin, P. Boulet, J. Rebizant, E. Colineau and G.H. Lander, *J. Phys. Condens. Matter*, 2003, **15**, S2279.
3 I. Opahle and P.M. Oppeneer, *Phys. Rev. Lett.*, 2003, **90**, 157001.
4 T. Maehira, T. Hotta, K. Ueda and A. Hasegawa, *Phys. Rev. Lett.*, 2003, **90**, 157001.
5 J.J. Joyce, J.M. Wills, T. Durakiewicz, M.T. Butterfield, E. Guziewicz, J.L. Sarrao, L.A. Morales, A.J. Arko and O. Eriksson, *Phys. Rev. Lett.*, 2003, **91**, 176401.
6 E.D. Bauer, J.D. Thompson, J.L. Sarrao, L.A. Morales. F. Wastin, J. Rebizant, J.C. Griveau, P. Javorsky, P. Boulet, E. Colineau, G.H. Lander and G.R. Stewart, *Phys. Rev. Lett.*, 2004, **93**, 147005.
7 I. Opahle, S. Elgazzar, K. Koepernik and P.M. Oppeneer, *Phys. Rev. B*, 2004, **70**, 104504.
8 Y. Bang, A.V. Balatsky, F. Wastin and J.D. Thompson, *Phys. Rev. B*, 2004, **70**, 104512.
9 A.B. Shick, V. Janiš and P.M. Oppeneer, *Phys. Rev. Lett.*, 2005, **94**, 016401.
10 N.J. Curro, T. Caldwell, E.D. Bauer, L.A. Morales, M.J. Graf, Y. Bang, A.V. Balatsky, J.D. Thompson and J.L. Sarrao, *Nature*, 2005, **434**, 622.
11 H. Sakai, Y. Tokunaga, T. Fujimoto, S. Kambe, R.E. Walstedt, H. Yasuoka, D. Aoki, Y. Homma, E. Yamamoto, A. Nakamura, Y. Shiokawa, K. Nakajima, Y. Arai, T.D. Matsuda, Y. Haga and Y. Ōnuki, *J. Phys. Soc. Jpn.*, 2005, **74**, 1710.
12 P. Söderlind, *Europhys. Lett.,* 2001, **55**, 525.
13 P. Söderlind, A. Landa and B. Sadigh, *Phys. Rev. B*, 2002, **66**, 205109.
14 J. Bouchet, B. Siberchicot, F. Jollet and A. Pasturel, *J. Phys. Condens. Matter*, 2000, **12**, 1723.
15 S.Y. Savrasov, G. Kotliar and E. Abrahams, *Nature*, 2001, **410**, 793.
16 A.B. Shick, V. Drchal and L. Havela, *Europhys. Lett.*, 2005, **69**, 588.
17 J.M. Wills, O. Eriksson, A. Delin, P.H. Andersson, J.J. Joyce, T. Durakiewicz, M.T. Butterfield, A.J. Arko, D.P. Moore and L. Morales, *J. Electron Spectrosc. Relat. Phenom.*, 2004, **135**, 163.
18 D.B. Ghosh, S.K. De, P.M. Oppeneer and M.S.S. Brooks, 2005, unpublished.
19 P. Söderlind, *Phys. Rev. B*, 2004, **70**, 094515.
20 A.B. Shick and W.E. Pickett, *Phys. Rev. Lett.*, 2001, **86**, 300.
21 J.F. Herbst, R.E. Watson and I. Lindgren, *Phys. Rev. B*, 1976, **14**, 3265.
22 D. van der Marel and G.A. Sawatsky, *Phys. Rev. B*, 1988, **37**, 10674.
23 K.T. Moore, M.A. Wall, A.J. Schwarz, B.W. Chung, D.K. Shuh, R.K. Schulze and J.G. Tobin, *Phys. Rev. Lett.*, 2003, **90**, 196404.
24 K. Izawa, H. Yamaguchi, Y. Matsuda, H. Shishido, R. Settai and Y. Ōnuki, *Phys. Rev. Lett.*, 2001, **87**, 057002.
25 W. Bao, G. Aeppli, J.W. Lynn, P.G. Pagliuso, J.L. Sarrao, M.F. Hundley, J.D. Thompson and Z. Fisk, *Phys. Rev. B*, 2002, **65**, 100505R.
26 F. Wastin, D. Bouexière, P. Boulet, E. Colineau, J.C. Griveau and J. Rebizant, 2003, unpublished. (Antiferromagnetic order in $PuIrGa_5$ was initially reported, but later investigations could not confirm its existence).

GROUND STATE AND SOLUTION BEHAVIOUR OF ACTINIDE IONS INVESTIGATED BY NUCLEAR MAGNETIC RESONANCE METHODS

J.F. Desreux, G. Vaast, A. Joassin and G. Gridelet

Coordination and Radiochemistry, University of Liège, Sart Tilman B16, B-4000 Liège, Belgium

1 INTRODUCTION

Nuclear magnetic resonance has long been established as a standard analytical method in chemistry but it is rarely used in actinide studies. Despite considerable improvements, the sensitivity of NMR remains low compared to most radioanalytical techniques and there are very few NMR instruments available for studying highly radioactive solutions. The present paper reports an unusual NMR technique called nuclear magnetic relaxation dispersion (NMRD) that is currently used in our laboratory for analyzing the magnetic properties and the dynamic behaviour of actinide ions in aqueous or organic solutions.[1] In contrast with NMR spectroscopy, the NMRD signal arises from the nuclei of the abundant solvent molecules and not from the dissolved substances. The relaxation properties of the solvent molecules are profoundly modified if paramagnetic metal ions are added to solutions. The solvent nuclei are then forced to relax faster than in diamagnetic conditions.[2] The modifications in the relaxation time T_1 of the solvent nuclei depend on the magnetic properties of the paramagnetic ions and thus yield information on these ions. In NMRD, the longitudinal relaxation rate of the solvent molecules ($1/T_1$, known as relaxivity) is measured *vs* the magnetic field, usually between 0.0003 and 1.9 T (corresponding to frequencies between 0.01 and 80 MHz for the ^{1}H nucleus, where $1/T_1$ is expressed in s^{-1} $mmol^{-1}$ of the metal ion).

The Solomon-Bloembergen-Morgan equations account for the dependence of relaxivity upon frequency for a 1 mmol water solution of a paramagnetic metal ion:

$$\frac{1}{T_1} = \frac{q_{solvent} \times 10^{-3}}{55.5} \frac{1}{(T_{1M} + \tau_m)} \tag{1}$$

$$\frac{1}{T_{1M}} = \frac{2}{15}\left(\frac{\mu_0}{4\pi}\right)^2 \frac{\gamma_H^2 \mu_B^2 g_e^2 S(S+1)}{r^6}\left[\frac{3\tau_c}{1+\omega_I^2\tau_c^2} + \frac{7\tau_c}{1+\omega_S^2\tau_c^2}\right] \tag{2}$$

$$\frac{1}{\tau_c} = \frac{1}{\tau_r} + \frac{1}{T_{1e}} + \frac{1}{\tau_m} \tag{3}$$

where $q_{solvent}$ is the number of solvent molecules directly coordinated to the metal ion and where τ_r and τ_m are respectively the rotational correlation time of the hydrated complex and the water exchange time. The other factors have their usual meanings. Relaxivity depends essentially on the smallest of the three factors in Equation (3). A simple S-shaped NMRD curve is obtained if a complex is tumbling rapidly (τ_r smallest) while slowly rotating species are characterized by high relaxivities at all frequencies with a maximum at 20-50 MHz because of a dependence on the τ_m and T_{1e} factors.[2]

Equation (1) holds for the Gd^{3+} ion which has a nearly pure ^8S ground state with a spherical distribution of the unpaired electronic spins and thus a poor coupling with the matrix. The electronic relaxation time is thus relatively long ($T_{1e} \approx 2x10^{-10}$ s) and the electrons maintain their spin state for a time sufficiently long to be coupled with solvent nuclei that are then forced to relax much faster than they normally do. Equation (1) also holds for all the other lanthanide ions provided the S and g_e terms are replaced by J and g_J respectively.[3] These ions have a non-spherical distribution of their unpaired electronic spins and their T_{1e} values are very small ($\approx 10^{-13}$ s). Gd^{3+} thus stands out against the other lanthanides because of its long T_{1e} and thus its strong effect on the proton relaxation time. Gd^{3+} complexes are used extensively in magnetic resonance imaging for improving the contrast of images.

2 RESULTS AND DISCUSSION

The principles presented above are applied here to the actinide ions in water. The Gd^{3+} and Cm^{3+} ions have the same f^7 configuration. Both ions give S-shaped NMRD curves characteristic of rapidly tumbling complexes but their relaxivities are quite different as illustrated in Figure 1. The much lower relaxivity of the Cm^{3+} solution can be interpreted with reference to electron paramagnetic resonance studies[4] at helium temperature that show that the high spin-orbit coupling constant of this ion brings about an admixture of higher energy levels to the ground state. Cm^{3+} is only 79-87 % pure $^8S_{7/2}$ and the resulting better coupling with the matrix leads to a shorter T_{1e} value and thus to a smaller relaxivity. A quantitative interpretation of the NMRD curves of Cm^{3+} at different temperatures yields a T_{1e} value of $\approx 10^{-11}$ s at 298 K and 0.47 T. The other parameters deduced from Equations (1) to (3) are similar to those obtained by the same technique for Gd^{3+} but a contact contribution could have to be taken into account.

What is perhaps most striking about the data in Figure 1 is the sensitivity of NMRD to small departures from a perfectly spherical distribution of the unpaired electronic spins of a metal ion. This technique also proved to be fruitful in an investigation of other actinides although very low relaxivities were expected for all non-S state ions as it is the case for the lanthanides. Low relaxivities with little magnetic field dependence are indeed observed for many actinide ions in various oxidation states. The NMRD curve of an aqueous solution of NpO_2^{2+} is included in Figure 1 as an example. However, the NpO_2^+ and PuO_2^{2+} ions exhibit unusually high relaxivities for non-S state ions, as shown in Figure 1, and as reported more than 25 years ago by Glebov in a single frequency study[5] (the NMRD curve of PuO_2^{2+} is not shown, the data are very similar to those of NpO_2^+). A $^3H_4$ ground state has been suggested for these two $5f^2$ ions on the basis of recent theoretical studies and electronic spectral analyses.[6,7] The relatively high relaxivities of NpO_2^+ and PuO_2^{2+} aqueous solutions are not in keeping with a pure $^3H_4$ state and admixtures with higher states must be taken into account. Glebov suggested that the $^3\Sigma_0$ level is the ground state at room temperature because the L_z quantum number is zero for this level. There would then be no interactions between the electronic spins and the crystal field through the spin-orbit coupling and thus little coupling with the matrix and long T_{1e} values.

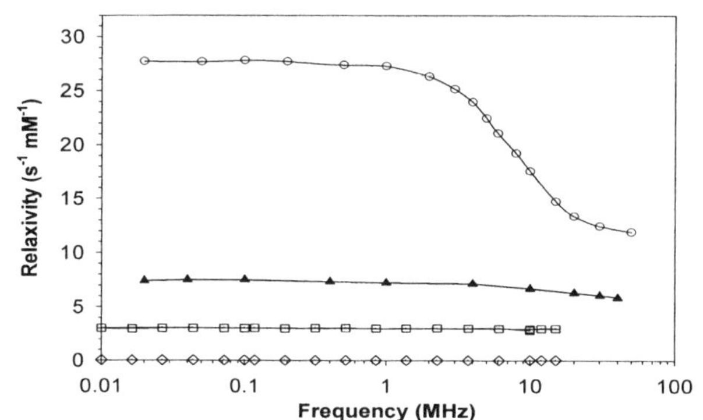

Figure 1 *Nuclear magnetic relaxation curve of Gd^{3+} (⊙), Cm^{3+} (▲), NpO_2^+ (⊡) and NpO_2^{2+} (◈). Relaxivities are corrected for the relaxivity of water, 1M $HClO_4$ solutions (Error bars are smaller than the symbol size, the relaxivity of NpO_2^{2+} is 1.8×10^{-2} s^{-1} $mmol^{-1}$)*

The exact nature of the energy levels involved in the ground state of the hydrated $5f^2$ ions is still an open question[6,7] but NMRD should shed light on the electronic structures of actinide ions. The dynamic properties of NpO_2^+ and PuO_2^{2+} have been deduced from Equations (1) to (3) despite the uncertainties on their ground states ($\tau_r \approx 85$ ps, $T_{1e} \approx 7 \times 10^{-11}$ s at 300 K and 0.47 T).

NMRD provides a new entry into the dynamics and the magnetic properties of the actinides but it also allows studies of processes in which a solvation change is involved. For instance, complexation, hydrolysis, and changes of oxidation states can be investigated by NMRD. Also noteworthy is the fact that the technique is not limited to aqueous solutions. Many organic solutions lend themselves to NMRD studies provided there is a direct interaction between ions and solvent molecules. Acetonitrile is particularly suitable for that purpose.[1]

Acknowledgements

The authors gratefully thank the FNRS of Belgium, the ACTINET NoE and the EUROPART IP for financial support.

References

1 J. F. Desreux, *Adv. Inorg. Chem.*, 2005, **57**, 381.
2 L. Banci, I. Bertini and C. Luchinat, 'Nuclear and electron relaxation', VCH, Weinheim, 1991.
3 I. Bertini, F. Capozzi, C. Luchinat, G. Nicastro and Z. C. Xia, *J. Phys. Chem.*, 1993, **97**, 6351.
4 N. M. Edelstein and W. Easley, *J.Chem. Phys.*, 1968, **48**, 2110.
5 V. A. Glebov, *Radiokhimiya,* 1979, **21**, 801.
6 S. Matsika and R. M. Pitzer, *J. Phys. Chem. A,* 2000, **104**, 4064.
7 L. Maron, T. Leininger, B. Schimmelpfennig, V. Vallet, J. L. Heully, C. V. Teichteil, O. Gropen and U. Wahlgren, *Chem. Phys.*, 1999, **244**, 195.

POLARIZATION EFFECTS IN EXAFS SPECTRA AT THE URANIUM L_I AND L_3 EDGE – A COMPARISON BETWEEN THEORY AND EXPERIMENT

C. Hennig

Forschungszentrum Rossendorf, The Rossendorf Beamline at the European Synchrotron Radiation Facility, P.O. Box 220, F-38043 Grenoble, France

1 INTRODUCTION

In anisotropic materials the amplitude of the EXAFS spectrum depends on the angle θ between the polarization vector $\vec{\varepsilon}$ and the vector $\vec{r}$ connecting the absorbing and backscattering atoms. Backscattering contributions of atoms along the polarization direction are accentuated, whereas the scattering contributions of atoms outside of the polarization direction are attenuated. The real or crystallographic coordination number N_{cryst} appears as an effective coordination number N_{eff} in the polarization dependent EXAFS measurements.[1] In this study, the polarization dependency at the L_I and L_3 edges of uranium was investigated using a single crystal of $Ca[UO_2PO_4]_2{\cdot}6H_2O$ with linearly polarized X-ray radiation.

1.1 U L_I Edge

For a single-scattering process and in plane–wave approximation, the angular dependence of the EXAFS amplitude is contained in N_{eff} and can be written for both L_I and K edges as:

$$N_{eff}(\theta) = \sum_{n=1}^{N} 3\left|\vec{\varepsilon}\cdot\vec{r}\right|^2 \quad (1)$$

In order to relate $\vec{\varepsilon}$ and $\vec{r}$ to crystallographic axes of the sample, θ is divided into two angles, where α is the angle between $\vec{\varepsilon}$ and a selected plane (defined e.g. by two crystallographic axes) and β is the angle between $\vec{r}$ and the normal vector $\vec{n}$ of this plane:[2]

$$\left|\vec{\varepsilon}\cdot\vec{r}\right|^2 = \cos^2\theta = \cos^2\beta\sin^2\alpha + (\sin^2\beta\cos^2\alpha)/2 \quad (2)$$

The effective coordination number N_{eff} is then related to the real crystallographic coordination N_{cryst} by:

$$N_{eff}(\theta) = N_{cryst}3\left(\cos^2\beta\sin^2\alpha + (\sin^2\beta\cos^2\alpha)/2\right) \quad (3)$$

A special case is defined by $\left|\vec{\varepsilon}\cdot\vec{r}\right|^2 = 1/3$, the "magic angle". In this situation the value N_{eff} is independent of β for an angle of $\alpha = 35.3$ °, and inversely, N_{eff} is independent of α at $\beta = 54.7$ °.[3]

1.2 U L_3 Edge

The polarization dependence of the EXAFS signal is more complicated at the L_2 and L_3 edges were the photoelectron is excited from a $2p$ core into s ($l = 0$) and d ($l = 2$) final states. The polarization dependent EXAFS expression of L_2 and L_3 edges contains s, d, and coupled s-d states.[4] Defining c as the ratio between the radial dipole matrix elements M_{01} and M_{21}, which couple the initial $2p$ wave function with the $l = 0$ and $l = 2$ final states, N_{eff} can be expressed as a sum of three effective partial coordination numbers:

$$N_{eff}^{d}(\theta) = 0.5\left[\frac{2}{2+c^2}\right]\sum_{i=1}^{N}(1+3|\vec{\varepsilon}\cdot\vec{r}|^2) \quad (4)$$

$$N_{eff}^{s} = 0.5\left[\frac{c^2}{2+c^2}\right] \quad (5)$$

$$N_{eff}^{sd}(\theta) = \left[\frac{2c}{2+c^2}\right]\sum_{i=1}^{N}(1-3|\vec{\varepsilon}\cdot\vec{r}|^2) \quad (6)$$

The value c is largely independent of k and has been approximated to 0.2 for elements with $Z > 20$.[5] Using $c = 0.2$ in Equations (4) to (6) the expression can be approximated by:[2]

$$N_{eff}(\theta) = \sum_{i=1}^{N}(0.7+0.9|\vec{\varepsilon}\cdot\vec{r}|^2) \quad (7)$$

The polarization dependency can be analyzed experimentally if the $[UO_2]^{2+}$ units have a single orientation in the crystal structure.

2 RESULTS AND DISCUSSION

In the crystal structure of $Ca[UO_2PO_4]_2{\cdot}6H_2O$ each uranium atom is coordinated by two opposing axial oxygen atoms (O_{ax}, R = 1.79 Å) and four equatorial atoms (O_{eq}, R = 2.28 Å) in a square planar arrangement (Figure 1). The four O_{eq} atoms are symmetry-equivalent in relation to the four-fold rotation axis aligned parallel to the $[UO_2]^{2+}$ unit, the axial oxygen atoms are not symmetry-equivalent.

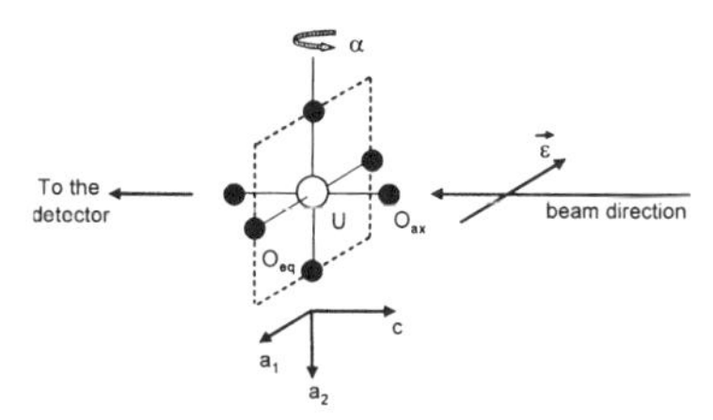

Figure 1 *Orientation of the U-O_{ax} and U-O_{eq} bonds with respect to the polarization vector $\vec{\varepsilon}$ shown for $\alpha = 0°$*

The U L_1 and L_3 edge EXAFS spectra are shown in Figure 2. The scattering contribution of the O_{ax} and O_{eq} shells were separated from the spectra by Fourier filtering. For the first shell fits the spectra collected at the magic angle (35 °) were chosen because there $N_{eff} = N_{cryst}$ is valid. In this fit the coordination numbers were fixed to the crystallographic values $N_{Oax} = 2$ and $N_{Oeq} = 4$. Free fit parameters were the distances R(U-O_{ax}) and R(U-O_{eq}), the associated Debye-Waller factors, σ^2, and ΔE. For all other angles α, N_{eff} was chosen as free fit parameter, while σ^2 was fixed at the values determined for the magic angle. The Debye-Waller factor was fixed for the EXAFS fit procedure in order to make the relation between N_{eff} and α comparable and to avoid the correlation problems between N and σ^2.

The fit results for the coordination numbers N_{eff} are summarized in Figure 3 as function of $cos^2\alpha$. A comparison between the N_{eff} values at the U L_1 edge and the theoretical function according to Equations (1) and (3) show a general agreement.

However, a systematic deviation is observed especially for N_{eff} of O_{eq}. This deviation may result from anisotropic σ^2, which were not considered here.

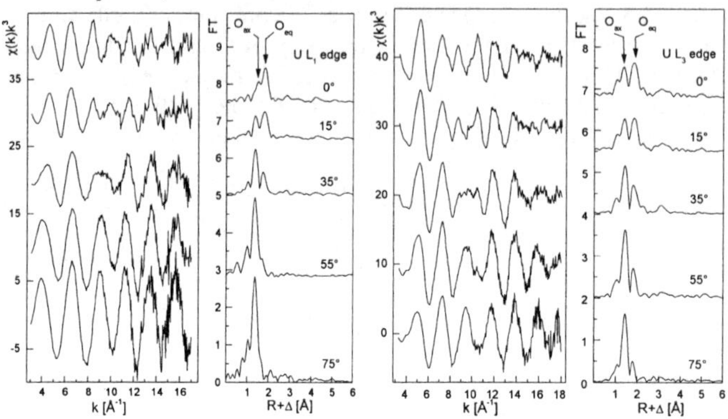

Figure 2 *U L_1-edge (left) and U L_3-edge (right) EXAFS spectra and their Fourier transforms at different angles α.*

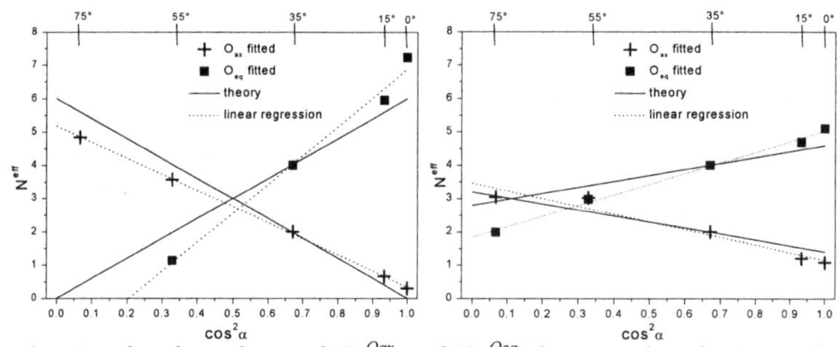

Figure 3 *Angular dependence of N_{eff}^{Oax} and N_{eff}^{Oeq} determined at the U L_1-edge (left) and at the U L_3-edge (right) in comparison with the theory.*

Correlating results at the U L_3 edge are shown in Figure 2 (right). The theoretical function was calculated according to Equation (7). The smaller polarization dependency at the L_3 edge over that at the L_1 edge is clearly visible. The dominant contribution at the L_3 edge, N_{eff}^{d}, running with $1+3cos^2\theta$, has a similar angle dependence as N_{eff} at the L_1 edge but its influence is attenuated by the N_{eff}^{sd} cross term running with $1\text{-}3cos^2\theta$ in the opposite direction.

References

1 E.A. Hudson, P.G. Allen, L.J. Terminello, M.A. Denecke, T. Reich, *Phys. Rev. B*, 1996, **54**, 156.
2 P.H. Citrin, *Phys. Rev. B*, 1985, **15**, 700.
3 R.F. Pettifer, C. Brouder, M. Benfatto, C.R. Natoli, C. Hermes and M.F. Ruiz López, *Phys. Rev. B.*, 1990, **42**, 37.
4 J. Stöhr and R. Jaeger, *Phys. Rev. B,* 1983, **27**, 5146.
5 B.K. Teo and P.A., Lee, *J. Am. Chem. Soc.*, 1979, **101**, 2815.

A STUDY OF THE COMPETITION BETWEEN DELOCALIZATION AND SPIN-ORBIT SPLITTING IN THE ACTINIDE 5*f* STATES

J.G. Tobin,[1] K.T. Moore,[1] B.W. Chung,[1] M.A. Wall,[1] A.J. Schwartz,[1]
G. van der Laan[2] and A.L. Kutepov[3]

[1]Lawrence Livermore National Laboratory, Livermore, California 94550, USA.
[2]Synchrotron Radiation Source, Daresbury Laboratory, Warrington, WA4 4AD, UK
[3]Russian Federation Nuclear Center, Institute of Technical Physics (VNIITF), Snezhinsk, Chelabinsk Region, Russia

1 INTRODUCTION

Synchrotron-radiation-based X-ray absorption, electron energy-loss spectroscopy in a transmission electron microscope, multi-electronic atomic spectral simulations and first principles calculations (Generalized Gradient Approximation in the Local Density Approximation, GGA/LDA) have been used to investigate the electronic structure of the light actinides: α-Th, α-U and α-Pu. It will be shown that the spin-orbit interaction can be used as a measure of the degree of localization of valence electrons in a material. The spin-orbit interaction in the light actinide metals, α-Th, α-U and α-Pu, has been determined using the branching ratio of the white line peaks of the $N_{4,5}$ edges, which correspond to $4d \rightarrow 5f$ transitions. Examination of the branching ratios and spin-orbit interaction shows that the apparent spin-orbit splitting is partially quenched in α-U, but is strongly dominant in α-Pu. These results are fully quantified using the sum rule.[1, 2] This picture of the actinide 5*f* electronic structure is confirmed by comparison with the results of electronic structure calculations for α-Th, α-U and α-Pu, which in turn are supported by a previous Bremstrahlung Isochromat Spectroscopy (BIS) experiment.

2 RESULTS AND DISCUSSION

The $N_{4,5}$ ($4d \rightarrow 5f$) white line spectra (XAS, EELS and spectral simulation) are shown in Figure 1 for α-Pu . For comparison, the $M_{4,5}$ ($3d \rightarrow 4f$) white line spectra (XAS, EELS and spectral simulations) for Ce are also included. Accompanying each set is the single crystal diffraction pattern taken in the TEM in concert with the EELS measurements, directly confirming the phase of each sample. The spin-orbit split initial states ($4d_{5/2}$ and $4d_{3/2}$ in Pu and $3d_{5/2}$ and $3d_{3/2}$ in Ce) are clearly resolved for each metal as a pair of white lines. While Ce exhibits a significant fine structure, none is resolved for either edge in Pu because the intrinsic lifetime broadening is about 2 eV for both core levels in Pu.

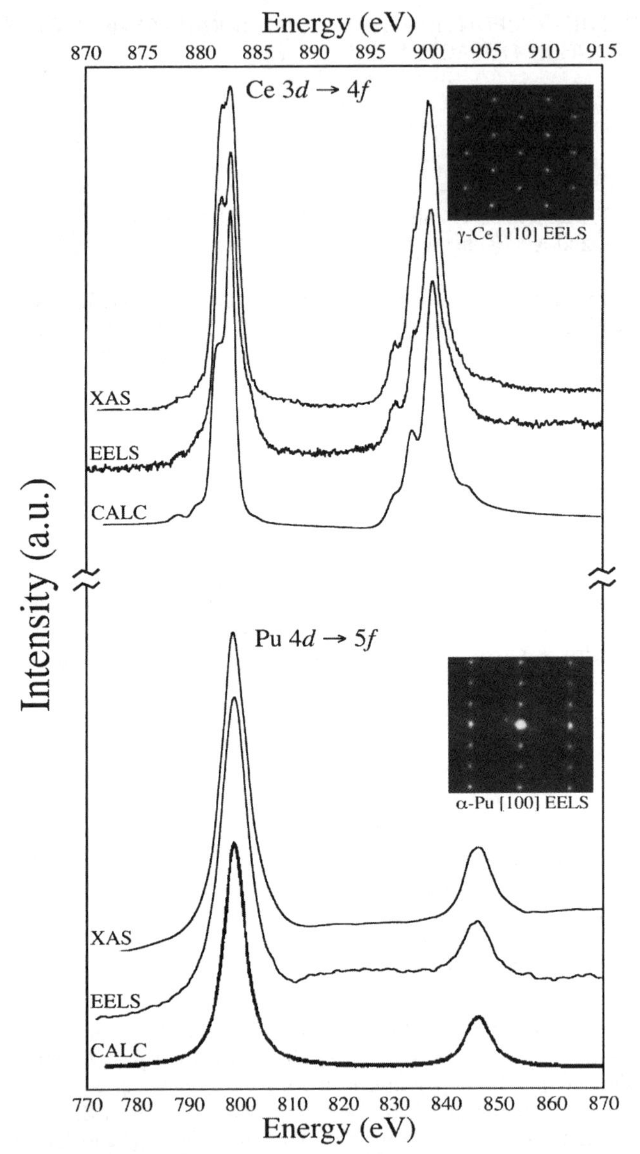

Figure 1 *White line spectra of α-Pu ($N_{4,5}$,4d→5f) and Ce ($M_{4,5}$,3d→4f) acquired by EELS in a TEM, XAS and spectral simulation are shown here. A single-crystal diffraction pattern from each metal is presented, confirming the phase being examined by EELS. For Ce, the $3d_{5/2}$ peak is near 884 eV and the $3d_{3/2}$ peak is near 902 eV. For Pu, the $4d_{5/2}$ peak is near 798 eV and the $4d_{3/2}$ peak is near 845 eV. Note the significantly different energy scales for the Ce and Pu*

Along the actinide series from Th to Pu the $4d_{3/2}$ peak progressively reduces in intensity relative to the $4d_{5/2}$ peak due to the fact that selection rules govern that a $d_{3/2}$ core-level electron can only be excited into an $f_{5/2}$ level.[1, 2] Because of the low $4f$ occupation in Ce, such a strong relative reduction in the $d_{3/2}$ peak intensity is not observed in Ce. (However, a similar but weaker trend is also observed for Rare Earths, where the ground state is still far enough from LS coupling to give a significant effect.[1, 2] In the case of Ce compounds, the trend in the branching ratio has been used to obtain the relative population of the spin-orbit split states.) Thus the decrease in intensity of the Pu $4d_{3/2}$ peak in Figure 1 illustrates a progressive filling of the $5f_{5/2}$ level along the actinide series, as observed previously.[1, 2]

Acknowledgements

This work was performed under the auspices of the U.S. DOE by University of California Lawrence Livermore National Laboratory under contract W-7405-Eng-48. The ALS and the Spectromicroscopy Facility (Beamline 7.0) have been built and operated under funding from the Office of Basic Energy Science at DOE. The authors wish to thank J. Lashley and M. Blau for synthesis of the Pu samples used at the ALS and D. Shuh, R. Schulze, J. Terry, J.D. Farr, T. Zocco, K. Heinzelman and E. Rotenberg for help with the data collection at the ALS.

References

1 G. van der Laan, K.T. Moore, J.G. Tobin, B.W. Chung, M.A. Wall and A.J. Schwartz, *Phys. Rev. Lett.*, 2004, **93**, 097401.
2 J.G. Tobin, K.T. Moore, B.W. Chung, M.A. Wall, A.J. Schwartz, G. van der Laan, and A.L. Kutepov, *Phys. Rev. B*, 2005, **72**, 085109.

A CROSSOVER EFFECT OF THE 5*f* ELECTRONS OF URANIUM COMPOUNDS: FROM ITINERANT TO LOCALIZED, WITH INCREASING TEMPERATURE

Y. Onuki,[1,2] A. Galatanu,[2] Y. Haga,[2] T.D. Matsuda,[2] S. Ikeda,[1,2] E. Yamamoto,[2] D. Aoki,[3] T. Takeuchi[4] and R. Settai[1]

[1]Graduate School of Science, Osaka University, Toyonaka, Osaka 560-0043, Japan
[2]Advanced Science Research Center, JAERI, Tokai, Ibaraki 319-1195, Japan
[3]Institute for Materials Research, Tohoku University, 2145-2 Narita Oarai, Ibaraki 311-1313, Japan
[4]Low Temperature Center, Osaka University, Toyonaka, Osaka 560-0043, Japan

1 INTRODUCTION

The 5*f* electrons of uranium-based intermetallic compounds are usually assumed to have a dual nature, both localized and itinerant. The reason for this property is that the 5*f* electronic states are more extended in real space than the 4*f* electronic states of the rare earths. The 5*f* electronic states thus hybridize with band levels from both the on-site states and also neighboring ligand states. From this point of view, the 5*f* electrons are close to the 3*d* electrons of the transition metals. A strong spin-orbit coupling, however, is present in the uranium compounds, which results in orbital moments and anisotropic properties.

To systematically assess the character of 5*f* electrons and to understand their properties, we have investigated magnetic properties over an extended temperature range, from 2 K to 800 K, by measuring the magnetic susceptibilities of uranium compounds.[1] Here we report the magnetic susceptibility data for ferromagnets of UGe_2 and UIr, heavy fermion compounds of UPt_3, UPd_2Al_3 and URu_2Si_2, and an antiferromagnet, $UPtGa_5$.

2 EXPERIMENTAL RESULTS

2.1 UGe_2 and UIr

UGe_2, with the orthorhombic crystal structure, is the first compound with the coexistence of superconductivity and ferromagnetism.[2] An itinerant nature was assumed for UGe_2 from both de Haas-van Alphen experiments and bulk investigations.[3] Above the Curie temperature, $T_C = 45$ K, the magnetic anisotropy is reduced but remains considerably, even at high temperatures. Above 450 K, the magnetic susceptibility shows the Curie-Weiss law for all the main crystallographic directions, with effective moments of 3.0, 3.6 and 3.4 μ_B/U for [100], [010] and [001] directions, respectively. The effective magnetic moments are close to a free ion value of $\mu_{eff} = 3.58\ \mu_B$/U for $5f^2(U^{4+})$ or $\mu_{eff} = 3.62\ \mu_B$/U for $5f^3(U^{3+})$. This implies a crossover effect from an itinerant nature at low temperatures to a localized one at high temperatures. A similar result is obtained for the ferromagnet, UIr,[4] which revealed a superconducting transition with a transition temperature $T_{SC} = 0.14$ K for pressures between 2.6 and 2.7 GPa.[5] The susceptibility in UIr follows the Curie-Weiss law above 500 K, with a free ion value of 3.6 μ_B/U.

2.2 UPt_3, UPd_2Al_3 and URu_2Si_2

UPt_3, UPd_2Al_3 and URu_2Si_2 are typical heavy fermion superconductors. For example, an itinerant nature of 5*f* electrons in UPd_2Al_3 was clarified from the de Haas-van Alphen experiments and spin (orbital)-polarized energy band calculations. The 5*f* electrons in UPd_2Al_3 also contribute the ordered moments of 0.85 μ_B/U at the uranium sites below a Néel temperature, T_N = 14.5 K. The magnetic susceptibility follows the Curie-Weiss law above 500 K, with μ_{eff} = 3.6 μ_B/U, 3.6 μ_B/U and 3.2 μ_B/U for H // [11$\bar{2}$0], [10$\bar{1}$0] and [0001] directions, respectively, which are close to a free ion value, indicating a localized 5*f* electron nature at high temperatures. The 5*f* electron nature of UPt_3 and URu_2Si_2 is similar to that of UPd_2Al_3. These heavy fermion compounds reveal a characteristic feature. Namely, the susceptibility has a peak at a characteristic temperature $T_{\chi max}$: $T_{\chi max}$ = 18 K in UPt_3, 30 K in UPd_2Al_3 and 60 K in URu_2Si_2, as shown in Figure 1. In Figure 1, the susceptibility of typical Ce-based heavy fermion compounds of $CeRu_2Si_2$ and $CeCu_6$ is also shown. In these compounds, the metamagnetic transition occurs at a critical magnetic field H_c: H_c=20 T in UPt_3, 18 T in UPd_2Al_3 and 40 T in URu_2S_{i2}.[6]

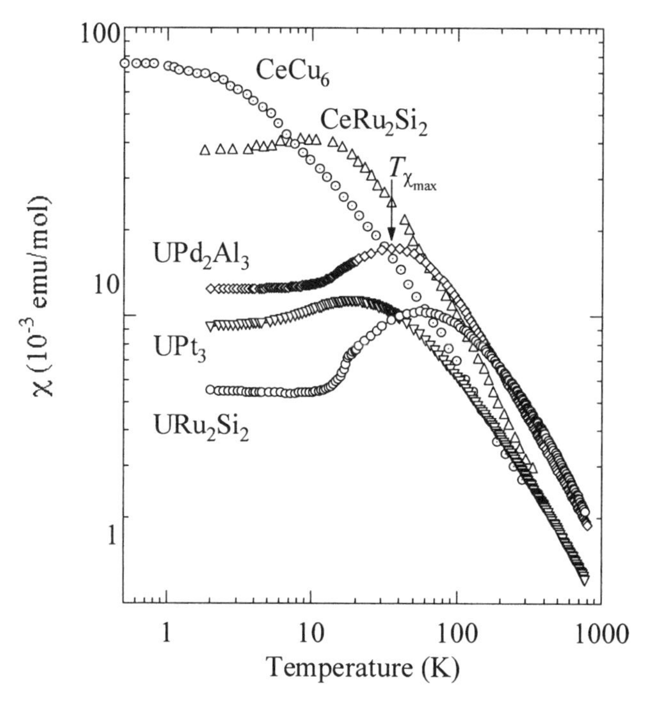

Figure 1 *Temperature dependence of the magnetic susceptibility for heavy fermion superconductors of UPt_3, UPd_2Al_3 and URu_2Si_2, together with typical cerium-based heavy fermion compounds of $CeRu_2Si_2$ and $CeCu_6$*

2.3 $UPtGa_5$

The quasi-two-dimensional electronic state with the 5*f* electron component was clarified in an antiferromagnet $UPtGa_5$ with the tetragonal structure.[7] On the other hand, the susceptibility follows the Curie-Weiss law above 500 K, with $\mu_{eff} = 3.2\ \mu_B$/U for H // [100] and $3.5\mu_B$/U for H // [001], close to a free ion value.

3 SUMMARY

We measured the magnetic susceptibility in the temperature range from 2 to 800 K for ferromagnets of UGe_2 and UIr, heavy fermion compounds UPt_3, UPd_2Al_3, URu_2Si_2 and an antiferromagnet $UPtGa_5$. The magnetic susceptibility follows the Curie-Weiss law at high temperatures. Surprisingly, the effective magnetic moment is close to a free ion value of 3.58 μ_B/U for $5f^2(U^{4+})$ or 3.62 μ_B/U for $5f^3(U^{3+})$. In these compounds, it is concluded that the itinerant nature of 5*f* electrons at low temperatures is changed into the 5*f* localized one at high temperatures.

Acknowledgements

This work was financially supported by the Grants-in-Aid for Scientific Research (A), Creative Research (15GS0213) and Scientific Research in Priority Area "Skutterudite" (No.16037215) from Japan Society for the Promotion of Science.

References

1 A. Galatanu, Y. Haga, T.D. Matsuda, S. Ikeda, E. Yamamoto, D. Aoki, T. Takeuchi and Y. Onuki, *J. Phys. Soc. Jpn.*, 2005, **75**, 1582.
2 S.S. Saxena, P. Agrawal, K. Ahilan, F.M. Grosche, R.K.W. Haselwimmer, M.J. Steiner, E.Pugh, I.R. Walker, S.R. Julian, P. Monthoux, G.G. Lonzarich, A. Huxley, I. Sheikin, D. Braithwaite and J. Flouquet, *Nature(London)*, 2000, **406**, 587.
3 Y. Onuki, S.W. Yun, I. Ukon, I. Umehara, K. Satoh, I. Sakamoto, M. Hunt, P. Meeson, P.-A. Probst and M. Springford, *J. Phys. Soc. Jpn.*, 1991, **60**, 2127.
4 E. Yamamoto, Y. Haga, A. Nakamura, Y. Tokiwa, D. Aoki, R, Settai and Y. Onuki, *J. Phys. Soc. Jpn.*, 2001, **70**, Suppl. A, 37.
5 T. Akazawa, H. Hidaka, T. Fujiwara, T.C. Kobayashi, E. Yamamoto, Y. Haga, R. Settai and Y. Onuki, *J. Phys.: Condens. Matter*, 2004, **16**, L29.
6 Y. Onuki, R. Settai, K. Sugiyama, T. Takeuchi, T.C. Kobayashi, Y. Haga and E. Yamamoto, *J. Phys. Soc. Jpn.*, 2004, **73**, 769.
7 S. Ikeda, Y. Tokiwa, Y. Haga, E. Yamamoto, T. Okubo, M. Yamada, N. Nakamura, K. Sugiyama, K. Kindo, Y. Inada, H. Yamagami and Y. Onuki, *J. Phys. Soc. Jpn.*, 2003, **72**, 576.

LOCALIZATION OF 5*f* ELECTRONS AND PHASE TRANSITIONS IN AMERICIUM

M. Pénicaud

Commissariat à l'Energie Atomique, DAM-Île de France, Département de Physique Théorique et Appliquée, BP12, 91680 Bruyères-le-Châtel, France

1 INTRODUCTION

Americium belongs to the series of actinide metals. It is known that the series of actinide metals, corresponding to the progressive filling of the 5*f* electronic sub-shell, must be split in two. In the first sub-series, from Pa to Pu, the 5*f* electrons bind in the same manner as *d* electrons in transition metals. In the other sub-series starting with Am, the 5*f* electrons are localized, similar to electrons in deep atomic layers, and like 4*f* electrons of lanthanides, they do not take part in metallic bonding.

The localization of the 5*f* electrons is due to strong correlations between them that are not accurately treated by density functional calculations (DFT) with a local density approximation (LDA) and many attempts have been made to solve this problem. In the case of Am the localization of the 5*f* electrons at atmospheric pressure has been simulated successively, by a ferromagnetic (FM) configuration,[1,2] by putting the 5*f* electrons in the atomic core,[3] by decoupling them from the other band electrons (Figure 1),[4,5] and by the self-interaction corrected (SIC) LDA.[6]

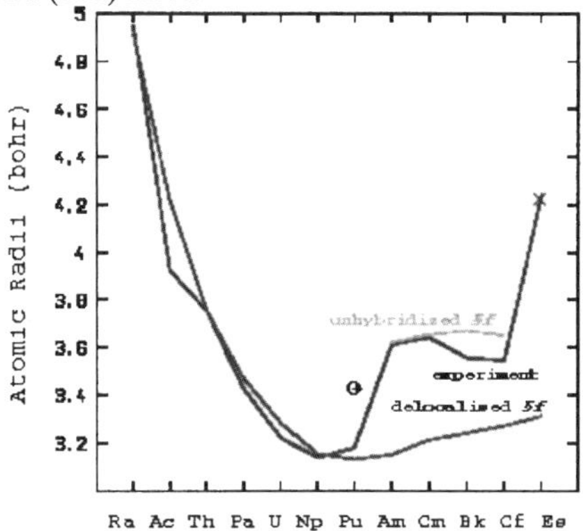

Figure 1 *Comparison of equilibrium atomic radii calculated*[4,5] *by the muffin orbital method in the fcc structure with experimental data. (+) Experimental and (o) calculated, with unhybridized $5f_{5/2}$ values for δ-Pu. (×) calculated value for Es with unhybridized 5f and 6d (unhybridized means decoupled)*

With the method used in Figure 1 of constrained LDA-type, it is not possible to determine *ab initio* localized-delocalized transitions because the total energies calculated in constrained-LDA and LDA are not comparable. Am has a nearly constant susceptibility and is non-magnetic (NM).[7] This is generally attributed to a configuration with six 5*f* electrons in which the spin (S) and the orbital (L) angular moments cancel (S = -L = 3, thus J = L + S = 0 in Russell-Saunders coupling). A similar conclusion is reached starting from the *jj* coupling scheme, with six 5*f* electrons. Am has a full $j = {}^5/_2$ shell and also J = 0. This cannot be reproduced theoretically,[2] but allowing magnetism in the theory permits a band splitting to simulate the localization of the 5*f* electrons. Here we allow for a FM or an anti-ferromagnetic (AFM) configuration (the latter is the configuration of the next elements, Cm and Bk) which was also used to simulate δ-Pu with some success.[8]

Allowing for magnetic configurations, in this study we aimed to reproduce the recent Am phase transitions obtained experimentally under pressure.[9] All calculations were performed using the "All Electrons Full Potential Linearized Augmented Plane Wave" (FPLAPW) program WIEN2k [10] in the generalized gradient approximation.

2 STRUCTURAL STABILITY UNDER PRESSURE

In a recent paper[5] we reported studies of the structural stability of Am by the comparison of total energies calculated in eleven crystal structures.[9,11] At high pressures the monoclinic α-Pu structure, followed by the orthorhombic Am(IV) structure, were found to be stable, unlike the face centred orthorhombic Am(III) (Figure 2). If we now allow for the magnetic configurations and relax the structures, which was not done previously, and we only consider the four structures obtained experimentally (dhcp Am(I), fcc Am(II), orthorhombic Am(III) and Am(IV)[9]), in addition the α-Pu structure[11] the results can be seen in Figure 3.

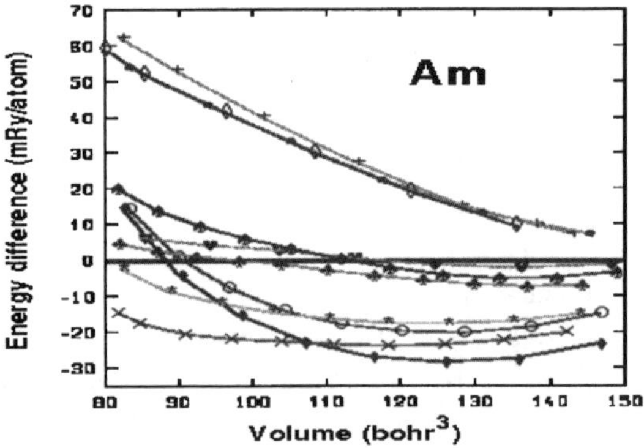

Figure 2 *Total energy differences for Am, calculated by the FPLAPW method in: α-Pu (♦), Am IV (×), α-Np (o), β-Np (*), Am III (♣), α-U (♠), α-Pa (♥), α''-Ce(◊), hcp (•) and fcc (+) crystal structures, relative to the bcc structure, as functions of volume. All structures were calculated in a NM configuration*

The phase transitions found now theoretically,[12] agree with the recent sequence obtained experimentally under pressure[9]: dhcp Am(I) → fcc Am(II) → face-centered orthorhombic Am(III) (same as the γ-Pu structure) → primitive orthorhombic Am(IV) (derivative of the α-U structure). The configurations of the first three phases are AFM, which is that with the lowest calculated energy compared with FM and NM. Only the fourth phase, Am(IV), has a NM configuration. The AFM configuration is a way of treating the localized $5f$ electrons. When the first three phases exist theoretically, they are in this configuration, i.e. with localized $5f$ electrons, but when the Am(IV) phase appears at a theoretical pressure of 198 kbar and with a theoretical volume of about 111 bohr^3, the magnetic moments of the AFM configuration collapse nearly to zero, which indicates a complete $5f$ delocalization.

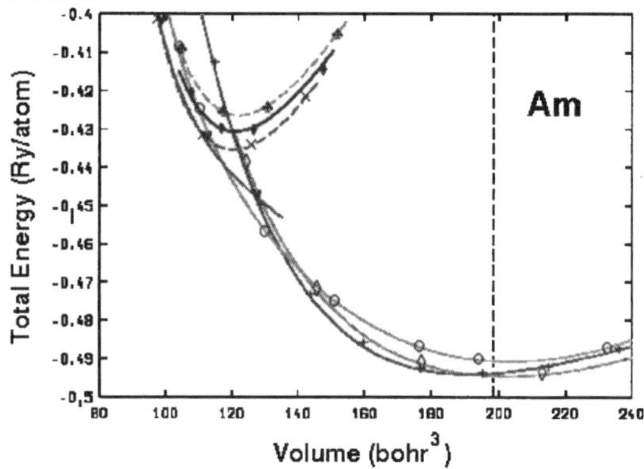

Figure 3 *Total energies as functions of volume for Am, calculated for different crystal structures: dhcp (◊), fcc (+), Am(III) (o) and Am(IV) (♥) with an AFM configuration and Am(III) (♣), Am(IV) (×) and α-Pu (♦) assuming spin degeneracy. All the structures are relaxed. The room temperature equilibrium volume for Am is denoted by a vertical broken line*

References

1 H.L. Skiver, O.K. Andersen and B. Johansson, *Phys. Rev. Lett.*, 1978, **41**, 42.
2 P. Söderlind, R. Ahuja, O. Eriksson, B. Johansson and J.M. Wills, *Phys. Rev. B*, 2000, **61**, 8119.
3 O. Eriksson, M.S.S. Brooks and B. Johansson, *J. Less Common Met.*, 1990, **158**, 207.
4 M. Pénicaud, *J. Phys.: Condens. Matter*, 1997, **9**, 6341.
5 M. Pénicaud, *J. Phys.: Condens. Matter*, 2002, **14**, 3575.
6 L. Petit, A. Svane, W.M. Temmerman and Z. Szotek, *Solid State Commun.*, 2000, **116**, 379.
7 J.M. Fournier and R. Troć, *Handbook on the Physics and Chemistry of the Actinides*, eds. A.J. Freeman and G.H. Lander, North-Holland, Amsterdam, 1985, Vol 2, p. 50.
8 Y. Wang and Y. Sun, *J. Phys.: Condens. Matter*, 2000, **12**, L311.
9 A. Lindbaum, S. Heathman, K. Litfin, Y. Méresse, R.G. Haire, T. Le Bihan and H. Libotte, *Phys. Rev. B*, 2001, **63**, 214101.
10 P. Blaha, K. Schwarz, G.K.H. Madsen, D. Kvasnicka and J. Luitz, *WIEN2k*, ed K. Schwarz, Vienna University of Technology, Vienna, 2001.
11 R.W.G. Wyckoff, *Crystal Structures*, Wiley, New York, 1963, Vol 1.
12 M. Pénicaud, *J. Phys.: Condens. Matter*, 2005, **17**, 257.

ORBITAL ORDERING IN ACTINIDE OXIDES: NEW PERSPECTIVES ON OLD PROBLEMS

S.B. Wilkins,[1,2] J.A. Paixão,[3] R. Caciuffo,[4] C. Detlefs,[2] J. Rebizant[1] and G.H. Lander[1]

[1]European Commission, Joint Research Center, Institute of Transuranium Elements, Postfach 2340, Karlsruhe, D-76125 Germany
[2]European Synchrotron Radiation Facility, BP 220, F-38043 Grenoble CEDEX, France
[3]Physics Department, University of Coimbra, Coimbra, P-3004-516, Portugal
[4]Dipartimento di Fisica ed Ingegneria dei Materiali e del Territorio, Università Politecnica delle Marche, I-60131 Ancona, Italy

1 INTRODUCTION

The behaviour of the actinide oxides has been of major interest since the 1940s. UO_2 is known to exhibit antiferromagnetism since neutron experiments of the 1960s. However, the puzzle of the lack of ordered magnetism in NpO_2 (also dating from the same period) was recently solved by the discovery[1,2] that the ordering phenomenon in NpO_2 is *not* connected with magnetic dipole moments (as in UO_2), but rather with the long-range ordering (at 25 K) of octupole magnetic moments which then induces ordering of the anisotropic $5f$ charge distribution surrounding the Np^{4+} ion. Such ordering of the $5f$ quadrupole charge distribution cannot be observed with neutrons (or muons) but can be measured by resonant x-ray scattering.

Following the work on pure NpO_2, we have re-examined single crystals of $(U_{0.75}Np_{0.25})O_2$ and shown that *both* the U and Np ions in this material exhibit dipole and quadrupole ordering with transverse symmetry.[3] To a lesser or greater extent all theoretical models of UO_2 emphasize the importance of the interplay between the Jahn-Teller and quadrupolar interactions. Allen[4] indeed proposed that the quadrupoles ordered and the subsequent internal strain would lead to a change in the position of the oxygen atoms without giving rise to an external change in the symmetry of UO_2. Although the internal distortion of the oxygen cage was observed by neutron diffraction,[5,6] no direct evidence for the ordering of electric quadrupole moments below T_N in UO_2 was obtained.

2 METHOD AND RESULTS

We present the results of an experiment involving resonant x-ray scattering that establish the *quadrupole* ordering in pure UO_2. The experiments consist of tuning the photon energy to the actinide M_4 absorption edge (at which photon energy core $3d$ electrons are promoted to the partially occupied $5f$ valence states). Then, once the repeat of the anisotropic charge distribution is known, the azimuthal distribution of the scattered resonant x-ray intensity is measured. This intensity distribution is related to the symmetry of the $5f$ anisotropic charge distribution. The experiments were performed at the ID20 beamline of the European

Synchrotron Radiation Facility (ESRF) in Grenoble, France, with the single crystals fabricated and mounted at ITU.

We show in Figure 1 the initial results of the study of the (112) reflection below T_N (~31 K). This reflection has the advantage that it is not sensitive to the internal distortion of the oxygen atoms observed in UO_2 below T_N.[5,6] The top frame shows the intensity of the $\sigma \rightarrow \pi$ scattered radiation which is normally associated with *dipole* magnetic scattering from E1 $F^{[1]}$ resonant scattering amplitudes.[7] The "dip" in the intensity near the resonant energy (dashed line) is characteristic of strong absorption at the maximum of the resonance and is a well-known phenomenon at these energies.[8] In the central panel is shown the $\sigma \rightarrow \sigma$ scattering from this reflection. This shows a sharp peak (in energy) that is displaced to lower energy by about 4 eV. Both effects are associated in actinide systems with an E1 $F^{[2]}$ electric quadrupole scattering amplitude.[1] The strong absorption in the top panel is supported by the dependence of the HWHM of the energy dependence shown in the bottom panel.

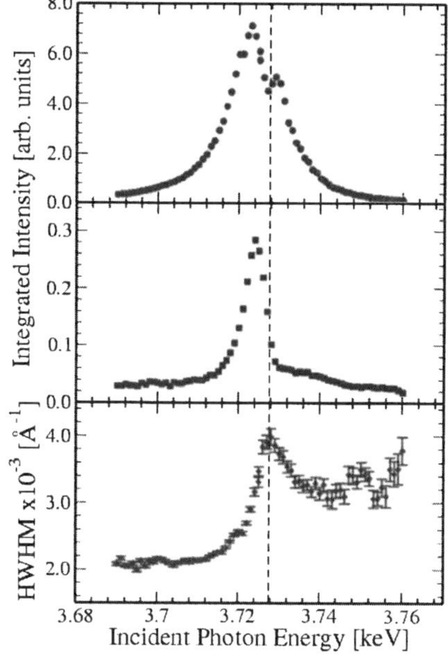

Figure 1 *Photon energy dependence of the (112) Bragg peak integrated intensity across the Np M_4 absorption edge in the σ–π polarisation channel (top panel) and in the σ–σ polarisation channel (central panel); the lower panel reports the half width at half maximum, showing a maximum at the dipole edge (vertical line). These data were collected at a temperature of 10 K*

Figure 2 shows the azimuthal dependence of the intensity in the σ–σ polarisation channel from the (112) reflection in UO_2 at 10 K, together with the predictions for an incoherent superposition of the two *transverse* quadrupole configurations. The excellent agreement between model predictions and experimental data proves the transverse nature of the quadrupolar ordering in UO_2, whereas that in NpO_2 is longitudinal.

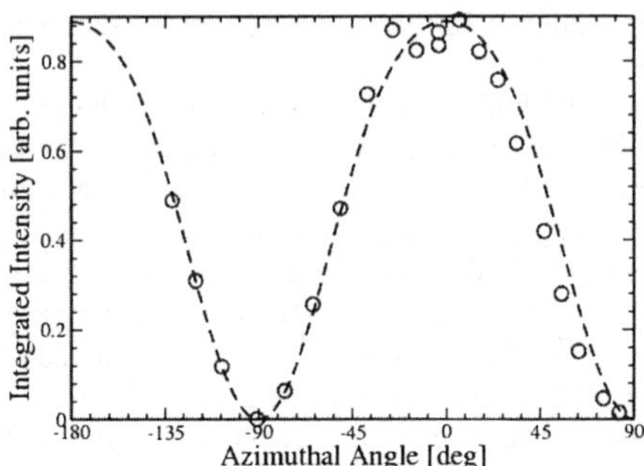

Figure 2 *Integrated intensity as a function of azimuthal angle for the (112) superstructure reflection in the σ–σ polarisation channel (open circles). The line is a fit to the model for quadrupolar ordering. In the model there is an incoherent superposition of the two transverse modes. These data were collected at a temperature of 10 K*

3 CONCLUSIONS

Although the presence of electric quadrupolar charge distributions has been theoretically invoked for almost 40 years in UO_2,[4] this study gives the first experimental proof of their long-range ordering below T_N. These new insights allow us to re-examine the phase diagram of $(U_{1-x}Np_x)O_2$. For example, neutron experiments[9] found that the dipole ordering at $x \sim 0.50$ shows an unusual short-range ordering with $\mathbf{q} \sim [^1/_2\ ^1/_2\ ^1/_2]$ and a much lower T_{ord} than the parent compounds. Given the presence of quadrupole ordering across the whole phase diagram we postulate that the ordering in the region $0.40 < x < 0.75$ is complex due to *quadrupole frustration.* For these values of x, neither the transverse nor the longitudinal quadrupole ordering can dominate, leading to frustration and only short-range ordering.

References

1 J.A. Paixão, C. Detlefs, M.J. Longfield, R. Caciuffo, P. Santini, N. Bernhoeft, J. Rebizant and G.H. Lander, *Phys. Rev. Lett.*, 2002, **89**, 187202.
2 R. Caciuffo, J.A. Paixão, C. Detlefs, M.J. Longfield, P. Santini, N. Bernhoeft, J. Rebizant and G.H. Lander, *J. Phys. Cond. Matter*, 2003, **15**, S2287, and references therein.
3 S.B. Wilkins, J.A. Paixão, R. Caciuffo, P. Javorsky, F. Wastin, J. Rebizant, C. Detlets, N. Bernhoeft, P. Santini and G.H. Lander, *Phys. Rev. B: Condens. Matter*, 2004, **70,** 214402.
4 S. J. Allen, *Phys. Rev.*, 1968, **166**, 530; *ibid*, 1968, **167**, 492.
5 J. Faber, G.H. Lander and B.R. Cooper, *Phys. Rev. Lett.,* 1975, **35**, 1770.
6 J. Faber and G.H. Lander, *Phys. Rev. B: Solid State,* 1976, **14**, 1151.
7 J. P. Hill & D. McMorrow, *Acta Crystallogr., Sect. A*, 1996, **52**, 236.
8 N. Bernhoeft, A. Hiess, S. Langridge, A. Stunault, D. Wermeille, C. Vettier, G.H. Lander, M. Huth, M. Jourdan and H. Adrian, *Phys. Rev. Lett.*, 1998, **81**, 3419.
9 A. Boeuf, R. Caciuffo, M. Pages, J. Rebizant, F. Rustichelli and A. Tabuteau, *Europhys. Lett.*, 1987, **3**, 221.

ELECTRONIC STRUCTURE OF δ-Pu: THEORIES AND PES EXPERIMENTS

L. Havela,[1] T. Gouder,[2] A.B. Shick[3] and V. Drchal[3]

[1]Charles University, Faculty of Mathematics and Physics, Ke Karlovu 5, Prague 2, Czech Republic
[2]European Commission, Joint Research Centre, Institute for Transuranium Elements, Karlsruhe, Germany
[3]Institute of Physics, Academy of Sciences of the Czech Rep., Prague 8, Czech Republic
Work supported by the Grant Agency of CR (Grant #202/04/1103) and by the Grant Agency of Academy of Sciences of CR (project IAA100100530), It is also part of the research program MSM 0021620834 of the Czech Ministry of Education

1 INTRODUCTION

The 5*f* localization occurs in the sequence of actinide elemental metals between Pu and Am. The cross-over regime can be vaguely characterized as dominated by strong correlations in narrow bands. In such a situation, diverse cooperative phenomena obscure any unified experimental picture, and theoretical calculations lose a true general predictive capability. In the sequence of actinide elements, additional information can be obtained from different allotropic phases. One of them is δ-Pu, which is located, as to its volume, roughly half way from still itinerant α-Pu to Am. As δ-Pu can be easily stabilized by dopants, it represents a very suitable case for testing the level of our understanding of actinides and strongly correlated systems in general.

2 PHOTOELECTRON SPECTROSCOPY

In the last decade, photoelectron spectroscopy (PES) techniques were applied to study Pu by three groups. Valence-band spectra of α- and δ-Pu surfaces cleaned by laser ablation were studied in Los Alamos.[1] A surprising degree of similarity was found for the two phases, with a satellite feature at 0.8 eV binding energy (BE, related to the Fermi energy E_F), and a narrow main peak in the close vicinity of E_F. A synchrotron experiment performed at LBNL on sputter-cleaned surfaces[2,3] yielded a similar picture in the resonant photoemission experiment. In addition, 4*f* core-level spectra corroborated the finding that the difference between the two phases is more quantitative than qualitative. The puzzle why α-Pu spectra are quite similar to δ-Pu and different from those in early studies in the 1980s, which exhibited only a broad triangular peak culminating at E_F, [4] was revealed by the study at ITU Karlsruhe,[5, 6] demonstrating that the surface of bulk α-Pu adopts a δ-Pu character, unless it is kept at low temperatures after cleaning by Ar ion sputtering. This effect has been considered before theoretically[7] and it became an illustration of the fact that diverse crystal structures of Pu are nearly degenerate in energy. Superior energy resolution and good statistics led to a recognition that a smaller feature at about 0.5 eV appears, besides the two peaks mentioned above. These three features were associated with those also seen before in PuSe.[8] Their occurrence in many other Pu-based systems[9,10] indicates that they do not relate to any particular features in one electron density of states. The following questions remain: what are those features which dominate the valence-band spectra of δ-Pu? Can they be compared with results of electronic structure calculations? In general, PES spectra reflect the imaginary part of electron self-energy, i.e., besides one-electron density of states, they also pick up potential many body features related to Kondo physics etc. In particular, the sharp resonance at E_F could be related to this kind of physics. On the other hand, final-state effects can dominate in photoemission from localized or nearly localized states. For example, a manifold of atomic-state multiplets appear in the spectra of lanthanides. Naturally, multiplets also do not appear in densities of state obtained from standard electron structure calculations.

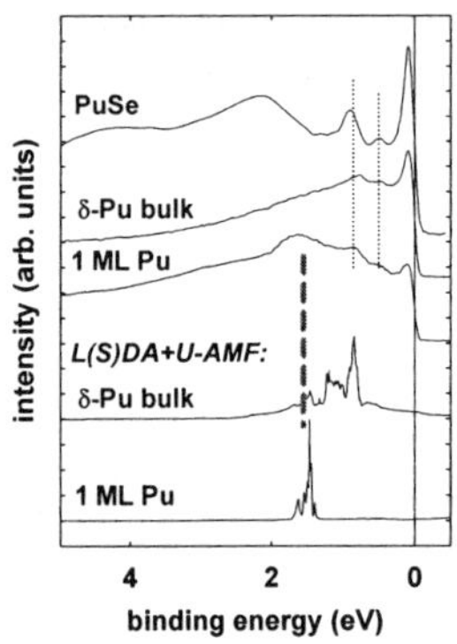

Figure 1 UPS *spectra* ($h\nu$ = *40.8 eV, yielding mainly the 5f emission), compared with results of L(S)DA+U calculations, which suggest that only high-energy features belong to ground state. The three-peak structure close to* E_F *(dotted lines), shared by many other Pu systems like PuSe, has to be due to final-state effects discussed in the text. Data have been arbitrarily re-scaled and shifted*

3 ELECTRONIC STRUCTURE CALCULATIONS

Density Functional Theory in Local Density (LDA) or Generalized Gradient (GGA) approximations do not reproduce the enhanced equilibrium volume (V_{eq}) of δ-Pu (compared to α-Pu), when the non-magnetic state is preserved. An increase of V_{eq} in LDA calculations of Pu occurs simultaneously with the occurrence of a magnetic ground state,[11] in contradiction to the experimentally-observed non-magnetic character of δ-Pu.[12] Similarly, the magnetic ground state was obtained by the LDA+U,[13] where the LDA is augmented by explicit correlations between *f* electrons in a narrow band regime. Another approach, the so called mixed-level model (MLM), considers a fraction of the 5*f* states as band-like, while keeping a certain integral number of 5*f* electrons still in localized states.[15,16] The MLM, as well as LDA and GGA, can produce a peak at E_F, which is observed in valence band spectra. On the contrary, LDA+U shifts the 5*f* states by a few eV below E_F, and cannot reproduce this feature. The question is: should the MLM reproduce the PES peak? One expects that spectra reflect not the ground state, but final-state multiplets ($5f^3$, originating from the considered $5f^4$ localized initial state). As those are not considered in the MLM model, a certain agreement with PES spectra can therefore only be coincidental. Moreover, the expectation that the same MLM model will lead to identical features in many other Pu systems does not seem realistic. On the other hand, dynamical electron correlations treated in the framework of the dynamical mean field theory (DMFT)[17] (beyond those in MLM and LDA/GGA) can lead to a high quasiparticle density of states at E_F for δ-Pu, even when starting from the LDA+U ground state, with the 5*f* states out of E_F. Recently we have shown[18] that a realistic description of δ-Pu is possible using a fully relativistic "around the mean field" formulation of the LDA+U method, which gives correct V_{eq} and bulk modulus, while converging to a non-magnetic (S =0, L = 0) state; robust with respect to considerable lattice expansion. In addition, this method correctly describes the shift of the 5*f* states to higher binding energy with reduced thickness (see Figure1 for one monolayer). As this method does not account for many-body or final-state effects, it insufficiently describes the spectra close to E_F. Further analysis reveals that the narrow band of 5*f* states is buried under the three-peak structure in bulk δ-Pu, and only its shift to higher BE in thin layers makes it conspicuous. The complete description of the spectra can be possible only by taking into account the physics beyond one-electron states.

References

1 A.J. Arko, J.J. Joyce, J. Wills, L. Morales, J. Wills, J. Lashley, F. Wastin and J. Rebizant, *Phys. Rev. B*, 2000, **62**, 1773.
2 J. Terry, R.K. Schulze, J.D. Farr, T. Zocco, K. Heinzelman, E. Rotenberg, D.K. Shuh, G. van der Laan, D.A. Arena and J.G. Tobin, *Surf. Sci.*, 2002, **499**, L141.
3 J.G. Tobin, B.W. Chung, R.K. Schulze, J. Terry, J.D. Farr, D.K. Shuh, K. Heinzelman, E. Rotenberg, G.D. Waddill and G. van der Laan, *Phys. Rev. B*, 2003, **68**, 155109.
4 J.R. Naegele, J. Ghijsen, L. Manes, *59/60 Structure and Bonding, Actinides-Chemistry and Physical Properties*, ed., L. Manes, Springer, Berlin, 1985, pp.197-262.
5 T. Gouder, L. Havela, F. Wastin and J. Rebizant, *Europhys. Lett.* 2001, **55**, 705.
6 L. Havela, T. Gouder, F. Wastin and J. Rebizant, *Phys. Rev. B*, 2002, **65**, 235118.
7 O. Eriksson, L.E. Cox, B.R. Cooper, J.M. Wills, G.W. Fernando, Y.G. Hao and A.M. Boring, *Phys. Rev. B*, 1992, **46**, 13576.
8 T. Gouder, F. Wastin, J. Rebizant and L. Havela, *Phys. Rev. Lett.*, 2000, **84**, 3378.
9 L. Havela, F. Wastin, J. Rebizant and T. Gouder, *Phys. Rev. B*, 2003, **68**, 085101.
10 T. Durakiewicz, J.J. Joyce, G.H. Lander, C.G. Olson, M.T. Butterfield, E. Guziewicz, A.J. Arko, L. Morales, J. Rebizant, K. Mattenberger and O. Vogt, *Phys. Rev. B*, 2004, **70**, 205103.
11 P. Söderlind and B. Sadigh, *Phys. Rev. Lett.*, 2004, **92**, 185702.
12 J.C. Lashley, A. Lawson, R.J. McQueeney and G.H. Lander, *Phys. Rev. B*, 2005, **72**, 054416.
13 J. Bouchet, B. Siberchicot, F. Jollet and A Pasturel, *J. Phys.:Condens. Matter*, 2000, **12**, 1723.
14 S.Y. Savrasov and G. Kotliar, *Phys. Rev. Lett.*, 2000, **84**, 3670.
15 O. Eriksson, J.D. Becker, A.V. Balatsky and J.M. Wills, *J. Alloys Compd.*, 1999, **287**, 1.
16 J.M. Wills, O Eriksson, A. Delin, P.H. Andersson, J.J. Joyce, T. Durakiewicz, M.T. Butterfield, A.J. Arko, D.P. Moore and L.A. Morales, *J. Electron Spectrosc.*, 2004, **135**, 163.
17 S.Y. Savrasov, G. Kotliar and E. Abrahams, *Nature*, 2001, **410**, 793.
18 A.B. Shick, V. Drchal and L. Havela, *Europhys. Lett.*, 2005, **69**, 588.

MAGNETIC PROPERTIES OF RADIATION DAMAGE IN Pu AND Pu ALLOYS

S. McCall,[1] M.J. Fluss,[1] B. Chung,[1] G. Chapline,[1] M. McElfresh,[1] D. Jackson,[1] N. Baclet,[2] L. Jolly[2] and M. Dormeval[2]

[1]Lawrence Livermore National Laboratory, Livermore CA 94552 USA
[2]CEA-Centre de Valduc, 21120 Is sur Tille, France
This work was performed under the auspices of the U.S. DOE by Lawrence Livermore National Laboratory, under contract W-7405-Eng-48

1 INTRODUCTION

Among the many exceptional properties of Pu[1] is its apparent lack of either local moments or cooperative magnetism. Lashley *et al.*,[2] have recently noted that little experimental evidence for the existence of local moments or collective magnetism has been found in over 50 years. The physical properties of Pu: resistance, magnetic susceptibility, and heat capacity,[2] all support a system with an enhanced electron density of states. Pu sits on the edge of both magnetism and superconductivity, possessing one of the highest elemental Pauli susceptibilities, consistent with a highly correlated electron system. The low-density δ-Pu eludes a first principles description and is an area of active investigation for theorists.

Recently, Griveau *et al.*[3] observed the variations in the resistance and superconducting properties of Am metal as a function of pressure to 27 GPa and T > 0.4K. They postulate that the pressure dependence of the superconducting temperature shows an *f*-electron Mott-like transition as pressure drives Am towards a Pu and then a U-like structure. Hence, we posit that dilating the Pu lattice will bring one to a similar transition. Experimental evidence supporting this point of view is given here.

2 PREVIOUS EXPERIMENTS

Recent experiments at Valduc and LLNL have pointed to the possibility of emerging magnetism with the accumulation of radiation damage. We briefly review these early experiments. Dormeval describes an observation of the consequential effects of room temperature radiation damage accumulation in a series of Pu(Am) alloys.[4] The Am concentration, which increased the damage rate, ranged from 4.9 % to 24 %. The damage accumulation periods were such that each specimen had the same number of alpha decays. When the magnetic susceptibility of these iso-damaged specimens was measured (Figure 1 *left*) a feature was seen below 75 K at fields below 2 Tesla that had not been present in the annealed specimen. What is interesting, however, is that this additional or excess magnetic susceptibility (EMS) could be significantly but not fully annealed at higher temperatures, indicating that radiation damage accumulation was indeed the source of the observed EMS. One expects that the accumulated damage consisted of helium-vacancy pairs, possibly small vacancy clusters, and the alpha decay products of Pu and Am, all of which may contribute to the EMS. In a related study, Fluss *et al.*[5] observed a temperature dependence for the resistivity of vacancies and vacancy clusters, ρ, in Pu(3.3at%Ga) that obeyed a Kondo impurity form ρ = a-b*ln*(T) (Figure 1 *right*). This suggests that vacancies, while possessing no moment of their own, induce a Kondo-like moment in the surrounding material. An effect similar to this has been studied in the hole-doped superconductors[6] where zero spin impurities, including vacancies, in the Cu-O plane result in both Kondo behaviour and a reduction in T_c.

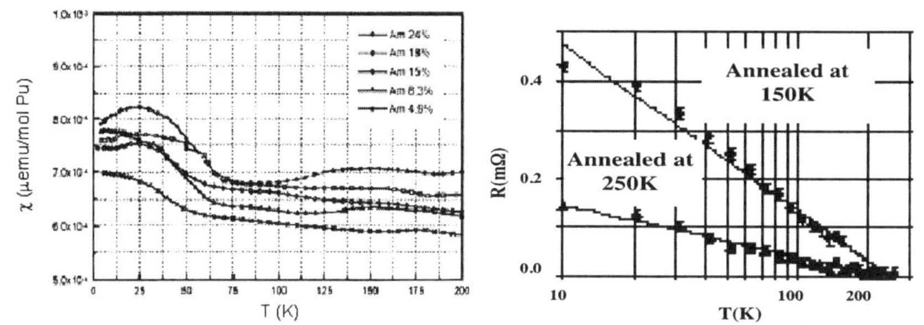

Figure 1 *Left: The appearance of EMS is seen in the data of Dormeval[4] for five iso-damaged specimens of Pu(Am). Right: The reported temperature dependence of the resistance of vacancy defects in Pu(3.3at%Ga) after 10K damage accumulation and either a 150K or 250K anneal[5]*

3 LOW TEMPERATURE DAMAGE ACCUMULATION

We report here preliminary data where we have measured the change in magnetic susceptibility as a function of time and temperature in α-Pu and δ-Pu(Ga). In terms of the atomic volume per atom α-Pu is close to Np while δ-Pu is nearer to Am. Hence, a comparison of the magnetic properties induced by accumulated radiation damage may track with this feature of Pu. Isochronal magnetization annealing defined the specimen Stage I temperature below which there is no defect mobility. Beginning with an annealed specimen lowered below the Stage I temperature, the magnetization was measured as a function of time and temperature, while cycling from 2-30 K, remaining safely below Stage I. In this way damage accumulation isotherms were determined for the two specimens. Some of this data is shown in Figure 2, where the EMS evolves substantial curvature as a function of time. The measured $\chi(t,T)$ fits a saturation model of the form in Equation (1), where

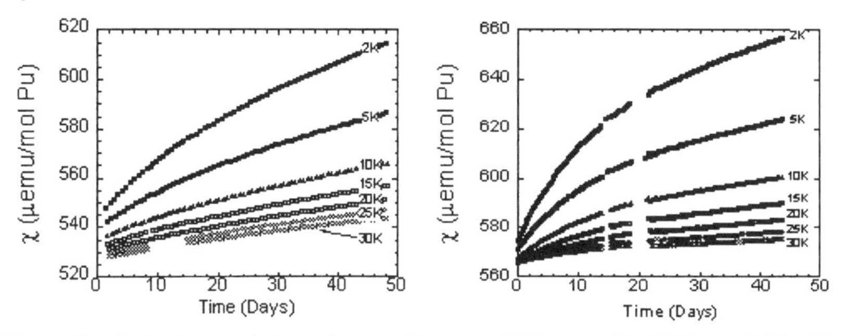

Figure 2 *Left: time evolution of magnetic susceptibility in α-Pu. Right: and δ-Pu(Ga), each measured in a 3T applied magnetic field*

$\chi_{Initial}(T)$ is the annealed susceptibility, χ_{EMS} is the maximum excess susceptibility, and $\tau(T)$ is a characteristic time.

$$\chi(t,T) = \chi_{Initial}(T) + \chi_{EMS}(T)(1 - e^{-t/\tau(T)}) \tag{1}$$

What is noteworthy is that $1/\tau(T)$ is proportional to an effective volume per alpha decay, which is the volume surrounding the U recoil damage cascade that has been modified; *i.e.* the volume in which the magnetic susceptibility has been increased by the local disorder. This volume for α and δ is 8 to 10 times greater than the damage cascade volume as estimated from molecular dynamics models. Delta-Pu reaches saturation sooner than α-Pu, meaning that the magnetically perturbed volume for δ-Pu is larger. This might be indicative of the concomitant increased localization from α– to δ–Pu.

4 CONCLUSION and FUTURE WORK

The exact structure-property relationships for the observed EMS described above remains as future work. However, we expect that vacancies, which create open regions in the lattice, should lead to increased *f*-electron localization *via* lattice dilation. Dormeval[4] noted that in the case of Pu(Am), Am is oversized in the Pu lattice (positive deviation from Vegard's law) exhibits a clear anomaly in both susceptibility and resistivity at Am 25 % (Figure 3). One might expect that low temperature damage accumulation studies close to this anomaly could lead to physical properties evidence for a delocalization to localization transition, analogous to that seen in Am under pressure.

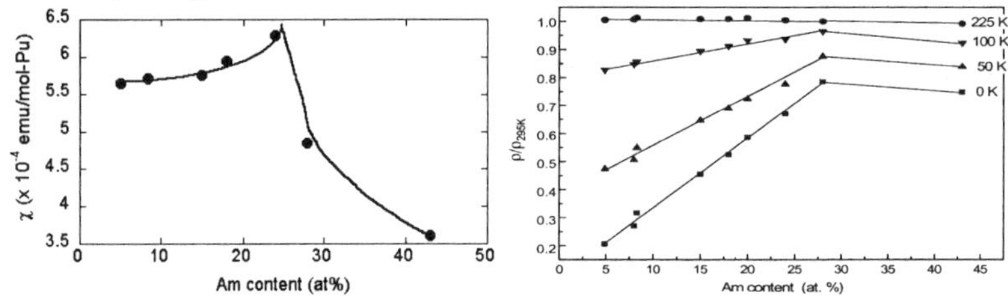

Figure 3 *Left: The Pauli susceptibility, χ_o, of Pu(Am) as a function of %Am, taken from Dormeval[4] and Fluss et al.[5] Right: Normalized resistive isotherms for Pu(Am) from Dormeval[4] and Fluss et al.[5] Both figures indicate an interesting discontinuity near 25 % Am suggesting increased localization. The lines are guides to the eye*

References

1 *Challenges in Plutonium Science*, ed., N. Cooper, Los Alamos Science, 2000, Vol. 26, I&II, LA-UR-00-4100, http://www.fas.org/sgp/othergov/doe/lanl/pubs/number26.htm.
2 J.C. Lashley, A. Lawson, R.J. McQueeney, and G.H. Lander, *Phys. Rev. B*, 2005, **72** 054416.
3 J.-C. Griveau, J. Rebizant, G.H. Lander and G. Kotliar, *Phys. Rev. Lett.*, 2005, **94,** 097002.
4 M. Dormeval, PhD Thesis, Universite de Bourgogne, Dijon, France, 2001.
5 M.J. Fluss, B.D. Wirth, M. Wall, T.E. Felter, M.J. Caturla, A. Kubota, T.D. de la Rubia, *J. Alloys Compd.*, 2004, **368**, 62.
6 F. Rullier-Albenque, H. Alloul and R. Tourbot, *Phys. Rev. Lett.*, 2003, **91,** 047001, (and references therein).

A PHOTOELECTRON SPECTROSCOPY INVESTIGATION OF THE EARLY STAGES OF NEPTUNIUM OXIDATION

P. Nevitt,[1] A. Carley,[1] P. Roussel,[2] T. Gouder[3] and F. Huber[3]

[1]School of Chemistry, Cardiff University, PO Box 921, Cardiff U. K. CF10 3TB
[2]AWE, Aldermaston, Berkshire U. K. RG7 4PR
[3]European Commission, Joint Research Centre, Institute for Transuranium Elements, Postfach 2340, D-76125, Karlsruhe, Germany

1 INTRODUCTION

The light actinide elements (thorium to americium) display evidence of 5*f* electron involvement in chemical bonding, an excellent example of this behaviour being the binary oxides of these elements. The Np-O phase diagram shows both tetra- and penta-valent oxides, NpO_2 and Np_2O_5, are stable.[1] Additionally, trivalent neptunium sesquioxide has been reported as a surface oxide,[2] prepared under ultra high vacuum conditions from the surface segregation of oxygen dissolved in the Np metal sample. However, this compound has not been synthesised as a pure material. We have therefore further investigated the possible existence of this oxide phase.

2 EXPERIMENTAL

All data were acquired at the Institute for Transuranium Elements using a Leybold-Heraeus LHS-10 spectrometer modified for work with radioactive materials, as described elsewhere.[3] The plasma triode deposition source has also been described elsewhere.[4] Typical parameters for neptunium deposition were: electron energy 50 – 100 eV, target bias –700 V and 4 x 10^{-3} mbar Ar. XPS spectra were acquired with non-monochromatic Al Kα radiation (1486.6 eV). UPS spectra were acquired using He^{II} (40.8 eV) and He^{I} (21.2 eV) radiation from a windowless UV rare gas discharge source.

3 RESULTS AND DISCUSSION

Thin film neptunium metal samples were prepared by *in situ* sputter deposition. X-ray photoelectron spectroscopy (XPS) of these films exhibited the characteristic asymmetric peaks attributable to Np 4*f* photoionisation at binding energies of 399.6 and 411.4 eV. No other elements were detectable in the XP spectra, confirming the purity of metal films produced by this method. The samples were exposed to increasing doses of oxygen at room temperature and the Np 4*f* and O 1*s* core level regions were followed using XPS (Figure 1). Initial exposure to oxygen produces a decrease in intensity of the Np $4f_{7/2}$ peak at 399.6 eV and slight asymmetric broadening. Following an 8 Langmuir exposure (1 Langmuir = 10^{-6} Torr-sec) two peaks become evident to the high binding energy side of the

Np $4f_{7/2}$ peak, centred at 401.2 and 403.5 eV. Further oxygen exposures, up to 50 Langmuir, result in a decrease in intensity of the Np $4f_{7/2}$ peaks centred at 399.6 and 401.2 eV, whereas the peak centred at 403.5 eV increases in intensity. At 50 Langmuir exposure we observe 'shake up' satellites at 421.9 eV indicative of NpO_2 formation.[5,6] The O 1*s* region displayed a single peak centred at 530.7 eV, which increases intensity with increasing oxygen exposure.

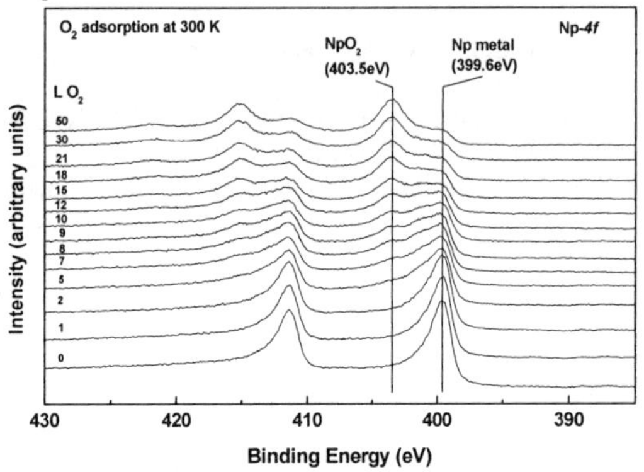

Figure 1 *Np (4f) core level spectra for increasing surface oxidation of Np metal*

Ultraviolet photoelectron spectroscopy was used to follow changes in the valence band. Consistent with the XPS observations, there is a loss of peak intensity attributable to Np metal at the Fermi level together with increasing intensities at approximately 1.5 and 3 eV below E_F (Figure 2). After 50 Langmuir exposure the He^{II} spectra is indicative of NpO_2 with the 5f band centred at 3 eV and an O 2*p* band as an asymmetric peak at 6 eV.[2,6]

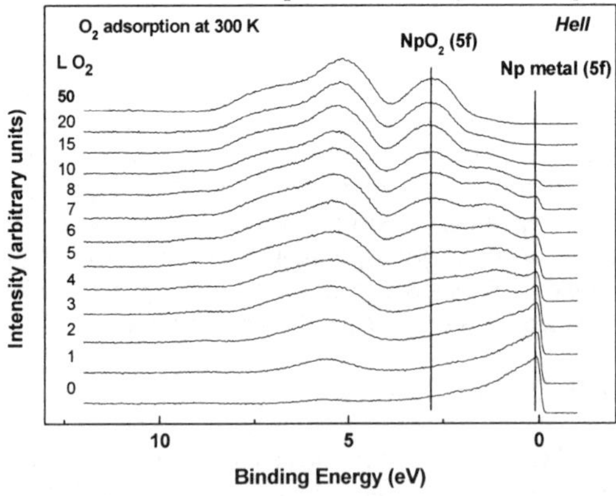

Figure 2 *Valance band photoelectron spectra for increasing surface oxidation at 300 K (He II, hυ = 40.8 eV)*

We attempted to isolate, as a single phase, the chemical species giving rise to the Np $4f_{7/2}$ XPS, and the valance band peaks centred at 401.2 and 1.5 eV, respectively. Controlled oxygen exposures at 80 K afforded similar data to that acquired at room temperature, albeit with an enhanced sticking coefficient. There was no change in the photoelectron spectra of a sample that had been pre-exposed to oxygen and heated to 473 K for short periods.

The data presented here agree well with those reported previously for Np_2O_3.[2] The original assignment of this surface sesqui-oxide phase was made by comparison with PES data obtained for Pu surface oxidation. However, it should be noted that according to the Pu-O binary phase diagram Pu_2O_3 can be prepared as a single phase material.[7] Furthermore, Pu_2O_3 is thermodynamically stable with respect to PuO_2 in the presence of Pu metal,[8] such that thin surface films of PuO_2 undergo a facile auto-reduction to Pu_2O_3 in UHV at room temperature.[9] As little change was observed for Np oxidation as a function of temperature, this would imply Np_2O_3 is thermodynamically stable. However, phase diagram studies for the system Np-O [1] exclude the existence of Np_2O_3 and a single phase Np_2O_3 material could not be prepared. This would suggest that the intermediate peaks observed, at a binding energy between the peaks due to Np metal and NpO_2, are due to surface Np(chem) states, associated with the chemisorption phase consisting of Np atoms bonded to less than two oxygen atoms. A plausible mechanism is that the chemisorbed state consists of a single layer of oxygen islands which grow to completely cover the surface at approximately 5 L. The onset of oxidation, evidenced by the appearance of Np^{4+} species in the XPS and UPS begins at around 5 L, only after the surface is completely covered by chemisorbed oxygen. From 5 L onwards oxide island nuclei grow to completely cover the surface by *ca* 50 L. Quantitative analysis of the curve-fitted Np $4f_{7/2}$ XP spectra suggest three different oxidation stages: a first stage up to ~5 L; a second stage up to ~20 L exposures; and a third stage characterized by a slow growth above 20 L. An unambiguous identification of Np_2O_3 cannot be made without surface structural measurements.

4 CONCLUSION

We have undertaken a study of the oxidation of neptunium thin films using XPS and UPS. The results are similar to those reported previously for the surface segregation of dissolved oxygen. We observe the formation of NpO_2 at high oxygen exposures and an intermediate species at low exposures. From the spectra collected, quantitative analysis of the XPS data, and thermodynamic data, we propose the existence of a chemisorbed phase consisting of a layer of oxygen islands which grow to completely cover the surface at ~5 L prior to oxide formation.

Acknowledgements

The authors would like to thank the EPSRC for funding PN, AWE for a CASE award to PN and ITU Actinide Userlab for supporting this work.

References

1 K. Richter and C. Sari, *J. Nucl. Mater.*, 1987, **148**, 266.
2 J.R. Naegele and L.E. Cox, *Inorg. Chim. Acta*, 1987, **139.** 327.
3 J.R. Naegele, *J. Phys. (Paris)*, 1984, **45**, 841.
4 T. Gouder, *J. Electro. Spectro. Rel. Phen.*, 1999, **101-103**, 419.
5 L.E. Cox and J.D. Farr, *Phys. Rev. B*, 1989, **39**, 11142.
6 B.W. Veal and D.J. Lam, *Phys. Rev. B*, 1976, **15**, 2929.
7 H.A. Wriedt, *Bull. Alloy. Phase Diag.*, 1990, **11**, 184.
8 J. Martz, J.M. Haschke and J. Stakebake, *J. Nucl. Mater.*, 1994, **210**, 130.
9 P. Morrall, S. Tull, J. Glascott, M. Pooley and P. Roussel, unpublished results.

ELECTRONIC STRUCTURE OF MOLECULAR ACTINIDE COMPOUNDS THROUGH SPECTROSCOPIC EXPERIMENTS COMBINED WITH THEORETICAL MODELLING

D.J.M. Meyer,[1] S. Hilaire,[1] D. Guillaumont,[1] S. Fouchard,[1] F. Wastin,[2] E. Colineau,[2] T. Gouder,[2] J. Rebizant,[2] J.C. Berthet,[3] M. Ephrethikhine[3] and E. Simoni[4]

[1]CEA, Marcoule, France
[2]EC-JRC-ITU, FZK, Karlsruhe, Germany
[3]CEA, Saclay, France
[4]IPN, Paris XI, Orsay, France

1 INTRODUCTION

The large variety of oxidation states observed for the light actinides (U to Am) is usually attributed to the contribution of the 5*f* states to the valence orbitals.[1-11] For the heavier actinides, where the 5*f* electrons are non-bonding, the actinides become rare-earth like with fewer oxidation states. However, it is still not understood what dictates the stability of a given oxidation state, and how it is dependent on the chemical environment (coordination sphere, nature of the counter-ion, etc.).

This communication focuses on the physical ground state investigations of molecular actinides species, using photoelectron spectroscopy (XPS), magnetic measurements (SQUID) and Mössbauer spectroscopy to aid in the understanding of the electronic structure. This work, undertaken by collaboration, and through the Actinide User Lab (ITU JRC), is described as a model for further interdisciplinary collaborations within the ACTINET network.

The actinyl structures of uranium and neptunium compounds have been analysed, from $5f_0$ through to $5f_2$ electronic configurations, by the above-mentioned techniques coupled with more classical UV-visible and infrared techniques. The representation of the molecular wave functions has been described by a theoretical approach using relativistic pseudopotential DFT modelling.

The overall results will not be discussed here in a way to know how far these techniques can be used for qualitative electronic speciation of actinides molecular compounds, especially for the stabilisation of high oxidation states.

2 EXPERIMENTAL METHODS

Photoelectron, SQUID and MÖSSBAUER experiments were undertaken at the JRC-ITU facility (Karlsruhe, Germany). XANES data were collected at the LURE synchrotron. All other characterisations, as well as neptunium compound syntheses, were undertaken at the ATALANTE facility (France).

For the uranyl compounds, to avoid ligand redistribution and in order to incorporate strong σ and π donation in the equatorial plane, non aqueous chemistry was considered. In the case of neptunium, due to higher radioactivity, the nature of the material and related experimental complication, synthesis was performed in air. Uranyl complexes were synthesized and structurally characterised at the "Service de Chimie Moleculaire" (CEA, Saclay) (Table 1). For the neptunyl complexes, thermogravimetric analyses, IPC-AES and α-counting were usc to obtain basic characterisation data (see Table 1).

Table 1 *Actinyl compounds studied*

Uranyl	Neptunyl	
$(UO_2Cl_2(THF)_3)_2$	$KNpO_2CO_3,H_2O$	(V)
$[UO_2(OPPh_3)_4][CF_3SO_3]_2$	NpO_2OH,H_2O	(V)
$UO_2(CF_3SO_3)_2$	NpO_2CO_3, H_2O	(VI)
	$NpO_2(OH)_2,H_2O$	(VI)

3 RESULTS FOR URANYL COMPOUNDS

L_{III} level XANES spectra were acquired for the uranyl compounds. The energy of the edge agrees with the oxidation state, but the width and the high energy of the peak do not permit the separation of the geometric effects from the electronic effects.

The Raman spectra show rather large differences in the UO_2 symmetric vibration between the two compounds (Table 2). For the actinyl species in aqueous solution, a difference of 50 cm^{-1} is usually observed in the case of an oxidation state change from AnO_2^+ to AnO_2^{2+}.

Table 2 *Raman data for uranyl compounds*

Compound	$[UO_2(OPPh_3)_4][CF_3SO_3]_2$	$UO_2(CF_3SO_3)_2$
Raman (cm^{-1})	993	897

Raman and U 4*f* XPS data exhibit a higher positive charge at uranium for the triflate than for the phosphineoxide. The experimental magnitude of these changes for both techniques is large as usually encountered for oxidation state changes in aqueous chemistry. This indicates that for these types of compounds, the equatorial ligands have greater importance in the overall electronic stabilisation than in the related aqueous compounds.

Relativistic pseudopotential DFT calculations on model compounds, $UO_2X_2(H_2O)_3$ (where X=Br^-, Cl^-, F^-, OH^-, CO_3^{2-}), show two types of interactions in the equatorial plane. One wherc the equatorial ligand disturbs the uranyl system, which is the case for Cl^- and Br^-, and another interaction observed for F^-, OH^-, CO_3^{2-}, strongly centred on the uranium atom interacting less with the axial uranyl system.

4 RESULTS FOR NEPTUNYL COMPOUNDS

Photoelectron data acquisition was performed on several Np(V) compounds, Np 4*f* levels are given in Table 3. These results show that the positive charge at the Np core for the carbonate is higher than for the hydroxide.

Table 3 *Photoelectron data for neptunyl compounds*

Compound	NpO_2OH,H_2O	$KNpO_2CO_3,H_2O$
Np $4f_{5/2-7/2}$ (eV)	404.3 / 415.9	405.8 / 416.6

Magnetisation measurements (SQID) were done on Np(V) and Np(VI) compounds and focused on the paramagnetic effective moment (Table 4).

Table 4 *Effective paramagnetic moments of neptunyl compounds (μ_B)*

$NpO_2(OH)_2,H_2O$	NpO_2CO_3, H_2O	NpO_2OH,H_2O	$KNpO_2CO_3,H_2O$
1.91	2.29	2.59	2.66

In a very simple way, the results suggest that a low magnetic moment can be correlated with a higher charge at the Np core, indicating that neptunium hydroxide has a higher positive charge than neptunium carbonate. A comparison with the XPS data highlights a contradiction, proving the need to integrate crystal and spin-orbital effects when interpreting magnetic data.

Relativistic pseudopotential DFT calculations on neptunyl halides, hydroxides and carbonates show the same results as for the uranyl model compounds.

5 CONCLUSIONS

To achieve direct charge investigations at the metal core, x-ray photoelectron spectroscopy will probably be the best tool available for molecular species, and shifting from x-ray to UV analysis will give a more accurate description of the valence shell. For magnetic measurements, physical models need to be developed for electronic structure investigations. As modelling tools, DFT calculations are a good starting point for a qualitative description of the ligand effects on the electronic structure of actinyl compounds.

References

1 B.D. Dunlap and G.M. Kalvius, *Handbook on the Physics and Chemistry of the Actinides*, ed., A.J. Freeman and G.H. Lander, 1985, vol 1-5.
2 G. Schreckenbach, P.J. Hay and R.L. Martin, *Inorg. Chem.*, 1998, **37**, 4442.
3 J.C. Eisenstein, M.H.L. Pryce, *Proc. R. Soc. London*, 1955, **A229**, 20.
4 R.E. Connick and Z.Z. Hugus, *J. Am. Chem. Soc.*, 1952, **74**, 6012.
5 P. Pyykkö, *Inorg. Chim. Acta*, 1987, **139**, 989.
6 K. Tatsumi and R. Hoffmann, *Inorg. Chem.*, 1980, **19**, 2656.
7 R.G. Denning, *Struct. Bonding*, 1992, **79**, 215.
8 Y.A. Teterin, V.A. Terehov, M.V. Ryzhkov, I.O. Utkin, K.E. Ivanov and A.S. Nikitin, *J. Electron. Spectrosc. Relat. Phenom.*, 2001, **114-116**, 915.
9 R.G. Denning, J.C. Green and T.E. Hutchings, *J. Chem. Phys.*, 2002, **117**, 8008.
10 E. Thibaut, J.-P. Boutique, J.J. Verbist, J.-C. Levet and H.J. Noël, *J. Am. Chem. Soc.* 1982, **104**, 5266.
11 G.R. Choppin, *Radiochim. Acta*, 1983, **32**, 43.

INFLUENCE OF SELF-IRRADIATION DAMAGE ON THE SUPERCONDUCTING BEHAVIOUR OF PLUTONIUM-BASED COMPOUNDS

F. Jutier,[1] J.-C. Griveau,[1] E. Colineau,[1] J. Rebizant,[1] P. Boulet,[1] F. Wastin[1] and E. Simoni[2]

[1]European Commission, Joint Research Centre, Institute for Transuranium Elements, Postfach 2340, Karlsruhe 76125, Germany
[2]Institut de Physique Nucléaire, Orsay cedex 91406, France

1 INTRODUCTION

External irradiation of materials creates defects and induces lattice disorders. Interestingly, defects are sometimes helpful for technological purposes hence, controlled defect creation has drawn prime attention to superconductors, especially A15 phases and high temperature superconductors. The aim is to make them economically viable by improving their critical current density (irradiation-induced defects act as pinning centres for vortices, which are quantized magnetic fluxes entering the material).[1]

In this context, the discovery of $PuCoGa_5$[2] and $PuRhGa_5$[3] superconductors is very interesting. Firstly, they represent two new cases of non-conventional superconductors[4,5] and secondly, they provide unique cases in which to study the impact of self-radiation damage (due to the α-decay of Pu atoms) on the superconducting critical parameters.

In this work, we report on the ageing effects on $^{239}PuCo_{1-x}Rh_xGa_5$ solid solutions to improve our understanding of the interactions between defects and vortices, without performing any extra irradiation experiments. The evolution of the critical parameters (T_c, H_{c2}, J_c) with time, has been followed by employing SQUID magnetometry and electrical resistivity measurements.

2 RESULTS

As a preliminary remark, we should highlight that although the evolution of the critical parameters (T_c, H_{c2}, J_c) has been followed experimentally as a function of time, we report our results versus displacements per atom (dpa). The dpa unit is the appropriate damage parameter to correlate the effects of radiation damage on different samples with various masses and containing different radioactive isotopes (i.e. different radiation doses received).

2.1 Evolution of the Critical Temperature T_c with Ageing

It was reported[6] recently that despite the same Pu batch being used for the synthesis of two Pu-based superconductor compounds, they were not equally sensitive to radiation damage (the decrease rate of the critical temperature (T_c) was twice as high in $PuRhGa_5$ compared to that of $PuCoGa_5$). In Figure 1a, we show the normalised variation of T_c from t = 0 versus displacements per atom (dpa) for various aged samples of $PuCo_{1-x}Rh_xGa_5$ solid solutions.

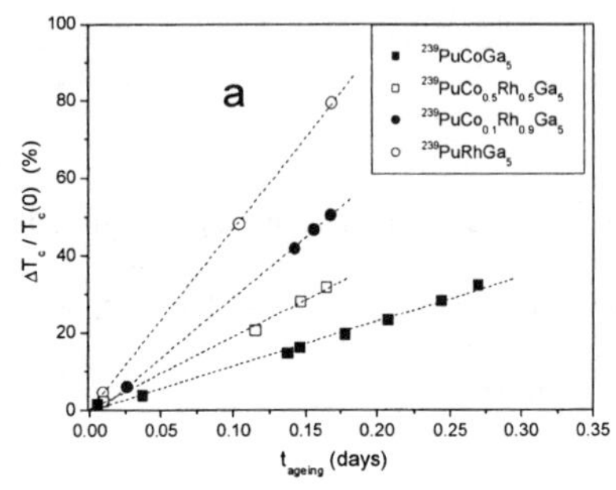

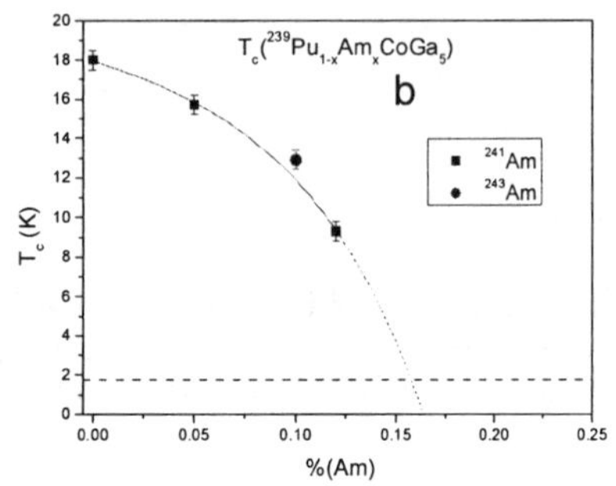

Figure 1 (a) *Evolution of the Normalized Decrease Rate (NDR) of T_c of $PuCo_{1-x}Rh_xGa_5$ solid solutions versus displacements per atom (dpa). (b) Critical temperature of $^{239}Pu_{1-x}Am_xCoGa_5$ solid solutions as a function of Am doping. The dashed line shows our low-temperature limit*

The normalised decrease rate (or NDR), ($\Delta T_c/T_c(0)$, is linear, as we are far from the regime of saturated defects, and it increases with the amount of Rh. The NDR difference observed between the Co- and Rh- Pu compounds can be explained by the coherence length (or ξ_0) of the Cooper pairs in these compounds. Calculation, using the Ginzburg-Landau theory, gives a value twice larger in $PuRhGa_5$ (4 nm) than in $PuCoGa_5$ (2 nm). A larger coherence length is expected to be more sensitive to self-radiation damage.

$Pu_{1-x}Am_xCoGa_5$ solid solutions have also been investigated to enhance the ageing phenomenon, utilising the higher specific activity of Am isotopes. The critical temperature is more strongly affected by Am substitution (Figure 1b). The decrease rates, obtained from T_c measurements, are in agreement with the assumption that the higher the self-radiation effect, the more the superconductivity is weakened by the creation of defects (~ -0.24 K.month^{-1} for $^{239}PuCoGa_5$, ~ - 0.28 K.month^{-1} for $^{239}Pu_{0.9}{}^{243}Am_{0.1}CoGa_5$ and ~ -1.13 K.month^{-1} for $^{239}Pu_{0.88}{}^{241}Am_{0.12}CoGa_5$) and this process finally destroys the Cooper pairs.

2.2 Evolution of the Upper Critical Field $H_{c2}(0)$ and Critical Current Density J_c with Ageing

Figure 2a shows that $H_{c2}(0)$ of $PuCoGa_5$, inferred from magnetoresistance measurements, increases up to a maximum of ~ 135 T at ~ 0.17 dpa and then decreases. On the contrary, the Rh compound shows an abrupt decrease of $H_{c2}(0)$ with ageing and already for ~ 0.10 dpa, is no longer superconducting. This different behaviour of the two compounds can be interpreted within the "clean-dirty" limit model[1] of superconducting materials. Considering the evolution of ξ_0 compared to the mean free path l (as illustrated in the case of $PuCoGa_5$ in Figure 2b), we observe that the values of ξ and l become comparable for an amount of damage equivalent to that corresponding to the $H_{c2}(0)$ decrease.

Sarrao *et al.*[2] reported a critical current density, $J_c > 10^4$ A/cm^2 for T > 16.2 K for a fresh sample, followed by an increase of nearly a factor 2 over 2 months ageing. The critical current density, J_c, inferred from the width of the hysteresis loops of magnetisation measurements according to the Bean critical state model, is in fair agreement with the value reported in Figure 2a.

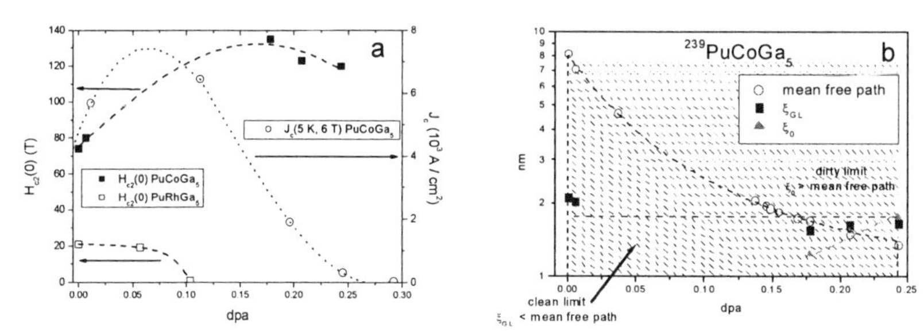

Figure 2 (a) *Evolution of H_{c2} (left hand axis) extrapolated at 0 K and J_c (right hand axis), and (b) Evolution of the superconducting coherence length ξ of the mean free path l of ^{239}PuCoGa$_5$ versus displacements per atom (dpa)*

Firstly, an enhancement of J_c is observed for $PuCoGa_5$ proving that self-radiation defects act as pinning centres. For longer ageing times, J_c decreases as the large defect density zones destroy the superconducting state and finally weaken the trapping forces which retain the vortices.

3 CONCLUSIONS

The evolution of the superconducting critical parameters in Pu-containing superconductors has been studied as a function of damage created by self-radiation effects induced from Pu α-decay. It was shown that the critical temperature is particularly sensitive with ageing time. $H_{c2}(0)$ and the critical current density, J_c, are at first, remarkably enhanced due to the very efficient properties of the pinning centres induced by self-radiation damage. It is also shown that the different behaviour observed for the $PuCoGa_5$ and its Rh-counterpart can be understood from the respective size of the coherence length of their Cooper pairs within the classical model of superconductivity. Further ageing weakens and then destroys the superconducting behaviour in these compounds.

Acknowledgements

The high purity Pu metals required for the fabrication of the compounds was made available through a loan agreement between Lawrence Livermore National Laboratory and ITU, in the frame of a collaboration involving LLNL, Los Alamos National Laboratory, and the US Department of Energy.

References

1 N.W. Ashcroft and N.D. Mermin, *Solid State Physics*, Saunders College Publishing, Cornell University, 1976.
2 J.L. Sarrao, L.A. Morales, J.D. Thompson, B.L. Scott, G.R. Stewart, F. Wastin, J. Rebizant, P. Boulet, E. Colineau, G.H. Lander, *Nature*, 2002, **420**, 297.
3 F. Wastin, P. Boulet, J. Rebizant, E. Colineau, G.H. Lander, *J. Phys.: Condens. Matter*, 2003, **15**, S2279.
4 T. Maehira, T. Hotta, K. Ueda, A. Hasegawa, *Phys. Rev. Lett.*, 2003, **90**, 207007.
5 I. Opahle, P.M. Oppeneer, *Phys. Rev. Lett.*, 2003, **90**, 157001.
6 F. Jutier, J.-C. Griveau, E. Colineau, J. Rebizant, P. Boulet, F. Wastin, E. Simoni, *Physica B*, 2005, **359-361**, 1078.

MARS BEAMLINE, A NEW FACILITY FOR STUDYING RADIOACTIVE MATTER FROM A SYNCHROTRON SOURCE

B. Sitaud[1] and S. Lequien[2]

[1]Synchrotron SOLEIL, Saint Aubin BP48, Gif sur Yvette, France, F-91192
[2]Laboratoire Pierre Süe, UMR 9956 CEA-CNRS, Gif sur Yvette, France, F-91191

1 INTRODUCTION

At the new third generation synchrotron SOLEIL[1] (Source Optimisée de Lumière d'Energie Intermédiare de Lure) and at the beginning of 2007, it will be possible to carry out high quality measurements on radioactive materials (α, β, γ and n emitters), including a large variety of actinides from MARS (MAtière Radioactive at Soleil) beamline. This will be the fourth beamline in Europe for analysing radionuclides, after ROBL at ESRF, INE beamline at ANKA, and the micro-focus beamline at SLS. The main characteristic of the MARS beamline is that its different experimental stations will accept radioactive matter with an activity of up to 18 GBq per sample and a dose rate limitation of 3 µSv/h at a distance of 1 m, values never offered before for any kind of emitters on a synchrotron (SR). This dedicated beamline has been designed mainly to bring together updated X-ray techniques (fluorescence, absorption, diffraction) available on high brilliant SR sources and all aspects of safety regulations which are generally encountered especially for "hot" laboratories. In this frame, new complementary fields in radionuclide sciences could be covered from the use of these different beamlines.

The needs of the scientific community, which are at the genesis of this beamline, are numerous and clearly multidisciplinary. In recent years, actinides and other radioactive elements have been the subject of important investigations related to environmental, geological and technological issues. In practice the technological field has been mainly concerned with different aspects of nuclear fuel processing and reprocessing which, for a large part, result in experimental proposals that could be developed on the MARS beamline. Most of them refer to results obtained from current beamlines on other SR facilities using ad-hoc sample confinements, considering low activity or dilute samples. These investigations[2-4] demonstrate the usefulness of synchrotron sources, with respect to laboratory ones, to solve physico-chemical problems. They emphasize the need for a complementary beamline for performing *in situ* characterisation under more or less drastic conditions of temperature, pressure, pH and/or Eh on highly irradiating or radioactive samples.

As a general rule the radioactive matter studied on the MARS beamline would be encapsulated within at least double containments and, when necessary, adapted shielding will be added around the samples to ensure the maximum dose rate limit is not exceeded.

Note that the eventual preparation of samples inside special glove boxes on the beamline will be possible and restricted to particular cases after safety approval.

2 MARS BEAMLINE LAYOUT

The MARS beamline is being built on the bending magnet port, D03-1, of the SOLEIL storage ring. The 1.71 T bending magnet field provides a continuous spectrum of photons with a critical energy of 8.6 keV. An overall layout of the MARS beamline is shown in Figure 1. The results of a global optimization design takes into account the experiment needs, safety regulations and the mechanical constraints of building inside the experimental hall of SOLEIL.

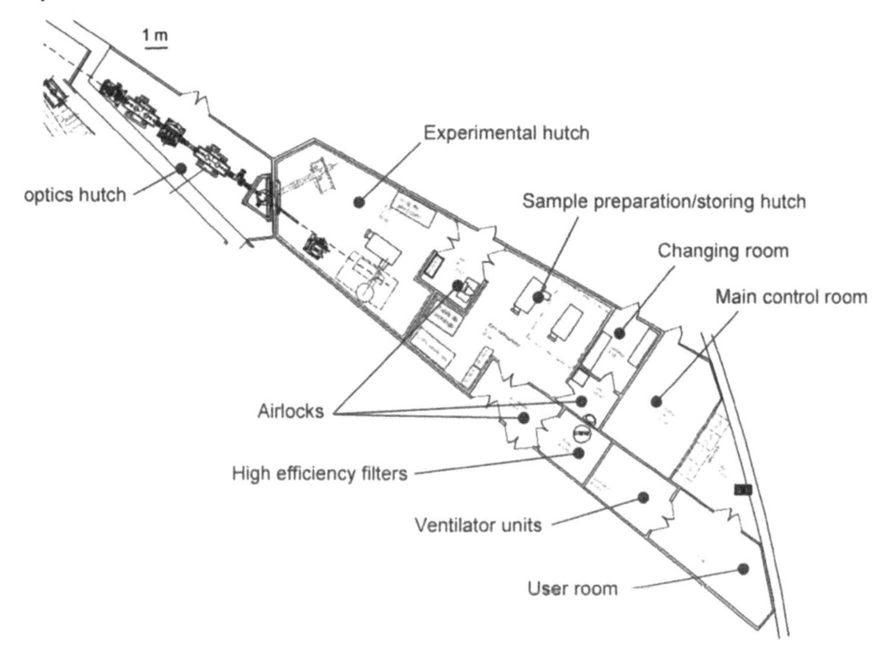

Figure 1 *Schematic view of the MARS beamline*

Basically the beamline is divided into three main functional areas. The first one is the Pb shield optics hutch in which the X-ray optical elements are implemented. The original design is based on the alternative use of two long mirrors and two monochromators, to allow both optical configurations, depending on the selected technique. All these elements have been defined to reach the maximum flux inside the energy range, from 3.5 to 35 keV, with the possibility of focusing the beam to a minimum size of around 100 x 100 μm^2. In addition, micro beams with dimensions below 20 x 20 μm^2 will be possible by the use of a set of mirrors in Kirkpatrick-Baez geometry. The second area includes the experimental and the sample preparation hutches, and will be classified as "Listed Installation for the Protection of the Environment" (or ICPE in French). Hence, in order to maintain a dynamic confinement, these hutches will be equipped with airlocks and kept at low pressure by a dedicated ventilation system. Very high efficiency filters located at the intake air point and at the blowing point will trap radioactive material particles in the case of failure of sample confinement barriers. Pressure, temperature and radioactivity shall be continuously controlled and recorded by a specific monitoring unit. In addition a fire-proof coating on the walls and on the ceiling will define two independent fire sectors having a fire resistance of 2 hours. Considering the regulation aspects, the fixed equivalent activity limits referring to radioisotopes of group 1 are 18.5 GBq per sample. A total activity of 185 GBq shall be stored inside the different beamline hutches. The last area is free of radiological (irradiation and contamination) safety regulations and is composed of a control room, ventilation and pump units room and a user room for preparation of experiments or online analysis of experimental results.

3 EXPERIMENTAL STATIONS

At least four complementary techniques of analysis will be proposed: high resolution powder diffraction (XRD), X-ray fluorescence (XRF), standard X-ray absorption spectroscopy (XAS) and dispersive XAS (DXAS) with the ability of focusing the X-ray beam over the whole energy range for all of them.

A robust 2-circle diffractometer will be available for high resolution powder diffraction measurements based on two coaxial rotary tables with high loading capacities for the use of heavy sample containers with weights up to 750 N. For particular studies (single crystal, stress and texture measurements) an open Eulerian cradle can be mounted on the omega circle to provide two more circles. In addition it would also be possible to perform quick measurements taking advantage of the angular resolution by using either an energy resolving Ge detector or a linear position-sensitive detector coupled to Söller slits attached on the detection arm. Because of the high background signal coming from γ radioactive decay of samples, experiments will be limited to α or β emitters in this latter case.

A general purpose X-ray absorption spectroscopy setup will be available 3 m downstream of the diffraction station. This second station will also be equipped for X-ray fluorescence analyses. Transmission and fluorescence detection will be available using photodiodes and multi-element solid-state detectors, respectively. However, as already mentioned above, additional equipment, like bandpass filter devices that remove unwanted energies (noise), will be necessary for optimizing the signal-to-noise ratio. Samples shall be directly analyzed inside either standard sample holders or a glove box depending on the risks estimated for each experiment (sample state, variations of external parameters, etc.). Measurements at low temperatures (e.g. 5 K) could be achieved using a special cryostat with additional shielding around the sample. Moreover, different options to reach high temperature (electrical resistance or laser radiation) will be considered to develop complementary devices. The sample confinement and the safety shall be guaranteed during the whole duration of the experiment and whatever the sample environment.

References

1 Synchrotron SOLEIL, *Synchrotron Radiation News*, 2003, **16**, 49.
2 P. Martin, M. Ripert, Th. Petit, T. Reich, Ch. Hennig, F. D'Acapito, J.-L. Hazemann and O. Proux, *J. Nucl. Mater.*, 2003, **312**, 103.
3 G. Rousseau, L. Desgranges, J.-C. Niepce, J.-F. Berar and G. Baldinozzi, *Mat. Res. Soc. Symp. Proc.*, 2004, **802**, DD1.2, 1.
4 Ch. Den Auwer, E. Simoni, S. Conradson and C. Madic, *Eur. J. Inorg. Chem.*, 2003, **21**, 3843.

SOLID STATE SYNTHESIS AND X-RAY DIFFRACTION CHARACTERIZATION OF $Pu^{3+}_{(1-2x)}Pu^{4+}_{x}Ca^{2+}_{x}PO_4$

D. Bregiroux,[1,2] R. Belin,[1] F. Audubert[1] and D. Bernache-Assollant[3]

[1]Commissariat à l'Energie Atomique, DEN/DEC/SPUA, Cadarache, 13108 Saint Paul Lez Durance, France
[2]SPCTS, Université de Limoges, 123 avenue Albert Thomas, 87060 Limoges, France
[3]Ecole Nationale Supérieure des Mines, 158 cours Fauriel, 42023 Saint Etienne, France

1 INTRODUCTION

In the framework of the 1991 French law concerning nuclear waste management, several studies have been carried out in order to elaborate crystalline matrices for specific immobilization of the radionuclides. In the case of high level and long-lived minor actinides (Np, Am and Cm), which are high level and long-lived radioactive elements, monazite, a light rare earth (*Re*) orthophosphate with general formula $Re^{3+}PO_4$ (with Re = La to Gd), has been proposed as a host matrix, due to its high resistance to self irradiation and its low solubility. Monazite crystallizes in the monoclinic space group $P2_1/n$. In this structure, trivalent cations (Re^{3+}) could be substituted by an equivalent amount of bivalent (A^{2+}) and tetravalent (B^{4+}) cations, allowing the simultaneous incorporation of Am^{3+}, Cm^{3+} and Np^{4+}. According to Podor's work,[1] the limit of incorporation of a tetravalent element in monazite is related to its size in the ninefold coordination (R^{IX}). $Re^{3+}_{1-2x}A^{2+}_{x}B^{4+}_{x}PO_4$ exists in the monazite structure if 1.216 Å $\geq R_{average} \geq$ 1.107 Å and 1.238 $\geq R_{ratio} \geq$ 1 with

$$R_{average} = (1-2x)R^{IX}_{Re^{3+}} + xR^{IX}_{A^{2+}} + xR^{IX}_{B^{4+}} \quad (1)$$

$$R_{ratio} = \frac{(1-2x)R^{IX}_{Re^{3+}} + xR^{IX}_{A^{2+}}}{(1-2x)R^{IX}_{Re^{3+}} + xR^{IX}_{B^{4+}}} \quad (2)$$

The present work deals with the incorporation of the Pu^{4+}/Ca^{2+} couple in the monazite structure by solid-state synthesis. According to the Equations (1) and (2), the maximum incorporation is x=0.43, with $Re^{3+}=Pu^{3+}$, leading to a compound with the formula $Pu^{3+}_{0.14}Pu^{4+}_{0.43}Ca^{2+}_{0.43}PO_4$.

2 MATERIALS AND METHODS

The monazite powders were prepared according to the following reactions:

Exp. A: $PuO_2 + NH_4H_2PO_4 \rightarrow Pu^{3+}PO_4 + NH_3\uparrow + {}^3/_2H_2O\uparrow + {}^1/_4O_2\uparrow$ (3)

Exp. B: ${}^1/_2PuO_2 + {}^1/_2CaO + NH_4H_2PO_4 \rightarrow Pu^{4+}_{0.5}Ca^{2+}_{0.5}PO_4 + NH_3\uparrow + {}^3/_2H_2O\uparrow$ (4)

Both compounds were prepared on the scale of a few milligrams. Starting materials were homogenized by manual grinding in an agate mortar and then fired in a platinum crucible at 1400°C for 2h under air atmosphere in an alumina tubular furnace. The whole process was repeated in order to obtain homogeneous materials. Powders were characterized by X-ray diffraction at room temperature using a high-resolution Siemens D5000 X-ray diffractometer with a curved quartz monochromator and copper radiation from a conventional tube source.

3 RESULTS AND DISCUSSION

3.1 Structure Analysis

X-ray diffraction patterns of the two resulting materials are shown in Figure 1. For both, it was found that the plutonium phosphate crystallizes in the monazite structure as expected. For experiment A, a secondary phase was identified as a tetravalent plutonium phosphate $Pu^{4+}P_2O_7$. This suggests that under air atmosphere, the Pu^{4+} is not completely reduced to Pu^{3+}. This result is in agreement with those of Bamberger[2] who showed that the apparent stability of PuP_2O_7 in air is due to its very slow rate of decomposition to $PuPO_4$. On the other hand, the result of experiment B is single-phased.

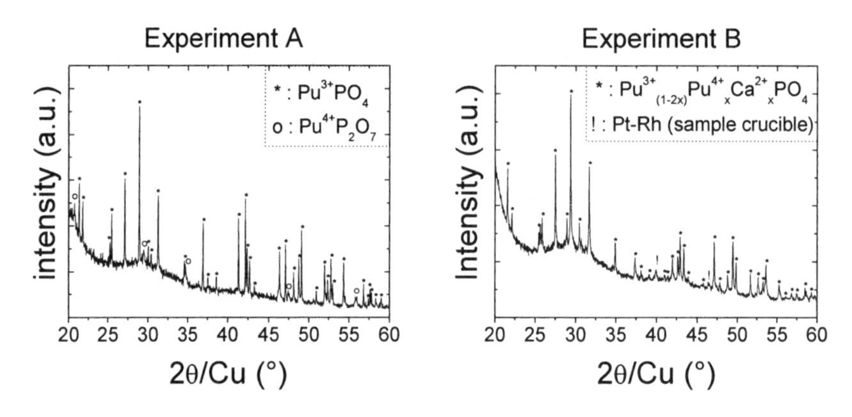

Figure 1 *X-ray diffraction diagrams of the synthesized powders*

Lattice parameters of the monazite structure are given in Table 1. They are lower for monazite B than for monazite A. Since Pu^{4+} and Ca^{2+} are smaller than Pu^{3+}, one can assume that Pu^{4+} was incorporated in the monazite structure.

Table 1 *Cell parameters of the synthesized monazite*

	a (nm)	b (nm)	c (nm)	β (°)
Monazite A	0.675	0.697	0.643	103.64
Monazite B	0.667	0.687	0.636	103.99

3.2 Determination of the Chemical Composition

Recently, Terra[3] showed a linear correlation between the cell parameters and chemical composition for $La^{3+}{}_{1-2x}Th^{4+}{}_xCa^{2+}{}_xPO_4$ and $Ca^{2+}{}_{0.5}Th^{4+}{}_{0.5-y}U^{4+}{}_yPO_4$. It is therefore possible to determine the Pu^{4+} incorporation ratio in our monazite B, from the cell parameters of $Pu^{3+}PO_4$ (monazite A) and those of $Ca^{2+}{}_{0.5}Pu^{4+}{}_{0.5}PO_4$. The cell parameters of $Ca^{2+}{}_{0.5}Pu^{4+}{}_{0.5}PO_4$ can be extrapolated from the two compositions of the $Ca^{2+}{}_{0.5}Np^{4+}{}_{0.5-y}Pu^{4+}{}_yPO_4$ reported by Tabuteau,[4] assuming a similar evolution (Figure 2 left: open circles). Figure 2 (right) shows the deduced Pu^{4+} incorporation ratio in the monazite B (squares). Results show that the composition of monazite B is $Pu^{3+}{}_{0.4}Pu^{4+}{}_{0.3}Ca^{2+}{}_{0.3}PO_4$ which has less Pu^{4+} than expected. According to our recent work on the incorporation of Ce^{4+} in the monazite structure,[5] residual Ca^{2+} should be incorporated in a secondary phase $Ca_2P_2O_7$. This compound was not observed by X-ray diffraction most likely because of the unfavourable signal-to-noise ratio.

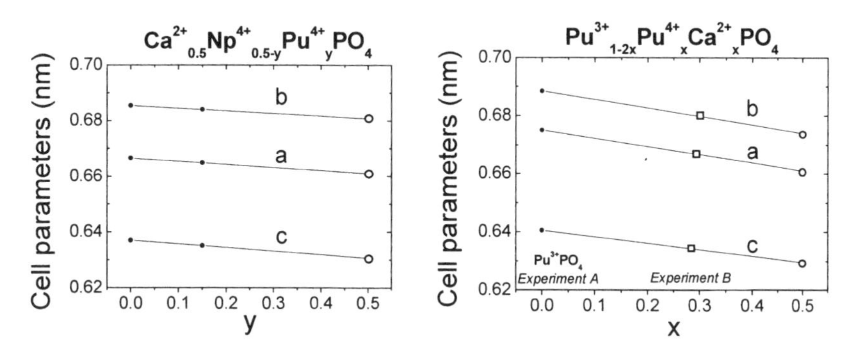

Figure 2 *Lattice parameters of $Ca^{2+}{}_{0.5}Np^{4+}{}_{1-y}Pu^{4+}{}_yPO_4$ and $Pu^{3+}{}_{1-2x}Pu^{4+}{}_xCa^{2+}{}_xPO_4$ solid solution versus y and x*

4 CONCLUSION

The solid-state synthesis of $Pu^{3+}{}_{(1-2x)}Pu^{4+}{}_xCa^{2+}{}_xPO_4$ was carried out in air. $Pu^{3+}PO_4$ was not obtained as a single phase and the maximum incorporation of Pu^{4+} in the monazite structure was found to be around x = 0.3. To complete the present work, the same experiments will be carried out under an inert atmosphere.

References

1 R. Podor and M. Cuney, *Amer. Miner.*, 1997, **82**, 765-771.
2 C.E. Bamberger, G.M. Begun and R.G. Haire, *J. Less-Common Met.*, 1984, **97**, 349-356.
3 O. Terra., PhD Thesis, Université d'Orsay, Paris XI, 2005.
4 A. Tabuteau, M. Pagès, J. Livet and C. Musikas, *J. Mater. Sci. Lett.*, 1988, **7**, 1315-1317.
5 D. Bregiroux, F. Audubert and D. Bernache-Assollant, *Study of tetravalent cerium incorporation in the monazite structure,* Atalante 2004 Conference, Nîmes, France.

CHEMILUMINESCENCE IN SOLID-PHASE REACTIONS OF URANIUM (IV) AND TERBIUM (III) COMPOUNDS

L.N. Khazimullina,[1] V.A. Antipin,[1] A.V. Mamykin,[1] I.G. Tananaev,[2] V.P. Kazakov[1] and B.F. Myasoedov[2]

[1]Institute of Organic Chemistry, Ufa Scientific Center of the RAS, Ufa
[2]Institute of Physical Chemistry of the RAS, Moscow

1 INTRODUCTION

Over the last ten years much of the attention of researchers has been directed towards the study of gas- and liquid-phase reactions involving the formation of excited molecules and ions. Solid-phase interactions possess interesting features, considerably differing from those in solutions, as they have the ability to maintain excited particles even in more stable intermediate stages of reactions.

Among the possible emitters of photons of chemiluminescence, the ions of lanthanide and uranyl attract a great deal of attention because of their typical emission spectra.[1-3]

2 EXPERIMENTAL METHOD AND RESULTS

2.1 The Reaction of Uranium (IV) Sulfate with Sodium Perxenate

We found that simple mixing of the $U(SO_4)_2{\cdot}4H_2O$ and $Na_4XeO_6{\cdot}8H_2O$ salts with a glass spatula and without trituration, resulted in greenish luminescence, visible by the unaided eye, even in only a slightly darkened room. At 18 °C and 26 °C, the time-dependence of the CL intensity (I) can be very well linearised in the $1/\sqrt{I}$ - t (time) coordinates (Figure 1 (*left*)). This type of dependence is typical of the CL of recombination processes.

The rate constants obtained from these curves are effective values, particularly since not only the elementary steps of formation of excited ions, but also the valence states of xenon compounds involved in the limiting steps, are unknown. An estimation of the activation energy, within such a small temperature range, gives a value of ≈ 22 (± 3) kcal. Presumably, the activation of redox processes themselves are overlaid by structural, relaxation or phase transitions either in the perxenate salt or in uranium sulfate. In the case of gradual heating, two maxima appear at 28 and 41 °C hence, dependence of the CL intensity on temperature exists. It should be noted that the thermal decomposition of pure perxenate occurs only above 300 °C, hence the changes in the curve are totally determined by the redox reactions of uranium ions with different valence forms of xenon compounds. The luminescence observed upon mixing $U(SO_4)_2{\cdot}4H_2O$ with $Na_4XeO_6{\cdot}8H_2O$ totally ceased after the complete decomposition of sodium perxenate at 300 °C.[4] It should be noted that no noticeable irradiation was detected during the gradual heating of pure sodium perxenate.

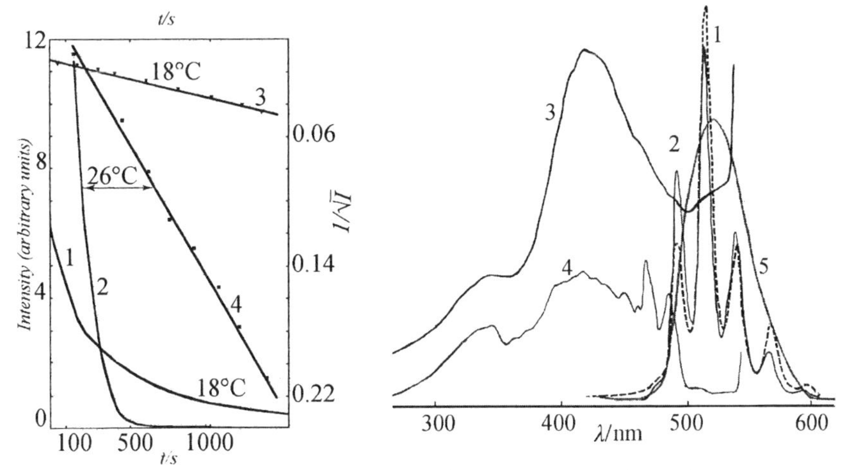

Figure 1 *(left) Kinetic curves showing the chemiluminescence decay for the reaction of $U(SO_4)_2{\cdot}4H_2O$ + $Na_4XeO_6{\cdot}8H_2O$ powders (5.2 and 1.4 mg, respectively) at 18 °C (1) and 26 °C (2); and their linear anamorphoses at their corresponding temperatures (3 and 4).*
(right) Photoluminescence (1, 2) and excitation spectra (3, 4) for the reaction products of $U(SO_4)_2{\cdot}4H_2O$ + $Na_4XeO_6{\cdot}8H_2O$ powders (7 and 1.5 mg, respectively) (1 and 3)(λ_{ex} = 380 nm, λ_{obs} = 570 nm) and for $UO_2SO_4{\cdot}3H_2O$ (2 and 4); (λ_{ex} = 350 nm, λ_{obs} = 560 nm); 18 °C. Chemiluminescence spectrum for the reaction of $U(SO_4)_2{\cdot}4H_2O$ + $Na_4XeO_6{\cdot}8H_2O$ powders recorded using light filters (5); 18 °C

Figure 1 (*right*) shows the photoluminescence and excitation spectra for reaction products of the $U(SO_4)_2{\cdot}4H_2O$ + $Na_4XeO_6{\cdot}8H_2O$ powders and for $UO_2SO_4{\cdot}3H_2O$. The CL spectrum obtained for the reaction of $U(SO_4)_2{\cdot}4H_2O$ and $Na_4XeO_6{\cdot}8H_2O$ powders confirms the fact that it is the excited uranyl ion $^*UO_2^{2+}$ that is the radiation emitter in the entire kinetic region studied.

The vigorous gas evolution is undoubtedly the main reason for the fast reaction between $U(SO_4)_2{\cdot}4H_2O$ and $Na_4XeO_6{\cdot}8H_2O$. This is accompanied by the decomposition of sodium perxenate crystals and renewal of the contact surface. Probably, rather an important role belongs to crystal water, which is also involved in the reaction with the perxenate ion; this also results in the destruction of crystals. The water molecules can form a water microphase on the surface of the reacting crystals, which considerably accelerates the redox reactions.

2.2 The Reaction of Terbium (III) Sulfate with Sodium Persulfate

The activation of persulfate salts with lanthanide ions was effected by the mixing and heating of terbium sulfate crystals with sodium persulfate, producing luminescence even at moderate temperatures (to 100 °C). With heating of the powdered mixture, green luminescence was observed with the unaided eye, in a slightly darked room (Figure 2). The emitter from this reaction was undoubtedly a terbium ion because the temperature of the reaction was too low for the occurrence of the diffusion of lanthanide ions into the crystal lattice of sodium persulfate, before its activation during the reaction. This testifies that the luminescence arises on the boundary of the crystal. During the reaction the surface is renewed continually due to the decomposition of persulfate ions, resulting in the formation of oxygen gas which, in turn, destroys the crystals. The luminescence ceases when the thermal destruction of persulfate is complete.

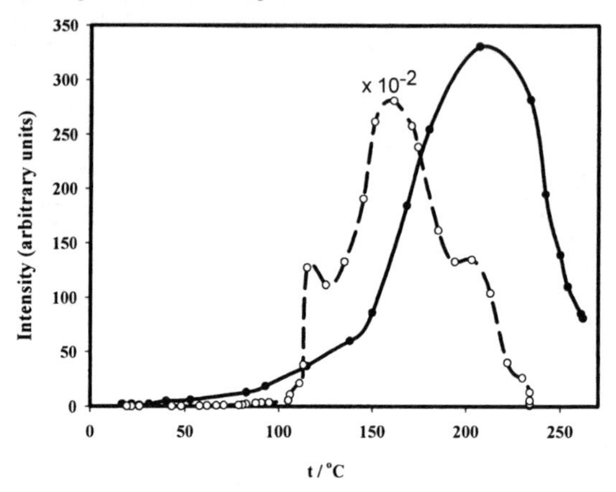

Figure 2 *Temperature dependence of CL intensity during gradual heating of (1) pure sodium persulfate(solid line) and (2) equimolar powder mixture of terbium (III) sulfate with sodium persulfate (broken line)*

3 CONCLUSION

The emitter of the luminescence during the solid-phase reaction of uranium (IV) sulfate with sodium perxenate is the uranyl ion. In the case of terbium (III) sulfate with sodium persulfate, it is the terbium ion which is responsible for the emission, confirmed by the correlation of photoluminescence and chemiluminescence spectra.

References

1 V.P. Kazakov, *Chemiluminescence of Uranium, Lanthanides and d-Elements*, Nauka, Moscow, 1980, p.176 (*in Russian*).
2 S.V. Lotnik, L.A. Khamidullina and V.P. Kazakov., *Dokl. Acad. Nauk*, 1999, **366**, 345 (*in Russian*).
3 S.V. Lotnik, L.A. Khamidullina and V.P. Kazakov. *Radiokhimiya*, 2001, **43**, 48.
4 A.H. Cockett, K.C. Smith, N. Bartlett and F.O. Sladky, *The Chemistry of the Monoatomic Gases*, Pergamon Press, Oxford, 1973, **4**, p. 338.

CHEMILUMINESCENT REACTIONS OF URANIUM AND THE ROLE OF UO_2^+

V.P. Kazakov and A.V. Mamykin

Institute of Organic Chemistry, Ufa Scientific Center of the RAS, Ufa

Since the discovery of electrochemiluminescence of the uranyl (UO_2^{2+}) and lanthanide ions,[1,2] these metals have found numerous applications as activators of chemiluminescence (CL). Moreover, CL has also been observed during the oxidation of uranium by various oxidants.[3]

At present, the formation of uranyl ions in the excited state, followed by their luminescence, is known to occur during many reactions. As shown in Scheme 1, these reactions can be classified into two types:

I. UO_2^{2+} is an energy acceptor[3]

$$\mathbf{Np(VII)} \xrightarrow[+UO_2^{2+}]{+H_20,\ HCOOH} \mathbf{Np(VI)} + \overset{*}{U}O_2^{2+} \longrightarrow \mathbf{hv} \quad \eta_{CL}=10^{-7}$$

II. Oxidation-reduction reactions of U (IV), U(V), U(VI)

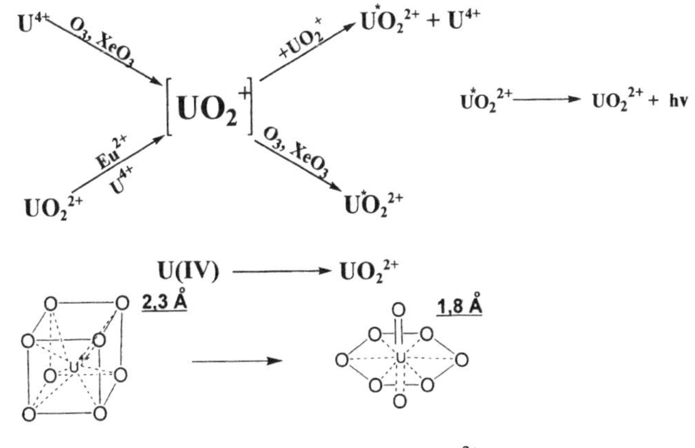

Scheme 1 Types of CL of UO_2^{2+}

UO_2^{2+} forms in the excited state because of the acceptance of the energy released during the reduction reaction of Np (VII). The excited intermediate is not always known as it takes place during the neptunium reaction mentioned above. However, there are reactions in which this intermediate can be identified precisely enough. For example, during the oxidation of elemental sulfur with ozone in uranyl solutions, in the presence of concentrated sulfuric acid, a short-lived molecule (SO*) has been identified as a primary excited particle, which penetrates into a coordination sphere of sulfate complexes of UO_2^{2+} and then transmits the excitation energy to this ion without emission.

$$S + O_3 \longrightarrow SO^* + O_2$$
$$SO^* + [UO_2(SO_4)_2(HSO_4)_2]^{4-} \longrightarrow [\overset{*}{U}O_2(SO_4)_2(HSO_4)SO]^{3-} + HSO_4^-$$
$$[\overset{*}{U}O_2(SO_4)_2(HSO_4)SO]^{3-} \longrightarrow [UO_2(SO_4)_2(HSO_4)SO]^{3-} + h\nu$$

Scheme 2

The transformation of chemical energy to light is a very efficient reaction and the yield of excited complex uranyl ions approximates 1.

During the oxidation-reduction reactions of uranium the yield of the excitation changes over a wide range. The oxidation reactions of low valency uranium species, with xenon compounds, are relatively more luminous.

Preceding the luminous stage, the formation of pentavalent uranium was found to be obligatory. The one-stage oxidation of tetravalent to hexavalent uranium, for example, via the transfer of an oxygen atom, does not occur because the distance between the atom of uranium and oxygen in the coordination sphere is 2.3 Å, while the distance of U=O in the uranyl ion is 1.8 Å, as seen in Scheme 1. That means that if this process took place, such a sharp shortening of the bond distance during a one-stage process should be equivalent to the production of a uranyl ion in the "hot" state, which inevitably leads to the deactivation of the electron excitation.

The disproportionation reaction of the pentavalent uranium ion (UO_2^+) is worthy of special mention. CL arises as a result of the reduction of the uranyl ion by divalent europium. The disproportionation reaction is the equibrium reaction.

In a perchloracidic solution the equilibrium (see Figure 1) is shifted strongly to the right, in the direction of the formation of tetravalent uranium and the uranyl ion. The most interesting feature of the reaction is that it is endothermic relative to the excited state of the uranyl ion. As seen from Figure 1, the sum of the activation energy and heat of reaction is less than the mean photon energy emitted by the excited ion ($^*UO_2^{2+}$) by 17 kcal. This is probably at the bottom of the relative low yield of CL of this reaction. At the same time, the equilibrium of the disproportion introduces also a new feature. Due to the reversible reaction between uranium (IV) and the uranyl ion, the equilibrium concentration of the pentavalent ion (UO_2^+) regenerates continuously in solution, in which following disproportionation, results in excited ions * UO_2^{2+} emitting photons.

A number of photons is not enough to affect the equilibrium appreciably, for example, by cooling the solution. However, this not great number of photons has to emit continuously, i.e. so much time as the direct and reversible reaction will proceed in a reaction vessel. No less than 100 photons/sec must be emitted from 1 ml of the solution, calculated using known values of the CL yield, rate and concentration constants at concentrations of U(IV) equal to that of the uranyl ion (10^{-3} M) and 0.1 M acidity. This amount must, at least, twice exceed the background radiation of radioluminescence by alpha particles emitted by U-238, which can be measured using modern techniques.

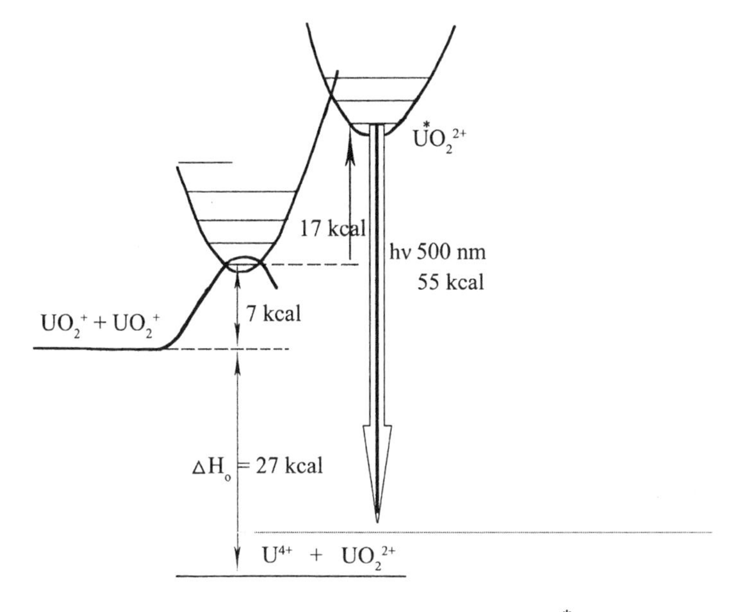

Figure 1 *Energetics of the reaction,* $2UO_2^+ + 4H^+ \rightleftharpoons U^{4+} + \overset{*}{U}O_2^{2+} + 2H_2O$

References

1 V.P. Kazakov, *Zh. Phys. Khim.*, 1965, **39**, 2936.
2 V.P. Kazakov, *Khemiluminestsentsiya uranila, lantanoidov i d-elementov (Chemiluminescence of uranyl, lanthanides and d-elements),* Nauka, Moscow, 1980, p. 176.
3 S.V. Lotnik, L.A. Khamidullina and V.P. Kazakov, *Radiochemistry (Russian),* 2004, **46**, 364.
4 A.B. Yusov, G.S. Parshin, A.V. Mamykin, V.P. Kazakov and N.N. Krot, *Radiochemistry (Russian),* 1983, **25**, 779.

MAGNETIC ANISOTROPY OF $U_2Co_{17-x}Si_x$ SINGLE CRYSTALS

A.V. Andreev,[1,2] E. Šantavá,[1] Y. Homma,[2] K. Koyama,[2] T. Yamamura,[2] Y. Shiokawa,[2] I. Satoh[2] and K. Watanabe[2]

[1]Institute of Physics, Academy of Sciences, Na Slovance 2, Prague, 18221 Czech Republic
[2]Institute for Materials Research, Tohoku University, Katahira 2-1-1, Sendai 980-8578

1 INTRODUCTION

Uranium and the 3*d* transition metals (T) do not form binary compounds with the hexagonal Th_2Ni_{17}-type of crystal structure. Nevertheless, this structure can be stabilized when the 3*d* metal is partly substituted by a third element, M. For T = Fe, M can be Si, Ge or Al.[1-4] For T = Co, it is found that the homogeneity range extends from $x = 1.3$ to 3 in the $U_2Co_{17-x}Ge_x$ system.[2] The related $U_2Co_{17-x}Si_x$ system has not been studied systematically, only the existence of $U_2Co_{15}Si_2$ with the Th_2Ni_{17}-type of crystal structure has been reported.[1] The magnetic properties of $U_2Co_{15}Si_2$ have not been studied. In this work, a single crystal with a higher Si content ($x = 3.4$) was grown and its magnetic properties were measured.[5,6] The goal of this work is a systematic study of the $U_2Co_{17-x}Si_x$ system. In particular, we wanted to determine the width of the homogeneity range, to try to grow single crystals of compounds within the homogeneity range and to study their magnetic properties.

2 EXPERIMENTAL

Ingots of 7 g each were prepared by arc melting stoichiometric mixtures of the pure elements (99.9% U and Co, 99.999% Si) in a tetra-arc furnace. Single crystals with a diameter of about 4 mm and a length of about 25 mm were grown by the Czochralski method using a tungsten wire as a seed with a 10 mm/hour pulling speed. After the homogeneity range was determined, the single crystals of the isostructural analogues of the terminal compositions of the homogeneity range were prepared for a comparison with non-magnetic Lu. The magnetization was measured by the Maglab-14 (Oxford) vibrating-sample magnetometer at 4.2-300 K in fields up to 6 T.

3 RESULTS AND DISCUSSION

In the $U_2Co_{17-x}Si_x$ system, the lattice parameter *c* does not depend on the Si content up to $x = 3.0$ whereas the parameter *a* decreases slightly with increasing *x* (Figure 1). However, in the terminal compound $U_2Co_{13.6}Si_{3.4}$, a sharp shrinkage along the *c* axis and a simultaneous expansion in the basal plane are observed. No other change in the crystal structure between $x = 3.0$ and 3.4 is indicated by powder x-ray diffraction. The unit-cell volume *V* has no anomaly and decreases monotonously with increasing x, because of the smaller atomic radius of Si compared to that of Co. Extrapolation of the a, $c(x)$ dependencies to $x = 0$ gives values for the hypothetical compound, U_2Co_{17}, which are in good agreement with similar results for the $U_2Co_{17-x}Ge_x$ system.[2]

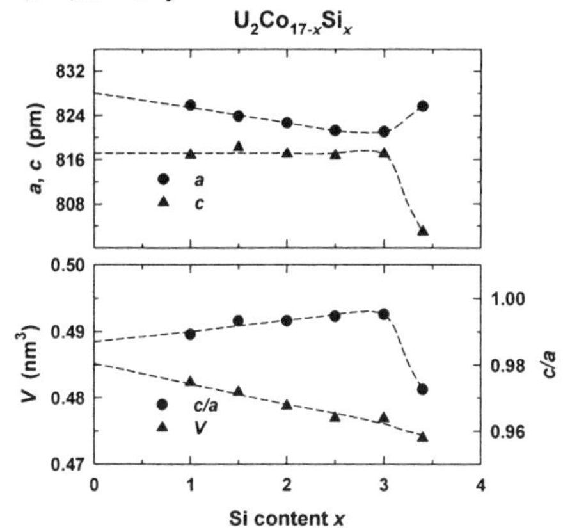

Figure 1 *Si-concentration dependence of the lattice parameters a and c, the unit-cell volume V and the c/a ratio of* $U_2Co_{17-x}Si_x$

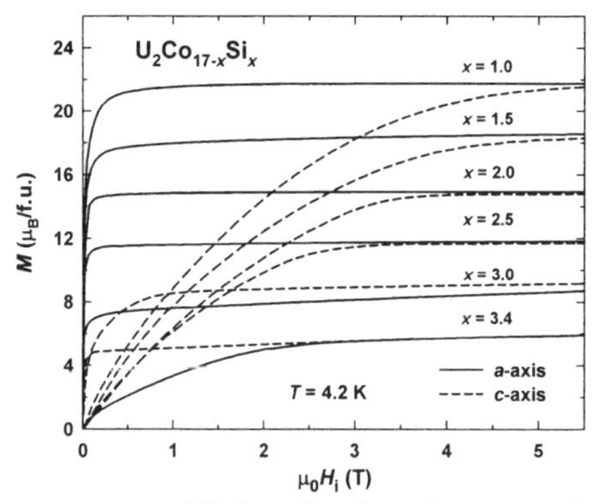

Figure 2 *Magnetization curves of* $U_2Co_{17-x}Si_x$ *along the principal axes at 4.2 K*

Figure 2 shows the magnetization curves along the principal axes for single crystals of $U_2Co_{17-x}Si_x$. Below $x = 3.0$, the compounds exhibit easy-plane magnetic anisotropy. $U_2Co_{14}Si_3$ is nearly isotropic and the compound with $x = 3.4$ has easy-axis anisotropy. The first anisotropy constant, K_1, of $U_2Co_{16}Si$ is larger by an order of magnitude than that of $Lu_2Co_{16}Si$, where the anisotropy field is only 0.3 T. This difference is attributed to a contribution from the U sub-lattice. Further proof for the magnetic state of U is a noticeable second anisotropy constant, K_2, which cannot be provided by the Co sub-lattice. Extrapolation of the anisotropy constants to $x = 0$ (hypothetical U_2Co_{17}) also gives values considerably larger than for Lu_2Co_{17}.[7] Since, with changing concentration, a spin re-orientation is observed for both the U and the Lu compound, this transition can be attributed to the Co sub-lattice. The easy-axis anisotropy of $U_2Co_{13.6}Si_{3.4}$ with the same K_1

as for $Lu_2Co_{13.6}Si_{3.4}$ points to a vanishing of the U sub-lattice anisotropy contribution with increasing Si content.

The spontaneous magnetic moment M_s decreases linearly with increasing x (Figure 3a). Extrapolation gives $M_s = 28.5\ \mu_B$ for hypothetic U_2Co_{17} which exceeds $M_s = 27.0\ \mu_B$ for Lu_2Co_{17}.[7] This may be interpreted as due to a ferromagnetic arrangement of the U and Co magnetic moments, with an estimated value for M_U of 0.7 μ_B/U atom. However, in the homogeneity range of $U_2Co_{17-x}Si_x$ the magnetic moment of the Lu compound is larger than that of the corresponding U compound. We suppose that this is due to a reduced Co moment in the U compounds rather than to anti-parallel arrangement of the U and Co magnetic moments, because in $U_2Co_{13.6}Si_{3.4}$, with presumably non-magnetic U, M_s is considerably lower than in the Lu counterpart.

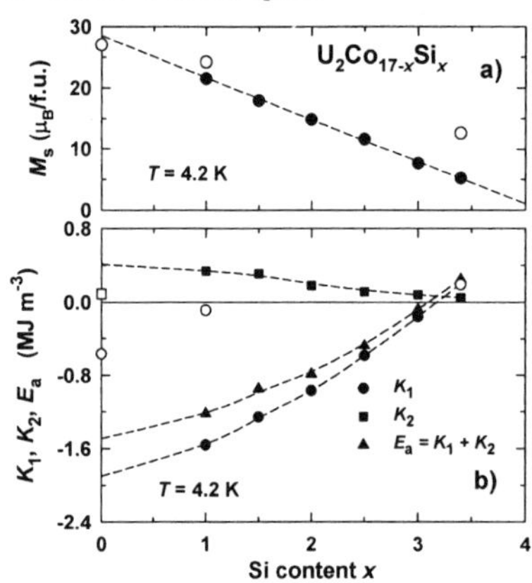

Figure 3 *Concentration dependences of the spontaneous magnetic moment M_s (a), the anisotropy constants K_1, K_2 and the anisotropy energy E_a (b) for $U_2Co_{17-x}Si_x$. The open symbols represent data for $Lu_2Co_{17-x}Si_x$*

Acknowledgements

This work is part of the AVOZ 10100520 research programs financed by the Academy of Sciences of the Czech Republic and was supported by grants: GACR 202/03/0550 and GAAV IAA100100530.

References

1 T. Berlureau, P. Gravereau, B. Chevalier and J. Etourneau, *J. Solid State Chem.*, 1993, **104**, 328.
2 B. Chevalier, P. Gravereau, T. Berlureau, L. Fournès and J. Etourneau, *J. Alloys Comp.*, 1996, **233**, 174.
3 B. Chevalier, P. Rogl and J. Etourneau, *J. Solid State Chem.*, 1995, **115**, 13.
4 A.P. Gonçalves, H. Noel, J.C. Waerenborgh and M. Almeida, *Chem. Mat.*, 2002, **14**, 4219.
5 A.V. Andreev, Y. Homma and Y. Shiokawa, *J. Alloys Comp.*, 2004, **383**, 195.
6 A.V. Andreev, A.V. Kolomiets, S. Daniš and T. Goto, *Physica B*, 2004, **348**, 134.
7 N.V. Kudrevatykh, A.V. Deryagin, A.A. Kazakov, V.A. Reimer and V.N. Moskalev, *Phys. Met. Metallogr.*, 1978, **45**, 38.

HIGH-FIELD MAGNETIZATION OF A UIrGe SINGLE CRYSTAL

S. Yoshii,[1] A.V. Andreev,[2] F.R. de Boer,[1,3] K. Kindo[1,4] and V. Sechovský[5]

[1]KYOKUGEN, Osaka University, Toyonaka, Osaka 560-8531, Japan
[2]Institute of Physics, Academy of Sciences, Na Slovance 2, Prague, 18221 Czech Republic
[3]Van der Waals-Zeeman Institute, University of Amsterdam, 1018 XE The Netherlands
[4]ISSP, University of Tokyo, 5-1-5 Kashiwanoha, Kashiwa, Chiba 277-8581, Japan
[5]Department of Electronic Structures, Faculty of Mathematics and Physics, Charles University, Ke Karlovu 5, Prague, 12116 Czech Republic

1 INTRODUCTION

UIrGe (orthorhombic TiNiSi structure) orders antiferromagnetically below the Neél temperature T_N of 16 K. Magnetic measurements reveal a large magnetic anisotropy in UIrGe with the hard magnetization axis along the *a* axis, i.e., along the direction of U-U zig-zag chains.[1] For magnetic fields along the *b* and the *c* axes, metamagnetic transitions are observed.[2] However, the measurements in Ref 2 were performed with a relatively low accuracy and only at 4.2 K. Recently, we have grown a new single crystal of UIrGe, which was used for magnetization, specific heat and resistivity measurements and for a neutron-diffraction study.[3,4] In this work, we present the results of high-accuracy, high-field magnetization measurements on a UIrGe single crystal in pulsed magnetic fields up to 51 T and in steady fields up to 14 T in the temperature interval 2-30 K and of DC magnetic-susceptibility measurements at 2-300 K. The single crystal was grown by a modified Czochralski method in a tri-arc furnace.

2 RESULTS AND DISCUSSION

The magnetization curves along the principal axes (Figure 1a) show that UIrGe has a huge magnetic anisotropy, which is typical for uranium intermetallics. Along the hardest *a*-axis, a linear magnetic response is observed up to the highest field. The magnetic susceptibility value, 2.2×10^{-8} m^3 mol^{-1}, corresponds well to the typical paramagnetic response in uranium compounds which do not exhibit magnetic ordering. Along the *b*- and the *c*-axis, the susceptibility is considerably higher and metamagnetic transitions are observed at 21 T and 14 T, respectively. The magnetization jumps of 0.36 and 0.28 μ_B are found to be considerably larger than those reported in Ref. 2 and agree better with the uranium magnetic moment, μ_U = 0.36 μ_B, found by neutron diffraction.[4] The strongly reduced μ_U value which, even in 51 T, does not reach 0.9 μ_B, points to a strong delocalisation of the 5*f* electrons (compared with μ_U = 3.3 μ_B for single U^{3+} or U^{4+} ions). A crystal-field splitting, the main mechanism of the moment reduction in the localised 4*f*-electron systems, could not play a considerable role in the actinides with a large spacing of the 5*f* electrons.

The temperature evolution of the metamagnetic transition (along the *b*-axis) is illustrated in Figure 1b. The transition becomes broader and its critical field, B_c (defined as inflection point of the $M(B)$ curve) decreases with increasing temperature. A similar temperature dependence of the transition is observed along the *c* axis as well.

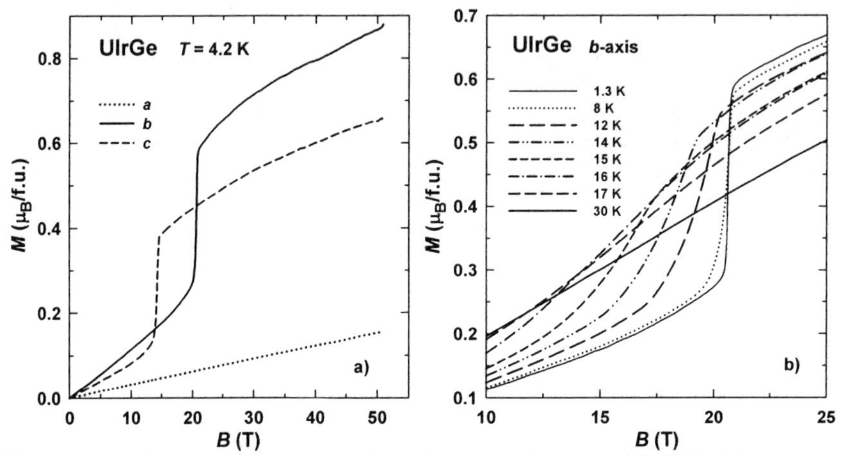

Figure 1 *a). Magnetization curves along the principal axes at 4.2 K. b). Metamagnetic transition in fields applied along the b-axis at different temperatures.*

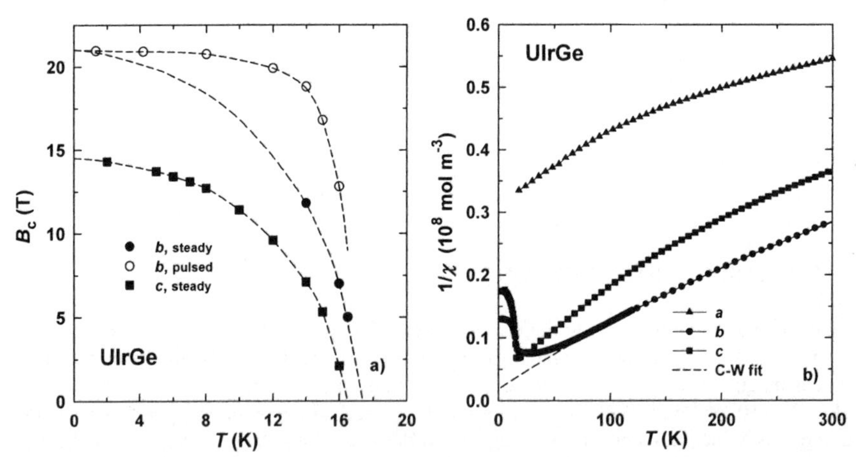

Figure 2 *a). Temperature dependence of the critical field, B_c, of the metamagnetic transition. b). Temperature dependence of the inverse magnetic susceptibility along the principal axes in a field of 4 T.*

Figure 2a shows the temperature dependence of B_c measured in steady magnetic fields along the *c* axis. Up to 12 K, the decrease of B_c with increasing temperature can be fitted by $B_c(T) = B_c(0) - a\,T^2$, typical for metamagnetic transitions. The shape of $B_c(T)$ along the *b* axis, which was measured in pulsed fields because B_c exceeds the maximum available steady field of 14 T, differs considerably from a quadratic fit. Moreover, there is a very large difference between the pulsed-field and the steady-field B_c values above 14 K, where B_c becomes low enough to be determined in steady fields. Since the magnetization process in pulsed fields is an adiabatic process, the observed difference with the isothermal steady-field results reveals a negative ΔT effect of about 4-6 K. This correlates well with specific-heat data.[3]

The strong magnetic anisotropy persists also in the paramagnetic range (Figure 2b). The magnetic susceptibility $\chi(T)$ along the *a* axis is much weaker than that along the *b* and the *c* axes. Similar to in the antiferromagnetic state, a noticeable anisotropy is found between the *b* and the *c* axis. The $\chi(T)$ curve along the *b*-axis can be described by a modified Curie-Weiss law with an effective magnetic moment μ_{eff} = 2.3 μ_B, a paramagnetic Curie temperature Θ_p = -16 K and a temperature-independent susceptibility of 0.9×10^{-8} m^3 mol^{-1}.

3 CONCLUSIONS

The antiferromagnet UIrGe (T_N = 16 K) exhibits metamagnetic transitions at fields of 21 T and 14 T applied along the *b* and the *c* axis, respectively. The magnetization jumps at the transition are 0.36 μ_B/U (*b*-axis) and 0.28 μ_B/U (*c*-axis). Along the *a*-axis, a weak linear magnetic response is observed up to 51 T, where the magnetic moment reaches only 0.16 μ_B/U. The metamagnetic-transition field, B_c, decreases with increasing temperature. Comparison of the temperature dependence of B_c measured in pulsed fields (almost adiabatic process) with isothermal steady-field results up to 14 T reveals a negative ΔT effect that correlates well with specific-heat results.

Acknowledgements

This work is a part of the research programs: AVOZ 10100520 (A.V.A.) and MSM 0021620834 (V.S.) that are financed by the Academy of Sciences of the Czech Republic and the Ministry of Education of the Czech Republic, respectively.

References

1 K. Prokeš, T. Tahara, T. Fujita, H. Goshima, T. Takabatake, M. Mihalik, A.A. Menovsky, S. Fukuda and J. Sakurai, *Phys. Rev. B*, 1999, **60**, 9532.
2 S. Chang, H. Nakotte, A.M. Alsmadi, A.H. Lacerda, M.H. Jung, M. Mihalik, K. Prokeš, J.C.P. Klaasse, E. Brück and F.R. de Boer, *Int. J. Modern Phys. B*, 2002, **16**, 3041.
3 V. Sechovský, J. Vejpravová, A.V. Andreev, F. Honda, K. Prokeš and E. Šantavá, *Physica B*, 2005 (in press).
4 K. Prokeš, H. Nakotte, V. Sechovsky, M. Mihalik and A.V. Andreev, *Physica B*, 2004, **350**, E199.

COMPARATIVE PHOTOEMISSION STUDY OF ACTINIDE METALS, NITRIDES AND HYDRIDES

T. Gouder, F. Wastin and J. Rebizant

European Commission, JRC, Institute for Transuranium Elements, Postfach 2340, 76125 Karlsruhe, FRG

1 INTRODUCTION

We present a comparative photoemission spectroscopic study of early actinide metals, hydrides and nitrides. In all three series, the *5f* electrons undergo the transition from itinerant to localized states, because of the *5f* contraction and the concomitant Mott-transition. But the different chemical environment of the actinide atoms in metal, nitride and hydride modulates this transition. In the nitride compounds, a nitride-band composed of N-*2p* and metal *6d* states forms. Because of the lower binding energy of the N-2*p* states the metal-nitride bond has mainly covalent character, and this favours hybridization with the *5f* states. In the hydrides, the H-1*s* states lie at higher binding energy (BE) – the bond is more polar, and energy overlap with the *5f* states is very small. There is thus a stronger tendency for *5f* localization in the hydrides than in the nitrides.

2 RESULTS AND DISCUSSION

Uranium is the most itinerant of the four actinides. In the metal, the *5f* states appear as a featureless triangular peak, cut by the Fermi-level. UN shows a very similar *f*-peak at the Fermi-level. *5f* itinerancy is also manifest in the band antiferromagnetism of UN. The N-2*p* band appears between 2-7 eV BE. It is close to the *5f* band, favouring hybridization and thus *f* delocalization. In UH_3, the shape of the *f*-emission changes. It is considerably broadened compared to UN. This is in sharp contrast with the specific heat, which is highest in UH_3 (α-U: 9 $mJmol^{-1}K^{-2}$, UN: 26 $mJmol^{-1}K^{-2}$, UH_3: 30 $mJmol^{-1}K^{-2}$) and the narrower *f*-bandwidth in the ground state.[1] Instead it fits well with a final-state multiplet of the $f^3 \rightarrow f^2$ transition (in the intermediate coupling scheme[2]). The lowest term of the multiplet (the ground state) lies at the Fermi-level: in the ground state, the *5f* states are itinerant. This is consistent with the large specific heat and the itinerant magnetism of UH_3. The appearance of the multiplets indicates a tendency towards localization: the photohole is no longer itinerant (as it would be in a broad band) but stays localized on the atom. This may even happen for systems, where in the ground state, the states are still essentially band-like. The more pronounced *f*-localization in the hydride, compared to the nitride, may be explained by the lower 1*s* overlap with the *5f* states, while the inter-An spacing is quite expanded compared to pure An metals.

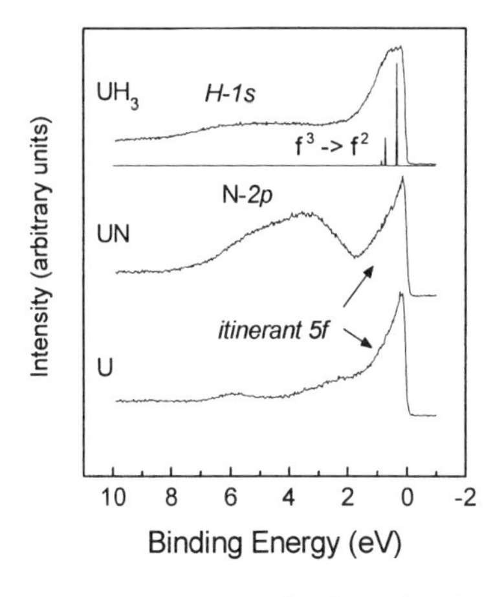

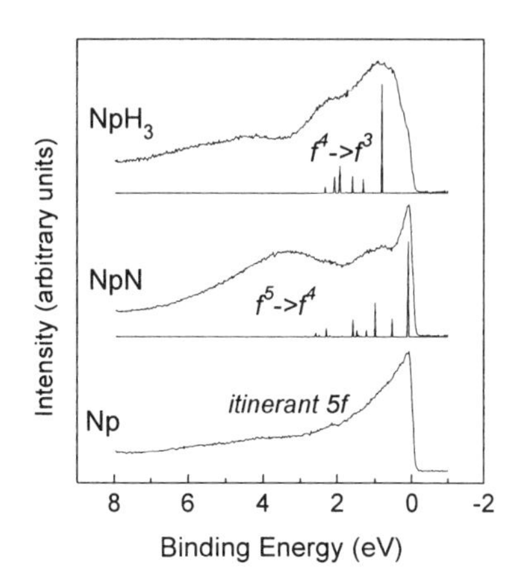

Figure 1 *He II excited valence band spectra of U (left) and Np (right): metal, nitride and hydride. Multiplet transitions, extracted from Gerken et al.*[2] *are superimposed. Appearance of multiplets indicates 5f localization. Np shows a more pronounced tendency for f-localization than U*

Neptunium metal has itinerant *5f* states, appearing as a featureless triangular peak at the Fermi-level, just as in uranium metal. But the spectrum of NpN no longer corresponds to the ground-state DOS but instead shows a final state multiplet structure. NpH_3 is also described by such a multiplet. However, the final state terms of hydride and nitride differ. While the nitride is well described by an f^4 configuration ($f^5 \rightarrow f^4$ transition), the hydride is described by an f^3 configuration ($f^4 \rightarrow f^3$ transition). However in both cases, the magnetic properties have been explained by an f^4 initial state configuration.[3,4] The solution to this apparent contradiction is provided by the final state screening concept. Photoemission from the same f^4 ground state leads to f^4 or f^3 final states, depending on whether the *f*-hole is refilled by an *f*-screening-electron, restoring the initial state configuration (f^4 final state, *f* – screening) or a *d*-screening-electron (f^3 final state, poor screening). Therefore the f^4 final state multiplet is a direct indication for *f*-screening. It shows that the *f*-states are still hybridized, while the f^3 final state appears if hybridization is too weak to permit *f*-refilling. So, the hydride is also more localized than the nitride in Np and this may again be related to a higher energy overlap between *f* and N-*2p* than *f* and H-*1s* band.

For Pu, the incipient 5*f*-localization is perceptible even in the metal spectrum. The overall triangular shape, as in U and Np, is preserved, but additional structures appear in the vicinity of the Fermi-level. The spectrum of PuN is well described by two multiplet transitions and the N-2*p* band. The multiplet transitions belong to a f^5 ground state configuration, which lead to the *f*-screened f^5 and the *d*-screened f^4 final states. The f^5 multiplet is observed in a multitude of Pu compounds (PuSe, PuTe, $PuSi_2$, $PuSb_{1-x}$, δ-Pu) and this general appearance in itself is a strong indication for the atomic (multiplet) rather than material specific (DOS) origin.[5] Indeed, the weak structures in the α-Pu spectrum are probably also due to this multiplet, starting to appear even in the relatively well itinerant metal. Finally, PuH_3 is again the most localized of all three Pu compounds. It is well described by the $f^5 \rightarrow f^4$ multiplet transition, which belongs to a weakly hybridized f^5 ground-state configuration.

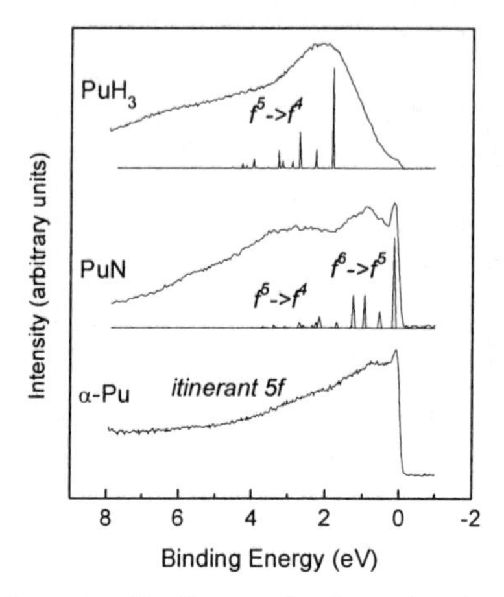

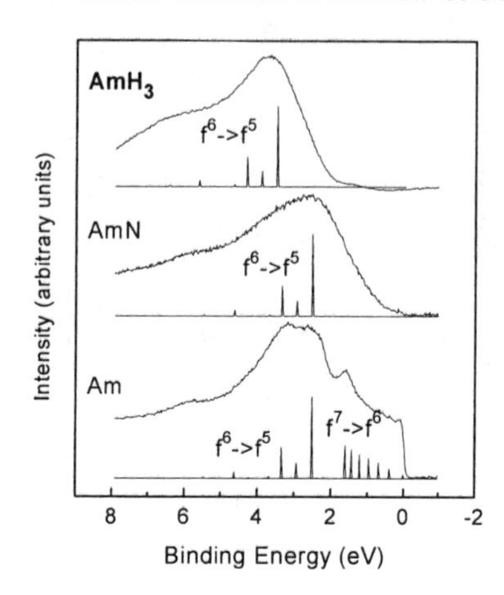

Figure 2 *He II excited valence band spectra of Pu(left) and Am (right): metal, nitride and hydride. Multiplet transitions, extracted from Gerken and Schmidt-May*[2] *are superimposed*

Am is the first actinide element, where in the ground-state the *f*-states are localized. Its non-magnetic character points to an f^6 ground-state configuration, where spin and orbital moment cancel ($L = S = 3$, $J = 0$). Photoemission from such configuration leads to the f^6 (*f*-screened) and f^5 (*d*-screened) final state multiplets. Indeed the spectrum of Am metal is well described by the ($f^7 \rightarrow f^6$) and ($f^6 \rightarrow f^5$) multiplet transitions.[6] The *f*-screened f^6 final state still witnesses some residual *f*-hybridization, persisting after localization. However, the main intensity between 0 and 2 eV BE belongs to the conduction band, which is composed of *d-s* states, like in the rare earth metals. In the nitride the *d-s* states are transferred in the nitride band, leaving a gap around the Fermi-level: AmN[7] is an insulator. The main signal can again be attributed to *f*-states in a multiplet structure, but the peak is strongly broadened due to lifetime and phonon effects. The same is true for AmH_3, which again is an insulator (with the *d-s* states in the hydride band) and the *f*-states appearing as an unresolved multiplet structure. Both in hydride and nitride, the *f*-screened final state (f^6) disappears, showing suppressed *f*-hybridization.

References

1 T. Gouder, R. Eloirdi, F. Wastin E. Colineau, J. Rebizant, D. Kolberg and F. Huber, *Phys. Rev. B*, 2004, **79**, 235108.
2 F. Gerken, J. and Schmidt-May, *J. Phys. F: Met. Phys*, 1983, **13**, 1571.
3 A.T. Aldred, G. Cinader, D.J. Lam and L.W. Weber, *Phys. Rev. B*, 1979, **19**, 300.
4 A.T. Aldred, B.D. Dunlap, A.R. Harvey, D.J. Lam, G.H. Lander and M.H. Mueller, *Phys. Rev. B*, 1974, **9**, 3766.
5 T. Gouder, R. Eloirdi, J. Rebizant, P. Boulet and F. Huber, *Phys. Rev. B*, 2005, **71**, 1.
6 N. Martensson, B. Johansson and J.R. Naegele, *Phys. Rev. B*, 1987, **35**, 1437.
7 T. Gouder, P. Oppeneer, F. Huber, F. Wastin and J. Rebizant, *Phys. Rev. B*, 2005, **72**, 115122.

CATHODOLUMINESCENCE OF ACTINIDE IONS IN CRYSTALLINE HOST PHASES

M.V. Zamoryanskaya and B.E. Burakov

Laboratory of Applied Mineralogy and Radiogeochemistry, the V.G. Khlopin Radium Institute, 28, 2-nd Murinskiy ave., St. Petersburg, 194021, Russia.

1 INTRODUCTION

The main objective of this paper is to summarize and discuss new results on the cathodoluminescent emission of actinide ions (U, Pu, Np and Am) which are incorporated into wide gap crystals. Luminescence experimental data of actinide ions are very limited. Experimentally, the intensive wide bands and weak and narrow lines are observed in the spectra of actinide and lanthanide ions. These features are related to two processes. There are allowed transitions (*5f-5f6d* for actinide ions or *4f-4f6d* for lanthanide ions) and forbidden transitions (*5f-5f* for actinide ions or *4f-4f* for lanthanide ions). As a rule, the narrow lines related to forbidden transitions have a small shift for different crystals. That is why such lines can be used as a diagnostic tool for the determination of ions and their valence state. Often, though, highly excited states belong to different overlapping configurations, which make interpretation difficult. Such configuration interaction is more important in the actinides than in the lanthanides. Previous studies of absorption properties of bulk crystals (AnF_3 and $LaCl_3$ doped with An) and actinide ions in solution showed several narrow lines in the spectra of Am^{3+} and Cm^{3+}. The absorption spectra of U^{3+}, Np^{3+} and Pu^{3+} are characterized by intensive bands in UV and visible region and narrow lines in red and IR region. We have very limited information about the luminescence of actinide ions in wide gap crystals, hence it may be very useful to interpret the electron structure of actinide-doped materials.

We studied the luminescent properties of actinide ions in different crystals and ceramics, utilised for the immobilization of actinides waste. These materials include: pyrochlore $(Ca,Gd,Hf,U,Pu,Am)_2Ti_2O_7$,[1] zircon $(Zr,Pu,Am)SiO_4$,[2] and garnet $(Y,Gd,Am..)_3(Al,Ga,Pu,...)_5O_{12}$.[3] Local cathodoluminescence (CL) excitation methods were used to study the excitation of actinide ions. *In situ* CL allows the investigation of single-phase grains in ceramics as small as 1-3 micrometers. The main objective of this paper is to summarize and discuss new CL emission results of U, Pu, Np and Am incorporated into zircon, garnet, and Ti-pyrochlore structures.

2 EXPERIMENTAL METHOD AND RESULTS

The following samples, obtained from previous investigations, were studied using the CL method:

- pyrochlore –10.0 % ^{239}Pu + 20.5 % ^{238}U;
- zircon – 1) un-doped synthetic zircon; 2) 0.2- 1.0 % ^{239}Pu; 3); 6.1 % ^{239}Pu, 0.05 % ^{241}Am;
- garnet – 1) $Y_3Al_5O_{12}$ doped with 0.01 wt % ^{241}Am; 2) $Gd_3Ga_5O_{12}$ doped with 3 % ^{238}U; 3) $Y_3Al_5O_{12}$ doped with 0.01 wt % ^{239}Pu.

All samples were mounted in epoxy, polished to reveal crystal interiors, and coated with aluminium or carbon. Cathodoluminescence spectra were acquired using the following parameters: accelerating voltage 15 kV; beam current of 10-50 nA. The in-house built CL spectrometer[4] was installed into the optical microscope port of a Cambax electron microprobe. The concentration of actinide ions in crystals and the composition of single-phase grains in ceramics were measured by electron microprobe spectroscopy.

2.1 Cathodoluminescence of Am^{3+}

The result obtained allowed for the identification of Am^{3+} ions in zircon and garnet. The CL of Am^{3+} is characterized in these host phases by narrow emission bands in the orange-red spectral region (see Figure 1) and showed similarities to the CL emission spectra of Eu^{3+}. The CL emission of Am^{3+} at 1.64 eV and 2.1 eV in garnet and zircon is caused by electron transitions between the levels $^5D_0$ to $^5F_{1,2}$ and at 2.3 eV – between $^5D_1$ to $^5F_1$. We assume that the spectral position of Am^{3+} in CL emission spectra is also similar to those of other wide gap materials.

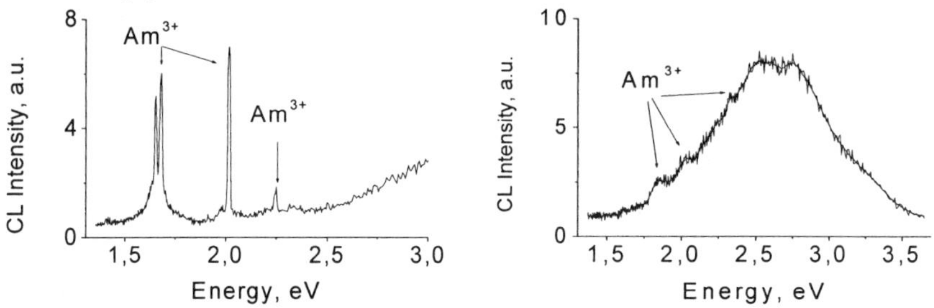

Figure 1 *The CL emission spectra of garnet, $Y_3Al_5O_{12}$ doped with Am^{3+} (left) and $ZrSiO_4$ doped with Pu and Am^{3+} (right)*

2.2 Cathodoluminescence of Pu, Np, U

No characteristic CL emissions bands of U and Np ions were observed in zircon and pyrochlore. The appearance of these ions leads to a decrease of CL intensity of the host phases.

The influence of Pu concentration on CL properties was observed in zircon and garnet. An increase in the Pu content in synthetic zircon and garnet causes the decrease of intensity of the intrinsic CL emission. The CL band in Pu-doped garnet appears at 3.2 eV. The band in Pu-doped zircon was observed at 2.8 eV. It may be related to a Pu^{3+} or Pu^{4+}emission band or to the formation of additional structural defects.

2.3 Cathodoluminescence of $[UO_x]^y$ Complexes

The CL study of U-doped garnet and pyrochlore allows for the identification of two forms of uranium: uranyl ion, $(UO_2)^{2+}$, and tetrahedral complex, $(UO_4)^{2-}$ (Figure 2).

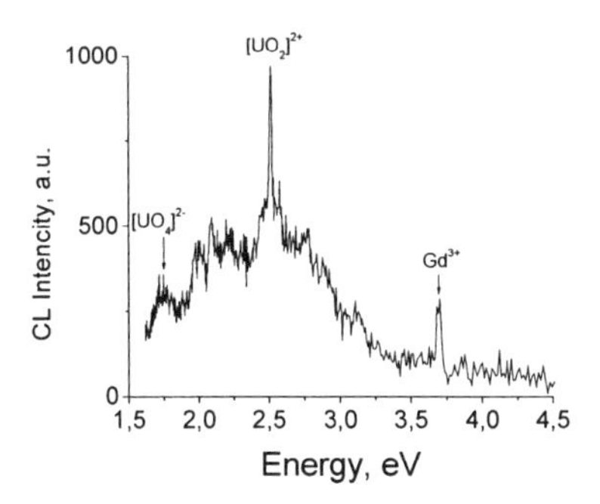

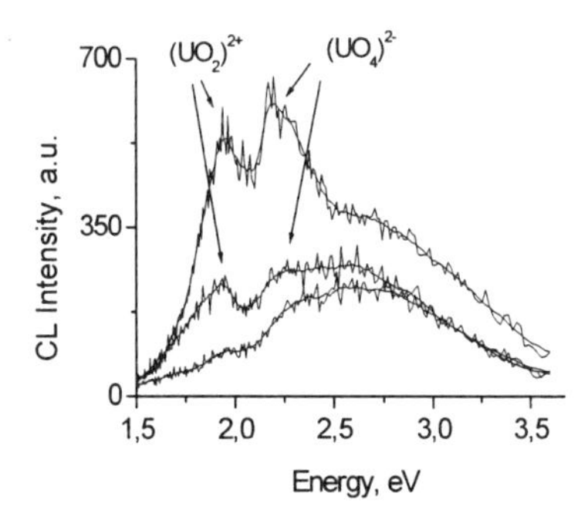

Figure 2 *The CL emission spectrum of garnet, $Gd_3Ga_5O_{12}$, doped with U (left) and of pyrochlore doped with ^{239}Pu and ^{238}U (right)*

The study of pyrochlore samples doped simultaneously with ^{238}Pu and ^{238}U has revealed that the CL intensity of uranyl ions increases depending on the cumulative dose of alpha-induced radiation damage (Figure 2 (*right*)). One explanation for this is that the self-irradiation from Pu causes the change of U valence state.

3 CONCLUSIONS

The CL emission of Am^{3+} is characterized by narrow characteristic bands in the orange-red spectral region that are similar to the CL emission of Eu^{3+} in synthetic zircon and garnet.

The presence of two forms of uranium: the uranyl ion, $(UO_2)^{2+}$, and the tetrahedral complex,[5] $(UO_4)^{2-}$ were identified in U-doped garnet and Pu-U-doped pyrochlore. The CL intensity of uranyl ions increases depending on the cumulative dose from alpha-induced radiation damage in the pyrochlore sample doped with ^{238}Pu and ^{238}U. The data suggest that self-irradiation might cause the change in actinide valence state.

References

1 M.V. Zamoryanskaya, B.E. Burakov, R.V. Bogdanov and A.S. Sergeev, in *Scientific Basis for Nuclear Waste Management XXV*, eds., B.P. McGrail and G.A. Cragnolino, *Mat. Res. Soc. Proc.*, 2002, **713**, Warrendale, PA, pp. 481-485.
2 B.E. Burakov, J.M. Hanchar, M.V. Zamoryanskaya, V.M. Garbuzov and V.A. Zirlin, *Radiochim. Acta,* 2002, **89**, 1.
3 M.V. Zamoryanskaya and B.E. Burakov, in *Scientific Basis for Nuclear Waste Management XXIV, Mat. Res. Soc. Proc.*, 2000, **608**, Warrendale, PA, pp. 437-442.
4 M.V. Zamoryanskaya, S.G. Konnikov and A.N. Zamoryanskii, *Instruments and Experimental Techniques*, 2004, **47**, 477.
5 A.M. Morozov, L.G. Morozova and P.P. Feofilov, *Opt. Spectrosc.*, 1972, **32**, 50.

ELECTRONIC STRUCTURE OF THE AMERICIUM MONOPNICTIDES: *AB INITIO* CALCULATIONS AND PHOTOEMISSION EXPERIMENTS

D.B. Ghosh,[1] S.K. De,[1] P.M. Oppeneer,[2] T. Gouder,[3] M.S.S. Brooks,[2,3] F. Wastin,[3] F. Huber[3] and J. Rebizant[3]

[1]Department of Materials Science, Indian Association for the Cultivation of Science, Jadavpur, Calcutta 700 032, India
[2]Department of Physics, Uppsala University, Box 530, S-751 21 Uppsala, Sweden
[3]European Commission, Joint Research Centre, Institute for Transuranium Elements, D-76125 Karlsruhe, Germany

1 INTRODUCTION

The *5f* electrons in the actinide elements are less localized than in the corresponding lanthanide (*4f*) elements, yet there is a trend of increasing *5f* localization with increase of atomic number. Americium, the element on the right-hand side of plutonium, "sits on the fence" of the *5f* delocalization-localization transition, i.e., the intriguing retraction of the *5f* states from bonding which occurs between Pu and Am.[1,2] This localization was proven by its photoemission spectrum,[3] which shows *5f* states at about 1–3 eV below the Fermi level.

The behaviour of *5f* electrons in Am compounds has practically not been studied. Here we report a combined experimental-computational study of the Am monopnictides. Experimental magnetization studies revealed temperature independent paramagnetism corresponding to a trivalent Am ion (i.e. $5f^6$, $J = 0$ state).[4-6] Thus far, a single electronic structure calculation of the Am monopnictides was reported, which employed the self-interaction corrected local spin-density approximation (SIC-LSDA).[7] These calculations predicted the Am monopnictides to be metallic, having a huge *5f* partial density of states (DOS) at the Fermi level.

2 CALCULATIONAL AND EXPERIMENTAL DETAILS

To elucidate the electronic structure of the Am monopnictides we have performed *ab initio* electronic structure calculations, employing both the LDA and LDA+U approaches. The calculations have been carried out using the full-potential linear muffin-tin orbital (FP-LMTO) method.[8] The LDA+U approach should especially be suited for situations where the *f* electrons are moderately localized. On the experimental side, we have investigated the electronic structures of AmN and AmSb (as well as, for comparison, Am and Am_2O_3) using ultraviolet photoemission spectroscopy (UPS). Thin films of AmN were prepared by reactive sputter deposition from an Am metal target in a N_2 atmosphere. AmSb films, on the other hand, were prepared by the co-deposition of Am and Sb from metallic targets.

3 RESULTS AND DISCUSSION

Our UPS spectra obtained for AmN and AmSb films reveal interesting electronic structures. For AmN, the valence-band signal smoothly vanishes for binding energies below the Fermi energy (E_F) and upon approaching E_F. Thus, AmN is either an insulator or a semiconductor (see Figure 1 (*left*)). The contributions of the 5*f* states have been separated by subtraction of the He I and He II signals, which reveals the 5*f* states to be positioned at a binding energy of 2.5 eV. No huge 5*f* DOS close to E_F could be observed, in contradiction with the prediction of the SIC-LSDA calculations.[7] Rather, the UPS spectra show the Am 5*f* electrons to be moderately localized, i.e., withdrawn from the Fermi energy, but still located in the energy region where the N *p*-states dominantly occur, and weakly hybridized with the N *p*-states. For AmSb, we find a small valence-band DOS remaining at E_F. This could imply that AmSb has a pseudo-gap at E_F. Alternatively, it could be that an inhomogeneity of the sample is responsible for this feature. The 5*f* electrons in AmSb are observed at binding energies of about 3 eV and no 5*f* intensity near E_F could be detected. In comparison, our UPS spectra for pure Am metal show a second peak at 1.8 eV binding energy, which is interpreted as good screening taking place through empty *f* states. This peak has disappeared for AmN, AmSb, indicating that the 5*f* electrons are more localized in the pnictides than in Am metal.

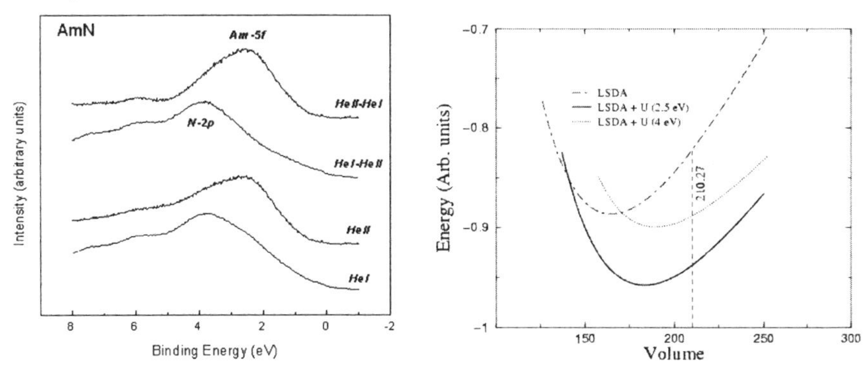

Figure 1 *He I and He II valence-band photoemission spectra of AmN (left). Total energy vs. volume for AmN (right). Vertical line denotes the experimental volume*

Our LDA band-structure calculations, predict all Am monopnictides (except AmN) to be narrow-gap semiconductors. In AmN the larger *f* overlap, owing to its smaller lattice parameter, results in a larger *f* bandwidth and hence, leads to a pseudogap only. In the LDA+*U* calculations a supplementary on-site Coulomb interaction of the *f* states is added. The Coulomb *U* parameter has been varied within the accepted range of 1–4 eV. The LDA+*U* approach yields calculated equilibrium lattice parameters closer to the experimental values than the LDA approach (see Figure 1 (*right*)). The inclusion of the Coulomb parameter shifts the occupied $5f_{5/2}$ states further downward and unoccupied $5f_{7/2}$ states further upward from the Fermi level. This opens up a gap in AmN (see Figure 2) and maintains the pseudo-gap or narrow gap behaviour in the other Am monopnictides. The insulating or pseudo-gap behaviour is fully consistent with our photoelectron experiments. The LDA+*U* calculated pnictogen *p* band and 5*f* band positions, respectively, correspond well with the available experimental results.

The Am monopnictides exhibit very high temperature-independent susceptibilities $\chi = 0.50–1.25x10^{-3}$ emu/mol.[4-6] This has been explained by Petit *et al.*[7] as Pauli paramagnetism arising from the high 5*f* density of states at E_F. However, the absence of 5*f* related emission near the Fermi energy, as detected by the photoemission experiments contradicts the above

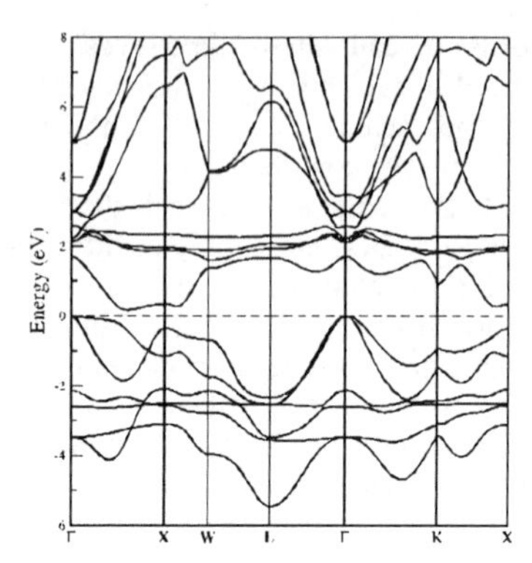

Figure 2 *Energy-bands of AmN, computed with the LDA+U (U=2.5 eV) approach.*

explanation. A consistent explanation of the high χ can be given in terms of a Van Vleck mechanism. Our calculations predict a $5f^6$ state for the Am ion (i.e. Am^{3+}). This being a J=0 state would permit magnetic excitations across the gap, leading to Van Vleck paramagnetism. Using the computed band gaps we estimate values for the susceptibility, χ_0, that are of the order of those obtained experimentally. Thus, the LDA+U approach provides a consistent electronic structure picture for the Am monopnictides.

4 CONCLUSION

We find the Am monopnictides to be semiconductors, with the $5f$ electrons withdrawn from the Fermi level to binding energies of a few eV. Our calculations and experiments suggest an electronic structure picture, which is qualitatively different from the high $5f$ DOS model predicted by SIC-LSDA calculations.[7] The gap formation in the Am monopnictides arises from the large spin-orbit interaction, which splits the $5f_{5/2}$ and $5f_{7/2}$ subbands, the Coulomb interaction, and the particular f—p—d hybridizations. A similar mechanism for gap formation was previously proposed for the Pu monochalcogenides.[9]

References

1 A.J. Freeman and D.D. Koelling, in *The Actinides: Electronic Structure and Related Properties*, eds., A.J. Freeman and J.B. Darby, Academic, New York, 1974, I, p. 51.
2 S.S. Hecker and L.F. Timofeeva, *Los Alamos Science,* 2000, **26**, 244.
3 J.R. Naegele, L. Manes, J.C. Spirlet and W. Müller, *Phys. Rev. Lett.*, 1984, **52**, 1834.
4 B.D. Dunlap, D.J. Lam, G.M. Kalvius and G.K. Shenoy, *J. Appl. Phys.*, 1971, **42**, 1719.
5 B. Kanellakopulos, J.P. Charvillat, F. Maino and W. Müller, in *Transplutonium-1975*, eds., W. Müller and R. Lindner, North-Holland, Amsterdam, 1976, p. 181.
6 O. Vogt, K. Mattenberger, J. Löhle and J. Rebizant, *J. Alloys Compd.*, 1998, **271-273**, 508.
7 L. Petit, A. Svane, W.M. Temmerman and Z. Szotek, *Phys. Rev. B*, 2001, **63**, 165107.
8 S.Y. Savrasov, *Phys. Rev. B*, 1996, **54**, 16470.
9 P.M. Oppeneer, T. Kraft and M.S.S. Brooks, *Phys. Rev. B*, 2000, **61**, 12825.

USING NANO-FOCUSSED BREMSTRAHLUNG ISOCHROMAT SPECTROSCOPY (nBIS) TO DETERMINE THE UNOCCUPIED ELECTRONIC STRUCTURE OF Pu: A PROPOSED STUDY

J.G. Tobin, M.T. Butterfield, N.E. Teslich Jr., R.A. Bliss, M.A. Wall, A.K. McMahan, B.W. Chung and A.J. Schwartz

Lawrence Livermore National Laboratory, Livermore, CA, USA
This work was performed under the auspices of the U.S. DOE by University of California, Lawrence Livermore National Laboratory under contract W-7405-Eng-48

1 INTRODUCTION

The details of the electronic structure of Pu remain undefined. While chemically toxic and highly radioactive, Pu may be the most scientifically interesting element in the periodic table. Its properties include the following: six different phases, close to each other in energy and sensitive to variations in temperature, pressure and chemistry; the face-centered-cubic phase (delta) is the *least* dense; Pu expands when it solidifies from a melt; and it is clearly the nexus of the actinide binary phase diagram of the actinides. In a sense, it is the boundary between the light (ostensibly delocalized $5f$ electrons) and heavy (ostensibly localized or correlated $5f$ electrons) actinide elements, but this is an over-simplification. The localized atomic $5f$ states are naturally correlated, but important regimes of correlated electron states are conceivable as extended states on the delocalized side of the possible Mott transition. The proximity to this crossover may be the driving force behind all these exotic properties. Pu remains of immense technological importance and the advancement to a firm, scientific understanding of the electronic structure of Pu and its compounds, mixtures, alloys and solutions, is a crucial issue. Moreover, while there are a number of ongoing experimental efforts directed at determining the occupied (valence band, below the Fermi Energy) electronic structure of Pu, there is essentially no experimental data on the unoccupied (conduction band, above the Fermi Energy) electronic structure of Pu.

2 PROPOSED EXPERIMENTAL METHODS

We have begun an experimental effort to help resolve the controversy regarding the electronic structure of Pu. The objective of this effort is to determine the conduction band (unoccupied) electronic structure of Pu and other actinides (and possibly rare earth metals as well), in a phase-specific fashion and emphasizing bulk contributions. Moreover, the conduction band (unoccupied) electronic structure is the missing link in studies of Pu. As illustrated in Figure 1, Bremstrahling Isochromat Spectroscopy (BIS) is the best way to determine the unoccupied electronic structure of the actinides. While experimental BIS data exist for Th and U, there are no such data for Pu! By using a nano- or micro-focussed beam of electrons as the excitation source (thus nBIS), we will be able to access single-crystalline regions of polycrystalline samples, permitting the determination in a phase-specific fashion.

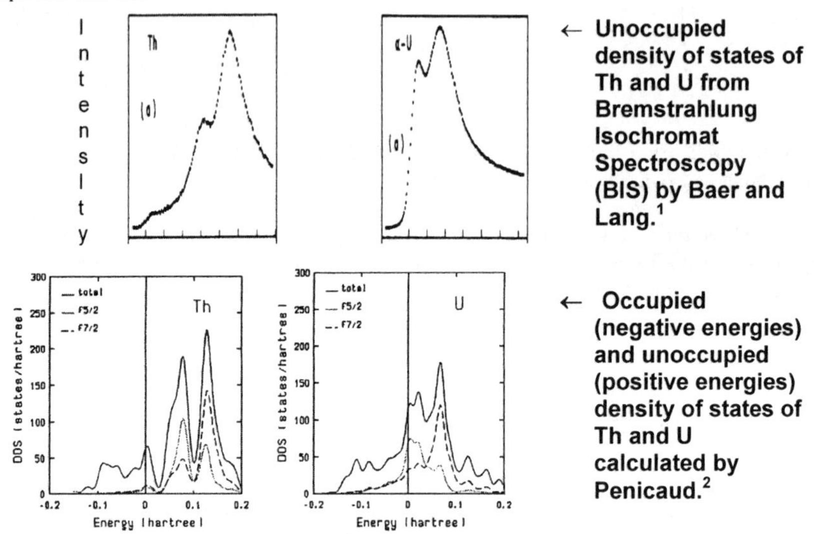

Figure 1 *The BIS and calculated electronic structure of Th and U. For each of these elements, there is strong agreement between the experimental and theoretical results, particularly the presence of the spin-orbit split* $5f_{5/2}$ *and* $5f_{7/2}$ *peaks. The detail is greater in the calculations than in the experiment: the BIS experiment is broadened but retains the essence of the spin-orbit split peaks. The BIS spectra also exhibit a Fermi cut-off at zero energy: the transitions cannot go into occupied states (as demonstrated below)*

The central technique is BIS, or high energy Inverse Photoelectron Spectroscopy. BIS is the high-energy variant of inverse photoelectron spectroscopy (IPES: electron in, photon out), which is essentially the time reversal of photoelectron spectroscopy (photon in, electron out), as illustrated in Figure 2 (*left*). IPES can be used to follow the dispersion of electronic states in ordered samples: an example of this is shown in Figure 2 (*right*). Owing to its low energies, IPES is usually very surface and band sensitive. However, by working at higher energies, we will sample preferentially for the bulk density of states, downgrading the impact of surface and band effects, following a philosophy similar to that of Mo *et al.*[3] Thus, from BIS, we would have a direct measure of the conduction band or unoccupied electronic structure of the bulk Pu.

Finally, by directly comparing the results of our nBIS studies to the predictions of Dynamical Mean Field Theory (DMFT) such as that of Pu by Savrosov *et al.*[6] and Ce by Held, McMahan and Scarlettar,[5] we hope to resolve issues regarding the 4*f* and 5*f* electronic structure and determine the nature of the electronic structure of Pu, including the effects of electron correlation.

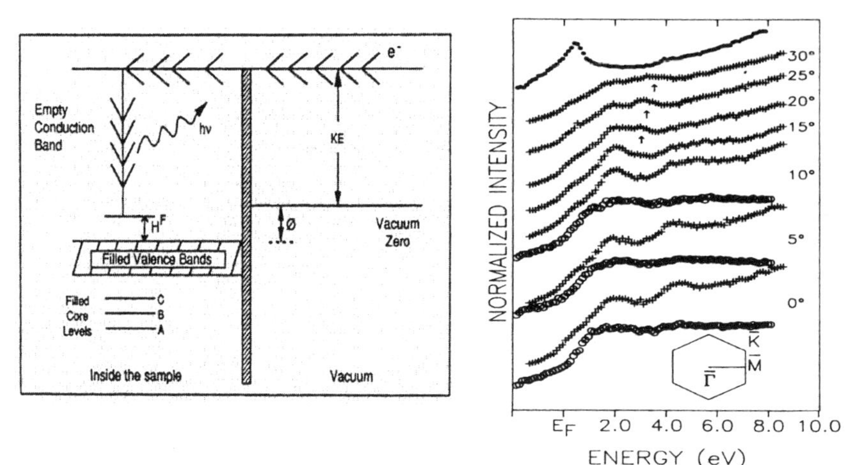

Figure 2 *(left) Inverse Photoelectron Spectroscopy (IPES). (right) IPES of Ag/Ge(111) from Knapp and Tobin,*[6] *where* O = Ge; + = Ag/Ge; ● = Ag

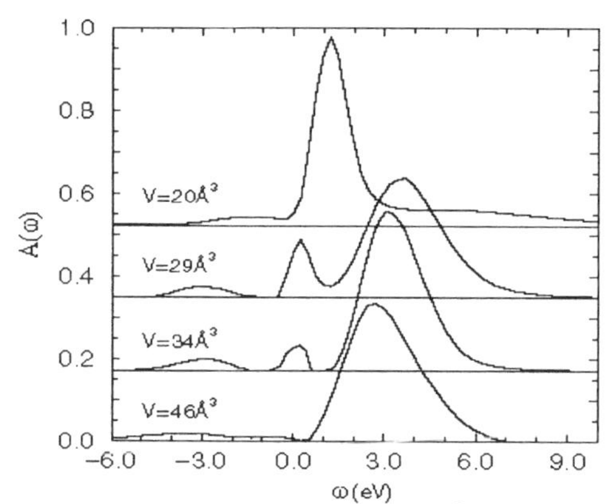

Figure 3 *Ce data from Held, McMahan and Scarlettar.*[5]

References

1 Y. Baer and J.K. Lang, *Phys. Rev B*, 1980, **21**, 2060.
2 M. Penicaud, *J. Phys.: Condens. Matter*, 1990, **9**, 6341.
3 S.-K. Mo, J.D. Denlinger, H.-D. Kim, J.-H. Park, J.W. Allen, A. Sekiyama, A. Yamasaki, K. Kadono, S. Suga, Y. Saitoh, T. Muro, P. Metcalf, G. Keller, K. Held, V. Eyert, V. I. Anisimov and D. Vollhardt, *Phys. Rev. Lett.*,2003, **90**, 186403.
4 S.Y. Savrosov, G. Kotliar and E. Abrahams, *Nature*, 2001, **410**, 793.
5 K. Held, A.K. McMahan and R.T. Scarlettar, *Phys. Rev. Lett.*, 2001, **87**, 276404.
6 B.J. Knapp and J.G. Tobin, *Phys. Rev. B*, 1988, **37**, 8656.

FLUORESCENCE SPECTROSCOPY OF PROTACTINIUM(IV)

C.M. Marquardt,[1] P.J. Panak,[1] C. Walther,[1] R. Klenze[1] and Th. Fanghänel[1,2]

[1]Forschungszentrum Karlsruhe GmbH, Institut für Nukleare Entsorgung, Postfach 3640, 76021 Karlsruhe, Germany
[2]Physikalisch-Chemisches Institut, Ruprecht-Karls-Universität
Im Neuenheimer Feld 253, D-69129 Heidelberg, Germany

1 INTRODUCTION

In a previous publication[1] we showed that fluorescence spectroscopy of tetravalent protactinium (Pa(IV)) in aqueous solution is possible. This might open up new vistas for spectroscopic speciation of tetravalent actinides at trace concentrations. In this paper we show that fluorescence spectroscopy is applicable for complexation studies with tetravalent protactinium. The fluoride and the hydroxide anion were chosen as complexing ligands.

2 EXPERIMENTAL

2.1 Chemicals

A stock solution of Pa-231 (3.3×10^4 y half-life) in 9.5 M HCl was used for all experiments with a concentration of 3.3×10^{-4} M. The protactinium was pentavalent in the stock. For the complexation studies aliquots were diluted with 1 M $HClO_4$ to final Pa concentrations between 8.6×10^{-6} and 1.3×10^{-5} M. The preparation of all Pa(IV) solutions was performed in a glove-box under oxygen-free (p_{O2}<10ppm) argon atmosphere. By adding 1-2 ml of liquid Zn-amalgam and vigorous shaking for 10 min all Pa(V) was reduced to Pa(IV). The Zn-amalgam was removed from the solutions before characterising by UV spectroscopy and TRLFS. After the spectroscopic measurements and before adding the next aliquot of NaF or NaOH, the sample was treated again with Zn-amalgam for two minutes to remove Pa(V), that had been produced meanwhile. Liquid scintillation counting was used for measuring the Pa concentration in solution.

The UV spectra of the solutions were recorded in 1 cm quartz cuvettes with a gas-tight screw cap (Hellma, Germany) by a Cary5 spectrophotometer (Varian).

TRLFS measurements were performed using a XeCl-excimer laser (Lambda Physics, LPX 202, 308 nm, 24 ns) for excitation. Details are described elsewhere.[1] The Pa(IV) was excited either at 308 nm by direct output of the laser light or at 278 nm by using a dye laser with a frequency doubling crystal. The absorption cross-section of Pa(IV) is about tenfold higher at 278 nm than at 308 nm. Emission spectra were recorded from 350 to 570 nm.

3 RESULTS AND DISCUSSION

3.1 Studies on Fluoride Complexation

The initial concentration of Pa(IV) used for complexation studies with F^- was 1.56 x 10^{-5} M in 1 M $HClO_4$. This sample was titrated with a 0.8 M NaF solution. Thus the total fluoride concentration varied from 0 to 0.11 M. In Figure 1 the absorption spectra of the samples are shown. It is discernable that with increasing fluoride concentration the absorption band at 279 nm of the free Pa^{4+}-ion is shifted to higher wavelengths. At higher fluoride concentrations (7 mM and higher) a new absorption band evolves at 340 nm. From thermodynamic calculations with complexing constants $\log\beta_1 = 7.91$ and $\log\beta_2 = 14.62$ from Guillaumont[2] at ionic strength of 3 and $\log\beta°_3 = 21.89$, $\log\beta°_4 = 26.34$, and the solubility product $\log K°_{s,0} = 29.4$ for the analogues metal ion U(IV),[3] four regions of dominant species (fractions > 25%) can be defined with increasing F^- concentrations: 1. Pa^{4+}, 2. ($PaF^{3+} + PaF_2^{2+}$), 3. ($PaF_2^{2+} + PaF_3^+$) 4. ($PaF_3 + PaF_4$(s)). It must be pointed out that the species distribution is only an estimation, because of the uncertainties in the complexing constants at given experimental conditions. But nevertheless, these regions comply qualitatively with the shifts of the absorption band in the spectra: 1. 279 nm, 2. 284 nm, 3. 286 nm, and 4. 340 nm.

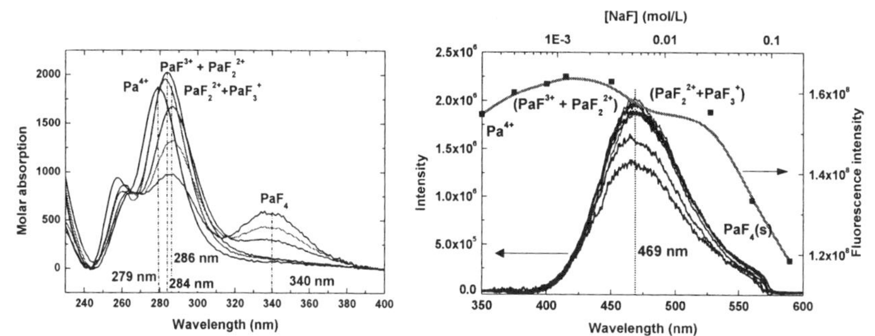

Figure 1 *Absorption (left) and fluorescence(right) spectra of Pa(IV) at various fluoride concentrations. In the graph on the right, the peak areas are plotted against the NaF concentration (grey line)*

However, the band at 340 nm may not solely be caused by the solved species PaF_4 (aq), because its concentration is too low for such high absorbances. We assume that it is rather caused by dispersed crystalline PaF_4 (s). This is in good agreement with the abrupt and enormous shift observed for NaF at concentrations higher than 0.025 M. In contrast to the absorption band the fluorescence spectra show no significant shift of the band at 469 nm with fluoride complexation. With increasing fluoride concentration the fluorescence intensity increases up to a concentration of 0.003 M and then decreases with an abrupt fall at 0.025 M NaF. This fall is correlated to an increase of the absorbance band at 340 nm and confirms the formation of PaF_4 (s).

3.2 Studies on the Hydrolysis Reaction

Small portions of concentrated NaOH solution were added stepwise to a solution of 1.4 x 10^{-4} M Pa(IV) in 1 M $HClO_4$ until pH 3.6. After each step the solution was characterised by spectroscopy. The absorption band of Pa(IV) shows a red shift from 280 nm at 1 M H^+

to 284 nm at pH 1.8. At higher pH it turns to a slight blue shift with a maximum at 281 nm at pH 3.6. This turn to shorter wavelengths is in accordance with a blue shift of the Pa(IV) fluorescence band at pH values higher than 1.6. A huge blue shift of 47 nm from 469 to 422 nm is observed at pH 3.6. The maximum of the fluorescence band remains constant at 469 nm at pH < 1.6. The absorbance declines steadily with increasing pH values, whereas the fluorescence increased until pH 0.38 and then dropped down at higher pH values. Only very low amounts of protactinium (< $1x10^{-7}$ M) were measured in solution by LSC in the last sample at pH 3.6. That means that almost all of protactinium is sorbed onto the quartz wall. An additional uncertainty arises about the amount of Pa(IV) that is oxidised to Pa(V). Also Pa(V) shows tremendous sorption behaviour at higher pH values.

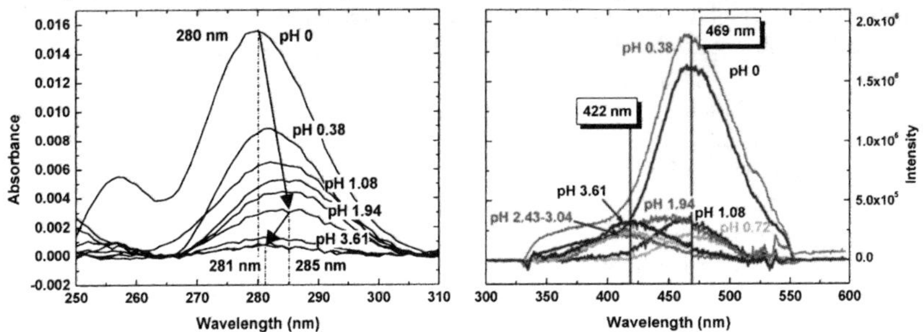

Figure 2 *Absorption (left) and fluorescence (right) spectra of Pa(IV) at various pH values in $HClO_4$ medium*

4 CONCLUSIONS

Fluorescence spectroscopy is a versatile tool for speciation of Pa(IV) in aqueous solution. The shifts of the fluorescence bands of Pa(IV) species among each other depends strongly on the ligand system. Speciation requires sufficiently large band shifts such that the bands of various species can be differentiated. For the fluoride system these shifts are very small contrary to UV absorption spectroscopy. Therefore, the latter is better applicable for fluoride complexation studies. For the hydrolysis reaction the shifts of the fluorescence bands are large for the complexes with OH^-, that is in contrast to the UV bands. Hence, fluorescence spectroscopy is the choice for hydrolysis studies. The most striking advantage of fluorescence spectroscopy is its high sensitivity, that allows the measurement of trace concentrations of Pa(IV) (< $1x10^{-7}$ M) in solution, as it was shown for the Pa(IV) hydrolysis species.

References

1 C.M. Marquardt, P.J. Panak, C. Apostolidis, A. Morgenstern, C. Walther, R. Klenze and Th. Fanghänel, *Radiochim. Acta,* 2004, **92**, 445.

2 S.H. Eberle, 'Chemie des Protactiniums in wässriger Lösung' in *Gmelin Handbuch der anorganischen Chemie. 8, System-Nr 51, Erg. Bd 2*, 1977, Chapter 13, p. 139.

3. I. Grenthe, J. Fuger, R.J.M. Konings, R.J. Lemire, A.B. Muller, C. Nguyen-Trung and H. Wanner, *Chemical Thermodynamics of Uranium,* Nuclear Agency Organisation for Economic Co-operation and Development (OECD), NEA Data Bank, 2004, p. 173.

Pu, Np AND U VALENCE STATES AND THE DETERMINATION OF THEIR MOLECULAR FORM BY CHEMILUMINESCENCE AND PULSED LASER SPECTROSCOPY

I.N. Izosimov, N.G. Gorshkov, L.G. Mashirov and N.G. Firsin

Khlopin Radium Institute, 194021 St.Petersburg, Russia. E-mail:izosimov@atom.nw.ru

1 INTRODUCTION

The detection of trace amounts of actinide elements in various samples is today of major importance for ecology; radioactive waste handling and control; the rehabilitation of contaminated areas; and risk assessment. Use of lasers open up new possibilities in the trace analysis of radionuclides in environmental samples.[1-3] However, the practical use of laser spectroscopic methods in the analysis of different samples encounters one essential difficulty; namely it is necessary to get the investigated element from the sample into a zone of interaction for laser irradiation. That is why the most attractive method, from a practical point of view, is to prepare solutions of the samples to be investigated.

The time-resolved technique (TR) allows the useful differentiation of long-lived fluorescence from short-lived fluorescence impurities and from scattered light, enabling a considerable increase in the sensitivity (down to 10^{-13} mol/l).[4-6] The valence forms of the actinides: UO_2^{2+}, Cm^{3+}, Am^{3+}, Cf^{3+}, Es^{3+}, and $Bk^{3+}$and some of their complexes can now be detected by Time Resolved Laser Induced Fluorescence (TRLIF).[5-9] The observation of direct luminescence from Np, Pu and many molecules containing U in solutions is practically impossible in most cases because of the very strong rate of radiationless deactivation[10] therefore, TRLIF cannot be applied.

We have found the possibility for the detection of Pu, Np and U in solutions, with high sensitivity, by using chemiluminescence. It is shown in this article that laser spectroscopic methods, with time resolution (TR), may be used for chemiluminescence spectroscopy and trace actinides detection in solutions. Chemiluminescence spectroscopy allows the determination of actinide valence states in solutions and the type of molecule.

2 CHEMILUMINESCENCE OF ACTINIDE SOLUTIONS INDUCED BY PULSED LASER RADIATION

In chemiluminescence, the energy is produced by a chemical reaction.[11] We study the process by which the excitation of actinides complexes produces the OH* radical by pulsed laser radiation. The OH* radical participates in chemical reactions with luminol, and chemiluminescence takes place.

A cuvette containing solution was irradiated with a 10 ns pulse laser. The time resolution of the start detector (p-i-n diode) was about 2 ns. Chemiluminescence was registered by photomultipliers operating in single photon counting mode connected to a monochromator. Chemiluminescence detection is performed after a laser pulse with a tuned delay time between 0.05 μs - 200 ms and a tuned gate time between 0.5 μs – 200 ms. Absorption spectra were measured using a Shimadzu UVI 3101 spectrophotometer. An ORION-611 pH-meter was used for solution pH measurements.

We selected an experimental scheme such that two actinides complexes, $AnO_2F_5^{3-}$ and $AnO_2F_4OH^{3-}$ (where An = U, Pu, Np), were formed in solutions, either containing luminol or not containing luminol.[12] OH* radicals were produced after actinide excitation by laser radiation in $AnO_2F_4OH^{3-}$. After OH* radical interaction with luminol typical chemiluminescence took place. The luminol concentration used was 10^{-4} mol/l. Absorption spectral measurements give us the possibility to control the influence of pH on the actinides complex speciation in solution. In our experiments, HF was used for pH regulation. A pH increase led to an increase in the $AnO_2F_4OH^{3-}$ complex concentration and a decrease in $AnO_2F_5^{3-}$ concentration. As a consequence we observed an increase in the chemiluminescence intensity. We thus studied the influence of pH on the chemiluminescence intensity in solutions and the kinetics of $AnO_2F_4OH^{3-}$ complex formation after laser radiation-induced excitation.[12]

From the kinetic curve of luminol chemiluminescence induced by excited plutonyl complexes (see Figure 1 (*left*)) one may determine the optimal delay time after the laser pulse for chemiluminescence detection. After a few microseconds the background, from luminescence impurities and scattered laser radiation, is practically absent and chemiluminescence (see Figure 1 (*right*)) may be detected by TR spectroscopic methods with high sensitivity.[13-15]

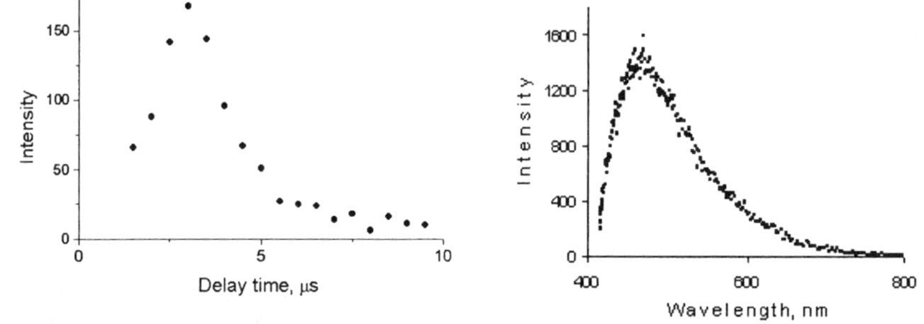

Figure 1 *Kinetic curve of luminol chemiluminescence (left) and chemiluminescence spectrumof luminol (right) induced by the excited $PuO_2F_4OH^{3-}$ complex*

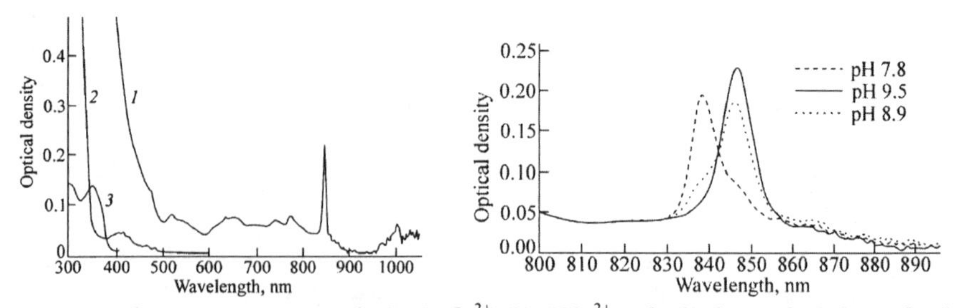

Figure 2 *Absorption spectra of (1) PuO_2^{2+},(2) $UO_2^{2+}$and (3) luminol (left) and of $PuO_2F_5^{3-}$ (band 839nm) and $PuO_2F_4OH^{3-}$(band 846.5nm) in 42% CsF+H_2O solutions at different pH values (right)*

With careful selection of laser radiation wavelengths (see Figure 2), one or multi-step actinides complexes excitation schemes, and chemiluminogenic labels, it is possible to induce chemiluminescence by the selective excitation of detectable actinide complexes.[15] The sensitivity of chemiluminescence methods[11,16] is higher than the sensitivity of other methods applied to determine Pu valence states, such as LIPAS.[17]

3 CONCLUSION

The behaviour of actinides in the environment is dictated by their valence states and their molecular form. Information about actinide valence states is essential to treat their emission source and determine propagation history. The combination of high sensitivity (up to 10^{-10}-10^{-13} mol/l) chemiluminescence effects with high selectivity laser spectroscopy TR and multi-step excitation methods[3,13] allows the possibility to effectively determine both luminescent and non-luminescent actinides species in different solutions, particularly those containing U, Pu and Np.

References

1 M. Nunnemann, N. Erdmann, H.-U. Hasse, G. Huber, J.V. Kratz, P. Kunz, A. Mansel, G. Passler, O. Stetzer, N. Trautmann and A. Waldek, *J. Alloys Compd.*, 1998, **271**, 4S.
2 C. Moulin, C. Beaucaire, P. Decambox and P. Mauchien, *Anal. Chim. Acta*, 1990, **238**, 291.
3 I.N. Izosimov, 'Laser spectroscopy on beams of radioactive nuclei' in *Proc. 4th Int. Workshop*, Poznan, Poland, ed., JINR, Dubna, , 1999, E15-2000-75, p. 169.
4 T. Berthoud, P. Decambox, B. Kirsch, P. Mauchien and C. Moulin, *Anal. Chim. Acta*, 1989, **220**, 235.
5 R. Klenze, J.I. Kim and H. Wimmer, *Radiochim. Acta*, 1991, **52-53**, 97.
6 G. Bernhard, G. Geipel, V. Brendler and H. Nitsche, *Radiochim. Acta*, 1996, **74**, 87.
7 C. Moulin, P. Decambox and P. Mauchien, *J. Phys. IV*, 1991, **C7**, 677.
8 C. Moulin, P. Decambox, P. Mauchien, D. Pouyat and L. Couston, *Anal. Chem.*, 1996, **68**, 3204.
9 P. Thouvenot, S. Hubert, C. Moulin, P. Decambox and P. Mauchien, *Radiochim. Acta*, 1993, **61**, 15.
10 C.P. Baird and T.J. Kemp, *Prog. React. Kinet.*, 1997, **22**, 87.
11 C. Dodeigne, L. Thunus and R. Lejeune, *Talanta*, 2000, **51**, 415.
12 N.G. Gorshkov, I.N. Izosimov, A.A. Kazimov, S.V. Kolychev, N.A. Kudryashev and L.G. Mashirov, *Radiochemistry*, 2001, **43**, 354.
13 I.N. Izosimov, N.G. Gorshkov, L.G. Mashirov, N.G. Firsin, A.A. Kazimov, N.A. Kudryshev and A.A. Rimski-Korsakov, 'Application of lasers in atomic nuclei research' in *Proc. 5th Int. Workshop,* Poland, ed., JINR, Dubna, 2001, E15-2002-84, p. 153.
14 N.G. Gorshkov, I.N. Izosimov, A.A. Kazimov, S.V. Kolychev, N.A. Kudryashev, L.G. Mashirov, A.A. Rimski-Korsakov and N.G. Firsin, *Radiochemistry*, 2001, **43**, 361.
15 N.G. Gorshkov, I.N. Izosimov, A.A. Kazimov, S.V. Kolychev, N.A. Kudryashev, L.G. Mashirov, A.V. Osokin and N.G. Firsin, *Radiochemistry*, 2003, **45**, 28.
16 T. Schlederer and P.G. Fritz, 'Method of detecting substances by chemiluminescence', 1998, United States Patent 5,736,320.
17 M.P. Neu, D.C. Hoffman, K.E. Roberts, H. Nitsche and R.J. Silva, *Radiochim. Acta*, 1994, **66-67**, 251.

AIR-OXIDATION BEHAVIOUR OF UO_2 AND Gd-DOPED UO_2 BY XAS

Y.-K. Ha, Y.-H. Cho, J.G. Kim, K.Y. Jee and W.H. Kim

Nuclear Chemistry Research Division, Korea Atomic Energy Research Institute, 150 Duk-Jin Dong, Yuseong, Daejeon, 305-353, Korea

1 INTRODUCTION

We previously reported[1,2] the effect of Gd^{3+} on the oxidation and phase transformation of UO_2 during air oxidation at temperatures up to 700 °C. We have further studied the phase transformations of uranium oxide during air oxidation from a different approach. In this investigation, the changes in oxidation state of U in oxidized phases of UO_2 and $(U_{0.844}Gd_{0.156})O_2$ solid solution were studied using X-ray absorption spectroscopy (XAS), since XAS is sensitive to changes in oxidation state.[3-4]

2 METHOD AND RESULTS

2.1 Experimental

The Gd-doped uranium oxides were prepared using a powder mixing technique.[1,2] The sample powders were oxidized at 250, 305, 340 and 420 °C for 4 hours under a continuous flow of air. The temperatures were chosen from the oxidation kinetic curves described in the previous report.[1] The data were collected from powder samples at Beamline 7C of the synchrotron radiation source at the Pohang Accelerator Laboratory, Korea. Absorption spectra were obtained by fluorescence detection with U L_{III} edge energy (17166.3 eV) at room temperature.

2.2 XANES

It is known that UO_2 has a cubic fluorite structure and transforms to the orthorhombic U_3O_8 phase upon oxidation.[5-7] The valence of U in UO_2 is tetravalent, U^{4+}, and that in U_3O_8 is mixed with tetravalent uranium and twice as many hexavalent ones, U^{6+}. Thus, the absorption edge for U_3O_8 will be shifted to a higher energy relative to that of UO_2. Figure 1 shows the XANES spectra for oxidized species of UO_2 and $(U_{0.844}Gd_{0.156})O_2$ at various temperatures. For comparison, the spectra of UO_2 and U_3O_8 were included as reference spectra. As can be seen in this figure, the edge position was shifted to a higher energy with increasing temperature for both UO_2 and $(U_{0.844}Gd_{0.156})O_2$. The shift of the peak position of $(U_{0.844}Gd_{0.156})O_2$ was less distinguishable than that of UO_2 and thus, the oxidation reaction was prohibited by the Gd atoms as described in the previous paper.

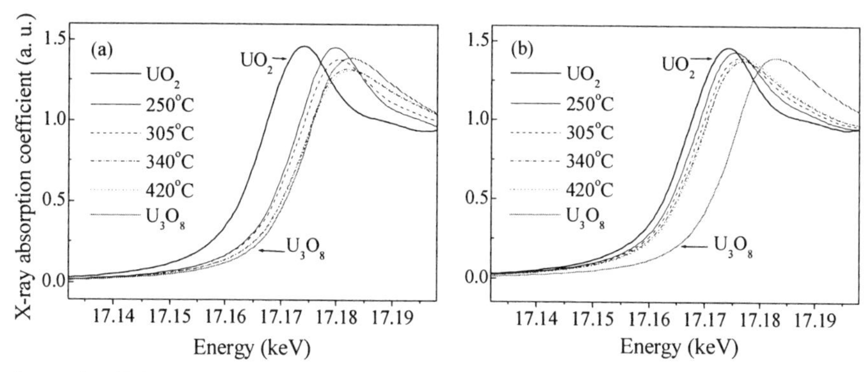

Figure 1 *U L_{III} X-ray absorption near edge spectra of (a) UO_2 and (b) $U_{0.844}Gd_{0.156}O_2$ oxidized at various temperatures*

The U L_{III} edge positions are expressed as the shifts (ΔE_0) from that of the UO_2 sample (Figure 2). In Figure 2, a lower energy is required to ionize a core electron from Gd-doped uranium oxide than from undoped uranium oxide. The E_0 increased as the average oxidation state of U increased, but it deviated from linearity.

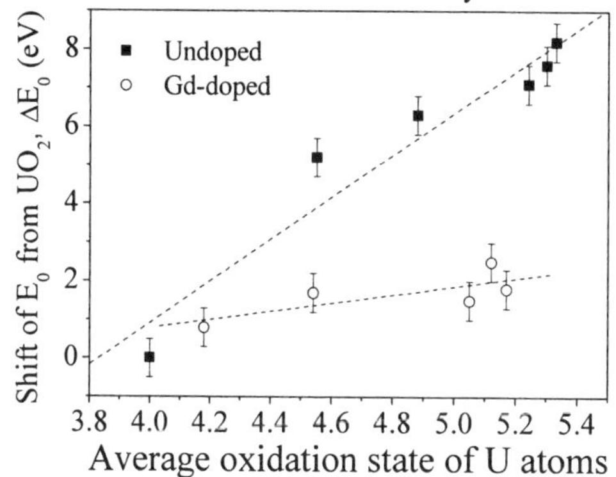

Figure 2 *Shifts of the U L_{III} edge energy with respect to UO_2 vs average formal charge of the U atoms in UO_2 and $U_{0.844}Gd_{0.156}O_2$ during air oxidation*

2.3 EXAFS

Figure 3 shows the k^2-weighted EXAFS spectra for oxidized species of UO_2 and $(U_{0.844}Gd_{0.156})O_2$. The spectra of UO_2 and U_3O_8 were also included as reference spectra. The spectral features at 250 and 305 °C seemed to be in between those of UO_2 and U_3O_8 (Figure 3a). These agreed well with XRD data, described as follows; UO_2 oxidized to: (1) U_4O_9 (70%)+ U_3O_7 (30%) at 250 °C, (2) U_4O_9 (25%) + U_3O_7 (35%) + U_3O_8 (40%) at 305 °C, (3) U_3O_7 (10%) + U_3O_8 (90%) at 340 °C and (4) U_3O_8 (100%) at 420 °C. Consulting the XRD data, it seems that the sharp peaks before peak I at 250 and 305 °C would be derived from U_3O_7. In the case of $U_{0.844}Gd_{0.156}O_2$, even at 420 °C, the spectral feature was similar to that of UO_2 rather than U_3O_8 (Figure 3b). It could be caused by the fact that Gd atoms inhibited the oxidation reaction, which also agrees well with XRD data described as follows; $U_{0.844}Gd_{0.156}O_2$ oxidized to (1) UO_2 (40%) + U_4O_9 (60%) at 250 °C, (2) U_4O_9 (40%) + U_3O_7 (40%) + U_3O_8 (20%) at 305 °C, (3) U_4O_9 (60%) + U_3O_8 (40%) at 340 °C and (4) U_4O_9 (65%) + U_3O_8 (35%) at 420 °C.

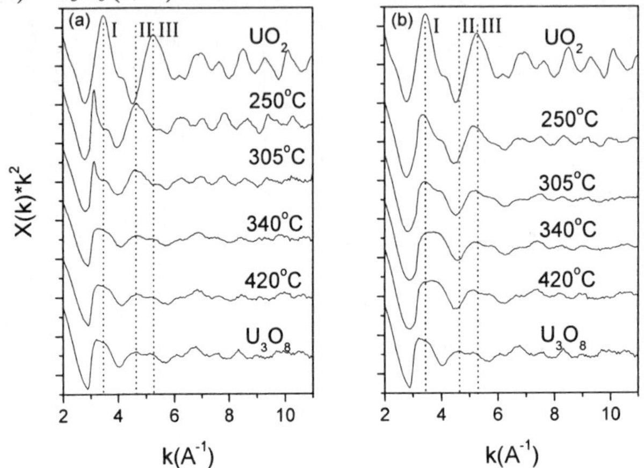

Figure 3 *U L_{III}-edge k^2-weighted EXAFS data of (a) UO_2 and (b) $U_{0.844}Gd_{0.156}O_2$ during air-oxidation at various temperatures*

3 CONCLUSIONS

The changes in the oxidation state of U for UO_2 and $(U_{0.844}Gd_{0.156})O_2$ solid solutions during air oxidation were studied by XAS. The shift of the U L_{III} edge energy showed the evidence of changes in the oxidation state and showed less oxidation by the Gd dopant.

Acknowledgements

We acknowledge the financial support of the Nuclear Development Fund from MOST. The XAS experiments at PLS were supported, in part, by MOST and POSCO in Korea.

References

1 J.G. Kim, Y-K. Ha, S.D. Park, K.Y. Jee and W.H. Kim, *J. Nucl. Mater.*, 2001, **297**, 327.
2 Y.-K. Ha, J.G. Kim, Y.J. Park and W.H. Kim, *J. Nucl. Sci. Technol.*, 2002, **Suppl. 3**, 772.
3 M.C. Duff, C. Amrhein, P.M. Bertsch and D.B. Hunter, *Geochim. Cosmochim. Acta*, 1997, **61**, 73.
4 N.J. Hess, W.J. Weber and S.D. Conradson, *J. Alloys Compds.*, 1998, **271-273,** 240.
5 I.J. Hastings, D.H. Rose, J.R. Kelm and D.A. Irvine, *J. Am. Ceram. Soc.*, 1986, **69,** C16.
6 G.C. Allen and N. R. Homes, *J. Nucl. Mater.*, 1995, **223,** 231.
7 L.E. Thomas, R.E. Einziger and H.C. Buchanan, *J. Nucl. Mater.*, 1993, **201,** 310.

SPECIATION OF RADIONUCLIDES WITH BIOLIGANDS USING TIME-RESOLVED LASER-INDUCED FLUORESCENCE (TRLIF) AND ELECTROSPRAY MASS SPECTROMETRY (ES-MS)

V. Lourenco,[1] E. Ansoborlo,[2] G. Cote[3] and C. Moulin[1]

[1]CEA Saclay DEN/DPC/SECR/Laboratoire de Spéciation des Radionucléides et des Molécules, 91 191 Gif-sur-Yvette cedex, F. E-mail : valerie.lourenco@cea.fr
[2]CEA Valrhô Marcoule DEN/DRCP/CETAMA, BP 17171, 30 207 Bagnols-sur-Cèze, F.
[3]ENSCP Laboratoire d'Electrochimie et de Chimie Analytique, 11 rue Pierre et Marie Curie, 75 231 Paris Cedex 05, F

1 INTRODUCTION

There is still great interest concerning the environmental and biochemical behaviour of radionuclides, in particular with respect to nuclear waste repository and to their removal (especially actinides) from contaminated workers. Nevertheless, studies on radionuclide complexation and uptake mechanisms, at the cellular and molecular levels, are very scarce. Hence, the chelating properties of these elements in biological media are not fully understood. Thus, this work was directed toward the study of the interactions between several bioligands of interest and radionuclides such as: uranium(VI) and several lanthanides(III) (also known as rare earths), the latter being chosen as analogues of trivalent actinides.

A literature survey has been carried out on several sequestering agents of heavy metals, present at the cellular scale, as well as in the cytoplasm, to select the most biologically relevant. This research revealed important discrepancies among the few existing complexation constants and led us to the choice of the following bioligands: amino acids (especially those constitutive of the binding sites of proteins) and some oligopeptides such as N-2-mercaptopropionyl glycine (MPG).

The selected systems were investigated using two major analytical techniques, namely electrospray mass spectrometry (ES-MS), as well as time-resolved laser-induced fluorescence (TRLIF). These two techniques are very powerful for such studies since, aside from their speciation capabilities, it is also possible to evaluate interaction constants as well as entropy and enthalpy, by undertaking experiments (for TRLIF) with varied temperatures. Moreover, these data acquisition can contribute to the improvement of thermodynamic data banks.

This work is intended to quantify the interactions between uranyl and rare earths, and previously quoted bioligands as a function of temperature (only for TRLIF) and pH, at negligible and biological ionic strengths. In this paper, only results between europium and cysteine at pH 3 and negligible ionic strength will be discussed.[1]

2 INTERACTIONS BETWEEN EUROPIUM AND CYSTEINE

The ligand, cysteine, is a thiol-containing amino acid. The complex, chemical significance of cysteine is determined by the mercaptosulfur donor atom, which is known to have a soft character. Via the mercapto group, this compound may thus participate in both redox and acid-base reactions. That is why cysteine is often constitutive of the binding sites of proteins and why the study of its interaction with metals is of such great biological importance.

Cysteine-rich proteins, such as metallothioneins or phytochelatins, have a heavy metal-scavenging role, since cysteine especially has the capacity to form stable complexes with heavy metals.[2] As a consequence, studying interactions between radionuclides and cysteine is of prime importance.

Europium was chosen as an analogue of trivalent actinides, not only because of its fluorescence properties but also for its isotopic pattern, easily identified by ES-MS. Cysteine contains three dissociable protons, with macroscopic protonation constants[3] $pK_{a\text{-}COOH} = 1.91$, $pK_{a\text{-}SH} = 8.14$ and $pK_{a\text{-}NH_3^+} = 10.28$.

The coordination chemistry of trivalent lanthanides is similar to that of calcium or strontium.[4] According to the Pearson's Hard and Soft Acids and Bases theory,[5] these elements are classified as "hard" (non polarizable) acids and are consequently expected to bind more strongly to hard bases, such as oxygen donor bases. Thus, the strongest binding site on the cysteine ligand is the carboxylate group. The amino group, even in the deprotonated form, is not expected to interact strongly with the metal ion.

The complexation equilibrium reaction and the global mononuclear complexation constants β_i are defined in Equation (1).

$$Eu^{3+} + i\,(H_2Cys)^0 \xleftrightarrow{\beta_i} Eu(H_2Cys)_i^{3+} \quad \text{with} \quad \beta_i = \frac{[Eu(H_2Cys)_i^{3+}]}{[Eu^{3+}].[(H_2Cys)^0]^i} \qquad (1)$$

3 EXPERIMENTAL

All chemicals used were reagent grade and Millipore deionised water (Alpha-Q, 18.2 MΩ cm) was used throughout the procedure. The mass spectrometric measurements were recorded in positive ion mode using a Quattro II tandem quadrupole mass spectrometer (Micromass, Manchester, England) equipped with an electrospray ionization source. The resolution of the quadrupole mass selector easily allows the observation of the isotopic mass distribution of Eu (i.e. ~48% ^{151}Eu and ~52% ^{153}Eu). TRLIF recordings were carried out using our "FLUO 2001" experimental set-up which is described elsewhere.[6] The excitation wavelength of the laser (266 nm quadrupled Brilliant Nd-YAG laser, coupled to an optical parametric oscillator system from Quantel, France) was tuned to 395 nm.

4 RESULTS AND DISCUSSION

4.1 TRLIF Results

Measurements of the variations of europium emission transition at 593 nm, and the hypersensitive transition at 618 nm, enabled the determination of: the fluorescence lifetime; the complex stoichiometry; and the complexation constant by using non-linear regression.

The complexation constant value found is consistent with that of calcium, whose charge/ionic radius ratio is comparable to that of trivalent europium.[4]

4.2 ES-MS Results

It has been previously observed that the ion abundance profile measured in ES-MS experiments is in good agreement with the aqueous solution speciation.[7-9]

To determine Eu speciation from mass spectrometry experiments, the europium species were divided into two categories: 1) complexed Eu species (with the ligand) and 2) "free" Eu species (solvated europium and inorganic complex (hydroxo and nitrate)). Relative amounts of complexed ligand were measured by assuming equal signal responses for charged complexes containing the same ligand and metal (as already observed for other similar systems investigated by ES-MS)[7-9] as well as for inorganic and organic species.

The main complexes observed are presented in Table 1.

Table 1 *ES-MS identification of main peaks*

Type	m/z	Species
(1:1) complex	414-6	$[Eu(H_2Cys)(NO_3)(HNO_3)(H_2O)]^+$
(1:2) complex	517-9	$[Eu(H_2Cys)_2(HNO_3)_2]^+$

Quantisation of the "free" Eu species was based on a calibration curve whose linear interpolation correlation coefficient R^2 was 0.997. The linearity range of the curve was $[Eu]_{free} = 1.10^{-5}$ M - 2.10^{-4} M . With peak identification and the quantification of the "free" Eu, it was possible to determine the formation constants of each complex. The complexation constant value found is in agreement with that obtained by TRLIF.

5 CONCLUSION

This study of the europium-cysteine system illustrates the complementarity of the two techniques used (TRLIF and ES-MS) for the determination of the stoichiometry and complexation constants, especially at very low metal concentrations.

Work is under progress for measuring complexation constants at higher pH values.

References

1 V. Lourenco, E. Ansoborlo, G. Cote and C. Moulin, unpublished results.
2 C.S. Cobbett, *Plant Physiol.*, 2000, **123**, 825.
3 R.M. Smith and A.E. Martell, 'Critically Selected Stability Constants of Metal Complexes Database, Version 3.0', 1997.
4 C.H. Evans, '*Biochemistry of the Lanthanides*', 1990, Vol. 8.
5 R.G. Pearson, *J. Amer. Chem. Soc.*, 1963, **85**, 3533.
6 G. Plancque, V. Moulin, P. Toulhoat and C. Moulin, *Anal. Chim. Acta*, 2003, **478**, 11.
7 C. Jacopin, M. Sawicki, G. Plancque, D. Doizi, F. Taran, E. Ansoborlo, B. Amekraz, and C. Moulin, *Inorg. Chem.*, 2003, **42**, 5015.
8 S. Colette, B. Amekraz, C. Madic, L. Berthon, G. Cote and C. Moulin, *Inorg. Chem.*, 2003, **42**, 2215.
9 G. Plancque, Y. Maurice, V. Moulin, P. Toulhoat and C. Moulin, *Appl. Spectrosc.*, 2005, **59**, 109.

MAGNETIC DIAGRAMS OF $PuGa_3$ UNDER PRESSURE

J.-C. Griveau,[1] P. Boulet,[1] E. Colineau,[1] P. Javorsky,[2] F. Wastin[1] and J. Rebizant[1]

[1]European Commission, Joint Research Centre, Institute for Transuranium Elements, P.O. Box 2340, 76125 Karlsruhe, Germany.
[2]Charles University, Faculty of Mathematics and Physics, Department of Electronic Structures, Ke Karlovu 5, 12116 Prague 2, The Czech Republic

1 INTRODUCTION

The recent discovery of superconductivity in $PuCoGa_5$ and $PuRhGa_5$[1,2] and their similarities with the cerium family have led to the study of binary systems Pu_xGa_y. $PuGa_3$ can exist with 2 structural/magnetic phases. The trigonal structure (R-3m) presents ferromagnetic ordering at T_C~21 K and the hexagonal structure (DO-19) an antiferromagnetic ordering at T_N~24 K. These 2 structures show heavy fermion features: γ~100 and 205 $mJ.mol^{-1}K^{-2}$, respectively.[3] By analogy to $CeIn_3$ behaviour under pressure,[4] which presents Quantum Critical Point at P_c~1.5 GPa where magnetism collapses and superconductivity occurs, we decided to study both structural phases under pressure and at low temperature by electrical resistivity.

2 RESULTS FOR THE TRIGONAL STRUCTURE

At ambient pressure, the resistivity presents a flat regime, from 300 K down to 25-30 K. A sharp decrease happens below the Curie Temperature, $T_C \approx 20$ K (Figure 1). At low pressure, the residual resistivity ratio, RRR=R(300K)/R(4.2K) ~1.5 to 2, is similar to that observed on the bulk sample at 0 GPa. After a change of regime around 75 K, we clearly observed the ferromagnetic transition at 21 K. Resistivity curves were slightly different from the pressure measurements on the bulk. The decrease of resistance under the T_C is smoother, and at higher temperature, we obtained a more pronounced "s-shape" curvature. This could have come from a possible anisotropic behaviour that we could not observe with the bulk and suggests strong anisotropic effects along the crystal directions, induced by the solid transmitting medium. T_C increased slightly with pressure at a rate of +0.45 K/GPa (see insert of Figure 1) and could not be determined above 4 GPa. At the maximum pressure achieved (6.7 GPa) we still observed the same resistive behaviour as at low pressure.

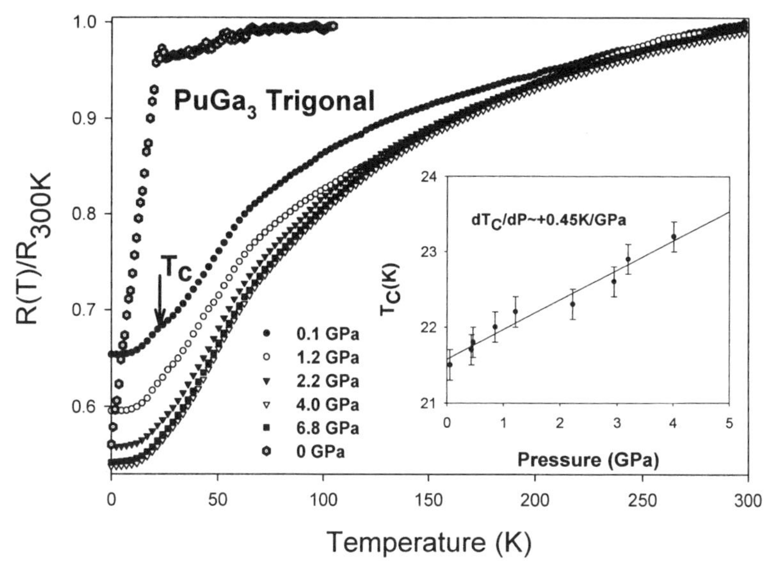

Figure 1 *Pressure evolution of $PuGa_3$ resistivity (trigonal structure) up to 6.8 GPa. We observe a slight increase of ferromagnetic transition temperature, T_C*

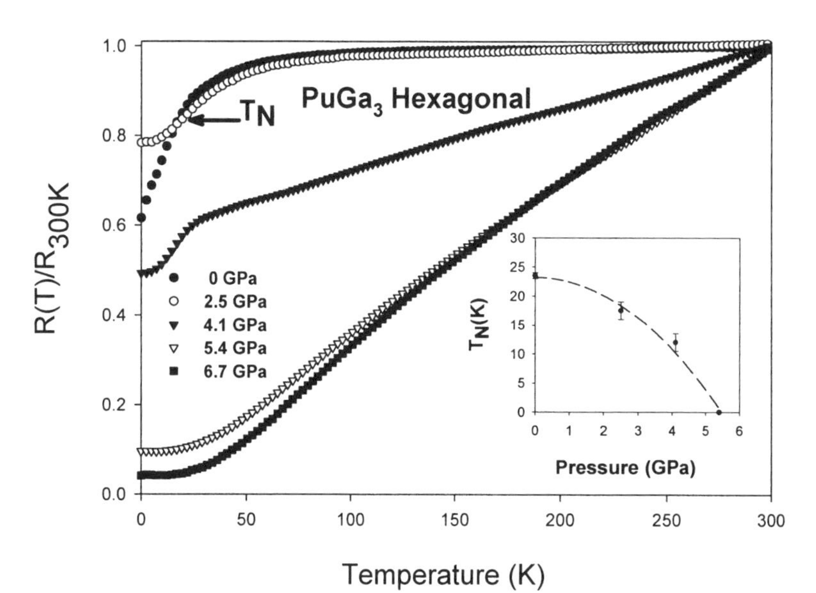

Figure 2 *Pressure evolution of $PuGa_3$ resistivity (hexagonal structure) up to 6.7 GPa. We observe a collapse of antiferromagnetic transition temperature, T_N, at around 5 GPa*

3 RESULTS FOR THE HEXAGONAL STRUCTURE

At ambient pressure, the hexagonal structure presented resistance curves similar to the trigonal structure with RRR=R(300K)/R(4.2K) ~1.6. However, the decrease of resistance was smoother at $T_N \approx 24$ K (Figure 2). At moderated pressure (2.5 GPa), the resistivity curve was still similar in shape to ambient pressure. Then, for higher pressures, dramatic changes were observed: a collapse of the resistance value was observed with a drastic increase of the RRR ratio from 1.3 at ambient pressure, up to 25 at the highest pressure achieved (6.7 GPa). The shape of the curve evolved to a metallic behaviour in the temperature range 300 K - 20 K. Below 20 K, T^3 behaviour was observed for p>4 GPa. This behaviour was clearly different from Fermi Liquid (T^2) or magnetic (Antiferromagnetic) behaviour. T_N decreased rapidly with pressure at –2K/GPa. At pressures higher than 4 GPa, we no longer observed any anomaly, and $PuGa_3$ showed metallic features over the temperature range studied.

4 DISCUSSION

Considering other $AnGa_3$ systems, we know that the inter-actinide distance exceeds the Hill limit without *f-f* overlap. UGa_3 ($d_{U\text{-}Ga}$=2.99 Å) is considered as a pure itinerant *5f* antiferromagnet with strong *5f-4p* hybridisation[5] while $NpGa_3$ ($d_{Np\text{-}Ga}$ =3.01 Å) shows AF / FM behaviour[6] with localised moment presenting a slight *5f* hybridization.[7] Pressure measurements reveal a fast collapse of AF magnetic order in UGa_3, while magnetism of $NpGa_3$ is reinforced.[7] For $PuGa_3$, we find $d_{Pu\text{-}Ga}$=3.02 Å and 2.89 Å for the trigonal and the hexagonal phases, respectively.[3] So, the trigonal phase properties of $PuGa_3$ should be close to $NpGa_3$, while the hexagonal phase should behave like UGa_3. Resistivity measurements under pressure confirm the following: for the trigonal phase, we clearly observed an increase in T_C with increasing pressure, while T_N collapsed for the hexagonal phase. By analogy with $CeIn_3$[4] this suggests a possible collapse of magnetism under pressure toward a quantum critical point (QCP) in $PuGa_3$. This description is correct for the hexagonal structure. However, its low temperature resistivity behaviour around 4 GPa (~T^3) is different from what is generally observed at QCP[4] where, for instance, a Fermi Liquid regime is induced from a Non Fermi Liquid with a resistivity behaviour fitting T^n (n<2).

References

1 J.L. Sarrao, L.A. Morales, J.D. Thompson, B.L Scott, G.R. Stewart, F. Wastin, J. Rebizant, P. Boulet, E. Colineau and G.H. Lander, *Nature (London)*, 2002, **420**, 317.
2 F. Wastin, P. Boulet, J. Rebizant, E. Colineau and G.H. Lander, *J. Phys.: Condens. Matter*, 2003, **15**, 2279.
3 P. Boulet, E. Colineau, F. Wastin, P. Javorský, J.C. Griveau, J. Rebizant, G.R. Stewart, and E.D. Bauer, *Phys. Rev. B*, 2005, **72**, 064438.
4 N.D. Mathur, F.M. Grosche, S.R. Julain, I.R. Walker, D.M. Freye, R.K. W. Haselwimmer and G.G. Lonzarich, *Nature (London)*, 1998, **394**, 39.
5 D. Mannix, A. Stunault, N. Bernhoeft, L. Paolasini, G.H. Lander, C. Vettier, F. de Bergevin, D. Kaczorowski, and A. Czopnik, *Phys. Rev. Lett.*, 2001, **86,** 4128.
6 E. Colineau, F. Bourdarot, P. Burlet, J.P. Sanchez and J. Larroque, *Physica B*, 1997, **230**, 773.
7 S. Zwirner, V. Ichas, D. Braithwaite, J.C. Waerenborgh, S. Heathman, W. Potzel, G.M. Kalvius, J.C. Spirlet, and J. Rebizant, *Phys. Rev. B*, 1996, **54**, 12283.

MAGNETIC ORDERING OF U AND Co MOMENTS IN (Th,U)Co_2X_2 (X = Si, Ge) SOLID SOLUTIONS

M. Kuznietz,* D. Li, T. Yamamura, K. Shirasaki and Y. Shiokawa

Institute for Materials Research, Tohoku University, Sendai, Miyagi 980-8577, Japan
*Present address: Nuclear Research Centre – Negev, P O Box 9001, 84190 Beer-Sheva, Israel

1 INTRODUCTION

The ternary 1:2:2 thorium and uranium compounds, $AnCo_2X_2$ (An = Th, U; X = Si, Ge), crystallize in the body-centered tetragonal $ThCr_2Si_2$-type structure (space group I4/mmm), with two formula units per tetragonal cell. The structure type was named after the lightest compound of the ThM_2X_2 series (M = 3*d* transition element), the first series of this large family to be prepared and characterized by x-ray diffraction (XRD) by Sikirica and Ban.[1,2]

The compounds, UCo_2Si_2 and UCo_2Ge_2, studied much later than the isostructural Th compounds, were found[3,4] to have the antiferromagnetic (AF) structure AF-I (+-+-) below a T_N of 85(3) K and 175(5) K, respectively. Neutron-diffraction studies indicated[3,4] magnetic ordering only of the uranium moments, aligned along the tetragonal axis perpendicularily to ferromagnetic basal planes, stacked alternately (+-+-) according to the AF-I sequence. Paramagnetic studies of both U compounds indicated[3,4] rather negative paramagnetic Curie temperatures, θ, of -285 K and -350(50) K, and high effective paramagnetic moments, μ_{eff}, of 4.85 μ_B and 4.5(5) μ_B, respectively. These paramagnetic moments are composed of U and Co moments, with (μ_{eff} /U) taken as those in the ferromagnetic copper compounds, UCu_2Si_2 (3.5 μ_B)[3] and UCu_2Ge_2, (2.1 μ_B).[5,6] The deduced (μ_{eff} /Co) values are: 2.3 μ_B in UCo_2Si_2, and 2.8 μ_B in UCo_2Ge_2. These paramagnetic Co moments do not order magnetically.

The paramagnetic properties of $ThCo_2Si_2$ and $ThCo_2Ge_2$ were studied initially in 1971 by Omejec and Ban.[7] From their magnetic susceptibility studies of all ThM_2X_2 compounds in the temperature range, 100-570 K, they predicted, for both Co compounds, ferromagnetic ordering at low temperatures (LT), quoting the paramagnetic values, θ, of 54 K and 14 K and (μ_{eff} /Co) of 3.82 μ_B and 1.42μ_B, respectively. Following some unclear magnetic data[7] of the $ThCo_2X_2$ compounds, we prepared both Th compounds, characterized them by XRD and determined their magnetic features. We recently reported[8] ferromagnetic ordering (of Co moments only) in $ThCo_2Ge_2$, below T_C of 120(20) K, with a positive θ of +30 K and μ_{eff} of 1.94 μ_B, or (μ_{eff} /Co)=1.37 μ_B, and ferromagnetism in $ThCo_2Si_2$ with $T_C > 320$ K.

The AF-I ordering of U moments in the UCo_2X_2 compounds and the ferromagnetic ordering of Co moments in the $ThCo_2X_2$ compounds led us to expect ordering of both U and Co moments in intermediate (Th,U)Co_2X_2 solid solutions. These form due to the moderate differences in lattice parameters between the end compounds, ~2 % in *a* and ~1

% in *c*. We report here on the preparation of (Th,U)Co_2X_2 solid solutions and studies of the magnetic ordering of U and Co moments in these newly-prepared materials.

2 EXPERIMENTAL DETAILS AND RESULTS

Polycrystalline $(Th_{1-x}U_x)Co_2Si_2$ solid solutions with x = 0, 0.17, 0.31, 0.40, 0.59, 0.69 and 0.86; and $(Th_{1-x}U_x)Co_2Ge_2$ solid solutions with x = 0, 0.13, 0.29, 0.40, 0.59, 0.71, and 0.86, were prepared by arc-melting under an argon atmosphere, followed by vacuum-annealing at 1023 K for 168 hrs and characterization by XRD at RT. All materials were studied by SQUID dc-magnetization measurements under zero-field-cooling (ZFC) and field-cooling (FC) procedures in an applied magnetic field of 0.01 T in the temperature range, 2-320 K, and their M *versus* H curves were followed at 5 K and several higher temperatures.

The newly-prepared solid solutions crystallize in the $ThCr_2Si_2$-type structure of the end compounds, $ThCo_2X_2$ and UCo_2X_2. Increasing the U-content, x, in the $(Th_{1-x}U_x)Co_2Si_2$ solid solutions reduces the ordering temperature, T_C, of the Co moments from above 320 K in $ThCo_2Si_2$ (x = 0) to 205 K for x = 0.40 (Figure 1) and finally to 165 K for x = 0.86. The ferromagnetic transition is followed by AF ordering of the U moments at 30 K for x = 0.40 (Figure 1) and 50 K for x = 0.86, being the only transition in UCo_2Si_2 (x = 1) at 85(3) K. In the $(Th_{1-x}U_x)Co_2Ge_2$ solid solutions AF ordering of the U moments is observed above a threshold composition 0.13< x <0.29, with a small increase in the ordering temperature, T_N, from 160 K for x = 0.29, to 165 K for x = 0.40 (Figure 2) and then to 175(5) K in UCo_2Ge_2 (x = 1), where it is the only transition. The AF transition is followed by the ferromagnetic ordering of the Co moments, with a stronger decrease from 120 K in $ThCo_2Ge_2$ (x = 0), being the only one there, through 80 K for x = 0.40 (Figure 2), down to 60 K for x = 0.86.

The totally different behaviour of the Si-based and Ge-based solid solutions is clear from the temperature dependence of the dc-magnetization and the inverse dc-susceptibility of the solid solutions $(Th_{0.60}U_{0.40})Co_2Si_2$ (Figure 1) and $(Th_{0.60}U_{0.40})Co_2Ge_2$ (Figure 2), that have similar U-contents. In the Si-based solid solution the initial ordering, due to the Co moments, is ferromagnetic, and it is followed by AF ordering of the U moments, with yet unknown magnetic structure. A rather negative θ (-1000 K) is observed, but μ_{eff} cannot be deduced. In the Ge-based solid solution the initial ordering due to U moments is AF, most probably AF-I, and it is followed by ferromagnetic ordering of the Co moments. For this solid solution the paramagnetic values, θ = -180 K and μ_{eff} = 3.37 μ_B, are easily determined and are compatible with the above values in the end compounds $ThCo_2Ge_2$ and UCo_2Ge_2.

3 DISCUSSION AND FUTURE PLANS

The coexistence of ordered U and Co magnetic moments is observed for the first time in intermediate $(Th_{1-x}U_x)Co_2Si_2$ and $(Th_{1-x}U_x)Co_2Ge_2$ solid solutions. From the magnetic features of the end compounds the ordering of the Co moments in the solid solutions is probably also ferromagnetic, while that of the U moments is AF, of yet an unknown type. These assumptions should be confirmed by the planned neutron-diffraction measurements.

In the system $(Th_{1-x}U_x)Co_2Si_2$ the ferromagnetic coupling of the Co moments is much stronger than the AF coupling of the U moments, as manifested by the corresponding ordering temperatures. In the system $(Th_{1-x}U_x)Co_2Ge_2$ the strengths of magnetic couplings are reversed, but they are closer to each other. Between the two transitions only one type of moment orders magnetically: Co moments in the former system and U moments in the latter. Coexistence of ordered moments of the two types occurs at LT, below the lower magnetic transition. Further studies of these magnetic features by neutron diffraction and other methods should lead to a better understanding of the absence of magnetic ordering of the Co moments in the UCo_2X_2 compounds,[3,4] mentioned above.

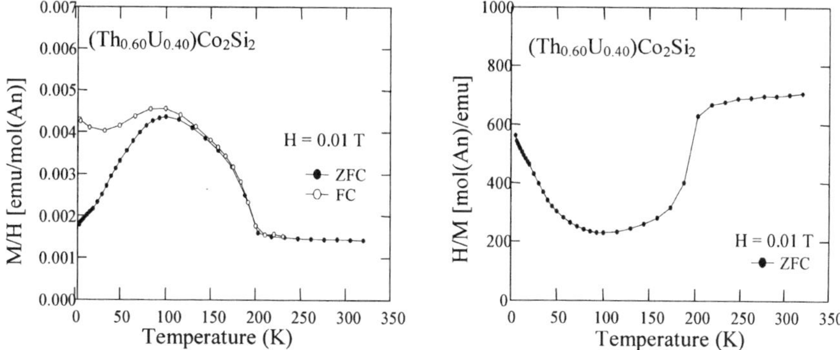

Figure 1 *DC-magnetization (in 0.01 T) measured under ZFC and FC procedures (left) and inverse dc-susceptibility (right) at 2-320 K in an annealed polycrystalline sample of* $(Th_{0.60}U_{0.40})Co_2Si_2$, *depicting initial ferromagnetic ordering of Co moments at 205 K, followed by AF ordering of U moments at 30 K*

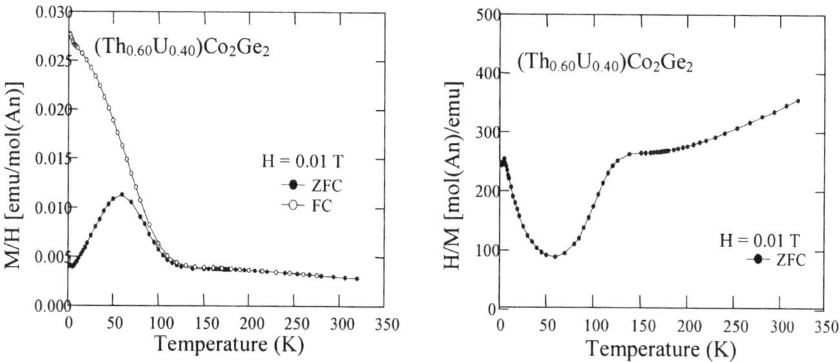

Figure 2 *DC-magnetization (in 0.01 T) measured under ZFC and FC procedures (left) and inverse dc-susceptibility (right) at 2-320 K in an annealed polycrystalline sample of* $(Th_{0.60}U_{0.40})Co_2Ge_2$, *depicting initial AF ordering of U moments at 165 K, followed by ferromagnetic ordering of Co moments at 80 K*

References

1 M. Sikirica and Z. Ban, *Croat. Chem. Acta*, 1964, **36**, 151.
2 Z. Ban and M. Sikirica, *Acta Cryst.*, 1965, **18**, 594.
3 L. Chełmicki, J. Leciejewicz and A. Zygmunt, *J. Phys. Chem. Solids*, 1985, **46**, 529.
4 M. Kuznietz, H. Pinto, H. Ettedgui and M. Melamud, *Phys. Rev. B*, 1989, **40**, 7328.
5 M. Kuznietz, H. Pinto, H. Ettedgui and M. Melamud, *Phys. Rev. B*, 1993, **48**, 3183.
6 M. Kuznietz, H. Pinto and H. Ettedgui, *J. Alloys Compd.*, 1995, **225**, 156.
7 L. Omejec and Z. Ban, *Z. Anorg. Allg. Chem.*, 1971, **380**, 111.
8 M. Kuznietz, D.X. Li, T. Yamamura and Y. Shiokawa, *J. Phys.: Condens. Matter*, 2005, submitted.

SPECTROSCOPIC INVESTIGATIONS OF URANIUM SPECIES IN ALKALI CHLORIDE MOLTEN SALTS

C.A. Sharrad,[1] I. May,[1] H. Kinoshita,[1] A.I. Bhatt,[1] V.A. Volkovich,[2] I.B. Polovov,[2] J.M. Charnock[3] and R.G. Lewin[4]

[1]Centre for Radiochemistry Research, School of Chemistry, The University of Manchester, Oxford Road, Manchester, M13 9PL, U.K.
[2]Rare Metals Department, Ural State Technical University, 19 Mira Street, Ekaterinburg, 620002, Russia.
[3]CCLRC Daresbury Laboratory, Daresbury, Warrington, Cheshire, WA4 4AD, U.K.
[4]Nexia Solutions, BNFL Technology Centre, Sellafield, Seascale, Cumbria, CA20 1PG, U.K.

1 INTRODUCTION

High temperature molten salts have many applications in actinide processing and separations. All plutonium electrorefining is currently undertaken in molten salt baths (*e.g.* equimolar NaCl:KCl). In addition, pyrochemical processes for the electrochemical separation of uranium (and plutonium) from irradiated nuclear fuel have been studied for several years. These processes have been developed to pilot plant scale at ANL, Argonne West, USA, which uses LiCl-KCl eutectic, and at RIAR, Dimitrovgrad, Russia, which is based on NaCl-KCl melts.[1] Current research is focussed on the next generation of pyrochemical processes for actinide separations.[2]

The development of electrochemical separation technologies for actinides in molten salts relies on a sound understanding of the key electrochemical processes, *i.e.* anodic dissolution, cathodic deposition and electrotransport. An increased awareness of *in situ* actinide speciation could help underpin such electrochemical developments. However, probing the actinide coordination environment in high temperature melts is extremely challenging, due to a limited number of spectroscopic techniques available.

We have used electronic absorption and X-ray absorption spectroscopy to study the *in situ* behaviour of actinide and transition metal species in high temperature molten salts.[3-5] The results of our latest investigations are reported here for uranium speciation in LiCl-KCl eutectic, focussing on different oxidation states and concentration dependence.

2 METHOD AND RESULTS

2.1 *In situ* Electronic Absorption Spectroscopy

The experiments were conducted using LiCl-KCl eutectic (Aldrich, 99.99 %, 58 mol% LiCl, mp 357 ^{o}C), as received. Chlorine (Aldrich, >99.5 %) and hydrogen chloride (Aldrich, >99 %) were dried by passing the gas through conc. H_2SO_4 before use.

High temperature electronic absorption spectra of uranium-containing melts were recorded using previously described methods[5] utilising an Avantes Avaspec 2048-2 fibre optic spectrometer with an Avantes AvaLight DHS deuterium-halogen light source. An Avantes collimating lens was attached to the end of the light source fibre optic to focus the beam. At the conclusion of the experiment the melt was quenched by extracting the melt into the silica tube used for chlorination, and allowed to cool to room temperature under an argon atmosphere.

Various soluble uranium species in molten LiCl-KCl eutectic at 450 °C were obtained by exposing UO_3, UO_2 and UCl_4 to HCl, and UO_2 and UCl_4 to Cl_2. The electronic absorption spectra of these species are presented in Figure 1(a). The spectra of UO_2 and UCl_4 exposed to HCl, are identical indicating a green melt characteristic for the U(IV) oxidation state. The spectral profiles of the UO_3 reacted with HCl, and UO_2 exposed to Cl_2, were assigned to uranyl species, which give a yellow colour. However, the characteristic fine structure observed for aqueous $\{UO_2\}^{2+}$ species in the region 400 – 500 nm, is obscured by an intense charge transfer band in the molten state. The spectrum of the yellow $\{U^{VI}Cl_6\}$ species (see Section 2.2) was also recorded. Spectra of U(III) species (Figure 1(a)) were obtained by reacting uranium metal with UCl_4 in molten LiCl-KCl eutectic, which resulted in an intense purple colour.

Concentration profiles of all uranium species in LiCl-KCl eutectic, by *in situ* EAS, indicate that the intensity of the absorbance at band maxima increases linearly with uranium concentration (see, for example, Figure 1(b)), thus adhering to Beer's Law. There are no changes to the positions of band maxima or the general spectral profiles with concentration. Hence, EAS can be a non-intrusive technique for the *in situ* determination of uranium in high temperature chloride melts.

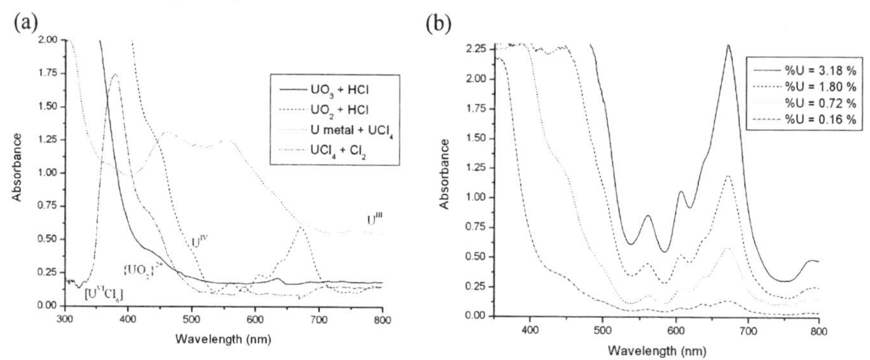

Figure 1 *Electronic absorption spectra of various uranium species (a) and of various concentrations of uranium from the reaction of UO_2 and HCl (b), in LiCl-KCl at 450 °C*

2.2 *In situ* X-ray Absorption Spectroscopy

Uranium L_{III}-edge XAS of high temperature melts were recorded in transmission mode on Station 9.3, and both transmission and fluorescence modes on Station 16.5 of the CCLRC Daresbury Radiation Source as previously described, with EXAFS data modelled using previously reported methods.[3]

In Table 1 are displayed the first coordination spheres and the range of bond lengths about uranium, modelled from the EXAFS data from the chlorination (HCl or Cl_2)

experiments, of various sources and concentrations of uranium in molten LiCl-KCl eutectic. The results show that uranium in the +VI oxidation state (see Section 2.1) has a preference to form $\{UO_2\}^{2+}$ in molten LiCl-KCl eutectic in the presence of oxygen. The coordination sphere is completed with four chloride ligands. However, when oxygen is eliminated from the system a U^{VI} species with a homoleptic chloride coordination sphere of six ligands can form by the oxidation of UCl_4 with Cl_2. Uranium (IV) species (UO_2 + HCl and UCl_4 + HCl) tend to undergo ligand exchange under mildly oxidising conditions, if necessary, to produce the $\{UCl_6\}^{2-}$ coordination shell.

Table 1 *First coordination spheres of uranium species in molten LiCl-KCl eutectic modelled using EXAFS analyses*[a]

1st Coord. Sphere	UO_2 + HCl	UO_2 + Cl_2	UO_3 + HCl	UCl_4 + HCl	UCl_4 + Cl_2	Distances (Å)[b]	
2 × O + 4 × Cl		✓	✓			U-O	1.70 – 1.80
						U-Cl	2.57 – 2.66
6 × Cl	✓			✓	✓	U^{IV}-Cl	2.57 – 2.68
						U^{VI}-Cl	2.62[c]

[a]Fits of all EXAFS experiments were refined such that $R \leq 45$. [b]Error ± 0.02 Å. [c]Single sample

3 CONCLUSIONS

The use of both electronic and X-ray absorption spectroscopic techniques has allowed for the *in situ* determination of both the coordination environment and the oxidation state of uranium in high temperature alkali chloride melts under chlorinating conditions.

Acknowledgements

We would like to acknowledge BNFL for funding, the EPSRC/CCLRC for the provision of synchrotron radiation facilities, and INTAS for a fellowship for I.B.P to work at the CRR. (Grant No. 03-55-1453).

References

1 J.J. Laidler, J.E. Battles, W.E. Miller, J.P. Ackerman and E.L. Carls, *Prog. Nucl. Energy*, 1997, **31**, 131; A.V. Bychkov, S.K. Vavilov, P.T. Porodnov, O.V. Skiba, G.P. Popkov and A.K. Pravdin, *Molten Salt Forum*, 1998, 525.
2 See, for example: J. Serp, R.J.M. Konings, R. Malmbreck, J. Rebizant, C. Scheppler and J.-P. Glatz, *J. Electroanal. Chem.*, 2004, **561**, 143; K. Uozumi, M. Iizuka, T. Kato, T. Inoue, O. Shirai, T. Iwai and Y. Arai, *J. Nucl. Mat.*, 2004, **325**, 34; K. Serrano and P. Taxil, *J. App. Electrochem.*, 1999, **29**, 497.
3 A.I. Bhatt, E.F. Kerdaniel, H. Kinoshita, F.R. Livens, I. May, I.B. Polovov, C.A. Sharrad, V.A. Volkovich, J.M. Charnock and R.G. Lewin, *Inorg. Chem.*, 2005, **44**, 2.
4 V.A. Volkovich, A.I. Bhatt, I. May, T.R. Griffiths and R.C. Thied, *J. Nucl. Sci. Technol.*, 2002, **Suppl. 3**, 595.
5 V.A. Volkovich, I. May, J.M. Charnock and B. Lewin, *Phys. Chem. Chem. Phys.*, 2002, **4**, 5753.

MAGNETIC AND ELECTRICAL PROPERTIES OF $UCu_3M_2Al_7$ ALLOYS

W. Suski,[1,2] K. Wochowski,[1] A. Gilewski,[2] T. Mydlarz[2] and D. Badurski[1]

[1]Institute of Low Temperature and Structure Research, Polish Academy of Sciences, P.O. Box 1410, 50-950 Wrocław 2, POLAND
[2]International Laboratory of High Magnetic Fields and Low Temperatures, P.O. Box 4714, 50-985 Wrocław 47, POLAND

1 INTRODUCTION

Our previous research on the $UCu_{4+x}Al_{8-x}$ derivatives (see e.g. Suski *et al.*),[1] is being extended now to the $UCu_3M_2Al_7$ alloys, where M = Cr, Mn and Fe. The aim of this project is to find the reason for the enhanced Sommerfeld coefficient:[1,2] the strongly correlated electrons or crystallographic disorder.

2 EXPERIMENTAL METHOD AND RESULTS

The investigated compounds were obtained by melting the components in stoichiometric quantities, in an arc furnace under an Ar protective atmosphere, and by annealing at 900 ^{0}C under vacuum for one week. The X-ray spectra of currently examined samples showed single-phase patterns. Note that we were not able to obtain the $UCu_3Fe_2Al_7$ alloy as a single phase and instead we succeeded only with obtaining $UCu_{3.5}Fe_{1.5}Al_7$. The lattice parameters and other magnetic parameters are listed in Table 1.

Table 1 *Crystallographic and magnetic data for the $UCu_3M_2Al_7$ alloys*

M	Lattice *a* (nm)	Parameters *c* (nm)	$T_{C,N}$ (K)	$\chi_0 \cdot 10^{-4}$ (cm^3/mol)	θ (K)	μ_{eff} (μ_B)
Cr	0.8779	0.5092		14.6	-83.9	2.20
Mn	0.8775	0.5086	320			
Fe(1.5)	0.8713	0.5052	32	5.3	46.7	3.98

The Cr alloy is paramagnetic, in the applied temperature and magnetic field range. The magnetic susceptibility roughly follows the modified Curie-Weiss law, with parameters listed in Table 1. The value of effective magnetic moment (2.20 μ_B), apparently lower than of the free U^{4+}/U^{3+}ion values, suggesting an influence of the crystal electric field (CEF). The magnetic flux density dependence of the magnetization at 1.8 K, is very weakly influenced by the applied field. The temperature dependence of the magnetization, M, of the Mn alloy (Figure 1 - *left*) exhibits a strong decrease at about 320 K (determined from the $^{dM}/_{dT}$ plot, not shown). This value is very close to the number previously obtained for the UCu_5MnAl_6 alloy[3] and proves the predominant role of the Mn atom in magnetic ordering. A low magnetic moment, and an absence of saturation, proves that magnetic order in this material has a ferrimagnetic character (Figure 1 - *right*). The anomaly observed at low temperature, existing at B = 5 kOe, and the shape of magnetization curve at 1.8 K, suggests that, at low temperature, there is a field-induced magnetic order in the uranium sublattice. The extension of the low temperature magnetization measurement in pulsed installation, up to a magnetic flux density of 350 kOe (35 T) does not reveal a change of the slope of the M(B) curve. In turn, the behavior of the iron compound is even more mysterious than the previous one. From Figure 1 (*right*) one can conclude that this compound is ferrimagnetic with a field-induced (B_{cr}~18 kOe) ferromagnetic contribution from clear remanence.

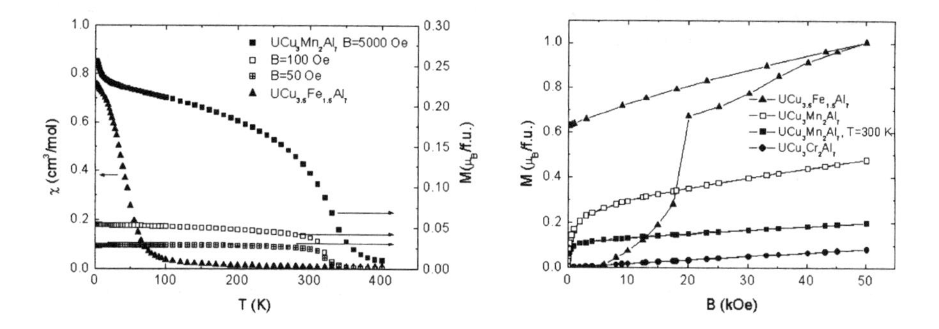

Figure 1 *Magnetization and magnetic susceptibility vs. temperature of the Mn and Fe alloys, respectively (left). Magnetization versus magnetic flux density at T = 1.8 K (right)*

However, even in the highest (pulsed) field of 35 T the M(B) plot, there is no saturation. Moreover, the magnetic susceptibility versus temperature measured in zero-field-cooling (ZFC) mode shows negative values (Figure 2). Such behavior was observed in the 4*f*-3*d* inter-metallics with iron and strongly anisotropic magnetic rare earth components[4] but also in the $YbFe_{4+x}Al_{8-x}$ alloys in which Yb ion is paramagnetic.[5] The subsequent electrical measurements excluded superconductivity as a reason for the negative magnetic susceptibility. For the field cooled (FC) sample, the $M(T)_B$ plots are reminescent of ferromagnetic materials. Above T = 60 K there is no difference between the ZFC and FC mode. The so called Arrott – Belov – Goryaga plot (not shown) provides the transition temperature to be 32 K. It is another indication that the iron compound is not a simple ferromagnet. Above about 100 K the susceptibility follows the modified Curie-Weiss law with the effective magnetic moment being 3.98 μ_B. Assuming that Fe contributes a moment of 1.73 μ_B and the paramagnetic moments of Fe and U add in quadrature, one obtains μ_{eff} = 3.58 μ_B, close to the value expected for localized U^{3+} or U^{4+}. This simplified reasoning can suggest that both the iron and uranium contribute to the effective moment.

In Figure 3 is plotted the electrical resistivity, ρ, versus temperature for all three samples. The ρ of all samples is weakly temperature dependent and exhibits relatively high values. All curves demonstrate, in principle, a decrease of the resistivity as the temperature increases; nevertheless, there are fine differences between them. However, at present we cannot propose any reasonable explanation for this behavior.

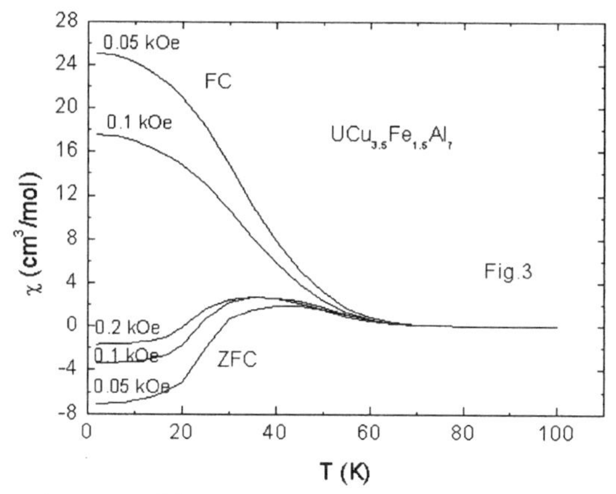

Figure 2 *Magnetic susceptibility versus temperature T < 100 K for the Fe alloy*

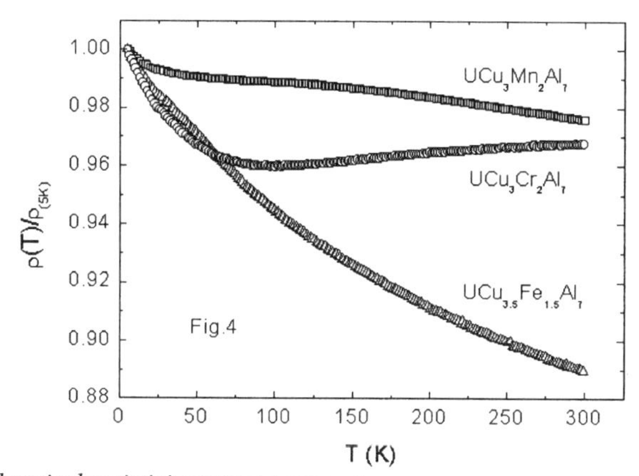

Figure 3 *Electrical resistivity versus temperature*

References

1 W. Suski, A. Czopnik, M. Sołyga, K. Wochowski, G. Bednarczyk, T. Mydlarz, *J. Alloys Compd.*, 2004, **383**, 126.
2 W. Suski, A. Czopnik, K. Gofryk, M. Sołyga, K. Wochowski, T. Mydlarz, *Solid State Sci.*, 2005, 7, 784.
3 W. Suski, A. Czopnik, M. Sołyga, K. Wochowski, T. Mydlarz, *Phys. B*, 2005, **359-361**, 1024.
4 I. Nowik and I. Felner, in *Proc. Int. Conf. Magnetism of Rare Earths and Actinides,* Bucharest, Sept.1983, Central Institute of Physics, p. 24.
5 H. Drulis, P. Gaczyński, W. Iwasieczko, W. Suski, B.Ya. Kotur, *Solid State Commun.*, 2002, **123**, 391.

X-RAY PHOTOELECTRON SPECTROSCOPY STUDY OF NEPTUNIUM-CONTAINING CERAMICS ON THE BASIS OF FERRITE AND TITANATE

Y.A. Teterin,[1] K.I. Maslakov,[1] A.Y. Teterin,[1] L. Vukcevic,[2] S.V. Yudintsev,[3] S.V. Stefanovsky,[4] K.E. Ivanov,[1] I.O. Utkin,[1] S.A. Perevalov,[5] T.S. Yudintseva[3] and A.V. Mohov[3]

[1]RRC "Kurchatov Institute", Moscow, 123182, Russia
[2]Faculty of Natural Sciences and Mathematics, University of Montenegro, Podgorica, Serbia and Montenegro
[3]Institute of Geology and Ore Depositions of RAS, Moscow, Russia
[4]Moscow NPO "Radon", Moscow, Russia
[5]Institute of Geology and Analytical Chemistry of RAS, Moscow, Russia
This work was supported by the RFBR grants 04-03-32892, 04-03-42902, 05-05-64005

1 INTRODUCTION

The most dangerous long-lived radionuclides in high-level radioactive wastes (HLW) are actinides. Their immobilization requires high stability matrices.[1] To understand chemical processes inside these matrixes it is necessary to know the physical and chemical states of the incorporated radionuclides (elemental and ionic compositions, oxidation states, number of uncoupled electrons on the metal ions, close environment structure, etc.).

The present work carried out X-ray diffraction (XRD), electron microscopy (EM) with energy-dispersion spectrometer SEM-EDS and X-ray photoelectron spectroscopy (XPS) analyses of murataite (M32) and garnet (G31) samples containing neptunium, as models of ceramic matrices for immobilization of actinides, in order to determine the structure and oxidation states of incorporated metal ions.

2 METHODS AND RESULTS

The samples were obtained after melting in a platinum ampoule at 1450 °C for 30 minutes following cooling, firstly to 1300 °C at a rate of 2 °C per minute, then to 20°C at a rate of 10 °C per minute. The ampoules were cut and the samples were studied by XRD, SEM-EDS and XPS. Calculated compositions of the samples in mass % are given in Table1.

XPS spectra of the studied samples were measured with electrostatic spectrometers (MK II VG Scientific) using non-monochromatized Al $K_{\alpha 1,2}$ and Mg $K_{\alpha 1,2}$ radiation.

XPS elemental and ionic quantitative analyses usually employ the most intense peaks from the elements present (Table 2). However, the fine XPS structural parameters are also very important for the determination of oxidation states of the M 3*d*, Ln 4*f* and An 5*f* transition elements (M – metal, Ln – lanthanide, An – actinide). For example, the Ti 3*s*, Mn 3*s*, Fe 3*s* multiplet splitting is proportional to the number of uncoupled 3*d* electrons in titanium, manganese and iron ions, which allow unambiguous determination of the oxidation states. The 'shake up' satellite structures in the Ln 3*d* and An 4*f* spectra allow the determination of lanthanide and actinide oxidation states and the ionic type of chemical bond.[2-4] For example, the Np 4*f* spectra from the studied samples exhibit the fine structure typical for Np^{4+} ions. Indeed, these spectra exhibit doublets, split at ΔE_{sl} = 11.6 eV and 11.8 eV. On the higher binding energy side of the basic peaks the typical 'shake up' satellites were observed at $\Delta E_{sat} \approx$ 6.8 eV with about 15 % intensity.

Table 1 *Calculated (Calc.) compositions for oxides (weight %), metals (at. %) and determined compositions for metals (at. %, XPS), murataite (Sample M32) and garnet (Sample G31)*

Muratite M32	*Calc. weight %, oxides*	*Calc. at.% metals*	*XPS, at. % metals*	*Garnet G31*	*Calc. weight % oxides*	*Calc. at.%, metals*	*XPS, at. %, metals*
TiO_2	56	54.0	54	Fe_2O_3	39	49.1	43
Mn_2O_3	12	11.7	13	Gd_2O_3	11	6.0	6
CaO	12	16.5	17	La_2O_3	10	6.2	8
NpO_2	5	2.2	2	CaO	10	17.9	24
ZrO_2	5	3.2	5	NpO_2	15	8.5	3
Al_2O_3	5	7.6	8	ZrO_2	15	12.3	16
Fe_2O_3	5	4.8	1				

The Mn 2*p* spectrum from murataite was observed as a low intense spin-orbit split doublet with ΔE_{sl} (Mn 2p) = 11.2 eV and Γ (Mn $2p_{3/2}$) = 2.1 eV. On the higher binding energy side of the basic Mn $2p_{3/2}$ peak, a 'shake up' satellite was observed with ΔE_{sat} (Mn 2*p*) $\approx$ 6 eV. Despite the low intensity, these spectra suggest that the manganese oxidation state in murataite is Mn^{2+} (Table 2).

In the La 4*d* region in the spectrum of garnet (G31) a complex structure was observed. Two low binding energy components at 93.4 eV and 99.1 eV are attributed to the Fe 3*s* electrons, whose spectrum, at first approximation, consists of two multiplet split peaks with ΔE_{ms} (Fe3s) = 5.7 eV. Since it is well known that one uncoupled electron leads to $\approx$ 1 eV multiplet splitting of the n*s* peak, the Fe 3*s* spectrum indicates that the incorporated iron ion has five uncoupled electrons (Fe $3d^5$), which is typical for the Fe^{3+} ion. It agrees with the Fe 2*p* and Fe 3*p* binding energies. The La 4*d* spectrum consists of a doublet with ΔE_{sl} (La 4*d*) = 3.2 eV, Γ (La $4d_{5/2}$) = 1.9 eV and 'shake up' satellites with ΔE_{sat} (La 4*d*) = 3.3 eV and a relative intensity I_{sat}/I_o = 27 % (Table 2).

The La 3*d* spectrum from garnet (Sample G31) consists of the spin doublet with ΔE_{sl} (La 3*d*) = 16.8 eV, Γ (La $3d_{5/2}$) = 1.7 eV and 'shake up' satellites with ΔE_{sat} (La $3d_{5/2}$) = 3.5 eV and a relative intensity, I_{sat}/I_o = 58 % (Table 2). The decrease in intensity of the satellites in the spectrum from garnet, compared to that from La_2O_3, can be explained by the increase in the ionic nature of the chemical bond.[4] These data unambiguously show that garnet contains La^{3+} ions.

The Gd 4*d* spectrum from garnet is typical for the Gd^{3+} ion.[4] By initial inspection, the Gd 4*d* spectrum could be interpreted as a spin-orbit split doublet, however, this is incorrect. The Gd^{3+} ion contains seven uncoupled Gd 4*f* electrons, which results in the complex Gd 4*d* multiplet splitting-related structure in the 40 eV range. Despite the difficulty in the interpretation of this structure, the gadolinium oxidation state can be determined as "fingerprints". The Gd $3d_{5/2}$ spectrum consists of the single-widened line at 1187.7 eV, which is also typical for the Gd^{3+} oxidation state.

On the basis of the XPS data, the relative atomic compositions of the studied samples were determined (Table 1):

$Ti_{1.00}Mn_{0.23}Ca_{0.32}Np_{0.03}Zr_{0.09}Al_{0.15}Fe_{0.01}OI_{3.58}O(II)_{0.57}$ (Sample M32 - murataite)

$Fe_{1.00}Gd_{0.15}La_{0.18}Np_{0.08}Ca_{0.56}Zr_{0.38}O(I)_{3.48}O(II)_{1.14}$ (Sample G31 - garnet)

Metal-oxygen inter-atomic distances were evaluated using the O 1*s* spectra. These spectra from murataite (Sample M32) and garnet (Sample G31) consist of two peaks at 530.1 and 532.2 eV. Taking into account the equations:

$$E_b\,(eV) = 2.27\ R_{M\text{-}O}^{-1}\ (nm) + 519.4 \quad (1)$$

derived previously elsewhere[5] and wherefrom:

$$R_{M\text{-}O}(nm) = 2.27\ (E_b - 519.4)^{-1}, \quad (2)$$

on the basis of the O 1*s* binding energy, one can evaluate in the studied samples, the inter-atomic metal-oxygen distance, $R_{M\text{-}O}$(nm). Thus, they are 0.212±0.002 nm and 0.177±0.002 nm, respectively. One can conclude that the value 0.177 nm is too low for ceramics and must be attributed to the hydroxide groups on the surface.

Table 2 *Binding energies, E_b (eV), and FWHMs, $\Gamma^{a)}$ (eV), of the outer (MO) and core electrons for murataite and garnet ceramics, given relative to E_b (C 1s)=285 eV*

Compound	*MO*	*$Np4f_{7/2}$*	*Ca2p*	*$Ti2p_{1.2}$*	*$Mn2p_{3/2}$*	*$Fe2p_{3/2}$*	*La4d, [$La3d_{5/2}$]*	*Gd4d [$Gd3d_{3/2}$]*	*$Zr3d_{5/2}$*	*$Al2p_{3/2}$*	*O1s*
Murataite M32	4.0; 21.6; 24.5;30.5; 36.8; 3.5; 48.1	403.0 (1.7) 6.2 sat	346.7 (2.3)	458.6 (1.1)	641.0 (2.1)	710.9 (2.6)			182.0 (1.1)	74.0 (1.1)	530.1 (1.4) 532.2 (1.3)
Garnet G31	2.7; 5.7; 7.7; 17.0; 21.4; 4.3; 30.3; 43.2	402.8 (1.4)	346.5 (1.5)			710.9 (2.5) 8.2 sat	101.7 104.9 108.2 [834.4] [3.5 sat]	141.2 143.4 146.9 [1187.7]	181.6 (1.0)		530.1 (1.4) 532.3

[a)]FWHMs are given in parentheses relative to the FWHM of the C 1*s* peak, accepted to be Γ (C 1*s*) = 1.3 eV

3 CONCLUSION

1. According to the XRD and SEM-EDS analyses, the titanate ceramic (Sample M32) is formed by the dominating murataite phase with a quintuple and octuple repeated basic fluorite cell, and the ferrite ceramic (Sample G31) is dominated by the garnet phase.
2. On the basis of the outer and core electron XPS structure parameters, in the binding energy range 0–1000 eV, quantitative elemental and ionic analyses were carried out. The oxidation states of the incorporated metal ions, in both samples, were determined to correspond to the following ions, as predicted: Ca^{2+}, Ti^{4+}, Mn^{2+}, Fe^{3+}, Zr^{4+}, Al^{3+}, Np^{4+} for murataite; and Ca^{2+}, Fe^{3+}, Zr^{4+}, La^{3+}, Gd^{3+}, Np^{4+} for garnet.
3. On the basis of the oxygen binding energies in the ceramic samples studied the metal-oxygen inter-atomic distances were evaluated to be 0.212±0.002 nm for murataite and garnet, and 0.177±0.002 nm for the impurity (hydroxyl groups) on the sample surface.

References

1 N.P. Laverov, S.V. Yudintsev, S.V. Stefanovsky and Y.N. Jang, *Dokl. RAN (Rep. Russ. Acad. Sci.)*, 2001, **381**, 399, (Russian).
2 Y.A. Teterin and A.S. Baev, *X-Ray Photoelectron Spectroscopy of Light Actinide Compounds*, TsNIIAtominform, Moscow, 1986, p. 104, (Russian).
3 Y.A. Teterin and A.Y. Teterin, *Russ. Chem. Rev.*, 2004, **73**, 541.
4 Y.A. Teterin and A.Y. Teterin, *Russ. Chem. Rev.*, 2002, **71**, 347.
5 M.I. Sosulnikov and Y.A. Teterin, *DAN SSSR*, 1991, **317**, 418, (Russian).

Subject Index